AF598583

Encyclopedia of Applied Physics

SPONSORS

AMERICAN INSTITUTE OF PHYSICS
DEUTSCHE PHYSIKALISCHE GESELLSCHAFT
JAPAN SOCIETY OF APPLIED PHYSICS
PHYSICAL SOCIETY OF JAPAN

DEUTSCHE PHYSIKALISCHE GESELLSCHAFT

JAPAN SOCIETY OF APPLIED PHYSICS

PHYSICAL SOCIETY OF JAPAN

ENCYCLOPEDIA OF APPLIED PHYSICS

VOLUME 16

Raman Spectroscopy Instrumentation to Schottky Barriers

Edited by

GEORGE L. TRIGG

Associate Editors

EDUARDO S. VERA

WALTER GREULICH

Managing Editor

EDMUND H. IMMERGUT

Assistant Managing Editor

CHRISTOPHER THOMAS MORAN

George L. Trigg
275 Beaver Dam Road
Brookhaven, New York 11719

Walter Greulich
Ziegeleiweg 30
D-69488 Birkenau
Federal Republic of Germany

Eduardo S. Vera
Science and Technology Information Center (ICT)
University of Chile
Beaucheff 850
Santiago, Chile

Edmund H. Immergut
Christopher Thomas Moran
2 Sidney Place
Brooklyn, New York 11201

Library of Congress Cataloging-in-Publication Data

Encyclopedia of applied physics.
(Revised for Vol. 16)

"Sponsors, American Institute of Physics . . . [et al]."
Includes bibliographical references and index.
1. Physics—Encyclopedias. 2. Engineering—Encyclopedias. I. Trigg, George L. II. Vera, Eduardo S. III. Greulich, Walter. IV. American Institute of Physics.
QC5.E543 1991 530'.05 91-8738
ISBN 1-56081-075-0 (VCH Publishers)

British Library Cataloguing in Publication Data

Encyclopedia of applied physics.
Vol. 16
I. Trigg, George L. (George Lockwood) II. Vera, Eduardo S. III. Greulich, Walter
621
ISBN 1-56081-058-0 (set)
ISBN 1-56081-075-0 (vol. 16)

Printed in the United States of America.

ISBN 3-527-28138-X (Volume 16) VCH Verlagsgesellschaft mbH
ISBN 3-527-26841-3 (set) VCH Verlagsgesellschaft mbH

Printing History:
10 9 8 7 6 5 4 3 2 1

Published jointly by:

VCH Publishers, Inc.
333 7th Avenue
New York, NY 10001

VCH Verlagsgesellschaft mbH
P.O. Box 10 11 61
69451 Weinheim
Federal Republic of Germany

VCH Publishers (UK) Ltd.
8 Wellington Court
Cambridge CB1 1HZ
United Kingdom

ADVISORY BOARD

EDITORIAL CONSULTANTS

MAIN ENTRIES

04-A	Raman Spectroscopy Instrumentation 1	16-B	Reinforcing Fibers 343
08-D	Random Processes 45	01-D	Relativity, Special 369
08-B	Rare Gases 71	06-C	Remote Sensing 385
08-D	Rarefied-Gas Dynamics 97	01-E	Research and Development 415
17-D	Reactions Induced by Light ...117	05-A	Resistors 423
02-A	Recording Devices 149	12-D	Resonant Tunneling 437
02-C	Recording, Photographic 185	09-E	Rheology 455
16-D	Recrystallization, Dynamic 205	09-D	Rheology Controlled by Magnetic Fields 487
16-D	Recrystallization, Static 227	02-E	Robotics 505
06-A	Reflectometers, Optical 247	20-A	Satellites, Artificial 519
06-A	Refractometers, Optical 255	07-A	Scanning Acoustical Microscopy 545
16-B	Refractory Materials—Ceramics 277	15-D	Schottky Barriers 573
16-B	Refractory Materials, Electronic Applications of 309		Contents of Previous Volumes 593
16-B	Refractory Metals 327		

The subject matter in the *Encyclopedia of Applied Physics* is presented in approximately 500 individual articles, arranged alphabetically. The topics can be classified into 20 sections, similar to the AIP Physics and Astronomy Classification Scheme (PACS):

01	General Aspects: Mathematical, Computational, and Information Techniques	11	Condensed Matter B: Thermal, Acoustic, and Quantum Properties
02	Measurement Science, General Devices and/or Methods	12	Condensed Matter C: Electronic Properties
03	Nuclear and Elementary Particle Physics	13	Condensed Matter D: Magnetic Properties
04	Atomic and Molecular Physics	14	Condensed Matter E: Dielectrical and Optical Properties
05	Electricity and Magnetism	15	Condensed Matter F: Surfaces and Interfaces
06	Optics (classical and quantum)	16	Materials Science
07	Acoustics	17	Physical Chemistry
08	Thermodynamics and Properties of Gases	18	Energy Research and Environmental Physics
09	Fluids and Plasma Physics	19	Biophysics and Medical Physics
10	Condensed Matter A: Structure and Mechanical Properties	20	Geophysics, Meteorology, Space Physics, and Aeronautics

Each article has been assigned a code number consisting of two digits which denotes the section, and a letter which gives the type of article. There are six types: A = Devices, Equipment; B = Materials; C = Methods, Processes; D = Phenomena, Effects; E = Scientific or Technological Fields; F = Institutions, Companies, Societies and other organizations.

CONTRIBUTORS

R. A. Aziz, Department of Physics, University of Waterloo, Waterloo, Ontario, Canada N2L 3G1
Rare Gases

A. R. Bunsell, Centre des Matériaux Pierre-Marie Fourt, Ecole Nationale Supérieure des Mines de Paris, B.P. 87, F-91003 Evry Cedex, France
Reinforcing Fibers

Bruce A. Campbell, Code 490, NASA/Goddard Space Flight Center, Greenbelt, MD 20771
Satellites, Artificial

John J. Craig, Adept Technology, Inc., 150 Rose Orchard Way, San Jose, CA 95134
Robotics

F. Cyrot-Lackmann, LEPES/CNRS, B.P. 166X, F-38042 Grenoble Cedex, France
Schottky Barriers

Ronald F. Dziuba, Electricity Division/NIST, Gaithersburg, MD 20899-0001
Resistors

Ralf Eck, Metallwerk Plansee GmbH, Reutte, Austria
Refractory Metals

Howell G. M. Edwards, Department of Chemistry and Chemical Technology, University of Bradford, Bradford, W. Yorkshire, 8D7 1DP UK
Raman Spectroscopy Instrumentation

L. Esaki, University of Tsukuba, Tsukuba, Ibaraki 305, Japan
Resonant Tunneling

Robert A. Frosch, John F. Kennedy School of Government, Harvard University, 79 John F. Kennedy Street, Cambridge, MA 02138
Research and Development

Hideo Fujiwara, Department of Physics and Astronomy, University of Alabama, P.O. Box 870324, Tuscaloosa, AL 35487-0324
Recording Devices

J. M. Ginder, Ford Motor Company Research Laboratory, 20000 Rotunda Drive, Mail Drop 3028 SRL, Dearborn, MI 48121-2053
Rheology Controlled by Magnetic Fields

Karl-Ernst Granitzki, DIFK—Deutsches Institut für Feuerfest und Keramik GmbH, An der Elisabethkirche 27, D-53113 Bonn, Germany
Refractory Materials—Ceramics

Ronald J. Holyer, Naval Research Laboratory, Code 7240, Stennis Space Center, MS 39529-5004
Remote Sensing

Ottó Horváth, Department of General and Inorganic Chemistry, Veszprém University, H-8201 Veszprém, Hungary
Reactions Induced by Light

F. J. Humphreys, Manchester Materials Science Centre, Grosvenor Street, Manchester M1 7HS, UK
Recrystallization, Static

Franklin C. Hurlbut, Department of Mechanical Engineering, Room 5138, Etcheverry Hall, University of California at Berkeley, Berkeley, CA 94720
Rarefied-Gas Dynamics

A. R. Janzen, Department of Physics, University of Waterloo, Waterloo, Ontario, Canada N2L 3G1
Rare Gases

J. J. Jonas, Department of Mining and Metallurgical Engineering, McGill University, 3450 University Street, Montréal, Québec, Canada H3A 2A7
Recrystallization, Dynamic

Elliot Kearsley, 10413 Englishman Drive, Rockville, MD 20852
Rheology

Butrus T. Khuri-Yakub, Edward L. Ginzton Laboratory, Stanford University, Stanford, CA 94305
Scanning Acoustical Microscopy

Nobuyoshi Kobayashi, Central Research Laboratory, Hitachi, Ltd., 1-280 Higashi-koigakubo, Kokubunji-shi, Tokyo 185, Japan
Refractory Materials, Electronic Applications of

L. Magaud, LEPES/CNRS, B.P. 166X, F-38042 Grenoble Cedex, France
Schottky Barriers

C. S. McCamy, 54 All Angels Hill Road, Wappingers Falls, NY 12590
Reflectometers, Optical

Samuel Walter McCandless, User Systems, Inc., Chesapeake Beach, MD 20732
Satellites, Artificial

E. E. Mendez, IBM Research Division, T. J. Watson Research Center, Yorktown Heights, NY 10598
Resonant Tunneling

Enrico P. Mercanti, NASA/Goddard Space Flight Center, Greenbelt, MD 20771
Satellites, Artificial

F. Montheillet, Department of Mining and Metallurgical Engineering, McGill University, 3450 University Street, Montréal, Québec, Canada H3A 2A7
Recrystallization, Dynamic

W. Nebe, Haselstrauchweg 31, D-07745 Jena, Germany
Refractometers, Optical

Stephen J. Paddack, Code 490, NASA/Goddard Space Flight Center, Greenbelt, MD 20771
Satellites, Artificial

Erwin Pink, Erich-Schmid-Institut für Festkörperphysik der Österreichischen Akademie der Wissenschaften, Jahnstrasse 12, A-8700 Leoben, Austria
Refractory Metals

Gerald Routschka, DIFK—Deutsches Institut für Feuerfest und Keramik GmbH, An der Elisabethkirche 27, D-53113 Bonn, Germany
Refractory Materials—Ceramics

Michael F. Shlesinger, 412 Green Pasture Drive, Rockville, MD 20852
Random Processes

R. O. Simmons, Department of Physics, University of Illinois, Urbana, IL 61801-3080
Rare Gases

Kenneth L. Stevenson, Department of Chemistry, Indiana University/Purdue University, Fort Wayne, IN 46805
Reactions Induced by Light

David M. Sturmer, Imaging Research Laboratories, Eastman Kodak Company, 343 State Street, Rochester, NY 14650-2103
Recording, Photographic

Romualdo Tabensky, Departamento de Física, Facultad de Ciencias Físicas y Matemáticas, University of Chile, Casilla 487-3, Santiago, Chile
Relativity, Special

Glen Wade, Department of Electrical and Computer Engineering, University of California at Santa Barbara, Santa Barbara, CA 93106-9560
Scanning Acoustical Microscopy

Albert C. Wey, Sonoscan, Inc., 530 E. Green Street, Bensenville, IL 60106
Scanning Acoustical Microscopy

RAMAN SPECTROSCOPY INSTRUMENTATION

HOWELL G. M. EDWARDS, *Molecular Spectroscopy Laboratories, Chemistry and Chemical Technology, University of Bradford, Bradford, West Yorkshire, United Kingdom*

	Introduction	1
1.	**Instrumentation for the Generation, Illumination, Dispersion, and Detection of Electromagnetic Radiation of Relevance to Raman Spectroscopy**	3
1.1	Electromagnetic-Radiation Sources	4
1.1.1	Lasers	4
1.2	Sample Illuminator and Cell Design	6
1.3	Spectrometers and the Dispersion of Radiation	9
1.3.1	Dispersion	9
1.3.2	Resolving Power	9
1.3.3	Brightness of Image	10
1.3.4	Speed of Spectrograph/ Spectrometer	12
1.3.5	Slit	13
1.3.6	Collimator	13
1.3.7	Diffraction Grating	14
1.3.8	Camera Mirror/Lens	16
1.3.9	Detectors	16
1.3.10	Problems	16
1.3.10.1	Light Leakage	16
1.3.10.2	Effect of Temperature, Pressure, and Humidity	17
1.3.10.3	Spectral-Line Tilt	17
1.3.11	Calibration	17
1.3.11.1	Fabry–Pérot Étalon	18
2.	**Raman Scattering**	19
2.1	Instrumentation for the Recording of the Pure Rotation and Vibration–Rotation Raman Spectra of Gases	19
3.	**Raman Microscopy and Imaging Techniques**	24
3.1	CCD Devices	27
3.2	Confocal Microscopy	28
3.3	Raman Imaging and Mapping	28
3.4	The UV Microscope	29
3.5	Raman Microspectroscopy with Near-Infrared Excitation	29
4.	**Nonlinear Spectroscopy**	31
4.1	Induced Electric Dipole-Frequency Mixing (Nonlinear Polarization)	31
4.2	Hyper-Rayleigh and Hyper-Raman Scattering	31
4.3	CARS	32
4.4	SRGS (SRLS)	34
5.	**Special Techniques in Raman Spectroscopy**	34
5.1	Surface-Enhanced Raman Scattering	34
5.2	Photoacoustic Raman Spectroscopy (PARS)	36
6.	**Fourier-Transform (FT) Methods in Vibrational Spectroscopy**	36
7.	**Intensity Measurements in Raman Scattering**	39
	Glossary	41
	Works Cited	42
	Further Reading	43

"I have kept my resolution. I have kept it strictly ever. I have allowed nothing to come between me and it."
Robert Graves, *Claudius the God*

INTRODUCTION

Molecular spectroscopy is primarily concerned with the qualitative and quantitative evaluation of the structures and transformation of materials at the molecular level. The

3-527-28138-X/96/$5.00 + .50

rotational, vibrational, and electronic transitions in systems that result in energy changes in the electromagnetic spectrum have long provided information about molecular species and chemical bonding and, thence, the kinetics of transformation from reactants to products in chemical reactions (Willard *et al.*, 1981).

The origin of Raman scattering is markedly different from that of infrared absorption; in Raman spectroscopy, the sample, which may be a gas, liquid, or solid, is irradiated with an intense beam of monochromatic radiation of wave number $\tilde{\nu}_0$ in the ultraviolet–visible–infrared regions of the electromagnetic spectrum. Most of the radiation is scattered without change in wave number; this is known as Rayleigh scattering. However, a very small part of the scattered radiation ($\sim 10^{-5}$ of the incident radiation intensity) has wave numbers $\tilde{\nu}_0 \pm \tilde{\nu}_m$, where $\tilde{\nu}_m$ is a characteristic vibrational wave number of the molecular species that is undergoing excitation. The $\tilde{\nu}_0 - \tilde{\nu}_m$ and $\tilde{\nu}_0 + \tilde{\nu}_m$ scattered-radiation components are known as Stokes and anti-Stokes, respectively; the Raman effect is often thought of as a vibrational- or rotational-frequency modulation of the electric field of the incident radiation. Normally the Stokes-scattered radiation, which is observed as a series of bands or lines shifted from the Rayleigh excitation, $\tilde{\nu}_0$, is more intense than the anti-Stokes radiation because of the Boltzmann distribution of molecular populations over the available energy levels of the molecular system. However, in certain cases, particularly in hot gases and jet-engine exhausts, the Stokes/anti-Stokes intensity ratios become comparable, and this has provided an application of Raman spectroscopy to temperature measurement in hostile environments.

In contrast, infrared absorption, as the name implies, requires the absorption of specific wave numbers defined by $\tilde{\nu} = \Delta E/hc = (E_2 - E_1)/hc$ for a molecular transition from energy level E_1 to E_2. Hence, infrared absorption can only occur in a precise region of the electromagnetic spectrum near $\tilde{\nu} = \lambda^{-1}$, but Raman scattering can be excited theoretically anywhere in the electromagnetic spectrum, dictated by the availability of suitable detectors and the intensity of the scattered radiation, which varies as $\tilde{\nu}^4$. For a given molecular system, therefore, and after accounting for instrumental detector difference and for similar laser-excitation radiation powers (strictly, irradiances), Raman scattering in the ultraviolet region at 300 nm (33.3×10^3 cm^{-1}) is 160 times more intense than that in the infrared region at 1064 nm (9.4×10^3 cm^{-1}). There are good reasons, however, for selectivity of Raman-spectroscopic excitation wavelengths, which are usually dictated by special considerations of sample stability, color, and molecular properties; the latter may give rise to the phenomenon of fluorescence, which can involve similar excitation processes, with radiationless transitions in an upper-molecular electronic state that can swamp the much weaker Raman scattering. Generally, onset of fluorescence diminishes toward the infrared, and this has been a powerful motivation for the excitation of Raman spectra at long wavelengths from fluorescent and colored materials.

It should not be inferred that because both Raman-scattering and infrared-absorption techniques give information about molecular vibrations and rotations, this information is identical from both sources. The selection rules for Raman scattering are based on molecular polarizabilities and induced dipoles, which change with molecular normal coordinates, whereas those that operate in infrared absorption are dependent on dipole-moment changes. The two techniques are complementary and, particularly in cases of molecular systems with high symmetries, provide very strong reinforcement for each other in the armory of the molecular spectroscopist. Even in molecules that have little or no symmetry, intensity differences in the observed infrared and Raman spectra often give valuable information; a good example of this is the polymerization of methyl methacrylate to poly(methyl methacrylate), where the consumption of monomer may be followed by Raman or infrared spectroscopy—in the former case using the ν(C=C) vibrational band, and in the latter the ν(C=O) band. In Raman spectroscopy, the ν(C=C) bond is much stronger than the ν(C=O), but the reverse is true in infrared spectroscopy since the C=O bond contains a dipole, whereas the C=C bond does not.

For a more specialized mathematical treatment of Raman scattering using natural, linearly, or circularly polarized radiation, the reader is referred to standard texts (Long, 1977; Ferraro and Nakamoto, 1994; Grasselli and Bulkin, 1991).

As indicated above, the absorption or emission of radiation by molecules occurs in precisely defined regions of the electromagnetic spectrum, such as the ultraviolet, visible, infrared, or microwave, and this has given rise to the molecular-spectroscopic techniques that are associated with these regions. Since the Raman-spectroscopic technique arises from a light-scattering phenomenon, Raman spectra may be observed in all of these spectral regions, i.e., the technique is not frequency or wavelength specific. Hence, improvements in spectroscopic instrumentation in recent years have had great impact in the area of Raman spectroscopy, where the excitation and detection of spectra may now be conventionally undertaken in the ultraviolet and infrared regions of the electromagnetic spectrum as well as in the visible region normally associated with this technique (Long, 1977, 1989; Loader, 1970).

The advent of continuous-wave and pulsed lasers operating in several wavelength regions of the electromagnetic spectrum has provided chemical physicists with new probes for the excitation and emission of molecular spectra. When these are coupled with interferometric instrumentation and detectors such as photomultipliers and charge-coupled diode arrays, new areas of molecular spectroscopy are now made available that hitherto were not accessible. For example, narrow spectral linewidths of visible and infrared lasers and extremely short pulse widths, e.g., picoseconds, femtoseconds, or smaller, have resulted in applications to ultrafast spectroscopy and real-time studies of chemical-bond breakage and reaction kinetics. The tunability of laser wavelength excitation has given rise to resonance-Raman–spectroscopic studies (Behringer, 1967) of molecular systems in the visible and ultraviolet regions and has particularly exciting applications in the fields of biological physics and chemistry. High-powered, giant-pulse lasers are used in the exploration of higher-order Raman scattering, such as coherent anti-Stokes Raman (CARS) and hyper-Raman (HRS) spectroscopies (Kiefer and Long, 1982), while the sonic effects of laser passage through a medium can be analyzed by photoacoustic Raman spectroscopy (PARS).

The presentation of samples in particular orientations is of critical application in surface-enhanced Raman-scattering studies (SERS) and for studies of liquid-crystalline and stretched-polymer behavior at the molecular level.

The recent achievements of Raman microscopy and laser microprobe spectroscopy for the molecular characterization of particles of material as small as 1 μm or for the mapping of heterogeneous surfaces at a microscopic level would not have been possible without the coupling of the microscope, perhaps the oldest scientific optical instrument still in use, to the laser, one of the newest scientific optical discoveries. Molecular microscopy has made an impressive impact in almost all scientific areas to which it has been applied, including geology, polymer science, biomolecular chemistry, and thin-film electronic devices (Siesler and Holland-Moritz, 1980). Recent application of Raman microscopy to the identification and characterization of molecules in single cells is witness to the rapid application of this technique.

This article provides an overview of the instrumentation available to the practicing Raman spectroscopist interested in the structure of molecules and the reasons why, for example, high-resolution gas-phase rotational spectroscopy of chlorine and studies of the microstructural composition of polybutadienes require different instrumental Raman techniques although both are problems in molecular spectroscopy.

1. INSTRUMENTATION FOR THE GENERATION, ILLUMINATION, DISPERSION, AND DETECTION OF ELECTROMAGNETIC RADIATION OF RELEVANCE TO RAMAN SPECTROSCOPY

The basic instrumentation for conventional molecular spectroscopy in the ultraviolet, visible, and infrared regions of the electromagnetic spectrum consists of a radiation source, a sample illuminator, a spectrometer dispersing device, a detector, and peripherals for the acquisition of data under controlled experimental conditions to include multiple scanning and facilities for the subtraction of spectra.

1.1 Electromagnetic-Radiation Sources

Raman spectra in the ultraviolet, visible, and infrared regions of the electromagnetic spectrum arise from electronic, vibrational, and rotational transitions, involved in scattering of incident radiation. As indicated earlier, Raman spectra are significantly different from infrared, microwave, or ultraviolet absorption spectra in that they arise from molecular-scattering phenomena, although the information contained therein is often considered to be complementary to infrared- or electronic-absorption spectra. Hence Raman spectra, scattered from an ideal, monochromatic radiation source, require different experimental arrangements from those of absorption spectra; thus, the impact of high-powered, monochromatic, coherent, laser sources has been observed most strongly in Raman spectroscopy where continuum sources are not usually necessary, unless one wishes to observe special Raman effects such as inverse-Raman or Raman-gain spectroscopies.

1.1.1 Lasers Laser action arises from stimulated emission processes (Andrews, 1990b) in which molecules in excited energy states with short decay lifetimes ($\sim 10^{-7}$ to 10^{-9} s) release energy and undergo relaxation to more stable, lower-energy states. Significant properties of lasers that give rise to their application in the field of molecular Raman spectroscopy are summarized as follows:

1. Collimated beams with divergences of the order of 200 μ radians.
2. High intensity, which can be increased further by focusing to give an irradiance of 10^{13} W m^{-2} for a 10-W argon-ion laser at 488 nm with an original cross-sectional area of 1 mm^2. This value should be compared with the mean irradiance of sunlight at the Earth's surface of 10^3 W m^{-2}.
3. Short pulse times for *Q*-switched lasers, of the order of 25 ns at 694.3 nm, giving a peak output of 1 GW in each pulse and an irradiance of 10^{12} W m^{-2}. Picosecond and femtosecond pulsed lasers are now being used for molecular Raman-spectroscopic studies of transient species and fast reactions of importance to biological processes.
4. Coherence: the emitted-laser photons are in phase with each other, unlike other light sources.
5. Monochromatic radiation. This term is something of a misnomer, an approximation since a typical plot of laser intensity vs wavelength output looks rather like Fig. 1, in which a number of closely separated wavelengths are actually involved in the laser action. With an étalon in the laser beam, mode selection occurs and the laser spectral linewidth is considerably reduced to typically 0.001 cm^{-1} or so; this linewidth is often necessary for very high-resolution studies of molecular gas-phase spectra.
6. Polarization characteristics, which can provide useful information about vibrational-mode assignment and molecular orientation.

For spectroscopic purposes, lasers can be broadly classified as fixed-wavelength and tunable, and there are lasers that cover the

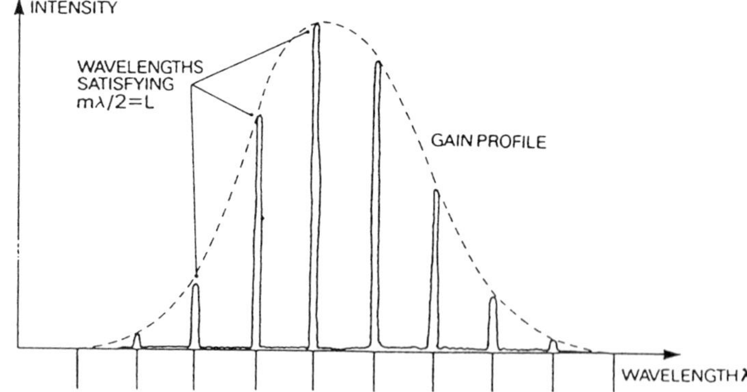

FIG. 1. Laser-emission intensity vs wavelength output plot; emission wavelengths given by the standing-wave condition $m\lambda/2 = l$, where m is an integer and l is the cavity length. Each division on the abscissa represents typically 0.001 nm.

electromagnetic spectrum from the far infrared (100 μm; 10^5 nm) to the vacuum ultraviolet (0.1 μm; 157 nm). Their mode of operation may be pulsed or continuous-wave (cw). The wavelength of operation, the tunability or otherwise, and the power and bandwidth of the lasers available as well as the type (pulsed or cw) have to be considered when Raman-spectroscopic applications are intended (Table 1). The most common sources of excitation in Raman spectroscopy are single-frequency lasers, which are commonly of the gas-ion laser (e.g., argon-ion, krypton, and helium–neon) or solid-state laser [e.g., Nd^{3+}/YAG (yttrium aluminum garnet)] type.

Tunable lasers are of special application particularly in vibrational Raman spectroscopy and may be classified as follows.

1. *Dye lasers,* which use the intense absorption and fluorescence of organic dyes, e.g., rhodamine 6G, to produce a population inversion between ground and excited electronic states. Although the spontaneous linewidth of such a laser is usually broad (~4 Å) it is reduced to 10^{-5} cm^{-1} by incorporating a Fabry–Pérot étalon in the laser cavity. The tuning range for any dye is ~600 Å, and use of suitable dyes gives a 2600–8000-Å overall tuning range, i.e., extending from the ultraviolet through the visible into the near infrared.
2. *Spin-flip Raman lasers (SFRLs),* which operate via stimulated Raman scattering from the electronic energy levels of an *n*-type semiconductor. Recent advances in this category will be described later.
3. *Semiconductor diode lasers (SDL),* in which stimulated emission of radiation arises from the recombination of electron–hole pairs in the valence band of a semiconductor. Electron injection at a *p-n* junction is used to achieve laser action. Typically, InAs, GaSb, PbS, PbSnTe, and PbSnSe are used as semiconductor materials for observation of SDL action. Maximum cw laser output is only 1 mW and typical powers are 10 μW. Linewidths are very narrow, e.g., 2×10^{-6} cm^{-1} for $Pb_{0.88}Sn_{0.12}$ operating at 10.6 μm with an output of 0.2 mW. Other types of semiconductor laser can readily achieve higher power levels, e.g., GaAlAs provides 100 mW cw at 810 nm.
4. *Nonlinear devices,* which depend on the interaction of laser radiation in nonlinear crystals; e.g., a $LiNbO_3$ parametric oscillator and a neodymium–YAG laser have been used with an $AgGaSe_2$ crystal to produce tunable radiation in the 7–12-μm (1400–850 cm^{-1}) region with a linewidth of 2 cm^{-1}.

Laser systems for spectroscopy are based on both fixed-wavelength and tunable-laser sources of radiation. However, fixed-wavelength sources that emit at only one or two discrete wavelengths are not as useful as the tunable lasers, which offer the facility of scanning across a continuous range of wavelengths. Some of the difficulties experienced in using these techniques in the visible and ultraviolet regions have been reviewed (Andrews, 1990a, 1990b), and the chemical applications of tunable lasers in absorption and scattering spectroscopies have been surveyed (Burdett and Poliakoff, 1974).

Table 1. Characteristics of some typical lasers used in Raman spectroscopy as radiation sources.

Region of spectrum[a]	Name of laser	Type	λ (nm)	Maximum power or energy	Pulse duration
V	Ruby (*Q*-switched)	Pulsed	694	20 J	20 ns
IR	Nd^{3+}/YAG	Pulsed	1 064	100 J	10 ns
UV	N_2	Pulsed	337	1 mJ	10 ns
UV	XeF (excimer)	Pulsed	351	80 mJ	10 ns
IR	CO_2	Pulsed/cw	10 600	2 kJ/50 W	5 μs
V	Ar^+	cw	500	30 W	...
V	Kr^+	cw	650	2 W	...
V	He/Ne	cw	633	0.3 W	...
Tunable IR-UV	Dye	Pulsed/cw	380–900	3 J/1 W	2 μs

[a]V, visible; UV, ultraviolet; IR, infrared.

Spin-flip Raman (SFR) scattering (Patel, 1974) arises from the inelastic scattering of an input photon by electrons in the presence of a magnetic field. The scattering process is described by the following expression:

$$\omega_s = \omega_0 \pm g\mu B, \qquad (1)$$

where ω_0 is the incident frequency, ω_s is the scattered frequency, g is the g value for electrons, μ is the Bohr magneton, and B is the magnetic-field strength. The negative sign corresponds to a Stokes spin-flip Raman process, and the positive sign to an anti-Stokes spin-flip Raman process. Both Stokes and anti-Stokes processes are tunable by changing the magnetic-field strength B.

A review of spin-flip Raman lasers and their application has been given (Patel, 1974). Generally, an InSb crystal pumped by a powerful laser is used as a "source" of electrons for the spin-flip Raman processes described above. A CO_2 laser (10.6 μm) or a CO laser (5.3 μm) is commonly used as a pump for the InSb semiconductor. With a 10^3-W pulsed CO_2 laser as a pump, the spin-flip Raman laser can be tuned over the 9.0–14.6-μm infrared region; with the CO laser (20 W pulsed or 0.5 W cw) the infrared region 5.2–6.2 μm is covered.

Large magnetic-field strengths are required for the operation of SFR lasers, e.g., 50–100 kG. Precise control of the magnetic-field strength is required for tuning purposes, since the wave number of the SFRL line changes by 0.002 cm^{-1} G^{-1}. The lowest magnetic field possible for the operation of an SFRL has been calculated to be ~10 kG at 20 K, which required a "carrier concentration" of 10^{15}–10^{16} electrons cm^{-3}.

Of great interest for rotational gas-phase molecular studies is the narrow linewidth of the SFRL. In the recording of the absorption spectrum of water vapor at a pressure of 30 Torr in the region of 1885 cm^{-1}, a spectral linewidth of 0.006 cm^{-1} has been observed, indicating that the SFRL linewidth must be <0.006 cm^{-1}.

The vibration–rotation spectra of NH_3 and of H_2O obtained using an InSb SFRL show great improvement in resolution over those spectra that were recorded using conventional infrared-absorption spectroscopy techniques. Further developments in SFRL technology involve the construction of an HF SFRL, which is tunable over the 14–20-μm region, and low-field operation of an InSb SFRL with magnetic-field strengths of 200 G.

1.2 Sample Illuminator and Cell Design

The presentation of the sample to be studied to the spectrometer and its orientation or otherwise with respect to the radiation source are of paramount importance in Raman spectroscopy (Long, 1977). A typical arrangement of a sample illuminator for Raman-spectroscopic purposes is shown diagrammatically in Fig. 2. The laser beam, with an electric vector $\mathbf{E}_x$, first passes through an interference filter of narrow spectral bandwidth to remove associated plasma wavelengths, a quartz half-wave plate that rotates the plane of polarization through 90°, i.e., to $\mathbf{E}_y$, an iris diaphragm to limit the physical beamwidth, and a focusing lens to increase the power density (W cm^{-2}) at the sample. The purpose of rotation of the plane of polarization of the incident electric vector is the determination of the depolarization ratio, ρ (see below), from consecutive measurements of the scattered-radiation intensity at $\mathbf{E}_x$ and $\mathbf{E}_y$ polarizations of the incident laser. The lower mirror serves to direct the laser beam along the z axis of the sample illuminator; the top mirror, set at focal distance from the focus of the incident laser beam in the sample, increases the power density at the sample focus.

The sample, which may be a liquid, solid, or gas, is placed at the focus of the laser beam, the optic axis of which (z) lies in the plane of observation of the scattered radiation (yx) containing the optic axis (x) of the spectrometer passing through the center of the entrance slit, collimator, grating, camera, and exit slit. Hence, scattered radiation from the sample focus passes through into the spectrometer, via a collection lens of high aperture (usually f/1.4 or better), which brings the focus to a magnified image on the entry slit of the spectrometer; the magnification is such that the angle subtended at the collection lens and at the collimator is sufficient only to fill the latter for the purpose of grating illumination. A quartz-wedge scrambler is normally sited before the entrance slit to randomize the polarization of the scattered radiation so that quantitative measurements are not affected by reflectivity changes

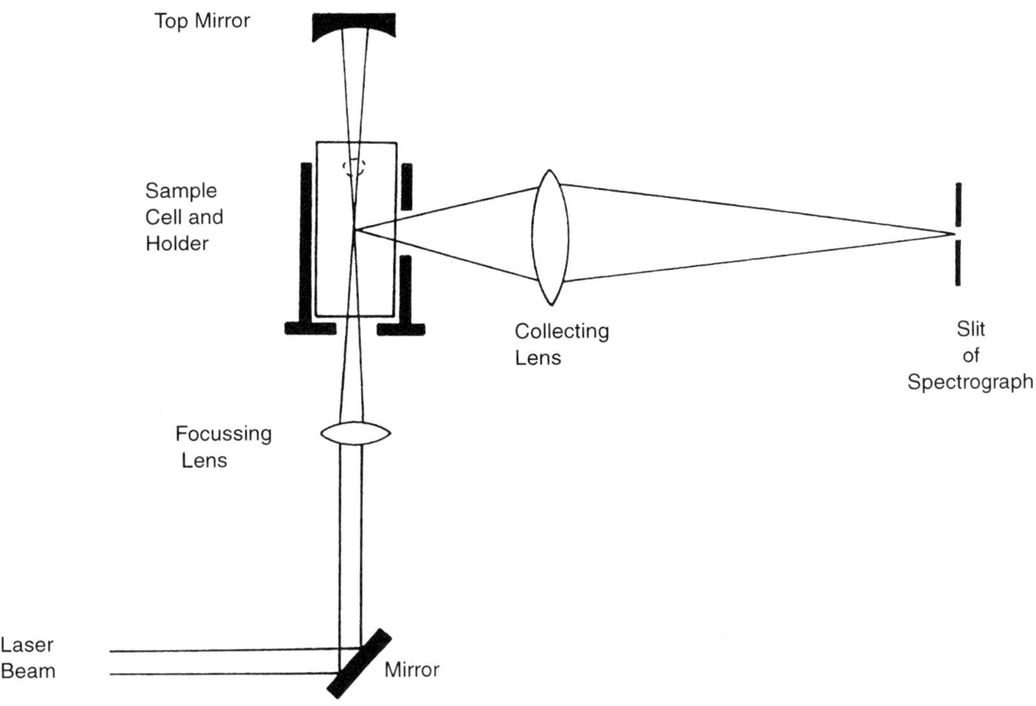

FIG. 2. Sample illuminator for conventional Raman spectroscopy.

at the grating surface. Also, an analyzer may be placed between the collecting lens and the scrambler to permit radiation of only $\|$ or $\perp$ polarization components to pass onto the slit. Here, $\|$ and $\perp$ refer to the polarization vectors and their orientation with respect to the plane of observation containing the spectrometer axis (x); so $\|$ is xz and $\perp$ is yz with the axes shown in Fig. 3. The example shown refers to observation of pure rotational and vibrational–rotational spectra from a gas sample; the pure rotational Raman lines (R and S family) are termed "depolarized," whereas the Q branch of the vibrational–rotational transition is "polarized." The depolarization ratios ρ of Raman lines or bands may be evaluated from band intensities I (Long, 1977), where $\rho = {}^{\perp}I_{\|}/{}^{\perp}I_{\perp}$ in the usual notation. From these polarization measurements, information may be provided about the symmetries of the modes of vibration of molecules, which can aid assignments in molecular-structure determinations.

In the gas-phase example cited here, the pure rotational transitions involve small energy changes within the $v = 0$ vibrational level (vibrationless ground state) in which the rotational quantum-number change is $\Delta J = +1$ or $+2$ for the R and S families of lines. For the vibration–rotation spectrum, $\Delta J = 0$ is a possibility, which gives rise to a Q-branch family of lines for the $v = 1 \leftarrow v = 0$ vibrational transition. For further explanation of the symbols and terminology of rotational and vibrational–rotational spectroscopy, the reader is referred to standard specialist texts (Long, 1977).

Orientation effects are particularly important for the study of samples such as single crystals (Garetz and Lombardi, 1986) and oriented polymer films (Bower and Maddams, 1989). A major problem in Raman-scattering experiments is the low signal-to-noise ratios experienced for the recording of spectra from the gas phase, and much ingenuity has gone into cell design to improve the situation (Weber, 1972, 1979). Using such cells the rotation and vibration–rotation spectra of thermally produced excited-state molecules in electric discharges have been obtained successfully. A pivotal development has been the construction of sample cells designed for intracavity use in lasers. These can normally only be used with transparent-

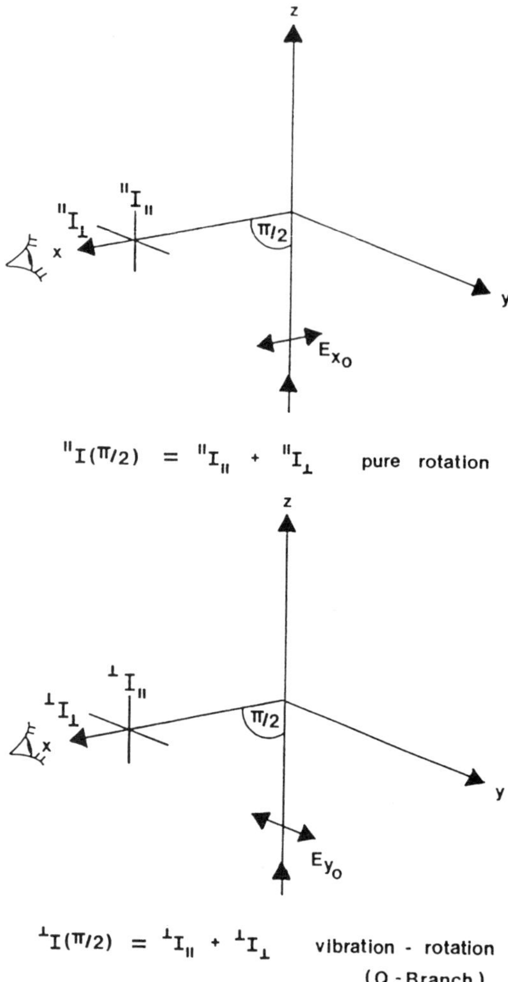

FIG. 3. Observed intensities and incident electric vectors for illumination of gas sample: pure rotational and vibrational–rotational spectra.

gas samples, since even small absorption by the cell windows or sample will be sufficient to negate lasing action. In one such cell (Fig. 4) of volume 10 mL, which forms an integral part of the laser cavity, one end of the cell is a planoconvex lens that focuses the laser beam to a diameter of about 50 μm, the other end being a high-reflectivity spherical laser mirror set at its radius of curvature (10 cm) from the focus. A pure rotational Raman spectrum of oxygen obtained using this device is shown in Fig. 5; the clean background should be noted, and a spin-flip transition of $^{16}O_2$ at only 1.9-cm^{-1} shift from the Rayleigh line can be clearly seen. Because of the high power densities at the laser focus in intracavity illumination, samples must necessarily be stable. For a sample of HCN gas irradiated with 488.0-nm argon-ion laser radiation in the intracavity cell described above, sample decomposition occurred and rotational Raman lines arising from the laser-induced synthesis of the trimer $(HCN)_3$ were assigned (Fig. 6).

There have been several designs for extracavity Raman sample cells for gas-phase studies in which the power density at the laser focus is increased by multipassing. Gains in effective illumination of up to 120 times that achieved with the normal laser intensity have been claimed. An example is shown in Fig. 7 and involves an ellipsoidal mirror and a plane mirror positioned between the two foci F_1 and F_2 of the ellipse; in such a "light-trapping illuminator" (LTI) the off-axis laser beam progressively approaches the illuminator axis with each reflection (Fig. 8), an optical property of this cell being that the laser beam is reflected repeatedly through the focus F_1, where the sample is placed. Because of the horizontal arrangement of the LTI an image rotator is required to focus the image in the vertical plane of the slit (Fig. 9); in such an arrangement an image rotation of 2θ is produced by an inclination of θ of the rotator to the vertical. An example of a vibration–rotation Raman spectrum of nitrogen in the interconal region (Fig. 10) of a hydrocarbon–air flame at 2000 K recorded using such a device is shown in Fig. 11. From such a spectrum, the temperature of the nitrogen gas in that part of the flame can be calculated; an advantage of this technique over that of gas pyrometry or thermocouple measurements is that the flame system is not physically disturbed by the laser beam. In this way, punctual measurements of temperature in flames of different fuel–air ratios (lean to rich flames) have been made using Raman spectroscopy.

The application of Raman-spectroscopic techniques to the remote analysis of air pollutants at the ppm level and to hot gases in flames over the temperature region 300–3500 K has received some attention in recent years. Excited vibrational and rotational molecular states can be studied from their increased population at higher temperatures. An interferometric method for the analysis of overlapping rotational Raman spectra

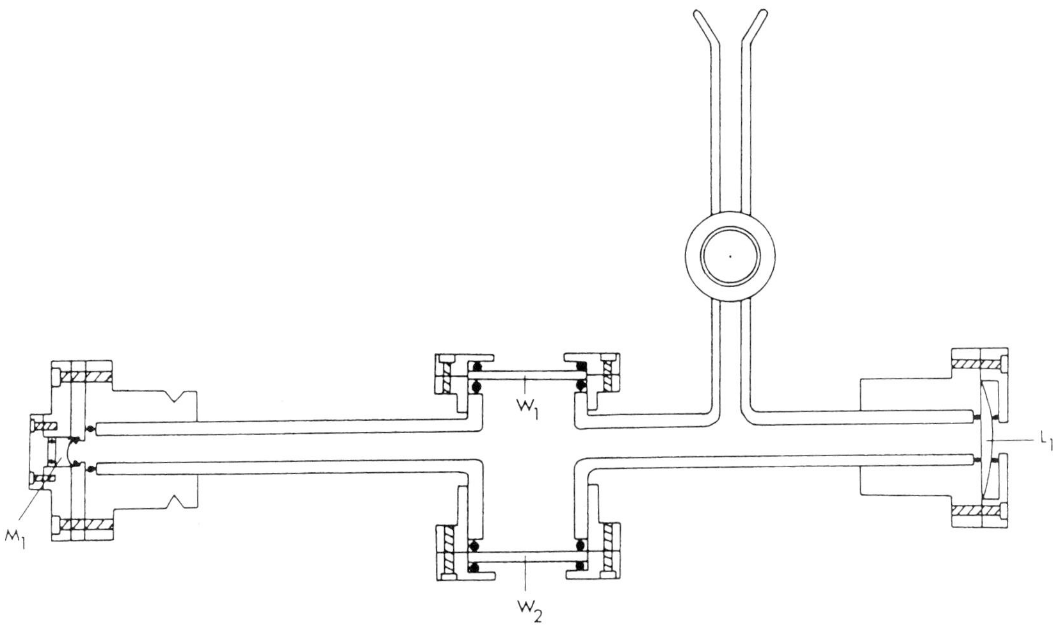

FIG. 4. Sample cell for the intracavity laser illumination of a gas sample; L is a planoconvex lens, M_1 is a spherical high-reflecting mirror, and W_1 and W_2 are antireflection-coated windows.

from atmospheric pollutants such as NO and SO_2 at a range of 300 m has been applied and a $10^4\times$ gain in intensity over that of a conventional spectrometer has been claimed, comprising a $10^2\times$ increase of rotational cross sections compared with the vibrational Q branches and a $10^2\times$ increase in light-collecting power of the Fabry–Pérot interferometer.

The coupling of a pressure-scanned Fabry-Pérot interferometer with a sample cell and a small-linewidth helium–neon laser has facilitated the recording of gas-phase Raman spectra of hydrogen with an effective resolution of 0.02 cm^{-1} and a free spectral range of 0.5 cm^{-1}. By use of forward-scattering conditions, the full width at half maximum intensity (FWHM) is reduced from 0.15 cm^{-1} at 90° scattering to a Doppler-limited linewidth of 0.04 cm^{-1}.

1.3 Spectrometers and the Dispersion of Radiation

A spectrograph is an instrument for producing a spectrogram, a photographic spectral image, whereas a spectrometer employs a photoelectric device for the quantitative measurement of dispersed radiation (Sawyer, 1963; Dodd, 1962). Generally, a spectrograph is a static system and a spectrometer is used in a scanning mode; however, scanning spectrographs may be used to drive the dispersing systems into a particular wavelength region, where they then remain statically controlled, and spectrometers may also be used in static arrangements when employed with photomultiplier (PMT) tube or charge-coupled diode (CCD) detector devices.

1.3.1 Dispersion An important characteristic of any spectrograph is its *dispersion*, or power to spread out different wavelengths so that they emerge from the dispersing system at different angles and are focused in different positions in the focal plane of the spectrograph. The difference in angles for different wavelengths is measured by the *angular dispersion* of the instrument, defined as $\Delta\theta/\Delta\lambda$, where $\Delta\theta$ is the difference in emergent angles corresponding to a difference in wavelength $\Delta\lambda$.

1.3.2 Resolving Power The *resolving power* is expressed as the ratio of the wavelength observed to the smallest difference between two wavelengths that can just be re-

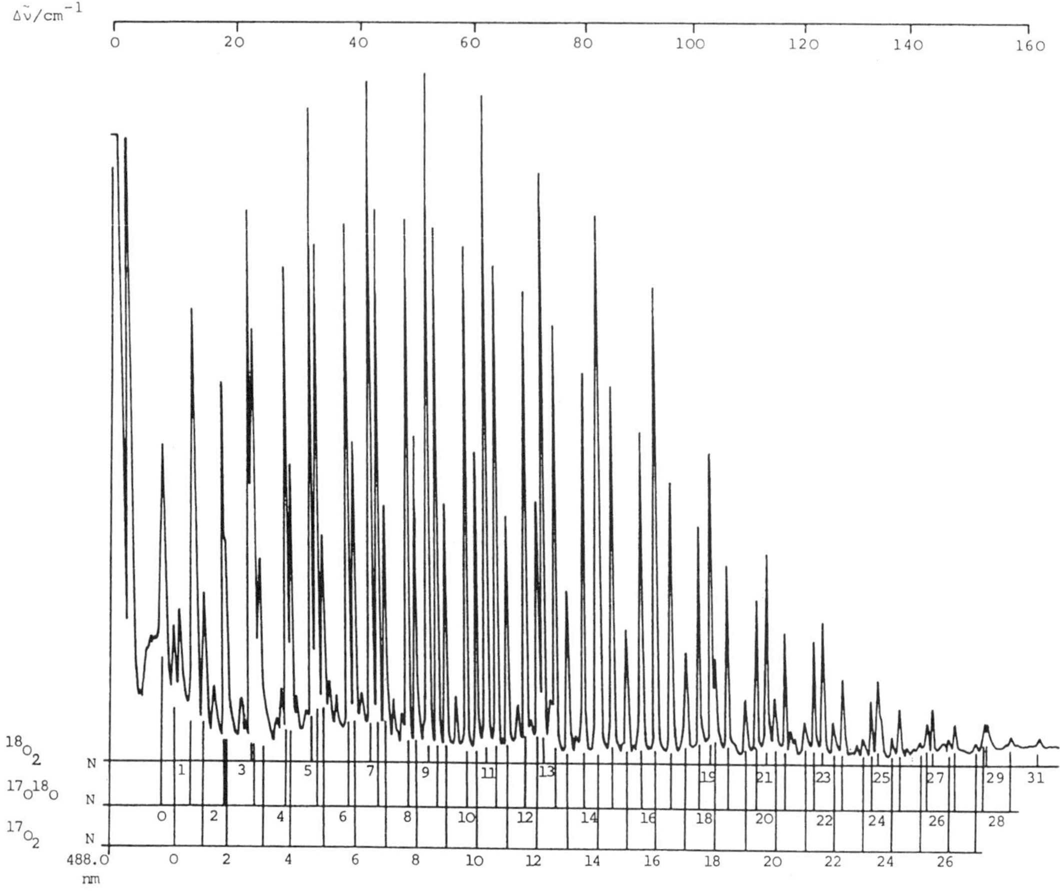

FIG. 5. Pure rotational Raman spectrum of $^{18}O_2$ enriched in ^{17}O recorded photographically using intracavity laser illumination and the cell shown in Fig. 4; argon-laser excitation, $\lambda = 488.0$ nm, nominal extracavity laser power 2 W, spectral slit width 0.14 cm^{-1}, reciprocal linear dispersion 1.4 cm^{-1} mm^{-1} in the twelfth order of diffraction at 488.0 nm.

solved or distinguished as two separate lines. The ratio $\lambda/\Delta\lambda$ is, then, a pure number without dimensions, and may vary widely for different instruments, from a few hundred for a small prism instrument to a million or more for the most powerful interference spectrographs or interferometers. For any spectrograph, the actual resolving power may be computed directly from observations made with the instrument; the theoretical resolving power may be calculated on the basis of suitable assumptions.

The resolving power may now be written in terms of angular dispersion to give a theoretical best resolving power

$$\frac{\lambda}{\Delta\lambda} = \frac{\lambda}{\Delta\theta}\frac{d\theta}{d\lambda}. \tag{2}$$

This theoretical resolving power is often attained in the best instruments. Actual resolution, however, is affected by numerous factors, such as the relative intensities of the two adjacent spectral lines, their form or structure, the width of the slit, the mode of illumination of the slit, the perfection of the optics and their adjustment, and the detector.

1.3.3 Brightness of Image The intensity of illumination of the spectra produced by a spectrograph depends upon such characteristics of the instrument as the focal length, the apertures of the lenses, and the reflection and transmission losses within the instrument. The illumination depends also upon the way in which the slit is illuminated by the external light source used.

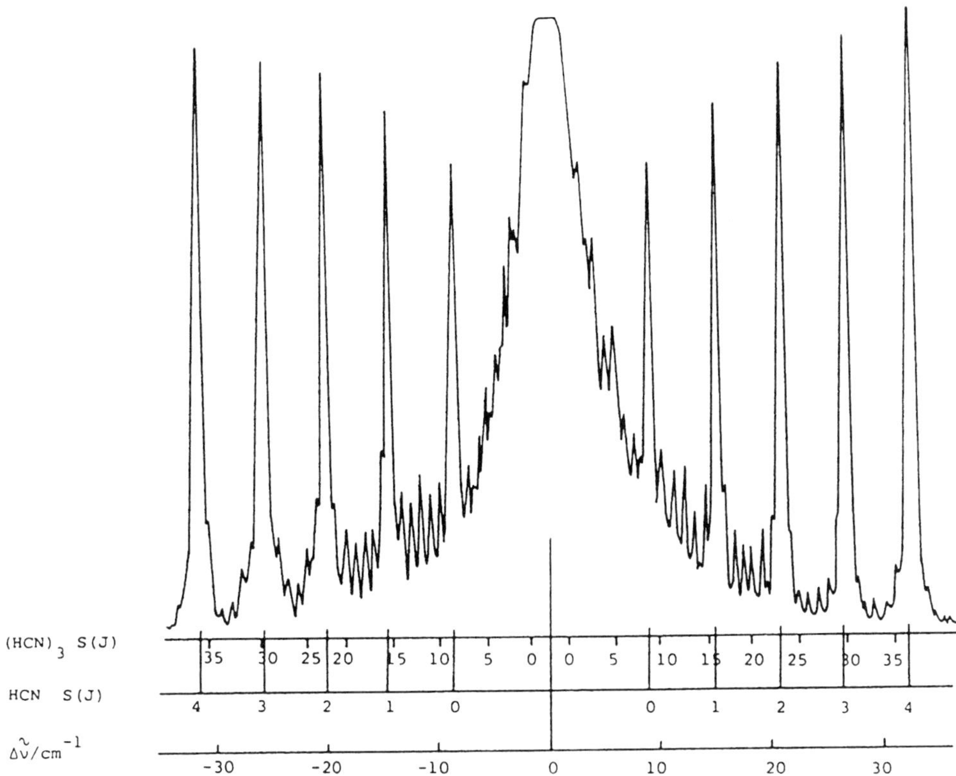

FIG. 6. Pure rotational Raman spectrum of $(HCN)_3$ trimer arising from the laser-induced photochemical decomposition of hydrogen–cyanide gas; the closely spaced rotational lines for the decomposition product are clearly seen between the stronger rotational lines for the monomer (J = 0 to J = 4).

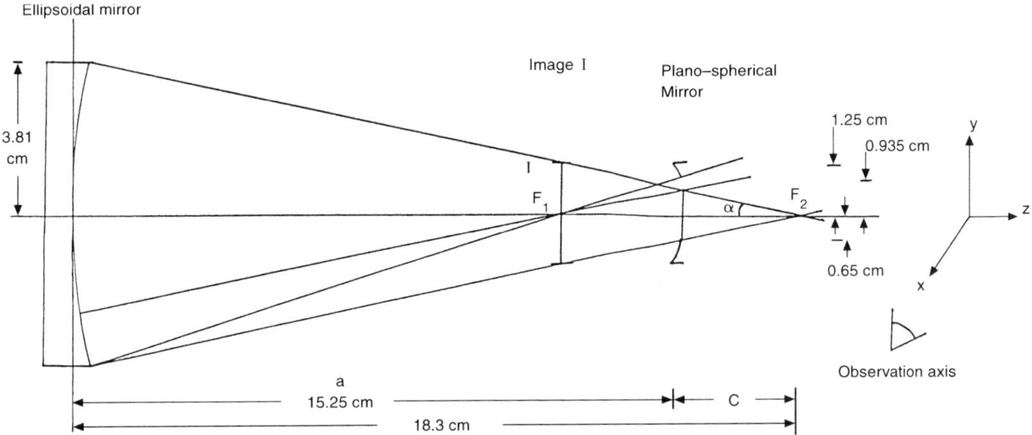

FIG. 7. Light-trapping illuminator (LTI) design.

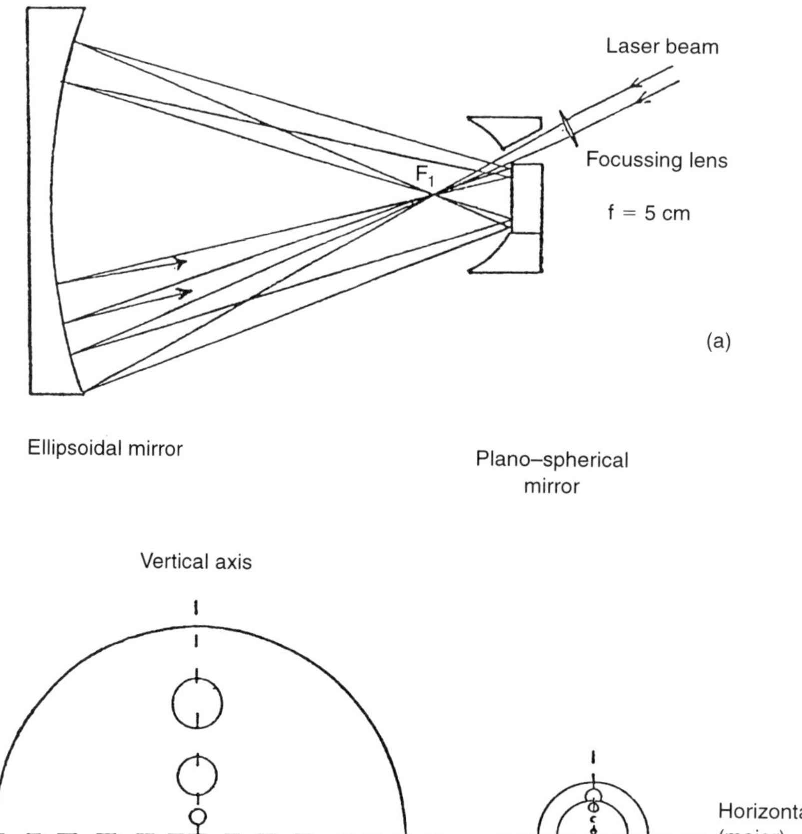

FIG. 8. (a) Ray diagram of LTI; off-axis laser beam progressively approaches the illuminator axis. (b) Observed beam spots.

1.3.4 Speed of Spectrograph/Spectrometer The brightness of a spectral-line image cannot be increased by altering the arrangement of the optical components used to couple with the spectrograph or spectrometer. However, the speed of the spectrograph or spectrometer (not to be confused with scanning speed!) depends on the product of the brightness B and the solid angle of illumination ω, $B\omega$. The total power per unit area in the spectral-line image, or illumination, can obviously be increased by increasing ω, the angular aperture of the radiant energy, as is done to obtain more rapid camera lenses. Spectrographs are often made symmetrical, that is, with the collimator lens/mirror and the observing or camera lens/mirror of the same diameter and focal length, a construction that results in a spectral line of the same size as the slit. High speed in a symmetrical spectrograph obviously requires both collimator and camera lenses or mirrors of large angular aperture, and as the cost of lenses and mirrors in-

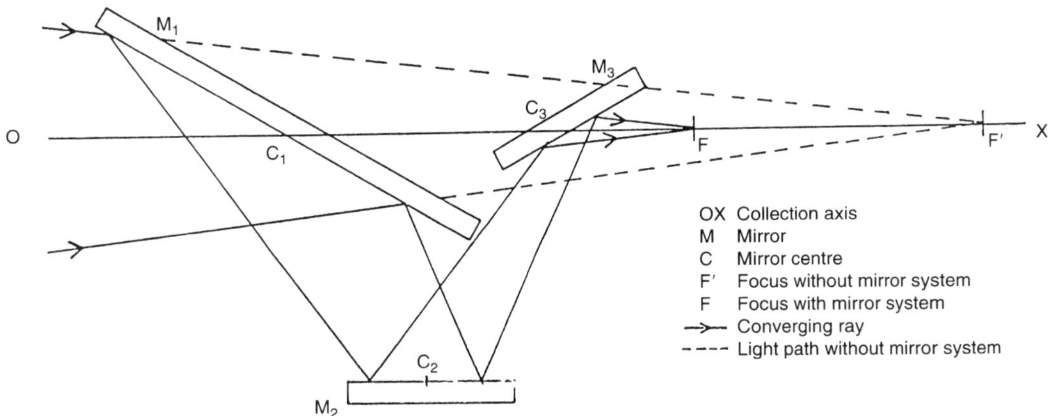

FIG. 9. Optical rotator of three-mirror type in vertical plane of 45° for the production of a vertical, focused slit image from a horizontally illuminated cylinder.

creases rapidly with the angular aperture, the cost of such high-speed, high-dispersion instruments becomes prohibitive.

1.3.5 Slit Fundamentally, the slit is a narrow, rectangular aperture that admits light to the spectrograph. The images of the slit formed by the spectrograph after the light has been dispersed are the spectral lines. It is essential, then, that the edges of the slit be straight and parallel, in order to give a clear image, and that they be sharp to avoid reflections from their edges into the spectrograph. Most spectrographs and spectrometers have adjustable slits, since the proper slit width for instruments of high dispersion will depend on the wavelengths used. The adjustable slit is formed by two jaws mounted in parallel ways and so designed as to insure parallelism of motion.

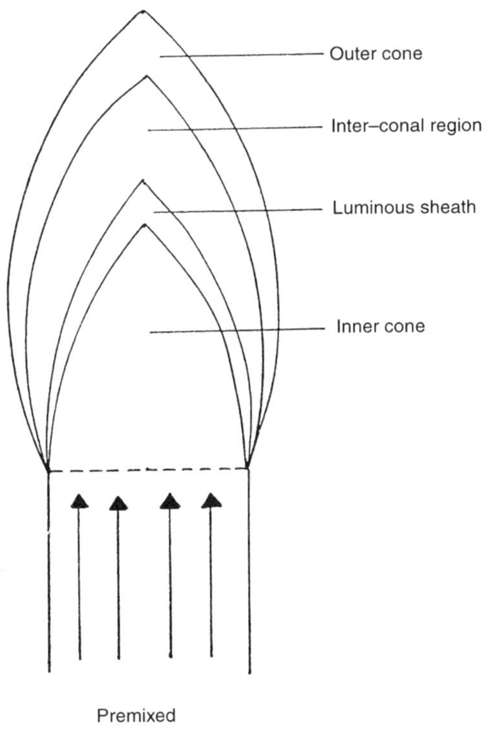

FIG. 10. Schematic diagram of a premixed laminar hydrocarbon–air flame, showing the interconal region.

It is important to recognize the difference between *physical* and *spectral* slit widths. The physical slit width as set by an operator is related to the spectral slit width through the *reciprocal linear dispersion;* hence, for a spectrometer working in the first order of diffraction with a reciprocal linear dispersion of 20 cm^{-1} mm^{-1} at 488 nm at the exit slit, a physical slit width of 100 μm corresponds to a spectral slit width (or spectral resolution) of 2 cm^{-1} at this wavelength. This is an important spectral concept and means that the spectral resolution changes through a spectrum unless computer-programmed slit-width changes are available to provide a constant spectral slit width over the wavelength range studied.

1.3.6 Collimator The function of the collimator lens or mirror is to render parallel the radiation from each point of the slit so

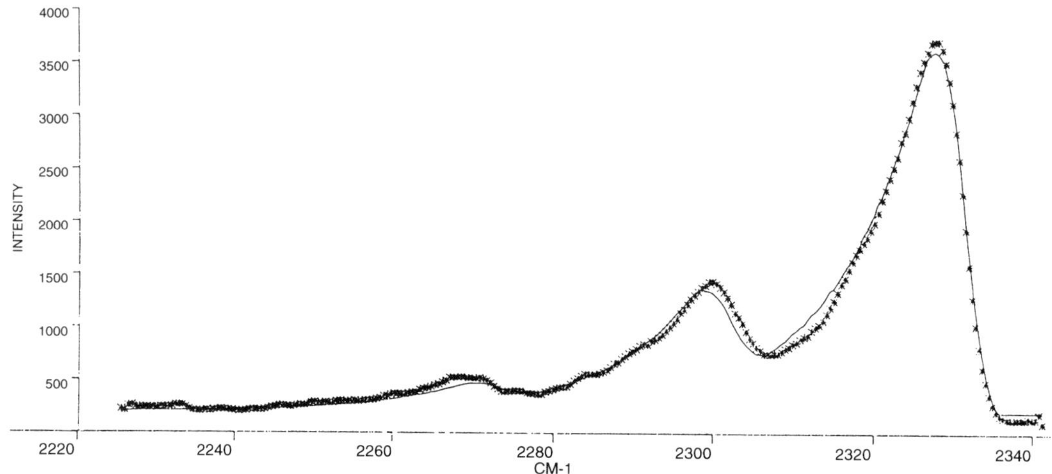

FIG. 11. Vibrational–rotational Raman spectrum of the Q branches for the $v = 1 \leftarrow v = 0$, $v = 2 \leftarrow v = 1$, and $v = 3 \leftarrow v = 2$ transitions of nitrogen in a methane–air flame; laminar flow, fuel–air ratio $\phi = 0.5$, interconal region 5 cm above burner orifice. Fitted temperature $T = 1615 \pm 10$ K. Unresolved rotational structure also present.

that the radiation of different wavelengths incident on the dispersing element (grating) all appears to emerge from an infinitely distant light source. Defects such as chromatic and spherical aberration and coma should all be considered in the design of optical instrumentation using lenses. Mirrors have distinct advantages over lenses in their reflectivity over a wide range of wavelengths and often can be coated to achieve high reflectivities in the visible or ultraviolet region, for example. Also, use of paraboloidal mirrors removes problems associated with chromatic and spherical aberration in lenses (Born and Wolf, 1980).

1.3.7 Diffraction Grating This is the most commonly used dispersing element of a spectrograph or spectrometer and is used from the far-ultraviolet spectral region to the far infrared, a wavelength range of 100–40 000 nm. The theoretical treatment of the diffraction grating has been given by several authors (e.g., Sawyer, 1963). The diffraction grating consists of a ruled surface, usually *blazed* at a particular angle, which receives parallel radiation from the collimator and produces a series of dispersed, parallel images that obey the equation

$$n\lambda = d(\sin\alpha + \sin\beta), \tag{3}$$

where n is the order of diffraction, λ is the wavelength of illumination of the grating surface, d is the groove (ruled) spacing, and α and β are the angles made by the incoming and emergent rays, respectively, to the grating normal (Fig. 12). The dispersion of a grating is given by

$$\frac{d\beta}{d\lambda} = \frac{n}{d\cos\beta}. \tag{4}$$

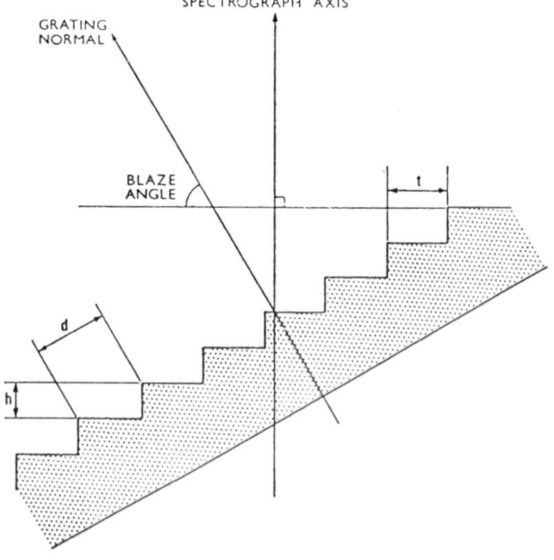

FIG. 12. The échelle grating; groove spacing d.

The dispersion is a minimum for $\beta = 0$, i.e., for the spectrum observed normal to the grating. A characteristic of gratings is the overlapping of spectra of different orders and the *free spectral range* decreases as n increases; for fixed values of α and β, if a wavelength λ is observed at $n = 1$, there will also appear at the same β, wavelengths of higher-order spectra such that $\lambda_1 = 2\lambda_2 = \cdots n\lambda_n$, i.e., wavelengths of 900 nm in the first order, 450 nm in the second order, and 300 nm in the third order are all observed at the same angle, β. Although in a normal diffraction pattern the diffracted intensity decreases with increase in n, selective increase in intensity for particular orders is accomplished using blazed gratings.

The free spectral range is of critical importance for échelle gratings, which can operate in high orders of diffraction with correspondingly small free spectral ranges. For incident radiation of wavelength λ, a grating groove spacing of d, and angles of incidence and reflection of α and β, respectively, with respect to the grating normal, $n\lambda = 2t$, where t is the step size of the grating (Fig. 12). The free spectral range is then given by

$$F_\lambda = \lambda^2/2t. \tag{5}$$

For the Czerny–Turner monochromator described below, in the $n = 12$ order of diffraction at $\lambda = 488.0$ nm, $F_\lambda = 40.7$ nm ($\equiv$ 1709 cm^{-1}).

The theoretical resolving power R_{theor} of a grating depends only on the product of the order and the number of ruled lines and not on the wavelength of illumination or grating spacing. Hence, it can be advantageous to use diffraction gratings in high orders to achieve good resolving powers—bearing in mind the free spectral range considerations outlined above. A 25.4-cm Bausch and Lomb ruled grating with 316 lines mm^{-1} operating in the twelfth order of diffraction will therefore have $R_{theor} = 10^6$. At 500 nm, this grating should resolve spectral wavelengths that are only 0.0005 nm or 0.02 cm^{-1} apart.

A useful concept for diffraction-grating instruments is the *reciprocal linear dispersion*, or rld,

$$\text{rld} = d\cos\beta/nf, \tag{6}$$

where f is the focal length of the camera mirror, which brings the dispersed wavelengths to a focused image in the plane of the exit slit. The rld is given as units of nm mm^{-1} or cm^{-1} mm^{-1}, and is a measure of the resolving capability of the spectrometer or spectrograph. The smaller the rld the higher the resolution that can be achieved by the instrument. For example, in the Czerny–Turner instrument at the University of Bradford, using the diffraction grating described above with $R_{theor} = 10^6$, the 3-m focal length gives an rld of 1.4 cm^{-1} mm^{-1} (0.002 nm mm^{-1}) at 500 nm. However, this instrument, although eminently suitable for rotational Raman-spectroscopic studies of gases with spectral lines perhaps only 1 cm^{-1} apart, is a factor of about 15× slower in speed (light-gathering power) than the general-purpose double-monochromator Spex Industries Model 1401, 0.8-m instrument (Fig. 13) used for conventional Raman-spectroscopic studies of liquids and crystals.

The blaze of the grating is selected to di-

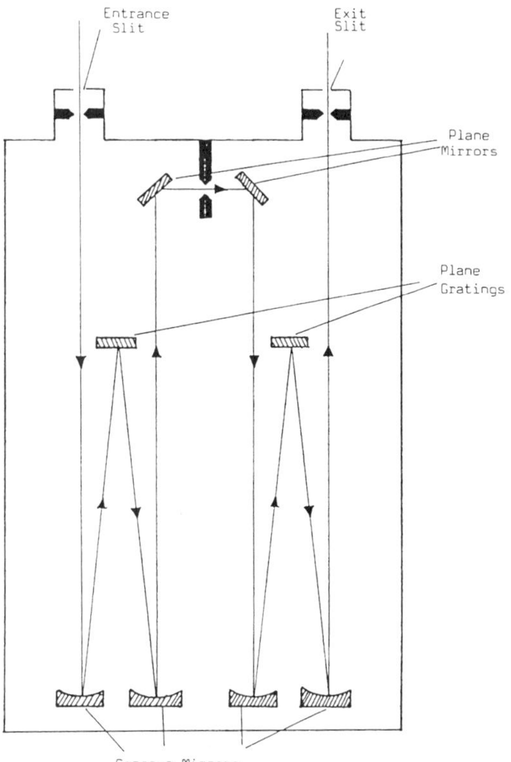

FIG. 13. Optical layout of double-monochromator, Czerny–Turner instrument.

rect the wavelengths diffracted in a particular order preferentially onto the camera. The luminosity of blazed-grating spectrometers is much greater than that of other instruments with equal resolving powers. Modern devices, such as holographic gratings, have contributed to increased spectral quality in Raman spectroscopy as periodic errors in ruling are eliminated and, thereby, grating artifacts are minimized.

1.3.8 Camera Mirror/Lens The function of the camera mirror/lens is to receive the dispersed radiation from the grating and to bring this to a focus in the plane of the exit slit; an extended image of a distant source is thus formed. High-speed camera lenses have been used up to apertures of *f*/0.4—for example, in the recording of spectra from astronomical nebulae in the wavelength range 360–500 nm in the Mt. Palomar observatory.

1.3.9 Detectors The detectors used for the scattered, dispersed radiation depend on the region of the electromagnetic spectrum being studied. Photographic plates, photomultipliers, and, more recently, charge-coupled diode (CCD) array detectors are commonly used in the visible and ultraviolet regions, while others such as the mercury cadmium telluride (MCT) and liquid-nitrogen–cooled GaAs are used for infrared excitation of Raman spectra.

Photographic plates are very insensitive beyond 900 nm but, nevertheless, are still being used with bright stellar objects up to 1100 nm. Photoelectron devices are insensitive beyond 1200 nm, where photoconductive materials must be used. Objective lens material also suffers from transparency problems; for example, quartz is transparent in the ultraviolet down to 200 nm but lithium fluoride is useful to 120 nm. Glass is transparent in the infrared to only 2.5 μm; beyond this alkali-halide crystals are used. Beyond about 40 μm the alkali halides, even cesium iodide, absorb infrared radiation, and then synthetic materials such as polyethylene are used (Dodd, 1962). Care must be taken in studies of molecular systems that interaction between the system under investigation and the window material does not occur. A major advantage of Raman spectroscopy is the use of material in windowless cells, especially in the near infrared.

Low sensitivity to the radiation is a problem in the infrared region, and thermocouples and bolometers have been used, depending on the response times required. A bolometer has a typical response time of 5 ms and, with a surface area of 0.004 cm^2, has a response of 0.8 $\mu V\ \mu W^{-1}\ cm^{-2}$. The infrared detector should have a small heat capacity to permit the rapid response to radiant energies. Detectors in the infrared have become important recently for FT-Raman spectroscopy at 1.06 μm (Nd^{3+}/YAG excitation), and Raman spectra with 1.3 μm excitation in the infrared have just been reported (Chase and Rabolt, 1994; Asselin and Chase, 1994).

Multichannel photon detectors consist of an array of small photosensitive devices (pixels) that convert an optical signal into a charge pattern. Several types of photodiode-array detectors are now available—e.g., silicon vidicon and silicon diode array. A typical diode-array detector contains 1024 diodes in an array about 1 in. (2.5 cm) long; dispersed radiation impinging on the multichannel detector gives rise to a charge pattern that is intensified by about 10^4 times.

The charge-coupled device (CCD) has been used increasingly in recent years, based on a two-dimensional optical-array detector, with each pixel having a dimension of about 10 μm. A major advantage of the CCD device compared with other multichannel detectors is the low readout noise and high quantum efficiency over a wide wavelength range (100–1100 nm). Hence, CCD detectors can be used with UV or IR excitation of Raman spectra as well as visible wavelengths. A disadvantage of CCD detectors, however, is the limited spectral coverage for general purposes; on the other hand, the sensitivity of these detectors is not surpassed—e.g., 75% quantum efficiency from 400 to 800 nm and 15% at 1000 nm is not atypical.

1.3.10 Problems

1.3.10.1 Light Leakage. Light may be lost in passage through a spectrograph or spectrometer by absorption in optical components such as lenses or prisms, and by reflection at tarnished surfaces, and stray light may enter the system from leakage and fluorescence within the optical train. This will result in poor signal-to-noise ratios for electronic detection, and fogging of photographic plates. Excess light caused by overfilling of the collimator or camera mirrors or from grating imperfections is a major cause of extraneous radiation reaching the exit slit. For this reason, masking of the optical train with baffles and painting of interior surfaces nonreflective (matte) black is recommended. Care should be taken with painting, particularly when infrared excitation is envisaged, because of transmission windows of visibly opaque materials in the IR.

1.3.10.2 Effect of Temperature, Pressure, and Humidity. Changes in temperature and pressure lead to serious problems in spectrographs and spectrometers through the resulting broadening and shifting of spectral lines caused by changes in focus (Weber, 1972). From the temperature coefficients of the refractive indices of the optical and metal components, quantitative measurements of the spectral-line shifts can be made, and these can amount to ~0.1 nm K^{-1}. Atmospheric pressure and humidity changes can also cause severe disruption to experiments, particularly over the long exposures required for stellar or high-resolution rotational spectroscopy. Humidity changes in the air contained within nonpurged or unevacuated spectrometers can also affect the optical paths of the dispersed radiation through refractive-index effects. In addition, at a temperature of 298 K, a barometric pressure of 760 Torr, and a relative humidity of 50%, the changes in the wave number of the 488.0-nm argon-ion laser line alone arising from changes in the temperature, pressure, and partial water-vapor pressure are -0.018 cm^{-1} K^{-1}, $+0.007$ cm^{-1} $Torr^{-1}$, and -0.001 cm^{-1} $Torr^{-1}$, respectively. These changes are nonlinear across the Stokes Raman region, and the temperature and pressure changes are -0.016 cm^{-1} K^{-1}, $+0.006$ cm^{-1} $Torr^{-1}$, respectively, at a wave number displacement of 3000 cm^{-1}; the response to the partial pressure of water vapor is much smaller than the temperature and pressure changes, namely -0.001 cm^{-1} $Torr^{-1}$ at 3000 cm^{-1} displacement. Special instrumentation has been devised to compensate for temperature and pressure changes during long photographic Raman exposures, using detectors and feedback loops to mirror-tuning drives on the laser.

1.3.10.3 Spectral-Line Tilt. With a Czerny–Turner spectrograph or spectrometer, spectral lines at the exit slit or the photographic plate will be tilted from the vertical by an angle θ given by $\cos\gamma/\sin 2\alpha$ where γ is the angle of incidence of the grating with the incoming ray and 2α is the vertical angle between the grating center and a line joining the centers of the entrance slit and collimator (Ebert angle). In the Czerny–Turner spectrograph described above, lines were tilted approximately 18° from the vertical for the $n = 12$ order at 500 nm. It should be noted that if the order of diffraction were changed, then $\cos\gamma$, and hence the tilt of the lines, would change. An example of this line tilt is given in Fig. 14 for the pure rotational Raman spectrum of carbon disulfide recorded in the twelfth order of diffraction using $\lambda = 488.0$ nm radiation. The rotational lines are only 0.8 cm^{-1} apart; the general clarity of the spectrum should be noted, with the absence of grating ghosts and interference from the exciting line. The first rotational line is only 0.8 cm^{-1} from the exciting line.

1.3.11 Calibration In the infrared, visible, and ultraviolet regions of the electromagnetic spectrum, calibration of the observed Raman spectra for intensity and wave numbers or wavelength is usually carried out by reference to primary standards such as neon-lamp emission lines, laser plasma lines, or secondary standards such as polyethylene and carbon tetrachloride (Robinson, 1991; Denny and Sinclair, 1987). However, the additional restraints imposed by accuracy and precision of wave-number measurements in the field of rotational and vibrational–rotational spectra require the observation of a large number of bands of accurately known wave number usually from species such as hydrogen chloride and carbon monoxide or from hollow-cathode lamps such as zirconium or thorium. In the pure rotational Raman spectrum of oxygen shown in Fig. 5 the calibration was effected by use of a hollow-cathode thorium lamp with 45 interferomet-

FIG. 14. Line-tilt effect at higher order of diffraction; pure rotational Raman spectrum of carbon disulfide at $n = 12$, $\lambda = 488.0$ nm, $R_{theor} = 10^6$.

rically measured standard lines in the ninth to thirteenth orders over a 300-cm^{-1} range around 488.0 nm. The constants a_i of constrained and unconstrained polynomial expressions of the type

$$\tilde{\nu} = a_0 + a_1x + a_2x^2 + a_3x^3 + a_4x^4 \qquad (7)$$

were obtained by a linear regression fitting procedure of the thorium wave numbers and measured positions with a standard error of ± 0.001 cm^{-1}.

1.3.11.1 Fabry–Pérot Étalon. A critical survey of the novel use of a Fabry–Pérot interferometer for the study of the rotation and vibration–rotation Raman spectra of gases has been produced (Jones, 1972). The greater part of the article outlines the manner in which a fixed–path-length étalon has been used for the precise measurement of the wavelengths of individual lines in the rotation Raman spectra of O_2 and N_2. The technique of "crossing" a Fabry–Pérot étalon having a small free spectral range of 0.4 cm^{-1} with a 1-m *f*/8 Czerny–Turner spectrograph equipped with a 1200-lines-mm^{-1} diffraction grating with a reciprocal linear dispersion of 11 cm^{-1} mm^{-1} was developed; the spectra are, in essence, pure rotation Raman spectra dispersed in the horizontal plane, but each Raman line displays a vertical interference pattern due to the étalon.

The spectrograph served simply to separate the individual rotation lines so that the set of fringes from one line did not overlap and mask the set of fringes from an adjacent line. The general experimental arrangement is shown in Fig. 15. The set of fringes observed on each rotation line is a section of a complete set of ring systems emerging from the étalon at angles θ to the normal given by the equation

$$n\lambda = 2d \cos\theta, \qquad (8)$$

where n is the order of interference, d the étalon spacing, and λ the air wavelength of the line in question. The radiation emerges from the étalon in a series of cones, which are imaged on the entrance slit of the spectrograph by lens L_2, so giving rise to the characteristic spectral pattern of Fig. 16. An analysis based on the twenty or so fringe diameters observed for a given line enabled the wave number of a Raman line to be determined to 0.003 cm^{-1}. The most precise determination of the rotational constants of $^{16}O_2$ by Raman spectroscopy has been achieved using this technique, namely,

$$B_0 = 1.437\,682 \pm 0.000\,009 \text{ cm}^{-1},$$
$$10^6 D_0 = 4.85 \pm 0.01 \text{ cm}^{-1}.$$

It is clear that this interesting Raman-spectroscopic technique is capable of providing highly accurate values for the rotational constants of simple molecules. The limitations on the method arise because of the finite spectral widths of the laser lines, typically 0.1 cm^{-1}, as well as the weakness of the scattered Raman lines. For obtaining higher resolution, and consequently greater accuracy, it would be necessary to reduce the width of the laser lines by mode selection and to decrease the pressure of the gas studied to avoid pressure-broadening effects. Both of these processes will undoubtedly decrease the intensity of the scattered light and hence lead to an increase in the exposure time necessary for the observation of the spectra. Such an increase in exposure time would be disadvantageous because of the dif-

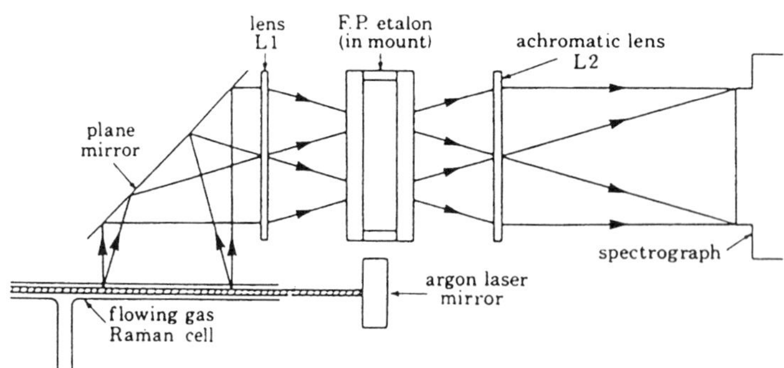

FIG. 15. Optical arrangement used for photographing the rotational Raman spectra of O_2 and N_2. [Reproduced with permission from Butcher *et al.* (1971).]

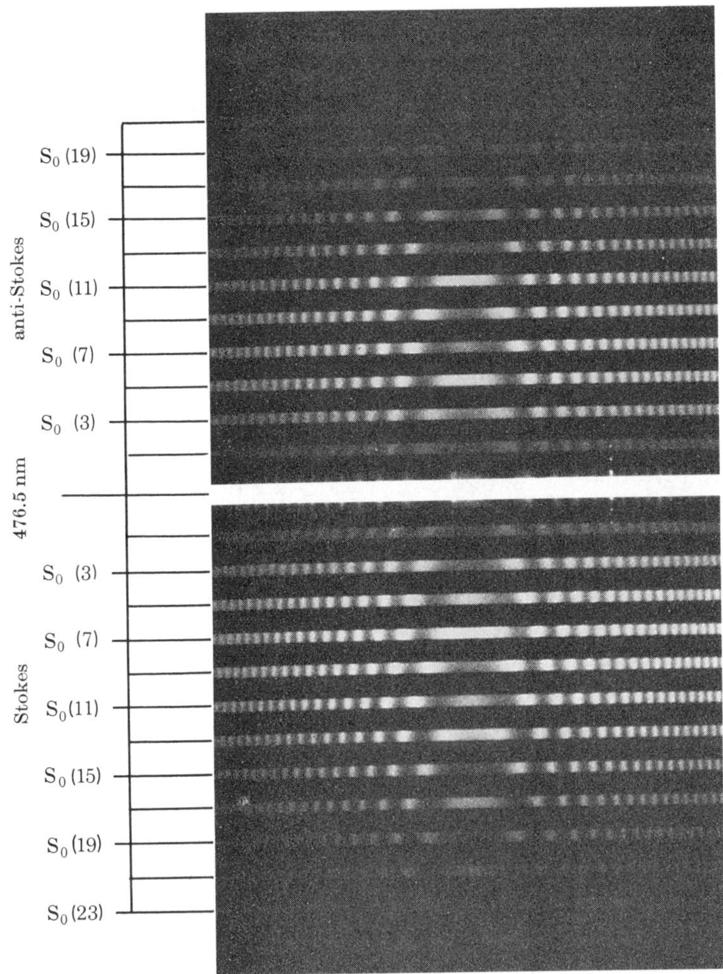

FIG. 16. Pure rotational Raman spectrum of O_2 excited by 476.5-nm argon laser radiation. Raman radiation dispersed by means of a Fabry–Pérot étalon with plate separation of 12 mm. Gas pressure 1 atm. [Reproduced with permission from Butcher *et al.* (1971).]

ficulties of controlling the temperature and pressure in the vicinity of the étalon. In spite of such difficulties, however, it should prove possible to study rotation Raman spectra at a resolution of ca. 0.05 cm^{-1} with simple extensions of the methods reported here (Weber, 1979).

2. RAMAN SCATTERING

2.1 Instrumentation for the Recording of the Pure Rotation and Vibration–Rotation Raman Spectra of Gases

A special illustration of the versatility of design of sample cells and instrumentation associated with the recording of molecular Raman spectra is provided by the rotational Raman spectra of gases (Weber, 1972; Brandmüller and Moser, 1962). These studies are undertaken to provide information about molecular structure and parameters and give data on moments of inertia, internuclear bond distances and angles, and nuclear-spin statistics.

Experimentally the investigation of the Raman spectra of gases has been fraught with difficulties because of the intrinsically low intensity of Raman scattering. When the rotational features are to be observed under conditions of high resolution then the following criteria must apply:

1. Fairly low pressures of gas must be used to minimize line broadening (and the Raman scattering is proportional to the concentration of scattering species).

2. Spectrographs (or spectrometers) of high resolving power are essential (and these are of low light-collecting power).
3. A light source of very high intensity must be used to excite the Raman spectra, in conjunction with efficient collecting optics.
4. The spectral linewidth of the exciting line must be as small as possible.
5. An efficient detection system, whether photographic or photoelectric, is required.

In the very early years (Long, 1988; Kohlrausch, 1972), Raman sources consisted of batteries of discharge lamps, and the means of detection was the photographic plate. Gas pressures were increased up to eighty atmospheres in an effort to observe Raman spectra from weakly scattering species. Exposure times were commonly of the order 20–100 hours for pure rotational spectra, and an order of magnitude greater for the even weaker rotation–vibration spectra. The major advances in instrumentation and apparatus (Stoicheff, 1959) that were so successfully applied to the rotational Raman studies of the 1950s consisted of the following:

1. the development of low-pressure, water-cooled, mercury arcs of low continuum and high intensity;
2. the use of multiple-reflection Raman cells, whereby the Raman scattering could be increased by as much as 40-fold using concave mirrors;
3. the construction of high-resolution spectrographs, such as the 21-ft concave grating and the faster Littrow spectrograph at the University of Toronto (Stoicheff, 1959).

The molecules listed in Table 2 were investigated with apparatus such as that described above. Few attempts were made to use discharge lamps other than those of the mercury-arc type, and this imposed a practical limit of resolution of about 0.2 cm^{-1}, the half-width of the natural mercury line. With the advent of the laser, its application to the field of Raman spectroscopy was quickly realized. The characteristics of lasers that distinguish them from other radiation sources are their very high powers, extreme directionality, polarization, and small spectral linewidth (Edwards, 1970). An intriguing situation developed when the new laser excitation was applied to the studies of gases; the narrow linewidth of the laser emission was favorable for high-resolution studies of rotational Raman spectra, but the very coherence and narrowness of the laser beam proved to be a disadvantage for the illumination of gases at low pressure. Insufficient quantities of gas could be illuminated for observation of the excitation of the rotational Raman spectra. Thus, the field of high-resolution Raman spectroscopy was denied its opportunity for expansion, until cell design met the requirements of the laser source. It was not until 1965 that the attempts at laser excitation of rotational Raman spectra were successful, when a He–Ne laser was used to produce the pure rotational spectrum of methylacetylene.

Table 2. Pure rotational Raman spectra.

Observer	Molecule	B_0 (cm^{-1})	I_0 (10^{-47} kg m^2)
Andrychuk	F_2	0.8828	31.70_5
Stoicheff	N_2	1.9897_3	14.067_0
	CO_2	0.3904_0	71.69_3
	CS_2	0.1091_0	256.54
	H_2	59.339_2	0.4716_8
	HD	44.667_8	0.6266_0
	D_2	29.910_5	09.357_6
Weber	O_2	1.4378	...
Stoicheff	C_2N_2	0.1575_2	177.69
	C_2H_2	1.1769_2	23.78_2
	C_4H_2	0.1468_9	190.54
	MeCCMe	0.1122_0	249.46
	$ZnMe_2$	0.1347_8	207.66
	$CdMe_2$	0.1140_5	245.41
	$HgMe_2$	0.1162_0	240.87
	C_6H_6	0.1896_0	147.59
Romanko	C_2H_4	0.10012	...
Stoicheff	HC≡CI	0.1022_2	263.50

With the discovery of more lasers operating in different regions of the visible spectrum, a new advantage of the laser as a source for the excitation of Raman spectra becomes apparent. For the first time Raman spectroscopists are now able critically to select an exciting line to suit their requirements, be these power or wavelength in the visible, infrared, or ultraviolet region (Demtröder, 1981). Laser physicists are still a long way from constructing the perfect laser, which should have reasonably high powers in quite a few lines, be stable, and be reasonably inexpensive to maintain. At present, there is a large selection of lasers available commercially. However, although the intensity of scattering in the Raman effect is proportional to ν^4, where ν is the frequency of the exciting radiation, the selection of a laser line of high ν (i.e., in or near the ultraviolet) is not always advantageous "chemically," since such radiation could initiate decomposition in many molecules that would be of interest to the Raman molecular spectroscopist. For this reason, a blue or green line represents a useful compromise; the argon-ion laser, with cw radiation at 488 and 514.5 nm, provides two such exciting lines.

A critical comparison of excitation with a laser and with a mercury-arc source is outlined in Table 3. Although the scattering intensity relative to Hg 435.8 nm is in all cases lower for laser excitation, the two factors that in themselves advocate the use of lasers as Raman sources are these:

1. The very much smaller sample volume needed for laser excitation. Thus, small quantities of expensive, isotopically substituted molecules may be studied. By this means, the amount of information obtained from the rotational Raman spectrum is increased. The first such isotopically substituted species other than ^2H studied by Raman spectroscopy was the radioactive β emitter, $^3H_2$.
2. The smaller spectral linewidth of a laser, relative to the mercury-arc Hg 435.8-nm line. This makes accessible the study of "heavy" molecules, in which the rotational lines are very closely spaced. Where the laser linewidth is the limiting factor for spectral resolution, a "maximum" moment of inertia I_{max} may be calculated. For molecules having an $I_0 > I_{max}$ rotational features will not be resolved using the laser excitation in question. At the foot of Table 3 is given the approximate value of I_{max} for each particular method of excitation.

There are, however, factors other than the width of the exciting line that can affect the resolution of rotational features in the Raman spectrum. Broadening of Raman lines can arise from two sources: namely, Doppler broadening and pressure broadening. Doppler broadening can be estimated from the expression

$$\Delta\tilde{\nu}_{1/2}/\tilde{\nu}_0 = 7 \times 10^{17}(T/M)^{1/2}, \tag{9}$$

where $\Delta\tilde{\nu}_{1/2}$ is the width at half-height of a spectral line of wave number $\tilde{\nu}_0$, where the sample of molecular weight M is at absolute temperature T. For benzene or carbon disulfide the Doppler linewidth is approximately

Table 3. Comparison of lasers and mercury arcs for the excitation of Raman spectra of gases.

	Laser			
	He–Ne	Ar	Pulsed ruby	Hg arc
Light source	Gas	Gas	Solid-state	Vapor
Exciting line(s), nm	632.8	488.0 514.5	692.3	435.8 (253.7, 541.6)
Power in line				
(a) outside cavity	80 mW	$1\frac{1}{2}$ W	10–100 MW	
(b) inside cavity	1 W	50–70 W	···	2 W
No. of sources needed	1	1	1	Battery of 2–4 lamps and light furnace
Maximum power density (W m^{-2})	10^6	10^8	10^{20}[a]	10^4
Linewidth (cm^{-1})	0.05	0.15	0.1[b]	0.20–0.25
Volume of sample irradiated		1–20 mL		1–51
Multiple-reflection cell		Unnecessary		Necessary
Relative scattering intensity (due to ν^4 law)	0.33	0.64 0.52	0.14	1
Maximum moment of inertia I_{max} resolvable with linewidth given (10^{-47} kg m^2)	1300	450	900	300

[a]Medium breakdown achievable.
[b]Normal linewidth of a ruby laser equipped with a Q switch and a sapphire resonator; with a Fabry–Pérot étalon for mode selection, linewidths as small as 0.005 cm^{-1} could be achieved.

0.04 cm^{-1} at room temperature, and for dimethylacetylene (which has one of the closest line spacings yet investigated in the rotational Raman effect) it is 0.06 cm^{-1}.

Pressure broadening is important, especially in cases where the resolution of lines closer than 0.4 cm^{-1} apart is to be achieved. The effects due to pressure broadening are rather difficult to estimate at lower pressures, about 0.1 atm, where most of the high-resolution work with heavy molecules would be done. Several recent investigations of pressure broadening at high pressures have been made and "broadening coefficients" have been calculated. The broadening of the rotational lines of N_2 is linear with pressure up to 25 atm, and the "broadening coefficient," $\Delta\tilde{\nu}_{1/2}/p$, is 0.043 cm^{-1} atm^{-1}. For isotropic scattering the broadening is less than for anisotropic scattering, and for the latter the broadening decreases with increasing rotational quantum number. A detailed impact theory of Raman scattering has been proposed in which both elastic and inelastic collisions contribute additively to the linewidth for anisotropic scattering, but isotropic scattering is affected only by inelastic collisions.

A few general remarks should be made concerning the use of high-resolution grating spectrographs or spectrometers with laser excitation. The three parameters resolving power, speed (i.e., light-collecting power), and signal intensity should be considered. Spectrographs (or spectrometers) of sufficiently high resolving power to utilize the extremely narrow spectral linewidth of the He–Ne laser could be constructed, but the optical arrangement would be such that to obtain the spectrum under conditions of high resolution would be a slow process. The luminosity of such a spectrograph would have to be decreased to the extent that the signal would be perhaps only one-twentieth of its intensity under less stringent conditions of resolution. Other parameters that warrant consideration include laser stability, the stability of the gas sample under laser irradiation, and thermostatting problems for constant-temperature control over very long periods of time. Laser stability is not merely an academic problem of maintaining the desired power levels for long "exposures"; it is also an economic one, since laser plasma tubes have a finite lifetime. Rather special rooms are required, ideally, to minimize the effects of temperature ambience or fluctuation on the spectrometer and detector. Artificial "line broadening," for example, on a photographic plate, can be caused by expansion or contraction of metal parts in the spectrograph. Temperature has some smaller effects on the detection system, too, and increase in grain size on the photographic plate or an alteration in the dark current of a sensitive photomultiplier can result from temperature ambience.

Resolution at the detector may, of course, be increased by using the grating (or gratings) in a higher order. Most modern Raman spectrometers incorporate double monochromators, being two- or three-grating instruments with a very high discrimination against stray-light intensity. The gratings are frequently used in higher orders to achieve the necessary resolution. However, signal intensity decreases markedly as one proceeds to the higher orders of diffraction of a grating, and selection of the required order is dictated by the blaze angle of the grating. Further, a change from blue to red excitation can provide a useful increase in dispersion at the detector. Thus, the pure rotational Raman spectrum of methylacetylene has been obtained with He–Ne laser excitation in the ninth order of a diffraction grating, giving 2.0 cm^{-1} per mm at the photographic plate. This dispersion was 30% greater than that achieved in the blue region (Hg 435.8 nm), assisting materially toward the resultant quality of the rotational Raman spectrum.

Apart from the effects of external parameters on detector efficiency, it is of interest to examine the detectors themselves. Basically, the detection system for high-resolution Raman spectroscopy may be photographic or photoelectric. With respect to the former, a photographic plate is selected that is sensitive at the wavelength of the laser line, or, what is more important, at that of the spectroscopic feature being investigated. An account of the difficulties inherent in photographing Raman spectra in the red is given by Stammreich *et al.* (1961). Increased speed of the photographic plate can be effected in several ways, but a common result of most sensitizing procedures is an increase in grain size—and this is most undesirable for high-resolution work (Weber, 1972).

In recent years Raman spectroscopists

have used more sensitive detection systems comprising a photomultiplier coupled to pulse-counting equipment or CCD detection. A pulse-counting detecting system for the photoelectric recording of Raman spectra gave a very high signal-to-noise ratio for discrimination against random noise levels and was achieved by cooling the photomultiplier tube to reduce the dark current. Although red-sensitive photomultipliers are available, their sensitivity falls off as the wavelength increases toward the red. Thus, such a photomultiplier is, typically, some three times more sensitive at 488.0 nm than it is at 632.8 nm.

It is in connection with the polarization properties of the laser beam that a very useful application to rotational Raman spectroscopy becomes evident. The bane of rotational Raman spectroscopy has always been the overwhelming intensity of the Rayleigh line. In photographic work this can result in a severely overexposed portion of the photographic plate, thereby masking such, if not all, of the region of interest. In earlier experiments using discharge lamps some work was carried out to design suitable post-filters with the characteristics of a very sharp absorption band at the Rayleigh frequency and a "clear" region on both the Stokes and anti-Stokes sides. Although this idea was fairly successful for helium-discharge lamps, no suitable post-filter could be found for mercury lamps. The mercury-discharge resonance line at 253.7 nm (Long, 1988) has been used to excite Raman spectra, since the Rayleigh portion of the scattered light could be extensively absorbed by passage through mercury vapor prior to its being focused on the slit of the spectrograph. In experiments using polarized laser excitation a simple solution to the problem is to observe the Raman effect in a direction parallel to the polarization vector in the incident laser beam. The Rayleigh line is due to isotropic scattering, so that, if the scattered radiation is observed in a direction coinciding with the direction of vibration of the electric vector of the exciting-laser radiation, the line will vanish to an extent that depends on its depolarization ratio (e.g., $\rho = 0.01$ for the Rayleigh line in O_2 or N_2). Rotational Raman lines are depolarized with $\rho = 0.75$ for linearly polarized exciting radiation, and so some loss of intensity is incurred by viewing their depolarized component. However, the improved spectral quality gained by such a maneuver is normally considered well worth while.

In Fig. 17 a diagrammatic example is given of a typical arrangement for a sample illuminator, calibration optics, fore-optics, and extracavity cell illumination for gas-

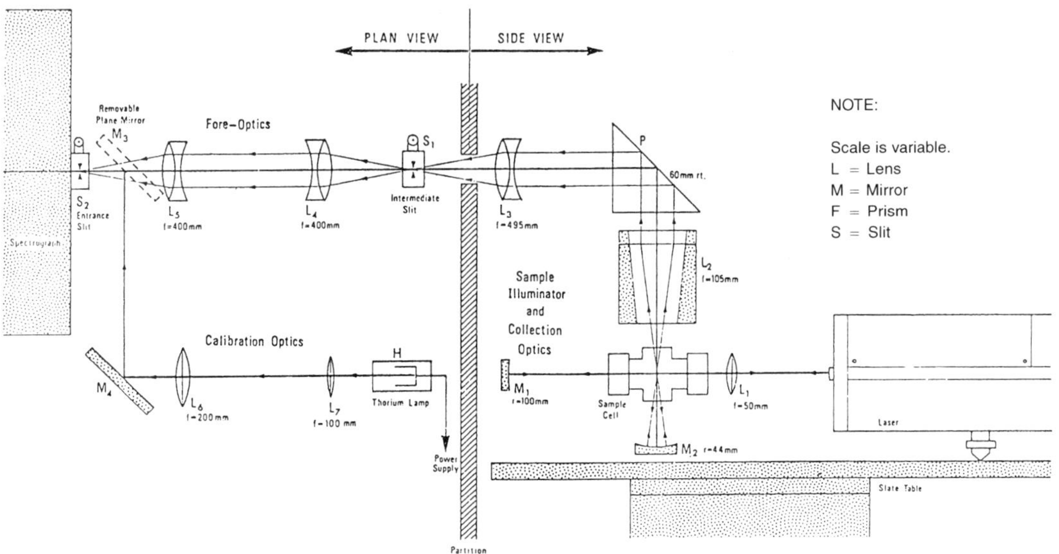

FIG. 17. Sample illuminator, fore-optics, and calibration optical arrangement for rotational Raman spectroscopy with extracavity laser illumination.

phase Raman spectroscopy. This reflects several of the necessary features mentioned above, including long-term stability of the optical system required by long exposures and the need to minimize stray-light scattering. The function of the fore-optical arrangement is similar to that of field lenses in that the parallel beam produced by lenses L_4 and L_5 gives flexibility in the relative siting of the monochromator and the sample illuminator. Note also the retroreflective mirrors M_1 and M_2, which are designed to increase the laser power density (W m^{-2}) at the focus in the sample cell.

3. RAMAN MICROSCOPY AND IMAGING TECHNIQUES

Optical microscopy depends on the spatial variations of refractive index, absorption, and reflectivity in analyzed specimens to produce the modulation of the transmitted light intensity that is required to form the magnified image (Messerschmidt and Hartcock, 1988). The ability to identify the materials in the specimen is severely limited, as the photographic viewed image is not specific to material properties, except perhaps in the case of fluorescence microscopy where light emitted by the specimen forms the recorded image. Thus, Raman microscopy, which is used to obtain molecular-structural information from microscopic samples, provides an attractive and novel development of molecular spectroscopy in which two possible modes of operation are envisaged, namely

1. the obtaining of molecular-vibrational data from microscopic samples;
2. the study of images formed by Raman-scattered radiation—an *imaging* technique.

Raman microscopy, which can be viewed as the blending of one of the oldest scientific instruments (the microscope) and one of the newest inventions (the laser), was first demonstrated in 1975. The principle is illustrated in Fig. 18; the laser-excitation source is brought onto the sample on the microscope stage using the epi-illuminator and microscope objective. The scattered Raman radiation is collected at a 180° scattering angle by the same microscope objective and imaged onto a spectrometer slit using a beam splitter. However, since the Raman-scattered light intensity is always much less, typically a factor of 10^6–10^{12}, than that of the incident radiation, and the sample volume is small, an efficient means of spectral filtering is normally required for the recording of suitable spectra. Offset against this disadvantage is the obvious improvement in light-collecting efficiency occasioned by the use of small–focal-length, short–working-distance objective lenses (e.g., 100×) with greater light-collecting ability.

Many applications followed upon the initial announcement of the Raman microscopic technique, and a novel example is illustrated here, namely, the Raman spectrum of a lichen-encrustation sample from a biodeteriorated 16th-century Renaissance fresco (Fig. 19). In this spectrum, the features due to calcium oxalate, incorporated stone, paint pigments, plaster, and lichen metabolic byproducts were identified; for example, the strong band at 1007 cm^{-1} is characteristic of calcium sulfate (gypsum) and that at 1086 cm^{-1} calcite, both of which occur as substratal materials. Calcium oxalate monohydrate has a doublet at ~1500 cm^{-1} and a weaker feature at 506 cm^{-1}. Other features at 1157, 660, 602, 480, and 400 cm^{-1} are ascribed to paint pigment and lichen material. The FT-Raman laser microprobe spectrum of the upper surface of the living lichen species (Fig. 20) excited at 1.06 μm demonstrates some remarkable features of the lichen itself rather than the biodeteriorated substratum and can be used to monitor environmental change in the microclimate generated by the plant colony. To this effect the distinguishability of the hydration states of mono- and dihydrates of calcium oxalate (Fig. 21) is important; all the spectra have been obtained from 2-μm-diam regions of the samples, representing a quantity of material in the nanogram to picogram range.

It must be realized that, despite concern in the literature for the degradative effects of laser radiation on biological samples, the high power densities generated using small–focal-length objectives and increased scattering collection capabilities could result in relatively low incident laser powers being used for the illumination of sensitive biological samples (Spiro, 1987). To illustrate the quality of the Raman microscope spectra ob-

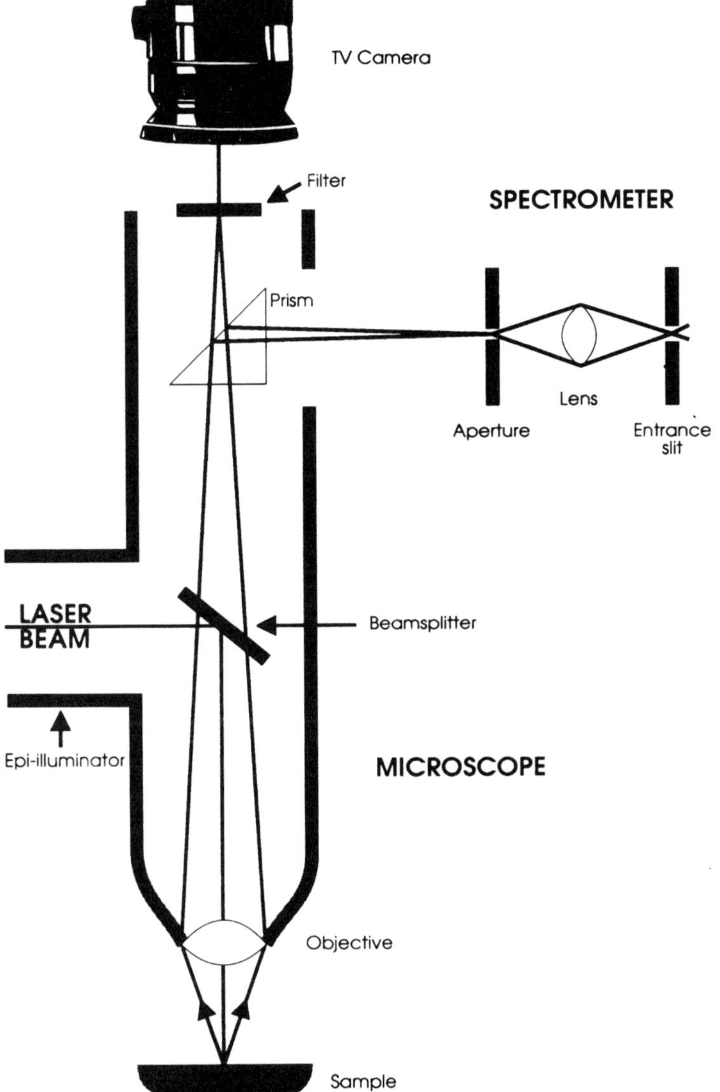

FIG. 18. Conventional Raman microprobe.

tained using microscopic laser techniques, the spectra of human skin (stratum corneum) and snake skin samples have been obtained; the latter is shown in Fig. 22 from a hinge area between the scales. The optical image of the snake hinge from which the spectrum in Fig. 22 was derived is shown in Fig. 23. It is stressed that no degradation of these biological samples was observed during the Raman microscopic experiments; and recently, Raman spectra of healthy and diseased human skin have been reported (Edwards and Williams, 1995). Some elegant examples of the use of Raman microscopy are provided by the vibrational analysis of paint fragments from colored medieval manuscripts and from atherosclerotic arteries. Such studies have now been applied to areas as diverse as electronic devices, interfaces between thin films and substrates, gemmology, geochemistry, and molecular biology. The natural progression to studies of healthy and diseased tissue has been accomplished, and biomedical application of Raman spectroscopy provides an exciting new development (Edwards and Williams, 1994; Schrader *et al.*, 1994).

However, several new approaches to Ra-

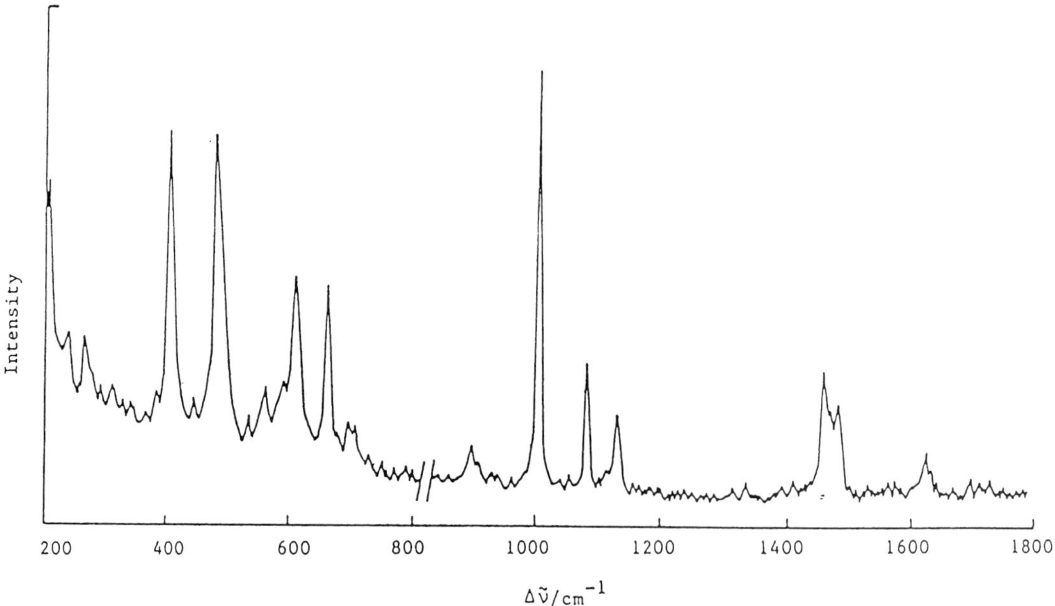

FIG. 19. Laser Raman microprobe spectrum of an encrustation of the lichen *Dirina massiliensis* forma *sorediata* on a 16th-century Renaissance fresco; λ = 488.0 nm excitation, 200 mW, $\Delta\tilde{\nu}$ = 200–1800 cm^{-1}.

man microscopy have recently been formulated and have resulted in instruments with greater light throughput and detector efficiencies, which makes them of importance in the area of Raman imaging. Hence, it is not only possible to identify the molecular species present in a specimen from Raman microscopy, but the location of a particular species within the specimen can also be obtained.

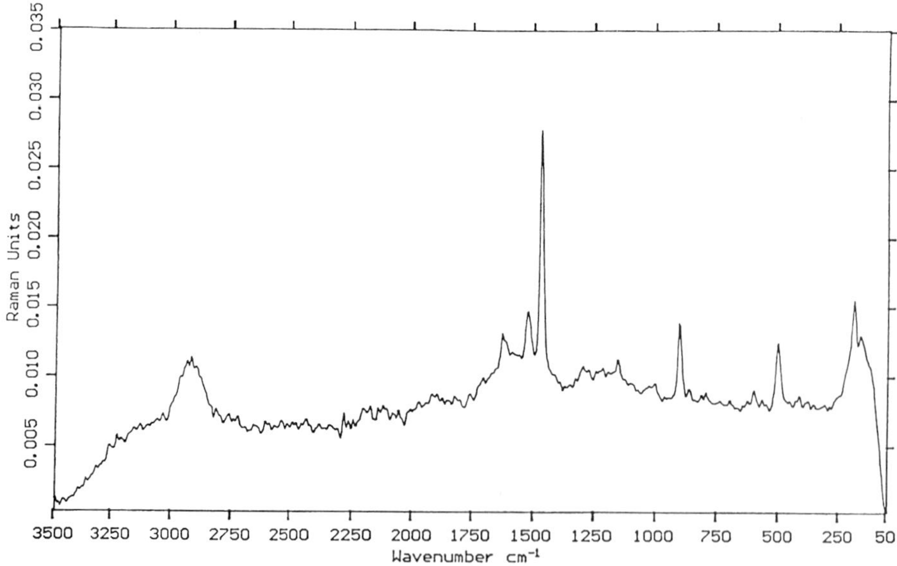

FIG. 20. FT-Raman laser microprobe spectrum of the upper surface of the *Dirina massiliensis* lichen species; λ = 1064 nm, 400 mW, $\Delta\tilde{\nu}$ = 50–3500 cm^{-1}.

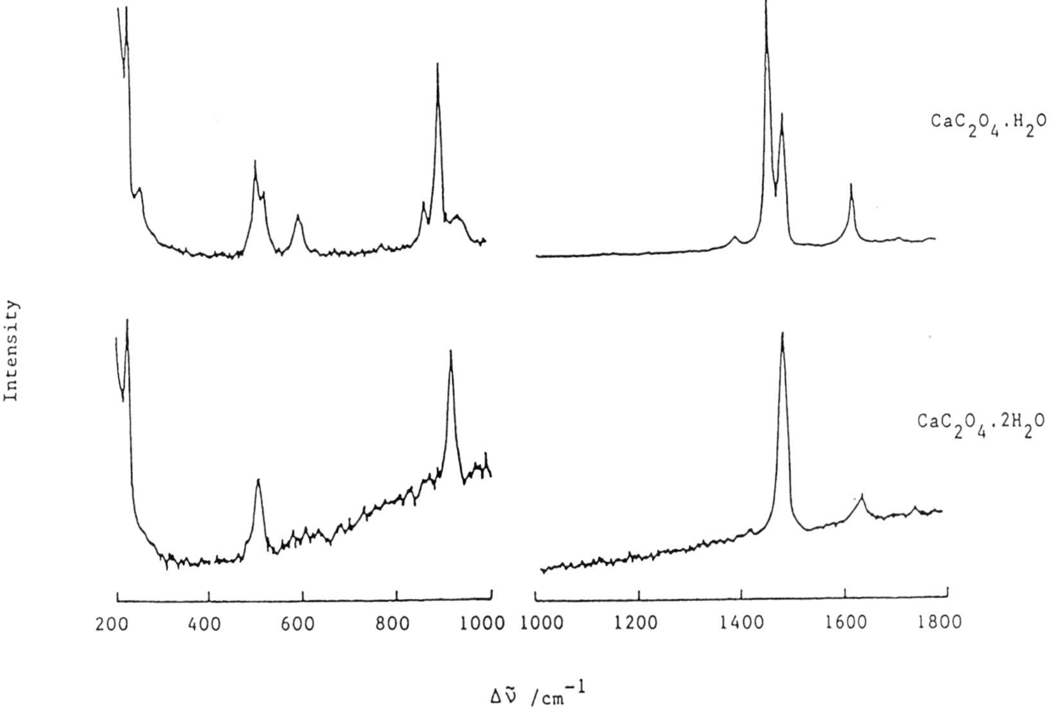

FIG. 21. Laser Raman microprobe spectra of $CaC_2O_4 \cdot H_2O$ and $CaC_2O_4 \cdot 2H_2O$; λ = 488.0 nm, 200 mW, $\Delta\tilde{\nu}$ = 200–800 cm^{-1}.

3.1 CCD Devices

Three technological developments have occurred that make possible significant improvements to the original Raman microscope. First, a cooled CCD detector provides an electronic photographic plate of very high quantum efficiency and extremely low noise. Second, multilayer dielectric filters can now be manufactured to a very high specification. These filters, which can be readily tuned by rotation, offer an attractive alternative to dispersive filtering of the image by a spectrometer; light throughput can be increased and optical complexity reduced without sacrificing stray-light rejection capabilities. Finally,

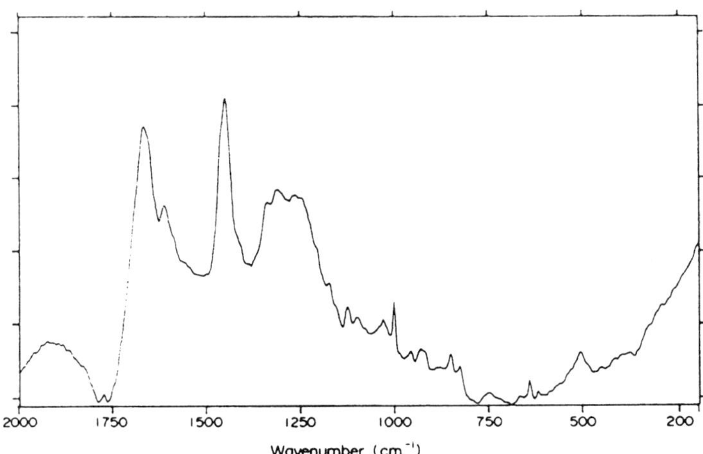

FIG. 22. FT-Raman spectrum of the shed dorsal skin of *Morelia argus* (carpet python); λ = 1064 nm, $\Delta\tilde{\nu}$ = 200–2000 cm^{-1}.

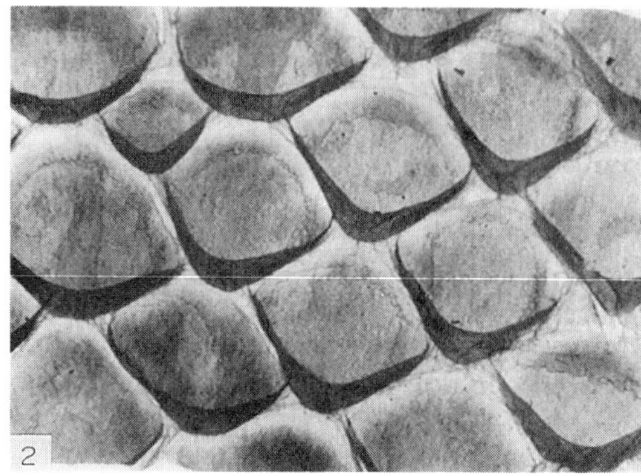

FIG. 23. Photograph of the shed dorsal skin of the snake *Morelia argus*, 8 × magnification, showing the hinge and scale construction.

a computer provides the means both for controlling the tunable filter and for processing the image produced on the CCD detector. The design of the Raman microprobe/microscope is similar to that of a fluorescence microscope. The major difference is that the optical filters used to select the wavelength with which the image is formed have a relatively narrow bandpass, which can be wavelength tuned by rotation of the filter. In the microscope mode the laser is defocused and the scattered light from the illuminated area of the sample is imaged onto the full area of the CCD detector. Exposures are taken with the filter fixed at the angle appropriate for the selected Raman band. The spatial resolution of the Raman image is essentially the same as that of the optical microscope since there is little perturbation by the optical filters. In the microprobe mode the scattered light from the focal spot of the laser is imaged onto an area of the CCD detector that is only a few pixels in diameter; the output from these pixels is combined to act as a single detector. Spectra are recorded similarly to those from a grating spectrometer by stepping the filter angle at a suitable rate.

3.2 Confocal Microscopy

It is not possible using conventional microscopy to obtain good resolution of the order of 1 μm along the optic axis, and radiation is collected equally from segments within the scattering volume of a particular sample. This depth resolution can be achieved by *confocal microscopy* and is determined by the objective used and the wavelength of the radiation. A simple diagram (Fig. 24) shows how a pinhole in the focal plane of the microscope (confocal arrangement) selectively removes radiation from the out-of-focus regions of the sample (Born and Wolf, 1980). Hence a *depth resolution* is made that, although dependent on sample optical properties, can be calculated from a step-response function described by the signal strengths from the focal segments. By use of such an arrangement the scattering from 2-μm segments has been studied in the imaged focal plane of thin polymer coatings deposited on substrates (Tabaksblat *et al.*, 1992). The applications of such a device for depth profiling of samples are now being described.

3.3 Raman Imaging and Mapping

Early attempts at obtaining images of surfaces exposed to laser radiation and studying the Raman scattering relied upon full-field il-

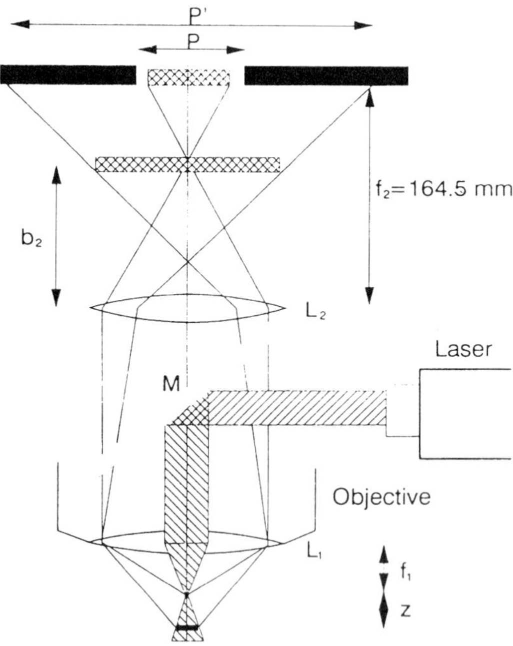

FIG. 24. Confocal Raman microscope. A laser spot in the focal plane passes through pinhole P, whereas a laser spot distant *Z* from this focal place forms a projected image P′ in the image plane and is excluded by the pinhole; L_1 and L_2 are lenses of focal lengths f_1 and f_2, respectively, M is a beam splitter, b_2 is the image distance of the out-of-focus laser spot.

lumination, selection of specific characteristic wavelengths, and use of vidicon detectors. These methods were severely limited by insensitivity and an inability to combat sample fluorescence. Recent advances in this area have centered on novel interference filters to act as "windows" for the scattered radiation.

However, it is simpler in practice to use the punctual illumination of the conventional microprobes described above and to couple the information received with *xy*-computerized microscope stage movements to provide a mapped image profile over the sample surface. This technique is aided by use of intensified linear diode-array detectors, sampling the experimental data in single exposures, or scanning charge-coupled devices (CCD) in a two-dimensional mode. The last has been further developed into a spectrometer using microline image focusing (MIFS) for data collection from all points along the line focus; hence, movement of the sample in only *one* direction gives two-dimensional data-collection parameters from the whole sample area under study. A high degree of speed and sophistication of collection has thereby been achieved, and data-collection times of 10^{-2} s for single-point images and 1 min for a 900-element map are now possible.

Wide-ranging examples of the use of these mapping techniques now exist in the literature in the field of materials analysis, including surface-corrosion product analysis of stainless steels, silicon devices, polymer surface-degradation studies, and stresses in lubricants on surfaces with applied shear (Grasselli and Bulkin, 1991; Chase and Rabolt, 1994; Ferraro and Nakamoto, 1994).

3.4 The UV Microscope

The adoption of the confocal principles outlined above with short-wavelength capabilities has recently resulted in the first UV microscope (Cannon, 1994). The enhanced resolution capability of such a system is impressive by visible standards and a resolution of 80 nm is claimed for radiation at 200 nm, currently the shortest wavelength attainable with calcium–fluoride optics, but already a system with lithium–fluoride optics is being designed to extend the range to 160 nm. At this wavelength the limiting factor will be absorption by water or water vapor in the sample. The advantages of tunability in the ultraviolet and high resolution are of immediate application to biological problems and biophysicists are already investigating the possibility of carrying out time-resolved studies using such systems, especially utilizing the natural fluorescence of key amino acids in living systems in the UV region. In this way, molecular passage through biological membranes can be studied and information about their environment obtained from time-domain spectroscopy. Another new area of research that is being opened up by this type of instrument arises because of molecular absorption in the UV; hence, animal or insect eyes, for example, which are transparent or translucent in the visible region, are now accessible to study. Similar work on semiconductor surfaces has also been undertaken.

3.5 Raman Microspectroscopy with Near-Infrared Excitation

Application of the Fourier-transform technique has found success in Raman microspectroscopy in the infrared region of the electromagnetic spectrum. A schematic diagram of a Raman microscope using infrared excitation is shown in Fig. 25. Until very recently the vibrational analysis of small particles (<25 μm) using infrared radiation posed considerable difficulties. However, the coupling of an FT-Raman spectrometer with an optical microscope has now achieved this objective (Messerschmidt and Hartcock, 1988). The limit of sample size for analysis using this technique is essentially that of the diffraction limit of the infrared radiation, and serious problems result when sample size or aperture size approaches the radiation wavelength, when loss of focal clarity is evident. For the near-infrared region using 1.06-μm excitation, the limit of resolution in an FT-Raman microscope arrangement is about 8 μm.

The infrared radiation is imaged onto the sample using objectives ranging from 5× to about 100×; the larger the objective magnification, the smaller the size of the illuminated spot. By means of the beam splitter, the scattered infrared radiation can be analyzed in the spectrometer and a comparison made with the visual image as viewed through the eyepiece or as recorded by video photography.

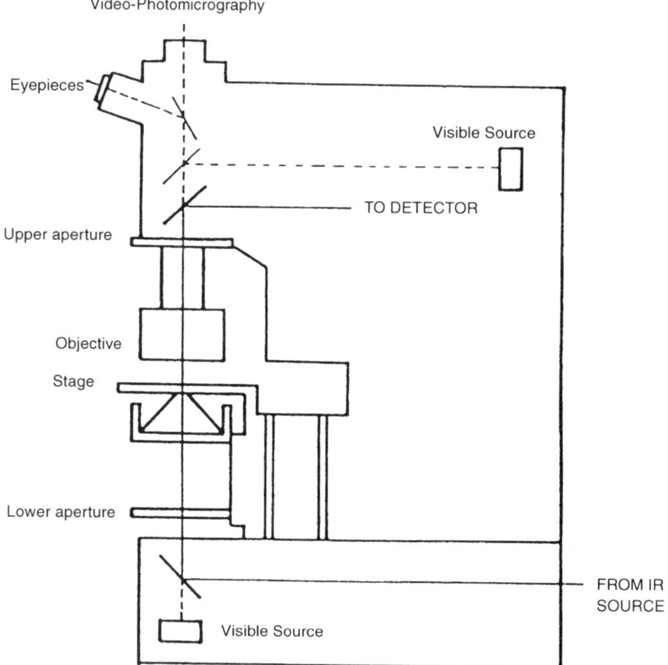

FIG. 25. FT-Raman microscope.

Generally, standard 15-μm objectives are appropriate for many applications, but special objectives for high focusing, for attenuated total reflection, and for grazing incidence angles are available for ultrathin films (2 nm) and coatings on reflective substrates. In addition, sensitive InGaAs or Ge detectors and motorized microscope stages for mapping experiments are now available.

New techniques for spectroscopic imaging have been introduced (Treado and Morris, 1992). A novel technique based on an acousto-optical tunable filter (AOTF) and a charge-coupled device (CCD) detector with infinity-corrected microscope has been used for chemical remote sensing and imaging for large areas (>20 km^2) and for micro-Raman spectroscopy. Other new techniques, such as Hadamard transform spectroscopy, are also being used for remote-sensing applications (Ferraro and Nakamoto, 1994).

Typical problems that have been addressed using Raman microscopy with infrared excitation are the characterization of particles on transparent or reflective substrates and the analysis of heterogeneous materials (Chase and Rabolt, 1994). Fiber analysis and polymer-degradation studies provide an extensive literature, especially novel in the area of biopolymers. For most applications, sample preparation for FT-Raman microscopy is relatively simple. For analysis in transmission mode, samples may be analyzed on IR-transparent supports, although it is preferable to avoid using thick supports since these may give rise to artifacts due to spherical aberrations. Nontransparent samples can be analyzed in a reflective mode of operation of the microscope. Samples can be mounted on the end of a needle, and thin films or fibers can be held in position on the microscope stage by fixing at each end with adhesive tape. The samples are then aligned in the IR beam with the *x-y* stage and adjustable apertures. Typically, the sample should have a thickness around 25–50 μm: thin films can be obtained from samples such as polymers, animal or plant tissue, or paint flakes by microtoming cross sections. Particles or small samples that are difficult to handle can be embedded in synthetic resins and then microtomed to obtain samples of optimal thickness, although it may sometimes be easier to crush or to sandwich them between transparent windows in a compression cell. Fibers also frequently need to be flattened, in order to decrease their thickness, but also to reduce the "lensing" effect caused by refraction of infrared radiation at the curved edges.

For analysis in the reflectance mode, virtually no sample preparation is needed, ordinary microscope slides can be used, and the only limitation is the size of both the sample holder and the stage.

Since macroscopic sampling techniques are generally not well adapted to the analysis of corrosion products or aged materials, especially where thin films are concerned, the advantages of FT-Raman microspectroscopy for the analysis of heterogeneous oxidation products are immediately apparent. A complete historical picture of the aging process in material surfaces can now be obtained by the spectroscopic analysis of a microtomed section cut perpendicular to the oxidation gradient. For this purpose, the embedding of the specimen in a resin is a requirement where sample thicknesses of 25–50 μm are to be studied. The thin film that is obtained is then positioned on the sample stage at the center of the field of view, and by means of the adjustable apertures, the area to be analyzed is delimited. Since the oxidation is homogeneous in, say, the y direction, the length Δy is chosen to be as large as possible (typically 600 μm), whereas Δx is as small as permitted by the diffraction limitation. The sample is then moved along the x axis in 5–10-μm steps and the spectrum of each area successively recorded. For example, Raman spectra in the 15-μm-wide layers of a polypropylene sample irradiated in artificial photo-aging conditions in the presence of air have been studied to monitor sample degradation.

In conclusion, it should be added that the profiling technique can be applied to polymer aging in order to measure any heterogeneous distribution of species, and permit, for example, the characterization of the diffusion or the migration of additives or the recognition of nonuniform aging effects that result from nonuniform processing.

4. NONLINEAR SPECTROSCOPY

The availability of pulsed lasers of high irradiance has given rise to the discovery of some new light-scattering phenomena, all of which are generically related to the Raman effect (Kiefer and Long, 1982). These phenomena originate in the nonlinear response of molecular systems to the large electric-field intensities associated with these lasers (Levenson and Kano, 1988). We will now consider some of these nonlinear Raman effects; hyper-Raman and second-order hyper-Raman scattering arise at $2\omega_0$ from one laser of frequency ω_0, whereas coherent anti-Stokes (CARS) and stimulated Raman-gain spectroscopy (SRGS) occur with two lasers, of frequencies ω_1 and ω_2.

4.1 Induced Electric Dipole-Frequency Mixing (Nonlinear Polarization)

The electric dipole **P** induced in a molecule by an electric field is related to the electric-field intensity **E** by the power series

$$\mathbf{P} = \chi^{(1)}\cdot\mathbf{E} + \chi^{(2)}:\mathbf{EE} + \chi^{(3)}\vdots\mathbf{EEE}, \tag{10}$$

where $\chi^{(1)}$, $\chi^{(2)}$, and $\chi^{(3)}$ are the first-, second-, and third-order bulk susceptibilities, respectively. The contribution from $\chi^{(1)}$ is dominant at low electric-field intensity and is responsible for processes such as spontaneous Raman scattering and absorption. $\chi^{(2)}$ and all other even-order terms are zero in isotropic media, but in the case of noncentrosymmetric crystals, $\chi^{(2)}$ can give rise to frequency doubling. At high electric-field intensities the contribution from $\chi^{(3)}$ becomes important and can produce a CARS signal. $\chi^{(3)}$ is a fourth-rank tensor.

In exponential notation, the resultant field **E** from three coherent fields may be expressed as

$$\mathbf{E} = \tfrac{1}{2}\sum_{1}^{3} (\mathbf{E}_{(\omega_i)}e^{i(\kappa_i - \omega_i t)} + c.c) \tag{11}$$

where ω_i and κ_i are the circular frequency component and wave vector of the ith electric field.

4.2 Hyper-Rayleigh and Hyper-Raman Scattering

When a laser of frequency ω_0 and irradiance I is incident on a sample, hyper-Rayleigh scattering is observed at $2\omega_0$ and hyper-Raman (Stokes and anti-Stokes) at $2\omega_0 \pm \omega_M$. The phenomenon is associated with the hyperpolarizability β (Long, 1989. The selection rules for hyper-Raman scattering differ from those of Raman scattering, and an important feature for vibrational spectroscopic applications is that transitions that are hyper-Raman active are often inactive in the Raman or infrared. However, since the intensity of hyper-Raman scattering is many orders of magnitude smaller than spontaneous linear Raman scattering, the technique has had only limited application because of instrumental problems.

Second-order hyper-Raman scattering is based on γ, the second hyperpolarizability, and gives scattering at $3\omega_0 \pm \omega_M$. The intensity is extremely low and has found no application. Other forms of third-order scattering do not suffer from these intensity limitations and are of spectroscopic importance, namely coherent anti-Stokes Raman scattering (CARS) and stimulated Raman-gain (or -loss) spectroscopy (SRGS or SRLS).

4.3 CARS

When two laser beams of sufficiently high irradiances interact nonlinearly in a medium one of the possible results is the production of a coherent beam of frequency ω_3 given by

$$\omega_3 = 2\omega_1 - \omega_2. \tag{12}$$

When $\omega_1 - \omega_2$ coincides with the frequency of a Raman-active mode ω_M then

$$\omega_3 = \omega_1 + \omega_M. \tag{13}$$

This special case is called coherent anti-Stokes Raman spectroscopy (CARS). The main advantage of CARS is the coherence of the emitted beam, which means that it can be detected easily without use of a monochromator; the rejection of fluorescence is also accomplished efficiently because of the small solid angle of deviation of the coherent CARS beam and its wavelength on the anti-Stokes side of the pump beam, ω_1.

Traditionally, CARS spectra were generated by using a narrow-bandwidth, high-powered pulsed laser as a source of ω_1 and a scanning dye laser for ω_2 (Stokes beam); varying ω_2, i.e., $\omega_1 - \omega_2$, generated the CARS spectrum. Phase matching of the three incident beams is crucial in the generation of an intense CARS signal. There are two main optical methods for producing correct phase matching, namely collinear CARS and BOXCARS. The difference between these can be seen in Fig. 26. Here, the K_i values are the wave vectors of the input and emitted beams.

The main disadvantage of collinear CARS, Fig. 26(a), is that poor spatial resolution is achieved. The crossed-beam phase-matching characteristic of BOXCARS, Fig. 26(b), however, gives excellent spatial resolution. In this arrangement, the generated CARS beam is separated from the other input beams by a filter.

A variation of BOXCARS, "folded BOXCARS," is also used, Fig. 26(c). This has the added advantage that the CARS beam emerges spatially separated from the input beams, thus facilitating the signal collection. In this arrangement, the polarization wave

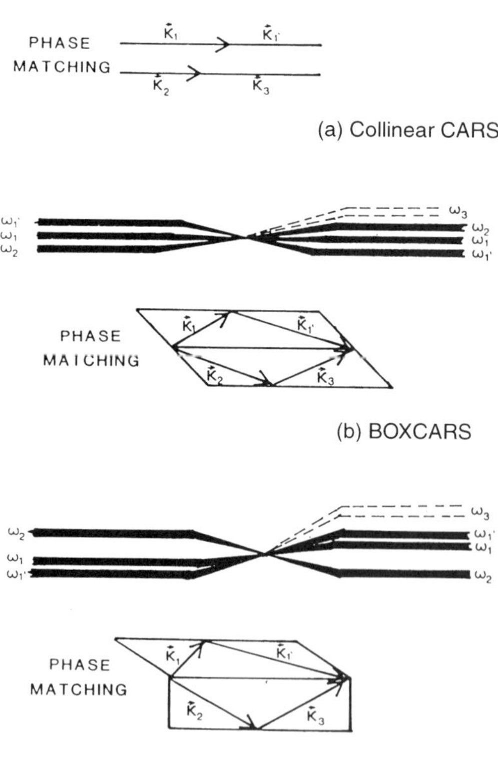

FIG. 26. Phase-matching characteristics of CARS: (a) collinear CARS, (b) BOXCARS, (c) folded BOXCARS.

vector remains equal to the anti-Stokes signal wave vector and the spatial resolution is better than 2 mm.

As the CARS signal emerges as a coherent beam, the intensity should be strictly regarded as an irradiance.

The efficiency of CARS signal generation is governed by the squared modulus of the third-order nonlinear electric susceptibility. As a result of this, the signal is dependent upon the square of the population difference between initial and final states. Thus it is favorable to maintain a reasonable population difference, as increasing pump-beam irradiance will eventually cause saturation and subsequent signal deterioration.

When phase matching is perfect, $\Delta K = 0$, then CARS signal strength will vary quadratically with the interaction length. In general, $\Delta K \neq 0$ and the CARS signal strength shows a succession of progressively weaker maxima. Thus, the CARS signal strength is proportional to

1. the square of the modulus of the third-order nonlinear susceptibility,
2. the square of the population difference between initial and final states,
3. the square of the irradiance of the pump beam,
4. the irradiance of the Stokes beam, and
5. the square of the beam interaction length.

As the CARS signal is generated through the third-order nonlinear molecular susceptibility, its magnitude is dependent on high input electric-field intensities.

In typical experimental CARS work the pump beams at frequency ω_1 were provided by a Nd:YAG laser operated in the Q-switched mode; see Fig. 27. In this system an oscillator and amplifier were employed to give increased power. The Nd:YAG oscillator rod, pumped coaxially by a 50-J xenon flash lamp, was placed between a Pockels-cell polarizer, which gave Q switching, and the output étalon. A telescope within the oscillator cavity reduced the beam diameter to cut down the Fresnel number of the cavity, enabling low divergency output to be attained. On entering the amplifier cavity the beam was passed through a telescope to fill a Nd:YAG rod that was pumped by two xenon flash lamps. The output beam was divided into two parallel beams by a coated 45° beam splitter. The Stokes beam was directed to be vertically above the two pump beams by means of a thick optical flat, the movement of which enabled the correct phase-matching angle of 2° to be achieved.

The spectrometer used in this experiment had a focal length of 0.9 m and incorporated a holographic grating ruled with 2400 lines per mm. As the technique used was broadband CARS, the detector was an intensified diode array, type B&M AK 500 with 512 channels. This array had an output efficiency

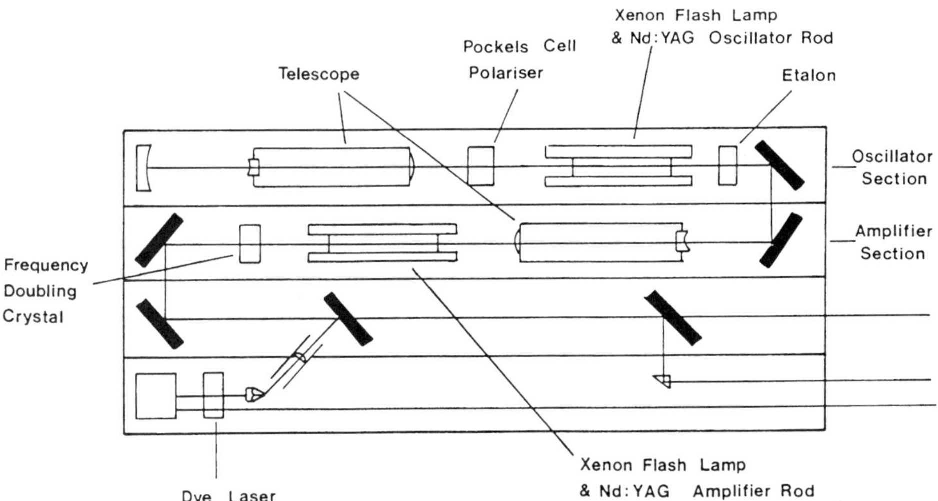

FIG. 27. Oscillator and amplifier experimental arrangement for CARS excitation.

of 10%. This combination of spectrometer and diode-array camera gave an overall dispersion of 0.00815 nm per channel number in the spectral region around the *Q* branch of hydrogen chloride. Calibration was effected by use of hollow-cathode emission lamps. Transference of data from diode array to the computer was carried out with a B&M STH 500 controller and direct memory interfacing. Each CARS spectrum recorded was averaged over 50 laser pulses.

As an example of the data from CARS spectra obtained with the above apparatus, the temperature dependence of the pressure-broadening coefficient for the *Q* branch of hydrogen chloride at 2990 cm^{-1} is shown in Fig. 28.

CARS has several advantages compared with normal Raman scattering, namely

1. high signal levels are achieved, often up to 10^5 times greater than normal Raman intensities;
2. the scattering is produced in a coherent beam whose divergence is small; signal-collection efficiency and spatial discrimination against fluorescence and thermal emission are favored;
3. sample quantities required are small, and spatial resolution is dependent only on the linewidths of the laser frequencies ω_1 and ω_2.

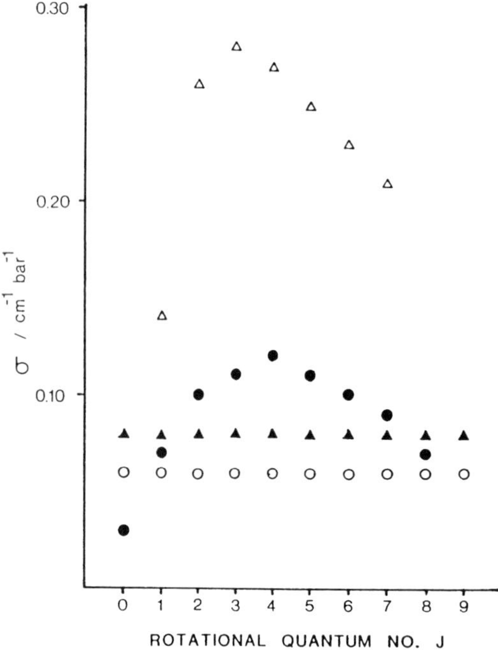

FIG. 28. Temperature dependence of the line-broadening coefficient σ for the *Q* branch of hydrogen chloride; Δ 298 K, ● 550 K, ▲ 798 K, ○ 1048 K.

Some disadvantages of CARS are that the intensity is dependent on the square of the species concentration and is not sensitive for species at low concentrations, the nonresonant background is often difficult to quantify, and samples are prone to damage in the focused laser beams. Experimentally, it is often difficult to maintain the necessary coherence of the input beams over little more than short distances.

Nevertheless, CARS finds application in flame-combustion diagnostic studies and in studies of photochemical reactions, e.g., the photolysis of formaldehyde (Andrews, 1990a).

4.4 SRGS (SRLS)

When two laser beams of frequencies ω_1 and ω_2, where $\omega_1 > \omega_2$, are coincident in a sample, a gain in intensity at ω_2 and a loss at ω_1 occurs, when $\omega_2 = \omega_1 - \omega_M$. With lasers of low power, the gain in intensity at ω_2 is linearly dependent on the laser power at ω_1 and a gain coefficient that depends on energy-level populations and a linewidth associated with ω_M. Similarly, the fractional loss in intensity at ω_1 is proportional to the laser power at ω_2 and a loss coefficient that is itself proportional to the gain coefficient.

Hence, by experimentally sweeping ω_2 through a range of frequencies below ω_1, all of the Raman-active frequencies are observed in a stimulated Raman-gain spectrum (SRGS) that lies on the Stokes side of ω_1, Fig. 29(a). Conversely, if ω_2 is swept through frequencies greater than ω_1, all of the Raman-active frequencies can now be observed as a stimulated Raman-loss spectrum (SRLS) on the anti-Stokes side of ω_1, Fig. 29(b). This phenomenon is also known as inverse-Raman spectroscopy.

Raman-gain and -loss spectra have similar selection rules and intensity dependence to normal Raman spectra.

5. SPECIAL TECHNIQUES IN RAMAN SPECTROSCOPY

5.1 Surface-Enhanced Raman Scattering

A special type of spectral enhancement can occur (Chang and Furtak, 1982) with molecules adsorbed on surfaces through the

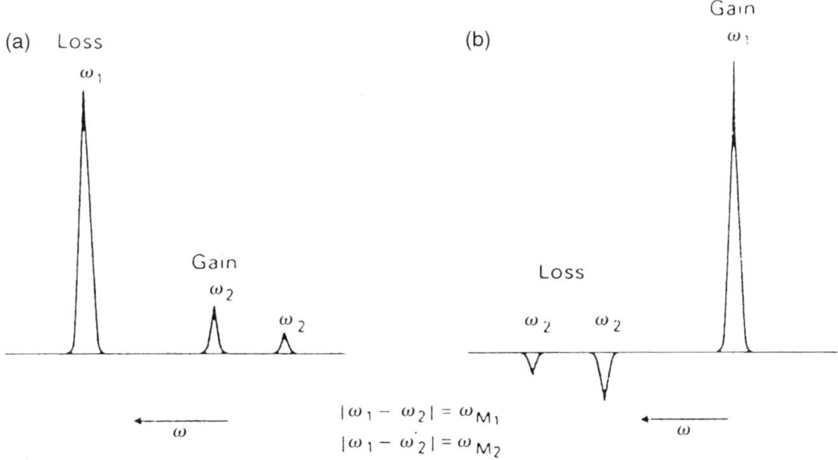

FIG. 29. Representation of (a) SRGS (b) SRLS.

phenomenon known as surface-enhanced Raman scattering (SERS). First observed in electrochemical studies using Raman scattering, SERS has now been adopted to study adsorption on catalysts, interfaces, colloids, corrosion, and metal–metal oxide systems (Weitz *et al.*, 1986). The maximum enhancements are noted with surfaces of roughened metals on the atomic scale and can approach 10^6 in Raman intensity for Cu, Ag, and Au electrodes. It is now accepted that the majority of the signal enhancement comes from an amplification of the electric-field strength ($10^4\times$) and the remainder from effects ascribed to surface roughness (Weitz *et al.*, 1986). The most widely tenable theory is that the incident photons used to excite Raman scattering also excite plasmons on the surface. In a resonance, which is dependent on the surface composition and particle size and on the electromagnetic properties of the base material, the electric fields at the surface are greatly enhanced—hence, the incident and Raman emitted photons benefit from the electric-field environmental enhancement.

The Raman spectra from SERS experiments are not identical with conventional Raman spectra since the enhancement of molecular-vibrational modes is strongly orientation dependent. The information is often difficult to interpret quantitatively, but the SERS effect is, nevertheless, seen to have potential as an analytical technique through the magnitude of the enhancements involved in its generation (Hendra *et al.*, 1991). Recent applications of SERS to studies of molecular structure include proteins adsorbed on silver sols (Grasselli and Bulkin, 1991; Ferraro and Nakamoto, 1994).

In an adaptation of this technique known as optical waveguiding (Chase and Rabolt, 1994), the enhancement of weak signals from interfaces is achieved. Single or multiple reflections of the incident laser beam from metallic surfaces covered with adsorbates normally produce unenhanced and very weak Raman signals. The detection of Raman spectra of thin films on a substrate or at an interface requires both a reasonable optical field at the surface and a long path length to increase the scattering volume. Optical waveguides can be used to produce strong Raman scattering from thin surface films. In a typical waveguide (Fig. 30) the optical field is contained in a narrow region near the surface.

A prism of high refractive index is used to couple the electric field of the incident radiation to the surface-coated waveguide. A small coupling gap of either air or vacuum exists and is generally caused by irregularities in the film. A guided wave can propagate in a film, provided its refractive index is larger than that of the substrate. Generally, this is the case for polymeric materials, which have reasonably high refractive indices ($n > 1.5$) relative to Pyrex or quartz sub-

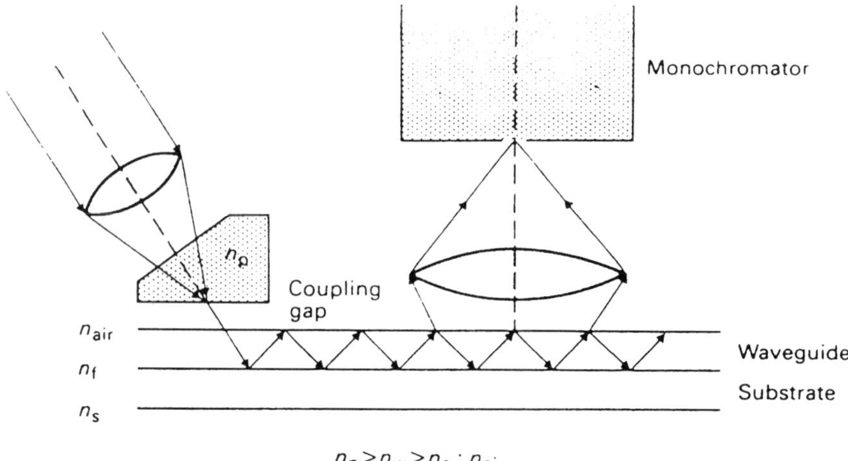

FIG. 30. Optical waveguiding: experimental arrangement for observation of Raman spectra; n_p, n_f, and n_s are the refractive indices of the prism, waveguide, and substrate, respectively.

strates. Raman spectroscopy may thus be used to study organic, inorganic, and biological films; typical improvements of 1500× in sensitivity using waveguiding techniques have been claimed.

5.2 Photoacoustic Raman Spectroscopy (PARS)

Unlike other spectroscopic infrared techniques, which depend on radiation attenuation due to absorption or scattering in a medium, PARS depends on the direct measurement of vibrational or rotational conversion into thermoclastic acoustic energy. The principle of PARS is similar to that of CARS in that a Stokes amplification arises at the expense of pump attenuation, with collisional deactivation of excited molecules resulting in a pressure change in the sample.

Absorption of electromagnetic radiation causes a temperature rise, which can be monitored thermometrically. This is a rather slow and inaccurate process, and a more efficient way of measuring the heating effect uses a gas-microphone cell, which operates through a periodic modulation of the incident radiation. Energy conversion to heat within the sample is fast ($\sim 10^{-8}$ s) and thermal waves originating in the sample cause expansion of a gas layer near the surface; the pressure wave created is observed as an acoustic signal by a microphone. The theory is complex and beyond the scope of the current article but is surveyed by Rosencwaig (1980); see also SPECTROSCOPY, PHOTOACOUSTIC.

The technique, which is closely related to stimulated Raman-gain spectroscopy, is applicable to solids, liquids, and gases but the effect in gases is much larger than in solids; this can cause problems where samples contain water vapor, for example, which can swamp the spectrum of interest. An excellent review is provided by Barrett (1981) with detailed applications in the area of gas-phase rotational spectra.

6. FOURIER-TRANSFORM (FT) METHODS IN VIBRATIONAL SPECTROSCOPY

The major advantage of FT methods in infrared or Raman spectroscopy arises from the multiplexing of the experimental system compared with the sequential elemental scanning of conventional spectroscopic methods (Hendra *et al.*, 1991). Simply, the multiplex advantage derives from the simultaneous measurement of signal from multiple spectral data points by analysis of a single signal from the detector. The first improvements in signal enhancement through this effect were made in the far-infrared region by replacement of the diffraction grating and the sequential scanning operation across an exit slit by a Michelson interferometer. This consists (Fig. 31) of a beam-splitter at 45° to the beam of radiation, which is thus split into mutually orthogonal beams. One of these is returned via a fixed mirror and the other by reflection at a moving mirror, the position of which affects the phase of the combined signal. An interferogram is produced that is composed of a series of cosine waves of different periodicities, whose amplitudes are dependent on signal intensities. Analysis of the interferograms can be accomplished by Fourier transformation. The resolution of such an interferogram is $1/x$, where x is the path length of the recorded interferogram; hence a 0.5-cm path difference gives an effective spectral resolution of 2 cm^{-1}.

Since radiation from all spectral elements over the desired wavelength range is measured in one interferogram scan, the signal-to-noise ratio (*S/N*) can be increased interferometrically (Fellgett advantage) by the rapid coaddition of scans; for example, modern interferometers can achieve 60 interferogram scans per second, and calculation reveals that a 30-min spectral accumulation for a 2000-cm^{-1} range therefore gives a *S/N* improvement of about 50× over the single-scan dispersive system at 1 cm^{-1} s^{-1} (Hendra *et al.*, 1991).

Additionally, because of the longer wavelengths in the infrared region the precision required for the mirror movements is readily achievable. However, because of the sensitivity of photomultiplier and CCD detectors in the visible region, FT methods are not so advantageous and UV/visible spectrometers still tend to be dispersive-grating systems.

For Raman spectroscopy, interferometers have several advantages over monochromators of similar aperture where radiation must pass through narrow slits (typically 50 to 100 μm cm^{-1} spectral bandpass) compared with the larger optical conductance of an interferometer (the Jacquinot advantage). For a spectral resolution of 4 cm^{-1}, for example, the étendue (radiation throughput) of an interferometer may be 100× better than that of the monochromator (Chase and Rabolt, 1994). Other interferometric properties such as the Connes advantage, which arises from the coaddition reproducibility and superimposability of data accumulation, means that in the infrared region the interferometer is

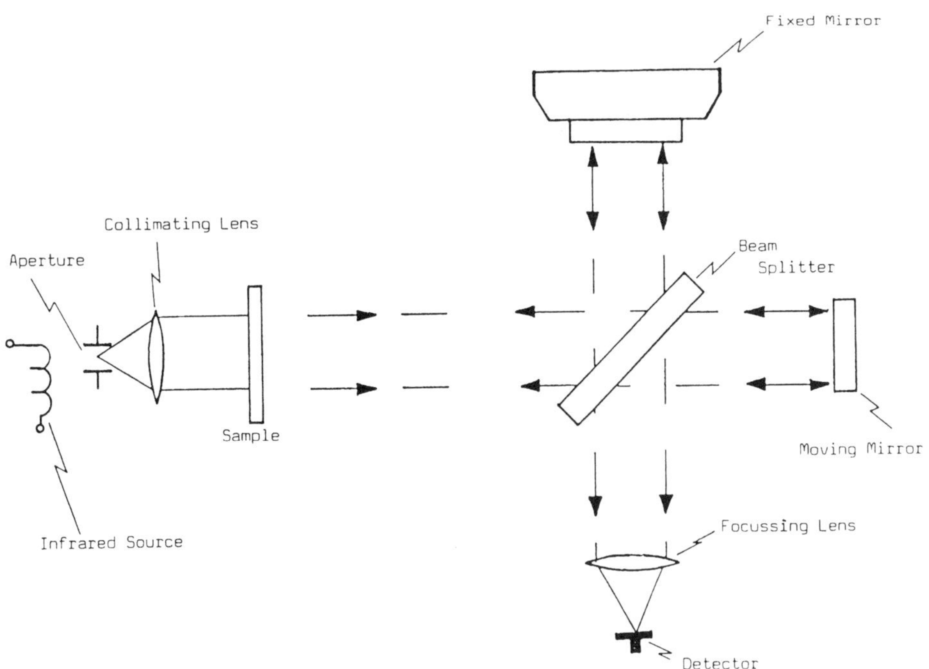

FIG. 31. Optical layout of a Michelson interferometer.

supreme for Raman-scattering experiments. The feasibility of using interferometry and near-infrared laser-excitation sources for Raman spectroscopy was suggested by Chantry *et al.* as long ago as 1964 but interest was renewed through the work of Chase and Hirschfeld (1986), which was quickly followed with the first commercial spectrometer in 1987/88.

For Raman-spectroscopic studies the operation of a system with a near-infrared excitation wavelength (e.g., Nd^{3+}/YAG laser at 1064 nm) obviates the problems associated with fluorescence excitation experienced in the blue and green regions of the visible (Grasselli and Bulkin, 1991; Walker and Straw, 1967). This is an inestimable advantage for the examination of the Raman spectra of fluorescent organic species such as anthracene and polyurethanes and inorganic materials such as uranium salts. In addition, Raman spectra of highly colored absorbing materials such as potassium permanganate, iodine, and metal phosphines have been obtained for the first time using FT-Raman spectroscopy with infrared excitation. On the negative side, however, the ν^4 intensity law for Raman scattering means that the intensity of Raman scattering decreases with increase in wavelength ($\lambda\nu = c$) from the visible to the infrared; hence, the Raman scattering from 435.8-nm mercury-arc excitation is about 40× as intense as that from a Nd^{3+}/YAG laser operating at 1064 nm, power for power. With generally lower powers for infrared lasers compared with the argon-ion system, this is compounded even further.

Despite this intrinsic disadvantage, however, the speed and ease of sampling occasioned by the FT spectrometers now make these attractive instruments for both Raman- and infrared-spectroscopic studies of molecular systems. The advent of FT-Raman spectroscopy and CCD-Raman spectroscopy has truly provided a renaissance for the Raman technique through the wider range of applications now available.

The detailed comparison of conventional disperse-Raman with the FT-Raman spectroscopic technique is difficult because of the correlation of the many instrumental factors involved in each technique. However, the major differences are as follows:

1. Moving to near-infrared excitation at 1.06 μm from the visible reduces the Raman-scattering intensity (dependent on the fourth power of the excitation wavelength) by up to 40× on 435.8-nm excitation.
2. An advantage of near-infrared excitation is the reduction in energy of the lasing transition from about 25 000 to 10 000 cm^{-1}, which therefore inhibits the onset of fluorescence; this is of supreme importance in the application of FT-Raman techniques to biological materials and to biomedical diagnostics.
3. In a scanning Raman monochromator system and to some extent also in spectrographs only a small portion of the dispersed spectrum is analyzed at any one time. With the interferometer, however, the Fellgett advantage applies; this is based on the acquisition of all spectral-resolution elements simultaneously. This greater "speed" of acquisition by an interferometer over a dispersing spectrometer is quantified as $\sqrt{r}$, where r represents the number of resolution elements in the spectrum.
4. The Jacquinot throughput is much greater for the interferometer than for scanning monochromators in any spectral range since the limitation of the spectral slit is removed. The J stop in a typical interferometer is about 8 mm diam compared with 0.1-mm slit width in a spectrometer system.
5. Internal calibration of the interferometer against a helium–neon laser (Connes advantage) provides exceptional wave-number reproducibility, which facilitates the superposition of spectral data (accumulation) and data subtraction (background, solvent, etc.). Hence, although all the spectral data in an FT-Raman experiment may be provided in one scan of about 2 s, multiple spectral-data accumulation to improve the signal-to-noise ratio is desirable. In cases of very weak Raman spectra, especially from biological materials for which the incident laser power has been reduced to less than 20 mW from 1 W to minimize sample degradation, accumulations of many hours duration are acceptably realistic; e.g., the first FT-Raman spectroscopic analysis of archaeological human tissue of the "Ice-man," a 5200-

year-old corpse, has been provided with up to 12 000 accumulated spectral scans at a resolution of 4 cm^{-1} from a 1-μg sample without deterioration of spectral resolution or quality or evidence of sample degradation (Edwards *et al.*, 1995).

A major advantage of Raman spectroscopy generally for sample investigation is the weak Raman scattering from water molecules; hence, for reactions in aqueous solution, biomolecules, and biopolymeric studies, little sample preparation in the form of desiccation or freeze drying is necessary. This is particularly crucial for the monitoring of living biological systems by vibrational spectroscopy, where infrared-spectroscopic absorption due to water is strong. However, as shown in Fig. 32, the move of excitation wavelength further into the infrared severely restricts the Raman data that can be acquired in the presence of water, because of increased absorption. This will present a severe limitation to longer-wavelength capability for FT-Raman excitation. Even at 1064 nm, the effect of higher-overtone water-based absorption on the higher–wave-number shifts will be manifest. For this reason, the substitution of heavy water (D_2O) is noted to alleviate partially the absorption problem with 1064-nm excitation in the near infrared.

The increased speed of data acquisition and ease of sampling have made the near-infrared excitation of FT-Raman spectra attractive for industrial remote-sensing applications. Figure 33 shows the FT-Raman spectrum of an aspirin tablet (acetylsalicylic acid) containing a starch bulking agent measured in 120 s using a fiber-optic sensing device.

Extension of FT-Raman spectroscopy into the realm of fast, real-time data acquisition in nanoseconds is limited by the relatively slow frequency response of the detectors, but time resolutions of 50 μs through step scanning have been shown to be possible. A critical parameter here, too, is the rather poor signal-to-noise ratio generally experienced from single-scan or single-shot Raman spectra (Chase and Rabolt, 1994).

7. INTENSITY MEASUREMENTS IN RAMAN SCATTERING

A major advantage of Raman spectroscopy, whether conventional or FT technique, is the linear dependence on concentration of molecular species and the observed band

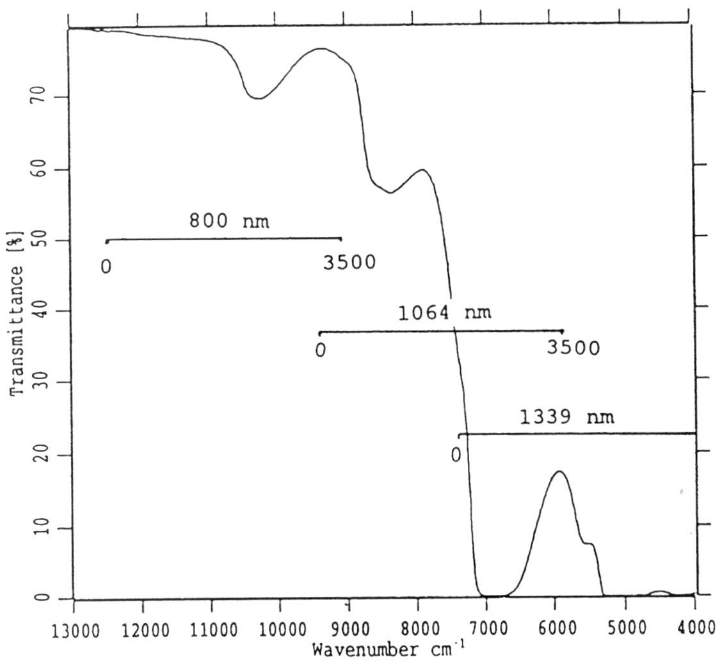

FIG. 32. Absorption spectrum of water in the near-infrared region, showing the wave-number shifts (0–3500 cm^{-1}) for Stokes Raman spectra excited with 800-, 1064-, and 1339-nm laser wavelengths.

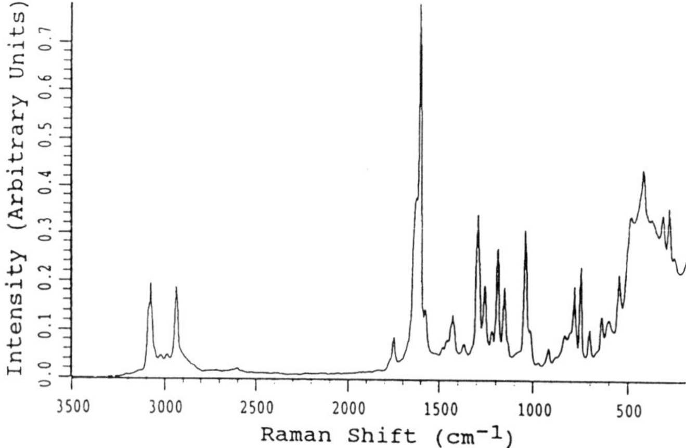

FIG. 33. FT-Raman spectra of a pharmaceutical tablet containing acetylsalicylic acid; resolution 8 cm^{-1}; 1064-nm excitation; 100-mW power; 120-s accumulation time.

intensity (Long, 1977). Unlike absorption spectroscopy, where the Beer–Lambert law dictates a logarithmic dependency of concentration on absorption intensity, which demands knowledge of the molar absorptivity ϵ for the quantification of spectral-intensity data, the linearity of concentration/intensity dependence in the Raman spectrum provides an excellent means of following the growth or depletion of molecular species during a reaction.

In this way, Raman spectroscopy has been developed recently as a technique for monitoring reaction kinetics of polymerizations, such as the living anionic polymerization of styrene and butadiene in hydrocarbon solvents by butyl–lithium initiator. The information about kinetic and thermodynamic quantities from Raman-spectroscopic measurements in solutions over various temperatures has thus provided detailed information about molecular and species equilibria, the presence of dimers or oligomers in a reaction system, hydrogen bonding in alcohol and water systems, nitrating species in acidic media, and tautomeric stability. All of these applications are of importance in commercial systems and synthetic procedures. Biologically, changes in distribution and composition of amino acid residues in large polypeptides and sulfur bridges in keratotic samples studied by Raman spectroscopy have provided an insight into drug-targeting systems and tissue degeneration, both of which are vital for biomedical diagnostic applications. Likewise, the absorption of drugs across the human skin barrier, transdermal drug delivery, of importance in pharmaceutical therapeutic formulations has been studied successfully by the FT-Raman spectroscopic technique and the influence of chemical enhancers on drug-delivery systems quantified.

However, the linearity between the observed intensity and the concentration of scattering species in the Raman spectrum is not always applicable, especially when the wavelength of laser excitation approaches that of an electronic absorption band in the scattering species under investigation. This is the situation that occurs in resonance Raman spectroscopy (RRS); the signal strength of Raman scattering from a molecular species is increased by as much as 10^8, and the spectrum is simplified because only the vibrations associated with the chromophore under excitation are able to generate the signals in resonance. For example, in the case of hemoglobin, laser excitation within the heme absorption (400–600 nm) produces a RRS that contains only the enhanced signals of the heme ring vibrational modes, and no bands arising from the protein component are observed.

The inherent advantage of RRS is an enhanced spectrum strength and simplification of the spectra; this has found particular application recently in biological studies of site-specific drug–DNA interactions and in the monitoring of the early stages of polymer degradation, which produces sites of Raman-active unsaturation. A critical limitation of RRS is the presence of an electronic absorption in the visible region that is accessible to probing by appropriate tunable dye lasers; of course, visible electronic absorption bands also confer color on chemical materials, and many systems worthy of study are colorless. To address the last point, the development of UV resonance Raman spectroscopy has been undertaken, since most molecules, particularly those of biological significance, have absorption bands in the 200–400-nm region of the electromagnetic spectrum. In addition, it is realized that for laser excitation at wavelengths less than about 270 nm fluorescence processes are reduced because the excited singlet states relax through fast radiationless transitions before onset of slower fluorescence occurs. Hence, UV RRS is seen to have great potential as a sensitive technique for the probing of the molecular structures of a range of interesting systems; the major drawback, however, is the lack of appropriate lasing frequencies, particularly in the 200–300-nm region. Most UV RRS studies reported in the literature have adopted a frequency-quadrupled Nd/YAG-pulsed laser, operating at 266 nm, but, because of small pulse rates and high peak pulse powers, there are associated problems related to sample exposure for unacceptably long times, which can result in photodegradation and chemical deterioration.

With the advent in 1994 of a new cw UV laser based on BBO (β barium borate), no fewer than six new wavelengths are now available for UV RRS studies, namely 229, 238, 244, 248, 257, and 264 nm; this range of new wavelengths and associated laser powers should provide a new platform for the next phase of UV RRS studies of a range of biological systems.

GLOSSARY

Collimation: The production of scattered, diffracted, or incident radiation in parallel wave form for illumination of optical devices.

Dispersion: A measure of the amount by which neighboring wavelengths are spread apart by a spectrograph or spectrometer. The greater the resolution required of an instrument, the greater the dispersion needed. Often quoted as reciprocal linear dispersion (rld), with units of cm^{-1} mm^{-1} at the exit slit of the **monochromator;** through the rld and the physical slit width of the instrument, the spectral **resolution** (in cm^{-1}) can be calculated.

Étalon: An optical device for the selection of individual wavelength components comprising a **monochromatic** radiation source.

Free Spectral Range: For diffraction studies, as the order of diffraction, n, increases the dispersion increases but the resulting spread of wavelengths gives rise to a situation where the orders then effectively overlap each other. The free spectral range (in cm^{-1} or nm) describes the absolute spectral separation of wavelengths generated in different orders.

Grating Ghost: The imperfections of a diffraction grating, particularly those in a mechanically produced or ruled blank result in periodic errors that become manifest as diffracted wavelengths. They can be identified in spectroscopic experiments through their wavelength dependence on measured positions.

Inertia Moment: The sum of the products of mass and distance squared for all atoms in the molecular system: $I = \Sigma mr^2$. As I increases, the **rotational constant** decreases; hence, heavier molecules have more closely spaced rotational lines that require better **dispersion** instruments for increased **resolution.**

Irradiance and Intensity: The irradiance of a source is measured in units of W m^{-2}, whereas the intensity has units of W sr^{-1}. This difference reflects the solid-angle dependence of illumination in the latter case through lack of **collimation** and the power/distance (collimated, nondivergence) relationship that typifies a laser beam with respect to the inverse-square law obeyed by normal, nonlasing sources.

Monochromatic: Greek "one color," which refers to radiation of a single wavelength λ. However, the concept of truly monochromatic radiation is a diffuse one since with instruments of high **dispersion** and **resolution** even monochromatic laser wavelengths have been found to contain many constituent wavelengths in the spectral band; e.g., the Ar^+ laser wavelength, 488.0 nm, contains up to 50 component wavelengths (each "monochromatic") in a measured band envelope of about 0.002 nm.

Pressure Broadening: Molecular interactions in the gas phase result in an increase in observed transition bandwidths through collisions. Hence, although this effect diminishes the capability of instrumental **resolution** of rotational transition, it also provides a mechanism for studying molecular short-range and long-range forces and the influence of environments upon these.

Resolving Power and Resolution: The efficiency of a spectrograph or spectrometer for the **dispersion** of radiation is related to the resolving power R, which is dependent in the widest sense on the diffraction order, grating-dispersion efficiency, and the focal length of the instrument. The resolving power is dimensionless, but application of this quantity to a particular wavelength or wave number gives rise to the spectroscopically useful resolution; e.g., for $R = 10^6$, a spectrometer is capable of resolving two wavelengths near 500 nm with a theoretical resolution of 0.0005 nm.

Rotational Constants: Termed B and D, with subscripts to indicate the vibrational state to which they belong, e.g., B_0, D_0 for the ground vibrational state. Technically, B is the rotational constant and D is a centrifugal distortion constant. B depends inversely on the **inertia moment;** hence heavier molecules have more closely spaced rotational lines. D, however, is dependent on a J^2 term and is more significant for heavier molecules at higher J levels.

Speed: The speed of an optical spectrograph or spectrometer is inversely proportional to the square of its aperture. Hence, for example, the speed of an $f/1$ spectrometer is only one-ninth of that of its $f/3$ counterpart. This is related to its light-gathering power (brightness) but the resolution capability is better for the $f/1$ system.

Works Cited

Andrews, D. L. (Ed.) (1990a), *Perspectives in Modern Chemical Spectroscopy,* Berlin: Springer-Verlag.

Andrews, D. L. (1990b), *Lasers in Chemistry,* 2nd ed., Berlin: Springer-Verlag.

Asselin, K. T., Chase, D. B. (1994), *Appl. Spectrosc.* **48,** 669–672.

Barrett, J. J. (1981), "Photoacoustic Raman Spectroscopy of Gases," in: A. B. Harvey (Ed.), *Chemical Applications of Nonlinear Raman Spectroscopy,* New York: Academic, Chap. 3.

Behringer, J. (1967), "Observed Resonance Raman Spectra," in: H. A. Szymanski (Ed.), *Raman Spectroscopy,* Vol. 1, New York: Plenum.

Born, M., Wolf, E. (1980), *Principles of Optics,* 6th ed., Oxford: Pergamon.

Bower, D. I., Maddams, W. F. (1989), *Vibrational Spectroscopy of Polymers,* Cambridge, UK: Cambridge Univ. Press.

Brandmüller, J., Moser, H. (1962), *Einführung in Die Ramanspektroskopie,* Darmstadt: Steinkoff.

Burdett, J. K., Poliakoff, M. (1974), "Tunable Lasers," *Chem. Soc. Rev.* **3,** 293–307.

Butcher, R. J., Willetts, D. V., Jones, W. J. (1971), "On the Use of a Fabry-Pérot Etalon for the Determination of Rotational Constants of Simple Molecules—the Pure Rotational Raman Spectra of Oxygen and Nitrogen," *Proc. R. Soc. London, Sect. A* **324,** 231–245.

Cannon, J. (1994), "Advances in UV-Resonance Raman Spectroscopy," *Int. Spectrosc. Lab.* **24,** 6–7.

Chang, R. K., Furtak, R. (1982), *Surface-Enhanced Raman Scattering,* New York: Plenum.

Chantry, G. W., Gebbie, H. A., Hilsum, C. (1964), "Interferometric Raman Spectroscopy Using Infra-Red Excitation," *Nature* **203,** 1052–1056.

Chase, D. B., Hirschfeld, T. (1986), "FT-Raman Spectroscopy: Development and Justification," *Appl. Spectrosc.* **40,** 133–137.

Chase, D. B., Rabolt, J. F. (1994), *Fourier Transform Raman Spectroscopy: From Concept to Experiment,* San Diego: Academic.

Demtröder, W. (1981), *Laser Spectroscopy: Basic Concepts and Instrumentation,* Berlin: Springer-Verlag.

Denney, R. C., Sinclair, R. (1987), *Visible and Ultraviolet Spectroscopy,* New York and London: Wiley Interscience.

Dodd, R. E. (1962), *Chemical Spectroscopy,* New York: Elsevier.

Edwards, H. G. M. (1970), "High-Resolution Raman Spectroscopy of Gases," in: D. A. Long, L. A. K. Staveley, A. J. Downs (Eds.), *Essays in Structural Chemistry,* London: Macmillan, Chap. 6.

Edwards, H. G. M., Williams, A. C. (1995), "Potential of FT-Raman Spectroscopy in Biomedical Diagnostics," *J. Mol. Struct.* **347,** 379–388.

Edwards, H. G. M., Williams, A. C., Barry, B. W. (1995), *Biochim. Biophys. Acta* **1246,** 98–105.

Ferraro, J. R., Nakamoto, K. (1994), *Introductory Raman Spectroscopy,* San Diego: Academic.

Garetz, B. A., Lombardi, J. R. (1986), *Advances in Laser Spectrometry,* Chichester: Wiley.

Grasselli, J. G., Bulkin, B. J. (1991), *Analytical Raman Spectroscopy,* New York: Wiley.

Hendra, P. J., Jones, C. H. Warnes, G. (1991), *Fourier-Transform Raman Spectroscopy: Instrumentation and Chemical Applications,* New York: Ellis Horwood.

Jones, W. J. (1972), *Contemp. Phys.* **13,** 419–439.

Kiefer, W., Long, D. A. (Eds.) (1982), *Non-Linear Raman Spectroscopy and Its Chemical Applications,* Dordrecht: Riedel.

Kohlrausch, K. W. F. (1972), *Ramanspektren,* London: Heyden.

Levenson, M. D., Kano, S. S. (1988), *Introduction to Non-Linear Laser Spectroscopy,* San Diego: Academic.

Loader, J. (1970), *Basic Laser Raman Spectroscopy,* London: Heyden.

Long, D. A. (1977), *Raman Spectroscopy,* New York: McGraw-Hill.

Long, D. A. (1988), *Int. Rev. Phys. Chem.* **7,** 317–349.

Long, D. A. (Ed.) (1989), Special issue on Renaissance of Raman Spectroscopy, *Chem. Brit.* **25,** 589–622.

Messerschmidt, R. G., Hartcock, M. A. (1988), *Infrared Microscopy: Theory and Practice,* New York: Marcel Dekker.

Patel, C. K. N. (1974), "Spin-Flip Raman Lasers," in: M. S. Feld, A. Javan, N. A. Kurnit (Eds.), *Fundamental and Applied Laser Physics, Proceedings of the Esfahan Symposium, August 29–September 5, 1971,* New York: Wiley.

Robinson, J. W. (1991), *Practical Handbook of Spectroscopy,* Boca Raton, FL: CRC Press.

Rosencwaig, A. (1980), *Photoacoustics and Photoacoustic Spectroscopy,* New York: Wiley.

Sawyer, R. A. (1963), *Experimental Spectroscopy,* New York: Dover.

Schrader, B., Keller, S., Hoffman, A., Schrader, W., Metc, K., Rehlaender, A., Pahnke, J., Ruwe, M., Budach, W. (1994), "Biomedical Applications of Near-Infrared Excited FT-Raman Spectra," *J. Raman Spectrosc.* **25,** 663–671.

Siesler, H. W., Holland-Moritz, K. M. (1980), *IR and Raman Spectroscopy of Polymers,* New York: Marcel Dekker.

Spiro, T. (1987), *Biological Applications of Raman Spectroscopy,* Vol. 2, *Resonance Raman Spectra,* New York: Wiley Interscience.

Stammreich, H., Forneris, R., Tavares, Y. (1961), "High-Resolution Raman Spectroscopy in the Red and Near Infra-Red II: Vibrational Frequencies and Molecular Interactions of Halogens and Diatomic Interhalogens," *Spectrochim. Acta* **17,** 1173–1178.

Stoicheff, B. P. (1959) in: H. W. Thompson (Ed.), *Advances in Spectroscopy,* Vol. 1, New York: Wiley Interscience, pp. 91–174.

Tabaksblat, R., Meier, R. J., Kip, B. J. (1992), "Confocal Raman Microspectroscopy: Theory and Application to Thin Polymer Samples," *Appl. Spectrosc.* **46,** 60–68.

Treado, P. J., Morris, M. D. (1992), in: M. D. Morris (Ed.), *Spectroscopic and Microscopic Imaging of the Chemical State,* New York: Marcel Dekker.

Walker, S., Straw, H. (1967), *Spectroscopy,* Vol. 2, London: Chapman and Hall.

Weber, A. (1972), *Dev. Appl. Spectrosc.* **10,** 137.

Weber, A. (Ed.) (1979), *Raman Spectroscopy of Gases and Liquids,* Berlin: Springer-Verlag.

Weitz, D. A., Moskovits, M., Creighton, J. A. (1986), *Chemical Structures of Interfaces,* New York: VCH.

Willard, H. H., Merritt, L. L., Dean, J. A., Settle, F. A. (1981), *Instrumental Methods of Analysis,* 6th ed., New York: Van Nostrand.

Further Reading

Elsley, G. L. (1981), *Coherent Raman Spectroscopy,* Oxford: Pergamon.

Garetz, B. A., Lombardi, J. R. (1986), *Advances in Laser Spectrometry,* Chichester: Wiley.

Hollas, J. M. (1993), *Modern Spectroscopy,* 2nd ed., Chichester: Wiley.

Long, D. A. (1977), *Raman Spectroscopy,* New York: McGraw-Hill.

Robinson, J. W. (1991), *Practical Handbook of Spectroscopy,* Boca Raton, FL: CRC Press.

Spiro, T. (1987/88), *Biological Applications of Raman Spectroscopy,* Vols. 1–3, New York: Interscience.

Strommen, D. P., Nakamoto, K. (1985), *Laboratory Raman Spectroscopy,* Chichester: Wiley.

RANDOM PROCESSES

Michael F. Shlesinger, *Office of Naval Research, Arlington, Virginia, U.S.A.*

	Introduction	45
1.	**A Brief History of Random Processes**	46
2.	**Gaussian Random Walks**	48
3.	**Langevin Equations**	49
3.1	Kubo Oscillator	50
4.	**Probability Equations**	51
4.1	The Chain Equation and the Master Equation	51
4.2	Fokker–Planck Equations	52
5.	**Random Walks**	53
5.1	Discrete-Time	53
5.2	Continuous-Time	55
5.3	Many Walkers	58
6.	**Reaction Kinetics**	59
6.1	Trapping	59
6.2	First- and Second-Order Kinetics	59
6.3	Reaction–Diffusion Equations	60
7.	**Fractal-Time Processes**	61
7.1	Fractal Waiting Times	61
7.2	Stretched-Exponential Relaxation	62
8.	**Fractal-Space Processes**	63
8.1	Lévy Flights	63
8.2	Lévy Walks	64
8.3	Random Walks on Fractals	65
9.	**Disordered Systems**	66
	Glossary	68
	Works Cited	69
	Further Reading	69

INTRODUCTION

Our main theme is the random motion of a particle. A random process that involves a time variable is also called a stochastic process. A physical approach to this problem, pioneered by Langevin, is to write a Newtonian equation of motion $F = Ma$ for a particle. One can include additive random forces, as well as noisy parameters. A frictional force and a force derived from a potential can be included, as well. This approach allows one to investigate random motion in a potential, where the force is written as minus the gradient of the potential. The force can be linear or nonlinear. Random motion in the linear force field of a harmonic potential is called an Ornstein–Uhlenbeck process. In the Langevin approach the particle is always in motion. In the random-walk approach, discussed below, a particle can become trapped and not move for a random length of time.

The variables in a Langevin equation, such as position and velocity, are random variables because they are determined, in part, by the noise terms in the equation. We describe how to find the time dependence of moments of these random variables. When calculating moments, the simplest case arises when the noise is uncorrelated with itself in time or space. The effect of noise correlation can, however, be of overriding importance. We will see in the Kubo oscillator example how time-correlated noise dramatically changes the high-frequency behavior of the moments of a noisy oscillator. Moments, however, only give limited information about a random process. All the possible information about a random process is contained in probability distributions. We will discuss the Fokker–Planck method for deriving an equation for the probability distribution for random variables in Langevin equations. Over all the Langevin and Fokker–Planck equations are the approach of choice when the random process is best described by an equation of motion with noisy terms.

A different approach to random processes is to work with the probabilities that govern changes in a particle's position, and not to write directly an equation for random variables. This has been called a random-walk or master-equation approach. The focus is on the elemental transition probability for the step of a particle. This is a powerful method

3-527-28138-X/96/$5.00 + .50

because it allows many complicated situations and features to be directly incorporated and analyzed. For example, we start with the case of a random walk where a particle makes jumps of random lengths in random directions, but then add several important generalizations. We let the time between jumps, and the jump velocity, become random. This even includes fractal probability distributions with infinite mean jump lengths and infinite mean waiting times between jumps. We can account for the random walker being restricted to a fractal space or placed in a disordered environment. Reaction processes that annihilate random walkers are also introduced. Over all the random-walk methods are the approach of choice when one wishes to model the elemental transitions in a random process.

1. A BRIEF HISTORY OF RANDOM PROCESSES

Man's conscious deliberation with random processes is as least as old as the ancient game of throwing the bones (David, 1962). The bones in question, called astrogali, are the heel bones of animals including dogs and sheep. These bones fit nicely in the hand and are the forerunners of dice. In fact, the French expression "jeu d'hasard" (games of hazard or chance) comes from the Arabic for dice, "al-zar" (Székely, 1986). In reality, gambling (from dice to actuarial tables to risk-benefit analysis) has always been in the forefront of expanding the frontiers of probability theory. The bones in question have only four faces upon which they come to rest. The four-sided spinning dreidel is the most similar, in this respect, to throwing of the bones. In ancient times a mathematics of probability did not arise. Perhaps this was because the early dice did not have equally likely outcomes and the occurrence of a rare event (landing on a less probable face) was ascribed to luck rather than sparking the development of the idea of the permanence of statistical ratios. On the other hand, a mathematics of probability also did not arise from the ancient Chinese *I Ching,* which has 64 equally likely outcomes. The *I Ching* was used as an oracle and perhaps has too many outcomes to provide the insight to inaugurate a new field of mathematics.

Despite some early writings on dice games, probability got its recognized start with a correspondence between Pascal and Fermat (Todhunter, 1865) in 1654 that cleared up a misconception between the probability to win and the expected winnings. Specifically, the expected number of sixes in four throws of a die is 4/6, which equals the expected number of pairs of sixes in 24 throws of a pair of dice. The probability to throw at least one six in four throws is $1 - (5/6)^4 \cong 0.5177$, and the probability to get at least one pair of sixes in 24 throws of a pair of dice is $1 - (35/36)^{24} \cong 0.4914$. Equal expectations do not imply equal probabilities. Having easily settled this matter, Pascal and Fermat wrestled with other problems such as how to split up the ante in games of chance that are not completed. With more than two players considered, these problems would tax many modern practitioners.

The Pascal–Fermat letters stimulated Huygens to write a treatise on probability theory. This, in turn, piqued the interest of Jacob Bernoulli, who turned his attention to questions of combinations and permutations, and games of chance. His works were published posthumously, in 1713, under the title *Ars Conjectandi* (Art of Conjecture). He introduced the Bernoulli process where a player has a probability p to win in each trial of a game, and probability $q = 1 - p$ to lose. We would today equate this to a random walk on a lattice by keeping track of a player's status along a one-dimensional axis, with success equated to a jump of one unit to the right, and failure a jump of one unit to the left. The probability $p_n(k)$ of k successes in n trials is given by $[n!/(n - k)!k!]p^k q^{n-k}$. In the limit of large n, and for $p = q$, DeMoivre showed, in 1756, that the probability to get $2h$ more successes than failures (or vice versa) in n Bernoulli trials is $(2\pi n)^{-1/2} \exp(-h^2/2n)$. This can be seen by expanding the factorials in $p_n(k)$ with $k = n/2 \pm h$. This probability function is today called a Gaussian because Gauss showed, in 1809, that DeMoivre's result holds in a more general situation. An intimate connection (Einstein, 1905) with these types of random-walk processes and Gaussian probabilities was later made to the phenomenon of diffusion called Brownian motion. For example, in Bernoulli's game (process) consider two players each starting with $R/2$ coins and with equal probability to win or lose a coin in each trial. The mean number of trials, $\langle N(R)\rangle$, before one player loses his fortune can be calculated to be $R(R + 1)/6$, and for large R, $\langle N(R)\rangle \approx R^2$. In Brownian motion, the mean-square distance a particle moves after N steps satisfies $\langle R^2(N)\rangle \approx N$.

Poisson, in the 1830s, found another possible limit for the Bernoulli process; let $p \to 0$ and $n \to \infty$ such that $np \to \lambda$, a constant. Then $p_n(k) = n(n - 1) \cdots (n - k + 1)\, p^k(1 - p)^{n-k}/k! \approx (np)^k e^{-np} e^{pk}/k! \approx \lambda^k e^{-\lambda}/k!$. This is called a Poisson probability distribution. Poisson wrote a book on probability, but devoted to social problems, such as the probability of a jury reaching the correct verdict.

Over in London, Bernoulli's contemporary De Moivre wrote, in 1718, the first edition of his famous treatise on probability, *The Doctrine of Chances,* which provided explicit odds for popular card games and, more importantly, introduced new concepts, including the use of generating functions, Sterling's formula for $\log(n!)$, the derivation of the formula $(\cos\vartheta + i \sin\vartheta)^n = \cos n\vartheta + i \sin n\vartheta$, and the proof that the Gaussian is the many-trial limit of a Bernoulli process (in the 1756 3rd edition). The Gaussian (also called the normal) is the probability distribution for a *sum* of identically distributed random variables with finite second moments. It was not until 1879 that MacAlister introduced the distribution for a *product* of random variables called the log normal distribution, as it is a Gaussian in the logarithm of the variable. De Moivre's other probability book, *Annuities upon Lives,* inaugurated actuarial science. Laplace in his 1815 treatise *Theorie Analytique des Probabilites* continued the tradition of introducing new tools in applied mathematics through probability. Before Laplace, most work in probability focused on the discrete, where all the possible outcomes of a trial could be enumerated. Card and dice games are good examples of this genre. Laplace brought the integral and continuum mathematics to the fore of probability theory with questions like: If an urn contained an infinite number of black and white tickets and one chooses A tickets of which p turn out to be white and q are black, then what is the probability that if an additional B tickets are chosen, m are white and n are black. Laplace's solution,

$$\frac{\int_0^1 x^{p+m}(1 - x)^{q+n}dx}{\int_0^1 x^p(1 - x)^q dx}, \tag{1}$$

led the way for calculating a wide variety of integrals, many of which are well known today.

Despite opposition to probability as a real mathematics, due to a number of seeming paradoxes, and prohibitions from church and state, due to its gambling connections, the ideas set in motion by the above-mentioned books would spread and be carried on by new generations. Poisson's book found its way to Russia where Chebyshev read it and did his 1846 Master's thesis on probability, and also started a school in this area. His star students were Markov and Lyapunov whose Markov chains and Lyapunov exponents are of great importance in modern work. Kolmogorov, in Russia, placed probability on a secure mathematical foundation. The early 1900s saw the introduction of Brownian motion by Einstein and Bachelier, and reaction kinetics by Smoluchowski. Langevin, in 1908, studied equations of motion with an additive random force. Jumping over potential barriers in noisy systems (Pontryagin *et al.*, 1933; Kramers, 1940) initiated the field of noise-induced transitions. Schottky (1922) studied the phenomenon of shot noise in diodes. Specifically, if electrons arrive at a detector with a rate ν and cause a decaying time-dependent current $i(t)$, then the total measured current is

$$I(t) = \sum_{j=1}^{\infty} i(t - T_j),$$

where T_j represents the past arrival time of the jth electron. This shot noise has an average value of $\nu\int_0^\infty i(t)dt$ and variance $\nu\int_0^\infty i^2(t)dt$. This is also known as Campbell's theorem to mathematicians, and it was introduced in 1909. Chandrasekhar's review of random processes (Chandrasekhar, 1943), including random walks, Brownian motion, escape over barriers, and stellar dynamics, dominated the field for many years. Rice (Wax, 1954) reviewed advances in analyzing noise in electrical devices. The analysis of telephone switching systems, by Fry at Bell Telephone in the 1920s and by Erlang in Denmark in the 1940s, produced advances in the statistical dynamics of congestion of networks. Fisher and Tippett (1928) introduced extreme-value theory to find the largest or the smallest value from a set of random variables. Statistical physics focused on noise, for its own intrinsic properties, and introduced the master equation, entropy, and fluctuation–dissipation theorems. Lévy, in the 1920s and 1930s, introduced scale-invariant random walks and distributions for random variables with infinite moments, which were an important precursor for the field of fractals. Such a wide variety of problems depend on probability that we cannot cover all the major topics and will omit many important problems including random-matrix theory, random fields, Ising-model dynamics, combinatorics, queuing theory, reliability theory, stochastic resonance, Monte Carlo techniques, signal processing, random-growth models, multifractals, and quantum probabilities.

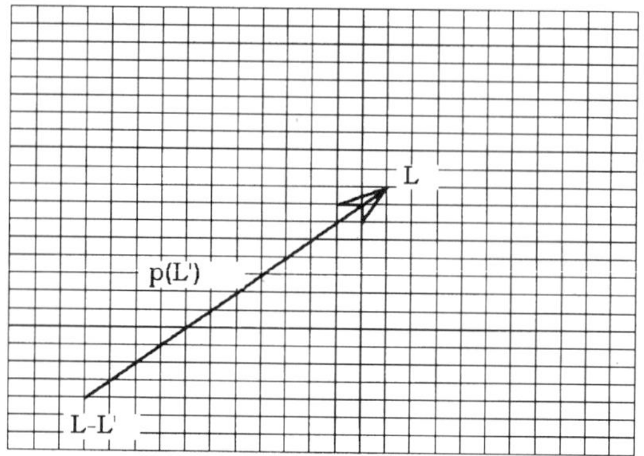

FIG. 1. A finite part of an infinite two-dimensional lattice. For a random walker on this lattice to reach site L after $n + 1$ steps, the walker first reaches some intermediate site $L - L'$ after n steps and then makes a transition with a displacement of L' at the next step with probability $p(L')$. A sum over all possible L' is necessary, as performed in Eq. (2), to obtain the probability to be at site L on the nth step.

2. GAUSSIAN RANDOM WALKS

Bernoulli's gambling problem allowed one's fortune to increase or decrease in discrete amounts. This is equivalent to a random walk on a lattice. Consider a random walk that starts at the origin of a lattice. We begin by choosing the probability $p(l)$ for the walker in one transition to change its position by l lattice spacings. All probabilities can be written in terms of $p(l)$ and the initial condition. The probability to be at site l after $n + 1$ steps is denoted by $P_{n+1}(l)$ (see Fig. 1 for a two-dimensional example), which we can write in one dimension as

$$P_{n+1}(l) = \sum_{l'} P_n(l - l')p(l'). \tag{2}$$

Let us introduce the discrete Fourier transform $\tilde{p}(k)$ defined by

$$\tilde{p}(k) \equiv \sum_{l} \exp(ikl)p(l). \tag{3}$$

Fourier transforming $P_n(l)$ we have, since Eq. (2) is in the form of a convolution,

$$\tilde{P}_n(k) = [\tilde{p}(k)]^n$$

or

$$P_n(l) = \frac{1}{2\pi}\int_{-\pi}^{\pi} [\tilde{p}(k)]^n \exp(-ikl)dk. \tag{4}$$

For an unbiased random walk with no preferred direction, $\tilde{p}(k)$ is real and has a maximum value of 1 at $k = 0$. For example, a random walk with equal probabilities of $\frac{1}{2}$ to jump to the right or left one lattice site has $\tilde{p}(k) = \cos(k)$. For large n, $[\tilde{p}(k)]^n$ is small, except at and near $k = 0$. This implies that for large n (many jumps of a random walk) the integral in Eq. (4) is dominated by small values of k, and so we use a small-k expansion of $\tilde{p}(k)$ to calculate the large-n behavior of $\tilde{P}_n(k)$. We start by expanding the exponential in Eq. (4) for small k,

$$\begin{aligned}\tilde{p}(k) &= \sum_{l} [1 + ikl - \tfrac{1}{2}k^2l^2 + \cdots]p(l) \\ &= 1 + ik\langle l\rangle - \tfrac{1}{2}k^2\langle l^2\rangle + \cdots,\end{aligned} \tag{5}$$

where the moments of the jump distributions are defined as $\langle l^n\rangle \equiv \Sigma_l l^n p(l)$. For the sake of simplicity let us assume that jumps occur with equal probabilities to the right and left so that $\langle l\rangle = 0$, and that $\langle l^2\rangle$ is finite. Then, for small k values,

$$[\tilde{p}(k)]^n \approx \exp\left(-\frac{n}{2}\langle l^2\rangle k^2\right), \tag{6}$$

which is Gaussian in k. We can now extend the limits on the integral in Eq. (4) from $-\infty$ to $+\infty$ without significant error because the integrand, for large n, has a sharp maximum around $k = 0$. This yields the important result

$$P_n(l) = \frac{1}{\sqrt{2\pi\langle l^2\rangle n}} \exp\left(-\frac{l^2}{2\langle l^2\rangle n}\right) \qquad (7)$$

for a many-step symmetric random walk with a finite $\langle l^2\rangle$ (see Fig. 2). This result of adding identically distributed random variables with finite second moments and arriving at a Gaussian distribution is known as the central limit theorem. This so impressed F. Galton that he wrote, "I know of scarcely anything so apt to impress the imagination as the wonderful form of cosmic order expressed by the Law of Frequency of Error. The law would have been personified by the Greeks and deified, if they had known of it. The larger the mob and the greater the apparent anarchy, the more perfect its sway." The Gaussian was originally applied to discuss errors in measurement, with its peak being the most probable value of the measurement and its variance related to typical errors in the measurement. The Maxwell–Boltzmann distribution is Gaussian in particle velocities, but the whole distribution has physical meaning, so that being away from the mean is not an error. In fact the whole distribution is a prediction. This was a major advance in thought.

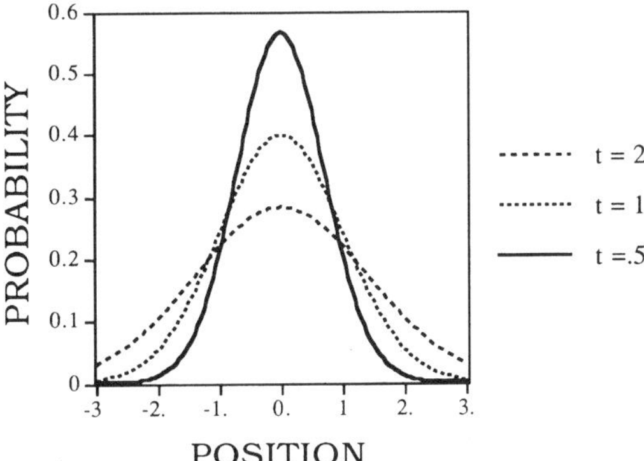

FIG. 2. Gaussian spreading. We let the number of steps become a continuous variable equivalent to the time and plot the spreading with time of a Gaussian probability. The variance of the distribution grows linearly with time.

It can be shown, via the Berry–Esséen theorem (Feller, 1971), that the value of n at which Gaussian behavior emerges is $n \approx (25/4)\langle |l^3|\rangle^2/\langle l^2\rangle^3$. The probability $P_n(l)$ satisfies the diffusion equation when we let the number of steps n and the distance l become continuous variables,

$$\frac{\partial P(l,n)}{\partial n} = D\,\frac{\partial^2 P(l,n)}{\partial l^2}, \qquad (8)$$

with $D = \langle l^2\rangle/2\langle t\rangle$ where $1/\langle t\rangle$ is introduced as a jump rate. This is also known as the heat equation, and Bachelier, in 1900, was amazed to discover that probability could radiate in the same fashion as heat.

While the central limit theorem is very powerful, there are many situations where it does not apply. One can ask questions about the first time a random walk reaches a certain distance from its origin, or the maximum of a set of random variables, or the conditional probability for a random walker to be at a site now, given that it was at another site at an earlier time, or the probability distribution for a random walker whose mean jump distance is infinite, etc. This means that, although the Gaussian distribution plays a central role in probability theory connected to summing random variables, it is only one of many important limit theorems.

The above analysis was for a simple walk on a one-dimensional regular lattice. We began by writing down directly the transition probability for a single step. A different approach for investigating the random motion of a particle is to employ a stochastic differential equation of motion for a particle's position or velocity (the Langevin approach) and then calculate via a partial differential equation (the Fokker–Planck equation) the probability for the particle being at position x at time t. We turn to the Langevin approach now, before returning to more complex random-walk models.

3. LANGEVIN EQUATIONS

One of the major themes in probability is the mathematics of Brownian motion and its variants. The original work of Einstein involved a large Brownian particle in a fluid being buffeted continuously by smaller fluid particles. If the Brownian particle has mass M and random velocity, denoted by v, then the equation of motion for the Brownian particle's dynamics is

$$M\frac{dv}{dt} = -fv + F(t), \tag{9}$$

where mass times acceleration is on the left-hand side (LHS) and the forces are on the right-hand side (RHS). The term f is the friction coefficient and $F(t)$ is a random force representing random collisions of the Brownian particle with the smaller fluid particles. An $F = Ma$ equation that involves random forces and/or parameters is called a Langevin equation. Although, one can readily solve for $v(t)$ as

$$v(t) = v(0)e^{-\beta t} + e^{-\beta t}\int_0^t e^{\beta\tau}A(\tau)d\tau, \tag{10}$$

where $\beta = f/M$ and $A = F/M$, one cannot directly evaluate $v(t)$ because it is a random variable, which is written in terms of $A(t)$, the random force per mass. One can, however, determine the probability distribution for values of v, or of related random variables such as the position $r(t)$, which is related by $v(t) = dr(t)/dt$. To proceed further, one must consider the statistical properties of the noise. A straightforward case, called white noise, is uncorrelated noise. A white-noise random force has the properties, where the brackets indicate a time average,

$$\langle F(t)\rangle = 0,$$
$$\langle F(t=0)F(t)\rangle = 2D\delta(t). \tag{11}$$

The choice of white noise yields $\langle v(t)\rangle = v(0)e^{-\beta\tau}$, and $\langle v^2(t)\rangle = \beta D + [v(0)^2 - \beta D] \times \exp(-2\beta t))$. In the long-time (equilibrium) limit $\frac{1}{2}M\langle v^2\rangle$ must be equal to $k_BT/2$, the energy of the system. Thus the choice of D is not arbitrary, but must be $D = k_BT/\beta M$. One can also solve for the moments of $r(t)$. Integrating Eq. (10) one obtains

$$r(t) = r(0) + \frac{v(0)}{\beta}(1 - e^{-\beta t}) - \frac{1}{\beta}e^{-\beta t}\int_0^t e^{+\beta\xi}A(\xi)d\xi + \frac{1}{\beta}\int_0^t A(\xi)d\xi. \tag{12}$$

Taking averages over values of A and initial velocities $v(0)$ one obtains $\langle r(t)\rangle = r(0)$. Squaring Eq. (12) and again taking double averages over the random forces and initial velocities one obtains the classic Brownian-motion result,

$$\langle r^2(t)\rangle = \frac{2k_BT}{M\beta}t = \frac{2k_BT}{f}t = 2Dt, \tag{13}$$

plus terms that decay exponentially with time. In his theory of Brownian motion, Einstein wrote specifically $\beta = 6\pi a\eta/M$, which is Stokes's law for a spherical particle of radius a and mass M, in a fluid with viscosity η.

The next challenge is to find the probability distribution of these stochastic variables, not just their moments. Before we address this matter, we mention that Eq. (9) can be easily generalized to govern Brownian motion in a potential by adding the force term $K(x)/M$ to the RHS. For a harmonic potential (the Ornstein–Uhlenbeck problem) $K(x)/M = -\omega^2x$, and this produces an oscillatory decay to the equilibrium value of zero for the mean position (and also the mean velocity), i.e.,

$$\langle r(t)\rangle = \frac{Mv(0)}{\omega}e^{-(1/2)\beta t}\sin\omega t + \frac{r(0)}{\omega}e^{-(1/2)\beta t}\left(\omega\cos\omega t + \frac{\beta}{2}\sin\omega t\right). \tag{14}$$

Expressions can be similarly calculated for the second spatial and velocity moments with oscillatory decays to equilibrium values, as opposed to the exponential decays when only the frictional force is present.

3.1 Kubo Oscillator

Consider randomness entering through a multiplicative factor in a stochastic differential equation. A classic problem is an oscillator (Kubo, 1962) with amplitude $x(t)$ and fluctuating frequency $\omega(t)$. This process satisfies the equation

$$\frac{dx(t)}{dt} = i\omega(t)x(t),$$

with the first moment given by

$$\langle x(t)\rangle = \left\langle \exp\left(i \int^{t} \omega(t')dt'\right)\right\rangle. \tag{15}$$

Because $\omega(t)$ is fluctuating, $x(t)$ is a random variable. We will not calculate the probability distribution for x, but will only concentrate on its first moment here, and the effect of colored noise on this behavior. For $\omega(t) = \omega_0 + \delta\omega(t)$, with $\langle\delta\omega(t)\rangle = 0$, the frequency fluctuates around the mean value ω_0. If $\delta\omega(t)$ possesses Gaussian statistics, then

$$\begin{aligned}\langle x(t)\rangle &= e^{i\omega_0 t} \exp\left(-\tfrac{1}{2}\int_0^t dt' \int_0^t dt''\langle\delta\omega(t')\delta\omega(t'')\rangle\right)\\ &= e^{i\omega_0 t} \exp\left(-\int_0^t (t - t')\langle\delta\omega(t')\delta\omega(0)\rangle dt'\right).\end{aligned} \tag{16}$$

If $\langle\delta\omega(t)\delta\omega(0)\rangle = 2\lambda\delta(t)$, then it follows that

$$\langle x(t)\rangle = e^{i\omega_0 t}e^{-\lambda t}. \tag{17}$$

The relaxation function, defined by $\Phi(t) = \exp(-\lambda t)$, leads to the line-shape spectrum $I(\omega)$, defined as

$$\begin{aligned}I(\omega - \omega_0) &= \frac{\text{Re}}{2\pi}\int_{-\infty}^{\infty} \exp[i(\omega - \omega_0)]\Phi(t)dt\\ &= \frac{1}{\pi}\frac{\lambda^2}{(\omega - \omega_0)^2 + \lambda^2}.\end{aligned} \tag{18}$$

The effect of the noisy frequency was to broaden the line shape around the peak at ω_0.

Temporally correlated "colored" noise can lead to other line shapes. If $\langle\delta\omega(t)\delta\omega(0)\rangle = (D/\tau)\exp(-|t|/\tau)$, then for $t \ll \tau$, $(\tau/D) \times \langle\delta\omega(t)\delta\omega(0)\rangle \approx 1$, and so Eq. (16) yields, for short times,

$$\langle x(t)\rangle = e^{i\omega_0 t} \exp\left(-\frac{D}{2\tau}t^2\right) \quad \text{for } t \ll \tau. \tag{19}$$

The line shape generated by this Gaussian relaxation at short times is a Gaussian at large frequencies. For long times $t \gg \tau$, $\langle\delta\omega(t)\delta\omega(0)\rangle \approx 0$, i.e., the correlation has decayed away, and so one can replace the upper limit in the last integral in Eq. (16) by infinity and recover the exponential decay of the delta-correlated noise case (effectively the $\tau = 0$ case) and Lorentzian line shape of Eq. (18) at low frequencies. This crossover from a Lorentzian to a Gaussian line shape at $\omega = 1/\tau$ illustrates the crucial role that correlated "colored" noise can play in random processes.

Equation (18) is also called a power spectrum. In general, take a random time series $h(t)$ whose mean is zero. Let $h(t)$ be zero outside a time interval T and represent $h(t)$ in terms of its Fourier transform

$$h(t) = \int_{-\infty}^{\infty} \exp(2\pi i f t)g(f)df.$$

The power spectrum $S(f)$ can be defined as

$$S(f) = \lim_{T\to\infty} \frac{2}{T}|g(f)|^2. \tag{20}$$

That also

$$4\text{Re}\int_{-\infty}^{\infty} \langle h(t + \tau)h(t)\rangle \exp(2\pi i f\tau)d\tau = S(f)$$

is called the Wiener–Khintchine theorem. If the correlation function is the product of an electrical current at two different times, then its power spectrum reflects the electric power as a function of frequency; hence the appellation of power. The Lorentzian power spectrum of Eq. (18) arises from a correlation function that decays exponentially. In many complex systems one finds a noise power spectrum, for low frequencies, of the form $S(f) \approx 1/f^{\alpha}$ where α is close to unity. This is called one-over-f noise. One method to derive $1/f$ noise is to have an exponentially decaying correlation $\exp(-t/\tau)$ with a distribution of correlation times $\rho(\tau) \approx \tau^{-1-\alpha}$. This produces $S(f) \approx 1/f^{1-\alpha}$. It is still an open question to determine the physics underlying $1/f$ noise in many complex systems (Weissman, 1988).

4. PROBABILITY EQUATIONS

4.1 The Chain Equation and the Master Equation

To describe fully a stochastic process $X(t)$ one must know all the joint probabilities,

$$P_n(x_1,t_1;\cdots;x_n,t_n) \quad \text{for all } n,$$

i.e., that at time t_i, the variable $X(t)$ has the value x_i, for $i = 1, \ldots, n$. Alternatively, one can describe $X(t)$ by defining joint conditional probabilities W_n for $X(t)$ to have the value x_n at time t_n when the earlier values x_i at t_i are given for $i = 1, \ldots, n - 1$. The probability P_n can be written as

$$P_n(x_1,t_1;\cdots;x_n,t_n) = W_n(x_n,t_n|x_1,t_1;\cdots;x_{n-1},t_{n-1}) \times P_{n-1}(x_1,t_1;\cdots;x_{n-1},t_{n-1}). \tag{21}$$

Knowing P_1 and all the W_n also completely specifies $X(t)$. A Markov process is defined by

$$W_n(x_n,t_n|x_1,t_1;\cdots;x_{n-1},t_{n-1}) = W_2(x_n,t_n|x_{n-1},t_{n-1}),$$

where the value of x_n depends only on x_{n-1} and the probability of transitioning from x_{n-1} to x_n. The earlier information of the x_{n-i} does not affect the outcome of the next step in a Markov process. For a Markov process the conditional probabilities satisfy the famous Bachelier–Smoluchowski–Chapman–Kolmogorov chain equation

$$W_2(x,t|x_0,t_0) = \int W_2(x,t|x',\tau)W_2(x',\tau|x_0,t_0)dx', \tag{22a}$$

where we considered going from x_0 to x in two steps, and we have integrated over all possible intermediate states x'. One can also write the chain equation as

$$\frac{\partial P(x,t)}{\partial t} = \int [K(x,x')P(x',t) - K(x',x)P(x,t)]dx', \tag{22b}$$

which is called a master equation (Oppenheim *et al.*, 1977). To derive the master equation from the chain equation all the initial conditions have been integrated out, so that $P(x,t)$ is the probability to be at state x at time t independent of the initial condition. The quantity $K(x,x')$ is the transition rate (it has the dimension of a frequency) for the system to change from state x' to state x. The derivative measures the rate of change with time of the probability to be in state x. The first term in the integral gives the rate of the system moving into state x from state x' (integrated over all x') times the probability that the system is in state x', while the second term calculates the rate for leaving state x times the probability that the system is in state x. In our discussion of continuous-time random walks we will study semi-Markovian (temporal, but no spatial, memory) processes where a trapping of a random walker occurs between each of its jumps. For these random-walk processes, an integro-differential master equation arises with an integral over all possible waiting times in a trap [see Eq. (52)].

4.2 Fokker–Planck Equations

From the chain equation, for only small transitions in x allowed, we can derive a second-order differential equation for $P(x,t)$. Let us denote W_2 in Eq. (22a) as P and look, following Wang and Uhlenbeck (Wax, 1954), at

$$\int dyR(y)\frac{\partial P(y,t|x,0)}{\partial t},$$

where $R(y)$ is an arbitrary function that asymptotically goes to zero at infinity. We can write the above derivative as

$$\lim_{\Delta t\to 0}\frac{1}{\Delta t}[P(y,t+\Delta t|x,t) - P(y,t|x,t)] = \lim_{\Delta t\to 0}\frac{1}{\Delta t} \times \left[\int [P(y,t+\Delta t|z,t')P(z,t'|x,t)dz - P(y,t|x,y)\right], \tag{23}$$

where we have used the chain equation (22a). Expanding $R(y)$ in a Taylor series up to the second derivative and using Eq. (23) for the derivative produces the following equation for $P(x,t)$, called the Fokker–Planck equation:

$$\frac{\partial P(x,t)}{\partial t} = -\frac{\partial}{\partial x}[A(x)P(x,t)] + \frac{1}{2}\frac{\partial^2}{\partial x^2}[B(x)P(x,t)], \tag{24}$$

where $A(x)$ is related to the first moment of $P(x,t)$ as

$$A(x) = \lim_{\Delta t\to 0}\frac{1}{\Delta t}\int P(y,\Delta t|x,t=0)(x-y)dy$$

and $B(x)$ is the same expression as above, except that the second moment of P appears through $(x - y)^2$ replacing $(x - y)$. These two terms on the RHS of Eq. (24) represent drift and diffusion in a force field. Although the expansion for $R(y)$ can be carried out to higher-order terms, they are usually small unless large jumps occur. In that case the full master equation should be used, as we do in our discussion of fractal processes. Setting the time derivative equal to zero, in Eq. (24), one has the following steady-state solution:

$$\lim_{t\to\infty} P(x,t) = \frac{\text{const.}}{B(x)} \exp\left(\int^x \frac{A(x')}{B(x')}\, dx'\right), \tag{25}$$

where the constant is chosen to insure normalization. When the first moment vanishes because there is no preferred direction, and the second moment is a constant, denoted by D, the Fokker–Planck equation reduces to the familiar diffusion equation.

The Ornstein–Uhlenbeck process for random motion in a harmonic potential has $A(x) = -\omega^2 x$, B = constant. The solution of the Fokker–Planck equation for $P(x,t)$ is then

$$P(x,t) = \frac{1}{\sqrt{2\pi\sigma^2(t)}} \exp\left\{-\frac{x^2}{2\sigma^2(t)}\right\}, \tag{26a}$$

where

$$\sigma^2(t) = \frac{D}{\omega^2}[1 - \exp(-2\omega^2 t)]. \tag{26b}$$

In the long-time, steady-state limit $\omega \to 0$, $\sigma^2(t) \to 2Dt$, which brings us back to the Brownian-motion limit.

In general, a Langevin equation of the form

$$\frac{dx(t)}{dt} = F(x(t)) + G(x(t))\eta(t), \tag{27}$$

with

$$\langle \eta(t)\eta(t')\rangle = 2D\delta(t - t'),$$

has the probability $P(x,t)$ governed by the Fokker–Planck equation

$$\frac{\partial P(x,t)}{\partial t} = -\frac{\partial F(x)P(x,t)}{\partial x} + D\frac{\partial}{\partial x} G(x) \frac{\partial G(x)P(x,t)}{\partial x}. \tag{28}$$

However, when a finite-correlation-time "colored" noise, e.g.,

$$\langle \eta(t)\eta(0)\rangle = (D/\tau) \exp(-|t|/\tau), \tag{29}$$

is required, a more complicated situation arises. Many techniques (Moss *et al.*, 1990) have been derived to treat stochastic differential equations with colored noise using expansions and infinite resummation in powers of D or τ.

5. RANDOM WALKS

5.1 Discrete-Time

Let us return to Eq. (2) for $P_n(l)$ and define the *generating function* $G(l;z)$ as

$$G(l;z) \equiv \sum_{n=0}^{\infty} P_n(l) z^n. \tag{30}$$

Multiplying both sides of Eq. (2) by z^{n+1} and summing over all n yields, for a walker that starts at the origin,

$$G(l;z) - z\sum_{l'} G(l - l';z)p(l') = \delta_{l,0}, \tag{31}$$

i.e., the random-walk generating function satisfies a Green's-function equation. Fourier transforming Eq. (31) gives an algebraic equation for $\tilde{G}(k;z)$, with the solution

$$\tilde{G}(k;z) \equiv \sum_l \exp(ikl)G(l;z) = \frac{1}{1 - z\tilde{p}(k)}. \tag{32}$$

For the case of nearest-neighbor transitions $p(l) = \frac{1}{2}[\delta_{l,-1} + \delta_{l,+1}]$, we have

$$\tilde{p}(k) = \cos(k) \tag{33}$$

and

$$P_n(l) = \frac{n!}{[\frac{1}{2}(n + l)]![\frac{1}{2}(n - l)]!} 2^{-n}. \tag{34}$$

This is the expression DeMoivre analyzed for $n/2 + l/2$ successes (jumps to the right) and $n/2 - l/2$ failures (jumps to the left) in n Bernoulli trials. He showed that $P_n(l)$ is Gaussian in the large-n limit. This is true for any symmetric random-walk process with finite mean squared jump lengths, as Gauss proved in 1809 in his central limit theorem.

The generating-function formalism is also good for calculating first-passage times. The probability $F_n(l)$ that a walker reaches site l for the first time on the nth step enters the following equation:

$$P_n(l) = \sum_{j=1}^{n} F_j(l)P_{n-j}(0) + \delta_{n,0}\delta_{l,0}, \qquad (35)$$

which accounts for being at l after n steps by first reaching l after j steps and then leaving and returning to l in $n - j$ steps (for all possible j values). Multiplying Eq. (35) by z^n, and summing over all n, gives the generating-function equation

$$G(l;z) = \delta_{l,0} + F(l;z)G(l = 0;z), \qquad (36)$$

where

$$F(l;z) \equiv \sum_{j=0}^{\infty} F_j(l)z^j.$$

This shows that knowledge of $G(l;z)$ allows the calculation of the first–passage-time generating function. The Green's function for return to the origin, which appears on the RHS of Eq. (36), has been calculated, for nearest-neighbor jumps, for z close to 1:

$$G(l = 0;z) = \left(\frac{1}{2\pi}\right)^D \int_0^{2\pi} \cdots \int_0^{2\pi} \frac{d^Dk}{1 - \frac{z}{D}[\cos(k_1) + \cdots + \cos(k_D)]}$$
$$= \begin{cases} (1 - z^2)^{-1/2} & \text{for } D = 1, \\ \approx -\frac{1}{\pi}\log(1 - z) & \text{for } D = 2, \\ \approx 1.516 - \frac{3}{\pi}\sqrt{\frac{3}{2}(1 - z)} & \text{for } D = 3. \end{cases} \qquad (37)$$

The 2D result is for a square lattice and the 3D result is for a cubic lattice. The manner in which $G(l = 0;z)$ diverges near $z = 1$ is important for calculating first–passage-time probabilities. Consider Polya's question of what is the probability for a symmetric random walker with $\langle l \rangle = 0$, on a regular lattice, to return to its origin. This probability of return, R, is given as

$$R = \sum_{n=0}^{\infty} f_n(0) = F(l = 0;z = 1) = 1 - \frac{1}{G(0;1)}, \qquad (38)$$

where we have used Eq. (36). We see from Eq. (37) that $G(0;1)$ diverges in one and two dimensions, and is finite in higher dimensions, yielding Polya's famous result that return to the origin is assured with probability 1 in one and two dimensions, but not in higher dimensions. The square-root divergence of the one-dimensional $G(0;1)$ in Eq. (37) implies that the probability that the walker has not returned by step n is

$$F_n(0) = \sum_{j=n+1}^{\infty} f_j(0) \approx \left(\frac{1}{n}\right)^{1/2}, \qquad (39)$$

causing, in one dimension, that the mean number of steps

$$\langle n \rangle = \sum_{n=0}^{\infty} nf_n(0)$$

for the first return of a random walker to the origin is infinite. The walker returns with probability 1, but the mean time to do this is infinite.

What if N walkers start at the origin? Is the mean time for any one of this set to return to the origin also infinite (Lindenberg *et al.*, 1980)? The probability $f_{N,n}(0)$ that any walker from a group of N walkers returns on the nth step is, using Eq. (39),

$$f_{N,n}(0) = F_{n-1}^N(0) - F_n^N(0), \qquad (40)$$

where the N on the LHS denotes the number of walkers and N is a power on the RHS.

The RHS is the difference between the probabilities that none of the walkers has returned by the $(n - 1)$st and the nth step. This difference is the probability that at least one walker returned on the nth step. The mean first time for a return to the origin from the set of N walkers is

$$\langle T_N \rangle = \sum_{n=1}^{\infty} n f_{N,n}(0) = \sum_{n=0}^{\infty} F_n^N(0). \tag{41}$$

In one dimension $F_n(0) \approx \sqrt{1/n}$, and so the divergence of the mean return time for the set of N walkers depends on the series

$$\sum_{n=1}^{\infty} \left(\frac{1}{n}\right)^{N/2}. \tag{42}$$

For three (or more) walkers, Eq. (42) determines that the mean time for return to the origin is finite. This further explores Polya's classic study of the effect of dimensionality on random-walk quantities.

There are several directions in which the generating-function approach to random walks can be naturally extended. One is to include finite memory of the walker's jump history. This can be accomplished through the use of internal states at each lattice site, say to keep track of whether a random walker came from the left or right. In one dimension if the probability for a walker to continue in the same direction is $p < 1$, and to change direction is $q = 1 - p$, then the generating-function method finds after many jumps that the probability density for the walker's position is Gaussian, as in Eq. (7), but with the extra factor p/q in the diffusion constant D.

Restricted reversal correlation can also be handled in terms of internal states. Suppose, in two dimensions, that a walker is not allowed to reverse its direction to go back immediately to its previous position. This is handled by having four states in a unit cell representing the four directions a walker could have come from, and not having a connecting path back to the reversed position. The mean squared displacement is doubled from the unrestricted walk, and for a periodic lattice with coordination number q, the mean squared displacement is enhanced by the factor $q/(q - 2)$ (Domb and Fisher, 1958). Correlation over a finite number of steps (Montroll, 1950) will only modify the diffusion constant, and not change the simple linear dependence on n (the number of steps) of the mean square displacement. One needs infinite correlation to go beyond simple asymptotic Gaussian probabilities for random walks with simple steps on a regular lattice. The self-avoiding random walk is one such example. This model forbids the walker to cross its own trajectory at any point. For a self-avoiding random walk on a three-dimensional cubic lattice one finds $\langle l^2(n) \rangle \approx n^{\gamma}$ with $\gamma \approx 6/5$.

5.2 Continuous-Time

So far we have treated random walks with steps taken at regular time intervals τ, 2τ, 3τ, ... We can generalize this by introducing a probability $\psi(t)dt$ for the interval between jumps to be between t and $t + dt$. The quantity $\psi(t)$ is called a waiting-time probability density, and it must be integrated over t to yield a probability. A quantity that we will find useful is $\Phi(t)$, the probability that a jump has not occurred by time t, since the previous jump. This is given by

$$\Phi(t) = 1 - \int_0^t \psi(\tau)d\tau. \tag{43}$$

The probability density $\psi_2(t)$ that the interval between two consecutive jumps is t is given by

$$\psi_2(t) = \int_0^t \psi(t - \tau)\psi(\tau)d\tau. \tag{44}$$

Similarly, we introduce $\psi_n(t) = \int_0^t \psi_{n-1}(t - \tau)\psi(\tau)d\tau$, and the Laplace transform of $\psi_n(t)$,

$$\psi_n^*(s) \equiv \int_0^{\infty} \psi_n(t)\exp(-st)dt = [\psi^*(s)]^n. \tag{45}$$

If $\psi(t) = \lambda\exp(-\lambda t)$ then $\Phi(t) = \exp(-\lambda t)$. This represents a pure random process with a constant jump rate λ. By a pure process it is meant that the next jump of a walker is independent of whether you start the random walker at $t = 0$ or observe it at a random time and do not know how much time has elapsed since the last jump of the walker. This is equivalent to waiting for a bus that goes in a loop. Do you have to wait longer for the bus to return if you see that the bus just left, or if you come and no bus is present? If the time between returns to your stop is governed by an exponential $\psi(t)$ then both cases have the same result. This is only true for the exponential $\psi(t)$, and for this case the probability $N_n(t)$ that n jumps have occurred by time t is $N_n(t) = \int_0^t \psi_n(t - \tau)\Phi(\tau)d\tau$. In Laplace space,

$$N_n^*(s) = [\psi^*(s)]^n \frac{1}{1 - \psi^*(s)} = \frac{\lambda^n}{(s + \lambda)^{n+1}}. \tag{46a}$$

Taking the inverse Laplace transform we find for $N(t)$ that

$$N_n(t) = \frac{(\lambda t)^n \exp(-\lambda t)}{n!}, \tag{46b}$$

which is the Poisson distribution.

It is useful to define the moments of $\psi(t)$,

$$\langle t^n \rangle \equiv \int_0^\infty t^n \psi(t)dt. \tag{47}$$

With these preliminaries, we can now write the probability density $Q(l,t)$ to reach site l *exactly* at time t, via the relationship

$$Q(l,t) = \sum_{l'} \int_0^t Q(l - l', t - \tau)p(l')\psi(\tau)d\tau + \delta_{l,0}\delta(t). \tag{48}$$

The RHS accounts for the walker reaching the site $l - l'$ at time $t - \tau$ and then a time τ later performing a jump of displacement l' that brings the walker to site l exactly at time t. The delta functions on the RHS take into account the initial condition of the walker being at the origin. In Laplace–Fourier space, Eq. (48) has the form

$$\tilde{Q}^*(k,s) = \frac{1}{1 - \tilde{p}(k)\psi^*(s)}, \tag{49}$$

which is equivalent to the Green's function for the discrete-time random walk when in Eq. (32) we set $z = \psi^*(s)$. The Green's function acting for a time t gives the probability evolution for that time interval. It is also called a propagator. The probability $P(l,t)$ to be at site l at time t is different from $Q(l,t)$ because the walker may have reached site l at an earlier time $t - \tau$, and then no jump occurs during the interval $t - \tau$ to t. We can write $P(l,t)$ as

$$P(l,t) = \int_0^t Q(l, t - \tau)\Phi(\tau)d\tau. \tag{50}$$

In Fourier–Laplace space the probability is

$$\tilde{P}^*(k,s) = \tilde{Q}^*(k,s)\frac{1 - \psi^*(s)}{s} = \frac{1}{1 - \tilde{p}(k)\psi^*(s)} \frac{1 - \psi^*(s)}{s}. \tag{51}$$

The above analysis describes a continuous-time random walk (CTRW) (Montroll and Weiss, 1965) (see Fig. 3).

We can write the probability in Eq. (51) in terms of a nonlocal-time master equation with a transition rate $p(r)K(t)$, i.e., which generalizes Eq. (22b) for the master equation:

$$\frac{\partial P(r,t)}{\partial t} = \int_0^t \left[-P(r,t) + \sum_{r'} P(r',t)p(r - r') \right] \times K(t - \tau)d\tau, \tag{52}$$

if we set in Laplace space

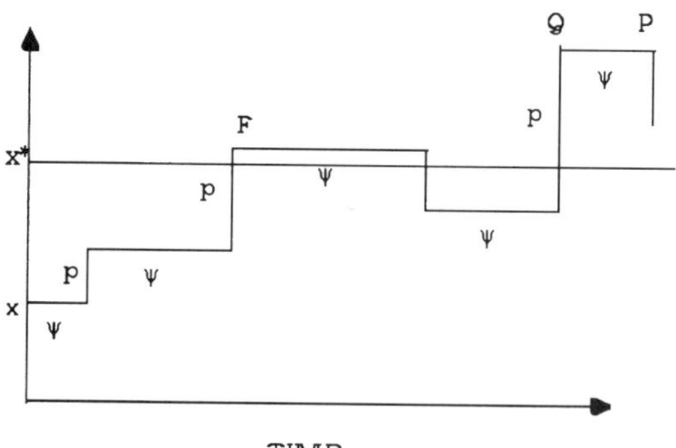

FIG. 3. A schematic plot of position x vs time t for a continuous-time random walk. The walker first waits with no movement for a time governed by $\psi(t)$, the waiting-time density. Next the walker jumps a distance r, governed by $p(r)$, then the walker again waits a random time. We choose a certain level x^*, and $F(x^*,t)$ denotes the first-passage time to this level. $Q(x,t)$ represents reaching position x exactly at time t, and $P(x,t + \tau)$ represents the probability of being at x at time $t + \tau$, which includes having arrived at an earlier time and not having left by time $t + \tau$.

$$K^*(s) = s\psi^*(s)/[1 - \psi^*(s)]. \tag{53}$$

An analysis can show that Eqs. (52) with (53) (the master equation) and (51) (the continuous-time random walk) are equivalent. The temporal memory relates how transitions (jumps) of the random walker at time t are related to the previous history of the random process. The case of $\psi(t) = \lambda \exp(-\lambda t)$ implies $K(t) = \lambda\delta(t)$. This $K(t)$, coupled with choosing $p(r)$ connecting only nearest neighbors on a one-dimensional lattice with equal probability, has the solution $P(l,t) = e^{-\lambda t} I_l(\lambda t)$, where I_l is the modified Bessel function. An analysis of the long-time behavior of the Bessel function demonstrates, in accord with the central limit theorem, that this probability asymptotically becomes Gaussian at long times.

Moments of $P(l,t)$ can be directly calculated from $\tilde{P}^*(k,s)$, e.g.,

$$\begin{aligned}\langle l(t)\rangle &= \sum_l lP(l,t) = \lim_{k\to 0}\left[-i\frac{\partial}{\partial k}\sum_l \exp(ikl)P(l,t)\right]\\ &= \lim_{k\to 0}\left[-i\frac{\partial}{\partial k}\mathcal{L}^{-1}\tilde{P}^*(k,s)\right]\\ &= \bar{l}\mathcal{L}^{-1}\left\{\frac{\psi^*(s)}{s[1-\psi^*(s)]}\right\},\end{aligned} \tag{54}$$

where the mean distance per jump is

$$\bar{l} = -i\frac{\partial}{\partial k}\tilde{p}(k = 0)$$

At long times if $\bar{l}$ is not zero, and $\bar{t}$ is finite, the walker has a drift,

$$\langle l(t)\rangle \approx (\bar{l}/\bar{t})t. \tag{55}$$

The variance, $\sigma^2(t) = \langle l^2(t) - \langle l(t)\rangle^2\rangle$, can be shown to behave asymptotically as

$$\sigma^2(t) \approx \left\{\frac{\overline{l^2}}{\langle t\rangle} + \left(\frac{2\bar{l}^2}{\langle t\rangle}\right)\left[\frac{1}{2}\frac{\langle t^2\rangle}{\langle t\rangle^2} - 1\right]\right\}t. \tag{56}$$

In the later sections we will explore ways in which this result can be modified either through infinitely long jumps or waiting times, or through walks on fractal lattices.

One of the strengths of the generating-function formalism is that many quantities can be expressed in terms of the Green's function. We now show how to calculate the first-passage time within the CTRW formalism. Let $F(l,t)$ be the probability density to reach site l for the first time at time t. This is related to the first passage in n steps (with a sum over all n) times the probability density for the nth step to occur at time t,

$$F(l,t) = \sum_{n=0}^{\infty} F_n(l)\psi_n(t). \tag{57}$$

In Laplace space, we can write $F^*(l,s)$ in terms of the first–passage-time generating function $\mathcal{F}$ as

$$\begin{aligned}F^*(l,s) &= \int_0^\infty F(l,t)\exp(-st)dt\\ &= \sum_{n=0}^{\infty} F_n(l)[\psi^*(s)]^n = \mathcal{F}(l;\psi^*(s)).\end{aligned} \tag{58}$$

From Eq. (36) we write

$$\mathcal{F}(l,\psi^*(s)) = \frac{G(l,\psi^*(s)) - \delta_{l,0}}{G(l = 0,\psi^*(s))} \tag{59}$$

or in Fourier space,

$$\tilde{\mathcal{F}}(k,\psi^*(s)) = \frac{[1 - \tilde{p}(k)\psi^*(s)]^{-1} - 1}{G(l = 0,\psi^*(s))}. \tag{60}$$

In one dimension, using Eq. (37) for $G(l = 0;z)$, we have for s, $k \to 0$

$$\tilde{F}^*(k,s) \approx \frac{1}{\sqrt{2\langle t\rangle s}\,(\frac{1}{2}k^2\overline{l^2} + s\langle t\rangle)}. \tag{61}$$

Taking the inverse Laplace and then the Fourier transform yields, for the long-time, large-distance limit,

$$F(l,t) = \frac{|l|}{\sqrt{4\pi D}}t^{-3/2}\exp\left(-\frac{l^2}{4Dt}\right), \tag{62}$$

where $D = \overline{l^2}/2\langle t\rangle$. The function $F(l,t)$ is a one-dimensional probability density and must be integrated over time to calculate a probability. This probability may superficially look similar to a Gaussian probability (except for the $|l|$ term), but for fixed time t, a Gaussian has a maximum at $l = 0$, while $F(l,t)$ has a maximum when $l^2 = 2Dt$. In fact, $F(l,t)$ is zero at $l = 0$. An important fact to notice is that the mean first-passage time, on an infinite lattice, to any lattice site is infinite. This is because the walker can drift out to infinity in a direction away from the chosen site to which the first passage is calculated. However, the average time $\langle T(R)\rangle$ to leave a region of length scale R (e.g., a circle of radius R) is finite because we then sum over all l on the boundary of the region. Since a Gaussian probability density spreads as $\langle R^2(T)\rangle \approx T$, one might suspect that $\langle T(R)\rangle \approx R^2$, and this is correct, as we show in Sec. 5.3.

An important quantity is $S(t)$, the number of distinct sites visited on a lattice in a time t. In the field of reactions it is important to know how much territory a diffusing reactant explores in a time t. This is related to $S(t)$. We begin with the quantity S_n, the number of distinct sites visited in n steps (and we count the origin as the first new site). We can write

$$S_n = 1 + \sum_l [F_1(l) + \cdots + F_n(l)]. \tag{63}$$

In continuous time the number of distinct sites visited in a time t is

$$S(t) = \sum_{n=0}^{\infty} S_n \int_0^t \psi_n(t-\tau)\Phi(\tau)d\tau, \tag{64}$$

where the integral accounts for the nth jump occurring at time $t - \tau$, and then no jump occurring from time $t - \tau$ to t. Using generating-function techniques one finds for the Laplace transform of $S(t)$ the form $\psi^*(s)/sG(l = 0;\psi^*(s))$. Using the explicit form of the Green's function, we find (assuming $\langle t\rangle$ is finite) that

$$S(t) \approx \begin{cases} t/\langle t\rangle G(0;1) & \text{for } D = 3, \\ \pi t/\langle t\rangle \log(t/\langle t\rangle) & \text{for } D = 2, \\ \sqrt{8t/\pi\langle t\rangle} & \text{for } D = 1. \end{cases} \tag{65}$$

Consider the number of distinct points visited in a set other than the entire lattice. The generating-function formalism still holds, with a sum over points in the set instead of over every lattice point. For the set being a plane in a three-dimensional cubic lattice $\langle S(t)\rangle \approx t^{1/2}$, so that after a time t the number of points in the plane visited by the walker (which can go anywhere in three dimensions) scales as $t^{1/2}$. For the special set being a line in three dimensions, $\langle S(t)\rangle \approx \log(t)$. The calculation of the probability distribution of $S(t)$, as opposed to its first moment, requires techniques beyond the generating-function approach.

5.3 Many Walkers

There are several ways to obtain noninteger exponents in random walks. We will later address the concepts of fractal space and time where fractional exponents reflect fractal properties. Here, we treat the case of more than one random walker. We consider N random walkers inside a sphere (see Fig. 4) of radius R, and ask the question for the mean first-passage time for the sphere to be empty of random walkers. This problem was posed in the context of the disentanglement time for N entangled polymer chains (Weiss *et al.*, 1988). If the walkers undergo Brownian motion we have, for the probability $p(\mathbf{r},t|\mathbf{r}_0,t = 0)$ for the walker to go between $\mathbf{r}_0$ and $\mathbf{r}$ in a time t,

$$p(\mathbf{r},t|\mathbf{r}_0,t = 0) = 4\pi\left(\frac{3}{4\pi Dt}\right)^{3/2} \exp\left[-\frac{3(\mathbf{r}-\mathbf{r}_0)^2}{4Dt}\right], \tag{66}$$

where D is the diffusion constant. Let $E(t)$ be the probability that a particle initially in the sphere is in the sphere at time t. We can express $E(t)$ as the integral of Eq. (66) over $\mathbf{r}$ and $\mathbf{r}_0$ for all points in the sphere. For N

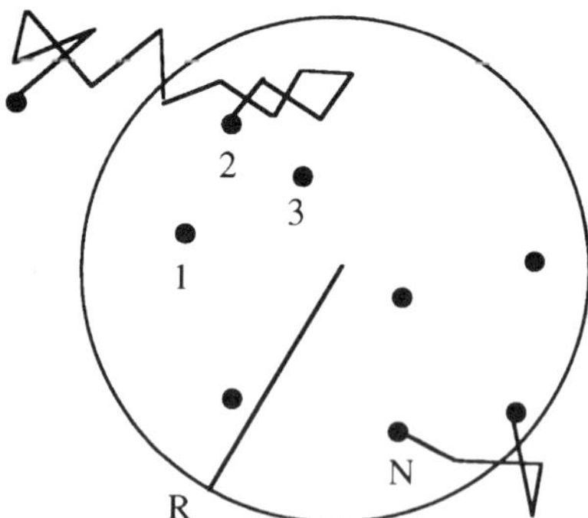

FIG. 4. N random walkers, denoted by ●, and labeled 1,2,3, . . ., N, start in random initial conditions inside a sphere of radius R. Each walker undergoes Brownian motion. A possible trajectory for walker 2 is shown. We are interested in the mean first-passage time for the sphere to be empty of walkers. Note that walker N left the circle, but reentered. This possibility complicates the calculation for the circle being empty of walkers.

noninteracting walkers the probability that at least one of them is in the sphere is given by $\Lambda(t) = \{1 - [1 - E(t)]^N\}$. The rate of change of $\Lambda(t)$ is the probability for the sphere to be empty. The first moment $\langle T_N \rangle$ for the sphere to be empty of N walkers (but not necessarily for the first time, although this is a good approximation in 3D for the first-time probability) is

$$\langle T_n \rangle = \int_0^\infty t \frac{d\Lambda(t)}{dt} dt = \int_0^\infty \Lambda(t) dt. \tag{67}$$

Now for t sufficiently large that $Dt \gg R^2$, then

$$E(t) \approx \sqrt{2\pi}\,(R^2/Dt)^{3/2}. \tag{68}$$

When N is large,

$$\begin{aligned}\langle T_N \rangle &\approx \int_0^\infty (1 - e^{-NE(t)})dt \\ &\approx \int_0^\infty \{1 - \exp[-N(T/t)^{3/2}]\}dt \\ &\approx TN^{2/3} \int_0^\infty [1 - \exp(-x^{3/2})]dx,\end{aligned} \tag{69}$$

where $T = cR^2/D$, with c a constant. If N, the number of walkers, is a function of R, say $N(R) = R$, then a fractional exponent results, i.e., $\langle T_N \rangle \approx R^{8/3}$ as opposed to the single-particle result of $\langle T_1 \rangle \approx R^2$.

Another example of a fractional exponent entering a many–random-walker problem deals with interacting walkers that annihilate upon contact. In one dimension, the probability that p walkers starting near each other survive for n steps scales as $n^{-p(p-1)/4}$ (Fisher, 1984). Fractional exponents also enter naturally in trapping and reaction problems, as discussed in Sec. 6.

6. REACTION KINETICS

6.1 Trapping

Consider Brownian motion of randomly spaced particles in the presence of randomly spaced trapping regions. One might as a first-order guess consider space to be broken up into cells of length R, each with a particle and a number of traps. The mean trapping time would be proportional to R^2/D, where R is the average distance to a trap and D is the particle's diffusion constant. The probability that a particle is not trapped would then be $\Phi(t) = \exp(-2nDT/R^2)$ where n is the dimension. This expression is, in fact, a good approximation for all but asymptotic times. On a finite lattice the survival time is always exponential (den Hollander, 1985).

The problem with this approach is that it ignores spatial fluctuations in the trap positions. Let us consider a particle that starts in a region of radius R without any traps. We must now average the lifetime of a particle over the probability of being in such a region of radius R (Donsker and Varadhan, 1979). If we assume a Poisson distribution for a region of volume V to be void of traps, i.e., $P(V) = (1/V) \exp(-V/V_0)$, with volume $V \sim R^d$ then the lifetime distribution of a walker is given by

$$\begin{aligned}\Phi(t) &\approx \int_0^\infty \exp\left(-\left[\frac{R}{R_0}\right]^d\right) \exp\left(-\frac{2dDt}{R^2}\right)dR \\ &\approx \exp(-\text{const.}t^{d/(d+2)}).\end{aligned} \tag{70}$$

This behavior is dominated by that small fraction of walks that start very far away from any traps, and Eq. (70) is only valid at very long times.

6.2 First- and Second-Order Kinetics

We now consider various reaction–diffusion schemes. In all cases we consider diffusion to be the rate-limiting step. This means the reactions are diffusion limited and not rate limited. In bimolecular reactions, if one species is unlimited in number, the population of the other species is said to follow a pseudo-unimolecular reaction.

For a unimolecular reaction if we start with N_0 reactants, we can write an equation for the probability $P(N,t)$ of having N reactants left at time t, where $K(t)$ is the reaction rate:

$$\frac{dP(N,t)}{dt} = K(t)[(N + 1)P(N + 1,t) - NP(N,t)]. \tag{71}$$

The equation for the average number of reactants $\langle N(t) \rangle$ is

$$\frac{d\langle N(t) \rangle}{dt} = -K(t)\langle N(t) \rangle, \tag{72}$$

with the solution

$$\langle N(t)\rangle = N_0 \exp\left(-\int_0^t K(\tau)d\tau\right). \tag{73}$$

Usually $K(t)$ is treated as a constant.

For a bimolecular reaction with species A and B annihilating upon contact, one writes for the average populations $A(t)$ and $B(t)$ the following rate equation:

$$\frac{dA(t)}{dt} = -kA(t)B(t) = \frac{dB(t)}{dt}, \tag{74}$$

with the solution

$$\frac{1 + C/A(t)}{1 + C/A(0)} = \exp(Ckt), \tag{75}$$

where $C = B(0) - A(0)$. If $B(0) \gg A(0)$ then $C \approx B(0)$ and

$$A(t) \approx A(0) \exp[-kB(0)t]. \tag{76}$$

If, however, $A(0) = B(0)$, then expanding Eq. (75) for small C yields

$$A(t) = \frac{A(0)}{1 + A(0)kt} \approx \frac{1}{kt}. \tag{77}$$

This solution is in fact not valid for dimensions $d \leq 2$ because it assumes that the rate constant *is* a constant. Actually, k is proportional to $dS(t)/dt$, where $S(t)$ is the number of new sites visited. Since $S(t) \approx \sqrt{t}$ and $t/\ln(t)$ in 1D and 2D, k is time dependent. A more serious matter involves spatial fluctuations. In the $A + B \rightarrow 0$ case, with $A = B$, fluctuations in the density will create regions with $A > B$ and vice versa. The number of particles in a region of radius R is proportional to R^d, with fluctuations of order $R^{d/2}$. After a time $t = R^2/D$, most particles of the lower density in a region have reacted, leaving mostly the excess majority species. So an A-rich region will have about $N(A) \sim R^{d/2} \sim (Dt)^{d/4}$ particles in a volume $V \sim (Dt)^{d/2}$. The density thus decays at long times as $N/V \sim (Dt)^{-d/4}$ for $d \leq 4$, and not $1/t$ as in Eq. (77). We derive this result in the next section.

6.3 Reaction–Diffusion Equations

The above rate-equation approach for average populations and constant reaction rates is useful and holds well at early times, but it lacks spatial information about the reactants, especially the segregation into A- and B-rich regions. This segregation phenomenon will slow down the reaction rate relative to the prediction of the rate equation. It is more accurate and interesting to start with a reaction–diffusion equation, where $P_A(r,t)$ is the probability that a particle of type A is at position $\mathbf{r}$ at time t (and similarly for particles of type B), and D is the diffusion constant for type A and type B particles. Type A and type B particles annihilate upon meeting with reaction rate k:

$$\frac{\partial P_A(\mathbf{r},t)}{\partial t} = D\nabla^2 P_A(\mathbf{r},t) - kP_A(\mathbf{r},t)P_B(\mathbf{r},t),$$
$$\frac{\partial P_B(\mathbf{r},t)}{\partial t} = DP_B\nabla^2(\mathbf{r},t) - kP_A(\mathbf{r},t)P_B(\mathbf{r},t). \tag{78}$$

Source and sink terms, plus boundary conditions, can also be added to the above equations. Further analysis proceeds by introducing two new variables (Ovchinnikov and Zeldovich, 1978),

$$s(\mathbf{r},t) \equiv \tfrac{1}{2}[P_A(\mathbf{r},t) + P_B(\mathbf{r},t)],$$
$$\delta(\mathbf{r},t) \equiv \tfrac{1}{2}[P_A(\mathbf{r},t) - P_B(\mathbf{r},t)]. \tag{79}$$

This method immediately produces the linear diffusion equation for the difference variable,

$$\frac{d\delta(\mathbf{r},t)}{dt} = D\nabla^2\delta(\mathbf{r},t). \tag{80}$$

We will treat the case of equal numbers of A and B particles, which are randomly distributed in an uncorrelated fashion. The average of δ vanishes identically. We need to calculate the second moment,

$$\begin{aligned}\Gamma(\mathbf{r} - \mathbf{r}',t) &\equiv \langle\delta(\mathbf{r},t)\delta(\mathbf{r}',t)\rangle \\ &= \frac{\rho(t = 0)}{2(8\pi Dt)^{d/2}} \exp\left(-\frac{[\mathbf{r} - \mathbf{r}']^2}{8Dt}\right),\end{aligned} \tag{81}$$

which is the d-dimensional form of the Gaussian, and $\rho(t = 0)$ is the initial particle density. The form for Γ follows from the Gaussian form for δ when the A and B initial positions are random and uncorrelated. For $\mathbf{r} = \mathbf{r}'$ we obtain the scaling $\delta(\mathbf{r},t) \sim t^{-d/4}$ for $d < 4$ and t^{-1} for $d \geq 4$. The bimolecular reaction–diffusion equation can produce other results for other conditions. For example, Lindenberg *et al.* (1990) find for geminate recombination with A's and B's created in initially close pairs (say, e.g., in electron–hole creation) that $\delta(\mathbf{r},t) \sim t^{-(d+2)/4}$.

7. FRACTAL-TIME PROCESSES

7.1 Fractal Waiting Times

In the continuous-time random walk let us now consider the Laplace transform of the waiting-time density, as this quantity appears in calculating many properties of a random walk, such as in determining the number $S(t)$ of distinct sites visited in a time t:

$$\psi^*(s) \equiv \int_0^\infty \psi(t)\exp(-st)dt = \sum_{n=0}^{\infty} (-1)^n s^n \langle t^n \rangle. \tag{82}$$

This expansion only holds if all the moments of $\psi(t)$ are finite. Let us consider a case where this does not occur. Choose $\psi(t)$ as

$$\psi(t) = \frac{1-q}{q} \sum_{j=1}^{\infty} q^j \lambda^j \exp(-\lambda^j t) \quad (\lambda < q < 1). \tag{83}$$

This has a scaling form, with an order-of-magnitude longer waiting time (in base $1/\lambda$) occurring an order of magnitude less often (in base $1/q$). This is like intervals in the triadic Cantor set where longer lengths (in base 3) occur two times less often. Our waiting-time distribution is easily checked to be normalized to 1, but the condition that $\lambda < q < 1$ leads to $\langle t \rangle$ being infinite, i.e.,

$$\langle t \rangle = \frac{1-q}{q} \sum_{j=1}^{\infty} \left(\frac{q}{\lambda}\right)^j = \infty, \tag{84}$$

and so the straightforward use of the moment expansion in Eq. (82) is invalid. How then can one proceed with a power-series expansion of $\psi^*(s)$ to determine its dominant behavior? We can write $\psi^*(s)$ as

$$\psi^*(s) = \frac{1-q}{q} \sum_{j=1}^{\infty} \frac{(q\lambda)^j}{[s+\lambda^j]}. \tag{85}$$

The scaling relation

$$\psi^*(s) = q\psi^*\left(\frac{s}{\lambda}\right) + \frac{\lambda(1-q)}{(s+\lambda)} \tag{86}$$

allows one to determine the noninteger exponent in the expansion of $\psi^*(s)$ in terms of powers of s. The inhomogeneous term in Eq. (86) involves only integer powers of s, and so any noninteger exponent must come from the homogeneous part of the equation $\psi^*(s) = q\psi^*(s/\lambda)$, which has a solution of the form

$$\psi^*(s) \approx s^\beta, \quad \text{with } \beta = \frac{\ln(q)}{\ln(\lambda)}. \tag{87}$$

This term is missing from the series expansion in Eq. (82) because that expansion assumed that all the moments of $\psi(t)$ were finite. Adding to Eq. (82) this missing term with its noninteger exponent yields (if β is less than 1)

$$\psi^*(s) \approx 1 - s^\beta + O(s), \quad \text{as } s \to 0,$$

and

$$\psi(t) \approx t^{-1-\beta}, \quad \text{as } t \to \infty. \tag{88}$$

If β is greater than unity, then for small s, the s term will dominate the s^β term in Eq. (82), and $\psi(t)$ will decay fast enough so that $\langle t \rangle$, the average time between transitions in the random process, is finite. This implies that Poisson statistics will asymptotically govern the number of transitions in the random process, i.e., the mean number of transitions will grow linearly with time as $t/\langle t \rangle$. This is similar to the central limit theorem where the important spatial quantity was $\langle r^2 \rangle$. If this was finite, Gaussian behavior ensued. The condition $\lambda < q < 1$, which makes $\langle t \rangle$ infinite, also keeps $\beta < 1$. When β equals 1, jumps occur at a well-defined constant rate and time flows forward (as measured by the number of jumps) as a one-dimensional process. The above $\psi(t)$, with β less than 1, is called a *fractal-time* random process because the mean number of jumps in a time t grows as t^β when $\beta < 1$, as opposed to linearly with t when $\beta > 1$. Time can be described as being β-dimensional for a fractal-time random process.

7.2 Stretched-Exponential Relaxation

A random process is said to be subordinated to a second random process if an event occurring in the first process depends on the probability of an event occurring in the second random process. We consider, specifically, the first random process to be a reaction that is subordinated to a second random process of random walkers reaching the reaction site. In this scenario, when the second process involves random walks moving in fractal time, the second process will be shown to be a stretched-exponential relaxation.

Consider the problem of dielectric relaxation of glasses involving a frozen-in dipole that can be relaxed only when it is hit by a mobile defect. This problem involves the first-passage time of random walkers (defects) to reach the origin (frozen-in dipole). The mathematical analysis of the problem is in the form of letting there be V lattice sites and letting N walkers be initially randomly distributed among these sites, not including the origin. The probability $\Phi(t)$ that none of the walkers has reached the origin by time t is given by

$$\Phi(t) = \left[1 - \frac{1}{V}\sum_{r}\int_0^t F(\mathbf{r},\tau)d\tau\right]^N, \tag{89}$$

where $F(\mathbf{r},\tau)$ is the probability density that a walker starting at site $\mathbf{r}$ will reach the origin for the first time at time τ. The integral allows for a first passage of a walker to the origin in the interval $(0,t)$. The $1/V$ enters as the probability of a walker starting at a site, and a sum over all possible starting points for a walker is performed. The quantity in brackets gives the probability that a particular walker has not reached the origin, and it is raised to the Nth power for the probability that none of the walkers has yet reached the origin. The problem is easier in the limit $N \to \infty$, $V \to \infty$, but with the ratio remaining constant, $N/V = c$. In this limit,

$$\Phi(t) = \exp\left[-c\sum_{\mathbf{r}}\int_0^t F(\mathbf{r},\tau)d\tau\right] = \exp[-cS(t)]. \tag{90}$$

The term in brackets was simplified by noting that any of the sites from which a walk can reach the origin in a time t are exactly the same sites a walker starting at the origin can reach in a time t. This is exactly $S(t)$, the number of distinct sites that a walker visits in a time t, as discussed in Sec. 5.3. Equation (90) has the following form for the random-walk jumps governed by finite and infinite $\langle t\rangle$ (see Fig. 5):

$$\Phi(t) \approx \begin{cases} \exp(-\text{const.}t/\langle t\rangle) & \text{for}\,\langle t\rangle\,\text{finite}, \\ \exp(-\text{const.}t^{\beta}),\ \beta < 1, & \text{for}\,\langle t\rangle\,\text{infinite}. \end{cases} \tag{91}$$

In addition to introducing subordination processes, we have found that the stretched exponential is a probability limit distribution. This means that the form of $\Phi(t)$ does not depend on the details of the underlying random process, but only on whether $\langle t\rangle$ for the time between jumps is finite or infinite.

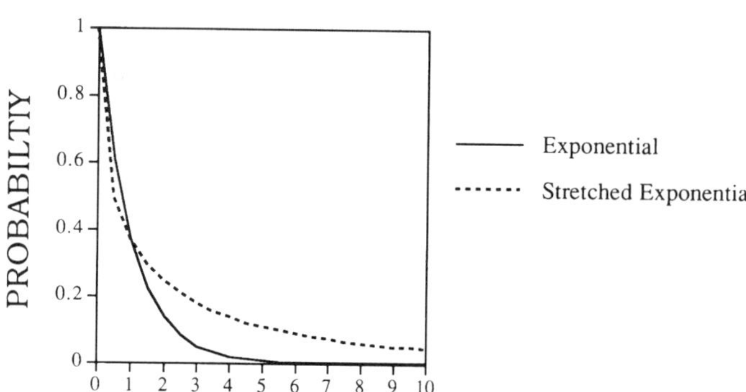

FIG. 5. The exponential decay (single time scale) compared to a stretched-exponential decay. The stretched-exponential decay is initially faster, but is slower after the $1/e$ decay of the exponential.

8. FRACTAL-SPACE PROCESSES

8.1 Lévy Flights

There are many ideas that have been introduced for generalizing Brownian motion, including fractional Brownian motion (Mandelbrot and van Ness, 1968). A fractional Brownian-motion time series would have a Gaussian distribution of values, but the variance would grow, in time t, as t^H with $0 < H < 1$. If $H > \frac{1}{2}$ the motion is called persistent and the variance grows at least as t^2, which is faster than Brownian motion, while the opposite is true for the antipersistent case of $H < \frac{1}{2}$.

A random-walk process that does not fall into the Brownian-motion class occurs when the variance of the jump lengths is infinite (Lévy, 1937). If the variance of each jump is infinite, then the variance of N jumps is also infinite. This in turn implies that the distribution of the sum of N steps should have similar properties to that of a single step. This is basically the question of fractals: When does the whole (the sum of N steps) look like its parts? Lévy found the answer that asymptotically for small k, the Fourier transform $\tilde{p}(k)$ of the probability $p(x)$ to make a jump of displacement x was of the form $\tilde{p}(k) = \exp(-|k|^\beta)$ with $\beta \leq 2$. The Gaussian is the case $\beta = 2$. The Cauchy is the case $\beta = 1$ (see Fig. 6). This is reminiscent of our analysis of fractal time where noninteger exponents entered and signaled self-similar behavior and infinite moments of a probability distribution. This time the moments are spatial, instead of temporal. Let us look at a particular pedagogical case, called the Weierstrass random walk, which illuminates the above discussion.

Begin with a random walk on a lattice with the following probabilities for jumps of length r:

$$p(r) = \frac{q-1}{2q} \sum_{j=0}^{\infty} q^{-j}[\delta_{r,b^j} + \delta_{r,-b^j}]. \tag{92}$$

In this scheme jumps an order of magnitude farther (in base q) occur an order of magnitude less often (in base b). About q jumps of size unity are made, and form a cluster of q sites, before a jump of length b occurs, and about q such clusters of visited sites are formed before a jump of length b^2 occurs, and so on until an infinite fractal hierarchy of clusters is formed. To see the mathematical condition necessary for this to occur, examine the Fourier transform of $p(r)$ to obtain

$$\tilde{p}(k) = \frac{q-1}{q} \sum_{j=0}^{\infty} q^{-j} \cos(kb^j). \tag{93}$$

This is the famous Weierstrass function, which is everywhere continuous, but nowhere differentiable, when $b > q$. Note that

$$\begin{aligned}\langle r^2 \rangle &= \sum_{r=-\infty}^{\infty} r^2 p(r) = \frac{q-1}{q} \sum_{j=0}^{\infty} \left(\frac{b^2}{q}\right)^j \\ &= -\frac{\partial^2}{\partial k^2} \tilde{p}(k=0) = \infty\end{aligned} \tag{94}$$

when $b^2 > q$. This divergence of the second moment (absence of a scale) is a sign of fractal self-similar properties. We have numerically graphed a two-dimensional version

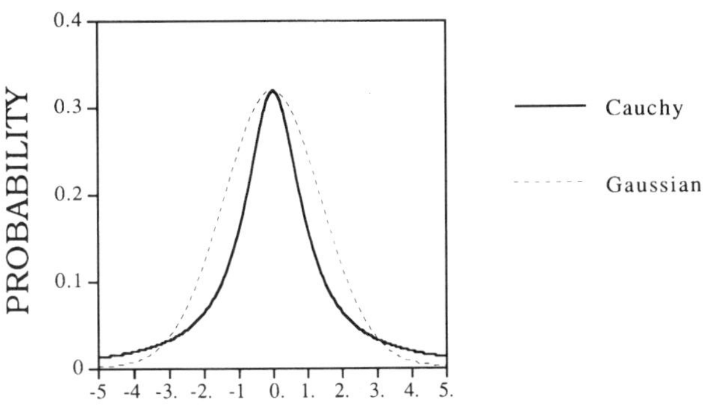

FIG. 6. A Gaussian distribution falls off rapidly enough that all of its positive moments are finite. The Cauchy distribution falls off slowly enough that its second and all higher moments are infinite. The probability in the tail is sufficiently strong to induce events far away from the typical Gaussian behavior. A random walker whose jumps are governed by a Cauchy distribution will visit in two or higher dimensions a fractal set of points with fractal dimension of 1. The traditional approach when encountering Cauchy distributions was to characterize them by their full width at half maximum, and to ignore the fractal consequences of a probability distribution with infinite moments.

FIG. 7. A 10 000-step random walk with nearest-neighbor transitions. Eventually an infinite-step walk would fill every site in the plane and achieve a trivial self-similarity of looking the same on every scale.

of a Gaussian random walk and a Weierstrass random walk in Figs. 7 and 8 to emphasize the difference between a Gaussian and a Lévy process. We cannot write a power-series expansion of $\tilde{p}(k)$, since the coefficient of the $\langle r^2 \rangle$ term is infinite. However, as in the case of the fractal-time distribution, a scaling equation can be employed to determine the small-k behavior of $\tilde{p}(k)$,

$$\tilde{p}(k) = \frac{1}{q}\tilde{p}(bk) + \frac{q-1}{q}\cos(k). \qquad (95)$$

The inhomogeneous term has no noninteger

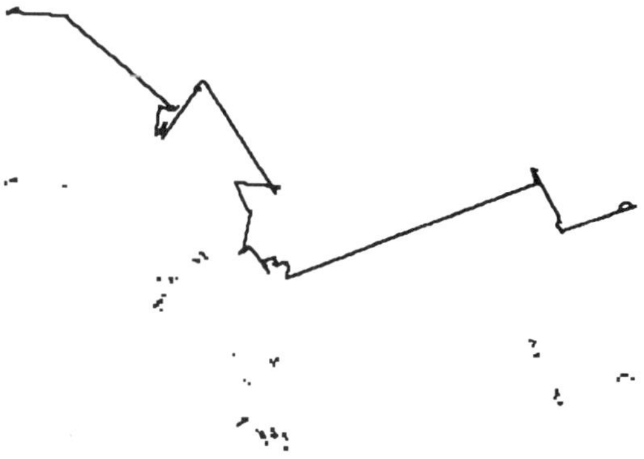

FIG. 8. A Weierstrass random walk of 1000 steps is shown with both the flight segments and the turning points (projected below) Note the self-similar nature of the cluster patterns. The more steps that are taken, the larger would be the probability that longer jumps would be made, thus starting new clusters at order-of-magnitude greater distances from the existing clusters. In this manner self-similar fractal clusters are generated.

powers, and so we focus on a solution to the homogeneous part of the equation, $\tilde{p}(k) = (1/q)\tilde{p}(bk)$, which has a solution of the form k^β with $\beta = \ln(q)/\ln(b)$. When $\beta < 2$ the small-k (large distances r) behavior of $\tilde{p}(k)$ involves the exponent β, while for $\beta > 2$ the moment $\langle r^2 \rangle$ is finite and a Taylor-series expansion of $\tilde{p}(k)$ will exist, i.e.,

$$\tilde{p}(k) \approx \begin{cases} 1 - \text{const.}|k|^\beta + O(k^2) \approx \exp(-|k|^\beta) & \text{for } \beta < 2, \\ 1 - \frac{1}{2}\langle r^2 \rangle k^2 \approx \exp(-|k|^2) & \text{for } \beta > 2. \end{cases} \qquad (96)$$

One can calculate $P_n(r)$, following Eq. (4), in terms of the inverse Fourier transform of $[\tilde{p}(k)]^n$. For large n, $[\tilde{p}(k)]^n$ will be dominated by small k values, which are what we have calculated in Eq. (96). For the case $\beta = 1$, $P_n(r)$, in the large-n limit, has the Cauchy form $n/[\pi(n^2 + r^2)]$. Lévy's work treats a more general case allowing for a bias and additional parameters, but the above discussion captures the general spirit.

8.2 Lévy Walks

Despite the beauty of Lévy flights, they have been largely ignored in the physics literature because of their infinite moments. This infinity can be tamed by associating a velocity with each flight-trajectory segment (Shlesinger *et al.* 1987). Still in the continuous-time random-walk framework, we consider $\Psi(r,t)$ to be the probability density to make a jump of displacement r in a time t. We write

$$\Psi(r,t) = \psi(t|r)p(r)$$

or

$$\Psi(r,t) = p(r|t)\psi(t), \qquad (97)$$

where $p(r)$ and $\psi(t)$ have the same meaning as before and $\psi(t|r)$ and $p(r|t)$ are conditional probabilities for a jump taking a time t given it is of distance r, and for a jump being of distance r, given it took a time t. A simple choice is

$$\psi(t|r) = \delta(t - r/V(r)). \qquad (98)$$

In the study of turbulent diffusion Kolmogorov assumed a scaling law implying $V(r) \sim r^{1/3}$. If we furthermore choose

$$p(r) \approx |r|^{1+\beta}, \tag{99}$$

which for small enough values of β produces a Lévy flight with $\langle r^2 \rangle = \infty$, one finds for the mean square displacement

$$\langle r(t) \rangle^2 = \begin{cases} t^3 & \text{for } \beta \le \frac{1}{3}, \\ t^{2+(3/2)(1-\beta)} & \text{for } \frac{1}{3} \le \beta \le \frac{5}{3}, \\ t & \text{for } \beta \ge \frac{1}{3}. \end{cases} \tag{100}$$

The t^3 case is called Richardson's law of turbulent diffusion. It corresponds to a Lévy walk with Kolmogorov scaling for $V(r)$ combined with a β such that the mean time spent in a segment of the trajectory is infinite. The reason we did not obtain the usual Lévy flight result of $\langle r^2(t) \rangle = \infty$ is that with the coupled space-time memory $\Psi(r,t)$ one does not calculate $\langle r^2 \rangle = -\partial^2 \tilde{p}(k = 0)/\partial k^2$, which would be infinite, but one uses $\int \exp(ikr)\psi(s|r)p(r)dr$ instead of $\tilde{p}(k)$. In effect, one is asking how far a walker has gone in a time t, instead of how long is the flight segment.

FIG. 9. The white squares represent a lattice that percolates (has continuous pathways for nearest-neighbor transitions) through the full square lattice. Many of the pathways, however, present an inefficient means for connecting two points, e.g., the nearby points A and B. A random walker on this tortuous percolation lattice only visits about $N^{2/3}$ new sites in an N-step random walk vs $N/\log(N)$ new sites on a perfect square lattice.

8.3 Random Walks on Fractals

Consider a random walk that is simple in the sense that it makes all of its jumps at a constant rate to a nearest neighbor. This can still lead to complex behavior if the set of sites is fractally distributed. One example is a random walk on a percolation cluster. The percolation cluster can be formed by connecting, with probability p, neighboring sites on a regular lattice. For low p values only finite-sized islands of connected sites are found. As p is increased there is a critical value when an infinite cluster of connected sites appears. Consider the two-dimensional case (see Fig. 9). If one draws a circle with radius R, then the number of sites on the connected cluster inside this circle grows as R^{d_f}, where d_f is the fractal dimension of the cluster. On a square lattice, with all sites accessible, $d_f = 2$, while on the critical percolation cluster $d_f \approx 1.89$. The cluster has many dead ends, which do not contribute to the mean square displacement, and so one may expect that $\langle R^2(t) \rangle$ grows more slowly in time than for Brownian motion in the plane. However, even if all the dead ends are excised and the random walk is restricted to the backbone of the percolation cluster, one still finds slower than linear growth of $\langle R^2(t) \rangle$.

For walks on fractal objects one defines the dimension of the walk, d_w, by

$$\langle R^2(t) \rangle \approx t^{2/d_w}. \tag{101}$$

Random walks on regular infinite lattices have $d_w = 2$. This is because by Polya's theorem a random walk will visit every site in two or fewer dimensions with probability 1, and so the walk trajectory dimension equals 2. Another dimension, called the spectral or harmonic dimension (Alexander and Orbach, 1982), associated with a random walker is defined via the equation for the number $S(t)$ of distinct sites visited, by

$$S(t) \approx t^{d_s/2}. \tag{102}$$

For a regular three-dimensional lattice, $d_s =$

$d_w = 2$. Using Eq. (101) for $\langle R^2(t)\rangle$ we see the scaling $R \approx t^{1/d_w}$. We can also write that after a time t the walker has explored a volume scaling as R^{d_f} that we can write as t^{d_f/d_w}. Equating the volume explored to the number of sites visited gives the following relationship between exponents:

$$d_s = 2d_f/d_w. \tag{103}$$

In general, d_w is not necessarily equal to d_f. For a percolation lattice $d_s \approx 4/3$. One can further write d_w in terms of scaling exponents of the percolation lattice (Gefen *et al.*, 1982) For percolation the probability P_∞ that any given lattice site belongs to the infinite cluster is $(p - p_c)^\beta$, and the linear size ξ of the cluster scales as $(p - p_c)^{-\nu}$. We rewrite P_∞ as $\xi^{-\beta/\nu}$, so that the fractal dimension is $d_f = d - \beta/\nu$. By the Einstein relation, the dc conductivity σ and diffusion D are linked by $\sigma = e^2nD/kT$, where e is the charge on an electron, n is the number of electrons, k is the Boltzmann constant, and T is the temperature in kelvins. The conductivity scales as $\sigma \approx (p - p_c)^\mu \approx \xi^{-\mu/\nu}$. Using that $n \approx P_\infty$, we have $D \propto \sigma/n \propto \xi^{-\mu/\nu+\beta/\nu}$. For $p < p_c$ the largest cluster is of size ξ, so that at long times $\langle R^2(t)\rangle \approx \xi^2 \approx t^{2/d_w}$, and so $\xi \approx t^{1/d_w}$. We also write $\langle R^2(t)\rangle \approx Dt$, which implies $D \approx t^{2/d_w-1} \approx \xi^{2-d_w}$. Now using the expression for D in terms of μ, ν, and β, we have

$$d_w = 2 + (\mu - \beta)/\nu. \tag{104}$$

The percolation lattice is a random fractal and only estimates of the various exponents governing random walks on this lattice have been achieved. For regular fractals, exact calculations of d_f, d_w, and d_s are possible. For example, on the Sierpinski gasket in d dimensions $d_w = \ln(d + 3)/\ln(2)$. For the 2D Sierpinski gasket (see Fig. 10), $2/d_w = 2 \times \ln(2)/\ln(5) = 0.862$, which is the exponent that describes the mean square displacement.

FIG. 10. A Sierpinski gasket. One performs a random walk on this structure by moving along the edges of the triangles. The jump length is the size of the side of the smallest triangle. The hierarchy of vacant triangular regions, in this fractal object, makes diffusion between widely separated sites more complicated and difficult than for a random walk on a regular square lattice. The mean square displacement of a random walker on the Sierpinski lattice grows with time as t^β with $\beta = 2\ \ln(2)/\ln(5) = 0.862$. On a regular lattice $\beta = 1$.

9. DISORDERED SYSTEMS

Consider a random walk of N steps on a straight line with only nearest-neighbor jumps allowed. The mean square displacement in this one-dimensional problem is the familiar $\langle r^2(N)\rangle \sim N$. Now take the straight line and deform it to follow a three-dimensional random-walk path. N units of the deformed line are separated from each other on the average only a distance of $N^{1/2}$, instead of the N units separating them on the straight line. The random walk of N steps on the deformed line has $\langle r^2(N)\rangle \sim N^{1/2}$ (Haus and Kehr, 1987). This is a simple example that provides results different from the standard random walk.

Sinai (1982) investigated the properties of a random walk on a one-dimensional lattice with fixed probabilities to move to the right and left at each lattice site. Although these probabilities are fixed, their values are chosen from a probability distribution such that there is a strong bias at each site for the walker to go in a preferred direction. Sinai found that the walker spends many jumps going between two sites: a leftmost site that when approached from the right sends the walker back toward the right with high probability, and similarly a rightmost site that sends the walker back to the left. Many such pairs of sites can exist, especially in a nested configuration. Sinai finds $\langle r^2(t)\rangle \sim [\ln(t)]^4$ for this random-walk process. A more complicated example involves a random walker in one dimension with a random distribution of symmetric transition rates to the right and left. [See the traps in Fig. 11(b)]. For this problem, one can find (Alexander *et al.*, 1982) the anomalous statistics $\langle r^2(t)\rangle \sim t^{2(1-\beta)/(2-\beta)}$ for a distribution $g(w) \approx w^{-\beta}$ of transition rates, between neighboring sites, for small w when $\beta \leq 1$. The difficulty in generalizing the above calculations beyond one dimension has led to other approaches for random walks on disordered lattices.

Consider a random walker faced with a distribution of transition rates $g(W)$. Effective-medium theory postulates an *unknown* time-dependent rate, $W(t)$, which represents the average behavior of all the possible transitions in the random system. One determines $W(t)$ through a self-consistency argument. Change one transition between two neighboring sites in the system from $W(t)$ to a random value chosen from the distribution $g(W)$. Next calculate the Green's function $G = \langle G\rangle + \delta G$, where $\langle G\rangle$ is the random-lattice–walk Green's function for the perfect lattice where all transitions are governed by $W(t)$, and δG involves transitions T that encounter the altered bond. The Green's function can also be written symbolically as $G = \langle G\rangle + \langle G\rangle T\langle G\rangle + \langle G\rangle T\langle G\rangle T\langle G\rangle + \cdots = \langle G\rangle + \langle G\rangle TG$ (Webman and Klafter, 1982), where the second term represents movement in the effective lattice governed by $W(t)$, then an interaction T with the defective bond, followed by movement under $\langle G\rangle$ again. Higher-order terms in T have more interactions with the defective bond. Taking the average of both sides, over the distribution $g(W)$, implies that $\langle TG\rangle = 0$. The assumption for the effective-medium theory is that $\langle TG\rangle = \langle G\rangle\langle T\rangle = 0$, i.e., $\langle T\rangle = 0$. This condition determines $W(t)$, usually through a numerical evaluation.

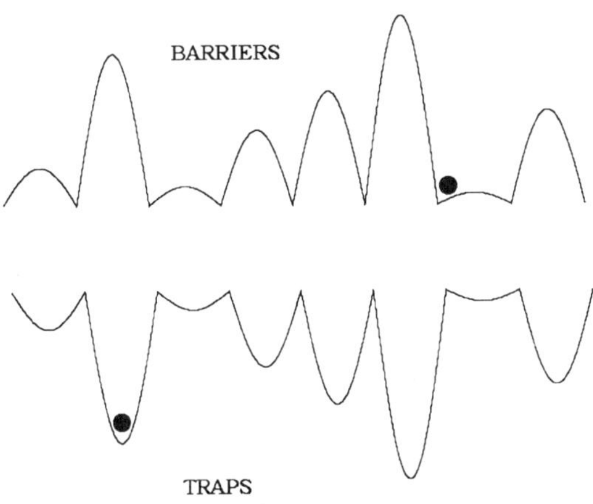

FIG. 11. The top figure represents random barriers, and the bottom one random traps. The dot, in each case, represents a random walker. In the barrier case (top), the walker has unequal probabilities at each site to go to the left or right. In this figure it is easier, from its present position, for the walker to go to the right. The walker tends to get trapped between high barriers. The walker's mean square displacement can grow logarithmically for this Sinai-type random walk. In the random-trap case (bottom), the walker always has equal probabilities to move to the right or left. The walker encounters a distribution of transition rates related to the depth of each trap. For a wide enough distribution of rates a sublinear growth of the mean square displacement with time occurs.

In the effective-medium approximation, one can let the rate to some sites be zero so that these sites are inaccessible. In this case a percolation critical probability p_c enters. If the probability is p that a site is inaccessible, and if $p < p_c$, then a walker is isolated on a finite island of connected sites, and so the diffusion constant would be zero. For a percolation model (Lax and Odagaki, 1982) the effective-medium calculation finds that the real and imaginary parts of the diffusion constant $D(\omega)$ vanish linearly and quadratically with frequency when $p < p_c$ and that $D(0)$ is proportional to $(p - p_c)/(1 - p_c)$ when $p > p_c$. Effective-medium theory was introduced in the 1950s to study scattering of waves in a random medium (Lax, 1952).

The continuous-time random walk discussed in Sec. 6.2 is an effective-medium theory. It was originally applied to electron and hole transport in disordered media (Scher and Lax, 1972). They derived the formula for the frequency-dependent diffusion constant $D(\omega)$ for a random-walk process,

$$\begin{aligned} D(\omega) &= -\frac{\omega^2}{6}\int_0^\infty \exp(-i\omega t)\langle[r(t) - r(0)]^2\rangle dt \\ &= \frac{\overline{r^2}}{6}\frac{i\omega\psi^*(i\omega)}{1 - \psi^*(i\omega)}. \end{aligned} \tag{105}$$

The function $\psi(t)$ represents the probability density for a walker (the electron or hole) to remain trapped between movements. The quantity $\psi(t)$ represents an effective medium and it is the same at each lattice site. An infinite first moment $\langle t\rangle$ of $\psi(t)$ was introduced (Montroll and Scher, 1973) to explain scaling behavior in a hole current set up by biased "fractal-time [$\psi(t) \approx t^{-1-\beta}$ for $\beta < 1$]" random walks. For fractal-time random walkers biased in one direction one finds that the first moment and its variance both grow as $\langle r(t)\rangle \approx \sqrt{\langle r^2(t)\rangle} \approx t^\beta$, and so the fluctuations (second moment) per distance traveled (first moment) are time independent. An approach to choosing $\psi(t)$ is to focus on the physics of the trapping process. If the hopping time t out of a potential well follows the Arrhenius law, $t^{-1} = \nu \exp(-E/kT)$, and the probability $\rho(E)$ of finding an energy barrier E falls off exponentially as $\exp(-E/E_0)$, then the randomness in energy induces a randomness of hopping times $\psi(t)dt = \rho(E)dE$, which produces the fractal time (see Sec. 8) $\psi(t) \sim t^{-1-\beta}$, where $\beta = kT/E_0$ when $kT < E_0$ and $\beta = 1$ if $kT > E_0$.

Random processes on random geometries is still an active area of research.

GLOSSARY

Brownian Motion: A random walk whose probability distribution is governed by the diffusion equation. The trajectory is fractal with the derivative (velocity) not well defined.

Central Limit Theorem: A theorem showing that a sum of identically distributed random variables with finite second moments converges to a Gaussian distribution as the number of variables becomes infinite.

Chapman–Kolmogorov–Smoluchowski–Bachelier Equation: An equation relating the probability for being at a site at a particular time, in terms of being at a different site at an earlier time and summing over all possible intermediate sites.

Characteristic Function: The Fourier transform of a probability function.

Colored Noise: Noise whose second-order correlation function is broader than a delta function.

First–Passage-Time Density: A probability density to reach a certain criterion (e.g., level, site) for the first time.

Fokker–Planck Equation: A second-order differential equation for a probability distribution based on an expansion relying on the existence of the first two moments of the distribution.

Fractal Brownian Motion: A fractal random process whose next step is built upon correlating its entire previous history. The uncorrelated case is Brownian motion.

Fractal Time: A random process where the mean interval between events is infinite. The set of event times looks like a random fractal set of dimension less than 1.

Gaussian Random Noise: Delta-correlated noise in time with an amplitude following Gaussian statistics.

Kubo Oscillator: An oscillator with a random frequency.

Langevin Equation: An equation of motion with an added noise term.

Lévy Process: A random walk in space with infinite second moment of the jump-size probability, or a random process with the first moment of the waiting-time distribution between events being infinite. A Lévy flight is a Lévy random walk jumping between points and a Lévy walk includes the whole trajectory as well as the end points of each jump.

Markov Process: A random process where the next change depends only on the present state of the random variable, and not on a more detailed prior history.

Master Equation: An equation relating the change with time of a probability for the value of a variable in terms of probabilities for the basic allowed transitions.

One-Over-*f* (1/*f*) noise: A power spectrum whose low-frequency behavior is described by a function that scales with frequency as $1/f$.

Power Spectrum: The Fourier transform of a second-order correlation function. The real part of the transform of a fluctuating current would correspond to the power as a function of frequency.

Random Walk: A sum of random variables. The walk can have correlated or uncorrelated steps.

Self-Avoiding Random Walk: A random walk whose steps may not overlap.

Shot Noise: The noise that arises from the arrival of randomly timed (usually Poisson arrival) distinct pulses each of which has an effect that decays away. Slow scale-invariant decays produce fractal shot noise.

Stochastic Differential Equation: A differential equation for the time evolution of random variables.

Stochastic Process: A random process that involves a time variable.

Subordinated Random Process: A process whose random event times depend on the outcome of a second random process.

Weierstrass Random Walk: A random walk whose generating function is the scaling Weierstrass function, and whose trajectory is a fractal set of hierarchical clusters.

White Noise: A noise whose second-order correlation function is a delta function, i.e., noise that is not correlated in space or time.

Wiener–Khintchine Theorem: A theorem that relates the second-order correlation function of a random process and its power spectrum as Fourier cosine transform pairs.

Works Cited

Alexander, S, Bernasconi, J., Schneider, W. R., Orbach, R. (1981), *Rev. Mod. Phys.* **53,** 175–198.

Alexander, S., Orbach, R. (1982), *J. Phys. Lett. (Paris)* **43,** L625–L631.

Chandrasekhar, S. (1943), *Rev. Mod. Phys.* **15,** 1–89.

David, F. N. (1962), *Games, Gods, and Gambling,* New York: Hafner.

den Hollander, W. Th. F. (1985), *J. Stat. Phys.* **40,** 201–204.

Domb, C., Fisher, M. E. (1958), *Proc. Cambridge Philos. Soc.* **54,** 48–59.

Donsker, M., Varadhan, S. (1979), *Commun. Pure Appl. Math.* **32,** 721–748.

Einstein, A. (1905), *Ann. Physik (Leipzig)* **17,** 549–560.

Feller, W. (1971), *An Introduction to Probability Theory and Its Applications,* 2nd ed., Vol. 2, New York: Wiley.

Fisher, R. A., Tippett, L. H. C. (1928), *Proc. Cambridge Philos. Soc.* **24,** 180–190.

Fisher, M. E. (1984), *J. Stat. Phys.* **34,** 667–729.

Gefen, Y., Aharony, A., Alexander, S. (1982), *Phys. Rev. Lett.* **50,** 77–80.

Haus, J. W., Kehr, K. (1987), *Phys. Rep.* **150,** 263–406.

Kramers, H. A. (1940), *Physica* **7,** 284–304.

Kubo, R. (1962), in: D. Ter Haar (Ed.), *Fluctuation, Relaxation, and Resonance in Magnetic Systems,* Edinburgh, U.K.: Oliver and Boyd.

Lax, M. (1952), *Phys. Rev.* **85,** 621–629.

Lax, M., Odagaki, T. (1982), in: M. F. Shlesinger, B. J. West (Eds.), *Random Walks and Their Applications in the Physical and Biological Sciences,* AIP Conference Proceedings No. 109, New York: American Institute of Physics.

Lévy, P. (1937), *Theorie de l'Addition des Variables Aleatoires,* Paris: Gauthiers-Villars.

Lindenberg, K., Seshadri, V., Shuler, K. E., Weiss, G. H. (1980), *J. Stat. Phys.* **23,** 11–25.

Lindenberg, K., West, B. J., Kopelman, R. (1990), in: F. Moss, L. A. Lugiato, W. Schleich (Eds.), *Noise and Chaos in Nonlinear Dynamical Systems,* Cambridge, U.K.: Cambridge Univ. Press.

Mandelbrot, B. B., van Ness, J. W. (1968), *SIAM Rev.* **10,** 422–437.

Montroll, E. W. (1950), *J. Chem. Phys.* **18,** 734–743.

Montroll, E. W., Scher, H. (1973), *J. Stat. Phys.* **9,** 101–135.

Montroll, E. W., Weiss, G. H. (1965), *J. Math. Phys.* **6,** 167–181.

Moss, F., Lugiato, L. A., Schleich, W. (Eds.) (1990), *Noise and Chaos in Nonlinear Dynamical Systems,* Cambridge, U.K.: Cambridge Univ. Press.

Oppenheim, I., Shuler, K. E., Weiss, G. H. (1977), *The Master Equation,* Cambridge, MA: MIT Press.

Ovchinnikov, A. A., Zeldovich, Ya. B. (1978), *Chem. Phys.* **28,** 215–218.

Pontryagin, L., Andronov, A. A., Vitt, A. (1933), *Zh. Eksp. Teor. Fiz.* **3,** 165–180.

Scher, H., Lax, M. (1972), *Phys. Rev. B* **7,** 4491–4501.

Schottky, W. (1922), *Ann. Phys. (Leipzig)* **68,** 157–176.

Shlesinger, M. F., West, B. J., Klafter, J. (1987), *Phys. Rev. Lett.* **58,** 1100–1103.

Sinai, Ya. G. (1982), *Theory Prob. Appl.* **27,** 256–268.

Székely, Gábor J. (1986), *Paradoxes in Probability Theory and Mathematical Statistics,* Dordrecht: Reidel.

Todhunter, I. (1865), *History of the Theory of Probability,* Cambridge, MA: Chelsea Pub.

Wax, N. (1954), *Selected Papers on Noise and Stochastic Processes,* New York: Dover.

Webman, I., Klafter, J. (1982), *Phys. Rev. B* **26,** 5950–5952.

Weiss, G. H., Bendler, J. T., Shlesinger, M. F. (1988), *Macromolecules* **21,** 521–523.

Weissman, M. (1988), *Rev. Mod. Phys.* **60,** 537–571.

Further Reading

Barber, M. N., Ninham, B. W. (1970), *Random and Restricted Walks: Theory and Applications,* New York: Gordon and Breach.

Blumen, A., Klafter, J., Zumofen, G. (1986), in: I. Zschokke (Ed.), *Optical Spectroscopy of Glasses,* Dordrecht: Reidel.

Bouchard, J.-P., Georges, A. (1990), *Phys. Rep.* **195,** 127–293.

Feder, J. (1988), *Fractals,* New York: Plenum.

Feller, W. (1973), *An Introduction to Probability Theory and Its Applications,* 3rd ed., Vol. 1, New York: Wiley.

Fox, R. F. (1978), *Phys. Rep.* **48,** 179–283.

Gardner, C. W. (1985), *Handbook of Statistical*

Methods for Physics, Chemistry and the Natural Sciences, Berlin: Springer-Verlag.

Gumbel, E. J. (1958), *Statistics of Extremes,* New York: Columbia Univ. Press.

Havlin, S., Ben Avraham, D. (1987), *Adv. Phys.* **36,** 695–798.

Havlin, S., Nossal, R., Shlesinger, M. (Eds.) (1991), special issue on Non-Classical Reactions, *J. Stat. Phys.* **65,** 837–1306.

Hughes, B. D. (1995), *Random Walks and Random Environments,* Oxford: Clarendon Press.

Kac, M. (1959), *Probability and Related Topics in Physical Sciences,* London: Interscience.

Lowen, S. B., Teich, M. C. (1989), *Phys. Rev. Lett.* **63,** 1755–1759.

Mandelbrot, B. B. (1982), *The Fractal Geometry of Nature,* San Francisco: Freeman.

Montroll, E. W., Shlesinger, M. F. (1984), in: E. W. Montroll, J. L. Lebowitz (Eds.), *The Wonderful World of Random Walks,* Studies in Statistical Mechanics Vol. VII, Amsterdam: North-Holland.

Montroll, E. W., West, B. J. (1979), in: J. L. Lebowitz, E. W. Montroll (Eds.), *On an Enriched Collection of Stochastic Processes,* Studies in Statistical Mechanics Vol. XI, Amsterdam: North-Holland.

Moss, F., McClintock, P.V.E. (1989), *Noise in Nonlinear Dynamical Systems,* Vols. 1–3, New York: Cambridge Univ. Press.

Moss, F., Bulsara, A., Shlesinger, M. (Eds.) (1993), special issue on Stochastic Resonance, *J. Stat. Phys.* **70,** 1–514.

Risken, H. (1984), *The Fokker-Planck Equation,* Berlin: Springer-Verlag.

Shlesinger, M. F., West, B. J. (1984), New York, AIP Conf. Proc. No. 109.

Shlesinger, M., Mandelbrot, B. B., Rubin, R. J. (Eds.) (1984), special issue on Fractals, *J. Stat. Phys.* **36,** 519–924.

Stratonovich, R. L. (1963), *Introduction to the Theory of Random Noise,* Vol. 1, New York: Gordon and Breach.

Stratonovich, R. L. (1967), *Introduction to the Theory of Random Noise,* Vol. 2, New York: Gordon and Breach.

Toussaint, D., Wilczek, F. (1983), *J. Chem. Phys.* **78,** 2642–2647.

van Kampen, N. G. (1981), *Stochastic Processes in Physics and Chemistry,* Amsterdam: North-Holland.

Weiss, G. (Chmn.) (1983), special issue on Random Walks, *J. Stat. Phys.* **30,** 249–561.

Weiss, G. H. (1994), *Aspects and Applications of the Random Walk,* Amsterdam: North Holland.

Weiss, G. H., Rubin, R. J. (1983), *Adv. Chem. Phys.* **52,** 363–505.

RARE GASES

R. A. Aziz and A. R. Janzen, *Department of Physics, University of Waterloo, Waterloo, Ontario, Canada*

R. O. Simmons, *Department of Physics, University of Illinois at Urbana-Champaign, Urbana, Illinois, U.S.A.*

	Introduction	72
1.	**Gaseous State**	73
1.1	Interatomic Interactions	73
1.1.1	General Considerations	73
1.1.2	Like Pairs	73
1.1.3	Unlike Pairs	74
1.2	Compounds	74
1.2.1	Chemical Compounds	74
1.2.2	Clathrates	75
1.2.3	Van der Waals Complexes	76
1.3	Adsorption	76
1.4	Thermodynamics	76
1.4.1	Equations of State	76
1.4.2	Heat Capacity and Vapor Pressure	77
1.4.3	The Universal Gas Constant	77
1.5	Transport Properties	77
1.5.1	Kinetic Theory	77
1.5.2	Measurement of Viscosity and Thermal Conductivity	78
1.5.3	Measurement of Ordinary and Thermal Diffusion	78
1.6	Law of Corresponding States (LCS)	78
1.7	Optical and Dielectric Properties	79
1.7.1	Refractivity and Dielectric Virial Coefficients	79
1.7.2	Spectroscopy of the Dimers	80
2.	**Liquid State and Dense Gases**	80
2.1	Interatomic Interactions in the Condensed State	80
2.2	The Structure of Liquids	81
2.3	Thermodynamic Properties	81
2.3.1	Quantum Corresponding States	82
2.3.2	Sound Velocity	82
2.4	Transport Properties	83
2.4.1	Transport Properties, Theory	83
2.4.2	Neutral and Charge Transport, Experiment	83
2.5	Microscopic Dynamics from Neutron-Scattering Evidence	83
2.6	Special Properties of Liquid Heliums and Their Isotopic Mixtures	84
3.	**Crystalline State**	84
3.1	Melting, Corresponding States	84
3.2	Thermoelastic Properties	85
3.3	Microscopic Properties	86
3.3.1	Phonons, Experiment and Theory	86
3.3.2	Momentum Distribution, Kinetic Energy	87
3.4	Dielectric and Electronic Properties	87
3.5	Crystal Defects	88
3.5.1	Point Defects	88
3.5.2	Line and Surface Defects, Thin Films, Clusters	89
3.6	Solid Heliums, Unique Properties	90
3.7	Defects and Excitations in Solid Heliums	90
4.	**Scientific and Technological Applications**	91
4.1	Arc Welding and Cutting	91
4.2	Passive Atmospheres and Inert Condensed Environments	91
4.3	Illumination	91
4.4	Lasers	92
4.5	Thermometry	92
4.6	Cryogenics and Refrigeration	92
4.7	Other Uses	92
	Glossary	93
	Works Cited	94
	Further Reading	96

3-527-28138-X/96/$5.00 + .50

INTRODUCTION

In 1904, the Nobel prize in Chemistry was awarded to Sir William Ramsay and the prize in Physics was awarded to Lord Rayleigh, in part for their work in the discovery of rare gases as a new family of elements, a whole new column in the periodic table. This family was unknown when they began precise work on the density of nitrogen in 1892, although there was earlier evidence for two of its members: Henry Cavendish isolated a small bubble of what is now known to be argon while studying the composition of air in 1785, and P. J. C. Janssen and J. N. Lockyer had observed a brilliant new yellow line in the solar spectrum in 1868, a line that later turned out to be that of helium.

The presence of argon occupying 1% of the Earth's atmosphere produced the discrepancy in the densities of "nitrogen" obtained from air and from chemical reactions. This provided the critical clue leading to the discovery of argon, the first of the family to be isolated. Forming successively smaller proportions of the Earth's atmosphere and crust are helium, neon, krypton, xenon, and radon. It has been estimated that helium forms one-quarter of the mass of the universe, although on Earth, gravity is too weak to retain helium from escape from the atmosphere, to which it and, for example, radon are released from the Earth's crust by natural processes.

The rare gases have important places both in the development of many branches of science and in numerous technological applications. Neon provided the first clear examples of stable nuclear isotopes when J. J. Thomson, studying liquid-air residues, discovered ^{22}Ne and ^{20}Ne in 1913. The nucleus of ^{4}He (the alpha particle of radioactivity) is the most stable known, whereas all radon isotopes are radioactive. In metrology, from 1960 to 1983 the SI standard of length was defined by an orange line in the ^{86}Kr spectrum; one meter is exactly 1 650 763.73 wavelengths of this line. In thermal science, the properties of helium and neon are used in establishing the temperature scale. General properties of rare gases are given in Table 1. Selected technical applications appear in Sec. 4.

Lennard-Jones (LJ) models for the denser phases of rare-gas atoms have considerable historical significance. Such a model represents the interatomic potential as a polynomial of only two terms in the interatomic separation, one each for the repulsive and the attractive parts, respectively. The computational simplicity of the LJ interaction provided the first qualitative confirmations of the usefulness of molecular dynamics and of Monte Carlo computational methods in representing the properties of fluids, typically the properties of fluid argon. Further, a multitude of scientific articles have been written about the static and dynamical properties of fcc solid models using an LJ 12-6 potential and the mass of a rare-gas atom, again typi-

Table 1. General properties of the rare gases. (The subscripts tp, bp, and c refer to triple-, boiling-, and critical-point coordinates, respectively. For the heliums, which have complicated phase diagrams, only the bcc–fcc–liquid triple point is given.)

	^{3}He	^{4}He	Ne	Ar	Kr	Xe
Molecular mass	3.016 05	4.002 60	20.183	39.948	83.804	131.30
Atomic number	2	2	10	18	36	54
T_{tp} (K)	17.8	14.9	24.56	83.806	115.78	161.40
P_{tp} (kPa)	13 700	3050	43	69	73	82
V_{tp} (cm^3 mol^{-1})	18.83	20.70	16.18	28.24	34.32	44.31
T_{bp} (K)	3.19	4.215	27.07	87.27	119.80	165.05
T_c (K)	3.32	5.20	44.45	150.73	209.29	289.74
P_c (Mpa)	0.1167	0.2274	2.72	4.865	5.4931	5.8400
V_c (cm^3 mol^{-1})	72.8	57.51	41.7	74.7	92.3	118.3
Reduced de Boer parameters[a]						
(A)	2.88	2.50	0.543	0.172	0.094	0.058
(B)	3.08	2.68	0.594	0.186	0.102	0.063

[a]Reduced de Broglie wavelength. See Sec. 2.3.1 for definition. (A) Based on HFD true pair-potential parameters ϵ/k_B and σ (see Sec. 1). (B) Based on Lennard-Jones "effective" pair-potential parameters ϵ/k_B and σ.

cally argon. Among these theoretical investigations of lattice dynamics and of the statistical mechanics of the condensed state are the introduction of many interesting concepts and ingenious computational methods. Because the models can be well defined, with common dependence upon a stated pair interaction with only two parameters, comparisons between models are thus facilitated. For most properties of actual rare-gas solids, however, such models frequently have only qualitative validity. As precision in the knowledge of the pair potential between rare-gas atoms has increased and computational power is enhanced, dependence upon such a poor approximation to the pair potential is no longer necessary, although it continues to have limited usefulness in some qualitative considerations.

1. GASEOUS STATE

1.1 Interatomic Interactions

1.1.1 General Considerations A rare-gas atom consists of a positively charged nucleus surrounded by a negatively charged spherically symmetric electron cloud and hence possesses no permanent dipole. When the atoms are separated by a finite distance, they interact, and the energy resulting from the interaction is the interatomic potential energy $U(r)$, which depends only on the separation between them. Qualitatively, for small separations, the electron clouds overlap. A repulsive force then arises between the atoms, as a consequence of the Pauli exclusion principle, from the increase of energy of the electrons. For large separations, these nonpolar atoms attract one another by means of the London dispersion energy. The electrons of each atom are moving, and at any instant, one atom may possess an instantaneous dipole, which induces an instantaneous dipole in the second atom. The dipoles interact to produce an attractive "dispersion" energy.

An *ab initio* determination of the forces from quantum mechanics has proved to be very difficult except in recent years in the case of helium, the simplest rare gas. An empirical approach has been consistently more successful. In this approach, a flexible potential form is assumed that is realistic at short and at long range and that lends itself to the inclusion of theoretical values in these regions if available. In order to ascertain the potential in the intermediate range, a multiproperty fit to a judiciously chosen set of properties is performed. The set is carefully chosen so that each property probes and constrains the potential in specific regions. The various dilute-gas properties that have been used are

1. macroscopic properties such as second virial coefficients and transport properties, and
2. microscopic properties such as spectroscopic data on the dimer and beam data (differential scattering cross sections and total, or integral, cross sections in the glory region, that relative velocity region where undulations in the cross sections occur as a consequence of the presence of repulsion and attraction in the interatomic interaction).

1.1.2 Like Pairs Researchers at IBM, Oxford University, and California Institute of Technology as well as others determined fairly accurate potentials in the early 1970s. A summary of the situation to 1982 was given by Aziz (1984), who described the fine tuning of potentials for the pure gases. As more accurate data became available and more sophisticated models were developed, state-of-the-art potentials, notably the multi-damped Hartree–Fock dispersion (HFD) and exchange Coulomb (XC), have been subsequently constructed. The problem of determining an accurate potential for helium from *ab initio* and semiempirical points of view has been especially difficult, and only recently has agreement been reached between the two approaches.

A particularly realistic but simple potential is the overall damped HFD in which the interaction energy is partitioned into uncorrelated (Hartree–Fock) and correlated parts and a suitable damping function to account for charge-overlap effects. The form is

$$U(r) = \epsilon U^*(x), \quad (1)$$

where

$$U^*(x) = A^* \exp(-\alpha^* x + \beta^* x^2) - F^*(x)[c_6/x^6 + c_8/x^8 + c_{10}/x^{10}] \quad (2)$$

with

$$F^*(x) = \begin{cases} \exp[-(D/x - 1)^2], & x < D, \\ 1, & x \geq D, \end{cases} \tag{3}$$

where $x = r/r_m$ and $c_n = C_n/\epsilon r_m^n$.

Parameters for the pure rare gases appear in Table 2. Figures 1 and 2 show a comparison between HFD and LJ potentials for argon.

Properties for which nonadditive many-body interaction energies are significant in pure rare gases are third virial coefficients, binding energies and structure of solids, pressure–volume relations, etc. The relative importance of many-body interactions is greater in the condensed phases.

1.1.3 Unlike Pairs Interactions for unlike pairs are not as well known as they are for the pure gases, largely because few spectroscopic data are available for these systems. Interatomic potentials determined from gas-phase properties for ten mixed systems are reviewed and critically assessed by Aziz (1984). Again, refinements to the knowledge of the interaction of these systems have subsequently been made as more accurate differential-scattering and diffusion data have become available.

1.2 Compounds

1.2.1 Chemical Compounds The term "chemical compound" is understood to mean a substance that is chemically stable at room temperature and pressure, held together by a chemical bond. This excludes hydrates and other clathrates as well as van der Waals complexes, in which the rare-gas atom is not held by a true chemical bond. Other substances that are excluded but that have been studied extensively include the short-lived rare-gas–cation complexes, some of which are important as lasing media.

The chemical reactivity of the rare gases increases with atomic number, as the outer electron shells get increasingly screened by the inner electrons. Nevertheless, the reactivity is always small compared with those of the other elements, because the outermost shell is filled. One would expect radon to be the most reactive rare gas, and this may indeed be the case. However, its rarity and radioactivity make it difficult to study, and its

Table 2. Parameters for HFD–B potentials (Sec. 1.1.2); σ is the value of r at which $U(r) = 0$. (Enough decimal places are displayed to avoid round-off errors.)

	He HFD–B3–FCI1[a]	Ne HFD–B[b]	Ar HFD–B3[c]	Kr HFD–B2[d]	Xe HFD–B1[d]
A^*	186 924.4	895 718.0	113 211.8	69 735.39	54 408.73
α^*	10.571 75	13.864 35	9.000 534	8.388 022	7.529 583
c_6	1.351 866	1.213 175	1.099 711	1.061 360	1.005 552
c_8	0.414 951	0.532 227	0.545 116	0.568 456	0.583 599
c_{10}	0.171 511	0.245 707	0.392 787	0.426 055	0.473 783
$10^{25}\ C_6(\mathrm{J} \times \mathrm{nm}^6)$	1.398 681	6.172 002	61.557 27	122.8311	271.7901
$10^{27}\ C_8(\mathrm{J} \times \mathrm{nm}^8)$	3.782 664	25.870 09	431.6150	1058.395	3006.292
$10^{28}\ C_{10}(\mathrm{J} \times \mathrm{nm}^{10})$	1.377 556	11.410 81	439.9169	1276.209	4651.408
β^*	−2.077 588	−0.129 938	−2.602 702	−2.796 115	−3.339 043
$\beta^*/r_m^2\ (\mathrm{nm}^{-2})$	−23.580 00	−1.360 000	−18.400 00	−17.380 00	−17.520 00
D	1.438 000	1.360 00	1.040 000	1.208 000	1.114 000
ϵ/k_B (K)	10.956	42.25	143.25	201.3	282.8
r_m (nm)	0.296 830	0.3091	0.3761	0.4011	0.436 56
σ (nm)	0.264 138	0.2759	0.3356	0.3571	0.389 10
Lennard-Jones 6–12 Parameters					
ϵ/k_B (K)	10.22	35.60	119.5	165.49	228.78
σ (nm)	0.2556	0.2749	0.3403	0.366	0.401

[a]Aziz *et al.*, 1995.
[b]Aziz and Slaman, 1989.
[c]Aziz and Slaman, 1990.
[d]Dham *et al.*, 1990.

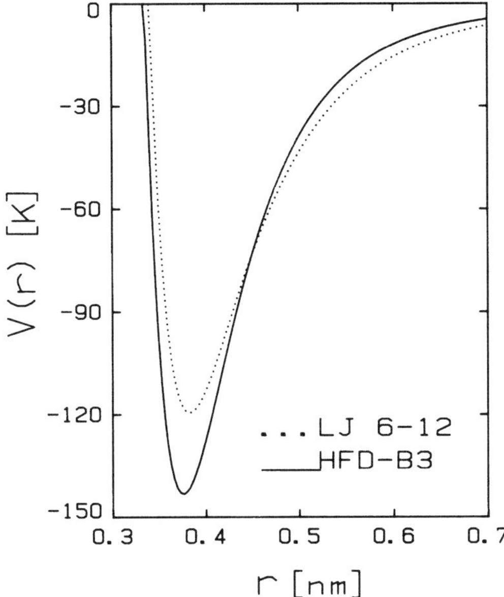

FIG. 1. The attractive region of the argon pair potential. LJ–6-12: the Lennard-Jones 6-12 argon potential of Table 2. HFD–B3: The Hartree–Fock–dispersion argon potential of Table 1.

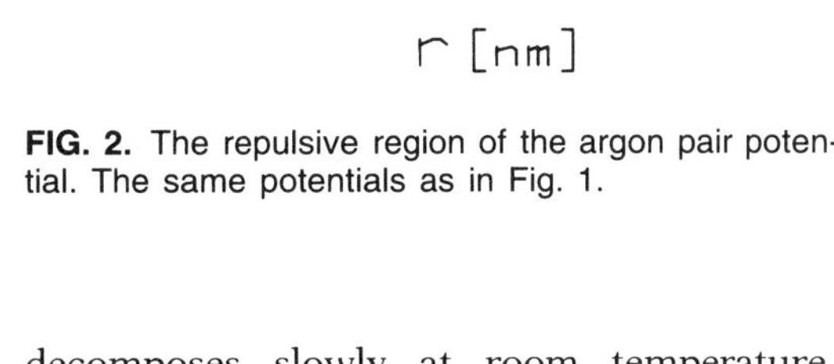

FIG. 2. The repulsive region of the argon pair potential. The same potentials as in Fig. 1.

only known compound is RnF_x, with x unknown.

Xenon is the most reactive of the nonradioactive rare gases. Xenon compounds were predicted by Linus Pauling as early as 1933, although the first one, $XePtF_6$, was not produced until 1962. Subsequently various binary compounds of xenon have been produced. The fluorides XeF_n, with $n = 2$, 4, and 6, are all stable, colorless crystals at room temperature. Evidence of XeF_8 has been reported. Xenon oxides have also been produced. XeO_3 is a white crystal and is dangerously explosive. XeO_4 has been detected but not produced in quantity and is the most volatile xenon compound known. The only other known binary compound of xenon is $XeCl_2$.

There is now a large catalog of more complex xenon compounds (Claassen, 1966). Besides the original $XePtF_6$, a yellow solid, they include $XeOF_4$, a clear, colorless liquid; metallic perxenates such as Na_4XeO_6 that, like the oxides, are very volatile; and mixed fluorides such as XeF_2SbF_5.

Krypton, with atomic number 36, is less reactive than xenon. Its only known binary compound with any stability is KrF_2, which decomposes slowly at room temperature. Some metallic krypton salts such as $KrFAuF_6$ have been made.

The three lighter gases argon, neon, and helium do not form any known stable compounds. However, calculations (Frenking *et al.*, 1990) suggest that all three may form stable compounds with beryllium and oxygen. Spectroscopic evidence for ArBeO has been reported (Thompson and Andrews, 1994), but the compound has not yet been produced in quantity.

1.2.2 Clathrates The weak van der Waals force that acts between two rare-gas atoms can also act between a rare-gas atom and an atom or molecule of another kind, resulting in a weak bond. The attachment of a rare-gas atom to other molecules is further enhanced if the molecules form a cage around the atom. Such a complex is called a clathrate. A familiar example of a clathrate is the hydrate.

Hydrates of argon, krypton, and xenon were prepared as early as 1896. All have a decomposition temperature below 0 °C, although for xenon hydrate it is just −3.4 °C. Many other rare-gas clathrates are now known including RG-hydroquinones (RG =

rare gas), RG-phenols, and the recent RG-fullerenes in which the rare-gas atom is trapped inside a C_{60} "buckyball" (Saunders *et al.*, 1993).

1.2.3 Van der Waals Complexes A physically bound two-atom or molecule-atom system, e.g., Ar_2, ArNe, H_2–Ar (X_2–Y), etc., may form under certain conditions. These dimers can arise when the temperature of the gas is reduced. Such dimers can also be formed in a pulsed supersonic jet. The concentration of dimers is small; for example, only about 0.5% of argon atoms are bound in the vapor phase near NPT (normal pressure and temperature) and are dissociated quickly by colliding atoms. A debate over whether diatomic helium could exist has persisted for many years. Dozens of potentials, *ab initio* and empirical, proposed since the 1939 Lennard-Jones (12-6) potential, produced contradictory conclusions about the existence of He_2. Generally, for potentials with reasonable long-range behavior, those with deeper wells predict the dimer while the shallower ones do not. To make an unequivocal prediction, one requires an accurate potential with, among other things, two key features: an accurate long range and an accurate well depth. The latter was determined from a knowledge of accurate low-temperature second virials provided by the standards laboratories for the ITS-90 temperature scale. Aziz *et al.* (1987) presented such a potential [HFD-B (HE); $\epsilon/k_B = 10.948$ K] from such data, which could predict a large number of other properties. This potential predicted the existence of a weakly bound dimer with a binding energy of 1.6 mK. While earlier *ab initio* potentials had shallower wells that would not predict the virials, recent *ab initio* methods have been refined to the point that *ab initio* and semiempirical potentials are in close agreement (Aziz and Slaman, 1991; Aziz *et al.*, 1995).

There have been a number of groups actively engaged in the pursuit of observing the very weakly bound He_2 dimer that would, by theoretical estimates, have the enormously long bond length of about 5.4 nm. The dimer was first detected in a supercooled beam at the University of Minnesota (Luo *et al.*, 1993) but perhaps the most convincing experimental confirmation lies in the diffraction experiments carried out in Göttingen (Schöllkopf and Toennies, 1994).

1.3 Adsorption

A great deal of experimental and theoretical work has been done on the adsorption of rare-gas atoms on surfaces of various kinds. Much of this work could be categorized as pure rather than applied, dealing, for example, with the general behavior of two-dimensional systems, but there are important applications as well.

Fundamental studies have focused upon using physisorbed layers on structured substrates such as graphite. The pattern of adsorbed rare gas depends upon its atomic size relative to the surface lattice parameter of the substrate, and the resulting layers are commensurate or not depending upon this as well as upon how many layers there are. The resulting systems are excellent prototypes for the investigation of the thermodynamic properties of two-dimensional systems (Nagler *et al.*, 1985). Both the static structures and the dynamics of motion of the adsorbed gas atoms have been studied, the latter using scattering probes such as slow neutrons and electrons.

Rare gases can be separated from gaseous mixtures by adsorption onto certain materials. For example, krypton and xenon can be adsorbed out of impure oxygen onto a certain kind of zeolite, a microporous material sometimes called a "molecular sieve." Conversely oxygen can be adsorbed out of impure argon by another type of zeolite. Neon has also been removed from helium–neon mixtures by adsorption onto activated carbon and silica gel.

1.4 Thermodynamics

1.4.1 Equations of State It has proved difficult to express the behavior of the rare gases over the whole range of pressure P, volume V, and temperature T in a single equation. The first success was the van der Waals equation

$$(P + a/V^2)(V - b) = RT, \tag{4}$$

which mimics, when used with the Maxwell equal-area construction, all the qualitative features of real equilibrium fluids. The quantity a/V^2 accounts for the attraction between atoms and b accounts for the finite atomic size. The parameters a and b differ from species to species. Many empirical relations have been proposed. Song and Mason (1989) found a compact, analytical equation of state that is simple, general, and accurate (see Sec. 2.3).

A useful equation of state from both a practical and a theoretical point of view is the so-called *virial equation*. In this form, the product PV of n moles of gas is expressed as a power series in $1/v$ where $v = V/n$:

$$Pv = RT[1 + B/v + C/v^2 + D/v^3 + \cdots]. \quad (5)$$

Here B, C, D, etc. are the second, third, and fourth virial coefficients, which depend only on T and the species of gas. The coefficients B, C, D, . . . represent deviations from ideal behavior when collisions involving two, three, four, . . . atoms, respectively, become significant in the gas. Hence B can be related to and provide information on the pair potential, C on the three-body potential, etc.

1.4.2 Heat Capacity and Vapor Pressure The heat capacities of the monatomic rare gases are nearly constant over a wide temperature range. The heat capacity at constant volume is nearly equal to the ideal-gas value of $\frac{3}{2}R$ while the heat capacity at constant pressure is very nearly equal to $\frac{5}{2}R$.

Very accurate values of the saturated vapor pressures P_σ for argon (Gilgen *et al.*, 1994) and other gases along the entire coexistence curve are available in the form

$$\ln(P_\sigma/P_c) = (T_c/T)(A_1\tau + A_2\tau^{1.5} + A_3\tau^{2.5} + A_4\tau^5), \quad (6)$$

where $\tau = (1 - T/T_c)$ where T_c and P_c are the critical temperature and pressure.

Accurate values of the condensed-phase enthalpy, entropy, and internal energy have been determined from accurate vapor-pressure data (Schwalbe *et al.*, 1977; Chen *et al.* 1975, 1978). In addition, accurate values of the cohesive energy at 0 K have been derived from triple-point solid enthalpies and integration of experimental heat capacity of the solid at saturated vapor pressure, C_{sat}, from T_{tp} to 0 K. Cohesive-energy data have been useful in studying many-body forces in solids.

1.4.3 The Universal Gas Constant A new value of the universal gas constant (R = 8.314 471 ± 0.000 014 J mol^{-1} K^{-1}) was determined from measurements of the speed of sound in argon as a function of pressure at the temperature of the triple point of water in a spherical acoustic resonator at the National Institute of Standards and Technology (NIST) (Moldover *et al.*, 1988). The error assigned to the measurement is some five times smaller than that assigned to the previous measurements. This reduces the uncertainty of values derived from R, such as the Boltzmann constant k_B and the Stefan–Boltzmann constant.

1.5 Transport Properties

1.5.1 Kinetic Theory Kinetic theory was developed mainly by Maxwell and Boltzmann in the last century. From the laws of mechanics and the theory of probability, it determines the macroscopic properties of gases in terms of the motion of molecules, their collisions with one another, and the forces of interaction. Central to kinetic theory is the time-dependent distribution function. If no gradients of velocity, temperature, or concentration exist, the distribution function reduces to the time-independent Maxwellian distribution. When nonequilibrium conditions prevail, then the distribution function satisfies the Boltzmann integrodifferential equation. When a dilute rare-gas system is not too far from equilibrium, a perturbation method due to Chapman and Enskog is used to solve the equation and derive expressions for the transport properties in terms of the properties of the atoms themselves and their interactions.

Transport phenomena in gases include viscosity, thermal conductivity, and ordinary and thermal diffusion. In terms of kinetic theory, the first three processes are equivalent to the transport through the gas of momentum, kinetic energy, and mass, respectively, when gradients in the gas are present. The viscosity of a gas results from a transport of momentum along a velocity gradient. Thermal conduction is the transport of kinetic energy along a temperature gradient and (ordinary) diffusion is the transport of mass along a concentration gradient. Thermal diffusion is a secondary effect in which an initially uniform gas mixture subjected to a temperature gradient results in a tiny separation of the components.

1.5.2 Measurement of Viscosity and Thermal Conductivity The viscosities of the rare gases and their mixtures are measured (Wakeham *et al.*, 1991) mainly in one of two techniques: capillary flow and oscillating-body viscometry. With the capillary-flow technique, the estimated error of the measurement ranges from $\pm 1\%$ to $\pm 2\%$ while accuracies up to $\pm 0.1\%$ have been claimed with the oscillating-disc method.

Data on the viscosity were measured using the oscillating-disc method at Rostock and Brown Universities (298 to 973 K) and using the capillary method at Oxford (114 to 1600 K) and at Los Alamos National Laboratory (1110 to 2200 K).

The thermal conductivity of the gases can be measured by either a steady-state or a transient method. In the former, the thermal gradient is measured when a heat flux is passed through gas between concentric cylinders or between parallel plates. In the latter, a wire is mounted along the axis of a vertical cylindrical cell and a constant heat flux is electrically generated in it. The thermal conductivity can be determined from temperature rise in the wire as a function of time. The transient hot-wire method seems to be the current method of choice. For very high temperatures, the shock-tube method is used, in which the thermal conductivity is deduced from the measurement of heat transfer rates from the shock-heated gas to the end wall of the shock tube.

1.5.3 Measurement of Ordinary and Thermal Diffusion The most common method of measurement of the diffusion coefficient is with the use of a two-bulb apparatus (Wakeham *et al.*, 1991). In this technique, gases of unequal composition residing in each bulb are allowed to mix through a connecting capillary. The diffusion coefficient is related to the geometry of the two-bulb system and the time required for the gas to approach uniform composition.

There are two basic means of determining the associated transport coefficient, viz., the thermal-diffusion factor. These are a two-bulb technique in which each bulb is maintained at a different temperature, and a thermal-diffusion column. In order to amplify the separation in the case of the former method, the so-called Trennschaukel (swing separator) was devised, in which many two-bulb systems are connected in series. In the column method, a temperature gradient is maintained between a concentric wire and the walls of a vertical tube. The natural convection of the gas in the column amplifies the separation by sweeping the lighter component adjacent to the hot wire up and the heavier component at the cold wall down the column. The column is the prototype apparatus employed for the separation of isotopes. It was the early method for separating uranium isotopes.

Diffusion data on mixtures of the rare gases have been measured at low temperatures at Leiden. More precise and extensive measurements have been performed at Adelaide University including some measurements on thermal diffusion. Self-diffusion and extensive thermal-diffusion measurements were also carried at the Mound Facility of the U.S. Department of Energy in Ohio.

1.6 Law of Corresponding States (LCS)

When expressed in reduced form, a property will have nearly the same value for all gases in a set. This law is based on the hypothesis that all gases in a set have the same reduced intermolecular potential. The reduced potential is defined as $U^* = U(r^*,\beta_i)/\epsilon$, where $r^* = r/\sigma$. The scaling factors ϵ (the energy parameter) and σ (the length parameter) differ from species to species in a set, and β_i are potential-shape parameters, which are identical for all members of the set.

An extended law of corresponding states for the dilute rare gases (helium excepted) and their binary mixtures was developed at Brown University (Kestin *et al.*, 1972). This LCS provided thermodynamically consistent correlations for second virial coefficients and transport properties. On the basis of limited experimental measurements on some of the species, the formulation provided universal functionals for these quantities that, with a set of scaling factors for each species and its binary mixtures, allow one to predict the properties of all the species in a set over an unusually large temperature range.

Realizing the deviation from congruence of reduced potentials for the various species at small separations, the group at Brown (Najafi *et al.*, 1983) improved their LCS principle by allowing some of the shape parameters β_i to be different for different rare-gas pairs. In addition, they appealed to quantum mechanics to determine the asymptotic behavior of the properties. As a result, they were then able to extend the temperature range of the correlation to high and low temperatures and to include helium and its mixtures. The formulation effectively synthesized the equilibrium and transport properties of the five rare gases and the 26 mixtures that can be formed with them.

A further improvement was achieved by the Rostock group (Bich *et al.*, 1987), who published accurate correlations for viscosity and thermal conductivity for the pure gases. It must be noted, however, that data based on accurate interatomic potentials that are more physically based cannot only correlate the data but can extend the range of temperature inaccessible to experiment.

1.7 Optical and Dielectric Properties

1.7.1 Refractivity and Dielectric Virial Coefficients If a rare-gas atom is placed in a uniform electric field, it acquires a dipole moment and is said to be polarized. If the moment **m** is proportional to the field **E** then $\mathbf{m} = \alpha_0 \mathbf{E}$ where α_0 is the polarizability of the isolated atom. In the presence of other atoms, e.g., in a gas, the polarizability α of an atom can be defined in terms of the index of refraction n and the density ρ by the Lorenz–Lorentz equation

$$(n^2 - 1)/(n^2 + 2)\rho^{-1} = A_R + B_R\rho + C_R\rho^2 + D_R\rho^3 \cdots. \qquad (7)$$

α will differ from α_0 because of the effects of atomic interactions. If ρ is in moles per unit volume, the left-hand side of Eq. (7) is the molar refraction R_m. As shown, it can be expanded as a power series in the density ρ, where A_R, B_R, C_R, D_R, . . ., respectively, are the first, second, third, fourth, . . . refractivity virial coefficients. A_R is the limiting value of R_m for a very dilute gas. That is, in the limit $\rho \rightarrow 0$, $A_R = N\alpha_0/3\epsilon_0$.

Determinations of A_R hence yield values for atomic polarizabilities α_0 for the isolated atom, while B_R, C_R, . . . are related to the excess contribution to R_m as a result of interacting pairs, triplets, . . . of atoms, respectively. Accurate values of A_R, B_R, C_R, . . . have been determined by a differential interferometric technique, and some of these are presented in Table 3 (Achtermann *et al.*, 1993).

The Clausius–Mosotti function for the dielectric constant ϵ may also be expanded as follows:

$$(\epsilon - 1)/(\epsilon + 2)\rho^{-1} = A_\epsilon + B_\epsilon r + C_\epsilon r^2 + D_\epsilon r^3 \cdots. \qquad (8)$$

The left-hand side is the molar polarization P_m and A_ϵ, B_ϵ, C_ϵ, D_ϵ, . . . are respectively the first, second, third, fourth, . . . dielectric virial coefficients. Again, these coefficients are related to the contributions from individual atoms, pairs, triplets, quadruplets, . . ., respectively. The limiting value of P_m at low density is $A_\epsilon = N\alpha_0/3\epsilon_0$. For the rare gases, accurate values of the dielectric virial coefficients have been determined by a differen-

Table 3. Measured refractivity and dielectric coefficients for the rare gases (Sec. 1.7.1).

Rare gas	T (K)	A_R (cm^3 mol^{-1})	B_R (cm^6 mol^{-2})	C_R (cm^9 mol^{-3})
He	303	0.5213	−0.068	...
Ne	303	1.0012	−0.11	...
Ar	303	4.1955	1.75	−85.2
Kr	303	6.414	5.96	−275.0
Xe	303	10.345	...	...
	348	10.344	28.5	−1802.0
Rare gas	**T (K)**	**A_ϵ (cm^3 mol^{-1})**	**B_ϵ (cm^6 mol^{-2})**	**C_ϵ (cm^9 mol^{-3})**
He	323	0.5196	−0.07	...
	3–18	0.006	−5.0	...
Ne	323	0.9969	−0.12	...
Ar	303	...	1.22	...
	323	4.1397	0.72	...
Kr	323	6.273	4.3	...
Xe	323	10.122	32	...

tial-capacitance technique (Huot and Bose, 1991; Gugan, 1984).

1.7.2 Spectroscopy of the Dimers Through the use of clever experimental techniques, absorption and emission bands in the vacuum ultraviolet have been observed that have provided information on the vibrational spacings (Ewing, 1975) for Ne_2, Ar_2, Kr_2, and Xe_2 dimers in the ground state and subsequently, with a spectrograph of higher resolution, on the rotational fine structure of Ar_2 and Kr_2 (Colbourn and Douglas, 1976; LaRocque *et al.*, 1986; Herman *et al.*, 1988). Such information has been invaluable for the accurate determination of ground-state constants and interatomic potentials. Results based on the potentials of Table 2 are given in Table 4. Energy levels for the argon dimer are shown in Fig. 3. Observation of vibrationally and isotopically resolved fluorescence excitation spectra has led additionally to information on some of the excited states involved in excimer lasing transitions. Infrared, microwave, radiofrequency, and Raman spectroscopy have been used to reveal the structure and properties of some of the dimers (Xu *et al.*, 1994).

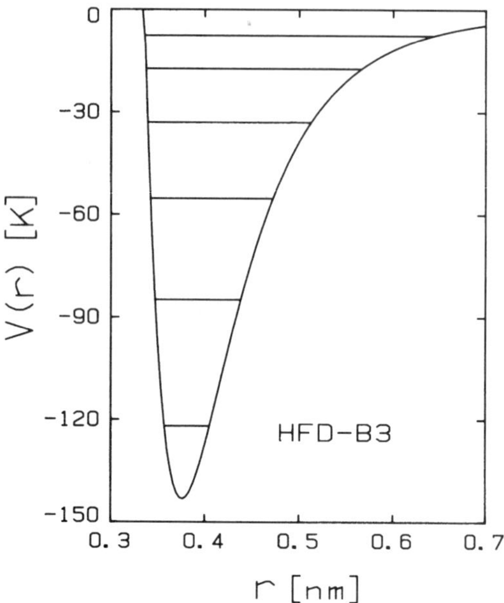

FIG. 3. Energy levels of the argon pair potential. HFD–B3: the Hartree–Fock–dispersion potential of Table 1. The energy levels are those of Table 4.

2. LIQUID STATE AND DENSE GASES

2.1 Interatomic Interactions in the Condensed State

Above the critical temperature, gases and liquids form one continuous state of matter. Coordinates of the critical points for the rare gases are given in Table 1. Experiments on rare-gas fluids can be carried out over very wide ranges of the state variables: pressure P, temperature T, and density (or molar volume V). Because the interatomic pair potentials are well understood, the rare gases form a prototype family of fluids upon which critical experiments can be compared with modern computations of the properties concerned. Such studies form a useful foundation upon which the complications of molecular fluids may be considered, including orientations of the constituents as well as their positions.

It cannot be said that many-body interactions in condensed rare gases are well understood. Quite often, for pressures that are not

Table 4. Vibrational levels of the rare-gas dimers calculated on the basis of potentials of Table 2 (Sec. 1.7.2). (Units in cm^{-1}.)

Level	He	Ne	Ar	Kr	Xe
$-\frac{1}{2}$	0	0	0	0	0
0 (ZPE)	7.613_6	12.3	14.85	11.35	10.23
1	…	26.0	40.57	32.52	29.84
2	…	29.1	61.09	52.20	48.29
3	…	…	76.62	69.40	65.57
4	…	…	87.50	84.51	81.66
5	…	…	94.31	97.52	96.54
6	…	…	…	108.49	110.25
Dissociation energy D_0	1.1×10^{-3}	17.1	84.71	128.56	186.32
Well depth D	7.6147_5	29.365	99.563	139.910	196.555

too large, a model of an accurate pair potential together with the Axilrod–Teller triple-dipole dispersion energy (DDD) is sufficient to explain properties of the fluids and binding energies of the solid phases. For crystals, the reason for this is that it has been assumed that contributions of the many-body dipole-interaction terms of fourth and higher order $(MBD)_{4\to\infty}$ cancel those of higher multipole third-order terms $(DDQ)_3$, $(QQD)_3$, $(DDO)_3$, and $(QQQ)_3$. D, Q, and O refer to dipole, quadrupole, and octupole, respectively, and subscripts 3 and $4 \to \infty$ refer to third and fourth to higher orders of quantum perturbation theory. In spite of the apparent success of the approach, however, it is worthwhile to point out that there are other nonadditive energies (e.g., nonadditive exchange energies) that could be of the same magnitude as (DDD) and that corrections must be incorporated when the simple approaches fail to yield results that agree with experiment, especially at high density (Barker, 1989; McLean *et al.*, 1988; Meath and Koulis, 1991).

Shock-wave experiments have been carried to high densities in dense-fluid rare gases, with compressions a factor of 3 or more (Yakub, 1994).

2.2 The Structure of Liquids

The structure of matter changes continually with time because its constituents are in motion. In the crystalline state, their displacements are small, and (except in solid helium) they are localized near their lattice sites. In the gas phase, the constituents move over such large distances that the term "structure" is rarely used. Since densities in liquids are similar to those of solids, the packings of the two states are somewhat similar. In fact, liquids possess considerable order or "structure." The static structure of a monatomic liquid may be expressed by a pair-distribution function $g(r)$, which is an equilibrium distribution of interatomic separations averaged over time and direction. Precise measurements of $g(r)$ in rare-gas liquids have been made by x-ray or neutron scattering where the intensity of scattering is obtained as a function of angle (Fig. 4). The measured distribution is then compared with one calculated using assumed potentials. Such calculations have advanced the development of theoretical methods and the understanding of the liquid state.

2.3 Thermodynamic Properties

The function $g(r)$ also plays a central role in determining the thermodynamic properties of the system. For example, the internal energy and the pressure can be expressed in terms of the pair-distribution function $g(r)$ and the potential $U(r)$ together with an ideal-gas term, if the interactions are assumed to be additive. For greater accuracy, a three-body distribution function and three-body

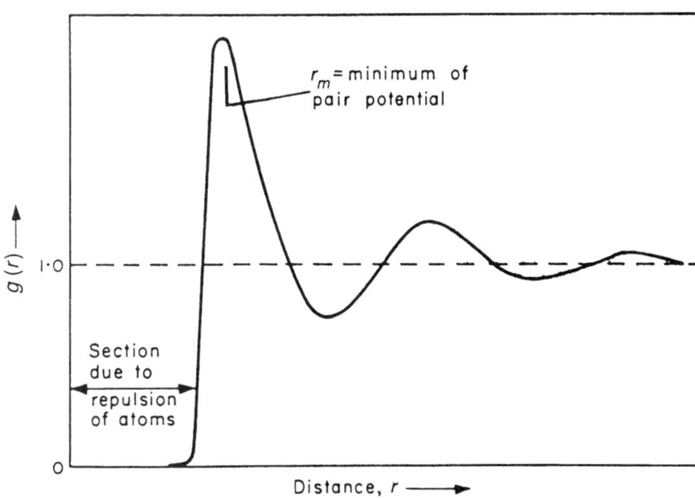

FIG. 4. A sketch of the typical form of the pair-distribution function $g(r)$ for a dense rare-gas liquid. The principal peak of $g(r)$ occurs at a position close to the minimum position r_m of the pair potential. For very low densities these two positions are coincident, and as the density is increased the maximum in $g(r)$ moves to slightly lower values of r than r_m (After Egelstaff, 1992).

potential must be introduced to reflect nonadditive contributions.

Computer-simulation methods can provide, in principle, exact values for most properties on the basis of a given potential model. There are two methods: the Monte Carlo (MC) and the molecular dynamics (MD). The MC method uses ensemble averages as in statistical mechanics and is used to calculate equilibrium properties. The MD method solves the Newtonian equations of motion of the particles and uses time averages. It is used to calculate both equilibrium and transport properties. MC can be applied to both quantum-mechanical and classical systems while MD is applicable only to classical systems. In each method, a small number of particles (less than a few hundred) are used with periodic boundary conditions to simulate the behavior of a large system. Usually an accuracy of 1% or better is attainable for many properties depending on computational speed. Such methods are useful in determining or testing interatomic potentials by comparing the results on the basis of an assumed potential with experiment. The simulated results can also provide a standard by which the results based on approximate methods may be compared.

There are a number of approximate methods to calculate thermodynamic properties that have been or are currently being employed. One of these is the so-called integral equation method. In this procedure, one introduces approximations into $g(r)$ in order to simplify the problem mathematically. One then determines whether such approximations are valid by comparing the results with experiment or computer simulation. Another approximate method meeting with success is the perturbation method (Barker and Henderson, 1976; Weeks *et al.*, 1971). In this approach, the pair potential is written as $U(r) = u_0(r) + \lambda u_i(r)$, where λ is a parameter that, when set equal to zero, gives a reference potential $u_0(r)$ (on the basis of which the properties of the reference system are well known from, say, computer-simulation studies) and, when set equal to unity, gives $U(r)$. The potential is introduced into the partition function and the thermodynamic properties are evaluated. Since the properties of the reference system are known, the properties of the real system can be determined.

Song and Mason (1989) developed a simple analytical equation of state based on perturbation theory for a hard-sphere fluid. By three separate integrations, they determine from a knowledge of the interatomic potential the second virial coefficient, an effective hard-sphere diameter, and a scaling factor as functions of temperature. Since the latter two are nearly universal functions when expressed in suitably chosen reduced units, it results that a knowledge of the second virial coefficient enables one to determine the entire equation of state for the rare gases (except helium; Song and Mason, 1993) up to 3 times the critical density. It allows one to predict such properties as the compressibility of the liquid, the vapor-pressure curve, and the critical constants.

2.3.1 Quantum Corresponding States The classical restriction of the law of corresponding states for translational states can be removed by introduction of a reduced Planck constant, a measure of quantum effects in the condensed state. For a Lennard-Jones interaction, the de Boer parameter $\Lambda^* = h/(m\epsilon\sigma^2)^{1/2}$ may be regarded as the reduced de Broglie wavelength of a particle of mass m. A property W in reduced form is then expressed by a relation including Λ^* as $W^* = f(P^*,T^*,\Lambda^*)$, where P^* is the reduced pressure. If Λ^* is less than about 0.2, then classical theories of the condensed state are often adequate.

2.3.2 Sound Velocity Measurements of sound velocity yield the most precise values for adiabatic compressibilities of rare-gas liquids. With knowledge that the heat-capacity ratio is a universal function of the reduced *PVT* coordinates, it is possible to summarize the sound velocity at saturated vapor v_σ for the heavier rare-gas liquids from their triple points to close to their critical points by a universal relation for the reduced velocity versus reduced temperature, $v_\sigma^* = f(T^*,\Lambda^*) \approx f(T^*)$. The reduction parameters for both velocity and T depend only on the energy parameter ϵ/k_B of some "effective" pair potential. By forcing corresponding states, accurate ratios of the energy parameters of krypton and xenon in terms of that of argon can be obtained (Aziz *et al.*, 1967). In fact, it is interesting to note that velocity appears to be almost perfectly linear with density.

2.4 Transport Properties

2.4.1 Transport Properties, Theory At low densities, transport properties can be accurately calculated at any temperature using exact kinetic-theory expressions. At high densities, however, there is at present no formal theory that provides an exact evaluation of transport properties in terms of a realistic characterization of the atomic interaction.

The MC method has provided significant progress toward a successful molecular theory of transport properties in dense fluids. It requires extensive numerical computation in order to determine the dependence of the transport coefficients on the experimental variables.

Enskog developed a kinetic theory of dense gases consisting of rigid spherical molecules of finite diameter. By considering only two-body collisions and the finite size of the molecule, Enskog modified the kinetic theory of the dilute gas. In this model there is an instantaneous transport of energy and momentum on collision from the center of one molecule to the center of the other; and because of the finite diameter, the frequency of collisions is altered. The density dependence of self-diffusion is given directly by the Enskog theory while viscosity and thermal conductivity require a knowledge of experimental data in order to bring theory and experiment into agreement. There is only one adjustable parameter, the molecular core size, which is determined from the data fit. In this modification as applied to the rare gases (smooth hard-sphere theory), the exact smooth hard-sphere transport coefficients $P_{\rm shs}$ are given by the product of the values from the Enskog theory $P_{\rm E}$ and computed corrections to it: $P_{\rm shs} = P_{\rm E}(P/P_{\rm E})_{\rm MD}$.

MD methods have been used to characterize the detailed time response in a model Lennard-Jones fluid (de Schepper *et al.*, 1988).

2.4.2 Neutral and Charge Transport, Experiment At the U.S. National Institute of Standards and Technology, extensive measurements have been obtained of the equation of state and the viscosity of liquid argon as a function of density, the latter using a torsional oscillating crystal. Thermal conductivity data for argon are also available. Since acceptable equations of state for krypton and xenon are unavailable, LCS can be used (Hanley *et al.*, 1974) to determine values of viscosity and thermal conductivity as follows. A reduced viscosity is defined as

$$\eta^* = \eta[r_{\rm m}^2(N/M)^{1/2}(k_{\rm B})^{-1/2}(\epsilon/k_{\rm B})^{-1/2}] \qquad (9)$$

and a reduced thermal conductivity is defined as

$$\lambda^* = \lambda[r_{\rm m}^2(M/N)^{1/2}(k_{\rm B})^{-3/2}(\epsilon/k_{\rm B})^{-1/2}] \qquad (10)$$

where $\epsilon/k_{\rm B}$ and $r_{\rm m}$ are potential parameters for an "effective" pair potential. A reduced pressure is given by

$$P^* = Pr_{\rm m}^3[(\epsilon/k_{\rm B})k_{\rm B}]^{-1} \qquad (11)$$

and a reduced temperature is given by

$$T^* = T/(\epsilon/k_{\rm B}). \qquad (12)$$

To obtain η and λ of krypton or xenon at a given temperature and pressure, a corresponding temperature and pressure are first found for argon using Eqs. (11) and (12). The equation of state for argon is used to find the density, and then the transport coefficients can be determined from Eqs. (9) and (10) and the appropriate potential parameters.

Charge-transport studies in rare-gas liquids show that injected electrons are in a quasifree state of high mobility. Carrier transport has typically been determined by time-of-flight methods using electrical shutters (Shikin, 1977). More unusual have been studies of the annihilation of positrons injected into rare-gas liquids (Stewart *et al.*, 1990). A localized bubble state is formed in liquid krypton, argon, neon, and helium (and in solid helium). The bubble radius is determined by a balance between shrinking forces of bubble surface tension and external pressure and the zero-point pressure exerted by the confined positronium. The lifetime for decay of each positronium mode is a measure of the electron charge density at the positron-annihilation site.

2.5 Microscopic Dynamics from Neutron-Scattering Evidence

Slow-neutron scattering has provided a powerful tool for extending studies to include the dynamics of liquid structure. That is, it can be used to measure various correlation functions appearing in dynamical models for liquids—for example, the self-correlation (the average motion of an atom among others) and the density–density correlation (the relative motion of atoms). In practice, argon provides a rare example of a liquid system in which both the coherent scattering (from a ^{36}Ar sample) and also the incoherent scattering (from a suitable mixed ^{36}Ar–^{40}Ar sample) can be determined (Skold *et al.*, 1972). For argon, analysis of the experimental results in view of molecular-dynamics simulations shows good agreement with a picture in which the atoms exhibit free-particle behavior at very short times with a smooth transition to simple diffusion already at times about 2 ps.

At very small scattering vectors Q, i.e., long wavelengths, observed fluid modes correspond to fluctuations in the density, at frequencies $\omega = cQ$, where c is the velocity of sound, and to fluctuations in the entropy, centered at zero frequency. These three peaks broaden and merge at larger Q, about 1 nm^{-1} for the rare gases. At very large Q the modes correspond to single-particle motion. At intermediate values of Q, interpretation of the response of rare-gas fluids is somewhat controversial, although results on argon and neon show interesting differences. Helium seems to be a special case (Montfrooij *et al.*, 1992).

2.6 Special Properties of Liquid Heliums and Their Isotopic Mixtures

Superfluidity in liquid-helium systems has been treated in many thousands of research papers (see SUPERFLUIDITY: LIQUID HELIUM SYSTEMS). The behavior of liquid ^4He (a boson) is qualitatively different from that of liquid ^3He (a fermion), although both exhibit superfluid phases at low temperatures. The only atom that forms an *anisotropic* liquid is ^3He.

Landau gave a phenomenological theory of normal Fermi liquids in 1956, only a few years after sufficient quantities of ^3He became available for its thermodynamic properties to be measured. The Landau theory uses the concept of elementary excitations to clarify the properties of strongly interacting quantum systems at low temperatures. It provides a useful way to correlate measured properties directly, such as the transport coefficients, with a limited number of parameters, without the need for a microscopic theory. Subsequent microscopic derivations of the Landau theories have been produced. In ^3He these excitations are quasiparticles, a concept that has proven useful in other forms of condensed matter.

Thermodynamic and physical properties of isotopic helium mixtures have an extraordinary range. For example, varying the ^4He content in ^3He produces an adjustment of the Fermi temperature with little change in number density of the liquid. Measurements of thermal conduction, viscosity, diffusion, etc. have been used to check theories of Fermi liquids over a very wide range of parameters. Work on nuclear matter and on nuclei is parallel, but only a much more limited range of parameters can be studied.

We note that the properties of these extraordinary liquids can be linked to those of the heavier rare gases when the heliums are brought to high density, at which internuclear distances are nearer the characteristic length of the pair potential, and at which the statistics obeyed by the atoms is insignificant (Boninsegni *et al.*, 1994).

3. CRYSTALLINE STATE

3.1 Melting, Corresponding States

The rare-gas solids form an interesting family. Among all interatomic interactions, van der Waals cohesion is the weakest in forming condensed systems, and the pure rare gases have low condensation temperatures (Table 1) and energies (Table 5). One kind of corresponding-states analysis simply uses the triple points as scaling parameters. The search for a critical point at the end of the solid–fluid line in a P-T diagram has been carried furthest in helium, for which one can access experimentally the largest range of V of any substance. No such point has been found.

In the rare-gas family, effects of different kinds occur in a smooth progression as one goes from one row of the periodic table to another. For example, Table 5 shows that with increasing mass comes a decrease in the importance of macroscopic quantum effects (compare the ratios E_z/E_s). The solid

Table 5. Binding energies $E(0)$ of rare-gas solids at 0 K based on the potentials of Table 2. [$E(0) = E_S + E_Z$ where E_S is the static energy of face-centered-cubic lattice and E_Z is the zero-point energy including anharmonic effects. Units of J mol^{-1}. "Complete" energies include $(DDD)_3$, $(DDQ)_3$, $(QQD)_3$, $(DDO)_3$, $(QQQ)_3$, and $(MBD)_{4\to\infty}$. (See Sec. 3 for definitions of these multibody interactions.)]

Energy	Interaction	Ne	Ar	Kr	Xe
E_S	Two-body	−2609.3	−9097.1	−12 776.8	−17 855.8
E_S	Triple dipole	63.0	571.6	1 000.1	1 686.7
E_S	Two-body + triple dipole	−2546.4	−8525.5	−11 776.7	−16 169.2
E_S	Complete many-body	−2537.0	−8444.8	−11 651.9	−15 883.6
E_Z	Two-body	612.7	781.7	597.7	534.0
E_Z	Triple dipole	3.1	7.2	6.9	7.0
E_Z	Two-body + triple dipole	615.8	789.0	604.6	541.0
E_Z	Complete many-body	617.0	793.5	609.4	547.2
$E(0)$	Two-body + triple dipole	−1930.6	−7736.5	−11 172.1	−15 628.2
$E(0)$	Complete many-body	−1920.0	−7651.3	−11 042.5	−15 336.4
$E(0)$	Experiment	−1933 ± 8[a]	−7722 ± 11[b]	−11 148 ± 13[b]	−15 784 ± 35[b]

[a]McConville (1974).
[b]Schwalbe *et al.* (1977).

heliums are a special case, because they require external pressure for solidification; for them, the ratio E_z/E_s is about unity. On the other hand, with increasing mass comes an increase in the relative importance of many-body interactions.

3.2 Thermoelastic Properties

For the heavier rare-gas solids under saturated vapor pressure, the heat capacity and thermal expansivity have been measured with high precision over the entire crystalline ranges above about 1 K. Heat capacity has been well determined by adiabatic calorimetry (Korpiun and Lüscher, 1977). Thermal expansion has been precisely measured by capacitance dilatometry at low temperatures and by x-ray diffraction at temperatures up to melting. X-ray measurements of lattice parameters of single crystals have approached the best accuracy available for usual solids (Losee and Simmons, 1968). On account of very large thermal expansion, the heat capacity at constant pressure for argon, krypton, and xenon rises to about 35 J mol^{-1} deg^{-1}, which is 40% above the Dulong–Petit value $3R$. This fact, together with the relatively large vibrational energies of these solids, has attracted the attention of theorists interested in the dynamics of crystals that are highly anharmonic. For example, because the isobaric coefficient of thermal expansion corresponds to a third derivative of the free energy, this is a most demanding test of a statistical-mechanical model.

Precise values have been obtained of the thermal properties of the solid neon isotopes ^{20}Ne and ^{22}Ne over the entire temperature range. Direct measurements of kinetic energy of crystalline neon, using eV neutron scattering, show that the vibrational energy is about 35% of the cohesive energy. The vibrational energy differs for the two isotopic masses. The isotopes therefore have different internal pressures and exhibit different equilibrium lattice parameters, vapor pressures, specific heats, etc. These can be correlated by a single-particle model in which the dominant agent is the vibrational energy (Batchelder *et al.*, 1968).

Elastic constants of the rare-gas solids are not as well determined. For these solids the equilibrium vapor pressures and thermal expansions are large and the melting temperatures low, which defeats ultrasonic pulse-echo techniques as usually applied to oriented and prepared single crystals. The solid heliums are an exception; for these, ultrasonic transducers can be kept in contact with the crystals confined at constant volume. For other crystals near melting, where single crystals could be kept in good optical contact with transparent cell walls, values have been obtained by Brillouin scattering, which is inelastic scattering of a photon by an acoustic phonon.

Anisotropy in the deduced elastic constants has been attributed to two effects: for the lighter rare gases, large zero-point energies, and for the heavier ones, many-body interactions. These attributions are qualitatively correct, but detailed interpretations remain somewhat inconclusive as long as the temperature dependence of the elastic constants is not well known over the entire temperature range.

Rare-gas solids are extraordinarily compressible compared with other solids. Unlike other materials, for them it is practical to study many properties under essentially isochoric conditions or over large changes in density. Brillouin scattering has been used to measure the velocity of sound (and hence the density) in argon compressed up to 30 GPa (Polian *et al.*, 1986). Isochoric studies of heat capacity have been carried out for the heliums, for neon, and for argon. For heliums, these have been correlated with precise measurements of pressure, using the relation $(\partial S/\partial V)_T = (\partial P/\partial T)_V$. For some phonon branches, dispersion has been studied in isochoric neon and argon samples using slow-neutron scattering.

Measurement of pressure–volume–temperature (*PVT*) relations by piston-displacement techniques have given the density dependence of the bulk modulus B of the heavier rare gases (Swenson, 1977). It has been found that a Birch type of relation often represents the results well at low temperature: $B = B_0 + bP$, where P is the pressure and b is a parameter having a value 5 to 6. It is remarkable that all members of this family that have a solid–liquid–vapor triple point fall on a common curve ($\pm 15\%$). At higher temperatures for the heavier rare-gas solids, the temperature dependence of the solid equation of state is approximately given by $(dP/dT)_V$ = constant, with the constant having a value about 2.7 to 2.9 MPa K^{-1}.

3.3 Microscopic Properties

3.3.1. Phonons, Experiment and Theory Measurements of phonon energies and dispersion, by coherent inelastic scattering of neutrons, provide information about the dynamics of interatomic interactions in rare-gas solids. All rare-gas solids except radon and ^{3}He have been studied in this way, radon because it is radioactive and ^{3}He because of its enormous neutron absorption.

It is conventional to parametrize results by fitting a Born–von Kármán model. This is an atomistic formulation of lattice dynamics that sets up the interaction forces in general fashion. Calculations of properties using the parameters of the model can then be compared with corresponding properties calculated from lattice-dynamical theories. Properties of interest within the harmonic approximation include the one-phonon density of states and its frequency moments, and the lattice specific heat. By examining the energy versus wave-vector relations at very small Q, the "zero-sound" elastic constants are obtained, i.e., those at frequencies large compared with a typical phonon–phonon collision frequency. Once again, the large compressibility of rare-gas solids has permitted measurement of dispersion in some phonon branches in isochoric helium, neon, and argon crystals, which exhibit considerable anharmonicity. This has the great advantage of displaying the intrinsic temperature dependence of anharmonic effects, which are often masked by changes in phonon frequencies due to changes in crystal density from thermal expansion.

Results for the fcc phases have provided a continual object of theoretical attention, as both experiment and theory were refined. For the heavier solids such as argon, very satisfactory agreement has been reached using computer simulations. Theories that use modern interatomic potentials are now able to account for many details of the phonon-excitation line shape as well as for the dispersion structure and hence the bulk thermodynamic properties. In favorable cases, observed agreement between theory and experiment tests the theory, the potential, and the experimental technique. On the other hand, results for solid ^{4}He are interesting, showing multiple-phonon interactions, but they are incomplete for higher phonon energies.

Thermal resistivity in perfect insulators arises from umklapp processes, those phonon-scattering processes for which the rule for conservation of wave vector is $\mathbf{Q}_1 + \mathbf{Q}_2 = \mathbf{Q}_3 + \boldsymbol{\tau}$, where $\boldsymbol{\tau}$ is a vector in the reciprocal lattice. Again, the large compressibility of rare-gas solids makes it possible to investigate thermal conduction in samples at constant volume, which removes the effects of thermal expansion from the analysis. Similarly, studies of the thermal conduction at different densities are possible; such studies are extremely rare in usual solids. At the highest temperatures, accessible for argon, krypton, and xenon, thermal conductivity at constant volume goes as T^{-1}, as given by a phonon-gas model, and it follows a law of corresponding states (Batchelder, 1977). At lowest temperatures, defects and boundary scattering limit thermal conduction. In highly pure and perfect helium crystals, an enhancement of thermal conduction has been observed, which can be attributed to a regime in which Poiseuille flow of the phonons is significant. Helium-isotope crystals have provided important information on mass-defect scattering processes, because the mass ratio is very large.

3.3.2 Momentum Distribution, Kinetic Energy The momentum distribution of individual atoms can be measured by analyzing the neutron recoil-scattering process in terms of neutron energy transfer at a given momentum transfer (Simmons, 1993). Neutrons of energy about 1 eV are required. From the momentum distribution, the kinetic energy can be deduced and compared with the results of simulations in order to test both the simulations and the experiments in a very direct fashion. For the heavier rare-gas solids the results have been according to expectation; for example, the equipartition law holds for translational states regardless of density of the system.

On the other hand, measurements on helium have shown that the actual kinetic energy is significantly larger than what would be expected from the Born–von Kármán (harmonic) analysis of neutron-measured phonon dispersion. Actually, for solid helium, the high-energy excitations corresponding to single phonons are generally so broad and weak that a Born–von Kármán analysis including these states is unjustified. For helium, path-integral Monte Carlo methods have proven most satisfactory (Ceperley, 1995). An illustration is provided by considering kinetic-energy results for samples of ^{4}He at fixed number density but in different phases, bcc or hcp or liquid (depending upon P and T); such a study is only possible with helium. The result is that the kinetic energies of these three phases are essentially the same, that is, helium is a quantum system dominated by the density, not by details of the local order (Blasdell *et al.*, 1993).

3.4 Dielectric and Electronic Properties

Experimental results on many dielectric and electronic properties of the rare-gas solids can be fruitfully interpreted using atomic characteristics. These solids have very large forbidden-energy gaps for electrons (Rössler, 1977). The gaps at low pressures range from about 9.3 eV in xenon to about 30 eV in helium. For these prototype insulators, optical properties at energies below the valence-band transitions can be succinctly expressed by the Lorentz–Lorenz function that relates the dielectric constant or index of refraction to the atomic polarizability [Eq. (7)]. Quantitative (phenomenological) descriptions of the dielectric properties can be constructed on the assumption that principal contributions arise from excitons (bound electron–hole pairs, the hole being a positive charge carrier) originating from the atomic doublet $^1S_0$-$^{3,1}P_1$.

Measurements of UV and XUV (extreme ultraviolet) reflectivity and absorption data on the heavier rare-gas solids have been correlated and extended by Kramers–Kronig analysis to give the complex optical constants (Sonntag, 1977). These have been further compared with electron energy-loss spectra. Again, information on atomic properties, and theoretical calculations of electron structure, are useful for interpretation of spectral features, although explanation of the observed intensities continues to provide a challenge to theorists. Data near the x-ray absorption edges are of special interest and have been collected for all rare gases xenon through helium (Schell *et al.*, 1995). Exciton effects are prominent, and the type of excitons observed depends upon the particular rare gas (Schwentner *et al.*, 1985; Varding *et al.*, 1993).

Drastic modification of the electron structure is possible at high pressures. For example, usual xenon (fcc structure at 34.7 cm^3 mol^{-1}) has a gap between the 5p valence band and the (empty) 6s conduction band. When compressed in diamond-anvil pressure cells solid xenon is found to transform at 17 GPa to a complex hcp structure, then at about 75 GPa to hcp, then finally to a metallic state, presumably through approach of the $5p$ and $5d$ electron bands. The observed pressure of transformation is 130–150 GPa, at a density of about 11 cm^3 mole^{-1}. In xenon, the metallization has no accompanying structural transition; x-ray diffraction shows a hcp structure from 75 to 175 GPa (Jephcoat *et al.*, 1987). For lighter rare-gas solids, expected pressures of transformation are higher, for example above 600 GPa in solid argon. The results of attempts to reach such pressures in shock-wave experiments are difficult to interpret because the accompanying large temperature rise also causes electronic excitations, apparently beginning at about 40 GPa in argon. Some insight into these processes has been gained through observations at high pressures on ionic compounds having an ion with the rare-gas–like electron configuration $(ns)^2(np)^6$ and on structural transformations in metals that involve promotion of electrons to the d band (Ross and Radousky, 1989).

Finite–sample-size effects were first seen in observations of the cutoff of excitonic series in thin films. More recently, as cluster physics and chemistry develop, measurements have been made of the soft x-ray spectra of rare-gas clusters of sizes from about 10^2 to about 10^4 atoms, produced in jets. Finally, electron energy-loss spectra have been collected both from thin films and from clusters. In films, evidence of plasma excitations has been seen.

Charge transport has been investigated using excess carriers injected by pulsed-electron excitation of the heavy rare-gas solids (Spear and le Comber, 1977) and by ionizing decay products of radioisotopes embedded in solid heliums. Charged-particle excitation has proven more convenient than photon excitation, because condensed rare gases exhibit photoemission more readily than photoconduction. For the solids formed from neon, argon, krypton, and xenon, low-field electron mobilities typically display a $T^{-3/2}$ temperature dependence, which is characteristic of quasifree-electron transport limited by phonon scattering. On the other hand, the drift mobility of excess holes (positive charge carriers) in these solids is typically several powers of ten smaller in magnitude than for electrons. In this case, it is more appropriate to describe the process as the formation and transport of localized entities, such as a polaron, a charge carrier trapped in its own polarization field. In this case the temperature dependence of the process is dominated by a term $\exp(-E_b/2k_BT)$, where E_b is a characteristic binding energy for the polaron. At high electric fields, electron velocity depends on essentially the square root of the applied field, an indication that the electron distribution becomes "hot." It saturates at a velocity near that of sound.

Altogether, rare-gas solids are providing rich sources of information about how electronic properties of condensed matter in various degrees of aggregation can be quantitatively related to the properties of the constituent atoms.

3.5 Crystal Defects

3.5.1 Point Defects The formation of equilibrium atomic vacancies has been measured in all rare-gas solids. The concentrations (n/N) are determined directly and absolutely by precise comparisons of changes in x-ray lattice parameter $(\Delta a/a)$ and of macroscopic expansion $(\Delta L/L)$ at a succession of temperatures. The applicable formula is

$$(n/N) = 3(\Delta L/L) - 3(\Delta a/a). \qquad (13)$$

From the concentration the free energy of formation of the vacancies, f, follows from the relation

$$(n/N) = \exp(-f/k_BT). \qquad (14)$$

The free energy has two parts: an energy e, and a nonconfigurational entropy s. From the magnitude of the entropy some idea of the local structure of the vacancy can be obtained. Table 6 displays some directly measured values for thermal-vacancy energies.

Other physical properties are affected by the presence of equilibrium defects—for example, the heat capacity and the compressibility. Inferences of vacancy content using these other properties depend, however, upon the assumption that one knows the behavior of a hypothetical perfect crystal, so that the vacancy contribution can be obtained by difference. Such assumptions have frequently proven unreliable, and so a direct method is preferred.

Rare-gas solids are composed of neutral atoms, so that comparison of their cohesive

Table 6. Some thermal-defect properties of rare-gas solids (Sec. 3.5.1). (The formation energies e correspond to the values shown for the entropies, s/k_B. The last column is a test for the influence of many-body interactions, present if the ratio is less than unity. For Ar an average value for e is used.)

Rare gas	f/k_BT_M	s/k_B	e (meV)	$e/(E_P - E_Z)$
Neon[a]	9.7	1.6	21	1.0
Argon[b]	8.5	0–4	62–93	0.9
Krypton[c]	7.8	2.8	86	0.8
Xenon[d]	7.0	~2	~100	0.6

[a]Schoknecht and Simmons, R.O. (1972).
[b]Schwalbe (1976).
[c]Losee and Simmons (1968).
[d]Granfors *et al.* (1981).

energies per atom with the energy of vacancy formation gives direct information about the presence of many-body forces and of local lattice relaxation around the vacancy site. Such a comparison is given in the last column of Table 6 for entropies near $2k_B$. The table shows that many-body effects become more important as one goes from solids composed of the less polarizable (neon) to the more polarizable atoms.

Self-diffusion in all rare-gas solids except ^{4}He has been measured by NMR methods or by exchange with radioisotopes in the vapor, because the usual sectioning of samples containing radiotracers is not practicable, on account of the low melting points and high vapor pressures of these solids. A consistent picture has emerged (Chadwick and Glyde, 1977). The dominant mechanism of self-diffusion in all solids except the heliums is the motion of single vacancies. In solid heliums, diffusion may occur both through direct interchange of atoms and through vacancy migration.

In principle, point defects analogous to some of the "color centers" in ionic crystals can be created in rare-gas solids by ionizing radiation. A likely example is a dimer–ion analogous to the V_K center found in alkali halides.

Extrinsic point defects such as foreign atoms, molecules, or radicals can be studied over a vast range of photon energies, microwave through x ray, because of the closed-shell character and large band gap of the rare-gas host. Low temperatures slow reaction kinetics so that ephemeral processes can be studied. Neutron scattering is also a useful tool to examine molecular excitations when the nearly (neutron) transparent ^{40}Ar is used as host (Langel, 1992).

3.5.2 Line and Surface Defects, Thin Films, Clusters Dislocations and surface defects in rare-gas solids have been studied extensively by transmission electron microscopy (Venables and Smith, 1977). Contrast methods for stacking faults (a planar defect in the crystal resulting from misstacking of successive planes of atoms) are applicable as in other thin films. Arrays of dislocations (a one-dimensional defect) are seen, and twinning as in other fcc crystals. Because of high equilibrium vapor pressures, grain boundaries that intersect a bulk surface can be thermally etched. In thin films of neon, both fcc and hcp structures are observed. The classic rare-gas structural question was "Why are most bulk structures fcc and not hcp" (as would be expected in a static lattice containing only pair interactions)? The answer is that experiments show that the energy differences are very small, and that relatively small amounts of zero-point motion, of many-body interactions, etc. are required to produce the observed fcc structures.

Clusters of rare-gas atoms can be formed in supersonic jets. Cluster size can be varied by varying the conditions of the jet. For argon, clusters of fewer than about 800 atoms exhibit a quasicrystalline (isocahedral) structure as analyzed by electron diffraction, whereas larger clusters exhibit the fcc structure of bulk argon (Kovalenko *et al.*, 1994; Schöllkopf and Toennies, 1994). Simulations suggest that in small neutral clusters there may be a dynamical coexistence between crystal and fluid structure that permits the investigation of melting mechanisms (Matsuoka *et al.*, 1992). For argon clusters, excitation by x rays or electron bombardment can produce effects such as x-ray fluorescence and fragmentation of the clusters. The phenomena can be consistently linked to atomic and bulk properties (Wörmer *et al.*, 1991). For helium clusters, however, considerable inconsistencies in such an interpretation remain (Joppien *et al.*, 1993). As usual, heliums are a special case in this rare-gas family.

3.6 Solid Heliums, Unique Properties

The two helium isotopes, ^{4}He and ^{3}He, are the only substances that do not solidify at atmospheric pressure even when their temperatures approach 0 K. External pressure is required, larger for ^{3}He (about 3.1 MPa) than for ^{4}He (about 2.5 MPa). The quantum zero-point motion of the lighter isotope is larger, producing a larger internal pressure to be overcome in order to bring the substance to a density at which the interatomic spacing is close to the distance at which the interatomic potential energy is a minimum. Table 1 indicates the remarkable differences in the properties of the helium solids from the other members of the family. These properties present a special challenge to lattice-dynamical theory. Max Born first addressed the possibility of a self-consistent phonon theory, and such theories, numerically computed, have been the most successful so far for phonons in solid heliums.

The isotope ^{3}He has a pronounced minimum in its melting curve; that is, the thermodynamic derivative along the solid–fluid coexistence line, $(\partial P/\partial T)_V$, is negative over a significant range of temperature. Qualitatively, the reason for this may be seen from the Maxwell relation, which can be used to equate the slope of the melting curve to the change in entropy with change in volume across the phase boundary: $(\partial S/\partial V)_T = (\partial P/\partial T)_V$. The liquid is less dense than the solid phase. However, in the temperature range concerned, the liquid phase is a Fermi liquid having a high degree of order in the spin system (although it has the usual positional disorder of a liquid) whereas the solid phase has much magnetic disorder (it is well above its magnetic transition temperature below 10 mK and hence has entropy about $R\ln 2$). This has led to very unusual studies such as the production of spin-polarized *solid* ^{3}He by sudden compression of the liquid, by cooling through sudden pressurization at temperatures below the minimum, and so on.

The magnetic structure of solid ^{3}He depends upon its density and upon the temperature. Magnetism in ordinary materials is accounted for by quantum-mechanical "exchange" of electron spins, which may appear to be a theoretical construct. In ^{3}He, which has only paired electrons, the spin concerned is nuclear, which, although linked to a very weak magnetic moment, becomes a dominant interaction at low enough temperatures and high enough densities. Because of the extraordinary mobility of the atoms, even in the solid state, there can be literal atomic exchanges (actually, in rings of three or of four atoms) to produce several observed magnetic phases (Roger *et al.*, 1983).

The properties of isotopic mixtures have been systematized by Edwards and Balibar (1989).

3.7 Defects and Excitations in Solid Heliums

By x-ray lattice-parameter measurements, thermal-vacancy concentrations have been directly measured in both hcp and bcc phases of both ^{3}He and ^{4}He (Simmons, 1994). The deduced activation energies have the expected variation with volume; that is, they are larger at smaller molar volumes. Nevertheless, the results are puzzling, because the concentrations are larger than those inferred indirectly from measurements (in isochoric samples) of pressure change and of specific heat. Because of the extremely high diffusivities measured by NMR in ^{3}He, explanations have been sought in vacancy excitations that are nonlocalized as a result of rapid tunneling processes. It has even been proposed, but not yet observed, that the ground states of these crystals contain vacancies. This would be one of the more bizarre consequences of the well-known nonlocalization of the atoms in solid heliums.

Negative-ion transport in the solid heliums can be attributed to motion of a hollow bubble surrounding a trapped electron (Dahm, 1986; Golov and Mezhov-Deglin, 1994). Such "bubble" states have been seen in both liquid neon and liquid helium, and in the latter have been used to probe excitations in superfluid helium samples. Bubble radii, determined by minimization of the total energy (electron confinement versus surface energy of the bubble), are thought to be about 0.7 nm in neon and from 1 to 2 nm in helium, depending upon the applied pressure.

The isotope ^{4}He also has a minimum in its melting curve, near 0.7 K, and the fact that at this temperature the adjacent liquid phase is superfluid has remarkable consequences. From the relation $(\partial S/\partial V)_T = (\partial P/\partial T)_V$ one sees that at and near this temperature the latent heat of freezing (melting) is zero or small. At the same time, the adjacent superfluid has extremely large thermal conduction, so that heat can be carried away from (or toward) the interface very rapidly. There is an interfacial energy, that is, a surface tension to be associated with the interface, so that "melting-freezing waves" can exist at this interface (Parshin, 1982; Chevalier *et al.*, 1994). Indeed, the observed frequencies of oscillation can be used to deduce information about the interfacial energy and structure. Observed velocities of propagation for these waves show the fastest freezing exhibited by any macroscopic condensed matter.

Fundamental optical studies of crystal growth have been made on ^{4}He and ^{3}He crystals growing in superfluid (the latter at 1 mK; Wagner *et al.*, 1994). Roughening transitions are seen, that is, temperatures at which the crystal–fluid surface changes from atomically smooth to rough.

4. SCIENTIFIC AND TECHNOLOGICAL APPLICATIONS

Scientific and technological applications for rare gases cover an extremely broad range. Many depend upon the inert property of the rare gases. The descriptions that follow are indicative, not comprehensive.

4.1 Arc Welding and Cutting

In volume of material, the largest use for argon and helium lies with gas-shielded welding and cutting of metals in an electric arc. They are used to exclude surrounding air from the molten weld metal. In this way, welds can be made in metals such as aluminum, magnesium, molybdenum, nickel, titanium, zirconium, and some stainless-steel alloys that are otherwise difficult to join. For some applications, one or more chemically active gases, such as hydrogen, may be added to control arc characteristics.

4.2 Passive Atmospheres and Inert Condensed Environments

The second-largest use for argon and helium, in terms of total quantity of gas, is as a passive atmosphere when processing or sparging reactive metals or other materials such as semiconductors. They are also used to regulate the rate of chemical reactions that depend on the concentration of the active gas. The excellent thermal conductivity of helium provides an efficient heat-transfer medium during the annealing of metals.

Condensed rare gases have often been used as inert matrices for spectroscopic studies of molecules and radicals that cannot otherwise be stabilized for extended study. Their transparency to electromagnetic radiation is extraordinary, and the low temperatures serve to retard kinetic processes. Spectroscopic properties of the guest molecules and radicals can therefore be studied from the microwave region through the far ultraviolet. This trapping can be carried to extreme selectivity: for examples, studies of the electron spin resonance of cesium atoms in solid helium and the influence on polarized light by rubidium implanted in solid xenon.

4.3 Illumination

A minor use of rare gases in terms of quantity, but all-pervasive in devices, is in sources of illumination. Electrical light sources fall into two groups: incandescent and gaseous discharge. In the case of incandescent lamps, a metal filament is heated to incandescence to produce light, usually in an atmosphere of argon gas together with a small amount of nitrogen. High-intensity incandescent bulbs are filled with krypton. Gaseous-discharge tubes produce light when an electric current is passed through an ionized gas between two electrodes. The color of the light produced depends on the specific gas present in the tube. Neon-discharge tubes, for example, emit light mainly in the red or orange region of the spectrum, while helium or argon or mixtures of these gases are used to produce other colors. Fluorescent lamps contain some argon, neon, or krypton to facilitate initiation of the discharge in the mercury vapor. Similarly, argon helps to start the electric discharge in mercury-vapor or sodium-vapor lamps. Xenon-filled flashlamps provide the photographer with economic, reproducible, brilliant, and brief illumination with a spectral distribution similar to that of daylight. A continuous source of extremely bright illumination for use in sports stadiums and in motion picture production is produced by an arc in high-pressure xenon placed in a quartz envelope.

4.4 Lasers

When highly collimated and monochromatic light is desired, rare-gas laser sources may be used. In the familiar helium–neon laser, laser emission results from excited levels of neutral neon atoms that have been populated by collision with excited helium atoms. Commercial versions are normally of a sealed-tube configuration. They typically have an output power of 1 mW and a lifetime about 10 000 h. Lasers based upon Ne–He mixtures are among the most widely used, including such technical applications as precise land-based surveying. Argon- and krypton-ion lasers provide high cw laser output power in the visible and near-ultraviolet regions of the spectrum but only at the expenditure of a considerable amount of electrical power. Other lasers involving rare gases are the He–Cd and pulsed rare-gas halide lasers. The latter class of "excimer" lasers produce stimulated emission as a result of transitions from a strongly bound excited state to a weakly bound ground state. For example, the laser transition occurs when $Kr^+ + F^-$ in a bound $^2\Sigma$ state undergoes a transition to the thermally unstable $X^2\Sigma$ state, which then dissociates into neutral ground-state atoms.

4.5 Thermometry

The International Temperature Scale of 1990 (ITS-90), adopted by the International Committee of Weights and Measures, is defined, in part, in terms of the properties of rare gases for its realization below 25 K. This scale (Preston-Thomas, 1990) supersedes previous scales, which have serious differences with the thermodynamic temperature. In the range from 0.65 to 5.0 K, ITS-90 is defined in terms of the *vapor pressure* of ^{3}He and ^{4}He. In the range from 3.0 K to the triple point of neon (24.5561 K), ITS-90 is defined in terms of a ^{3}He or ^{4}He gas thermometer of the constant-volume type calibrated at three experimentally realizable temperatures (the triple point of neon, the triple point of equilibrium hydrogen, and a temperature between 3.0 and 5.0 K determined in a prescribed way using a ^{3}He or ^{4}He vapor-pressure thermometer). The nonideality of the gas must be accounted for explicitly using the appropriate and accurate virial coefficient. To this end, many standards laboratories proceeded to measure these coefficients with the highest possible degree of accuracy.

4.6 Cryogenics and Refrigeration

The liquid heliums serve as cryogenic fluid baths at temperatures that can be precisely controlled through control of the vapor pressure. With readily available pumping the limiting temperatures are about 1.2 and 0.3 K for ^{4}He and ^{3}He, respectively. The isotopic mixture of ^{4}He and ^{3}He can be used to produce a continuously operating refrigerator at temperatures down to about 3 mK. ^{3}He displays about 6% solubility in the liquid mixture, even at the lowest temperatures. Concentrated ^{3}He floats on a mixture at higher temperatures, providing a source of ^{3}He, which dissolves into the dilute ^{3}He phase. This is analogous to ordinary evaporation, with accompanying absorption of heat. On the other hand, ^{3}He has a higher vapor pressure, which permits it to be distilled out of the resulting mixture and recirculated continuously.

Large-scale use of liquid ^{4}He occurs in the cooling of superconducting magnetic structures in accelerators and detectors used by high-energy physicists. This use will be important until high-T_c superconductors can be used in these structures, which experience extremely large magnetic forces and must operate with changing currents.

4.7 Other Uses

Rare gases have found an extremely wide variety of other applications. Among these for helium are as buoyant filling for nonrigid airships and balloons, and as a gas for wind tunnels at extremely high Mach numbers. A "wind" tunnel for unusual hydrodynamic regimes could use liquid helium. Perhaps most familiar to scientists are uses for highly sensitive vacuum leak detection with mass spectrometers and as a carrier gas in chromatography. Rare gases have been a common filling for gas detectors of radiation. High-energy particle physicists continue to develop the use of liquid Xe, Kr, and Ar as vacuum-ultraviolet scintillation materials that have high quantum efficiency, uniformity, and resistance to radiation damage.

An important application of rare-gas adsorption is the measurement of the total surface area of a microporous material. The area can be determined if the pores are large enough to allow the gas to enter and if its adsorption behavior on the material is well known. Conversely the nonadsorption of helium makes it useful in volume measurements of microporous materials. Helium penetrates the finest pores of a substance without being adsorbed appreciably. The volume of displaced helium then provides an accurate measurement of the volume of the substance.

All the rare gases have been investigated for biological effects and possible medical use. Helium is useful in breathing atmospheres in deep diving or in sealed environments, and it and xenon have been investigated for use as nonexplosive anesthetics. The radioactive isotopes ^{85}Kr and ^{87}Kr have found some use, for example as tracers in studies of circulation and in radiotherapy.

It is remotely possible that xenon oxides and perxenates could find application as specialized explosives or oxidizing agents (Asimov, 1966). Another possible application of solid or liquid rare-gas compounds would be as a storage method for the gas. Such compounds, if sufficiently stable, might eliminate the need for the high-pressure and cryogenic facilities currently used to store the fluids.

GLOSSARY

Chemical Compound: A substance that is chemically stable at room temperature and pressure, held together by a chemical bond.

Clathrate: A **van der Waals force** can act between a rare-gas atom and an atom or molecule of another kind, resulting in a weak bond. This attachment is further enhanced if the molecules form a cage around the rare-gas atom. Such a complex is called a clathrate.

Correlation Function: In dynamical models for liquids, for example the self-correlation (the average motion of an atom among others) and the density–density correlation (the relative motion of atoms).

Equation of State: For a gas, a relation over a range of pressure P, volume V, and temperature T in an equation, such as $PV = RT[1 + B/V + C/V^2 + D/V^3 + \cdots]$. Here B, C, D, etc. are the second, third, and fourth **virial coefficients,** which depend only on T and the species of gas. For condensed systems, the relation $P(V,T)$ generally has no such simple analytical form.

Extrinsic Point Defects: Examples are foreign atoms, molecules, or radicals.

HFD Potential: In the overall damped HFD potential the interaction energy is partitioned into uncorrelated (Hartree–Fock) and correlated parts and a suitable damping function to account for charge-overlap effects.

Kinetic Theory: From the laws of mechanics and the theory of probability, it determines the macroscopic properties of gases in terms of the motion of molecules, their collisions with one another, and the forces of interaction.

Landau Theory: A phenomenological theory of normal Fermi liquids, which are formed of spin-$\frac{1}{2}$ particles. It uses the concept of elementary excitations to clarify the properties of strongly interacting quantum systems at low temperatures. It provides a useful way to correlate measured properties directly, such as the transport coefficients, with a limited number of parameters, without the need for a microscopic theory. Subsequent microscopic derivations of the Landau theories have been produced. In ^3He these excitations are quasiparticles, a concept that has proven useful in other forms of condensed matter.

Law of Corresponding States: A physical relation based on the hypothesis that all gases in a set have the same reduced intermolecular potential. The reduced potential is defined as $U^* = U(r^*,b_i)/e$, where $r^* = r/s$, where e and s are scaling parameters for the energy and length, respectively. One kind of corresponding-states analysis simply uses the triple points as scaling parameters.

Many-Body Interactions: Many-body multipole interaction terms of third order are $(DDQ)_3$, $(QQD)_3$, $(DDO)_3$, and $(QQQ)_3$ and those of fourth and higher order $(MBD)_{4\to\infty}$. D, Q, and O refer to dipole, quadrupole, and octupole, respectively, and subscripts 3 and 4→∞ refer to third and to fourth to higher orders of quantum perturbation theory. Short-range many-body interactions may result from charge overlap.

Monte Carlo (MC) and Molecular Dynamics (MD) Methods: The MC method uses ensemble averages as in statistical mechanics and is used to calculate equilibrium properties. The MD method solves the Newtonian equations of motion of the particles and uses time averages. It is used to calculate both equilibrium and transport properties.

Pair-Distribution Function $g(r)$: An equilibrium distribution of interatomic pair separations averaged over time and direction, used to describe the static structure of a liquid or amorphous material.

Path-Integral Monte Carlo (PIMC) Method: A simulation technique employing Feynmann's path-integral approach to extend the applicability of the **Monte Carlo** method to quantum systems.

Phonon: In the vibrations of crystal lattices, a collective mode with well-defined energy and wave vector. Generally, there is phonon dispersion: The energy depends upon the wave vector.

Polaron: A charge carrier trapped in its own polarization field.

Quantum Corresponding States: The classical restriction of the law of corresponding states for translational states can be removed by introduction of a reduced Planck constant, a measure of quantum effects in the condensed state. For a Lennard-Jones interaction, the de Boer parameter $L^* = h/(mes^2)^{1/2}$ may be regarded as the reduced de Broglie wavelength of a particle of mass m.

Umklapp Processes: Those **phonon**-scattering processes for which the rule for conservation of wave vector is $\mathbf{Q}_1 + \mathbf{Q}_2 = \mathbf{Q}_3 + \boldsymbol{\tau}$, where $\boldsymbol{\tau}$ is a vector in the reciprocal lattice.

Vacancy: A vacant atomic site in a crystal.

Van der Waals–London Interaction: An attraction that varies as the inverse sixth power of the separation of two electric dipole oscillators.

Van der Waals Complex: A two-atom system physically bound by the van der Waals force, e.g., Ar_2, ArNe, H_2–Ar(X_2–Y), etc.

Virial Coefficients, Refractivity and Dielectric: Those pertaining to optical refraction and to dielectric properties.

Works Cited

Achtermann, H. J., Hong, J. G., Magnus, G., Aziz, R. A., Slaman, M. J. (1993), *J. Chem. Phys.* **98,** 2308–2318.

Asimov, I. (1966), *The Noble Gases,* New York: Basic Books.

Aziz, R. A., Bowman, D. H., Lim, C. C. (1967), *Can. J. Chem.* **45,** 2079–2086.

Aziz, R. A. (1984), in: M. L. Klein (Ed.), *Inert Gases: Potentials, Dynamics, and Energy Transfer in Doped Crystals,* Berlin: Springer-Verlag.

Aziz, R. A., McCourt, F. R., Wong, C. W. (1987), *Mol. Phys.* **61,** 1487–1511.

Aziz, R. A., Slaman, M. J. (1989), *Chem. Phys.* **130,** 187–194.

Aziz, R. A., Slaman, M. J. (1990), *J. Chem. Phys.* **92,** 1030–1035.

Aziz, R. A., Slaman, M. J. (1991), *J. Chem. Phys.* **94,** 8047–8053.

Aziz, R. A., Janzen, A. R., Moldover, M. R. (1995), *Phys. Rev. Lett.* **74,** 1586–1589.

Barker, J. A. (1989), in: A. Polian, P. Loubeyre, N. Boccara (Eds.), *Simple Molecular Systems at Very High Density,* New York: Plenum, pp. 331–351.

Barker, J. A., Henderson, D. (1976), *Rev. Mod. Phys.* **48,** 587–672.

Batchelder, D. N., Losee, D. L., Simmons, R. O. (1968), *Phys. Rev.* **173,** 873–880.

Batchelder, D. N. (1977), in: M. L. Klein, J. A. Venables (Eds.), *Rare Gas Solids,* Vol. 2, London: Academic, Chap. 14.

Bich, E., Millat, J., Vogel, E. (1987), *Wiss. Z. WPU Rostock* **36** (8), 5–11.

Blasdell, R. C., Ceperley, D. M., Simmons, R. O. (1993), *Z. Naturforsch. A* **48,** 433–437.

Boninsegni, M., Pierleoni, C., Ceperley, D. M. (1994), *Phys. Rev. Lett.* **72,** 1854–1857.

Ceperley, D. M. (1995), *Rev. Mod. Phys.* **67,** 279–355.

Chadwick, A. V., Glyde, H. R. (1977), in: M. L. Klein, J. A. Venables (Eds.), *Rare Gas Solids,* Vol. 2, London: Academic, Chap. 19.

Chen, H. H., Aziz, R. A., Lim, C. C. (1975), *J. Chem. Thermodynam.* **7,** 191–199; *ibid.* (1978), **10,** 649–659.

Chevalier, E., Guthmann, C., Rolley, E., Balibar, S. (1994), *Physica B* **194–196,** 919–920.

Claassen, H. (1966), *The Noble Gases,* Boston: Heath.

Colbourn, E. A., Douglas, A. E. (1976), *J. Chem. Phys.* **65,** 1741–1745.

Dahm, A. J. (1986), in: D. F. Brewer (Ed.), *Pro-*

gress in Low-Temperature Physics, Vol. 10, Amsterdam: North-Holland, pp. 73–137.

de Schepper, I. M., Cohen, E. G. D., Bruin, C., van Rihs, J. C., Montfrooij, W., de Graaf, L. A. (1988), *Phys. Rev. A* **38,** 271–287.

Dham, A. K., Allnatt, A. R., Meath, W. J., Slaman, M. J., Aziz, R. A. (1990), *Chem. Phys.* **142,** 173–189.

Dobbs, E. R. (1994), *Solid Helium Three,* Oxford: Clarendon.

Edwards, D. E., Balibar, S. (1989), *Phys. Rev. B* **39,** 4083–4097.

Egelstaff, P. A. (1992), *An Introduction to the Liquid State,* 2nd ed., Oxford: Clarendon.

Ewing, G. F. (1975), *"Structure and Properties of van der Waals Molecules," Accounts Chem. Res.* **8,** 185–192.

Frenking, G., Koch, W., Reichel, F., Cremer, D. (1990), *J. Am. Chem. Soc.* **112,** 4240–4256.

Gilgen, R., Kleinrahm, R., Wagner, W. (1994), *J. Chem. Thermodynam.* **26,** 399–413.

Glyde, H. R. (1995), *Excitations in Liquid and Solid Helium,* Oxford: Clarendon.

Golov, A. I., Mezhov-Deglin, L. P. (1994), *Physica B* **194–196,** 951–952.

Granfors, P. R., Macrander, A. T., Simmons, R. O. (1981), *Phys. Rev. B* **24,** 4753–4763.

Gugan, D. (1984), *Metrologia* **19,** 147–162.

Hanley, H. J. M., McCarty, R. D., Haynes, W. M. (1974), *J. Phys. Chem. Ref. Data* **3,** 979–1017.

Herman, P. R., LaRocque, P. E., Stoicheff, B. P. (1988), *J. Chem. Phys.* **89,** 4535–4549.

Huot, J., Bose, T. K. (1991), *J. Chem. Phys.* **95,** 2683–2687.

Jephcoat, A. P., Mao, H. K., Finger, L. W., Cox, D. E., Helmley, R. J., Zha, C. S. (1987), *Phys. Rev. Lett.* **59,** 2670–2673.

Joppien, M., Karnbach, R., Möller, T. (1993), *Phys. Rev. Lett.* **71,** 2654–2657.

Kestin, J., Ro, S. T., Wakeham, W. A. (1972), *Physica* **58,** 165–211.

Korpiun, P., Lüscher, E. (1977), in: M. L. Klein, J. A. Venables (Eds.), *Rare Gas Solids,* Vol. 2, London: Academic, Chap. 12.

Kovalenko, S. I., Solnyshki, D. D., Verkhovtseva, E. T., Eremenko, V. V. (1994), *Low-Temp. Phys.* **20,** 758–763.

Langel, W. (1992). *Spectrochim. Acta, Part A* **48,** 405–428.

LaRocque, P. E., Lipson, R. H., Herman, P. R., Stoicheff, B. P. (1986), *J. Chem. Phys.* **84,** 6627–6641.

Losee, D. L., Simmons, R. O. (1968), *Phys. Rev.* **172,** 944–957.

Luo, F., McBane, G. C., Kim, G., Giese, C. F., Gentry, W. R. (1993), *J. Chem. Phys.* **98,** 3564–3567.

Matsuoka, H., Hirokawa, T., Matsui, M., Doyama, M. (1992), *Phys. Rev. Lett.* **69,** 297–300.

McConville, G. T. (1974), *J. Chem. Phys.* **60,** 4093.

McLean, A. D., Liu, B., Barker, J. A. (1988), *J. Chem. Phys.* **89,** 6339–6347.

Meath, W. J., Koulis, M. (1991), *J. Mol. Struct. (Theochem.)* **226,** 1–37.

Moldover, M. R., Trusler, J. P. M., Edwards, T. J., Mehl, J. B., Davis, R. S. (1988), *Phys. Rev. Lett.* **60,** 249–252.

Montfrooij, W. de Graaf, de Schepper, I. M. (1992), *Phys. Rev. B* **45,** 3111–3114.

Nagler, S. L., Horn, P. M., Rosenbaum, T. F., Birgenau, R. J., Sutton, M., Mochrie, S. G. J., Moncton, D. E., Clarke, R. (1985), *Phys. Rev. B* **32,** 7373–7383.

Najafi, B., Mason, E. A., Kestin, J. (1983), *Physica* **119A,** 387–440.

Parshin, A. Ya. (1982), *Physica* **110B,** 1819–1829.

Polian, A., Loubeyre, P., Grimsditch, M. (1986), *Phys. Rev. B* **33,** 7192–7200.

Preston-Thomas, H. (1990), *Metrologia* **27,** 3–10 and 107 (erratum).

Roger, M., Hetherington, J. H., Delrieu, J. M. (1983), *Rev. Mod. Phys.* **55,** 1–64.

Ross, M., Radousky, H. B. (1989), in: A. Polian, P. Loubeyre, N. Boccara (Eds.), *Simple Molecular Systems at Very High Density,* New York: Plenum, pp. 47–84.

Rössler, U. (1977), in: M. L. Klein, J. A. Venables (Eds.), *Rare Gas Solids,* Vol. 1, London: Academic, Chap. 8.

Saunders, M., Jimenez-Vasquez, H., Cross, R., Poreda, R. (1993), *Science* **259,** 1428–1430.

Schell, N., Simmons, R. O., Kaprolat, A., Schülke, W., Burkel, E. (1995), *Phys. Rev. Lett.* **74,** 2535–2538.

Schoknecht, W. E., Simmons, R. O. (1972), in: N. G. Graham, H. E. Hagy (Eds.), *Symposium on Thermal Expansion,* New York: American Institute of Physics, pp. 169–182.

Schöllkopf, W., Toennies, J. P. (1994), *Science* **266,** 1345–1348.

Schwalbe, L. (1976), *Phys. Rev. B* **14,** 1722–1732.

Schwalbe, L. A., Crawford, R. K., Chen, H. H., Aziz, R. A. (1977), *J. Chem. Phys.* **66,** 4493–4502.

Schwentner, N., Koch, E.-E., Jortner, J. (Eds.) (1985), *Electronic Excitations in Condensed Rare Gases,* Berlin: Springer-Verlag.

Shikin, V. B. (1977), *Sov. Phys. Usp.* **20,** 226–248.

Simmons, R. O. (1993), *Z. Naturforsch. A* **48,** 415–424.

Simmons, R. O. (1994), *J. Phys. Chem. Solids* **55,** 895–906.

Skold, K., Rowe, J. M., Ostrowski, G., Randolph, P. D. (1972), *Phys. Rev. A* **6,** 1107–1131.

Song, Y., Mason, E. A. (1989), *J. Chem. Phys.* **91,** 7840–7853.

Song, Y., Mason, E. A. (1993), *Phys. Rev. E* **47,** 2193–2196.

Sonntag, B. (1977), in: M. L. Klein, J. A. Venables (Eds.), *Rare Gas Solids,* Vol. 2, London: Academic, Chap. 17.

Spear, W. E., le Comber, P. G. (1977), in: M. L. Klein, J. A. Venables (Eds.), *Rare Gas Solids,* Vol. 2, London: Academic, Chap. 18.

Stewart, A. T., Briscoe, C. V., Steinbacher, J. J. (1990), *Can. J. Phys.* **68,** 1362–1376.

Swenson, C. A. (1977), in: M. L. Klein, J. A. Venables (Eds.), *Rare Gas Solids,* Vol. 2, London: Academic, Chap. 13.

Thompson, C. A., Andrews, L. (1994), *J. Am. Chem. Soc.* **116,** 423–424.

Varding, D., Becker, J., Frankenstein, L., Peters, B., Runne, M., Schröder, A., Zimmerer, G. (1993), *Low-Temp. Phys.* **19,** 427–436.

Venables, J. A., Smith, B. L. (1977), in: M. L. Klein, J. A. Venables (Eds.), *Rare Gas Solids,* Vol. 2, London: Academic, Chap. 10.

Wagner, R., Andreeva, O. A., Ras, P. J., Remeijer, P., Steel, S. C., van Woerkens, C. M. C. M., Frossati, G. (1994), *Physica B* **194–196,** 961–962.

Wakeham, W. A., Sengers, J. V., Nagashima, A. (Eds.), (1991), *Experimental Thermodynamics III. Measurement of the Transport Properties of Fluids,* London: Blackwell.

Weeks, J. D., Chandler, D., Andersen, H. C. (1971), *J. Chem. Phys.* **54,** 5237–5247.

Wörmer, J., Joppien, M., Zimmerer, G., Möller, T. (1991), *Phys. Rev. Lett.* **67,** 2053–2056.

Xu, Y., Jäger, W., Gerry, M. C. L. (1994), *J. Chem. Phys.* **100,** 4174–4180.

Yakub, E. S. (1994), *Low-Temp. Phys.* **20,** 579–598.

Further Reading

Allen, M. P., Tildesley, D. J. (Eds.) (1993), *Computer Simulation in Chemical Physics,* Dordrecht: Kluwer.

Barker, J. A., Auerbach, D. J. (1985), *Surf. Sci. Rep.* **4,** 1–99.

Bennemann, K. H., Ketterson, J. B. (Eds.), (1976/8), *The Physics of Liquid and Solid Helium,* Vols. I and II, New York: Wiley.

Cook, G. A. (Ed.) (1961), *Argon, Helium, and the Rare Gases,* Vols. 1 and 2, New York: Interscience.

Dobbs, E. R. (1994), *Solid Helium Three,* Oxford: Clarendon.

Duke, C. B. (Ed.) (1994), *Surface Science: The First Thirty Years,* Amsterdam: North-Holland.

Duley, W. W. (1995), *UV Lasers: Effects and Applications in Material Science,* Cambridge, U.K.: Cambridge Univ. Press.

Dymond, J. H., Smith, E. B. (1980), *The Virial Coefficients of Pure Gases and Mixtures,* Oxford: Clarendon.

Egelstaff, P. A. (1992), *An Introduction to the Liquid State,* 2nd ed., Oxford: Clarendon.

Glyde, H. R. (1995), *Excitations in Liquid and Solid Helium,* Oxford: Clarendon.

Glyde, H. R., Svensson, E. C. (1987), "Solid and Liquid Helium," in: D. L. Price and K. Skold (Eds.), *Methods of Experimental Physics,* Vol. 23, Pt. B, London: Academic, Chap. 13.

Gray, C. G., Gubbins, K. E. (1984), *Theory of Molecular Fluids,* Vol. 1, Oxford: Clarendon.

Hirschfelder, J. O., Curtiss, C. F., Bird, R. B. (1964), *Molecular Theory of Gases and Liquids,* New York: Wiley.

Horton, G. K., Maradudin, A. A. (Eds.), *Dynamical Properties of Solids,* especially Vol. 2 (1975) and Vol. 7 (1995), Amsterdam: North-Holland.

Jorgenson, C. K., Frenking, G. (1990), *Structure Bonding* **73,** 1–15.

Kauffman, G. B. (1988), *J. College Sci. Teaching* **17,** 264–268.

Klein, M. L., Venables, J. A. (Eds.) (1976/7), *Rare Gas Solids,* Vols. 1 and 2, London: Academic.

Klein, M. L. (Ed.) (1984), *Inert Gases: Potentials, Dynamics, and Energy Transfer in Doped Crystals,* Berlin: Springer-Verlag.

Maitland, G. C., Rigby, M., Smith, E. B., Wakeham, W. A. (1981), *Intermolecular Forces,* Oxford: Clarendon.

Polian, A. Loubeyre, P., Boccara, A. (Eds.) (1989), *Simple Molecular Systems at High Density,* New York: Plenum.

Price, D. L., Skold, K. (Eds.), (1987), *Methods of Experimental Physics,* Vol. 23, Pt. B, London: Academic.

Saha, G. (1992), *Fundamentals of Nuclear Pharmacy,* 3rd ed., New York: Springer-Verlag.

Schwentner, N., Koch, E.-E., Jortner, J. (Eds.) (1985), *Electronic Excitations in Condensed Rare Gases,* Berlin: Springer-Verlag.

Smith, B. L. (1971), *The Inert Gases: Model Systems for Science,* London: Wykeham.

Varotsos, P. A., Alexopoulos, K. D. (1986), *Thermodynamics of Point Defects and Their Relation with Bulk Properties,* Amsterdam: North-Holland.

Wilks, J., Betts, D. S. (1987), *An Introduction to Liquid Helium,* 2nd ed., Oxford: Oxford Univ. Press.

RAREFIED-GAS DYNAMICS

Franklin C. Hurlbut, *Department of Mechanical Engineering, University of California at Berkeley, Berkeley, California, U.S.A.*

	Introduction	97
1.	**Historical Summary**	98
1.1	Fundamental Beginnings, to 1939	98
1.2	Second Phase, the Expansion Period, to 1978	98
1.3	Contemporary Phase; the Growing Sophistication of Tractable Problems	99
2.	**Theoretical Descriptions**	99
2.1	Preliminary Discussion	99
2.2	The Boltzmann Equation	100
2.3	Momentum and Energy Transfers in Free-Molecule Flow	101
2.3.1	Background	101
2.3.2	Momentum- and Energy-Transfer Coefficients	102
2.3.3	Forces of Lift and Drag on a Flat Plate, General Expressions	103
2.4	Direct Molecular Flow Simulation Using Monte Carlo Methods	104
2.5	Gas–Surface Interactions	105
3.	**Experimental Methods**	108
3.1	Ground-Based Experimental Systems	108
3.1.1	Wind Tunnels and Shock Tunnels	109
3.1.2	Flow Production	109
3.1.3	Molecular and Ion Beams	109
3.2	Instruments and Diagnostic Methods	110
3.3	Rocket and Satellite Experimental Systems	111
4.	**Applications**	111
	Glossary	112
	Works Cited	112
	Further Reading	114

INTRODUCTION

Rarefied-gas dynamics rests upon the theoretical disciplines of statistical mechanics and the molecular theory of gases and derives essential parameters from several areas of the physical sciences. In this subject we are concerned with those dynamic processes in which the discrete particle nature of gases plays a critical role in the modes of analysis and the prediction of behavior. The gas is said to be "rarefied" wherever the ratio of the mean distance between molecular collisions (mfp) to a characteristic length of the flow exceeds a very small value, usually of the order of 0.01. This ratio, the Knudsen number, Kn, expresses the departure of the flow from a near-continuum regime at low values of Kn to one at large values of Kn in which intermolecular collisions occur infrequently.

At one time it was thought that rarefied gases were found principally in laboratory vacuum systems or at extreme altitude. Rarefied gases were identified as being of "low density" as compared with the "normal density" of our familiar atmosphere. One sees, however, that the phenomena of rarefaction are a matter of scale, independent of the absolute number density of molecules or the physical size of the field, so long as binary molecular collisions predominate. Flows about microscopic filaments, through small orifices, or in particle beds may be subject to the effects of rarefaction, just as are flows about orbiting bodies. Regimes of rarefied flow are customarily defined in terms of the Knudsen number, as follows:

$$\text{Near-continuum flow, } 0.01 < Kn < 0.1; \tag{1a}$$

$$\text{Transition flow, } 0.1 < Kn < 10; \tag{1b}$$

$$\text{Nearly free-molecule flow, } 10 < Kn < 100. \tag{1c}$$

3-527-28138-X/96/$5.00 + .50

Discussions of the analysis and applications of rarefied flows in these regimes and in the regime of free-molecule flow in which no intermolecular collisions occur comprise the material of this article.

The physical phenomena associated with the rarefied-gas–dynamic regimes are so singularly numerous and so varied that it is not possible to categorize them in compact form. This is true in part because of the wide range of energy transfers that lie within these regimes, and in part because of the extreme variation in scale of the phenomena that lie within its domain. It is not surprising that scientists from so many areas of the physical sciences and engineering have contributed. A possible unifying factor may be found in the attractions that lie beyond the limits of continuum fluid mechanics. In addition there is always the promise of exploration, to learn, as we have learned in the past few years, of our Earth and its relation to the universe. Each period in the history of rarefied-gas dynamics has been filled with a sense of exploration, but each has had its particular nature. It is useful to distinguish among these as we approach this subject.

The formative years, which began with the rise of classical physics, were marked by the great works of Maxwell and Boltzmann during which the mathematical consolidation of the molecular hypothesis was achieved. Pioneering studies continued during this period until its close in about 1939. There next came a period of expansion and discovery in which, aided by rapidly developing research technology and by man's incursion into space, our knowledge grew with unprecedented speed. In the third and present period there exists a new departure in inquiry; we are enabled to look more deeply into gas-dynamic processes and we now have computational tools allowing the exploration of problems that once seemed unsolvable. This article examines these perceptions and skills as they relate to present applications.

1. HISTORICAL SUMMARY

1.1 Fundamental Beginnings, to 1939

Natural scientists had agreed over many years on the atomic nature of matter, but it was not until the first of the twentieth century, following Dunoyer's (1911) demonstration using a molecular beam of sodium, that the discrete particle structure of dilute gases became fully accepted. Still, the model of the particulate gas had been used by kinetic theorists for at least ninety years prior to that experiment. Maxwell had formulated the law for the distribution of molecular velocities in such a gas in 1859, and by 1866 had made substantial progress in establishing the basis for modern theories of nonuniform gases, that is, gases in which gradients of temperature, velocity, and density are present. Boltzmann (1872) published the celebrated equation that would guide the acquisition of much of our present knowledge of gas transport properties and that provided the link between continuum fluid motion and that of dilute gases. At the same time, he also presented his H theorem, which demonstrated the inevitable relaxation of nonuniform gases toward the Maxwellian distribution of velocities. Efforts to determine general expressions for the distributions of velocity in nonuniform gases met with little success until Chapman in 1916 and Enskog in 1917 published work leading to equivalent terms in expressions for the general velocity distribution function f and, thus, provided the background for the modern theory of transport coefficients. This phase in the development of rarefied-gas–dynamic theory was essentially completed in 1935 by Burnett, who introduced Sonine polynomials as terms in the expansion for f, and extended the order of expansion to the second. A readable, more detailed account of this period may be found in the well-known treatise by Chapman and Cowling (1939).

1.2 Second Phase, the Expansion Period, to 1978

The close of World War II marked the beginning of a period of an extraordinary increase in interest in the field of rarefied-gas dynamics. Attention was now given to simplification of the Boltzmann equation, as in the work of Bhatnagar *et al.* (1954), and to alternative modes of solution for the full equation, as is seen in the work of Grad (1949). Experimental laboratories were set up in departments of physics, chemistry, engineering, and aeronautics throughout the world and also in the research arms of defense departments and aeronautics agencies such as NACA, the National Advisory Committee for Aeronautics, precursor to NASA. Hirschfelder *et al.* (1954) published a monumental resource incorporating theoretical background and experimental results over a broad range of topics relating to collision processes and transport properties, covering much of importance to that year. Sputnik electrified the world in 1957 and many nations turned more directly to the application of rarefied-gas dynamics in space. An important series of symposia was begun in 1958 by Devienne at Nice, France. These meetings, the International Symposia on Rarefied Gas Dynamics, have brought together a significant number of the world's specialists every two years. The Proceedings of the symposia serve to define the field and to provide a compact entrée to it. A list of volume titles appears elsewhere in this article.

New experimental techniques were developed or adapted to a new environment. These included optical methods involving the excitation of fluorescence radiation, making possible noninvasive determinations of temperatures and velocities in monatomic gases and rotational temperatures in diatomic gases. Cryogenic pumping was introduced to enhance the extreme low-density capability of rarefied-gas wind tunnels. Several sources of flow were developed, notably the free-jet expansion. Plasma torch and arc-jet stagnation heaters were introduced to elevate the stagnation enthalpy. Space does not permit enumeration of the large number of investigators, but the record may be found in the proceedings of the RGD symposia. Rapid growth was also seen in the number and sophistication of studies on the interaction of gas molecules and surfaces. These were undertaken in numerous laboratories throughout the world by several thermal and dynamic methods. A few of these are discussed here in Sec. 3. It is particularly worth noting that the experimental study and parametrization of free-jet expansions by Ashkenas and Sherman (1966), which led to the frequent application of these flows in work with rarefied-gas wind tunnels, also led to the revolutionary method of aerodynamic intensification for molecular beams proposed by Kantrowitz and Gray (1951), now very widely used.

This second phase was one of great experimental activity, resulting in enormous extensions of our understanding of molecular collision parameters and of the aerodynamics and thermodynamics of bodies in high-speed rarefied-gas flows, and in our knowledge concerning the processes of momentum and energy exchange in gas–surface interactions. It was also a period of important theoretical activity, much of which was directed to the further exploration of the Boltzmann equation and its derivatives. Other interesting work was undertaken on "first-collision" departures from the free-molecule regime.

By the end of the 1970s there began a general reduction in activity, due, as one immediate cause, to changes in funding patterns, and we may mark the close of the second period as corresponding with this time.

1.3 Contemporary Phase; the Growing Sophistication of Tractable Problems

Beginning in the middle years of the 1980s the content of rarefied-gas dynamics broadened substantially in many aspects but most particularly in connection with the interest in more realistic treatments of molecular collisions, inelastic processes, and chemical reactions. The method of direct simulation using Monte Carlo techniques had been introduced by Bird in the late 1960s. During subsequent years the development of this method had been continued by Bird and by others, but it was not until several years had passed that the method of DSMC reached the current level of acceptance in the scientific community. This important topic is discussed more carefully here in Sec. 2.4. The contemporary period has just begun, but an important feature is visible; it is marked by a significant growth, both experimental and theoretical, in attention to inelastic and reactive collision processes, a growth coincident with and dependent upon the growing capability of DSMC.

2. THEORETICAL DESCRIPTIONS

2.1 Preliminary Discussion

Modern descriptions of rarefied gases rest upon concepts that are well verified by physical observation; gases are composed of discrete molecular particles that are small by comparison with their mutual separation and with the mean distance between collisions. Molecules are constantly in motion; their trajectories are interrupted by collisions with other molecules or with surfaces; the collisional potentials of interaction are relatively short ranged; in general analytical representations it is assumed that only binary collisions occur. The distribution of molecular velocities for such a gas in thermal equilibrium translating at velocity U with respect to a certain coordinate system is known as the translating or drifting Maxwellian and is written as follows:

$$f(\xi) = (n/2\pi RT)^{3/2} \exp\{-(\boldsymbol{\xi} - \mathbf{U})^2/2RT\}. \qquad (2)$$

In equation (2), n is the number density, T is the thermodynamic temperature, R is the species gas constant, $\boldsymbol{\xi}$ is the molecular velocity, and $\mathbf{U}$ is the velocity of translation. The difference between $\boldsymbol{\xi}$, the total velocity, and $\mathbf{U}$ gives the local velocity of thermal motion, the *peculiar* velocity. With the help of equation (2) and standard concepts of the kinetic theory and thermodynamics relations, we can find most of the expressions applied in free-molecule aerodynamic calculations. This matter will be taken up in subsequent paragraphs.

2.2 The Boltzmann Equation

It should be noted that the macroscopic velocity $\mathbf{U}$ is determined by averaging the total velocities of all molecules in a certain region, whether or not the gas is rarefied. Values of other quantities such as the flux of momentum across a control surface are determined by taking the average of appropriate velocity components over the velocity distribution function. However, where values of field properties such as density alter significantly within lengths of the order of the mean distance between collisions, the gas may be slightly or perhaps greatly out of equilibrium, in which event the Maxwellian distribution cannot be used in performing the weighted average. When the departures from equilibrium are small we are enabled to find velocities, temperatures, densities, and so on, through application of the Boltzmann equation. The Boltzmann equation is a conservation equation expressed in terms of the complete molecular velocity distribution, f. The equation equates the time rate of change in the number of particles of a certain velocity class within a certain small volume due to the physical movement of those particles to the change in the number within that same velocity class in that same volume due to molecular collisions. Solutions of the Boltzmann equation are found in terms of the distribution function f.

The Boltzmann equation is an integro-differential equation of considerable complexity. The several approaches to solutions are discussed in Schaaf and Chambré (1958) and in the monograph for the *Encyclopedia of Physics* (Schaaf, 1963) in which Schaaf has provided an extensive entrée to the literature to 1960. In their monumental work, Chapman and Cowling (1939) set forth a detailed and authoritative coverage. Excellent accounts are also found in Hirschfelder *et al.* (1954) and in Vincenti and Kruger (1965).

Solutions for f are often sought as a series expansion in powers of a small number, essentially the Knudsen number. If f_0 is the Maxwellian then f may be written

$$f = f_0[1 + \phi_1(\lambda/L) + \phi_2(\lambda/L)^2 + \cdots]. \qquad (3)$$

In Eq. (3) λ is the mean distance between collisions and L is a characteristic length of the flow. Clearly λ/L must be small. At normal densities, gradients of velocity and temperature are usually sufficiently small to permit the assumption that the system is in local equilibrium within each elementary region. In this event $f = f_0$. From the zeroth order of the expansion, Eq. (3), we are led directly to the Euler equations. The Navier-Stokes equations result from the retention of the first-order term in λ/L while the Burnett equations (1935) result from the retention of the second-order term. It is sometimes useful to think of small departures from this condition in terms of the ratio of Mach number M to Reynolds number. Since the kinematic viscosity ν is proportional to $a\lambda$, where a is the sound speed, and the Reynolds number is given by $Re = VL/\nu$, in which V is a velocity and L is a characteristic length, we find that $\lambda/L \approx M/Re$. Thus M/Re is a measure of rarefaction equivalent to the Knudsen number for small departures from continuum flow.

Tests of the validity of the Navier-Stokes equations and of the Burnett equations have been made under conditions that minimize the influence of wall–gas collisions. Sherman (1955) and Talbot and Sherman (1959) examined shock-wave profiles in weak shock waves of helium and air, and later in argon. It was found that the Navier-Stokes computations produced shock-wave profiles in fair agreement with experiment, but except for the very low–Mach-number calculations, the Burnett terms did not yield stable solutions. This matter is discussed again in subsequent paragraphs.

In applications of the Navier-Stokes equations in the presence of boundaries within regimes of near-continuum flow, it is necessary to apply slip-flow and temperature-jump boundary conditions. These boundary conditions express an important effect of rarefaction. Within a distance λ from the surface, the gas cannot be considered exactly at rest with respect to the surface, nor, where a gradient of temperature exists, can the gas be exactly at the surface temperature. Thus, there will be a "slip" of velocity and a "jump" in temperature at the wall, although these will be of small magnitude where the Navier-Stokes equations remain valid.

2.3 Momentum and Energy Transfers in Free-Molecule Flow

2.3.1 Background During the years when the foundations of the kinetic theory were taking shape, little thought was given to the idea that man's own satellites would orbit the Earth. However, collisionless flows were of practical and theoretical interest to physicists and chemists, so that by the early 1900s an increasing number of persons were measuring momentum and thermal transfers and attempting to understand the relation between these effects and the nature of the bounding surface. The Maxwellian distribution of molecular velocities was experimentally verified by Stern (1920), wall–gas interaction studies were conducted by Millikan (1923), and heat transfers in rarefied gases were examined by Knudsen (1911). By the end of the first quarter of the twentieth century, a great deal had become known about processes that were of low thermal order and that involved modest molecular velocities. The context broadened greatly with the flight of Sputnik, but much of the mathematical background for the description of flows about satellites had already been well established. There still remained the problem of describing the aerodynamic and thermal effects of wall–gas collisions.

Free-molecule flows are considered to occur for values of the Knudsen number greater than about 100. The flow is collisionless only in the sense that intermolecular collisions do not alter the distribution of velocities incident upon a body; however, the Maxwellian distribution assumed for the incident flow would normally have arisen from intermolecular collisions at remote distances or from diffusely scattering surroundings. Densities and molecular velocities in the vicinity of the body, on the other hand, become greatly modified by wall collisions. This point is elaborated in paragraphs to follow.

A body in free-molecule flow is assumed to be subject to the collisions of large numbers of independently moving molecules. Our knowledge of molecular energy, momentum, and number flux incident upon an element of surface or across any arbitrary elementary plane, and our knowledge of energy density and other densities, must be reported in terms of averaged values. These averages are found by integrating appropriate quantities, usually products of velocities, over the molecular velocity distribution function. Coordinates for an element of surface are illustrated in Fig. 1. As is customary in aerodynamic calculation, the attack angle θ is defined to be taken between the surface tangent and the macroscopic velocity vector **U**, and the axes are oriented to place **U** in the (x_1,x_2) plane. The number flux to an elementary surface dA is

$$dN_i = \int_{-\infty}^{\infty}\int_{-\infty}^{\infty}\int_{0}^{\infty} f\xi_1 d\xi_1 d\xi_2 d\xi_3, \tag{4}$$

in which f is the Maxwellian distribution from Eq. (2). It is useful to define a molecular speed ratio S as the ratio of the macroscopic speed of flow, U, to the most probable molecular speed, $(2RT)^{1/2}$. Integration of equation (4) leads to

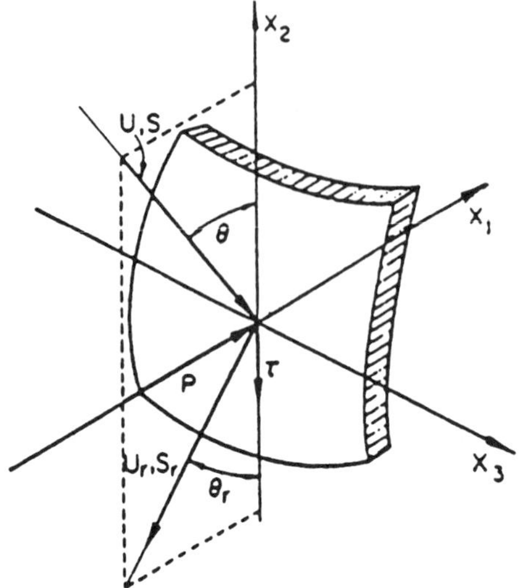

FIG. 1. Schematic diagram of coordinates for molecular flow at the surface.

$$dN_i = n\sqrt{RT}/2\pi\{\exp[-(S\sin\theta)^2] + \sqrt{\pi}(S\sin\theta)[1 + \text{erf}(S\sin\theta)]\}dA. \quad (5)$$

Conservation in number density at the surface is assumed so that dN_r, the reemitted number flux, is equal to that incident, $dN_r = dN_i$. Expressions for the flux of momentum incident to the surface in the normal direction, the flux of momentum in the tangential direction, and the flux of energy are found in a similar fashion. Let p_i be the flux of normal momentum to the surface as given by the integration of $m\xi_1^2 f d\xi_1 d\xi_2 d\xi_3$, and let τ_i be the flux of tangential momentum given by the integration of $-m\xi_1\xi_2 f d\xi_1 d\xi_2 d\xi_3$. Let $S\theta = S\sin\theta$. We may then write

$$p_i = \frac{\rho U^2}{2\sqrt{\pi}S}\{(S\theta)e^{-(S\theta)^2} + \sqrt{\pi}[0.5 + (S\theta)^2][1 + \text{erf}(S\theta)]\}, \quad (6)$$

$$\tau_i = \frac{\rho U^2 \cos\theta}{2\sqrt{\pi}S}\{e^{-(S\theta)^2} + \sqrt{\pi}(S\theta)[1 + \text{erf}(S\theta)]\}. \quad (7)$$

The flux of incident energy, which may be found in an analogous manner, must include that due to the translational energy, $\frac{1}{2}m\xi^2$, and that of the internal energy, $\frac{j}{2}RT$, in which j is the number of internal degrees of freedom:

$$E_i = E_{i,\text{tran}} + E_{i,\text{int}}. \quad (8)$$

We find

$$E_{i,\text{tran}} = \rho RT\sqrt{\frac{RT}{2\pi}}\{(S^2 + 2)e^{-(S\theta)^2} + \sqrt{\pi}(S^2 + \tfrac{5}{2})(S\theta)[1 + \text{erf}(S\theta)]\}, \quad (9a)$$

$$E_{i,\text{int}} = \frac{\rho(5 - 3\gamma)}{\sqrt{\pi}(\gamma - 1)}\left(\frac{RT}{2}\right)^{3/2} \times \{e^{-(S\theta)^2} + [1 + \text{erf}(S\theta)]\}. \quad (9b)$$

2.3.2 Momentum- and Energy-Transfer Coefficients Our next task is to develop expressions for the fluxes of momentum and energy that leave the surface through molecular scattering. At this point we encounter one of the least understood and most speculative areas of rarefied-gas dynamics, since our knowledge of these exit fluxes is extremely limited. Few molecular exit-velocity distributions have been obtained from laboratory experiment and none at all from observations aboard orbiting satellites. Much more is known about angular distributions of number flux, particularly in connection with laboratory studies of scattering from clean, regular surfaces at energies of thermal order. While speculations based upon reasonable conjectures can be made in light of present knowledge, there are insufficient experimental data to permit accurate performance predictions for extended shapes. A short account of our present knowledge of wall–gas interactions is given in Sec. 2.5.

The standard kinetic theory does, however, include useful formulas for drag, lift, and heat transfer for bodies in free-molecule flow, which treat the effects of gas collisions at the wall in terms of momentum- and energy-transfer coefficients. These coefficients, in essence, are measures of the degree to which the net reflected momentum or energy approaches that which would be emitted in Maxwellian equilibrium at the wall temperature. These coefficients can be connected with Maxwell's nonphysical conjecture dividing reflection into two kinds, specular and diffuse, but that is not their meaning, although some confusion persists. In the current and standard usage, thermal energy transfer is discussed in terms of α, the thermal accommodation coefficient, following von Smoluchowski (1911) and Knudsen

(1911), where

$$\alpha = (E_i - E_r)/(E_i - E_w). \quad (10a)$$

The subscript i identifies the energy carried to the surface by incident molecules, the subscript r identifies the energy carried away from the surface, and the subscript w identifies the energy that would be carried away by molecules in diffuse Maxwellian scattering.

In similar fashion we define σ, the transfer coefficient for tangential momentum, as

$$\sigma = (\tau_i - \tau_r)/(\tau_i - \tau_w), \quad (10b)$$

in which τ_w represents the tangential momentum carried from the wall by molecules scattered in Maxwellian reemission from it. However, these molecules are scattered diffusely and hence isotropically, in which event the value of τ_w is 0.

The coefficient of normal momentum transfer, σ', was introduced by Schaaf to complete the triad of surface interaction coefficients. It is seen as an essential feature in the current analysis of aerodynamic forces:

$$\sigma' = (p_i - p_r)/(p_i - p_w). \quad (10c)$$

The energy carried from the wall in Maxwellian reemission is given by

$$E_w = (4 + j)mRT_w \frac{dN_i}{dA} = \frac{\gamma + 1}{2(\gamma - 1)} mRT_w \frac{dN_i}{dA}. \quad (11)$$

We write the scattered normal momentum flux p_w in equation (10c) as

$$p_w = \tfrac{1}{2} m\sqrt{2\pi RT_w}\frac{dN_i}{dA}. \quad (12)$$

The coefficients σ, σ', and α are, in principle, independently determinable by physical measurement in particular physical systems. Through these empirical coefficients one may express the net shear stress τ, or the net pressure p, or the heat flux Q, acting on a surface, since these coefficients link the incoming with the outgoing quantities.

The quantities p_i and p_r are added to form the total normal pressure, giving

$$p = \frac{\rho U^2}{2S^2}\left\{\left(\frac{(2 - \sigma')}{\sqrt{\pi}}(S\theta) + \frac{\sigma'}{2}\sqrt{\frac{T_w}{T}}\right)e^{-(S\theta)^2} + \left[(2 - \sigma')[(S\theta)^2 + 0.5] + \frac{\sigma'}{2}\sqrt{\frac{\pi T_w}{T}}(S\theta)\right][1 + \mathrm{erf}(S\theta)]\right\} \quad (13)$$

and in a similar manner, letting $\chi(S\theta) = \exp[-(S\theta)^2] + \sqrt{\pi}S\theta[1 + \mathrm{erf}(S\theta)]$, we find

$$\tau = \frac{\sigma\rho U^2 \cos\theta}{2S\pi}\chi(S\theta) \quad (14)$$

and

$$Q = \alpha\rho U^2 RT\sqrt{\frac{RT}{2\pi}}\left\{\left(S^2 + \frac{\gamma}{\gamma + 1} - \frac{\gamma + 1}{2(\gamma - 1)}\frac{T_w}{T}\right)\chi(S\theta) - 0.5e^{-(S\theta)^2}\right\}. \quad (15)$$

Expressions (13), (14), and (15) provide the essential key to free-molecule analysis. It should be noted that each expression depends solely upon a single transfer coefficient, thus permitting a clear separation of effects in experimental studies. Both the pressure and the net heat flux, but not the shear stress, depend upon the wall temperature. No complication arises where T_w is known, but where it is not it may be necessary to solve for T_w with Q set equal to zero. Terms for radiation exchange will be needed.

2.3.3 Forces of Lift and Drag on a Flat Plate, General Expressions Where the transfer coefficients are known, solutions to problems of external free-molecule aerodynamics are accessible in principle but may be difficult to solve in analytic form. In general the transfer coefficients will be dependent on the angle θ, measured between the incoming gas molecules and the surface tangent. This feature may add substantially to the complication of solution. A number of authors have calculated values of aerodynamic coefficients and heat transfer for various conventional shapes, but have treated the coefficients as independent of the incident angle. For cases of high-speed flows over vehicles in orbit, where it is probable that the elementary assumpical are inadequate, it seems advisable to treat complex shapes through numerical integration over elementary areas. To this end the expressions for drag and lift per unit area for a flat plate at angle of attack θ may prove useful. These are listed as follows:

$$
\begin{aligned}
F_D = \frac{\rho_\infty U_\infty^2}{\sqrt{\pi}S} A \Big\{ & [(2 - \sigma')\sin^2\theta \\
& + \sigma\cos^2\theta] e^{-(S\sin\theta)^2} + \frac{\sigma'\pi}{2}\sqrt{\frac{T_W}{T}}\sin^2\theta \\
& + \sqrt{\pi}(S\sin\theta)\operatorname{erf}(S\sin\theta)[(2 - \sigma') \\
& \times (\sin^2\theta + \tfrac{1}{2}S^2) + \sigma\cos^2\theta]\Big\}, \qquad (16)
\end{aligned}
$$

$$
\begin{aligned}
F_L = \frac{\rho_\infty U_\infty^2}{\sqrt{\pi}S} A \Big\{ & (2 - \sigma - \sigma')\cos\theta\sin\theta\, e^{-(S\sin\theta)^2} \\
& + \frac{\sigma'\pi}{2}\sqrt{\frac{T_W}{T}}\sin\theta\cos\theta \\
& + \frac{\sqrt{\pi}}{2} S\cos\theta\Big[2(2 - \sigma - \sigma')\sin^2\theta \\
& + \frac{(2 - \sigma')}{S^2}\Big]\operatorname{erf}(S\sin\theta)\Big\}. \qquad (17)
\end{aligned}
$$

2.4 Direct Molecular Flow Simulation using Monte Carlo Methods

During the late 1950s scholars turned to the examination of computational models that exploited directly the dynamics of molecular motion and collisions and that left aside immediate concern with time-varying fluid behavior. Such directions were made possible by the increasing capability of high-speed computers. Alder and Wainwright (1958) used exact dynamical calculation to study the behavior of one hundred hard spheres. These were started into motion with uniform speeds and with randomized directions within a cubical enclosure. Velocities were found to approach the Maxwellian distribution following a few collisions. It was recognized, however, that the approach would quickly impose unrealistically large demands on computer capability as the number of molecules increased. Davis (1960), in what proved to be very promising directions, studied the free-molecule conductance of a tube having a right-angle bend. Randomized direction and position were given to each of a sequence of test particles started from the tube entrance. Trajectories were followed and statistical decisions were made for next directions of flight. These directions were based on diffuse scattering at the wall. The probability of success, as measured by the ratio of the number of molecules emerging to those starting, may be seen as a measure of the conductivity of the system. Haviland (1965) introduced the calculation of intermolecular collisions as an extension of the test-particle method in application to heat conduction at intermediate Knudsen numbers.

It was a natural step to move from these beginnings to the method of direct simulation of molecular dynamic processes using Monte Carlo methods (DSMC), a step that has most profoundly affected the landscape of rarefied-gas dynamics. The conceptual basis for the method and much of its operational detail are due to Professor Graeme Bird (1976). DSMC rests upon two critical assumptions. In the first of these Bird assumed that the calculation of collisional outcomes occurring within a very short period of time in a certain region of the flow may be decoupled from the positional changes of molecules occurring within the same period. It is also assumed that the outcomes of intermolecular collisions and molecule–wall collisions can be determined using statistical methods when appropriate physical parameters are given. A number of enabling assumptions are made. Molecules move within a field divided into a large number of cells, each containing 15 to 20 numbered particles; the motions and collisions of these are followed. Each particle represents a very large number of “real” gas molecules. The cell dimension is small as compared with the molecular mean free path, so that only small changes in gas properties can occur within the cell. The motion of the particle is followed through successive time intervals Δt_m, each of which is followed by calculations of collisional outcomes. The basic time interval is smaller than the mean collision time per molecule. Collision partners are selected from within the cell. Since the change in gas properties is small over cell dimensions, the exact coordinates of collision partners are immaterial, each partner being representative of molecules anywhere within the cell. A cell-by-cell record of densities and velocities is maintained for the later calculation of flow properties. The method of DSMC is developing swiftly. Numerous comparisons have been made between the results of experiment and the predictions of DSMC, leading now to general acceptance of molecular flow simulation as the mode of solution for rarefied-gas problems. Calculations incorporating chemical reactions and electron production are often made. Recent work, notably by Harvey (1989) and later by Boyd (1991) and others, has focused on the examination of phenomenological methods to guide calculation for inelastic collisions. As a result of these several advances, it would appear that some of the problems that had imposed an apparently impenetrable barrier to the advancement of understanding and application have been removed.

It is useful to consider again the determinations of temperature and density distributions within a shock wave. Experiments by Sherman (1955) and by Talbot and Sherman (1959) were conducted in the Berkeley tunnel using a free-molecule temperature probe. Argon was used in this work, in which the flow Mach numbers were low, always below 1.61. Experimental results were compared with Navier-Stokes solutions and also, as permitted, those of the Burnett equations. In addition comparisons were made with solutions to the 13-moment equations of Grad (1949) and with those from the two-temperature equations of Mott-Smith (1954). Experimental profiles of temperature compared well with density profiles from Navier-Stokes and with those from the Burnett equations (1935) at these low Mach numbers but not with those from the 13-moment solutions nor those from the two-temperature model. In subsequent work Schmidt (1969) used electron-beam fluorescence (EBF) to measure density profiles for Mach-8 flows of argon. At a later date remarkable agreement was found between Schmidt's experimental results and those from DSMC (Bird, 1976). Robben and Talbot (1965) using EBF-measured rotational temperature and density profiles for a shock wave of nitrogen at Mach 1.71. Again, results obtained by Bird (1976) using DSMC were found to be in excellent agreement with experiment. Much more recently, Lumpkin and Boyd (1993) examined flow properties for a detatched shock resulting from high–Mach-number air flow over a half-sphere. Comparisons of DSMC simulations with results of a Navier-Stokes–based computation agree well except in prediction of the flow angle at the rim, DSMC showing a much greater deflection of flow toward the central axis than given by Navier-Stokes. The Burnett equations did not yield stable solutions. Clearly, physical experiment is now required.

The foregoing material leads to the following important observations. Rarefied-gas experimentation can provide information concerning dynamic processes in high-speed flows. Where boundary conditions are simple, as, for example, for a plane shock held at the entrance to a shock holder, and the Knudsen number is small enough, Navier-Stokes solutions model the physical process fairly well. On the other hand, Monte Carlo simulations, which are also good in these cases, are not limited to low–Knudsen-number flows and accommodate more readily to complex geometries. With DSMC, as is also the case for Navier-Stokes, we are not always sure of the values of essential parameters or of the validity of physical assumptions. Physical experimentation is needed, to validate codes and to assist in the determination of essential parameters.

2.5 Gas–Surface Interactions

Collisions of gas molecules with walls occur in gas flows of any density, but at normal densities the frequency of collisions in the stream greatly exceeds that for collisions of gas molecules with solid surfaces. Under conditions of low Knudsen number, where knowledge of velocity slip and temperature jump is required, the determination of these quantities has been formulated in terms of a coefficient σ for velocity slip and α for temperature jump. In the context of the conventional formulation, σ is defined as the number of molecules diffusely reflected per unit time and surface area, $1 - \sigma$ is the number specularly reflected, while α is the standard thermal accommodation coefficient. Both α and σ are assumed equal to 1 in the absence of well-documented information.

At larger Knudsen numbers, wall–gas collisions become relatively more numerous until, in the limit of free-molecule flow, only wall–gas collisions remain. Throughout the rarefied regimes it is critically important to incorporate wall–gas collision data into computations. Where free-molecule assumptions can be made, a knowledge of the transfer coefficients, Eqs. (1a)–(1c), will permit the calculation of forces and heat transfer. In all other regimes it is essential to know the exit distributions of velocity, that is, to know the statistical description of postcollision velocities, in order to proceed in computation with DSMC.

In current conventional computions, exit distributions follow directly from the assumption that the scattering is diffuse or that it is specular. This nonphysical approximation is used in simulations of high-speed flows simply because there has been no generally accepted alternative in the aerospace context. Accordingly, it is useful to examine the understanding that has come to us from physical experiment and to look, too, at models for exit distributions that are currently being studied.

The first studies of thermal accommodation in wall–gas collisions were made using the loss of heat from a central wire through molecular transport to a cylindrical enclosure. The system could be freed of adsorbed gas molecules by baking under high vacuum. Much was learned from these experiments regarding the formation of adsorbed gas films, and values of thermal accommodation coefficients were obtained for gases of many sorts on various surfaces, both clean and with gas layers. The method was brought to its present level of development through the work over many years of Thomas (1981) and associates. Almost all gas–surface combinations exhibited incomplete thermal accommodation, i.e., $\alpha < 1$, for clean surfaces, the lowest values being found for helium in contact with clean tungsten. Values of α close to unity were found for complex molecules interacting with gas-covered surfaces. Values of α ranged from 0.0164 for helium on tungsten to 0.89–0.97 for air on etched aluminum. It seems safe to conclude that for many gas–surface combinations, some degree of incomplete thermal accommodation will be found.

Somewhat similar results were obtained in experimental studies of the tangential momentum-transfer coefficient. These studies were usually conducted in rotating-cylinder devices or by the measurement of flow through tubes. Values of σ were found to lie near to but often less than 1.0, although very infrequently, values have been found to exceed 1.0. There is no anomaly here; the result can be produced by backscattering due to roughness. Representative values for σ and for α are shown in Tables 1 and 2.

Table 1. Representative values of the tangential momentum-transfer coefficient σ, obtained from rotating-cylinder experiments.

Gas	Surface	σ	Ref.
Air	Machined brass	1.00	a
Air	Oil	0.90	a
Air	Glass	0.89	a
Air	Oil film on Al	0.90	b
N_2	Air on oil film on Al	0.8–0.93	c

Investigators: a. Millikan (1923). b. Chiang (1951). c. Hurlbut (1960).

Measurements using heat-conductivity cells or rotating cylinders, as in the foregoing, do not yield velocity or flux distributions of scattered molecules. On the other hand, from the earliest experiments using molecular-beam methods, it had been observed that distributions of scattered molecular flux were often of strikingly lobal form, although for many systems the flux patterns followed the cosine distribution to close approximation. In cosine scattering the flux of molecules scattered from the point of impingement of a collimated beam into a small element of solid angle above the surface is proportional to the cosine of θ, where θ is the angle from the surface normal to the center of the element of solid angle. Once the Kantrowitz–

Table 2. Representative values of the thermal accommodation coefficient α, obtained from measurement in thermal conductivity cells.

Gas	Surface	α	Ref.
N_2	Pt	0.50	a
N_2	W	0.35	a
Air	Machined bronze	0.89–0.93	b
Air	Machined cast iron	0.87–0.88	b
He	W 30 °C	0.0164	c
Ne	W 30 °C	0.0412	c
Ar	W 30 °C	0.271	c
Xe	W 30 °C	0.773	c
He	H_2 adsorbed on W	0.041	d
He	N_2 adsorbed on W	0.064	d
He	W, clean	0.02	d
CO_2	CO_2 adsorbed on W	0.990	d

Investigators: a. Oliver (1950). b. Wiedmann and Trumpler (1946). c. Silvernail (1954). d. Wachman (1957).

Gray sources came into common use, it was soon found that fast beams of heavier molecules could be produced by the expansion of a mixture of light and heavy species. The heavier species in beams skimmed from such mixture expansions might reach kinetic energies of one or more electron volts. The production of such "seeded" beams became standard in the art and is now used by molecular physicists and physical chemists in many research applications.

Molecular-beam methods have been used in a number of studies of the scattering of common gases from many kinds of surface including regular crystals, machined surfaces, and foils. The observations of Hinchin and Sheppard (1967) on the angular density distributions of argon scattered from platinum foil surfaces are represented in part in Fig. 2. One may see in these results certain patterns that are commonly observed and that merit discussion. It is first apparent that all distributions in this instance exhibit a strong lobality and show a remarkable regularity in the location of the lobe maxima. These positions are related to the beam reservoir temperature so that the most energetic beams scatter in patterns nearest to the surface tangent. Scattering patterns such as these, and in fact other scattering patterns that show a departure from the pure diffuse or cosine pattern, imply that the energy- and momentum-transfer coefficients also must depart from the value 1. The scattering from rough surfaces or those heavily laden with adsorbed gas layers is most often diffuse. One may summarize the results of much observation by noting that patterns of molecular scattering depend upon the nature of gas and surface, upon the surface smoothness, upon the cleanliness of the surface, and upon the energy of the gas molecule relative to the surface. Other factors may also be important if molecule internal energies are transferred or reactive events occur. Lobality, when present, tends to increase with surface cleanliness, with surface regularity, with incident velocity, and with a reduction in the angle of attack. Scattering patterns from suitably degassed and adequately crystalline surfaces are, with few exceptions, of distorted elliptical form having maxima near the exit specular angle.

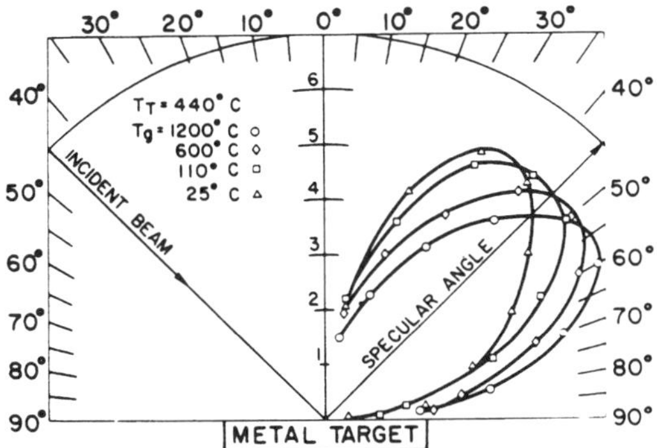

FIG. 2. Scattering of argon from degassed platinum surface (Hinchen and Shepherd, 1967).

Technologic change again intervened to make possible the measurement of distributions of velocity for the scattered molecular flux. Results from two of the relatively few studies completed are shown in Fig. 3. Both studies, Subbarao and Miller (1973) and Jih and Hurlbut (1977), employed argon beams of energy up to 1.3 eV on (111) surfaces of silver. Jih and Hurlbut observed exit velocities increasing from directions near to the surface normal to those near the surface tangent except for a pronounced dip in the region of the density peak. Subbarao and Miller, measuring only in the vicinity of the peak, observed a less pronounced dip. At higher energies, Tenner *et al.* (1986) found calculated energy distributions for 35-eV potassium ions scattered from the tungsten (110) surface to reproduce remarkably the observed experimental distributions in density and energy of exit particles. The reader is referred to surveys by Hurlbut (1986, 1989), for more detailed reviews of these and related observations.

We are also instructed by measurements of the transfer of momentum to common surfaces from high-speed beams of neutral molecules or of ions. Studies by Doughty and Schaetzle (1969) using neutral argon or nitrogen molecules of 25-eV energy on surfaces of aluminum or fresh varnish gave values for the normal momentum-transfer coefficient that decreased from values of 0.9 or smaller at normal incidence to values of 0.5 or smaller at incidence angles of 60° from the normal. Knechtel and Pitts (1969), using nitrogen ion beams incident on aluminum surfaces at 5 to 15 eV, observed coefficients for normal momentum transfer over the same range of incident angles that differed

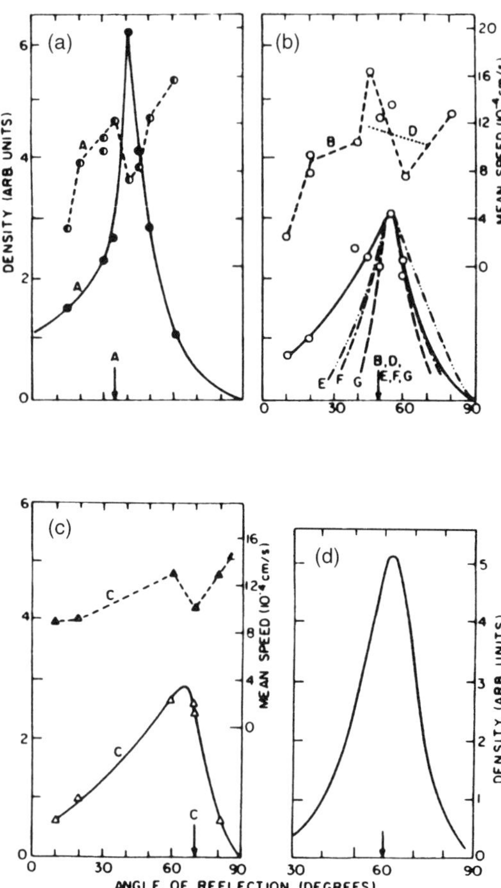

FIG. 3. Scattering of argon on silver (111) surfaces: (a)–(c) 0.8-eV incident argon; solid lines show density and broken lines show mean speed (Jih and Hurlbut, 1977). In (b), curve E is 1.23-eV argon, and curve F is 0.3-eV argon; both show density. Curve D, 0.3 eV, shows mean speed (Subbarao and Miller, 1973), and curve G, 0.8 eV, is from DSMC (Hurlbut, 1974); (d) Hays *et al.* (1972).

slightly in numerical value, as one might expect, but were qualitatively very similar to the observations of Doughty and Schaetzle. These slender fragments of data suggest very strongly that much more experimental observation is needed, some of it in the orbital environment, before one can predict with confidence in rarefied regimes.

Data from detailed measurements of velocity distributions for scattered particles must be parametrized or modeled in order to make them accessible for use in DSMC. Two general models are identified here as having merit, since each is based on its own substantive technical premise. The first derives from a conjecture due to Nocilla (1961) that lobal exit distributions could be modeled by use of suitable drifting Maxwellians consisting of a macroscopic velocity upon which a random thermal velocity is superimposed. The macroscopic velocity vector makes an angle θ_r with respect to the surface normal. Subsequently, molecular-beam experiments have shown the best representation of reflected distributions of velocity along observed directions to be the drifting Maxwellian. Hurlbut and Sherman (1968) used Nocilla's conjecture to develop a model in which parameters for the exit Maxwellian distribution are connected with those for the incident molecular flow. For a given system, experimental values of the thermal accommodation coefficient are required and a functional relation between the angle θ_i and the magnitude of the exit molecular speed ratio S_r is also required. The model is easily applied in DSMC calculations and has been used by a number of authors.

The second of these general models is based on the conjecture that the principle of reciprocity must apply between incident and exit velocity distributions in scattering processes if thermal equilibrium is to be preserved. It is further asserted that the same exit distributions found for a given set of incident conditions should correctly apply to the nonequilibrium flow. Detailed formulation and application to surface scattering appears in Cercignani and Lampis (1971). Formulas for flux distributions and for aerodynamic forces have been given in subsequent work. Both models have created substantial interest, but to the time of the present writing, neither is fully explored.

3. EXPERIMENTAL METHODS

3.1 Ground-Based Experimental Systems

Experimental systems used in the study of rarefied-gas aerodynamics and of related molecular processes are of four major types: wind tunnels, shock tunnels, molecular- and ion-beam systems, and ballistic ranges. There are also specialized systems developed for the study of momentum and energy transfers.

3.1.1 Wind Tunnels and Shock Tunnels Rarefied-gas wind tunnels are designed to produce low-density, high-speed flows of gas through a diagnostic region in which models may be placed for test. Low-density flows are frequently produced by oil-vapor booster pumps backed by lobe-type blowers that are exhausted to the atmosphere by rotary vacuum pumps. A few tunnels have been cryogenically pumped, some have been pumped by mechanical systems alone, and some have been pumped by steam-jet ejectors. All are characterized by offering an extended period of continuous testing. They are usually of large size and serve as the test bed for a wide variety of experiments. From the beginnings of orbital flight, however, it was acknowledged that limitations imposed on the magnitude of the stagnation enthalpy would make full flight simulation impossible. This notwithstanding, rarefied-gas wind tunnels are valuable research tools.

Experiments may be conducted over a wide range of flow conditions, permitting the comparison of flows in highly rarefied regimes with those in regimes of smaller Knudsen number. Long-duration flows are of particular advantage in studies involving the detection of low-level electrical or optical signals.

Many features of tunnel construction, modes of operation, instrumentation, force balance design, etc., may be found in Maslach and Schaaf (1959). Accounts of more recent design and practice may be found in the RGD Symposium Volumes 9 (1974) through 12 (1980).

Shock tunnels are, in essence, wind tunnels that derive reservoir supply gas at high temperatures from a shock tube and yield flows of substantially larger stagnation enthalpy than can be obtained in steady-state devices. While stagnation temperatures can reach several thousand kelvins, the flow duration is short, typically a few milliseconds.

3.1.2 Flow Production In continuous-flow wind tunnels and in shock tunnels, gas flows from the reservoir through a converging region and then emerges through a throat into a nozzle. Following expansion in the nozzle the flow passes into the test section. Alternatively, gas emerges from the throat directly into a test section of size sufficient to permit free-jet expansion. Such expansions contain a region of supersonic flow that is terminated by a plane shock and is bounded at its circumference by a shock barrel, as we see in the schematic, Fig. 4. Whereas the analysis of flow in nozzles is taken up in standard texts, the most satisfactory guide to free-jet expansions remains the original paper by Ashkenas and Sherman (1966).

It is useful to consider certain characteristic features of free-jet expansions. The gas, which is at sonic speed as it leaves the throat, rapidly expands to supersonic velocities along straight flow lines. These diverge from a single point of origin, as shown. Properties of the flow at any value of the nondimensional distance x/D may be fully characterized in terms of γ, the ratio of specific heats, the stagnation temperature T_0, and the ratio of the stagnation pressure p_0 to the downstream tunnel pressure p. As the Mach number increases with larger values of x/D, the local density and temperature of the flow decrease and the Knudsen number, based on the size of a diagnostic probe, can become large. Rarefied conditions are easily achieved, the flow properties can be calculated for any molecular species, and every region of the flow field is physically and optically accessible. For some purposes, these advantages are offset by the divergence of the flow.

3.1.3 Molecular and Ion Beams Molecular-beam systems are powerful tools for the study of molecular collision processes and for the study of gas collisions at the surface. A collimated stream of molecules traverses a region of high vacuum in which it can intersect a second beam or impinge on a solid test surface. Measurements of densities,

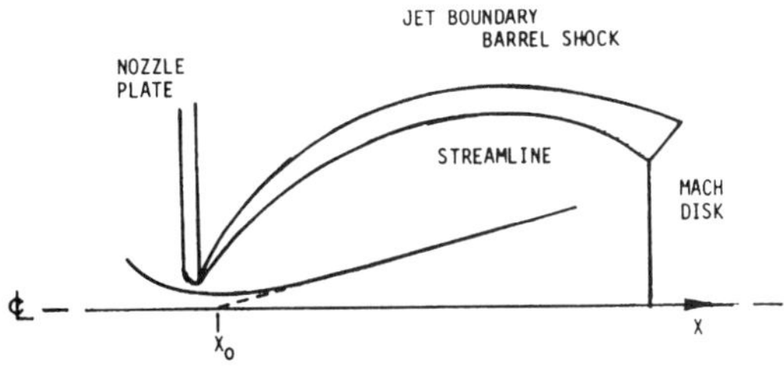

FIG. 4. Schematic diagram for free-jet expansion. (See Ashkenas and Sherman, 1966).

of composition, and of distributions of velocity and of electronic state may be obtained. In most current systems beams are formed by "skimming" from the rarefied region of a free-jet expansion, a method introduced by Kantrowitz and Gray (1951). The skimmer is simply a hollow conical shell with a small sharp-edged hole at the tip. Flow external to the skimmer is diverted to a large pump, while the skimmed beam passes into and through a high-vacuum region and thence to the target. Beam flux is higher than that from effusive sources, and because of the mechanics of expansion, the mean speed of the beam is higher and the thermal spread of velocities is much smaller. These features offer critical opportunities.

To reach beam energies above a very few electron volts it is necessary to accelerate ionized molecules. These may be used directly, or as neutral molecules following neutralization by charge exchange. Space charge limits the beam current to very low values at beam energies of the order of 10 eV. Unfortunately, energies of approximately this value are of particular interest to the experimenter interested in the effects of gases at the velocities encountered by satellites.

Beam systems are designed for operation under conditions of very high vacuum through careful attention to materials, welds, seals, and pumping equipment. A typical experiment might require scattered particles to be velocity selected, then ionized and accelerated into an electron multiplier detector. The reader is referred to the RGD Symposium proceedings for accounts of developments and applications of beams in rarefied-gas research.

3.2 Instruments and Diagnostic Methods

Wind tunnels and other experimental systems for the study of rarefied gases are, in essence, vacuum systems that must be equipped with the standard instrumentation for vacuum measurement. Balance systems are required that are suitable for the measurement of aerodynamic forces of perhaps a few hundredths of a newton. Advantage is often taken of the relatively large mean free path to position noninvasive physical probes in the field of flow. This point is illustrated by the pioneering measurement of shock-wave profiles by Sherman (1955) using a free-molecule thermocouple probe, and at an earlier time by Stalder *et al.* (1950), who measured the recovery temperature of a thermocouple probe in free-molecule flow.

Most of today's noninvasive measurement systems are based in some way upon the production and detection of light. The familiar methods of interferometry and of Schlieren flow visualization are not suitable in rarefied regimes because of insufficient optical depth. However, certain methods that employ the emission or absorption of light through the alteration of molecular electronic states have proved to be very useful. The most prominent of these methods has used an energetic electron beam to excite fluorescence radiation from well-defined regions of a flowing gas. A slender electron beam of 10- to 30-keV energy and a few milliamperes of current excites fluorescence radiation specific to the species and its state. Collection optics, focused on a portion of the electron beam, connect through a fiber optic to analyzing equipment. Detection is usually accomplished by photomultiplier and pulse counter or electrometer. Spectroscopic data from molecular gases, frequently N_2, lead to density measurement and to determinations of rotational temperatures and vibrational states. Atomic fluorescence radiation is used in density determinations, and, with the help of very high-resolution instruments such as the Fabry–Pérot interferometer, translational velocities and translational temperatures can be determined through the measurement of Doppler shifts. The method of electron-beam–induced fluorescence (EBF) in its various aspects is largely to be credited to its pioneering development by Muntz (1966) and to his continuing productive research.

During the past decade the quality and availability of tunable lasers have facilitated more general use of these systems in diagnostic applications. A valuable survey due to Lewis (1989) compares the utility of EBF, laser-induced fluorescence (LIF), Rayleigh scattering, and Raman scattering (RRS and VRS), and rates the relative difficulty of measurement over a wide range of gas densities. A bar graph from the Lewis paper is reproduced in Fig. 5. In the production of fluorescence, changes in electronic state are precisely determined by the controlled laser wavelength. Experience has shown that the laser beam is more easily focused into the

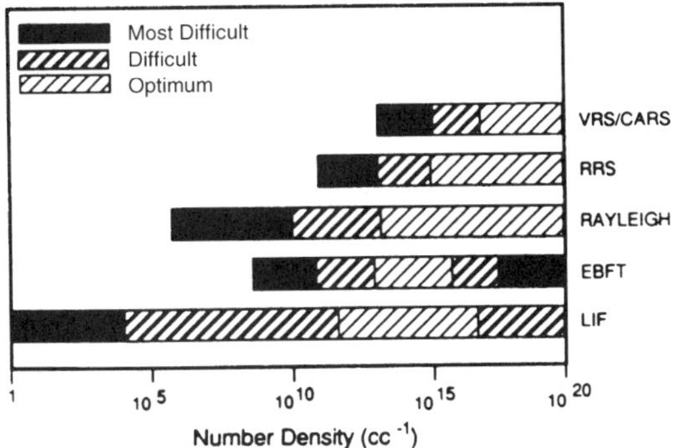

FIG. 5. Optical diagnostic comparisons (Lewis, 1989).

flow and is more stable in operation than the electron beam. Moreover, laser excitation avoids side effects due to secondary electrons, which compromise the effectiveness of EBF under some circumstances. One sees that the range of usable densities for LIF is much greater than for EBF. These advantages must be weighed against the potential for economical exploratory investigation using the electron beam as a broad-spectrum exciter.

3.3 Rocket and Satellite Experimental Systems

Once rockets were sufficiently developed to carry instruments to extreme altitudes, explorations of the thermosphere were conducted to measure atmospheric density and composition, charged-particle concentration, solar energy flux, and so on. These studies by sounding rocket have been of the greatest importance to the development of atmospheric models and to the understanding of atmospheric processes. Orbiting satellites, the next level of development, soon served as valuable experimental platforms. These vehicles may be instrumented for mass spectrometry, optical spectrometry, the detection of charged particles of various kinds, or the detection of electromagnetic fields, and with high-resolution systems for Earth viewing. Issues of astrophysical interest have largely displaced studies of the Earth's atmosphere or the gas dynamics of vehicle interactions with the flow, in part because of the altitude required for long-lived orbits. It should be noted, however, that the understanding of rarefied-gas dynamics always is important in the design of the satellite and its instrumentation.

In the late 1980s the tethered satellite became increasingly discussed as a potential research tool for atmospheric and aerodynamic studies. A research satellite, tethered to a parent satellite by a small-diameter line, may be deployed perhaps 100 km Earthward from the parent vehicle, where it will find an appropriate environment for the study of thermal and aerodynamic effects on the vehicle. The potential for aerodynamic testing using an orbiting "wind tunnel" is very attractive. By the date of this writing, two tethered satellites have been launched, the second demonstrating fully the feasibility. Dynamic and aerodynamic analysis are in progress. For further information relating to applications in space sciences and to related materials, the reader is referred to the *Journal of Geophysical Research* (see, e.g., Jacchia *et al.*, 1977, on atmospheric modeling) and to recent volumes of the *Journal of Spacecraft and Rockets*.

4. APPLICATIONS

The world's technologic life has been extraordinarily transformed through the effects of our knowledge of the behavior of rarefied gases. The influences are pervasive, touching our lives at every moment. Consider, for example, this short list of industrial processes that derive help from vacuum science and engineering: vacuum film deposition, chip manufacture, ion implantation, the production of vacuum electronic devices from x-ray sources to klystrons, diagnostic industrial control. One must also consider the essential aids to laboratory research, including electron microscopes, spectrometers, mass spectrographs, particle accelerators, and lasers.

Probably the greatest change of our times has come from the growth in our facilities for communication. This change, with its uncountable societal consequences, must be attributed to the rise of Sputnik and to the subsequent development of a vast communication umbrella based upon satellites in low Earth orbit. Every part of the technical and operational background for this enterprise rests upon aspects of rarefied-gas dynamics, just as will our continuing exploration of the Earth and its solar planetary environment.

At present much of the focus of aerospace work is on thermodynamic and thermochemical processes in very high-speed flows. This work is motivated by a need for more efficient space transportation. Specific emphasis is placed in two areas, the development of the aerospace plane and of the lifting aeropass and aerobraking shapes for reentry or orbit changes. The aerospace plane will depend upon airbreathing propulsion for much or all of its ascent and requires much new research before this concept of propulsion wherein the vehicle is an integral part of the propulsion system can be implemented. The aerobraking, lifting configurations also require much new research in order to resolve the many questions of chemistry, heat transfer, and wake flow that surround these new shapes at very high speeds and in unfamiliar atmospheres.

Both of the foregoing examples involve vehicles that will experience the full range of flow conditions from continuum to free molecule. There are in these efforts fascinating technical challenges; meeting these will result in new discoveries, and create new knowledge the impact of which cannot be predicted.

GLOSSARY

Cosine Law: Refers to molecular scattering in which the number flux of molecules scattered from a small element of surface into a small solid angle at θ with respect to the surface normal is proportional to the cosine of that angle.

Degassing: In vacuum systems, the process of removing adsorbed gas films on surfaces by heating to elevated temperatures.

Dilute Gases: A model of discrete particle gases in which only binary collisions occur.

Direct Simulation Monte Carlo Method (DSMC): A method for the simulation of the flow of rarefied gases in which the collisions and trajectories of representative particles are treated using statistical techniques.

Free-Jet Expansion: The expansion to supersonic speed of a gas issuing from an orifice into a suitably large region maintained at low pressure.

Free-Molecule Flow: Flow of very dilute gases in which the number of intermolecular collisions within or near to the system is negligibly small.

Knudsen Number: The ratio of the mean distance between molecular collisions to a characteristic length of the system. In external flows the characteristic length may be a diameter, but the choice is particular to the system.

Noninvasive Diagnostic Methods: Methods, usually optical, for determining the properties of flowing gases without altering the flow field.

Normal Momentum-Transfer Coefficient: Defined in Eq. (10c). Characterizes the effectiveness of the wall in absorbing the normal component of momentum carried to it by the incident molecular flux.

Phenomenological Methods: In the present context, statistical methods for selecting the transformation of excited states in molecular collisions from a range of alternatives.

Rarefied Gases: Dilute gases in which the ratio of the mean distance between molecular collisions to a characteristic length of the system lies between 0.01 and 100, approximately.

Specular and Diffuse Molecular Scattering: By specular scattering of molecules at surfaces it is implied that the full normal component of momentum is reversed and that the full component of tangential momentum is preserved. Diffuse scattering of molecules implies scattering according to the **Cosine Law.**

Tangential Momentum-Transfer Coefficient: Defined in Eq. (10b). Characterizes the effectiveness of the wall in absorbing the tangential component of momentum carried to it by the incident molecular flux.

Thermal Accommodation Coefficient: Defined in Eq. (10a). Characterizes the effectiveness of the wall in absorbing the thermal energy carried to it by the incident molecular flux.

Vehicle: Used in the aerospace context to refer to the entire rocket or satellite or other space system designed for flight in the upper atmosphere or beyond.

Works Cited

Alder, B. J., Wainwright, T. (1958), in: *Proceedings of the International Symposium on Transport Processes in Statistical Mechanics,* New York: Interscience, pp. 97–131.

Ashkenas, H., Sherman, F. S. (1966), "The Structure and Utilization of Supersonic Free Jets in Low Density Wind Tunnels," in: J. H. de Leeuw (Ed.), *Rarefied Gas Dynamics, Proceedings of the Fourth International Symposium,* New York: Academic Press, Vol 2, pp. 84–105.

Bhatnagar, P. L., Gross, E. P., Krook, M. (1954), "A Model for Collision Processes in Gases. I. Small Amplitude Processes in Charged and Neutral One-Component Systems." *Phys. Rev.* **94,** 511–525.

Bird, Graeme A. (1976), *Molecular Gas Dynamics,* Oxford: Clarendon Press.

Boltzmann, L. (1872), "Further Studies on the Thermal Equilibrium among Gas Molecules," *Wien. Ber.* **66,** 275.

Boyd, I. D. (1991), "Analysis of Vibrational-Translational Energy Transfer Using the Direct Simulation Monte Carlo Method," *Phys. Fluids A* **3,** 1785–1791.

Burnett, D. (1935), "The Distribution of Molecular Velocities and the Mean Motion in a Non-Uniform Gas," *Proc. London Math. Soc.* **40,** 382–400.

Cercignani, C., Lampis, M. (1971), "Kinetic Models for Gas-Surface Interactions," *Transport Theory Statist. Phys.* **1,** 101–114.

Chapman, S., Cowling, T. G. (1939), *The Mathematical Theory of Non-Uniform Gases,* 2nd ed, Cambridge, U.K.: Cambridge Univ. Press; (1952).

Chiang, S. F. (1951), Ph.D. Thesis, University of California, Berkeley.

Davis, D. H. (1960), "Monte Carlo Calculation of Molecular Flow through a Cylindrical Elbow and Pipes of Other Shapes," *J. Appl. Phys.* **31,** 1169–1176.

Devienne, F. M. (Ed.) (1960), *Rarefied Gas Dynamics; Proceedings of the First International Symposium,* Oxford: Pergamon.

Doughty, R. G., Schaetzle, W. J. (1969), "Experimental Determination of Momentum Accommodation Coefficients at Velocities up to and Exceeding Earth Escape Velocities," in: L. Trilling, H. Y. Wachman (Eds.), *Rarefied Gas Dynamics; Proceedings of the Sixth International Symposium,* New York: Academic, Vol. 2, pp. 1035–1054.

Dunoyer, L. (1911), *Compt. Rend.* **152,** 592–595.

Grad, H. (1949), "Lectures on the Kinetic Theory of Gases," *Commun. Pure Appl. Math.* **2,** 331–407.

Harvey, J. (1989), "Inelastic Collision Models for Monte Carlo Computation," in: E. P. Muntz, D. Weaver, D. Campbell (Eds.), *Rarefied Gas Dynamics; Proceedings of the 16th International Symposium,* Vol. 2, Progress in Astronautics and Aeronautics Vol. 117, Washington, DC: American Institute of Aeronautics and Astronautics, pp. 3–24.

Haviland, J. K. (1961), "Monte Carlo Application to Molecular Flows," MIT Dynamics Research Laboratory Report No. 61.5.

Hays, W. J., Rogers, W. E., Knuth, E. L. (1972), "Scattering of Argon Beams with Incident Energies up to 20 eV from a (111) Silver Surface," *J. Chem. Phys.* **56,** 1652–1657.

Hinchen, J. J., Shepherd, E. F. (1967), "Molecular Beam Scattering from Surfaces of Various Metals," in: C. L. Brundin (Ed.), *Rarefied Gas Dynamics; Proceedings of the 5th Symposium,* London: Academic, Vol. 1, pp. 239–252.

Hirschfelder, J. O., Curtis, C. F., Bird, R. B. (1954), *Molecular Theory of Gases and Liquids,* New York: Wiley.

Hurlbut, F. C. (1960), "The Influence of Pressure History on Momentum Transfer in Rarefied Gas Flows," *Phys. Fluids* **3,** 541–544.

Hurlbut, F. C. (1974), "Current Experiments and Open Questions in Gas-Surface Scattering," in: M. Becker, M. Fiebig (Eds.), *Rarefied Gas Dynamics; Proceedings of the 9th Symposium,* Porz-Wahn, Germany: DFVLR-Press, Vol. 2, pp. E-1–E-23.

Hurlbut, F. C. (1986), "Gas/Surface Scattering Models for Satellite Applications," in: J. N. Moss, C. D. Scott (Eds.), *Thermophysical Aspects of Re-Entry Flows,* Progress in Astronautics and Aeronautics Vol. 103, New York: American Institute of Aeronautics and Astronautics, pp. 97–119.

Hurlbut, F. C. (1989), "Particle Surface Interactions in the Orbital Context," in: E. P. Muntz, D. Weaver, D. Campbell (Eds.), *Rarefied Gas Dynamics; Proceedings of the 16th International Symposium,* Vol. 1, Progress in Astronautics and Aeronautics Vol. 116, Washington, DC: American Institute of Aeronautics and Astronautics, pp. 419–450.

Hurlbut, F. C., Sherman, F. S. (1968), "Applications of the Nocilla Wall Reflection Model to Free-Molecule Kinetic Theory," *Phys. Fluids* **11,** 486–496.

Jacchia, L. G., Slowey, J. W., Von Zahn, V. (1977), "Temperature, Density and Composition in the Disturbed Thermosphere from ESRO 4 Gas-Analyzer Measurements: A Global Model," *J. Geophys. Res.* **22,** 684–688.

Jih, C. T. R., Hurlbut, F. C. (1977), "Time of Flight Studies of Argon Beams Scattered from a Silver (111) Crystal Surface," in: L. Potter (Ed.), *Rarefied Gas Dynamics; Proceedings of the 10th International Symposium,* Progress in Astronautics and Aeronautics Vol. 51, New York: American Institute of Aeronautics and Astronautics, Part II, pp. 539–554.

Kantrowitz, A., Gray, J. (1951), "A High Intensity Source for the Molecular Beam, Part I," *Rev. Sci. Instrum.* **22,** 328–332.

Knechtel, E. D., Pitts, W. C. (1969), "Experimental Momentum Accommodation on Metal Surfaces of Ions near and above Earth-Satellite Speeds," in: L. Trilling, H. Y. Wachman (Eds.), *Rarefied Gas Dynamics; Proceedings of the 6th International Symposium,* Vol. 2, New York: Academic, pp. 1257–1266.

Knudsen, M. (1911), "Die Moleculäre Wärmleitungen der Gase und der Akkommodations-koeffizient," *Ann. Phys. (Leipzig)* **31,** 539–XXX.

Lewis, J. W. L. (1989), "Optical Diagnostics of Low Density Flow Fields," in: E. P. Muntz, D. Weaver, D. Campbell (Eds.), *Rarefied Gas Dynamics; Proceedings of the 16th International Symposium,* Vol. 1, Progress in Astronautics and Aeronautics Vol. 116, Washington, DC: American Institute of Aeronautics and Astronautics, pp. 107–132.

Lumpkin, F. E., Boyd, I. D., Venkatapathy, E. (1993), "Comparison of Continuum and Particle Simulations of Expanding Rarefied Flows," presented at 31st Aerospace Sciences Meeting, 11–14 January, AIAA Paper 93-0728.

Maslach, G. J., Schaaf, S. A. (1959), "Slip Flow Testing Techniques," *AGARDograph* 39, July.

Maxwell, J. C. (1867), "On the Dynamical Theory of Gases," *Philos. Trans. R. Soc.* **157,** 49. Reprinted in: W. D. Niven (Ed.), *The Scientific Papers of James Clerk Maxwell,* Vol. 2, Cambridge, U.K.: Cambridge Univ. Press, pp. 26–78.

Millikan, R. A. (1923), "Coefficients of Slip in Gases and the Law of Reflection of Molecules from Surfaces of Liquids and Solids," *Phys. Rev.* **20,** 217–238.

Mott-Smith, H. M. (1954), "A New Approach to the Kinetic Theory of Gases," MIT Lincoln Laboratory Group Report No. V-2.

Muntz, E. P. (1966), "The Direct Measurement of Velocity Distribution Functions," in: J. H. de Leeuw (Ed.), *Rarefied Gas Dynamics; Proceedings of the 4th International Symposium,* Vol. 2, New York: Academic, pp. 128–150.

Nocilla, S. (1961), "On the Interaction between Stream and Body in Free-Molecule Flow," in: L. Talbot (Ed.), *Rarefied Gas Dynamics; Proceedings of the 2nd International Symposium,* New York: Academic, pp. 169–208.

Oliver, R. N., Farber, M. (1950), Ph.D. Thesis, California Institute of Technology.

Robben, F., Talbot, L. (1965), "Measurement of Rotational Temperatures in Non-equilibrium Free-Jet Expansions of Nitrogen," *Phys. Fluids* **9,** 644–652.

Schaaf, S. A. (1963), "Mechanics of Rarefied Gases," in: S. Flügge (Ed.), *Encyclopedia of Physics,* Vol. VIII/2, Berlin: Springer, pp. 591–624.

Schaaf, S. A., Chambré, P. L. (1958), in: H. W. Emmons (Ed.), *Fundamentals of Gas Dynamics,* High Speed Aerodynamics and Jet Propulsion Series Vol. 3, Princeton, NJ: Princeton Univ. Press, pp. 687–741.

Schmidt, B. (1969), "Electron Beam Density Measurement in Shock Waves in Argon," *J. Fluid Mech.* **39,** Pt. 2, 361–373.

Sherman, F. S. (1955), "A Low Density Wind Tunnel Study of Shock Wave Structure and Relaxation Phenomena in Gases," NACA Technical Note 3298, Washington, DC: U.S. GPO.

Silvernail, W. L. (1954), Ph.D. Thesis, University of Missouri.

Stalder, J. R., Goodwin, G., Creager, M. O. (1950), "A Comparison of Theory and Experiment for High Speed Free-Molecule Flow," NACA Ames Aeronautical Laboratory Technical Note No. 2244.

Stern, O. (1920), "Eine direkte Messung der Thermischen Molekulargeschwindigkeit," *Z. Phys.* **2,** 49–56.

Subbarao, R. B., Miller, D. R. (1973), "Velocity Distribution Measurements of 0.06-1.4 eV Argon and Neon Atoms Scattered from the (111) Plane of a Silver Crystal," *J. Chem. Phys.* **58,** 5247–5257.

Talbot, L., Sherman, F. S. (1959), "Structure of Weak Shock Waves in a Monatomic Gas," NASA Memo 12-14-58W.

Tenner, A. D., Gillen, K. T., Horn, T. G. M., Los, J., Kleyn, A. W. (1986), "Energy and Angular Distribution for Scattering of K^+ from W (110) at Normal Incidence," *Surf. Sci.* **172,** 90–120.

Thomas, L. B. (1981), "Accommodation of Molecules on Controlled Surfaces—Experimental Developments at the University of Missouri, 1940–1980," in: S. Fisher (Ed.), *Rarefied Gas Dynamics; Proceedings of the 12th International Symposium,* Progress in Astronautics and Aeronautics Vol. 74, New York: American Institute of Aeronautics and Astronautics, Part I, pp. 83–108.

Vincenti, W. G., Kruger, C. H. (1965), *Introduction to Physical Gas Dynamics,* New York: Wiley.

von Smoluchowski, M. (1911), "Zur Theorie der Wärmeleitung in verfeinten Gasen und der dabei auftretenden Druckkräfte," *Ann. Phys. (Leipzig)* **35,** 983–1004.

Wachmann, Y. H. (1957), Ph.D. Thesis, University of Missouri.

Wiedmann, M. L., Trumpler, P. R. (1946), *Trans. Am. Soc. Mech. Eng.* **68,** 57–64.

Further Reading

Volumes of the International Symposia on Rarefied Gas Dynamics

Devienne, F. M. (Ed.) (1960), *Rarefied Gas Dynamics; Proceedings of the 1st International Symposium,* Paris: Pergamon.

Talbot, L. (Ed.) (1961), *Rarefied Gas Dynamics; Proceedings of the 2nd International Symposium,* New York: Academic.

Laurman, J. A. (1963), *Rarefied Gas Dynamics; Proceedings of the 3rd International Symposium,* Vols. 1 and 2, New York: Academic.

de Leeuw, J. H. (Ed.) (1965), *Rarefied Gas Dynamics; Proceedings of the 4th International Symposium,* Vols. 1 and 2, New York: Academic.

Brundin, C. L. (Ed.) (1967), *Rarefied Gas Dynamics; Proceedings of the 5th International Symposium,* Vols. 1 and 2, London: Academic.

Trilling, L., Wachman, H. Y. (1969), *Rarefied Gas Dynamics; Proceedings of the 6th International Symposium,* New York: Academic.

Dini, D., Cercignani, C., Nocilla, S. (Eds.)

(1971), *Rarefied Gas Dynamics; Proceedings of the 7th International Symposium,* Pisa: Editrice Tecnico Scientifica.

Karamcheti, K. (Ed.) (1974), *Rarefied Gas Dynamics; Proceedings of the 8th International Symposium,* New York: Academic.

Becker, M., Fiebig, M. (Eds.) (1974), *Rarefied Gas Dynamics; Proceedings of the 9th International Symposium,* Porz-Wahn, Germany: DFVLR-Press.

Potter, L. (Ed.) (1977), *Rarefied Gas Dynamics; Proceedings of the 10th International Symposium,* Progress in Astronautics and Aeronautics Vol. 51, Parts I and II, New York: American Institute of Aeronautics and Astronautics.

Campargue, R. (Ed.) (1979), *Rarefied Gas Dynamics; Proceedings of the 11th International Symposium,* Vols. 1 and 2, Paris: Commissariat à l'Energie Atomique.

Fisher, S. (Ed.) (1981), *Rarefied Gas Dynamics; Proceedings of the 12th International Symposium,* Progress in Astronautics and Aeronautics Vol. 74, Parts I and II, New York: American Institute of Aeronautics and Astronautics.

Belotserkovskii, O. M., Kogan, M. N., Kutateladze, C. S., Rebrov, A. K. (Eds.) (1985), *Rarefied Gas Dynamics; Proceedings of the 13th International Symposium,* New York: Plenum.

Ogucho, H. (Ed.) (1984), *Rarefied Gas Dynamics; Proceedings of the 14th International Symposium,* Tokyo: Univ. of Tokyo Press.

Boffi, V., Cercignani, C. (Eds.) (1986), *Rarefied Gas Dynamics; Proceedings of the 15th International Symposium,* Stuttgart: B. G. Teubner.

Muntz, E. P., Weaver, D., Campbell, D. (Eds.) (1989), *Rarefied Gas Dynamics; Proceedings of the 16th International Symposium,* Progress in Astronautics and Aeronautics Vols. 116, 117, and 118, Washington, DC: American Institute of Aeronautics and Astronautics.

Beylich, A. E. (Ed.) (1991), *Rarefied Gas Dynamics; Proceedings of the 17th International Symposium,* Weinheim: VCH.

Shizgal, B. (Ed.) (1994), *Rarefied Gas Dynamics; Proceedings of the 18th International Symposium,* Progress in Astronautics and Aeronautics Vols. 158–160, Washington, DC: American Institute of Aeronautics and Astronautics.

REACTIONS INDUCED BY LIGHT

KENNETH L. STEVENSON, *Department of Chemistry, Indiana University-Purdue University Fort Wayne, Fort Wayne, Indiana, U.S.A.*

OTTÓ HORVÁTH, *Department of General and Inorganic Chemistry, Veszprém University, Veszprém, Hungary*

	Introduction	117
1.	**Photochemical Laws**	118
1.1	Light and Energy	118
1.2	Light-Absorption Principles	118
1.3	Quantum Yields	119
2.	**Fundamental Principles of Photophysics and Photochemistry**	120
2.1	Quantum-Mechanical Principles of Absorption	120
2.2	Energy Dissipation Pathways	122
2.3	Photophysical Pathways: Luminescence and Nonradiative Decay	122
2.4	Photochemical Pathways	124
3.	**Types of Photochemical Experiments**	125
3.1	Continuous (cw) Irradiation	125
3.1.1	Preparative Photochemistry	125
3.1.2	Quantum-Yield Determination and Product Identification	125
3.2	Laser Flash Photolysis	126
3.3	Ultrafast Intermediates	127
3.4	Laser Spectroscopy	128
4.	**Photochemistry of Organic Compounds**	130
4.1	Electronic Transitions and Spectra of Organic Molecules	130
4.2	Electronic Energy Transfer	131
4.3	Reactions of Excited Species	131
4.3.1	Reactivity of Excited Species	131
4.3.2	Correlation Rules and Symmetry Conservation	132
4.3.3.	Intramolecular Processes	132
4.3.3.1	Primary Dissociation Processes	132
4.3.3.2	Photoisomerization	133
4.3.3.3	Hydrogen Abstraction	134
4.3.4	Intermolecular Processes	134
4.3.4.1	Hydrogen Abstraction	134
4.3.4.2	Addition Reactions	134
5.	**Photochemistry of Inorganic and Coordination Compounds**	136
5.1	Spectra and Photochemistry	136
5.1.1	Ligand-Field (LF) Transitions	136
5.1.2	Ligand-to-Metal (LMCT) and Metal-to-Ligand (MLCT) Charge-Transfer Transitions	138
5.1.3	Charge-Transfer-to-Solvent (CTTS) Transitions	139
5.1.4	Other Charge-Transfer Transitions	139
6.	**Applications of Photochemistry**	140
6.1	Photography	140
6.2	Photopolymerization	140
6.3	Photoimaging	142
6.4	Phototherapy	142
6.5	Photochromism	142
6.6	Photosynthesis	143
6.7	Photochemistry of Vision	144
7.	**Acknowledgments**	145
	Glossary	145
	Works Cited	146
	Further Reading	147

INTRODUCTION

The study of chemical reactions induced by light is usually termed photochemistry. It is concerned specifically with the ways that light can bring about changes in materials. It involves a wide variety of topics, to be found in other articles, such as CHEMICAL REACTIONS; QUANTUM MECHANICS; MOLECULAR SPECTROSCOPY; LASER PHOTOCHEMISTRY;

Encyclopedia of Applied Physics, Vol. 16

3-527-28138-X/96/$5.00 + .50

CHEMICAL KINETICS; and RADIATION INTERACTIONS WITH MOLECULES.

The interaction of light with matter is of fundamental importance. Indeed, all of the energy we derive from fossil fuels was accumulated over vast periods of time by photochemical processes that converted energy from the sun into chemical energy (see SOLAR ENERGY). But not until the late nineteenth century did the study of photochemistry begin in earnest. At that time a number of simple organic reactions, such as dimerizations, reductions, cycloadditions, decompositions, and isomerizations, that could be promoted by the action of sunlight were observed. With the development of high-intensity nonsolar light sources in the early part of this century, the quantitative and systematic study of photochemistry began. Gas-phase photochemistry became a common area of study in the 1930s and 1940s; in the early 1940s the importance of the triplet state in organic chemistry was confirmed; and in the 1950s spectroscopic correlations, theoretical progress, and the development of flash irradiation were important advances. During the 1960s a great number of photochemical reactions were discovered and probed with the rapidly expanding list of new instruments and techniques. Through the 1970s, 1980s, and into the 1990s, the development of instrumentation and theory, coupled with the strongly perceived need to change the world energy economy from one that depletes nonrenewable resources to one that relies on a virtually inexhaustible energy source such as solar energy, placed photochemistry into the realm of highly interesting and technologically significant branches of science.

The discussion of photochemistry has been organized such that the basic theory of the physics of light absorption is presented first, followed by a description of types of experiments that can be performed, sections on organic photochemistry and inorganic photochemistry, and finally a review of some of the more important applications of photochemistry.

1. PHOTOCHEMICAL LAWS

Changes in chemical structure caused by light are the subject of *photochemistry,* whereas light-induced changes in the ways electrons are distributed within a chemical entity while the atomic nuclei remain intact are the realm of *photophysics.* In order to understand these processes, it is necessary to describe some of the properties of light (see ELECTROMAGNETIC RADIATION).

1.1 Light and Energy

On the one hand, it is convenient to think of light as an electromagnetic wave, summarized by the expression

$$\lambda = c/\nu \tag{1}$$

relating the wavelength λ to the frequency ν through the speed of propagation c. On the other hand, light consists of quanta, or photons, each of which has energy given by

$$\mathscr{E} = h\nu, \tag{2}$$

where h, known as Planck's constant, can be interpreted as the proportionality constant between the energy and frequency of the wave associated with the photon.

When chemical systems absorb light, individual molecules or atoms absorb only those photons with the right amount of energy to effect an excitation of the molecule from a lower to a higher energy level, if such a transition is allowed. Energy is distributed in molecules in the various levels of molecular motions, i.e., translation, rotation, vibration, electronic, etc., and the kind of electromagnetic radiation absorbed will depend on the kind of motion that is affected. Table 1 (Hollenberg, 1970) outlines the relationship between light energy and the kinds of molecular-energy transitions that can give rise to absorption.

1.2 Light-Absorption Principles

The first law of photochemistry from Grotthus (1817) and Draper (1843) states that only the light that is absorbed by a molecule can produce photochemical change within the molecule. The amount of light absorbed in a homogeneous medium is defined by the Beer–Lambert law:

$$I = I_0 10^{-\epsilon c l}, \tag{3}$$

Table 1. Light energies and molecular-energy transitions.

Type	Wavelength	Energy (kJ/mol)	Kind of transition
Microwave	33–0.33 cm	0.0004–0.04	Rotation of heavy molecules
Far IR	0.33–0.0033 cm	0.04–4	Rotation of light molecules and heavy-atom vibrations
Infrared	33–3.3 μ	4–40	Most vibrational transitions
Near IR	3300–1000 nm	40–120	Light-atom vibrations and small electronic transitions
Visible	1000–360 nm	120–330	Lower-energy electronic transitions (e.g., $d \rightarrow d$, $\pi \rightarrow \pi^*$, charge transfer)
UV	360–200 nm	330–600	Higher-energy electronic transitions (e.g., $d \rightarrow s$, charge transfer)
Far UV	200–150 nm	600–800	Bond-breaking electronic transitions

where I_0 and I are the light intensities at the front of the absorber, and at distance l within the absorbing medium, respectively; and ϵ is the *extinction coefficient* in units consistent with the path length l (usually in centimeters) and the *concentration c* (usually in mol L^{-1}).

If one is to do quantitative continuous photochemical studies, it is important that an accurate value of the light absorbed by a photoactive sample can be measured, since this determines the rate of the reaction. From Eq. (3) one can show that the *absorbed light intensity,* or *radiant flux J,* is given by

$$J = I_0(1 - 10^{-\epsilon cl}). \qquad (4)$$

Photochemists generally measure the incident light intensity I_0 falling on a sample, and then use this equation to calculate J.

The accurate measurement of light intensities, or actinometry, is very important in determining the photokinetics of the system. One convenient method for doing this is the irradiation of a chemical actinometer that has been calibrated for the wavelength of the photoreaction under study. Almost any photoreaction system that has a well-defined quantum yield at the irradiating wavelength can be used as an actinometer, but for continuous irradiation work in the wavelength range 250 to 550 nm, the potassium trioxalatoferrate (III) system, developed by Hatchard and Parker (1956), is the most widely used and reliable actinometer available. For longer wavelengths to 750 nm, the photoaquation (i.e., the photoinduced substitution of water molecules into the coordination sphere) of aqueous Reinecke's salt, $KCr(NH_3)_2(NCS)_4$, works quite well (Wegner and Adamson, 1966). The unit used by photochemists for J and I_0 in the equations above is *einstein* s^{-1}, where an einstein is one mole or 6.02×10^{23} photons. Occasionally this is expressed per unit volume of reactant, such as E s^{-1} L^{-1}, analogous to the method for expressing rates of reactions.

1.3 Quantum Yields

The quantum yield or quantum efficiency of a photochemical reaction is normally defined as the fraction of photoexcited molecules that undergo some chemical change. If there are a number of parallel pathways resulting in different products, then each pathway will have its own quantum yield, ϕ. Sometimes the primary step in a photoreaction is the initiator of subsequent thermal reactions, resulting in quantum yields that may exceed unity. The second law of photochemistry states that *the absorption of light by a molecule is a one-quantum process, so that the sum of the primary process quantum yields* ϕ *must be unity, i.e.,* $\Sigma\phi_i = 1$, *where* ϕ_i *is the quantum yield of each of the primary processes,* which may include but are not limited to dissociation, isomerization, charge separation, luminescence (see PHOTOLUMINESCENCE), or radiationless deactivation.

For a reaction

$$A \xrightarrow{h\nu} B + C + \cdots, \qquad (5)$$

the rate can be expressed as

$$-\frac{d[A]}{dt} = \phi_A J_A, \qquad (6)$$

where $[A]$ is the concentration of photoactive

species. One thus determines quantum yields by measuring the ratio of the reaction rate to the absorbed light intensity J_A.

2. FUNDAMENTAL PRINCIPLES OF PHOTOPHYSICS AND PHOTOCHEMISTRY

The initial event in any photochemical or photophysical process is the absorption of a photon by an atom, molecule, or complex ion, resulting in an electronically excited state. What happens after this event determines whether or not a photochemical reaction will occur, or if some other photophysical process occurs such as fluorescence, phosphorescence, or the simple conversion of the light energy to heat. The following discussion describes the fundamental principles governing these processes.

2.1 Quantum-Mechanical Principles of Absorption

If A and A^* represent the ground and excited states, respectively, the process can be written as the reaction

$$A \xrightarrow{h\nu} A^*. \tag{7}$$

In order for the absorption to occur, two criteria must be met by the absorbing species:

1. the difference between the energy of the excited state, E_2, and the energy of the ground state, E_1, must be exactly equal to the energy of the photon, i.e.,

$$h\nu = E_2 - E_1, \tag{8}$$

and

2. there must be a finite probability for the transition, as defined by the electronic structure of the molecule.

The Einstein transition probability for absorption, B_{12}, is given by the expression

$$B_{12} = \frac{8\pi^3}{3h^2c} g_2(\mu_{12})^2, \tag{9}$$

where g_2 is the degeneracy of the excited state and μ_{12} is the transition moment given by the integral

$$\mu_{12} = \int \Psi_1(e\,\Sigma_i\, r_i)\Psi_2 dq. \tag{10}$$

In this equation, Ψ_1 and Ψ_2 are the wave functions of the ground and excited states, respectively; the sum is over all of the dipole-moment vectors r_i of the electrons; e is the charge on the electron; and the integral is taken over the three-dimensional coordinates q of the molecule.

While the exact solution of Eq. (10) is impossible for anything but the hydrogen atom, approximate wave functions can be used to derive for various transitions qualitative selection rules that reveal when the transition moment μ_{12} is nonzero and thus results in an allowed transition, as opposed to one that is forbidden.

Two selection rules are particularly important and relate directly to photochemical processes: the *spin-conservation rule* and the *Laporte rule.* For systems containing atoms with relatively few electrons (i.e., light atoms) a reliable approximation allows for the separation of the spin wave functions, S_1 and S_2, from the total wave functions Ψ_1 and Ψ_2, such that the transition moment can be written as the product of two integrals,

$$\mu_{12} = \int \phi_1(e\,\Sigma_i\, r_i)\phi_2 dq \times \int S_1S_2 dq, \tag{11}$$

where ϕ_1 and ϕ_2 are the orbital wave functions of ground and excited states, respectively. The spin part of the transition moment vanishes when $\Delta S \neq 0$, i.e., when the total spin quantum numbers, or spin multiplicities, of the excited and ground states are not the same, and the transition is said to be "spin-forbidden." The Laporte rule applies to species with a center of symmetry, such as octahedral complexes, and requires that transitions between orbitals must be either u → g or g → u (g = *gerade* or even, and u = *ungerade* or uneven, i.e., symmetric and antisymmetric, respectively, for inversion of the orbital about the center of symmetry), but not g → g or u → u, such transitions being "parity forbidden." The methods of group theory are often necessary in applying the Laporte rule.

In practice, not only allowed transitions,

but sometimes forbidden transitions may be observed because of subtle perturbations in the system, such as spin–orbit coupling, not taken into account in the approximation procedures used to derive the selection rules. The allowed transitions, of course, are much stronger, as indicated by the bandwidth and extinction coefficient of the absorption band. The classical oscillator strength f of the band is related to the Einstein absorption probability by the expression

$$f = \left(\frac{m_e hc^2 \omega_{12}}{\pi e^2}\right) B_{12}, \quad (12)$$

where m_e is the mass of the electron and ω_{12} is the wave number of the transition in cm^{-1}. Experimentally, the oscillator strength can be determined from the integral of the absorption band, i.e.,

$$f = 4.32 \times 10^{-9} F \int_{\omega 1}^{\omega 2} \epsilon d\omega, \quad (13)$$

where the integration is over all frequencies from the onset to the disappearance of the band in question and the value of F is related to the index of refraction n through $F = 9n/(n^2 + 2)^2$. As a very crude approximation, the oscillator strength can be estimated from

$$f \cong 4.32 \times 10^{-9} F \epsilon_{max} \Delta\omega_{1/2}, \quad (14)$$

where $\Delta\omega_{1/2}$ is the width of the band when the extinction coefficient is $\epsilon = \frac{1}{2}\epsilon_{max}$ (half-bandwidth). Since the oscillator strength for a single electron in a fully allowed transition should be unity, the highest value of ϵ_{max} one should observe is about 5×10^4 L mol^{-1} cm^{-1}, for a half-bandwidth of 5000 cm^{-1}. Parity- and spin-forbidden transitions have much smaller oscillator strengths, resulting in values of ϵ on the order of 10^2 or less, but they are easily measured with the crudest of spectrometers. Thus, measurement of the oscillator strengths of various bands observed in a spectrum can be of help in making the correct assignments to the transitions.

The foregoing discussion has been based on the assumption that nuclear motions in ordinary molecular vibrations are so slow that they do not affect the electronic energy levels of the molecule, an approximation known as the Born–Oppenheimer principle. This means that the total wave function for the molecule can be separated into the product of an electronic part and a nuclear part,

$$\Psi_{total} = \Psi_{elec} \times \Psi_{nucl}, \quad (15)$$

and each part treated independently. This permits the construction of the familiar energy diagrams of diatomic molecules that show the electronic potential energy as a function of internuclear distance, a general example of which is shown in Fig. 1. From a purely classical viewpoint, one would expect the molecule to reside at the exact bottom of either well, but in fact this is not the case because it would imply that the atoms would be stationary, i.e., not vibrating. The quantum-mechanical treatment of the simple harmonic oscillator (which approximates a diatomic molecule to the extent that the bottom parts of the potential wells are approximated

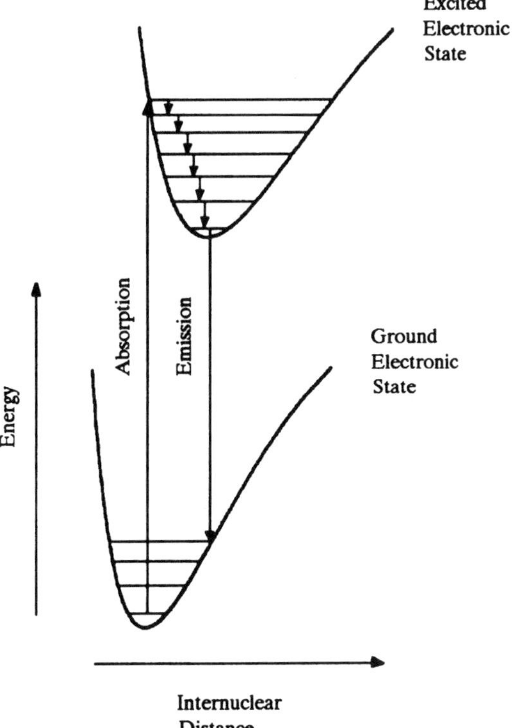

FIG. 1. Typical ground and excited electronic state potential-energy wells for a molecule, showing vibrational relaxation and emission.

by a parabola) requires that the zero-point energy of the vibrating molecule must be $\frac{1}{2}h\nu_0$ above the minimum [a result of the uncertainty principle (see QUANTUM MECHANICS), which does not allow both the position and momentum to be completely defined as would be the case if the zero-point energy were 0] where ν_0 is the fundamental frequency of vibration in this ground state. Several vibrational levels for each electronic state are represented in the wells by horizontal lines, which show not only the value of vibrational energy but also the turning points, i.e., limits of compression and extension of the atoms during the vibrations.

Figure 1 shows what happens during an electronic excitation by absorption of light. Because nuclear motions are vastly slower than electronic transitions, the latter occur vertically, i.e., the excited molecule has the same internuclear separation as the ground-state molecule, a rule known as the Franck–Condon principle. At temperatures in the vicinity of 25 °C most molecules in a thermally equilibrated system are in the ground vibrational state as a result of the Boltzmann distribution law (see STATISTICAL MECHANICS, CLASSICAL); therefore, nearly all of the electronic transitions will arise out of this ground vibrational state, as shown in the diagram. However, because the excited electronic potential well is wider and somewhat displaced to the right as a consequence of having a weaker bond than in the ground electronic state, the vertical transitions end in excited vibrational levels in the upper well. Because of the necessity for optimum overlap of the initial and final vibrational wave functions in the transitions, those with highest probability occur between vibrational levels in which the molecule is near a turning point in its vibration (except for lowest energy vibrations of an electronic state where this highest probability occurs at the center, or equilibrium internuclear distance). This results in a series of transitions to excited vibrational levels in the upper well, giving rise to a set of closely spaced lines in the absorption spectrum of a gas at low pressure, or to a broad absorption band in a condensed medium in which the lines are broadened and overlap.

In polyatomic systems there are numerous vibrational modes and internuclear distances, requiring a multidimensional analog of Fig. 1 to describe the energy states. Nevertheless, one can still imagine minimum electronic potential-energy wells with superimposed vibrational levels in the hyperspace of n dimensions, where n is the number of atoms in the system. Thus the simple principles of the energetics of diatomic molecules are routinely conceptually applied to more complex systems.

For a more thorough treatment of this topic see QUANTUM MECHANICS and MOLECULAR SPECTROSCOPY.

2.2 Energy Dissipation Pathways

What happens to the photoactive species after initial electronic excitation by absorption determines the eventual outcome of the photophysical or photochemical process. Since the excited system has a large number of its molecules in excited vibrational levels, it usually undergoes a very rapid thermal equilibration with its surroundings, resulting in a Boltzmann population distribution, which means that most molecules drop back to the lowest vibrational level of the excited electronic state, a process called *vibrational relaxation* and shown by the short, downward arrows in Fig. 1.

2.3 Photophysical Pathways: Luminescence and Nonradiative Decay

After the vibrational relaxation of the excited electronic state, the system may return to the ground-state potential well by spontaneously emitting a photon, also shown in Fig. 1. Note, however, that in this case, since the excited-state potential well is shifted somewhat to the right, the most likely electronic transitions are from the lowest vibrational level in the excited electronic state to vibrationally excited states in the ground potential well. This gives rise to an emission band, said to be Stokes-shifted, which is at lower energies than, and almost an exact mirror image of, the absorption band. This form of luminescence is often referred to as *fluorescence* (see PHOTOLUMINESCENCE), a term that is generally applied to emissions of very short lifetimes between singlet energy levels. From quantum mechanics, the Einstein probability for spontaneous emission,

A_{21}, is related to the oscillator strength of the absorption band, through

$$A_{21} = \left(\frac{8\pi^2\omega_{21}^2}{m_e c}\right)\left(\frac{g_1}{g_2}\right)f, \tag{16}$$

where g_1 and g_2 are the degeneracies of the ground and excited electronic states, respectively, and the other variables have been previously defined. If emission is the only mode of decay, then

$$\tau_0 = 1/A_{21}, \tag{17}$$

where τ_0 is the *mean radiative lifetime* for the emission, a quantity that is thus inversely proportional to the oscillator strength of the absorption.

Nonradiative decay from excited electronic states may occur if there is sufficient overlap between the excited vibrational levels of the lower electronic level and the lowest vibrational states of the excited electronic state. If this occurs between electronic levels of the same multiplicity it is called *internal conversion;* if electronic levels of different multiplicities are involved it is termed *intersystem crossing.* Figure 2 illustrates the situation in which there are two excited electronic states of differing multiplicities, e.g., singlet and triplet, which have considerable overlap in their vibrational manifolds with each other. Intersystem crossing from the uppermost potential well to the one that crosses it should be relatively easy from the viewpoint of vibrational relaxation, but forbidden because of the spin-conservation rule. However, such processes occur readily because of spin–orbit coupling. Decay of either of the two excited electronic states to the ground state is less likely from the viewpoint of vibrational relaxation because of the poor mixing of vibrational levels, but it can occur through *collisional deactivation* in which other molecules absorb the energy difference and behave as quenchers of the excited states without actually undergoing chemical change. Since this process is controlled by the collision rate in the system, it is easy to see why this kind of deactivation increases in liquid media at higher temperatures, causing competing luminescence processes to be decreased.

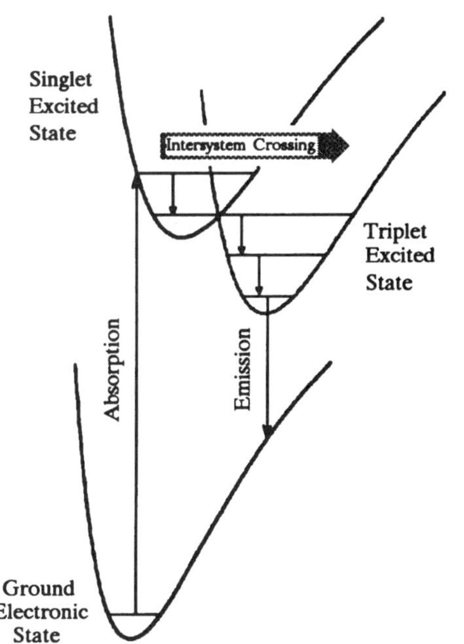

FIG. 2. Potential-energy diagram showing excited singlet and triplet states.

Decay from the excited state of different multiplicity back to the ground state may also occur by emission or nonradiative processes. In both cases, however, the transition has a low probability because of the different spin multiplicities of the two states and the wide separation between the two electronic levels; thus the excited state has a rather long lifetime. Emission from such a state is often called phosphorescence and is characterized by lifetimes as long as seconds, minutes, or even hours after the absorption of light.

Table 2 summarizes the various photo-

Table 2. Photophysical processes, typical rate constants, and lifetimes in liquid media.

Process	Rate constants (s^{-1})	Lifetimes
Absorption of photon	$\sim 10^{15}$	1 fs
Vibrational relaxation	10^{11}–10^{12}	1–10 ps
Fluorescence	10^{8}–10^{9}	1–10 ns
Intersystem crossing between excited states	0–10^{10}	>0.1 ns
Phosphorescence	0–10^{3}	>1 ms
Radiationless deactivation		
1) Internal conversion	0–10^{9}	>1 ns
2) Intersystem crossing	0–10^{3}	>1 ms

physical processes that may occur in liquid media and the typical ranges in rate constants (see CHEMICAL KINETICS) and lifetimes for each process. Not all emission processes fall neatly into the definitions for fluorescence and phosphorescence, and therefore all types of emission can be generally referred to as luminescence.

2.4 Photochemical Pathways

The extent to which photochemical products are formed will depend on how well the photochemical pathways compete with the various photophysical decay processes described above. This, in turn, will depend on the reactivity of the excited states, which are usually chemically very different from the ground states. Figure 3 is another way of outlining the many choices of pathways through which the excited system can dissipate the energy it gained in absorbing a photon. Nearly all types of reactions can be classified as either unimolecular or bimolecular. The excited states that are reactive may be those that were originally excited in the absorption of light, or they may be other states, labeled "reactive intermediate" in Fig. 3, accessible by radiationless decay. There are two special types of reactive intermediates that are formed by bimolecular processes: *excimers*, which are formed by reaction of a molecule in an excited state with an otherwise identical one in the ground state, and *exciplexes*, which are formed by a complexation reaction of an excited state with ligands. In any case, the potential-energy well of the excited reactive intermediate relative to the ground and excited states of the photoreactant can also be represented by the diagram in Fig. 2. Photophysical processes are those that result in a return of the system to the ground state, while photochemical processes result in different products.

Many of these processes occur simultaneously and compete with each other such that one usually observes a mixture of outcomes, each with its own quantum yield. Each of the various decay pathways leading away from the excited photoreactant has its own rate (see CHEMICAL KINETICS), and the quantum yield for a specific pathway, ϕ_j, is the ratio of an individual rate of decay for that pathway to the sum of the rates of all of the decay pathways, such that (assuming first-order reactions)

$$\phi_j = k_j/\Sigma_i\, k_i, \tag{18}$$

where k_j is the rate constant for the pathway of interest and k_i represents the rate constant for each of the other pathways.

An often-cited quantity is the lifetime τ of an excited state or reactive intermediate. This is defined as the time required for a sample to decay to $1/e$ of its original concentration, and can be shown to be equal to the reciprocal of the sum of the first-order rate constants for all decay pathways,

$$\tau = (\Sigma_i\, k_i)^{-1}. \tag{19}$$

It is often the case that excited states may react through bimolecular collisions with other species, thus causing the excited states to be "quenched." Such reactions offer additional pathways for the decay of the excited states, and they must be accounted for in

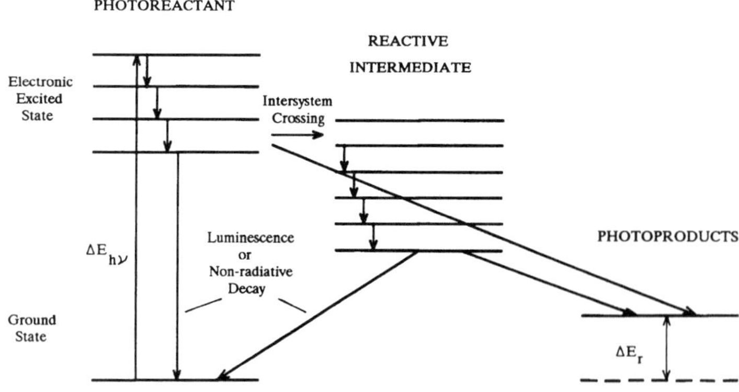

FIG. 3. Energy diagram summarizing the various pathways available to an excited molecule or complex.

any quantum-yield expression. Since these reactions are second-order processes, their rates will be proportional to the concentrations of both the excited state and the quencher; thus the quantum-yield expression, Eq. (18), will contain a second-order rate constant times quencher concentration for each of these quenching pathways. For example, suppose the only pathways for decay of an excited photoreactant are luminescence, nonradiative decay, or reaction with a quencher, Q, to form a product, as in

$$A \underset{k_L, k_{NR}}{\overset{h\nu}{\rightleftarrows}} A^* \xrightarrow{k_P[Q]} \text{products}, \tag{20}$$

where k_L and k_{NR} are the first-order luminescence and nonradiative decay constants and k_P is the second-order constant for the bimolecular quenching that forms product. The quantum yield of product formation, ϕ_P, would then be

$$\phi_P = \frac{k_P[Q]}{k_L + k_{NR} + k_P[Q]}, \tag{21}$$

where $[Q]$ is the concentration of the quencher. One can establish that a second-order quenching process occurs by obtaining a linear plot of $1/\phi_P$ vs $1/[Q]$ since the reciprocal of Eq. (21) is

$$\frac{1}{\phi_P} = 1 + \frac{k_L + K_{NR}}{k_p[Q]}. \tag{22}$$

The slope of such a plot, known as a *Stern–Volmer plot,* can be used to calculate the second-order quenching constant if the lifetime in the absence of quencher is known. An alternative method of obtaining the second-order rate constant directly is from the slope of measured pseudo first-order decay constants vs $[Q]$.

Many photoreaction schemes are considerably more complex than that shown in Fig. 3. For an excellent detailed analysis of photochemical kinetics the reader should consult the book by Demas (1983).

3. TYPES OF PHOTOCHEMICAL EXPERIMENTS

There are usually two kinds of experiments performed by photochemists: continuous-wave or cw irradiation, and flash irradiation. Each type is useful for gaining certain kinds of information. A brief description of each type of experiment is given below.

3.1 Continuous (cw) Irradiation

In cw irradiation, a continuous source of light such as a halogen or an arc lamp is directed or focused onto a sample in a cuvette or photochemical reactor. The goal of such experiments may be to use light as a chemical reactant to prepare usable quantities of some chemical product, or to determine certain parameters such as quantum yield, the effective wavelength of light needed to carry out the reaction, or the identity of one or more products of the reaction.

3.1.1 Preparative Photochemistry In order to prepare materials using light, one needs a photochemical reactor optimally designed to produce the desired amount of product. A laboratory reactor usually consists of a borosilicate glass vessel for containing the reacting chemicals, and a quartz or borosilicate immersion well that fits into the reaction vessel and contains the lamp (usually mercury medium-pressure arc lamp) and a jacket for circulating water or some other coolant. The lamp produces a range of wavelengths in the visible and near UV. For more information on preparative photochemistry the reader should consult Kopecky's excellent book (Kopecky, 1992).

3.1.2 Quantum-Yield Determination and Product Identification In this kind of experiment the goal is to determine the ratio of a reaction rate to the rate of absorption of light (see quantum-yield definition in Sec. 1.3). To do this one generally utilizes a well-collimated beam of light impinging on the sample in a cuvette. The beam may first be passed through a monochromator or filter if wavelength information is desired. The rate of reaction may be determined by monitoring at various time intervals properties that are functions of concentration, such as absorbance, pH, conductance, luminescence, or volume of gas emitted or absorbed. The light-absorption rate can be determined from actinometry (see Sec. 1.2). The kinds of information obtained in such experiments might be

1. the nature of the products of the photoreaction,
2. the net quantum yield, and the dependence of quantum yield on various reactants, solvents, temperature, or
3. the action spectrum, i.e., the dependence of quantum yield on wavelength.

This type of experiment is comparatively simple and requires inexpensive equipment. Since the photoreaction occurs in a *stationary-state* mode one cannot directly "observe" the intermediates in the reactions, as they may have very short lifetimes and thus very low concentrations compared with the net reactants and products. Thus, one can only make measurements of the overall or net reaction.

Since photochemical processes are sensitive to specific wavelengths of light, careful attention must be paid to the spectral output of the light source selected for experiments. For quantitative cw experiments the usual type of lamp used is a high-pressure arc lamp containing mercury, xenon, or a mixture of mercury and xenon. These types of lamps are efficient converters of energy; they have a wide spectral response from the UV to the infrared; and because of their short arc lengths it is easy to focus their outputs through monochromators and onto a cuvette. The energy output of a high-pressure mercury arc lamp is concentrated primarily on the series of emission lines of the mercury spectrum, excluding the resonance line at 254 nm, whereas xenon arc lamps tend to give a more continuous output from the UV to the IR (see Fig. 4). The features of both types of lamp can be obtained in an arc-lamp envelope filled with a mixture of xenon and mercury. Figure 4 also shows the visible portion of the output from a quartz–halogen lamp, which behaves as a blackbody radiator.

3.2 Laser Flash Photolysis

In flash photolysis, a pulsed light source delivers a short burst of light to the sample. This generates short-lived intermediates whose properties, such as absorption or lu-

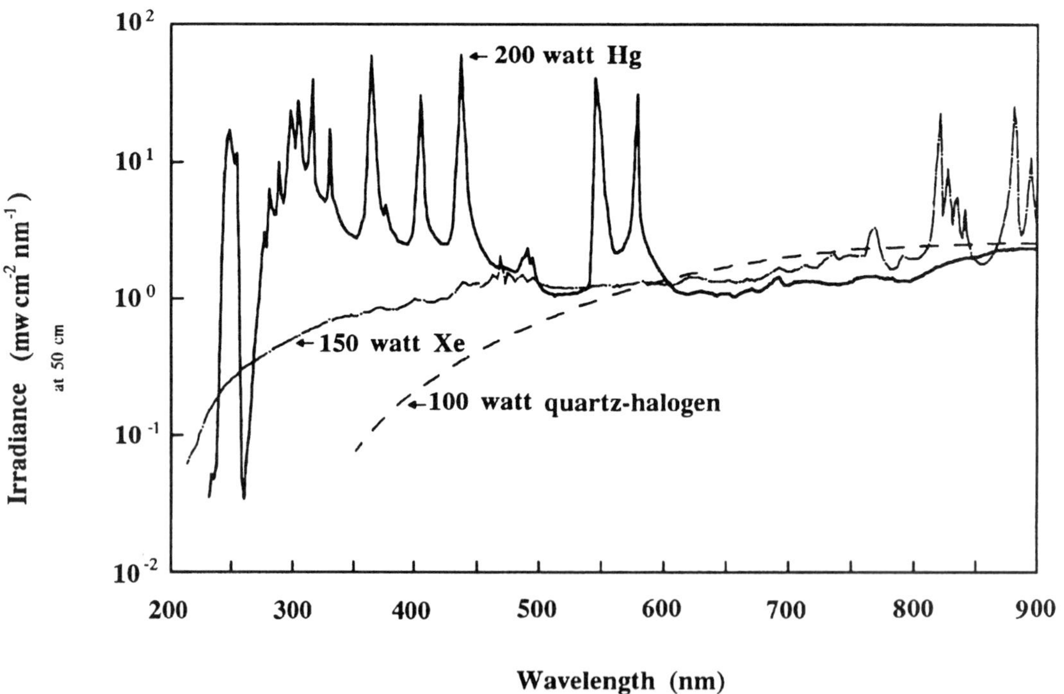

FIG. 4. The energy-output spectra for (solid curve) 200-watt mercury and (dot-dashed curve) 150-watt xenon high-pressure short-arc lamps, and (dashed curve) 100-watt tungsten–halogen lamp.

minescence, can be monitored as a function of time. The kinds of information obtained are

1. lifetimes of intermediates,
2. absorption and luminescence spectra of intermediates, and
3. quantum yields of intermediate formation,

all of which lead to an understanding of the intermediate steps and the mechanism of the photochemical reaction. Thus, flash photolysis can "observe" intermediates directly, even those of very fast lifetimes—on the nanosecond or even femtosecond time scale. Both cw and flash photolysis yield important complementary information about photochemical systems.

Modern flash experiments rely almost exclusively on pulsed lasers (see LASER TECHNOLOGY; SPECTROSCOPY, LASER) to produce the required high-energy and short-pulse-width flash. Lasers may operate in either the pulsed or cw mode, depending on the type. In all cases they have extremely narrow bandwidths, and virtually nonexistent beam divergences compared with noncoherent light sources. The energy output is centered at only one wavelength and depends on the material in the optical cavity. Availability of a wide selection of wavelengths and the ability to tune lasers over a limited wavelength range have emerged in the past several decades by the use of a variety of cavity materials, dye lasers, frequency-modulating crystals, and optical parametric oscillators. Another type of laser, the *free-electron laser,* produces coherent light at wavelengths from the far IR to the deep x-ray region, by the relativistic acceleration of a cluster of electrons in a storage ring. Such radiation is called *synchrotron radiation* (*q.v.*), and can be a very useful source of tunable laser light at almost any given pulse width, provided one has access to the rather expensive equipment required for its generation. For a complete description of the many laser configurations possible for laser flash photolysis the reader is referred to the book *Applied Laser Spectroscopy* (Andrews, 1992).

A simple laser flash setup is diagrammed in Fig. 5. When the sample is excited by absorption of the laser pulse, the oscilloscope is triggered to receive the signal coming from the photomultiplier (PM) attached to the monochromator. The monochromator selects the wavelength of light of interest, and if the transient produced by the flash absorbs at this wavelength a signal like that shown may appear as the transient decays over a selected time span. This signal can be computer-processed to yield plots of absorbance vs time, and the decay constant and lifetime can be calculated. Alternatively, if the monochromator has a photodiode-array detector, or if the sample is exposed to series of flashes, each one at a different wavelength, then it is possible to obtain a time-resolved spectrum, i.e., an absorption spectrum of the short-lived transient at a given time after the flash. The setup also allows the user to obtain analogous luminescence decay information with the analyzing lamp turned off.

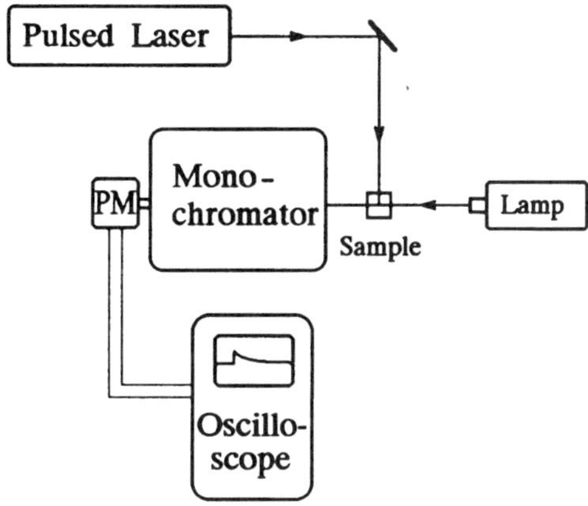

FIG. 5. A typical laser flash-photolysis experimental setup.

3.3 Ultrafast Intermediates

Such fundamental processes in chemical reactions as energy dissipation, molecular motions, and structural changes occur on the picosecond (10^{-12} s) to femtosecond (10^{-15} s) time scale. With the advent of femtosecond lasers, it has become possible to measure events that occur on a time scale shorter than a single vibrational oscillation of a molecule. Thus, ultrafast "snapshots" can be taken of molecules, crystal lattices, unstable intermediates, or other structures in various stages of distortion, yielding valuable information about the structures of molecules and the dynamics of crystal lattices, liquids, individual molecules, and biological systems (see ULTRAFAST SPECTROSCOPY).

An example of the application of picosecond spectroscopy to the study of ultrashort intermediates is given by the photoisomerization of *cis*-stilbene to *trans*-stilbene (the stilbenes are 1,2 diphenylethenes; see Sec. 4.3.3.2 for more information about this type of reaction). Excitation by a 70-fs, 312-nm laser pulse into the lowest singlet excited state of the *cis* complex results in the photoisomerization reaction, along with formation of a by-product, dihydrophenanthrene (DHP) (Anfinrud *et al.*, 1992). Ultrafast transient decays at 330 and 480 nm (Fig. 6), where the *trans*-stilbene and DHP, respectively, have maximum absorptions, show the disappearance of the electronically excited state within 2 ps of excitation, followed by a slower decay in the case of *trans*-stilbene, or no decay in the case of the DHP. The change in the spectrum of *trans*-stilbene from a broad band to a fairly narrow band after about 50 ps [Fig. 7(a)] is attributed to the fact that the newly formed product has an excess of rotational and vibrational energy that it discharges during that time span. The spectrum of the DHP [Fig. 7(b)], on the other hand, remains essentially unchanged from 5 ps after formation to the relatively long time of 50 ms, indicating that not much vibrational energy must be disposed of after the DHP forms.

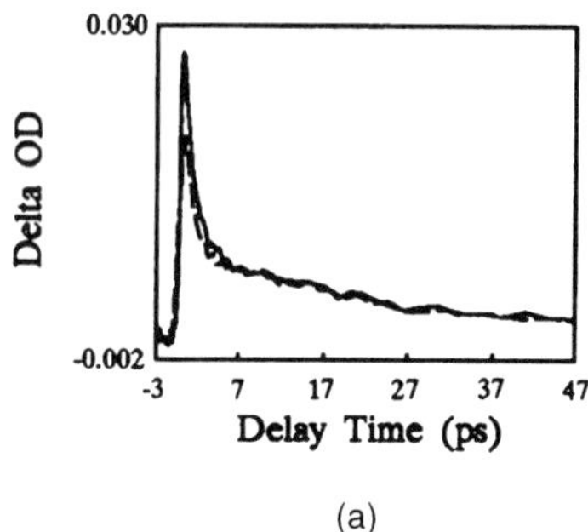

(a)

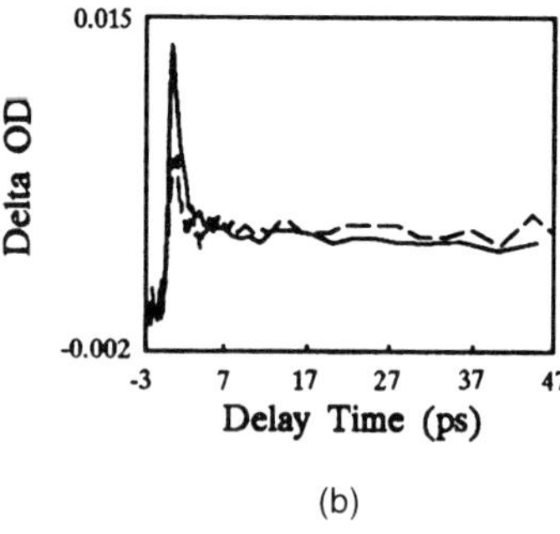

(b)

FIG. 6. Magic-angle time-resolved absorption (given as a change in optical density, or DOD) of *cis*-stilbene in hexadecane. (a) Data obtained using a 312-nm pump (solid line) or a 250-nm pump (dashed line) and a 330-nm probe wavelength. The primary photoproduct observed at 330 nm is *trans*-stilbene. (b) Data obtained using a 312-nm pump and a 480-nm probe wavelength (solid line) or a 250-nm pump and a 460-nm probe wavelength (dashed line). The primary photoproduct observed at 480 and 460 nm is DHP.

Many systems have recently been examined via femtosecond time-scale experiments by Zewail and others (Polanyi and Zewail, 1995). For example, the dissociation of the ionic compound NaI through a covalent intermediate to the atomic products, Na + I, upon absorption of energy, occurs in less than 10 ps. During this time, the activated intermediate oscillates on the average about eight times, and through an ultrafast series of probe laser pulses the amounts of activated intermediate, ionic $Na^{+}I^{-}$ species, and product sodium atoms can be monitored, with a time resolution down to about 7 fs. The results have given direct information about the reaction time, the probability of dissociation, and the extent of ionic and covalent character in the NaI bond, and have confirmed many theoretical predictions about reaction dynamics.

3.4 Laser Spectroscopy

The extremely high spectral purity of a laser beam, coupled with the advances in tunability of lasers, has made them indispensable tools for obtaining high-resolution spectral information of materials. Figure 1, above, shows two electronic energy levels depicted as potential-energy wells, with several vibrational levels superposed on each one. Not shown are the more closely spaced molecular rotational levels that should be superposed on top of each vibrational level. In normal spectroscopy the exciting source for the various transitions between energy levels is some kind of lamp–monochromator combination that has a fairly wide bandwidth and thus limits the degree of resolution of the energies of the transitions. Lasers, on the

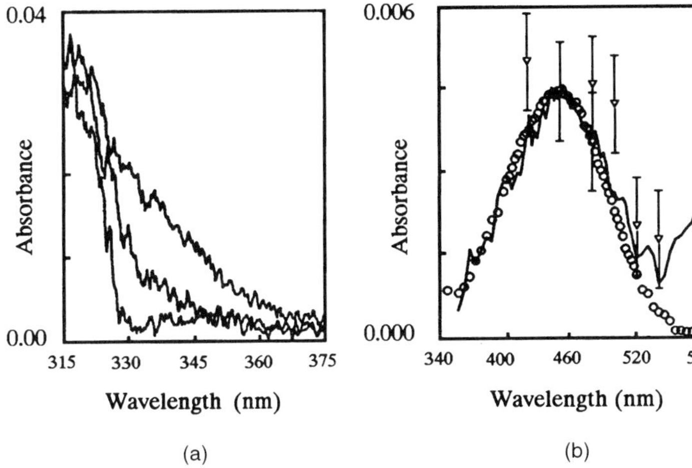

FIG. 7. (a) Absorption spectrum of hot *trans*-stilbene obtained at 6, 15, and 50 ps. The pump wavelength is 312 nm. The 50-ps spectrum is indistinguishable from the spectrum of ground-state *trans*-stilbene obtained using the same apparatus. (b) DHP absorption at ~5 ps (data points with error bars), 200 ps (solid line), and 50 ms (circles). The pump wavelength is 312 nm.

other hand, have an extremely narrow bandwidth as an excitation source, so that high-resolution spectra can be obtained. Several important laser-spectroscopy techniques will be briefly described. For more details on these techniques the reader is referred to the book *Applied Laser Spectroscopy* (Andrews, 1992).

In *electronic photoabsorption spectroscopy* a sample is excited by a visible or UV laser such that electronic excited states can be probed in the absorption mode. Determining the absorption spectrum of an excited state can yield important information about photochemical intermediates, and the various reaction pathways available to a photoexcited species. Alternatively, because the bandwidth of a laser may be considerably smaller than the linewidth of the electronic transition that absorbs the laser beam, replacing the conventional monochromator of a spectrometer with a tunable laser can result in highly resolved spectra of the ground state, yielding precise energies and transition probabilities of the excited electronic states. This is especially applicable to gaseous systems at low temperatures, where the rotational and vibrational features of the spectra become minimized and the electronic transitions become "purer." One common method of significantly lowering the temperature of a gas under study is by supersonic jet-cooling, i.e., ejecting the gas from a nozzle at supersonic speeds.

Laser-induced fluorescence spectroscopy (LIF) is a technique that produces a fluorescence spectrum by exciting the sample with a laser. In conventional fluorescence spectroscopy one can obtain two types of spectra: an *excitation spectrum* in which the intensity of emission at a fixed wavelength is measured as a function of excitation wavelength; or an *emission spectrum* in which intensity of emission is measured as a function of emission wavelength under excitation at a fixed wavelength. Figure 1 shows how the fluorescence process occurs and indicates that the excitation spectrum gives the energy changes occurring in the absorption process whereas the emission spectrum gives the energy released when the excited species returns to the ground state. The use of a tunable laser as the excitation source dramatically increases the resolution of an excitation spectrum so that vibrational and rotational fine structure can be observed in samples such as jet-cooled gases, crystals, and molecular beams. Important temporal information can also be obtained by use of ultrashort pulsed lasers as the excitation source.

Infrared (IR) spectroscopy has been an important tool of chemists for most of the twentieth century. It is used to probe the energy transitions between vibrational levels in the ground electronic state (i.e., the vibrational levels shown in the lower potential-energy well of Fig. 1). Such information has been used to understand the nature of molecular bonds, geometry, and thermodynamic properties, as well as serving as a powerful tool for the identification and characterization of molecular systems. The development of infrared lasers has given rise to high-resolution infrared spectroscopy, which routinely measures rotational details of IR transitions. Because of the high resolution and sensitivity of this technique, useful information has been obtained in understanding the properties of free radicals such as $CH_3\cdot$, molecular ions (i.e., electrically charged molecules), and van der Waals molecules (species held together by forces much weaker than chemical bonds).

In order for a molecule to absorb infrared radiation the molecule must undergo a change in its dipole moment during a vibration. Many vibrational transitions, particularly in large molecules, are thus infrared active, but for many simple molecules, especially homonuclear diatomic molecules like H_2 or Cl_2, this is not the case, and IR spectroscopy cannot be used to study them. Even in complex molecules there may be many vibrational modes that are not infrared active. Fortunately a complementary technique, *laser Raman spectroscopy,* has been developed that probes most of the IR-inactive vibrational transitions in molecules. It is based on the phenomenon of Raman scattering, the process of excitation of a molecule from a specific vibrational level in the ground electronic state to a "virtual" electronic energy level, followed by return to a different vibrational level in the ground electronic state. If the final vibrational state is higher in energy than the initial state, the resulting spectral line is called a *Stokes* line; if it is lower in energy it is called *anti-Stokes.* The requirement for a Raman-active vibrational transition is that there must be a change in the polarizability of the molecule during the transition; thus molecules like N_2, Cl_2, and the symmetric stretching vibration of CH_4 are Raman active. Normal Raman scattering is a very inefficient process, with quantum yields of 10^{-7} or less, and early Raman spectra required hours of exposure to conventional UV or visible lamps. However, the high power output of lasers has brought about the development of the modern laser Raman spectrometer. Even greater enhancement of Raman spectra can be achieved by resonance-enhanced Raman scattering, in which the exciting wavelength of the laser is at or near an electronic absorption band in the molecule, or by *surface-enhanced Raman scattering* (SERS), in which the molecules to be probed are adsorbed on the surface of a metal, usually silver.

The absorption of laser light can sometimes occur in a multistep process, because of the very high light intensities of a laser beam. For example, a molecule can be excited to an electronic energy level that has a value of E above the ground state by the absorption of two photons of energy $E/2$ at precisely the same time. This nonlinear effect gives rise to multiphoton absorption spectroscopy. One important advantage of this technique over conventional spectroscopy is that the electronic transitions do not obey the Laporte rule; in two-photon absorption, for example, the transitions must be $g \leftrightarrows g$ or $u \leftrightarrows u$, rather than $g \leftrightarrows u$ or $u \leftrightarrows g$. This increased access to excited states results in a greater understanding of the electronic and vibrational structure of molecules.

4. PHOTOCHEMISTRY OF ORGANIC COMPOUNDS

4.1 Electronic Transitions and Spectra of Organic Molecules

Absorption of a photon by an organic molecule causes a change in its electronic state in a variety of ways that depend on the type of substrate and the energy of the photon. The *h*ighest-energy *o*ccupied *m*olecular *o*rbitals (or HOMOs) of an alkene are the π orbitals in the C–C bond. Absorption of a photon of wavelength around 180 nm promotes an electron from this HOMO to the LUMO (*l*owest *u*noccupied *m*olecular *o*rbital) π^* orbital. The excited state is termed the (π,π^*) excited state, and the electronic transition leading to it is a $\pi \rightarrow \pi^*$ transition. While simple alkenes absorb at wavelengths below 200 nm, conjugation of double bonds can shift the absorption maximum into the visible range; e.g., λ_{max} = 476 nm for $CH_3(CH{=}CH)_{10}CH_3$.

The HOMOs of molecules containing heteroatoms (such as N, O, or halogen) are usually nonbonding, and the lowest-energy transitions for such molecules are $n \rightarrow \pi^*$ (in the presence of multiple bonds) or $n \rightarrow \sigma^*$ (for saturated molecules). Since $n \rightarrow \pi^*$ transitions are forbidden, ketones and aldehydes have their lowest-energy absorption maxima at about 280 and 290 nm, respectively. Acids, anhydrides, and esters have absorption bands at shorter wavelengths (<250 nm), while aromatic substitution causes a considerable redshift (e.g., $\lambda_{max} \approx 340$ nm for benzophenone). Less important are the allowed $\sigma \rightarrow \sigma^*$ transitions of alkanes, giving rise to bands in the vacuum UV.

Most organic molecules are singlets in their ground states; thus, their first excited triplet state is of lower energy than the corresponding singlet state because of the repulsive nature of interactions between electrons of the same spin.

4.2 Electronic Energy Transfer

An important overall mechanism of quenching in which the excitation energy is transferred to the quencher in a bimolecular process,

$$M^* + Q \rightarrow M + Q^*, \tag{23}$$

is also called photosensitization, where Q^* is some excited state of Q usually not directly accessible by optical absorption. The quencher, Q, is said to be sensitized by M; i.e., although Q is by itself normally transparent at this wavelength, it is able to become excited by the excited sensitizer, M^*. For this kind of reaction to occur, the energy level of the excited sensitizer, M^*, should be greater than (or equal to) that of Q^*.

Energy transfer can be either radiative or *nonradiative*. In the former, the emission from M^* is reabsorbed by Q. (Of course, the absorption spectrum of Q should overlap the emission spectrum of M^*.) Since the S_0–T_1 transition is spin-forbidden, the dominant radiative energy-transfer processes are from a singlet to a singlet or from a triplet to a singlet state.

Nonradiative energy transfers involve the mutual perturbation of the electronic structures of M^* and Q. In long-range Coulombic energy transfer (Förster, 1959) the donor and acceptor molecules may be coupled through an electrostatic interaction at large (up to 100 Å) intermolecular separations. The energy-transfer processes favored by the Coulombic mechanism are from a singlet to a singlet or from a singlet to a triplet state.

Nonradiative energy transfer can also take place by an electron-exchange mechanism (Dexter, 1953), which requires closer contact between M^* and Q (at encounter distances of 6–10 Å). This mechanism is thought to involve the formation of a short-lived exciplex intermediate (between the energy donor and acceptor molecules). It is worth mentioning that the ground-state O_2 molecule (as ^3Q) is an effective quencher of both singlet and triplet excited states. In the latter case very reactive singlet molecular oxygen, $^1O_2$, may be produced.

In many sensitization reactions, triplet–triplet energy transfer plays the key role. One of the earliest examples is the reaction between triplet excited-state benzophenone and ground-state naphthalene using 370-nm irradiation (Terenin and Ermolaev, 1956). Benzophenone can also be used for the photosensitization of dimerization of buta-1,3-diene (to form cyclobutene; see also Sec. 4.3.3.2).

4.3 Reactions of Excited Species

4.3.1 Reactivity of Excited Species

The apparent reactivity of an excited species may be determined by the intrinsic reactivity of its electronic arrangement, by the excitation energy, and by the lifetime of the particular excited state.

The intrinsic reactivity of a species largely depends on the way in which its electrons are distributed in the available orbitals. Electronic excitation alters this arrangement and thus influences the geometry, the dipole moment, the electron-donating or -accepting ability, and the related acid–base properties.

If excitation affects the nature of the bonding, the shape of the molecule may also be changed. For example, while the ground state of an alkene (e.g., ethene) is planar, in the equilibrium structure of the (π,π^*) excited state the two CH_2 groups lie in perpendicular planes and only a σ bond remains between the carbon atoms.

Excited species are generally both better electron donors and better electron acceptors than the corresponding ground-state species. Therefore, excitation both decreases the ionization energy and increases the electron affinity of the molecule.

The contrasting reactivities of the corresponding singlet and triplet states are well demonstrated by the photoisomerization of *trans*-penta-1,3,-diene. Direct excitation (to S_1) leads to cyclization,

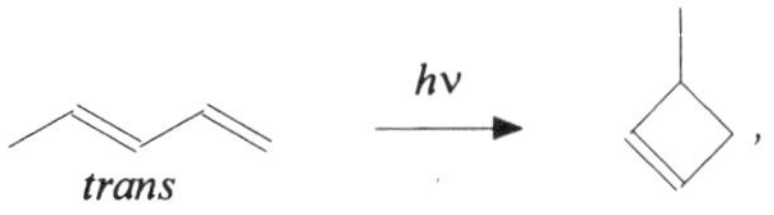

$$\text{trans} \xrightarrow{h\nu} \text{(cyclobutene product)}, \quad (24a)$$

while sensitization (to T_1) results in the formation of the *cis* isomer,

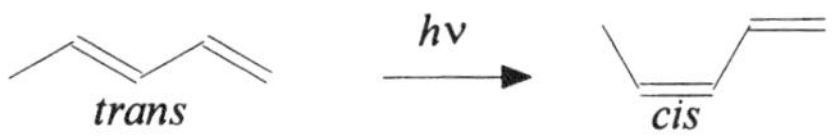

$$\text{trans} \xrightarrow{h\nu} \text{cis}. \quad (24b)$$

Generally, the excited singlet may be more reactive than the triplet. But, since for most organic molecules the ground state is a singlet, the triplet excited state may survive radiative and nonradiative quenching much better than the more reactive singlet. Thus, generally, the triplet is the more important excited species in terms of the overall reactivity in photoinduced organic reactions.

4.3.2 Correlation Rules and Symmetry Conservation Electron spin-correlation rules are also valid for reactions where chemical change occurs. Thus, in permitted adiabatic reactions the total electron spin is conserved, i.e., the sum of the spins of the reactants should be the same as that for the products. Conservation of symmetry is especially important to the understanding of what are termed *pericyclic reactions,* or reactions that pass through a cyclic transition state in a concerted manner, i.e., with bond-breaking and -forming processes occurring essentially simultaneously. The well-known Woodward–Hoffmann rules (Woodward and Hoffmann, 1970), using the concepts of orbital symmetry, give a good compass for prediction or interpretation of these reactions. Without going into a detailed explanation of these rules, the experimental observations regarding the pericyclic reactions show that the stereochemical course of these processes occurring photochemically is in general opposite to that for thermal reactions. For example, the cyclization of the polyene,

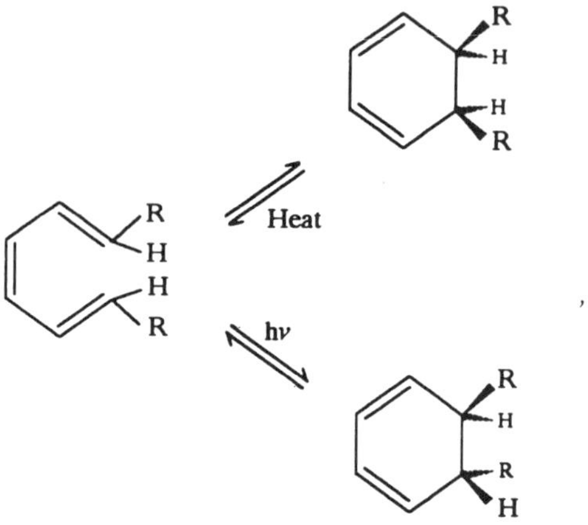

(25)

occurs thermally such that the terminal *R* and *H* groups rotate in opposite directions (*disrotation*) whereas the photochemical process causes them to rotate in the same direction (*conrotation*).

4.3.3 Intramolecular Processes

4.3.3.1 Primary Dissociation Processes. While dissociation of a diatomic molecule can only give two atomic fragments with unambiguous chemical identity, that of a polyatomic molecule can sometimes yield many sets of products. For a detailed discussion of the methods used to elucidate primary dissociative mechanisms the reader can consult the books by Calvert and Pitts (1966) and by Okabe (1978). Primary photodissociation in organic species can be demonstrated by reference to two classes of compound: hydrocarbons and carbonyl compounds.

Photodissociation of alkanes mostly results in elimination of molecular hydrogen, but bond ruptures giving a variety of free radicals can also occur as minor processes. Photofragmentation of alkenes yields hydrogen, alkynes, and a variety of radicals, while cleavage of aromatic hydrocarbons gives polymers, carbon, and traces of volatile products.

Two main classes of photodissociation of carbonyl-containing compounds (aldehydes and ketones) are distinguished by Norrish and Bamford (1937): the radical-forming processes (known as Norrish Type I reactions),

$$RCOR' \xrightarrow{h\nu} R + (COR')^* \quad \text{or} \quad (RCO)^* + R' \rightarrow R + CO + R', \qquad (26)$$

and the intramolecular fission of the C–C bond α–β to the carbonyl group (called Norrish Type II process),

$$R_2CHCR_2CR_2COR' \xrightarrow{h\nu} R_2C{=}CR_2 + CR_2{=}C(OH)R'(\rightarrow CHR_2COR'). \qquad (27)$$

Aliphatic aldehydes, for example, can photodissociate simultaneously according to both mechanisms. The ratio of the processes is wavelength dependent. Norrish Type II photolyses of ketones generally proceed via a six-membered intermediate (as shown in Sec. 4.3.3.3).

Cyclic ketones undergo photodecomposition in a way involving biradical formation in the first step. For example, photodissociation of cyclopentanone in the gas phase can yield ethene, cyclobutane, 4-pentenal, and CO, which can be interpreted in terms of initial formation of $\cdot CH_2CH_2CH_2CH_2O\cdot$.

4.3.3.2 Photoisomerization. The perpendicular configuration of ethene and its derivatives in the lowest-energy (π,π^*) excited state is geometrically equivalent whether it is derived from a *cis* or *trans* ground-state molecule. Thus, the probability ratio for the formation of *cis* and *trans* isomers depends on the conversion rates of the excited state to the two (isomeric) ground-state forms.

In accordance with the Franck–Condon principle, the significant difference between the geometries of the ground and excited states forbids emission from the latter one. Therefore, in the case where the quantum yields are similar for both *cis* → *trans* and *trans* → *cis*, prolonged irradiation of either isomer results in the composition of the isomeric mixture at the photostationary state being predominantly determined by the individual absorbances of each isomer at the excitation wavelength. For simple alkenic systems generally *trans* isomers possess higher extinction coefficients at longer wavelengths. Thus, irradiation in that range results in the predominance of the *cis* isomer in the photostationary state. Electrocyclic ring-closure reactions of dienes and trienes are well-known examples of valence and structural isomerizations. These processes leading to the formation of cycloalkenes require the *s-cis* conformation. Starting with the *s-trans* conformation, cyclization yields, e.g., bicyclo products:

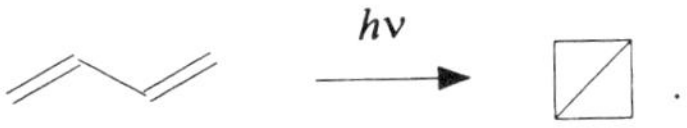

hν (28)

This product can also be formed in the triplet sensitized reaction, indicating that ring closure may involve a diradical in a nonconcerted reaction.

The reactions of polyenes in the excited singlet state often differ from those in the triplet state. While direct excitation (to S_1) results in internal cyclization, triplet sensitization usually leads to (cyclic) dimerization via processes similar to those in Eqs. (24a) and (24b). Photocyclization of aromatic compounds often results in nonaromatic products. Cyclic conjugated enones and dienones can also undergo structural photoisomerizations with ring rearrangement. These concerted stereospecific reactions proceed via a triplet (π,π^*) state.

Another important type of photoisomerization is the di-π-rearrangement, which is formally a 1,2-shift with ring closure:

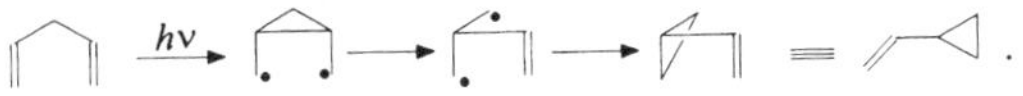

hν (29)

This reaction, however, is probably a nonconcerted process, and promoted via the excited singlet.

Photochemical *sigmatropic shift* is also a significant type of photoisomerization. In these pericyclic reactions a σ bond migrates with respect to a system of π electrons, resulting in a switching of double and single bonds as shown in the following example:

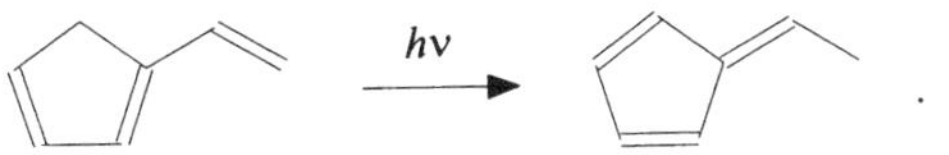

hν (30)

A more detailed discussion of these topics is given by Wagner and Hammond (1967).

4.3.3.3 Hydrogen Abstraction. Intramolecular hydrogen-atom abstractions, in contrast to photoisomerization processes, involve molecules almost exclusively in the (n,π^*) excited state. This type of reaction is especially significant in the photochemistry of most carbonyl compounds, because it plays an important role in processes leading to the Norrish Type II fragmentation (Wagner, 1971):

$$CH_3CH_2CH_2COCH_3 \xrightarrow{h\nu} \begin{array}{c} H \quad O \\ H_2C \qquad C-CH_3 \\ H_2C-CH_2 \end{array}$$

intramolecular H abstraction

$$\begin{array}{c} OH \\ H_2C\bullet \quad \bullet C-CH_3 \\ H_2C-CH_2 \end{array} \longrightarrow \begin{array}{c} CH_2{=}COH \\ CH_3 \\ + \\ CH_2{=}CH_2 \end{array} \longrightarrow CH_3COCH_3 \qquad (31)$$

The hydroxy diradical intermediate can also undergo a ring-closure reaction leading to cyclobutanol formation. Since only 1,4-diradicals undergo the cleavage reaction, photocyclization is more important in systems where H abstraction cannot occur via the six-membered transition state, i.e., where no hydrogen atom can be found in the γ position of the carbonyl compound.

4.3.4 Intermolecular Processes

4.3.4.1 Hydrogen Abstraction. Hydrogen abstraction can also take place from (n,π^*) excited states [especially of carbonyl compounds; see Scaiano (1973)] in intermolecular processes. Benzophenone undergoes such a reaction with an appropriate hydrogen donor:

$$C_6H_5COC_6H_5 + RH \xrightarrow{h\nu} C_6H_5C(OH)C_6H_5 + R. \qquad (32)$$

If the reagent RH has a low ionization potential, as is the case for amines or unsaturated hydrocarbons, charge transfer or electron abstraction may be the favored process, yielding the ketyl radical anion, $(C_6H_5)_2COH$. Such a radical can undergo subsequent secondary reactions such as dimerization (e.g., photopinacolization) or further hydrogen abstraction. Substituents of aryl ketones can, however, significantly change the nature of the photoinduced reaction.

4.3.4.2 Addition Reactions. Both homoaddition and heteroaddition reactions can be photochemically induced. The latter type of reaction can take place between unsaturated hydrocarbons and water, alcohols, and carboxylic acids, or between aromatic compounds and amines, e.g.,

benzene + cyclohexylamine (NH_2) $\xrightarrow{h\nu}$ N-H (cyclohexadienyl)(cyclohexyl)amine + N-H (cyclohexadienyl)(cyclohexyl)amine. (33)

Of special importance are the homo and hetero cycloaddition reactions. Two common classes of these types of reaction are the (2 + 2) addition of two alkenes to give cyclobutane,

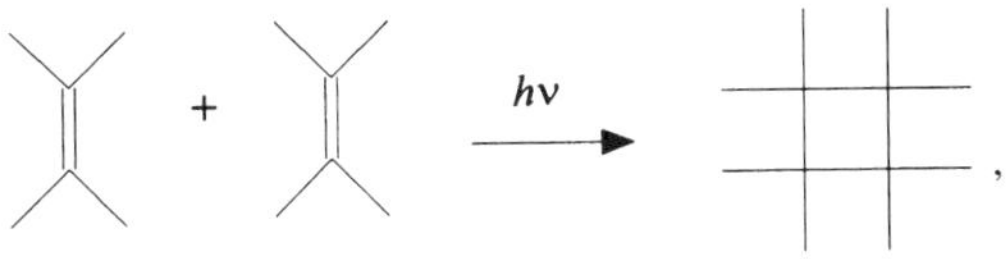

(34)

and the (4 + 2) addition of a conjugated diene and an alkene to give a cyclohexene,

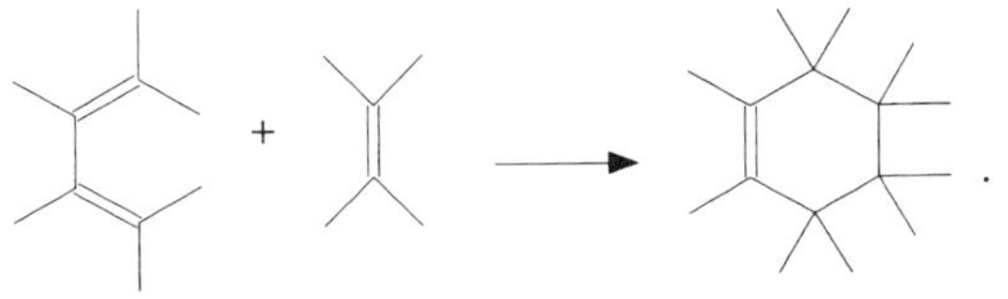

(35)

Cycloaddition of benzene and its derivatives can take place across 1,2-, 1,3-, or 1,4-positions. The 1,3-addition predominates with alkenes having only alkyl substituents on the double bonds:

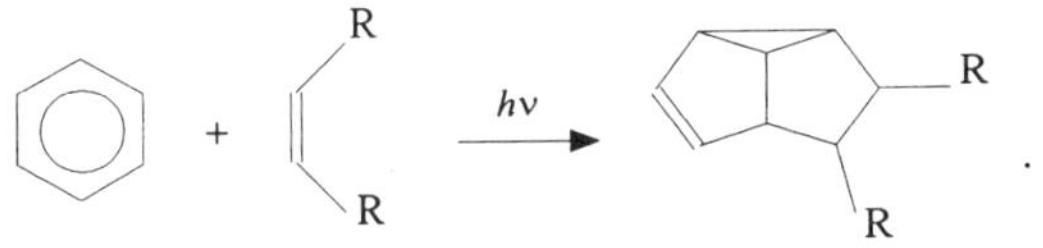

(36)

1,2-addition is the main mode of reaction if the electron donor–acceptor properties of the aromatic species and the alkene are significantly different:

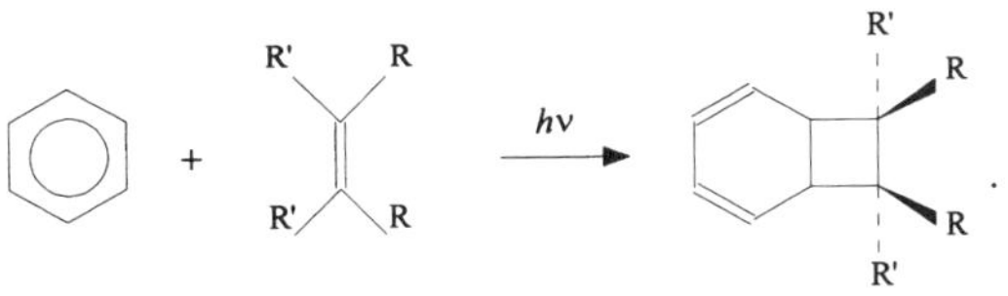

(37)

The 1,4-cycloaddition occurs least frequently and may take place through an exciplex interaction, as do the 1,2-additions as well.

Cyclobutane derivatives are the main products in the photoaddition of alkenes to the unsaturated bond in α,β-unsaturated ketones. Cyclic enones undergo photocyclic addition more readily than their acyclic analogs because their reduced flexibility inhibits intersystem crossing from the excited triplet state to the ground state. The first step in the interaction of the excited-state enone and ground-state alkene gives rise to an exciplex, the geometry of which might affect the stereochemistry of the final products.

Addition of the carbonyl group itself (in the excited ketones or aldehydes) to suitable alkenes can also occur, producing oxetanes (Arnold, 1968), e.g.,

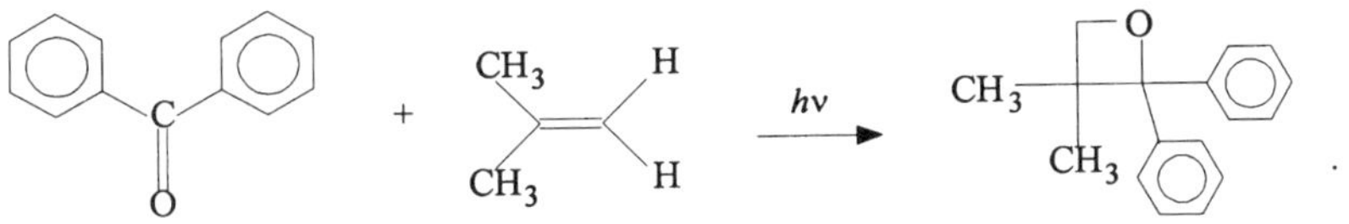

(38)

An alternative isomer can form with the CH_3 and phenyl groups reversed, and the ratio of these isomeric oxetane products is determined by the stability of the 1,4-diradical intermediate formed in the first step after the triplet (n,π^*) state aryl ketone attacks the ground-state alkene. Oxetane formation from alkyl ketones requires the first excited singlet (n,π^*) state of the precursor compounds.

Oxygen is one of the most important reactants in photochemical addition reactions of unsaturated compounds. Very efficient photosensitized oxidations are observed with alkenes, dienes, dienoid heterocycles, and polycyclic aromatic compounds. The first products of these reactions are often peroxides—as in the case of dimethylfuran,

$$H_3C\text{-(furan)-}CH_3 + O_2 \xrightarrow[sens]{h\nu} H_3C\text{-(endoperoxide)-}CH_3 \qquad (39)$$

—or hydroperoxides,

$$(H_3C)_2C{=}C(CH_3)_2 + O_2 \xrightarrow[sens]{h\nu} (H_3C)_2C(OOH){-}C(=CH_2)CH_3 \; , \qquad (40)$$

which may undergo subsequent oxidation steps. Alkenes that have electron-rich double bonds or no allylic hydrogen can undergo a (2 + 2) cycloaddition giving a dioxetane:

$$C_2H_5O{-}CH{=}CH{-}OC_2H_5 + O_2 \xrightarrow[sens]{h\nu} C_2H_5O{-}\underset{H}{C}H(O{-}O)\underset{H}{C}{-}OC_2H_5 \; . \qquad (41)$$

In most of these photosensitized oxidations the very reactive excited singlet state of oxygen, $O_2(^1\Delta_g)$, is favored via energy transfer from the triplet state of the sensitizers (Foote, 1968). Hence, carotenoids (e.g., β-carotene) are very important photobiological reagents, because as efficient quenchers of singlet oxygen they can protect photosynthetic organisms against photooxidation sensitized by their own chlorophyll.

5. PHOTOCHEMISTRY OF INORGANIC AND COORDINATION COMPOUNDS

To illustrate the kinds of electronic transitions that occur in inorganic photochemical reactions Fig. 8 shows a generic molecular-orbital diagram for an octahedral transition-metal coordination complex and the various kinds of electronic transitions that may occur between the orbitals. These are categorized as follows:

1. intraligand or IL bands, which are transitions between energy levels of the ligands (see discussion of organic photochemistry, above) and are relatively unaffected by the metal centers;
2. ligand-field or LF bands, which are the low-intensity transitions, usually in the visible region, caused by the splitting of the metal *d* orbitals (hence the other often-used term, *d–d* bands) by the field generated by the coordination of ligands to the metal;
3. *ligand-to-metal-charge-transfer* or LMCT bands, which are transitions in which electronic charge is essentially transferred from the ligands toward the coordinating metal;
4. metal-to-ligand-charge-transfer or MLCT bands, in which electronic charge is transferred from the coordinating metal to the ligands; and
5. charge-transfer-to-solvent or CTTS transitions, in which electronic charge moves to the solvent.

Charge-transfer bands are characterized by their tendency to produce charge separations, which may result in oxidation–reduction reactions.

5.1 Spectra and Photochemistry

5.1.1 Ligand-Field (LF) Transitions The transition metals are in the middle part of the periodic table where the elements have partially filled $(n-1)d$ electron subshells (where n is the period or main shell number). Zn, Cd, and Hg are also considered to be transition metals even though their *d* subshells are filled. There are five *d* orbitals in a given subshell, each with its own geometrical orientation in space, as shown in Fig. 9. In a neutral isolated atom, the five *d* orbitals of Fig. 9 are of the same energy, but in a metallic ion coordinately bound to electron-donor ligands there is a splitting of the energies, termed crystal-field splitting,

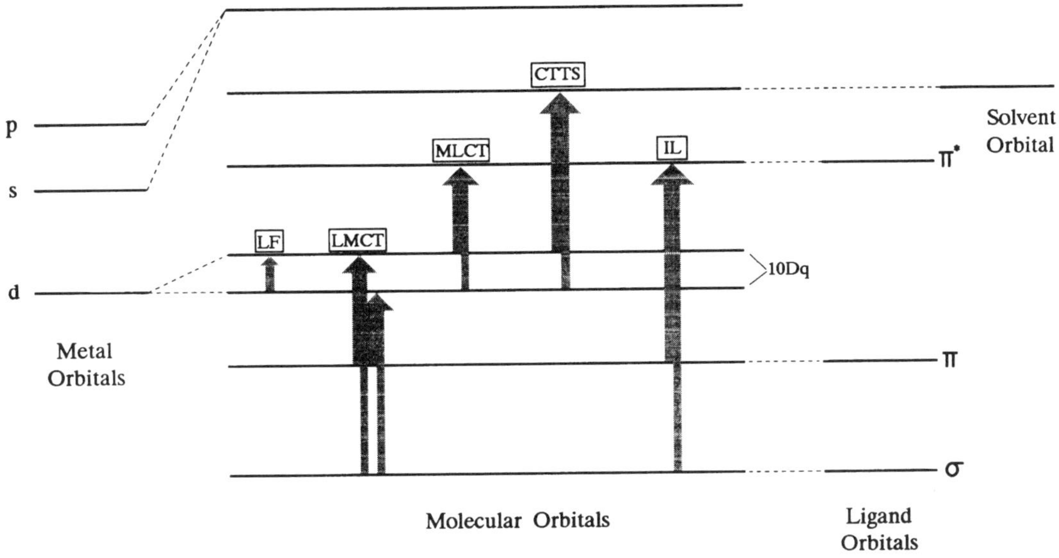

FIG. 8. Molecular-orbital diagram for an octahedral coordination complex and the various kinds of electronic excitation transitions that may occur between the orbitals.

caused by the interaction of the ligand electrons with the metal *d* orbitals. The amount of energy splitting depends on the geometry and electron-donating ability of the ligand, giving rise to the *spectrochemical series,* which roughly ranks some of the common ligands in their capacity to split the *d*-orbital energies as follows: $CN^- > NO_2^- >$ phenanthroline > dipyridine > $SO_3^{2-} > NH_3 > H_2O > OH^- > SCN^- > F^- > Cl^- > Br^- > I^-$. Which atomic orbitals undergo the strongest interaction with the ligands depends on the geometry of the complex: For an octahedral complex the two e_g orbitals point in the directions of the incoming ligands, thus raising their energy, whereas the three t_{2g} orbitals point away from the ligands, resulting in no change in energy. For a tetrahedral complex the situation is reversed, and for other geometries the splitting becomes more complicated. A molecular-orbital picture gives the same splitting patterns between metal orbitals, and these are shown for an octahedral complex in Fig. 8 by the commonly used symbol for this difference, *10Dq*.

The split atomic orbitals of the metal ion give rise to electronic transitions resulting in ligand-field or *d–d* bands. There are two general features of such bands:

1. Because these transitions are Laporte-for-

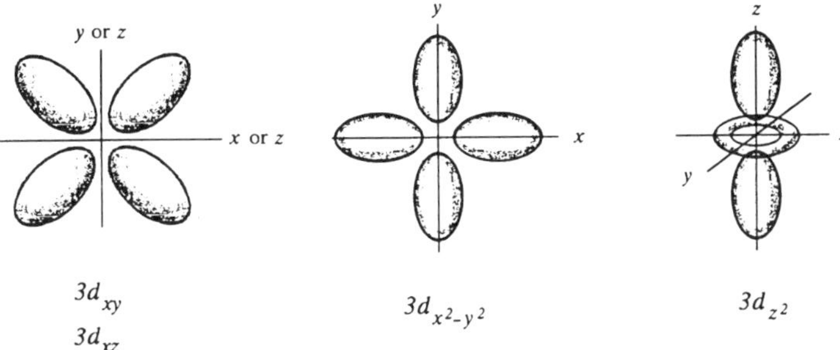

FIG. 9. Electron probability pictures of the 3*d* orbitals of an atom.

bidden, they are rather weak such that extinction coefficients rarely exceed several hundreds; and

2. the energies of the transitions are usually in the visible to near-IR region of the spectrum, giving a noticeable color to many transition-metal complexes.

In order for a transition to occur the coordinating metal ion must have at least one *d* electron and there must be at least one unfilled orbital in the higher-energy state to which an electron can be excited; therefore LF transitions are ruled out for such coordinating metals as Zn(II), Cu(I), Ag(I), Sc(III), and Ti(IV).

Excitation of a metal center changes the geometrical distribution of electrons and ligands and hence may change the strength of the metal–ligand bond. For example, in an octahedral complex, promoting an electron from one of the lower t_{2g} orbitals to an e_g orbital may cause a repulsion between a ligand and the metal, thus weakening the bond. At the same time, it might open a pathway for attack of the metal by another ligand. Thus, LF transitions may result in ligand-exchange, isomerization, or rearrangement reactions without changing the oxidation state of the metal ion. An example of a photosubstitution reaction of a complex ion with ligands arranged in an octahedral configuration is that of the Reineckate ion, *trans*-$Cr(NH_3)_2(NCS)_4^-$:

$$Cr(NH_3)_2(NCS)_4^- + H_2O \xrightarrow{h\nu(315\text{–}600\ nm)} Cr(NH^3)_2(NCS)_3(H_2O) + NCS^-. \quad (42)$$

This reaction is particularly useful as an actinometer in the visible region (Wegner and Adamson, 1966). A great many photochemical reactions induced by ligand-field transitions have been identified and may be reviewed in the books by Balzani and Carassiti (1970), Adamson and Fleischauer (1975), Geoffroy and Wrighton (1979), and Ferraudi (1988).

5.1.2 Ligand-to-Metal (LMCT) and Metal-to-Ligand (MLCT) Charge-Transfer Transitions Normally, charge-transfer bands are observed in the UV and short-wavelength visible region in complexes consisting of a donor and acceptor of electronic charge. If the donor is an electron-rich ligand such as NH_3 or a halide ion (such as Cl^- or Br^-), and the acceptor a readily reduced metal such as Co(III) or Fe(III), the transition is LMCT; if the donor is an oxidizable metal such as Cu(I) or Fe(II), and the acceptor an electron-poor ligand such as 1,10-phenanthroline or bipyridine, the transition is MLCT. Such bands are Laporte- and (usually) spin-allowed, yielding absorption coefficients on the order of 10^4. Since the transition is regarded as a displacement of charge, the initial absorption step can be represented by the following:

$$ML \xrightarrow{h\nu} {}^*M^-L^+ \quad (43)$$

for LMCT, or

$$ML \xrightarrow{h\nu} {}^*M^+L^- \quad (44)$$

for MLCT, where *M* and *L* are the metal center and a ligand, respectively, in the absorbing complex.

One should expect that the excited species following charge-transfer absorption is at once a stronger oxidizing and reducing agent than the ground state because of the shift in electron density between metal and ligand. Thus, the photochemistry resulting from such transitions should involve redox processes of metal and ligand. This certainly seems to be the case following MLCT absorption, as exemplified by the well-known tris(bipyridyl) ruthenium (II) system (Gafney and Adamson, 1972):

$$Ru(bpy)_3^{2+} \xrightarrow{h\nu} {}^*Ru(bpy)_3^{2+}, \quad (45)$$

where bpy represents the bipyridyl ligand. Whereas the normal reduction potentials E^0 in water for the couples $Ru(bpy)_3^{3+}$/$Ru(bpy)_3^{2+}$ and $Ru(bpy)_3^{2+}$/$Ru(bpy)_3^{1+}$ are +1.26 V and −1.26 V, respectively, the reduction potentials for the couples involving the excited Ru(II) species, i.e., $Ru(bpy)_3^{3+}$/${}^*Ru(bpy)_3^{2+}$ and ${}^*Ru(bpy)_3^{2+}$/$Ru(bpy)_3^{1+}$, are −0.84 V and +0.84 V, respectively. Thus, the excited state is both a stronger reductant and stronger oxidant than the ground state of this complex by 2.10 V in either case. This, of course, is a familiar example, which has been heavily researched as a possible photocatalytic agent for water splitting (Creutz and Sutin, 1976) since the transitions occur in the visible portion of the spectrum, but in principle it represents a common situation for MLCT photochemistry. In addition to redox processes, MLCT transitions often result in reductive elimination.

In general, MLCT transitions usually result in processes that maintain the integrity of the metal–ligand bonds because they originate in filled or nearly filled metal *d* orbitals that do not have much effect on the bonding, and they terminate in ligand-localized orbitals, which also are not usually involved in the metal–ligand bond.

By contrast, LMCT transitions often occur between ligand orbitals involved in the ligand–metal bond and a metal in a higher oxidation state with unfilled *d* orbitals, resulting in a coordinatively unstable, reduced metal–radical ion pair, which usually decomposes further to a complicated array of products. A classic example of the complexity that can arise from LMCT excitation is that of the irradiation of $Fe(C_2O_4)_3^{3-}$, the net reaction for which is

$$2Fe(C_2O_4)_3^{3-} \xrightarrow{h\nu} 2Fe(C_2O_4) + 3C_2O_4^{2-} + 2CO_2. \tag{46}$$

Flash studies of this reaction (Cooper and Degraff, 1971, 1972) suggest that there may be two competing primary steps in this reaction, one of which is a charge transfer resulting in the formation of the radicals $C_2O_4^-$ and/or CO_2^-, and the other of which is a photoaquation resulting from the change in the ligand field caused by the excitation. A useful list, prepared by Endicott (1975), tabulates the primary reduced metal and radical formed from the LMCT excitation of each of some 39 complexes of Co(III), Rh(III), Ir(III), Pt(IV), Fe(III), and Ru(III) in aqueous solution.

5.1.3 Charge-Transfer-to-Solvent (CTTS) Transitions Charge-transfer-to-solvent transitions are favorable when the irradiated species is easily oxidized and electron-rich, and there is some low-lying unoccupied orbital on the solvent molecule in the first or second coordination sphere of the complex. Thus, a representation of the absorption process would be

$$M\text{–Sol} \xrightarrow{h\nu} {}^*M^+\text{–Sol}^-, \tag{47}$$

where M represents the metal complex and Sol is the solvent molecule into whose molecular orbital the electron has been transferred. Since the first criterion for such transitions is the same as that which favors MLCT transitions, it is common for the lower oxidation states of such metals as Cu, Ru, and Fe to exhibit both MLCT and CTTS bands, sometimes with both types of transitions occurring in the same complex when an acceptor ligand is present, as has been observed, for example, in the case of the cyano complexes of copper(I) and iron(II).

The most obvious indicator of a CTTS transition is the ejection by the excited species of a solvated electron into the bulk of the solvent. In aqueous systems, this can be shown by the reaction

$$M(\text{aq}) \xrightarrow{h\nu} [{}^*M^+\text{–}H_2O^-] \rightarrow M^+(\text{aq}) + e^-_{\text{aq}}, \tag{48}$$

where $[{}^*M^+\text{–}H_2O^-]$ is the CTTS excited state of the complex M(aq) and e^-_{aq} is the hydrated electron. Hydrated electrons are easily detected in cw experiments by their reaction with scavengers such as N_2O(aq),

$$e^-_{\text{aq}} + N_2O \rightarrow N_2 + OH + OH^-, \tag{49}$$

which leads to oxidation of substrate by OH radical; or H^+,

$$e^-_{\text{aq}} + H_3O^+ \rightarrow H_3O, \tag{50}$$

resulting in evolution of dihydrogen by eventual combination of hydrogen radicals. If nanosecond flash photolysis is utilized, the spectrum of the hydrated electron is easily identified by its absorption maximum at 720 nm.

5.1.4 Other Charge-Transfer Transitions The number of dinuclear and polynuclear complexes exhibiting metal–metal bonding has increased dramatically in the past few years. In such compounds, it is expected that there will be significant orbital overlap such that new metal–metal delocalized molecular orbitals will be formed. These molecular orbitals can then take part in transitions resulting in charge transfer between the metals themselves, i.e., metal–metal charge transfer (MMCT), or between the metal–metal orbitals and ligand orbitals (perhaps labeled M_2LCT or LM_2CT, depending on the direction of the charge transfer). Most of the observations of such bands have occurred in dinuclear complexes [see Geoffroy and Wrighton (1979)], but there is intrinsically no reason why clusters of more than two metals should not exhibit such bands. The photochemical result of excitation is often metal–metal bond cleavage, which is exemplified by the reaction

$$Re_2(CO)_{10} \xrightarrow{h\nu} 2Re(CO)_5. \tag{51}$$

For a recent compilation of charge-transfer photochemistry of coordination compounds see the book by Horváth and Stevenson (1993).

6. APPLICATIONS OF PHOTOCHEMISTRY

6.1 Photography

Photography is the process of recording on a two-dimensional matrix such as paper, glass, or polymer an accurate optical image containing brightness information, such as in black-and-white photography, as well as some wavelength information as in color photography (see PHOTOGRAPHIC RECORDING). Although a variety of photochemical systems have been used during the 150-year history of photography, nearly all modern photographic processes are based on the photoreduction of silver-halide salts,

$$AgX(c) \xrightarrow{h\nu} Ag(c) + \tfrac{1}{2}X_2, \tag{52}$$

where X can be Cl, Br, or I. A suspension of silver-halide crystals in gelatin can be coated onto a polymeric film to produce a light-sensitive emulsion, which produces a negative *latent image* (i.e., one that is too faint to see with the naked eye) of photoreduced silver when exposed inside a camera. The latent image is made up of silver-atom centers in microscopic crystallites (grains) of silver halide, and they can be intensified by a factor of several orders of magnitude by development with a suitable water-soluble organic reducing agent. During this dark development the silver halide in close proximity to the latent-image silver is preferentially and catalytically reduced, producing the highest density of opaque silver on those areas on the film that had the highest light flux. The developed negative image can be reversed by transmitting light through it onto another sheet of silver emulsion, in a paper matrix if a fine-quality reflecting print is desired, and developing that in the same way. After development, emulsions must be "fixed" by dissolving the excess, undeveloped light-sensitive silver halide with a complexing agent such as the thiosulfate ion. The "speed" of films and papers is determined by the size of the silver-halide grains; in general, the larger the grain the faster the emulsion because the development process can result in a much more intensified image.

Silver halides vary in their sensitivity to the different wavelengths of light; all are sensitive to ultraviolet, but the sensitivity to visible wavelengths drops off from iodide to bromide to chloride. The spectral sensitivity can be enhanced by the addition to the emulsion of photosensitizers. A color film is made by sandwiching three layers of silver emulsions containing sensitizers, separated by suitable filters if necessary, such that each layer is sensitive to only one of the three primary colors, red, blue, and green. In color-reversal film, i.e., transparency film that gives a positive color image, the latent silver images in all three layers are developed, and then the unexposed silver-halide grains are developed successively in a chromogenic developer, i.e., one that forms a dye in the developing process. The dye formed is opaque to the primary color to which the emulsion layer was sensitive; therefore after all of the developed silver is dissolved from the film there remain three layers of subtractive filters that recreate the color and light-intensity information of the original image when illuminated. A color-negative film or paper uses dye-forming couplers in each of the three layers so that chromogenesis occurs in the original development step.

6.2 Photopolymerization

There are generally two types of photopolymerization processes: photoinitiated polymerization and photocrosslinking (see POLYMERIZATION AND POLYMER REACTIONS).

Photoinitiation of polymerization is used, for example, for drying or hardening of thin coatings, a process often termed photocuring. Most applications of photoinitiated polymerization involve a free-radical mechanism using the monomer based on acrylate esters, ($CH_2{=}CHCOOR$). Photoinitiators are generally aromatic carbonyl compounds with absorption spectra matching available UV sources. Upon irradiation they undergo α cleavage (Norrish Type I; see Sec. 4.3.3.1) or hydrogen abstraction (with an appropriate donor; see Sec. 4.3.4.1) producing radicals.

However, oxygen inhibits radical polymerization since it is an efficient triplet quencher. Therefore, cationic polymerization is more advantageous in aerobic processes. Photoinitiation of polymerization can be accomplished through charge-complex formation, using mixtures of aromatic diazonium or iodonium and sulfonium salts and nonnucleophilic anions such as PF_6^-.

For example, the generally accepted mechanism for photoinitiation using diphenyliodonium hexafluorophosphate, $(C_6H_5)_2I^+$-PF_6^-, results in the formation of a Brönsted acid (see CHEMICAL REACTIONS), ArI^+H, as a primary initiator:

$$Ar_2I^+X^- \xrightarrow{h\nu} [Ar_2I^+X^-]^* \rightarrow ArI^+ + Ar + X^-, \tag{53}$$

$$ArI^+ \xrightarrow{SH} ArI^+H \rightarrow ArI + H^+.$$

Here, SH designates a monomer or solvent molecule. Cationic photopolymerization is generally sensitized because the iodonium salts themselves absorb poorly in the visible or near-UV range.

Photocrosslinking is commonly used for the production of photoresist materials for the production of negative images. It involves formation of crosslinks between preexisting polymer chains. For example, after preliminary controlled cyclization of a polymer such as polyisoprene, the original methylene groups and double bonds remain intact. If aromatic diazide is introduced to the polymer as a photosensitizer, then during irradiation it yields a nitrene, which crosslinks the polymer molecules:

$$N_3\text{–Arom–}N_3 \xrightarrow[-N_2]{nh\nu} \ddot{N}\text{–Arom–}\ddot{N} + \begin{matrix} \sim CH_2 \sim \\ \sim C(CH_3){=}CH \sim \end{matrix} \longrightarrow \begin{matrix} \sim CH(NH\text{–}) \sim \\ \text{Arom} \\ \sim C(CH_3)\text{–}CH \sim \ (\text{aziridine ring via } N) \end{matrix} \tag{54}$$

Photocycloaddition or photodimerization can also be the key reaction of crosslinking. A typical example is the photopolymerization involving poly(vinyl cinnamate), which, upon irradiation, undergoes a dimerization reaction involving the double bonds of neighboring cinnamoyl groups, as

$$\begin{matrix} \sim CH_2\text{—}CH(\text{O–CO–CH{=}CH–}C_6H_5) \sim \\ \sim CH(\text{O–CO–CH{=}CH–}C_6H_5)\text{—}CH_2 \sim \end{matrix} \xrightarrow{h\nu} \begin{matrix} \sim CH_2\text{—}CH(\text{O–CO–}) \sim \\ \text{cyclobutane: } C_6H_5\text{–CH—CH–CO / CH—CH–}C_6H_5 \\ \sim CH(\text{O–CO–})\text{—}CH_2 \sim \end{matrix} \tag{55}$$

Since the cinnamate chromophore and related compounds do not absorb above 320 nm, triplet sensitizers are generally used for the photocrosslinkings.

6.3 Photoimaging

The formation of images by photochemical reactions not involving silver is called *photoimaging.* Such processes usually involve polymers, either by photoinitiation of polymerization, or by photolytic modification of an existing polymer. Photoimaging finds many applications in the printing and electronics industries, in the production, for example, of lithographic plates or printed circuit boards.

The most common polymer imaging systems use photoinitiation of polymerization to produce the image. Through chain reactions in the polymerization process a latent image can be intensified many times, causing changes in hardness, tackiness, solubility, or chemical reactivity, properties that develop and enhance the image. The ideal photoinitiator is one that has a high absorption coefficient and a high quantum yield, does not fade upon irradiation, and is photoactive in a desirable (for example, visible) wavelength region.

Among the many usable systems, a recently described one utilizing dye sensitizers is that involving triaryl alkyl borate anion activators (Chatterjee *et al.,* 1988, 1990). The basic free-radical–forming photoreaction is

$$(\mathrm{Dye})^+ \mathrm{Ph_3BC_4H_9}^- \xrightarrow{h\nu} (\mathrm{Dye})\cdot + \mathrm{Ph_3BC_4H_9}\cdot$$
$$\mathrm{Ph_3BC_4H_9}\cdot \rightarrow \mathrm{Ph_3B} + \mathrm{C_4H_9}\cdot, \qquad (56)$$

where Ph represents the phenyl group, C_6H_5. The butyl radical $C_4H_9\cdot$ formed in the second step is the polymerization initiator. Although a variety of dyes can be used, it is interesting that the class of dyes known as *cyanines* works quite well. This is significant because these same dyes are used as photosensitizers in color photography, and much is known about them. They are available in a wide selection of wavelength sensitivity so that the photoimaging process can be tuned to almost any visible-wavelength region.

A recent important technological development is the process of *stereophotolithography* (Brulle *et al.,* 1994), in which precision three-dimensional objects are formed layer by layer by computer-controlled laser photopolymerization. This technology may be of great significance to medicine—for example, for noninvasive modeling from CT scans a part of the body to aid in preparing for surgery, or for fabricating precise artificial joints or other tissue. Other applications are the duplication and fabrication of machine parts and art objects, and the creation of any kind of three-dimensional model from digitized information.

6.4 Phototherapy

Phototherapy is the use of photochemical reactions in treating disease. It involves the absorption of light of selected wavelengths by chromophores within tissue cells, or in drugs that have been administered specifically for the phototherapy.

A number of diseases may be successfully treated by phototherapy. Hyperbilirubinemia, a disease of infants in which bilirubin builds up in the serum causing neurological problems, can be treated by exposure of the skin to sunlight or simulated sunlight resulting in the photoisomerization of bilirubin to a more readily eliminated form. In photodynamic therapy a photosensitive drug is used to target tumor cells or diseased tissue, which is then irradiated to activate the drug. *Photochemotherapy* utilizes the combined effects of light and a secondary drug to treat the disease, and this has proved useful in the treatment of skin diseases such as psoriasis, vitiligo, and other disorders. Lasers have come into use in certain surgical procedures, especially those involving the eyes, and these rely on the ability of tissue to absorb light and undergo chemical or physical change. The book by Grossweiner (1994) is a good source of information on these techniques.

6.5 Photochromism

While photography is based on irreversible photochemical processes, photochromism is the general term for the production of reversible photoinduced color changes. Irradiation of photochromic compounds drastically alters their absorption spectra, and the return to the original state of the substrate can take place simply by removal of the light source or by irradiation at a different wavelength. The processes responsible for photochromic behavior are isomerization, dissociation, and redox reactions. Generally, photochromic materials utilizing isomerization are more durable, because mechanisms involving bond cleavage often lead to chemical decomposition in side reactions, causing a rapid "fatigue."

As indicated in Sec. 4.3.3.2, *cis* and *trans* isomers of many substituted alkenes and azo compounds are interconverted by light. Generally, the *trans* isomers are more stable and absorb at longer wavelengths. Photoinduced valence isomerization such as tautomerism (via intramolecular hydrogen transfer) is of great importance in photochromic systems. Many aromatic nitro compounds can be converted by irradiation from the colorless nitro form to the colored aci form:

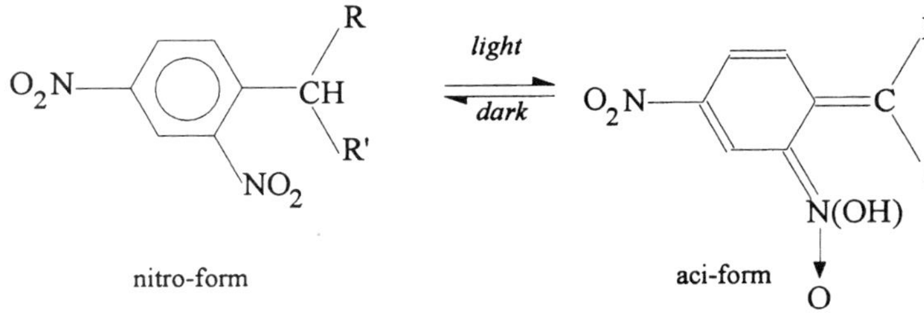

light / dark (57)

nitro-form aci-form

where R = H, CH_3, C_6H_5, etc.; R' = electron-withdrawing group.

The photochromism exhibited by the so-called fulgides (Heller, 1991),

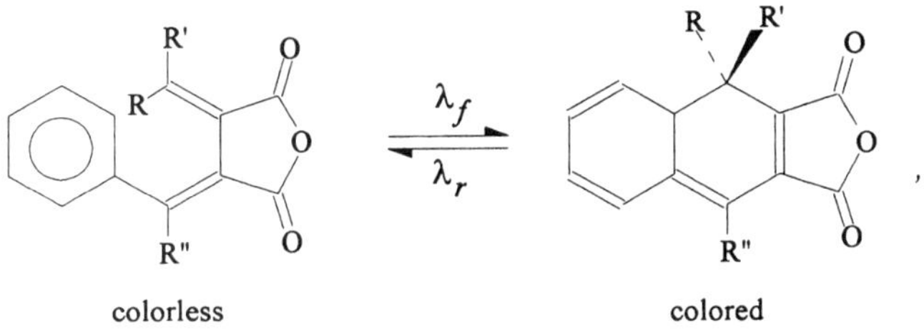

λ_f / λ_r (58)

colorless colored

requires irradiation for the reverse reaction as well. The wavelength of the light used for this process is usually longer than that for the forward reaction, because of the absorbance of the colored product.

Inorganic photochromic systems are also known. In a mixture of Hg_2I_2 and AgI a typical reversible redox reaction takes place:

$$\underset{\text{green}}{Hg_2I_2} + \underset{\text{yellow}}{2AgI} \underset{\text{dark}}{\overset{\text{light}}{\rightleftharpoons}} \underset{\text{red}}{2HgI_2} + \underset{\text{black}}{2Ag}\ . \qquad (59)$$

6.6 Photosynthesis

The overall reaction for the natural photosynthesis (*q.v.*) occurring in green plants and some other systems except for photosynthetic bacteria is

$$H_2O + CO_2 \underset{\text{chlorophyll}}{\overset{\text{light}}{\rightarrow}} (CH_2O) + O_2, \qquad (60)$$

where (CH_2O) designates carbohydrate (e.g., starch or sugar). This is the result of a very complex sequence of photochemical and thermal reactions. The essential photochemical reaction is water splitting, which produces gaseous oxygen, accompanied by electron and proton transfer to reduce (indirectly) carbon dioxide to carbohydrate:

$$2H_2O \rightarrow O_2 + 4e^- + 4H^+, \qquad (61)$$

$$4e^- + 4H^+ + CO_2 \rightarrow (CH_2O) + H_2O. \qquad (62)$$

Experimental measurements have shown that two photons are needed for each electron transfer, which implies a two-step process involving rather long-lived intermediates. In fact, two separate photosystems (PS I and PS II) operate in photosynthesis. The scheme (known as the *Z* scheme, Fig. 10) demonstrates the mechanism of this two-photon excitation. The position of any component in the diagram represents the redox potential of an oxidant/reductant couple (e.g., +0.81 V for the $\frac{1}{2}O_2/H_2O$ couple, for the two-electron reduction of $\frac{1}{2}O_2$). According to the scheme, a quantum of light absorbed by the pigment of PS II excites a molecule called P690 (because it absorbs at wavelengths up to 690 nm). This mediates an uphill transfer of an electron from species *Z* to *Q*, and from Q^- the electron is transferred in spontaneous dark reactions involving plastoquinone (PQ) and cytochromes (C) to the reaction center of PS I (called P700). Absorption of another photon by P700 mediates the second uphill transfer of the electron received from the PS II system to the strongly reducing substrate *X*. At this point a second series of dark reactions takes place via the protein ferredoxin (FD), yielding NADP (the reduced form of nicotinamide adenine dinucleotide diphosphate), and then finally giving NADPH (nicotinamide adenine dinucleotide diphosphate) as a result of two-electron reduction. This thermally reduces carbon dioxide through a series of catalyzed steps called the Calvin cycle (Calvin, 1962). This cycle also requires ATP (adenosine triphosphate), which is synthesized in a reaction parallel to the *Z* scheme and utilizing some energy from the downhill electron-transfer steps. The oxidized form of the manganese-containing *Z* oxidizes water to oxygen (in four steps to give a molecule of O_2).

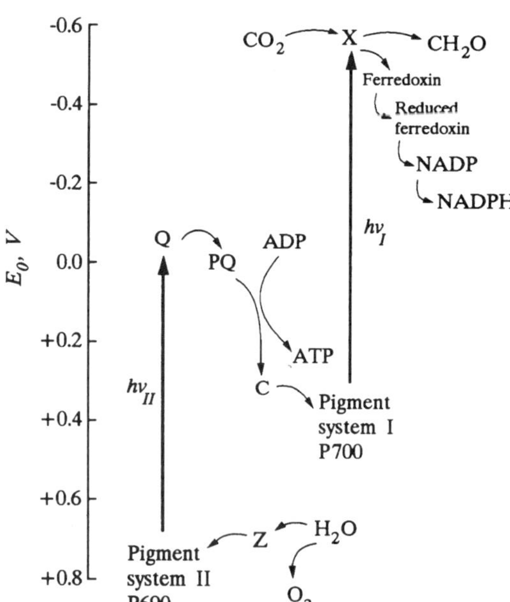

FIG. 10. The *Z* scheme of photosynthesis.

Experimental observations indicate that about 300 pigment molecules associate with each reaction center of PS I and PS II, and excitation of any of these can result in the excitation of the reaction center. This array of pigment molecules is called the light-harvesting antenna. Most of the pigments in the photosynthetic systems absorb at wavelengths up to 650–700 nm. Chlorophyll-*a* is usually the major pigment in green plants and algae, whereas chlorophyll-*b* and carotenes are the principal auxiliary pigments.

A different method of photosynthesis is utilized by bacteria. Unlike plant photosynthesis, bacterial photosynthesis uses H_2S rather than water as the reducing agent. The principal pigment in these bacteria is bacteriochlorophyll (BChl), a reduced form of chlorophyll. Reduction of carbon dioxide occurs in these systems too, but only one photon is used for each electron transfer. The electron-transport chain involves the protein cytochrome *c* (Cyt-*c*), BChl, bacteriopheophytin (BPh) (which is BChl without its central magnesium ion), and ubiquinone (*Q*), arranged in that sequence. The electron-transfer series following the excitation of BChl leads to the formation of Cyt-c^+ and Q^- (which is complexed to iron).

6.7 Photochemistry of Vision

In the photochemistry of the visual process, transformation of a pigment containing a moiety related to vitamin A (retinol) plays the key role. The receptors of the retina of the eye consist of "rods" and "cones"; the rods possess high sensitivity for low intensity of light, while the less sensitive cones are used for color vision.

The visual pigments of the retina are carotenoid substances combined with a protein. For example, rhodopsin ("visual purple"), discovered by Böll (1876), contains

11-*cis*-retinal (Fig. 11) as the carotenoid chromophore and the protein scotopsin found in rods.

Irradiation of rhodopsin triggers a series of conformational changes involving intermediates of different colors:

$$\begin{array}{l} \text{rhodopsin} \xrightarrow{h\nu} \text{bathorhodopsin} \\ \quad\text{(red)} \qquad\qquad \text{(red)} \\ \quad\rightarrow \text{lumirhodopsin} \rightarrow \text{metarhodopsin I} \\ \qquad\text{(orange-red)} \qquad \text{(orange)} \\ \quad\rightarrow \text{metarhodopsin II} \rightarrow \text{retinal + opsin.} \\ \qquad\quad\text{(yellow)} \qquad\qquad \text{(colorless)} \end{array}$$

The straight all-*trans*-retinal cannot be accommodated on the surface of the opsin, while the bent 11-*cis*-retinal "fits" into the protein. In other words, absorption of a photon results in a progressive photoisomerization toward more strained structures and, thus, the cleavage of the protein–chromophore bond. At each step the protein backbone also rearranges. The cycle is completed by a slow thermal isomerization back to the 11-*cis*-retinal that combines spontaneously with opsin.

Color vision, associated mostly with the cones, involves three different pigments absorbing in the blue, the green, and the red wavelength regions, and closely related to rod rhodopsin. The photochemical changes in the retina are then converted to electrical impulses stimulating the brain.

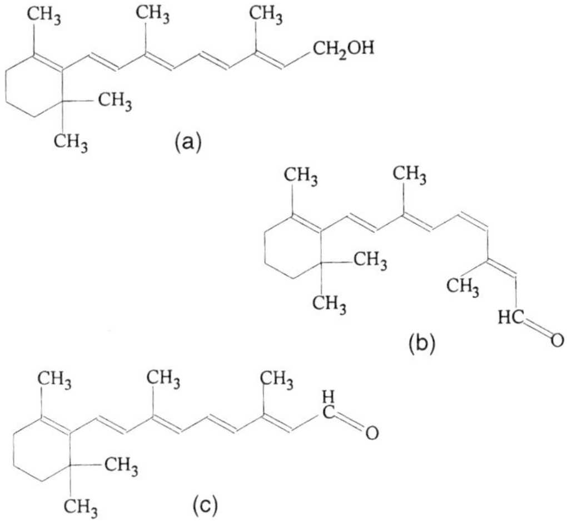

FIG. 11. Formulas of (a) retinol, (b) 11-*cis*-retinal, and (c) all-*trans*-retinal.

7. ACKNOWLEDGMENTS

This work was supported by the Division of International Programs, U.S.–Eastern Europe Cooperative Science Programs, of the National Science Foundation, and by the Hungarian Academy of Sciences.

GLOSSARY

Absorbance: $\log_{10}(I_0/I)$, where I_0 is the incident light intensity falling on an absorbing sample and I is the transmitted light intensity.

Actinometry: Measurement of light intensities for quantum-yield determinations.

(2 + 2), (2 + 4) Additions: Reactions of unsaturated hydrocarbons; the terms in the parentheses indicate the position numbers of the π electrons involved.

Chromophore: Light-absorbing chemical group.

Cis: A molecular structural configuration in which two like functional groups are adjacent to each other on the molecular skeleton.

Concerted Reaction: A reaction in which bond-breaking and forming occur essentially simultaneously.

Conjugation: Alternating double and single bonds within a molecular chain or ring, resulting in bond stabilization via electron delocalization.

Dipole Moment: A measure of the separation between centers of positive and negative charges in polar molecules.

Excimer: A complex formed from the excited state of a molecule with an identical molecule in the ground state.

Exciplex: A complex formed from an excited molecule and a different molecule in the ground state.

Extinction Coefficient: Specific absorbance of 1-cm pathlength of solution at 1 mole/L concentration.

Flash Photolysis: Experimental study of photochemical or photophysical processes initiated by a short pulse of light.

HOMO: Highest occupied molecular orbital.

Internal Conversion: Energy loss within a given electronic excited or ground-state manifold.

Intersystem Crossing: Transition of an excited electron from a given multiplicity (such as the singlet) to a different one (such as the triplet).

Lifetime: The time required for an excited species to decay to 1/*e* of its initial concentration.

LUMO: Lowest unoccupied molecular orbital.

Monomers: Low–molecular-weight molecules or atoms that can join together to form a polymer.

Nonradiative Energy Transfer: Loss of energy from an excited state by any process that does not produce light.

Norrish Type I and II Reactions: Typical classes of photodissociation of carbonyl compounds.

Oscillator Strength: A measure of the area under an absorption band, or the degree to which a transition is allowed.

Pericyclic Reactions: Concerted reactions proceeding via a cyclic transition state.

Polarizability: The measure of charge displacement by an externally applied electric field.

Radiative Energy Transfer: Loss of energy of an excited state by emission of light.

Sensitizer: Compound that absorbs light at the irradiation wavelength, then transfers its excited-state energy to a substrate that cannot be directly excited.

Singlet: An electron configuration in which the sum of the electron-spin quantum numbers is 0, i.e., all the electrons are paired with respect to their spins.

Spin–Orbit Coupling: Interaction of the orbital and spin angular momenta of an electron in an atom or polyatomic species.

Trans: A molecular structural configuration in which two like functional groups are opposite to each other on the molecular skeleton.

u → g or g → u Transitions: Laporte-allowed transitions between wave functions that are *even* (gerade or g) or *odd* (ungerade or u) with respect to inversion about a center of symmetry.

Triplet: An electron configuration in which the sum of electron-spin quantum numbers is 1, i.e., there are two unpaired electrons of parallel spin.

Works Cited

Adamson, A. W., Fleischauer, P. D. (1975), *Concepts of Inorganic Photochemistry,* New York: Wiley.

Andrews, D. L. (1992), *Applied Laser Spectroscopy,* New York: VCH.

Anfinrud, P. A., Johnson, C. K., Sension, R., Hochstrasser, R. M. (1992), in: D. L. Andrews (Ed.), *Applied Laser Spectroscopy,* New York: VCH, p. 410.

Arnold, D. R. (1968), *Adv. Photochem.* **6,** 301–423.

Balzani, V., Carassiti, V. (1970), *Photochemistry of Coordination Compounds,* New York: Academic.

Böll, F. (1876), *Monatsber. Akad. Wiss. Berlin,* 23 November, 783–787.

Brulle, Y., Bouchy, A., Valance, B., André, J. (1994), *J. Photochem. Photobiol. A: Chemistry* **83,** 29–37.

Calvert, J. G., Pitts, J. N., Jr. (1966), *Photochemistry,* Chichester and New York: Wiley, Chaps. 3–5.

Calvin, M. (1962), *Angew. Chem. Int. Ed.* **1,** 65–75.

Chatterjee, S., Gottschalk, P., Davis, P. D., Schuster, G. B. (1988), *J. Am. Chem. Soc.* **110,** 2326–2328.

Chatterjee, S., Davis, P. D., Gottschalk, P., Durz, M. E., Sauerwein, B., Yang, X., Schuster, G. B. (1990), *J. Am. Chem. Soc.* **112,** 6329–6338.

Cooper, G. D., DeGraff, B. A. (1971), *J. Phys. Chem.* **75,** 2897–2901.

Cooper, G. D., DeGraff, B. A. (1972), *J. Phys. Chem.* **76,** 2618–2625.

Creutz, C., Sutin, N. (1976), *J. Am. Chem. Soc.* **98,** 6384–6385.

Demas, J. N. (1983), *Excited State Lifetime Measurements,* New York: Academic.

Dexter, D. L. (1953), *J. Chem. Phys.* **21,** 836–850.

Endicott, J. F. (1975), in: A. W. Adamson, P. D. Fleischauer (Eds.), *Concepts of Inorganic Photochemistry,* New York: Wiley, Chap. 3.

Ferraudi, G. J. (1988), *Elements of Inorganic Photochemistry,* New York: Wiley.

Foote, C. S. (1968), *Acc. Chem. Res.* **1,** 104–110.

Förster, Th. (1959), *Discuss. Faraday Soc.* **27,** 7–17.

Gafney, H. D., Adamson, A. W. (1972), *J. Am. Chem. Soc.* **94,** 8238–8239.

Geoffroy, G. L., Wrighton, M. S. (1979), *Organometallic Photochemistry,* New York: Academic.

Grossweiner, L. I. (1994), *The Science of Phototherapy,* Boca Raton, FL: CRC Press.

Hatchard, C. G., Parker, C. A. (1956), *Proc. R. Soc. London, Ser. A* **235,** 518–536.

Heller, H. G. (1991), "Photochromics for the Future," in: L. S. Miller, J. B. Mullin (Eds.), *Electronic Materials: From Silicon to Organics,* New York: Plenum, pp. 471–483.

Hollenberg, J. L. (1970), *J. Chem. Educ.* **47,** 2–14.

Horváth, O., Stevenson, K. L. (1993), *Charge Transfer Photochemistry of Coordination Compounds,* New York: VCH.

Kopecky, J. (1992), *Organic Photochemistry: A Visual Approach,* New York: VCH, Chap. 14.

Norrish, R. G. W., Bamford, C. H. (1937), *Nature* **140,** 195–196.

Okabe, H. (1978), *Photochemistry of Small Molecules,* Chichester and New York: Wiley, Chaps. 2, 4–7.

Polanyi, J. C., Zewail, A. H. (1995), *Acc. Chem. Res.* **28,** 119–132.

Scaiano, J. C. (1973), *J. Photochem.* **2,** 81–118.

Terenin, A. N., Ermolaev, V. L. (1956), *Trans. Faraday Soc.* **52,** 1042–1052.

Wagner, P. J. (1971), *Acc. Chem. Res.* **4,** 168–177.

Wagner, P. J., Hammond, G. S. (1967), *Adv. Photochem.* **5,** 21–156.

Wegner, E. E., Adamson, A. W. (1966), *J. Am. Chem. Soc.* **88,** 394–404.

Woodward, R. B., Hoffmann, R. (1970), *The Conservation of orbital Symmetry,* Deerfield Beach, FL: VCH.

Further Reading

Adamson, A. W., Fleischauer, P. D. (1975), *Concepts of Inorganic Photochemistry,* New York: Wiley.

Andrews, D. L. (1992), *Applied Laser Spectroscopy,* New York: VCH.

Balzani, V., Carassiti, V. (1970), *Photochemistry of Coordination Compounds,* New York: Academic.

Barltrop, J. A., Coyle, J. D. (1975), *Excited States in Organic Chemistry,* Chichester and New York: Wiley.

Barltrop, J. A., Coyle, J. D. (1978), *Principles of Photochemistry,* Chichester, and New York: Wiley, Chap. 6.

Cowan, D. O., Drisko, R. L. (1976), *Elements of Organic Photochemistry,* New York: Plenum.

Coyle, J. D., Hill, R. R. (Eds.) (1982), *Light, Chemical Change and Life: a Source Book in Photochemistry,* Walton Hall, Milton Keynes, U.K.: The Open University Press.

Coyle, J. D. (1986), *Introduction to Organic Photochemistry,* Chichester and New York: Wiley.

Dartnall, H. J. A. (Ed.) (1972), *Photochemistry of Vision,* Berlin: Springer-Verlag.

Demas, J. N. (1983), *Excited State Lifetime Measurements,* New York: Academic.

De Mayo, P. (Ed.) (1980), *Rearrangements in Ground and Excited States,* Vol. 3, New York: Academic.

Ferraudi, G. J. (1988), *Elements of Inorganic Photochemistry,* New York: Wiley.

Fleming, G. R. (1986), *Chemical Applications of Ultrafast Spectroscopy,* New York: Oxford Univ. Press.

Frimer, A. A. (Ed.) (1985), *Singlet Oxygen,* Vols. 1–4, Boca Raton, FL: CRC Press.

Geoffroy, G. L., Wrighton, M. S. (1979), *Organometallic Photochemistry,* New York: Academic.

Gilbert, A., Bagott, J. (1991), *Essentials of Molecular Photochemistry,* Oxford, U.K.: Blackwell Scientific.

Gilchrist, T. L., Storr, R. C. (1979), *Organic Reactions and Orbital Symmetry,* Cambridge, U.K.: Cambridge Univ. Press.

Grossweiner, L. I. (1994), *The Science of Phototherapy,* Boca Raton, FL: CRC Press.

Horspool, W. M. (1976), *Aspects of Organic Photochemistry,* New York: Academic.

Horspool, W. M. (Ed.) (1984), *Synthetic Organic Photochemistry,* New York: Plenum.

Horváth, O., Stevenson, K. L. (1993), *Charge Transfer Photochemistry of Coordination Compounds,* New York: VCH.

Klessinger, M., Michl, J. (1995), *Excited States and Photochemistry of Organic Molecules,* New York: VCH.

Kopecky, J. (1992), *Organic Photochemistry: A Visual Approach,* New York: VCH.

Mees, C. E. K., James, T. H. (1966), *The Theory of the Photographic Process,* New York: Macmillan.

Monroe, B. M., Weed, G. C. (1993), *Chem. Rev.* **93,** 435–448.

Neblette, C. P. (1962), *Photography, Its Materials and Processes,* New York: Van Nostrand.

Padwa, A. (Ed.) (1979–87), *Organic Photochemistry,* Vols. 4–9, New York: Marcel Dekker.

Rabek, F. F. (1982), *Experimental Methods in Photochemistry and Photophysics,* New York: Wiley.

Roffey, C. G. (1982), *Photopolymerization of Surface Coatings,* Chichester: Wiley.

Shipley, T., Crescitelli, F. (1979), *Visual Photochemistry: The Beginnings,* Oxford, U.K.: Pergamon.

Turro, N. J. (1978), *Modern Molecular Photochemistry,* Menlo Park, CA: Benjamin/Cummings, Chaps. 10–14.

Wayne, R. P. (1988), *Principles and Applications of Photochemistry,* Oxford, U.K.: Oxford Univ. Press.

RECORDING DEVICES

HIDEO FUJIWARA, *Center for Materials for Information Technology, The University of Alabama, Tuscaloosa, Alabama, U.S.A.*

	Introduction	149
1.	**Magnetic Recording**	150
1.1	Principles and Modes of Magnetic Recording	150
1.2	Magnetic Recording Devices	152
1.2.1	Audio Tape Recorders	152
1.2.2	Video Tape Recorders	153
1.2.3	Digital-Data Recording	154
1.3	Magnetic Heads	155
1.3.1	Inductive Heads	156
1.3.2	Magnetoresistive (MR) Heads	160
1.4	Magnetic Recording Media	163
1.4.1	Particulate Media	163
1.4.2	Thin-Film Media	166
1.4.3	Perpendicular-Recording Media	169
2.	**Optical Recording**	170
2.1	Concept of Optical Recording	170
2.2	Optical Recording Devices	170
2.3	Optical Heads	172
2.4	Optical Recording Media	173
2.4.1	Read-Only Disks	173
2.4.2	Write-Once (W-O) Disks	174
2.4.3	Rewritable (Erasable) Disks	175
2.4.3.1	MO (magneto-optic) Disks	175
2.4.3.2	Phase-Change–Type Disks	179
2.4.4	Multiplex Recording	180
2.4.4.1	Hole Burning	180
2.4.4.2	Holographic Recording	181
3.	**Conclusion**	181
	Glossary	181
	Works Cited	182
	Further Reading	183

INTRODUCTION

People since prehistoric ages have been managing to create various methods of recording to store and communicate information. Varieties of media were utilized such as rocks, bones and skins of animals, trunks and skins of trees or plants including the pith of the papyrus cut and pressed into sheets, etc. Recording was performed by putting a mark upon them by carving or painting. The inventions of paper and typography were indeed epoch-making in the history of recording technology. However, with the progress of electronics, recording technology has made a huge advance in its performances such as quality, capacity, speed, etc., and its progress is still being accelerated in accordance with the ever-increasing demand of the present information-technology era. This article focuses on magnetic and optical recording devices that will play a decisive role in the information processing, although there exist several key technologies in recording such as pen recorders, phonograph-type records, and photographic recording, about which the readers may be referred to the articles LABORATORY INSTRUMENTATION, SOUND REPRODUCTION, and RECORDING PHOTOGRAPHIC, respectively.

Magnetic recording has a history of almost one century, during which enormous development has been made in all aspects, starting from a steel wire recorder. The inventions of magnetic-powder–coated medium and the ring-type head especially have promoted surprisingly great progress in this technology, resulting in various kinds of systems such as audio and video tape recorders, disk memories of both rigid and flexible types, card systems, etc., which are all now indispensable in human life. On the other hand, it has not been much longer than fifteen years since optical recording systems were brought into market as video disk and compact disk (CD) players right after the mass-production technology for laser diodes had been accomplished. In the following, brief concepts of both magnetic and optical recordings will be given with some emphasis on the materials used recently and some idea

3-527-28138-X/96/$5.00 + .50

about future trends rather than historical ones.

1. MAGNETIC RECORDING

1.1 Principles and Modes of Magnetic Recording

Magnetic recording devices basically consist of two parts: a medium on which information is written in the form of magnetic patterns corresponding to the information, and a head by which the information is written and/or read out. The medium, therefore, has a magnetic layer, which is usually coated on a substrate—rigid or flexible depending on the devices in which it is used. The write head is composed of a magnetic core and a coil wound around it and is designed to generate a magnetic flux in a restricted tiny portion, the strength of which is controlled by an electric current applied to the coil. The read head can be the same write head, or a specifically designed one that is sensitive to a magnetic flux coming out of the recorded medium.

In Figs. 1(a) and 1(b) are illustrated schematically the write processes of the two typical modes (longitudinal and perpendicular; Iwasaki and Nakamura, 1977) of magnetic recording. A magnetic head having (a) a small gap or (b) a thin main pole accompanied by an auxiliary pole is put close to a recording medium, with the gap or the pole tips facing it. During writing, the head moves relative to the medium, and a current modulated according to the information to be written is applied to the coil wound around the head core (or pole), making magnetic flux flow out of the gap (or the pole tip). This flux creates a track of magnetic pattern carrying the information in the form shown in the figures, with the head width determining the track width. For readout, the track is traced by the head. This time, the leakage flux from the magnetic pattern flows into the head core or the poles passing through the coil inducing a voltage in the coil, from which the information embedded in the magnetic pattern is reproduced.

Figure 2 illustrates the principle of the magnetic recording process by use of hysteresis curves (curves of magnetization M vs field H) of the magnetic layer of a medium, taking a pulse recording as an example. Here, H is the sum of all the fields applied to the relevant spot, i.e., the head field, the leakage field from the surroundings H_{srd}, and

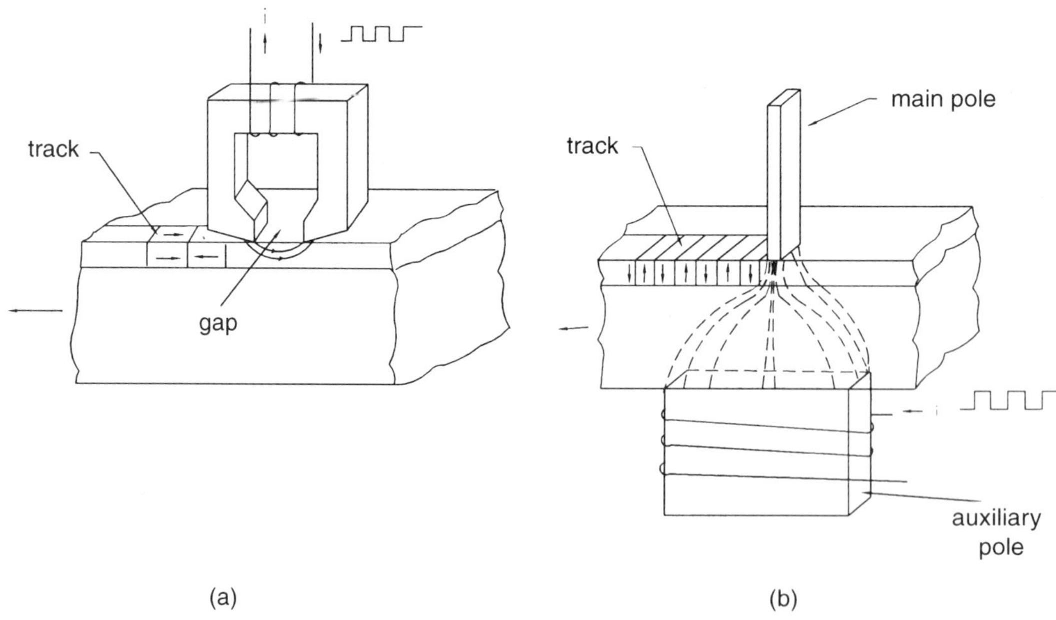

FIG. 1. Write process of magnetic recording: (a) longitudinal and (b) perpendicular recording.

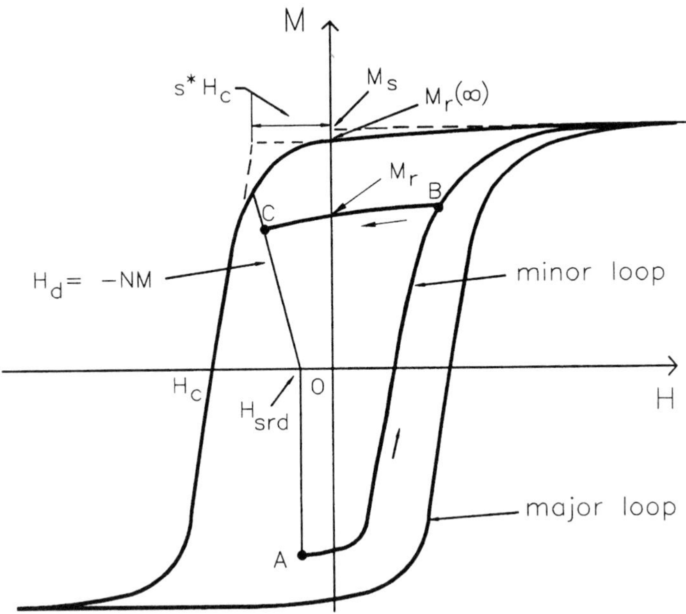

FIG. 2. Recording process on the *M–H* curves (pulse recording). *A*, initial state; *B*, maximum point (corresponding to the maximum write field), *C*, final state (recorded state); H_{srd}, field from the surroundings; H_d, demagnetization field; *N*, demagnetization factor. The saturation magnetization M_s, remanence M_r, squareness ratio M_r/M_s, coercive field H_c, and s^* (which is defined as shown in the figure) are the fundamental factors determining the recording characteristics.

the demagnetization field H_d, which is expressed as

$$H_d = -NM, \tag{1}$$

where N is a coefficient determined by the shape of the written bit and called "demagnetization factor." The recording starts from the initial point A and is completed at the final point C as is indicated by the arrows. The hysteresis curves represent the magnetic properties inherent to the medium and depend on the materials and fabrication conditions, and proper properties should be chosen for each specific device.

The modes of writing are classified basically into longitudinal and perpendicular recording and also into analog and digital recording according to the direction of the magnetization upon which signals are loaded and the nature of the signals to be recorded, respectively.

Although perpendicular recording is expected to show an excellent characteristic especially in the very high recording density, all practical systems have historically employed the longitudinal mode, with the exception of the evaporated tape case, which can be called "intermediate mode." Figures 3(a) and 3(b) show examples of the cross-sectional view of the recorded pattern of three different recording densities obtained by simulations for each mode of recording upon media having typical magnetic properties suitable for high-density recording for each case (for instance, the thickness for longitudinal recording is chosen much thinner than for perpendicular recording). The scaling ratio of the magnetization vectors for the longitudinal and perpendicular recordings is set as 3. Thus, it is seen that for the highest recording density shown here [300 kbpi (bpi: bits/inch)] the maximum recorded magnetization at the surface in perpendicular recording is much greater than that in longitudinal recording. Analog signals are recorded either by the ac bias method or by the frequency-modulation (FM) method, while digital signals are recorded by the pulse-coding method. The ac bias method was invented in order to improve the linearity of the recording response to the input head current, in which a high-frequency current (bias current) is added to the signal current. The FM method is rather similar to digital recording in the nature of the current wave form. The only difference is that in the FM method, the zero-crossing points are modulated in the analog manner, while in digital recording, they appear with intervals of integers times the predetermined minimum interval. The digital recording method

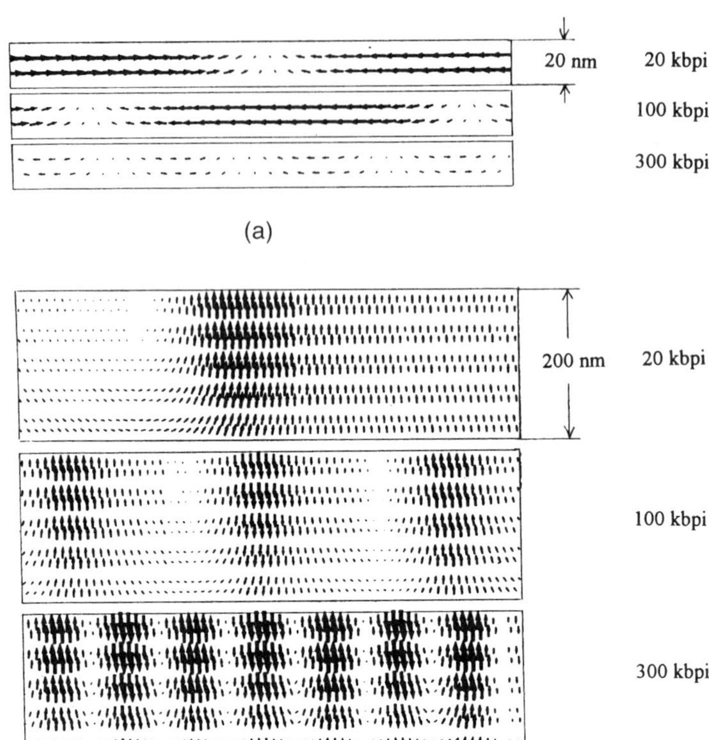

FIG. 3. Cross sections of recorded patterns for (a) longitudinal and (b) perpendicular recording at various recording densities obtained by simulation. The scaling ratio of the magnetization vectors for (a) and (b) is 3:1. (Courtesy of Dr. K. Yoshida and Mrs. F. Akagi, Hitachi.)

was first started from digital computer use. However, it is now being applied widely for all kinds of devices because of its potential perfectness in the reconstruction of the original signals by use of error-detection and -correction schemes.

1.2 Magnetic Recording Devices

A variety of devices exist that are designed to match various applications in the forms of a tape, drum, disk, card, sheet, etc. Among them, audio-video recording and digital-data storage for computers and instrumentation are the major applications. In Fig. 4, some typical magnetic recording devices are shown schematically.

1.2.1 Audio Tape Recorders Originally, audio tape recorders started from an open-reel type, in which two reels were set on a deck equipped-with a fixed read/write head and the tape was transported from one reel to the other during the read/write processes. However, they have been almost completely replaced by the two-reel "compact cassette" type, especially for consumer use, in which two stereo tracks are recorded on a 3.8-mm-wide tape.

In order to provide better sound quality and duplicability, digital tape recorders have been developed, which are divided into two categories; rotary-head type and stationary-head type. The rotary-head type recorder of the standard type is called R-DAT (3.8-mm tape), which is a modification of the VCR (video cassette-tape recorder) in which a helical-scan rotary head and a cassette tape are employed. The format called DASH (digital audio stationary head) was the first adopted for the stationary type. It is of an open-reel type carrying 2-in. tape and uses 24 or 48 heads in parallel. The plurality of heads allowed reducing the tape speed, which must have otherwise been made greater than 10 m/s. Another format of 32 channels using 1-in. tape is now also available. Recently, a stationary-head–type digital tape recorder

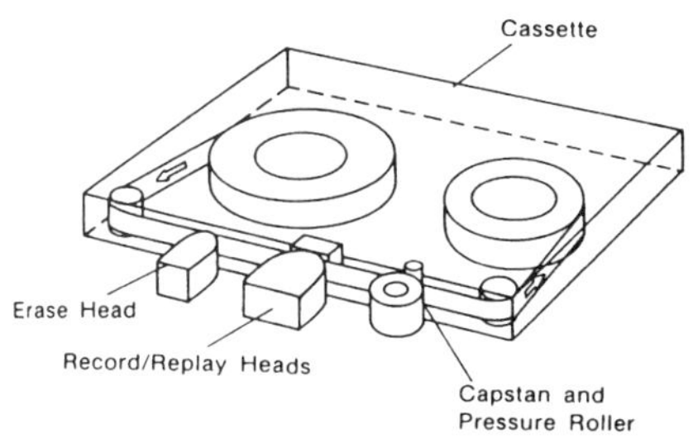

(a) Audio cassette recorder.

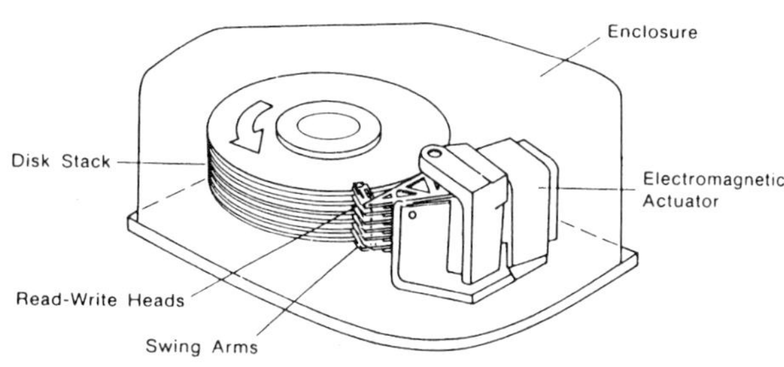

(c) Rigid disk drive.

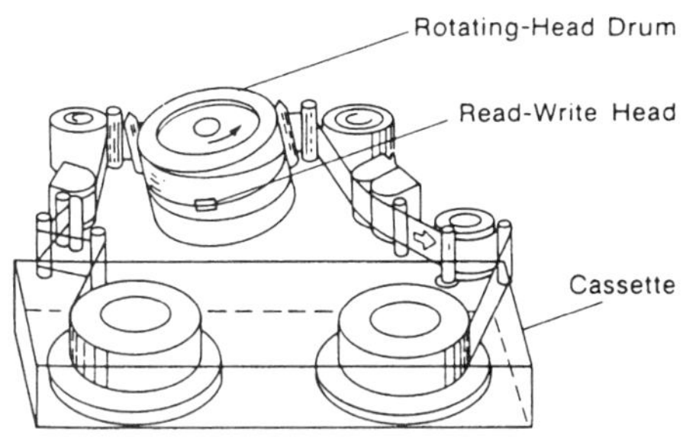

(b) Helical-scan video recorder.

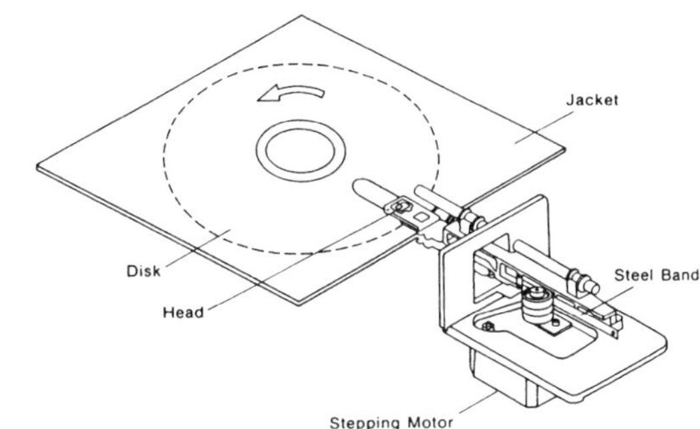

(d) Floppy disk and head accessing system.

FIG. 4. Schematic illustration of typical magnetic recording devices (after Mee and Daniel, 1989).

named DCC (digital compact cassette) equipped with 18 parallel thin-film heads has been developed for consumer use. The cassette has the same basic dimensions as the conventional "compact cassette" providing read compatibility with the conventional one. The heads used for each of the above devices are designed differently from each other as will be seen in Fig. 7. In Table 1, recording characteristics and some other parameters are listed for typical audio tape recorders described above.

1.2.2 Video Tape Recorders Video signals usually carry more than two orders of magnitude wider frequency bandwidth (several megahertz) than audio ones. The frequency response of magnetic recording is not good enough to record directly such wide-band signals, and several modulation methods have been explored. Among them frequency modulation (FM) was proved to be most appropriate because of its insensitivity to the low-frequency noise inherent to electronics.

The first video recording system for broadcasting was accomplished in the form of the so-called "quadruplex head" scheme, in which a rotary drum carrying four heads that scanned a 2-in.-wide tape transversely was employed. One of its drawbacks was that a picture field was divided into 16 tracks making the electronics for reconstructing the field exceedingly complex. The "helical-scan" rotary head was thus invented to solve this problem, first using a 1-in. tape, and subsequent downsizing has continued via $\frac{3}{4}$-in. (U format) to $\frac{1}{2}$-in. (L and M formats).

On the other hand, the consumer-use video tape recorders started with the helical-scan type from the beginning. Although some of the U format were used for some time, they were soon replaced by $\frac{1}{2}$-in. video cassette-tape recorders (VCR) of β format and VHS format. The 8-mm video tape recorder was later developed for portable video cameras.

Digital video tape recorders called D-1 to D-5 have been successively developed for professional use within the past ten years. A number of experimental works have also been performed for the application of the

Table 1. Audio tape recorders.

	Open reel (Open reel)	Compact cassette (Cassette)	R-DAT (Cassette)	Digital micro (Cassette)	DCC (Cassette)
Recording scheme	Analog	Analog	Digital	Digital	Digital
Head type	Stationary	Stationary	Rotary	Rotary	Stationary (MR)
Relative speed	38, 19, 9.5 cm/s	4.76 cm/s	3 m/s		4.76 cm/s
Tape speed	Same	Same	8.15 mm/s	6.35 mm/s	Same
Track width	2.03 cm (mono) 1.05 cm (stereo)	1.6 cm (mono) 0.61 cm (stereo)	15 μm	13 μm	185 μm
Gap length (μm)	2	1.6	0.2		1.6
Tape width (mm)	12.7, 6.3	3.78	3.81	2.5	3.78
Playing time (min)	30 (7-in. reel)	120 (max.)	120 (max.)	120 (max.)	120 (max.)
Kind of tape	γ-Fe_2O_3	γ-Fe_2O_3, CrO_2, MP[a]	MP[a]	ME[b]	MP[a]
First delivery	1947	1966	1986	1992	1992

[a]Metal particle tape (acicular Fe particles).
[b]Metal evaporated tape [Co(Ni)].

digital technology to consumer use, where an 8-mm cassette is expected to be employed. However, in order to make it compatible with high-definition television (HD-TV), a bit rate as high as 130–160 Mbits/s is required, claiming the bit size as small as 1 μm^2, which is almost one order of magnitude smaller than that for D-5 (6 μm^2).

In Tables 2 and 3, the characteristics of typical analog and digital video tape recorders are listed, respectively.

1.2.3 Digital-Data Recording Tape recorders are also used for the recording of digital data. However, the demands for very quick access to the stored data gave birth to a magnetic drum, followed by the magnetic rigid disk that is now employed for the major on-line file memories for almost all computers. The head is mounted on a slider supported by a suspension that is carried by an arm. The slider is designed to float extremely close (a few thousand angstroms) to the

Table 2. Video tape recorders—professional broadcasting/home use (analog).

	2-in. type (Transverse scan)	1-in. type (Helical scan)	U (Helical)	M-II (Helical)	β-cam-sp (Helical)	UNIHI (HD-TV) (Helical)	VHS (Helical)	8-mm video (Helical)
Tape								
Width (mm)	50.8	25.35	19.0	12.65	12.65	12.65	12.56	8.0
Materials	γ-Fe_2O_3	γ-Fe_2O_3/Co-γ-Fe_2O_3	Same	MP	MP	MP	Co-γ-Fe_2O_3	MP/ME
Coercive field (Oe)	250/270	270/700	500/630	1500	1500	1500	600–700	1500/1050
Head								
Core materials	Sendust	Ferrite	Ferrite	Amorphous	Sendust (sputtered)	Sendust (multilayered)	Ferrite	Sendust (sputtered)
Number of heads	4	1 + sub.	2	2	2	4	2	2
Track width[a] (μm)	250	130	85	Y:44, C:36	Y:86, C:36	24.8	58	20.5
Relative speed (m/s)	39.5	25.59	10.26	7.09	6.9	21.4	5.8	3.8
Playing time (min)	96	180	60	95	94	63	120/540	120
Delivery	1956	1976	1970[b]	1985	1986	1989	1976	1983

[a]Y for brightness signals, C for color signals.
[b]Announcement of standardization.

Table 3. Video tape recorders—professional broadcasting (digital).

	D-1 (Transverse scan)	D-2 (Helical scan)	D-3 (Helical)	D-5 (Helical)	1-in. open (HD-1) (Helical)
Tape					
Width (mm)	19.01	19.01	12.65	12.65	25.4
Materials	Co–γ-Fe_2O_3	MP	MP	MP	MP
Coercive field (Oe)	850	1500	1600	1600	1500
Head					
Core materials (write/read)	Ferrite/ferrite	MIG/ferrite	Co-based multilayer/ ferrite	Fe–N thin film	Amorphous/ ferrite
Number of write heads	4	4	4	8	8
Track width (μm)	40	39.1	20	20	37
Relative speed (m/s)	35.6	27.4	21.4	21.3	10.26
Bit length (μm)	0.45	0.43	0.38	0.30	0.35
Playing time (min)	76	208	245	120	90
Delivery	1987	1988	1990	1993	1988

smoothly finished magnetic surface of the disk. The data tracks are recorded circumferentially, and high-speed access to each track is achieved by moving the arm quickly either by a linear or by a rotary actuator bestowed with a precise servoing accuracy. The disks at the early stages were as large as 24-in. in diameter, with their capacities being only 10 MB (megabytes)/disk. Downsizing has also been performed in this field through 14-, 10.8-, 8.3-, 5.25-, 3.5- to 2.5-in., and now even a 1.8-in. disk is also available. Their typical characteristics are listed in Table 4. The rigid disks are in most cases used in a stack of 2 to 10 disks set on a spindle, to each surface of which one or a plurality of heads is assigned. This allows the achievement of much higher access speed for high-capacity file memories than a single disk drive.

The flexible disk called "floppy disk" is another major digital-data storage means. The medium is a flexible plastic disk coated with magnetic paint and is put in a soft jacket or a hard cartridge removable from the disk drive. The head comes in contact with the disk surface, except for the Bernoulli-type floppy-disk drive, and is driven by a step motor. The track density is designed much lower than that for the rigid disk. These design concepts have resulted in low-cost media and drives meeting a demand in the low-end uses. The disk size started as 8-in. followed by 5.25- and 3.5-in, successively, and the capacity has been increased from 0.4 to 2 MB. Even 3.5-in. disks of 4 and 12.5 MB have already been developed. In order to attain more than 20-MB capacity on a 3.5-in. disk, an optical-servoing–type floppy disk called "floptical disk" has been developed, in which the magnetic coating is set upon a PET (polyethylene terephthalate) substrate the surface of which is embossed so as to facilitate optical tracking. Very recently, a 3.5-in. floppy-disk drive of 100 MB in which a flexible disk of ultrathin metal particle coating (several hundred nanometers thick) is employed has been announced. In Table 4, the characteristics of typical floppy disks are also listed.

1.3 Magnetic Heads

A magnetic head must be endowed with "write" and/or "read" functions. The basic structures of write heads are seen in Fig. 1. The longitudinal recording head is essentially composed of a high-permeability magnetic core that forms a ring having a narrow gap and a coil wound around the core, thus being called a "ring head." For the perpendicular recording, on the other hand, a specific head consisting of a main pole with a narrow tip at one end and an auxiliary pole wound by a coil has been proposed, which may best be called a "probe head," although a ring head is also applicable. Since both types of heads are inductive in their nature, they are called "inductive heads." Inductive heads are also used as read heads, but some

Table 4. Rigid disks, floppy disks, and data recorders.

	Rigid disks						Floppy disks			
	14 in.	8 in.	5.25 in.	3.5 in.	2.5 in.	8 in.	5.25 in.	3.5	Floptical (3.5 in.)	Digital data recorder QIC
Medium	γ-Fe_2O_3	Co–Ni–P (plated)	Co-based alloy on Cr	Same	Same	γ-Fe_2O_3	Co–γ-Fe_2O_3	Co–γ-Fe_2O_3	Ba–ferrite	Co–γ-Fe_2O_3
Head[a]	NiFe thin film (ferrite)	Ferrite	NiFe thin film	Fe–N thin film MR	Same MR	Ferrite	Ferrite	Ferrite	Ferrite	NiFe thin film MR
Rotation speed (rpm)	3600 (1500)	3000	5400	5400	3800	360	360	360	720	Tape speed 120 ips
Linear density (kbpi)	15.9 (1)	14	50	87	74.8	6.8	9.9	14.2	17.7	80
Track density (tpi)	2100 (50)	1100	2900	4100	4300	48	96	135	1245	400
Capacity (MB)	7500 (6)	400	3700	1050	344	1.6	1.6/2.0	1.6/2.0	25	10 000
Delivery	1987 (1963)	1981	1991	1993	1993	1977	1981	1985	1991	1994

[a]MR: magnetoresistive.

read-only heads, such as a Hall head and a MR (magnetoresistive) head, have been developed to obtain higher field sensitivity. The Hall* or MR element is set at the front end of a head tip or in a gap of the magnetic core (either the front gap or a gap especially made at a recessed portion of the core in order to provide a wider space for it) or to avoid direct contact of the sensor with the medium.

1.3.1 Inductive Heads In order to obtain a distribution of the head field quantitatively in the vicinity of the head gap or the pole tip, it is generally required to solve a Poisson's equation regarding a vector potential, which can only be solved numerically in most cases (see Fig. 5). However, analytical expressions obtained under some simplified but reasonable assumptions are often useful. For an infinite-permeability ring head having a sufficiently high ratio of the track width to the gap length, the field components H_x and H_y along the center of the track width are well expressed by

$$H_x = \frac{H_0}{\pi}\left(\tan^{-1}\frac{g/2 + x}{y} + \tan^{-1}\frac{g/2 - x}{y}\right), \tag{2a}$$

$$H_y = -\frac{H_0}{2\pi}\ln\frac{(g/2 + x)^2 + y^2}{(g/2 - x)^2 + y^2}, \tag{2b}$$

where the head is supposed to be placed with respect to the coordinate as shown in Fig. 5(a); g is the gap length and H_0 is the field at the center of the gap at $y = 0$. These formulas are well known as the Karlqvist expressions. For the probe head, no such simple expressions have been given, but qualitatively, H_x and H_y of the probe head have the nature of $-H_y$ and H_x, respectively, of the ring head, as is seen in Figs. 6(a) and 6(b) in which they are plotted as functions of x.

The readout voltage e to be obtained by using an inductive head is given, by applying the reciprocity law,† as follows, in terms of

*For increasing the recording density, narrowing the gap length is essential. However, the semiconductor film of the Hall element is difficult to make thin enough to put into the narrow gap. Therefore, Hall heads have almost been discarded from present-day magnetic recording devices.

†The reciprocity law between two coils can be applied for the system of the coil of the head and the magnetization in the medium because a magnetic moment is equivalent to a circular current.

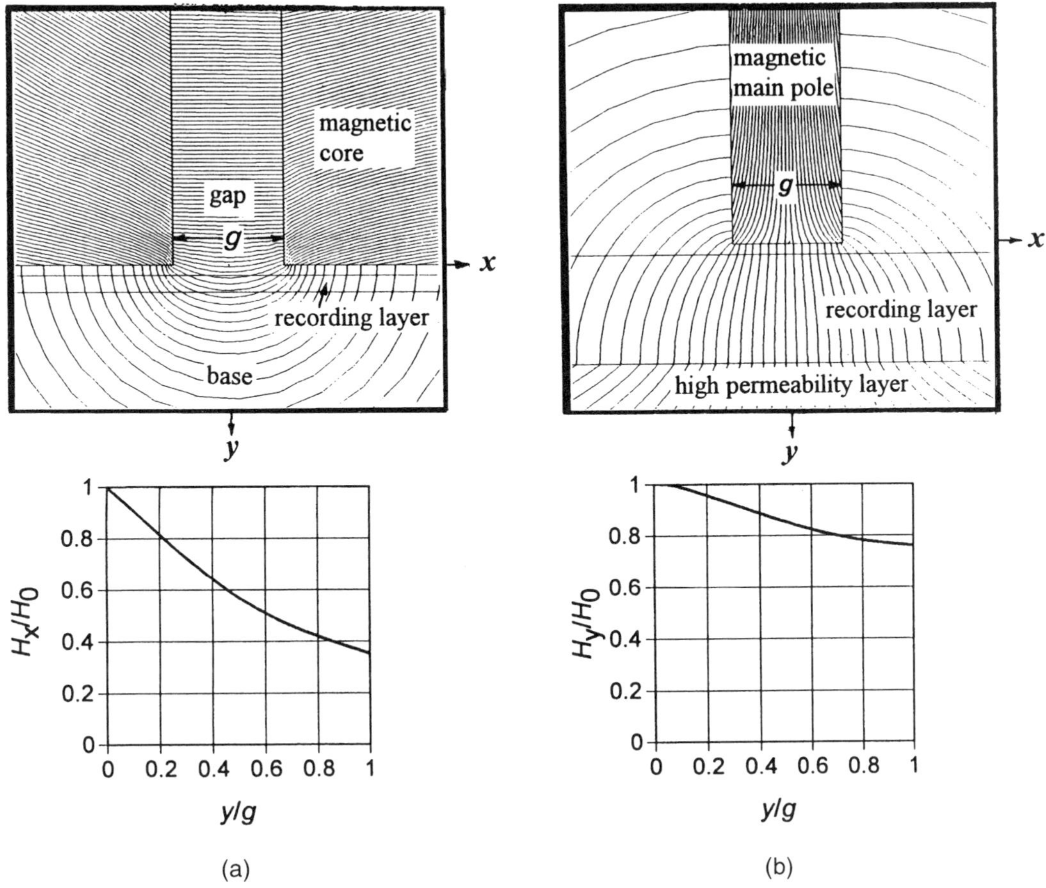

FIG. 5. Magnetic flux distribution: (a) at the head gap of a ring head and (b) at the pole tip of a probe head, calculated by simulation. The head fields at the center of the gap or the pole tip are also plotted as a function of the distance along the *y* direction. (Courtesy of Dr. Y. Yoshida and Mrs. F. Akagi, Hitachi.)

the magnetization distribution $M(x,y,z)$ in the medium and the head field distribution $H(x,y,z)$ that would be generated in the medium if a unit current be applied to the coil:

$$e = \mu_0 v \int H(x,y,z) \frac{dM(x,y,z)}{dx} d\tau, \tag{3}$$

where μ_0 is the permeability of the space, v is the relative speed of the head and the medium in the x direction, and the integration is performed for the whole volume of the medium. Since the permeability of the core or the pole materials has a finite value, the efficiency η of the head is generally reduced from that expected for the infinite-permeability case. Thus it is highly required that the core (pole) materials possess a high permeability. The following expression regarding the efficiency explains this situation:

$$\eta = \left[1 + \frac{g/\mu_0 S_g + l_c/\mu S_c}{R_l}\right]^{-1}, \tag{4}$$

where g and S_g are the length and area of the gap, l_c, S_c, and μ are the length, area, and permeability of the core, respectively, and R_l is the reluctance of the flux leakage paths.

The attainable maximum H_0 that appears in Eq. (2a) or Eq. (2b) is proportional to the saturation induction B_s of the pole material, especially in the vicinity of the head gap or the pole tip. Therefore, high B_s is another

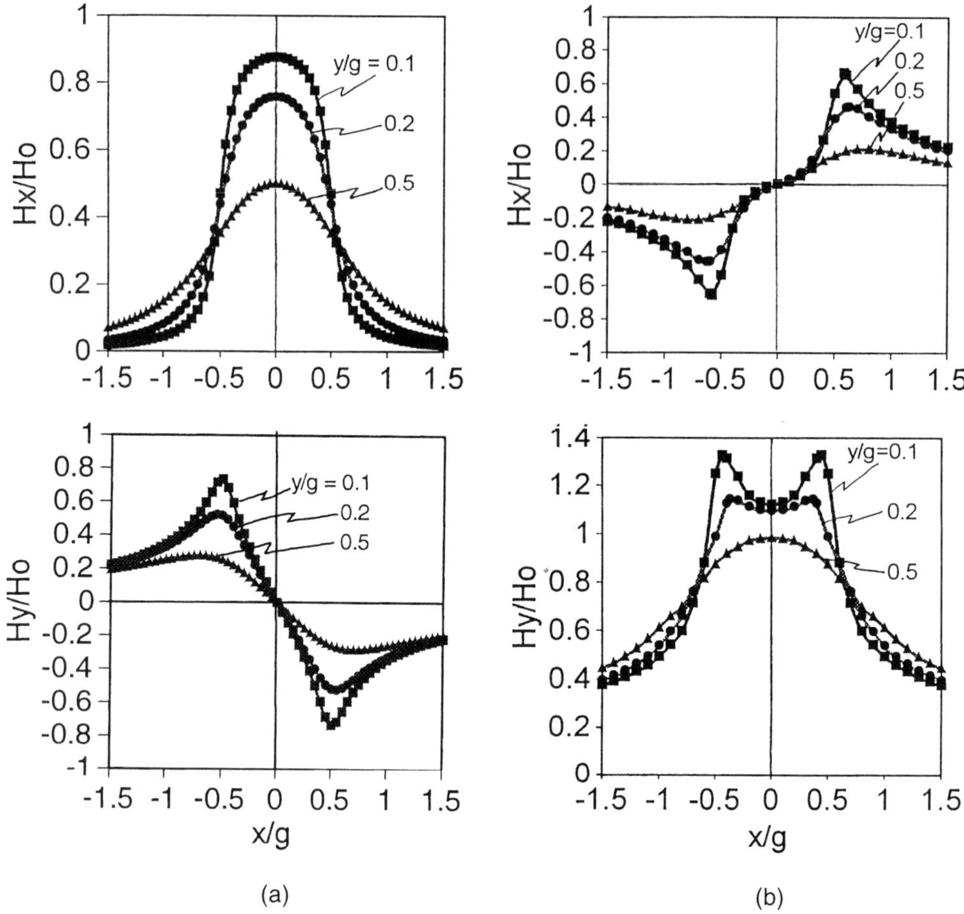

FIG. 6. Head field components H_x and H_y as functions of x (track direction) with the distance from the head surface y as a parameter: (a) of a ring head calculated by use of Karlqvist expressions, (b) of a probe head calculated assuming that the magnetizations at the main pole edge are all equal and parallel to the y direction.

important characteristic that the core/pole materials are required to assume.

Inductive heads are constructed in many ways by using bulk magnetic materials and/or thin films. In Fig. 7, the structures of typical heads are schematically illustrated. The audio heads are mostly composed of laminated Permalloy cores, the layers of which are insulated from each other to avoid eddy-current losses at high frequencies. The video heads for professional use were first composed of two parts, the back core and the front core. In order to match the video frequencies, sintered ferrite was chosen for the back core. For the front core, a mechanically hard Al–Fe–Si alloy called "Alfesil" or "Sendust"* was chosen because of its excellent precise machinability and abrasion durability. When the VCRs for consumer use were developed, single-crystalline Mn–Zn ferrites were available and were utilized as the core materials. The choice of specific crystalline orientations has been confirmed essential for obtaining both high read/write performances and a high wear resistance (Fujiwara *et al.*, 1982). A typical choice of the crystalline ori-

*This name came from the fact that an Al–Fe–Si alloy with a composition of Fe with Al 12%, Si 5%, which shows especially high permeability, was discovered in Tohoku University in Sendai (Japan), and was applied to making "dust" cores.

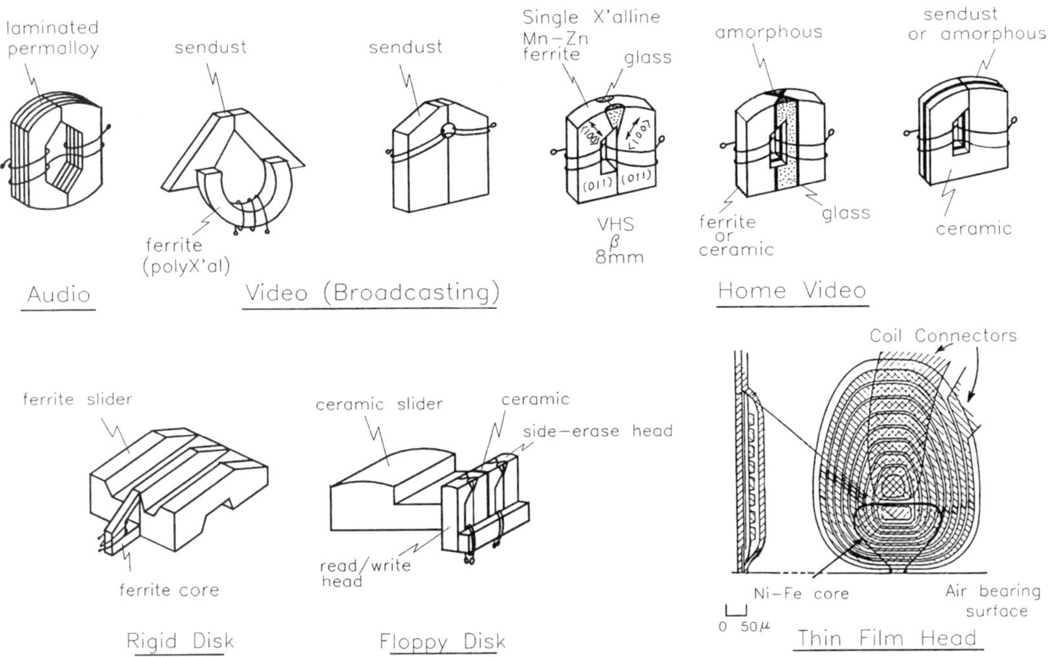

FIG. 7. Schematic illustration of typical magnetic heads (thin-film head: after Mee and Daniel, 1989).

entations is shown in Fig. 7. For rigid disk heads, polycrystalline Ni–Zn ferrites had long been used. This was because Ni–Zn ferrite had been believed to possess better characteristics at high frequencies than Mn–Zn ferrite on account of its very high resistivity. However, it was later confirmed that the machining effect improves the high-frequency characteristics of Mn–Zn ferrite and its higher saturation induction is more beneficial than Ni–Zn ferrite. The head cores are imbedded in a slider designed to float upon the disk surface hydraulically, the slider itself often making a part of the magnetic flux path. Floppy-disk heads are usually made of sintered polycrystalline ferrite and are imbedded in a slider.

In order to match the higher coercivity of the medium, heads having a metal gap have been developed for video tape recorders (see Fig. 7). Some of them are called MIG (metal-in-gap) or composite head as the core is mostly constructed by a ferrite and the gap portion is made of magnetic metals such as Permalloy, Sendust, and amorphous magnetic materials.

The thin-film head was introduced into disk systems first. Its chief merits are in the simplicity of its manufacturing process beneficial to mass production and in the characteristics of high permeability at high frequencies and high B_s (saturation induction). A typical structure of the thin-film head is shown in Fig. 7. Permalloy (a Ni–Fe alloy) is usually employed as a core material. The patterning of the head core and coils is performed by utilizing various fine lithography methods such as photolitography, electron lithography, and ion milling. Magnetic domain-wall stabilization or elimination is one of the principal concerns, which makes it possible to avoid the Barkhausen noise.

The demand for higher recording density, in which very high coercivity is to be employed, has stimulated the development of high-B_s head core materials. Fe-based alloys had been known as high-B_s materials, but their high crystalline anisotropy was preventing them from becoming magnetically soft enough to be employed as head core materials. The control of the grain size small enough to average out the crystalline anisotropy, which is recently referred to as nanocrystalline technology, was the breakthrough for that. This was first tried by multilayering and then by adding nitrogen and other ele-

Table 5. Head materials.

Material	Composition (wt%)	μ (dc/5 MHz)	H_c [Oe (A/m)]	B_s [kG (T)]	ρ ($\mu\Omega$ cm)	Hardness, Vickers
Mo Permalloy	4% Mo, 17% Fe, 79% Ni	11 000/–	0.025 (2.0)	8 (0.8)	50	120
Hardperm	9% Nb, 12% Fe, 79% Ni	8 000/–	0.005 (0.4)	6 (0.6)	80	350
Sendust	5.4% Al, 9.6% Si, 85% Fe	8 000/–	0.025 (2.0)	10 (1.0)	85	480
Amorphous	15% Si, 10% B, 5% Fe, 70% Co	80 000/500	0.006 (0.5)	6.6 (0.66)	134	910
	7% Mo, 8% Zr, 85% Co	–/620	0.6 (50)	10 (1.0)		1000
MnZn ferrite (single crystal)	30% MnO, 20% ZnO, 50% Fe_2O_3	4 000/500	0.05 (4)	3.8 (0.38)	10^3	640
MnZn ferrite (polycrystal)	19% MnO, 11% ZnO, 70% Fe_2O_3	3 000/800	0.1 (8)	4.8 (0.48)	10^4	650
NiZn ferrite (polycrystal)	11% NiO, 22% ZnO, 67% Fe_2O_3	850/550	0.4 (32)	3.9 (0.39)	10^7	600

ments that prevent excessive grain growth. In Table 5, properties of typical head core materials are listed.

1.3.2 Magnetoresistive (MR) Heads

The key element in MR heads is a MR element that consists of a narrow magnetoresistive thin-film stripe having a uniaxial magnetic anisotropy and current leads connected at each end of the stripe through which a current is supplied (see Fig. 8). The resistivity ρ changes as a function of the angle θ be-

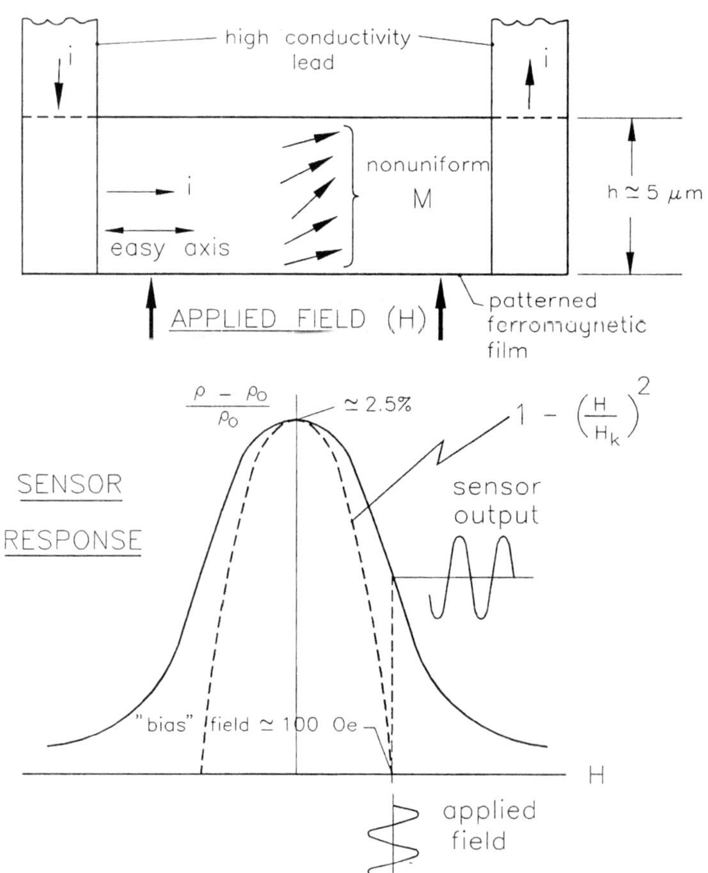

FIG. 8. Response of the MR head to the applied field (after Markham and Jeffers, 1990).

tween the current and the magnetization of the element, according to the expression

$$\rho = \rho_0 + \Delta\rho \cos^2\theta, \tag{5}$$

where ρ_0 and $\Delta\rho$ depend on the material of the element. In most cases the easy axis of the element is set parallel to the stripes. Therefore ρ is expected to change quadratically with the field H as

$$\rho = \rho_0 + \Delta\rho[1 - (H/H_k)^2], \tag{6}$$

since

$$\sin\theta = H/H_k, \tag{7}$$

with H_k denoting the anisotropy field. However, because of the demagnetization field the magnetizations lying close to the edge do not deflect so much as the center, giving rise to the characteristic inflection points and the skirts in the response curve as is illustrated in Fig. 8 (Markham and Jeffers, 1990). Thus, in order to obtain a linear response, various kinds of biasing methods have been explored, typical examples of which are shown schematically in Fig. 9. The elimination of Barkhausen noise is another important issue in practical use. Therefore, some measure has to be taken to avoid the generation of reverse domains at the edges, such as putting hard magnetic or antiferromagnetic films at the ends of the element or making the element double-layered, etc.

A magnetic shield is usually used to obtain a fine resolution as shown in Fig. 10. The reciprocal relation in this case is formally expressed as

$$e \sim \phi = \mu_0 \int \mathbf{H}(x,y,z) \cdot \mathbf{M}(x,y,z) d\tau, \tag{8}$$

where ϕ is the flux entering the MR element and $\mathbf{H}(x,y,z)$ is the field distribution in the space when a unit current is applied to an imaginary coil supposed to be wound around the MR element. Analytical expressions for $\mathbf{H}(x,y,z)$ can be obtained for the infinite track width almost in the same way as the Karlqvist expressions:

$$H_x = H_{rx}(x + (b + t)/4, y) - H_{rx}(x - (b + t)/4, y), \tag{9a}$$

and

$$H_y = H_{ry}(x + (b + t)/4, y) - H_{ry}(x - (b + t)/4, y), \tag{9b}$$

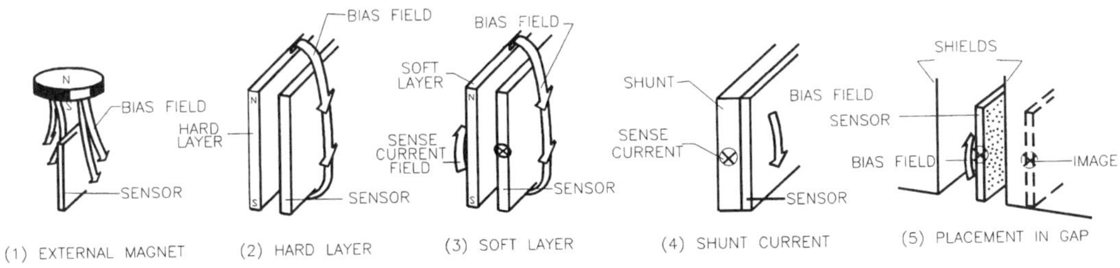

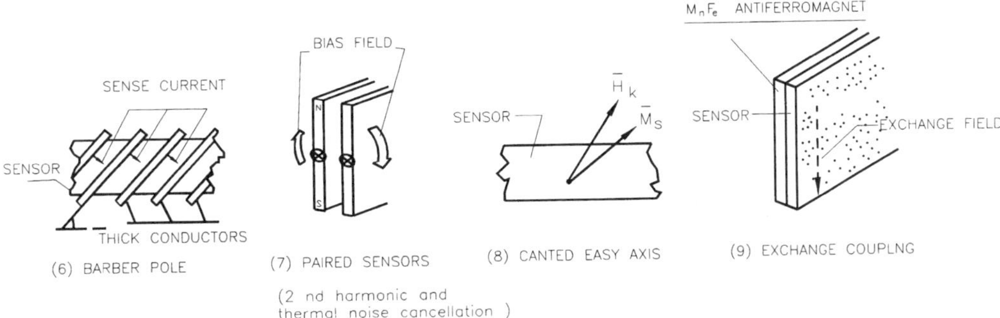

FIG. 9. Various biasing methods for MR heads (Jeffers, 1986; © 1986 IEEE).

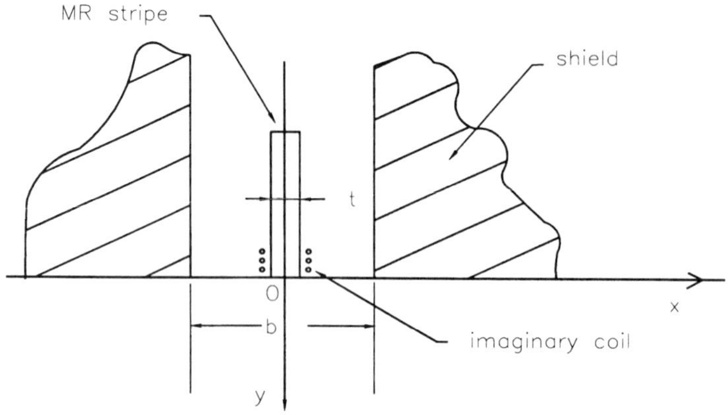

FIG. 10. Cross section of a shielded MR head, showing the coil that is imagined to exist for the purpose of computing the sensitivity function (after Potter, 1974).

where by H_{rx} and H_{ry} the Karlqvist's expressions for the ring head are denoted, with b and t denoting the gap length between the shield cores and the thickness of the MR element, respectively. The problem is, however, that the relation between e and ϕ is not usually simple. Therefore, the theoretical evaluation of the sensitivity requires some numerical calculations.

For MR heads, a high magnetoresistive ratio ($\Delta\rho/\rho_0$) is one of the basic requirements for the material. Although Permalloy has $\Delta\rho/\rho_0$ of only 2% to 3%, it has been the only material employed for practical use. The ρ_0 of Permalloy is about 20 $\mu\Omega$ cm and the applicable current density is almost 10^7 A/cm^2, leading to the maximum electric field of about 200 V/cm. This implies that for a track width of 10 μm, the signal output of 2–3 mV can be obtained taking into account the reduction of the usable amplitude to half of the maximum obtainable amplitude because of the nonlinearity in the transfer curve (Markham and Jeffers, 1990). This is almost an order of magnitude higher than the output obtainable by the inductive head at a bit density of 60 kfci (fci: flux changes per inch) and a relative speed of 1 m/s (MR heads are relatively speed insensitive by their nature). Some Ni–Co alloys show 6% to 7% $\Delta\rho/\rho_0$ ratio, but because of their high crystalline anisotropy and high magnetostriction, they have not been made applicable to magnetic heads.

In 1988, a new phenomenon called "the giant magnetoresistive (GMR) effect" was discovered in antiferromagnetically coupled multilayered films composed of a number of magnetic layers with nonmagnetic layers in between, each having the thickness of a few nanometers (Baibich *et al.,* 1988). This phenomenon comes from the spin-dependent resistivity at the interface and in the bulk of the magnetic layers and is isotropic with respect to the current direction, and is expressed as

$$\rho = \rho_0 + \Delta\rho \cos\phi, \tag{10}$$

where ϕ is the angle between the magnetizations of the adjacent layers, in contrast to the case for the conventional MR, which is now called anisotropic MR because of its dependence on the angle of the magnetization and the direction of the current.

The GMR was first found in Fe/Cr systems at very low temperature, but soon it was observed even at room temperature in Co/Cu systems, showing MR ratios as high as 65% (Parkin *et al.,* 1991). At first, to obtain this high ratio, a magnetic field of the order of 10 000 Oe was needed, but a lot of efforts have made it possible to reduce the field drastically. Recently, a structure that gives a substantial sensitivity in fields as low as several oersteds has been proposed in which a granular nature is introduced into the magnetic layers (Hylton *et al.,* 1993). It has been made clear that the antiferromagnetic coupling between the adjacent layers is not essential to the GMR effect. Thus, any metallic magnetic bilayer structures in which the magnetization of one layer is pinned and that of the other is free exhibit the GMR effect (Shinjo and Yamamoto, 1990; Dieny *et al.,* 1991). They are now called spin valves and are expected to be applied for read heads before long.

1.4 Magnetic Recording Media

It may safely be said that the history of magnetic recording media is finding ways of increasing the recording density. This is especially true with digital recording. For longitudinal recording, assuming the magnetization function at the bit transition with a transition length a_T as

$$M_x(x) = (2/\pi)M_r \tan^{-1}(x/a_T), \tag{11}$$

where x refers to the track direction, the output voltage with use of a ring head is given by

$$e_x(x) = -2\mu_0 vwM_r tH_x(x, a_T + h), \tag{12}$$

where v is the relative velocity of the head and medium, w is the track width, M_r is the saturation remanence of the medium, and H_x is given by Eq. (2a) (h: head-to-medium spacing). Then the output pulse width p_{50} (the width at 50% of peak amplitude) is given by

$$p_{50} = 2[(g/2)^2 + (a_T + h)(a_T + h + t)]^{1/2}, \tag{13}$$

where t is the thickness of the medium. This tells that a small gap g, spacing h, thickness t, and transition width a_T are essential to obtain a high resolution. Among those factors, a_T, t, and h in part are related to the medium properties. It should be noted that the above expressions are derived by assuming that the recording is perfect from the top to the bottom, which is the case only for very thin-film media relative to g. A lot of experiments have been performed and an empirical relationship between p_{50} and M_r, H_c, and t of the medium has been obtained for the longitudinal recording as follows:

$$p_{50} \sim (M_r/H_c)^a t^b, \quad a = 0.3\text{–}0.5, \quad b = 0.3\text{–}0.8, \tag{14a}$$

while Nakamura (1991) has obtained by simulation

$$p_{50} \sim (M_r/H_c)^{0.4} t^{0.4}. \tag{14b}$$

Thus, it is concluded that in order to obtain a high resolution, H_c of the medium should be made as high as possible within the limit of the writing capability of the head, although some compromise is necessary between the resolution and output in terms of M_r and t. The surface roughness is another critical factor that should be minimized from the viewpoint of both output and resolution.

The magnetic characteristics of the medium are mostly expressed by its B–H curves, schematic illustration of which was given in Fig. 2. Other than M_r and H_c, the squareness ratio, which is defined as shown in the figure, is also critical to the resolution.

Noise reduction is another important issue. The key point is to reduce the factors that cause inhomogeneity in the medium. Among them, the reduction of the particle size (or the effective size of magnetization switching unit), their homogeneous dispersion, and the smoothing of the surface are most critical.

In accord with the guidelines stated above, varieties of recording media have been developed and brought into practical use. Historically, the first medium was a steel wire, meanwhile replaced by tapes constituted of a plastic base and iron-oxide particles coated upon it with an organic binder. A plated medium was introduced for drum devices but they were soon replaced by rigid disks of particulate coating and now thin-film media are beginning to replace them, starting with disks such as 5.25-in., 3.5-in., or smaller-sized ones. Floppy disks are again of particulate coating.

1.4.1 Particulate Media A particulate medium is basically composed of a base, with a magnetic coating and lubricant coated upon it. As base materials, biaxially oriented polyethylene terephthalate (PET), in some cases with some fillers embedded to reduce the friction coefficient, is mostly employed for flexible media, and mirror-finished Al–Mg(4%–5%) alloys for rigid disks. Other than these, polyimides, polyamides, hard glass, ceramics, etc., are now under development. For most particulate media, acicular magnetic particles are used on account of their high coercivity. For tapes and disks, the particles are aligned in the track direction by applying a magnetic field during the coating process in order to give a higher squareness ratio, while in floppy disks, they are not. This is because the floppy disks are produced by punching out of a wide plastic sheet with a magnetic coating on it. As a lubricant, fluoro resins such as perfluoropolyether are used. The typical thickness of the coatings is a few micrometers for flexible media and several hundred nanometers for rigid disks, respectively, with some exceptions such as double-coated media in which less than 300-nm thickness has been attained for flexible media (Inaba *et al.*, 1995).

The magnetic paint (or dispersion) for coating is prepared by mixing the magnetic powder, a small amount of nonmagnetic powder (α-Al_2O_3, Cr_2O_3, etc., for strengthening the durability, graphite for preventing electrostatic charging), an organic binder, and some organic solvent by use of a sand mill and/or a kneader, though some water-borne paints are under development from the environmental point of view. The binder is designed to be composed of hydrophilic and hydrophobic segments so that it may prevent the particles from aggregation. Figure 11 illustrates the function of the binder. The hydrophilic segments work as an anchor and the hydrophobic ones as a repeller. For flexible media, thermoplastic compositions such as vinyl resin and/or urethane resin are employed, while for rigid disks, thermosetting ones such as epoxy resins or phenolic resins are used. Those have been chosen to facilitate each coating process: roll coating for flexible media and spin coating for rigid disks. The volume fraction of the magnetic particles in the coatings is limited to not much greater than 35% (25% for rigid disks) in order to endow the media with sufficient durability and smooth surfaces.

Magnetic particles in practical use are listed in Table 6. Acicular γ-Fe_2O_3 particles elongated in the [110] direction are prepared according to the processes shown in Fig. 12. The magnetic anisotropy of the particles thus prepared is the sum of the crystalline and shape anisotropies, the contribution of each being almost of the same order. The dehydration and reduction involved in the process are accompanied by a substantial material transport resulting in a more or less irregular shape containing many pores, which leads to a lower coercivity than expected. In order to overcome this drawback, a new process has been developed in which acicular γ-Fe_2O_3 particles are directly precipitated without passing through α-FeOOH (goethite). Thus their coercivity ranges from 250 to 450 Oe according to their preparation methods. CrO_2 is synthesized by hydrothermal synthesis (~400 °C) from Cr_2O_5 by using seeds having a rutile structure. The particles are mostly a single crystal of well-defined shape, which is beneficent for obtaining a coercivity as high as 700 Oe. However, there was a drawback to CrO_2 that toxic hexavalent Cr ions were to be treated during the preparation process. For its replacement, Co-

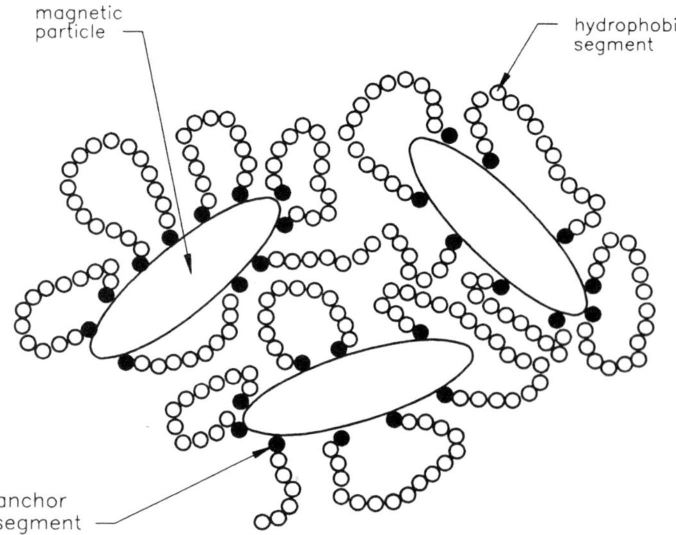

FIG. 11. Schematic model of the function of binder. The binder is composed of hydrophobic segments and hydrophilic segments located at the ends. The hydrophilic segments act as anchors to the magnetic particles.

Table 6. Magnetic particles (after Mee and Daniel, 1989).

Material	σ_s[a] (emu/g = A m^2/kg)	ρ (g/cm^3)	H_c [Oe (kA/m)]	$\langle \nu \rangle_{av}$ (10^{-4} μm^3)	SSA[b] (m^2/g)
α-Fe	125–170	5.8	940–1600 (75–130)	<1	35–60
Fe(70 at %)Co(30 at %)	140–190	5.8	1100–2000 (90–160)	<1	35–60
CrO_2	76–84	4.8	380–730 (30–58)	0.8–10	18–40
γ-Fe_2O_3	73–75	4.8	250–450 (20–36)	1–25	12–35
Co–γ-Fe_2O_3	70–75	4.8	380–940 (30–75)	0.7–10	16–50
$BaFe_{2-2x}Co_xTi_xO_{19}$					
$x = 0.7$	58	5.3	1200 (95)	<1	20–55
$x = 0.8$	58	5.3	700 (56)	<1	20–55

[a]Saturation moment per gram.
[b]Specific surface area.

modified γ-Fe_2O_3 has been developed. At first, replacement of a fraction of the Fe^{2+} by Co^{2+} was tried, which was expected to increase the crystalline anisotropy substantially resulting in an increase in coercivity. However, because of the drastic temperature dependence of its anisotropy, this trial was not successful. Instead, methods in which only the surface of the particle is modified, including epitaxial growth of Co-ferrite upon the surface, have been developed. Thus, coercivity as high as 900 Oe has become available. Figure 13 demonstrates the epitaxially grown Co-ferrite upon the surface of a γ-Fe_2O_3 particle. Acicular α-Fe particles were developed to obtain higher B_s particles than oxides. They are now being produced by the reduction of acicular precursors of α-FeOOH or γ-Fe_2O_3. In order to obtain chemical stability, the surfaces are passivated by mild oxidation, and some stable oxides such as Al_2O_3 and SiO_2 are coated, although the recently developed amine-quinone polymer

$FeSO_4$ water solution
NaOH →
$Fe(OH)_2$
air →
α–FeOOH (acicular) <0 1 0>
H_2O ←
α – Fe_2O_3
H_2 ,heating(400°C) →
Fe_3O_4
air(H_2O), 250°C →
γ –Fe_2O_3 < 110 >

FIG. 12. Typical production process of acicular γ-Fe_2O_3 particles.

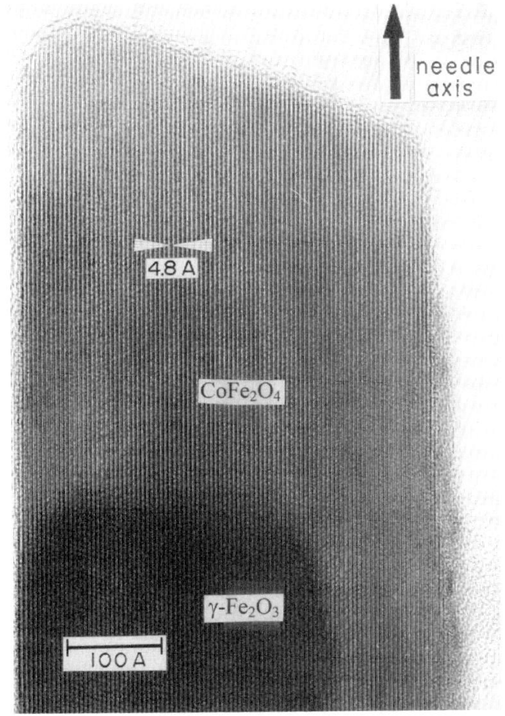

FIG. 13. Lattice image of a Co (epitaxial)–γ-Fe_2O_3 particle taken by a high-resolution electron microscope (courtesy of Hitachi Maxell).

binder may relieve the necessity of protection coating through their corrosion-protecting capability (Nikles *et al.*, 1994). Typical magnetic properties are seen in Table 6. Barium ferrite ($BaFe_{12}O_{19}$) has a hexagonal crystalline structure and has a strong uniaxial anisotropy (3.3×10^5 ergs/cm^3) with its easy axis parallel to the *c* axis. This high anisotropy results in a very high coercivity (~2000 Oe), and as a material for magnetic recording, the use was limited up to now only to master tapes or magnetic cards for specific applications. However, the idea of its application to perpendicular recording promoted its adaptability to more general use. The primary concern was the reduction of the coercivity, although it is estimated that coercivity as high as 3000 Oe will be necessary for future high-density recording. The reduction was accomplished by replacing some fraction of Ba or Fe ions by divalent and tetravalent elements such as Co–Ti or Zn–Ti. The particles are prepared either by precipitation in glass flux or by hydrothermal synthesis.

Figure 14 shows electron microscope images of typical magnetic particles mentioned above.

1.4.2 Thin-Film Media As is evident from the expressions (14a) and (14b) for p_{50}, thinness of the medium is essential to high-density recording especially for longitudinal recording. This is the reason why thin-film technology has received strong attention. Another advantage inherent to the thin-film media is the binderlessness that enables achieving higher saturation magnetizations than the particulate ones. However, obtaining a sufficiently high coercivity in thin films was the primary challenge, which was eventually achieved by utilizing crystalline anisotropy, with the aid of shape anisotropy in some cases, of the grains constituting the films. Thus, in most of the thin-film media, cobalt-based alloys having a hexagonal structure, in which a high uniaxial crystalline anisotropy is inherent, have been used, although there have been some exceptions such as gamma ferric oxide and barium ferrite. In thin magnetic films, zigzag walls as shown in Fig. 15 are apt to be formed at the recorded bit boundaries where the magnetizations point head to head (or tail to tail) against each other, reducing the magnetostatic energy at the expense of exchange and anisotropy energies. Their amplitude should be made as small as possible in recording. This can be attained by the reduction of the exchange interaction between the grains. Thus, by taking advantage of the ease of obtaining much smaller grains than in the particulate coatings, very low-noise thin-film media are now available. Protection from abrasion or mechanical damage is also important, a typical example of which is a carbon coating of the order of 10 nm thick.

Thin-film media are produced by plating, sputtering, or evaporation, the appropriate one of which is chosen according to its application: for tapes, evaporation because of its mass productivity and the ease of obtaining a high coercivity and high squareness ratio along the tape direction by making use of the oblique incident evaporation method, and for rigid disks, plating or sputtering because of its ease in producing films with isotropic high coercivity (in some cases, circumferential anisotropy is attached to the

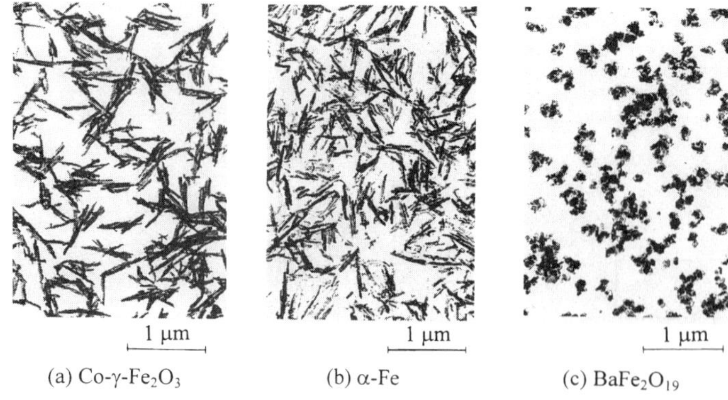

FIG. 14. Electron micrographs of typical magnetic particles for recording (courtesy of Hitachi Maxell).

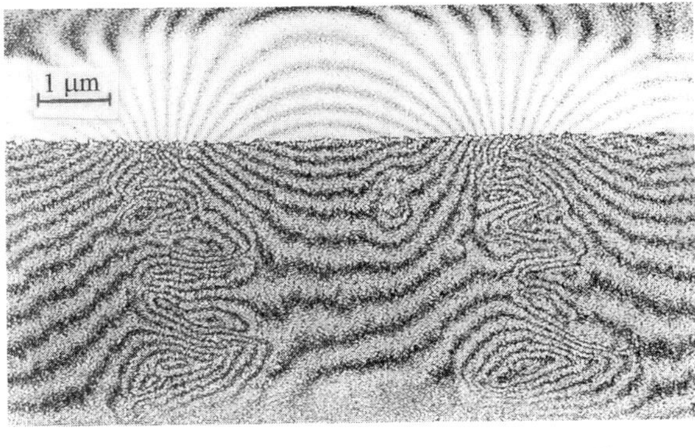

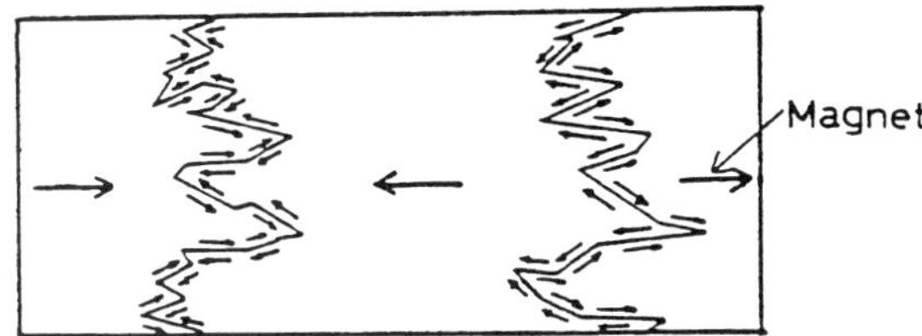

FIG. 15. Zigzag walls at the transition of the magnetization of a Co film observed by electron holography (after Yoshida *et al.*, 1983).

disk by making scratched trails on the substrate in that direction). In Table 7, the properties of some typical thin-film disks are also listed.

Plated disks are produced by electroplating or electroless plating of Co–Ni(P) upon an Al-alloy substrate covered with an electroless-plated nonmagnetic Ni(P) layer. The Ni(P) layer here facilitates obtaining a smooth finish by virtue of its hardness. The coercivity can be controlled by the composition of Co–Ni, additives, the *p*H of the plating solution, and the current density in the case of electroplating. For films having a rel-

Table 7. Properties of magnetic recording media.

Medium	H_c [Oe (kA/m)]	B_s (kG = 0.1 T)	B_r/B_s	Thickness (μm)	Areal density (Mb/in.^{2})
		Tapes			
Audio cassette tape					
γ-Fe_2O_3	360–410 (28–33)	1.7–1.9	0.85–0.91	5	
CrO_2	450–650 (36–52)	1.6–2.3			
Co–γ-Fe_2O_3	610–690 (49–55)	1.7–1.8	0.89–0.91	5	
MP (α-Fe)	~1200 (95)	~3.6	0.91	4	
Video tape					
Co–γ-Fe_2O_3	650–900 (52–72)	1.7–2.0	0.82–0.88	3.5–4.5	17
MP (8 mm)	1550 (122)	3.0	0.83–0.85	2.5	90
Double-coated MP (8 mm)	1550 (122)	4.0	0.82	0.25 2.5 (nonmag.)	
ME (8 mm)	910–1400 (72–110)	4.6–6.4	0.6–0.78	0.17–0.24	128
		Rigid disks			
γ-Fe_2O_3	330 (26)	1.3	0.85	0.34	
Co (Ni) (plated)	790 (63)	2.2	0.8	0.055	40
	600 (48)	8.8	0.8	0.08	15
γ-Fe_2O_3 (sputtered)	700 (56)	8	0.8	0.16	40
Co(Ta)/Cr (sputtered)	1300 (103)	9	0.8	0.05	150
Co(Pt)/Cr (sputtered)	1800 (173)	10	0.85	0.045	235
		Floppy disks			
γ-Fe_2O_3	300 (24)	1.2	0.57	2.5	0.054
MP	1550 (122)	2.5	0.6	2.5	1.5
Ba-ferrite	700 (56)	1.25	0.47	0.6	2.8

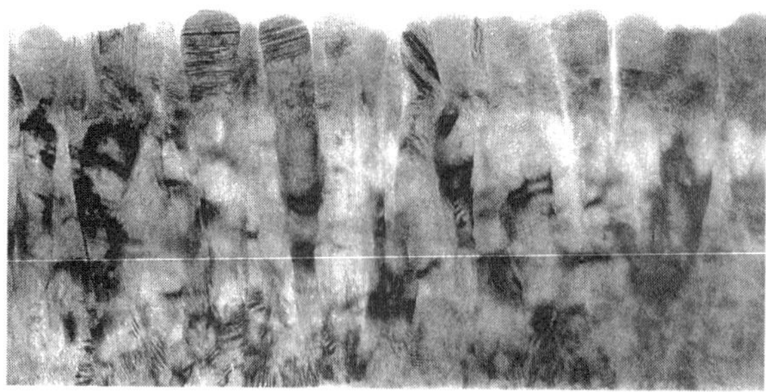

FIG. 16. Cross-sectional view of a typical Co-based–alloy sputtered disk (courtesy of Dr. M. Futamoto, Hitachi).

atively high coercivity it has been confirmed that they have an hcp structure with its *c* axis preferentially aligned in the in-plane direction forming a networklike structure or that they are composed of a mixture of hcp and fcc structures. It has been also made clear that the control of *p*H and addition of some elements such as Zn and Mn facilitate the reduction of grain sizes, leading to obtaining low-noise media (Goto and Osaka, 1993).

The γ-Fe_2O_3 sputtered disk is mostly made by reactive sputtering of Fe in an Ar–O_2–H_2 atmosphere to obtain a Fe_3O_4 film and then heat-treating it in H_2. Addition of Co increases the crystalline anisotropy. In some films of the media that show good high-density recording characteristics, it has been confirmed that the films are composed of two kinds of columns (several tens of nanometers in diameter); one is of a single crystal with the (111) plane oriented parallel to the plane of the substrate and the other is an assembly of fine grains separated by amorphous materials (Takagaki *et al.*, 1987). Co-based–alloy sputtered disks are made by depositing the alloy upon a polycrystalline Cr underlayer mostly with the (110) or (100) planes preferentially oriented parallel to the substrate surface, upon which Co alloys are epitaxially grown resulting in the *c* axis of the Co-alloy grains oriented parallel to or at a small angle with the disk surface. Circumferential trails are sometimes introduced to obtain a circumferential easy-axis orientation. As Co-based alloys, first CoNi and then CoCr, CoCrTa, and CoCrPt have been explored to obtain a higher and higher coercivity. It is now understood that Cr and Ta are likely to precipitate at the grain boundaries and give a desirable effect of the reduction of the exchange coupling between the grains. Sputtering conditions that give a low mobility to the deposited atoms upon the substrate, such as high Ar pressure, low substrate temperature, and no biasing, are also effective to reduce the noise. Figure 16 shows a cross-sectional view of a typical Co-based–alloy sputtered disk. By utilizing Co-based thin-film media, recording densities of 1–3 Gb/in.^{2} have been successfully demonstrated.

Evaporated media are prepared by such a method as is illustrated schematically in Fig. 17. A Co(Ni) alloy is electron-beam deposited obliquely onto a 6–20-μm-thick PET base film running at a speed of 10–20 m/min over a rotating drum. The incident angle of the depositing atom beams is controlled by a shutter. During the deposition, air is blown

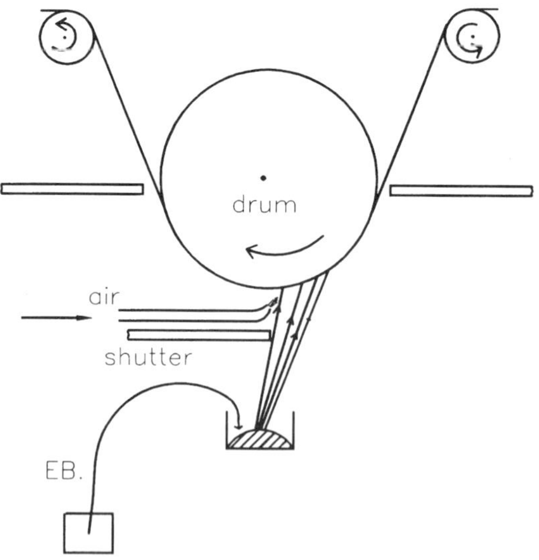

FIG. 17. Schematic illustration of the production process of evaporated metal (ME) tapes.

FIG. 18. Electron micrograph of a cross section of an evaporated metal (ME) tape (courtesy of Hitachi Maxell).

onto the substrate in order to form an oxide around each grain growing on the substrate. Figure 18 shows a cross section of a typical evaporated tape (ME tape). The oxide acts as a reducer or inhibitor of the exchange coupling between the grains and a protector against both corrosion and abrasion. The inclination of the grain growth with respect to the substrate surface makes the role of the perpendicular component of the magnetization greater, especially in high–recording-density regions. The optimization of the inclination angle is one of the present primary concerns.

1.4.3 Perpendicular-Recording Media Investigations on perpendicular recording were first started by using CoCr films (Iwasaki and Nakamura, 1977). Although various other Co-based alloys (CoV, CoMo, CoW, etc.) and barium ferrite have been explored, no better recording characteristics have been obtained than with CoCr media. One of the important criteria the media should satisfy is that the quality factor Q, which is defined as the ratio $K_{\perp}/2\pi M_s^2$, be almost equal to or greater than unity, with $K_{\perp}$ denoting the intrinsic perpendicular anisotropy. For CoCr, this criterion can be satisfied in the Cr-content region greater than 15 at. %, with the limit at about 25% at which M_s tends to vanish. CoCr films with a definite perpendicular anisotropy can be formed either by sputtering or by evaporation. They are composed of very fine columnar crystallites of hcp structure with their c axes aligned almost perpendicular to the film surface. Morphological analysis has revealed that each column has a microstructure such that there exist Co-rich areas and Cr-rich areas if observed from the top of the film, an example of which is shown in Fig. 19, which facilitates high-density recording. Co-based–alloy films with a good quality factor can also be made by electroless plating in a Co–Ni–Zn–P bath with a relatively high concentration of $NiSO_4$ compared with that for the longitudinal recording media (Goto and Osaka, 1993). Barium-ferrite films have been reported successfully prepared on ceramic substrates by the sputtering/heat-treating method. A drawback of this method is that they require a high-temperature (almost 500°C or higher) heat treatment, although some breakthrough is being developed by using laser flash annealing.

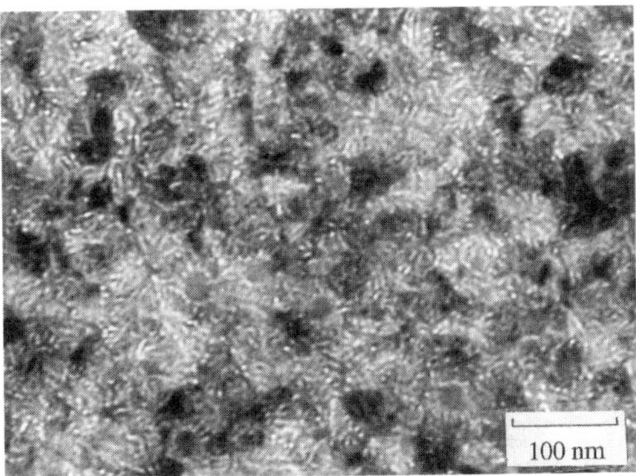

FIG. 19. Transmission electron micrograph of an etched Co–Cr film. Bright area, Co-rich; dark area, Cr-rich (courtesy of Drs. Y. Maeda and S. Hirono, NTT).

In order to match the medium to the one-sided probe-type head system, the recording layer should be formed upon a several hundred nanometers thick soft magnetic (under)layer, which acts as a flux path from the main pole to the auxiliary pole or vice versa. In most cases, Permalloy is used for that purpose. It is noteworthy that even for the ring-type head system, a very thin, rather soft magnetic underlayer (several to several tens of nanometers) improves the recording characteristics substantially (Kitakami *et al.*, 1988; Iwasaki and Ouchi, 1989)

2. OPTICAL RECORDING*

2.1 Concept of Optical Recording

The concept of optical recording is not much different from that of magnetic recording. The only primary difference between the two is that a light beam is utilized to write and read a bit in the former, a magnetic flux in the latter. Therefore, the setup for optical recording devices is quite similar to that for magnetic ones. Optical recording devices basically consist of a medium upon which signals are written and an optical head that applies an optical beam to write and/or read the signals. One of the prominent characteristics of optical recording is its extremely high-density recording capability due to the fact that the light can be focused down to the limit of the wavelength, say less than 1 μm. Other features especially to be mentioned will be the facts that the spacing between the medium and the head can be set wide enough to keep them from contact with each other and that each bit is accessible by the light beam through the transparent substrate at the surface of which the beam diameter can be made as large as about 1 mm. This provides the system with high endurance against dust, enabling the media to be made detachable from the drive and portable, with a much higher recording density than floppy disks.

Figure 20 illustrates the optical recording system schematically. An optical disk is rotated at a high speed and a head is set under the disk and is driven by an actuator so that the light beam may access each track of the disk transversely. The head system is provided with both autofocusing and autotracking functions in order to take full advantage of the high-density recording capability.

2.2 Optical Recording Devices

Optical recording is classified into three categories: read only, write once, and rewritable (erasable), according to the mechanism for recording the information. Just like magnetic recording, the medium can assume any

*A substantial part of this section is based on "Optical Recording" (in Japanese) (Terao *et al.*, 1990).

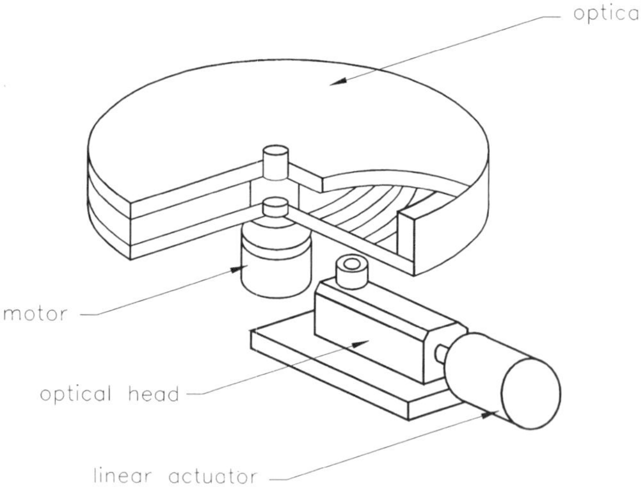

FIG. 20. Conceptual illustration of an optical disk system (after Terao *et al.*, 1990).

style—a disk, card, or tape, to meet various usages, disk being the most popular.

Among the read-only devices are video disk (VD), CD (compact disk), and its derivatives CD-ROM, CD-I (CD-interactive), CD-ROM.XA, and CD-V (CD-video). Video disks employ the FM method to record images. They are either double-sided or single-sided with diameters 200 mm (8 in.) or 300 mm (12 in.), with the playing time ranging from 20 to 120 min. The CD has a diameter of 120 mm and has been developed for recording music by the PCM (digital) method with playing time more than 60 min. An 80-mm CD has also been developed recently with the name "CD single." The CD is rotated at a constant angular velocity (CAV) whereas the VD has two types, CAV and CLV (constant linear velocity). The CD-ROM is a "read-only memory" of any data or software for digital information processors, having the same geometrical, mechanical, and optical specifications for the read procedures as those for the CD. The data structures are also made similar to each other as shown in Fig. 21. On CD-I and CD-ROM.XA are recorded both image and sound by means of a time-division multiplexing method, the latter being standardized for PC use. The CD-V has also both image and sound recordings but they are recorded separately from each other, image data on the outside and sound data inside. All the VD's and CD's have a single spiral track per surface.

Write-once [W-O or WORM (write once and read many)] and rewritable types are both intended to be used for digital information files of a huge capacity, having either a single spiral track or a large number of concentric circular tracks on each surface. Various sizes of W-O–type disks are now on the market such as 356 mm (14 in.), 300 mm (12 in.), 200 mm (8 in.), and 130 mm (5.25 in.), with various kinds of recording media. As for the rewritable ones, 130-mm (5.25-in.) and 85-mm (3.5-in.) magneto-optic

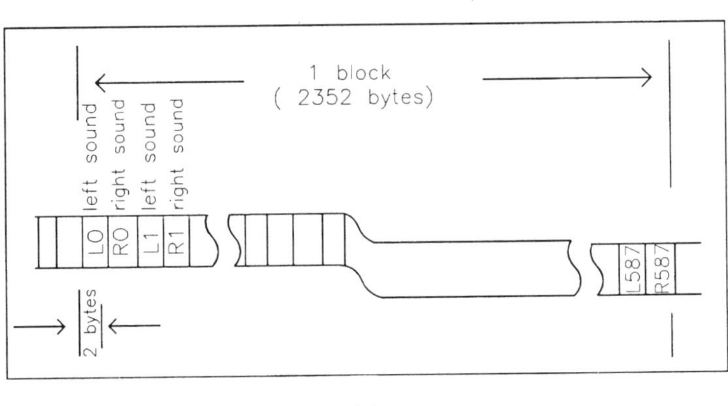

(a)

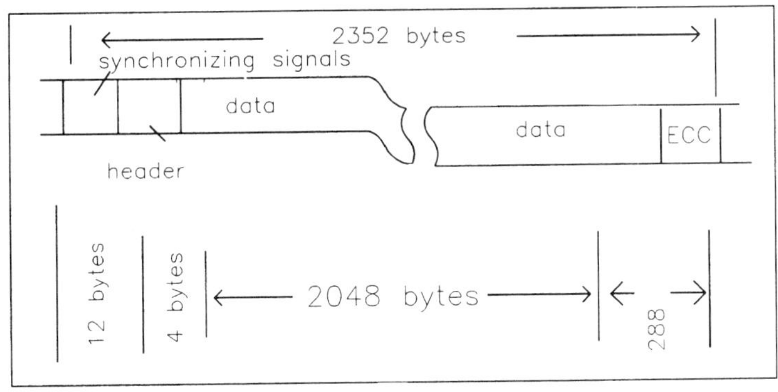

(b)

FIG. 21. Data recording schemes of (a) CD and (b) CD-ROM (after Otomo, 1988).

(MO) disks are now commonly used, although a 300-mm (12-in.) MO disk has also been developed for special applications such as video editors for broadcasting uses. The number of possible write repetitions exceeds 10^7. Phase-change–type optical disks are now also beginning to be put on the market, which will provide disks of lower cost with lower write repeatability. In Table 8 are listed the specifications and characteristics of typical optical disks and drives.

2.3 Optical Heads

In Fig. 22 is illustrated the basic system of an optical head. The beam from the laser is made parallel through the coupling lens and is shaped into a circular beam by the first prism. After passing through the polarized beam splitter, the linearly polarized light emitted from the laser is transformed into a circularly polarized beam through the $\frac{1}{4}\lambda$ (wavelength) plate, reflected by the mirror toward the medium, and focused upon the surface of the recording layer of the medium by the condenser lens mounted upon a fine positioner (not shown). The reflected beam from the medium comes back to the beam splitter after being changed into linearly polarized light by the $\frac{1}{4}\lambda$ plate and forwarded to the light detectors of the autofocusing and autotracking signals as well as the data signals. For MO recording, the $\frac{1}{4}\lambda$ plate is omitted because the linearly polarized light is utilized to read the signals.

As a laser, GaAlAs laser with 780- to 830-nm wavelength is now adopted, although, for a further increase of the recording density, light sources with shorter wavelengths are under development such as the InGaAlP laser (670 nm), the ZnSe laser (600 nm), and second-harmonic generators that generate the light of a wave length $\frac{1}{2}$ of that of the incident beam (when using a YAG laser, 532-nm light is generated).

It is well known that the diameter of the focal point (measured between points of $1/e^2$ times maximum intensity) is expressed by λ/NA and the focal depth by λ/NA^2, where λ is the wavelength of the light and NA is the numerical aperture of the condenser lens. In order to make the focal diameter small, it is necessary to employ a lens of a large NA, which makes the focal depth extremely small. Therefore, in the optical recording

Table 8. Optical disks.

	Video disk		CD	WORM disk			MO disk			MD
Disk size	200/300 mm	300 mm	120/80 mm	14 in.	12 in.	5.25 in.	12 in.	5.25 in.	3.5 in.	2.5 in.
Type of disk	Read-only	WORM	Read-only	Write-once read-many			Video	Rewritable	Rewritable	Rewritable
Capacity/side (MB)	Analog	Same	640/180	4500	1300/3500	300	11 500	340/1000	128/1230	140
Linear density (kbpi)	White peak: 9.3 MHz	Same	25		19.5/33.2	24	74.7	24/45.6	24.4/34.1	
Track density (ktpi)	16	16	16		16/17	16	21.2	16/19	16/18	16
Playing time										
CLV (min)	20/60	60	74/20	…	…	…	32 (both sides)	…	…	60/74
CAV (min)	–/30	30								
Rotation speed										
CLV (m/s)	10.0–11.3	Same	1.2–1.4							1.2–1.4
CAV (rpm)	1800	Same			600/1000	1800	1800	1800/3000	1800/3600	
Laser power (mW)	<1	11	<1		6	6	<10	5	5	5
Delivery year	1978	1990	1982	1991	1984/1990	1987		1987/1993	1990/1993	1993

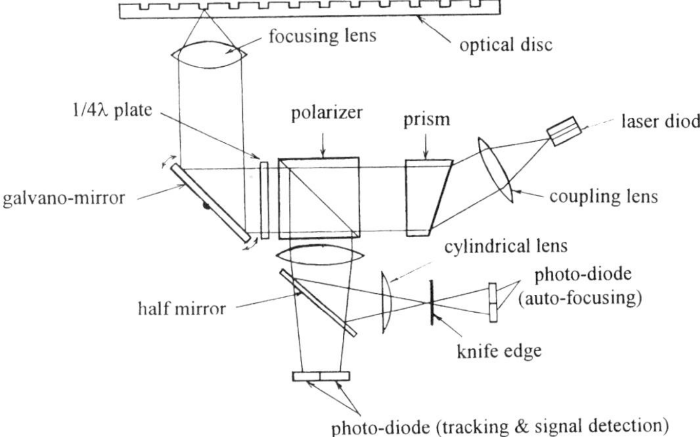

FIG. 22. Basic system of an optical head (after Terao *et al.*, 1990).

system, autofocusing is inevitable. Autofocusing is performed by detecting the change of the focus of the reflected beam condensed through a condenser system set in front of the light detectors. Several methods are proposed for it:

1. two detectors, set before and behind the nominal focal point;
2. a knife-edge shutter set at the nominal focal point and detection of the change of angular distribution of the light;
3. astigmatism method; and
4. astigmatism method with a knife edge.

In the astigmatism methods, a cylindrical lens is combined with a condenser lens in the condenser system to provide an elongated beam cross section in the straight and curved direction of the cylindrical lens right before and after the average focal point of the system, respectively. In method **3**, four detectors are set at the average focal point for the normal position, and by adding/subtracting the four outputs, the amount and the direction of the defocusing are sensed. In method **4**, a knife-edged shutter is set at the spot where the detectors are set in **3**, having its edge at 45° with the principal axis of the cylindrical lens, providing a semicircular beam cross section behind the shutter. The semicircle rotates clockwise or counterclockwise according to the defocusing direction. Therefore, by sensing the rotation direction, the current to be applied to the electromagnetic coil is determined to control the position of the condenser lens.

Autotracking methods are classified into two: continuous servoing [differential diffraction (push-pull) method and three-beam method] and sample servoing (SS). The differential diffraction method is employed when disks having guide groves are used. In this case, a two-division detector is set to receive equal zeroth- and first-order diffracted light when the beam is right on the track, and the off-tracking is sensed by the difference between the first-order diffracted beams. In most of the read-only disk drives, the three-beam method is adopted, in which the beam is split into three by a grating before entering the condenser lens and the light spot is divided into three, one main spot and two subspots aligned in a line tilted a little with respect to the data track direction. The off-track condition is sensed by the difference between the two signals detected through the two subbeams. In the sample servoing method, a number of marks are put periodically on each track (say 1000/track), each mark typically having a pair of dots aligned across the center of the track located a little before and behind in the track direction, and the data are recorded between the marks. Thus the off-track signals are obtained through the read beam.

2.4 Optical Recording Media

2.4.1 Read-Only Disks Read-only disks are produced by stamping. Figure 23 illustrates schematically the stamping process. The first three steps are those for preparing a stamper. First, an original disk in which the information is recorded as a series of

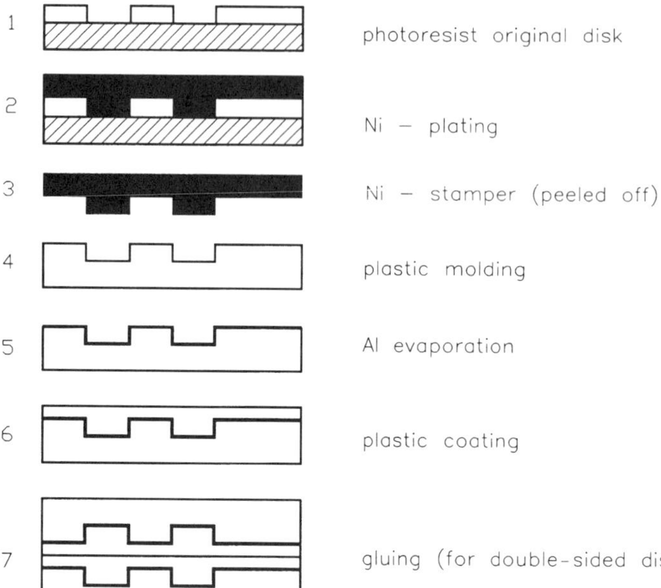

FIG. 23. Production process of W-O–type optical disks (after Terao *et al.*, 1990).

pits is prepared by means of a laser-beam recording upon a photoresist-coated glass disk, then nickel is first sputtered (several micrometers) and then electrodeposited (about 0.2 mm thick), and finally the deposited nickel layer is peeled off making a mother stamper. Although not shown in the figure, actually, working stampers are made by repeating the Ni deposition and peeling similarly to the above with the aid of peeling agent, using the mother stamper. Then, plastic materials, such as polycarbonate (PC) and polymethyl methacrylate (PMMA), are compressed or injection-molded upon the stamper. After being peeled from the stamper, the mold is finished with aluminum coating for reflection and resin coating for protection (and adhered face to face when making a double-sided disk).

The signals are recorded as a series of pits having a width smaller than the read beam diameter ($\sim\lambda$/NA) and a depth equal to $\lambda/4n$ (n: refractive index of the substrate). In the read process, the reflected light from a pit and its surrounding land cancel each other through interference, which gives a weaker return beam than from the land area, providing the pit signal.

2.4.2 Write-Once (W-O) Disks There are various kinds of W-O disks in terms of the mechanism of recording: deformation (hole, dip, or bubble formation), phase change, alloying, and color developing, with the deformation type, especially the hole-formation type, being the most common.

The mechanism of hole or dip formation by irradiation with a laser beam is explained as follows: When the disk is irradiated, the viscosity of the recording layer at the spot is decreased because of the elevation of the temperature, and a viscosity gradient appears with a minimum at the center, which in turn generates a gradient in the surface tension. The surface tension increases toward the perimeter, causing material to gather toward the perimeter, which makes a banklike bulge (or mound) leaving a dip inside as is schematically shown in Fig. 24; and when the viscosity becomes sufficiently low, a hole is formed at the center. There-

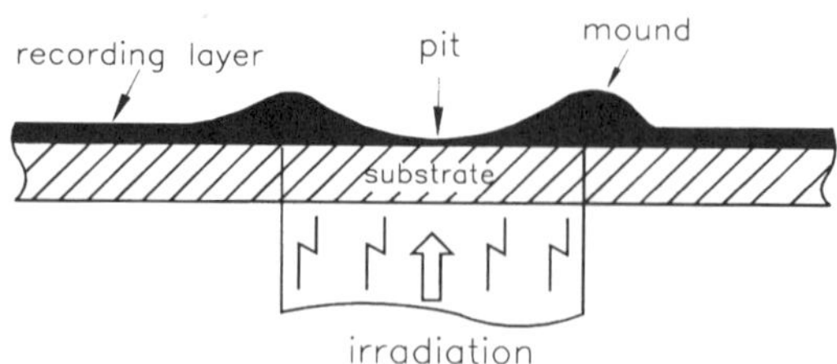

FIG. 24. Pit formation in W-O–type recording (after Oba *et al.*, 1985; Terao *et al.*, 1990).

fore, it is desirable that the materials for the recording layer possess a steep temperature dependence of the viscosity as well as a low melting point, high light-absorption coefficient, and low thermal conductivity in order to obtain high resolution and high light sensitivity. Of the many materials explored thus far, tellurium alloys such as (Pb–)Te–Se and Te–C are outstanding among inorganic materials, although there have been developed a lot of potential organic materials such as cyanine dyes, azulenium dyes, phthalocyanines, naphthoquinone dyes, etc., which have a high absorption coefficient for the light emitted from the laser diode. Generally, those dye polymers are rather apt to be deteriorated by the read beam or even by natural light, and their disk life cannot be expected to be as long as that of the inorganic ones. However, some of them show a read durability of more than 10^6 times and they are acceptable for low-cost disks.

There is another type of deformation by irradiation, in which the surface of the substrate itself is deformed through the heating up of the light-absorbing layer coated directly upon it. Au–Sn alloy on PC (polycarbonate) substrate is one of the most noticeable examples because it can provide a reflectivity contrast that is compatible with the CD's.

Te–O–Pd is known as an excellent material for phase-change–type media, in which fine Te particles of several nanometers are dispersed in the amorphous TeO_2 matrix. Without Pd, the heat/cool process under irradiation causes the recrystallization of Te in the form of long chains, which makes the spot shape irregular; Pd prevents this phenomenon. As alloying-type media, the $Sb_2Se_3/Bi_2Te_3/Sb_2Se_3$ triple-layered film coated with an Al reflection layer is well known. By irradiation, the Bi_2Te_3 layer absorbs the light and the temperature is elevated, which causes selective mutual diffusion of Bi and Se because of their strong affinity. A color-developing–type recording layer consists of a colorless dye layer and a developer layer mostly having a light-absorbing layer in between, an example being the combination of crystal violet lactone (dye), phenolphthalein (developer), and vanadyl phthalocyanine (absorber).

2.4.3 Rewritable (Erasable) Disks Although several methods have been investigated for rewritable optical recording, such as magneto-optic (MO), phase change, deformation, photochromic, liquid crystal, etc., most rewritable disks now on the market are MO disks, with some of the phase-transition types being also put into practical use.

2.4.3.1 MO (Magneto-optic) Disks. Figure 25 illustrates the principle of the write process. The recording layer is composed of a magneto-optic material of several tens of nanometers thickness, having a magnetic anisotropy with its easy axis perpendicular to the plane. Initially, the magnetization of the layer is saturated in one direction, say upward. When a spot is irradiated by a laser beam (<10 mW), the temperature of the spot is elevated and the coercive force decreases to almost zero near the Curie temperature. Thus the magnetization of the spot is reversed by the effect of the reverse field H_{srd} from the surroundings or, in most cases, by a field H_a (typically several hundred oersteds) applied through a coil set close to the condenser lens through which the beam is applied. After irradiation, the spot begins to cool accompanied by an increase in the coercive force, which sustains the small reverse domain (reversed magnetization region) even if the applied field is removed. Erasing is

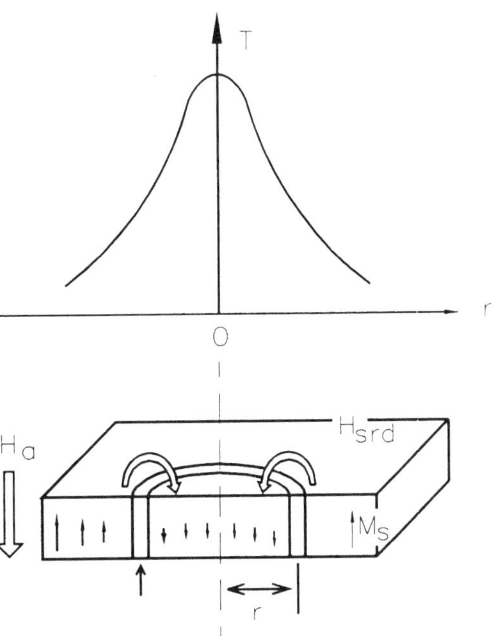

FIG. 25. Mechanism of magneto-optical (MO) recording (after Terao *et al.*, 1990).

performed before the rewrite process by applying an opposite field while irradiating. For reading, a magneto-optic effect called the polar Kerr effect is utilized. When linearly polarized light (2–3 mW) is reflected at the surface of a magnetic body, the polarization plane rotates clockwise or counterclockwise (a few tenths of a degree) according to the direction of the magnetization, which enables the sensing of the magnetization direction. Thus the magneto-optic materials for a recording layer should have

1. a high Kerr rotation angle,
2. appropriate Curie temperature (150 to 250 °C), and
3. low thermal conductivity and heat capacity, as well as sufficiently high perpendicular anisotropy $K_\perp$ to keep the magnetization perpendicular to the film plane, and high coercive force H_c, which sustains the written bit in the ambient conditions including reading.

More specific conditions that $K_\perp$ and H_c should satisfy are

$$K_\perp > 2\pi M_s^2, \tag{15}$$

$$H_c > (AK_\perp)^{1/2}/2\, M_s d, \tag{16}$$

where A is the exchange constant of the material and d is the diameter of the written spot. The readout signal-to-noise ratio is given by

$$S/N \sim R^{1/2}\theta_k \tag{17}$$

where R is the reflectivity and θ_k is the Kerr rotation angle. Therefore, the reflectivity and absorption should be appropriately compromised.

A lot of materials have been explored, and at present amorphous RE–TM (rare-earth–transition-metal) alloys are regarded as the most favorable (Imamura and Ota, 1980), although a lot of work has been devoted to manganese chalcogenides such as Mn(Cu)Bi, and to rare-earth iron garnets. This is because the RE–TM alloys have much flexibility in material designing, and their amorphousness is thought to be advantageous in making the written-spot perimeter smoother, leading to less noise than polycrystalline materials like MnBi and garnets.

Among RE–TM alloys, TbFeCo alloy is the most common, although GdTbFe with a greater Kerr effect and low-cost DyFeCo and TbDyFeCo are now also in practical use. In Fig. 26, curves of the saturation magnetization M_s and coercive force H_c vs temperature

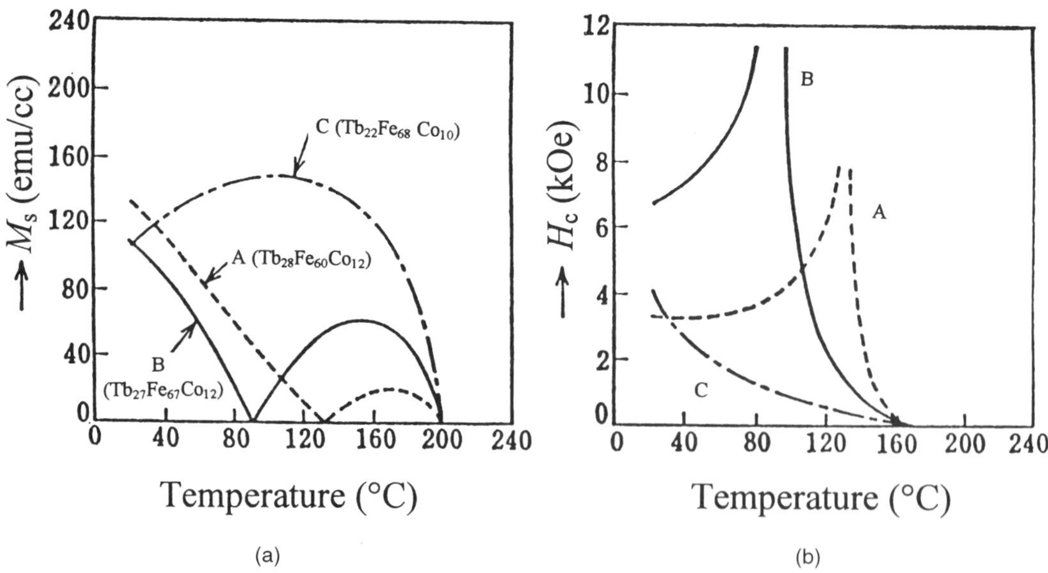

FIG. 26. Temperature dependence of the saturation magnetization M_s and the coercive field H_c of typical TbFeCo magneto-optical (MO) recording films (after Takahashi *et al.*, 1988).

are shown for three typical kinds of TbFeCo alloys with different compositions. Figure 27 shows the written spots on each of the three kinds of disks for various applied fields during recording. It is seen that disk B is the most insensitive to variation of the field between 100 and 400 Oe, giving an almost constant size to the written spots. This behavior is explained by considering the balance of the forces acting on the domain wall during the write process: the forces through the applied field H_a, leakage flux of the surroundings H_{srd}, the surface tension and potential energy of the wall, and the coercive force taking into account the temperature distribution during recording:

$$H_c(r) = H_a + H_{srd} - \frac{\sigma_w}{2rM_s} - \frac{\partial M_s}{\partial r}(2M_s)^{-1}, \tag{18}$$

where r is the radius of the written domain and σ_w is the wall energy.

For the next-generation MO disks increases in speed and recording density are now being explored. Higher speed will be attained by introducing overwritability, lighter heads, and multichannel heads. Typically, two schemes are under development to obtain overwritability: the field-modulation type and the beam-modulation type. In the former scheme, a small magnetic head is used for providing high-frequency modulation fields. To make the head small enough, the field necessary to write/erase is required to be weak. The coating of a very thin (several nanometers or less) ferromagnetic capping layer upon the recording layer is known to be extremely effective for field reduction (Ohnuki *et al.,* 1991). On the other hand, in the latter, exchange-coupled double-layered films consisting of a memory layer and an auxiliary layer or their modifications are being investigated (Saito *et al.,* 1987). Figure 28 illustrates how it works. One way to decrease the head weight to be moved is to divide the head into two parts, one consisting of the condenser-lens systems and the other the light-source and detector systems. Another is making it up with thin films. The former is almost at a practical use level, although the latter is still at a research stage. Four typical methods are being explored for increasing the recording density;

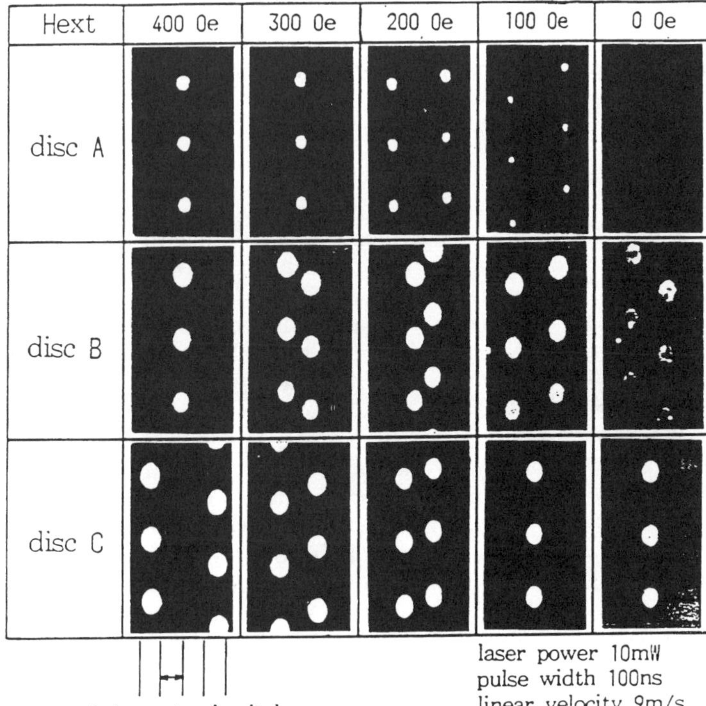

FIG. 27. Recorded patterns on the disks A, B, and C the magnetic properties of which are shown in the previous figure. The applied field was varied from 0 to 400 Oe, keeping the other conditions constant. (After Takahashi *et al.,* 1988).

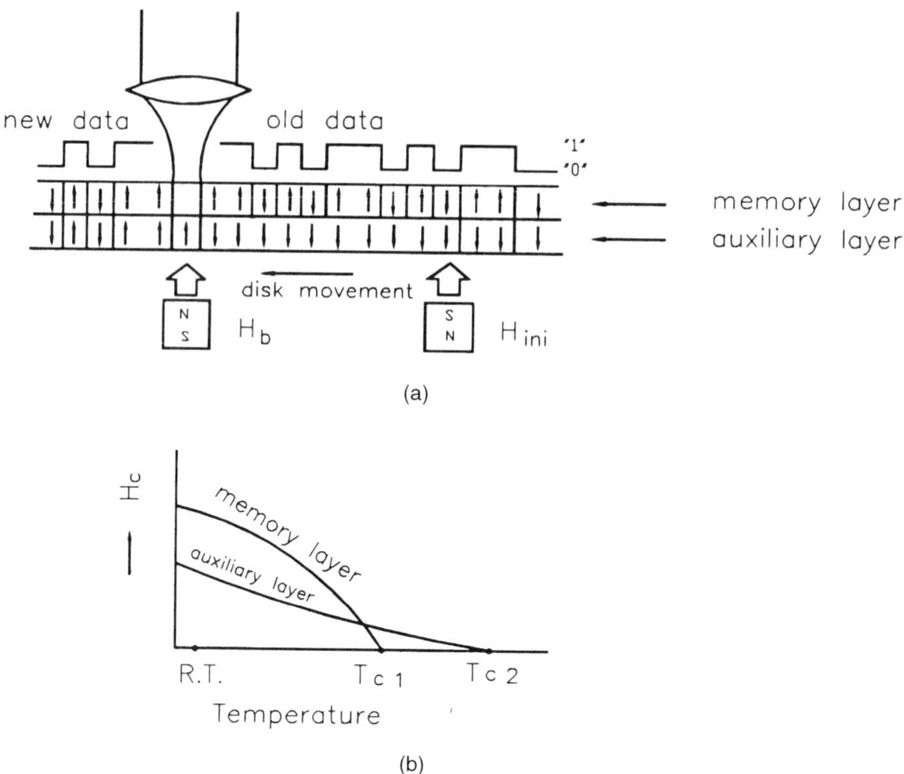

FIG. 28. Mechanism of overwritability in double-layered films for light (beam) modulation mode. The recording film consists of a memory layer and an auxiliary layer. The H_c vs temperature curve of each layer is shown in (b). Before overwriting, the auxiliary layer is initialized by a field H_{ini}. For overwriting, H_b is applied in the reverse direction. For the high-level beam ("1"), both magnetizations are aligned upward; for the low-level ("0"), the magnetization of the auxiliary layer does not change and the memory layer follows it by the exchange coupling. (After Saito *et al.*, 1987.)

1. reduction of the light wavelength λ,
2. mark-edge recording,
3. removal of track guard band, and
4. MSR (magnetically induced super resolution), described later (Ohta et al., 1991).

The reduction of λ basically depends on the development of short-wavelength laser diodes. However, materials for short wavelength are also being investigated, amorphous NdFeCo, garnets, and Co/Pt(Pd) multilayers (Zeper, *et al.,* 1989) being typical ones. Mark-edge recording, in which information is stored by the position of the edge of a recorded mark, can increase at least 1.5 times the conventional mark-position recording, and it is almost at hand. The three-beam method has been proposed for removing the guard band that enables cancellation of the crosstalk from the neighboring tracks. There are two typical types in the MSR method: front-aperture detection (FAD) and rear-aperture detection (RAD). The FAD method utilizes trilayered films exchange coupled when no beam is applied; the recording layer, a switching layer, and a readout layer (see Fig. 29). At room temperature, the recording layer has the highest coercivity and the coercivities of the readout layer and the switching layer are low enough for the magnetization of each layer to align in the same direction as that of the memory layer through exchange coupling. The Curie temperature of the switching layer is made lower than that of the other two layers. Thus, when a read beam is applied to a rotating disk while an erase field H_r is applied, the film is heated up to make a part of the

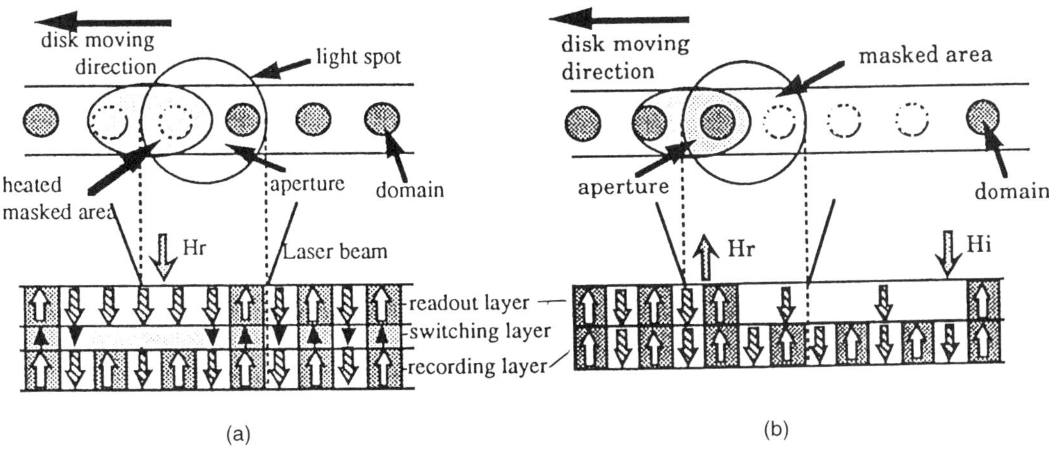

FIG. 29. Schematic illustration of the magnetically induced super-resolution (MSR) methods: (a) front-aperture detection (FAD) method, (b) rear-aperture detection (RAD) method (after Ohta *et al.*, 1991).

irradiated switching layer nonmagnetic, decoupling the readout layer from the recording layer and resulting in the erasure of the readout layer, which acts as a kind of mask in the spot providing an extremely high readout resolution. Regarding the mechanism of RAD, see Fig. 29(b).

Very recently, multilevel (four-level) recording has been achieved by using multilayered MO media, which will also be a candidate for a future high-density recording scheme (Shimazaki *et al.*, 1995).

2.4.3.2 Phase-Change–Type Disks. Amorphous materials have a glass point T_g above which atoms become rather easy to move around, and a melting point T_m above which they become liquids. If they are kept at a temperature between T_g and T_m for a sufficient time, they turn into crystallites, and if they are cooled rapidly from their liquid state, they become amorphous. The write/erase or overwrite in the phase-change–type disks utilizes this phenomenon. Figure 30 illustrates the recording process schematically. A series of light pulses corresponding to the signal to be written, having a sufficient height to melt the recording layer, are applied with a constant biasing power that elevates the temperature just above the crystallization temperature, thus leaving amor-

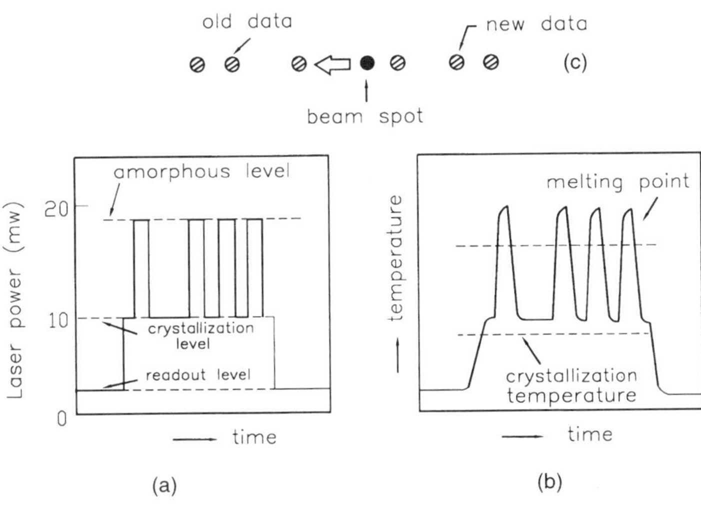

FIG. 30. Mechanism of phase-change–type recording (one-beam overwrite mode). (a) Form of three-level beam (amorphous, crystallization, and readout levels) applied. (b) Temperature profile generated by the irradiation. The spots where the temperature has exceeded the melting point become amorphous and the spots where the temperature has been raised a little higher than the crystallization temperature become crystallites. Thus the pattern as shown in (c) is recorded. (After Terao *et al.*, 1990.)

phous spots corresponding to the pulse peaks and crystalline spots corresponding to the valleys. Therefore, one of the crucial points to design the recording layer is to choose an appropriate rate of the phase change from amorphous to crystalline state in the range of the temperatures that the recording layers will experience.

The crystallization ratio x as a function of time t is generally given by the Avrami's formula,

$$x = 1 - \exp(-kt^n), \qquad (19)$$

where n is an index called the Avrami index that takes on values from 1 to 4 according to the dimension of the crystal growth and whether the number of nuclei is kept constant or increases with time, and k is a function of the nucleation speed and the crystal-growth rate. The average crystallization rate $\langle \nu_0 \rangle$ from the beginning to the time when x becomes x_0 is given by

$$\langle \nu_0 \rangle = [-k/\ln(1 - x_0)]^{1/n}. \qquad (20)$$

Reading the written bit is performed by detecting the change of the reflectivity as in the case of the write-once type.

The Ge–Sb(Co,Tl,Ag)–Te system is known as one of the most excellent material systems in terms of overwrite repeatability and erase characteristics (Akahira *et al.*, 1988; Ohta *et al.*, 1989). GeTe possesses a crystallization time as short as 30 ns and a high stability in the amorphous state because of its high crystallization temperature (Chen *et al.*, 1986). However, it is easily cracked during crystallization, as a substantial volume change takes place during the phase transition. On the other hand, Sb_2Te_3 has a long crystallization time of several microseconds but shows an excellent resistance to crack formation (Yagi *et al.*, 1987). The Ge–Sb–Te system is endowed with the compromised characteristics of those two. The addition of Co, Tl, and Ag controls the crystallization speed (Terao *et al.*, 1989). In–Se–Tl–Co, In–Sb–Te, and In–Sb have also been investigated extensively. However, none of them have surpassed the Ge–Sb–Te system in erasability and overwrite repeatability.

2.4.4 Multiplex Recording In order to obtain much higher recording density, multiplex recording methods have been investigated, such as multiwavelength recording and holographic recording, although a great deal of work is left to be done before they are put into practical use.

2.4.4.1 Hole Burning. Several media consisting of pluralities of dyes, such as multilayered LB (Langmuir–Blodgett) films of spiropirane compounds, have been shown to have the potential of being applied to multiwavelength recording. However the multiplexity number must be limited. “Hole burning,” on the other hand, has demonstrated the possibility of attaining multiplexity of almost 1000 (Bjorklund *et al.*, 1988). The recording medium is composed of a host material with a guest material dispersed in it. The guest must be or must possess a light-absorption center, the spectrum of which is modulated by the inhomogeneity of the surroundings, resulting in a broad dispersion in the absorption spectrum of the medium. Recording is performed by applying a monochromatic light with a sharp spectrum making some of the absorption centers immune to the light of the same wavelength by exciting the electrons away. This generates a “hole” in the absorption spectrum of the irradiated spot, which enables reading out later. By changing the wavelength, a multiple of recordings can be done on the same spot.

A lot of materials have been investigated including both organic and inorganic materials, such as porphyrin, phthalocyanine, quinizarin, carbazole–boric acid, single-crystalline alkali halide with color centers or divalent ions, single-crystalline transition-metal oxide, glass doped with trivalent ions (Pr^{3+}, Nd^{3+}), etc. In terms of the multiplexity in recording, quinizarin/PHEMA(poly-2-hydroxyethylmethacrylate) and ABDAQ [4-amino-2,6-bis (4-butylphenoxy)-1,5-dihydroxy-anthraquinone]/PHEMA are known as outstanding. For a double-layered film of these two, 600 multiplex recording has been demonstrated successfully (Yoshimura *et al.*, 1989).

In order to obtain a reasonable signal-to-noise ratio and speed, however, the guest must simultaneously fulfill conditions with respect to both quantum efficiency and absorption cross section, which has not yet been attained. All the successful experiments have been performed at very low temperatures such as liquid He or N_2 temperatures. This is another drawback left to be overcome. At present, by reading, the record is

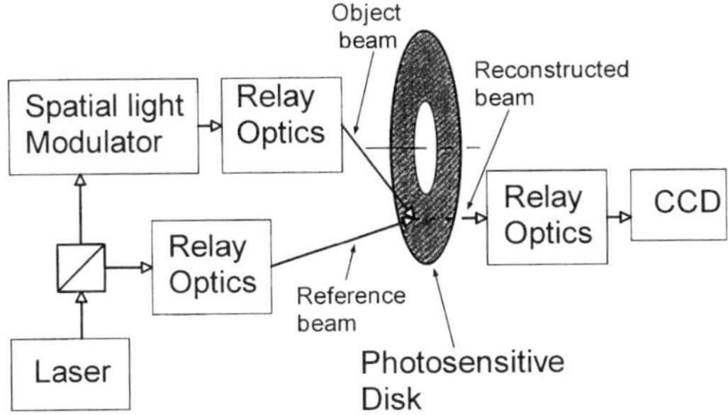

FIG. 31. Diagram of a holographic recording system (after Curtis, 1994).

substantially faded out, which is also one of the crucial issues to be improved.

2.4.4.2 Holographic Recording. Holography is a well-known image-recording technology in which both intensity and phase are recorded simultaneously in the form of interference patterns of the signal and reference beams. Figure 31 illustrates a possible system approach (Curtis, 1994). The spatial light modulator (SLM), which is either a liquid-crystal–based device or a deformable mirror, creates pages of data and multiple pages are recorded in the form of holograms (images of interference pattern) on the same area of the photosensitive recording medium by changing the wavelength or the incident angle of the reference beam. Reading is performed by applying the reference beam to generate a reconstructed beam, which is led to the light detector (CCD). Thus more than 10^5 parallel channel processing can be achieved. As recording materials, several photosensitive materials such as $LiNbO_3$, photorefractives, photopolymers, photorefractive polymers, and glasses are now under investigation. By using those materials, a recording density as high as 10^{10} pixels/cm^3 is expected to be available, 3×10^9 having been achieved experimentally by using $LiNbO_3$ (Heanue *et al.*, 1994). Although the necessary components now seem to be available, full evaluation of the total systems is still to be performed.

3. CONCLUSION

The history of recording technology has been and will continue to be, so to speak, a history of the increase in the recording density and rate. The basic trend in magnetic recording is the utilization of thin films both for heads and media. For the head, the development of high-permeability materials with a high saturation induction and the application of giant magnetoresistance will be major concerns, while for the media, thin-film media constituted of fine grains with high saturation moment and high anisotropy and the utilization of the inclined anisotropy will be. Perpendicular recording will also participate in ultrahigh-density recording in the long run. Although it was not mentioned above, it is worth pointing out that some recent studies imply that the switching speed of recording media will be a matter of serious concern when we go into the frequency range of a few hundred megahertz (He *et al.*, 1995). In optical recording, the development of shorter-wavelength laser diodes including SHG, and the media matching for them, and multilevel recording will be the main issues for increasing the recording density, although other multiplex recording technologies, such as holographic and hole-burning recording, give us an unlimited dream for the future.

GLOSSARY

Barkhausen Noise: In the magnetization process in which domain-wall motion takes place, the change in the magnetization becomes rather discontinuous because there exist a lot of wall pinning sites originating from various kinds of inhomogeneity such as holes, inclusions, crystalline boundaries, etc., generating noise. This effect was first observed by H. Barkhausen in 1919, who made an indirect verification of the presence of ferromagnetic domains in a magnetic body.

Demagnetization Field: A magnetic field that appears when a magnetic body is magnetized, due to the appearance of magnetic free poles mostly at the two ends of the body. As this field opposes the magnetization direction, resulting in some degree of demagnetization, it is called the demagnetization field. When a magnetic body is saturated in one direction with a saturation magnetization M_s, the demagnetization field H_d is expressed as $H_d = -NM_s$, where N is called a demagnetization factor.

GMR (Giant Magnetoresistance): A phenomenon found in magnetic multilayers each layer of which couples with the neighboring magnetic layers antiferromagnetically through a nonmagnetic layer in between, each layer having a thickness of a few nanometers (Baibich *et al.*, 1988). The resistance of the multilayers is changed drastically by applying a field that changes the angle of the magnetizations of the neighboring layers. This phenomenon comes from the spin-dependent resistance at the interface and in the bulk of the magnetic layers and is isotropic with respect to the applied current direction.

Inductive Head: A magnetic head consisting of a magnetic core and a coil wound around it. Because it operates through the phenomenon of inductance, it is called "inductive head." This term began to be used to distinguish the one from noninductive heads such as Hall head or MR (magnetoresistive) heads, which are used for reading only.

MR (Magnetoresistance): A phenomenon in which the resistance of a magnetic substance changes as a function of the magnetization direction with respect to the direction of the current applied. Since it is anisotropic in this sense in contrast to the giant magnetoresistive effect (GMR), it is called "anisotropic magnetoresistance."

Spin Valve: The metallic magnetic bilayer structure in which the magnetization of one layer is pinned and that of the other is free exhibiting the GMR effect. The magnetization in the free layer acts as if it were a valve for the electric current.

Works Cited

Akahira, N., Yamada, N., Kimura, K., Takao, M. (1988), in: D. B. Carlin, Y. Tsunoda, A. A. Jamberdino (Eds.), *Optical Storage Technology and Applications,* SPIE Proceedings No. 899, Bellingham, WA: SPIE.

Baibich, M. N., Broto, J. M., Fert, A., Nguyen Van Dau, F., Petroff, F. (1988), *Phys. Rev. Lett.* **61,** 2472–2475.

Bjorklund, G. C., Haarer, D., Hayes, J. M., Jankowiak, R., Lenth, W., MacFarlane, R. M., Sievers, A. J., Small, G. J. (1988), "Persistent Spectral Hole-Burning," in: W. E. Moerner (Ed.), *Science and Application,* Topics in Current Physics Series, New York: Springer-Verlag.

Chen, M., Rubin, K. A., Barton, R. W. (1986), *Appl. Phys. Lett.* **49,** 502–504.

Curtis, K., Psaltis, D. (1994), "Volume Holographic Data Storage," in *Proceedings of the ITA Sixth Annual Magnetic and Optical Media Seminar.*

Dieny, B., Speriosu, V. S., Parkin, S. S. P., Gurney, B. A., Wilhoit, R., Mauri, D. (1991), *Phys. Rev. B* **43,** 1297–1300.

Fujiwara, H., Kudo, M., Tamura, T., Sugishita, N., Shiroishi, Y., Kimura, T., Shinagawa, K., Kumasaka, N. (1982), U.S. Patent No. 4,316,228.

Goto, F., Osaka, T. (1993), "Plated Media for High Density Magnetic Recording," *Proc. Electrochem. Soc.* **93-20,** 410–422.

He, L., Doyle, W. D., Varga, L., Fujiwara, H., Flanders, P. J. (1995), presented at MRM '95 in Oxford (to be published).

Heanue, J. F., Bashaw, M. C., Hesselink, L. (1994), *Science* **265,** 749–752.

Hylton, T. L., Coffey, K. R., Parker, M. A., Howard, J. K. (1993), *Science* **261,** 1021–1024.

Imamura, N., Ota, C. (1980), *Jpn. J. Appl. Phys.* **19,** Suppl. **L,** 731.

Inaba, H., Saitoh, S., Kitahara, T., Kashiwagi, A. (1995), *IEICE Trans. Electron.* **E78-C,** 1536–1542.

Iwasaki, S., Nakamura, Y. (1977), *IEEE Trans. Magn.* **13,** 1272–1277.

Iwasaki, S., Ouchi, K. (1989), *J. Magn. Soc. Jpn.* **13,** Suppl. **S1,** 21–26.

Jeffers, F. (1986), *Proc. IEEE* **74,** 1540–1556.

Kitakami, O., Ogawa, Y., Fujiwara, H., Kugiya, F., Suzuki, M. (1988), *IEEE Trans. Magn.* **25,** 4177–4179.

Markham, D., Jeffers, F. (1990), "Magnetoresistive Head Technology," *Proc. Electrochem. Soc.* **90-8,** 185–204.

Mee, C. D., Daniel, E. D. (1989), *Magnetic Recording Handbook,* New York: McGraw-Hill.

Nakamura, Y. (1991), *J. Magn. Soc. Jpn.* **15,** S2, 497–506.

Nikles, D. E., Cain, J. L., Chacko, A. P., Webb, R. I. (1994), *IEEE Trans. Magn.* **30,** 4068–4070.

Oba, H., Abe, M., Umehara, M., Sato, T., Ueda, Y., Kunikane, M. (1985), "Organic Dye Materials for Optical Recording Medium," *Digest of Topical Meeting on Optical Data Storage,* WDD, Washington, DC: Optical Society of America, pp. 1–15.

Ohnuki, S., Shimazaki, K., Ohta, N., Fujiwara, H. (1991), *J. Magn. Soc. Jpn.* **15,** Suppl. **S1,** 399–402.

Ohta, M., Fukumoto, A., Aratani, K., Kaneko, M., Watanabe, K. (1991), *J. Magn. Soc. Jpn.* **15,** Suppl. **S1,** 319–322.

Ohta, T., Uchida, M., Yoshioka, K., Inoue, K., Akiyama, T., Furukawa, S., Kotera, K., Nakamura, S. (1989), in: G. R. Knight, C. N. Kurtz (Eds.), *Optical Data Storage Topical Meeting,* SPIE Proceedings No. 1078, Bellingham, WA: SPIE.

Otomo, Y. (1990), *Optical Disks* (in Japanese, Frontier Technology Series, No. 31), Tokyo: Maruzen.

Parkin, S. S. P., Li, Z. G., Smith, D. J. (1991), *Appl. Phys. Lett.* **58,** 2710–2712.

Potter, R. I. (1974), *IEEE Trans. Magn.* **13,** 502–508.

Saito, J., Sato, M., Matsumoto, H., Akasaka, H. (1987), *Jpn. J. Appl. Phys.* **26,** Suppl. **26-4**, 155–159.

Shimazaki, K., Yoshihiro, M., Ishizaki, O., Ohta, N. (1995), in: G. R. Knight *et al.* (Eds.), *Optical Data Storage '95,* SPIE Proceedings No. 2514, Bellingham, WA: SPIE.

Shinjo, T., Yamamoto, H. (1990), *J. Phys. Soc. Jpn.* **59,** 3061–3064.

Takagaki, T., Takaoka, H., Furusawa, K., Abe, K., Kamei, T., Sano, M. (1987), *IEEE Trans. Magn.* **23,** 3420–3422.

Takahashi, M., Niihara, T., Ohta, N. (1988), *J. Appl. Phys.* **64,** 262–269.

Terao, M., Miyauchi, Y., Ando, K., Yasuoka, H., Tamura, R. (1989), "Progress of Phase-Change Single Beam Overwrite Technology," in: G. R. Knight, C. N. Kurtz (Eds.), *Optical Data Storage Topical Meeting,* SPIE Proceedings No. 1078, Bellingham, WA: SPIE.

Terao, M., Ohta, N., Horigome, S., Ojima, M. (1990), *The Fundamentals of Optical Memories* (in Japanese), Tokyo: Corona.

Yagi, S., Fujimori, S., Yamazaki, H. (1987), *Jpn. J. Appl. Phys.* **26,** Suppl. **S26-4**, 51–54.

Yoshida, K., Okuwaki, T., Osakabe, N., Tanabe, H., Ouchi, Y., Matsuda, T., Shinagawa, K., Tonomura, A., Fujiwara, H. (1983), *IEEE Trans. Magn.* **19,** 1600–1604.

Yoshimura, M., Nishimura, T., Tsukada, N. (1989), "Ultrahigh Density Optical Memory by Photochemical Hole Burning (PHB) and Multilayered PHB System," in: G. R. Knight, C. N. Kurtz, (Eds.), *Optical Data Storage Topical Meeting,* SPIE Proceedings No. 1078, Bellingham, WA: SPIE.

Zeper, W. B., Greidanus, F., Carcia, P. F. (1989), *IEEE Trans. Magn.* **25,** 3764–3766.

Further Reading

Magnetic Recording

Bate, G. (1981), "Recent Developments in Magnetic Recording Materials," *J. Appl. Phys.* **52,** 2447–2452.

Bertram, H. N. (1994), *Theory of Magnetic Recording,* New York: Cambridge Univ. Press.

Fert, A., Bruno, P. (1994), "Interlayer Coupling and Magnetoresistance in Multilayers," in: B. Heinrich, J. A. C. Bland (Eds.), *Ultrathin Magnetic Structures II,* New York: Springer-Verlag.

Fujiwara, H. (1992), "Emerging Magnetic Technologies for Consumer Audio/Video," *J. Appl. Phys.* **73,** 5757–5762.

Guerst, J. A. (1963), "The Reciprocity Principle in the Theory of Magnetic Recording," *Proc. IEEE* **51,** 1573–1577.

Iwasaki, S. (1984), "Perpendicular Magnetic Recording—Evolution and Future," *IEEE Trans. Magn.* **20,** 657–668.

Jorgensen, F. (1980), *The Complete Handbook of Magnetic Recording,* Blue Ridge Summit, PA: TAB Books.

Karlqvist, O. (1954), "Calculation of the Magnetic Field in the Ferromagnetic Layer of a Magnetic Drum," *Trans. R. Inst. Tech.,* reprinted (1985), in: R. M. White (Ed.), *Introduction to Magnetic Recording,* New York: IEEE.

Levy, P. M. (1994), "Giant Magnetoresistance in Magnetic Layered and Granular Materials," *Solid State Phys.* **47,** 367–463.

Mallinson, J. C., Bertram, H. N. (1984), "A Theoretical and Experimental Comparison of the Longitudinal and Vertical Modes of Magnetic Recording," *IEEE Trans. Magn.* **20,** 416–467.

Mallinson, J. C. (1987), *The Foundations of Magnetic Recording,* San Diego: Academic.

Mee, C. D., Ciureanu, P., Daniel, E. D. (1989), *Magnetic Recording Handbook,* New York: McGraw-Hill.

Middelhoek, S. (1992), *Thin Film Resistive Sensors,* London: Institute of Physics Publishing.

Miura, Y. (1983), "Thin Film Head," in: Y. Sakurai (Ed.), *Recent Magnetics for Electronics,* Japan Annual Reviews in Electronics, Computers and Telecommunications Vol. 10, Tokyo: Ohmsha, Ltd., p. 77.

Potter, R. I. (1975), "Analytical Expression for the Fringe Field of Finite Pole-Tip Length Recording Head Fields," *IEEE Trans. Magn.* **11,** 80–81.

Sharrock, M. P. (1990), "Particulate Recording Media," *MRS Bull.* **15,** (3), 53–61.

White, R. M. (Ed.) (1985), *Introduction to Magnetic Recording,* New York: IEEE.

Williams, M. L., Comstock, R. L. (1971), "An Analytical Model of the Write Process in Digital Magnetic Recording," in: *Seventeenth Annual AIP Conference Proceedings,* New York: AIP, pp. 738–742.

Optical Recording

Bell, A. E. (1987), "Materials for High Density Optical Data Storage," in: M. J. Weber (Ed.), *Handbook of Laser Science and Technology,* Vol. 5, New York: CRC, Part 3.

Bouwhuis, G., Braat, J., Huijser, A., Pasman, J., van Rosmalen, G., Schouhaumer-Immink, K. (1985), *Principles of Optical Disc Systems,* Bristol, U.K.: Adam Hilger.

Brady, D., Psaltis, D. (1992), "Control of Volume Holograms," *J. Opt. Soc. Am. A* **9,** 1167–1182.

Curtis, K., Pu, A., Psaltis, D. (1994), "Method for Holographic Storage Using Peristrophic Multiplexing," *Opt. Lett.* **19,** 993–994.

Curtis, K., Psaltis, D. (1994), "Characterization of the DuPont Photopolymer for Three-dimensional Holographic Storage," *Appl. Opt.* **33,** 5396–5399.

Emmelius, M., Pawlowski, G., Vollmann, H. W. (1989), "Materials for Optical Data Storage," *Angew. Chem., Int. Ed. Engl.,* **28,** 1445–1600.

Heanue, J. F., Bashaw, M. C., Hesselink, L. (1994), "Volume Holographic Storage and Retrieval of Digital Data," *Science* **265,** 749–752.

Isailovic, J. (1985), *Videodisc and Optical Memory Systems,* Englewood Cliffs, NJ: Prentice-Hall.

Ito, F., Kitayama, K., Oguri, H. (1992), "Holographic Image Storage in $LiNbO_3$ Fibers with Compensation for Intrasignal Photorefractive Coupling," *J. Opt. Soc. Am. B* **9,** 1432–1439.

Johnson, G. H. (1989), "Information Storage Materials" in: G. E. Thomas (Ed.), *Ullman's Encyclopedia of Industrial Chemistry,* Vol. A 14, New York: VCH.

Mok, F. H., Tackitt, M. C., Stoll, H. M. (1991), "Storage of 500 High-Resolution Holograms in a $LiNbO_3$ Crystal," *Opt. Lett.* **16,** 605–607.

Nagao, Y., Sakata, H., Mimura, Y. (1993), "Improvements in Bi12SiO20 Thin Spatial Light Modulators," *Appl. Opt.* **32,** 5036–5042.

Niihara, T., Takahashi, M., Miyamoto, H., Kirino, F., Ogihara, N., Ohta, N. (1990), "Thermomagnetic Recording Mechanism on TbFeCo Disks," *J. Magn. Magn. Mater.* **88,** 177–182.

Qiao, Y., Psaltis, D., Gu, C., Hong, J., Yeh, P., Neurgaonkar, R. R. (1991), "Phase-Locked Sustainment of Photorefractive Holograms Using Phase Conjugation," *J. Appl. Phys.* **70,** 4646–4648.

Rakuljic, G. A., Leyva, V., Yariv, A. (1992), "Optical Data Storage by Using Orthogonal Wavelength-Multiplexed Volume Holograms," *Opt. Lett.* **17,** 1471–1473.

Van Heerden, P. J. (1963), "Theory of Optical Information Storage in Solids," *Appl. Opt.* **2,** 393–400.

Yu, F. T. S., Wu, S., Mayers, A. W., Rajan, S. (1991), "Wavelength Multiplexed Reflection Matched Spatial Filters Using $LiNbO_3$," *Opt. Commun.* **81,** 343–347.

RECORDING, PHOTOGRAPHIC

David M. Sturmer, *Eastman Kodak Company, Rochester, New York, U.S.A.*

	Introduction	185
1.	**Principles of Photographic Recording**	186
1.1	Silver-Halide Photography	186
1.2	Silicon-Based Photography	189
1.3	Photographic Quantum Efficiency	189
1.4	Image Structure: Silver Halides and Silicon	190
1.5	Color Reproduction	191
1.6	Holographic Image Recording	192
2.	**Imaging Materials: From Silver Halides to Phosphors to Silicon**	192
2.1	Silver Halides	192
2.2	Silicon	194
2.3	X Rays and Radiography	196
2.4	Infrared Image Detectors	197
3.	**Applied Photographic Recording**	198
3.1	Clinical and Scientific Photography	198
3.2	Holography	199
3.3	HDTV Sensors and High-Resolution Still Sensors	201
	Glossary	201
	Works Cited	201
	Further Reading	203

INTRODUCTION

Photographic recording provides capture and storage of images. The capture–storage sequence places high demands on materials, costs, transportability, ease of use, image resolution, color reproduction, high photosensitivity, etc. The photographic industry has grown from a fledgling enterprise (Mees, 1961) of a century ago to an extraordinary cultural medium for recording daily life in several present-day cities and countries (Cohen, 1989). George Eastman convinced early consumers to expect "You push the button, we do the rest," and many large and small imaging companies have been running fast to approach this goal ever since. Photographic imaging has, of course, expanded greatly over the past hundred years. The decade of the 1990s began with an estimated 50 billion consumer images being printed worldwide. News photography is global. Scientific and engineering imaging extend from laboratory instrumentation and commercial production lines to outer space. Countless images are a part of the daily routine for medical clinics, cinematographers, quality-control engineers, astronomical observatories, scientific laboratories, space exploration, and holographers. Diverse sets of needs among these customers require imaginative combinations of materials and environment for the high-quality, easy-to-use, cost-effective imaging that we have come to expect.

Today, conventional silver-halide photography is complemented by digital imaging (Khosla, 1992). Capture, storage, manipulation, transmission, soft display, and hard display, the typical elements of an imaging chain (Fig. 1), are realistically accomplished by either technology. Conventional photography produces high-quality monochrome or color images where the image is stored in a visually recognizable analog form. Digital imaging, which traces its modern high-quality implementation to the invention of charge-coupled devices (CCDs) in 1970, captures images in arrays of picture elements (pixels) and (after proper readout) stores images as bits in computer mass storage (Tredwell, 1995).

Gateways between the conventional and digital imaging chains are common, after the initial capture and storage of an image. Many gateways occur between the two imaging chains, representing opportunities to convert analog to digital or digital to analog images. The specific gateway shown in Fig. 1 represents the conversion process for a developed and processed film image, which can

3-527-28138-X/96/$5.00 + .50

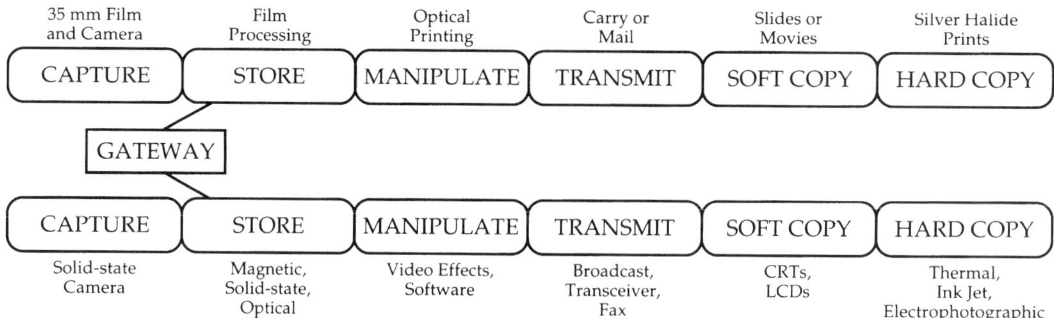

FIG. 1. Hybrid imaging chain for analog (film/paper) imaging and digital (electronic) imaging. Photographic recording can start with film and camera in the analog imaging chain, followed by chemical development of the exposed film to "store" the image in a usable form. The digital imaging chain can start with a solid-state camera containing a CCD array, which is subsequently read out and stored in a usable digital form. The gateway shown is one of many, consisting of the steps necessary to convert an analog image to a digital image.

be scanned and stored in digital form. Reverse gateways (not shown) would convert a digital image to a film or paper image by utilizing the digital image to expose the appropriate photographic film or paper. Thus, high-speed conventional movie films containing high-quality color images are digitized for video editing and reassembled onto film for distribution to theaters. Digital photonews images are compressed and then transmitted with sufficient detail for use in both black-and-white and color newsprint.

Images are also desired from many segments of the electromagnetic spectrum. Short-wavelength, high-energy regions (x ray and ultraviolet) are important in medical imaging, circuit-board photolithography, and quality control for metallic structural parts. Commercial and consumer photography focuses on the blue–green–red visible spectrum. Long-wavelength, lower-energy regions (various segments of the infrared) are important for environmental survey mapping and other specialized uses. Selecting a capture–storage technology for an imaging task requires a good knowledge of imaging principles and an awareness of the behavior of imaging materials.

1. PRINCIPLES OF PHOTOGRAPHIC RECORDING

The appearance and usefulness of images require matching the spectral and image-structure properties of imaging recording materials to the characteristics of objects being photographed (overall scene wavelengths and brightness, range of brightness levels in scene, resolution required in image, etc.). Technical data that define these properties, as provided by most major manufacturers, help select the appropriate materials, environment, and exposure time for a specific task.

1.1 Silver-Halide Photography

Silver-halide photography relies on spectrally sensitized silver-halide microcrystals (grains), dispersed in a gelatin binder, and coated on plastic or paper supports to give layers with random arrangements of grains within the layers. A brief overview of the image characteristics of these materials is given below. More detail may be found in specialized texts (James, 1977; Sturge *et al.*, 1989).

Exposure and chemical development provides either a black-and-white or color image that can be measured optically. In the exposure step, the silver halide or the spectral sensitizing dyes on its surface absorb selected regions of the exposing radiation, converting a photon to a hole and electron within the silver halide or at its surface. In most commercial imaging materials, the hole is trapped or chemically reacted at a variety of sites, including reaction with gelatin. The electron becomes mobile in the silver-halide conduction band and either combines with a trapped hole or reacts with an interstitial silver ion to form a silver atom. Single silver atoms are unstable at room temperature, so that formation and decay (to electron + interstitial silver ion) occur until the electron either recombines with a hole (no image) or a sufficient number of additional photons are absorbed to induce stepwise aggregation to form silver-atom clusters (Hamilton, 1991). These silver-atom clusters are more stable, and the three- or four-atom–sized clusters are catalytic centers for the process of chemical development. The silver-atom clusters are the photographic latent images formed during the exposure (or image-capture) step in the imaging chain (Fig. 1). The subsequent chemical development amplifies this image and effectively stores it in a visible form. A black-and-white image is silver metal, formed by the catalytic chemical reduction of the silver ions in microcrystals that have latent image. The chemical developers, which are oxidized during development, are not part of the image. A color image can be formed by utilizing the oxidized organic chemical developer, formed when silver ions are catalytically reduced. The oxidized developer reacts with organic coupler molecules to produce stable colored dyes as the visible photographic image. Quantitative color photography, the chemistry of color photography, image structure in color photography, and color electronic cameras are often subjects of updated reviews (see, for example, Hunt, 1995).

There are many variations on the basic steps to capture and store photographic images. Surface and internal sites for latent-image formation exist in most microcrystals, but commercial chemical developers are catalyzed primarily by the surface latent image. Preformed dyes can replace the organic coupler molecules, react with oxidized developer to affect their diffusibility through gelatin layers, and thereby lead to the basis for instant photography. The details of such variations are covered elsewhere (Sturge *et al.*, 1989).

Photographic speed (or sensitivity) is inversely related to the exposure required to produce an image. Because imaging materials are used to perform many tasks, international standards (ISO speed definitions) are specified differently for pictorial, medical, microfilm, and other uses. In Fig. 2, idealized responses for fast, medium, and slow films indicate that more exposure [higher ratio of input signal to noise, Fig. 2(d)] is required to obtain a specific film density for slower-speed films [Fig. 2(c)]. As is often the case for well-sensitized commercial (pictorial) imaging, the three films could be similar in photographic efficiency [similar detective quantum efficiency, Fig. 2(b)]. The faster film would be the result of precipitating and sensitizing larger silver-halide microcrystals to respond to about the same number of incident photons per microcrystal as the smaller microcrystals in slower films.

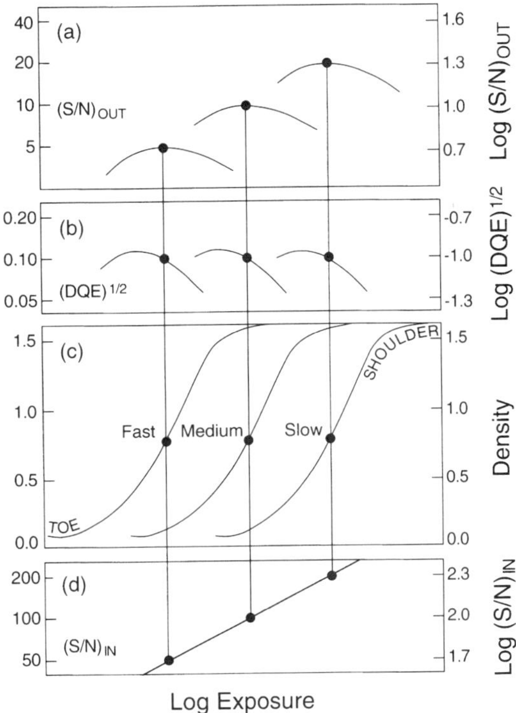

FIG. 2. Relationship of (a) output signal-to-noise ratio, (b) detective quantum efficiency, (c) photographic speed, and (d) input signal-to-noise ratio for three materials with different photographic speeds (fast, medium, slow). The curves in (c) illustrate the typical nonlinear relation between logarithm of exposure and optical density.

Image quality is, however, not a simple function of photographic speed. Comparison of photographic imaging with a $(1\text{-}\mu m)^2$ grain area and electronic imaging with a $(10\text{-}\mu m)^2$ grain area (Shepp, 1989) indicates ISO speeds for threshold responses near 1000 and 10 000, respectively, if no significant loss processes occur. A standard 35-mm color negative contains 2400 × 3600 picture elements (pixels) per color or about 24 million pixels for a three-color negative. Electronic imaging devices for pictorial photography or video are approaching the three-color information requirement of high-definition television (>6 million pixels). Additionally, a simple expression for image quality is still being debated (Jacobson, 1995). For both silver-halide films and electronic imaging devices, several physical measurements are often required to describe the optical processes, chemical pathways, and sources of noise that degrade image quality. The mathematical relationships in photographic science (Saunders, 1994) attempt to treat some of these factors, such as the influence of randomness in collections of microcrystals in films and coupler droplets involved in forming color-image dyes on macroscopic responses like optical density curves [Fig. 2(c)] and image-quality factors like granularity and sharpness.

The multiple photons necessary for a single microcrystal to form a latent image cause conventional silver-halide films to show optical densities that are nonlinear functions of exposure. Collections of multiphoton receptors like the silver-halide microcrystals can be shown to exhibit S-shaped photographic response curves [Fig. 2(c)], where the density and exposure terms are defined as follows:

$$\text{Density} = D = \log_{10}(1/T),$$

where D = optical density, T = transmittance of film sample;

$$H = Et,$$

where H = total exposure, E = irradiance, t = time for which radiation acts on photosensitive layer(s);

$$\text{Contrast} = \text{slope of } D \text{ vs } \log H \text{ curve};$$

$$\text{Sensitivity} \propto 1/H.$$

The middle section of each photographic response curve is often close to linear, and the macroscopic contrast of a film is easiest to measure by the slope of this middle section. Increasing the number of photons required to produce latent image in a microcrystal causes the toe and shoulder segments to be more sharply defined and the middle section to exhibit a higher slope (higher contrast). Very high-contrast films are important in printing and publishing, where well-controlled exposures shift film response from minimum density to maximum density over a short exposure range. Moderate-contrast films (applicable to pictorial, medical, and scientific imaging) will retain useful density changes across several orders of magnitude of exposure.

Note that the total exposure (Et) alone does not determine photographic optical density. For a constant value of Et, E and t are not always interchangeable, and dramatic decreases in density often occur at high intensity (high E, short t) or very low intensity (low E, long t) as compared with intermediate values for both quantities. This effect is termed reciprocity failure.

Other measures of photographic sensitivity are detective quantum efficiency (DQE) and quantum sensitivity. DQE relates input signal/noise to output signal/noise. The detective quantum efficiency for photographic materials can be expressed in terms of measurable quantities (Bird, 1989): $(0.434)^2\gamma^2/HA\sigma_D^2$, where γ = contrast of photographic response curve at density D, H = exposure (ergs/cm^2), A = aperture area at which noise (granularity) is measured, and σ_D = granularity at density D. Photographic films show maximum DQE values of about 1% as manufactured, since other factors like long-term storage of unexposed film must be considered.

For imaging very weak signals as, for example, in astronomical surveying, higher-speed films are typically pretreated before use (Babcock *et al.*, 1974) to increase output signal/noise, or a slow-speed film (long exposure time) is selected so that input signal/noise will be enhanced. As the idealized relationships in Fig. 2 show, films of varying speed with DQE values near 1% will exhibit a higher output signal/noise for the slower photographic speeds. Experimental studies have quantified these relationships. Signal-to-noise ratios for color and black-and-white films are proportional to the reciprocal of the films' sensitivities, (sensitivity)$^{-a}$ where a = 0.28 for commercial color films, 0.44 for commercial black-and-white films, and 0.35 for a series of model films where the dye-forming chemicals (couplers) varied considerably in their reactivity (Topfer *et al.*, 1994).

Important sources of noise for these light-exposed films are the statistical distribution of silver-halide microcrystals in the coated film layers, the color-image–dye yields, and random coupler-droplet distribution. For radiation-exposed systems, quantum mottle (statistical photon arrival for a limited number of x-ray photons) is an additional source of noise, as would occur in x-ray–exposed phosphor screens or films. The noise contribution due to the finite number of x-ray photons sufficient to expose a screen–film combination is a necessary component of noise needed to explain experimental data (Bunch *et al.*, 1987).

1.2 Silicon-Based Photography

Digital photography using silicon devices relies on a regular array of solid-state sensors that absorb incident photons, collect the resulting electrons, and participate in subsequent readout steps so that the image may be stored digitally. DQE for solid-state charge-coupled device (CCD) arrays depends largely on two factors:

1. the "fill factor" or percentage of surface area available to the photosensitive silicon, and
2. the optical (exposure) enhancement features and the electron-loss processes (exposure and readout) that result from a specific CCD design.

Manufacturer's descriptions of CCD arrays include considerable information related to integral DQE for the array (e.g., sensitivity, noise vs signal level, saturation, uniformity). Nonetheless, other (unspecified) factors affecting DQE will vary significantly from pixel to pixel [noisy pixels, stray response, aliasing/moire (spatial quantization), banding (amplitude quantization), nonuniformities like shading, blemishes, and blocked channels]. (See also CHARGE-COUPLED DEVICES.)

Visible-wavelength radiation is almost completely absorbed by CCD arrays, so that in a practical sense the detective quantum efficiency and quantum efficiencies are nearly interchangeable. Typical quantum efficiencies for an interline-transfer CCD are quite high (Fig. 3) and are enhanced by an optical lens array (above the CCD) (Nichols *et al.*, 1995). CCD arrays can be specifically designed for 100% fill factor, including front-illuminated CCDs and back-illuminated CCDs, as noted below in Sec. 2.2. Electron generation per absorbed photon is essentially one electron per photon in the visible spectral region. Image-readout effectiveness is compromised by exposure-unrelated factors (e.g., dark-current shot noise, dark-current pattern noise, output amplifier noise) and exposure-enhanced factors (e.g., photon shot noise). Estimates for the nonexposure factors are about 30 rms electrons per pixel ($14 \times 12\ \mu m$) in camera- or video-type sensors at 40 °C (Tredwell, 1995). Use of sensors at lower temperatures and lower readout rates can significantly lower noise levels.

1.3 Photographic Quantum Efficiency

Quantum efficiency in photographic recording is usually defined as useful image derived from the photons absorbed by the sensitized material. Reflection and transmission losses of incident photons, for example,

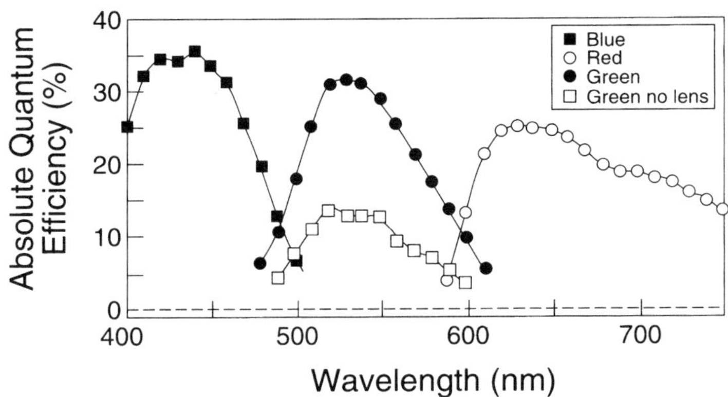

FIG. 3. Measured quantum efficiency for a KODAK KAI-2090 Image Sensor with color-filter and microlens arrays. Only quantum efficiencies above about 5% are shown.

are not included. Larger differences between DQE and quantum efficiency are noted for silver-halide films than for silicon, since spectrally sensitized silver halides absorb only about 5%–10% of incident photons. After correction for this low absorption, silver halides can provide a latent image to catalyze chemical development with as few as three photons per grain. These values can be obtained for slower-speed films when exposure is carried out under nitrogen atmosphere after evacuation and hydrogen treatment (Babcock *et al.*, 1974). The optimal three-photon response is experimentally observed for spectroscopic plates, which are typically evacuated and treated with forming gas (95% nitrogen, 5% hydrogen) before astronomical use to reduce exposure times from several hours to near 40 min. Higher-speed consumer films contain larger grains (larger volume and larger surface area), are exposed in the presence of oxygen and water (no evacuation), and consequently use photons considerably less efficiently.

The approach of silver halides to their theoretical limit is presently a strong function of preparation technique, inadvertent impurity and defect incorporation, and the final sensitizing process. These factors are harder to control for larger grains, thus leading to experimental results that show larger grain sizes to be less efficient. For grain sizes of interest to image recording, the smaller-sized grains (near 0.2 μm) do exhibit the three-photon sensitivity described above for astronomical plates and films. Octahedra near 0.4-μm edge length have also exhibited these optimal sensitivities. Data summaries that include these and other grains (Hailstone, 1994) indicate that ten photons or more are typically required for developable latent image for grain sizes between 0.5 and 3 μm.

1.4 Image Structure: Silver Halides and Silicon

Two additional and measurable image characteristics are granularity and resolving power. Granularity measures image-density variations from an average for an image area of constant density. Because most silver-halide images are composed microscopically of small silver or colored dots, enlargements of a high-granularity image will make these discrete elements of the image too visible. Representative granularity values (defined as 1000 × the rms deviation of diffuse density measured at an average optical density of 1.0 with a 48-μm circular aperture) are 4–5 for slow-speed (50 ISO) color-negative film, 6 for black-and-white microfilm, and 5–7 for higher-speed (400 ISO) color-negative film. The appropriate measuring aperture for granularity measurements is a function of the intended use for an image.

The most universal measure of granularity is the noise–power spectrum, often indexed in photographic texts as the Wiener spectrum (Jones, 1955). If a (constant average density) photographic sample is scanned at constant speed with a small slit of light, the transmitted light intensity can be transformed to squared density fluctuations per two-dimensional spatial-frequency interval as a function of spatial frequency (and exposing-radiation wavelength, if appropriate). The Wiener spectrum is the representation of these data as an equivalent set of sine waves, plotting sinusoidal power (amplitude squared) as a function of sine-wave frequency. The rms granularity of a negative can be tracked through a printer system in terms of the Wiener spectrum of the photographic-negative image and the modulation-transfer functions of the printer system elements (Doerner, 1962).

The ability to record the details in an object can be quantified in several ways. Resolving power, measured for film following procedures in ANSI method PH2.33 1969, is the ability to resolve images of closely spaced lines in an object. Modulation transfer measures the ability of optical components (e.g., lenses, films) to record detail in an object. Modulation transfer is measured with test objects that vary sinusoidally in transmittance. When the peak/valley ratio in the test object (sine function) is decreased by an optical component (influence of a lens or image recorded on film), then modulation-transfer values decrease. The modulation-transfer function (MTF) is an expression of modulation-transfer values as a function of the frequency of the optical transmittance input. Representative MTF values for films are typically cited as the frequencies for which the modulation-transfer value is measured as 50%: 40–60 cycles/mm (slow color), >200 cycles/mm (microfilm), and 25–40 cycles/mm (higher-speed color). The measurement procedure is described in ANSI PH2.39-1984 for continuous-tone photographic films. For determining modulation-transfer functions of experimental film samples, the exposing camera, MTF-test object, microdensitometer, and evaluation methods have been described in detail (Lamberts *et al.*, 1965).

Image sharpness at the edges of objects within an image can show microscopic but measurable enhancement from chemical effects during development of photographic films. A high-density region within an image will deplete development chemicals rapidly (development by-products restrain development), but near the edge of this region extra developer can diffuse into the image region from lower-density (less depleted) regions. The net effect of these factors is an enhanced visual perception of the edge of an object resulting from

1. somewhat lower than expected density near the center of a high-density region;
2. enhanced density at the edge of the region; and occasionally
3. somewhat restrained (lower) density immediately adjacent to the high-density region (development by-products diffuse out of a high-density region and restrain development in adjacent areas).

Granularity and resolving power for CCD arrays are limited by pixel size, pixel uniformity, and wavelength. Each pixel can have a different gain than a neighboring pixel, resulting in an output signal that amplifies a constant-input signal differently at each pixel. Photoresponse uniformity is approximately $\pm 5\%$ (maximum) across a full array. For pixels illuminated at 70% of saturation, a pixel that deviates by more than 6% from neighboring pixels (the surrounding 100×100 pixels) is a point defect in the sensor array. A C1-class sensor would have fewer than 23 of these point defects over the total array and fewer than 10 point defects in the central 1024×1024-pixel region of the array. Typically, a specific sensor design may be produced at two or three levels of defect classes by a CCD-array manufacturer (Eastman Kodak Company, 1993). Applying gain corrections to each pixel can be critical for low-contrast images, such as those encountered in aerial photography (Maver *et al.*, 1989).

Pixel size and exposing-radiation wavelength limit resolving power. At short wavelengths, the MTF closely matches a function with a cutoff frequency equal to the reciprocal of the pixel dimensions (e.g., 100 cycles/mm for a 10-μm pixel). At longer wavelengths, photons penetrate deeper into the silicon before being absorbed, and the resulting photoelectrons have more chances to drift laterally into neighboring pixels. A particularly instructive computational approach (Lavine *et al.*, 1983) to this image-degradation problem considered spectral wavelength and depletion depth (amount of exposure vs saturation) as variables. As a measure of image degradation, the degree of crosstalk from the exposed pixel to the nearest-neighbor pixel was computed to vary from 4.5% to 13.5% as a function of depletion depth in the exposed pixel. The next-nearest-neighbor pixel experienced 1%–2% crosstalk. In the computation, the exposed pixel was characterized by 12-μm diameter, a variable (5.0–2.0 μm) depletion depth, and 700-nm exposing-radiation wavelength. CCDs designed with deep depletion layers can improve MTF to useful values for wavelengths up to about 800 nm.

The modulation-transfer function is also degraded by transfer inefficiency, smear, and lag during the readout process. The primary degradation is transfer inefficiency, which occurs during transfer of charge from one phase to the next in the CCD shift register. Although charge-transfer inefficiencies are less than 1×10^{-5} per transfer, readout processes with nearly 10 000 transfers can result in significant loss in signal charge. Image smear and lag are presently less important, since CCD designs have improved to overcome these problems.

1.5 Color Reproduction

Image structure in color photographic recording includes not only granularity and sharpness, but also the accurate reproduction of colors onto a final print or transparency. Since the entire imaging chain affects the color reproduction (Hunt, 1995; Krause, 1989), only a brief commentary on color-reproduction issues is given here. The spectral sensitivities that record various colors during image capture must be matched with the subsequent imaging-chain steps (e.g., printing) for spectral response. Early color-reproduction studies focused on matching the spectral distribution for each color record in color film with the spectral distributions for both the exposing-light sources (Schwan and Graham, 1972) and the output material (color paper) (Evans *et al.*, 1953). Computerized color models have improved the simulation of color reproduction and now include not only spectral factors, but adjustment and matching of curve shape as well (Buhr and Franchino, 1994).

1.6 Holographic Image Recording

(See also HOLOGRAPHY, OPTICAL.) Holographic recording captures and stores three-dimensional images of an object. The stored images can reconstruct the original wave fronts from the object, since the hologram contains both phase and amplitude information. (A conventional photograph only contains amplitude information.) The increased availability of lasers has provided the monochromatic (temporally coherent) and unidirectional (spatially coherent) light sources needed for the practical use of holography (Conley and Robillard, 1992). Industrial uses of holography include cataloging, training, remote inspection, and nondestructive testing (Lin, 1992; Brownell *et al.*, 1992). Artistic uses of holography span fields like advertising, archaeology, and (traveling) museum exhibitions.

Image quality in holograms is related to the resolution (lines/mm) of the holographic medium, its sensitivity (particularly for recording dynamic objects), and the image-viewing efficiency afforded by the final hologram. Approximate values for the first two quantities vary with the recording medium:

1. silver-halide (very small grain size) emulsion (microcrystals dispersed in gelatin) on film or glass supports, 5000 line pairs per mm, 5×10^{-5} J/cm^2;
2. dichromated gelatin, 3000 line pairs per mm, 4×10^{-2} J/cm^2;
3. photo-thermoplastics, 4000 line pairs per mm, 5×10^{-5} J/cm^2.

Reconstructing a holographic image is most quantitatively accomplished using the same spectral wavelength that exposed the holographic material. A reconstructing beam of shorter wavelength will produce a smaller object, whereas a longer wavelength gives a larger object. Additionally, dimensions of the object parallel to the plane of the hologram scale linearly with wavelength, but depth dimensions are distorted (square of the wavelength).

2. IMAGING MATERIALS: FROM SILVER HALIDES TO PHOSPHORS TO SILICON

2.1 Silver Halides

The image-capture step in the photographic imaging chain (Fig. 1) often requires both high efficiency and high photographic speed. For silver halides silver iodobromide [Ag(Br,I)] is the most widely used of these photographic image-capture materials. The large microcrystals for high speed are conveniently formed by precipitation in aqueous gelatin solution. A range of 2%–20% iodide occurs in the microcrystals. After exposure, chemical development of Ag(Br,I) microcrystals proceeds at rates that allow organic materials to interact with developing crystals; these organic materials modify the rate or extent of development to improve the granularity, sharpness, or color balance of the final image.

Silver iodobromide itself is a solid-state, rock-salt–structured crystal that is naturally sensitive to x-ray, ultraviolet, and short-wavelength visible (blue) radiation. It can be spectrally sensitized by adsorbed dyes to wavelengths throughout the visible and infrared up to about 1.2 μm. The image-capture step is successfully modeled (Hamilton, 1991; Hailstone, 1994) by assuming that incident photon absorption produces mobile electrons and somewhat less mobile holes. These species equilibrate between free, trapped, and atom states [Fig. 4(a)]. Since most silver-halide microcrystals contain a multiplicity of trapping sites for electrons and holes throughout the crystal, many trapping events can occur. The electrons proceed to the image-forming step in regions of crystals where intentional sensitivity sites have been introduced, and where sufficient interstitial silver ions exist to form a silver atom at an electron-trap sensitivity site. Subsequently, preferential capturing of additional

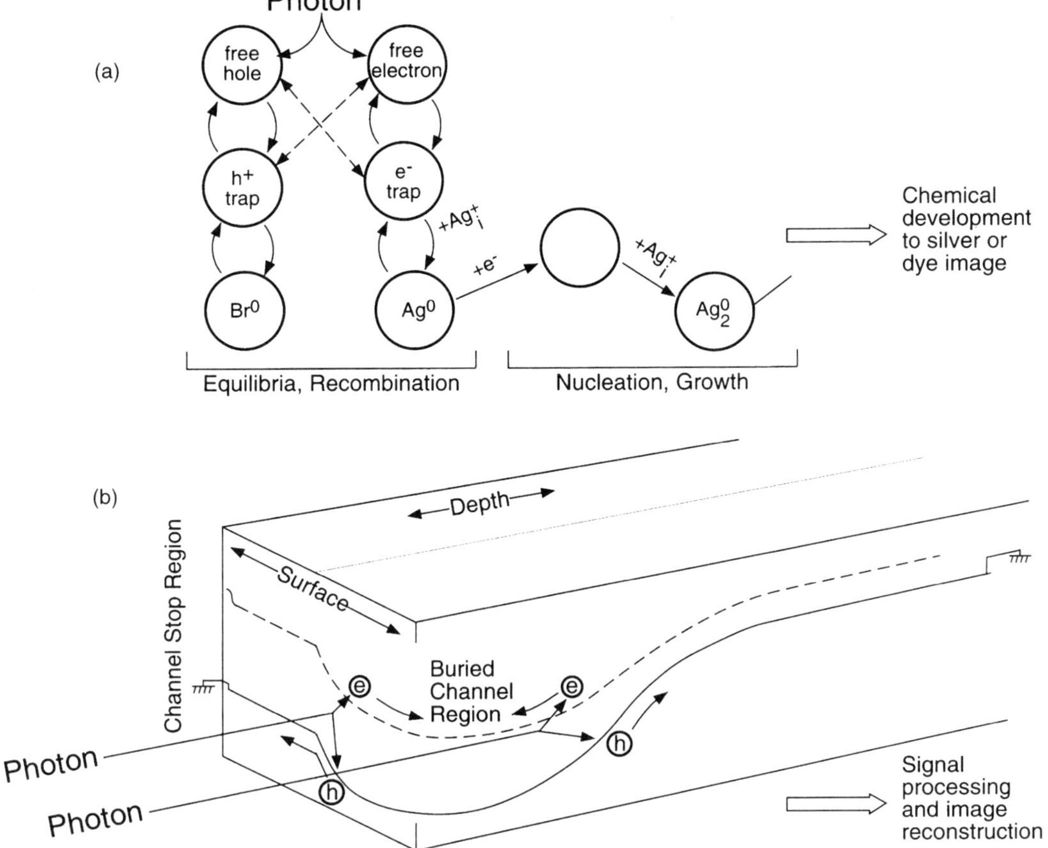

FIG. 4. Solid-state equilibria, recombination, and accumulation of photoelectrons for image capture. (a) Microcrystalline silver halides contain a multiplicity of traps throughout the crystal, making an equilibrium model appropriate for electrons and holes. Intentional sensitization sites are preferential electron traps with sufficient interstitial silver ions (Ag_i^+) nearby to favor nucleation and growth of silver-atom clusters (latent image = catalyst for chemical development). (b) Crystalline silicon, doped to favor electron accumulation in a buried-channel region, has sufficiently low impurity levels (about $10^{16}/cm^3$) to allow millisecond mean free lifetimes for electrons and holes. Conduction (dashed curve) and valence (solid curve) band energies are shown as a function of depth into the silicon and also across the surface of a single-pixel element. Potential fields separate the photocarriers, and little recombination occurs. Holes are neutralized deep in the interior region [or move to the silicon surface and then to the edge of the pixel (channel stop region) and are neutralized]. Signal processing (pixel-to-pixel transfer and then charge-to-voltage conversion) and image reconstruction (especially for color CCD arrays) provide a usable digital image.

photoelectrons and interstitial silver ions occurs at sites where silver atoms have already formed. The alternating sequence of electron trapping and addition of interstitial silver ions leads to the growth of a silver-atom cluster as the latent image [Fig. 4(a)].

The dominant commercial silver-halide films require the latent image to be on the surface of the silver-halide crystal because the photographic developers (which chemically amplify the latent image) do not dissolve the crystals sufficiently to uncover internal latent image. Surface latent-image stability is affected by chemical constituents of the intentional sensitivity sites, the near-monolayer of sensitizing dye(s) on the silver-halide–crystal surfaces, the crystallographic surfaces of the silver-halide microcrystals, and the presence of oxygen and moisture. Consequently, the technology to improve image capture by silver halides must achieve a complex balancing of factors to improve sensitization, but also eliminate trapping sites for photoelectrons in the dye layers, improve trapping sites for holes, minimize recombination tendencies (trapped hole + free electron; trapped electron + free hole), and control instabilities of latent image prior to development.

The ideal conditions produced by hydrogen sensitization and removal of oxygen/moisture allow three absorbed photons to create a developable latent image. Photographic development converts approximately 10^9 additional silver atoms to silver metal at full development. Color development is less complete with approximately 25%–50% silver development, primarily to improve image-structure attributes like granularity.

Silver chloride, silver bromochloride, and silver bromide are also suitable as imaging materials. All can be dye sensitized efficiently, but significant incorporation of transition-metal dopants like iridium [Ir(III)] is required for satisfactory responses to high-intensity, short-duration exposures. The silver-chloride–containing materials are most frequently applied to photographic papers (printing step) and graphic-arts films. Silver bromides are used in high-speed x-ray films (see below). Much of the quantitative latent-image theory is derived from silver-bromide experimentation (Hamilton, 1991). Other silver salts (fluorides, iodides, organic salts) are not widely used for image capture.

2.2 Silicon

Silicon-based image-capture materials employ four basic structures: the photodiode, the photocapacitor, the pinned ($p + np$) photodiode, and the photoconductor. The last uses an amorphous silicon layer deposited over the image-readout array. The first three types of sensors are fabricated in the single-crystal silicon as part of the image-capture device and utilize front-side illumination and charge collection. Some image sensors consist of a crystalline silicon wafer thinned to 10 μm for back-side illumination (with a front-side, charge-coupled device for charge collection and readout). Since the important camera-speed, charge-coupled devices (CCDs) are arrays of metal-oxide-semiconductor (MOS) photocapacitors designed for front-side illumination, the features of silicon image-capture materials will be illustrated with a typical buried-channel MOS photocapacitor (Burkey *et al.*, 1984). A schematic band diagram for a buried-channel CCD is shown in Fig. 4(b) with both depth-wise and surface-potential gradients for a single-pixel element (see also CHARGE-COUPLED DEVICES).

The MOS capacitor is a p-type silicon substrate with a thin layer (200–1000 Å) of silicon dioxide and an electrode (heavily phosphorus-doped polycrystalline silicon) subsequently deposited on the surface. These surface layers, especially the polycrystalline silicon, modify the fraction of incident photons reaching the silicon imaging sensor. For a 3000-Å-thick polysilicon layer, less than 30% of the blue photons and less than 70% of the green photons reach the silicon substrate where they can be absorbed and generate useful image charge. More transparent electrode materials, such as indium-tin-oxide (ITO), are sufficiently conductive, but until recently were limited by materials and processing complexities. Potential wells in the single-crystal silicon are created by lightly doping the subsurface region by diffusion or implantation early in the fabrication process. Photoelectrons and photoholes, which result from photon absorption, are separated into subsurface and interior regions, respectively. A buried-channel photocapacitor has a potential well for electrons just below the silicon–silicon-dioxide interface [Fig. 4(b)]. The primary purpose of this buried-channel design is to maintain the image (electron) charge below the surface, and thus prevent electrons from being trapped by interface states during exposure (charge collection) and readout (charge transfer).

Within crystalline silicon, photon absorption occurs at various wavelength-dependent depths, in contrast to dye-sensitized silver halides where photons are initially absorbed within the surface monolayer of the sensitizing dye. In the ultraviolet, photon absorption occurs very near the silicon surface, and surface recombination is significant. For visible wavelengths, absorption depths range from 0.55 μm (blue, 450 nm) through 1.5 μm (green, 550 nm) to 3 μm (red, 640 nm). Since p-type silicon can exhibit electron-diffusion lengths of 150 μm, there is a high probability that the electrons from visible exposures will be collected in the buried-channel potential well (Lavine *et al.*, 1983). Further into the infrared, absorption depths are 10 μm (at 800 nm) or larger. A photon that is absorbed too deeply in the silicon creates an electron that may diffuse laterally to a neighboring pixel or may be trapped by defects at the 20–50-μm depths, which arise from impurity gettering (a process step in manufacturing that removes many metal ions from the subsurface region where the photocapacitor is fabricated).

CCDs used for image capture are designed as full-frame photocapacitors, which use 100% of the silicon-surface area for capturing image photons. An array of these photocapacitors with up to 6 million pixels is now manufacturable. In addition to their 100% fill factor, full-frame CCDs have low dark current, but designs with electrodes from polysilicon have low blue-light sensitivity. The photocapacitors also function as elements of the CCD used during the readout of the captured image, making full-frame image sensors most suitable for scientific or still electronic photography where a single exposure can be followed by a readout step. The layer architecture for the three-color, full-frame photocapacitors in a CCD array is shown in Fig. 5, as compared with traditional color-film layers.

Much lower fill factors are typical of interline-transfer image sensors. A photodiode accomplishes the image capture, and separate elements of the vertical CCD provide charge transport, so that sequential exposures for typical video applications (30 frames/s) are possible. The 20%–30% fill factors for interline-transfer sensors are not the sole determinants of DQE because a microlens array on top of the device directs light away from the photoinactive regions into the photodiode areas of each pixel. As noted above in Fig. 3, microlens arrays increase photosensitivity by factors of 2.5–3 (Nichols *et al.*, 1995).

In image capture, the photocapacitor must not only collect photoelectrons, but also store electrons until the charge is read out. If too many electrons are generated, the potential on the storage well can drop to zero, and no more electrons will be stored. This is known as saturation. A photocapacitor that is too easily saturated will not exhibit enough exposure latitude to record the range of brightness levels in typical image scenes. The diffusion of excess charge into neighboring pixels is called blooming, again causing image degradation of the high-intensity features of an image. Antiblooming designs in image sensors incorporate either lateral potential wells for excess charge accumulation or vertical potential wells where excess charge moves deeper into the silicon substrate. The former design results in less than 100% of the silicon surface being used for photon absorption (fill factor <100%), whereas the latter design leads to lower quantum efficiencies beyond 500 nm (photons absorbed too deeply within the silicon are also lost to the substrate).

Accomplishing the electron-readout pro-

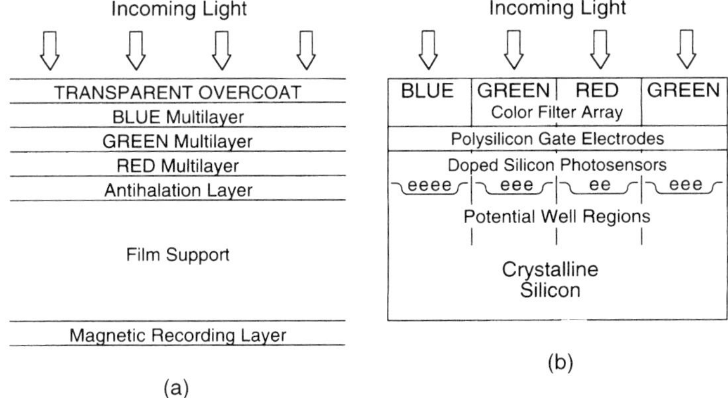

FIG. 5. Color-selective imaging. Color photographic films (a) contain several light-sensitive layers, each of which is dye-sensitized to a single color region. The other colors are transmitted, allowing each color record to have access to 100% of the area exposed by the incoming light. Color electronic sensors (b) are solid-state imagers sensitive to many wavelengths, from the blue to the near infrared. They can be overcoated with a color-filter array pattern, so that incoming light exposes, for example, only 50% of sensor elements to green light and 25% for each of red and blue. Consumer color films may also contain a magnetic layer to record information about the exposure (duration, flash, date, time) and subsequently about photofinishing (e.g., reprint number).

cess also influences the photocapacitor design in CCD image sensors. Briefly, a typical CCD is the four-phase CCD where two levels of electrode are fabricated (see $\phi 1$ and $\phi 3$ in Fig. 6; also $\phi 2$ and $\phi 4$). The process of charge transfer (readout) in this four-phase CCD is illustrated by the sequences t_0, t_1, t_2, t_3 (Fig. 6). To hold a group of electrons, two adjacent gates (for example, $\phi 2$ and $\phi 3$) would be held at a high positive potential (+5 V) while the other two phases would be at a low potential (0 V). The depletion layer or well under $\phi 2$ and $\phi 3$ holds the electrons, while $\phi 4$ and $\phi 1$ serve as a potential barrier. To transfer the electrons through the silicon, the electrode ahead of the charge packet ($\phi 4$) is clocked positive (to +5 V) and the electrode behind ($\phi 2$) is clocked negative (to 0 V). The electrons move parallel to the silicon surface following the positive potential. Repeating the sequence through all four phases moves the charge forward one pixel.

Because the million-pixel CCD image sensors will need to move certain charge packets across a thousand or more pixels, buried-channel CCDs are designed to avoid electron trapping by silicon–silicon-dioxide surface states, and consequently move charge between pixels with high efficiency. Nearly all image sensors use buried-channel CCDs with typical transfer rates of 20 megapixels/s. After sufficient numbers of transfers, the CCD output converts the charge packets to a voltage signal, described in more detail elsewhere (Tredwell, 1995). In the sequences shown in Fig. 6, the potential at the last phase before the output gate is clocked low (time t_3, potential $\phi 4$ = 0 V). The packet of electrons is transferred over the output gate onto the output diffusion region of the array. The voltage of the floating diffusion changes by an amount $V = Nq/C$, where N is the number of electrons, C is the total capacitance of the floating diffusion itself, and q is the charge per electron. The capacitance C is in the 10–50-fF range. Because the capacitance is small, there is a large voltage change for a small change in charge. A 10-fF capacitance allows a charge-to-voltage conversion of 16 μV per electron. After this charge is sensed by the output amplifier, the charge packet is removed via the reset drain before the next charge packet arrives.

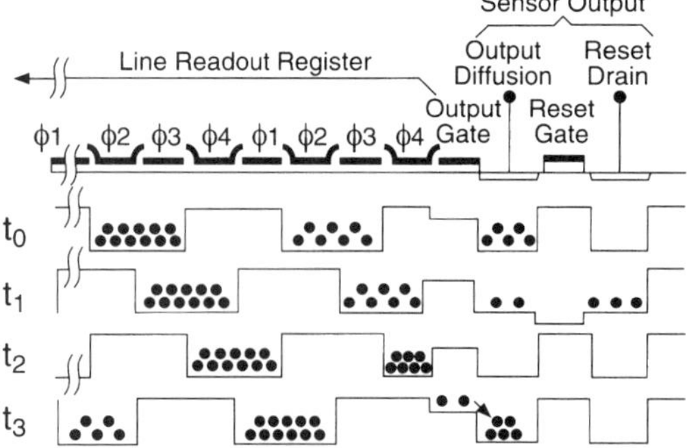

FIG. 6. An illustration of charge transfer along a four-phase CCD, showing both transfer of groups of electrons (●) along a register for voltage changes at the electrodes (gates $\phi 1$, $\phi 2$, $\phi 3$, $\phi 4$) and the sensor output region. The t_0, t_1, t_2, t_3 are arbitrary sequential times during the readout steps. (Reproduced with permission of McGraw-Hill, Inc., from Bass, 1995.)

2.3 X Rays and Radiography

Photographic recording of high-energy radiation encompasses imaging of x rays, short-wavelength ultraviolet, and autoradiography. X-ray imaging, such as used in medical applications, is an important and technologically growing imaging field (Wagner *et al.*, 1989). Conventional medical x-ray imaging employs a phosphor screen as the primary x-ray detector, and subsequently the phosphor emits near-ultraviolet, blue, or green light to be recorded by a silver-halide film (adjacent to the phosphor screen). Phosphor-based screen–film systems show detective quantum efficiencies above 20% (Bunch *et al.*, 1987; Van Metter and Dickerson, 1994), compared with the 1% or less for silver-halide films.

Blue-emitting phosphors such as calcium tungstate ($CaWO_4$) have been supplanted by rare-earth–activated materials, which reduce x-ray doses by factors of 4 or more. The following combinations are useful phosphors: GdOS:Tb(III), LaOBr:Tb(III), $HfZrO_2$:TiIn, $Ba(F,Cl)_2$:Eu(II), and $Ba(F,Br)_2$:Eu(II). The oxysulfides emit in the green (onto green-sensitized x-ray films), whereas the halide materials emit in the blue or near ultraviolet. The intensity of prompt emission from an x-ray–stimulated phosphor needs to be as high as possible, whereas long-lived phosphorescence (also described as persistent luminescence or afterglow) is disadvantageous. Prompt emission exhibits lifetimes in the microsecond range or shorter; afterglow can persist for several minutes. Altering methods of synthesis for the phosphors can reduce phosphorescence and decrease optical scattering of the promptly emitted light by controlling phosphor-particle size. Compositional changes also markedly affect phosphorescence; the addition of indium to titanium-activated hafnium/zirconium oxides reduced afterglow sufficiently to make these useful x-ray phosphors (Bryan *et al.*, 1991). Optical scattering by phosphor particles in phosphor screens influences the sharpness of a medical x-ray on the silver-halide x-ray film (adjacent to the phosphor screen). Quantum mottle is an important source of noise for x-ray exposures, significantly reducing image quality in either phosphor-screen–film systems or direct x-ray–exposed film (Saunders, 1994; Bunch *et al.*, 1987).

Industrial x-ray and autoradiographic imaging have conventionally used direct exposure of silver-halide films. For example, nucleic-acid sequencing for DNA as in the Human Genome Project attains the highest-resolution images for the radioactive ^{35}S or ^{32}P isotopes using direct film-exposure methods.

Medical x ray, x-ray crystallography, and autoradiography can now utilize storage phosphors for image detection (Whiting *et al.*, 1988), based on systems that incorporate materials developed earlier. When high-energy radiation or particles impinge on a storage phosphor screen, materials like $Ba(F,Br)_2$:Eu(II) absorb the energy and store photocarriers in traps within the crystal. The storage phosphors form stable image centers by trapping an electron at a fluoride- or bromide-ion vacancy and trapping a hole at (unspecified) sites within the crystal. Red-light excitation can subsequently excite and detrap the electron, allowing it to diffuse and recombine with the trapped hole. Energy from this recombination is transferred to Eu(II), which gives an emission at 390 nm. If laser scanning is used, the stored image can be accurately read out and imaged onto film or converted by photomultipliers to a digital form suitable for image processing and/or transmission.

2.4 Infrared Image Detectors

Aerial imaging and other applications include a wide range of infrared wavelengths in the spectral regions of interest: ultraviolet, visible, near-infrared, short-wavelength infrared, mid-wavelength infrared, long-wavelength infrared, thermal infrared, and microwave. Early silver-halide films adapted their three-color (visible) structure to include the near infrared by changing spectral sensitizations from blue/green/red to (blue + green)/red/(near infrared), but still used yellow/magenta/cyan dyes as image dyes to display a visible image. These "false-color" films showed the near-infrared image as cyan dye, the red image as magenta dye, and the (blue + green) image as yellow dye. Other silver-halide films are sensitized out to wavelengths of 1.2 μm. Silicon also absorbs wavelengths to about 1.1 μm, so that infrared imaging by band-gap absorption is limited to wavelengths similar to silver halide.

Substantial technological interest continues for infrared imaging at wavelengths well beyond 1 μm. Although a long list of materials are sufficiently sensitive as detectors (Koslowski and Kosonocky, 1995; Slater, 1986), fabrication into infrared sensors in an array format with a useful number of pixels is accomplished with PtSi, HgCdTe, InSb, AlGaAs/GaAs, and Si:As. Many of these are incorporated into focal-plane arrays to operate at temperatures from 4 K to room temperature. Historically, HgCdTe and InSb provided high–quantum-efficiency infrared imaging by intrinsic photoconduction, and compositional changes pushed limiting spectral responses from near 3 μm to beyond 10 μm for the HgCdTe. The development of larger array sizes slowed in the mid-1980s (at about 10^4 pixels), but recent progress approaches 10^6-pixel arrays.

The most advanced technology for imaging short-wavelength infrared (SWIR, 1–3 μm) and mid-wavelength infrared (MWIR, 3–5 μm) uses the photovoltaic internal photoemission response of platinum silicide (PtSi). When fabricated onto CCD image arrays, these yield low–readout-noise infrared-CCD imagers with quantum efficiencies of 0.5% to 1% at 4.0 μm and with capabilities for 300-K thermal imaging (noise equivalent temperature of 0.04–0.15 K) at 30 frames per second and *f*/1.2 to *f*/2.8 optics (Kozlowski and Kosonocky, 1995). A PtSi focal-plane array design of 640 × 486 pixels (25 × 25-μm^2 pixel size; 54% fill factor) with two-phase CCDs has been described (Nelson *et al.*, 1990). Array-size fabrication is clearly led by PtSi, for which 10^6-pixel arrays were achieved in approximately 1992. Technology estimates place PtSi, with its fabrication compatibility with VLSI methods, approximately two years ahead of other materials with respect to array sizes.

Wavelengths in the long-wavelength infrared (8–14 μm and beyond) utilize the well-developed technology around liquid-phase–epitaxy fabrication of HgCdTe/CdZnTe arrays, more specialized materials like arsenic-, gallium-, or germanium-doped silicon, and quantum-well composites (AlGaAs/GaAs). The first are detectors of choice for arrays to be used above 40 K. The doped-silicon detectors as small arrays operate in the 4–18-K range by impurity-band conduction. Their advantages include responses at very long wavelengths. Arsenic-doped silicon detectors exhibit rather high quantum efficiencies throughout the longer wavelengths up to 28 μm. Comparisons of quantum efficiencies, mechanisms for additional detectors like solid-state photomultipliers and thermal detectors, array sizes, and infrared-detector array architectures were published recently (Kozlowski and Kosonocky, 1995).

3. APPLIED PHOTOGRAPHIC RECORDING

Imaging materials, lighting conditions, wavelength shifting, and environmental factors all contribute to successful photographic recording onto films, plates, and electronic sensors. The adaptations required are illustrated below by some examples.

3.1 Clinical and Scientific Photography

Clinical photography is photography related to patients and is primarily visible-light photography. Optics, optical fibers, lenses, lighting angles, and exposure times are important. Subjects of interest include dermatology, plastic or reconstructive surgery, orthopedics, veterinary science, ophthalmology, archaeological and/or specimen photography, optical microscopy, ultraviolet (direct and fluorescence) photography, and television/computer-screen photography. Scientific photography includes similar needs for visible-light detection, as in astronomy (see previous section) and spectroscopy, but also extends to the short-wavelength (vacuum) ultraviolet and to the infrared up to wavelengths of 1 μm. Three illustrative examples are micrographic, medical, and underwater photography.

Optical photomicrography requires both black-and-white and color images. Color imaging, which provides the most challenge to microscopic recording, was historically accomplished by a variety of color films or video methods. High-resolution, color-adapted CCD sensors for still images and high-definition video now provide satisfactory image capture at reasonable cost. The specific image-recording method(s) employed will also depend on characteristics of the image and the optical light source. For example, bright-field images of stained tissues can be recorded on color-adapted CCD sensors or slower-speed color films (ISO 40–50). Typical tungsten–halogen lamps in many microscopes match well with tungsten-balanced color films; manufacturers provide lamp-voltage recommendations to help match the lamp color temperature to their films or to the color-correction algorithms of CCDs. Fluorescence micrography captures weak, transient signals and necessitates higher speed (ISO 400–1000). Reversal (slide) films offer some versatility, since these can be selectively processed to higher speeds by altering the first development step. Reflection micrography, used traditionally for metals, now applies to other crystals, circuit boards, ceramics, painted or corroded surfaces, cosmetics, and microbiological cultures.

Electron microscopy in both scanning and transmission modes integrates the image-capture, manipulation, and display functions of the imaging chain. The capture step has traditionally relied on scintillators (e.g., Eu-doped CaF_2) combined with photomultiplier tubes, or, alternatively, electron-image film. Recent developments broaden the options for this image-capture step to include silicon-based thin-film detectors for scanning (Goldstein *et al.*, 1992), and either improved phosphor screens (analogous to medical x ray) or yttrium aluminum garnet (YAG) crystals as primary image receivers for transmission. The output from these primary image-capturing materials is often being coupled to digital-imaging technology like CCD-based cameras for both visualization and image transmission.

Newer medical x-ray films are designed to enhance specific imaging tasks. Since the screen–film system has a complete film-plus-phosphor-screen imaging system on each side of the plastic film support, obtaining improved images of high-density objects (dense tissue) and low-density objects (lung tissue) can be achieved (Dickerson and Bunch, 1994; Van Metter and Dickerson, 1994). The film is designed so that visible light emitted by either phosphor screen does not cross over from one side of the film to the other. A slow-film–screen combination with high photographic contrast is nearest the x-ray source, and a faster-speed (thicker phosphor screen) film–screen combination with low photographic contrast is on the opposite side of the film support. The lower intensity of x rays transmitted by dense tissue requires the faster film–screen combination to produce an image (low contrast). Less dense tissue transmits sufficient x rays to produce a high-contrast image in the slower film–screen combination nearest the x-ray source. This high-contrast image is readily discernible even with additional (low-contrast) density from the emulsion on the opposite side.

Underwater photographic recording requires special attention to light sources and to enhancing the red speed of color films for accurate color reproduction. Daylight-balanced color films, when used at underwater depths or distances of more than 1 or 2 m, capture images that have a cyan appearance. Electronic flash (daylight) illumination for photographing an object at about 3 m from the camera/flash unit (or 6 m total light-travel distance) will detect four times less red light (relative to blue and green) than in the daylight balance of the flash. The lack of red light causes the cyan-appearing image. To attain highest image quality (i.e., to avoid filtering out blue/green light also), some conventional daylight color films are now altered during manufacture to increase their red speed by more than four times, in order to compensate for a total (flash to subject to camera) distance of about 8 m. Intermediate distances are color corrected by the use of cyan-colored filters on the light source.

3.2 Holography

Capture, storage, and subsequent display of three-dimensional information from an original scene or object have attracted renewed artistic and industrial attention since the advent of lasers. For example, there is an immense amount of technology surrounding the editing, printing, transfer/duplication, and viewing of holographic and lenticular image display. The photographic recording of these three-dimensional images likewise coalesces multiple technologies. Only capture and storage of holograms are the focus of the following examples. Subsequent steps (e.g., manipulation, transmission, soft display, hard copy), including computer-generated holograms and holographic data storage, are treated elsewhere (see HOLOGRAPHY, OPTICAL).

Holograms are coherent-light versions of photographs. Analogous steps to conventional photography are used to record and view the hologram. Recording a hologram captures the interference pattern produced by two coherent beams of light (same wavelength, same polarization, different propagation directions), where one of the beams is reflected or transmitted by the object of interest. Light sources for holography are most frequently lasers, although monochromatic (filtered) light (or other electromagnetic radiation) emanating from a point source (pinhole) can also attain enough temporal and spatial coherence. In the laboratory, vibrational isolations for the laser, object, and holographic recording material are required to avoid positional shifts by even a fraction of the light wavelength during exposure. Small cw lasers (HeNe, 632.8 nm, 1–25 mW) are suitable for small objects, whereas several-watt lasers (argon ion, 514.5- and 488-nm main lines) allow larger objects to be imaged with reasonable exposure times. Nonlaboratory work requires large pulsed lasers (ruby, 694.3 nm, up to 10-J output) to achieve satisfactory imaging in the 20–50-ns pulse-duration times.

Holographic recording media include silver halides, dichromated gelatin, photoresists, photopolymers, photothermoplastics, photochromics, and photorefractives. The more advantageous materials for industry, advertising, art, and science are described below (as well as in recent texts) (Conley and Robillard, 1992).

1. **Dichromated** Gelatin—light exposure crosslinks gelatin, which is subsequently washed and then dehydrated with alco-

hol. A preparative formula for dichromated gelatin along with the coating and processing method is published (Saxby, 1991). Usually sensitive only to blue light, red sensitization (with long exposure) is achievable. Primary limitations: low sensitivity and physical defects (cracking, tearing) that form in the remaining noncrosslinked gelatin when dehydration causes it to shrink.

2. Photoresists—more commonly used for lithography in integrated-circuit fabrication. The surface relief pattern (which remains after holographic exposure and photoresist development steps) can be nickel electroplated. This provides a master hologram to mass produce (via embossing heat-softened plastic) copies as holograms for large-volume applications. Primary limitation: low photosensitivity.
3. Silver-Halide Emulsions—the most common recording material, very high sensitivity, spectrally sensitizable to many wavelengths, and commercially available in both large and small sizes on film supports or glass plates (Bjelkhagen, 1993).
4. Photothermoplastics—actually multilayer structures on glass plates, starting with a conductive metal film, a photoconductor, and then a thermoplastic material. The recording process is an electrophotographic one: Apply a static charge, discharge the photoconductor by (holographic) exposure, apply a second charge to the surface, and heat until the plastic layer softens. Deformation of the thermoplastic occurs, being largest in regions of highest charge, and a relief, phase hologram is formed on the thermoplastic surface.

Holographic interferometry sets new image-capture requirements on the recording medium and the light source, since it compares two different holograms, each of which represents a different state of the object under observation. Since it is also possible to compare a previously obtained holographic wave front with an actual wave front from an object, the interference pattern can monitor changes in objects as a function of real time. Vibrations in both fixed and rotating objects, for example, are readily observed because the change in distance between the exposing source and a vibrating object changes the hologram as a function of time. Optically comparing two different holograms results in interference fringes (Brownell *et al.*, 1992). Temperature gradients in the air can also cause interference fringe fluctuations of the order of a wavelength or more. In a different area of application, defects as small as 0.5 μm occuring on 6-in. patterned silicon wafers can be detected (Lin, 1992).

Table 1. Two-phase CCD image sensors.

Full-Frame Still-Image Sensor	
Photosensitive size	18.48 × 27.65 mm; 2048 × 3072 pixels
Pixel dimension	9 μm (square)
Quantum efficiency	15% (450 nm), 35% (550 nm), 40% (620 nm)
Signal linearity	Less than 1% rms nonlinearity up to 80% of saturation
Saturation	85 000 electrons (no antiblooming) 45 000 electrons (with antiblooming)
Color-filter transmission (peak)	65% (blue), 75% (green), 85% (red)
Readout rate	20 megapixels/s
Dynamic range	12 bits (higher at lower readout rates)
Noise sources	
Output amplifier	15 electrons rms
Dark current	10.5 pA/cm^2 at 25 °C
Shot and pattern (dark)	Less than four electrons rms
Shot and pattern (photon)	Less than 1% of signal (550 nm)
Partial-Frame Video-Image Sensor	
Photosensitive size	14.0 × 7.9 mm; 1084 × 1928 pixels
Pixel dimension	7.3 μm (square)
Quantum efficiency	35% (440 nm), 32% (540 nm), 25% (620 nm) (with 2.5-fold gain from microlenses)
Fill factor	25%
Signal linearity	As above
Saturation	—
Color-filter transmission (peak)	as above
Readout rate	To yield 30 frames per second
Noise sources	
Photodiode dark current	100 pA/cm^2 at 40 °C

3.3 HDTV Sensors and High-Resolution Still Sensors

Commercial high-definition television (HDTV) sensors with 1- or 2-megapixel CCD arrays will be able to accomplish the simultaneous image capture and readout required by video, as well as by applications in microscopy, medical imaging, and machine and robotic vision. Video imaging requires image integration simultaneous with image readout steps, high readout rates (greater than 72 megapixels per second), and often interlaced readout. Dynamic range, color reproduction, linearity, and sharp antiblooming are not as critical as for still-image arrays. [The latter can use a mechanical shutter and a lower readout rate (150 ms) but does require high resolution, progressive (noninterlaced) scan, square pixels, wide dynamic range, high sensitivity, low noise, very linear response, antiblooming protection with very sharp turn-on, and excellent color reproduction.] Both HDTV-level video and still-image CCD arrays now utilize a simpler, more manufacturable two-phase CCD design that requires only one polysilicon electrode per phase (Nichols *et al.*, 1995). However, each application makes different performance compromises. A 6.4-megapixel CCD sensor/color-filter array can provide single-chip color imaging using square 9-μm pixels with the two-phase CCD design and achieve very low noise. Other data are given in Table 1 to compare still-image array features with those for video imagers.

GLOSSARY (see also Stroebel and Zakia, 1993)

Contrast: Slope of photographic response curve [density vs log (exposure)].

Density (*D*): Optical density.

Detective Quantum Efficiency (DQE): For photographic film, $(0.434)^2\gamma^2/HA(\sigma_D)^2$, where γ = contrast of photographic response curve density D, H = total exposure, A = granularity measuring aperture, σ_D = rms density deviation at density D.

Exposure (*H*): Total exposure (erg cm^{-2}).

Granularity: Rms deviation of density values (based on a microdensitometer trace) with respect to the average density of a uniformly exposed and processed area.

Irradiance (*E*): Light intensity per unit time (erg cm^{-2} s^{-1}).

Latent Image: Metal-atom aggregates (silver and other metals) formed by light exposure of silver halides; catalytic centers for chemical development.

Modulation-Transfer Function (MTF): Measurement of contrast reduction on images of sinusoidal targets of various spatial frequencies.

Moire: Artifact of imaging systems resulting from interference between two regularly spaced sets of lines.

Noise: Granularity in light-exposed photographic images; granularity plus quantum mottle in x-ray–exposed photographic images; unwanted energy (dark current, shot noise, etc.) in electronic images.

Pixel: Picture element in an array; smallest individually processable element in electronic image.

Power (Wiener) Spectrum: Result of representing complex wave form as an equivalent set of sine waves; plot of sinusoidal power (squared amplitude) vs frequency; used to represent grain-size distribution in processed film.

Reciprocity: Interchangeability of irradiance and time to produce a constant photographic speed, density, or contrast; failure of interchangeability is reciprocity failure.

Resolving Power: Visual determination of the resolution limit of imaging material for images of various spatial frequencies.

Sensitivity: Inverse exposure; also expressed as ISO speed.

Shot Noise: Current fluctuations due to random emission of carriers or random injection of carriers through space-charge regions.

Signal-to-Noise Ratio: Signal (mean average density of an image minus the mean background density) divided by noise (granularity).

Transmittance (*T*): Fraction of light transmitted by film sample, collected for 10° cone angle (diffuse).

Works Cited

Babcock, T. A., Sewell, M. H., Lewis, W. C., James, T. H. (1974), "Hypersensitization of Spectroscopic Films and Plates Using Hydrogen Gas," *Astron. J.* **79** 1479–1487.

Bass, M. D. (1995), *Handbook of Optics*, New York: McGraw-Hill.

Bird, G. R. (1989), "A Critique of Imaging Systems," in: J. Sturge, V. Walworth, A. Shepp (Eds.), *Imaging Processes and Materials,* New York: Van Nostrand Reinhold, pp. 587–636.

Bjelkhagen, H. I. (1993), *Silver Halide Recording Materials for Holography and Their Processing,* Springer Series in Optical Sciences, Vol. 66, Berlin: Springer-Verlag.

Brownell, J., Parker, R., Jones, D. (1992), "Recent Developments in Holography at Rolls-Royce," in: E. Conley, J. Robillard (Eds.), *Industrial Applications for Optical Data Processing and Holography,* Boca Raton, FL: CRC pp. 3–32.

Bryan, P. S., Lambert, P. M., Towers, C. M., Jarrold, G. S. (1991), "X-Ray Intensifying Screen Including a Titanium-Activated Hafnium Dioxide Phosphor Containing Indium," assignors to Eastman Kodak Company, U.S. patent 4,983,847 (January 8).

Buhr, J. D., Franchino, H. D. (1994), "Color Image Reproduction of Scenes with Preferential Tone Mapping," assignors to Eastman Kodak Company, U.S. patent 5,300,381 (April 5).

Bunch, P. C., Huff, K. E., Van Metter, R. (1987), "Analysis of the Detective Quantum Efficiency of a Radiographic Screen-Film Combination," *J. Opt. Soc. Am. A* **4,** 902–909.

Burkey, B. C., Lubberts, G., Trabka, E. A., Tredwell, T. J. (1984), "Channel Potential and Channel Width in Narrow Buried Channel MOSFETs," *IEEE Trans. Electron Devices,* **ED-31,** 423–429.

Cohen, D. (1989), "A Day in the Life of China," San Francisco, CA: Collins.

Conley, E., Robillard, J. (1992), "Industrial Applications for Optical Data Processing and Holography," Boca Raton, FL: CRC.

Dickerson, R. E., Bunch, P. C. (1994), "Minimal Crossover Radiographic Elements Adapted for Flesh and Bone Imaging," assignors to Eastman Kodak Company, EP patent application 611,226 (August 17).

Doerner, E. C. (1962), "Wiener-Spectrum Analysis of Photographic Granularity," *J. Opt. Soc. Am.* **52,** 669–672.

Eastman Kodak Company (1993), "Performance Specifications for KAF-6300 3072 (H) × 2048 (V) Pixel Full-Frame CCD Image Sensor," Rochester, NY, pp. 3, 13.

Evans, R. M., Hansen, W. T., Jr., Brewer, W. L. (1953), *Principles of Color Photography,* New York: Wiley.

Goldstein, J. I., Newbury, D. E., Echlin, P., Joy, D. C., Romig, A. D., Lyman, C. E. (1992), *Scanning Electron Microscopy and X-Ray Microanalysis, A Text for Biologists, Material Scientists, and Geologists,* 2nd ed., New York: Plenum.

Hailstone, R. K. (1994), "Transputer-Based Simulations of Image Recording in Silver Halide Materials," *Comput. Phys.* **8,** 205–214.

Hamilton, J. F. (1991), "Conventional Photographic Materials," in: A. S. Diamond (Ed.), *Handbook of Imaging Materials,* New York: Dekker, pp. 1–42.

Hunt, R. W. G. (1995), *The Reproduction of Colour,* Kingston-upon-Thames, England: Fountain, pp. 280–318, 361–388, 389–432, 463–482.

Jacobson, R. E. (1995), "An Evaluation of Image Quality Metrics," *J. Photogr. Sci.* **43,** 7–16.

James, T. H. (Ed.) (1977), *The Theory of the Photographic Process,* New York: MacMillan.

Jones, R. C. (1955), "New Method of Describing and Measuring the Granularity of Photographic Materials," *J. Opt. Soc. Am.* **45,** 799–808.

Khosla, R. P. (1992), "Photons to Bits," *Phys. Today* **45** (12), 42–49.

Kozlowski, L. J., Kosonocky, W. F. (1995), "Infrared Detector Arrays," in: M. D. Bass, (Ed.), *Handbook of Optics,* Vol. 1, New York: McGraw-Hill, Chap. 23.

Krause, P. (1989), "Color Photography," in: J. Sturge, V. Walworth, A. Shepp (Eds.), *Imaging Processes and Materials,* New York: Van Nostrand Reinhold, pp. 110–134.

Lamberts, R. L., Straub, C. M., Garbe, W. F. (1965), "Equipment for Routine Evaluation of the Modulation Transfer Function of Photographic Materials: I. The Camera, II. The Microdensitometer, III. Evaluating and Plotting Instrument," *Photogr. Sci. Eng.* **9,** 331–342.

Lavine, J. P., Trabka, E. A., Burkey, B. C., Tredwell, T. J., Nelson, E. T., Anagnostopoulos, C. (1983), "Steady-State Photocarrier Collection in Silicon Imaging Devices," *IEEE Trans. Electron Devices,* **ED-30,** 1123–1134.

Lin, L. (1992), "A Holographic Patterned Semiconductor Wafer Defect Detection System," in: E. Conley, J. Robillard (Eds.), *Industrial Applications for Optical Data Processing and Holography,* Boca Raton, FL: CRC, pp. 47–55.

Maver, L. A., Bisbee, J., Erdman, C. D., Scarff, L. A., Clark, J. A. (1989), "Aerial Imaging Systems," in: J. Sturge, V. Walworth, A. Shepp (Eds.), *Imaging Processes and Materials,* New York: Van Nostrand Reinhold, pp. 423–464.

Mees, C. E. K. (1961), *From Dry Plates to Ektachrome Film: A Story of Photographic Research,* New York: Ziff-Davis.

Nelson, E. T., Wong, K. Y., Yoshizumi, S., Rockafellow, D., DesJardin, W., Elzinga, M., Lavine, J. P., Tredwell, T. J., Khosla, R. P., Sorlie, P., Howe, B., Brickman, S., Refermat, S. (1990), in: E. L. Dereniak, R. E. Sampson (Eds.), *Wide Field of View PtSi Infrared Focal Plane Array,* SPIE Proceedings Vol. 1308, Bellingham, WA: SPIE, pp. 36–37.

Nichols, D. N., Stevens, E. G., Burkey, B. C., Stancampiano, C. V., Lee, Y.-R., Kosman, S. L., Losee, D. L., Lavine, J. P., Torok, G. R., Khosla, R. P. (1995), "High Resolution Interline Image Sensors Using Two-Phase CCD Technology," personal communication.

Saunders, A. E. (1994), "Mathematical Quirks

of Photographic Science," *J. Photogr. Sci.* **42,** 142–148.

Saxby, G. (1991), *Manual of Practical Holography,* London: Focal, pp. 154, 171.

Schwan, J. A., Graham, J. L. (1972), "Multicolor Silver Halide Photographic Materials and Processes," assignors to Eastman Kodak Company, U.S. patent 3,672,898 (June 27).

Shepp, A. (1989), "Introduction to Images and Imaging," in: J. Sturge, V. Walworth, A. Shepp (Eds.), *Imaging Processes and Materials,* New York: Van Nostrand Reinhold, pp. 1–38.

Slater, P. N. (1986), "Survey of Multispectral Imaging Systems for Earth Observations," in: A. Cracknell, L. Hayes (Eds.), *Remote Sensing Yearbook, 1986,* London: Taylor and Francis, pp. 141–163.

Stroebel, L., Zakia, R. (1993), *Encyclopedia of Photography,* 3rd ed., Boston: Focal.

Sturge, J., Walworth, V., Shepp, A. (1989), *Imaging Processes and Materials,* New York: Van Nostrand Reinhold.

Topfer, K., Jarvis, J. R., Hertel, D. (1994), "Physical and Chemical Properties of Colour Couplers in Relation to the Image Quality of Colour Negative Films," *J. Photogr. Sci.* **42,** 149–156.

Tredwell, T. J. (1995), "Visible Array Detectors," in: M. D. Bass (Ed.), *Handbook of Optics,* Vol. 1, New York: McGraw-Hill, Chap. 22.

Van Metter, R., Dickerson, R. (1994), "Objective Performance Characteristics of a New Asymmetric Screen-Film System," *Med. Phys.* **21,** 1483–1490.

Wagner, L. K., Naragara, P., Murry, P. C. Jr. (1989), "Medical Imaging," in: J. Sturge, V. Walworth, A. Shepp (Eds.), *Imaging Processes and Materials,* New York: Van Nostrand Reinhold, pp. 497–566.

Whiting, B., Owen, J., Rubin, B. (1988), "Storage Phosphor X-Ray Diffraction Detectors," *Nucl. Instrum. Methods A* **266,** 628–635.

Further Reading

Crowe, D., Norton, P. R., Limperis, T., Mudar, J. (1993), "Performance and Configuration of IR Image Sensors," in: W. D. Rogatto (Ed.), *Electro-Optical Components,* Bellingham, WA: SPIE, pp. 258–273.

Diamond, A. S. (Ed.) (1991), *Handbook of Imaging Materials,* New York: Marcel Dekker.

Hunt, R. W. G. (1995), *The Reproduction of Colour,* Kingston-upon-Thames, England: Fountain.

James, T. H. (Ed.) (1977), *The Theory of the Photographic Process,* New York: MacMillan.

Leith, E. N. (1993), "Holography" in: S. N. Robinson (Ed.), *Emerging Systems and Technologies,* Bellingham, WA: SPIE, pp. 483–516.

Stroebel, L., Compton, J., Current, I., Zakia, R. (1986), *Photographic Materials and Processes,* Boston: Focal.

Sturge, J. (1977), *Neblette's Handbook of Photography and Reprography (Materials, Processes, and Systems),* 7th ed. New York: Van Nostrand Reinhold Company.

Sturge, J., Walworth, V., Shepp, A. (Eds.) (1989), *Imaging Processes and Materials,* 8th ed., New York: Van Nostrand Reinhold.

Theuwissen, A. J. P. (1995), *Solid-State Imaging with Charge-Coupled Devices,* Solid-State Science Library, Vol. 1, Boston: Kluwer.

Thomas, W. Jr. (1973), *SPSE Handbook of Photographic Science and Engineering,* New York: Wiley.

RECRYSTALLIZATION, DYNAMIC

F. MONTHEILLET* AND J. J. JONAS, *Department of Metallurgical Engineering, McGill University, Montréal, Québec, Canada*

	Introduction	205
1.	**Discontinuous Dynamic Recrystallization**	206
1.1	Flow Curves Associated with Discontinuous DRX and the Critical-Strain Model	206
1.2	Grain-Size Evolution and the Relative–Grain-Size Model	209
1.3	Nucleus-Density Condition	211
1.4	Effect of Alloying Elements on DRX	213
2.	**Dynamic Recovery and Continuous Dynamic Recrystallization**	213
2.1	Flow Curves Associated with Continuous DRX	213
2.2	Microstructural Changes Associated with Continuous DRX	215
3.	**Texture Changes Associated with DRX**	218
4.	**Postdynamic Softening**	221
5.	**Implications for Industrial Processing**	223
	Glossary	223
	Works Cited	224
	Further Reading	225

INTRODUCTION

The recrystallization that takes place *during* the high-temperature deformation of metals is known as dynamic recrystallization (DRX) and was initially a subject of considerable controversy. At the Sheffield conference on Deformation under Hot Working Conditions, for instance, it was still being claimed by Stüwe (1968) that the nucleation and growth of new grains (which involve grain-boundary migration) can only take place *after* straining, while the microstructural changes occurring *during* deformation are restricted to the rearrangement and removal of dislocations (dynamic recovery). It was, indeed, rather difficult to decide whether the new grains present in deformed and cooled samples were generated dynamically (i.e., during deformation) or statically, during the interval between hot working and quenching. As will be seen below, *postdynamic* restoration processes generally take place very rapidly, which makes it difficult to interpret metallographic data without ambiguity.

Because of these difficulties, the early references to DRX were based on the special shapes of the stress–strain curves sometimes observed under metal-working conditions. (Similar remarks apply to the unusual strain–time curves occasionally determined under creep conditions.) These were characterized by considerable amounts of strain softening, and more particularly by oscillations indicating that successive cycles of work hardening and flow softening were taking place. An early example of such wavy flow behavior is presented in Fig. 1, obtained from compression tests carried out on lead specimens at room temperature (Thomsen *et al.*, 1954). Since then, similar observations have been made on a large number of metals and alloys (copper, carbon steels in the austenite range, austenitic stainless steels, nickel, etc.) that have been subjected to hot-compression, tension, or torsion tests. DRX has also been reported to occur in nonmetallic materials, such as ice (Steinemann, 1958), and some minerals (Poirier, 1972).

Although the occurrence of DRX under laboratory testing conditions has now been amply assessed and analyzed, some doubt still remains about its importance during in-

*Permanent address: Ecole des Mines de Saint-Etienne, Centre Science des Matériaux et des Structures, URA CNRS 1884, Laboratoire Microstructures et Mise en Forme, Saint-Etienne, France.

3-527-28138-X/96/$5.00 + .50

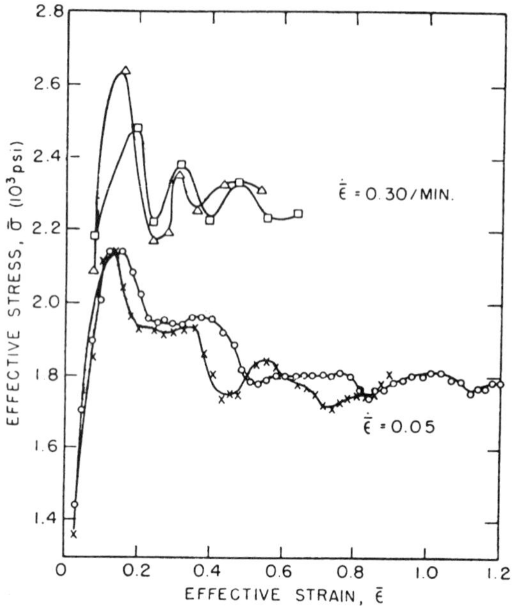

FIG. 1. Room-temperature stress–strain curves determined on lead specimens; the wavy flow behavior is an early example of the periodic softening produced by dynamic recrystallization (after Thomsen *et al.*, 1954).

dustrial hot-forming operations, as recently discussed by Jonas (1994). In processes where large strains are applied in a single operation (as in seamless tube piercing, extrusion, and planetary hot rolling), the material grain size during processing is clearly controlled by DRX. This also appears to be the case for multiple-pass processes such as hot rolling when the interpass times are short ($\ll 1$ s) so that interpass-restoration effects are negligible or absent, and the strain can accumulate. It will be seen later that such industrial forming conditions, which usually involve high strain rates (10 to 1000 s^{-1}), generally lead to *grain refinement* as a result of DRX and, therefore, to increases in yield stress and hardness, as well as in the resistance to cracking and fracture. No longer a "scientific curiosity," dynamic recrystallization is now becoming an industrial tool (Jonas, 1994).

Careful analysis of the stress–strain curves obtained over various ranges of strain rate and temperature, together with that of the associated microstructures, has shown that two broad classes of DRX can be distinguished. The first, usually referred to simply as "dynamic recrystallization" in the literature, involves the nucleation and growth of new grains within a strained matrix; it will be categorized here as *discontinuous dynamic recrystallization*. The second type results solely from the operation of dynamic recovery and grain-boundary migration; i.e., it does *not* involve nucleation. This mechanism can also produce "new" grains at large strains. Because of the absence of nucleation, it will be referred to here as *continuous dynamic recrystallization*.

1. DISCONTINUOUS DYNAMIC RECRYSTALLIZATION

The most striking microstructural feature of discontinuous DRX is the formation of dislocation-free nuclei, which grow during further straining. This leads to the appearance of equiaxed grains that, unlike those produced by *static* recrystallization, contain dislocations as a result of the continued straining.

1.1 Flow Curves Associated with Discontinuous DRX and the Critical-Strain Model

Some stress–strain curves that are typical of DRX are illustrated in Fig. 2(a); these were obtained from hot-compression tests of OFHC (oxygen-free high-conductivity) copper specimens at various temperatures (Blaz *et al.*, 1983). Here, the shape of the curve changes from the *multiple-peak* variety to the *single-peak* type when the temperature is decreased. A similar transition is observed when the strain rate is increased, as shown in Fig. 2(b) for the case of a 0.25% C steel (Rossard and Blain, 1959). Such equivalence between the effects of decreasing temperature and increasing strain rate is associated with the occurrence of thermally activated processes and has long been recognized in the study of hot working. It arises because an increase in strain rate decreases the time available for thermal activation (and therefore the probability of successful activation), while a decrease in temperature has an equivalent effect on successful activation by decreasing the amplitude of the thermal fluctuations. The equivalence permits use of the

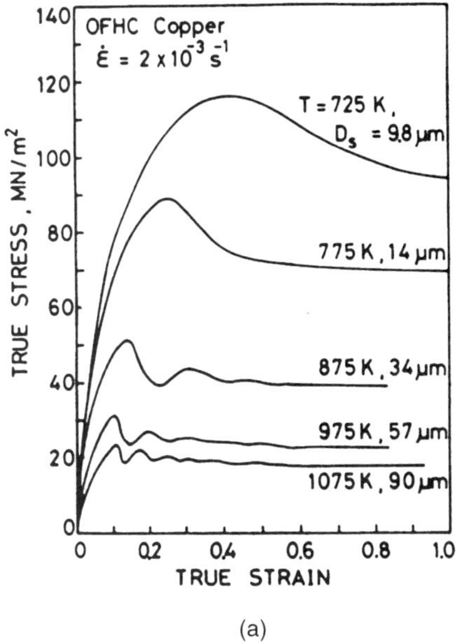

(a)

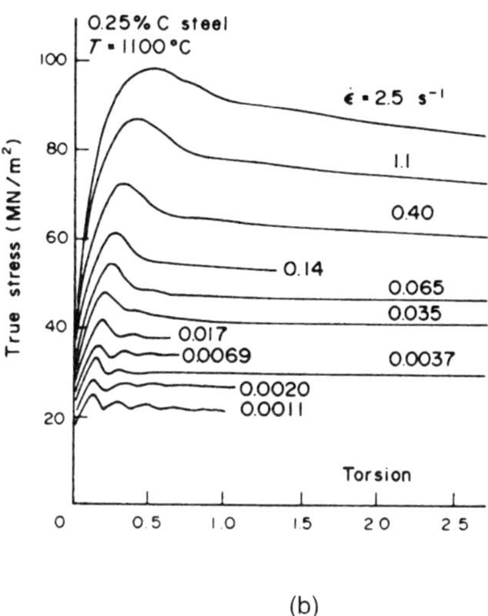

(b)

FIG. 2. Stress–strain curves typical of discontinuous dynamic recrystallization obtained on (a) OFHC copper at various temperatures (after Blaz *et al.*, 1983) and (b) 0.25% C steel at various strain rates (after Rossard and Blain, 1959).

temperature-corrected strain rate, or Zener–Hollomon parameter (Zener and Hollomon, 1944):

$$Z = \dot{\epsilon} \exp(Q/RT), \tag{1}$$

where $\dot{\epsilon}$ is the strain rate, T the absolute temperature, R the gas constant, and Q the apparent activation energy of deformation. Thus when Z is less than a critical value Z_c, multiple peaks are present, whereas when $Z > Z_c$, single-peak flow occurs.

The positive-slope portions of the oscillations (when $Z < Z_c$) can be intuitively understood as resulting from strain hardening, whereas the negative-slope regions can be associated with softening by recrystallization. Such softening can only be reflected in the flow curve when recrystallization is "synchronized" throughout the sample. Conversely, in the single-peak type of curve ($Z > Z_c$), work hardening and softening by recrystallization take place concurrently in different parts of the specimen, so that recrystallization is now "asynchronized." Starting from this very simple idea, a first model of discontinuous DRX was proposed by Luton and Sellars (1969). They introduced two characteristic strains associated with the flow curves (Fig. 3): ϵ_p, the strain to the first stress peak,* and ϵ_x, the recrystallization

*A more physical quantity, viz, the critical strain ϵ_c, has also been used. It refers to the strain required to initiate DRX, which is less than ϵ_p [$\epsilon_c \approx (0.6$ to $0.8)\epsilon_p$]. As ϵ_c cannot be readily determined from stress–strain curves, the "nucleation" strain is frequently specified in terms of ϵ_p rather than ϵ_c.

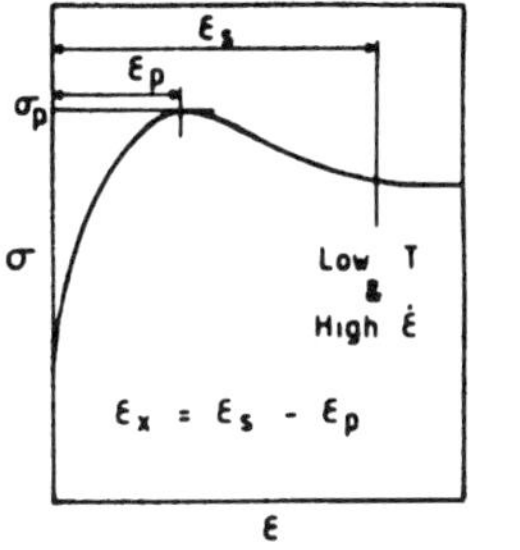

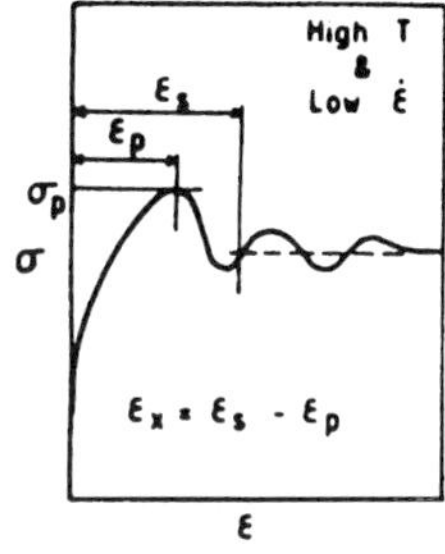

FIG. 3. Schematic definitions of the peak and recrystallization strains, ϵ_p and ϵ_x, respectively, in the single- (left) and multiple-peak (right) cases. ϵ_s is either the minimum strain required to attain steady-state flow or the strain at which the mean value of the flow stress is attained during the second cycle of work hardening. (After Luton and Sellars, 1969.)

strain. The latter was defined as $\epsilon_s - \epsilon_p$, where ϵ_s is the minimum strain required to attain steady-state flow in single-peak curves, and the strain at which the mean value of the flow stress is attained during the second cycle of work hardening in multiple-peak curves. On the basis of their results for the torsion testing of nickel, the authors showed that when $\epsilon_p > \epsilon_x$ (low Z), DRX is synchronized, whereas when $\epsilon_p < \epsilon_x$ (large Z), asynchronized DRX takes place. By means of computer simulations, they were able to reproduce the two alternative shapes of stress–strain curves. This *critical-strain model* is consistent with the data obtained from numerous *torsion*-test investigations.

It has been shown more recently, however (Sakui *et al.*, 1977; Weiss *et al.*, 1984), that the critical-strain comparison does not apply to *tension* or *compression* experiments, at least in unmodified form. This is illustrated in Fig. 4, which shows the peak stress dependences of ϵ_p and ϵ_x for three plain carbon steels deformed in torsion, tension, and compression (Sakai and Jonas, 1984). In the case of *torsion* [Fig. 4(a)], the intersection of the two curves identifies the peak stress that corresponds to the critical value Z_c of the Zener–Hollomon parameter introduced above. The latter defines the transition from multiple-peak to single-peak flow. However, unlike the behavior displayed by the torsion results, the ϵ_p and ϵ_x curves do *not* intersect in the case of the tension [Fig. 4(b)] or compression [Fig. 4(c)] experiments, although the flow-curve–shape transition is still observed within the σ_p range investigated.

Such an unexpected disagreement between the results of torsion tests on the one hand, and those obtained from tension or compression tests on the other, has been analyzed in detail by Sakai and Jonas (1984). Their main conclusions can be summarized as follows.

1. The determination of material stress–strain curves (σ-ϵ) from torque (T) versus twist (N) data determined on solid-bar torsion specimens is subject to error, because of the radial strain and strain-rate gradients that are present. Although the problem was solved by Fields and Backofen (1957) for a wide class of materials that undergo work hardening and display rate sensitivity, their method involves calculation of the partial derivatives of T with respect to the number of twists, N, and the rate of twisting, $\dot{N}$. This requires the processing of large amounts of experimental data, and is rarely carried out. The algebraic shortcuts taken instead by most experimenters lead to the overestimation of ϵ_p.
2. Because necking develops fairly rapidly in tension after the stress peak has been attained, ϵ_x values determined by this

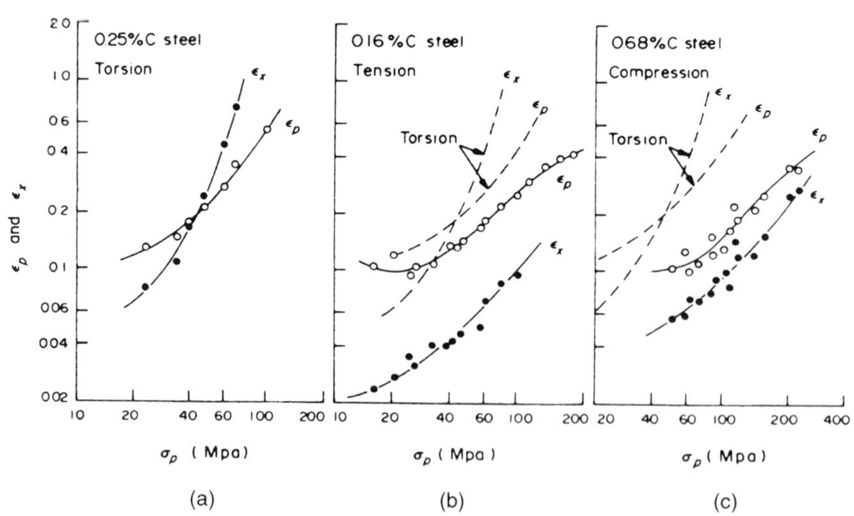

FIG. 4. Peak stress dependences of ϵ_p and ϵ_x for three plain carbon steels deformed in (a) torsion, (b) tension, and (c) compression (after Sakai and Jonas, 1984).

means will generally be lower than those measured using other test methods [cf. Fig. 4(b) with Figs. 4(a) and 4(c)].

3. There is also an effect of *strain path* on the rate of work hardening and therefore on the strain required to attain ϵ_p. This is because the flow stress of a material can be written in the form

$$\sigma(\epsilon) = \bar{M}(\epsilon)\bar{\tau}_c(\epsilon), \quad (2)$$

where $\bar{M}$, the average Taylor factor, is associated with the current crystallographic texture and $\bar{\tau}_c$ is the mean value of the critical resolved shear stress, a quantity that is closely linked to the current dislocation configuration. The strain dependences of these two factors are distinctly different for specimens deformed along different strain paths (Jonas *et al.*, 1982). In particular, the flow stress of samples strained at room temperature is generally lower in torsion than in tension or compression. Such strain-path effects also contribute to the different ϵ_p and ϵ_x dependences displayed in Figs. 4(a)–4(c).

The errors and discrepancies involved in determining ϵ_p and ϵ_x from mechanical test data thus led Sakai and Jonas (1984) to conclude that the critical condition $\epsilon_p = \epsilon_x$ associated with the transition from single- to multiple-peak flow only applies, strictly speaking, to solid-bar torsion experiments and then only in an approximate and perhaps fortuitous way.

1.2 Grain-Size Evolution and the Relative–Grain-Size Model

The critical-strain approach was principally concerned with the mechanical or macroscopic aspects of DRX. Little attention was paid to the characteristics of the actual microstructural processes taking place within the material. More recently, these have been studied in some detail, principally by Sakai and co-workers (Sakui *et al.*, 1977; Sakai and Jonas, 1984). To avoid the occurrence of postdynamic restoration, they used special tensile machines in which the specimen could be quenched at a cooling rate of about 2000 K s^{-1} at the end of the deformation. By this means, these authors were able to investigate the evolution of the average grain size during DRX. In particular, they compared the dynamic "equilibrium" grain size D_s present in the steady state (at large strains) with the initial grain size D_0.

The results obtained on a 0.16% C steel tested at 940 °C and a range of strain rates are presented in Fig. 5. The strains at which stress peaks were observed are identified as P_1, P_2, etc. It can be seen that, at low strain rates, multiple-peak behavior is displayed and either grain coarsening ($\dot{\epsilon} = 2.6 \times 10^{-4}$ and 2.0×10^{-3} s^{-1}) or slight refinement ($\dot{\epsilon} = 2.0 \times 10^{-2}$ s^{-1}) takes place. Conversely, at high strain rates ($\dot{\epsilon} = 1.48 \times 10^{-1}$ and 18 s^{-1}), single-peak behavior is associated with a large decrease in grain size. The critical D_0/D_s ratio above which single-peak behavior occurs can be estimated to be about 2. This can be shown more quantitatively as follows.

It is well known that the equilibrium grain size D_s is a power-law function of the steady-state flow stress σ_s for a number of metals and alloys (Derby, 1991):

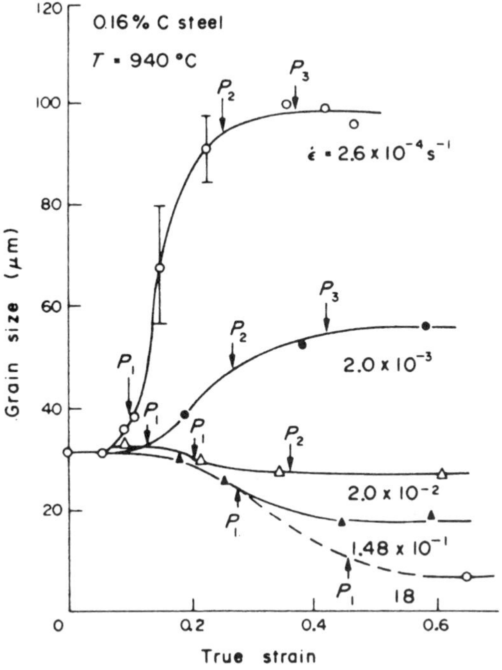

FIG. 5. Strain dependence of the average grain size during the tensile deformation of a 0.16% C steel at 940 °C. Whether grain coarsening or grain refinement takes place depends on the imposed strain rate. The strains at which stress peaks were observed are identified as P_1, P_2, etc. (After Sakui *et al.*, 1977.)

$$\sigma_s = k_1/D_s^{\alpha}. \tag{3}$$

Here, $\alpha \approx 0.66$–0.75 is an empirical constant and k_1 is a constant. Since σ_s is itself related to the Zener–Hollomon parameter Z by a power-law-relationship (provided that the strain rate and stress are not too high), viz.,

$$\sigma_s = k_2 Z^m, \tag{4}$$

Z can be expressed in the form

$$Z = k/D_s^{\alpha/m} \tag{5}$$

In the above equations, k_2 and k are further constants, and $m \approx 0.1$ to 0.2, so that $\alpha/m \approx 4$ to 8. Equation (5) essentially states that D_s depends solely on the temperature-corrected strain rate Z. This can be interpreted as signifying that the growth of new grains stops when the driving force for grain-boundary migration (the dislocation-density difference) decreases to zero, or drops below a critical level. The loss of driving force is caused by strain hardening, the rate of which must be balanced against the velocity of grain-boundary migration. It is the latter quantity that is directly related to the parameters of straining, via Z.

The relationship described by Eq. (5) is illustrated in the semilogarithmic plot of Fig. 6 for a 0.16% C steel. Here it can be seen that the dependence of the critical value of the Zener–Hollomon parameter Z_c on the initial grain size D_0 is *nearly parallel* to the Z-D_s dependence. Furthermore, the above authors showed that the Z_c vs $2D_s$ relation approximately coincides with the one pertaining to Z_c vs D_0 (Fig. 6). This means that $D_0 = 2D_s$ corresponds to $Z = Z_c$ and is thus associated with the transition from multiple- to single-peak behavior.

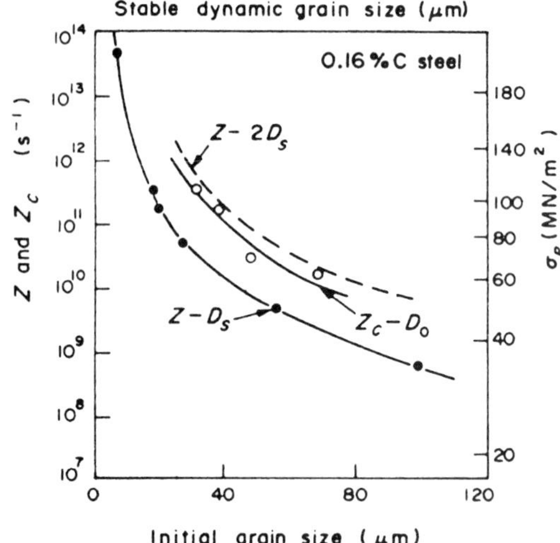

FIG. 6. Relationship between the steady-state grain size D_s and Zener–Hollomon parameter Z. The initial grain-size dependence of the critical value Z_c associated with the transition from single- to multiple-peak flow (and vice versa) is shown to be *parallel* to the latter curve. (After Sakui *et al.*, 1977.)

These experimental observations led to the *relative–grain-size model* described in Fig. 7. In this double-logarithmic plot, the Z_c-D_0 and Z-D_s relations are now represented by two parallel straight lines of slope $-\alpha/m$. The first divides the diagram into two parts, associated with single-peak (large initial grain size, high Z value) and multiple-peak (small initial grain size, low Z value) behavior, respectively. The Z-D_s line in turn distinguishes between the grain-coarsening and grain-refinement domains. From the diagram, the outcomes of two different types of tests can be predicted. "Vertical" experiments are carried out with a fixed initial grain size D_0 but at various values of Z, as in Figs. 2(a) and 2(b). According to Fig. 7, given an initial grain size D_{01}, stress oscillations will occur when a combination of relatively low testing strain rate and high tem-

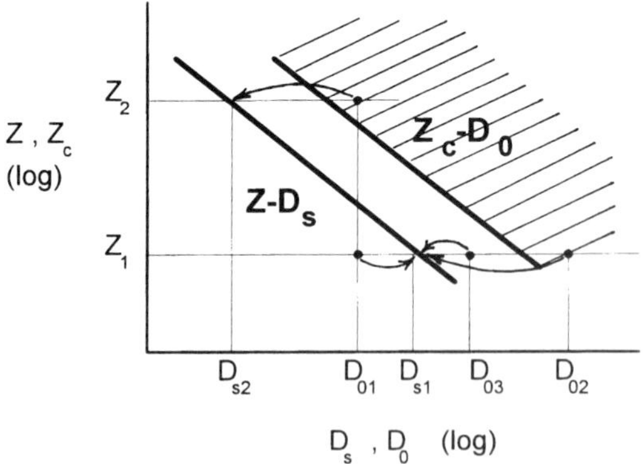

FIG. 7. Double-logarithmic plot illustrating the relative–grain-size model (see text). The Z_c-D_0 and Z-D_s relations are represented here by two parallel straight lines. The cross-hatched area corresponds to the temperature-corrected strain rate and initial grain-size conditions associated with single-peak dynamic recrystallization.

perature, e.g., $Z_1 < Z_{c1}$, is used. Each cycle of recrystallization then produces *grain coarsening* until the stable grain size D_{s1} pertaining to Z_1 is attained. Conversely, if a combination of a relatively high testing strain rate and low temperature, e.g., $Z_2 > Z_{c1}$, is used, a single peak is observed and *grain refinement* occurs until the stable grain size D_{s2} is reached.

The "horizontal" tests are carried out, in turn, at a fixed value of Z (generally with both the strain rate *and* temperature held constant) but at various values of D_0, as illustrated in Fig. 8. Thus for $Z = Z_1$, when the initial grain size D_{01} is finer than D_{s1}, multiple peaks and grain coarsening are called for, persisting until D_{s1} is attained (Fig. 7). Alternatively, when the initial grain size D_{02} or D_{03} is larger than D_{s1}, the occurrence of grain refinement is predicted; this can be associated with either single- (for $D_{02} > 2D_{s1}$) or multiple-peak (for $D_{s1} < D_{03} < 2D_{s1}$) flow. The grain size decreases in this case until it attains D_{s1} and a steady state of flow is achieved.

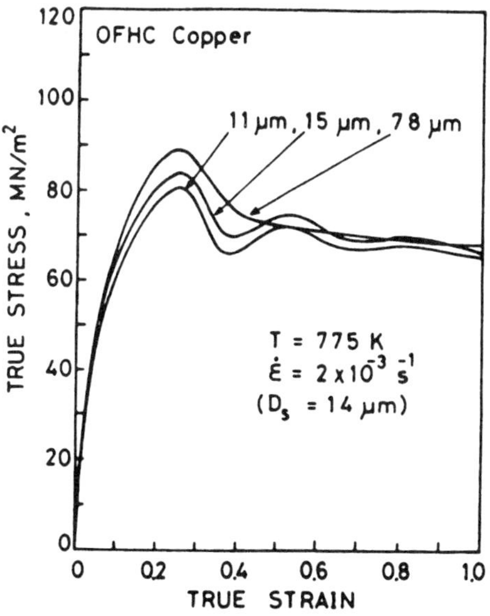

FIG. 8. Hot-compression stress–strain curves for OFHC copper specimens with various initial grain sizes, illustrating the transition from multiple- to single-peak dynamic recrystallization (after Blaz *et al.*, 1983).

1.3 Nucleus-Density Condition

A physical basis for the microstructure-mechanism map of Fig. 7 will now be considered. It is first assumed that the nucleation of new grains occurs only at or near existing grain boundaries. This is because the stored energy (dislocation density) is greater in these parts of the crystals, and also because less additional surface area need be created when a nucleus is formed. The relations between nucleus density N and grain size D are thus

$$N_0 = \beta P_0/D_0 \quad \text{and} \quad N_s = \beta P_s/D_s \tag{6}$$

in the initial and steady states, respectively, where β is a geometrical constant. P_0 and P_s are the probabilities per unit surface area that a grain-boundary site is activated to form a nucleus. Since some recently recrystallized parts are unable to supply nuclei in the steady-state microstructure, P_s is lower than P_0; the experimental data in fact suggest that $P_0/P_s \approx 2$. Then, from Eqs. (6),

$$\frac{N_0}{N_s} = \frac{P_0}{P_s}\frac{D_s}{D_0} \approx 2\,\frac{D_s}{D_0}\,. \tag{7}$$

Thus the relative–grain-size condition $D_0 = 2D_s$ is associated with the *nucleus-density condition*

$$N_0 = N_s. \tag{8}$$

The transition between single- and multiple-peak behavior can now be rationalized with the help of the above criterion.

When $D_0 > 2D_s$ or $N_0 < N_s$, there is a *deficit* in the initial nucleus density with respect to its steady-state value. Because nucleation occurs at the existing grain boundaries, a "necklace" mechanism operates as illustrated in Fig. 9. Since $D_s < D_0/2$, the new grains reach their equilibrium size D_s before the whole structure has recrystallized. After the formation of the first strand of new grains, DRX continues to spread progressively toward the interiors of the deformed initial grains (Fig. 9). According to this model, at any instant in time, the different volume fractions of the sample are at different stages of recrystallization. DRX is thus a very heterogeneous process and the occurrence of softening is not at all synchronized at any

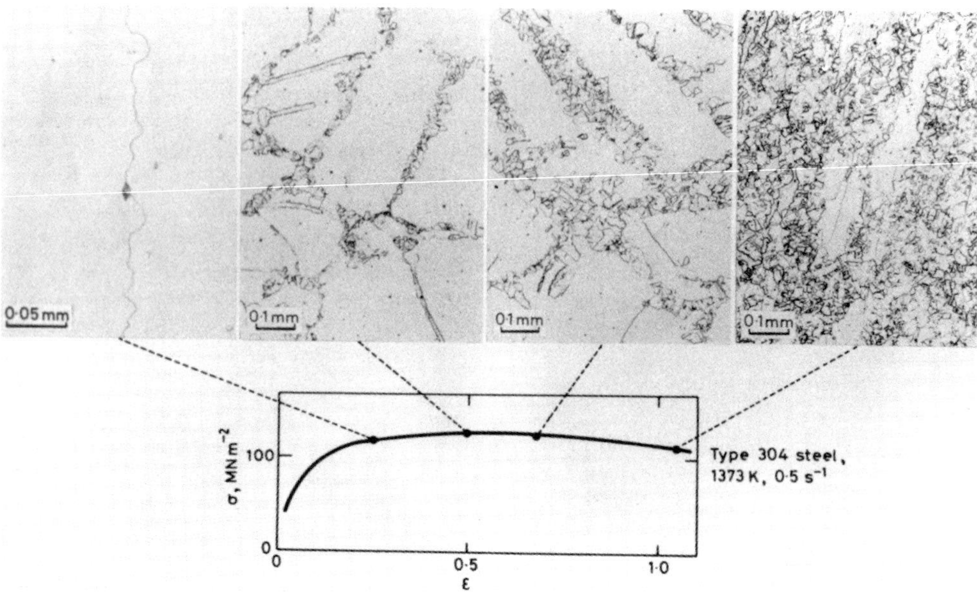

FIG. 9. Microstructures illustrating the progress of necklace dynamic recrystallization in type 304 austenitic stainless steel with a coarse initial grain size. The associated single-peak stress–strain curve is also shown. (After Roberts *et al.*, 1979.)

particular moment. Nevertheless, there is overall flow softening until the equilibrium grain size has been attained everywhere within the material, i.e., single-peak DRX occurs. The limiting case for this mechanism is a necklace containing a single strand, which corresponds to the relative-grain-size condition $D_0 = 2D_s$ or to the nucleus-density condition $N_0 = N_s$.

When $D_0 < 2D_s$ or $N_0 > N_s$, there is an *excess* in nucleus density. Since $D_s > D_0/2$, the growth of recrystallized grains is stopped by mutual impingement before they have reached their steady-state size D_s. Once the first cycle of recrystallization is complete, work hardening begins again, followed by a second cycle of recrystallization, and so on, until grain coarsening finally produces the stable grain size D_s. Such coarsening also progressively reduces the excess nucleus density until the stable level given by N_s is attained. It should be noted that, although this process generally leads to an increase in grain size (i.e., when $D_0 < D_s$), it can also be associated with a small amount of grain refinement (i.e., when $D_s < D_0 < 2D_s$), as illustrated in Fig. 5. Under these conditions, the progress of DRX is highly synchronized and the macroscopic multiple-peak flow curve readily reflects the local cycles of work hardening and recrystallization.

Unresolved Issues. Since the steady state is only attained at large strains ($\epsilon > 1$) and is rarely achieved during industrial processing, its characteristics have only been investigated experimentally in a few materials, such as Cu, Ni, and stainless steels (Ryan and McQueen, 1990). With regard to mathematical treatments, only a few tentative models have been proposed (Stüwe and Ortner, 1974; Sandström and Lagneborg, 1975). Although these can account for the stress dependence of the steady-state grain size, in their present states of development they are unable to distinguish between the grain-refinement and grain-coarsening modes of recrystallization. More recently, numerical models based on the Monte Carlo simulation technique have been used to study DRX (Rollett *et al.*, 1992). Although many of the essential features of the latter are faithfully represented, the relationship between flow-stress oscillations and grain coarsening has not yet been clearly reproduced.

During steady-state flow, both the flow stress and the microstructural parameters (e.g., the distributions of grain size and dislocation density) remain stationary (Fig. 10).

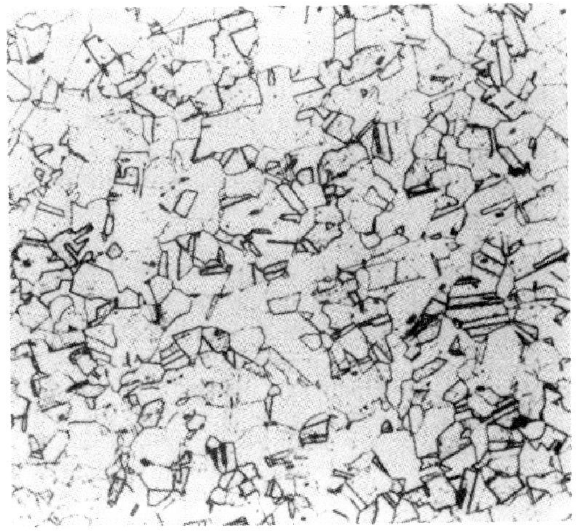

FIG. 10. Micrograph illustrating the steady-state microstructure associated with dynamic recrystallization in type 304 austenitic stainless steel deformed in torsion to $\epsilon \approx 5$ at 900 °C and 5 s^{-1}. The latter conditions correspond to single-peak flow and to grain refinement. (After Ryan and McQueen, 1990.)

Similar remarks apply to the crystallographic texture. The mechanism of nucleus formation is another topic that has been insufficiently explored. Nuclei can be created within grains from dislocation cells or subgrains (e.g., by "coalescence") or alternatively at grain boundaries, as assumed in the relative–grain-size model (e.g., by "grain-boundary bulging"). The role of oriented nucleation in the formation of nuclei remains to be clarified, as does that of selective growth. With the increasing availability and ease of use of equipment for EBSD (electron backscattering diffraction) measurements, considerable progress is expected in this area over the next few years.

1.4 Effect of Alloying Elements on DRX

The four steels compared in Fig. 11(a) all had the same initial grain size (110 μm). Their flow curves differ, however, because of differing amounts of V, Nb, or Mo in solid solution. The solutes decrease the mobility of grain boundaries in the course of DRX. This, in turn, increases the peak strain ϵ_p and decreases the steady-state grain size D_s, as shown in the figure. These effects are still more pronounced when precipitation occurs during DRX, as shown in Fig. 11(b), which pertains to the plain carbon, V, and Nb steels of Fig. 11(a) when deformed at a lower temperature. The possibility of delaying the onset of DRX by the addition of very small amounts of niobium or vanadium (e.g., 0.035% Nb or 0.115% V in the above examples) is what makes it possible to carry out the controlled rolling of HSLA (high-strength low-alloy) steels. Provided the accumulated strain in a given pass is less than the current value of ϵ_p, no DRX occurs and postdynamic recrystallization is also avoided (see below).

2. DYNAMIC RECOVERY AND CONTINUOUS DYNAMIC RECRYSTALLIZATION

The behavior described in the previous section concerned metals and alloys of medium or low stacking-fault energy (e.g., 10 to 50 mJ/m^2), such as copper, γ-iron, and austenitic steels. In these materials, the cross-slip of dislocations is relatively difficult, so that the ease of the rearrangement and removal of dislocations during deformation, i.e., *dynamic recovery*, is reduced (although it still plays an important role in the nucleation of new grains). By contrast, in materials of high stacking-fault energy (e.g., ≈100 mJ/m^2), such as aluminum, α-iron, and ferritic steels, dynamic recovery is so efficient that its operation can preclude the occurrence of discontinuous DRX. As conventional nucleation is unable to take place, the "new grains" are generated instead in a more gradual way by dynamic recovery followed by grain growth, as described in more detail below.

2.1 Flow Curves Associated with Continuous DRX

An example of such flow-stress behavior is presented in Fig. 12, where the results of hot-torsion experiments on grade 1100 aluminum carried out at 400 °C and various strain rates are reproduced (Perdrix *et al.*, 1981). The shapes of the stress–strain curves are similar at all strain rates: The strain-hardening stage leads to a smooth maximum, followed by a very slow decrease in the flow stress. Quite large strains are necessary in order to attain a true steady state, e.g., $\epsilon \approx 20$ in Fig. 12; these levels cannot be reached by other testing techniques such as

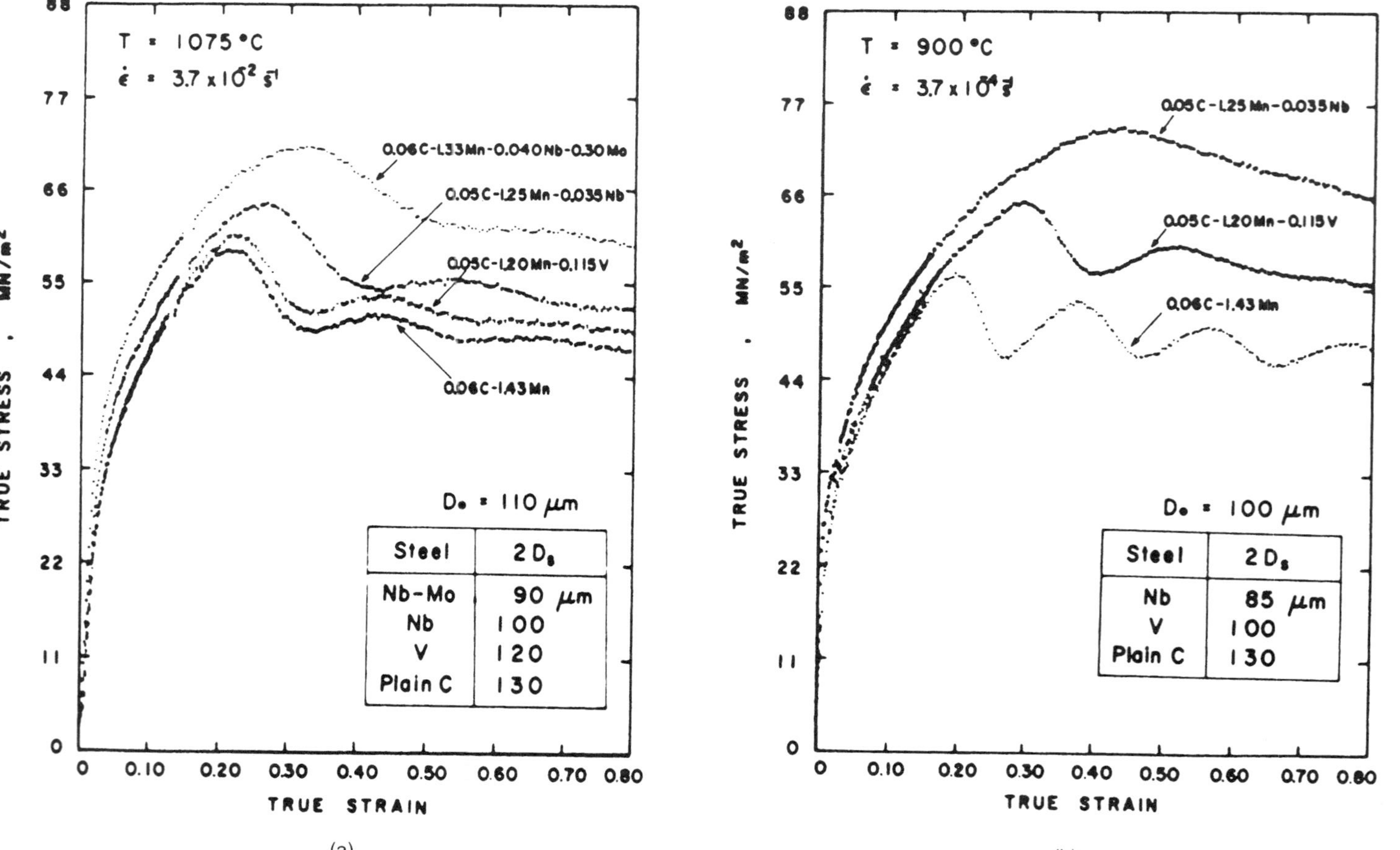

FIG. 11. (a) Effect of the presence of Nb, Mo, and V in solution on the stress–strain curves associated with dynamic recrystallization of four steels with identical initial grain sizes, during deformation at the same strain rate and temperature (after Sakai *et al.*, 1982). (b) Effect of precipitation occurring during dynamic recrystallization on the stress–strain curves of the plain carbon, V, and Nb steels of (a), when deformed at a lower temperature (after Sakai and Jonas, 1986).

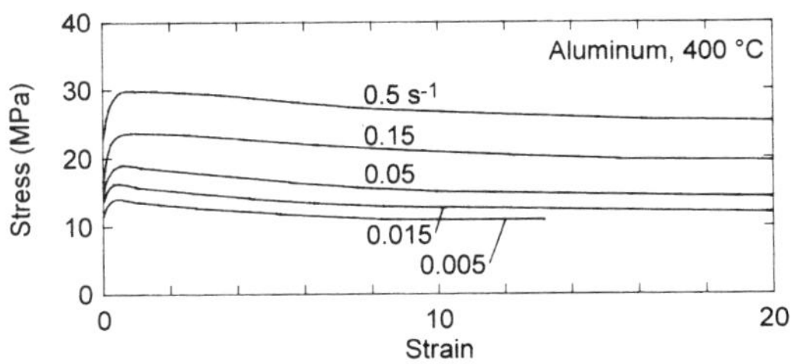

FIG. 12. Stress–strain curves typical of continuous dynamic recrystallization obtained on grade 1100 aluminum at 400 °C and various strain rates (after Perdrix *et al.*, 1981).

compression or tension. Similar behavior has been observed in α-iron and more recently during the testing of the high-temperature β phase (bcc) of titanium alloys. The smooth shape of the stress–strain curves, as well as the absence of oscillations at all temperatures and strain rates, clearly distinguishes the continuous from the discontinuous type of DRX process.

2.2 Microstructural Changes Associated with Continuous DRX

Dynamic recovery has been extensively investigated at low and moderate strains ($\epsilon < 1$), i.e., up to the maximum of the stress–strain curve. The essentially equiaxed *subgrains* formed by this process within the initial grains are illustrated in Fig. 13, obtained from an aluminum specimen tested in torsion at 400 °C to a final strain of 0.85 and rapidly quenched (Montheillet, 1981). The

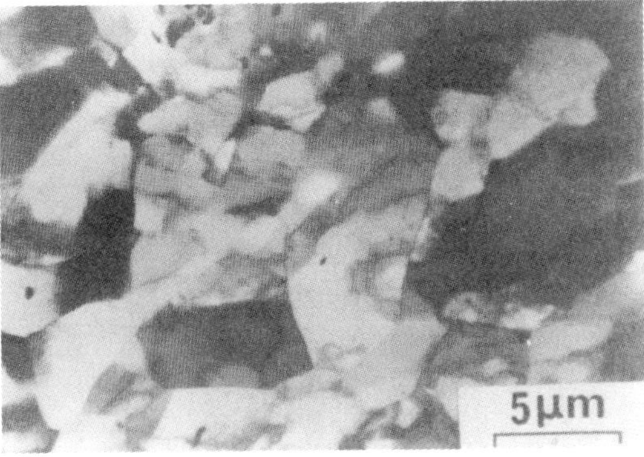

FIG. 13. Subgrain microstructure of an aluminum specimen tested in torsion at 400 °C and 4.2×10^{-2} s^{-1} to a strain of 0.85 (after Montheillet, 1981).

low misorientations that apply to these subgrains can be verified by using electron microdiffraction (or by the recently developed EBSD technique). The orientation of each subgrain with respect to the specimen axes can also be determined in this way. There are 12 different rotations θ_i (i.e., axes and angles of rotation) that can cause the orientations of two sets of rectangular coordinates to coincide. By convention, the misorientation that applies to two crystals is $\theta = \min(\theta_i)$. It can be shown that θ falls between 0 and ψ, where $\tan(\psi/2) = (\sqrt{2} - 1)(5 - 2\sqrt{2})^{1/2}$ in cubic structures, so that $\psi \approx 62.8°$ (Mackenzie, 1958). When $\theta < 15°$, the boundary is made up of one, two, or three sets of parallel dislocations and is called a *subgrain* or *low-angle boundary*. However, when $\theta > 15°$, the misorientation can no longer be accommodated by planar arrays of dislocations and is then referred to as a *grain* or *high-angle boundary*.

A schematic view of part of the specimen of Fig. 13 that was investigated in this way is presented in Fig. 14(a). The thick line represents a high-angle boundary, which was present in the starting material and has become somewhat serrated as a result of straining. The narrow lines in turn denote low-angle, i.e., subgrain, boundaries. The pole figure of Fig. 14(b) displays the orientations of the ⟨100⟩ (i.e., cube face) directions of the various crystals. These clearly form two families associated with the two original grains, A and B. Within each grain, the misorientations between neighboring subgrains remain small ($\theta < 15°$), a finding that is also illustrated in the misorientation-distribution diagram of Fig. 14(c).

Less work has been carried out on the dynamically recovered microstructures formed at large strains ($\epsilon > 1$) (e.g., Perdrix et al., 1981). An example of the microstructure developed at a strain of 40 in a specimen of grade 1100 aluminum twisted at 400 °C is presented in Fig. 15 (Montheillet, 1981). Comparison with the low-strain microstructure depicted in Fig. 13 shows that straight boundaries as well as 120° triple points (arrows) are much more frequent. The large-strain microstructure is further described in Fig. 16(a), from which it can be seen that numerous high-angle boundaries have formed within the material; as a result, the original 80-μm-diam grains can no longer be

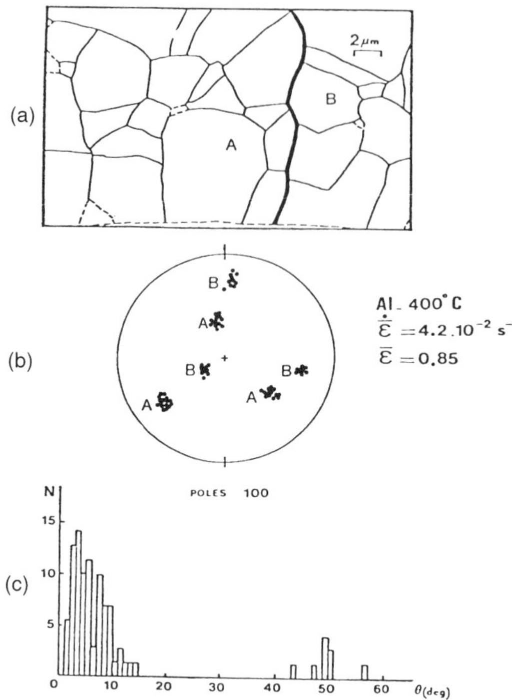

FIG. 14. (a) Schematic view of part of the specimen of Fig. 13 showing a high-misorientation boundary (thick line) and several low-angle, i.e., subgrain boundaries (narrow lines). (The dashed lines correspond to boundaries whose misorientations were not measured.) (b) (100) pole figure pertaining to the same area of the specimen. (c) Associated misorientation-distribution diagram. (After Perdrix *et al.*, 1981.)

identified. This is evident from Fig. 16(b) as well, where the ⟨100⟩ directions are seen to be located in several parts of the pole figure. Finally, Fig. 16(c) shows that the misorientations are more or less continuously distributed over the possible range.

More recently, similar results were obtained from compression tests carried out on grade 1200 aluminum specimens (Gourdet and Montheillet, 1994). In the latter case, a significant increase in the misorientation of the subgrains was observed at much lower strains ($\epsilon \approx 1$); the earlier increase in misorientation can be partly attributed to the different strain path that was employed (i.e., compression instead of torsion). It should be noted, however, that physical models remain to be developed both for the transition from low- to high-angle boundaries and for the evolution and motion of high-angle boundaries during straining.

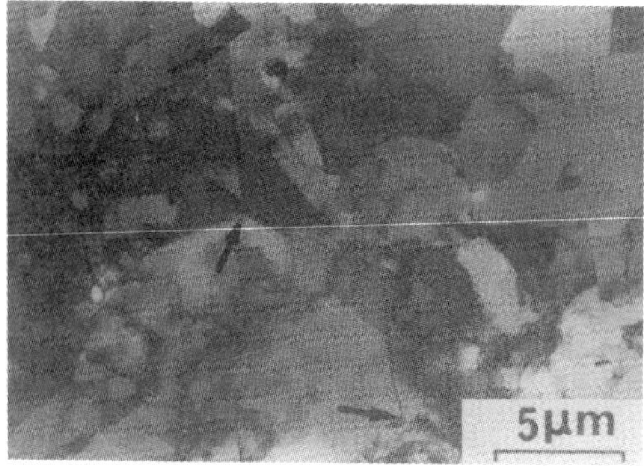

FIG. 15. Microstructure developed at a strain of 40 in a specimen of grade 1100 aluminum twisted at 400 °C and 4.2×10^{-1} s^{-1} (after Montheillet, 1981).

Two alternative models have been proposed to rationalize the above observations. Perdrix *et al.* (1981) first suggested that new high-angle boundaries form by a continuous process involving the absorption of mobile dislocations by the subgrain boundaries present at low strains. More recently, a different mechanism, referred to as *geometric dynamic recrystallization,* has been proposed by McQueen *et al.* (1989). According to these authors, the high-angle boundaries observed at large strains originate from the elongation and deformation of the boundaries present in the starting material. When the distance between two neighboring boundaries is reduced to the equilibrium size of the subgrains, pinched-off regions appear, which break up the elongated grains (Fig. 17). The repetition of this mechanism can lead eventually to the formation of an equiaxed microstructure. In contrast to the first mechanism, the second does not involve the formation of *new* high-angle boundaries. By using a simple geometrical model, it can be shown that at large torsional strains, the grain-boundary surface area is multiplied by a factor close to $\bar{\epsilon}$, the von Mises equivalent strain. Thus, starting from an initial grain size of 50 μm, an equivalent strain of about 10 is large enough to produce "geometrically recrystallized" grains of 5 μm.

In order to decide whether the first or the second of the above mechanisms operates during the hot deformation of materials of high stacking-fault energy, large-strain torsion experiments have been conducted on aluminum *single crystals* (Kassner, 1989).

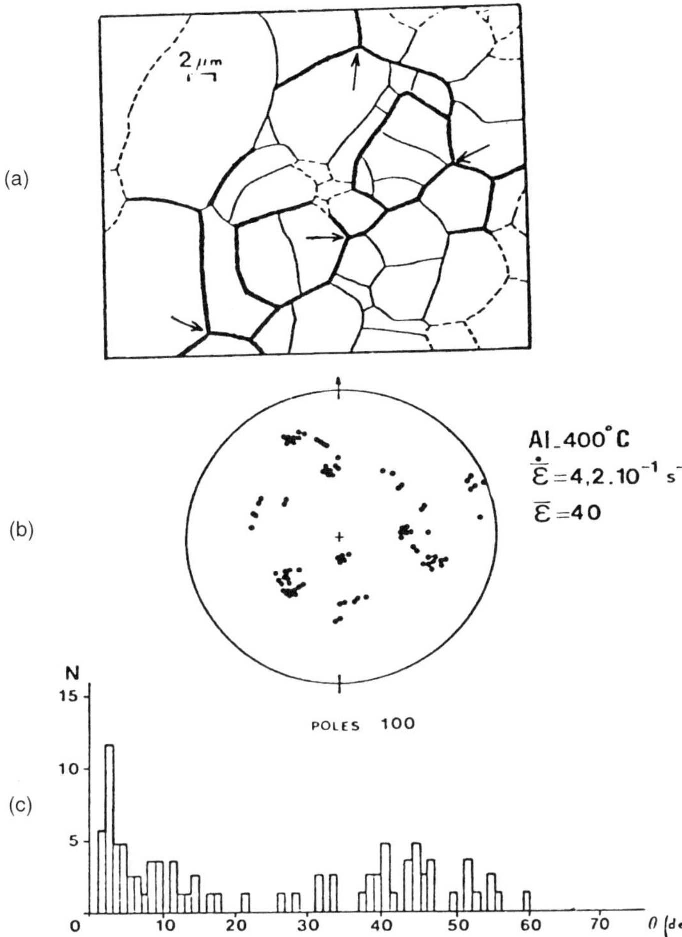

FIG. 16. (a) Schematic view of part of the specimen of Fig. 15 exhibiting numerous high-angle boundaries (thick lines) in addition to the low-angle boundaries (narrow lines). (The dashed lines correspond to boundaries whose misorientations were not measured.) (b) (100) pole figure corresponding to the same area of the specimen. (c) Corresponding misorientation distribution diagram. (After Perdrix *et al.*, 1981.)

Since only a limited number of high-misorientation interfaces was observed after straining, it was concluded that the much larger increase in the number of high-angle boundaries in deformed *polycrystals* was the result of geometric rather than continuous DRX. However, this interpretation is invalid at lower strains, in particular when the starting

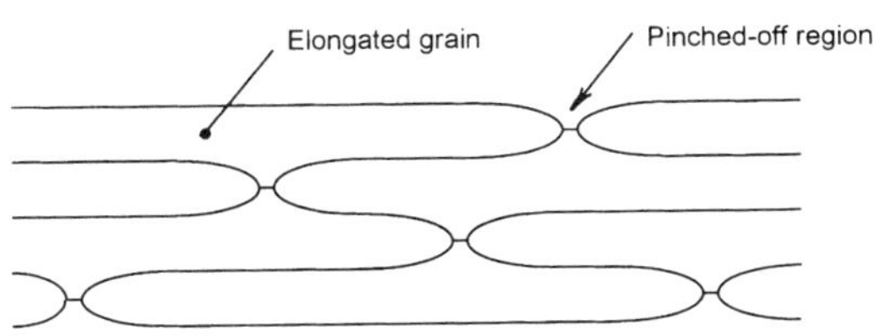

FIG. 17. Simplified schematic representation of the geometric dynamic-recrystallization mechanism.

grain size is large, and new grains are still observed under these conditions (Gourdet and Montheillet, 1994). It therefore seems likely that solely the first mechanism operates at the strains involved in commercial processing, but that both mechanisms operate simultaneously at very large strains (e.g., $\epsilon > 10$) in polycrystalline materials.

Finally, it is of interest to note that the steady-state flow stress σ_s and stable grain or subgrain size D_s are related by a power law:

$$\sigma_s = k/D_s^{\alpha}, \qquad (9)$$

where k and α are constants. Note that this relationship has the same form as in the case of discontinuous DRX [see Eq. (3)]. At low strains (i.e., when a subgrain structure is present), a number of investigations have shown that $\alpha \approx 1$. By contrast, at large strains (mixture of low- and high-angle boundaries), the available results (Montheillet, 1981) indicate that $\alpha \approx 0.75$; this suggests that a different mechanism is controlling the formation of the high-angle boundary microstructure.

By combining Eqs. (1) and (4) above, the following relationship is obtained between the steady state flow stress σ_s, strain rate $\dot{\epsilon}$, and absolute temperature T:

$$\sigma_s = A\dot{\epsilon}^m \exp(mQ/RT), \qquad (10)$$

where A is a constant. Although the strain-rate sensitivity exponent m generally increases with temperature, it is often considered to be constant to a first approximation. In such cases, the material is considered to be *power-law viscoplastic*. For both types of DRX, experimental values of m generally fall within the range 0.1 to 0.2. The temperature dependence is in turn a special form of the more general Arrhenius law, according to which Q denotes the apparent activation energy of deformation. In the case of dynamic recovery and continuous recrystallization, Q is close to the activation energy for self-diffusion (e.g., $Q \approx 150$ kJ/mol in aluminum). This can be attributed to the occurrence of dislocation climb in the subgrain boundaries, which involves self-diffusion, and which is considered to control the evolution of the microstructure. However, when discontinuous recrystallization occurs, appreciably greater values of Q are generally observed (e.g., $Q \approx 300$ kJ/mol in copper). Such a difference in temperature dependence of the flow stress has been associated with an accommodation process controlled by solute drag and by grain-boundary migration.

3. TEXTURE CHANGES ASSOCIATED WITH DRX

The changes in preferred orientation that take place during discontinuous or continuous DRX have long been of interest, in part because their control could have useful applications in deformation processing. Nevertheless, experimental data on the development of such crystallographic "textures" are scarce, particularly in the form of detailed ODFs (orientation distribution functions), in part because the rapidity of postdynamic recrystallization makes it difficult to determine the exact state of the metal at the moment when straining was completed.

Early results obtained from torsion experiments by Montheillet *et al.* (1984) indicate that, in the case of aluminum, where only continuous DRX takes place, a sharp texture develops at 400 °C [Fig. 18(a)]. The latter consists of the "twin symmetric" $\{\bar{1}12\}\langle 110\rangle/\{1\bar{1}\bar{2}\}\langle\bar{1}\bar{1}0\rangle$ ideal orientation referred to as the $B/\bar{B}$ component, where $\{hkl\}$ and $\langle uvw\rangle$ are the shear plane and shear direction, respectively [Fig. 18(d)]. In the case of copper deformed at 200 °C, although the $B/\bar{B}$ component is still evident, the $A/\bar{A}$ component, viz., $\{1\bar{1}\bar{1}\}\langle 110\rangle/\{\bar{1}11\}\langle\bar{1}\bar{1}0\rangle$, is considerably sharper [Figs. 18(b) and 18(d)]. At higher temperatures (300 to 500 °C), the $A/\bar{A}$ and $B/\bar{B}$ orientations continue to be present, but they appear to be more and more scattered [Fig. 18(c)]. This was attributed by the authors to the effect of more and more rapid discontinuous DRX, which tends to consume the deformation texture, although some amount of postdynamic recrystallization may also have occurred after straining. More recently, Tóth and Jonas (1992), Jonas and Tóth (1992), and Tóth *et al.* (1992) have characterized the textures developed during the hot torsion of copper bars in considerably more detail. For this, they employed relatively low deformation temperatures (200 and 300 °C) and fairly rapid quenches. Analysis of the ODFs confirmed that there is a considerable intensification of the $A/\bar{A}$ component with increasing temperature [Figs. 19(a) and 19(b)].

In order to account for the texture changes produced by high-temperature deformation, the above authors considered that two complementary recrystallization mechanisms take place, viz., oriented nucleation and selective growth. In this case, *oriented nucleation* concerns the preferred nucleation of particular orientations among the deformed grains; these are the ones that recover more easily than the others, and thus form many of the recrystallization nuclei. These grains (or parts of grains) have the lowest possible Taylor factor M. This is because, for a given von Mises equivalent strain increment $\Delta\epsilon$, the lowest M grains have experienced the least total microscopic slip $\Delta\Gamma$, since

$$\Delta\Gamma = M\Delta\epsilon. \qquad (11)$$

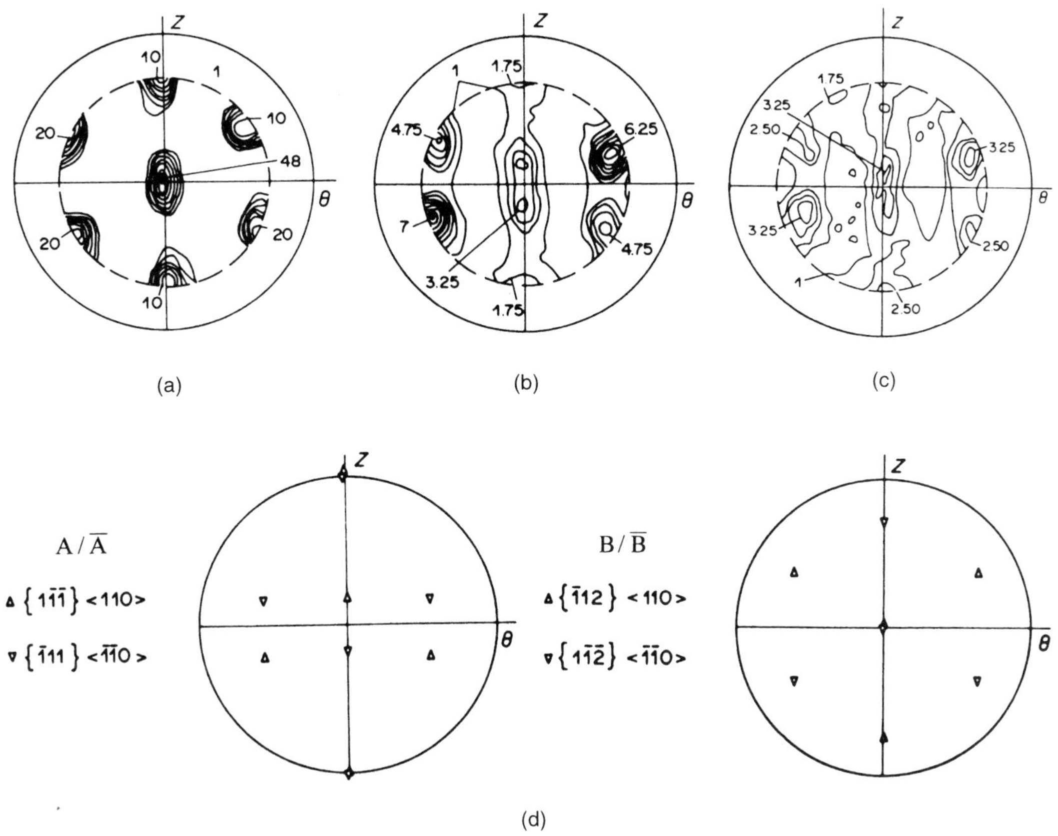

FIG. 18. (111) pole figures illustrating the crystallographic textures obtained from large-strain hot-torsion experiments: (a) aluminum, 400 °C; (b) copper, 200 °C; (c) copper, 400 °C; (d) $A/\bar{A}$ and $B/\bar{B}$ ideal orientations. (After Montheillet *et al.*, 1984.)

They are therefore likely to have undergone the least work hardening or increase in dislocation density, and so to be the most highly recovered at any particular strain. For the torsion (i.e., simple shear) of fcc materials, the lowest possible value of M (1.73) corresponds to the $A/\bar{A}$ component, whereas for the $B/\bar{B}$ component, $M = 2.45$.

Selective growth in turn involves the rapid increase in size of nuclei characterized by particular misorientations with respect to the matrix. In fcc metals, the orientations of these nuclei generally differ by ±40° rotations around appropriate ⟨111⟩ directions. In their simulations, Jonas and Tóth (1992) assumed that the reorientation axis is perpendicular to the most active slip plane. This choice is based on the high mobility of grain boundaries perpendicular to active slip planes, which generally contain planar arrays of dislocations and subboundaries. The high mobility is caused by impurity diffusion away from these boundaries by pipe diffusion along the planar arrays.

Comparison of the experimental and simulated textures allowed the relative importance of oriented nucleation and selective growth to be estimated. In the case of copper twisted at 300 °C, it was found that the first mechanism was the more important, in a ratio of about 2:1 [Fig. 19(c)]. Oriented nucleation can thus account for the high intensity of the minimum–Taylor-factor $A/\bar{A}$ component in copper twisted at 300 °C. By contrast, when the temperature is increased to 400 °C, as in Fig. 18(c), selective growth is likely to become more important, thus accounting for the more scattered texture evident in the latter figure. Finally, the development of the $B/\bar{B}$ (medium–Taylor-factor)

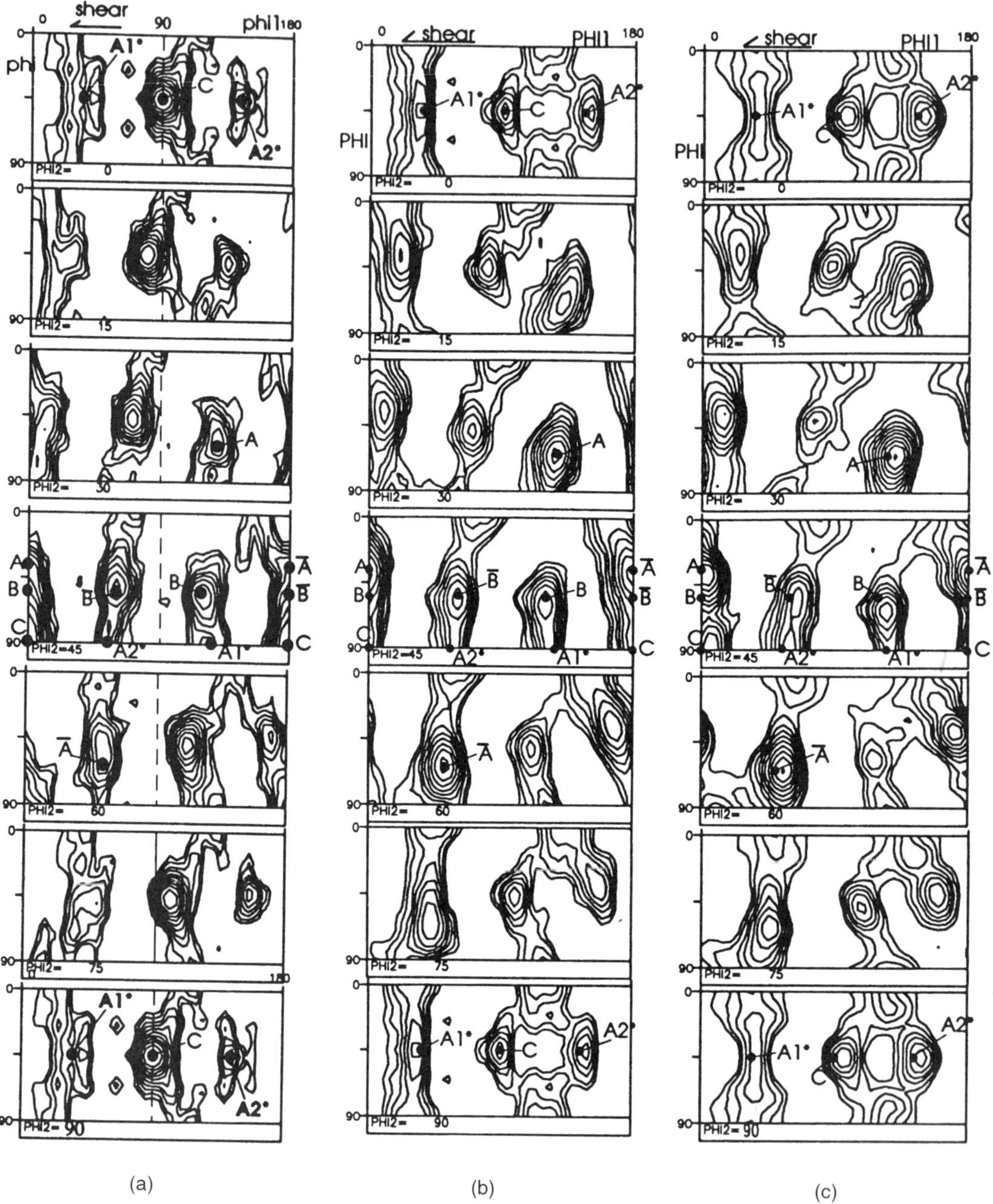

FIG. 19. Orientation-distribution functions characterizing the textures developed during the torsion testing of copper. (a) Experimental texture developed at room temperature at a strain of about 3.2. (b) Experimental texture developed at 300 °C and a strain of about 2.3. (c) Texture obtained from computer simulation under conditions corresponding to (b). (After Jonas and Tóth 1992.)

component in aluminum deformed in torsion at 400 °C still remains unexplained.

4. POSTDYNAMIC SOFTENING

As already mentioned, postdynamic restoration often takes place very rapidly in the interpass intervals of multiple-pass industrial processing, or during the final cooling stage. As a result, it can strongly influence the microstructure (e.g., grain size, texture), and therefore the properties (e.g., yield strength, fracture toughness, plastic anisotropy), of the final product. In fact, it is the characteristics of postdynamic, rather than dynamic, recrystallization that are often more relevant to industrial hot working.

The postdynamic softening mechanisms were extensively studied by Petkovic *et al.* (1975, 1979) by means of *interrupted-compression tests* carried out on several kinds of steel and copper specimens. In such experiments, a sample is first deformed at a given strain rate and temperature. It is then unloaded and held for a time t at the same temperature. Finally it is reloaded, thus allowing the softening produced by the postdynamic restoration to be measured. Samples can also be quenched for subsequent metallographic examination (although these are not easy to interpret in steel specimens because of the martensitic transformation).

Whereas only two distinct softening mechanisms can take place in a specimen held at temperature after cold deformation, *three* have been identified when hot deformation is involved:

1. static recovery;
2. conventional static recrystallization; and
3. a type of recrystallization termed *metadynamic* or *postdynamic*.

The microstructural aspects of the first two processes are exactly those of their counterparts following cold deformation. By contrast, metadynamic recrystallization takes place only if discontinuous DRX has occurred during the previous interval of hot deformation. It is not preceded by an appreciable incubation period, since it relies on further growth of the nuclei already formed during DRX. Furthermore, it proceeds much more quickly than classical recrystallization. The kinetic equations for these three mechanisms, however, are all of the Avrami type, i.e.,

$$X_i = X_i^{\infty}[1 - \exp(-k_i t^{n_i})]. \qquad (12)$$

In this equation, $X_i = X_r$, X_R, or X_M is the fractional softening that can be associated with static recovery, static recrystallization, or metadynamic recrystallization, respectively; t denotes the time measured from the onset of the restoration process; and X_i^{∞} and k_i are parameters that depend on the detailed conditions of prior straining and holding. Experimental measurements of the exponent n_i for the hot working of copper led to the following values (Luton *et al.*, 1980): $n_r = 1.33$ (whereas $n_r = 1$ after cold working), $n_R = 2.2$, and $n_M = 1$ (to be compared with $n = 3$ to 4 in the original Avrami theory).

The progress of softening as described by Eq. (12) is illustrated in Figs. 20(a) and 20(b), which show the time dependence of the softening occurring in copper after deformation at 500 °C. In Fig. 20(a), prior straining was interrupted *before* the onset of discontinuous DRX. Accordingly, the subsequent softening is produced solely by static recovery (X_r) and static recrystallization (X_R), as shown in the figure. In Fig. 20(b), the initial compression test was stopped *after* the initiation of DRX, so that metadynamic recrystallization took place during the subsequent holding interval. In this case, static softening is described by the sum of the three components X_r, X_R, and X_M.

Metadynamic recrystallization leads to some coarsening (by a factor of about 1.5) of the dynamically recrystallized grains, as shown in Fig. 21 (Hodgson *et al.*, 1992). This diagram is based on data obtained from torsion tests carried out on 0.15% C steel specimens. It can be seen that the sizes of both the dynamically and metadynamically recrystallized grains decrease as the strain rate of the prior deformation is increased and/or the temperature is lowered (i.e., as Z is increased). It is of practical importance that, under normal hot-working conditions, the grain size produced by metadynamic recrystallization is 2 to 3 times finer than that resulting from conventional static recrystallization.

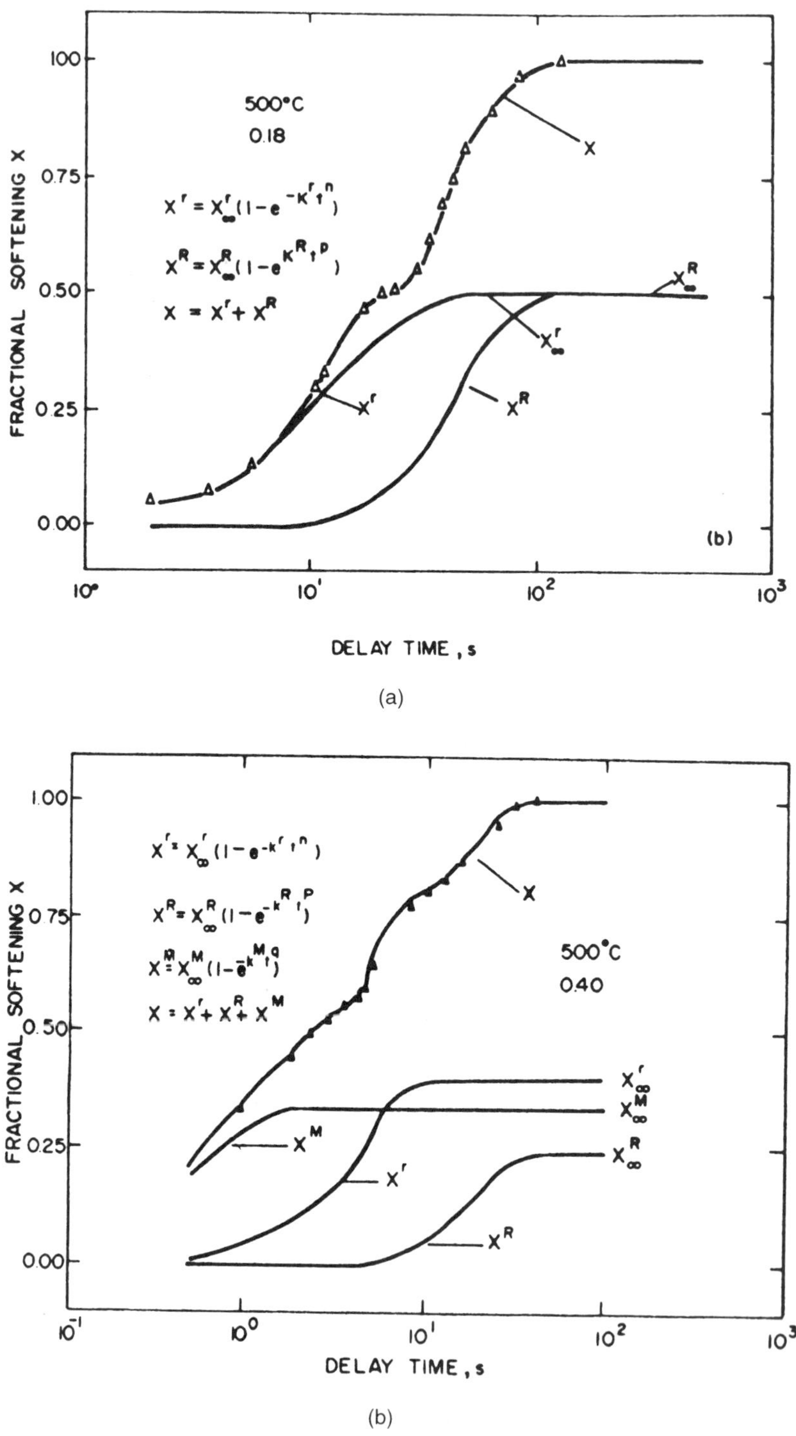

FIG. 20. Progress of softening following the hot compression of copper specimens at 500 °C to a strain of (a) 0.18, i.e., before the onset of DRX, and (b) 0.40, i.e., after DRX has started. The curves that follow the experimental data points correspond to the sum of the X_r and X_R components in the first case, and of the X_r, X_R, and X_M components in the second. These are associated with static recovery, static recrystallization, and metadynamic recrystallization, respectively. (After Luton *et al.,* 1980.)

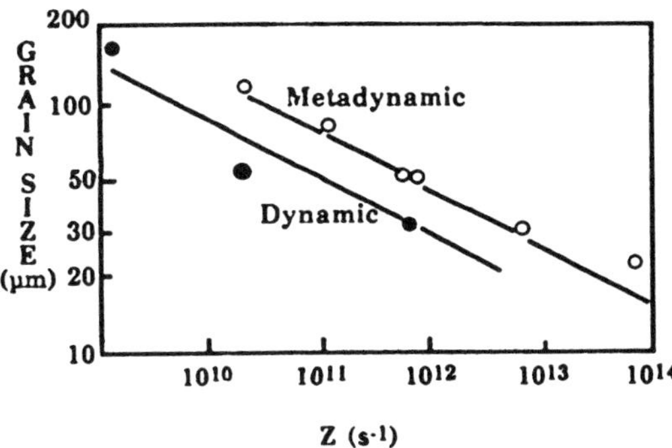

FIG. 21. Dependence of the grain sizes produced by dynamic and metadynamic recrystallization on the temperature-corrected strain rate Z (after Hodgson *et al.*, 1992).

5. IMPLICATIONS FOR INDUSTRIAL PROCESSING

In industrial practice, DRX plays a role in two types of processes. In the first, large reductions are applied in a single uninterrupted pass or operation. This description applies to planetary hot mills and extrusion presses. In the second, although multipass operations are involved, the interpass times are relatively short (i.e., <1 s). Under these conditions, there is insufficient time for interpass softening, so that strain hardening is accumulated at first, until the critical strain for DRX is attained. During subsequent straining, the flow stress remains relatively low, on account of continued dynamic recrystallization during deformation, supplemented by the occurrence of metadynamic (postdynamic) recrystallization during the interpass intervals.

The interpass softening produced by metadynamic recrystallization increases not only with the length of the interpass interval, but also with the strain rate of the previous pass. This dependence can be described by the following relationship:

$$t_{50} = A\dot{\epsilon}^{-p} \exp(Q/RT), \tag{13}$$

where t_{50} is the hold time for 50% softening, A and p are material constants ($p \approx 0.6$), and Q is an apparent activation energy (Roucoules *et al.*, 1994). Although interpass times are progressively reduced during typical rolling operations, the strain rate simultaneously increases from pass to pass. Thus the rate of postdynamic softening also increases from pass to pass.

When little postdynamic recrystallization takes place, such low-temperature finishing has been referred to as dynamic-recrystallization–controlled rolling (DRCR). Under these conditions, the final grain size is controlled by DRX (Roucoules *et al.*, 1994; Jonas, 1994). Conversely, when there is nonnegligible postdynamic softening, it has been referred to as metadynamic-recrystallization–controlled rolling (MDRCR), in which case the final microstructure is a result of postdynamic recrystallization. The grain sizes produced by both DRX and DRX followed by metadynamic recrystallization are much finer than those attributable to the static recrystallization that takes place in conventional rolling, i.e., when longer interpass times are involved. Finally, when the rate of softening is so fast that appreciable postdynamic softening takes place, the continued occurrence of DRX is no longer possible, and the process of strain accumulation is reinitiated, as in the first few passes of DRCR and MDRCR rolling.

GLOSSARY

Apparent Activation Energy: Parameter (expressed in kJ/mol) that specifies the temperature sensitivity of the flow stress in an Arrhenius law.

Arrhenius Law: An exponential relationship between the rate of a process and the inverse absolute temperature. This is used in the present context to relate the flow stress to the inverse temperature when constant testing strain rates are used.

Avrami Equation: Relationship between the recrystallized volume fraction and time used to describe the kinetics of recrystallization.

Bulging (Grain Boundary): Local migration of a high-misorientation grain boundary, which moves into a neighboring grain containing a higher dislocation density. Such migration can lead to nucleus formation, particularly in lightly deformed materials.

Coalescence (of Subgrains): The conversion of two or more neighboring subgrains into a single subgrain of higher misorientation, by the annihilation of their common (low-misorientation) subgrain boundary. This can lead to nucleus formation and applies particularly to highly deformed materials.

Continuous Dynamic Recrystallization: DRX occurring by a progressive increase in misorientation of the subgrains formed by dynamic recovery.

Critical Resolved Shear Stress: Shear stress required to activate a slip system, i.e., to produce dislocation glide on that system.

Cross-Slip: Glide of the screw component of a dislocation when it leaves its original slip plane to travel along another plane inclined at an angle to the first.

Dislocation Climb: Movement of a dislocation in a direction perpendicular to its slip plane. Since it requires vacancy migration, rapid climb only takes place at elevated temperatures.

Dynamic Recovery: The formation of lower-energy configurations by the partial removal and rearrangement of dislocations during straining at elevated temperatures. This leads to the formation of fairly well-defined subgrain boundaries.

Dynamic Recrystallization: Nucleation and growth of initially dislocation-free new grains; this occurs *during* straining at elevated temperatures.

Geometric Dynamic Recrystallization: DRX occurring at very large strains by the elongation and subsequent pinching off and fragmentation of elongated grains.

Ideal Orientation: Low-index crystallographic orientation in the vicinity of which is located a significant volume fraction of the grains in a given specimen. Also referred to as a *texture component.*

Metadynamic Recrystallization: Type of static recrystallization that follows discontinuous DRX after the cessation of straining. Also referred to as *postdynamic recrystallization.*

Miller Index: Triplet of integer numbers used to specify the crystallographic orientation of a lattice plane or direction.

Nucleation: Generation of a nucleus, which is a small dislocation-free grain surrounded by mobile high-angle boundaries. The three most common mechanisms are grain-boundary bulging, subgrain coalescence, and particle-stimulated nucleation.

Nucleus: Small dislocation-free grain bounded by a high-misorientation boundary that is able to grow into the strain-hardened matrix.

Orientation-Distribution Function: Scalar function specifying the volume fraction of grains possessing the various possible orientations pertaining to "orientation space." Here the grain orientations of interest are expressed in terms of the Euler angles that relate a given orientation to the orientation of the specimen axes.

Self-Diffusion: Thermally activated process of bulk (or matrix) diffusion in a metal; it involves the migration of vacancies.

Stacking-Fault Energy: Parameter associated with the mobility (ease of cross-slip) of dislocations. When the stacking-fault energy is low, dislocations are widely dissociated and their mobility is low, and vice versa.

Steady-State Flow: Dynamic equilibrium that can be achieved at large strains at elevated temperatures. It is associated with both a steady flow-stress level and a statistically time-independent microstructure.

Strain Path: Type of deformation imposed on or experienced by a material, e.g., uniaxial tension or compression, plane-strain compression, simple shear (torsion).

Strain-Rate Sensitivity: Dimensionless parameter that specifies the strain-rate sensitivity of the flow stress; it involves the ratio of the logarithm of the flow-stress increase to the logarithm of the strain-rate increase.

Taylor Factor: Ratio of the macroscopic flow stress to the critical resolved shear stress for dislocation glide within a grain. It also relates the total microscopic slip taking place within a grain to the macroscopic strain.

Zener–Hollomon Parameter: Function of strain rate and temperature that accounts for the equivalent effects of these two parameters on the flow stress at high temperatures. Also referred to as the temperature-corrected strain rate.

Works Cited

Blaz, L., Sakai, T., Jonas, J. J. (1983), *Metal Sci.* **17,** 609–616.

Derby, B. (1991), *Acta Metall. Mater.* **39,** 955–962.

Fields, D. S., Jr., Backofen, W. A. (1957), *Proc. ASTM* **57,** 1259–1272.

Gourdet, S., Montheillet, F. (1994), in: "Microstructures et recristallisation," 37ème Colloque de Métallurgie de l'INSTN, Saclay, *J. Phys. (Paris)* **5,** Suppl. JPIII, No. 4, 255.

Hodgson, P. D., Jonas, J. J., Yue, S. (1992), in: A. J. DeArdo (Ed.), *Processing, Microstructure and Properties of Microalloyed and other Modern HSLA Steels,* Warrendale, PA: Iron and Steel Society of the Metallurgical Society of AIME, p. 41.

Jonas, J. J. (1994), *Mater. Sci. Eng. A* **184,** 155–165.

Jonas, J. J., Canova, G. R., Shrivastava, S. C., Christodoulou, N. (1982), in: E. H. Lee, R. L. Mallett (Eds.), *Plasticity of Metals at Finite Strain: Theory, Experiment and Computation,* Stanford University, CA, p. 206.

Jonas, J. J., Tóth, L. S. (1992), *Scripta Metall. Mater.* **27,** 1575–1580.

Kassner, M. E. (1989), *Metall. Trans. A* **20,** 2182–2185.

Luton, M. J., Petkovic, R. A., Jonas, J. J. (1980), *Acta Metall.* **28,** 729–743.

Luton, M. J., Sellars, C. M. (1969), *Acta Metall.* **17,** 1033–1043.

Mackenzie, J. K. (1958), *Biometrika* **45,** 229–240.

McQueen, H. J., Solberg, J. K., Ryum, N., Nes, E. (1989), *Philos. Mag. A* **60,** 473–485.

Montheillet, F. (1981), in: P. Costa, A. Herpin, J. C. Jousset, P. Lacombe, A. Le Bon, G. Pomey, J. L. Strudel (Eds.), *Les traitements thermomécaniques,* 24ème Colloque de Métallurgie de Saclay, Paris, France: INSTN, pp. 57–70.

Montheillet, F., Cohen, M., Jonas, J. J. (1984), *Acta Metall.* **32,** 2077–2089.

Perdrix, Ch., Perrin, M. Y., Montheillet, F. (1981), *Mém. Sci. Rev. Metall.* **78,** 309–320.

Petkovic, R. A., Luton, M. J., Jonas, J. J. (1975), *Can. Metall. Quart.* **14,** 137–145.

Petkovic, R. A., Luton, M. J., Jonas, J. J. (1979), *Acta Metall.* **27,** 1633–1648.

Poirier, J.-P. (1972), *Philos. Mag.* **26,** 713–725.

Roberts, W., Boden, H., Ahlblom, B. (1979), *Met. Sci.* **13,** 195–205.

Rollett, A. D., Luton, M. J., Srolovitz, D. J. (1992), *Acta Metall. Mater.* **40,** 43–55.

Rossard, C., Blain, P. (1959), *Mém. Sci. Rev. Metall.* **56,** 285–299.

Roucoules, Ch., Hodgson, P. D., Yue, S., Jonas, J. J. (1994), *Metall. Trans. A,* **25,** 389–400.

Ryan, N. D., McQueen, H. J. (1990), *Can. Metall. Quart.* **29,** 147–162.

Sakai, T., Akben, M. G., Jonas, J. J. (1982), in: A. J. De Ardo, G. A. Ratz, P. J. Wray (Eds.), *Thermomechanical Processing of Microalloyed Austenite,* Warrendale, PA: AIME, pp. 237–252.

Sakai, T., Jonas, J. J. (1984), *Acta Metall.* **32,** 189–209.

Sakai, T., Jonas, J. J. (1986), in: N. Hansen, D. Juul Jensen, T. Leffers, B. Ralph (Eds.), *Annealing Processes—Recovery, Recrystallization and Grain Growth,* 7th Risø International Symposium on Metallurgy and Materials Science, Roskilde, Denmark: Risø National Laboratory, p. 143.

Sakui, S., Sakai, T., Takeishi, K. (1977), *Trans. Iron Steel Inst. Jpn.* **17,** 718–725.

Sandström, R., Lagneborg, R. (1975), *Acta Metall.* **23,** 387–398.

Steinemann, S. (1958), *Beiträge Geol. Schweiz (Hydrol.)* **10,** 1–72.

Stüwe, H. P. (1968), in: W. J. McG. Tegart, C. M. Sellars (Eds.), *Deformation under Hot Working Conditions,* ISI Special Report 108, London: Iron and Steel Institute, p. 1.

Stüwe, H. P., Ortner, B. (1974), *Met. Sci.* **8,** 161–167.

Thomsen, E. G., Yang C. T., Bierbower, J. B. (1954), in: *An Experimental Investigation of the Mechanics of Plastic Deformation of Metals,* Berkeley and Los Angeles: University of California Press, p. 89, quoted by Thomsen, E. G., Yang, C. T., Kobayashi, S. (1965), *Mechanics of Plastic Deformation in Metal Processing,* New York: The MacMillan Company, p. 113.

Tóth, L. S., Jonas, J. J. (1992), *Scripta Metall. Mater.* **27,** 359–363.

Tóth, L. S., Jonas, J. J., Daniel, D., Bailey, J. A. (1992), *Textures and Microstructures,* **19,** 245–262.

Weiss, I., Sakai, T., Jonas, J. J. (1984), *Met. Sci.,* **18,** 77–84.

Zener, C., Hollomon, J. H. (1944), *Trans. Am. Soc. Met.* **33,** 163–235.

Further Reading

Jonas, J. J., Bieler, R., Bowman, K. (Eds.), (1994), *Advances in Hot Deformation Textures and Microstructures,* Warrendale, PA: The Minerals, Metals, and Materials Society.

Krauss, G. (Ed.) (1984), *Deformation, Processing and Structure* Metals Park, OH: American Society for Metals, especially Chap. 4, "Dynamic Changes That Occur during Hot Working and Their Significance Regarding Microstructural Development and Hot Workability" (W. Roberts), Chap. 5, "A New Approach to Dynamic Recrystallization" (J. J. Jonas and T. Sakai), and Chap. 6, "Static Recrystallization and Precipitation during Hot Rolling of Microalloyed Steels" (C. M. Sellars).

H. J. McQueen, Jonas, J. J. (1975), "Recovery and Recrystallization during High Temperature Deformation," in: R. J. Arsenault (Ed.), *Plastic Deformation of Materials,* New York: Academic, pp. 393–493.

Yue, S. (Ed.) (1990), *Mathematical Modelling of Hot Rolling of Steel,* Montréal: Canadian Institute of Mining and Metallurgy.

RECRYSTALLIZATION, STATIC

F. J. HUMPHREYS, *Manchester Materials Science Centre, Manchester, United Kingdom.*

Introduction 227
1. **The Driving Force for Recrystallization** 228
2. **The Fundamental Mechanisms of Annealing** ... 229
2.1 Recovery 229
2.1.1 The Mechanisms of Dislocation Recovery 229
2.1.2 Subgrain Growth 232
2.2 Recrystallization of Single-Phase Alloys 232
2.2.1 Kinetics of Primary Recrystallization 232
2.2.2 The Nucleation of Recrystallization 233
2.2.2.1 Strain-Induced Grain-Boundary Migration 233
2.2.2.2 Preformed-Nucleus Model 234
2.2.2.3 Nucleation Sites 234
2.2.3 Computer Modeling of Recrystallization 234
2.3 Migration of High-Angle Grain Boundaries 235
2.3.1 Experimental Observations 235
2.3.2 The Effect of Variables on Boundary Mobility 235
3. **Recrystallization of Two-Phase Alloys** 236
3.1 Particle-Stimulated Nucleation of Recrystallization 236
3.2 Pinning Effects of Particles (Zener Drag) 238
3.3 Bimodal Alloys and the Prediction of Grain Size 238
3.4 Interaction of Precipitation and Recrystallization 239
4. **Grain Growth after Primary Recrystallization** 239
4.1 Factors Affecting Grain Growth 240
4.2 Computer Simulation of Grain Growth 240
4.3 Secondary Recrystallization ... 241
5. **Recrystallization Textures** ... 241
5.1 Orientation Effects in Nucleation 241
5.1.1 Transition Bands 241
5.1.2 Shear Bands 242
5.1.3 Prior Grain Boundaries 242
5.1.4 Second-Phase Particles 242
5.2 Orientation Effects in Growth 242
5.3 Competition between Texture Components 242
Glossary 243
Works Cited 243
Further Reading 244

INTRODUCTION

During plastic deformation of a crystalline solid, dislocations and point defects are generated in the material. Although thermodynamics would suggest that these defects should spontaneously disappear, in practice, the kinetics of the diffusion processes by which these defects migrate are often very slow at low homologous temperatures, with the result that the defects are retained after deformation. However, if the material is subsequently heated to a high temperature (or annealed), then solid-state diffusion provides a mechanism whereby the defects may be removed or alternatively arranged in lower energy configurations.

The material may be partially restored to its original state by recovery, in which annihilation and rearrangement of the dislocations occur. The changes in microstructure are shown schematically in Figs. 1(a) and 1(b). Similar softening processes (dynamic recovery) may also occur during the deformation, particularly at high temperatures.

There is a further annealing process called recrystallization in which new dislocation-free grains are formed within the de-

3-527-28138-X/96/$5.00 + .50

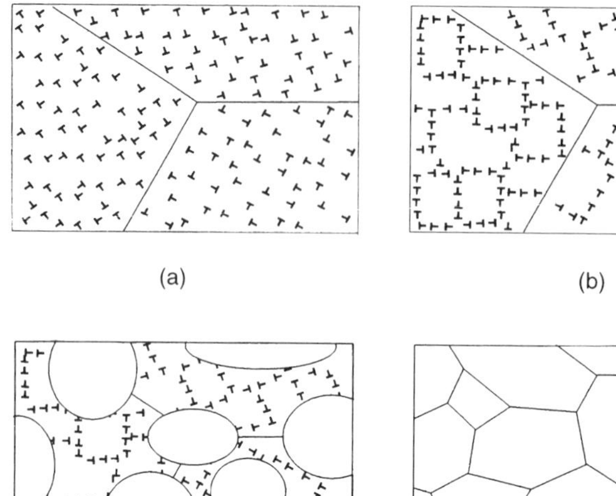

FIG. 1. Schematic sketch of recovery and recrystallization. (a) Deformed microstructure, (b) recovered structure, (c) partial recrystallization, (d) fully recrystallized specimen.

formed or recovered structure. These then grow and consume the old grains, resulting in a new grain structure with a low dislocation density [Figs. 1(c) and 1(d)].

Although recrystallization removes the dislocations, the material still contains grain boundaries, which are thermodynamically unstable. Further annealing may result in grain growth, in which smaller grains are eliminated, larger grains grow, and the grain boundaries assume a lower-energy configuration. In certain circumstances this uniform grain growth may give way to the selective growth of a few large grains, a process known as abnormal grain growth or secondary recrystallization.

The understanding and control of recrystallization is of particular importance in the mechanical processing of alloys. To a large extent the mechanical properties and behavior of a metal depend on the dislocation structure, the size of the grains, and the orientation or texture of the grains. The first of these parameters is dependent on the amount of deformation, the temperature of deformation, and the subsequent annealing treatment as well as other material and microstructural parameters. The grain size and texture are controlled by the recrystallization process.

1. THE DRIVING FORCE FOR RECRYSTALLIZATION

The nature of the deformed material is a key element in determining the annealing behavior, because the stored energy of deformation provides the driving force for recovery and recrystallization, while the mechanisms of recrystallization are determined by the microstructure of the deformed material, and in particular the microstructural heterogeneities.

Most of the work expended in deforming a metal is given out as heat, with typically only some 1% being stored in the form of dislocations and, at low deformation temperatures, point defects. In a single-phase single crystal, dislocation storage depends on the trapping of dislocations by others, and this is not easy to predict from first principles. The density of such statistically stored dislocations (ρ_S) is typically 10^{12} m^{-2} for a lightly deformed metal and 10^{16} m^{-2} for a heavily cold-rolled metal. If the metal is constrained to deform inhomogeneously by the presence of nondeformable second-phase particles, then the plastic incompatibility creates additional geometrically necessary dislocations (ρ_G) (Ashby, 1970). The density of these dislocations depends on the strain and the particle size, shape, and spacing. For example, for a volume fraction F_V of spherical particles of diameter d, Ashby shows that ρ_G is given by

$$\rho_G = 8F_V\gamma/bd, \tag{1}$$

where b is the Burgers vector and γ the shear strain. For large volume fractions or small particles, the predicted geometric dislocation density may be much larger than ρ_S. However, this may in reality be significantly lowered by dynamic recovery, even at ambient temperatures. Grain boundaries also lead to inhomogeneous deformation and hence to geometrically necessary dislocations (Ashby, 1970).

The line energy of a dislocation per unit length is given approximately by Gb^2 where G is the shear modulus. Therefore the stored energy of the deformed metal (E) is given approximately by

$$E = Gb^2(\rho_G + \rho_S). \tag{2}$$

A highly deformed metal typically has a stored energy of about 10^6 J m^{-3}. Note that this is very much smaller than the energies associated with phase transformations, e.g., only about 1% of the latent heat of fusion of the metal.

2. THE FUNDAMENTAL MECHANISMS OF ANNEALING

2.1 Recovery

During recovery, many physical and mechanical properties change, and recovery may be measured by a variety of experimental techniques. Figure 2, from the work of Schmidt and Haessner (1990), shows the annealing of high-purity aluminum that was deformed at 77 K and maintained at that temperature until the calorimetry was performed. The peak at 200 K is due to the recovery of point defects, and the peak at 260 K is due to recrystallization. A separate peak due to dislocation recovery is not detected. The high-purity aluminum recrystallizes at such a low temperature that either no dislocation recovery occurs prior to recrystallization or else the peak is too close to the recrystallization peak to be detected. The data from calorimetry are compared in Fig. 2 with those from other techniques.

One of the most important parameters in determining the amount and rate of recovery is the stacking fault energy, which, by affecting the extent to which dislocations dissociate, determines the rate of dislocation climb, which is usually the rate-controlling process during recovery. In metals of low stacking fault energy such as copper, α-brass, and austenitic stainless steel, climb is difficult, and little or no recovery of the dislocation structure occurs prior to recrystallization. However, in metals of high stacking fault energy such as aluminum and α-iron, climb is rapid, and significant recovery may occur.

2.1.1 The Mechanisms of Dislocation Recovery During recovery the stored energy of the material is lowered by dislocation movement. There are two primary processes, these being the annihilation of dislocations and the rearrangement of dislocations into low-energy configurations. Both processes are achieved by glide, climb, and cross-slip of dislocations.

A schematic drawing of a crystal containing an array of edge dislocations is shown in Fig. 3. The forces between the dislocations will depend on the Burgers vectors and relative positions of the dislocations. For example, dislocations of opposite sign on the same glide plane, e.g., A and B, may annihilate by gliding toward each other. Such processes can occur even at low temperatures, lowering the dislocation density during deformation and leading to dynamic recovery. Dislocations of opposite Burgers vector on different glide planes, e.g., C and D, can annihilate by a combination of glide and climb. As climb requires thermal activation, this can only occur at high homologous temperatures (homologous temperature is actual temperature normalized by the melting point).

If unequal numbers of dislocations of the two signs are produced during deformation then the excess cannot be removed by annihilation [Fig. 3(b)]. On annealing, these excess dislocations will arrange into lower-energy configurations in the form of regular arrays or low-angle grain boundaries. The simplest case is that shown in Fig. 3(c) in which dislocations of only one Burgers vector are involved. This process is often known

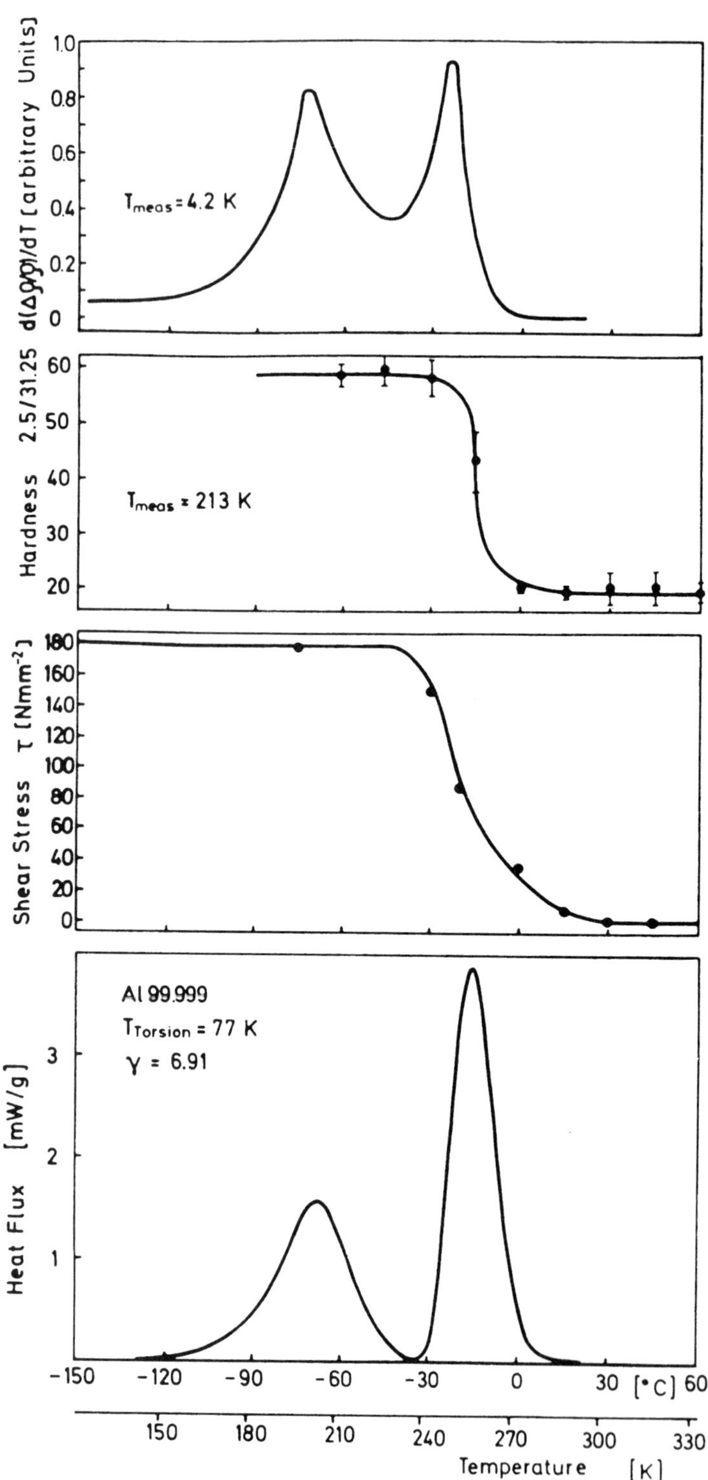

FIG. 2. Comparison of the heat flux, stress, hardness, and differentiated isochronous resistance $d(\Delta\rho/\rho)/dt$ as a function of the annealing temperature of Al 99.999, deformed to a true strain of 6.91 at 77 K (from Schmidt and Haessner, 1990).

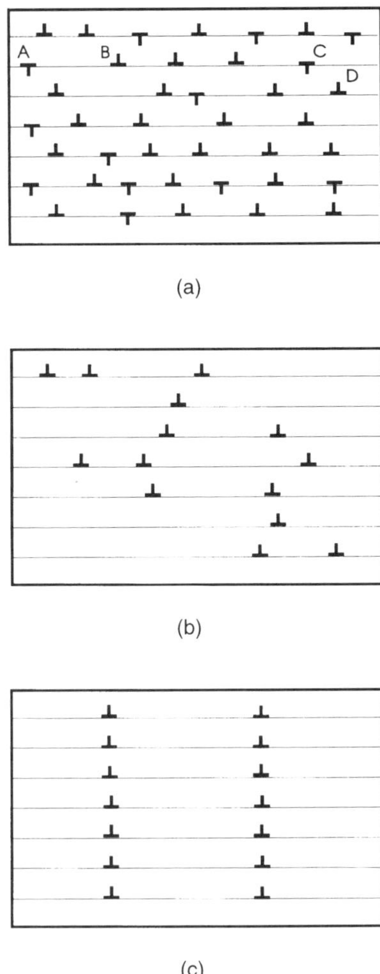

FIG. 3. (a) Crystal containing dislocations of two types. (b) The crystal after annihilation of dislocations has occurred. (c) Rearrangement of the remaining dislocations into low-angle tilt boundaries.

as polygonization, and the excess dislocations form tilt boundaries. In the case of a polycrystalline material subjected to large strains the dislocation structures produced on deformation and on subsequent annealing are more complex than the simple case shown in Fig. 3 because dislocations of many Burgers vectors are involved.

In an alloy of medium or high stacking fault energy, the dislocations are typically arranged after deformation in the form of a three-dimensional cell structure, the cell walls being complex dislocation tangles. The transmission electron micrograph of Fig. 4(a)

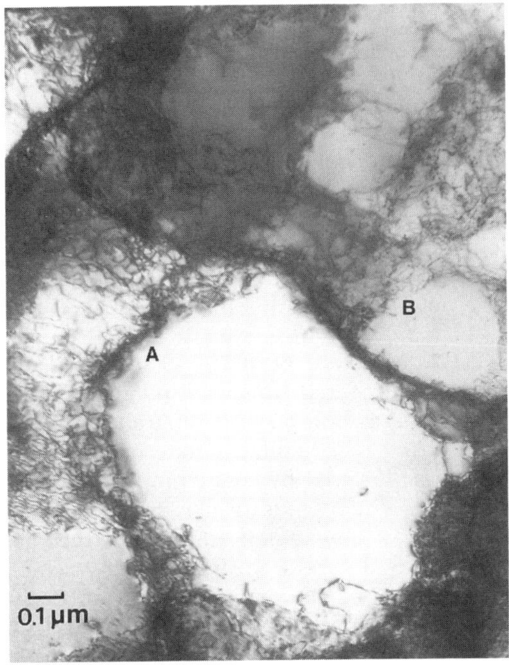

(a)

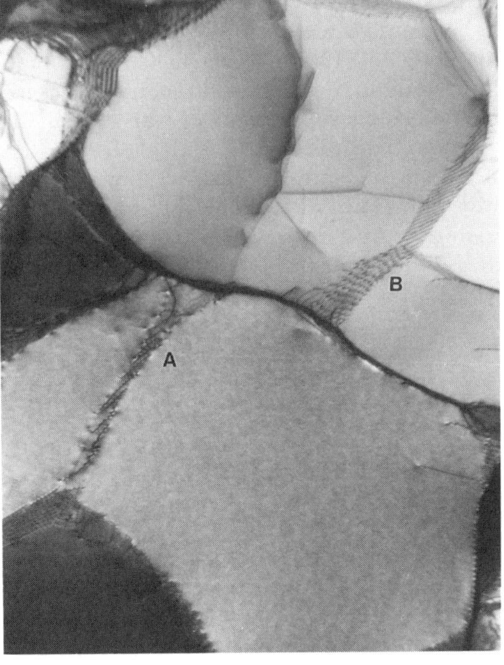

(b)

FIG. 4. Transmission high-voltage electron micrographs of aluminum deformed 10% and annealed *in situ*. (a) Deformed structure, (b) same area after 2 min anneal at 500 K.

shows equiaxed cells of diameter about 1 μm in deformed aluminum. By annealing this specimen *in situ* in a high-voltage electron microscope, the recovery processes can be followed directly. Figure 4(b) shows the same area after annealing. The tangled cell walls, e.g., at A and B, have become more regular dislocation networks or low-angle grain boundaries and the number of dislocations in the cell interiors has diminished. The cells have now become subgrains.

2.1.2 Subgrain Growth The stored energy of a recovered substructure such as that of Fig. 4(b) is still large compared with that of the fully recrystallized material, and can be further lowered by coarsening of the substructure, which leads to a reduction in the total area of low-angle boundary in the material. Two different mechanisms by which coarsening of the substructure can occur have been proposed: low-angle boundary migration and subgrain rotation. Subgrain boundary migration has been frequently observed experimentally and the mobility of a low-angle boundary is known to depend on the type and misorientation of the boundary. An alternative recovery mechanism was proposed by Li (1966). He suggested that subgrains might rotate by diffusional processes until adjacent subgrains were of similar orientation. The two subgrains involved would then coalesce into one larger subgrain with little boundary migration, the driving force arising from a reduction in boundary energies. Although it has been shown that coalescence is thermodynamically feasible, there is little direct evidence of its occurrence and the importance of this mechanism during recovery has yet to be established.

2.2 Recrystallization of Single-Phase Alloys

Recrystallization involves the formation of new strain-free grains in certain parts of the specimen and the subsequent growth of these to consume the deformed or recovered microstructure [Figs. 1(c) and 1(d)]. It is convenient to divide recrystallization into two regimes, the nucleation of recrystallization, which is the formation of the new grains, and grain growth during recrystallization, which follows it.

2.2.1 Kinetics of Primary Recrystallization The progress of recrystallization with time is commonly represented by a plot of the fraction of material recrystallized (X) as a function of time. This plot usually has a characteristic sigmoidal form which is shown schematically in Fig. 5. This typically shows an incubation time before recrystallization is detected, followed by an increasing rate of recrystallization, a linear region, and finally a decreasing rate of recrystallization. Such a curve is typical of many transformation reactions (see, e.g., Shewmon, 1969), and can be described phenomenologically in terms of the constituent nucleation and growth processes. Nuclei are formed at a rate $\dot{N} = dN/dt$ and grains grow into the deformed material at a rate $\dot{G}$. The volume of spherical grains varies as the cube of their diameter, and the recrystallized fraction rises rapidly with time. However, the new grains will eventually impinge on each other and the rate of recrystallization will decrease, tending to zero as X approaches 1. Such a model leads to a kinetic equation of the form

$$X = 1 - \exp(-Bt^k), \qquad (3)$$

which is generally known as the Avrami (1939) or Johnson–Mehl (Johnson and Mehl, 1939) equation.

The constants B and k depend on the details of the model—for example on how the nucleation and growth rates vary during recrystallization—and on the shapes of the growing grains. Although this equation gives

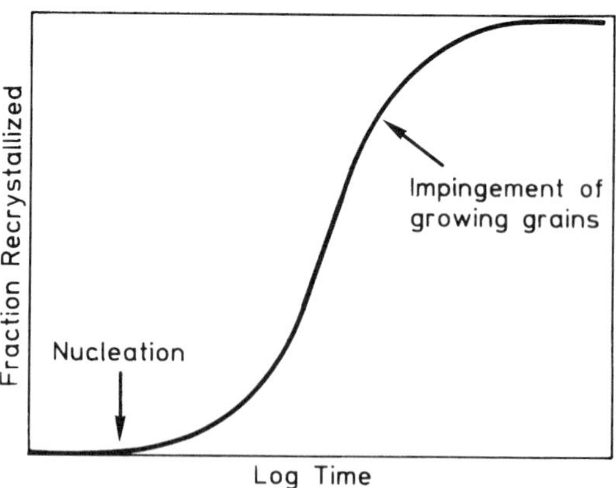

FIG. 5. The effect of time on the fraction recrystallized.

a broad description of recrystallization, careful experiments have shown that it is rarely accurately obeyed in practice. The reason for this is that the recrystallization process is far too complex to be described by an equation with two variables, and, in particular, the inhomogeneity of the distribution of stored energy and of the nucleation events has been shown to cause significant deviations from Avrami kinetics. This has led to the application of computer modeling to the problem (see Sec. 3.2.3).

Because recrystallization is a thermally activated process, the temperature of annealing has a profound influence on the rate of recrystallization as shown in Fig. 6. For isothermal annealing, the rate of recrystallization is often expressed in terms of $t_{1/2}$, the time at which the material is 50% recrystallized. For material which is isochronally annealed, it is often convenient to measure the ease of recrystallization in terms of a recrystallization temperature, the temperature at which recrystallization is complete (or 50% complete) in a fixed time, e.g., 1 h.

The amount and, to some extent, the type of deformation affect the rate of recrystallization, because the deformation alters both the amount of stored energy and the number of effective nuclei. There is a minimum amount of strain, generally a few percent, below which recrystallization will not occur. Above this strain the rate of recrystallization increases, leveling out to a maximum value at true strains of around 2–4.

The initial grain size may affect the stored energy of deformation as well as the number of nucleation sites, and it is found that a fine-grained material will recrystallize more rapidly than a coarse-grained material, as shown in Fig. 6.

The magnitude of the grain size after recrystallization can be rationalized in terms of the effects of various parameters on the nucleation and growth processes. Any parameter favoring a larger number of nuclei or a rapid nucleation rate, such as a high strain or a small initial grain size, will lead to a small final grain size.

2.2.2 The Nucleation of Recrystallization A considerable research effort has been aimed at understanding the processes by which recrystallization originates. The importance of recrystallization nucleation is that it is a critical factor in determining the size and orientation (texture) of the resulting grains. In order to control recrystallization effectively, it is necessary to understand the mechanisms and the parameters that control it. Although the term nucleation of recrystallization is universally used, it is known that recrystallization does not occur by a true nucleation event such as occurs in phase transformations. Recrystallization starts by the growth of a small crystallite that was present in the deformed microstructure.

2.2.2.1 Strain-Induced Grain-Boundary Migration. This mechanism, an example of which is seen in Fig. 7, was first reported by Beck and Sperry (1950). It involves the bulging of part of a pre-existing grain boundary leaving a dislocation-free region behind the migrating boundary. The driving force

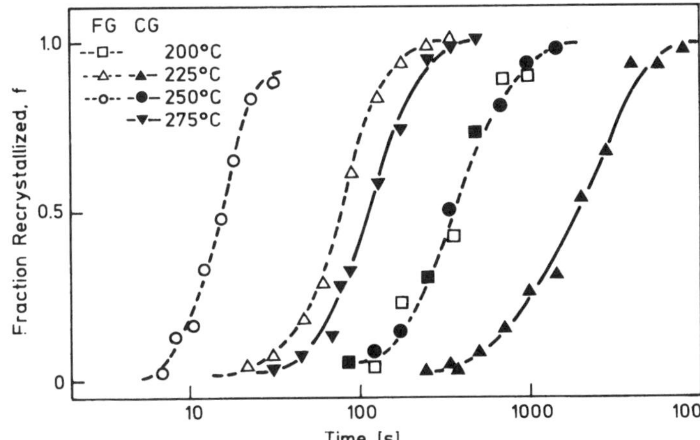

FIG. 6. The effect of temperature and initial grain size on the recrystallization kinetics of copper. The dotted curves show the behavior of material with a 15-μm grain (FG) size and the full curves show the behavior of material with a 50-μm grain (CG) size (Ryde *et al.*, 1990).

FIG. 7. Strain-induced grain-boundary migration in aluminium deformed 40% by compression (Bellier and Doherty, 1977).

for such an event arises from a difference in dislocation density on opposite sides of the grain boundary, and the process has been analyzed by Bailey and Hirsch (1964). This mechanism is particularly important after low strains. The characteristic feature is that the new grains have similar orientations to the old grains from which they have grown.

2.2.2.2 Preformed-Nucleus Model. In the process of strain-induced grain-boundary migration discussed above, the high-angle grain boundary, which is a prerequisite for recrystallization, was already in existence. However, in many cases, recrystallization originates in other regions of the material, and we need to consider how a nucleus is formed in such circumstances. It is now established beyond reasonable doubt that recrystallization originates from dislocation cells or subgrains that are present after deformation. Although there are still uncertainties about how these pre-existing crystallites become nuclei, several points are now clear:

1. The orientation of the nucleus is present in the deformed structure. There is no evidence that new orientations are formed during or after nucleation, except by twinning.
2. Nucleation occurs by the growth of subgrains by the mechanisms discussed in Sec. 2.1. Although all direct TEM observations have shown the mechanism to be one of subboundary migration, subgrain coalescence may also occur. A characteristic of nucleation is that the recovery in that area is significantly faster than in the majority of the material.
3. In order for a high-angle grain boundary to be produced by this rapid recovery, there must be an orientation gradient present.

2.2.2.3 Nucleation Sites. The site of the nucleus is very important in determining its viability, and it is clear that the general mechanism discussed above can occur at a variety of sites. The grains of a polycrystal do not deform homogeneously, but tend to split into regions of different orientations. This is a consequence of the constraining effects of neighboring grains and of the different slip systems which operate in different grains. Following the terminology used by Hatherly (1982) we note that the regions of different orientation are known as deformation bands. The boundary of a deformation band is a small region of continuous orientation change which is called a transition band. In certain cases the orientation is identical on either side of a deformation band, and in that case it is called a kink band. The transition bands at the edges of the deformation bands are ideal sites for nucleation of recrystallization, having a large dislocation density and a large orientation gradient. During rolling, another form of deformation heterogeneity called a shear band is formed. This is a planar region oriented at around 30° to the rolling plane, in which extensive shear has occurred locally. These bands are another favored site for nucleation, particularly in metals of medium and low stacking fault energy (Duggan *et al.*, 1978). These nucleation sites are shown in the sketch of Fig. 8.

2.2.3 Computer Modeling of Primary Recrystallization There have recently been several attempts to simulate primary recrystallization by computer modeling (see, e.g., Saetre *et al.*, 1986; Doherty *et al.*, 1986; Furu *et al.*, 1990). The main advantage of such models is that they can allow for realistic spatial distribution of nuclei and for complex variations of nucleation and growth rates. Such models are capable of predicting grain size distributions as well as recrystallization kinetics. However, the models are not yet sophisticated enough to give any detailed insight into the mechanisms of recrystallization.

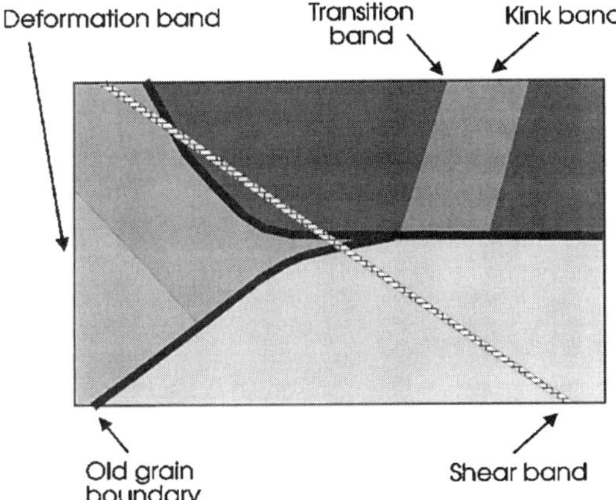

FIG. 8. Sketch of deformation inhomogeneities that may act as sites for recrystallization.

2.3 Migration of High-Angle Grain Boundaries

In the nucleation stage of recrystallization discussed above, a high-angle boundary, capable of migrating into the deformed material, is produced. In order to complete the recrystallization process, the deformed material is consumed by these migrating boundaries.

2.3.1 Experimental Observations It is generally accepted that the velocity (V) of a migrating boundary is related to the boundary mobility (M) and the driving force (P) by

$$V = MP. \tag{4}$$

If the driving force is known, then the boundary mobility can be determined. Experiments in which the velocity of a boundary during recrystallization is measured are very difficult to carry out because the driving force, arising from the stored energy, typically 10 MPa, is not easy to measure accurately, varies throughout the microstructure, and does not remain constant with time, decreasing as recovery proceeds. For this reason, many measurements of boundary mobility have been carried out on materials with a better characterized driving force, e.g., an as-cast substructure. However, it is not clear to what extent measurements of boundary mobility in materials with such low driving forces are directly applicable to migrating boundaries in recrystallizing material.

2.3.2 The Effect of Variables on Boundary Mobility As was the case for low-angle boundaries, the mobility of high-angle boundaries is very temperature dependent. The activation energy for pure metals is typically about half that for self diffusion (see Haessner and Hofmann, 1978) but the activation energy is higher for impure materials.

As boundary migration involves diffusion processes in and across the boundary, it is to be expected that the structure of the boundary should affect its mobility. Boundaries may be classified in part by reference to the number of coincidence sites, which are atom sites common to the grains on either side of the boundary. Special boundaries with a large number of coincidence sites are known to have properties different from those with few coincident sites. Aust and Rutter (1959) measured the migration rate of boundaries in zone-refined lead under a low driving force, as a function of both orientation and impurity level. Their results, shown in Fig. 9, indicate that in very pure lead, the boundary

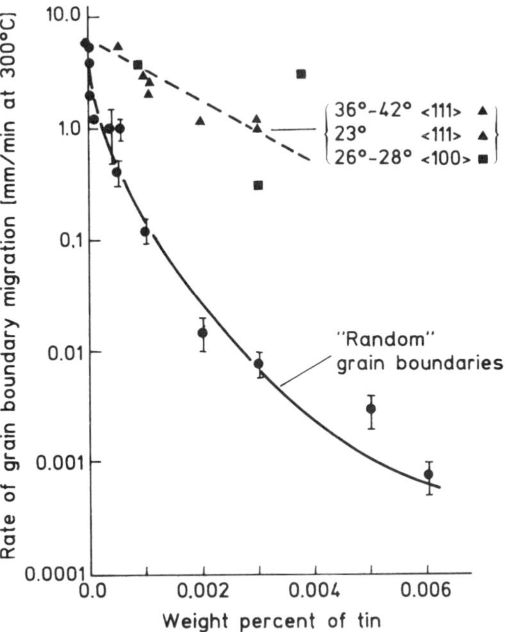

FIG. 9. The rate of grain-boundary migration at 573 K into zone-refined lead crystals doped with small amounts of tin (after Aust and Rutter, 1959).

mobility is almost independent of orientation. However, as the impurity level rises, the mobilities of the randomly oriented boundaries fall much more than those of the coincident-site-lattice boundaries, which are denoted in Fig. 9 by the angle and axis of rotation between the adjacent grains, suggesting that impurities are adsorbed more easily on the more open and disordered structure of the random boundaries.

Liebmann *et al.* (1956) found that during recrystallization of an aluminum single crystal, the boundary migration rate was highest for boundaries misoriented about a ⟨111⟩ axis by around 40° (Fig. 10). Rapid growth of grains of certain orientations will lead to the development of a recrystallization texture, and the results above have often been cited in favor of the model of oriented growth of textures (Sec. 5).

As may be seen from Fig. 9, small amounts of solute have a very large effect on the mobility of boundaries. Lücke and Detert (1957) were the first to formulate a quantitative theory of the effect of solute atoms on boundary mobility. This theory has been extended by others, and the developments are reviewed by Grant *et al.* (1984).

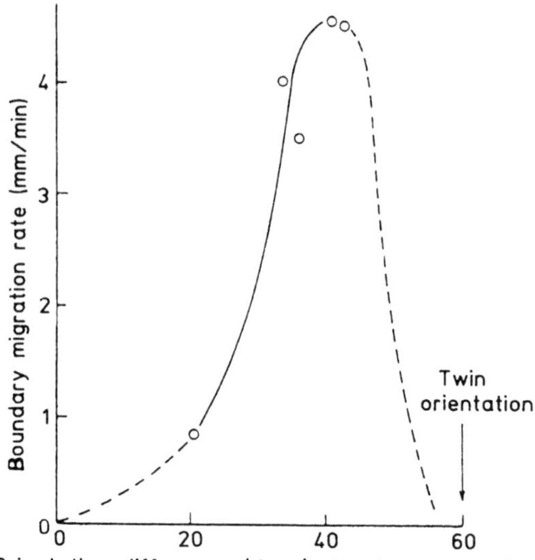

FIG. 10. Growth rates of new grains into a deformed aluminium crystal at 888 K (after Liebmann *et al.*, 1956).

3. RECRYSTALLIZATION OF TWO-PHASE ALLOYS

As most alloys of commercial importance are multiphase, an understanding of the recrystallization behavior of such materials is of practical as well as of scientific interest. Second-phase particles have three important effects on recrystallization:

1. Particles may increase the stored energy and hence the driving force for recrystallization.
2. Large particles may act as nucleation sites for recrystallization.
3. Particles, particularly if closely spaced, may exert a significant pinning effect on grain boundaries.

The first two effects tend to promote recrystallization, whereas the last effect tends to prevent recrystallization. Thus the recrystallization behavior, particularly the kinetics and the resulting grain size, will depend on which of these effects dominates. The final grain size tends to be small when recrystallization is accelerated and coarse when recrystallization is retarded. The recrystallization kinetics strongly depend on both the particle size and the interparticle spacing, as was first clearly demonstrated by Doherty and Martin (1962), and it is clear that by comparison with a single-phase alloy, recrystallization is retarded or even completely inhibited by closely spaced particles, and is accelerated by widely spaced particles, as shown in Fig. 11.

3.1 Particle-Stimulated Nucleation of Recrystallization

If the particles do not deform, then geometrically necessary dislocations are generated at the particles [Eq. (1)]. The form and distribution of these dislocations are primarily functions of strain and particle size, although other factors such as shape, interface strength, and matrix are known to be important (see, e.g., Humphreys, 1985). At particles larger than around 0.2 μm, plastic relaxation results in the formation of deformation zones near the particles. These are regions of high dislocation density that are misoriented from the matrix (Humphreys, 1977, 1979). Nucleation of recrystallization may occur

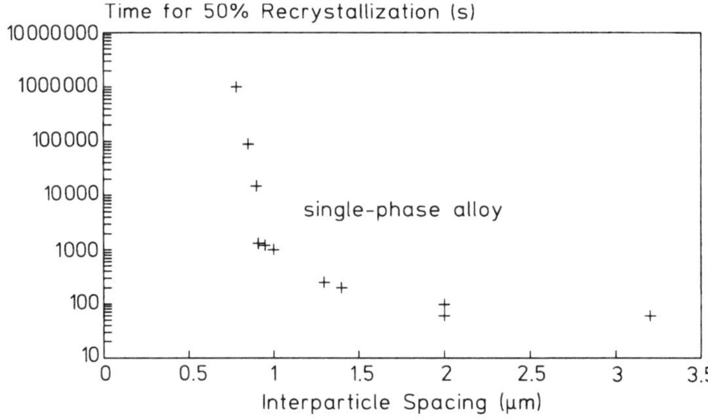

FIG. 11. The effect of interparticle spacing on the time for 50% recrystallization in Al-Cu single crystals (data from Doherty and Martin, 1964).

within these deformation zones and this is commonly referred to as particle-stimulated nucleation (PSN).

The characteristics of PSN are that nucleation occurs by rapid sub-boundary migration, and that the grain may stop growing when the deformation zone has been consumed. A sequence of electron micrographs from an *in situ* annealing sequence, Fig. 12, shows the occurrence of particle-stimulated nucleation in aluminum. There is no evidence that PSN occurs by a mechanism different from nucleation of recrystallization at heterogeneities in single-phase materials, and

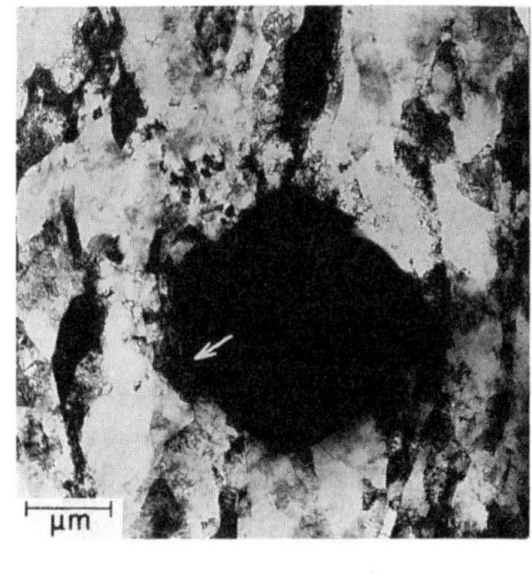

(a)

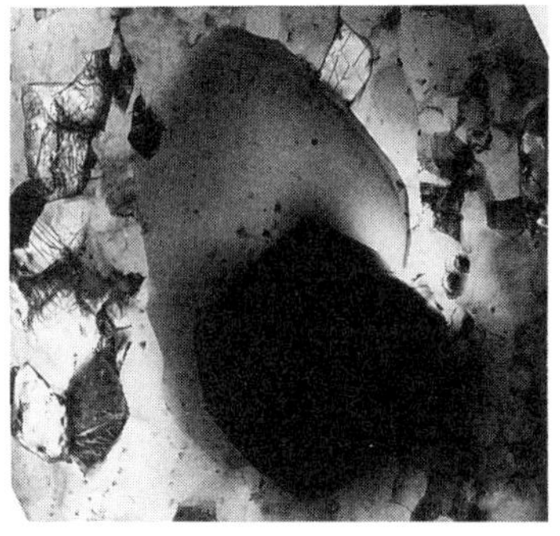

(b)

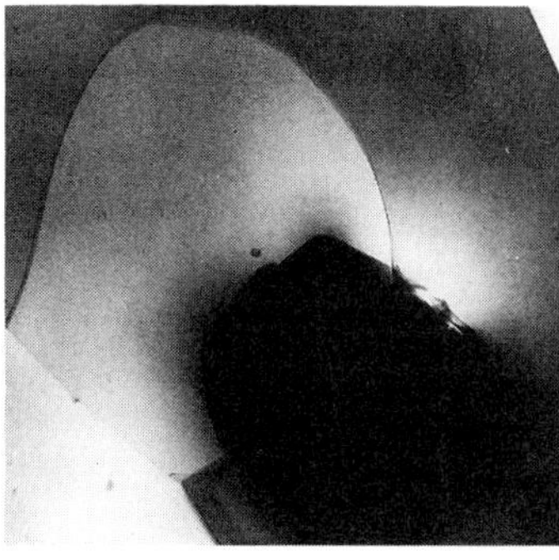

(c)

FIG. 12. *In situ* high-voltage electron microscope annealing of Al-Si. Recrystallization originated in the deformation zone (arrowed).

the models of PSN are consistent with the preformed-nucleus models discussed in Sec. 2.2.2.

The main parameters that determine whether or not PSN occurs are the strain and the particle size. This is shown in Fig. 13 for rolled aluminium containing Si particles. Two criteria must be fulfilled for growth of a nucleus beyond the deformation zone (Humphreys, 1977). First, a deformation zone with sufficient misorientation to create a high-angle boundary must be created on deformation, and second, the growing grain must also be able to grow into the surrounding matrix, which has a stored energy per unit volume of E. Humphreys showed that this condition approximated to

$$d_c = 4\sigma/E, \qquad (5)$$

where d_c is the critical particle diameter and σ is the grain-boundary energy.

It is difficult to predict values of E in deformed alloys accurately. However, if reasonable values are taken (Humphreys, 1977) or if experimental subgrain data are used, then this equation gives agreement with the results of Fig. 13, showing that the growth criterion is the critical one.

3.2 Pinning Effects of Particles (Zener Drag)

A dispersion of particles will exert a retarding force on a grain boundary. The effect is known as Zener drag after the original analysis by Zener (Smith, 1948).

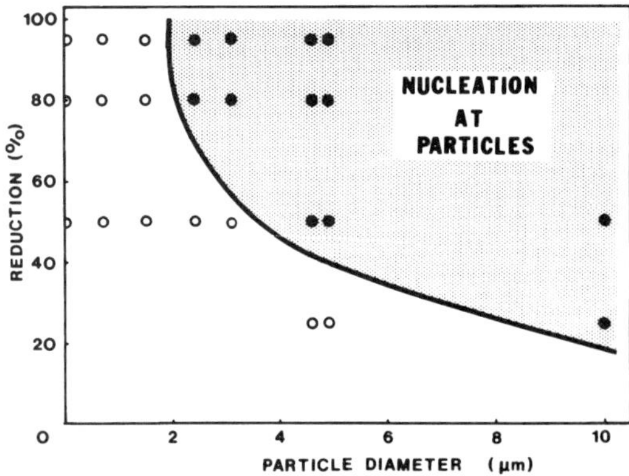

FIG. 13. The effect of rolling reduction and particle size on the occurrence of particle-stimulated nucleation (Humphreys, 1977).

When a grain boundary, of specific energy σ, migrates onto a spherical incoherent particle, when it can be assumed that the interfacial energy of the particle is unchanged by the passage of the grain boundary, then the particle, of radius r, effectively removes an area of the boundary equal to πr^2, and the energy is reduced by an amount $\pi r^2\sigma$. To move the boundary past the particle a force F must be applied, the maximum value of which is $\pi r\sigma$. If the boundary is planar and randomly intersects particles, then the pinning force exerted on the boundary in an alloy with a volume fraction F_v of particles of radius r is given by

$$p_z = 3F_v\sigma/2r. \qquad (6)$$

Coherent particles or nonspherical particles may be more effective in pinning boundaries. The Zener drag force has been examined by several authors, and the reader is referred to the reviews by Nes *et al.* (1985) and Hillert (1988) for further details. Particle pinning is undoubtedly the cause of retarded recrystallization in alloys containing closely spaced small particles, and plays a dominant role in grain growth after recrystallization (Sec. 4).

3.3 Bimodal Alloys and the Prediction of Grain Size

Many commercial alloys contain distributions of both large (>1 μm) particles, which will act as nucleation sites, and small particles, which will pin the migrating boundaries. In such a situation, the driving force is offset by the Zener pinning force and the critical particle size for nucleation [Eq. (5)] now becomes

$$d_c = \frac{4\sigma}{E - 3F_v\sigma/2r}. \qquad (7)$$

Thus, as the Zener pinning force increases, the critical particle diameter for PSN increases. As there will be a distribution of particle sizes in a real alloy, this means that fewer particles are able to act as nuclei, and that the recrystallized grain size will increase. The number of particles capable of acting as nuclei (N) is the number of particles of diameter greater than d_c. The grain size will be given approximately by

$$D_N = N^{-1/3}. \quad (8)$$

There is considerable interest in being able to predict the grain size of particle-containing alloys as a function of the particle parameters and the thermomechanical processing route, Nes (1976), Nes and Hutchinson (1989), and Wert and Austin (1988) have developed models, based on the mechanisms discussed above, for predicting grain size in commercial aluminum alloys.

3.4 The Interaction of Precipitation and Recrystallization

The mutual interaction of precipitation and recrystallization has been extensively studied by Hornbogen and colleagues in Al-Cu and Al-Fe alloys, and has been comprehensively reviewed by Hornbogen and Köster (1978). If a supersaturated solid solution is deformed and annealed, then unless recrystallization is complete before precipitation begins, the two processes affect each other. Precipitation on low-angle or high-angle boundaries will hinder recovery and recrystallization, and dislocations will themselves promote the nucleation of precipitates.

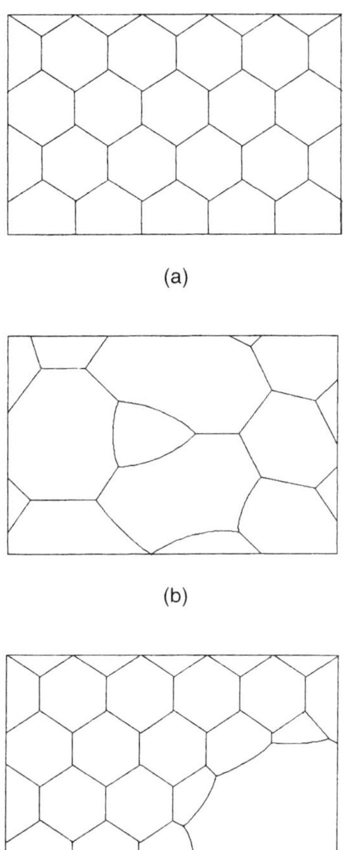

FIG. 14. Growth of a two-dimensional grain structure. (a) An array of regular hexagonal grains is stable. (b) In an irregular grain structure the boundaries are curved. (c) Secondary recrystallization.

4. GRAIN GROWTH AFTER PRIMARY RECRYSTALLIZATION

When primary recrystallization, which is driven by the stored energy of cold work, is complete, the structure is not yet stable, and further growth of the recrystallized grains may occur. The driving force for this is a reduction in the energy stored in the material in the form of grain boundaries. The driving pressure for grain growth is some two orders of magnitude less than that for primary recrystallization, and consequently, grain growth will be slower than during primary recrystallization and will be more affected by solutes and particles that pin grain boundaries. General reviews of grain growth have been undertaken by Higgins (1974), Doherty and Martin (1976), and Atkinson (1988).

Consider first a two-dimensional grain structure as shown in Fig. 14. If the grain boundaries are assumed to have equal energies, then the triple points are in equilibrium when the boundaries make an angle of 120° with each other. This is possible for an array of regular hexagonal grains [Fig. 14(a)] and in this case we have achieved a metastable structure that will not coarsen. However, if we take a more realistic, less regular array of grains as shown in Fig. 14(b), then the boundaries will become curved to achieve the required triple-point angles. Curved boundaries will be unstable and tend to migrate in the direction toward their centers of curvature, so as to shorten their lengths. The result is that grains of fewer than six sides will tend to shrink and eventually disappear, while those of more than six sides will tend to grow, and the average grain size therefore increases with time. Extension of these ideas to a three-dimensional grain structure leads to similar results except that a truly stable three-dimensional array, equivalent to Fig. 14(b), does not appear to exist.

The kinetics of grain growth are generally found to obey the relationship

$$D^2 - D_0^2 = Kt^n, \tag{9}$$

where D is grain diameter at time t and D_0 the grain size at $t = 0$. The constant n typically lies between 0.3 and 0.5.

4.1 Factors Affecting Grain Growth

Grain growth is inhibited by a number of factors, and the effect of solutes on boundary migration has already been discussed in Sec. 2.3.

The rate of grain growth diminishes when the grain size becomes greater than the thickness of a sheet specimen (Burke, 1949). This is because the grains are now curved only in one direction rather than two, and thus the driving force is diminished. Thermal etching grooves may also be formed on the sample surface by diffusion, and these will also impede grain growth.

Grain growth may also be affected by the presence of a sharp crystallographic texture (Beck and Sperry, 1949). This arises at least in part from a large number of grains of similar orientation leading to more low-angle and hence low-energy boundaries. Thus the driving force for growth is reduced. The texture may also alter during grain growth (Abbruzzese and Lücke, 1986), thereby affecting the kinetics.

Undoubtedly the most important features affecting grain growth are second-phase particles. During grain growth of a fully recrystallized alloy, the driving force is much smaller than during primary recrystallization. Therefore the pinning effects of particles on grain boundaries discussed in Sec. 3 are relatively more important. The pinning force (p_z) exerted on a boundary by an array of particles is given by Eq. (6). During grain growth, the driving force (P_R) arises solely from the curvature of the boundaries; the magnitude of this driving force is given by $2E/R$. On average, we expect the grains to stop growing at a diameter D_z (the limiting grain size or Zener limit) when $p_Z = p_R$:

$$D_z = 4r/3F_v. \tag{10}$$

The pinning effect of even very low volume fractions of small particles is sufficient to limit the grain size significantly. For example, a volume fraction of 5×10^{-3} of particles of diameter 50 nm gives $D_Z = 67$ μm. If the interparticle spacing is of a similar magnitude to the grain size then the particles will lie predominantly on the grain boundaries and the pinning force will be greater (Hellman and Hillert, 1975; Hutchinson and Duggan 1978).

4.2 Computer Simulation of Grain Growth

In a recent series of papers, Anderson and colleagues (e.g., Anderson *et al.*, 1984; Anderson, 1986) have described Monte Carlo simulation of grain growth in two and three dimensions. The material is divided into a number of discrete points each of which is given a number, corresponding to a grain orientation as shown in Fig. 15. The grain-boundary energy is then specified by the interaction between the grid sites. For example, a certain value of energy might be assigned to a 4–6 boundary and the same or a different one to a 3–9 boundary. Pairs of adjacent points are then swapped and the energy change (ΔE) measured. If the energy is lower, then there is a probability of

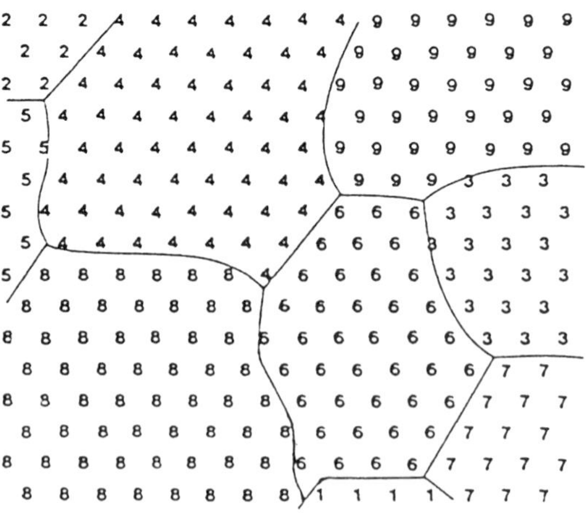

FIG. 15. The type of "microstructure" used for atomistic simulation of grain growth. The integers denote orientations and the lines are grain boundaries (Anderson, 1986).

$\exp(-\Delta E/kT)$ that the change will be accepted. Transitions occur at grain boundaries and the grains grow, showing most of the features of grain growth. If this technique is applied to three dimensions, the size of the array that can be used is limited to around $100 \times 100 \times 100$ points.

4.3 Secondary Recrystallization

In the previous section we discussed the relatively uniform growth of grains after recrystallization. However, in certain circumstances, a few grains, such as the large grain in Fig. 14(c), may grow excessively, consuming the recrystallized grains. This may lead to a grain structure of the order of several millimeters or greater. This process is known as secondary recrystallization. Detailed discussions of this phenomenon are given by Detert (1978) and Cahn (1983). The rapidly growing grains may be larger than the average or they may have an orientation which is more favorable for growth. If the latter is the case then a pronounced texture may result on secondary recrystallization. The best-known examples of this are the iron-silicon alloys used for transformer cores (e.g., Luborsky *et al.*, 1983). In these a pronounced (110)[001] texture (a {110} plane parallel to the rolling plane and an ⟨001⟩ direction parallel to the rolling direction) is produced on secondary recrystallization. The driving force for secondary recrystallization is the energy of the grain boundaries and therefore the process is more likely to occur when pinning of boundaries by second-phase particles restricts normal grain growth (Sec. 4.1). If, during a subsequent anneal, the particles coarsen or dissolve, then secondary recrystallization commonly occurs (Detert, 1978).

5. RECRYSTALLIZATION TEXTURES

When a metal is recrystallized, the crystallographic orientations of the new grains may be quite different from those of the old ones. As textures have a significant effect on the mechanical and physical properties of a material, the understanding of the origin of textures is of great importance. The published literature on the subject of textures is extensive, and there are several contentious issues that continue to arouse much debate. For further general information the reader is referred to the reviews of Hatherly and Dillamore (1975), Grewen and Huber (1978), Nes and Hutchinson (1989), and Hatherly (1990).

The texture after annealing is primarily determined by two factors: the orientations that are produced during nucleation, and the relative growth velocity of grains of different orientations. For many years there has been heated debate as to which of these two factors controls the final texture. It has been claimed that the recrystallization texture has its origin in either the preferred nucleation of grains with the required orientation (oriented nucleation theory) or the preferred growth of such grains from a more randomly oriented array (oriented growth theory).

Recent advances in techniques for determining local textures, which enable detailed correlation of microstructure and local texture, are now providing the types of information necessary for obtaining a better understanding of the formation of annealing textures. It is clear that there are some instances in which nuclei of a limited range of orientations are produced, and this factor then will predominantly determine the final texture. Similarly, there are cases when grains of certain orientations clearly grow more rapidly. In general we must expect the final texture to be determined by both factors in proportions that will vary for different materials, deformations, and annealing conditions.

5.1 Orientation Effects in Nucleation

As discussed in Sec. 2.2, a recrystallized grain originates in a small region of the deformed material and will therefore have the orientation of that region. Although this orientation is present in the deformed material, it may not be present in sufficient quantity to be detected by bulk texture measurements and therefore is unlikely in general to be an important component of the deformation texture. In Sec. 2.2 we discussed several different sites for recrystallization, and each of these may lead to one or more texture components.

5.1.1 Transition Bands A transition band separates different parts of an old grain which has split during deformation and where the individual new parts have rotated toward different, but stable, end orientations (Hu, 1962). A transition band therefore is a region with a sharp orientation gradient bridging the two neighboring texture components. Dillamore and Katoh (1974) predicted that in fcc metals, transition bands containing the cube orientation (001)[100] would develop. This was confirmed experimentally by Ridha and Hutchinson (1982) who showed that thin pancake-shaped regions of this orientation developed into grains of the cube orientation. The success of these nuclei in developing was undoubtedly enhanced by their shape. Similar cube nuclei have also been found in aluminum by Hjelen and Nes (1986), although only in transition bands within (112)[111] grains.

5.1.2 Shear Bands Shear bands are also important sites for recrystallization. However, it is more difficult to predict the resulting textures. For example in steels, it has been shown (Haratani *et al.*, 1984) that the Goss component (110)[001] originates at shear bands. However, in copper, the nuclei that form from shear bands are of widely spread orientations (Ridha and Hutchinson, 1982).

5.1.3 Prior Grain Boundaries Strain-induced grain-boundary migration, which is an important mechanism after lower amounts of deformation, results in texture components that are present in significant amounts in the deformed material. For example, the (001)[110] and (112)[110] orientations in steel are thought to originate in this way (Inokuti and Doherty, 1977).

5.1.4 Second-Phase Particles Nucleation in the vicinity of large second-phase particles gives rise to grains whose orientations reflect those in the deformation zones. In single crystals, where a very restricted range of orientations is produced, sharp textures may be formed on recrystallization (Humphreys, 1977). However, in polycrystals, the large spread of orientations available near the particles results in a severe weakening of the deformation texture with no new major components being formed (Humphreys and Juul Jensen, 1986; Humphreys and Kalu, 1990).

5.2 Orientation Effects in Growth

The effects of orientation on grain growth were discussed in Sec. 2.3. In summary, there is clear evidence, particularly from lightly deformed fcc single crystals, that the growth rate of grains is a maximum when there is a misorientation of ~40° about a ⟨111⟩ axis. For highly deformed polycrystals there is less information, and the situation is complicated because the orientation of the deformed material changes over short distances and so no precise orientation relationship exists between the old and new grains.

Another factor which must be taken into account is that the deformation structure is heterogeneous and different regions of the microstructure contain different components of the deformation texture. Therefore a model in which oriented growth is taken into account must consider the local nuclei and their environment. Microtexture work such as that by Hjelen *et al.* (1990) is now beginning to provide the information which will enable more complete theories of texture development to be formulated.

Twinning will provide new orientations, and may also be important in producing grains of suitable orientation for rapid growth. There is still some debate about the conditions under which annealing twinning is important in determining textures, particularly in metals of higher stacking fault energy. A review by Hatherly (1990) gives a good summary of the current situation.

5.3 Competition between Texture Components

The final texture is the result of the selection of orientations during nucleation and growth. In many materials several components are formed, originating from different microstructural sites. Nucleation and growth rates for the different components may also alter with time, making it therefore very difficult to predict the final texture.

As an example, we can consider the recrystallization of commercial-purity aluminum, which contains some large second-phase particles. The growth of various texture components was followed during the annealing process and the results are shown in Fig. 16. Initially, microstructural observation showed nucleation to be predominantly at particles (random component). However, during the anneal, these grains grew slowly, whereas grains of the rolling texture (formed at the old grain boundaries) grew more rapidly. Cube grains grew very rapidly, and the

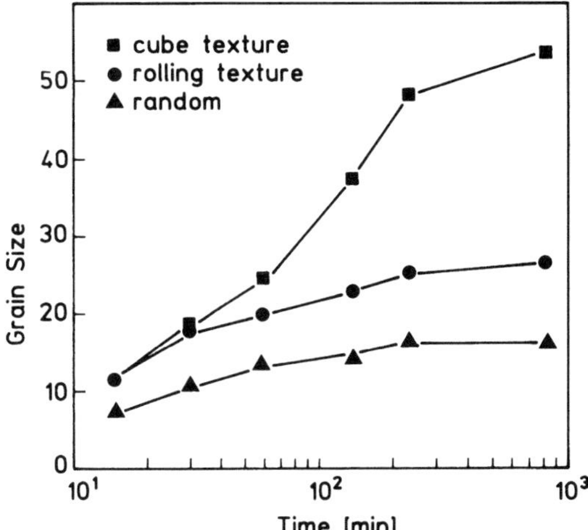

FIG. 16. The average size of grains with different orientations, measured for a series of partly recrystallized specimens of aluminium of commercial purity (Juul Jensen *et al.*, 1985).

final texture contained a high percentage of this component. This example illustrates the way in which both selection of orientations, different growth rates, and competition between the various components play a significant role in determining the final texture.

GLOSSARY

Cell: A small region, typically about one micrometer in diameter, surrounded by a tangled wall of dislocations.

Dislocation: A line defect in a crystal. Dislocations produced during plastic deformation provide the driving force for the various annealing phenomena.

Grain Growth: The growth of grains following recrystallization. The driving force for this process is a reduction in the energy of the grain boundaries.

Grains: Crystals comprising a polycrystalline aggregate.

High-Angle Grain Boundary: A grain boundary between grain misoriented by more than about 15°. In general the energy and mobility of such a boundary are independent of misorientation, although special or coincidence boundaries with low energies are found for specific misorientations.

Low-Angle Grain Boundary: A grain boundary between crystals misoriented by a small angle, typically taken to be 15°. A low-angle boundary is made up of an array of dislocations. The energy and mobility of these boundaries are dependent on the misorientation.

Particle-Stimulated Nucleation: Nucleation of recrystallization at large second-phase particles.

Recovery: The lowering of the stored energy by the annihilation and rearrangement of dislocations into low-energy configurations.

Recrystallization: The formation and growth of essentially dislocation-free grains within a deformed or recovered material.

Secondary Recrystallization: If grain growth is restricted, then a few grains may grow discontinuously thereby producing a very large grain size.

Stored Energy: The energy stored in a deformed metal, primarily in the form of dislocations.

Subgrain: Similar to a cell, but surrounded by low-angle boundaries. Recovery may convert a cell structure to a subgrain structure.

Texture: A term describing the crystallographic orientation of a polycrystalline material. Textures range from strong, if the grains have similar orientations, to weak if the grains are almost randomly oriented. A texture is produced both during deformation and during recrystallization.

Zener Drag: The pinning of boundaries by a distribution of small particles.

Works Cited

Abbruzzese, G., Lücke, K. (1976), *Acta Metall.* **34,** 905.

Anderson, M. (1986), in: N. Hansen (Ed.), *Proceedings of the 7th International Riso Symposium,* Riso, Denmark, p. 15.

Anderson, M. P., Grest, G. S., Srolowitz, D. J., Sahni, P. S. (1984), *Acta Metall.* **32,** 783.

Ashby, M. F. (1970), *Philos. Mag.* **21,** 399.

Atkinson, H. V. (1988), *Acta Metall.* **36,** 469.

Aust, K. T., Rutter, J. W. (1959), *Trans AIME* **215,** 119 and 608.

Avrami, M. J. (1939), *Chem. Phys.* **7,** 103.

Bailey, J. E., Hirsch, P. B. (1964), *Proc. R. Soc. London, Ser. A* **267,** 11.

Beck, P. A., Sperry, P. R. (1949), *Trans. AIME* **180,** 240.

Beck, P. A., Sperry, P. R. (1950), *J. Appl. Phys.* **21,** 150.

Bellier, S. P., Doherty, R. D. (1977), *Acta Metall.* **25,** 521.

Burke, J. E. (1949), *Trans AIME* **180,** 73.

Cahn, R. W. (1983), in: R. Cahn, N. Haasen (Eds.), *Physical Metallurgy,* 3rd ed., Amsterdam: North-Holland, 1595.

Detert, K. (1978), in: F. Haessner (Ed.), *Recrystallization of Metallic Materials,* Stuttgart: Dr. Riederer Verlag GmbH, 97.

Dillamore, I. L., Katoh, H. (1974), *Metal Sci.* **8,** 73.

Doherty, R. D., Martin, J. W. (1962), *J. Inst. Met.* **91,** 332.

Doherty, R. D., Martin, J. W. (1964), *Trans. ASM* **57,** 874.

Doherty, R. D., Martin, J. W. (1976), Stability of Microstructure in Metallic Systems, Cambridge, U.K.: Cambridge University Press.

Doherty, R. D., Rollett, A. R., Srolovitz, D. J. (1986), in: N. Hansen (Ed.), *Proceedings of the 7th International Riso Symposium,* Riso, Denmark, p. 53.

Duggan, B. J., Hutchinson, W. B., Hatherly, M. (1978), *Scripta Metall.* **12,** 1293.

Furu, T., Marthinsen, Nes, E. (1990), *Mater. Sci. Tech.* **6,** 1093.

Grant, E., Porter, A., Ralph, B. (1984), *J. Mater. Sci.* **19,** 3554.

Grewen, J., Huber, J. (1978), in: F. Haessner (Ed.), *Recrystallization of Metallic Materials,* Stuttgart: Dr. Riederer Verlag GmbH, p. 111.

Haessner, F., Hofmann, S. (1978), in: F. Haessner (Ed.), *Recrystallization of Metallic Materials,* Stuttgart: Dr. Riederer Verlag GmbH, p. 63.

Haratani, T., Hutchinson, W. B., Dillamore, I. L., Bate, P. (1974), *Metal Sci.* **18,** 57.

Hatherly, M. (1982) in: R. Gifkins (Ed.), *Proceedings of the 6th International Conference on Strength of Metals and Alloys,* Pergamon: Melbourne.

Hatherly, M. (1990), in: T. Chandra (Ed.), *Recrystallization '90,* Warrendale, OH: The Metals Society, p. 59.

Hatherly, M., Dillamore, I. L. (1975), *J. Australian Inst. Met.* **20,** 71.

Hellman, P., Hillert, M. (1975), *Scand. J. Metall.* **4,** 211.

Higgins, G. T. (1974), *Met Sci. J.* **8,** 143.

Hillert, M. (1988), *Acta Metall.* **36,** 3177.

Hjelen, J., Nes, E. (1986), in: N. Hansen (Ed.), *Proceedings of the 7th International Riso Symposium,* Riso, Denmark, p. 367.

Hjelen, J., Orsund, R., Nes, E. (1990), *Acta Metall.* **39,** 1377.

Hornbogen, E., Köster, U. (1978), in: F. Haessner (Ed.), *Recrystallization of Metallic Materials,* Stuttgart: Dr. Riederer Verlag GmbH, p. 159.

Hu, H. (1962), *Trans. Met. Soc. AIME* **224,** 75.

Humphreys, F. J. (1977), *Acta Metall.* **25,** 1323.

Humphreys, F. J. (1979), *Acta Metall.* **27,** 1801.

Humphreys, F. J. (1985), in: M. Loretto (Ed.), *Dislocations and Properties of Real Materials,* London: Institute of Metals, 175.

Humphreys, F. J., Juul Jensen, D. (1986), in: N. Hansen (Ed.), *Proceedings of the 7th International Riso Symposium,* Riso, Denmark, p. 93.

Humphreys, F. J., Kalu, P. N. (1990), *Acta Metall.* **38,** 917.

Hutchinson, W. B., Duggan, B. J. (1978), *Metal Sci.* **12,** 372.

Inokuti, Y., Doherty, R. D. (1977), *Texture* **2,** 143.

Johnson, W. A., Mehl, R. F. (1939), *Trans. AIME* **135,** 416.

Juul Jensen, D., Hansen, N., Humphreys, F. J. (1985), *Acta Metall.* **33,** 2155.

Li, J. C. M. (1966), in: H. Margolin (Ed.), *Recrystallization Grain Growth & Textures,* Warrendale, OH: The Metals Society, p. 45.

Liebmann, B., Lücke, K, Masing, G. (1956), *Z. Metallk.* **47,** 57.

Luborsky, F. E., Livingston, J. D., Chin, G. Y. (1983), in: R. Cahn, P. Haasen (Eds.), *Physical Metallurgy,* Amsterdam: North-Holland, 1698.

Lücke, K., Detert, K. (1957), *Acta Metall.* **5,** 628.

Nes, E. (1976), *Acta Metall.* **24,** 391.

Nes, E., Hutchinson, W. B. (1989), in: Bilde-Sorensen (Ed.), *Proceedings of the 10th International Riso Symposium,* Riso, Denmark, p. 233.

Nes, E., Ryum, N., Hunderi, O. (1985), *Acta Metall.* **33,** 11.

Ridha, A. A., Hutchinson, W. B. (1982), *Acta Metall.* **30,** 1929.

Ryde, L., Hutchinson, W. B., Jonsson, S. (1990), in: T. Chandra (Ed.), *Recrystallization '90,* Warrendale, OH: The Metals Society, p. 313.

Schmidt, J., Haessner, F. (1990), *Cond. Matter* **81,** 215.

Saetre, T. O., Hunderi, O., Nes, E. (1986), *Acta Metall.* **34,** 981.

Shewmon, P. G. (1969), *Transformations in Metals,* New York: McGraw-Hill.

Smith, C. S. (1948), *Trans. AIME* **175,** 47.

Wert, J. A., Austin, L. K. (1988), *Metall. Trans. A* **19,** 617.

Further Reading

Byrne, J. G. (1965), *Recovery, Recrystallization and Grain Growth,* New York: McMillan.

Cahn, R. W. (1983), in: R. Cahn, P. Haasen (Eds.), *Physical Metallurgy,* 3rd ed., Amsterdam: Elsevier Science Publishers, 1595.

Cotterill, P., Mould, P. R. (1976), *Recrystalliza-*

tion and Grain Growth in Metals, Surrey, U.K.: Surrey Univ. Press.

Doherty, R. D., Martin, J. W. (1976), *Stability of Microstructure in Metallic Systems,* Cambridge, U.K.: Cambridge University Press.

Haessner, F. (Ed) (1978), *Recrystallization of Metallic Materials,* Stuttgart: Dr. Riederer Verlag GmbH, p. 111.

Humphreys, F. J., Hatherly, M. (1995), *Recrystallization and Related Annealing Phenomena,* Oxford, U.K.: Pergamon.

RECYCLING

See WASTE MANAGEMENT AND RECYCLING

REFLECTOMETERS, OPTICAL

C. S. McCamy, *Wappingers Falls, New York, U.S.A.*

	Introduction	247
1.	**Physical and Technical Principles**	247
1.1	Basic Components of Reflectometers	247
1.2	Conditions of Measurement	247
1.3	Physical Quantities Measured	248
2.	**Design, Function, and Operation**	248
2.1	Geometric Design	248
2.2	Spectral Design	249
2.3	Light Sources	250
2.4	Sensors	250
2.5	Standards	250
2.6	Quality of Measurements	250
2.7	Operation	251
3.	**Manufacture**	251
4.	**Applications**	251
4.1	Chemical Analysis	251
4.2	Densitometers	251
4.3	Colorimetry	252
4.4	Gloss, Distinctness of Image, and Haze	253
5.	**Economic Aspects**	253
	Glossary	253
	Works Cited	253
	Further Reading	253

INTRODUCTION

Reflectometers measure light (or other electromagnetic radiant flux) reflected from a surface, as a percentage of the flux incident on the surface or as a percentage of the flux reflected by a highly reflecting reference standard. They are used to characterize materials and to control the manufacture of colored products.

1. PHYSICAL AND TECHNICAL PRINCIPLES

1.1 Basic Components of Reflectometers

All reflectometers have the same basic components, though they differ widely in details of design, depending on the nature of the specimens to be measured and the intended use of the data. An illuminator (or irradiator in the case of radiant flux outside the visible spectrum) provides incident radiant flux, simply called the influx. The illuminator consists of a source, which radiates flux, and a director, which directs the flux to a sampling aperture, where the specimen is placed for measurement. The source may be a lamp filament, for example. The director might simply be the lamp envelope, but more often it includes various optical components called the influx optics.

Light reflected by the specimen is collected and measured by a receiver. The receiver consists of a collector, which collects flux reflected by the specimen, and a sensor, which senses the collected flux and produces a corresponding signal. The collector usually includes optical components, comprising the efflux optics. The sensor requires supporting electronics and data display. Usually, a separate or built-in computer records and processes data and interprets them in useful terms.

1.2 Conditions of Measurement

Values obtained by reflectometry depend on the angles at which the surface is illuminated and the angles at which light is collected for measurement. These angular distributions are called the geometry or geometric conditions of measurement. The values also depend on the spectral distribution of the incident light and the spectral sensitivity of the receiver, known as the spectral conditions. The geometric and spectral conditions characterize the measurement system and characterize applications, includ-

3-527-28138-X/96/$5.00 + .50

ing visual observation. To be useful, measurements must be made under conditions appropriate to the way the data will be applied. Thus the principle of simulation: The geometric and spectral conditions of measurement must simulate the geometric and spectral conditions of the application.

1.3 Physical Quantities Measured

Reflectance is the ratio of the amount of reflected flux to the amount of incident flux, under prescribed geometric and spectral conditions. Reflectance factor is the ratio of the amount of reflected flux to the amount of flux reflected by a reflection standard, under the same geometric and spectral conditions. These quantities are usually quite different. For example, white paper may reflect over 80% of incident light and absorb the rest. The eye collects a very small fraction of the light reflected in all directions, so the reflectance is very small. However, the reflectance factor (ratio of reflected light to that reflected by a white standard) might exceed 80%. If the flux reflected in all directions is collected, the adjective "total" is applied. The application dictates which quantity is appropriate. A quantity measured within a narrow wavelength band is a spectral quantity, such as spectral reflectance. Concepts, terminology, symbols, notation, and methods of describing geometry delineated by McCamy (1966) have been standardized by the American Society for Testing Materials (ASTM), the International Organization for Standardization (ISO), and the Commission Internationale de l'Éclairage (CIE).

Reflectivity is the reflectance or the reflectance factor of a generic material, such as polished gold. The reflectance factor of paper, plastics, and textiles may depend on reflections from the surface behind them. In this case, reflectivity is the reflectance factor of a specimen or stack of specimens so thick that greater thickness does not change the value.

Radiance is the radiant flux emanating from a surface, in a given direction, per unit of projected area, per unit of solid angle. Radiance factor is the ratio of the radiance from a point on a specimen to that from an ideal diffuser, under the same conditions. This quantity is the same as reflectance factor, if only reflection is involved, but if the material is fluorescent, phosphorescent, or otherwise self-luminous, the flux emanated may be partly reflected and partly emitted. Both are included in radiance factor.

2. DESIGN, FUNCTION, AND OPERATION

2.1 Geometric Design

Geometry is designed with regard for two different kinds of reflection. Incident light may be partly reflected by the first surface encountered, but some may penetrate the surface and be reflected by underlying matter. Reflection at the first surface of a glossy material is called specular reflection. The reflected light is the color of the incident light. If the incident light is white, the reflected light is white. Such highlight reflections are seen on glossy ceramics, plastics, and paints. Light that penetrates the surface is scattered and diffusely reflected by underlying pigments or other colorants and has the color imparted by them. Light reflected by a metal surface has a color characteristic of the metal.

When we examine the color of a nonmetal surface, we avoid any specular reflection of the light source. If a lamp is 45° off the normal to a surface and we observe along the normal, the specular reflection is avoided. This is one standard geometry for measurements. The colors of some materials, such as wood and some textiles, observed this way, vary if they are rotated in their own plane. This variation is eliminated by illuminating at 45°, from all azimuthal directions. Such illumination is called annular.

In another standard form of illumination, hemispherical illumination, the specimen is uniformly illuminated from nearly all directions by an integrating sphere. A hollow sphere, usually around 150 mm in diameter, is painted matte white inside. Light is introduced through an entrance port. Multiple diffuse reflections cause the whole sphere wall to be uniformly bright. A specimen placed at the sample port is diffusely illuminated. The light reflected by the specimen is measured through the exit port, a few degrees off the specimen normal. White baffles in the sphere block direct rays from the entrance port to the sample port and from the entrance port to the exit port.

Such a measurement includes light specularly reflected at the surface of a glossy specimen. To avoid including the specular component, a specular port is made in the sphere at the same angle to the normal as the exit port, but on the opposite side. A glossy specimen reflects that dark open port to the exit port, and so no light is specularly reflected to the exit port. The specular component is often excluded from measurements of glossy nonmetal specimens, because it characterizes the source, not the specimen. If desired, a white cover is placed on the specular port to include the specular component. If the specimen has a matte surface, light is reflected at the first surface in all directions and cannot be separated and excluded.

The measurement of the color of metals and the evaluation of gloss require measuring reflection in the specular direction. The illuminator and receiver are placed symmetrically about the specimen normal. Specular reflection depends on the angle of incidence, and so that angle is critical.

The sampling aperture is the area on which measurements are made. It may be limited by the influx or the efflux optics. The influx sampling aperture and efflux sampling aperture are usually not the same size. That avoids misalignment errors. Furthermore, many specimens scatter light internally, so that it is necessary to illuminate a small area and collect flux from a larger area, to include light re-emitted outside the influx sampling aperture. Alternatively, one can illuminate a large area and collect flux from a small one.

Any ray of light that passes through an optical system can pass through the same system in the opposite direction. Though this principle of reversibility pertains to rays, it is applied to the design of reflectometers and is generally valid. Illumination at 45° and collection at 0° gives the same results as illumination at 0° and collection at 45°. Likewise, the direction of rays in sphere geometry may be reversed.

2.2 Spectral Design

The spectral conditions of measurement or application are specified in terms of the spectral power distribution of the illuminator and the spectral sensitivity of the receiver, which are called the influx spectrum and efflux spectrum, respectively. These spectra are specified relative to their maximum values. The influx spectrum depends on the spectral power distribution of the source and the spectral effect of the director. The efflux spectrum depends on the spectral sensitivity of the sensor and the spectral effect of the collector. The product, at each wavelength, of the influx spectrum and the efflux spectrum is called the spectral product of the instrument or application.

A spectrophotometer measures reflected flux as a function of wavelength. The wavelength of the light used for the measurement is selected by filters, which may be dyed gelatin, glass, or interference types, or by a dispersive system. A dispersive system employs a diffraction grating or prism to disperse light into a spectrum, so that a desired wavelength can be isolated. Such an optical system is called a monochromator. Gratings are generally preferred to prisms, because they disperse a spectrum with wavelength variation proportional to distance. A monochromator may be placed in the illuminator, to illuminate the specimen with monochromatic light, or in the receiver, to sense a narrow wavelength band. In the absence of fluorescence, either can be used.

Fluorescence is the absorption of radiant energy of a given wavelength and emission of part of that energy at a longer wavelength. Characterization of the excitation spectrum and emission spectrum of a fluorescent specimen requires a monochromator in both the illuminator and receiver. Fluorescent dyes, called fluorescent whitening agents, are added to white paper, plastics, and textiles, to absorb ultraviolet energy and emit blue light, counteracting the undesirable tendency of such materials to look yellowish.

To measure the appearance of a specimen that may be fluorescent, it must be illuminated with simulated daylight or other illuminant of interest, including the ultraviolet component. Light is reflected and emitted by the specimen. The monochromator must be placed in the receiver, so that both components are measured at the designated wavelength.

To measure how light a surface appears to the eye, the reflectance factor must be spectrally weighted by the sensitivity of the human eye. Filters may be used to simulate that sensitivity, or the measured spectral reflectance factor may be mathematically integrated with respect to the standard spectral luminous efficiency function. Measured quantities of this kind are distinguished by the adjective "luminous." We differentiate radiant flux, which may have any spectral distribution, from luminous flux, which is integrated as described, and we speak of luminous reflectance. These quantities are called photometric quantities.

2.3 Light Sources

Incandescent lamps are inexpensive, simple to operate, and provide a smooth spectral power distribution over the whole visible spectrum, but are deficient in ultraviolet power. Xenon-arc lamps are powerful continuous sources with ample ultraviolet. Xenon-flash lamps, introduced in the mid-1970s, provide enough visible and ultraviolet power to simulate daylight and, having a flash duration of about 50 μs, they may be used to measure moving materials, such as paper on a web press. The momentary illumination is so high that, if measurement is made only during the flash, extraneous light, including full sunlight, on the specimen does not interfere. Reflecting properties may depend on temperature, a phenomenon known as thermochromism, and the use of a flash source avoids heating the specimen before measurement. Photodiodes are used in simple inexpensive instruments.

2.4 Sensors

Thermosensitive devices, such as thermocouples, thermopiles, and bolometers, were used as sensors before photocells were invented. Photocells were replaced by phototubes and photomultipliers and, more recently, by solid-state sensors, such as silicon diodes. An array of silicon diodes can record a number of spectral intervals simultaneously.

2.5 Standards

A documentary standard is a document recording national or international agreements on terminology, symbols, physical quantities to be measured, or methods of measurement. Standards are prepared by experts in the field and are reviewed periodically. Standard symbols and terminology facilitate concise and unambiguous communication. Standard methods of measurement assure that values obtained in different places or at different times have the same meaning. Standards on reflectometry are published by ASTM, the American National Standards Institute (ANSI), ISO, and CIE.

A physical standard is a material object, having a physical property, the quantity of which is known, used as the basis of calibrating a measuring instrument. The true value of the physical quantity is assigned by measurement in a standardizing laboratory.

Reflectance factor depends on which standard is chosen. By international agreement, unless otherwise specified, the reflection standard is taken to be an ideal diffuse reflecting surface. That condition is sometimes part of the definition of the term "reflectance factor." No ideal surface actually exists. Working standards are calibrated by standards laboratories, so that measured values can be reported as though they were obtained with an ideal standard. The first widely used white standard was calcium carbonate (chalk), then came magnesium oxide condensed from the fumes of burning magnesium, and then pure barium-sulfate powder pressed into a flat plaque. These have generally been replaced by pressed plaques of powdered polytetrafluoroethylene, sold under the registered trademark "Halon." Such standards are used to transfer primary calibrations. White ceramic tiles are generally used as working standards. For measurement of specular reflection, polished black glass or polished or deposited pure metal surfaces are used.

2.6 Quality of Measurements

The quality of measurements depends on the degree to which the method of measurement meets the standard specifications, on the precision, and on the accuracy. Precision is the degree to which repeated measurements agree, based on statistical analysis of a number of measurements. Accuracy is the closeness of the measured value to the true value, assessed by comparing measured values with those assigned by a standards laboratory. Inaccuracy is generally categorized as failure to meet geometric or spectral specifications (design inaccuracy), failure to indicate values linearly and equal to true values (photometric inaccuracy), or failure to isolate the desired spectral wavelength (wavelength inaccuracy).

2.7 Operation

Reflectometers vary so widely that only a few basic operating rules can be given. Samples must be representative and be prepared and presented for measurement in a standard way. Ideally, specimens should be plane. Options, such as inclusion or exclusion of the specular component, the spectral range, the source, and various standard computational parameters, must be chosen carefully and recorded with measured values. Instruments must be operated within the required temperature range. Some have sensors to detect temperature out of range. Some instruments require calibration with a working standard for each measurement; others require calibration a few times a day. Regularly scheduled battery checks and cleaning of optics are recommended. (Integrating spheres may be degraded by atmospheric pollutants, such as tobacco smoke.) Standard samples are measured to assure the accuracy of measurements and repeated measurements are statistically analyzed. Statistical process control is applied to this measurement process.

3. MANUFACTURE

Manufacture of reflectometers requires careful design, conformance to documentary standards, precise part manufacture (or specification and purchase), and precise assembly. Tasks are of five kinds: mechanical, optical, electronic, computational, and, where analysis is the objective, chemical. Since their introduction, personal computers have been used for computations and data storage. Portable instruments have built-in computers. Most instrument manufacturers assemble components made by firms specializing in certain components. Holographic gratings and sensor arrays have greatly reduced costs.

4. APPLICATIONS

4.1 Chemical Analysis

Infrared reflection spectra correlate with molecular structure, and so many instruments are made to perform chemical analyses. Some are general-purpose laboratory spectrophotometers; others are special-purpose field instruments to measure, for example, the moisture content of grain or the fat content of meat. Many analyses are performed by placing a specimen of blood or urine on treated paper and measuring the spectral reflectance factor of the resulting spot. Simple pocket-sized instruments are used by patients to monitor blood sugar or detect pregnancy. In chemistry, the term "colorimeter" refers to a device, sometimes a simple visual device, to compare colors of chemicals, as part of an analytical procedure.

4.2 Densitometers

Reflection densitometers are used to control photographic and color printing processes. Measurements may be made on images or on control patches printed for the purpose. A "measuring head" includes the illuminator and receiver. There are bench-type instruments, hand-held instruments, and "on-line" instruments mounted on printing presses, to measure during a run. Microdensitometers measure density on a microscopic area, to scan and quantify images and analyze image structure (Dainty and Shaw, 1974; Shaw, 1977; Swing, 1995). Measurements on very small areas involve the optical theory of partial coherence (Born and Wolf, 1965).

The standard geometry for densitometry is 45° annular. Reflectance factor is measured, but a derived quantity, reflection density, is indicated. Reflection density is the logarithm, to base ten, of the reciprocal of the reflectance factor. It increases with the amount of dye or ink present. The yellow, magenta, and cyan dyes or inks are measured in the spectral region of maximum absorption, with narrow-band blue, green, and red light, respectively. Densitometry is used to control amounts of colorants, but the colors produced depend on the colorants, and so densitometry is not colorimetry. Densitometry is standardized by ANSI and ISO.

4.3 Colorimetry

Color is a psychophysical quantity; that is, it depends not only on the spectral quality of the light source and the spectral reflectance factor of the specimen, but also on the nature of the human visual system. Data specifying the normal human observer and methods of color measurement have been standardized by the CIE and ASTM.

The spectrum of light reflected by a surface depends on the spectrum of the illumination, and so the CIE has standardized the spectra of common illuminants. An incandescent lamp, some fluorescent lamps, and daylight have been standardized. Phases of daylight are identified by correlated color temperature. (The correlated color temperature of a light is the temperature in kelvins of a thermal radiator having the same color.) Sunlight, average daylight, and north-sky daylight have correlated color temperatures of 5500, 6500, and 7500 K, respectively, and are given the standard symbols D55, D65, and D75.

CIE standardized the nature of normal human color vision, for use in colorimetry, by adopting three color-mixture functions, which are the amounts of CIE primary colored lights, in a mixture of three primaries, required to match the color of light of a given wavelength.

Usually, a spectrophotometer is used to measure the spectral reflectance factor. That quantity must be mathematically integrated with respect to a chosen illuminant and the three CIE color-mixture functions to determine tristimulus values, the amounts of the three primaries required to match the color of the given surface.

In a filter colorimeter, the specimen is illuminated by a simulated standard illuminant and reflected light is measured with sensors fitted with filters designed to simulate the three color-mixture functions, so that tristimulus values are read directly. These instruments are simpler and less expensive than a spectrophotometer, but less stable, precise, and accurate. In color technology, the term "colorimeter" usually refers to this type of instrument.

The tristimulus values are mathematically transformed in many ways to correlate with colors and color differences as they are perceived (Wyszecki and Stiles, 1982). The derived quantities are used to specify colors and color differences, compute colorant formulations to produce desired colors, correct off-color batches, and control processes that produce colors.

Color-measuring instruments may be bench type, hand-held, or on-line. Bench-type instruments are in the form of a box with an aperture on which the specimen is placed for measurement. A personal computer controls the operation of the spectrophotometer, records data, performs computations, displays results, and communicates with process-control instruments. Hand-held instruments perform the same kinds of functions, but with limited capacity. On-line instruments mounted on production machines must be resistant to vibration, moisture, heat, and chemical fumes. They are used to control the paper-making process without human intervention, a system known as closed-loop process control.

The colors of metallic automobile paints are characterized by measuring color with three different geometries. The specimen is illuminated at 45° to the normal and reflected light is measured at three widely different angles. Hand-held multiangle spectrophotometers have come into widespread use in the automotive industry. This is an example of "gonioreflectometry," the measurement of reflection as a function of angles. Standardization of such measurements is addressed by ASTM. The luster of textiles can be characterized by gonioreflectometry.

Color may be indicative of some aspect of quality, even when color itself is of little or no concern. For example, the measured color of toasted corn flakes indicates how well done they are and predicts the flavor. Sheet materials to package milk and juices for shelf storage have several layers to provide strength, retain contents, prevent influx of oxygen and other gases, block ink odors, etc. They are coextruded and to measure the thickness of each layer, the methodology of quantitative chemical analysis by infrared reflectometry is used. Very thin transparent films present interference colors, and the thickness may be gauged by colorimetry.

4.4 Gloss, Distinctness of Image, and Haze

Gloss is evaluated on a 100-point scale by measuring the specular reflectance factor, relative to highly polished black glass of a specified refractive index. The angle of incidence may be 20°, 30°, 45°, 60°, 75°, or 85°. The instrument is called a gloss meter. The spectral product must be the product of the luminous efficiency function and the spectrum of daylight. Some instruments have sensors to assure precise alignment with the specimen. Gloss meters are widely used in the paint, plastics, ceramics, and paper industries. Illuminating at 30° and receiving at 30° is called 30° gloss, receiving at 30.3° evaluates distinctness of image (DOI), and receiving at 32° or 35° evaluates 2° haze or 5° haze. Methods are standardized by ASTM.

5. ECONOMIC ASPECTS

All industries concerned with appearance depend on reflectometry to quantify and specify appearance, predict colorant formulations, control production processes, and assure quality. These industries include photography, printing, paints, plastics, and textiles, as well as all other industries, such as the automobile, clothing, housing, packaging, and advertising industries, that are supplied by these industries. Appearance is the most obvious aspect of quality perceived by the potential buyer of most products. Appearance may be a crucial factor in the decision to buy, and a variety of colors or finishes promotes sales. Reflectometry and formulation software put custom color formulation in neighborhood paint stores. Chemical analysis by reflectometry takes a small fraction of the time and effort required by gravimetric wet chemistry, and thickness measurements, easily made by reflectometry, are virtually impossible by any other nondestructive means.

GLOSSARY

Aperture: An opening permitting the flow of radiant energy.

Appearance: The way a surface looks, as characterized by color, gloss, texture, grain, etc.

Emission Spectrum: In fluorescence theory, the spectral power distribution of flux emitted by a material as a result of fluorescence.

Excitation Spectrum: The spectral distribution of energy that causes a fluorescent material to absorb energy and emit at a longer wavelength.

Filter: An optical window that transmits a selected spectrum.

Flux: The time rate of flow radiant energy.

Port: An opening in an integrating sphere.

Specimen: A sample of a class of objects, chosen to be measured.

Spectral: Pertaining to a particular wavelength, a narrow band of wavelengths, or a distribution of radiant power with respect to wavelength.

Spectrum: The spatial arrangement of components of radiant power in order of wavelength.

Specular: Reflection, as by a mirror, with the angle of reflection equal to the angle of incidence.

Wavelength: The distance in the direction of propagation between nearest maxima of an electromagnetic wave, such as a light wave. In reflectometry, the nanometer is the unit most often used.

Works Cited

Born, M., Wolf, E. (1965), *Principles of Optics*, New York: Pergamon.

Dainty, J. C., Shaw, R. (1974), *Image Science*, New York: Academic.

McCamy, C. S. (1966), *Photogr. Sci. Eng.* **10**, 314–325.

Shaw, R. (1977), *Image Analysis and Evaluation*, Washington, DC: Society of Photographic Scientists and Engineers.

Swing, R. E. (1995), *Selected Papers on Microdensitometry*, Bellingham, WA: SPIE Optical Engineering Press.

Wyszecki, G., Stiles, W. S. (1982), *Color Science: Concepts and Methods, Quantitative Data and Formulae*, 2nd ed., New York: Wiley.

Further Reading

Andrews, D. L. (1990), *Current Topics in Modern Chemical Spectroscopy*, New York: Springer-Verlag.

Burgess, C., Mielenz, K. D. (Eds.) (1987), *Ad-*

vances in Standards and Methodology in Spectrophotometry, New York: Elsevier.

Grum, F., Bartleson, C. J. (Eds.) (1980), *Optical Radiation Measurements,* Vol. 2, *Color Measurement,* New York: Academic.

Hunter, R. S., Harold, R. W. (1987), *The Measurement of Appearance,* 2nd ed., New York: Wiley.

Stearns, E. I. (1969), *The Practice of Absorption Spectrophotometry,* New York: Wiley.

REFRACTOMETERS, OPTICAL

W. Nebe, *Carl Zeiss Jena GmbH, Jena, Germany*

	Introduction	255
1.	**Physical Foundations**	256
1.1	Refraction of Light	256
1.2	Prism Combinations	258
1.3	Critical-Angle Refractometry	258
1.4	Intensity Relations	258
1.5	Interference of Light	259
1.6	Refractive-Index Gradients	260
2.	**Refractometers**	261
2.1	Deviation-Measuring Refractometers	261
2.1.1	Spectrometer	261
2.1.2	Autocollimation Spectrometer	261
2.1.3	Double-Prism Refractometer	261
2.1.4	Vee-Prism Refractometer	263
2.2	Critical-Angle Refractometers	263
2.2.1	Abbe Refractometer	263
2.2.2	Pulfrich Refractometer	263
2.2.3	Dipping Refractometer	264
2.2.4	Hand Refractometer	265
2.3	Automatic Refractometers	265
2.3.1	Deviation-Measuring Automatic Refractometers	265
2.3.2	Intensity-Measuring Refractometers	265
2.3.3	Integrated-Optics Refractometers	266
2.4	Interference Refractometers	266
2.4.1	Rayleigh–Loewe Interferometer	266
2.4.2	Jamin Interferometer	267
2.4.3	Photoelectric Interferometers	268
2.5	Refractive-Index-Gradient–Measuring Refractometers	268
2.5.1	Schlieren Apparatus	268
2.5.2	Interferometers	269
3.	**Refractometry of Homogeneous Substances**	269
3.1	Glasses and Other Solids	269
3.2	Solutions and Binary Mixtures of Liquids	270
3.3	Ternary Mixtures of Liquids	271
3.4	Continuously Varying Liquids	272
3.5	Binary Mixtures of Gases	272
3.6	Ternary Mixtures of Gases	273
4.	**Refractometry of Inhomogeneous Substances**	273
4.1	Transparent Solids	274
4.2	Schlieren in Liquids	274
4.3	Stratified Solutions	274
4.4	Inhomogeneous Gases	274
	Glossary	275
	Works Cited	275
	Further Reading	276

INTRODUCTION

Refractometry means the investigation of substances by measuring their refractive index, which is a characteristic property of any homogeneous substance, and by observing variations of refractive index in inhomogeneous substances.

After the fundamental discoveries made by Snell (1591–1626) and Huygens (1629–1695)—the law of refraction and the nature of light as a wave phenomenon, respectively—the first purpose-oriented measurements of refractive index can be dated to the beginning of the 19th century. Fraunhofer (1787–1826) determined the refractive indices of glasses with the aid of the light deflection by a prism, and Arago (1786–1853) was able to ascertain the refractive index of air by means of interference methods. The decisive impulse for developing the rather sporadic measurements into a standard analytical method, known as refractometry, was Abbe's (1840–1905) idea to use the refractive index of a solution as a measure of its concentration (1869). Subsequently, the company of Carl Zeiss in Jena, Germany, developed various refractometers tailored for specific measurement of solids, liquids, or gases. Technical advancements such as the development of powerful spectral light

3-527-28138-X/96/$5.00 + .50

sources and the use of new glass types improved the precision of the instruments and expanded the scope of their application.

Since the 1950s, refractometry has been significantly boosted by the development of electronics. Whereas before, observations were purely visual, i.e., subjective, the use of electronic devices has created capabilities not only for objective measurement, but also for automation, which is important for the checking and control of continuously running industrial processes as well as for the monitoring of chemical reactions in the laboratory. Recent efforts are directed at miniaturizing the instruments and, above all, at minimizing the sample volumes required, by making use of integrated optical devices, which provide new possibilities of analytical applications.

This article describes instruments and methods of refractometry that are widely applied in many fields of chemical research, industrial control, and medical diagnostics.

1. PHYSICAL FOUNDATIONS

The refractive index (RI) of a transparent medium is defined as the relation of velocity of light in a vacuum (c) to that in this medium (c_M):

$$n = c/c_M. \tag{1}$$

From this there follow various relations for the behavior of light that are employed in refractometry.

1.1 Refraction of Light

If light passes the boundary area of two substances with different RI (n_A,n_B) it will be refracted according to Snell's law:

$$\sin(\alpha)/\sin(\beta) = n_B/n_A, \tag{2}$$

where α = angle of incidence, β = angle of refraction. The angles α and β and the normal to the boundary surface between the two media lie in one plane.

A light beam passing a glass prism is bent from its original direction by an angle δ that is dependent on α, the apex angle (φ), and the RI (N) of the prism and the RI of the surrounding media (n,n') (Fig. 1):

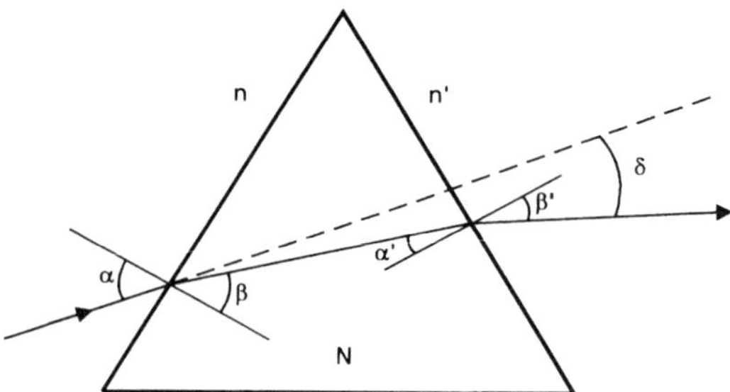

FIG. 1. Refraction by a prism.

$$\sin(\alpha)/\sin(\beta) = N/n, \quad \alpha' = \varphi - \beta;$$
$$\sin(\alpha')/\sin(\beta') = n'/N, \quad \delta = \alpha - \varphi + \beta'. \tag{3}$$

If the angle of incidence is $\alpha = 0$ then also $\beta = 0$ and

$$\delta = \beta' - \varphi, \quad \sin(\beta') = \sin(\varphi)N/n'. \tag{4}$$

Another special case will happen if the angle of deviation reaches a minimum. Then, supposing $n' = n$, the light passes the prism symmetrically, and from Eq. (3) follows

$$N = n\,\frac{\sin[(\delta + \varphi)/2]}{\sin(\varphi/2)}. \tag{5}$$

Generally measurements of N are made in relation to air so that $n' = n = 1$.

The RI of a medium is dependent on the wavelength of light. A beam of white light passing a prism is split up according to Eq. (3), the refracted beams forming a colored spectrum with different wavelengths, a phenomenon known as dispersion. In optical refractometry certain wavelengths in the visible region of the spectrum are preferred; they are emitted by spectral lamps, light-emitting diodes (LEDs), or lasers (Table 1). Like the RI, the dispersion of light is a characteristic parameter of a medium. It is quantified by the difference of the RI at two wavelengths:

$$D = n_F - n_C, \quad \text{or} \quad D = n_{F'} - n_{C'}. \tag{6}$$

A substance without an absorption peak in the visible region shows a normal dispersion, i.e., its RI decreases with increasing wavelength (Fig. 2). If the RI is known for three different wavelengths, the RI for any

Table 1. Light sources and preferred wavelengths for refractometry.

Light source	Color	Symbol	Wavelength (nm)
Spectral lamps			
Helium (He)	Red	···	706.5
Hydrogen (H)	Red	C	656.3
Cadmium (Cd)	Red	C′	643.8
Sodium (Na)	Yellow	D	589.3
Helium (He)	Yellow	d	587.6
Mercury (Hg)	Green	e	546.1
Cadmium (Cd)	Blue	···	508.6
Hydrogen (H)	Blue	F	486.1
Cadmium (Cd)	Blue	F′	480.0
Mercury (Hg)	Violet	g	435.8
Mercury (Hg)	Violet	h	404.7
LED	Red	···	640–660
HeNe laser	Red	···	632.8

other wavelength is represented approximately by a formula by Cauchy:

$$n = A + B/\lambda^2 + C/\lambda^4, \tag{7a}$$

where A, B, C are constants characteristic of a substance. In the region of an absorption band, Eq. (7a) must be replaced by Sellmeier's formula

$$n^2 = 1 + A\lambda^2/(\lambda^2 - \lambda_0^2), \tag{7b}$$

where A, λ_0 are constants, the latter corresponding to the wavelength of the absorption peak. In this region anomalous dispersion appears, i.e., RI increases with increasing wavelength.

If light passes an anisotropic medium, e.g., a crystal plate of calcite or quartz, double refraction (birefringence) may occur. The light beam is split up into two refracted beams that are polarized perpendicularly to each other and that have different angles of refraction. One beam is refracted according to Eq. (2), irrespective of the position of the crystal axis ("ordinary ray"), while the refraction of the other is dependent on this position and the angle of incidence ("extraordinary ray").

A birefringent crystal is characterized by two RI: n_o (ordinary ray) and n_e (the RI that deviates extremely from n_o, for the extraordinary ray) (Table 2). Most crystals occurring in nature possess not only one axis (uniaxial) but two crystal axes (biaxial). Therefore, they are characterized by three principal RI.

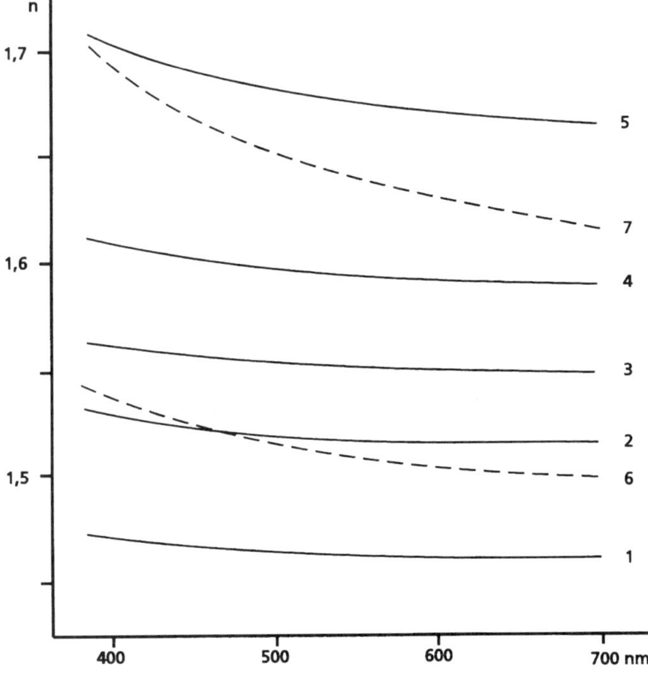

FIG. 2. Dispersion curves of some solids and liquids. (1) Fused quartz; (2) borosilicate crown; (3) crystalline quartz (o); (4) light flint; (5) dense flint; (6) benzene; (7) carbon disulfide.

Table 2. Refractive indices, dispersions, and Abbe numbers of some optical materials. (*o*) ordinary ray, (*e*) extraordinary ray.

Material	n_D (20°)	$n_F - n_C$	Abbe number
Quartz, fused	1.45845	0.00678	67.6
Crystalline (*o*)	1.54425	0.00778	70.0
Crystalline (*e*)	1.55336	0.00805	68.7
Calcite			
(*o*)	1.65836	0.01347	48.9
(*e*)	1.48641	0.00615	79.1
Borosilicate crown	1.5170	0.00802	64.5
Light barium crown	1.5410	0.00907	59.7
Light flint	1.5760	0.01398	41.2
Dense flint	1.6490	0.01913	33.9
Dense barium flint	1.6705	0.01409	47.5
Extra dense flint	1.7200	0.02477	29.1

1.2 Prism Combinations

Two prisms may be in contact so that the contact faces are inclined to the outer faces at a known angle φ, while the outer faces are parallel to each other. If the angle of incidence at the first prism is $\alpha = 0$ (Fig. 3) and the RI of the surrounding medium $n = 1$, the angle of deviation can be calculated from

$$\sin(\delta) = n_B \sin(\beta - \varphi), \quad \sin(\beta) = \sin(\varphi) n_A/n_B. \tag{8}$$

Let n_B be the known RI of a reference medium; then n_A is calculated from δ.

A combination of three prisms is of some importance for refractometers to avoid disturbing effects of dispersion. Prisms of different glasses (crown and flint) are arranged in contact with each other, their angles being exactly calculated so that light of the central wavelength (*D*) emerges from the triplex prism parallel to the incident light while the red (*C*) and the blue (*F*) beams are deflected in opposite directions (Fig. 4). This type is known as a direct-vision or Amici prism. The angle of deflection may reach a few degrees.

1.3 Critical-Angle Refractometry

In Eq. (2) we assume n_B to be greater than n_A. Then the largest angle of incidence, $\alpha = 90°$, results in a refractive angle γ with

$$\sin(\gamma) = n_A/n_B. \tag{9a}$$

No light will enter the second medium at an angle larger than γ [Fig. 5(a)]. γ is called the "critical angle of refraction." By means of a telescope it is possible to observe the boundary line between a bright and a dark field and to read the critical angle off a circular scale.

If, on the other hand, we assume $n_{B'}$ to be smaller than $n_{A'}$, there also exists an angle γ' such that

$$\sin(\gamma') = n_{B'}/n_{A'}. \tag{9b}$$

While a light beam with an incident angle smaller than γ' will enter the second medium, each light beam with α greater than γ' will be totally reflected back into the first medium [Fig. 5(b)]. γ' is called the "critical angle of total reflection." Here again the border line between a bright and a less bright area can be observed in the eyepiece of a telescope. The position of this border line is a measure of the critical angle and thus of the RI. If $n_A/n_B = n_{B'}/n_{A'}$, then $\gamma' = \gamma$.

1.4 Intensity Relations

With each refraction of light at the interface of two substances, a portion of light will be reflected back into the first medium. This portion is dependent on the parameters α, n_A, n_B according to Eq. (2) and to Fresnel's formulas

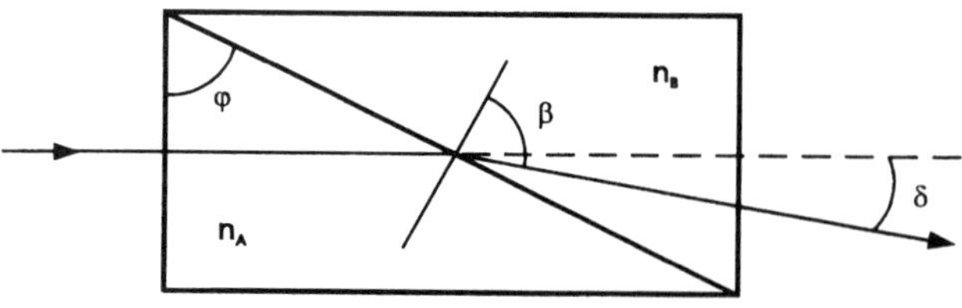

FIG. 3. Deviation of light by a double prism.

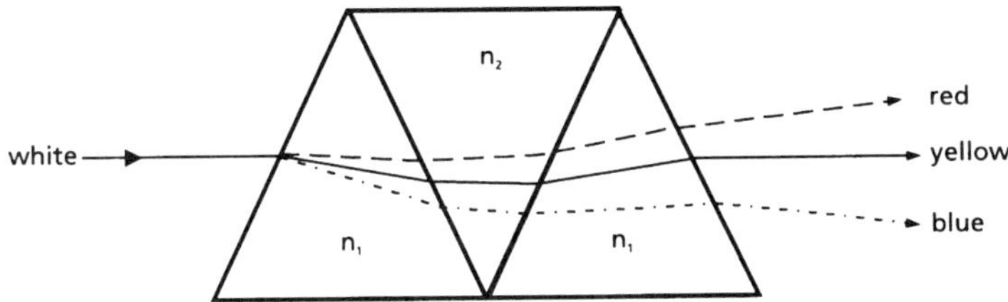

FIG. 4. Dispersion by a direct-vision prism.

$$R_s = -\frac{\sin(\alpha - \beta)}{\sin(\alpha + \beta)} I_s, \quad R_p = \frac{\tan(\alpha - \beta)}{\tan(\alpha + \beta)} I_p,$$
$$R'_s = \frac{2\cos(\alpha)\sin(\beta)}{\sin(\alpha + \beta)} I_s,$$
$$R'_p = \frac{2\cos(\alpha)\sin(\beta)}{\sin(\alpha + \beta)\cos(\alpha - \beta)} I_p, \tag{10}$$

in which I, R, R' are amplitudes of the electric vectors of the incident, the reflected and the refracted beam; indices s and p denote the state of polarization of light perpendicular and parallel to the plane of incidence, respectively. Intensities are the squares of amplitudes. Although Eqs. (10) are not used directly in refractometry, they form the background of intensity measurements based on RI relations.

In this mode of measurement the geometrical form of the contact surface between the two substances is irrelevant: It may be plane, cylindrical, spherical, or otherwise. According to Eq. (2) n_B/n_A only depends on α and β. If the parameters α (i.e., the geometrical conditions) and I of an arrangement are kept constant, the relation n_B/n_A is found by measuring the resultant intensity of the refracted or of the reflected light.

Like the geometry of the contact face the wavelength of the incident light is not relevant either. It is necessary, however, that each refractometer is calibrated individually.

1.5 Interference of Light

Light may be understood as a propagation of waves of electromagnetic energy. The velocity of propagation in a medium is

$$c_M = \lambda\nu, \tag{11}$$

in which λ = wavelength, ν = frequency of vibration.

If a light beam passes along a path of length s through a medium of RI n, the term sn is called "optical path length," and the difference from another optical path length is measured in terms of a multiple of the wavelength:

$$sn_2 - sn_1 = k\lambda \quad \text{or} \quad n_2 = n_1 + k\lambda/s. \tag{12}$$

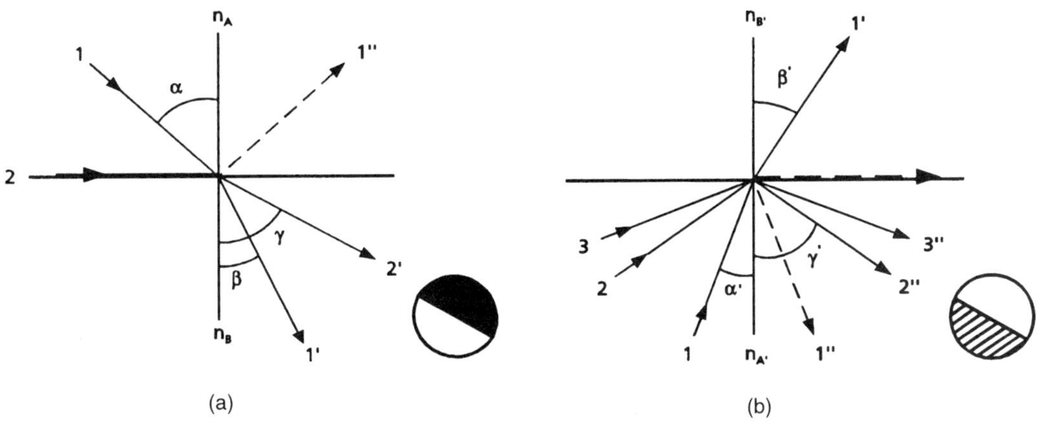

FIG. 5. Refraction (1′,2′) and reflection (1″,2″,3″) at the interface of two media; (a) n_A smaller than n_B; (b) $n_{A'}$, larger than $n_{B'}$. To the right of it, are the fields observed by a telescope.

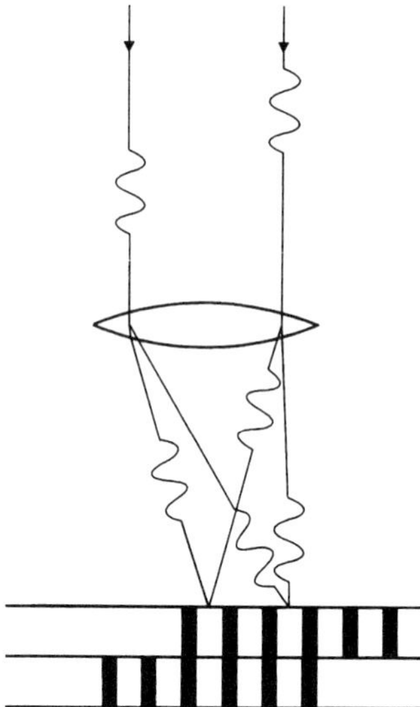

FIG. 6. Interference of two coherent light beams, with a phase difference of two wavelengths, in the focal plane of an objective. The position of the lower fringe pattern would be valid for a phase difference of zero. In the figure, only six fringes of the extended system are shown. In reality, the fringes have no sharp edges, but intensity changes continuously between minimum and maximum.

As light is emitted in quanta, it propagates in wave trains of finite length (coherence length). Two light waves are called "coherent" if they propagate in the same manner of vibration (equal wavelengths, amplitudes, and state of polarization) and if the difference of their optical paths is smaller than their coherence length. Two coherent waves superimposed on each other at one point may interfere. The superposition of the vibrations will result in an increase or decrease of the intensity at this point, so that under special phase relations complete extinction takes place at this point.

An interferometric arrangement using a *monochromatic* light source produces a system of many interference fringes. The number of fringes depends on the degree of coherence of the interfering beams and increases with the coherence length. A phase difference of the light beams causes a displacement of the fringes (Fig. 6). As all fringes look alike it is not possible to distinguish the fringe of zeroth order ($k = 0$), which marks the start position.

A *white* light source, however, because of dispersion of light, produces a superposition of fringe systems of each wavelength of the spectrum. A white zero-order fringe is seen between two black half-order fringes while the fringes of the first and higher orders show colored zones. Consequently the zero-order fringe is an excellent indication mark.

1.6 Refractive-Index Gradients

RI gradients, i.e., variations of RI within a medium, appear in the form of striae, also called "schlieren" or streaks. Striae are inhomogeneous zones within an otherwise homogeneous substance; they have no sharp borderlines, and their RI changes continuously from the margin to the center.

The most convenient method for determination of RI gradients by aid of geometrical optics is the "schlieren method" by Toepler (1864) (Fig. 7). A light beam passing striae will be deviated, the magnitude of deviation depending on the RI gradient dn/dx as well as on the path length (s) within the striae. Approximately we have

$$\delta = \frac{s}{n}\frac{dn}{dx}, \quad n = \text{RI of the surrounding.} \tag{13}$$

A knife edge ("Foucault edge") is arranged in the focal plane of the objective with its edge parallel to the image of the slit. A part of the light that is deviated by striae will fall on the knife edge and is thereby prevented from reaching the screen. Consequently,

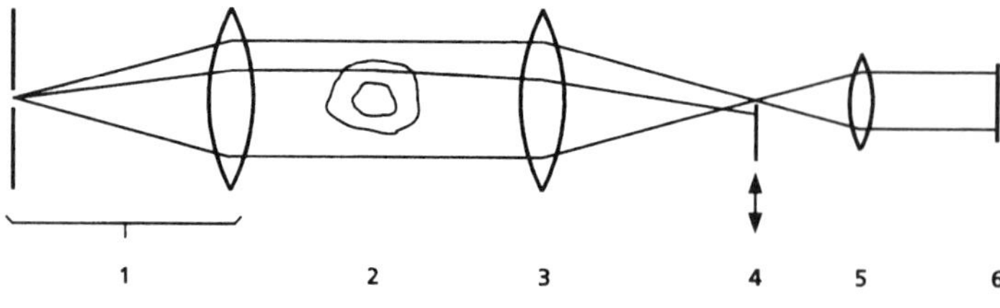

FIG. 7. Toepler arrangement for the observation of striae. (1) Collimator with entrance slit; (2) striae; (3) objective; (4) knife edge; (5) lens system; (6) screen.

striae are recognized on the screen as dark (or bright) zones in a bright (or dark) field; see Schardin (1942).

A disordered distribution of inhomogeneities generally allows only a qualitative analysis by observation of the light deviation. Special techniques allow a partially quantitative determination of the magnitude and direction of the RI gradients (Sec. 2.5.1).

For the quantitative analysis of weak inhomogeneities, interferometers of the Michelson, Twyman, or Mach–Zehnder type are applied; see INTERFEROMETERS AND INTERFEROMETRY.

In stratified liquids, there exists a constant dn/dx in each layer, and all gradients have the same direction. In a cell with stratified solutions, the deviation (regarding refraction at the exit window) will be

$$\delta = s\, dn/dx, \quad s = \text{length of the cell.} \tag{14}$$

Interferometric investigations of stratified solutions are also used. They are based on the measurement of changes of the optical path length according to Eq. (12).

2. REFRACTOMETERS

2.1 Deviation-Measuring Refractometers

According to Eq. (3) these instruments measure the deflection of a defined light beam passing a homogeneous prismatic object.

2.1.1 Spectrometer By means of a collimator, a parallel bundle of light is produced and directed onto the prismatic sample, where it is deflected (Fig. 8). In the eyepiece of a telescope, the position of the image of the entrance slit is observed before and after inserting the prism in its position with predetermined angle α; e.g., for $\alpha = 0$ Eq. (4) holds. The angle of rotation of the telescope is read off at a scale. To determine the RI, the apex angle φ of the prism must be known.

The entrance slit is illuminated by a monochromatic light source. Several light sources may be fitted interchangeably with one another (Table 1).

High-precision spectrometers allow the determination of the absolute RI of glasses and other transparent solids at different wavelengths with an accuracy of some 10^{-6} RI units (RIU). The sample, however, must fulfill strong conditions: It must be transparent, homogeneous (without any striae or scattering effects), well thermostatted; its optical faces must be exactly plane, its diameter must be large, and the prism angle must be known to 1″. Moreover the RI of the surrounding medium must be controlled exactly.

The measuring range is not limited. An exact determination of dispersion is possible.

2.1.2 Autocollimation Spectrometer An "autocollimation telescope" generates a parallel bundle of light and, after reflection of the bundle back toward its original direction, allows the observation of the entrance slit (Fig. 9). This instrument is very well suited for measuring RI *and* prism angle by turning the prism around a vertical axis and reading the angles in three positions of rotation. The controlling telescope remains in a fixed position, thus providing greater convenience of measurement.

2.1.3 Double-Prism Refractometer The determination of the RI of a substance in relation to that of a reference medium is

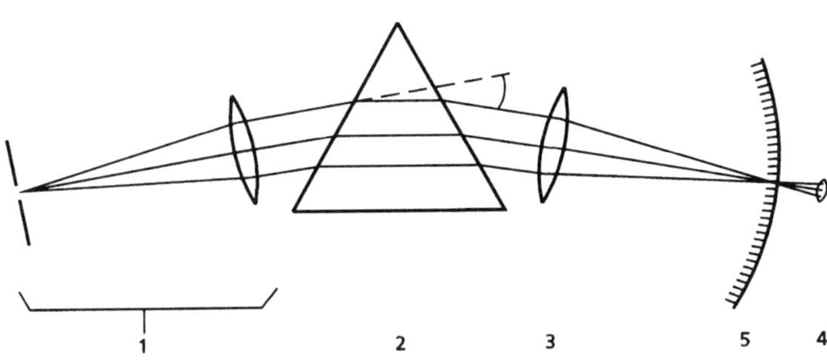

FIG. 8. Spectrometer arrangement (illumination with monochromatic light). (1) Collimator; (2) prism (sample); (3 and 4) objective and eyepiece of the telescope, respectively; (5) circular scale.

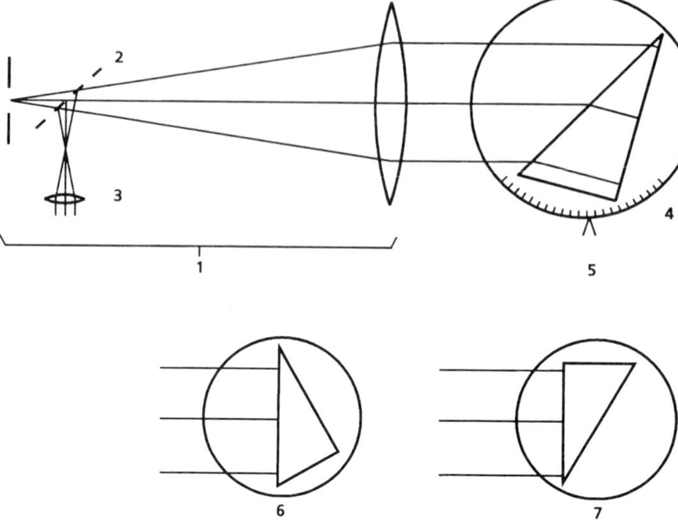

FIG. 9. Autocollimation spectrometer for the measurement of prism angle and deviation of light. (1) Autocollimation telescope with (2) semitransparent glass plate and (3) eyepiece; (4) rotating table with prism and circular scale; (5) reading device with indicative mark; (6 and 7) two further positions of the table for measuring the angle of the prism.

established if both substances take the form of prisms. For the investigation of liquids a double cell consisting of two hollow prisms is illuminated by a collimator. One prism is filled with a known reference liquid, and the other with the sample. An objective projects an image of the entrance slit onto a divided scale. The deviation of light by different substances in the double prism changes the position of the slit image; from the displacement and the focal length of the objective, the RI difference is calculated according to Eq. (8). The scale is nowadays usually replaced by a detector, such as a differential photocell or a position-sensitive detector, e.g., a charge-coupled device (CCD) array. A built-in computer calculates the RI of the sample [Fig. 10(a)].

This type of refractometer is excellently suited for the continuous measurement of

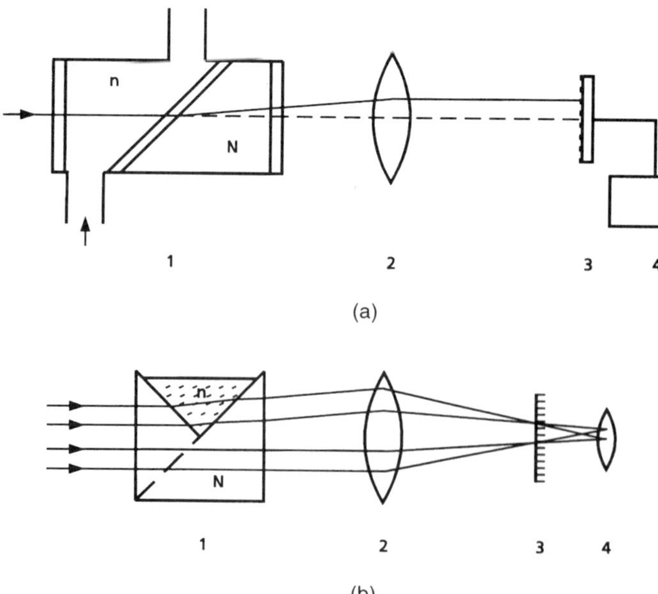

FIG. 10. Measurement of deviation with (a) flow-through refractometer with (1) double-prism cell for sample (n) and reference medium (N), (2) objective, (3) position-sensitive detector, and (4) electrical unit; (b) refractometer with (1) Vee block prism (N) and sample (n, solid or liquid), (2) objective, (3) scale, and (4) eyepiece.

liquids flowing through the sample cell. This arrangement has many advantages for continuous industrial control:

no moving parts in the instrument;

high accuracy better than 1×10^{-5} RIU dependent on the prism angle, the focal length of the objective, and the resolution of the detector;

the transmission of the sample may be rather low;

sedimentation of sample particles at the interior faces of the windows is largely negligible;

automatic temperature compensation by the close proximity of the liquids and, moreover, by the sample running around the reference cell before entering the flow-through cell.

2.1.4 Vee-Prism Refractometer In this refractometer a vee block consisting of two prisms of the same high-refracting glass holds the sample [Fig. 10(b)], which may be liquid or solid. The vee-block angle is exactly 90°. Light passing the lower part remains undeviated while that traversing the upper part is deviated according to the RI difference between the sample (n) and the vee block (N). From Eq. (2) there results

$$\sin(\delta) = (\sqrt{1.5\,N^2 - n^2} - \sqrt{n^2 - 0.5\,N^2})/\sqrt{2},$$
$$n^2 = N^2 - \sin(\delta)\sqrt{N^2 - \sin^2(\delta)}. \tag{15}$$

The very small entrance slit is illuminated by an interchangeable spectral lamp combined with filters to select the required monochromatic radiation. The telescope is rotated about a horizontal axis. The circular scale is divided into degrees and allows the determination of RI to an accuracy of up to 1×10^{-5} RIU absolute and nearly 3×10^{-6} RIU for the measurement of RI differences and for dispersion.

The prism block is surrounded by a hollow jacket containing a thermostatted liquid.

Solid samples must have two polished faces at right angles to each other. With two samples arranged side by side in the vee block, a very exact comparison is possible.

2.2 Critical-Angle Refractometers

The special advantage of this type of refractometer compared with deviation measurement lies in the simpler construction, as no collimating device is necessary, and in the very easy change of samples.

However, the measuring range is limited by the RI of the measuring prism. Determination of RI difference in a direct way is not possible. The sharpness of the black–dark borderline is decisive for accuracy.

2.2.1 Abbe Refractometer The original type Abbe refractometer provides a rotation of the prism relative to the telescope [Fig. 11(a)]. The measuring prism is complemented by a prism confining the sample space on the illumination side. Only a few mm^3 of sample are required. The prisms are contained in hollow jackets for thermostatting. By rotating the prism assembly, the bright–dark boundary is brought to coincidence with the measuring mark of the telescope. The angle of rotation is read off a circular scale through a microscope. The eyepieces of telescope and microscope are situated side by side for convenient observation.

Nowadays most Abbe refractometers have a fixed measuring prism with a horizontal measuring face. Rotation is effected by a rotatable mirror arranged between prism and telescope. In the eyepiece, the borderline and the circular scale are seen simultaneously. Most instruments contain two scales, one each for RI and for percentage sugar dissolved in water. The measuring range for RI extends from 1.3 to 1.7 and is limited by the RI of the measuring prism. The accuracy is about 1×10^{-4} RIU.

Under illumination with white light—tungsten lamp or sun light—the bright–dark borderline appears colored because of dispersion. To avoid this disturbing effect, a built-in color compensator is used, which consists of a pair of direct-vision prisms (Sec. 1.1) rotatable about the optical axis.

The Abbe-type critical-angle refractometer allows measurement with both transmitted and reflected light, so that also opaque liquids and solids as well as plastics may be observed.

2.2.2 Pulfrich Refractometer The Pulfrich refractometer in its original form provides critical-angle measurements with grazing incidence of light, $\alpha = 90°$ [Fig. 11(b)]. The measuring prism (N) has a fixed position with a horizontal measuring face and is illuminated by spectral lamps (Table 1) in com-

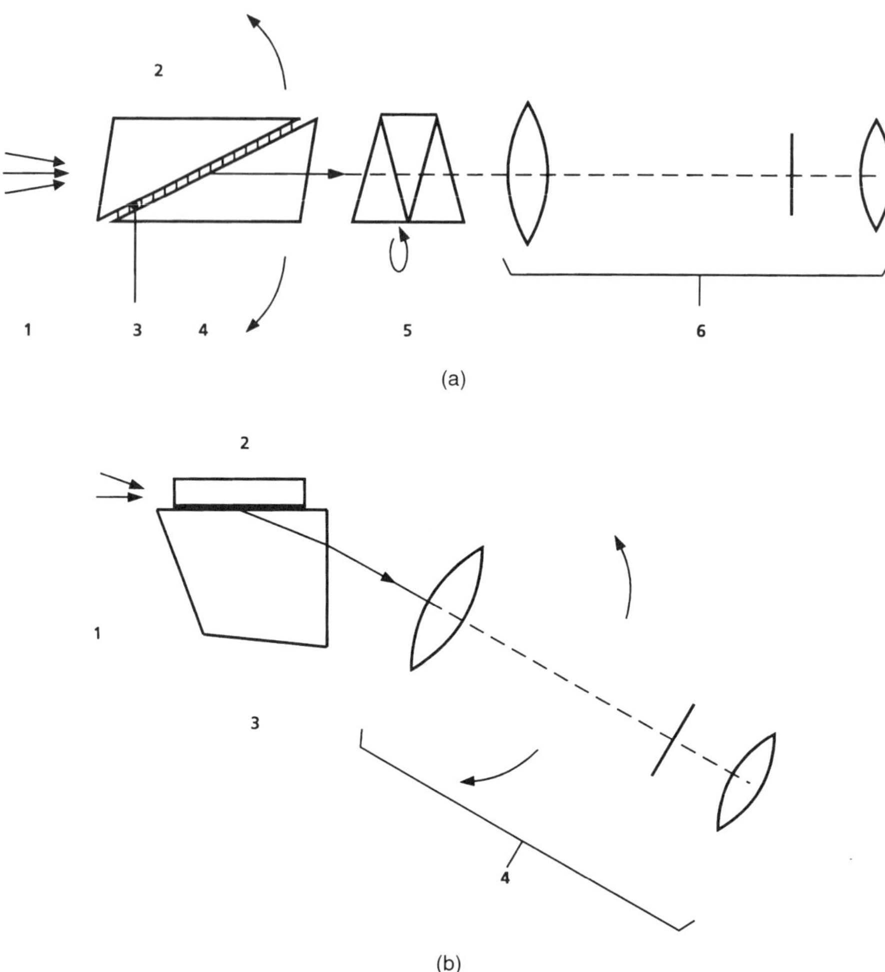

FIG. 11. Measurement of the position of the critical angle borderline. (a) Abbe refractometer: (1) diffuse illumination with white light, (2) prism for light entrance, (3) liquid sample, (4) measuring prism, (5) direct vision prism, and (6) telescope. (b) Pulfrich refractometer: (1) diffuse illumination with monochromatic light, (2) solid sample, (3) 90° measuring prism, and (4) rotatable telescope.

bination with monochromatic filters. As the prism angle is 90°, Eq. (3) results in

$$n = \sqrt{N^2 - \sin^2(\delta)}. \quad (16)$$

The borderline of the bright and dark fields is brought to the center of diagonal cross wires by rotating the telescope about a horizontal axis. The PR2 Pulfrich refractometer (Carl Zeiss Jena) has interchangeable prisms and illumination units to enable both critical-angle and deviation measurement (Sec. 2.1.4).

The measuring range of critical-angle measurement is limited by the RI of the measuring prism. The accuracy is nearly 2×10^{-5} RIU, an exactly constant temperature provided.

A special type of Pulfrich refractometer is a crystal refractometer, which has a hemisphere of highly refracting glass instead of the measuring prism with two polished plane faces.

2.2.3 Dipping Refractometer To avoid inaccuracies arising from rotatable parts, some refractometer types hold measuring prism and telescope in a fixed position. The bright–dark boundary is observed in the eyepiece and its position is directly read off a fixed scale.

To permit quick and convenient measurement and to provide an exactly grazing incidence of light at the prism surface, the prism end of the refractometer is dipped into the liquid sample contained in a small cup, which is immersed in a thermostatting water bath. More commonly, however, one uses a prism in a temperature-controlled jacket.

The measuring range is small, only about 0.03 RIU, but it may be extended by a set of interchangeable prisms with different RI.

A color-compensating prism is built in just as in the Abbe refractometer so that white-light illumination is possible. For high-precision measurements, a monochromatic light source such as a Na spectral lamp is preferred. A long focal length of the telescope objective makes a high accuracy of 2×10^{-5} RIU possible. The scale is divided into equal intervals, and RI is found from tables.

2.2.4 Hand Refractometer This type is designed for special applications in mobile use. The main characteristics of visual instruments are the rigid connection of prism and telescope; small measuring range about 0.03 RIU; scale divided into percentage of a special solution; no thermostatting. Some types of hand refractometers are temperature controlled by way of a bimetal-mounted scale so that a separate temperature measurement is superfluous.

The accuracy is up to 2×10^{-4} RIU. Light-weight instruments weigh less than 100 g.

For automatically indicating hand refractometers see Sec. 2.3.2.

2.3 Automatic Refractometers

To avoid errors caused by the subjective observation of a measuring mark, automatic refractometers have been designed on the basis of light deviation by a prism (Sec. 1.1) or of the changes in light intensity (Sec. 1.4). The major parts of these arrangements are an illuminating system with a white or monochromatic light source, a space for the interaction between the light and the sample, a detector to receive the light signal, an amplifier, and a display. Many instruments have electrical output ports for a control signal, which can be linked to recording or process-control equipment.

2.3.1 Deviation-Measuring Automatic Refractometers In these instruments, the illumination system is a collimator with an entrance slit. As far as the measurement is concerned with liquids, a prismatic differential flow-through cell is used so that the RI difference from a reference substance results in a deviation of light. The slit-image position on the detector is recognized by the signals of a differential photocell, by the modulated signal of a simple photocell, or by a semiconductor detector.

Most instruments employ a compensation method: The displacement of the slit image is reduced to zero by a rotating glass plate arranged between cell and objective or by a rotating mirror. Modern instruments use a CCD array that makes compensation superfluous, so that all movable parts can be omitted [Fig. 10(a)].

Deviation-measuring automatic refractometers are temperature compensated and yield a high accuracy of 1×10^{-5} RIU. However, there are restrictions as to the flow rate, viscosity, and absorption of the sample.

2.3.2 Intensity-Measuring Refractometers These refractometers use the variation of the intensity of light that is partially or totally reflected at the contact face between the sample and a body of glass. The form of the body is irrelevant; it may be a prism, a rod, a fiber, a spherical lens, or other shape (Fig. 12). For illumination no collimator is necessary; the reflected light is collected on the sensor by means of a simple lens. As the light does not enter the sample, opaque media can also be tested. The electrical signal is amplified and converted into RIU or percentage concentrations and displayed and/or recorded.

Hand refractometers for nonstationary applications use reflection at a prism near the critical angle and give a direct indication of RI or of percentage sugar in water.

The accuracy of intensity-measuring refractometers is less high than that of deviation-measuring instruments, reaching only close to 1×10^{-4} RIU. But they are robust in construction, of small size, and easily handled. They are applicable also for opaque and highly viscous liquids, and flow-through types allow a high flow rate. The contact face must be meticulously clean, and sample temperature must be measured separately.

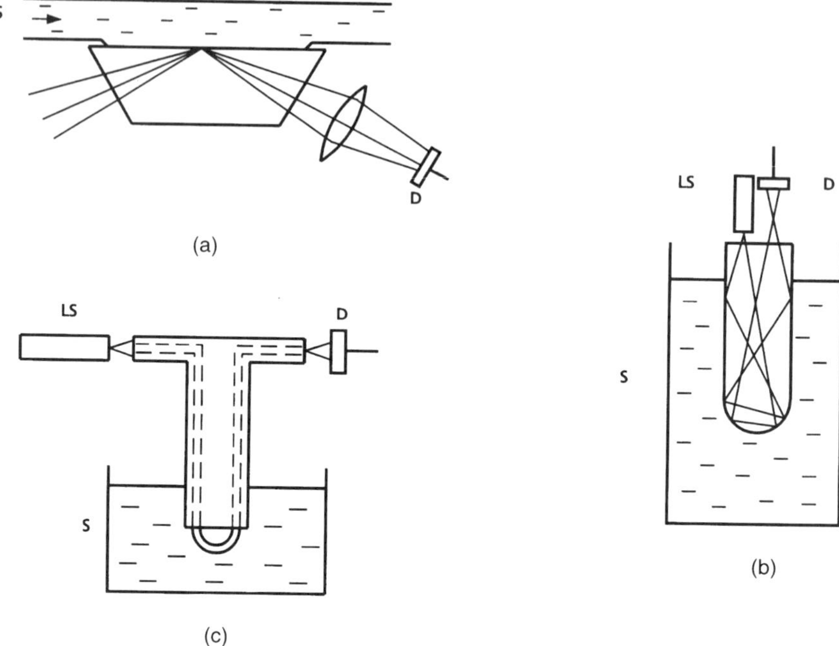

FIG. 12. Measurement of intensity of reflected light. LS: light source; S: sample; D: detector. (a) Prism in contact with a flowing sample, diffuse illumination; (b) a rod dipped in the liquid; (c) the bare zone of a glass fiber dipped in the liquid; the upper region of the fiber has a protective cover.

2.3.3 Integrated-Optics Refractometers In recent years the development of lasers has brought forth a new branch in optics: integrated optics. Integrated optical devices make it possible to reduce the size of optical instruments and to minimize sample volume. This can be achieved by the principle of total reflection of light at the boundary of a medium with lower RI.

The main parts of an integrated-optics refractometer are a laser, an optical waveguide, and a detector. The laser is usually of the HeNe or the semiconductor type. The radiation is coupled into the waveguide via an optical fiber. The waveguide is a thin metallic layer of a thickness smaller than the wavelength of light. One surface of the layer is in contact with a substrate, e.g., a glass plate or a prism, while the other surface of about 1–2 mm^2 contacts the sample. The radiation of the laser generates waveguide modes with vibration states perpendicular to each other, which vary when the RI of the sample changes.

Different arrangements for the detection of the RI-dependent signals are described, some of them based on interferometric principles. Clerk and Lukosz (1993, 1994) inserted a diffraction grating as an output coupler and measured the outcoupling angles of the modes with position-sensitive detectors. RI differences of about 5×10^{-5} can be recognized.

2.4 Interference Refractometers

The highest accuracy in refractometric measurements is reached by refractometers based on interference of light. For analytical applications two types are commonly employed, which use geometrical or amplitude division of the wave front, respectively.

2.4.1 Rayleigh–Loewe Interferometer Two coherent beams are generated by a double slit intercepting a parallel bundle of light [Fig. 13(a)]. Each beam traverses a chamber filled with the sample or a reference medium, respectively. The interference fringe pattern in the focal plane of an objective is observed through an eyepiece with a cylindrical lens. Two further beams passing below

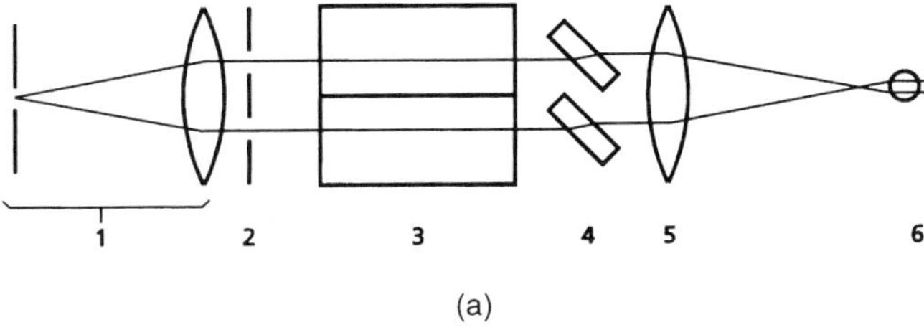

(a)

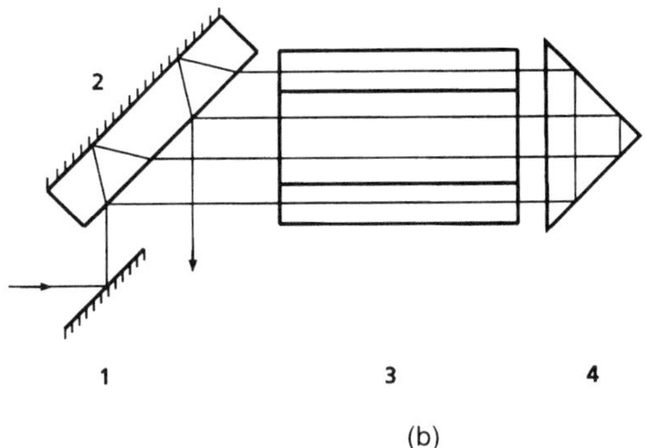

(b)

FIG. 13. (a) Rayleigh-Loewe interferometer: (1) collimator, (2) double slit, (3) double cell, (4) compensator, (5) objective, (6) eyepiece with cylindrical lens. (b) Jamin-Doi interferometer: (1) plane mirror, (2) glass plate with a semitransparent front face, (3) three-cell system with inner cell for the sample and both outer cells for a reference medium, (4) reflection prism.

the chambers generate a second pattern of interference fringes, which serves as a reference. Media with different RI (n_1,n_2) in the chambers cause a shift of the fringe pattern that, by a rotatable glass plate of a compensation device, is brought back to coincidence with the reference pattern.

White-light illumination allows an unambiguous recognition of the zero-order fringe. Because of the different wavelengths contained in white light, the spaces between the fringes differ in proportion to the wavelength. Therefore, approximate coincidence at all wavelengths exists only for fringes up to the $\pm$ first order, producing black half-order minima, while the spacings of higher-order fringes differ increasingly, so that the fringes appear colored.

According to Eq. (12), the accuracy of RI-difference determination is dependent on the length of the chambers. While in portable instruments, which use an autocollimation device, chamber length will not exceed 10 cm, in laboratory instruments it may be up to 1 m or, by internal reflections within the chamber, up to 3 m, so that an accuracy of better than 1×10^{-8} RIU will be reached in gas analyses.

2.4.2 Jamin Interferometer The coherent rays are generated by transmission and reflection of light at a semitransparent layer on a plane-parallel glass plate. After passing the chambers and a compensating device, the rays are reunited at the semitransparent layer of a second glass plate, where they interfere with one another.

Under illumination with white light, a system of interference fringes with two black fringes in the middle is observed. A RI difference of the media in the chambers causes a shift of the fringe pattern, which is either compensated (as in the Rayleigh–Loewe interferometer) or read directly off a scale.

The accuracy is nearly that of Rayleigh–Loewe interferometers.

In portable interferometers of the Jamin–Doi type, a reflecting 90° prism is arranged behind the chambers so that the light twice passes the sample chambers of, e.g., 10-cm length [Fig. 13(b)]. The recombination of the coherent rays takes place at the same layer that splits up the incoming bundle of light. The fringe pattern appears together with a scale where its position is read off directly. By a small displacement of the eyepiece, the center of the fringe pattern is shifted from the zero-order maximum to a half-order minimum, so that *one* small black fringe constitutes the zero mark.

As the chambers are short and a compensation device is omitted, these instruments are small. Moreover, by threefold reflection of light within the chambers, the size of such an instrument can be minimized. The accuracy is about 2×10^{-7} RIU.

2.4.3 Photoelectric Interferometers While the above-mentioned visual interferometers are based on the identification of the zero-order fringe in white light, photoelectric interferometers commonly use a laser source. As this monochromatic source generates an extended fringe pattern without any prominent fringe, the procedure is to vary the optical path length continuously and to determine the shift of the fringe pattern by counting the number of fringes passing a detector. This is made possible by placing the cell filled with the sample into one beam of the interferometer and rotating the cell about a defined angle.

Photoelectric interferometers are based on an amplitude division of the wave front, so that no intensity-limiting entrance slit is required. To have a large lateral separation ("shear") of the interfering beams, interferometers of the Mach–Zehnder or Michelson type are preferred (Hassan *et al.*, 1995).

2.5 Refractive-Index-Gradient–Measuring Refractometers

Inhomogeneous media show RI gradients (dn/dx), i.e., a variation of RI within the substance.

2.5.1 Schlieren Apparatus For the *qualitative* investigation of RI gradients (schlieren, striae) in a transparent medium a schlieren apparatus is employed. A Toepler arrangement according to Fig. 7 provides a high sensitivity so that very low gradients can be recognized. The visible diameter of the object is limited by the diameter of the objectives. For extended objects one prefers a Z-shaped arrangement with large mirrors instead of the objectives so that the visible field may be up to 300 mm (Fig. 14).

Instead of the knife edge a wire [Fig. 15(b)] that is as broad as the image of the slit may be inserted so that all the light deviated by striae in a direction perpendicular to the slit will produce bright zones on a dark background on the screen.

To give a higher accuracy, the wire may be replaced by a phase plate [Fig. 15(c)], i.e., a glass plate one-half of which is coated with a transparent layer of a thickness of $\lambda/2$, so that interference effects will arise that mark the presence also of very weak striae. In this way, a variation of the phase-contrast technique is achieved (Wolter, 1967).

A rough quantitative determination of striae is possible by illumination with white light and using stripes of transparent filters in different colors [Fig. 15(d)] instead of the knife edge. Thus, the object image shows strongly deflected light in a different color than weakly deflected light; light deflected in the opposite direction appears in still other colors.

For the investigation of stratified solutions, the original Toepler arrangement can easily be completed as described by Philpot

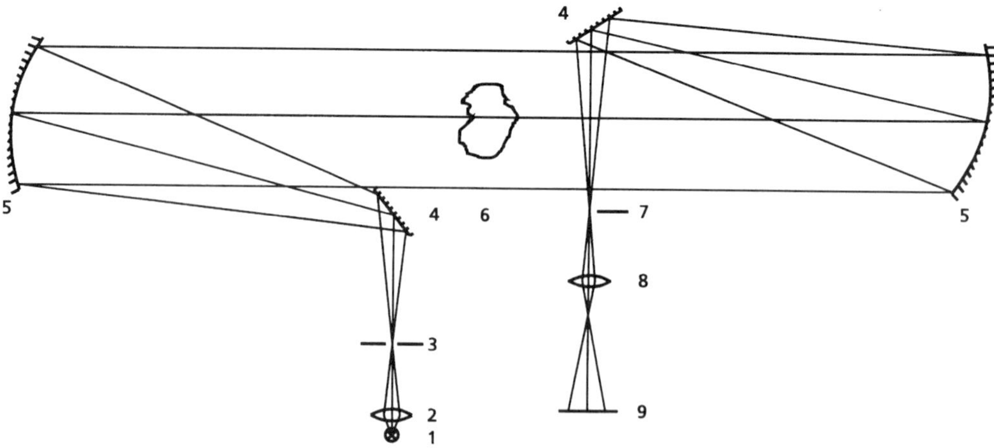

FIG. 14. Z-shaped schlieren arrangement with mirrors. (1) Light source, (2) lens, (3) entrance slit, (4) plane mirrors, (5) concave mirrors, (6) striae (deviation of light is not shown), (7) knife edge, (8) lens system, (9) screen.

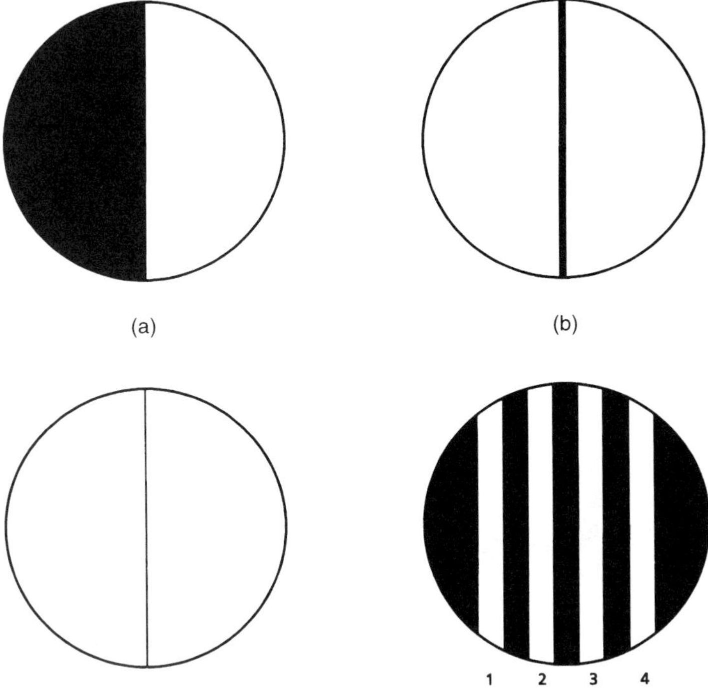

FIG. 15. Some typical sorts of insets in the focal plane of a schlieren arrangement: (a) knife edge, (b) wire, (c) phase edge, (d) colored stripes (1, 2, 3, and 4 indicate different colors).

(1938) and Svensson (1939): The knife edge is replaced by a diagonal slit, and a cylindrical lens projects an image onto the screen, which shows vertically the cell and horizontally the plane of the slit. The deviation of light caused by the gradients results in a lateral displacement of light on the screen, and a gradient curve is obtained.

2.5.2 Interferometers For a *quantitative* determination of disordered inhomogeneities, interferometers are used. Based on Eq. (12), the interference fringe pattern of an amplitude-dividing interferometer (Mach–Zehnder or Michelson type) is changed by schlieren, and the magnitude of the fringe displacement at one point allows the RI gradient to be determined. These instruments are well suited for the exact analysis of very weak inhomogeneities.

High-accuracy measurements of stratified solutions are performed by Rayleigh-type interferometers, which have a grid instead of the entrance slit, and a large vertically extended double cell (Svensson, 1939). The uniform direction of gradients allows interference fringes to be generated, which are shifted according to the magnitude of the RI difference between the sample and a homogeneous reference liquid at the respective height:

$$n_x - n_{o,x} = k_x \lambda / s; \tag{17}$$

x marks the height in the cell.

The gradient curve is obtained by two beams of light both passing through the sample cell with a very small vertical separation (shear) between them.

The RI and the RI-gradient curve can be imaged simultaneously using a biprism in the interferometer.

3. REFRACTOMETRY OF HOMOGENEOUS SUBSTANCES

3.1 Glasses and Other Solids

Optical glasses are mainly characterized by their RI and their dispersion. The high accuracy required for close inspection is reached only by measurement of the deviation of light at a prismatic object with a spectrometer. Samples must have exactly plane, polished surfaces of several centimeters in diameter. Table 2 presents the RI and Abbe numbers of some transparent optical materials.

For routine measurements a refractometer with a vee-block prism is well suited. The sample only needs to have two unpolished plane faces approximately at right angles to each other, which are put into the vee block with a contact liquid of nearly the same RI. With this arrangement it is also possible to control the process of hardening of adhesives or other synthetics by RI measurement.

To gain comparable values, temperature measurement is necessary. While the temperature coefficients dn/dt of glasses are low, i.e., smaller than $\pm 2 \times 10^{-5}$ RIU/K, those of plastics may reach some 10^{-4} RIU/K.

Routine RI measurements of lower precision can be made by critical-angle methods with a Pulfrich, Abbe, or dipping refractometer. The sample must have a plane, polished face, which is brought into contact with the face of the measuring prism, and a contact liquid with a RI higher than that of the sample is needed, e.g., monobromonaphthalene.

The RI of a small piece of a transparent solid may be measured by an immersion method: A sample immersed in a liquid of the same RI is invisible. By immersing the sample into a liquid, changing the RI of the liquid (by substitution, mixing with another liquid, or change of temperature), and observing the disappearance of the contour, it is possible to identify the immersion liquid with exactly the same RI as that of the sample. The RI of that liquid can then be measured separately.

3.2 Solutions and Binary Mixtures of Liquids

Values of RI, Abbe numbers, and temperature coefficients of RI for several liquids and solutions are given in Table 3. The RI of solutions and binary mixtures depends on their concentrations. Refractometry is an excellent method of determining concentration c because of the simple and quick procedure. For each solution or mixture there exists a specific n–c relation, which must be found by calibration. Calibration data for most

Table 3. Refractive indices, Abbe numbers, and temperature coefficients of some liquids.

Liquid	n_D (20°)	Abbe number	$dn/dt \times 10^4$
Water	1.33299	55.8	−0.9
Methanol	1.3286	59.9	−4
50% in water	1.3424	...	...
Ethanol	1.3613	59.3	−4
50% in water	1.3612	...	...
Ethylene glycol	1.4319	59.8	−3
Glycerol	1.4740	59.0	−2.3
Benzene	1.5014	30.2	
Tetralin	1.5449	32.7	
Carbon disulfide	1.6277	18.4	−8.1
Monobromonaphthalene	1.6582	20.3	−4.5

commonly used solutions and binary mixtures are compiled in standard tables.

The most typical example is the analysis of sugar content in water, a test widely applied in agriculture (sugar cane, sugar beet, fruit), wine growing, sugar factories, the production of jams, fruit juices, etc. Precise data of sugar concentration and its dependence on temperature have been established by the International Commission for Unified Methods of Sugar Analysis (ICUMSA, 1974).

The gradient of RI with concentration is called "refractive-index increment" (RI increment):

$$dn/dc = (n - n_o)/c, \quad (18)$$

where c = concentration in g/mL, n, n_o = RI of solution and solvent. The RI increment has proved a characteristic constant of polymer solutions, dependent only on temperature and wavelength of light (Huglin, 1965; Lorimer, 1972). The values of dn/dc lie between −0.24 and 0.25. They are used for the determination of molecular weights of polymers (Baltog *et al.*, 1970).

The calibration curve of most liquid mixtures is bent. For some mixtures it has a maximum (Fig. 16), so that two different concentrations have the same RI. To get unambiguous results, the mixture is diluted 1:1 with one of its components, and the RI of the new mixture is measured.

In organic chemistry and pharmacy the terms "specific refraction" r and "molar refraction" R have become important for the identification of substances and the investigation of two- or three-component systems.

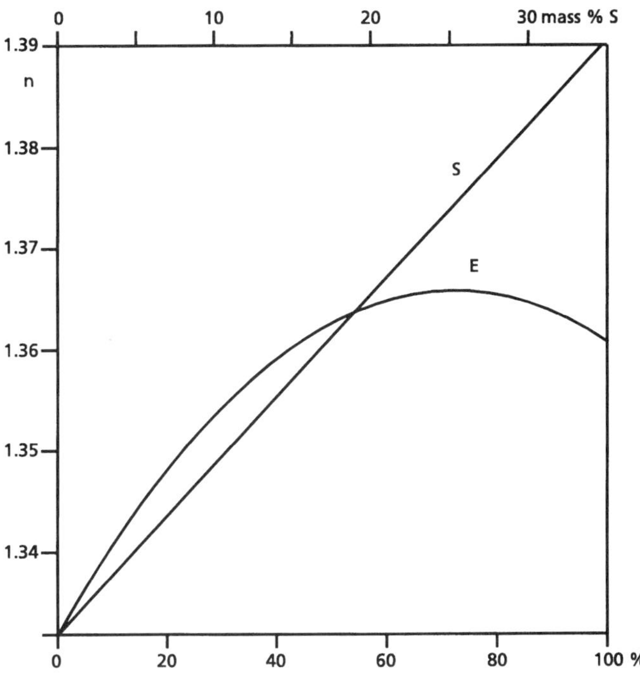

FIG. 16. Calibration curves of sugar in water (S) and of ethanol in water (E).

According to Lorenz and Lorentz the terms

$$r = \frac{n^2 - 1}{n^2 + 2}\frac{1}{d}, \quad R = rM,$$
$$M = \text{molecular weight,} \qquad (19)$$

are specific constants for each substance independent of their state of matter. Besides Eq. (19) there exist some approximations given by other authors; see Joffe (1983).

The dependence of RI of liquids on pressure is negligible, whereas that on temperature is considerable. Comparable results are gained only if the temperature t is thermostatted and noted together with the RI, the same as the wavelength.

Routine laboratory tests are commonly carried out with the Abbe refractometer. Higher precision (nearly 2×10^{-5} RIU) is reached with the dipping refractometer; differential measurements to some 10^{-6} are possible with the Pulfrich refractometer, and to about 1×10^{-7} with an interferometer. With interferometric measurements of liquids, the double cell is immersed in a water bath. Interferometers serve, e.g., for determination of the solubility of gases in water or for observation of a course of titration.

Applications outside the laboratory are widespread—e.g., for the determination of salt contents in ocean water, for testing of antifreeze agents, or for the above-mentioned measurements of sugar contents. The refractometers must be equipped with a thermometer or an automatic temperature-compensation device. They allow a direct reading of the percentage content of sugar or other substances.

3.3 Ternary Mixtures of Liquids

If in a mixture of three components the relation of concentrations c_1, c_2 of two components is a known constant, these components may be regarded as one ($c_{1,2}$) so that the problem is reduced to a binary mixture of $c_{1,2}$ and c_3. Generally, however, concentrations of each of the three components change independently so that a second parameter besides RI must be determined. Preferably this is density d, because it can be measured as easily as RI; this method is known as "refractodensitometry." The measurement of other parameters (e.g., optical activity, viscosity, velocity of sound) is also possible.

The correlations between RI, density, and concentrations c_i must be found experimentally, and calibration tables must be erected. All measurements must be made at a constant temperature.

A typical and important example is the analysis of beer. In the brewing process, part of the extract changes into alcohol, so that in the ternary system of alcohol, extract, and water the concentrations as well as the original extract are to be determined. Similar examples are the testing of wine, liqueur, and spirits.

Commonly used instruments are a dipping refractometer, a densitometer, and a thermostat. Brewery processes can be monitored with an automatic flow-through refractometer and an automatic densitometer, both equipped with temperature-controlling devices.

3.4 Continuously Varying Liquids

The development of photoelectric techniques has opened up new and highly useful applications of refractometry. This applies especially to the continuous measurement of flowing liquids or of liquids that change with time. One of the most important applications is the monitoring and control of processes in industrial production plants. Automatic process refractometers (Sec. 2.3) are well suited for the indication and control of concentration in the food and beverage industries (sugar content), of fractionating and synthesizing processes in the petrochemical industries, of spinning and cleaning processes in the textile industry, of distillation and extraction processes, etc. These instruments supply electric signals that actuate control valves to correct out-of-tolerance concentrations in the product.

Process control is the most important field of application of refractometers today.

Automatic refractometers with extremely small cells have been designed for application as detectors in liquid chromatography with columns. The necessary high accuracy of better than 1×10^{-5} RIU requires differential refractometers with a prismatic double cell. Some types of RI detectors use an autocollimation device so that the light passes the differential cell twice to gain a higher accuracy. Though in HPLC (high-performance liquid chromatography) generally a UV detector is preferred, a RI detector must be taken if the eluate and the solvent show no UV absorption.

In immunology, the amount of sample available (e.g., body fluids) is extremely small. The refractive index of a sample, which changes, e.g., in the course of reaction between antigens and antibodies, is determined with refractometers built around integrated-optics devices (Sec. 2.3.3). It is also possible with this method to trace chemical reactions in minute substance volumes.

3.5 Binary Mixtures of Gases

Refractometry is a useful tool for controlling the purity of gases or determining their concentrations by measuring RI, or the RI difference from a reference gas. The RI of gases are very low, smaller than 1.003, so that only interference refractometers are eligible. For simplification one sets $m = (n - 1) \times 10^6$ (Table 4). The RI of gases and vapors are strongly dependent on temperature and pressure. Therefore, care must be taken that these parameters are identical for sample and reference medium.

Published tables list the RI of gases for "standard conditions," i.e., temperature t_0 = 0 °C and pressure p_0 = 101.325 kPa (= 760 Torr). These are calculated from the param-

Table 4. Refractive indices of gases and vapors at standard conditions (Sec. 3.5).

Gas/Vapor	$m_e = (n_e - 1) \times 10^6$	$m = m_F - m_C$
Air, dry and CO_2 free	293.11	3.27
Helium	34.95	0.18
Neon	67.25	0.28
Hydrogen	139.37	1.81
Oxygen	272.27	3.83
Nitrogen	299.14	3.33
Carbon dioxide	450.6	5.79
Chlorine	784.0	13.9
Carbon disulfide	1477	70
Methane	443.3	6.7
Methanol	563.0	8.5
Ethanol	870	10.8
Chloroform	1412	19.9
Halothane	1582	23.8
Benzene	1759	43
Carbon tetrachloride	1781.9	30.6
Toluene	2092	...

eters of measurement, t and p, by

$$m(t_0,p_0) = Fm(t,p), \tag{20}$$

where $F = (1 + t/273)\ p_0/p$. With this, Eq. (12) yields

$$n_2(t_0,p_0) - n_1(t_0,p_0) = Fk\lambda/s. \tag{21}$$

The great dependence of the RI of air on pressure and temperature makes it possible to monitor the condition of ambient air by interferometry, which is important, e.g., in high-precision metrology.

Concentration measurements require calibration. Most mixtures show a straight calibration curve. But through the effect of dispersion (Sec. 2.4.1) the colors within the interference fringes change with increasing concentration so that the zero-order fringe becomes colored while that of a higher order becomes black. In the calibration curve this change of colors results in "jumps" by exactly one interference fringe spacing (Fig. 17).

The accuracy of a laboratory interferometer with a chamber of 1-m length is about 2×10^{-8} RIU for differential measurement; this corresponds, for example, to 0.01% CH_4 in air or to 0.001% benzene in air.

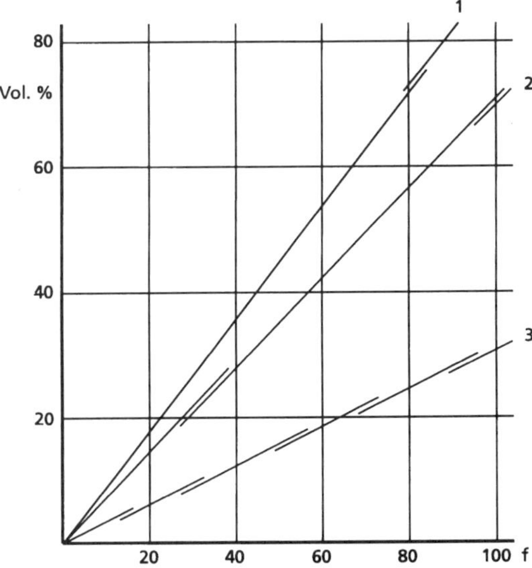

FIG. 17. Calibration curves for interferometric determination of (1) hydrogen, (2) carbon dioxide, (3) acetylene in air; Rayleigh-Loewe interferometer with cell length 1 m. f is the measured number of interference fringes.

Applications of interferometric analysis of gaseous mixtures are many and varied; examples are the calibration and testing of anesthetic vaporizers, testing of lung function, and the monitoring of air in rooms with explosion hazards. An important field of application is the monitoring of air in underground coal mines to prevent fire-damp explosions. For this purpose the small Jamin–Doi type interferometer, equipped with cells for the absorption of water vapor and CO_2 from air and with its scale divided in percentage CH_4, is used worldwide. A similar task is the testing of air in factories where organic solvents are used for production or cleaning. Most solvent vapors can generate explosive mixtures in air or are harmful to health.

The accuracy of portable interferometers is approximately 0.1% CH_4 or 0.01% benzene in air.

3.6 Ternary Mixtures of Gases

The methods employed for the investigation of ternary mixtures of gases are similar to those used for liquids (Sec. 3.3). If the relative concentration of two components is a constant these may be regarded as one gas. The problem is thus reduced to a binary system. A typical example is determination of CO_2 (or another gas) in air, as air is a mixture of essentially 79% N_2 and 21% O_2. If, however, the composition of air changes, the procedure is to measure the RI of the original sample first, and then that of the sample after absorption of CO_2, both relative to normal air. This method is applied in investigations of respiration in medical and biological tests, also in animal stables. Useful sorption media are listed by Nebe (1977).

4. REFRACTOMETRY OF INHOMOGENEOUS SUBSTANCES

Investigations of inhomogeneous substances are widespread in technical, medical, environmental and other fields. Depending on whether a more vivid visualization or an exact quantitative analysis is required, a schlieren apparatus or an interferometer is preferred. Cited below are only the most important investigations.

4.1 Transparent Solids

In glass factories and in the optical industry, a close inspection of the homogeneity of glasses is necessary. Also, optical surfaces can be tested for flatness.

In the same way, plastics are inspected. As the temperature coefficient of RI of plastics is high, heat-transfer processes can be observed. Moreover, plastics subjected to stress show effects of double refraction (photoelasticity). If such an object is brought between crossed polarization filters the stress distribution is visible. The photoelastic study of models made of transparent plastics, above all PMMA, is an important method of forecasting the behavior of products under stress in use.

4.2 Schlieren in Liquids

Within liquids striae are examined in many fields of research. Some examples are solvation processes, electrochemical processes in electrolytes, characterization of polymeric materials by measuring disperse solutions, heat current and heat transfer in liquids, and ultrasonic effects in the bulk and at the surface of liquids.

RI gradients in liquids are generally large. Therefore, interferometers are difficult to use. Instead, Toepler schlieren arrangements are preferred. Their precision may be varied by changing the focal length of the imaging objective.

4.3 Stratified Solutions

For the investigation of stratified liquids, refractometric methods have proved very useful. The optical arrangements are based on the Toepler schlieren device, with special adaptions, or on interferometric devices (Sec. 2.5). The most important measurements are those of electrophoretic, diffusion, and sedimentation processes (Weissberger, 1972).

Electrophoretic investigations are mainly aimed at the identification of colloidal substances, particularly proteins, which differ only by their rate of migration in an electric field. An important application in medical diagnostics is the differentiation of serum proteins.

Examination of diffusion and measurement of the diffusion coefficient are a useful tool for the characterization of macromolecules in chemical research. As solutions to be measured are highly diluted, very precise methods of detection, preferably interferometric methods, are applied.

Sedimentation processes are mainly observed in a high-speed ultracentrifuge because a very strong gravitational field is a prerequisite. Interferometric techniques based on Rayleigh–Svensson or Jamin-type interferometers are employed. From the measured rate of sedimentation, the sedimentation coefficient is calculated, which in combination with the diffusion coefficient allows the determination of molecular weight.

4.4 Inhomogeneous Gases

Striae in gases generally have no preferred direction of RI gradient. Very weak striae may be determined interferometrically. Strong striae, however, disturb the pattern of interference fringes, and therefore, the use of the Toepler schlieren arrangement is preferred.

The most important applications are aerodynamic investigations of all sorts of quickly moving objects by a simulation procedure in a wind tunnel; investigation of shock waves and pressure waves (Fig. 18); investigation of flames, combustions, and explosions; obser-

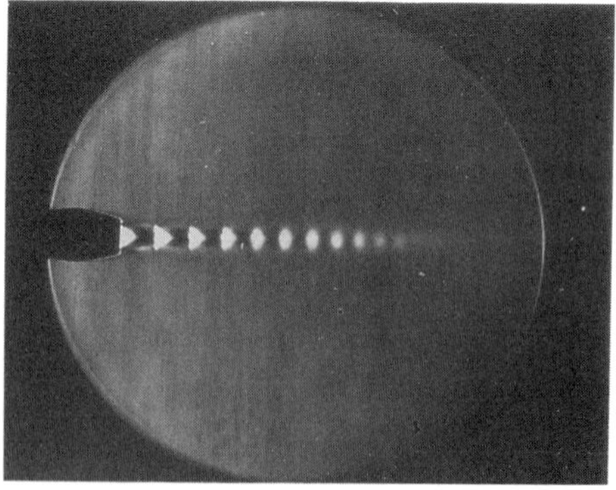

FIG. 18. Pressure waves in air escaping from a nozzle. Photo taken with a Toepler schlieren arrangement.

vation of electrical discharges and plasma research (electron concentration in plasma); heat exchange processes; and surface effects in vapor-liquid two-phase processes.

GLOSSARY

Abbe Number: Dispersive power; an expression characterizing the dispersion of a medium.

Amici Prism: See **Direct-Vision Prism.**

Beam Splitter: Glass plate with a semitransparent face that splits an incident beam of light into a refracted and a reflected one.

Birefringence: See **Double Refraction.**

Coherence: Coincidence of two electromagnetic waves in certain parameters; a condition for the interference of light beams.

Collimator: A device for the generation of a bundle of parallel light beams by illuminating a pinhole (or slit) in the focal plane of an objective.

Critical Angle of Refraction: Largest angle of refraction at the face of two substances.

Critical Angle of Total Reflection: Smallest angle of incidence for which total reflection at the face of two substances occurs.

Critical-Angle Refractometer: Refractometer based on the measurement of the critical angle of refraction/reflection.

Direct-Vision Prism: Combination of two or more prisms so that light of a selected wavelength will leave the prism undeviated.

Dispersion: $dn/d\lambda$, dependence of RI on the wavelength of light.

Double Refraction: A light beam is split into two beams; one (ordinary ray) is refracted independently of its angle to the optical axis of the medium, while the refraction of the other (extraordinary ray) depends on this angle.

Integrated-Optics Refractometer: Refractometer containing integrated-optics constituents.

Interference Fringe Pattern: A system of interference fringes generated by an interferometer.

Interference of Light: Superposition of two (or more) light waves so that they may either attenuate or amplify each other; also extinction may occur.

Refractodensitometry: Analytical method based on the measurement of RI and density.

RI = Refractive Index: Quotient of velocity of light in a vacuum and that in a medium.

RI Gradient: dn/dx; local change of RI within a medium.

RI Increment: dn/dc; change of RI of a solution with the concentration of the solute.

Schlieren: Striae, streaks; inhomogeneities (RI gradients) within a medium, caused by differences of temperature, pressure, or concentration.

Schlieren Apparatus: Instrument for recognition of striae.

Total Reflection: Reflection of light at a surface without loss of intensity.

Works Cited

Baltog, I., Ghita, C., Ghita, L. (1970), *Eur. Polym. J.* **6,** 1299–1303.

Clerk, D., Lukosz, W. (1993), *Sensors Actuators B* **11,** 461–465.

Clerk, D., Lukosz, W. (1994), *Sensors Actuators B* **18–19,** 581–586.

Hassan, G. E., El-Kashef, H., El-Baradie, B. Y., El-Labban, M. (1995), *Rev. Sci. Instrum.* **66,** 38–42.

Huglin, M. B. (1965), *J. Appl. Polym. Sci.* **9,** 3963–4024.

ICUMSA (1974), *Refractive Index,* ICUMSA Report, Subject 12, 144–156.

Joffe, B. W. (1983), *Refractometric Methods of Chemistry* (Russian), 3rd ed., Leningrad: Chimia.

Lorimer, J. W. (1972), *Polymer* **13,** 46–56 and 274–276.

Nebe, W. (1977), "Analytical Applications of Interferometry," in: G. Svehla (Ed.), *Comprehensive Analytical Chemistry,* Vol. 8, Amsterdam: Elsevier, pp. 391–546.

Philpot, J. St. L. (1938), *Nature* **141,** 283.

Schardin, H. (1942), "Die Schlierenverfahren und ihre Anwendungen," *Ergebn. Exakt. Naturwiss.* **20,** 303–439.

Svensson, H. (1939), *Kolloid Z.* **87,** 181; **90,** 141.

Toepler, A. (1864), *Beobachtungen nach einer neuen optischen Methode,* Bonn: Cohen.

Weissberger, A. (Ed.) (1972), *Techniques of Chemistry,* Vol. 1, *Physical Methods of Chemistry,* New York: Wiley Interscience.

Wolter, H. (1967), *"Schlieren-, Phasenkontrast- und Lichtschnittverfahren,"* in: S. Flügge (Ed.), *Handbuch der Physik,* Vol. 24, Berlin: Springer, pp. 555–645.

Further Reading

Born, M., Wolf, E. (1991), *Principles of Optics,* Oxford, U.K.: Pergamon.

Creath, K. (1988), "Phase-Measurement Interferometry Techniques," in: E. Wolf (Ed.), *Progress in Optics,* Vol. 26, pp. 349–393.

International Conference on Refractometry (1994), Warszawa, Abstracts; Polish Chapter of the International Society for Optical Engineering (SPIE).

Jenkins, F. A., White, H. E. (1976), *Fundamentals of Optics,* 4th ed., New York: McGraw-Hill, 746 pages.

Malacara, D. (Ed.) (1978, 1991), *Optical Shop Testing,* New York: Wiley, 523 pages.

Nebe, W. (1970), *Analytische Interferometrie,* Leipzig: Geest and Portig.

Schmitz, E. H. (1981–1993), *Handbuch zur Geschichte der Optik,* Vols. 1 to 5, Bonn and Ostende: Wayenborgh.

Weissberger, A. (Ed.) (1972), *Techniques of Chemistry,* Vol. 1, *Physical Methods of Chemistry,* New York: Wiley Interscience.

REFRACTORY MATERIALS—CERAMICS

GERALD ROUTSCHKA, *DIFK—Deutsches Institut für Feuerfest und Keramik GmbH (formerly Forschungsinstitut der Feuerfest-Industrie), Bonn, Federal Republic of Germany*

KARL-ERNST GRANITZKI, *Königswinter, Federal Republic of Germany (Sec. 5.2)*

	Introduction	277
	Definitions	277
1.	**General Classification and Nomenclature**	278
2.	**Production**	278
3.	**General Composition, Properties, and Testing**	279
3.1	Porosity	280
3.2	Bulk Density, Density	280
3.3	Cold Mechanical Properties	281
3.4	Thermal Expansion	281
3.5	Reheat Change	281
3.6	Refractoriness under Load (R.U.L.)	281
3.7	Creep in Compression	282
3.8	Thermal Conductivity, Specific Heat, and Emission	283
3.9	Hot Mechanical Properties	283
3.10	Elasticity and Stress–Strain Behavior	283
3.11	Thermal-Shock Resistance	285
3.12	Electrical Resistance	286
3.13	Corrosion Resistance	286
4.	**Types of Refractory Ceramics**	286
4.1	Dense, Shaped Refractory Products	286
4.1.1	Silica Bricks and Fused-Silica Ceramics	286
4.1.2	Fireclay and High-Alumina Bricks	288
4.1.2.1	Fireclay Bricks	288
4.1.2.2	High-Alumina Bricks	289
4.1.3	Basic Bricks	290
4.1.3.1	Magnesia Bricks	290
4.1.3.2	Doloma Bricks	291
4.1.3.3	Basic Bricks Containing Chrome	291
4.1.3.4	Spinel-Containing Bricks	292
4.1.3.5	Forsterite Bricks	292
4.1.4	Zircon- and Zirconia-Containing Bricks	292
4.1.5	Carbon and Graphite Bricks	293
4.1.6	Carbon- and Graphite-Containing Refractories	294
4.1.6.1	Fired Graphite-Containing Refractories	294
4.1.6.2	Pitch- and Resin-Bonded Unfired Refractories	294
4.1.7	Silicon Carbide Bricks	295
4.1.8	Fine-Grained Oxide and Nonoxide Ceramics	296
4.1.9	Fusion-Cast Bricks	297
4.1.10	Refractory Ceramics with Low Thermal Expansion	297
4.2	Unshaped Refractory Materials	297
4.2.1	Ramming Mixes	299
4.2.2	Plastic Mixes	299
4.2.3	Refractory Castables	299
4.2.4	Gunning Materials	301
4.2.5	Prefabricated Shapes	301
4.2.6	Jointing and Coating Materials	301
4.3	Heat-Insulating Ceramic Materials	301
4.3.1	Insulating Bricks	302
4.3.2	Insulating Refractory Bricks	302
4.3.3	Refractory Ceramic Fibers	302
5.	**Selection and Use**	304
	Glossary	304
	Works Cited	306
	Further Reading	307

INTRODUCTION

Definitions

According to ISO/R 836, refractory materials are nonmetallic materials or products (including those containing a proportion of metals) having a minimum pyrometric-cone equivalent of 1500 °C. Determination of the pyrometric-cone equivalent (refractoriness) is described in ISO/R 528. Test pieces of the refractory material in the form of small cones are raised in temperature alongside reference

3-527-28138-X/96/$5.00 + .50

pyrometric cones (blunt-tipped, skewed, triangular pyramids) with a defined temperature of collapse (softening).

Certain products from the refractory industry do not comply with this definition, however, but still fall under refractory classifications and are therefore conventionally termed refractories.

In the common sense, refractories are ceramic construction materials capable of withstanding temperatures above 800 °C. Numerous refractory compositions in a wide variety of shapes and forms are used in a broad range of industrial applications. The three main groups of refractory ceramics are shaped refractory products (bricks), unshaped refractory products (monolithics), and ceramic-fiber products. Their three main applications are the construction of reaction chambers (e.g., furnaces), thermal insulation, and recycling of heat in regenerators or recuperators. In some cases refractory ceramics are used as functional or mechanical parts of hot-working aggregates, especially for metal melt processing, like sliding gates and filters.

The temperature of use for "hot face" linings ranges from ca. 800 to 1800 °C, and sometimes to 2000 °C or higher.

In addition to high temperature resistance, other essential properties for refractories are resistance to temperature change (thermal shock) and to mechanical and corrosive influences.

1. GENERAL CLASSIFICATION AND NOMENCLATURE

Classifications of refractory products generally make reference only to characteristic limits for specific property values. The nomenclature and forms of designation are derived from features such as raw material base, manufacturing process, nature of bond, and typical application. The main classification criteria are the following:

1. Chemical and/or mineralogical constitution (see Table 1).
2. Total porosity: dense refractory products, <45 vol%; insulating refractory products, ≥45 vol%.
3. Form of delivery: shaped products (bricks and shapes), unshaped products (monolithics), or fibrous products (ceramic-fiber products).
4. Type of bonding: bonding at <150 °C, unfired products—ceramic bond (clay bond), inorganic chemical bond (e.g., phosphate), hydraulic bond, or organic bond (e.g., pitch, resin); bonding at 150 to 800 °C, thermally treated (tempered) products—inorganic chemical bond, hydraulic bond, organic bond (e.g., carbon bond); bonding at >800 °C—firing (ceramic bond), fusion and solidification.
5. Application field.

2. PRODUCTION

The refractories industry consumes a wide variety of raw materials, binders, and special additives (*Am. Ceram. Soc. Bull.*, 1992; Coope and Dickson, 1991).

Most of the raw materials are based on naturally occurring beneficated, upgraded, crushed minerals, usually acquired in captive production by the refractories manufacturer. Several high-purity synthetic raw materials have become important in recent years because of the increasing demand for high-performance refractories. Synthetic raw materials are produced by special suppliers.

Various methods are employed to produce refractory ceramics of different types:

1. refractory products with large grain sizes, normally ≤6 mm, in special cases ≤20 mm;
2. special fine-grained refractory products (fine ceramics);
3. products formed via a melt (i.e., fused cast shapes);
4. ceramic-fiber products.

The production process includes the following steps: preparation of mixes, shaping, drying, thermal treatment up to 800 °C, or firing, aftertreatment (if necessary), and packing.

During firing the characteristic ceramic microstructure develops as a result of reactions, recrystallization, and transformation or partial formation of a liquid phase or melt, respectively (at room temperature this phase consolidates to a glassy phase).

The main groups of refractory ceramics roughly share in total production as follows:

Table 1. Classification of refractory products (ISO 1109–1975).

Products	Limiting content of principal constituent	Criteria of subdivision and general observations
High-alumina products Group 1	$Al_2O_3 \geq 56\%$	A complete designation of these products shall include an indication of either the raw material actually used or the mineralogical constitution of the final product. In the latter case, the method of ascertaining this constitution shall be stated
High-alumina products Group 2	$45\% \leq Al_2O_3 < 56\%$	
Fireclay products	$30\% \leq Al_2O_3 < 45\%$	
Low-alumina fireclay products[a]	$10\% \leq Al_2O_3 < 30\%$ $SiO_2 < 85\%$	
Siliceous products[a]	$85\% \leq SiO_2 < 93\%$	
Silica products	$SiO_2 \geq 93\%$	Quality specifications according to use
Basic products		In view of recent and possible future developments in basic products, new subdivisions and new criteria of classification may become necessary
Magnesite (magnesia)	$MgO \geq 80\%$	Products in which the principal constituent is magnesite (magnesia)
Magnesite–chrome (magnesia chrome)	$55\% \leq MgO < 80\%$	Products in which the principal constituents are magnesite (magnesia) and chromite
Chrome–magnesite (chrome magnesia)	$25\% \leq MgO < 55\%$	Products in which the principal constituents are chromite and magnesite (magnesia)
Chromite	$Cr_2O_3 \geq 25\%$ $MgO \leq 25\%$	Products in which the principal constituent is chromite
Forsterite		Products in which the principal constituent is forsterite
Dolomite (doloma)		Products in which the principal constituent is dolomite (doloma)
Special products		Products based on carbon, graphite, zircon, zirconia, silicon carbide, carbides (other than silicon carbide), nitrides, borides, spinels (other than chromite)
		Products based on several oxides (other than those of basic products)
		Products based on pure oxides, including alumina, silica, magnesia, zirconia; products of high purity

[a]Products designated in certain countries by the name "semi-silica products" may belong either to the low-alumina fireclay class or to the siliceous products class.

silica bricks, 1%–2%; fireclay bricks, 15%; high-alumina bricks, 12%; basic bricks, 20%–30%; other bricks, 6%; and unshaped refractories, 35%–50% (mainly castables and gunning mixes).

3. GENERAL COMPOSITION, PROPERTIES, AND TESTING

The properties of refractory products are determined primarily by their chemical compositions, mineral compositions, and structure. Structure here refers to the type, quantity, size, form, and arrangement of phases in the substance involved. Refractory ceramics generally comprise three structural elements: coarse grain, binding matrix, and pores. The binding matrix is of special importance with respect to refractoriness and corrosion resistance. It is usually associated with a higher degree of porosity than the grains, which have well-defined properties.

Fired bricks possess fixed, measurable properties. By contrast, unfired and unshaped ceramic products acquire their properties through use, in the course of which a temperature gradient is usually established. This temperature gradient gives rise to considerable property gradients between the hot face (sintered at the high-temperature end) and what is often a loose mass at the low-temperature face.

Properties are ascertained by applying various tests, most of which are standardized (Annual Book of ASTM Standards, 1995; *Refractory Materials, Recommendations,* 1990). Values depend on the test method employed and, in the case of unshaped products, also on the method of preparation of test pieces. Even standardized test methods are of limited value because of problems with reproducibility and comparability, a problem especially prevalent with destructive testing.

Table 2 summarizes important property tests conducted on refractory materials.

3.1 Porosity

Porosity has a considerable effect on properties (Fig. 1). Three types of porosity can be distinguished: total, open (apparent), and closed porosity. Open porosity is normally used to characterize porosity of refractory ceramics. The amount of closed pores in refractory ceramics with total porosity exceeding 15% is small (ca. <3%.)

Relationships between open porosity and other properties can be established most accurately by determining pore-size distribution (Carswell, 1977) and gas permeability.

3.2 Bulk Density, Density

Bulk density characterizes the weight of a product per unit volume, including all pores. It is needed to calculate porosity, refractory requirement, weight of a refractory lining, stored heat, etc. Bulk density is influenced by the material density (weight per unit volume, excluding pores), which is a constant for each material, and total porosity. Densities for refractory ceramics range between 2.2 g/cm^3 (fused silica) and 5.9 g/cm^3 (ZrO_2).

Table 2. Tested characteristics of refractory ceramics.

Chemical, mineralogical	Texture	Cold mechanical	Hot mechanical	Thermophysical	Thermochemical
Chemical analysis[a]	True density	Cold compressive strength[a]	Hot compressive strength	Thermal expansion[b]	Refractoriness[b] (pyrometric-cone equivalent)
Microprobe analysis	Bulk density[a]	Cold modulus of rupture[b]	Hot modulus of rupture[b]	Thermal conductivity[b]	Corrosion resistance (slag, melts, vapors)
Microscopy	Porosity[a]	Abrasion resistance		Specific heat	CO resistance
X-ray diffraction	Permeability[b] to gas (air)			Electrical resistance	Oxidation resistance
	Pore-size distribution	Modulus of elasticity (stress–strain behavior)			Hydration resistance
	Nondestructive testing (mainly ultrasonic)		Change in dimensions on heating[b]		Acid resistance
			Refractoriness under load[b]		Bubble formation (only for glass-melt–contacting materials)
			Creep in compression		
			Thermal-shock resistance[b]		
		Special and additional tests and test conditions for			
Unshaped refractories	Joining and gunning mixtures		Carbon-containing refractories		Ceramic fibers[d]
Grain-size distribution Consistency, workability Moisture content[b]	Bonding or adhesive strength		Tests in neutral or reducing atmosphere,[c] special chemical analysis methods, coking behavior, residual carbon content[b]		Compressibility Shot content (nonfibrous particles) Tensile strength

[a]For statistical quality control.
[b]Additional information for data sheets.
[c]Le Doussal and Prieur, 1989.
[d]Dietrichs, 1987.

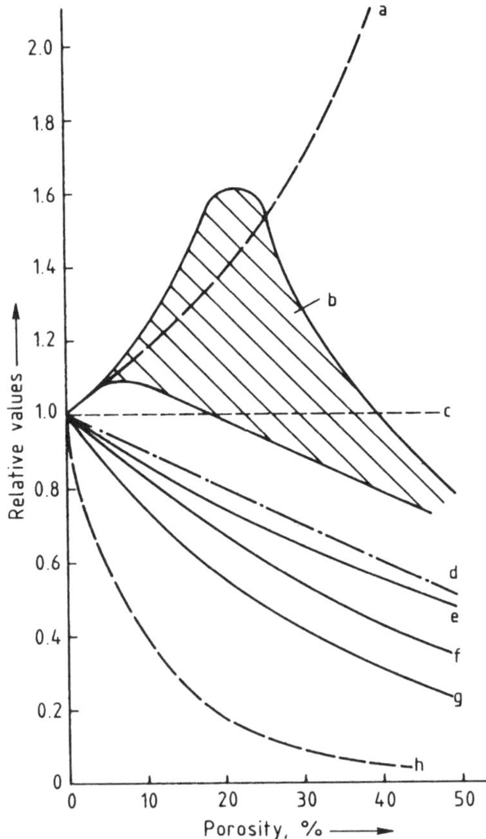

FIG. 1. Porosity effects on the properties of refractory ceramics: Curve a, corrosion; curve b, thermal-shock resistance; curve c, thermal expansion; curve d, bulk density; curve e, strength (fine ceramics); curve f, thermal conductivity; curve g, modulus of elasticity; curve h, cold compressive strength (coarse refractories).

3.3 Cold Mechanical Properties

Cold crushing and modulus of rupture (bending) tests provide information about the structure and structural uniformity of a ceramic and assist in statistical quality control.

3.4 Thermal Expansion

Thermal expansion is a material constant determined primarily by the basic component of the ceramic (e.g., SiO_2, MgO). With the exception of silica- and zirconium-oxide–containing products, fired ceramic materials have smooth thermal-expansion curves (Fig. 2). During initial heating, unfired bricks and unshaped refractories show overlap between reversible thermal expansion and irreversible expansion/shrinkage effects, which are influenced even by small loads. Reversible and permanent changes in length are of importance for construction of a refractory lining, e.g., when measuring joints. Because thermal expansion cannot be suppressed even with high loads, some fraction (usually 0.5–0.8) of the laboratory expansion value is used for dimensioning joint thickness.

3.5 Reheat Change

Heating to high temperatures and subsequent cooling often causes a permanent, time-dependent change in dimensions. If a refractory lining shows substantial shrinkage, joints or cracks may form. The opposite case, after-expansion, is also dangerous, because the resulting pressure may destroy the refractory lining. Dense bricks should not show more than 1%–1.5% linear shrinkage at maximum service temperature.

The classification-temperature criterion for insulating and unshaped refractory products is limited linear heat shrinkage at a defined temperature and holding time (5–24 h). The following shrinkage limits apply: insulating fire bricks, max. 2%; ceramic-fiber products, max. 4%; castables, max. 2%; ramming mixes, max. 2% drying and firing shrinkage; and moldable mixes, max. 3% drying and firing shrinkage.

3.6 Refractoriness under Load (R.U.L.)

Determination of the refractoriness under load (differential, with rising temperature in oxidizing atmosphere; ISO 1893) is a characteristic test for refractory products. The sample is normally a cylinder (height 50 mm, diameter 50 mm) with an internal axial bore (diameter 12.5 mm). This is heated in air under a load of 0.2 N/mm^2 at a rate of 5 °C/min until it becomes soft. Characteristic temperatures (e.g., T_{05}, the temperature at which the sample is compressed by 0.5%) are read from a height-change–temperature curve (Fig. 3). Some products display a broad softening range (e.g., fireclay bricks), while others show a sudden collapse (e.g., silica and magnesia bricks).

The precursor of ISO 1893, described for

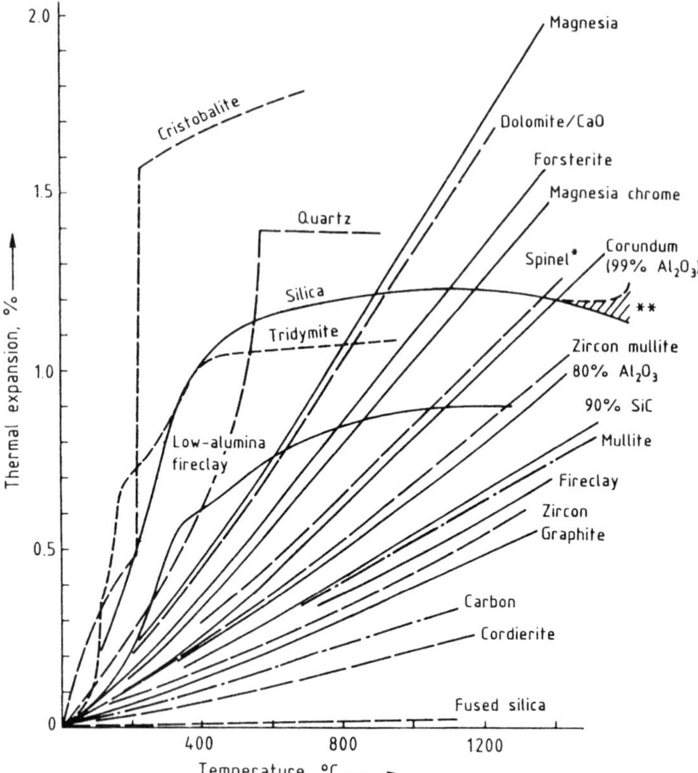

FIG. 2. Thermal-expansion curves of fired refractory materials (VDI-Richtlinie 3128, Blatt 4, 1977). *$MgO \cdot Al_2O_3$, $MgO \cdot Cr_2O_3$, chrome ore. **After expansion when residual quartz is present. For thermal expansion of ZrO_2 ceramics see Fig. 8.

instance in DIN 51064, is still used to some extent. The test is conducted in a reducing atmosphere and the measured expansion includes the expansion of the loading columns (carbon). The characteristic temperature is t_a, the temperature at which the sample is compressed by 0.6%. Generally higher values than ISO 1893 are found, especially with materials that are sensitive to redox conditions (e.g., iron-rich magnesia).

Softening behavior under pressure is influenced by the type of refractory material, its mineralogical and pore structure, and the quantity and the viscosity of the melts formed at testing temperature. Softening behavior improves with increasing firing temperature of the refractory material.

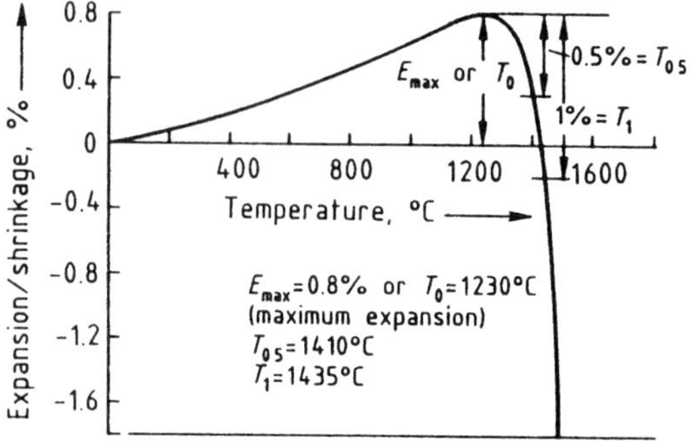

FIG. 3. Refractoriness under load curve (example according to DIN 51053, Part 1).

3.7 Creep in Compression

Refractory products are often exposed to high temperature and pressure for extended periods of time (e.g., the checkerwork of regenerators). This results in slow shrinkage and creep processes that influence refractory-lining lifetimes. Such behavior is characterized by the creep test (ISO 3187), which is performed analogous to ISO 1893 but at constant temperature and pressure over 25 h or longer (Fig. 4).

Creep over longer periods can be estimated by extrapolation. For rough calcula-

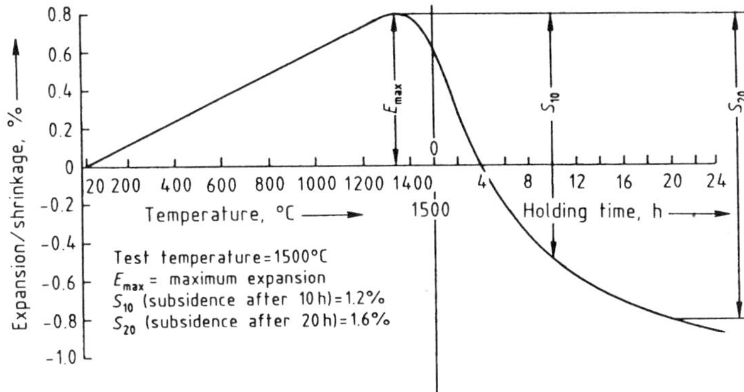

FIG. 4. Creep curve (example according to DIN 51053, Part 2).

tion, the creep increases proportional to the applied load (Konopicky *et al.*, 1971).

3.8 Thermal Conductivity, Specific Heat, and Emission

Thermal-conductivity values are used primarily to calculate the temperature distribution in a refractory wall and the heat loss of a furnace. In some cases products with high conductivity are used to keep the temperature of the hot side of the lining low, thereby reducing wear. The thermal conductivity of a refractory material depends on its chemical and mineralogical composition, and especially on its porosity and pore-size distribution. The atmosphere is also of great importance, most notably in the case of porous products (Young *et al.*, 1964).

There are two standard methods for measuring thermal conductivity: the hot-wire test (ISO 8894) and the calorimeter method (ASTM C 201). Values obtained by the two methods may differ by up to 25%. The method used should therefore be specified whenever thermal-conductivity values are given. This is particularly important for unshaped products; in such cases, special attention must also be given to pretreatment of the sample (Routschka, 1988). The thermal conductivity of refractory products ranges from ca. 0.05 (ceramic fibers) to 100 W m^{-1} K^{-1} (graphite) (Fig. 5).

Specific heat is used to calculate the heat stored by a lining, and for determining the temperature diffusivity that results from thermal conductivity in order to calculate nonstationary temperature fields in, for example, regenerators. The specific heat for conventional refractory products is ca. 1 kJ kg^{-1} K^{-1} (VDI-Richtlinie 3128, 1977).

The thermal emission of a refractory construction material depends on the type of material and the nature of the surface. Refractory products can be regarded as graybody radiators, with emissions 50%–85% as great as blackbody radiators (Glazman *et al.*, 1989).

3.9 Hot Mechanical Properties

Thermomechanical behavior is important in the application of a refractory product, and it can be characterized by high-temperature strength testing. Hot strength is a complex parameter that depends on the structure, quantity, and viscosity of the formed melts at high temperature. The temperature dependence of thermal behavior is a function of the type of loading involved (e.g., bending, compression, tension, or torsion).

One common measure of the thermomechanical character is the hot modulus of rupture test (ISO 5013), which is frequently incorporated into the quality control of refractory products. Examples of the hot modulus of rupture of different brick types are shown in Fig. 6.

3.10 Elasticity and Stress–Strain Behavior

A modulus of elasticity value is determined in order to estimate the critical stress on a refractory material brought about by mechanical and thermal loading. The static modulus of elasticity is established by measuring deformation under compression, bending, and tension. For simplicity, dy-

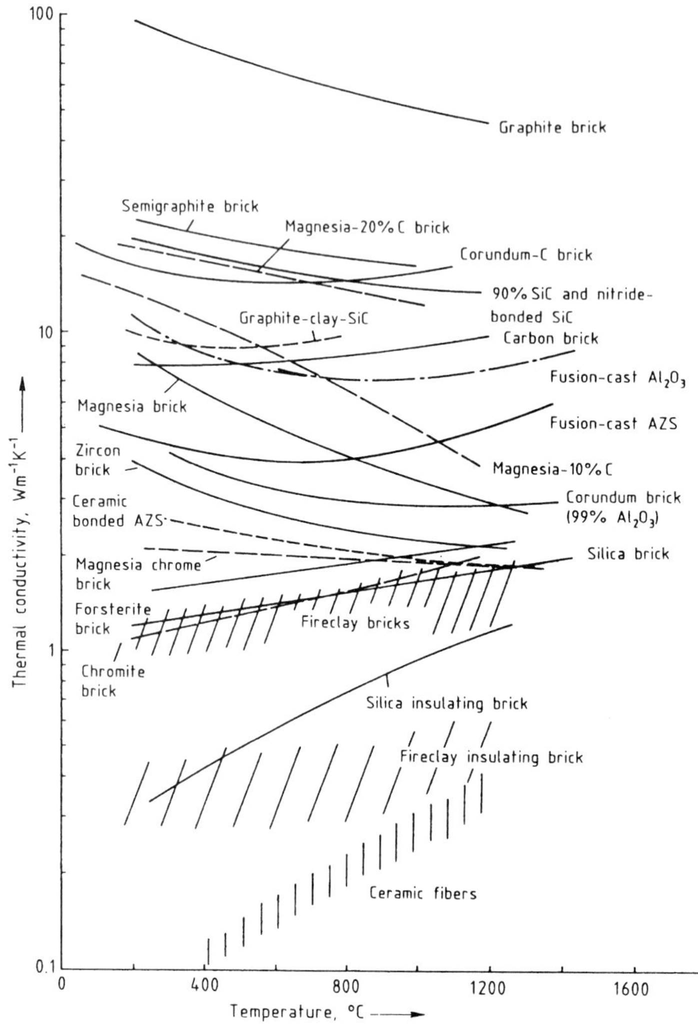

FIG. 5. Thermal conductivity of refractory products (VDI-Richtlinie 3218, Blatt 4, 1977, completed)

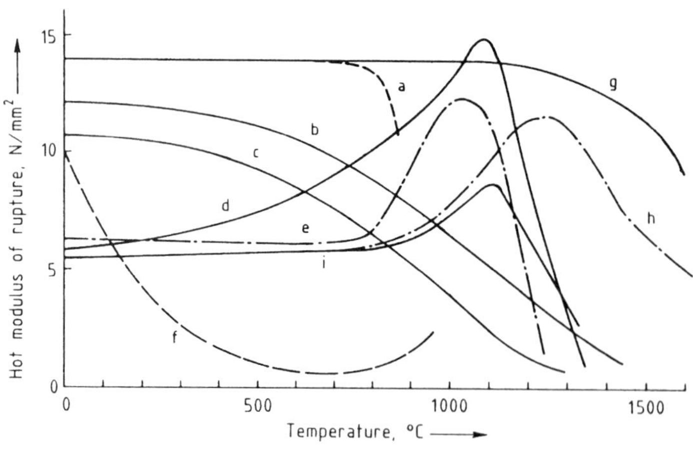

FIG. 6. Hot modulus of rupture of refractory bricks: Curve a, low-iron magnesia with 0.2% B_2O_3; curve b, corundum (99% Al_2O_3); curve c, iron-rich magnesia; curve d, low-alkali fireclay; curve e, alkali-rich fireclay; curve f, chemically bonded magnesia chrome; curve g, low-iron magnesia (C_2S bond); curve h, high-fired magnesia chrome; curve i, sillimanite.

namic elasticity is determined by means of ultrasonic velocity or resonance frequency. Values obtained by different methods can differ substantially, and they also depend on the type and structure of the refractory material under investigation (Table 3; Das and Jeschke, 1975). Refractory bricks often show strictly elastic behavior only within a narrow deformation range. Increasing stress and temperature lead to irreversible deviations from linearity. This is a consequence of deformations caused by mechanical or creep factors at higher temperatures. Irreversible changes are an indirect result of the porous, inhomogeneous textures of refractory materials.

The temperature dependence of stress–strain curves for a refractory material provides an indication of the elasticity of the microstructure (relaxation) and the onset of plastic deformation.

3.11 Thermal-Shock Resistance

Thermal-shock resistance is an important property that describes the behavior of a material during a more or less rapid change in temperature. Cracks can lead to considerable wear due to spalling, and they also increase susceptibility to corrosion. The test procedures are quite diverse (water or air quenching, dipping in metal melts, flame-impingement tests, cycling between two temperatures, etc.). Thermal-shock resistance is highly dependent on sample shape and the test conditions. Densification, uptake of foreign material, and the formation of zones of differing thermal expansion during refractory-product use can also lead to substantial changes in thermal-shock resistance. Thus, only a test procedure adapted to suit specific service conditions has any chance of success. Another way of circumventing the difficulties with thermal-shock tests is to invoke thermal-shock parameters. These represent an attempt to classify materials in terms of other properties (e.g., strength, modulus of elasticity, thermal expansion, thermal conductivity, work of fracture, and fracture toughness) (Hasselmann, 1978; Uchida *et al.*, 1990).

Two important parameters have been defined to assess the thermal-shock resistance of refractories:

Table 3. Strength and elasticity properties of refractory bricks (Das and Jeschke 1975).

Brick type	Microstructure	Cold strength (N/mm^2)			Static modulus of elasticity (10^4 N/mm^2)				Dynamic modulus of elasticity (10^4 N/mm^2)		
		Compression	Bending	Tension	Compression	Bending	Tension	Torsion	Young's modulus	Shearing modulus	Poisson's ratio
Magnesia	Crystalline, homogeneous	50	6	3.5	4.5	2.0	2.2	0.8	7.0	2.9	0.2
Magnesia–chrome	Crystalline, inhomogeneous	65	4	1.5	3.0	1.0	0.8	0.3	2.9	1.2	0.2
Fireclay	Crystalline + glassy phase	30	7	4.5	1.0	0.8	0.9	0.4	2.3	1.0	0.1
Fused silica	Glass, homogeneous	400	20	10	4.0	3.0	1.6	1.2	4.8	2.1	0.14

$R'''' = WE/S^2$

(situation involving a fracture of a catastrophic nature), and

$R_{ST} = [W/\alpha^2 E]^{0.5}$

(situation in which crack propagation is stable or quasistable), where W = work of fracture, E = modulus of elasticity; S = modulus of rupture, and α = thermal-expansion coefficient.

3.12 Electrical Resistance

The electrical resistance of a refractory material is important with respect to the insulation and operation of electrical furnaces. Conventional refractory ceramics display very high electrical resistance at room temperature ($>10^5$ Ω m), but this decreases with increasing temperature, reaching values like 1 Ω m at 1400 °C (VDI-Richtlinie 3128, 1977). Even low concentrations of foreign components can substantially reduce electrical resistance.

3.13 Corrosion Resistance

The corrosion resistance of a refractory product is influenced primarily by the chemical compositions of the corrosive agents and the refractory material. Porosity, pore-size distribution, and mineral structure of the refractory material determine the reaction kinetics. Corrosion is greatly affected by specific operating conditions (furnace atmosphere, temperature, temperature gradient in the lining, crack formation due to additional temperature changes, static or dynamic action of the corrosive media), and so conclusions drawn from laboratory slagging experiments are of limited value unless these have been carefully related to real operational conditions. The diverse character of corrosive attack on refractory materials is reflected in the numerous corrosion tests proposed (Kobayashi *et al.*, 1982).

Refractory products are subject to three main types of corrosive stress:

1. effects of the atmosphere and changes in the atmosphere,
2. attack by vapors, flue dusts, and condensates, and
3. attack by slag and other fluid media.

Crucible tests are usually performed in cavities prepared in cubes of the test material. The corrosive material (e.g., glass, slag, or a salt) is placed in the prepared cavity and the crucible is then heated. This test provides only a rough classification of materials.

The rod-immersion method has achieved importance for testing the attack of melts on refractory materials. Bars of the refractory ceramics are dipped into the melt. By rotating the bars a dynamic test is achieved.

Semi-industrial corrosion tests are somewhat extravagant in terms of time and equipment, and are a response to the demand for a test procedure that provides assessments as nearly equivalent as possible to actual operating conditions. This in turn implies that each procedure must be adapted to a specific application. Special emphasis is placed on the testing of large samples (e.g., a brickwork), which allows one also to test the joining materials.

4. TYPES OF REFRACTORY CERAMICS

4.1 Dense, Shaped Refractory Products

Characteristics of selected types of conventional refractory bricks are summarized in Tables 4 and 5.

4.1.1 Silica Bricks and Fused-Silica Ceramics Silica bricks ($>93\%$ SiO_2) (Jeschke *et al.*, 1989; Lepère *et al.*, 1992) consist mainly of silicon dioxide (SiO_2), generally described as silica. Silica occurs in a variety of crystalline modifications (e.g., quartz, cristobalite, tridymite), each with a high- and low-temperature modification, and also as an undercooled melt called fused silica. The crystal structures of the SiO_2 modifications differ widely, and so distinct density (volume) changes and steps in the thermal expansion are seen during transformation. This is of great importance during heating and cooling (i.e., during firing and use of silica refractories).

The raw materials for silica bricks are naturally occurring quartzite and other quartz materials with low Al_2O_3 and low alkali-oxide content. The grained material is

Table 4. Properties of conventional refractory bricks of the system SiO_2–Al_2O_3; for maximum service temperatures see Fig. 7.

Type	Al_2O_3 (%)	SiO_2 (%)	Fe_2O_3 (%)	$Na_2O + K_2O$ (%)	Density (g/cm^3)	Bulk density (g/cm^3)	Open porosity (%)	CCS[a] (N/mm^2)	T_{05}, °C
Fused silica, compact		99.5			2.2	2.16	0	200	
Fused silica, slip cast		>99			2.2	1.90	13–16	50–80	1640–1690[b]
Silica	<1.5	>95	<0.5	<0.2	2.34–2.36	1.75–2.0	15–23	20–100	
		CaO < 3							
Low-alumina fireclay	20–30		<2.5	1–3	2.5–2.65	1.9–2.3	15–25	>35	1250–1300
Fireclay	30–45		0.6–2.5	0.5–2.5	2.55–2.75	1.8–2.4	12–25	20–85	1300–1550
Andalusite, sillimanite	55–65		<1.5	<1	3.0	2.4–2.6	15–20	50–90	1470–1600
Bauxite	80–85	8–14	1.3	0.5	3.6	2.7–2.9	18–22	65–75	1400–1500
Corundum (80% Al_2O_3)	80–85		<1.3	<1	3.6	2.8	18	70	1550
Corundum (97% Al_2O_3)	>97		0.3	<0.4	3.96	3.1	<20	>50	>1650

[a]Cold compressive strength.
[b]Fused silica, when devitrified to cristobalite. In the glassy state T_{05} ca. 1350 °C.

Table 5. Properties of conventional ceramic-bonded basic bricks.

Type	MgO (%)	CaO (%)	SiO_2 (%)	Fe_2O_3 (%)	Al_2O_3 (%)	Cr_2O_3 (%)	Density (g/cm^3)	Bulk density (g/cm^3)	Open porosity (%)	CCS[a] (N/mm^2)	T_{05} °C
Magnesia (range)	85–95	0.8–3.9	0.2–6	0.2–6	0.1–5			2.8–3.1	14–18	40–110	1500 – >1700
Forsterite bonded	94	1.7	3.8	0.2	0.2		3.5–3.8			90	1700
C_2S bonded	96	2	1	0.5	0.2			3.0	15	60	>1700
99% MgO	99	0.1	0.1	0.1			3.58			55	>1740
Magnesia–spinel	86		0.3	0.8	13			2.95	15	55	>1600
Magnesia–chrome	55–80	<3	<5	5–12	<11	6–20	3.7–3.85	2.8–3.2	14–21	>30	>1500
Chrome–magnesia	25–25	<2.5	<5	7–15	<22	15–35	3.8–4	2.8–3.35	16–21	>40	>1500
Chromite	<25	<2	<7	9–18	<28	>25	3.8–4	3–3.3	17–21	>30	>1500
Forsterite	50–60	<1.5	<40	<8	<8	<1.5	3.35	2.6–2.8	15–21	>25	>1600
Doloma	>36	<61	<1.5	<1	<1		3.38	2.6–2.9	14–21	30–80	>1700

[a]Cold compressive strength.

normally bonded with about 2%–3% slaked lime and some sulfite lye solution as a temporary binder.

Fired silica bricks contain cristobalite, tridymite, perhaps some residual quartz (today usually <6%), some wollastonite ($CaSiO_3$), and about 2%–10% glassy phase. Silica bricks are characterized by very good thermal-shock resistance above 600 °C and high-temperature load-bearing capacity up to about 1600 °C.

Applications: coke ovens, hot-blast stoves, glass-melting furnaces (crowns), and furnaces for the ceramics industry.

Silica bricks in the applications cited usually show no deformation or critical wear. Acquired impurities migrate with melts formed in the bricks themselves deep into the colder part of the lining, leading to a zoned texture.

Fused-silica products are fabricated as fused blocks (manufactured with a special sink fuse method) or in special shapes starting from crushed fused grain, formed (often by slip casting), and fired at about 1050 °C. A characteristic feature of fused silica is its very low thermal expansion (i.e., high thermal-shock resistance). At about 1150 °C cristobalite crystallizes without any large volume change, but this crystallized material is sensitive to cooling below 300 °C (like silica bricks).

Applications: mainly in the glass industry, for hot repairs; submerged nozzles and shrouds for the casting of steel; foundries.

4.1.2 Fireclay and High-Alumina Bricks

These refractory products, also called alumosilicate bricks, belong to the system SiO_2–Al_2O_3 (Pask, 1990). Only three phases exist in this system: crystalline SiO_2 modifications, corundum (α-Al_2O_3), and mullite. The composition of the mullite ranges from $3Al_2O_3 \cdot 2SiO_2$ with about 72% Al_2O_3 (formed on sintering above about 1000 °C) to $2Al_2O_3 \cdot SiO_2$ with about 78% Al_2O_3 (crystallizes from melt).

The morphology of mullite crystals may vary considerably depending on the type and composition of the raw materials, the formation conditions during firing, and subsequent thermal treatment. Mullite shapes range from fine-needled crystals formed at low temperature in the presence of a melt to coarse crystals in materials with low flux content, fired at high temperatures, or in fused mullite. The shape of the mullite crystals and their arrangement in the texture of the brick (e.g., "mullite felt") have a pronounced influence on the properties of fireclay and high-alumina bricks, especially on hot strength.

An overview of refractoriness and maximum application temperatures for fireclay and high-alumina bricks is given in Fig. 7. Refractoriness in the field of fireclay refractories falls considerably below the liquidus line of the pure system SiO_2–Al_2O_3 mainly because of their high content of alkali oxides and alkaline earths (up to 4.5%). Maximum application temperatures are distinctly lower because of time-dependent shrinkage or creep, which significantly affects the life of a refractory lining. It is also necessary to distinguish between application under a thermal gradient (as in a refractory wall) and under quasi-isothermal conditions (checkers in regenerators). The broad range of maximum application temperatures for high-alumina bricks shows clearly the influence of the diverse chemical and mineralogical compositions of these bricks.

4.1.2.1 Fireclay Bricks. Fireclay bricks have an Al_2O_3 content of 10%–45%. They are manufactured from fireclay grog (see Glossary) and bonding clay mixtures. The amounts of bonding clay and added water vary, a factor that largely determines both the subsequent forming process (plastic, semidry, or dry shaping) and the properties of the resulting fired bricks. Low-alumina or siliceous fireclay bricks may contain as additives granular SiO_2 raw materials. Variation of the clays, grogs, and grain sizes permits the production of numerous brick types (see Stahl-Eisen-Werkstoffblatt 915, 917, 1984). Fireclay bricks contain up to ca. 3% Fe_2O_3 (for low-iron bricks, <1%), ≤3% TiO_2, <1% CaO + MgO, and alkali oxides, mainly K_2O (≤3.5%) and Na_2O (usually <0.25%). High-quality bricks include <0.5% Na_2O + K_2O.

Mullite, cristobalite, and/or quartz and a glassy phase (composition: ca. 80% SiO_2, 10% Al_2O_3, the rest fluxes, mainly alkalies) are the usual components of fireclay bricks. The mullite content is generally proportional to the Al_2O_3 content of the bricks. The amounts of cristobalite (0%–25%) and of the glassy phase (20%–60%) depend largely on

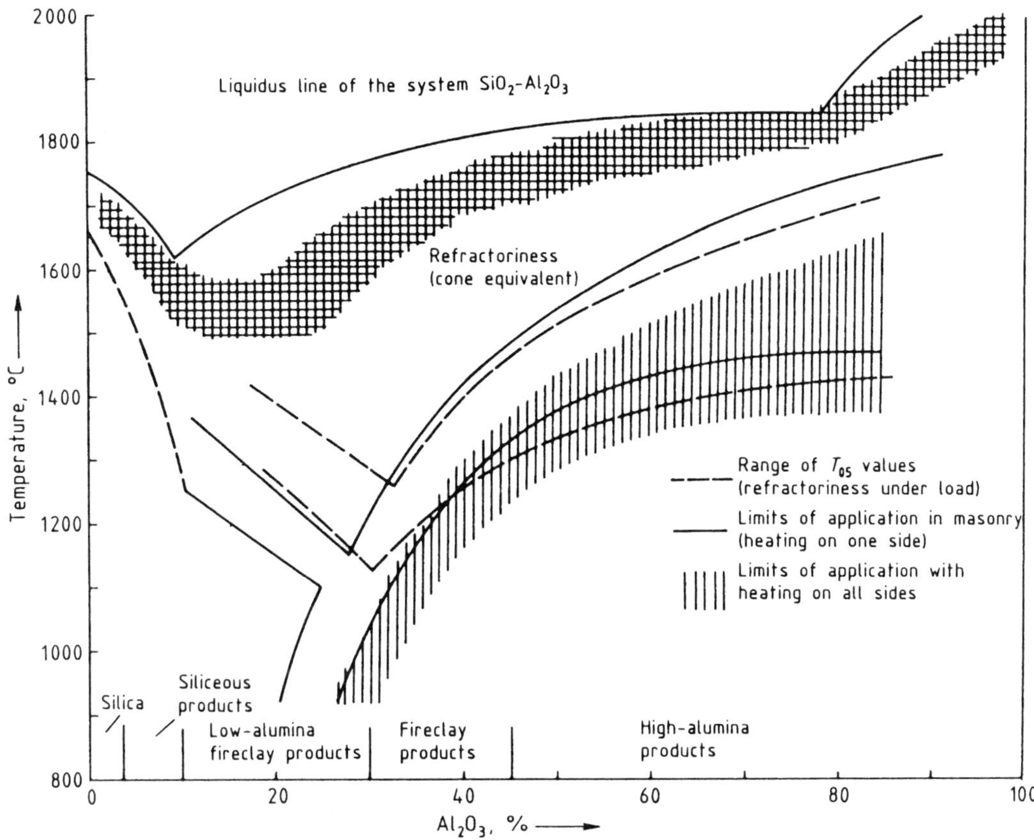

FIG. 7. Refractoriness, refractoriness under load, and maximum application temperature of fireclay and high-alumina bricks (Routschka, 1981).

the content of alkali oxides, which can be deduced from the phase diagram of Al_2O_3–SiO_2–K_2O.

Attack by slag on fireclay bricks normally occurs in relatively small reaction zones; wear progresses by erosion of the resulting highly viscous reaction zone. In contact with glass melts, alkali migrates into the brick texture and the bricks are slowly dissolved in the glass melt. Other forms of damage include alkali bursting (for bricks with >ca. 35% Al_2O_3), CO disintegration, and thermal spalling (dense bricks).

Main applications: general furnace construction, blast furnaces and hot-blast stoves, foundries, furnaces for the nonferrous-metal industry, coke ovens, glass industry, cement industry.

4.1.2.2 High-Alumina Bricks. Bricks with >45% Al_2O_3 are required for increased thermal and chemical stresses. These are produced from various high-alumina raw materials (sillimanite, andalusite, sintered and fused mullite, sintered alumina or fused alumina, often called corundum, or sintered bauxite), alone, in combination, or together with other raw-material types, mainly fireclay grog. The nomenclature applied to the products depends on the primary high-alumina raw material.

A small amount of clay (<15%) is normally used to produce ceramic-bonded fired bricks. Use of a mixture of clay and fine alumina leads to a mullite-bonded brick. A chemical bond on an aluminum–phosphate base has also proven successful regarding corrosion resistance.

Addition of Cr_2O_3 powder (5%–15%) to corundum bricks, either alone or in combination with ZrO_2-containing raw materials, leads to a highly corrosion-resistant corundum–chrome brick.

The chemical and mineralogical compositions of high-alumina bricks vary over wide ranges because firing of the bricks normally fails to produce phase equilibrium, and significant changes may occur in the properties of the bricks in service. Bauxite bricks, which constitute the majority of high-alumina bricks, show more or less after-expansion in service caused by reactions between the clay bond and corundum and resulting in mullitization.

The main mineralogical components are corundum and mullite. Sillimanite and andalusite bricks normally contain residual amounts of these minerals. Compared with fireclay refractories, the alkali-oxide content and the extent of glassy phase are low (<20%).

Properties of the different types of high-alumina bricks are compiled in Stahl-Eisen-Werkstoffblatt 912 (1984).

The corrosion resistance of such bricks to glass, slag, and iron and steel melts is good. They are prone to alkali attack: high-alumina bricks with low Al_2O_3 content react to produce low-melting alkali aluminum–silicate phases or glasses. With higher alumina content alkali bursting occurs (due to crystallization of, for example, nepheline, $Na_2O \cdot Al_2O_3 \cdot 2SiO_2$), as does the formation of β-Al_2O_3 ($Na_2O \cdot 11Al_2O_3$).

Main applications: iron and steel industry, nonferrous-metal industry, glass and cement industry, ceramic industry (also as high-temperature kiln furniture).

4.1.3 Basic Bricks (Alper, 1970) Basic bricks belong mainly to the system MgO–CaO and consist of magnesia (periclase), doloma (dolomite), and CaO without or in combination with chrome ore, chrome oxide, spinel ($MgO \cdot Al_2O_3$), and carbon (graphite). Forsterite, chrome ore, and spinel bricks are included as well, since these materials are often used in place of or in contact with magnesia bricks. An overview is provided in Table 6.

4.1.3.1 Magnesia Bricks. According to ISO 1109, magnesia bricks contain at least 80% MgO. A variety of magnesia raw materials are used to manufacture magnesia and magnesia-containing products, e.g., natural magnesite sinters, seawater magnesia, or fused magnesia with various chemical and mineralogical compositions. The magnesias contain varying amounts of Fe_2O_3 (<8%), Al_2O_3 (<0.3%), CaO (<4%), SiO_2 (<4%), and B_2O_3 (<0.2%). Besides the main mineralogical component periclase, silicates and ferrites occur as secondary components. The CaO/SiO_2 ratio and the B_2O_3 content (formation of low-melting components) determine the mineral compositions and properties (hot strength, corrosion resistance) of magnesia bricks. Monticellite ($CaO \cdot MgO \cdot SiO_2$, CMS) with a CaO/SiO_2 mass ratio of 0.94 has the lowest melting point (1495 °C) of the silicates.

High-performance bricks are highly fired, have a low iron content (<1.5% Fe_2O_3), and contain dicalcium silicate $2CaO \cdot SiO_2$ (short

Table 6. Classification and types of basic bricks.

Brick type	Groups' MgO content (%)	Remarks
Magnesia	80–98	Including impregnated fired and pitch- or resin-bonded (untempered and tempered) bricks (carbon content after coking <7%)
Magnesia–doloma	50–80	
Dolomite	40	
Magnesia–carbon	30–70	Addition of natural graphite
Magnesia–doloma–carbon	C ≥7	Pitch- or resin-bonded untempered and tempered bricks
Doloma–carbon	(up to 25 C)	With or without antioxidants
Magnesia–chrome	30–80	Subgroups: direct-bonded, rebonded, and chemically bonded bricks
Chrome ore	<30	
Magnesia–spinel	60–80	Addition of sintered or fused magnesium aluminum spinel
Magnesia–zirconia	80, 90	Addition of zirconia and/or zircon
Spinel	20, 30	$MgO \cdot Al_2O_3$, $MgO \cdot Cr_2O_3$
Forsterite	40, 50	Fired and unfired bricks
CaO	CaO 90	

formula: C_2S, m.p. 2130 °C) or forsterite $2MgO \cdot SiO_2$ (short formula: M_2S, m.p. 1890 °C) between the periclase crystals. These bricks show first subsidence at temperatures ≥1600 °C. Magnesia bricks display the highest thermal expansion of refractory ceramics and are therefore prone to thermal shock and thermally induced stress in furnace linings. Thermal-shock resistance can be increased by a coarse-grain distribution or the addition of Al_2O_3 (spinel formation on firing, microcrack formation) or chrome ore (texture elasticity). Special products include magnesia–forsterite (olivine) bricks and zircon- or zirconia-containing magnesia bricks.

High refractoriness, good resistance to attack by basic slags, and alkali resistance characterize magnesia refractories. Slagging causes diffusion and migration of silicate or ferritic melts deep into the brick, producing a zoned structure prone to peeling (structural spalling). A pitch or resin, or carbon bond, reduces the infiltration of metal and slags under reducing atmosphere service conditions.

Magnesia bricks are sensitive to hydration (humid atmosphere between 40 and 120 °C), SO_3 (magnesium–sulfate formation), and SiO_2 attack (forsterite bursting). Excessive growth of magnesia crystals during service, promoted by acquired impurities and high temperature, can also loosen brick texture.

Applications: steel-melting and -refining furnaces and vessels, regenerators for glass-melting furnaces.

4.1.3.2 Doloma Bricks. These consist of sintered dolomite (doloma), the main components of which are evenly distributed crystals of CaO and MgO. The principal impurities are calcium compounds like dicalcium ferrite, $2CaO \cdot Fe_2O_3$ (C_2F); brownmillerite $4CaO \cdot Al_2O_3 \cdot Fe_2O_3$($C_4AF$); tricalcium aluminate, $3CaO \cdot Al_2O_3$ (C_3A); tricalcium silicate (C_3S); and dicalcium silicate (C_2S). The SiO_2 content should be <1%.

Their free-lime content makes these products susceptible to hydration. For this reason, manufacture, transport, storage (packed in shrink foils), and installation require virtually moisture-free environments and early use or heating. Hydration resistance is increased by pitch or resin bonding and tempering or pitch dipping and impregnation of the fired bricks.

Doloma bricks are sensitive to attack by iron-rich and SiO_2-rich slags. The periclase component is more resistant than the CaO component. Fluid melts are formed with slags rich in FeO. MgO absorbs the FeO in a solid solution.

Applications: converters and ladles for the steel industry (pitch- or resin-bonded, tempered bricks) and the cement and lime industry (fired carbon-free bricks) in competition with magnesia and magnesia–spinel bricks.

4.1.3.3 Basic Bricks Containing Chrome. Most bricks of this type are fired mixtures of magnesia and varying amounts of chrome ore, usually in coarse to medium grain sizes, and sometimes also amounts of chrome oxide. The refractory chrome ores consist of a chromite, a spinel $(Mg,Fe^{II})O \cdot (Cr,Al,Fe^{III})_2O_3$ of variable composition depending on the ore deposit, and greater or lesser amounts of a magnesium silicate. The content of the silicate phase should be as low as possible. High-quality magnesia–chrome bricks have a SiO_2 content of <2.5% and a CaO + SiO_2 content of <3.5%. In contrast to magnesia bricks, the CaO/SiO_2 ratio and the B_2O_3 content have little influence on hot strength.

Special brick types are high-fired bricks, with low SiO_2 content and a predominant direct bond between magnesia, residual chrome ore, and so-called secondary spinels (direct-bonded bricks); bricks containing prereacted sintered or fused magnesia–chrome grain (rebonded basic bricks), the grain in this case containing a large amount of directly bonded phases; and chromite bricks, consisting mainly of chrome ore and a small amount of magnesia, which are quite inert and very resistant to acidic and basic slags.

The advantages of adding chrome ore to magnesia mixtures are better thermal-shock resistance, increased corrosion resistance to slags with a CaO/SiO_2 ratio <1.5 (acidic slags) due to the chromite, lower thermal expansion, and lower thermal conductivity.

Applications: ladles and vessels for vacuum treatment of steel and argon–oxygen decarburization (AOD) converters; rotating cement furnaces and glass-furnace regenerators (in these applications formation of toxic hexavalent chromium occurs and chrome-free alternative bricks like magnesia–spinel or magnesia bricks are used increasingly);

open-hearth furnaces; limited applications for chromite bricks.

4.1.3.4 Spinel-Containing Bricks. The magnesium–aluminum spinel $MgO \cdot Al_2O_3$ has recently attracted greater attention in the refractory industry (Dal Maschino *et al.*, 1988). Spinel (71.7% Al_2O_3, 28.3% MgO) is the only compound in the system $MgO \cdot Al_2O_3$, and it has a melting point of 2135 °C. First formation of a melt occurs at 2020 °C, so spinel can be classed among the high-temperature materials. Sintered spinel raw materials are available with 66%–90% Al_2O_3, as are fused spinels.

The following brick types are currently in active service:

Magnesia–spinel bricks consist mainly of very low-iron and low-SiO_2 magnesia together with sintered spinel (up to 25%). This brick type was developed as a chrome-free alternative to magnesia–chrome brick in rotating cement furnaces.

A special type of spinel brick is the picrochromite brick, containing fused chrome–spinel ($MgO \cdot Cr_2O_3$) grain, which shows extraordinary resistance to blast-furnace slags and acidic coal ashes.

4.1.3.5 Forsterite Bricks. The principal component is magnesium orthosilicate, forsterite ($2MgO \cdot SiO_2$). Olivine is the main raw material, containing <8% FeO.

Applications: regenerators for glass-melting furnaces, safety linings in the iron and steel industry (restricted use).

4.1.4 Zircon- and Zirconia-Containing Bricks For characteristics of these bricks, see Table 7 (Begley and Herndon, 1971).

The primary raw material for ZrO_2-containing refractories is zirconium silicate, $ZrSiO_4$ (mineral name: zircon), with a theoretical composition of 67.2% ZrO_2 and 32.8% SiO_2. It is stable to ca. 1800 °C, but the temperature for thermal decomposition into ZrO_2 (baddeleyite) and SiO_2 may be far lower depending on the type and amount of accompanying components. Chemical processing of zircon results in powdered zirconium oxide, which is worked up to various types of refractory raw materials (sintered and fused material, stabilized or partially stabilized materials).

Pure ZrO_2 (e.g., the mineral baddeleyite) has a monoclinic lattice, which is transformed reversibly beginning at about 950 °C into a tetragonal modification. This is stable to ca. 2300 °C, at which point it is transformed into a cubic modification. The monoclinic–tetragonal transformation is accompanied by a volume change of ca. 7% together with considerable hysteresis (Fig. 8), resulting in stress and cracking in pure ZrO_2 ceramics. Consequently, such ceramics cannot be manufactured. Addition of suitable metal oxides (for refractory purposes, mainly CaO, MgO, and Y_2O_3 in solid solution with ZrO_2) permits the cubic modification to be stabilized down to room temperature. A fully stabilized ZrO_2 can be achieved by the addition

Table 7. Ceramic-bonded zircon- and zirconia-containing refractories.

Type	ZrO_2 (%)	SiO_2 (%)	Al_2O_3 (%)	Bulk density (g/cm^3)	Open porosity (%)	CCS[a] (N/mm^2)	Thermal expansion, 1000 °C (%)	Thermal conductivity, 1000 °C (W m^{-1} K^{-1})
Zircon–silicate								
Pressed or slip cast	62–65	33–34	0.5–1	3.6–3.8	17–22	20–100	0.4	2
Isostatically pressed				3.6–4.2	1–17	>70		
Zircon–mullite (AZS)[b]								
Normally pressed	11–23	10–15	65–73	2.9–3.2	15–24	60	0.65	1.9–2.5
Isostatically pressed				3.2–3.5	1–14			
Zircon–zirconia (nozzle for steel casting)	85			4.25	19			
Stabilized zirconia								
MgO stab.	95			5.0	10	400	0.75	1.5
Dense (CaO stab.)				5.6	0		1.1	

[a]Cold compressive strength.
[b]Mineral composition: zircon 0%–30%, baddeleyite 0%–20%, mullite 25%–30%, glassy phase <10%.

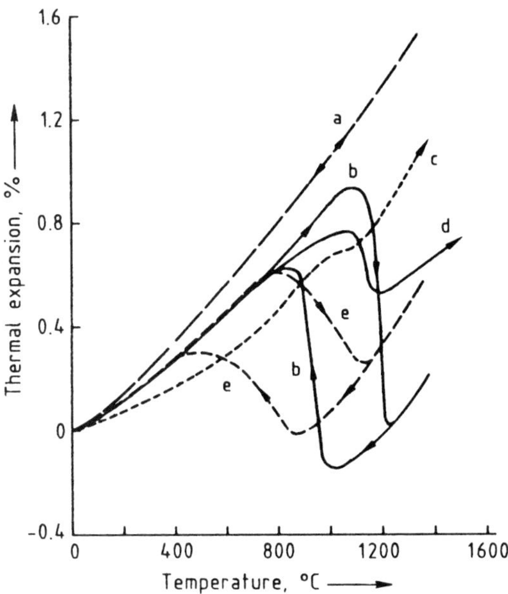

FIG. 8. Thermal expansion of ZrO_2 ceramics: Curve a, fully stabilized; curve b, baddeleyite; curve c, ceramic-bonded AZS brick; curve d, fusion-cast AZS brick; curve e, partially stabilized (Begley and Herndon, 1971, completed).

of about 5%–10% stabilizer. Partially stabilized zirconia raw materials are now used in most cases (3%–6% stabilizer added). This leads to smaller volume changes on heating and provides the additional advantage of creating microcracks, which increases the thermal-shock resistance of the brick.

The following types of ZrO_2-containing products can be distinguished:

1. ceramic-bonded zirconium–silicate bricks;
2. ceramic-bonded Al_2O_3–ZrO_2–SiO_2 (AZS) products, manufactured from mixtures of zircon and high-alumina raw material;
3. fusion-cast ZrO_2-containing bricks, such as fusion-cast AZS bricks (see Sec. 4.1.9); and
4. stabilized or partially stabilized sintered ZrO_2 products (zirconia ceramics).

Applications: zircon-containing refractories are used in glass-melting furnaces (superstructure, bottom, feeders) and in the steel industry as ladle bricks and nozzles. Zircon bricks are resistant to oil residues and especially to V_2O_5. ZrO_2 materials are used as nozzles and sliding nozzle plates for the extremely corrosive conditions associated with the teeming of special steel. Other refractory applications of ZrO_2 ceramics are in the ceramic industry as high-temperature kiln linings and kiln furniture.

4.1.5 Carbon and Graphite Bricks (Table 8) Carbon is unmeltable and not wetted by melts of metals, slag, or salts under oxygen-free conditions, but it is soluble in pig-iron melt. Carbon resists the attack of many acids. High thermal conductivity (especially of graphite) and low thermal expansion impart high thermal-shock resistance. Applications of these bricks are restricted because of susceptibility to attack by oxygen, steam, and CO_2 above 400 °C.

Raw materials with the required grain size (low-ash cokes, anthracite, natural and artificial graphite) are mixed together with tar or pitch in heated mixers and, while the mix is still hot, compacted by ramming or vibration. The cold, hardened shapes are subsequently fired at temperatures ≤1400 °C under the exclusion of air. Pure graphite shapes are produced by a second heat treatment (resistance heating) at ca. 3000 °C.

Depending on the crystallinity of the carbon or graphite and mixtures of the various raw materials, products can be fabricated with widely varying properties (thermal con-

Table 8. Carbon and graphite refractory bricks.

Type	Ash content (%)	Bulk density (g/cm^3)	Open porosity (%)	CCS[a] (N/mm^2)	Thermal expansion, 1000 °C (%)	Thermal conductivity, 1000 °C ($W\,m^{-1}\,K^{-1}$)
Carbon	4.5–6.5	1.5–1.6	13–19	25–65	0.4	8–14
Semigraphite	1–3	1.6–1.7	13–21	25–48	0.38	14–24
Graphite	1	1.55	26	15	0.27	51

[a]Cold compressive strength.

ductivity and oxidation, erosion, and alkali resistance). Addition of, e.g., Al_2O_3 or SiC increases corrosion resistance (Piel *et al.*, 1989).

Applications: as large carbon blocks in the lower parts of blast furnaces; linings for electric reduction furnaces (electrolytic aluminum cells, FeSi production); acid-resistant linings for the chemical industry. Because graphite-containing bricks display high thermal conductivity they are used as a heat-abducting medium in conjunction with water-cooling systems in furnaces.

4.1.6 Carbon- and Graphite-Containing Refractories (Table 9) Combinations of different refractory materials with carbon (originating from pitch or resin bond or impregnation and/or graphite, mainly natural graphite) play a dominant role in the metallurgy of ferrous and nonferrous metals, and above all in the casting of steel.

4.1.6.1 Fired Graphite-Containing Refractories. Three types of carbon-containing fired refractory ceramics are produced.

1. Chamotte–clay–silicon carbide–graphite shapes, used mainly as crucibles for the melting and casting of nonferrous metals. The added SiC and silicon metal increase oxidation resistance and strength. To minimize oxidation during firing under reducing conditions (≤1350 °C) and during use, such shapes are glazed, or a glaze-forming additive is introduced into the mix. Silicon–carbide crucibles are manufactured with a pitch bond; i.e., a carbon bond is formed during firing.
2. Chamotte–graphite and high-alumina–graphite shapes for steel casting such as stoppers and nozzles.
3. Pitch- and resin-impregnated fired bricks (high alumina, magnesia, and doloma products).

4.1.6.2 Pitch- and Resin-Bonded Unfired Refractories. When manufacturing pitch- or resin-bonded products, preheated refractory grain and binder (usually <10%) are hot mixed (temperatures up to 160 °C depending on the viscosity of the binder) and hot

Table 9. Graphite-containing refractory products.

Type	Application	Al_2O_3 (%)	SiO_2 (%)	C (%)	ZrO_2 (%)	SiC (%)	Others[a] (%)	Bulk density (g/cm^3)	Open porosity (%)	CCS[b] (N/mm^2)
Chamotte and high-alumina graphite (fired)	Steel casting nozzles, stoppers	30–70		7–20				2–2.3	17–24	
Pitch- or resin-bonded Al_2O_3–SiO_2–SiC–C	Torpedo ladle bricks	68–90	19–30	10–17		5–8	0–6	2.7–2.8	4–10	55–80
Al_2O_3–C base[c]	Continuous steel casting (shrouds, nozzles, stopper rods)	42–56	14–24	24–33	0–3		0–5	2.2–2.45	12–18	
		60	0.9	24	0–9	4		2.7	17	
ZrO_2–C		1–2	5–7	7–10	75–80		0–2	3.55–4.2	12–16	
				20–25	67–74			3.2–3.6	15–18	
MgO–C			0–2	5–9		MgO 85–92	0–5	2.45–2.5	15–18	
			12–16	10–20		58–78		2.25–2.5	14–17	
	Converter, ladle, arc furnace bricks[c]			10–25		70–89	0–6		3–6	30–50

[a]Mainly antioxidants (metal powder, B_2O_3-containing compounds, etc.).
[b]Cold compressive strength.
[c]Thermal conductivity at 1000 °C, 6–13.5 W m^{-1} K^{-1}.

pressed. Liquid resins or solutions containing curing agents (hexamethylenetetramines) allow cold mixing. Resin hardening occurs at 200 °C (thermosetting). Impregnation of bricks with pitch or resin is carried out under vacuum and at the appropriate temperature (with resin solutions, room temperature).

Heating causes volatile matter to be released, and the "carbon bond" develops. The carbon yield depends on the type of binder and the coking conditions. The release of volatile organic compounds requires environmental precautions during the production and use of organic-bonded refractories.

The following types can be distinguished:

1. unfired pitch- or resin-bonded or tempered high-alumina bricks (after coking, ca. 5% C);
2. Al_2O_3–SiO_2–SiC–C bricks: combinations of high-alumina raw materials, SiC (5%–15%), and graphite (5%–15%);
3. pitch- or resin-bonded, unfired or tempered magnesia, magnesia–doloma, and doloma bricks and shapes (Zednicek, 1988; Zoglmeyer and Romai, 1991) (after coking, <6%, at which point additional carbon black is added up to 10%);
4. pitch- or resin-bonded or tempered oxide–graphite products from mixtures of mainly pure oxides (sintered and fused grain) together with ≤ca. 30% graphite, with or without the addition of so-called antioxidants (metal or nonoxide powders), usually <6%.

The group **4** includes corundum–carbon and zirconia–carbon products, used mainly as nozzles and shrouds for the continuous casting of steel, and magnesia–carbon bricks and shapes, used for lining steel converters, for the slag area in ladles and steel-refining vessels, and in electrical arc furnaces, and also as nozzles for continuous steel casting. Addition of fused silica or ZrO_2-containing aggregates to corundum–graphite products increases thermal-shock resistance. Special additives like BN reduce for instance the clogging of nozzles.

Especially the oxide–graphite products combine the nonwettability of graphite crystals by metal melts and slag (no infiltration into the bricks) with the high corrosion and erosion resistance of dense oxide grain. The graphite content additionally increases thermal-shock resistance. The products have low open porosity (<6%) in the delivered state, but porosity after coking increases to ca. 10%–12%.

The addition of metal powders (e.g., aluminum, magnesium, silicon) or alloys and/or nonoxide compounds like SiC, B_4C, or ZrB_2 inhibits or lowers carbon oxidation.

The effects of graphite content on essential properties of graphite-containing refractories are shown in Fig. 9.

During service, carbon-containing refractories normally show only a small oxidized working zone, where the effects of corrosion mechanisms known from comparable non–carbon-containing refractories are evident. Carbon still present behind this zone minimizes a deeper infiltration of slag or metal.

4.1.7 Silicon Carbide Bricks For characteristics of these bricks, see Table 10 (Wecht, 1977; Fickel and Völkel, 1988; Röttenbach and Pögl, 1989).

These can be divided into two classes according to their SiC content: SiC-containing refractories with ca. 40%–70% SiC, usually in combination with fireclay grog and a clay bond; and SiC products with 70%–100% SiC and different types of binding (and therefore different properties). The bonding types encountered are as follows:

1. Silicate or oxide bond, where the SiC

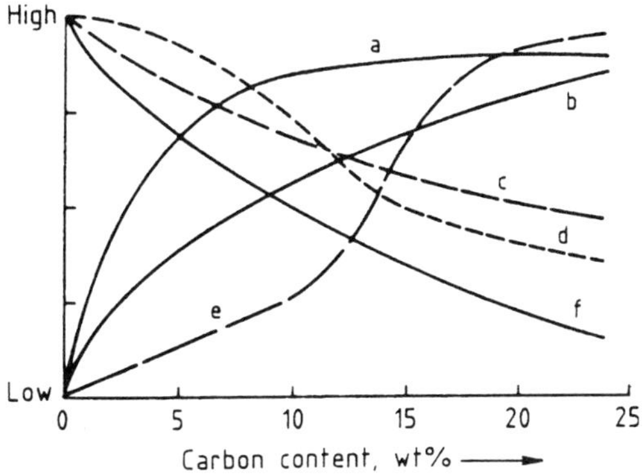

FIG. 9. Effect of graphite content on the properties of magnesia–graphite refractories: Curve a, slag-penetration resistance; curve b, thermal-shock resistance; curve c, oxidation resistance and hot strength with antioxidants; curve d, thermal expansion; curve e, thermal conductivity; curve f, oxidation resistance and hot strength without oxidants (Rathner *et al.*, 1992; Yamaguchi and Tanaka, 1992).

Table 10. Silicon–carbide refractories.

Type	Al_2O_3 (%)	SiC (%)	Si_3N_4/ SiAlON (%)	Bulk density (g/cm^3)	Open porosity (%)	CCS[a] (N/mm^2)	CMR[b] (N/mm^2)	Thermal expansion, 1000 °C (%)	Thermal conductivity, 1000 °C ($W\ m^{-1}\ K^{-1}$)	Max. service temperature (°C)
Silicate-bonded	27	50		2.4	21	70	18	0.45	5	1450–1550
	8	90		2.5–2.7	12–20	100	30		10	
Nitride-bonded		70–85	≤25	2.6	14–20	130–150	50–70	0.45	10–15	1500–1600
Recrystallized		>99		2.6	15–18	300–700	100	0.48	22	1600 ox. 1700 red.
SiSiC		>80		3.05	0	1250	300	0.45	60	1350

[a]Cold compressive strength.
[b]Cold modulus of rupture.

grains normally are bonded with about 10%–20% bonding clay, alone or in combination with fine Al_2O_3, leading to a mullite bond; hot strength is determined largely by the composition of the formed glass phase.

2. Silicon–nitride and sialon (SiAlON) bond, produced by firing under a nitrogen-containing atmosphere a green mixture composed of SiC and Si, which may also include Al_2O_3 or other alumina-bearing compounds.
3. Self-bond, whereby a SiC bond is achieved by adding silicon and carbon and firing in an oxygen-free atmosphere.
4. Recrystallized SiC, where the bond phase (SiC) forms during high-temperature sintering (at ca. 2000 °C) by vaporization and condensation reactions of fine SiC particles in the presence of C.

A special type is the dense SiSiC ceramic, a self-bond SiC product with an infiltration of silicon metal. The exceptional properties of SiSiC ceramics include high thermal conductivity, low thermal expansion, very good resistance to thermal shock, high hot strength, up to 1350 °C (melting point of Si 1410 °C), and abrasion resistance.

SiC is sensitive to oxygen and such oxygen-containing gases as water vapor. In an oxidizing atmosphere SiC begins to decompose to SiO_2 and CO or CO_2 at ca. 900 °C. Above ca. 1200 °C the SiO_2 layer that forms on the SiC grains becomes glassy and retards further oxidation. SiC is also sensitive to alkali melts, basic slags, chlorine gas, and metal melts other than lead, zinc, and copper. Alkali resistance can be increased by using a nitride bond.

Applications: blast furnaces, kiln furniture in the ceramic industry (maximum service temperature 1500 °C), municipal incinerators and special refuse incinerators, retorts for zinc distillation, aluminum industry (submerged burner tubes), tubes for ceramic regenerators.

4.1.8 Fine-Grained Oxide and Nonoxide Ceramics (Kriegesmann, 1989) These normally very dense and highly refractory materials have not yet found appreciable application in the classical markets for refractories (exceptions include ZrO_2, Sec. 4.1.4, and SiC, Sec. 4.1.7) because of high price, a restriction to small or thin shapes, and often low thermal-shock resistance. The production of nonoxide shapes demands special fabrication methods not yet adapted to the refractory industry. Nonoxide refractory ceramics (carbides, nitrides, borides) are prone to oxidation at relatively low temperatures. They are normally not wetted by metal melts.

One of the most promising areas of potential development for fine ceramics is their use as engineering or functional refractory materials and in very special applications in industries such as the steel industry.

Examples of current uses for special fine ceramics include

1. chromium oxide bricks: used in highly corroded areas of glass tank furnaces (fiberglass production);
2. silicon nitride–based materials: skid rails in reheating furnaces for the steel industry, anchors for fixing ceramic-fiber linings, aluminum industry (not wetted by aluminum melt).

3. boron nitride–based composites with Si_3N_4 and AlN, tested also in combination with ZrB_2 and TiB_2 or mullite: break rings for horizontal continuous steel castings, reinforcement of the slag zones of submerged nozzles in continuous steel casting.

Most special ceramics are very dense, but porous materials based on Al_2O_3, ZrO_2, and SiC with defined pore-size distribution are increasingly being used for filtering steel and nonferrous-metal melts, as well as for high-temperature filtration of gases.

4.1.9 Fusion-Cast Bricks (Cichy, 1990) Fusion-cast (fused-cast) bricks are produced by melting of finely ground raw materials (alumina, zircon, zirconia, chrome ore, chrome oxide, special additives) in an electric-arc furnace and casting the melt into sand or graphite forms. The shapes produced range from small parts to blocks with lengths up to 1 m or more. During the slow cooling process (for larger blocks, 6–20 d) major cracks are prevented and the characteristic texture of a fused-cast brick develops: a dense structure (practically free of open pores) with intercrystalline bonding of the mineral phases.

The dense structure imparts high strength and both corrosion and erosion resistance. Disadvantages are low thermal-shock resistance and structural inhomogeneities caused by contraction and segregation during cooling. Location of casting voids may be partially controlled by special casting and cutting away an excess part of the block. Cutting and grinding are inherent parts of the technology of larger fusion-cast bricks. The types of fusion-cast bricks are described in Table 11. Most fusion-cast products belong to the system Al_2O_3–SiO_2–ZrO_2(–Cr_2O_3).

The different alumina types of fusion-cast bricks are produced by adding controlled amounts of soda to the raw-material alumina. The β type has a coarse friable structure with no glassy phase. The β'''' type contains additionally MgO and shows high stability to alkali attack (special development as checker bricks in glass tank regenerators). The corundum–mullite type contains coarse mullite crystals in a glassy matrix.

Fusion-cast corundum–zircon blocks (also called fusion-cast AZS material) are the predominant refractory ceramics for glass-melt contact. The structure contains a corundum–zirconia eutectic phase with segregated modular and dendritic zirconia (baddeleyite) in about 20% of glassy matrix. A special feature of AZS fused-cast bricks is draining out of glassy phase during first heating up, which may lead to bubble formation in contact with glass melt. The addition of chrome oxide to corundum or AZS material considerably increases corrosion resistance caused by the formation of a solid solution of chrome–alumina.

Fused magnesia–chrome bricks have a high corrosion resistance due to its chrome–spinel phase (see Sec. 4.1.3.3.)

A newly developed fusion-cast ZrO_2 brick contains a special SiO_2-rich glassy phase that acts as a buffer against the transformation of baddeleyite. The bricks show high corrosion resistance to some special glasses (Zanoli, 1991).

Applications: primarily in glass tank furnaces and regenerators for glass furnaces (cruciform checker bricks), reheating furnaces in the steel industry (skid rails).

4.1.10 Refractory Ceramics with Low Thermal Expansion As in the case of glassy SiO_2, ceramics with low thermal expansion display extraordinarily high thermal-shock resistance, and they are especially qualified for service conditions involving frequent changes in temperature (e.g., kiln furniture). These ceramics are products based on cordierite ($2MgO \cdot 2Al_2O_3 \cdot 5SiO_2$), lithium–aluminium silicates (eucryptite, spodumene, petalite), and aluminum titanate ($Al_2O_3 \cdot TiO_2$).

Cordierite products have acquired considerable industrial significance. Fireclay grog or cordierite grog is bonded with cordierite-forming mixtures of clay, kaolin, talc, or magnesia through firing at ca. 1150–1350 °C. The cordierite content of the fired products is ca. 40%–80%. The maximum application temperature of normal cordierite shapes is 1250 °C. Addition of mullite and corundum increases both hot strength and service temperature.

4.2 Unshaped Refractory Materials (Plibrico Japan Comp., 1984; Taikabutsu Overseas, 1989)

Unshaped refractory materials, also called monolithics, are mixtures consisting of an aggregate (grains) and one or more bonds,

Table 11. Properties of fusion-cast bricks.[a]

Type	Al_2O_3 (%)	SiO_2 (%)	ZrO_2 (%)	Na_2O (%)	MgO (%)	Cr_2O_3 (%)	FeO (%)	Bulk density (g/cm^3)	Thermal expansion, 1000 °C (%)	Thermal conductivity, $W\ m^{-1}\ K^{-1}$ 300 °C	800 °C	1400 °C	Mineral phases
Corundum (α-Al_2O_3)	99.5			0.2									Corundum
α/β-Al_2O_3	95	1.2		3.5				3.2	0.8	5	4.2	7	Corundum, $Na_2O \cdot 11Al_2O_3$
β-Al_2O_3	94.5			5.2				2.8					$Na_2O \cdot 11Al_2O_3$
β'''-Al_2O_3[b]	87.5	0.3		4.5	7.7			2.8		2.3	2	6.5	$15Al_2O_3 \cdot 4MgO \cdot Na_2O$
Corundum–mullite	74	20	2	0.8					0.75	5	4	6.5	Corundum, mullite, 15% glass
Corundum–zircon	50	16	33	1.1					0.75[c]	4.5	4	6.5	Corundum, baddeleyite,
(AZS)	45.5	12	41	1						4.8	3.9	5.9	16%–21% glass
Zirconium oxide	0.8	4.5	94	0.4				5.3	0.7	3.6	2.5	3	Baddeleyite, 6% glass
Corundum–chrome	60	2			12	28	4	3.7					Ruby, spinel
Corundum–chrome–zircon	32	13	26			26	3	4.0	0.7	5.5	4	6	Ruby, baddeleyite, 20% glass
Magnesia–chrome	7	2.5			57	20	12	3.2[d]	1	4	4.5	5.5	Periclase, spinel
Chromoxide–corundum	14	0.8				82		4.6	0.75				Spinel, ruby

[a]Open porosity <2%; cold crushing strength >200 N/mm^2.
[b]Microporosity ca. 12%; cold crushing strength 60 N/mm^2.
[c]See also Fig. 8.
[d]Porosity 15% (block).

prepared ready for use either directly or after the addition of one or more suitable liquids (usually water). These mixtures are either dense or insulating.

The ENV 1402-1 defines such products according to

1. intended application: materials for monolithic constructions, materials for repairs, materials for jointing;
2. product type and placement: castables, gunning materials, moldable materials (mainly ramming mixes, plastics, and taphole mixes), refractory cements and mortars, others (mainly dry mixes, injection mixes, and coatings, each corresponding to various types of placement), and prefabricated shapes made from unshaped materials (mainly castables and moldables).

Classification within the various categories is based on the temperature below which the material does not shrink more than a specified amount (linear shrinkage 2%–3%).

Four main types of unshaped products can be identified according to chemical composition:

1. alumina–silica refractories, including products consisting mainly of aggregates of alumina, silica, and/or alumosilicates;
2. basic refractories, mainly aggregates of magnesia, doloma, chromite, and spinel;
3. refractories containing carbon (i.e., graphite or carbon together with other nonwetting agents), including mixes for troughs of blast furnaces;
4. special products containing oxide or nonoxide aggregates different from the above (e.g., silicon carbide, zircon, zirconia).

The raw materials and compositions of monolithics correspond essentially to those of shaped products.

An overview of materials for monolithic construction is provided in Table 12.

4.2.1 Ramming Mixes These are noncoherent before use and made up of refractory aggregates, bond(s), and (if necessary) liquid(s). Processing is accomplished by hand or, more frequently, with the help of some mechanical device (pneumatic rammer, slinger, etc.) or vibration. Hardening generally occurs under the influence of heat.

Dry ramming mixes are a specialty product, often vibrated behind steel forms after mixing with a low-melting fritting agent (e.g., boric acid) or a thermosetting resin (hardening at ca. 200 °C).

4.2.2 Plastic Mixes These are ready-for-use products with a moldable, plastic consistency. The principal bonding agent is plastic clay. These mixes are supplied in a soft or preformed state (extrusion-pressed blocks). They are placed by ramming, vibration, pressing, or "extrusion."

4.2.3 Refractory Castables These are supplied dry and used after the addition of water or some other liquid followed by mixing. They are placed by casting with or without vibration (self-flowing), or by rodding, pumping, injecting, and sometimes tamping. Bond formation and hardening occur at ambient temperature. Hydraulic- or cement-bonded castables are the most common type of monolithics. Low-alumina cements with ca. 40% Al_2O_3 and 12% Fe_2O_3, and calcium–aluminate cements with 50%–80% Al_2O_3 and

Table 12. Types of unshaped refractory materials for monolithic constructions.

Type	Subdivision	Water content upon application (%)	Nature of bond
Moldable mixes	Dry ramming mixes (fritting mixes)	0–1	Ceramic
	Moist friable ramming mixes	3–6	Ceramic }
	Plastic friable mixes	5–8	Chemical }
	Preformed plastic mixes	6–10	Organic }
Refractory castables	Dense castables	5–12	Hydraulic }
	Insulating castables	10–50	Chemical }
Gunning mixes	Moist friable mixes		Hydraulic }
	Different types of castables	5–50	Chemical }
	Special hot gunning mixes		Ceramic (special binders) }

19%–38% CaO, are used as hydraulic binder (Kopanda and MacZura, 1990).

Two main families are the dense and the insulating castables, with a limit of 45 vol% porosity. The various types of dense castables are listed in Table 13. Figures 10 and 11 describe properties of castables, classifying them within the broader context of unshaped materials and comparing them with bricks.

As a result of a drastic decrease in cement content together with the presence of ultrafine powders (microsilica SiO_2, reactive Al_2O_3) and dense packing, dense castables require little mixing water and display low porosity and high cold and hot strength up to 1500 °C (due to, e.g., mullitization of the matrix). Low-cement castables without addition of microsilica show high hot strength at 1500 °C and higher (Kriechbaum *et al.*, 1994).

Insulating castables contain porous grains (like porous fireclay grog, thermally treated perlite, or vermiculite) and normally a high content of refractory cement (up to 40%).

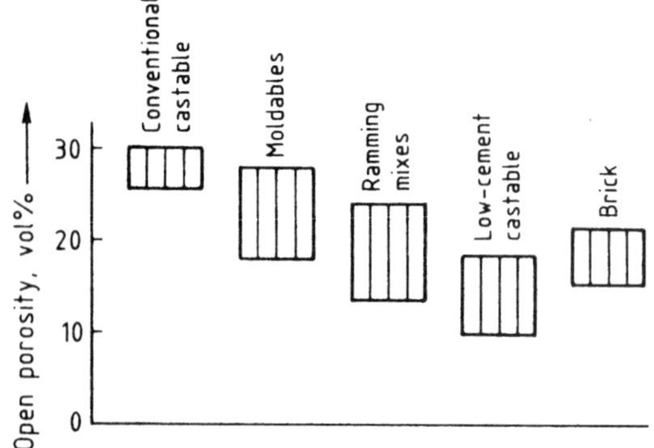

FIG. 10. Comparison of open porosities of various fired unshaped refractory materials (60%–90% Al_2O_3).

Prior to heating a castable must be cured for a sufficiently long time (i.e., 24 h) to ensure full hydration and strength. Drying and

Table 13. Types of dense refractory castables.

Type	Classification ASTM C 401-89 and ENV 1402-1	Mixing water (approx.)	Essential components, bonding
Conventional castable (CC)	>2.5% CaO	9–12	Main hydraulic bonding: CA[a] + 10H[b] → CAH_{10}[c] (<20 °C), 2CA + 11H → C_2AH_8[d] + AH_3[e] (20–35 °C)
Low-moisture conventional castable[f]		7–9	Addition of ultrafine powders (high strength and abrasion resistance)
Low-cement castable (LCC)	1–2.5% CaO	4.5–7	Deflocculated castables containing aluminous cement, a minimum of 2% ultrafine particles (SiO_2, Al_2O_3; <1 μm), and at least one deflocculating agent
Ultralow-cement castable (ULCC)	0.2–1% CaO		Thixotropic rheological behavior (vibrational casting)
Clay-bonded castable	See LCC and ULCC	7	Contains clay (flocculated) and aluminous cement as a deflocculant
Self-flowing castable[g]		6	Consistency (rheopex rheological behavior), which allows flow and deaeration without vibration
No-cement castable	<0.2% CaO		Bonding: phosphate, alkali silicate, pitch or resin, Si or Al sol
Mixerless castable[h]			Delivered ready-mixed with water for casting (vibrating); contains 0.5% sodium silicate, thermosetting at 130–200 °C by gelation

[a]CA = $CaO \cdot Al_2O_3$.
[b]H = H_2O.
[c]CAH_{10} = $CaO \cdot Al_2O_3 \cdot 10H_2O$.
[d]C_2AH_8 = $2CaO \cdot Al_2O_3 \cdot 8H_2O$.
[e]AH_3 = $Al_2O_3 \cdot 3H_2O$ (gel).
[f]Weavers *et al.*, 1985.
[g]Brachet *et al.*, 1992.
[h]Tabata *et al.*, 1992.

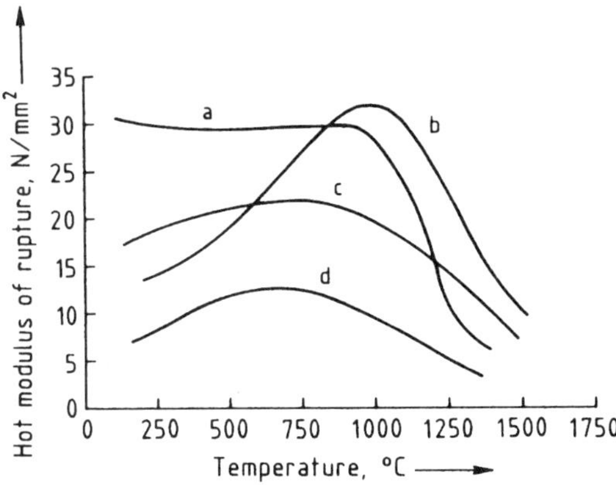

FIG. 11. Hot modulus of rupture of low-cement microsilica-containing castables, ramming mix, and dense brick (corundum-based materials) (Fletcher, 1989): Curve a, LCC; curve b, ULCC; curve c, brick; curve d, dry ramming mix. Low-cement castables without addition of microsilica achieve high hot strength at 1500 °C and higher.

first heat-up steps require careful monitoring to prevent explosive effects caused by excessive vapor pressure inside the lining.

4.2.4 Gunning Materials These are specially prepared for placement by pneumatic or mechanical projection means in cold and also hot application. Various special bonds have been developed for achieving a dense and adhesive gunned layer. Such materials are used mainly for repairs, especially in the iron and steel industry.

4.2.5 Prefabricated Shapes These are rammed or cast and may be pretreated up to 600 °C by the manufacturer as a way of providing blocks that are ready for use. Relatively large-volume parts are fabricated. The advantages are optimal manufacture, simple lining, and short installation times.

4.2.6 Jointing and Coating Materials These are essential for establishing a safe refractory brickwork or lining surface. Such products are fine-grain materials whose characteristics and composition must correspond to those of the associated bricks. Table 14 describes such materials.

Applications: principal applications of materials for monolithic construction are in the iron and steel industry, foundries, the petrochemical industry (especially insulating gunning mixes), and incinerators.

Table 14. Joints and coatings.

Material	Mixing liquid or water (%)	Type of bonding
Refractory mortars	10–25	Chemical, hydraulic, ceramic
Refractory mastics	12–30	Chemical
Adhesives	30–80	Chemical
Troweling materials	15–30	Chemical, hydraulic, ceramic
Coatings	30–60	Chemical

4.3 Heat-Insulating Ceramic Materials

Ceramic materials with a high pore volume (i.e., low thermal conductivity) are used extensively for thermal insulation of furnaces and auxiliary equipment. Apart from the effect of decreased thermal wall losses, the low weight of the insulating ceramic material allows lighter furnace construction and low heat-storage losses in discontinuous service.

Thermal conductivity depends not only on total porosity (overall gas volume) but also to a considerable extent on pore size and pore-size distribution, as well as texture ("thermal bridges") and mineralogical composition. Normally, the finer the pores, the lower the thermal conductivity. A maximum pore size of <1 mm is the target. Microporous materials have a thermal conductivity lower than that of air. Above 1000 °C thermal radiation must be taken into account as a heat-transfer mechanism.

High porosity implies low mechanical strength, high gas permeability, and low corrosion resistance, all of which must be considered in the use of insulating refractory materials. Furthermore, a porous texture leads to lower volume stability at high temperatures than in the case of dense materials (greater sensitivity to overheating) and a greater susceptibility to reducing atmospheres (reduction of, e.g., Fe_2O_3; increased shrinkage).

The main groups of insulating ceramic products are

1. insulating bricks,
2. insulating refractory bricks,
3. ceramic fibers,
4. porous or dense grains and granules of ceramic materials used as poured backup insulation,

5. insulating castables (see Sec. 4.2), and
6. special insulating products based on calcium silicate and microporous high-SiO_2 materials, with application temperatures ≤1100 °C.

4.3.1 Insulating Bricks Insulating bricks are based on diatomite (moler), vermiculite, and perlite, with bulk densities of ca. 0.45–0.8 g/cm^3, cold crushing strength of ca. 1–7 N/mm^2, and a thermal conductivity at 200 °C of 0.1–0.2 W m^{-1} K^{-1}. They are normally used up to 800 °C, but low-flux diatomite bricks may be applicable to 1100 °C.

4.3.2 Insulating Refractory Bricks According to official definitions, insulating refractory bricks have a total porosity of at least 45% and an application temperature of at least 800 °C. Porosity normally lies between 60% and 80%. The criteria for classification (ISO 2245) are as follows:

1. the temperature at which the material does not show a permanent linear shrinkage greater than 2%, and
2. the bulk density (which for a given product is directly related to porosity, and consequently to thermal conductivity).

Classification temperatures are not to be taken as acceptable service temperature limits, which depend on the particular service conditions.

Insulating refractory bricks are usually fabricated from fireclay, sillimanite or mullite, corundum, and silica as raw materials, together with a combustible material.

Types and properties of insulating refractory bricks are summarized in Table 15.

4.3.3 Refractory Ceramic Fibers (Cooke, 1991; Smith, 1990) A proposed definition describes ceramic fibers as synthetic mineral fibers suitable for use as heat-insulating materials and containing less than 2 wt% alkali-metal or alkaline-earth oxides. Most ceramic fibers are based on glassy or crystalline aluminosilicates and alumina.

Glassy (amorphous) refractory ceramic fibers (Al_2O_3 content ca. 45%–60%) are produced by fusing an aluminosilicate raw material (e.g., calcined kaolin) and/or sand and calcined alumina. Addition of ZrO_2, Cr_2O_3, B_2O_3, and other materials leads to changes in the properties of the fiber. The molten stream is attenuated into fiber form either by impact with a high-velocity jet of air and/or steam or by downward flow onto rotating wheels.

Polycrystalline fibers (62%–100% Al_2O_3) are sol-gel–derived fibers from, e.g., aluminum salts, colloidal silica, organosilane, and modifiers. Rotary spinning (short-staple fibers) is also practiced, as is the production of continuous filaments. After formation the fiber must be dried and calcined. Other types of oxide fibers (e.g., ZrO_2 fiber) are produced in the same way. This type of fiber is relatively expensive. To improve cost effectiveness, polycrystalline fibers are often blended with low-cost glassy fiber.

Properties and application temperatures for fibers and fiber products are shown in Table 16.

Table 15. Properties of insulating refractory bricks.

Type	Classification temperature (°C)	Al_2O_3 (%)	SiO_2 (%)	Fe_2O_3 (%)	Bulk density (g/cm^3)	Porosity (%)	CCS[a] (N/mm^2)	Thermal expansion, 1000 °C (%)	Thermal conductivity, 1000 °C (W m^{-1} K^{-1})
Silica	1500–1600	1	92 (3.5% CaO)	<0.5	0.8–1.2	50–65	3–4	1.2	0.5–0.8
Anorthite fireclay	1100–1250	36	45 (15% CaO)	0.7	0.5	85	1	0.6	0.2
Fireclay	1250–1400	35–42		<1.7	0.55–1.3	55–75	2–8	0.5–0.6	0.3–0.5
High-alumina 50–75% Al_2O_3	1400–1700	55–75		<1.5	0.8–1.5	45–75	2–12	0.6–0.7	0.35–0.55
Corundum hollow sphere	1800	98	0.6	0.1	1.2–1.4	60	8–18	0.8	0.8–1.1
Forsterite	1500	MgO 57	34	6	1.7	45	2.5	1.2	0.8

[a]Cold compressive strength.

Table 16. Ceramic fibers and fiber products.[a]

Type of fiber	Al_2O_3 (%)	SiO_2 (%)	Cr_2O_3 (%)	ZrO_2 (%)	Liquidus temp. (°C)	Classification temp. (°C)	Fiber diameter (μm)	Density (g/cm^3)	Maximum service temperature (°C): Oxidizing atmosphere, continuous	Oxidizing atmosphere, discontinuous	Reducing atmosphere
Glassy, amorphous	47	52			1760	1250	2–5	2.58	1100	1250	<1100
	63	36			1800	1400	3–6	2.63	1200–1400	1400 (1500[b])	Ca. 200° lower than in oxidizing atmosphere
	41	55	3		1785	1400	2–5	2.61			
	35	50		15	1810	1450	≈3	2.60			
Polycrystalline	85	15			1950	1600	≈7	3.40	1500	1600 (1700)	Ca. 100° lower than in oxidizing atmosphere
	95	5			2000	1600	≈3	3.58			

[a]For fiber products in general, the bulk density is 0.05–0.3 g/cm^3 (special products, ≤1.3 g/cm^3). Thermal-conductivity ranges are 0.08–0.12 $W\,m^{-1}\,K^{-1}$ (400 °C) and 0.16–0.40 $W\,m^{-1}\,K^{-1}$ (1000 °C). Fiber tensile strength is 400–1500 N/mm^2.
[b]Mixed fiber from high-alumina glassy fiber and alumina fiber.

The fiber is used in loose form or processed into any of a variety of flexible or essentially rigid forms. Blankets and textiles (ropes, yarns) are examples of products made by an air-laid system. Papers, felts, and boards are produced by wet-laid systems (like the traditional paper-forming process). Boards and special shapes are processed via vacuum casting. The ceramic fiber is slurried in water with organic and/or inorganic binders (e.g., SiO_2 sol) and fillers. The slurry is dewatered onto molds via a vacuum. Fiber is also used in unshaped refractory materials like gunning mixes and castables. Products include ceramic-fiber veneering elements, folded modules, and ready-for-use system blocks with anchoring.

Ceramic-fiber products have very low bulk density (0.05–0.3 g/cm^3; i.e., a porosity >90%) and unusually low thermal conductivity. In contrast to insulating bricks, thermal conductivity of ceramic-fiber products decreases with increasing bulk density. Thermal conductivity increases more with increasing temperature than in the case of comparable insulating bricks (Fig. 12). Thermal-shock resistance is extraordinarily high.

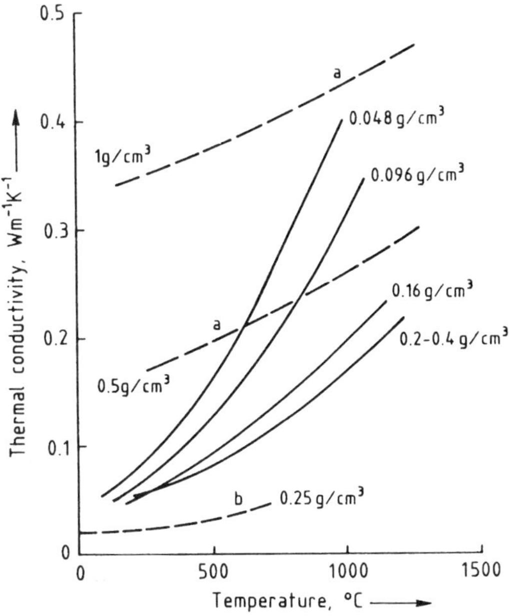

FIG. 12. Thermal conductivities of ceramic-fiber products compared with other insulating materials: Curves a, fireclay insulating bricks; curve b, microporous material.

Ceramic-fiber products are classified into groups according to the temperature (in 50 °C steps) at which the material does not show more than 4% shrinkage. Long-time application of ceramic fibers at maximum service temperature is limited significantly by shrinkage and crystallization and/or recrystallization (crystal growth). Impurities acquired from a furnace atmosphere (e.g., alkalies) may greatly increase shrinkage and crystal transformation. Ceramic-fiber products are sensitive to mechanical stress as well as, e.g., to high-velocity gas streams. Improvement is achieved by brushlike orientation of the fibers perpendicular to abrasion stresses, or by impregnating the surface with SiO_2 sol or the use of rigid ceramically bonded and prefired shapes.

Applications: Ceramic fibers have found application in all types of low- and high-temperature thermal processes, from use as a filler for joints or seals and insulation over veneering of furnace walls up to complete fiber-lined furnaces (e.g., in the ceramics industry, and heating furnaces for the steel and glass industries).

In the nonferrous-metal industry preshaped parts made from ceramic fibers have found application in direct contact with molten aluminum and copper alloys.

Proper construction and the use of appropriate joining techniques (gluing procedures, anchoring systems) are important factors in the lifetime of fiber furnace linings.

5. SELECTION AND USE

The service performance of a refractory material depends in roughly equal measure on

1. the properties of the refractory material,
2. the construction, lining and brick design, installation technique, drying, and initial heating, and
3. service conditions.

Most data sheets accompanying refractory products specify a service-temperature limit or maximum service temperature. This refers to the maximum temperature to which the refractory can be exposed during operation without corrosive influences (heating on one side only). No test yet exists for determining the maximum service temperature. Consequently, the indicated temperatures are based on practical experience supported by suitable hot tests (refractoriness under load, hot strength, reheat change). Classification temperatures for refractory products are not to be considered as service-temperature limits, which are usually lower.

In most cases, refractory ceramics are subject to a multitude of thermal, thermomechanical, mechanical, and corrosive stresses. Corrosion is the factor most likely to change the service properties of a refractory material in a significant way. A refractory material must be selected in accordance with those properties that are most important for a given service application. This requires close consultation between the refractory producer, the furnace builder, and the user. Often a refractory ceramic is "tailor made" for a specific application.

The output (tonnage) of refractory products is distributed among various industries in approximately the following way:

Iron and steel industry	55%–60%
Cement and lime industry	5%–8%
Glass industry	5%–8%
Ceramic industry	5%–8%
Nonferrous-metal industry	2%–5%
Chemical and petrochemical industry	2%–4%
Energy-producing industry	2%–4%

GLOSSARY

Aggregate: As applied to refractories, a ground mineral, consisting of particles of various sizes, used with much finer sizes (normally under 10 mm) for making formed or monolithic bodies.

Alumina: Aluminum oxide, Al_2O_3, in any of its many forms. As a general term in the refractory industry the α form (mineral name corundum), including calcined alumina, low-fired fine Al_2O_3 types, which may contain residual hydroxide or other oxides, high-fired alumina grain, called sintered (or tabular) alumina, fused alumina (often named corundum), and bubble alumina, hollow spheres of Al_2O_3. Beta alumina is not an alumina but a family of alkali-metal polyaluminate, e.g., $Na_2O \cdot 11Al_2O_3$, often a minor impurity in aluminas.

Bulk Density: The ratio of the mass to its bulk volume (including volume of total pores).

Bursting: Disintegration of refractories characterized by having the exposed face swell and grow until it breaks away from the brick mass, caused by a permanent increase in volume. Typical for chromite-containing material when exposed to iron oxide at high temperature. Alkali attack on aluminosilicate refractories leads to alkali bursting by formation of crystalline alkali aluminosilicates (e.g., nepheline, $Na_2O \cdot Al_2O_3 \cdot 2SiO_2$).

Calcine, Calcines: Refractory material, often fireclay or alumina (calcined alumina), that has been heated to eliminate volatile constituents and to produce desired physical changes.

Chamotte: See **Grog.**

CO Disintegration: A carbon deposition resulting from the decomposition of carbon monoxide gas into carbon within a critical temperature range. When deposited within the pores of refractory brick, carbon can build up sufficient pressure to destroy the bond and cause the brick to disintegrate.

Consistency: A measure of the flow, softness, or wetness of castables.

Corrosion: Destruction of refractory surfaces by the chemical action of external agencies.

Creep: Isothermal deformation of a stressed product as a function of time.

Direct-Bonded Brick: A fired refractory in which the grains are joined predominantly by a solid-state diffusion mechanism. The term "direct bond" was initially applied to fired magnesia–chrome refractories.

Erosion: Wearing away of refractories' surfaces by the washing action of moving liquids or particles in flowing gases.

Firebrick: Refractory brick of any type.

Fireclay: An earthy or stony mineral aggregate that has as the essential constituent hydrous silicates of aluminum with or without free silica, plastic when sufficiently pulverized and wetted, rigid when subsequently dried, and of suitable refractoriness for use in commercial refractory products.

Fireclay Brick: A refractory brick manufactured substantially from fireclay and fireclay grog.

Firing (Burning): The final heat treatment in a kiln to which refractory brick and shapes are subjected in the process of manufacture for the purpose of developing bond and other necessary physical and chemical properties.

Flux: A material that, when present in or added to a refractory material (even in very small amounts), appreciably lowers the temperature of formation of a liquid phase (e.g., promoting densification or sintering).

Fused Grain: Made ordinarily by arc-melting and refreezing; crushed and sized.

Fused Silica: Silica in the glassy or vitreous state. Synonyms include vitreous silica and silica glass.

Fusion Point: The temperature at which melting takes place. Most refractory materials have no definite melting points, but soften gradually over a range of temperatures.

Grog: A granular product produced by crushing and grinding calcined or burned refractory material, usually of alumina–silica composition. Grog from fireclay is sometimes named chamotte.

Gunning: An application technique that uses a pneumatic means to transport a refractory material and place it onto a cold or hot surface.

Kiln Furniture: Various supports, shapes, plates, or boxes (saggars) used to hold or contain ceramic ware during firing.

Matrix: The medium in which grains or crystal are embedded.

Modulus of Rupture: A measure of the transverse or "crossbreaking" strength of a solid body.

Monolithic Refractory: An unshaped refractory that is normally to be installed *in situ,* without joints, to form an integral structure.

Permeability: The capacity of a refractory for transmitting a fluid (gas or liquid).

Plastic Clay: A fireclay that has sufficient natural plasticity to bond together other materials that have little or no plasticity (also ball clay).

Plasticity: That property of a material that enables it to be molded into desired forms, which are retained after the pressure of molding has been released.

Pyrometric Cones: Elongated trigonal pyramids of standard size made from specified mixtures of ceramic materials, which when under stated conditions tip the supporting plaque at a specified temperature. May be used as an index of heat treatment or for testing of pyrometric-cone equivalent (refractoriness).

Pyrometric-Cone Equivalent (PCE): The number or temperature equivalent of that standard pyrometric cone whose tip would touch the supporting plaque simultaneously with a cone of the refractory material being investigated.

Rebonded Brick: Fired refractory bricks or other shapes made predominantly or entirely from fused grain.

Refractoriness: The capability of maintaining a desired degree of chemical and physical identity at high temperatures and in the environment and conditions of use. Synonym for pyrometric-cone equivalent.

Refractoriness under Load: The resistance (softening) of a refractory subjected to a constant load (normally 0.2 N/mm^2) under progressively rising temperature (5 °C/min). Temperatures corresponding to characteristic deformations are determined.

Reheat Change: The permanent changes in length (or volume) taking place in a fired refractory upon reheating.

Secondary Expansion: The property exhibited by some fireclay and high-alumina refractories of developing permanent expansion at temperatures within their useful range.

Shrinkage: The fractional reduction in dimensions or volume of a material or object when subjected to drying, calcining, or firing (sintering).

Sintering: A heat treatment that causes adjacent particles of material to cohere at a temperature below that of complete melting. Used especially when firing dense fine-grained ceramics.

Slip: A suspension or slurry (concentrated suspension) of normally finely divided ceramic materials in a liquid.

Spalling: The cracking or rupturing of a refractory material or unit caused by differential expansion due to thermal shock (thermal spalling) or crystalline inversions (structural spalling)

Tempering: Preheating at 200–600 °C of pitch- or resin-bonded refractory bricks and prefabricated blocks from castables to eliminate volatile materials or of chemically bonded bricks to gain strength.

Texture: Macroscopic relationship existing between the various shapes and sizes of grains and pores.

Thixotropy: A reversible process by which certain materials in suspension (e.g., castables) become more fluid on vibrating or mixing and less fluid or stiff on standing.

Workability: Measure of the facility with which plastic refractories can be rammed or pounded into place.

Works Cited

Alper, A. M. (Ed.) (1970), *High Temperature Oxides, Magnesia, Lime, and Chrome Refractories,* New York: Academic.

Am. Ceram. Soc. Bull. (1992), **71,** 799–824.

Annual Book of ASTM Standards (1995), Vol. 15.01, *Refractories, Carbon and Graphite Products, Activated Carbon.*

ASTM C201-93, Standard test method for thermal conductivity of refractories.

Begley, E. R., Herndon, P. O. (1971), in: A. M. Alper (Ed.), *High Temperature Oxides,* Part 4, New York: Academic, pp. 185–208.

Brachet, F. C., Avis, R., Clavaud, B. (1992), in: *Unitecr '91 Congress,* Düsseldorf: Verlag Stahleisen, pp. 264–266.

Carswell, G. P. (1977), *Refractories J.,* **52** (6), 7–19.

Cichy, P. (1990), in: *Alumina Chemicals Science and Technology,* Westerville, OH: American Ceramic Society Inc., pp. 393–426.

Cooke, T. F. (1991), *J. Am. Ceram. Soc.* **74,** 2959–2978.

Coope, B. M., Dickson, E. M. (Eds.) (1991), *Raw Materials for the Refractory Industry,* IM Refractories Survey, Metal Bulletin Ltd., London.

Dal Maschino, R., Fabbi, B., Fiori, C., (1988), *Ind. Ceram. (Faenza Italy)* **8** (3), 121–126.

Das, T. K., Jeschke, P. (1975), *Ber. Dtsch. Keram. Ges.* **52,** 126–130.

Dietrichs, P. (1987), *Silikattechnik* **38,** 349–353.

DIN 51064 (1981), Bestimmung der Druckfeuerbeständigkeit (DFB) an feuerfesten Steinen.

ENV 1402-1 (1995), (European Standard). Unshaped refractory products. Part 1: Introduction and definitions.

Fickel, A., Völkel, W. (1988), *InterCeram, Aachen Proceedings,* pp. 38–40.

Fletcher, W. (1989), *Steel Metals Mag.* **27** (5), 389–390.

Glazman, M. S., Landa, Ya. A., Litovskii, E. Ya., Puchkelevich, N. A. (1989), *Refractories (Eng. Trans.)* **30** (5), 312–319.

Hasselmann, D. P. M. (1978), *Ceramurgia Int.* **4,** 147–150.

ISO 528 (1983), Refractory products—Determination of pyrometric cone equivalent (refractoriness).

ISO/R 836 (1968), Vocabulary for the refractories industry. Trilingual edition.

ISO 1109 (1975), Refractory products—Classification of dense shaped refractory products.

ISO 1893 (1989), Refractory products—Determination of refractoriness under load (differential—with rising temperature).

ISO 2245 (1990), Shaped insulating refractory products—Classification.

ISO 3187 (1989), Refractory products—Determination of creep in compression.

ISO 5013 (1985), Refractory products—Determination of modulus of rupture at elevated temperatures.

ISO 8894-1 (1987), Refractory materials—Determination of thermal conductivity—Part 1: Hot-wire method (cross-array).

Jeschke, P., Bertling, H., Breidenbech, D. (1989), in: *Unitecr '89 Proceedings*, Vol. 1, Westerville, OH: American Ceramic Society Inc., pp. 263–283.

Kobayashi, M., Nishi, M., Miyamoto, A. (1982), *Taikabutsu Overseas* **2** (2), 5–13.

Konopicky, K., Routschka, G., Hagemenn, L. (1971), *Archiv Eisenhüttenwesen* **42,** 433–437.

Kopanda, J. E., MacZura, G. (1990), in: *Alumina Chemicals Science and Technology,* Westerville, OH: American Ceramic Society Inc., pp. 171–183.

Kriechbaum, G. W., Gnauk, V., Routschka, G. (1994), *Stahl u. Eisen Speciel Okt.,* Düsseldorf: Verlag Stahleisen, pp. 150–159.

Kriegesmann, J. (Ed.) (1989), *DKG-Technische Keramische Werkstoffe,* Köln: Lose-Blatt-Sammlung, Wirtschaftsdienst.

Le Doussal, H. L., Prieur, C. (1989), *Ind. Ceram.* no. 837, 255–262.

Lepère, K. E., Overkott, E., Beckmann, R., Eschner, L. (1992), in: *Unitecr '91 Congress,* Düsseldorf: Verlag Stahleisen, pp. 450–453.

Pask, J. A. (1990), in: *Ceramic Transactions,* Vol. 6, *Mullite and Mullite Matrix,* Westerville, OH: American Ceramic Society Inc., pp. 1–13.

Piel, K. T., Wilkening, S., Lechthaler, K., Santowski, K. (1988), *InterCeram Special Issue, Aachen Proceedings,* pp. 22–25.

Plibrico Japan Comp. (1984), *Technology of Monolithic Refractories,* Tokyo.

Rathner, R., Knauder, J., Weissensteiner, H. (1992), in: *Unitecr '91 Congress,* Düsseldorf: Verlag Stahleisen, pp. 12–18.

Refractory Materials, Recommendations (1990), Brussels: Fédération Européene des Fabricants de Produits Réfractaires (PRE).

Röttenbach, R., Pögl, P. (1989), *InterCeram, Aachen Proceedings,* pp. 72–74.

Routschka, G. (1981), *Ziegelind. Intern.,* pp. 690–700.

Routschka, G. (1988), *InterCeram* **37,** 24–28, 33.

Smith, R. D. (1990), in: *Alumina Chemicals Science and Technology,* Westerville, OH: American Ceramic Society Inc., pp. 385–394.

Stahl-Eisen-Werkstoffblatt 912 (1984), *Tonerdereiche Steine,* Düsseldorf: Verlag Stahleisen.

Stahl-Eisen-Werkstoffblatt 915 (1984), *Schamottesteine für Hochöfen und Winderhitzer,* Düsseldorf: Verlag Stahleisen.

Stahl-Eisen-Werkstoffblatt 917 (1984), *Schamottesteine für allgemeine industrielle Einsatzzwecke,* Düsseldorf: Verlag Stahleisen.

Tabata, K., Yamamura, T., Katayama, M., Enoki, K. (1992), in: *Unitecr '91 Congress,* Düsseldorf: Verlag Stahleisen, pp. 46–53.

Taikabutsu Overseas (1989), *Recent Progress on Castable Refractories* **9** (1), 2–75.

Uchida, S., Tchikawa, K., Matsuo, A., Miyoshi, S. (1990), *Taikabutsu Overseas* **10** (2), 74–84.

VDI-Richtlinie 3128, Blatt 4 (1977), Induktive Erwärmung. *Physikalische Stoffeigenschaften nichtmetallischer Werkstoffe (Ofenbaustoffe),* Düsseldorf: VDI-Verlag.

Weaver, E. P., Tally, R. W., Engel, A. J. (1985), in: *Advances in Ceramics,* Vol. 13, *New Developments in Monolithic Refractories,* Westerville, OH: American Ceramic Society Inc., pp. 219–229.

Wecht, E. H. P. (1977), *Feuerfestsiliciumcarbid,* Wien: Springer-Verlag.

Yamaguchi, A., Tanaka, H. (1992), in: *Unitecr '91 Congress,* Düsseldorf: Verlag Stahleisen, pp. 32–38.

Young, R. G., Hartwig, F. J., Norton, C. L. (1964), *J. Am. Ceram. Soc.* **47,** 205–210.

Zanoli, A., DuVierre, G., Sertain, E. (1991), *Ceram. Eng. Sci. Proc.* **12,** 496–517.

Zednicek, W. (1988), *Silikattechnik* **39,** 385–390.

Zoglmeyer, G., Romai, D. (1991), *InterCeram* **40,** 81–89.

Further Reading

Brook, R. J. (1991), *Concise Encyclopedia of Advanced Ceramic Materials,* Oxford: Pergamon, p. 588.

Carniglia, S. C., Barna, G. L. (1992), *Handbook of Industrial Refractories Technology. Principles, Types, Properties and Application,* Park Ridge, NJ: Noyes, p. 627.

Ceramics and Glasses (1991), *Engineering Materials Handbook,* Vol. 4, The Materials Information Society, p. 1217.

Deutsche Gesellschaft Feuerfest- und Schornsteinbau e.V. (1994), *Feuerfestbau. Stoffe-Konstruktions-Ausführung,* Essen: Vulkan Verlag, p. 358.

Didier-Werke AG (1990), "Refractory Materials and Their Properties," company brochure, Wiesbaden, p. 104.

Fisher, R. E. (Ed.) (1985), *Advances in Ceramics,* Vol. 13, *New Developments in Monolithic Re-*

fractories, Westerville, OH: American Ceramic Society, p. 424.

Granitzki, K. E., Krönert, W., Müller, E. (1989), *Feuerfeste Stoffe im Gießereibetrieb,* Düsseldorf: Gießerei-Verlag, p. 247.

Harbison-Walker (1992), *Modern Refractory Practice,* 5th ed., Pittsburgh: Harbison Walker Refractories.

Hart, L. D. (Ed.) (1990), *Alumina Chemicals Science and Technology Handbook,* Westerville, OH: American Ceramic Society Inc., p. 617.

O'Bennon, L. S. (1984), *Dictionary of Ceramics Science and Engineering,* New York: Plenum, p. 303.

Proceedings Unitecr '95 Congress [Unified International Technical Conference on Refractories (1995), Fourth Biennial Worldwide Conference on Refractories], Kyoto, Nov. 19–22, 1995, Vols. 1–3, Düsseldorf: Verlag Stahleisen.

Routschka, G., Granitzki, K. E. (1993), *"Refractory Ceramics,"* in: *Ullmann's Encyclopedia of Industrial Chemistry,* Vol. 23, Weinheim: VCH, pp. 1–48.

Schulle, W. (1990), *Feuerfeste Werkstoffe. Feuerfestkeramik—Eigenschaften, prüftechnische Beurteilung, Werkstofftypen,* Leipzig: Deutscher Verlag für Grundstoffindustrie, p. 494.

Taikabutsu Overseas, Quarterly English Journal of The Technical Association of Refractories, Japan, Tokyo, Vol. 1, 1989, and current.

REFRACTORY MATERIALS, ELECTRONIC APPLICATIONS OF

NOBUYOSHI KOBAYASHI, *Central Research Laboratory, Hitachi, Ltd., Japan*

	Introduction	309
1.	**Refractory-Metal Films**	310
1.1	Formation Method	310
1.1.1	Magnetron Sputtering	310
1.1.2	Chemical Vapor Deposition (CVD)	311
1.1.3	Surface Cleaning	312
1.2	Characterization of Film Properties	312
1.2.1	Film Properties	312
1.2.2	Characterization Method	313
1.3	Patterning	313
1.3.1	Dry Etching	313
1.3.2	Etching Using Wet Chemicals	314
2.	**Applications to Si ULSIs**	314
2.1	Tungsten or Molybdenum Gate Metallization for MOS LSIs	314
2.1.1	Requirements for MOS Gate Material	314
2.1.2	Tungsten or Molybdenum Gate Process	315
2.2	Refractory-Metal Barrier for Aluminum Interconnection	317
2.2.1	Barrier Properties	317
2.2.2	Layered Aluminum Interconnects with Refractory Metals	318
2.3	Contact Filling using Chemical Vapor Deposition of Tungsten	318
2.3.1	Requirements for Contact Filling	318
2.3.2	Selective Chemical Vapor Deposition of Tungsten	320
2.3.3	Blanket Chemical Vapor Deposition of Tungsten	323
2.4	Solid-Phase Reaction of Refractory Metal with Silicon	323
2.4.1	Silicide Formation	323
2.4.2	Use of Silicides in MOS Source and Drain Metallization	324
3.	**Application to GaAs MESFETs**	324
4.	**Acknowledgments**	325
	Works Cited	325

INTRODUCTION

Industrial applications of refractory materials are very extensive, and they range from propulsive systems through electric power generation systems to nuclear systems. The maximum temperature at which a material can be used is limited by its melting point, and refractory materials that have high melting points are widely used in these applications. Besides the above applications, refractory materials have begun to be widely used in microelectronics. In particular, their use in silicon large-scale integrated circuits (Si LSIs) has become the most pervasive application. In microelectronics, refractory metals are used in the form of "thin" films, and so their technological development has been quite different from applications as bulk material. This article describes electronic applications of selected refractory metals mainly in the field of metallization for Si LSIs.

As was pointed out by Moore (1980), silicon device integration has continued to increase over the last 40 years, and it now achieves 100 megabit to gigabit integration of devices on a single silicon chip. The level of megabit to gigabit integration is usually called ULSI (ultralarge-scale integration). With this rapid technological progress in ULSI, device interconnect technology is becoming more and more significant in determining integrity and performance. Conventional interconnect technology uses aluminum (Al) for the conductive material and silicon dioxide (SiO_2) for the dielectric. The

3-527-28138-X/96/$5.00 + .50

SiO_2 film is still used for the dielectric, but new conductive materials and their processes have been developed in order to obtain reliable high-performance devices. Use of refractory metals has made great progress in metallization of Si ULSI. Figure 1 shows schematically the cross section of multilevel metallization in Si ULSI. The demand for closely packed devices has pushed the metallization system into such a multilevel structure, and in some cases, more than five-level metallization has been realized. This figure shows three-level metallization. Important features of refractory materials are the resistance to high-temperature processes, mechanical hardness, barrier properties to Al diffusion, and relatively low electrical resistivity. As shown in this figure, refractory metals are used for gate electrodes of MOSFETs (metal-oxide-semiconductor field-effect transistors), source and drain metallization, interconnects, contact and via filling, barrier metals to Al diffusion, and so on. Among refractory metals, tungsten (W) has low electrical resistivity and is widely used for these applications. This article discusses these electronic applications of selected refractory metals in the following aspects:

1. thin-film formation and characterization,
2. applications to Si ULSIs, and
3. applications to GaAs MESFETs.

1. REFRACTORY-METAL FILMS

In Si ULSIs, typical dimensions of devices are reduced to below 1 micron (μm), and films used for the device fabrication become very thin (e.g., less than a few hundred nm). Under the usual deposition conditions used in manufacture, thin refractory-metal films are mostly polycrystalline or amorphous rather than single-crystal. Refractory compound material is also used. It is very important for practical uses to determine how to form a refractory thin film and to characterize it.

1.1 Formation Method

Table 1 shows vapor pressures of refractory metals as a function of temperature. In practice, vapor pressures of 1–10 Pa are necessary for evaporation of refractory metals with deposition rates more than 0.5 μm/min. As shown in this table, refractory metals have high boiling points, and they are not usually deposited using the conventional resistance-heat evaporation or electron-beam evaporation for ULSI applications. Instead, sputtering or chemical vapor deposition (CVD) has been used. In particular, magnetron sputtering, where large plasma density is sustained by magnets for high deposition rate, is widely used in manufacture. A metal-film formation method that uses a gas-phase chemical reaction [CVD (Chemical Vapor Deposition), *q.v.*], has been developed to obtain the deposition of films with good step coverage (conformal deposition) onto a substrate with complicated structure.

1.1.1 Magnetron Sputtering Thornton and Penfold (1978) havc proposed a magnetron sputtering method. In practice, a planar magnetron cathode is used, in which magnets are attached to the back side of the

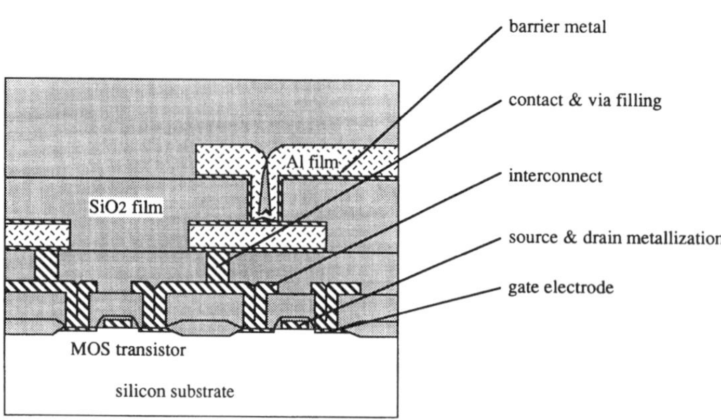

FIG. 1. Cross-sectional view of Si ULSI multilevel metallization.

Table 1. Vapor pressures of refractory metals at various temperatures.

Material	Vapor pressure (Pa)			
	1.33×10^{-4}	1.33×10^{-3}	1.33	133
	Temperature (K)			
Titanium, Ti	1508	1726	2020	2464
Chromium, Cr	1229	1403	1637	1968
Cobalt, Co	1095	1247	1454	1744
Nickel, Ni	1061	1207	1402	1681
Zirconium, Zr	1975	2265	2662	3227
Molybdenum, Mo	2090	2386	2797	...
Palladium, Pd	1336	1539	1820	2246
Tantalum, Ta	2507	2862	3329	3978
Tungsten, W	2667	3029	3502	4141
Rhenium, Re	2480	2844	3333	4036

sputtering target. It is important to obtain a high-purity refractory-metal film, and so sputtering is done with a high deposition rate at low background pressures. With advances in vacuum pumping technology, base pressures lower than 10^{-6} Pa are achieved in commercially available sputtering machines. In the sputtering, an inactive gas such as argon (Ar) is used to form the plasma, and a small number of gas atoms (around 1 at.%) can be incorporated in the metal films. Impurities have a large effect on the electrical resistivity and mechanical stress of the films.

In reactive sputtering, an active gas is intentionally added to Ar to deposit chemical compounds of refractory metals. Refractory nitrides such as titanium nitride (TiN) and zirconium nitride (ZrN) can be deposited by reactive sputtering in nitrogen (N_2) atmosphere. In addition, refractory oxides such as tantalum oxide (Ta_2O_5) and tungsten oxide (WO_3) can be obtained by adding oxygen (O_2).

1.1.2 Chemical Vapor Deposition (CVD) Chemical vapor deposition has been used in microelectronics processes to form Si and SiO_2 films. However, it has only recently been used for metal-film deposition. This is because technological problems such as the choice of suitable source gases for CVD and the design of the CVD reactor are rather difficult compared with Si or SiO_2 CVD using silane (SiH_4) or silane and oxygen (O_2). Metal precursors used for the CVD of refractory metals were summarized by Bernard *et al.* (1989). Among them, CVD of refractory metals such as tungsten (W), tungsten silicide (WSi_2), molybdenum (Mo), molybdenum silicide ($MoSi_2$), titanium nitride (TiN), and titanium silicide ($TiSi_2$) have been extensively investigated, and some are used for Si ULSI applications. An example of a CVD reactor, which is used for the CVD of these materials, is shown in Fig. 2. This figure shows a bell jar reactor that is shown in Sherman (1987). In this configuration, source gases enter at the top of the bell jar and flow between the bell jar and susceptor before exiting at the bottom. Silicon wafers are set on the susceptor. The susceptor can be heated by halogen lamps or resistance heaters located inside the bell jar. In CVD, a gas-phase chemical reaction proceeds at low pressures on a hot substrate surface the temperature of which is controlled. Typically, CVD temperature ranges from 200 to 700 °C, and total pressure from 10 to 100 Pa. The CVD reactor wall is usually cooled by water to prevent metal deposition on the wall. Moreover, some commercially available CVD

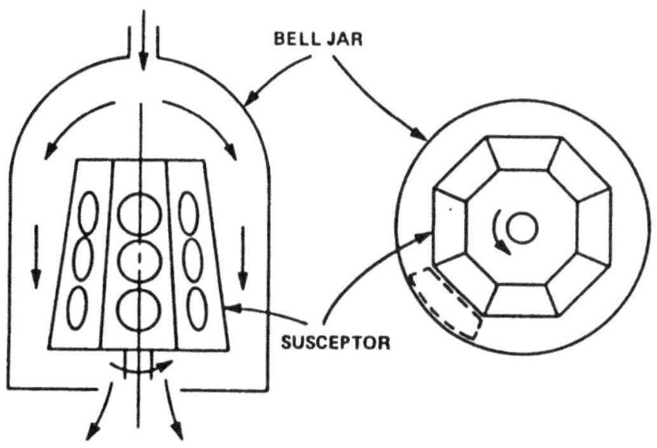

FIG. 2. Schematic figure of a bell jar CVD reactor.

machines are equipped with a plasma cleaning system for dry etching the metal deposited on the wall. With an increase of wafer size (e.g., 8-inch wafers currently in production), the single-wafer system has been widely accepted for CVD machines. A single-wafer reactor is schematically shown in Fig. 3, which is taken from Schmitz (1992). Film properties depend strongly on CVD parameters: temperature, pressure, and gas flow of source gas.

1.1.3 Surface Cleaning Film properties also depend on the physical and chemical conditions of the surface: surface roughness, surface oxidation, and contamination. Thus, it is important to clean the surface before CVD to achieve reliable metal deposition. Native oxide on a Si surface is usually removed by an HF dip. To remove native oxide on a metal surface (e.g., oxide on an aluminum surface), *in situ* Ar sputter cleaning is carried out before sputtering deposition of metal. Much work has been done to characterize refractory metals in terms of CVD parameters and surface conditions.

1.2 Characterization of Film Properties

The physical and chemical properties of thin films often differ from those of the bulk material. Characterization of thin films is important for scientific research as well as for electronic applications. General descriptions of characterization have appeared in textbooks—for example, Sze (1983)—and so this section focuses on refractory thin films.

1.2.1 Film Properties Electrical resistivity, Schottky-barrier height, mechanical stress, mass density, film composition, and microstructure (preferred orientation, grain size, subgrain structure, etc.) are important factors in determining the effectiveness of refractory metals for ULSIs. Low electrical resistivity is required for refractory thin films to be used for gates and interconnections. Low Schottky-barrier height to both n^+ and p^+ Si is preferable for Ohmic contact. Table 2 shows typical values of electrical resistivity and Schottky-barrier height of bulk refractory metals. Resistivity and barrier-height data are from the American Institute of Physics Handbook (Gray, 1972) and Rhoderick (1978). The resistivities of Mo and W are the lowest among refractory metals, and so they are widely used for metal interconnects. The electrical resistivity of refractory thin films is larger than that of the bulk material (almost twice in W and Mo) even if sputtering deposition is done at very low pressures ($<10^{-6}$ Pa). This might be because defects in the crystalline lattice of thin refractory-metal films, which are deposited at temperatures much lower than the melting point, are more numerous than those of the bulk material. Mechanical tensile stress of refractory metals is generally high, for example, greater than 1 GPa. So peeling of a refractory-metal film from the substrate is a concern in microelectronics processes. Film stress also has an influence on device characteristics such as hot-electron reliability of MOS devices and leakage current at junctions.

In general, the contact between refractory metal and Si substrate is characterized as rectifying or Ohmic, depending on dopant concentration of the Si substrate. In Ohmic contact, the most important feature is low

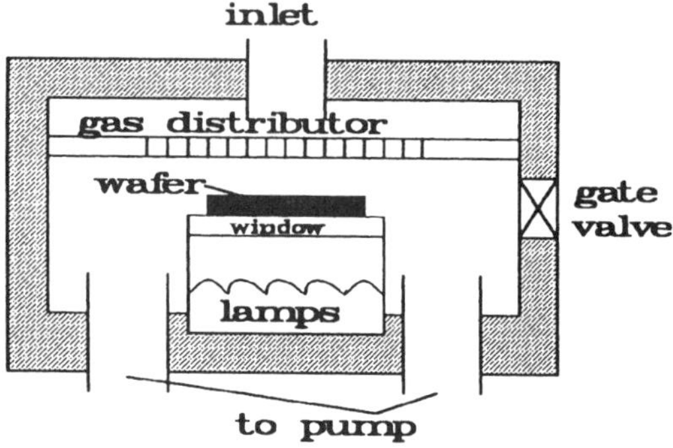

FIG. 3. Schematic figure of a single-wafer reactor.

Table 2. Resistivities and Schottky-barrier height of refractory metals.

Material	Electrical resistivity ($\mu\Omega$ cm) (bulk)	Schottky-barrier height (eV) (to n^+ Si)
Ti	47	0.5
Cr	12.9	0.6
Co	5.8	...
Ni	7.04	0.65
Zr	45	...
Mo	5.33	0.60
Hf	30.6	...
Ta	13.1	0.58
W	5.44	0.65
Re	18.6	...

resistance. For this purpose, Si substrate is heavily doped with arsenic (Ar) or phosphorus (P) for n^+ contact and with boron (B) for p^+ contact. In the case of heavy doping ($>10^{19}$ cm^{-3}), Sze (1981) discussed that tunneling current through the Schottky barrier becomes important, and the specific contact resistance (R_c) may be written as

$$R_c \sim \exp[4\pi\epsilon m^*/h(\Phi_B/\sqrt{N_D})], \quad (1)$$

where ϵ is the dielectric permittivity of silicon, N_D is the doping concentration, h is Planck's constant, m^* is the effective mass of the charge carrier (electron or hole), and Φ_B is the Schottky-barrier height. In order to obtain low specific resistivity, refractory metal having low Φ_B is preferable.

1.2.2 Characterization Method Characterization of film properties is carried out using refractory-metal films that are formed on a substrate. In Si LSI processing, refractory-metal films are usually deposited on silicon wafers. Electrical resistivity is calculated from the sheet resistance of the film multiplied by its thickness according to the formula

$$\rho = R_s t, \quad (2)$$

where ρ (Ω m) is electrical resistivity, R_s ($\Omega/\square$) is sheet resistance, and t (m) is metal thickness. Sheet resistance is usually measured by a four-point terminal method, and film thickness is measured from the step height that is formed by removing a metal film by means of wet chemicals and an etching mask. Other electronic properties such as electrical contact resistivity are measured by using a series-chain-contact pattern device or a Kelvin pattern device, as was reported by Loh *et al.* (1985). Mechanical stress is measured from the curvature of the substrate on which the refractory-metal film is deposited. A formula for the stress, developed assuming the substrate to be a cubic plate, is expressed as

$$\sigma = (\delta/r^2)[E/3(1 - \nu)](t_s^2/t), \quad (3)$$

where σ is stress (N/m^2), δ is disk deflection (m), ν is substrate Poisson's ratio, E is substrate Young's modulus, t_s is substrate thickness (m), t is metal thickness, and r is disk radius (m).

In binary-metal systems such as titanium–tungsten (Ti–W) and titanium nitride (TiN), film composition is strongly related to physical and chemical film properties. Film composition is usually determined by Rutherford backscattering. The profile of the impurity distribution in films is measured by sputtering Auger electron spectroscopy. Impurities affect the microstructures and the electrical properties of films. In particular, oxygen impurity is easily absorbed in refractory-metal films, and this increases electrical resistivity. A little impurity that cannot be detected by Rutherford backscattering and Auger electron spectroscopy (less than 0.1 at.%) in films is measured by secondary-ion mass spectroscopy. Crystal orientation is determined by x-ray diffraction, and grain size is estimated from width of the x-ray diffraction peak. The microstructure of films is usually observed by SEM (scanning electron microscopy) and TEM (transmission electron microscopy). The cross section of the device structure can be observed by TEM by thinning a sample that is cut from a test wafer. Nanometer-level surface roughness is evaluated using scanning tunneling microscopy and atomic-force microscopy.

1.3 Patterning

1.3.1 Dry Etching In microelectronics applications, refractory-metal films are patterned to fine lines using photolithography and dry etching. Dry etching is a plasma-assisted pattern transfer technique using partially ionized gases consisting of ions, electrons, and neutral in a low-pressure ($\sim 10^{-2}$–10^{-3} Pa) electric discharge. In the dry-etching process, other material such as SiO_2 film and masking material exist under a refractory-metal film, and so selectivity of etching is a parameter of considerable importance. Typical etching gases used for dry etching of refractory metals are shown in Table 3. The

Table 3. Etching gases used in dry etching of refractory metals.

Material	Gases used for dry etching
Cr	$CCl_4(+O_2)$, Cl_2
Ti	CCl_4, BCl_3, $SiCl_4$
Mo	$CF_4(+O_2)$, $CBrF_3$
W	$SF_6(+O_2)$, $CF_4(+O_2)$, NF_3

vapor pressures of tungsten (W) and molybdenum (Mo) fluorides are relatively high, so that fluorine-based chemistry is used in dry etching for them. Selectivity values of 3–5 to SiO_2 can be obtained using $CF_4 + O_2$ or SF_6, although it strongly depends on the etching conditions. Dry etching of titanium (Ti) and titanium compounds uses chlorine-based chemistry such as CCl_4, BCl_3, and $SiCl_4$. On the other hand, there are no volatile compounds for nickel (Ni) and cobalt (Co), and they are physically etched using ion milling or sputtering. However, it is difficult for these metals to be used in LSI because etching using ion milling or sputtering shows little selectivity to an underlying SiO_2 film. However, Ni and Co have been investigated as metals for self-aligned silicided source and drain, where these metals are removed by wet chemicals listed in Table 4. The self-aligned silicide technology is described in Sec. 2.4.

1.3.2 Etching Using Wet Chemicals Etching of refractory metals using wet chemicals is not suitable for fine patterning. However, it is a very convenient way to remove an unnecessary metal film, or to form a relatively large pattern. As is shown in Sec. 2.4.2, wet etching of refractory metal is used for self-aligned silicide formation. In addition, several kinds of wet chemicals are used to clean Si wafers during device fabrication, and so it is important to know the resistance of refractory metals to wet chemicals when they are used in ULSI devices. Wet chemicals that are often used for patterning refractory metals in microelectronics fabrication are shown in Table 4. It should be noticed that Ti and Ta can be dissolved in HF solutions when these metals are used in the Si ULSI process.

Table 4. Wet chemicals for refractory metals.

Material	Chemical solution
Cr	HCl/H_2O_2 = 1/1
Co	HCl/H_2O_2 = 1/1
Ni	HCl/H_2O_2 = 1/1
Ta	HF/HNO_3 = 1/1
Ti	$HF/HNO_3/H_2O$ = 1/1/50, NH_4OH/H_2O_2 = 1/2
Mo	$H_3PO_4/HNO_3/CH_3OOH/H_2O$ = 5/2/4/150
	$HNO_3/H_2SO_4/H_2O$ = 1/1/3
W	$KH_2PO_4/KOH/K_3Fe(CN)_6/H_2O$ = 34/13.4/33 (in weight)
	H_2O_2

2. APPLICATIONS TO Si ULSIs

2.1 Tungsten or Molybdenum Gate Metallization for MOS LSIs

2.1.1 Requirements for MOS Gate Material We use a conductive thin film for the gate electrode, which is a key component of MOS transistors. It is used to control the current under the gate by applying voltage to it. The details of MOS device operation are explained in textbooks—for example, Muller and Kamins (1986). In terms of the electrical properties of gate material, its electrical resistivity and work function are important factors for MOS device operation. Saraswat and Mohammadi (1982) calculated the *RC* time delay for gate and interconnection in ULSI circuits and pointed out the importance of low-resistivity gate and interconnection. The work function of the gate material determines the threshold voltage of the MOS device, where the current under the gate begins to flow. On the other hand, the gate material is required to have these characteristics to make it suitable for the manufacturing processes: stable interfacial electrical and mechanical properties, fine patternability, resistance to high-temperature process (800–900 °C), use as a mask for ion implantation, and so on. Figure 4 schematically shows cross sections of MOS transistors. It describes the method of fabricating MOS transistors. Gate material is deposited on the Si substrate, whose surface is covered by thin Si oxide (gate oxide), and patterned to form a gate electrode. Highly doped regions (source and drain) are formed on the substrate by ion implantation using the gate material as a mask. This is followed by annealing at 800–950 °C to activate implanted dopants. These processes constitute what is called self-aligned gate technology. In the self-aligned gate processes, the gate material is required to stop the accelerated ions and to have thermal stability during annealing.

Conventionally, polycrystalline silicon (poly Si) has been used for the gate material, because it satisfies these requirements quite well. However, a gate material with a resis-

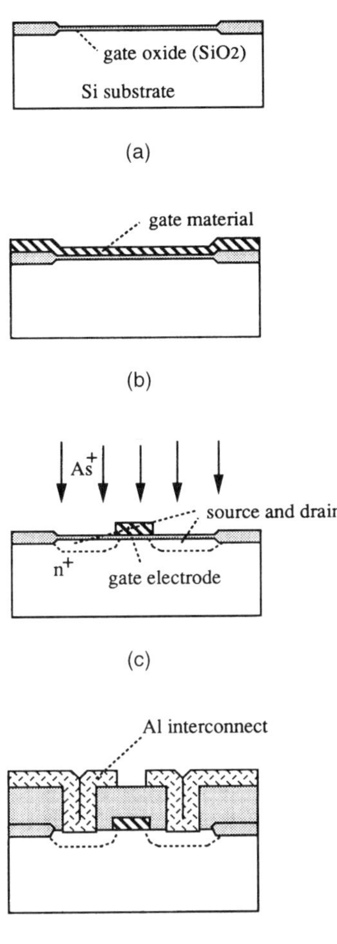

FIG. 4. Fabrication process flow of MOS transistors using the self-aligned gate technology.

tivity lower than that of poly Si is greatly needed to reduce the *RC* delay of the device as the dimensions of the gate electrode decrease. Low-resistivity refractory-metal silicide has therefore been used in manufacturing. In this application, a layered gate structure where silicide is put on poly Si is used because of its compatibility with the poly Si gate process. This gate structure is called the "polycide" gate. Wang *et al.* (1982) reported application of a polycide gate using titanium silicide ($TiSi_2$) for low-resistivity gate and interconnection. The polycide gate is used in manufacture; however, refractory metals such as W and Mo have gathered much interest for research and development as an alternative to poly Si and silicides because of their low resistivity.

2.1.2 Tungsten or Molybdenum Gate Process Tungsten (W) and molybdenum (Mo) are the most promising candidates for low-resistivity gate materials, and they have been studied for Si LSI applications. In Table 5, the electrical resistivities of refractory-metal films are compared with those for poly Si and silicides. Brown *et al.* (1971) investigated the self-aligned Mo gate technology, although impurities were doped into the Si using doped glass diffusion sources instead of ion implantation. Iwata *et al.* (1984) compared Mo and W, and pointed out that W is more suitable for gate-electrode material because of the thermal stability of the W/SiO_2 interface. There are, however, several problems in using W (or Mo) for gate electrodes.

Generally, the adhesion between W (or Mo) and gate SiO_2 is not strong compared with that between poly Si and SiO_2. In most cases, the W (or Mo) is deposited by sputtering because it cannot be stably formed on SiO_2 by CVD. Therefore it is necessary to reduce mechanical film stress by optimizing the sputtering conditions. Figure 5 shows the internal stress change of Cr, Mo, Ta, and Pt films as a function of argon pressure during sputtering, as reported by Hoffman and Thornton (1982). Although the threshold Ar pressures, where internal stress changes from compressive to tensile, differ among these metals, the film stress changes from compressive to tensile as Ar pressure is increased. At low Ar pressures, the mean free path is a few centimeters and Ar ions easily hit the substrate surface to be incorporated in the film. The compressive stress is thought to come from interstitial argon incorporation in the W lattice. Yamamoto *et al.* (1987) reported that the film stress has an influence

Table 5. Comparison of film resistivities for MOS gate materials.

	Material	Thin-film resistivity ($\mu\Omega$ cm)
Silicon	Phosphorus-doped poly Si	700–1000
Refractory silicide	WSi_2	70–120
	$MoSi_2$	60–110
	$TiSi_2$ (after annealed at 900–1000 °C)	15–30
Refractory metal	W	8–12
	Mo	7–11

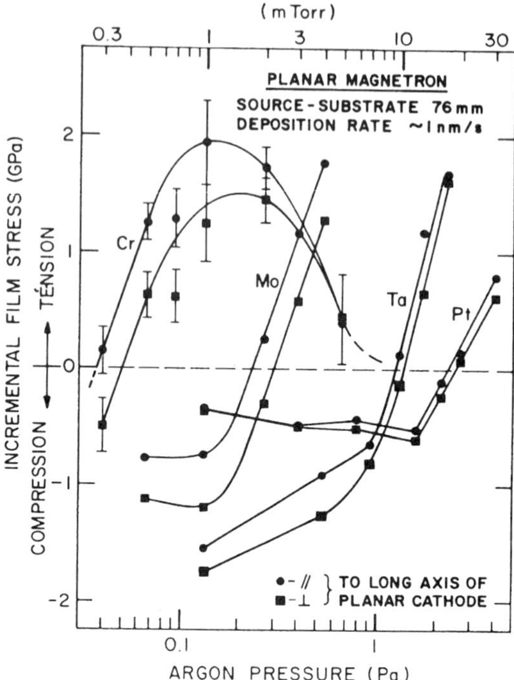

FIG. 5. Mechanical stresses as a function of Ar pressure for Cr, Mo, Ta, and Pt metal films sputter deposited from dc planar magnetron source onto glass at 1 nm/s to thickness of 50–350 nm.

on the hot-electron reliability of MOS devices. Thus, the film stress of W (or Mo) is controlled to be nearly zero by optimizing the argon pressure and other sputtering conditions.

The second problem is poor mask properties of W (or Mo) film against ion implantation to form source and drain regions. This is due to the fine columnar structure of W films, although the relatively large W atom itself has a large stopping power against the accelerated ions. In addition, ion channeling occurs through (110)-oriented W films to the Si substrate. W (or Mo) has the bcc (body-centered cubic) structure, and so its preferred orientation is (110). In practice, a 10–50-nm-thick SiO_2 film is deposited onto the W gate to prevent ion penetration through the W film.

Oxidation of W (or Mo) is also a serious problem. In the course of Si LSI fabrication, Si is oxidized in the presence of gate material. During oxidation, poly Si is covered with SiO_2, which is very stable and protective against contamination. On the other hand, W (or Mo) is easily oxidized to form unstable oxides. Figure 6 shows the temperature dependence of oxidation of sputter-deposited W films in dry O_2 ambient. SEM observation and step-height measurement of W oxides are utilized to determine the W oxide thickness. The oxidation rate is a thermally activated process, and it proceeds as $(\text{time})^{1/2}$. The oxidation rate is larger than that of poly Si by more than two orders in magnitude. Moreover, the W (or Mo) oxide is volatile at high temperatures (800–1000 °C), and its stoichiometry strongly depends on the oxidation conditions. To prevent W oxidation, an oxygen-free process is preferable. For example, measures should be taken to prevent oxygen incorporation from outside when W (or Mo) is annealed in inert gases using a quartz furnace tube. For the case that oxidation of Si is inevitable in the presence of gate material, Kobayashi *et al.* (1984) developed a wet hydrogen process, where an appropriate amount of H_2O is added to H_2, which has been developed to oxidize Si without oxidizing W.

The W/SiO_2 interface is thermally stable for obtaining reliable MOS characteristics. The interfacial reaction of refractory metal (M) with SiO_2 is expressed as

$$SiO_2 + xM = Si + M_xO_2. \quad (4)$$

This reaction favorably occurs when the

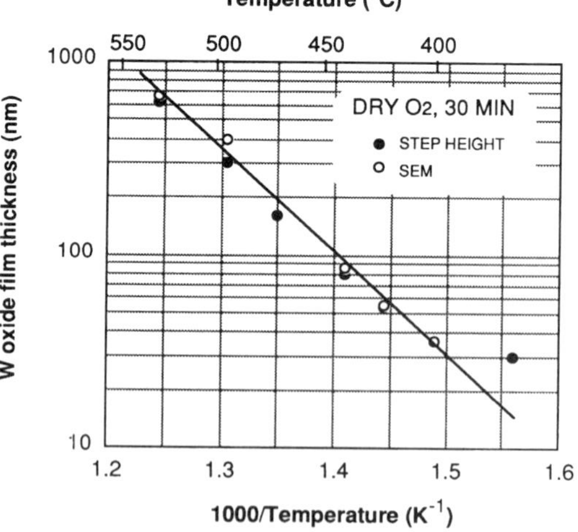

FIG. 6. Tungsten-oxide thickness of sputter-deposited tungsten films as a function of temperature.

Gibbs free energy change (ΔG) is negative. In other words, this reaction proceeds spontaneously when the absolute value of the Gibbs free energy of the metal oxide (M_xO_2) is greater than that of SiO_2. For W or Mo, ΔG is positive, and so W (or Mo) is not expected to react with SiO_2. No metallurgical reaction of the W/SiO_2 interface was observed by XPS up to 1000 °C annealing, as reported by Iwata *et al.* (1984). However, Ti or Ta reacts with SiO_2 because ΔG is negative. This reaction can degrade the electric characteristics of the gate oxide. On the other hand, it helps to reduce the contact resistance when metal can reduce the native oxide of the Si surface. Thus, Ti or Ta is used not for the gate electrode but for a low resistive contact. Although W (or Mo) is thermodynamically stable on SiO_2, its deposition method is important to obtain reliable MOS characteristics. As the gate oxide thickness decreases to less than 10 nm in the deep-submicron ULSI, damage induced by film deposition becomes serious. In sputtering deposition, plasma damage can cause degradation of the gate oxide. In CVD, the tungsten source gas (WF_6) reduces thin gate SiO_2 at high temperatures (400–500 °C). Further study is necessary in depositing W directly on thin gate SiO_2 in order to achieve good electrical performance.

Finally, avoidance of mobile-ion (Na^+, K^+, etc.) and radioactive-element (U, Th, etc.) contamination and damage induced by the sputtering deposition is critical in achieving reliable MOS characteristics. Mobile ions are easily diffused to the W/SiO_2 interface, and change the threshold voltage. May and Woods (1970) found a soft error mechanism in Si LSIs induced by alpha particles emitted in the radioactive decay of U and Th. Since then, these radioactive elements have been reduced to less than a few ppb level for interconnect and packaging materials. For this purpose, a high-purity sputtering target (99.999%–99.9999%) has been developed for use in microelectronics fabrication. Moreover, high-purity film formation by CVD has been investigated.

2.2 Refractory-Metal Barrier for Aluminum Interconnection

2.2.1 Barrier Properties Direct contact of Al to Si suffers from Al spikes caused by local dissolution of Si. The problem of Al spikes is solved by depositing Al mixed with Si. In practice, more than 1 wt.% Si is required to prevent the spikes. However, the excess Si present in the Al precipitates on the Si after a heating cycle at around 450 °C, which causes a non-Ohmic contact to Si, especially to n^+ Si. Thus, in current Al metallization, a refractory-metal barrier is used between the Al and the Si contact. Refractory-metal barriers have been used in microelectronics fabrication. Among these refractory-metal compounds, Ti–W and TiN are widely used. Ghate *et al.* (1978) reported application of Ti–W barrier metallization for Si integrated circuits. Ti–W is usually deposited by sputtering using a Ti–W compound target. On the other hand, TiN is usually deposited by reactive sputtering, and a CVD method has recently been developed.

It is known that TiN prevents Al diffusion into Si, and does not react with Al at up to 500 °C. This is why TiN is used as a barrier metal especially for high-temperature processing. Barrier properties strongly depend on the morphology and stoichiometry of the film. Pramanik and Jain (1991) reported that titanium nitride can exist as Ti_2N and TiN and nitrogen can form a solid solution with these compounds. Stoichiometric TiN has excellent barrier properties and low resistivity (50–100 $\mu\Omega$ cm). Thus, the nitrogen (N_2) gas pressure and applied voltage are key parameters in determining the film properties in reactive sputtering. However, the sputter-deposited metal barrier can be too thin to prevent the Al diffusion into Si at the bottom of a contact when the contact becomes deep. Thus, TiN CVD using $TiCl_4$ and NH_3 has been investigated for conformal deposition. Figure 7 shows the excellent step coverage of TiN film deposited in contacts with various sizes, as was reported by Yokoyama *et al.* (1991). The contact diameter is varied from 0.5 to 0.8 μm, and the substrate material is Si. In this CVD, the gas-flow ratio of $TiCl_4$/NH_3 and the CVD temperature are important parameters, and conformal TiN film can be deposited in deep contacts. High-temperature deposition (>700 °C) is necessary to reduce Cl incorporation in the films because incorporated Cl causes an increase of film resistivity. However, these deposition temperatures are too high to apply this CVD for Al metallization, and so low-temperature CVD using metal organic precursers such

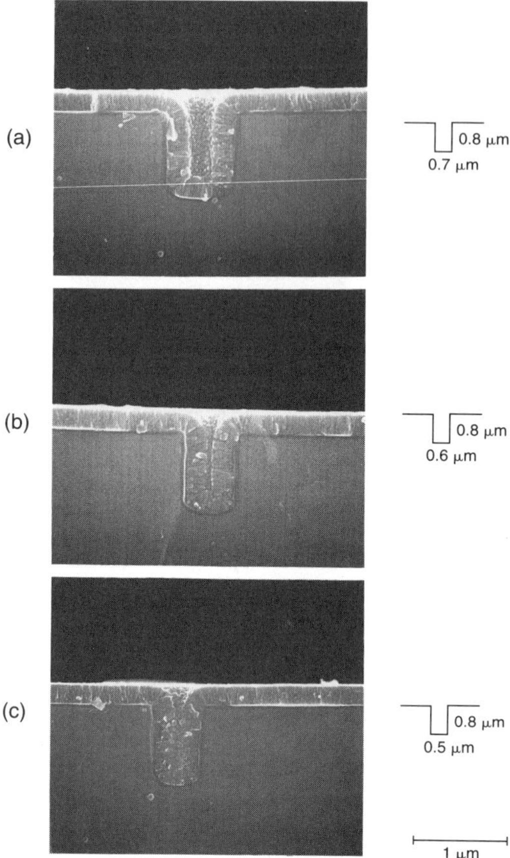

FIG. 7. SEM micrographs of 0.25-μm-thick CVD-TiN films in high-aspect-ratio contact holes with various diameters. (a) 0.7-μm diameter and 0.8-μm depth; (b) 0.6-μm diameter and 0.8-μm depth; and (c) 0.5-μm diameter and 0.8-μm depth.

as *tetrakis*-dimethylamino-titanium (TDMAT) and *tetrakis*-diethylamino-titanium (TDEAT) with NH_3 is being pursued. Moreover, a plasma-excited process using electron cyclotron resonance is being investigated.

Besides Ti–W and TiN, other refractory metals are often used as barrier metals. Holloway *et al.* (1992) reported that Ta or tantalum nitrogen alloy (Ta_2N) is a suitable barrier metal for Cu interconnect because of small reactivity with Cu and thermal stability in metal interdiffusion at the Cu/Ta structure.

2.2.2 Layered Aluminum Interconnects with Refractory Metals In device reliability, electromigration is a great concern for Al fine interconnects. Electromigration is observed as a material transport in conductive material, and it is caused by momentum transfer from electrons, moving under the influence of the electric field applied along the conductor. In addition, stress-induced migration (stress migration) has been found to cause failure of Al interconnects. In general, electromigration and stress-migration phenomena are related to Al self-diffusion along the grain boundaries, and easily occur in conductive materials with low melting points. Refractory metals have large electromigration and stress-migration resistance due to their high melting points, and this has been confirmed for W interconnects. To overcome these migration problems, layered Al interconnects with refractory metals have been widely used in microelectronics fabrication. In layered structures, refractory metal is deposited beneath the Al, and if necessary on top of the Al. Metals such as Ti–W, TiN, W, and refractory silicides are used. The use of the refractory metal beneath (and on) the Al causes an increase in resistivity because the refractory metal reacts with the Al to form high-resistivity compounds such as WAl_{12} during heat treatment at around 450 °C. The reaction of refractory metals with Al has been extensively studied.

2.3 Contact Filling Using Chemical Vapor Deposition of Tungsten

Chemical vapor deposition (CVD) of tungsten (W) achieves conformal deposition in deep contacts. Moreover, selective deposition on a conductive layer without deposition on an insulator layer can be achieved when appropriate CVD conditions are selected, and this has been widely investigated for filling submicron contacts.

2.3.1 Requirements for Contact Filling Figure 8 shows the aspect ratios of contact holes in Si LSI plotted as a function of the linewidth of the interconnects. The dotted line represents the dielectric layer thickness. Aspect ratio is defined as the dielectric layer thickness (= contact depth b) divided by contact diameter (a). The contact diameter decreases with scaling of device dimensions, whereas interdielectric layer thickness remains almost constant because a certain minimum thickness is necessary for electrical insulation between interconnects and for

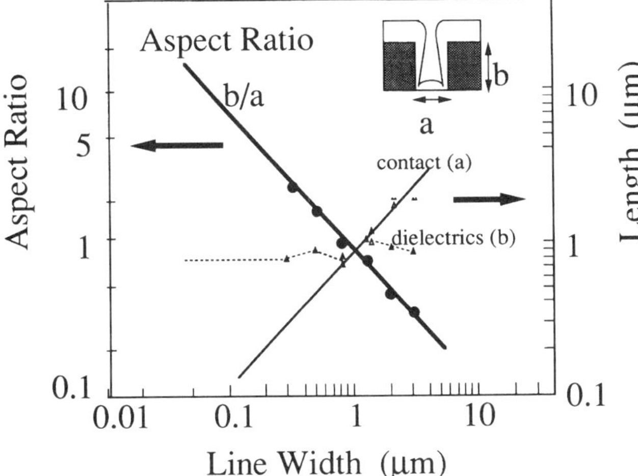

FIG. 8. Aspect ratio of contact holes as a function of linewidth of interconnect. Dotted line represents the dielectric thickness.

planarization of the multilevel metallization structure. As a result, the contact aspect ratio increases as the linewidth decreases, and it reaches around 2 in 0.5-μm interconnects. Conventional sputtering cannot achieve conformal metal deposition in high–aspect-ratio contacts. Figure 9 shows an SEM micrograph of W interconnects deposited by planar magnetron sputtering. The step coverage factor, defined as the ratio of film thickness at the bottom of the contact to that at the flat surface, is below 5% even at aspect ratio of 0.8, and it will cause electrical open-circuit failure of thin W interconnects when the aspect ratio exceeds 2. Low-pressure CVD is expected to achieve conformal deposition compared with the sputtering. For comparison, W interconnects formed by CVD are shown in Fig. 9 along with sputter-deposited W film. The step coverage factor is almost 100% for CVD-W interconnects. Because of conformal film deposition and high reliability against electromigration and stress migration, the CVD-W interconnect has begun to be used in the Si ULSI as an alternative for sputter-deposited Al interconnects. The film properties are compared between W and Al in Table 6. However, the characteristics of films deposited by CVD strongly depend on the CVD chemistry. The metals that can be deposited by CVD and that are widely used in microelectronics fabrication are restricted to W and WSi_2. CVD of Al, Cu, and TiN is now being widely investigated for LSI applications.

Figure 10 schematically shows the cross section of multilevel metallization using W-CVD. There are two approaches to fill the high–aspect-ratio contact and via using W-

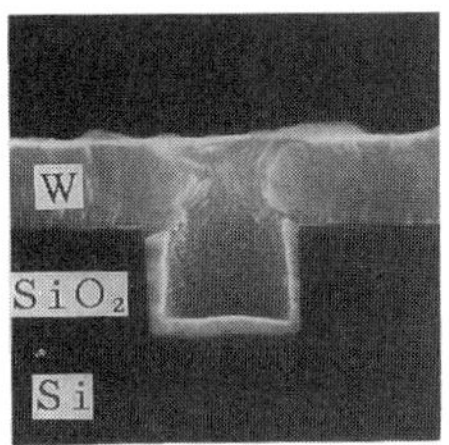

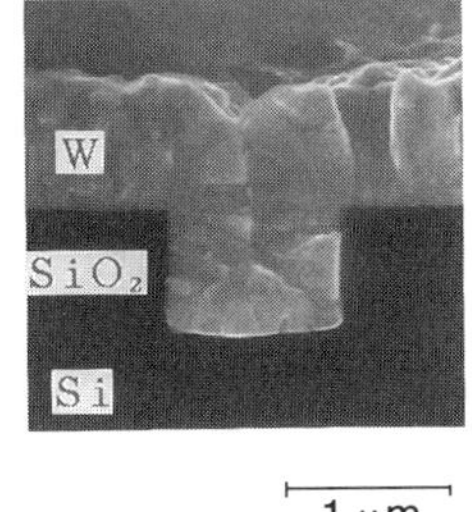

1 μm

(a) (b)

FIG. 9. SEM micrograph of W interconnects deposited by planar magnetron sputtering in Si contact. (a) Sputter-deposited W interconnect and (b) CVD-W interconnect.

Table 6. Comparison of tungsten with aluminum for interconnect application.

	W	Al (containing 1 wt.% Cu,Si)
Resistivity (μΩ cm)	8–10	3–3.5
Melting point (°C)	3380	660
Schottky barrier (eV) n^+ Si	0.5–0.6	0.65–0.75
Dry etching	SF_6	BCl_3
CVD gas	WF_6	TIBA[a], DMAH[b]

[a]TIBA: $Al(iC_4H_9)_3$
[b]DMAH: $(CH_3)_2AlH$

Conventional Al interconnect (by sputtering method)

Tungsten (W) interconnect (by CVD method)

Al interconnect

SiO2

Si - LSI substrate

selective W-CVD

blanket W-CVD

W

problem: poor film step-coverage low migration resistance

good film step-coverage high migration resistance

FIG. 10. Cross-sectional view of multilevel metallization using tungsten CVD.

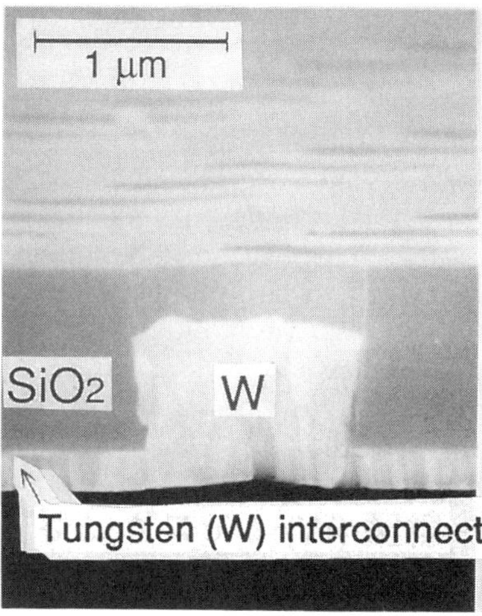

FIG. 11. SEM micrograph of W selectively deposited on W interconnect. Tungsten is deposited using WF_6 and SiH_4.

CVD: One is selective deposition of W in contact and via holes, and the other is blanket deposition of W for W interconnect formation. In selective deposition, W is deposited only onto Si and metal without being deposited on dielectrics. Figure 11 shows SEM micrographs of W selectively deposited on W interconnects. After filling the contact and via with W, the interconnect metal (typically Al) is usually deposited by sputtering. In blanket deposition, the sputter-deposited metal film is usually deposited on the interdielectric SiO_2 film because adhesion between the W-CVD film and the SiO_2 is poor. Source gases for W-CVD that have been investigated are compared in Table 7. For Si LSI applications, tungsten hexafluoride (WF_6) has been mainly used because the vapor pressure of WF_6 is high enough for use of a conventional gas supply system and it causes relatively little damage to Si substrates. Tungsten hexachloride (WCl_6) has rarely been used in manufacture because it etches Si heavily and corrodes the stainless steel of the CVD reactor. Tungsten carbonyl [$W(CO)_6$] is a solid source with a low vapor pressure, but it is not suitable for high-quality W film deposition because of carbon incorporation in the film. When WF_6 is used for the source gas, a reducing gas for WF_6 is necessary because WF_6 itself is not decomposed at temperatures around 300–500 °C. It is necessary to point out that the characteristics of W-CVD depend strongly on the choice of reducing gas as well as the source gas. Thus, we should choose a gas to reduce WF_6 according to the application.

2.3.2 Selective Chemical Vapor Deposition of Tungsten Although the history of selective W-CVD is long, it has been exten-

Table 7. Source gases for W-CVD.

Source gas	Vapor pressure (Pa)	CVD temperature (°C)	Chemical reaction
WF_6	1.3×10^5 at 20 °C	250–500	$WF_6 + 3H_2 = W + 6HF$
WCl_6	(solid)	550–1250	$WCl_6 + 3H_2 = W + 6HCl$
$W(CO)_6$	(solid, 1330 at 60 °C)	400–500	$W(CO)_6 = W + 6CO$

sively investigated for Si LSI application for the past 10 years. Cuomo (1972) first reported selective W-CVD using WF_6 and H_2 for Si LSIs. The global chemical reaction is

$$WF_6 + 3H_2 \rightarrow W + 6HF. \quad (5)$$

McConica and Krishnamari (1986) studied growth kinetics and found that the growth rate of W film [R (nm/s)] is given by

$$R = 6.8 \times 10^4 \exp(-8000/T) P(WF_6)^0 P(H_2)^{0.5}, \quad (6)$$

where T is the deposition temperature (K) and P is the partial pressure (Pa). The origin of selectivity is thought to be the dissociative adsorption of H_2 on a conductive layer, which is expressed by

$$H_2 \rightarrow 2H. \quad (7)$$

Atomic hydrogen can react with adsorbed WF_6 to form W and HF. Broadbent (1984) explained that an activation energy in Eq. (6) corresponds to the surface migration of H on a growing W surface. When W is deposited on Si by this chemistry, the reaction of WF_6 with Si,

$$WF_6 + \tfrac{3}{2}Si \rightarrow W + \tfrac{3}{2}SiF_4, \quad (8)$$

precedes reaction (5) because reaction (8) is very rapid. Reaction (8) is called a Si reduction process, and it shows self-limiting growth after which no further W growth occurs. The self-limiting thickness depends on the CVD conditions and Si surface conditions, and it ranges from 10 to 1000 nm. In Si contacts, highly doped shallow junctions are formed to obtain an Ohmic contact, and the Si reduction process should be suppressed as much as possible because "encroachment" of W into the shallow junctions degrades leakage current characteristics. Moreover, a by-product gas of reaction (5), HF, can etch Si at relatively high temperatures and form a "worm hole" at the interface of W/Si, as was observed by Stacy *et al.* (1985).

To overcome this problem of Si damage, Kusumoto *et al.* (1987), Ohba *et al.* (1987), and Kotani *et al.* (1987) developed W-CVD using monosilane (SiH_4) as a reducing gas for WF_6. In this SiH_4-reduced W-CVD, the growth rate is high because of the high reactivity of SiH_4 with WF_6, and Si damage is greatly suppressed. The most important parameter of this chemistry is the gas-flow ratio of SiH_4/WF_6. Tungsten silicide (WSi_2) film can be deposited with ratios greater than 20. Tungsten can be selectively deposited with ratios less than 1–1.3. Film properties such as surface morphology and mechanical stress also depend on the flow ratio. Although the chemistry is not yet fully understood for SiH_4-reduced W-CVD, Kobayashi *et al.* (1991) observed two reaction pathways using *in situ* FT-IR (Fourier-transform infrared spectroscopy) of reaction gases during the selective W-CVD:

$$WF_6 + 2SiH_4 \rightarrow W + 2SiHF_3 + 3H_2, \quad (9)$$

$$WF_6 + \tfrac{3}{2}SiH_4 \rightarrow W + \tfrac{3}{2}SiF_4 + 3H_2. \quad (10)$$

Yu and Eldridge (1989) also observed reaction (10) using mass spectroscopy. A gas-phase reaction was observed even at room temperature in SiH_4/WF_6 gas mixtures when the flow ratio was larger than 1.3, and this causes loss of selectivity during deposition. Kobayashi *et al.* (1993) reported that SiH_4 is selectively dissociated at the W surface at temperatures higher than 150 °C, using the FT-IR reflection absorption spectroscopy, and this can be the cause of selective deposition on W without W deposition on SiO_2.

The growth kinetics has been studied by many researchers; however, there are some discrepancies among their data. The growth rate (R) is summarized as

$$R \propto \exp(-E/kT) P(WF_6)^a P(SiH_4)^{1+b}, \quad (11)$$

where E is the activation energy = 0.29–0.30 eV, P is the partial pressure, a is the exponent for $P(WF_6)$ = −0.5 to 0, and b is the exponent for $P(SiH_4)$ = 0 to 0.5. Suzuki *et al.* (1990) reported that the Si content incorporated in W films is a function of the gas-flow ratio as shown in Fig. 12. The Si content slightly increases with increasing gas-flow ratio, and decreases with increasing deposition temperature in selective deposition. On the other hand, in nonselective deposition where the gas-flow ratio exceeds 1.3, the Si content greatly increases, and it does not strongly depend on the deposition tem-

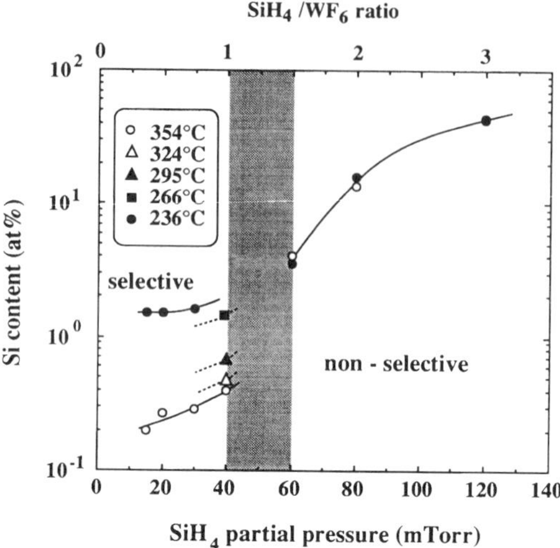

FIG. 12. The Si content in CVD-W films as a function of the gas flow of SiH_4/WF_6 at various temperatures.

perature. X-ray deflection shows that the film consists of polycrystalline α-W with gas-flow ratios less than 1, and it also involves β-W with ratios around 1.3. When the ratio exceeds 3, a broad x-ray diffraction peak of WSi_2 phase was also observed. The electrical resistivity of the film is almost proportional to the Si content and is determined by impurity scattering. This is confirmed by electrical resistivity measurement at liquid He temperatures. The superconducting transition at around 10 K was observed with W films containing 13–25 at.% Si. Kondo (1992) explained that a great increase of the superconducting transition temperature was ascribed to lattice softening in the amorphous W films. The film morphology also depends on the Si content in the film as shown in Fig. 13. The film texture changes from equiaxed to columnar surface features with increasing Si content.

In the ideal selective W-CVD, W can be deposited without any deposition of W on SiO_2. However, unnecessary W nuclei were often observed on the SiO_2 surface in the manufacturing process. These W nuclei are considered to be harmful particles produced in the process, and they should be reduced as much as possible to increase the device yield. Empirically, the cause for selectivity loss can be classified into two categories; one is the cause associated with chemical reactions during W-CVD, and the other is related to the surface condition of the SiO_2 surfaces. As for the former, besides the gas-phase reactions, reaction by-products such as tungsten subfluoride (WF_x) and silicon subfluoride (SiF_x) are reported by Creigton (1987) and Hirase *et al.* (1987) to cause selectivity loss. In the case of the latter, the SiO_2 surface is generally contaminated and damaged by conventional photolithography and dry etching, and this can cause loss of selectivity. It is known that Ar sputtering can break the Si–O chemical bond, and this provides a nucleation site for W deposition. In order to maintain the selectivity, nucleation sites on

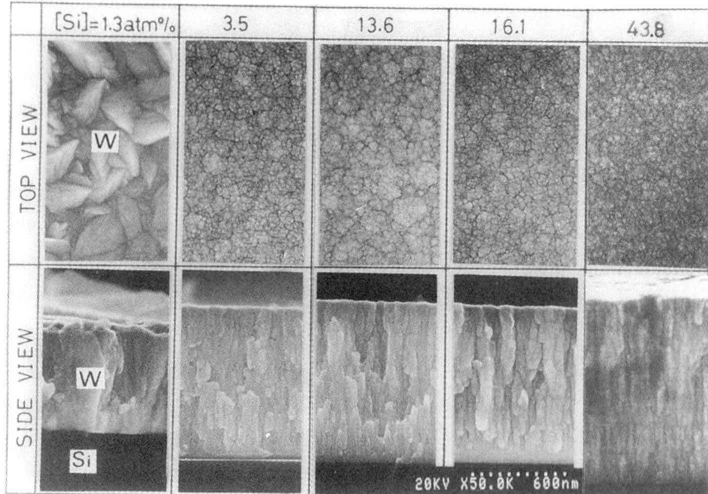

FIG. 13. Film morphology of CVD-W films as a function of the Si content.

the SiO_2 surface should be stably terminated with chemisorbed species. For this approach, *in situ* RIE (reactive ion etching) using Cl-based chemistry is being investigated and is shown to be effective to maintain the selectivity, as was reported by Tamaru *et al.* (1992).

2.3.3 Blanket Chemical Vapor Deposition of Tungsten In the formation of W interconnects, W-CVD has been widely used to fill submicron contacts. Because H_2- or SiH_4-reduced W-CVD provides selective deposition, W is not deposited on SiO_2. Thus, an adhesion metal layer is usually sputter deposited before the W-CVD when W film is formed on SiO_2. Titanium nitride (TiN), titanium–tungsten compound metal (Ti–W), and W are usually used for the adhesion layer. In the choice of adhesion layer, fine patternability, reactivity with WF_6, and contact resistivity (Schottky-barrier height) should be considered along with the adhesion to SiO_2. After deposition of an adhesion metal layer, W is deposited on the entire surface of the metal by CVD. We call this blanket W-CVD in contrast to selective W-CVD. Blanket W-CVD is used for forming W interconnection.

In blanket W-CVD, the H_2-reduced process has been used because it can provide conformal, low-resistivity W deposition. SiH_4-reduced W-CVD is not suitable for blanket deposition because of the low step coverage factor and high resistivity of the W film. The problem of H_2-reduced W-CVD is its rather high deposition temperature (450–475 °C) and low growth rate (<50 nm/min). A high-pressure process, in which H_2-reduced W-CVD is carried out at high pressures around 10^4 Pa, has been developed to increase the growth rate and obtain a smooth film surface. On the other hand, a new W-CVD chemistry has been proposed to overcome the problem. Van der Jeugd *et al.* (1991) developed W-CVD using WF_6 and germane (GeH_4), and Goto *et al.* (1991) proposed difluoro-silane (SiH_2F_2) as a reducing gas for WF_6 for low-temperature and conformal blanket deposition. The problems associated with H_2-reduced W-CVD, such as "encroachment" and "worm holes," are not crucial in blanket deposition because an adhesion layer acts as a barrier against the WF_6/Si reaction. However, the adhesion layer can be too thin to prevent the reaction at the bottom of the deep contact if the contact aspect ratio increases. So, the conformal deposition of the adhesion layer by CVD is being studied. CVD of TiN, W_2N, and WSi_2 have been reported for the formation of adhesion layers.

2.4 Solid-Phase Reaction of Refractory Metal with Silicon

As mentioned in the previous section, refractory metal is applied to interconnect material or barrier metal. In these applications, a refractory-metal film is formed directly on Si, and so solid-phase reactions of refractory metal with silicon during high-temperature processing is a great concern regarding the thermal stability of the interface between the refractory metal and Si. The refractory metal reacts with Si to form silicide, and it is associated with volume shrinkage as well as dopant redistribution. On the other hand, silicide formation is intentionally used to get a clean metal/Si interface during LSI processing.

2.4.1 Silicide Formation Typical silicidation temperatures for refractory metals, at which solid-phase reactions form stable silicide, are listed in Table 8. These data are from Murarka (1983). In case of W interconnects to n^+ Si, the silicide reaction begins at around 600 °C, and it changes the contact resistance. A sudden increase in contact resistance is observed as a result of the dopant redistribution caused by the silicide reaction of W with Si. In general, solid solubility of arsenic, phosphorus, and boron is large in silicide, and so dopant concentration at the interface will decrease after the silicide reaction. In addition, fast dopant diffusion in silicides can cause the mixing of dopants when silicides are used for interconnects connected to diffused Si areas. However, some silicides such as platinum silicide and palladium silicide are known to push the dopant out during silicidation.

A lot of mechanical stress occurs at the interface because of the volume shrinkage during silicidation, and this can degrade the electrical characteristics of the contacts. LSI processes should therefore be designed to avoid the problems associated with silicidation, when refractory metals are used in Si LSIs.

Table 8. Silicidation temperatures and resistivities for refractory-metal silicides.

Silicide	Formed by starting from	Sintering temperature (°C)	Resultant resistivity ($\mu\Omega$ cm)
$TiSi_2$	Metal on polysilicon	900	13–16
	Cosputtered	900	25
$ZrSi_2$	Metal on polysilicon	900	35–40
$HfSi_2$	Metal on polysilicon	900	45–50
	Cosputtered alloy		60–70
VSi_2	Metal on polysilicon	900	50–55
$NbSi_2$	Metal on polysilicon	900	50
$TaSi_2$	Metal on polysilicon	1000	35–45
	Cosputtered alloy	1000	50–55
$CrSi_2$	Metal on polysilicon	700	~600
$MoSi_2$	Metal on polysilicon	1100	~90
	Cosputtered alloy	1000	~100
WSi_2	Cosputtered alloy	1000	~70
$FeSi_2$	Metal on polysilicon	700	>1000
$CoSi_2$	Metal on polysilicon	900	18–20
	Cosputtered alloy	900	25
$NiSi_2$	Metal on polysilicon	900	~50
	Cosputtered alloy	900	50–60
PtSi	Metal on polysilicon	600–800	28–35
	Cosputtered alloy		
Pd_2Si	Metal on polysilicon	400	30–35

2.4.2 Use of silicides in MOS Source and Drain Metallization A silicide reaction is intentionally used to obtain the low-resistivity source and drain in MOS transistors. Shibata *et al.* (1981) showed an improvement in device performance of submicron MOS transistors by forming silicides on source and drain. We call this self-aligned silicide (SALICIDE) technology. Figure 14 shows a typical process flow for SALICIDE using titanium silicide ($TiSi_2$), as was reported by Lau (1982). After n^+ or p^+ source and drain formation, sidewall SiO_2 is formed on the gate [Fig. 14(a)]. A Ti film is deposited on the Si substrate by sputtering [Fig. 14(b)], and this is followed by annealing at silicidation temperature (650–700 °C) to form titanium silicide on the gate and source/drain. Unreacted Ti film on SiO_2 is removed by wet chemicals (H_2O_2 + NH_4OH + H_2O). After these procedures, low-resistivity $TiSi_2$ film is formed on the gate as well as the source and drain regions. This process greatly decreases the parasitic resistance of MOS transistors, and it is especially effective in obtaining high-speed performance in submicron MOS transistors. Similar processes have been investigated for reducing the base resistance of bipolar and other kinds of transistors. For SALICIDE technology, Co and Ni have also been used to form silicides on account of their low resistivity.

To obtain a reliable silicide contact, it is important to clean the Si surface before refractory-metal deposition. Native oxide on a Si surface is known to greatly suppress silicidation, and should therefore be removed to get good reproducibility of the silicidation process. Native oxide on the Si surface is usually removed using buffered HF dip. Much work has been done to investigate the silicide reaction focusing on Si surface conditions and refractory-metal deposition conditions.

3. APPLICATIONS TO GaAs MESFETs

Refractory metals are used for the self-aligned MESFET (MEtal-Semiconductor Field-Effect Transistor) gate in GaAs devices just as they are used for the MOSFET gate in Si ULSIs. The metal gate technology was reviewed by Weitzel and Doane (1986). It is important that the criteria for MESFET gate material selection are different from those for Si MOSFET. Thermal stability of the Schottky contact between the gate material and GaAs is crucial because the gate material is directly deposited on the n-type GaAs substrate in the MESFET structure. For this reason, refractory metals such as W and Mo are not suitable for the MESFET gate; however, alternatively, tungsten silicide (W_5Si_3)

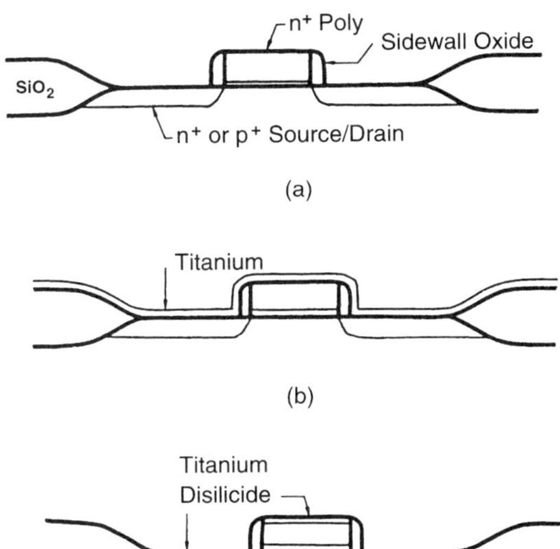

Gate oxide thickness: 250 Å
Gate polysilicon: 5000 Å
Phosphorus doped to 20 Ω/□
Source/drain n^+ implant dose: 1×10^{16} arsenic
p^+ implant dose: 1.5×10^{15} BF_2
Silicide thickness: 350–700 Å

FIG. 14. Fabrication process flow of the self-aligned silicide (SALICIDE) devices using titanium silicide ($TiSi_2$).

or tungsten nitride (W_2N) are adopted for the GaAs MESFET fabrication. Tungsten silicide (W_5Si_3) is the most suitable gate material in terms of thermal stability of the Schottky contact barrier height at up to 800 °C as compared with TiW or TiW silicide. Besides tungsten silicide, WN has been reported to be used for the MESFET gate by Uchitomi *et al.* (1984). These refractory metals are usually deposited by magnetron sputtering.

In the GaAs devices, Au is used for interconnects because Al can be an active dopant in GaAs. When Au interconnects make a contact to substrate, Ti or Mo films with a thickness of around 50 nm are usually sputter deposited under the Au film. They work as barrier metals against Au diffusion into GaAs. At the same time, these refractory metals play the role of an adhesion layer between the Au interconnect and the SiO_2.

Besides the applications described above, refractory compound metals such as WSi_xN_y, Ta_3N_4, and NiCr have been used as electrical resistors in GaAs devices. These metals are usually deposited by reactive sputtering. For example, WSi_xN_y films are deposited by sputtering a W_5Si_3 target in N_2/Ar mixed atmosphere.

4. ACKNOWLEDGMENTS

The author is much obliged to Dr. Seiichi Iwata, Dr. Naoki Yamamoto, and Dr. Masayoshi Saitoh for their collaboration in the study of refractory-metal applications to Si ULSIs. Above all, encouragement from Dr. Shojiro Asai in writing this article is highly appreciated.

Works Cited

Bernard, C., Madar, R., Pauleau, Y. (1989), *Solid State Technol.* **32,** (2), 79–84.

Broadbent, E. K., Ramiller C. L. (1984), *J. Electrochem. Soc.* **131,** 1427–1432.

Brown, D. M., Cady, W. R., Sprague, J. W., Salvagni D. J. (1971), *IEEE Trans. Electron Devices* **ED-18,** 931–940.

Creigton J. R. (1987), *J. Vac. Sci. Technol.* **A5,** 1739–1740.

Cuomo, J. J. (1972), in: F. A. Glaski (Ed.), *Proceedings of the 3rd International Conference on Chemical Vapor Deposition,* Hinsdale, IL: American Nuclear Society, pp. 270–291.

Ghate, P. B., Blair, J. C., Fuller, C. R., McGuire, G. E. (1978), *Thin Solid Films* **53,** 117–128.

Goto, H., Kobayashi, N., Homma, Y. (1991), in: *Extended Abstracts of International Conference on Solid State Devices and Materials,* Tokyo: The Japan Society of Applied Physics, p. 183.

Gray, D. E. (Ed.) (1972), *American Institute of Physics Handbook,* 3rd ed., New York: McGraw-Hill.

Hirase, I., Sumiya, T., Schack, M., Umishima, S., Rufin, D., Shinshikura, M., Matsuura, M., Ito, A. (1987), in: V. A. Wells (Ed.), *Tungsten and Other Refractory Metals for VLSI Applications III,* Pittsburgh: Materials Research Society, p. 133.

Hoffman, D. W., Thornton, J. A. (1982), *J. Vac. Sci. Technol.* **20,** 355–358.

Holloway, K., Fryer, P. M., Cabral, C., Jr., Harper, J. M. E., Bailey, P. J. (1992), *J. Appl. Phys.* **71,** 5433–5443.

Iwata, S., Yamamoto, N., Kobayashi, N., Terada, T., Mizutani, T. (1984), *IEEE Trans. Electron Devices* **ED-31,** 1174–1179.

van der Jeugd, C. A., Leusink, G. J., Janssen, G. C. A. M., Radelaar, S. (1991), *Appl. Phys. Lett.* **57,** 354–356.

Kobayashi, N., Yamamoto, N., Iwata, S., Mizutani, T., Yagi, K. (1984), in: *International Electron Devices Meeting Technical Digest,* New York: IEEE, p. 122.

Kobayashi, N., Goto, H., Suzuki, M. (1991), *J. Appl. Phys.* **69,** 1013–1019.

Kobayashi, N., Nakamura, Y., Goto, H., Homma, Y. (1993), *J. Appl. Phys.* **73,** 4637–4643.

Kondo, S. (1992), *J. Mater. Res.* **7,** 853–860.

Kotani, H., Tsutsumi, T., Komori, J., Nagao, S. (1987), in: *International Electron Devices Meeting Technical Digest,* New York: IEEE, p. 213.

Kusumoto, Y., Takakuwa, K., Hashinokuchi, H., Ikuta, T., Nakayama, I. (1987), in: V. A. Wells (Ed.), *Tungsten and Other Refractory Metals for VLSI Applications III,* Pittsburgh: Materials Research Society, p. 103.

Lau, C. K., See, Y. C., Scott, D. B., Bridges, J. M., Perna, S. M., Davies, R. D. (1982), in: *International Electron Devices Meeting Technical Digests,* New York: IEEE, p. 714.

Loh, W. H., Swirhun, S. E., Crabbe, E., Saraswat, K., Swanson, R. M. (1985), *IEEE Electron Devices Lett.* **EDL-6,** 441–443.

May, T. C., Woods, M. H. (1970), in: *Proceedings IEEE Reliability Physics,* New York: IEEE, p. 33.

McConica, C. M., Krishnamani, K. (1986), *J. Electrochem. Soc.* **133,** 2542–2548.

Moore, G. (1980), *Electron. Aust.* **42,** p. 14.

Muller, R. S., Kamins, T. I. (1986), in: *Device Electronics for Integrated Circuits,* 2nd ed., New York: Wiley, Chap. 9.

Murarka, S. P. (1983), *Silicides for VLSI Applications,* New York: Academic Press, p. 30.

Ohba, T., Inoue, S., Maeda, M. (1987), in: *International Electron Devices Meeting Technical Digest,* New York: IEEE, p. 213.

Pramanik, D., Jain, V. (1991), *Solid State Technol.* **34** (5), 97–101.

Rhoderick, E. H. (1978), *Metal-Semiconductor Contacts,* Monographs in Electrical and Electronic Engineering No. 19, Oxford, U.K.: Clarendon Press, p. 53.

Saraswat, K. C., Mohammadi, F. (1982), *IEEE J. Solid-State Circuits* **SC-17,** 275–280.

Schmitz, E. J. (1992), *Chemical Vapor Deposition of Tungsten and Tungsten Silicides for VLSI/ULSI Applications,* Park Ridge, NJ: Noyes Publications, p. 143.

Sherman, A. (1987), in: *Chemical Vapor Deposition for Microelectronics,* NJ: Noyes Publications, p. 35.

Shibata, T., Hieda, K., Sato, M., Konaka, M., Dang, R. L. M., Iizuka, H. (1981), in: *International Electron Devices Meeting Technical Digest,* New York: IEEE, p. 651.

Stacy, W. T., Broadbent, E. K., Norcott, M. H. (1985), *J. Electrochem. Soc.* **132,** 444–448.

Suzuki, M., Kobayashi, N., Mukai, K., Kondo, S. (1990), *J. Electrochem. Soc.* **137,** 3213–3218

Sze, S. M. (1983), in: Sze, S. M. (Ed.), *VLSI Technology,* New York: McGraw-Hill, Chap. 12.

Sze, S. M. (1981), in: *Physics of Semiconductor Devices,* 2nd ed., New York: Wiley, p. 304.

Tamaru, T., Iwata, S., Kobayashi, N., Tokunaga, T. (1992), in: *Proceedings of the Dry Process Symposium,* Tokyo: The Japan Society of Applied Physics, p. 68.

Thornton, J. A., Penfold, A. S. (1978), in: V. L. Vossen, W. Kern (Eds.), *Thin Film Processes,* New York: Academic Press, p. 75.

Uchitomi, N., Kitaura, Y., Mizoguchi, T., Ikawa, Y., Toyoda, N., Hojo, A. (1984), in: *Extended Abstracts International Solid State Devices and Materials Meeting,* Tokyo: The Japan Society of Applied Physics, p. 383.

Weitzel, C. E., Doane, D. A. (1986), *J. Electrochem. Soc.: Reviews and News* **133,** 409C.

Wang, K. L., Holloway, T. C., Pinizotto, R. F., Sobczak, Z. P., Hunter, W. R., Tasch, A. F. (1982), *IEEE Trans. Electron Devices* **ED-29,** 547–553.

Yamamoto, N., Iwata, S., Kume, H. (1987), *IEEE Trans. Electron Devices* **ED-34,** 607–614.

Yokoyama, N., Hinode, K., Homma, Y. (1991), *J. Electrochem. Soc.* **138,** 190–195.

Yu, M. L., Eldridge, B. N. (1989), *J. Vac. Sci. Technol.* **A7,** 625–629.

REFRACTORY METALS

ERWIN PINK, *Erich-Schmid-Institut für Festkörperphysik der Österreichischen Akademie der Wissenschaften, Leoben, Austria*

RALF ECK, *Metallwerk Plansee GmbH, Reutte, Austria*

	Introduction	327
1.	**Materials and Their Properties**	328
1.1	The Pure Metals	328
1.1.1	General Physical Properties	328
1.1.2	Chemical Properties	328
1.1.3	Strength and Ductility	329
1.1.4	Essential Differences between V*a* and VI*a* Metals	329
1.2	Refractory Alloys	329
1.2.1	Systematics of Alloys	329
1.2.2	Commercial Alloys	330
1.2.2.1	Substitutional Alloys	330
1.2.2.2	"Doped" Refractory Metals	330
1.2.2.3	Dispersion-Strengthened Alloys	330
1.2.2.4	Combinations	330
1.2.2.5	Dual-Phase Alloys	332
1.2.2.6	Refractory-Metal Cermets	333
1.2.2.7	Alloys Reinforced by Refractory-Metal Fibers	333
1.2.2.8	Amorphous Refractory Alloys	334
2.	**Resources**	334
3.	**Production Technologies**	334
3.1	Powder Compacting	334
3.2	Sintering	335
3.3	Melting	335
3.4	Forming	335
3.5	Heat Treatments and Thermomechanical Treatments	335
3.6	Recycling	336
4.	**Further Processing**	337
4.1	Fastening and Joining	337
4.1.1	Mechanical Fastening	337
4.1.2	Welding	337
4.1.3	Brazing	337
4.2	Coatings	337
4.2.1	Pack and Slurry Cementation	337
4.2.2	Electrolytic Coating	338
4.2.3	Flame and Plasma Spraying	338
4.2.4	Chemical and Physical Vapor Deposition (CVD and PVD)	338
5.	**Applications**	338
5.1	Main Use of Refractory Metals and Alloys	338
5.2	Special Requirements	339
5.2.1	High-Purity Materials	339
5.2.2	Porous Metals	339
5.2.3	Alloys for Thermocouples	339
5.2.4	Refractory Metals for Superconductors	339
5.2.5	Use in Thermonuclear Reactors	339
6.	**Economic Aspects**	339
7.	**Toxicological Aspects**	340
	Glossary	340
	Works Cited	341
	Further Reading	342

INTRODUCTION

By convention, commercial refractory, or high-melting, metals are those that belong to groups V*a* and VI*a* of the periodic table of elements. They are vanadium, niobium, and tantalum (V*a*), and chromium, molybdenum, and tungsten (VI*a*). Their melting points exceed roughly 1900 °C, culminating at 3410 °C in the case of tungsten, which is a melting temperature comparable to those of the most heat-resisting carbides. There exist other metals with high melting points; some of them are, however, classified appropriately as precious metals (see SILVER, GOLD, AND OTHER PRECIOUS METALS). The other high-melting metals, rhenium and hafnium, are used today as alloying elements for the refractory metals, and only rhenium in exceptional cases for its own metallic merits.

3-527-28138-X/96/$5.00 + .50

1. MATERIALS AND THEIR PROPERTIES

1.1 The Pure Metals

1.1.1 General Physical Properties For high-temperature use, a high melting point, a low vapor pressure, low thermal expansion, and high thermal conductivity are of advantage. Other important physical properties of these metals are compiled in Table 1. All metals except for rhenium and hafnium (whose lattices are hexagonal close-packed) have a body-centered cubic crystal structure, and only hafnium exhibits a crystallographic phase change upon variation of the temperature. The microstructure of all pure refractory metals consists of one phase only.

In commercially pure metals, hydrogen is contained in amounts less than 1 μg/g. The oxygen, nitrogen, and carbon concentrations in V*a* metals are smaller than 50 μg/g, in VI*a* metals today mostly smaller than 10 μg/g. Substitutionally dissolved impurity elements (nickel, iron, or the other refractory metals) amount to 5–100 μg/g. The purity levels routinely available are better than 99.95%.

1.1.2 Chemical Properties Refractory metals have, in comparison with other metals, superior resistance to many chemical agents (Lambert, 1990). In general, they are resistant to corrosion in various media, but those of the V*a* group are outstanding in their performance. The stability of refractory metals in dilute nonoxidizing acids is good, but in oxidizing acids they corrode severely, especially molybdenum. All refractory metals are attacked by sodium and potassium hydroxides, especially when oxidizing substances are present. The resistance to solids and metal or glass melts is good to excellent. In molten vanadium pentoxide, a normally strongly corrosive medium (present in the ashes of all fossil fuels, with a melting point of 675 °C), the resistance of chromium is unsurpassed.

Refractory metals are to various extents prone to destruction through gases (Fromm and Gebhardt, 1976). This behavior, it will be shown, has consequences for the manufacture, use, and recycling methods of metals of the two groups. V*a* metals absorb large quantities of all gases (except those that are inert) and are then seriously embrittled. VI*a* metals remain unaffected. The effect of oxygen is opposite. Molybdenum and tungsten start to oxidize in air around 300–400 °C. Volatile oxides form and evaporate catastrophically above 600 °C in the case of molybdenum, above 800 °C for tungsten, and al-

Table 1. Important physical properties and mechanical room-temperature properties of pure refractory metals.

	Group V*a*			Group VI*a*			VII*a*
Property	V	Nb	Ta	Cr	Mo	W	Re
Structure	bcc	bcc	bcc	bcc	bcc	bcc	hcp
Density (g/cm^3)	6.1	8.6	16.6	7.2	10.2	19.3	21.0
Melting point (°C)	1920 ± 20	2470 ± 20	3000 ± 30	1920 ± 20	2620 ± 20	3380 ± 20	3180
Coefficient of linear thermal expansion 20 °C, (μm/m pd K)	8.3	7.6	6.6	6.2	5.3	4.6	8.4
Specific electrical resistivity at 20 °C (μΩ m)	0.19	0.14	0.12	0.12	0.05	0.05	0.17
Approximate interstitial solubility (μg/g), 20 °C, for C to H	10^3 to 10^4	10^2 to 10^4	70 to 4000	0.1 to 1	0.1 to 1	<0.1	<3000
Start of recrystallization (°C) (1-hour annealing)	650	750	1100	700	800	1150	900
Modulus of elasticity (GPa)	130–150	100–120	180–190	220–280	320–360	390–410	470
Poisson's ratio	0.37	0.40	0.34	0.21–0.28	0.30	0.30	0.29
Hardness HV10							
Recrystallized (min.)	70	80	80	100	150	350	250
As-worked (max.)	200	200	200	300	500	650	580
Minimum UTS (MPa) (recrystallized)	300	300	300	300	550	(480 °C) 600	1070
DBTT[a] (°C) bending (recrystallized)	−100	−200	−269	+200	−20	+400	

[a]Ductile–brittle transition temperature.

ready above 300 °C for rhenium. The oxidation behavior of chromium is exceptional, insofar as the progress of oxidation, once the oxide layer is formed, is much slower than even for niobium or tantalum.

Resistance to gaseous hydrocarbons and carbon oxides is defined by the temperatures of carbide formation. Nitrides form, at elevated temperatures, on surfaces of molybdenum and tungsten under high-pressure nitrogen and under ammonia at 1 bar (Martinz and Prandini, 1993–1994). In air, brittle layers grow very slowly at the usually employed temperatures below 1400 °C.

The metals niobium and tantalum have excellent biocompatibility (Zitter and Plenk, 1987).

1.1.3 Strength and Ductility The temperature dependences of strength values and fracture elongations of the recrystallized pure metals are shown in Fig. 1 in relation to each other. The strength of metals in the worked state is higher, but this gain can only be exploited in high-temperature applications as long as the recrystallization temperature is not exceeded. These limiting temperatures are also listed in Table 1, together with other mechanical characteristics.

1.1.4 Essential Differences between V*a* and VI*a* Metals The most notable feature of VI*a* metals—aside from other differences between the metals of the V*a* and VI*a* groups apparent from Table 1—is the severe embrittlement at low temperatures [below the "ductile–brittle transition temperature"; see Fig. 1(b)]. In principle, this is a common inherent feature among body-centered cubic metals (Honeycombe, 1984). However, the embrittlement temperature in the VI*a* metals is augmented and reaches, or even exceeds, room temperature when oxides and large carbides cover the grain boundaries and weaken their cohesion, especially in the recrystallized state. This is a consequence of the interplay of low interstitial solubility and actual content of impurities (see Table 1 and Sec. 1.1.1). The ductile–brittle transition temperature may be high even when low deformation rates, as in bending, are applied (see Table 1). Various means can be employed to reduce this handicap: purifying, alloying (cf. Sec. 1.2.2), and mechanical working (in order to neutralize the effect of contaminated grain boundaries).

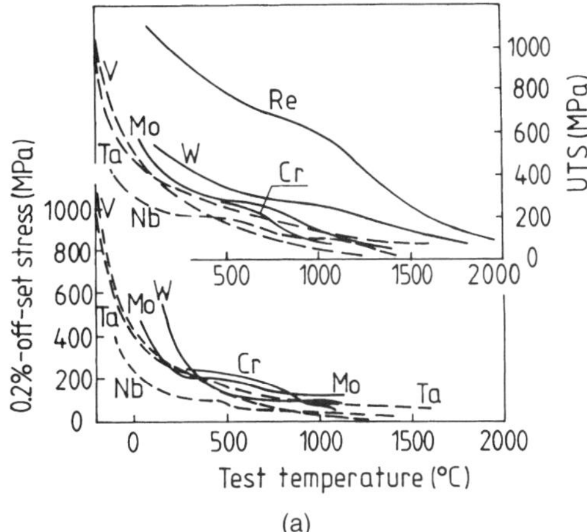

(a)

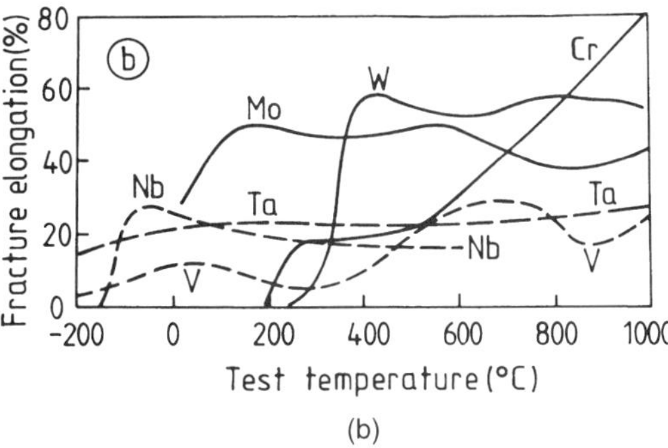

(b)

FIG. 1. The temperature dependences of (a) the ultimate tensile (UTS) and yield stresses and (b) the fracture elongations of recrystallized refractory metals, for medium strain rates (Tietz and Wilson, 1965).

1.2 Refractory Alloys

1.2.1 Systematics of Alloys In the case of refractory metals, the term "alloy" comprises—in addition to its standard meaning—also types commonly not produced from other metals. Some of these are not even "alloys" in the same sense the word is used otherwise, and so the variety of refractory alloys exceeds that of other metals. The purpose of alloying is manifold (Pink and Eck, 1996): It ranges from influencing the recrystallization behavior to raising the strength and ductility by various means, at low as well as at high temperatures, or to combining special physical properties of the alloy constituents. The common strengthening methods can equally well be employed (with the exception of transformation, i.e., martensitic, strengthening, which is not possible because of the absence of phase transformations). In addition, certain alloys are designed for special purposes to make use of mechanisms based on principles that are rarely exploited otherwise.

The alloy types are substitutional alloys, "doped" refractory metals, dispersion-strengthened alloys, combinations, dual-phase alloys, metal-ceramic composites, alloys reinforced by refractory-metal fibers, and amorphous refractory alloys.

1.2.2 Commercial Alloys Table 2 lists the most important types of alloys for the six group V*a* and VI*a* refractory metals, and the improvements that are achieved. The classes introduced in the previous section are described in the following.

1.2.2.1 Substitutional Alloys. Most solution hardening has limited importance. But rhenium additions to molybdenum and tungsten reduce or even eliminate their brittleness completely (this is not caused by a solution effect; the reasons are not yet quite explored). In certain substitutional solutions, the corrosion resistance is improved.

1.2.2.2 "Doped" Refractory Metals. Tungsten wires for high-temperature application (e.g., in incandescent lamps) need a specially modified recrystallized grain structure (Fig. 2) (Snow, 1989). This "nonsag" structure with coarse elongated and overlapping recrystallized grains with a high "grain aspect ratio" (grain length-to-width ratio) will not sag, i.e., deform along grain boundaries, as a normally recrystallized tungsten wire with a "bamboo" grain structure would. This doped or nonsag (NS) tungsten is produced through doping with aluminum chlorides and potassium silicates ("AKS" tungsten is a further denomination). The concentration of potassium that remains in the wire is smaller than 100 μg/g, but sufficient to form bubbles less than 0.1 μm in diameter, filled with metallic potassium. In the course of the severe deformation to wire diameters less than 0.5 mm, the bubbles become aligned, less than 1 μm apart, in the direction of the wire axis. They prevent the migration of the grain boundaries in the direction perpendicular to the wire axis during recrystallization, and raise the recrystallization temperature. The nonsag structure is not only creep resistant, but strengthens the matrix also at low temperatures since the bubbles serve as dispersed obstacles against dislocation motion (Pink and Eck, 1996). The overlapping grains also increase to a certain extent the bending ductility at the lower temperatures. A disadvantage is that bubbles migrate and coalesce when the wire is overheated. The effective cross section may then be reduced, and the wire becomes prone to melting when heated by direct current. This kind of doping has also been applied to produce rods, wires, sheets, and foils of molybdenum that do not embrittle as severely as the pure metals when they are recrystallized.

1.2.2.3 Dispersion-Strengthened Alloys. Commercial alloys are based on the effects of dispersed carbides that form from additions of alloying metals and carbon. Strengthening through nitrides is not technically exploited, although research has proved it equally effective. A prominent example is the molybdenum alloy TZM, containing titanium ("T") and zirconium ("Z"). "M" stands for molybdenum. At present, the strongest alloy is ZHM with hafnium ("H") carbide, which has the second-highest melting point of all carbides. In current practice, alloying is complemented by thermal treatments such as aging or strain aging only in the case of HfC-containing alloys (see Sec. 3.5).

A recent successful development is sheets and rods made from ODS (i.e., oxide-dispersion-strengthened) molybdenum. The additions are oxides of rare earths, especially of lanthanum, of high thermal stability. These materials increasingly replace the doped molybdenum because, aside from being dispersion hardened, they also develop an elongated grain structure upon recrystallization, and exhibit thus all the advantages offered by a dopant without having its drawbacks: There is no danger of bubble blistering by potassium evaporation during exposure to high temperatures, and no need for the extreme degree of deformation as is necessary for producing the bubble rows. A similar development concentrates on the production of ODS-tungsten wires.

1.2.2.4 Combinations. A further improvement is achieved by combinations of the measures described in Secs. 1.2.2.1–1.2.2.3. For instance, V*a*-group metals are strengthened by hafnium carbides, nitrides, and oxides, and their effect is enhanced by solution strengthening through tungsten, zirconium, and hafnium. Complex tungsten al-

Table 2. Properties and applications of commercial alloys (with additions in wt%).

Alloy	Oustanding properties	Applications
Vanadium alloys:		
V–(5–30)Ti–(5–15)Cr	High strength, good strength/ specific-weight ratio, good corrosion resistance	Aircraft and space applications; in liquid heat exchangers, nuclear engineering
Niobium alloys:		
Nb–1Zr Nb–10W–10Hf–15C Nb–10W–2.5Zr Nb–28Ta–10W–1Zr	High strength, good strength/ specific weight ratio	Space applications; nuclear engineering
Tantalum alloys:		
Ta–(2.5; 7.5; 10)W Ta–8W–2Hf Ta–10W–2.5Hf–0.01C	Good corrosion resistance; high strength	Chemical process equipment
Chromium alloys:		
Cr–(40–75)Cu	Good resistance against electrical wear; good electrical conductivity	Vacuum contacts
Molybdenum alloys:		
Doped (KS) Mo	High recrystallization temperature, elongated recrystallized grain; creep resistant, ductile at 20 °C	Wires, sheets, foils for exposure at 1600–1800 °C
Mo–(5–30)W	Resistant against liquid zinc	Zinc-producing industry
Mo–(5–41)Re	Highest ductility, high strength; electromotoric force	Ductile thermocouples for high temperatures
Dispersion alloys:		
"TZM" Mo–0.5Ti–0.08Zr–0.05C "MHC" Mo–1.2Hf–0.06C "ZHM" Mo–(0.4–0.7)Zr–(1.2–2)Hf–(0.15–0.25)C	High strength, good ductility, age hardenable (MHC, ZHM)	Construction materials for 900–1600 °C, forging tools
ODS alloy Mo–0.7La_2O_3	High recrystallization temperature, elongated recrystallized grain; creep resistant; ductile at 20 °C	Construction material for use at 1200–2200 °C
Complex alloys:		
Mo–25W–1.2Hf–0.06C ODS-ZHM-1Y_2O_3/1La_2O_3	Best strength at highest temperatures	Construction material for use at 1200–2200 °C
Tungsten alloys:		
Doped (AKS) W	High recrystallization temperature, elongated recrystallized grain; high creep resistance at high temperatures	Incandescent lamp filaments; metallizing wires
Doped W–(1–2) ThO_2	Also: low work function	Electronic emitters
W–(3–26)Re	Best low-temperature ductility, high-temperature strength; electromotoric force	X-ray targets, ductile thermocouple wires for highest temperatures
Dispersion (ODS) alloys:		
W–(1–4)ThO_2	Low work function; good machinability	Welding electrodes; plasma generators
W–(1–2)ZrO_2/CeO_2/La_2O_3	The same, as a nonradioactive alternative material for W–ThO_2	The same
Heavy alloys:		
W–(2–7)Ni–(1–3)Fe/Cu	Ductile at room temperature, excellent machinability; high specific weight	X- and γ-radiation shields; counter weights and high-inertia material; artillery penetrators
Electrical contact materials:		
W–(10–50)Cu/Ag W–20Cu–1Ni	Ductile at room temperature, excellent machinability; good electrical conductivity, good resistance against electrical wear	High-voltage contacts; welding electrodes

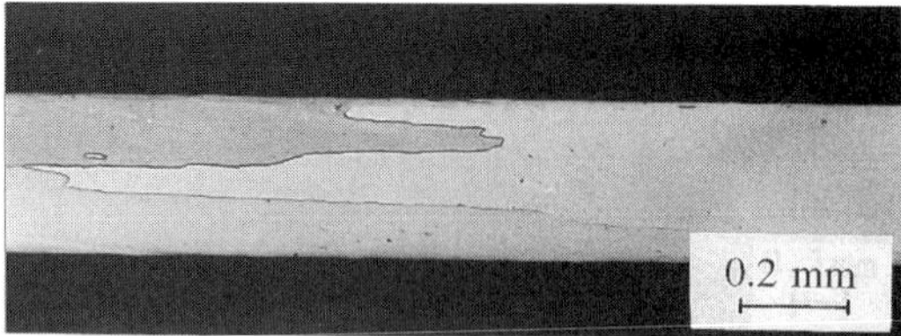

FIG. 2. The microstructure of a doped-tungsten wire with recrystallized elongated grains (courtesy of H. Yamamoto).

loys combine the advantages of rhenium and of hafnium carbides. Tungsten alloyed with rhenium can be AKS-doped to develop an elongated grain structure after recrystallization.

1.2.2.5 Dual-Phase Alloys. *Tungsten Heavy Metals.* In the early days of producing tungsten, attempts were made to create an easily machinable material by embedding the tungsten grains in a binder phase (e.g., nickel). Further development led to alloys containing 90–98 wt% tungsten and a face-centered cubic binder phase of nickel and iron. The powder mixture is heated above the melting temperature of the binder. Aside from general densification, the tungsten grains are partly or even completely dissolved during sintering (German, 1985). During cooling, the reprecipitation of the dissolved portions creates tungsten "spheroids" that are embedded in the binder phase (Fig. 3).

The Ni/Fe ratios in heavy metals are 1:1 to 4:1 (preferably 7:3, because this composition allows precipitation hardening of the matrix and is less likely to embrittle as a result of intermetallic phases at interfaces). Alloys in which copper is substituted for iron (in Ni/Cu ratios 3:2 to 4:1) have better electrical conductivity and are not ferromagnetic. Typical values of strength and fracture elongation are evident from Fig. 4. They vary with composition and with the amount of contiguity (which is a measure of the ratio of tungsten/tungsten to tungsten/matrix contact areas—a parameter of great importance for the ductility). The temperature dependence of the fracture energy obtained from Charpy tests shows a transition to brittle behavior as the temperature decreases (German, 1985). Heavy alloys can be coldworked to achieve a high ultimate tensile strength at the cost of toughness and fracture elongation.

Infiltrated (Pseudo) Alloys. Chromium, molybdenum, or tungsten is combined with silver or copper to create materials for electrical contacts (Shen *et al.,* 1990). Certain properties are combined that are incompatible in normal alloys: high electrical and thermal conductivity, high resistance to chemical reactions, a high melting point for good resistance to burn-off and spark erosion, and good mechanical strength and easy machinability. A skeleton of the refractory metal with pores of optimum size is first produced. The interconnecting pores are then infiltrated by placing the necessary amount of silver or copper in contact with the outer surface, and heating the whole above the melting point of the infiltrant. With the exception of the chromium-base alloys, the

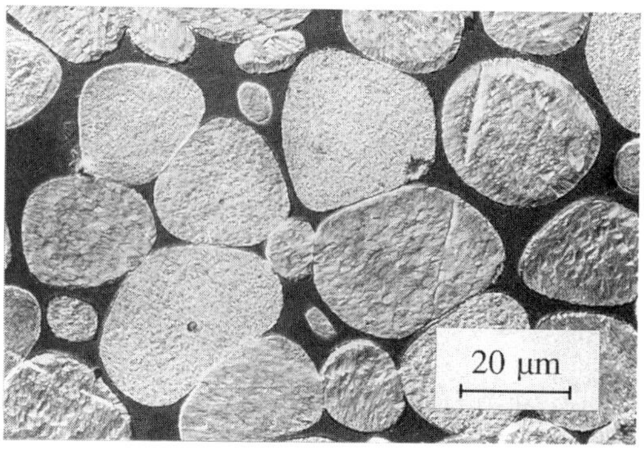

FIG. 3. The microstructure of a tungsten heavy alloy with 7 wt% nickel and 3 wt% iron (courtesy of H. Danninger). The binder phase has been etched away.

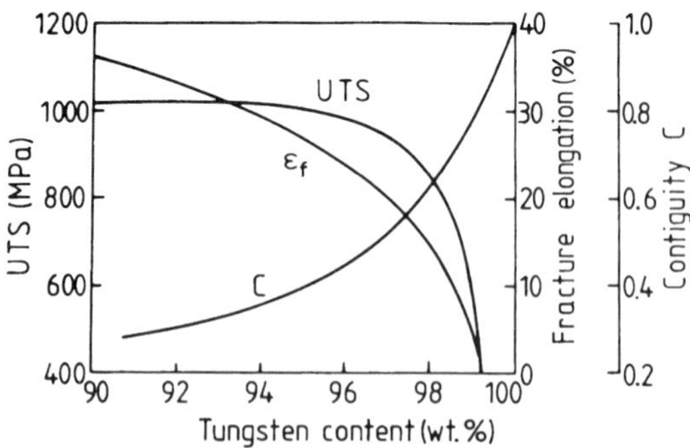

FIG. 4. Strength and fracture elongation of heavy alloys with varying tungsten contents, and their correlation with the contiguity according to Edmonds (Pink and Eck, 1996).

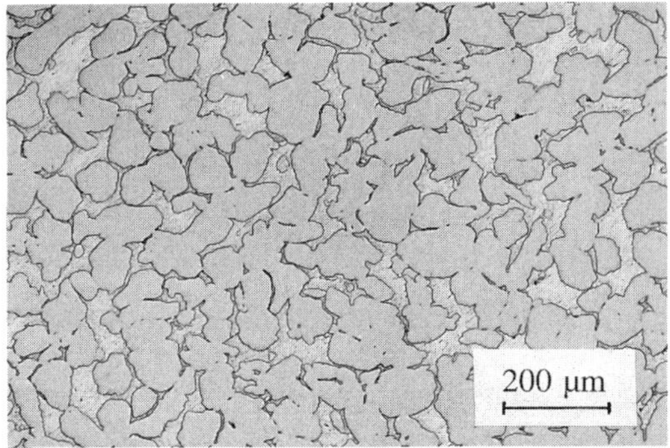

FIG. 5. The microstructure of tungsten (the larger component) infiltrated with 30 wt% copper.

components do not react with each other, and a typical microstructure of an infiltrated material (Fig. 5) therefore does not consist of the same rounded grains as the tungsten heavy alloy. Electrical contact materials with large or excess amounts of copper and silver (>50 vol%) are produced differently. A mixture of the two powder components is pressed and solid-state sintered at temperatures below the melting point of the low-melting constituent. In such structures, the refractory-metal component does not form a continuous skeleton. Upon heating above the melting point of copper or silver, the alloy would be destroyed. Such solid-state sintered tungsten-copper can be produced to densities near 90% of the theoretical. Further densification is possible by plastic forming at elevated temperatures. Some specific properties of W–Cu/Ag and Cr–Cu alloys are collected in Table 3. The strength of the composite material can be improved by cold rolling to obtain work-hardened sheet material, or—rarely done—by solution strengthening of the soft phase.

1.2.2.6 Refractory-Metal Cermets. The name designates a heterogeneous combination of a metal and a ceramic. The volume fraction of the metallic binder can be 40%–85%. Cermets with high concentrations of refractory metals are materials that are strong and wear resistant, and capable of withstanding chemical attack. In general, they can be easily machined. A combination of molybdenum with zirconia is used as sheath material for thermocouples for measuring the temperature of molten steel (Heitzinger, 1974). The strength (about 300 MPa at room temperature) varies little (or even increases slightly) with temperature up to about 1000 °C.

The following alloy variants are not standard materials.

1.2.2.7 Alloys Reinforced by Refractory-Metal Fibers. Niobium alloys reinforced with tungsten wires have been proposed as potential materials for long-term high-temperature applications in space-power systems. Wires of refractory metals (especially tungsten) strengthen other metals (primarily nickel) and superalloys intended for use at high temperatures (Lambert, 1990; see also COMPOSITE MATERIALS). Embedded within the matrix, the problem with the poor oxidation resistance of some of the refractory metals is reduced. In comparison to ceramic fibers, their thermal expansion and ductility are closer to those of the matrix materials so that they are better suited. The strength of all refractory-metal fibers is high in relation to that of most matrix metals at the applied temperatures, which are in general "low" from the point of view of tungsten fibers. Fiber–matrix interactions are possible, especially at the higher temperatures. Chemical reactions lead to the formation of brittle intermetallic phases. Traces of certain metals (e.g., nickel) that are deposited on the fiber surface or come from the matrix metal itself may diffuse into the wire and activate recrystallization of the reinforcement. This may be avoided by ductile coatings yet to be devel-

Table 3. Properties of contact alloys.

Alloy	Mean linear coefficient of thermal expansion (10^{-6}/K)	Specific electrical resistivity ($\mu\Omega$ m)	Heat conductivity (W/m K)
W–20Cu	8.5	0.05	134
W–20Ag	8.7	0.05	138
Cr–75Cu	13.5	0.04	235

oped, or by recrystallization before embedding.

1.2.2.8 Amorphous Refractory Alloys. Refractory metals are quenched into the glassy state by vapor deposition in the form of thin films onto substrates that are cooled down to temperatures below 10 K (Johnson, 1990). The pure amorphous refractory metals crystallize easily, but amorphous alloys can retain their structure upon annealing up to 1000 K. The alloys are of the metal–metal or metal–metalloid type—for instance, $(Nb,Mo)_{50}Ni_{50}$, $Mo_{30}Re_{70}$, $Mo_{52}Ru_{32}B_{16}$, or $W_{40}Re_{40}B_{20}$. Yield strength and hardness are typically high: With the Vickers hardness $HV = 2400$ of $W_{40}Re_{40}B_{20}$, for instance, and with a hardness-to-yield-point ratio of 3, common for metallic glasses, the yield stress should be 8000 MPa, which places these materials among the strongest known. Sputter deposition of amorphous layers a few microns thick delays, for instance, wear and corrosion of steels because amorphous metals have no grain boundaries.

2. RESOURCES

Specific mineral deposits of vanadium do not exist; vanadium is mainly extracted from iron, titanium, or uranium ores occurring in South Africa, Russia, China, and Finland. Increasing amounts of vanadium are obtained from petroleum. Niobium and tantalum are frequently found in close association in the mineral $(Fe,Mn)(Ta,Nb)_2O_6$, called either tantalite or columbite, depending on which metal is in excess. Cassiterite (SnO_2) deposits found in Thailand, Malaysia, Australia, etc. also contain significant amounts of tantalum and niobium. Slags from tin smelting are frequently used as tantalum feedstock. The highest-grade deposits of niobium, consisting of the mineral pyrochlore, $(Na,Ca)_2(Nb,Ta,Ti)_2O_4(OH,F) \cdot H_2O$, are found in Brazil.

Almost all chromium is obtained from deposits of chromite, $(Fe,Mg)O \cdot (Cr,Al,Fe)_2O_3$, principally in southern Africa, and in Russia, Finland, Turkey, and the Philippines. Ninety-five percent of the world production of molybdenum is from deposits of the disulfide MoS_2 that occur in the eastern Rocky Mountains of the U.S.A., in Chile, and in the Caucasus and the Ural; the most important of these are the so-called porphyry molybdenum deposits, in which molybdenum occurs with a grade of about 1% MoS_2 in conjunction with copper. Of the many tungsten minerals, only two groups are of commercial importance: the wolframite group, which consists of a solid solution of ferberite ($FeWO_4$) and huebnerite ($MnWO_4$) and occurs mainly in China, Korea, and the U.S.A.; and the scheelite group ($CaWO_4$) found primarily in Canada and Austria. Most ores contain 0.25%–2.5% WO_3.

Rhenium is obtained as a by-product from the mining of molybdenum-bearing porphyry copper ores. Major deposits occur in Chile, China, and Russia. Rhenium is particularly expensive because ores containing only 10–20 g/t have to be processed.

3. PRODUCTION TECHNOLOGIES

Since the product extracted from the ore and chemically treated is in many cases a metal powder, it is convenient to continue the further production by means of "powder metallurgy." In addition, this method is technically and economically preferable because melting at high temperatures is avoided. It needs, in the beginning, special production steps.

3.1 Powder Compacting

Powders of all commercially fabricated refractory metals (with typical particle sizes of 1–10 μm) and alloys are consolidated before sintering to reach densities averaging 60%–65% of the theoretical, mainly by two methods:

1. *Pressing of powders in rigid dies* on hydraulic presses using pressures of about 3000 bar. Dimensional precision is high. Molybdenum blanks can be machined without sintering; tungsten has to be presintered.
2. *Isostatic pressing* of powders in flexible containers. For very large parts, isostatic wet-bag pressing is the most common method, leading to homogeneous densities.

3.2 Sintering

Presintering of billets compacted from powder serves to facilitate handling and machining. The primary goal of sintering is densification to at least 90%–95% of the theoretical density, which is required to make further working operations successful. In addition to ensuring high density, purification occurs during sintering. The grain sizes after sintering are 10–30 μm.

Sintering is carried out in different ways:

1. Self-resistance heating or "direct sintering" has been used extensively for tungsten because of the extremely high temperatures required for practical sintering. Rods are heated to nearly 3000 °C in a hydrogen atmosphere. Tantalum and its alloys are sintered at 2600 °C in vacuum.
2. Ceramic-free high-temperature furnaces with vertical heating elements of tungsten sheet and heat shielding made of molybdenum alloys reach 2700 °C.
3. Induction radiant heating uses a cylindrical susceptor ring of tungsten or molybdenum that is inductively heated. The compact is heated up to 2400 °C by radiation.
4. For liquid-phase sintering and infiltration treatments, pusher-type horizontal furnaces operating up to 1600 °C are in use.

Among the many factors that influence sintering, the following are important:

1. Time and temperature are fundamental determinants of the sintering kinetics. Mass transport by diffusion is very slow for refractory metals at the sintering temperatures that are economically or technically feasible.
2. The purity of the compact greatly affects the sintering. Because molybdenum and tungsten are sintered at high temperatures, most impurities are molten and may have high vapor pressures and prevent achievement of high density.
3. Thermal gradients and heating rates are particularly important when large furnace loads are sintered.

3.3 Melting

Melting leads to a higher degree of purity in alloys, but the resulting coarse grain structure makes further processing difficult. To break up the large crystals in the ingots of molybdenum, they must be extruded further with additional cost. For vanadium, niobium, tantalum, and some alloys, arc and electron-beam melting are the most important methods.

3.4 Forming

Once a sintered ingot has been obtained, the further working follows the conventional path. Heating is mostly performed in hydrogen furnaces to avoid oxidation of molybdenum or tungsten and sublimation of their oxides.

The initial heating temperature necessary for molybdenum is 1300–1400 °C, and 1500–1600 °C for tungsten. During mechanical forming, the intermediate heating temperatures are successively lowered. After severe deformation (to produce thin wires, sheets, or foils), molybdenum and alloys can be processed at room temperature, and tungsten and alloys are rolled and drawn at temperatures of 500–200 °C without protective atmosphere. Normally, molybdenum alloys, tungsten, and tungsten alloys do not recrystallize during processing. Only pure molybdenum may recrystallize during heating at standard temperatures.

All forming processes that are common in other metal-producing industries are equally suited for refractory metals: hot and cold rolling, forging in dies, rotary swaging, extruding, flow turning, spinning, deep drawing, and wire drawing through dies. Drawing is also suited for the fabrication of long seamless tubes.

In contrast to standard metal-forming operations, the initial forming has to remove the pores that are present in the sintered condition. Figure 6 illustrates how the basic properties change during these procedures: Note how the ductile–brittle transition temperature is continuously decreased. Extremely high degrees of cold deformation can be achieved (for certain products, the diameters can be reduced by means of wire drawing to 10 μm). The production volume of certain products ranges from a few milligrams to maximum pieces weighing 5 tons.

3.5 Heat Treatments and Thermomechanical Treatments

Annealing of worked metals below their recrystallization temperature may relieve residual stresses and improve the ductility to a

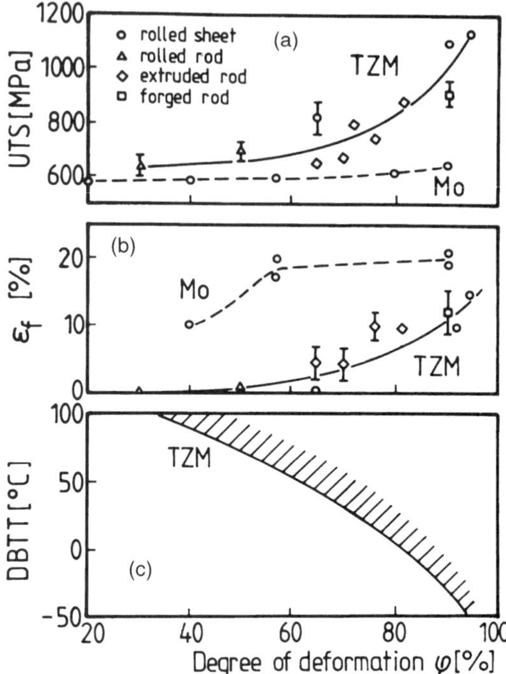

FIG. 6. The evolution of (a) the strength, (b) the fracture elongation, and (c) the ductile–brittle transition temperature (DBTT) in bending with the degree of working (Pink and Eck, 1996).

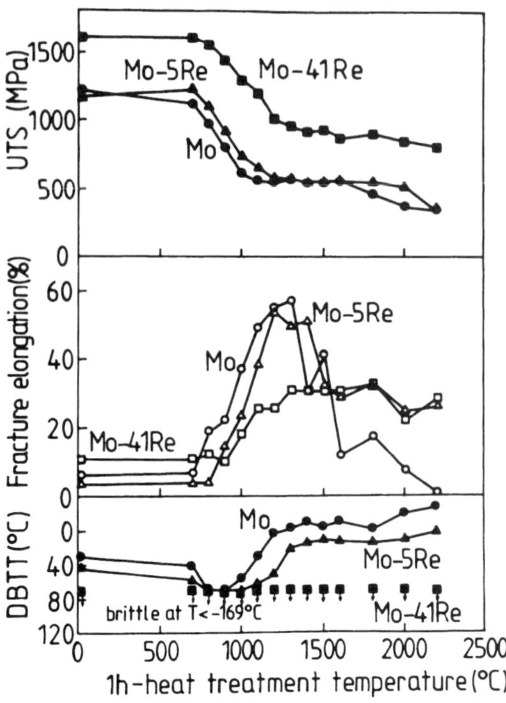

FIG. 7. The effect of annealing on room-temperature strength and elongation, and on the ductile–brittle transition temperature (DBTT) in bending, as demonstrated for the case of molybdenum and molybdenum–rhenium alloys (Pink and Eck, 1996).

certain extent (Fig. 7). Internal cracking and lamination of highly cold-worked wires and sheets can be avoided by applying intermediate partial recrystallization. In alloys, the precipitation of phases that strengthen or embrittle may be controlled by selected time-temperature cycles. Heat treatments between certain stages of forming are also performed to control final mechanical properties, e.g., in ZHM.

Heat treatments are performed under hydrogen or protective atmosphere or in vacuum. A problem of heat-treating alloys is the oxidation of alloying elements such as titanium, zirconium, and hafnium, and the depletion of carbon. In contrast to unalloyed molybdenum and tungsten, such alloys must not be annealed in nitrogen-containing atmospheres, because nitrides of the alloying elements formed on the surfaces have an embrittling effect.

Deformation in connection with the application of heat ("strain aging") may create special dislocation structures or influence the precipitation behavior, improving the strength of alloys (Trefilov and Milman, 1989).

3.6 Recycling

Most of the scrap is converted by similar chemical methods as applied for the powder production from the ore. High-purity scrap, which can be collected during the production, is recycled the easiest way, as high-purity alloying additions to superalloys and steels. Cleaned and recrystallized scrap can be recycled by comminution milling at temperatures below −40 °C using liquid nitrogen as coolant. Tungsten powder of such origin is further processed to tungsten heavy metal; molybdenum powder can be used for plasma spraying. Another process using the cooling effect of adiabatic expansion of nitrogen is called *coldstream process* and can be applied to tungsten. Pure tungsten–copper and tungsten heavy metal can be recycled by oxidation in air and successive reduction in hydrogen to the respective alloy powders, which can be reprocessed directly. Vacuum distillation could be used to recover infiltrated tungsten, but is expensive. Contaminated scrap of tantalum is pretreated mechanically and chemically, and arc-cast or electron-beam melted.

It is essential that rhenium as a "precious" metal be recovered. Chemical processing is an established procedure, as is also sublimation of the oxide.

4. FURTHER PROCESSING

4.1 Fastening and Joining

4.1.1 Mechanical Fastening For construction parts, refractory metals are preferably fastened by riveting, or with screws and nuts made of the same material. The threads of screws and nuts sinter at service temperature to the parts they are joining, and cannot be dismounted again. Rivets of molybdenum and tungsten are locally heated by torches to 800–1000 °C before pressing.

4.1.2 Welding Only "clean" welding methods are applicable (Lison *et al.*, 1989; Lambert, 1990). Welding under vacuum or pure inert gas prevents embrittling contaminations. Electron-beam (EB) and tungsten inert-gas (TIG) welding give best results. For thin parts, low-energy methods such as laser welding are also performed. Methods that do not create a liquid phase are friction welding (where temperatures of 2000 °C are reached at the contacting faces for a few seconds) and diffusion bonding. If not geometrically restricted, all joints of refractory metals and alloys can be made in this way, even when the parts to be joined are different materials.

Unalloyed tungsten and molybdenum embrittle as a result of excessive grain growth during welding. For perfect welding, the expensive filler alloys containing rhenium (Mo– or W–Re) can be used.

The welding behavior of V*a*-group metals is different from that of the group VI*a*. At high temperatures, protection from gases such as oxygen, nitrogen, hydrogen, water vapor, and carbon oxides is necessary, because the solubility for these gases is high, and even low contents have an embrittling effect (Buckman, 1988).

4.1.3 Brazing The disadvantage of brazing is the limitation imposed by the melting point of the braze, and in many cases by the temperature of the eutectic formed; another severe restriction is often the corrosion of the braze or the brazed area. Surfaces must be free of contaminations. The cleanest possible method is vacuum brazing. Brazing is important for the fabrication of refractory-metal composites where no other method is economically feasible: for brazing to ceramics, graphite, or semiconductor material (silicon).

A variety of braze materials are in use (Lambert, 1990) of which many are commercially available as foils. The temperatures where they can be used range from 700 °C in the case of silver–copper to 2200 °C for tantalum–hafnium alloys. Nickel, platinum, and rhodium are excellent brazes because of their good wetting power. Brazing materials for still higher temperatures are refractory metals or their alloys. For brazing of tungsten heavy alloys and contact materials, low-temperature brazing is sufficient: Elements contained in the binder phase such as nickel, copper, or silver are the base for brazing alloys.

4.2 Coatings

Great efforts, not too successful, have been made to find coating materials that protect molybdenum, tungsten, and rhenium against the formation of volatile oxides, and tantalum and niobium against taking atmospheric gases into embrittling solution, when these metals are exposed to elevated temperatures (Wang and Webster, 1982). The limiting temperature for creating coating under isothermal conditions is the recrystallization temperature, in order to prevent softening of the substrate material. Protective coatings develop microcracks when heated, and the smallest crack causes catastrophic failure at high temperatures.

Coatings are also deposited to protect surfaces against wear and to form layers for brazing.

4.2.1 Pack and Slurry Cementation Pack cementation based on silicides and aluminides has been developed. An improvement is to let a homogeneous slurry of complex silicides react with the refractory-metal surface (Lambert, 1990). Si–20Cr–20Fe prepared from elemental powders suspended in organic liquids is the standard coating for niobium and its alloys. Pack cementation of silicides developed for molybdenum has recently achieved commercial status. Layers may be up to 250 μm thick.

4.2.2 Electrolytic Coating

Aqueous electrolytic processes to form layers of nickel, silver, gold, platinum, rhodium, tin, and copper are more widely used for molybdenum than for tungsten. Very often a nickel flash improves adherence of subsequent layers. These coatings are necessary for brazing operations. Instead of depositing a very thin layer on wire, it turned out convenient to coat, for instance, molybdenum rods that are subsequently reduced in diameter to coated wires. Molybdenum electroplated with platinum or platinum–rhodium is widely used in the glass industry for operation in oxidizing atmospheres.

4.2.3 Flame and Plasma Spraying

One of the largest areas of consumption of molybdenum is for thermal spray coatings (Clare and Crawmer, 1982). Flame spraying (during which molybdenum wire is passed through an acetylene flame) is used for wear-resistant coatings on steel; their thickness may amount to millimeters. The wear-resistant layer contains pores that in service act as reservoirs for the lubricating oil. Plasma spraying in air (APS) and plasma spraying in vacuum (VPS, which avoids the oxidation of the layers and gives better densities up to 98%) of molybdenum starts from powder granules. Tungsten can be plasma-sprayed equally well, but is of limited importance.

Layers of iron- and/or nickel-based alloy powders containing 20 wt% chromium, 5 wt% aluminum, and up to 1% yttria protect molybdenum against oxidation. Electrodes for glass melting can be successfully protected against burner gases during the start of the furnace operation.

4.2.4 Chemical and Physical Vapor Deposition (CVD and PVD) Oxidation-resistant coatings up to 20 μm thickness can be deposited in dense layers by CVD and PVD processes. Coating compositions are based on silicides of refractory metals. Rhenium, platinum, and iridium layers are deposited by PVD processes such as magnetron sputtering and ion plating. These coatings serve as intermediate layers against carbide formation, for instance for refractory-metal–graphite composites.

CVD of layers (1 mm thick and more, with densities over 99%) of tungsten and tungsten–rhenium alloys on graphite discs serves to produce x-ray targets, large crucibles, and other shapes.

5. APPLICATIONS

5.1 Main Use of Refractory Metals and Alloys

The use of refractory metals is determined by their exceptional (both positive and negative) properties. Many applications of V*a* metals are based on their good resistance against chemical attack, which makes them a favorite material for equipment of the chemical industry. The VI*a* metals can only be applied in inert atmospheres or in vacuum. Under such conditions, their properties are exploited in the lighting industry, in the low- and high-energy electronic industry, for purposes of heat production (windings for furnaces, electrodes), and for high-temperature applications (construction material, heat shielding, jet propulsion).

Table 2 provides examples of uses of refractory alloys, based on the special properties of the materials. Up to service temperatures of 1600 °C, strain-aged ZHM and MHC are the superior refractory alloys (better than the strongest tantalum and niobium alloys or ODS nickel), especially when the data are compared in terms of a strength-to-density ratio, which is interesting for applications in aviation. From 1600 to above 2000 °C, tungsten alloys (e.g., with rhenium and hafnium carbides) are not surpassed by any other metallic materials, and are, or rather may become, important even for the aerospace industry despite their high specific weight. They are alternatives to the ceramic- and carbon-based materials.

Aside from conventional, though advanced, applications, pure refractory metals of normal quality are sometimes utilized for peculiar or exceptional purposes, e.g., chromium for sputtering targets, molybdenum for flame spraying (see Sec. 4.2.3), or rhenium as filament in carburizing atmospheres.

5.2 Special Requirements

5.2.1 High-Purity Materials Refractory metals, primarily tungsten, in use for microelectronics (microchips) need purity levels of 99.999% and 99.9999%. Lithium, sodium, and potassium contents can be reduced to below 0.1 μg/g. Sputtering targets of W–10Ti, used to deposit diffusion barriers on semiconductors, are presently produced by powder metallurgy with a purity of 99.99%, keeping the concentrations of the radioactive traces of thorium and uranium below 5 ng/g.

5.2.2 Porous Metals The most important use of tantalum is for capacitors for application in electronics (Belz, 1959). The metal surface is electrolytically oxidized to generate dielectric Ta_2O_5 layers. It is desirable to press the powders to as low a density as possible, because the capacitance is directly proportional to the surface area accessible to an electrolyte. Electron-beam melting purifies the tantalum, thereby allowing electrolytic capacitor devices to operate at higher voltages.

5.2.3 Alloys for Thermocouples For long-term use up to 2000 °C and for short-term use above, ductile Mo–41Re/Mo–5Re couples have been developed. The voltage of this type of thermocouple is equal to that of the tungsten–rhenium couples, in the range of Pt/Pt–10Rh.

5.2.4 Refractory Metals for Superconductors Pure metals of the V*a* and VI*a* groups are transition-element superconductors. Some of their alloys and compounds belong to the group of high-field type-II superconducting materials having simultaneously favorable critical transition temperatures for superconductivity as well as high upper critical magnetic fields (Warnes, 1990). NbTi alloys have the lowest superconducting characteristics (Kreilick, 1990). The A15 superconductors (Smathers, 1990) are better; they have a cubic crystal structure of type A_3B (A standing for the metals of the groups V*a* and VI*a*). In contrast to NbTi superconductors they are, as intermetallic phases, brittle. The third group of superconductors with high upper critical magnetic fields are the ternary molybdenum chalcogenides or "Chevrel" phases (Le Lay, 1990) of type $M\mathrm{Mo}_6X_8$, where M is a cation and X one of the chalcogenes sulfur, selenium, or tellurium. Amorphous modifications of refractory metals, e.g., $Mo_{30}Re_{70}$ or $Mo_{52}Ru_{32}B_{16}$, also belong to the group of high-field type-II superconductors, but their transition temperatures are below 9 K (Johnson, 1990).

It is interesting to mention that, for physical reasons, the NbTi superconductor has to be surrounded with a suitable metal like copper. The diameter of the superconducting wire must be extremely small (10 μm). Wire packages containing several hundreds or thousands of individual superconducting filaments are produced by sophisticated sequences of extruding and drawing, which can be done thanks to the good malleability of the material.

5.2.5 Use in Thermonuclear Reactors Refractory metals are favored because the operations are carried out at high temperatures, because they may retain their strength upon neutron irradiation, because of their low neutron-absorption cross section and short induced radioactivity, and because of their good corrosion resistance (e.g., to liquid alkali metals) at high temperatures (Mazey and English, 1984). Irradiation-induced void swelling and the elevation of the ductile–brittle transition are disadvantages. Refractory metals have also been considered for application in fusion reactors as "first-wall" materials, i.e., as materials situated closest to the plasma.

6. ECONOMIC ASPECTS

The production volume of refractory alloys is small compared to steel or nonferrous metals. Table 4 shows the total consumption of refractory elements in recent years. Table 5 presents a more detailed list giving the percentages of the different applications. An increase in the consumption of tungsten and molybdenum metal and alloys of 2%–3% per year is predicted. Chromium is a relatively new metal on the market. While its consumption in various forms is the highest of

Table 4. Worldwide consumption of refractory elements (in t/y).

Year	V	Nb[a]	Ta	Cr[b]	Mo	W	Re
1972	16 000	5 200					
1976		6 000	800		90 000	36 000	4
1981		20 000		2.5×10^6	100 000	40 000	5
1988			1 000		97 000	44 000	8.4
1990					105 000		15.9

[a]Nb (1989): 250–300 t metal products.
[b]Cr (1990): 75 t metal products.

all refractory elements (see Table 4), only a small fraction is produced in metallic form. The rhenium consumption will grow as new applications are found.

While prices for refractory metals fluctuate, the relations between the prices remain much the same. Molybdenum is the cheapest, followed by chromium, tungsten, vanadium, tantalum, and rhenium, in that order. Tantalum is about 10 times more expensive than molybdenum, and rhenium about 10–50 times more expensive than tantalum, depending on the market.

7. TOXICOLOGICAL ASPECTS

A certain amount of the elements of the V*a* and VI*a* groups is necessary for the human body. Of the refractory metals, only chromium, molybdenum, and vanadium appear in presently available tables of toxic materials (Goyer, 1990), yet the toxic effects appear not to be a consequence of the elementary metals, but of their chemical compounds.

Vanadium compounds as by-products of petroleum refining are present in urban atmospheres. Excess exposure can damage the respiratory tract, the kidney, liver, and bone marrow, and can cause gastrointestinal distress and affect the nervous system. Sources of chromium-compound pollution are coal-fired power plants, the metal-finishing industries, textile industries, and tanneries. Compounds of chromium are carcinogenic and provoke immune reactions. An overdose of a molybdenum compound picked up by cattle from pastures in the vicinity of molybdenum factories is responsible for anemia, diarrhea, retarded growth, and deformities of the joints. The symptoms can, in some cases, disappear when the intake is terminated.

GLOSSARY

Aging: Annealing of an alloy at a certain temperature for a certain time in order to create precipitates that strengthen the matrix (see **Dispersion-Strengthened Alloys** and **Heat Treatment**).

AKS Tungsten: Tungsten with minute additions of Al–K–Si compounds, which cause—after severe deformation—the formation of elongated grains upon recrystallization. The same effect can be also accomplished in other metals, e.g., in molybdenum.

Amorphous Metals/Alloys: Metals and alloys that could not assume their normal crystal structure during extremely rapid cooling from the melt.

Contiguity: A measure for the amount of direct contacts in a dual-phase alloy between phases (grains) of the same crystal structure.

Table 5. Percentage of refractory elements used for main products and industries.

Product	V (1991)	Nb (1972)	Ta (1988)	Mo (1978)	W (1990) Euro.	W (1990) Japan	W (1990) USA	Re (1990)
Metal products	<2	3	83	3–4	15	17	23	10–20
In steels and (super)alloys	94	79	8	86	20	23	5	10–20
In hard metals			6		55	58	62	
For chemicals etc.		18	3	10	10	2	10	60–70

Dispersion-Strengthened Alloys: Metals that are strengthened by dispersoids (intermetallic or nonmetallic inclusions) of 5–500-nm diameter, which obstruct the movement of dislocations. The dispersoids in refractory alloys are introduced either to the metal powder or by aging treatments.

Doped Tungsten/Molybdenum: See **AKS Tungsten.**

Dual-Phase Alloys: Alloys consisting of grains of two different metals or alloys with different crystal structures, arranged in well-defined distributions (see **Contiguity**).

Ductile–Brittle Transition Temperature: Temperature below which a metal or an alloy with a body-centered cubic lattice structure loses its capability for plastic deformation, and breaks in a brittle manner.

Heat Treatment: Heating and slow or fast cooling of a metal or an alloy in order to improve its workabilities and/or its mechanical properties.

Liquid-Phase Sintering: Sintering under conditions that favor the appearance of one liquid phase which then surrounds the solid phase, thereby dissolving part of it in accordance with the solubility limits.

Nonsag (NS) Tungsten/Molybdenum: See **AKS-Tungsten.** The "sagging" is the deformation that a wire, heated to high temperatures, experiences when it is horizontally suspended.

ODS Alloys: Alloys strengthened by oxide dispersions.

Precipitates: Intermetallic or nonmetallic coherent or incoherent particles of 5–50-nm diameter that precipitate out of the crystal lattice of a supersaturated solid solution during aging.

Refractory Metals and Alloys: Metals and their alloys with high melting points (higher than 1900 °C), mainly of the V*a* and VI*a* groups of the periodic table of elements.

Sintering: Heating of a compressed metal powder at high temperatures (below the melting point), in order to achieve consolidation by diffusional material transport.

Solid Solution: An alloy in which the minor constituent is completely dissolved (and not precipitated) within the primary metal lattice.

Strain Aging: Increasing the strength by aging after a certain amount of cold work, due to a segregation of alloying atoms and/or precipitates at dislocations.

Thermomechanical Treatments: Mechanical working of an alloy in combination with a heat treatment, in order to improve the mechanical properties.

Tungsten Heavy Metal: Tungsten alloy consisting of spheroidal tungsten grains embedded in a low-melting binder phase (of iron and nickel, mostly).

Works Cited

Belz, L. H. (1959), in: W. R. Clough (Ed.), *Reactive Metals,* Metallurgical Society Conference Vol. 2, New York: Interscience, pp. 525–539.

Buckman, R. W., Jr. (1988), in: J. L. Walter, Ch. Jackson, T. Sims (Eds.), *Alloying,* Metals Park, OH: ASM International, pp. 419–445.

Clare, J. H., Crawmer, D. E. (1982), in: *Metals Handbook,* 9th ed., Vol. 7, Metals Park, OH: ASM, pp. 361–374.

Eck, R., Pink, E. (1992), *Int. J. Refract. Met. Hard Mater.* **11,** 337–341.

Fromm, E., Gebhardt, E. (1976), *Gase und Kohlenstoff in Metallen,* Berlin: Springer-Verlag.

German, R. M. (1985), *Liquid Phase Sintering,* New York: Plenum Press.

Goyer, R. A. (1990), in: *Metals Handbook,* 10th ed., Vol. 2, Metals Park, OH: ASM International, pp. 1233–1269.

Heitzinger, F. (1974), in: H. H. Hausner, W. E. Smith (Eds.), *Modern Developments in Powder Metallurgy,* Vol. 8, Princeton, NJ: American Powder Metallurgical Institute, pp. 371–390.

Honeycombe, R. W. K. (1984), *The Plastic Deformation of Metals,* London: E. Arnold.

Johnson, W. L. (1990), in: *Metals Handbook,* 10th ed., Vol. 2, Metals Park, OH: ASM International, pp. 804–821.

Kreilick, T. S. (1990), in: *Metals Handbook,* 10th ed., Vol. 2, Metals Park, OH: ASM International, pp. 1043–1059.

Lambert, J. B. (1990), in: *Metals Handbook,* 10th ed., Vol. 2, Metals Park, OH: ASM International, pp. 557–585.

Le Lay, L. (1990), in: *Metals Handbook,* 10th ed., Vol. 2, Metals Park, OH: ASM International, pp. 1077–1081.

Lison, R., Ambroziak, A., Horn, H. (1989), in: H. Bildstein, H. M. Ortner (Eds.), *Proceedings of the 12th Plansee Seminar,* Vol. 1, Reutte, Austria: Metallwerk Plansee, pp. 75–105.

Martinz, H. P., Prandini, K. (1993–1994), *Int. J. Refract. Met. Hard Mater.* **12,** 179–186.

Mazey, D. J., English, C. A. (1984), *J. Less-Common Met.* **100,** 385–427.

Pink, E., Eck, R. (1996), in: K. H. Matucha (Ed.), *Structure and Properties of Non-Ferrous Alloys,* Materials Science and Technology, Vol. 8, Weinheim: VCH-Verlag, Chap. 10, pp. 589–641.

Shen, Y. Sh., Lattari, P., Gardner, J., Wiegard, H. (1990), in: *Metals Handbook,* 10th ed., Vol. 2, Metals Park, OH: ASM International, pp. 840–868.

Smathers, D. B. (1990), in: *Metals Handbook,* 10th ed., Vol. 2, Metals Park, OH: ASM International, pp. 1060–1076.

Snow, D. B. (1989), in: E. Pink, L. Bartha (Eds.), *The Metallurgy of Doped/Non-Sag Tungsten,* London: Elsevier Science Publishers, pp. 189–202.

Tietz, T. E., Wilson, J. W. (1965), *Behavior and Properties of Refractory Metals,* London: E. Arnold.

Trefilov, V. I., Milman, Yu. V. (1989), in: H. Bildstein, H. M. Ortner (Eds.), *Proceedings of the 12th Plansee Seminar,* Vol. 1, Reutte, Austria: Metallwerk Plansee, pp. 107–131.

Wang, Ch. T., Webster, R. T. (1982), in: *Metals Handbook,* 9th ed., Vol. 5, Metals Park, OH: ASM, pp. 659–666.

Warnes, W. H. (1990), in: *Metals Handbook,* 10th ed., Vol. 2, Metals Park, OH: ASM International, pp. 1030–1042.

Zitter, H., Plenk, H., Jr. (1987), *J. Biomed. Mater. Res.* **21,** 881–896.

Further Reading

ASM-International Handbook Committee (Eds.) (1990), *Metals Handbook,* 10th ed., Vol. 2, Metals Park, OH: ASM International.

Benesovsky, F. (1979), *Wolfram, Gmelins Handbuch der anorganischen Chemie,* Ergänzungsband A1, 8 Auflage, Berlin: Springer-Verlag.

Bildstein, H., Ortner, H. M., Eck, R. (Eds.) (1985,1989,1993), *Proceedings of the 11th–13th Plansee Seminar,* Reutte, Austria: Metallwerk Plansee.

Kieffer, R., Braun, H. (1963), *Vanadin, Niob und Tantal,* Berlin: Springer-Verlag.

Kieffer, R., Jangg, G., Ettmayer, P. (1971), *Sondermetalle,* Wien: Springer-Verlag.

Pink, E., Bartha, L. (Eds.) (1989), *The Metallurgy of Doped/Non-Sag Tungsten,* London: Elsevier Sci. Publ.

Sully, A. H., Brandes, E. A. (1967), *Chromium,* London: Butterworths.

Trefilov, V. I., Mil'man, Yu. V., Firstov, S. A. (1975), *Fizicheskie Ocnovy Prochnosti Tugoplavkikh Metallov,* Kiev: Naukova Dumka.

Trefilov, V. I., Moiseev, V. F. (1978), *Dispersnie Chastitsy b Tugoplavkikh Metallakh,* Kiev: Naukova Dumka.

Yih, S. Y. H., Wang, C. T. (1979), *Tungsten: Sources, Metallurgy, Properties and Applications,* New York: Plenum Press.

REINFORCING FIBERS

A. R. BUNSELL, *Centre des Matériaux P. M. Fourt, Ecole Nationale Supérieure des Mines de Paris, Evry, France*

	Introduction	343
1.	**Fiber Development**	344
2.	**Reinforcement by Fibers**	345
3.	**Fiber Processing, Structure, and Properties**	349
4.	**Statistics of Fiber Reinforcement**	352
5.	**Organic Fibers**	354
5.1	Molecular Structure and Mechanical Properties	354
5.2	Melt-Spun Fibers	354
5.3	Aramid Fibers	356
5.4	Theoretical Performance of Organic Fibers	357
5.5	Ultrahigh-Performance Polyethylene Fibers	358
6.	**Glass Fibers**	359
7.	**Boron and Silicon-Carbide Fibers on a Substrate**	360
8.	**Carbon Fibers**	361
9.	**Alumina Fibers**	363
10.	**Ceramic Fibers from Organic Precursors**	365
11.	**Whiskers**	367
12.	**Conclusion**	367
	Glossary	367
	Works Cited	368
	Further Reading	368

INTRODUCTION

A fiber is a long fine filament of matter with a diameter generally of the order of 10 μm and with an aspect ratio of length to diameter between a hundred and virtually infinity for continuous fibers. Fibers are one of the most extraordinary forms of ordinary matter. They are often much stronger and stiffer than the same material in bulk form. They have remarkable specific properties so that for the same strength, for example, a steel filament would be three or four times heavier than most fibers. The fineness of fibers allows them to be woven and draped into complex shapes. Cloth, woven or nonwoven, is a means of converting a one-dimensional filament into a two-dimensional component. Many and perhaps most fiber-reinforced composite structures are essentially two dimensional whether they are carbon-epoxy skins over a honeycomb structure, a filament-wound tube, or a car-body panel made by press molding. Three-dimensional weaving and needle punching are recent techniques of the textile industry that are also used for composites. Mechanical properties are but one aspect in the choice of a material, and others, such as thermal transfer or insulation behavior, water retention, and cost, have determined the use of fibers in the past as they do today in composite materials.

Natural fibers have long found applications in clothing, ropes, and sails; however, they were largely ignored for more structural applications up to the beginning of the twentieth century when artificial fibers were made by reprocessing the natural polymeric structure of cellulose. These viscose-rayon fibers were used to reinforce the rubber that was used for tires. The second half of the twentieth century has seen an ever widening use of synthetic fibers to reinforce plastics, metals, and ceramics. These are the subjects of this article on reinforcing fibers. The methods of production of the fibers are sketched and, most importantly, the relationships between the fibers' microstructures and their mechanical properties are treated. The mechanics of reinforcement by these long and thin filaments are explained, so revealing the importance of the rigidity of the fibers in fiber-reinforced composites but also the shear of the surrounding matrix material on the ultimate properties of the composites. The presence of often millions of fibers in

3-527-28138-X/96/$5.00 + .50

the cross section of any fibrous structure underlines the importance of a statistical approach to analysis of fiber properties. This is also put into context in the article.

1. FIBER DEVELOPMENT

The reinforcement of a medium by a network of fibers has evolved in the living world to be the natural solution for structures that respond to and resist mechanical stresses. The combination of two very different phases, the matrix medium and the fibers, allows a great degree of flexibility in the development of microstructural arrangements. In addition, the long fine form of the fibers, coupled with their mechanical strength and stiffness, allows natural structures to be preferentially reinforced in areas of greater stress (see COMPOSITE MATERIALS).

Although surrounded by naturally occurring fiber-reinforced materials and the wide use of wood for construction, man, with very few exceptions, only began to create fiber-reinforced structural materials at the beginning of the twentieth century. Up to that time there were no artificial fibers produced, but their development at the end of the nineteenth century coincided with the beginning of the automobile industry. Fiber-reinforced rubber forms the basis of the tire industry. Fiber-reinforced resins, which form the basis of the composite industry, have developed in the second part of the twentieth century and owe their existence to the production of organic resins and the beginning of the synthetic-fiber industry in the 1930s.

Fibers can be divided into three groups: natural, regenerated, and synthetic. Naturally occurring fibers can themselves be divided into three subgroups—vegetable, animal, and mineral fibers—as described by Moncrieff (1979). Vegetable fibers such as flax, cotton, and hemp are short or at least discontinuous and based generally on cellulose. Mineral fibers such as asbestos are also short, being rarely longer than a few centimeters, and their fully crystalline structure places them apart from the others. Animal fibers, which are based on animal proteins, can be discontinuous, as in the case of wool, or they can be virtually continuous, as with silk. The natural fibers all share a common trait of possessing molecular structures that themselves are arranged in a fibrillar manner. A cotton fiber, for example, has a microfibrillar structure arranged in a helical fashion around the fiber axis. The helix reverses in direction at points along the fiber length and is reminiscent of the structure of cotton threads. Threads consisting of discontinuous fibers owe their integrity to the intertwining of the individual fibers and the frictional forces developed between them that lock the structure together and allow a continuous thread to be produced, as indicated by Morton and Hearle (1975). The load transfer developed between the fibers within a thread is exactly analogous to the load transfer between matrix and fiber that accounts for the properties of fiber-reinforced composites; see Sec. 2.

Regenerated fibers benefit from the fibrillar nature of the molecular structure of plants, which is processed to form continuous filaments. In this way, this first class of artificial fiber to be produced avoids the difficulty of constructing its basic molecular structure.

Artificial fibers are only just one hundred years old, and the earliest produced were based on the naturally occurring cellulose molecule. Toward the end of the nineteenth century workers in Great Britain and in France had found means of dissolving natural cellulose and then extruding it through holes to produce filaments. The first commercial exploitation of these regenerated fibers was in France, where a patent to produce the earliest form of rayon was awarded to Count Hilaire de Chardonnet in 1885. This was the beginning of a revolution in fiber technology. The regenerated fibers did not exploit the total potential of the cellulose molecules, as they were broken up to shorter lengths during the fiber-manufacturing process. The cellulose was obtained from wood, usually spruce, made up of molecules containing about 1000 glucose residues, but these were broken down to about a fifth of their original length in the manufacture of rayon. The intention of the first fiber producers was to produce an artificial silk for textile purposes; however, they found a major application in the carcasses of tires for the developing automobile industry. Rayon fibers have evolved significantly over the years, and the descendants of these earliest artificial fibers are still to be found in some car tires today. Rayon fibers were used as precursors for the early production of carbon fibers, mainly in the U.S.A., and there are still some carbon fibers produced in this way; however, they have now mainly been superseded by other fibers for this purpose.

Truly synthetic organic fibers for which the long molecular structures, necessary for most fiber production, are made from relatively simple molecules have existed only since the late 1930s. The first organic truly synthetic fibers were developed independently in the U.S.A. and Germany in the late 1930s. These were polyamide or "nylon" fibers and these, together with polyester fibers developed in Great Britain in the 1940s, have become the most widely produced commodity synthetic fibers. Organic fibers have since been developed with elastic moduli rivaling that of steel but with only a fifth or less of the weight. Two approaches have been taken to achieve this remarkable behavior. First, the molecular structures of organic fibers have been made more complex and consequently stiffer. The second approach has been to align the molecular structure of simple polymers to achieve greatly improved tensile properties.

Glass fibers, first commercially produced in the U.S.A. in the early 1930s, were the first artificial fibers with a relatively high Young's modulus. The manufacture of glass fibers by drawing the molten glass through a spinneret is similar to the process used to produce fine thermoplastic fibers. However, the act of drawing aligns the molecular structure of the thermoplastic fiber producing an anisotropic structure. This is not the case with the glass fibers, which remain isotropic, but in both instances the act of drawing and plastically deforming the material during fiber production removes surface defects and accounts for the very high strength of glass fibers compared with bulk glass.

Although glass fibers possess Young's moduli considerably higher than those of conventional textile fibers, still greater tensile stiffness is required for structural composites in many applications and notably in the aerospace industry. Boron fibers produced in the early 1960s, first in the U.S.A. and then in France, allowed a leap from 70 GPa, which is the elastic modulus of glass, to over 400 GPa. These remarkable fibers were technically a success, but their use has been limited by their extremely high cost.

Carbon fibers, first produced in Great Britain in the middle 1960s and produced commercially for the first time there in 1967, were also developed in parallel in Japan, where commercial production began in 1970. These carbon fibers, which are made by the pyrolysis of polyacrylonitrile precursor fibers, have revolutionized the aerospace industry and are finding increasing applications in other industrial areas. An alternative route for producing carbon fibers is from pitch, which is the residue from oil refining or the coking of coal. Fibers with very high Young's moduli, approaching that of diamond, can be made from pitch.

Large-diameter ceramic fibers of silicon carbide on a core have existed since the early production of boron fibers; however, fine ceramic fibers started to appear at the end of the 1970s and in the 1980s. One group of these fibers, developed in Great Britain and in the U.S.A., is based on fine-grained alumina whereas others, developed in Japan, are silicon carbide or silicon nitride. This class of fibers allows light-metal alloys to be reinforced and, probably more importantly, the production of fiber-reinforced ceramics for high-temperature structural applications. They are made by the pyrolysis of precursor fibers in an analogous way to the production of carbon fibers.

Single-crystal continuous fibers can also be made. Such fibers, based on α-alumina and first commercialized in the U.S.A., usually have large diameters around 100 mm; however, the absence of grain boundaries offers the possibility of maintaining useful mechanical properties to 1500 °C and above.

The range of fibers available for reinforcement has been described in detail by Bunsell (1988).

2. REINFORCEMENT BY FIBERS

Any shape of inclusion could be put into a matrix, but a long thin form provides the most efficient reinforcement. This can be seen if a single fiber of length l and circular cross section is considered embedded in a cylindrical block of radius r_1 as illustrated in Fig. 1. Consider the case of an elastic fiber having a Young's modulus much greater than that of the matrix, which itself is elastic. In the absence of the fiber any applied

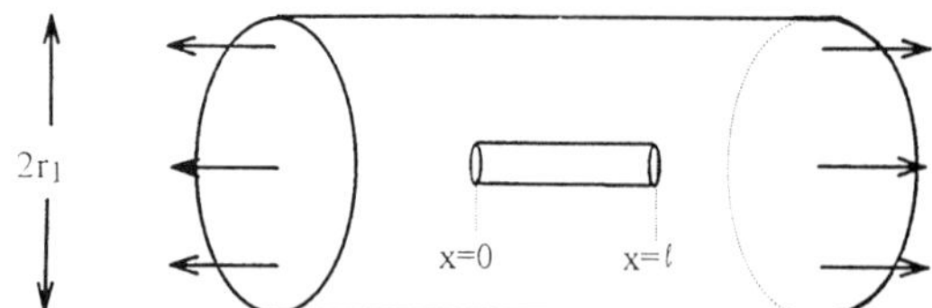

FIG. 1. Consider a short stiff elastic fiber embedded in a cylinder of elastic matrix material that is less rigid and more deformable than the fiber.

load on the specimen would produce a uniform deformation of the matrix. If the specimens were viewed from the side and divided up into identical square zones these would be transformed into identical rectangular areas. In the case in which a high-modulus fiber is embedded in the matrix and there is good adhesion between the two, a different situation occurs under load. This matrix near the fiber is limited in deformation by the lower strain produced in the fiber, whereas further away the matrix is free to deform as in the unreinforced case. Figure 2 shows that the deformations produced in the presence of a higher-modulus fiber are far from simple, as in this case the displacements produced near to and far from the fiber are different and induce shear stresses in the matrix around the fiber. The shear stresses transfer load from the matrix to the fiber, which supports a load that increases from zero at the ends, assuming no load transfer across the ends, to a maximum in the center.

This model of fiber reinforcement was first proposed by Cox (1952).

The load P generated in the fiber produces in it a deformation of U so that we can describe the load transfer due to the shear of the matrix in mathematical form by

$$\frac{dP}{dx} = H(U - V), \tag{1}$$

where V is the overall composite displacement, H is a factor of proportionality, and x is the distance along the fiber from the end.

We can also write

$$P = E_f A_f \left(\frac{dU}{dx}\right), \tag{2}$$

where E_f is the Young's modulus of the fiber, A_f is the fiber cross-sectional area, dU/dx is the strain in the fiber at x, and $P/A_F = \sigma_X$, the stress supported by the fiber at a distance x from the origin taken as one end of the fiber.

We also know that the strain e produced in the composite, considered as a continuum, would be the same all over the specimen so that

$$\frac{dV}{dx} = e = \text{const.} \tag{3}$$

From Eq. (1),

$$\frac{d^2P}{dx^2} = H\left(\frac{dU}{dx} - \frac{dV}{dx}\right).$$

So from the above three equations we have

$$\frac{d^2P}{dx^2} = H\left(\frac{P}{E_f A_f} - e\right), \tag{4}$$

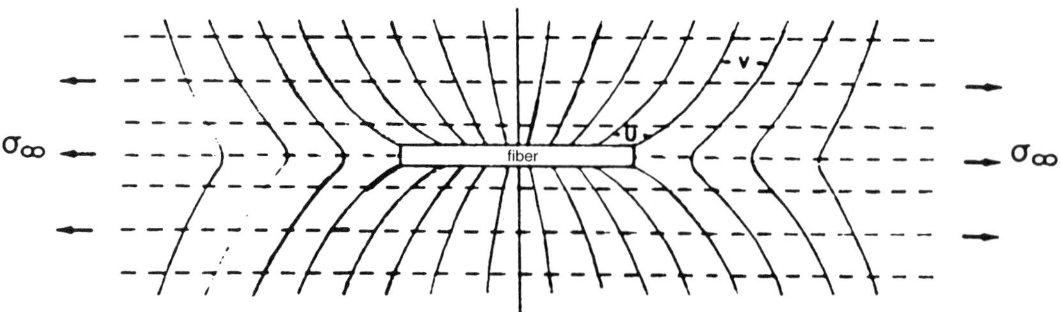

FIG. 2. A uniform tensile load parallel to a short fiber embedded in a deformable matrix produces a nonuniform strain of the matrix. The stress σ_∞ applied to the specimen produces an overall deformation of V; however, the deformation in the neighborhood of the fiber is U.

which can be solved by the substitution

$$P = E_f A_f e + R \sinh\beta x + S \cosh\beta x, \qquad (5)$$

in which R, S, and β are parameters that can be determined making use of straightforward but elaborate algebra involving hyperbolic functions and by noting that $P = 0$ where $x = 0$ and l. This gives

$$\beta = \sqrt{H/E_f A_f}$$

and

$$\sigma_x = E_f e\left[1 - \frac{\cosh\beta(l/2 - x)}{\cosh\beta l/2}\right]. \qquad (6)$$

Equation (6) reveals that the load supported by the fiber increases from each end to attain a maximum at the midpoint of the fiber, as is shown in Fig. 3(a). In order that the average strain in the fiber be that of the composite so that $e = \sigma_x/E_f$, the function inside the brackets in Eq. (6) must become equal to unity. This is only possible when the fiber is continuous and l tends to infinity.

The average stress in a short fiber of length l is given by

$$\sigma_m = \frac{E_f e}{l}\int_0^l \left[1 - \frac{\cosh\beta(l/2 - x)}{\cosh\beta l/2}\right]dx.$$

Simplifying gives

$$\sigma_m = \frac{E_f e}{l}\left[l - \frac{2}{\beta}\tanh\frac{\beta l}{2}\right].$$

Dividing through by l we obtain

$$\sigma_m = E_f e\left[1 - \frac{\tanh\beta l/2}{\beta l/2}\right].$$

The effective modulus of a short fiber embedded in a matrix is therefore given by

$$\frac{\sigma_m}{e} = E_{fM} = E_f\left[1 - \frac{\tanh\beta l/2}{\beta l/2}\right]. \qquad (7)$$

Equation (7) shows that a short fiber does not have the same reinforcing efficiency as does a long fiber. The longer the fiber, however, the closer it gets to being as efficient as a continuous fiber, as the load-transfer lengths at each end remain the same. A practical difficulty that this analysis does not consider is that of aligning short fibers in the direction of the applied load.

The shear forces around the fiber can be calculated by considering the equilibrium state in which the shear forces in the matrix are balanced by the load in the fiber so that

$$\frac{dP}{dx} = -2\pi r_0 \tau(r_0), \qquad (8)$$

in which $\tau(r_0)$ is the shear stress at the interface and r_0 is the fiber radius.

From Eq. (1) we can write

$$H = -2\pi r_0 \tau(r_0)/(U - V) \qquad (9)$$

If w is the actual displacement of the matrix and r is the radial distance from the fiber axis, then at $r = r_0$, $w = U$, and at $r = r_1$, $w = V$. For equilibrium,

$$2\pi r \tau(r) = 2\pi r_0 \tau(r_0). \qquad (10)$$

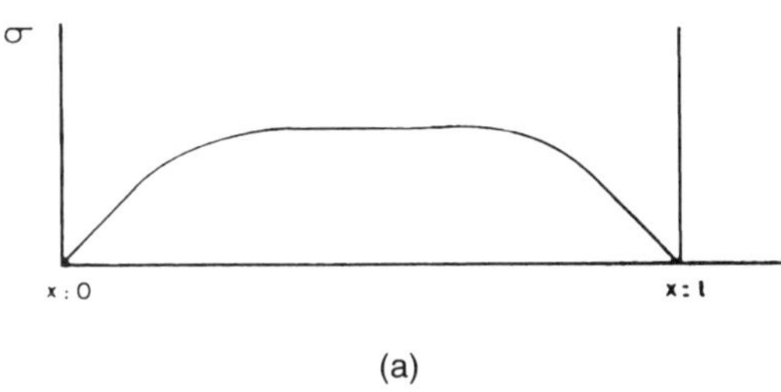

(a)

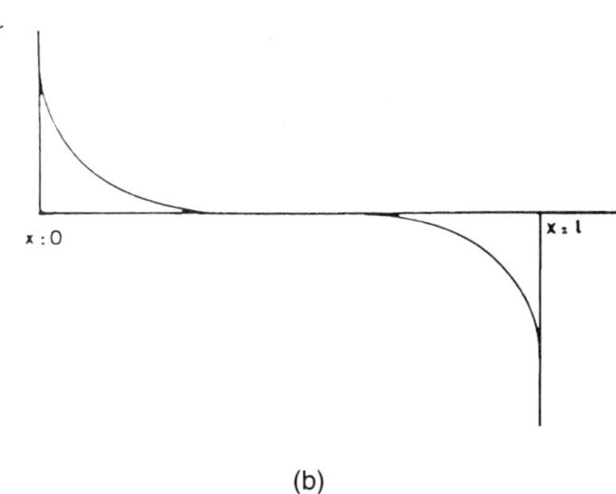

(b)

FIG. 3. (a) The stress transferred to the fiber tends to a plateau in the center of the fiber. (b) The shear of the matrix around the fiber is a maximum at the fiber ends and tends to zero at the center of the fiber. The shear in the matrix is in opposite directions at the two ends of the fiber.

This shear deformation produced in the matrix is given by

$$\frac{dw}{dr} = \frac{\tau(r)}{G_m} = \frac{\tau(r_0)r_0}{G_m r}, \qquad (11)$$

where G_m is the matrix shear modulus.

Integrating Eq. (11) between r_0 and r_1 given that Δw is the difference in displacements produced in the matrix at r_0 and r_1, which is $U - V$, we obtain

$$U - V = \frac{\tau(r_0)}{G_m} r_0 \ln\left(\frac{r_1}{r_0}\right), \qquad (12)$$

so that from Eq. (9),

$$H = \frac{2\pi r_0 G_m}{\ln(r_1/r_0)}. \qquad (13)$$

Since

$$\beta = [H/E_f A_f]^{1/2},$$

we can see that

$$\beta = \left[\left(\frac{G_m}{E_f}\right)\left(\frac{2\pi}{A_f \ln(r_1/r_0)}\right)\right]^{1/2}. \qquad (14)$$

The significance of this result is that the larger G_m/E_f, the more rapidly the stress in the fiber increases with distance from the fiber ends.

Equations (6) and (8) allow us to calculate the shear stress $\tau(r_0)$ in the matrix at the interface so that, as

$$P = \pi r_0^2 \sigma_x$$

and

$$\tau(r_0) = -\frac{dP}{dx}\frac{1}{2\pi r_0},$$

it follows that

$$\tau(r_0) = e\left[\frac{G_M E_F}{2\ln(r_1/r_0)}\right]^{1/2}\left[\frac{\sinh\beta(l/2 - x)}{\cosh\beta l/2}\right]. \qquad (15)$$

Figure 3(b) shows the shear-stress distribution described by Eq. (15). It can be seen that the shear stresses are a maximum at the ends of the fiber and decrease to zero at the midpoint. The shear stresses are in opposite directions at opposing ends of the fiber; hence their opposite signs.

Equations (6) and (15) reveal that for the case of a long fiber we can write

$$\frac{\tau_{\max}}{\sigma_{\max}} = \left[\frac{G_m}{2E_f \ln(r_1/r_0)}\right]^{1/2}. \qquad (16)$$

As indicated the above analysis is for an elastic fiber embedded in an elastic matrix. Now let us consider an elastic fiber embedded in an ideally plastic matrix. The plastic case is simpler than the elastic situation considered above, as it can be assumed that the shear stress in the matrix is constant and equal to the yield stress τ_y, so that Eq. (8) can be directly integrated giving

$$P = -2\pi r_0 \tau_y x|_0^x,$$

which divided through by the cross-sectional area of the fiber, πr_0^2, can be written as

$$\sigma_x = 2\tau_y x/r_0. \qquad (17)$$

Equation 17 shows that there is a linear buildup of stress from the fiber ends when it is embedded in a plastic matrix. In the case of a long fiber this buildup of stress is limited by the overall stress applied to the composite, as shown in Fig. 4. In this case the central section of the fiber experiences a constant stress. If the stress applied to the composite is increased the load-transfer length is increased and the central constant-stress region diminishes. Fracture will occur in the central section if the applied stress exceeds the fiber failure stress. If the fiber is long or continuous multiple fractures can occur. There clearly exists a length of fiber for which the maximum stress that can be attained at the midpoint is equal to the fiber failure stress. This is known as the critical length $2l_c c$. At the critical length Eq. (17) can be rewritten so that

$$l_c/r_0 = \sigma_f/2\tau_y, \qquad (18)$$

where σ_f is the failure stress of the fiber.

It can be seen therefore that the aspect ratio l_c/r_0 is the important factor in determining the efficiency of a reinforcement. A long fine reinforcement, in other words a fiber, is clearly the ideal reinforcement.

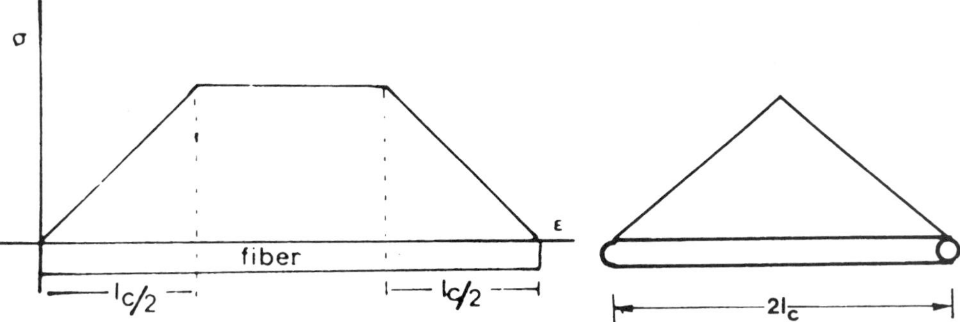

FIG. 4. An elastic fiber embedded in a perfectly plastic matrix leads to a uniform buildup of stress in the fiber. If the fiber is long this leads to a constant stress borne by the fiber in the midsection. A fiber of critical length $2l_c$ is just long enough that the maximum stress than can be attained at its midpoint is equal to its fracture stress.

3. FIBER PROCESSING, STRUCTURE, AND PROPERTIES

Many fibers, particularly organic and glass fibers, are produced by being drawn or stretched between leaving the spinneret and being wound onto a bobbin. The drawing process reduces the filament diameter from 1 or 2 mm to the order of 10 μm. This change of form has immediate consequences on the strength of the fiber even if effects of induced orientation, such as occur with organic fibers, are ignored.

Glass fibers can be several hundred times stronger than the same glass in bulk form, and this is due to the drawing of the fiber, which occurs during its manufacture. Glass is a brittle material, and its strength is determined by the size of defects it contains. Inglis (1913) first showed that irregularities such as a small crack in a brittle solid induce stress concentrations under load and that these localized high stresses σ precipitate failure. The finer the crack, the greater the stress concentration and the lower the strength of the body.

Most of the defects on glass are on the surface, as it is there that it is most easily damaged. The process of drawing the glass into a fine filament greatly deforms it and in particular modifies the surface. The movement of matter from the interior of the filament to its surface eliminates the surface defects that weaken the bulk material and so greatly increases its strength. If the filament is to retain its strength the surface must be protected from damage by abrasion and environmental effects. For this reason glass fibers receive a protective surface coating at an early stage in their manufacture.

Strength can be seen therefore not as an intrinsic quality of the material but as determined by the probability of finding defects in the body. A statistical approach to analyzing fiber failure is necessary, as is explained below.

Glass fibers are generally considered to be isotropic. The effect of drawing on polymers, however, is to orient the macromolecular structure more or less in the drawing direction, which, in the case of a fiber, is parallel to its axis. The greater degree of orientation increases both the Young's modulus and failure stress of the fiber and reduces its strain to failure. This is because the applied load is supported by an increasing number of covalent atomic bonds in the polymer chains. As a direct consequence, however, the structure of drawn organic fibers becomes anisotropic with the transverse strength being more influenced by the secondary weak atomic bonds linking the macromolecules. The alignment of a polymer structure does not always necessitate mechanical drawing, as by inducing a mesophase or liquid-crystal phase the polymer may be induced to possess an inherent alignment while still in the liquid phase, which enables very high-modulus fibers to be spun. Fibers such as the aramid fibers, as well as some pitch-based carbon fibers, are produced in such ways as to take advantage of the intrinsic anisotropy of the liquid mesophase.

The first organic fibers were based on a

Table 1. Molecular structure of several organic fibers showing the development of increasingly rigid molecules.

Fiber type	Repeat unit in the macromolecule	Maximum elastic modulus (GPa)
Polyamide 6/6 (nylon 6/6)	$—NH—CH_2—CH_2—CH_2—CH_2—CH_2—CH_2—NH—CO—CH_2—CH_2—CH_2—CH_2—CO—$	5
Polyamide 6 (nylon 6)	$—NH—CH_2—CH_2—CH_2—CH_2—CH_2—CO—$	4
Polyethylene terephthalate	$—O—CO—C_6H_4—CO—O—CH_2—CH_2—$	18
Poly (*m*-phenylenediamine-isophthalamide) (Nomex)	$—[—N(H)—C_6H_4—NH—CO—C_6H_4—C(=O)—]—$	17
Poly(*p*-phenylene terephthalamide) (Kevlar)	$—[HN—C_6H_4—NH - CO—C_6H_4—CO]—$	160

structure made up of linear macromolecules. Table 1 shows that polyamide fibers consist of macromolecules made up of repeat units that determine their name. The first structure shown is of polyamide 6/6, which consists of repeat units containing two series of six carbon atoms. The molecule is flexible, as it is able to bend at each bond. The structure of the fiber consists of microfibrils aligned more or less parallel to the axis and composed of regions of molecules folded into well-ordered crystallites as shown ideally in Fig. 5. The crystallites may be rectangular as indicated or rhomboidal. This type of structure is considered likely for both polyamide and polyester fibers. The folded molecules and the amorphous regions of the structure, in which the molecules are randomly organized, lead to a low Young's modulus.

The polyester fiber, however, contains an

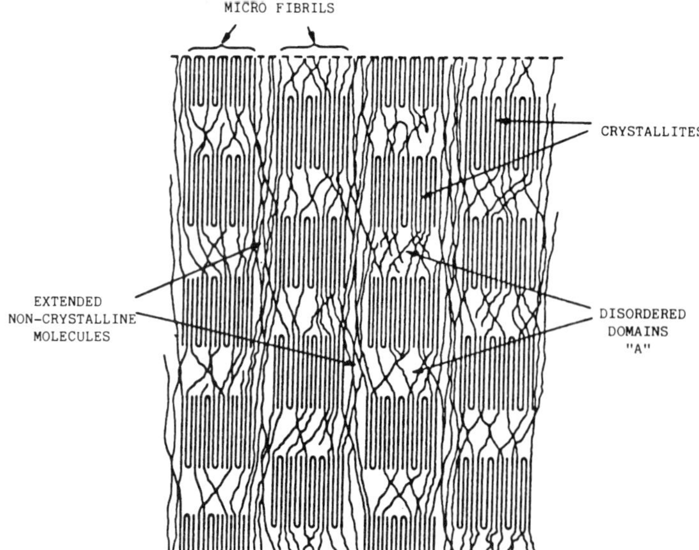

FIG. 5. Schematic and idealized representation of the microfibrillar structure of polyester and nylon fibers.

aromatic ring, which makes it less flexible than the polyamide molecule. The consequence of this increased molecular rigidity is an increase in the tensile modulus of polyester fibers compared with polyamide fibers.

Fibers with increasingly rigid molecular structures have been developed, and this has brought about better thermal stability and much higher Young's moduli. The Nomex fiber described in Table 1 has much improved thermal properties when compared with polyamide fibers, giving it an increase in useful temperature range of 100 °C, and as such is used for protective clothing where the possibility of fire may be encountered. The Nomex fiber is not produced for high mechanical properties and fails with a breaking strain of about 22% similar to that of polyester. It is used as a paper to form the core of honeycomb structures. It has an initial Young's modulus of 17 GPa.

The production of fibers from already anisotropic liquid-crystal solutions allowed a leap in initial modulus as well as final properties to be made, as described by Black and Preston (1973).

The aromatic polyamide PPT, described in Table 1 and produced by Du Pont under the name Kevlar, was the first commercial fiber of its type and launched the aramid family of fibers for composite and other applications. Table 1 shows that the high-modulus form of this fiber has a much higher tensile modulus than other organic textile fibers.

An alternative approach to producing high-modulus organic fibers is to ensure that the macromolecules are aligned parallel to the fiber axis. In order to be able to disentangle the molecular structure relatively simple polymers have so far been used such as polyethylene. Fibers of this type are produced either by spinning them from dilute solutions or by optimizing the melt-spinning process. Very high Young's moduli can be achieved, approaching that of high-performance aramid fibers. The polymer retains its low melting point, however.

The ultimate tensile properties of a solid depend on the interatomic bonds. These depend on the interaction of valence electrons associated with neighboring atoms and not on the atomic weight. This has encouraged attempts to produce high-modulus fibers from the lightest elements in the periodic table. The lightest element with which it is conceivable to make fibers is the fourth in the periodic table, beryllium. It is possible to make filaments from beryllium but the properties are not exceptional and above all the material is toxic. The fifth element, boron, however, has been made into fibers by vapor deposition onto a tungsten substrate. The properties of boron fibers, which were first produced in the early 1960s, are very good with a Young's modulus twice as high as that of steel and excellent properties in compression. The specific gravity of boron fibers is 2.45 and their diameter is normally 140 μm. The slow rate of fiber production and the fact that it is impossible to produce more than one fiber at a time with each production unit leads to an unacceptably high cost of the fiber for most applications. As a consequence almost all of the applications for which boron fibers had been considered have used carbon fibers as reinforcements.

Carbon fibers, which were developed in the middle and late 1960s, can be made with a range of properties depending on the temperature used during production. Some of the very first artificial fibers to be produced at the end of the nineteenth century were of carbon. Cotton fibers were pyrolyzed to make carbon fibers and used as elements in some of the first electric lamps produced by Edison. These fibers had little mechanical strength, and the high-performance fibers produced in the 1960s were made by the pyrolysis of polyacrylonitrile (PAN) fibers. High modulus is achieved by preventing shrinkage of the fiber from occurring during pyrolysis, which produces a 45% loss in weight. The inherent anisotropy of the precursor fiber is preserved in this way to give carbon fibers made up of polycarbon basic structural units arranged in the form of ribbons parallel to the fiber axis. This longitudinal organization is not reflected in the radial direction, in which the structure is completely random.

Most of the carbon fibers used in composites are now of the high-strength type. Remarkable progress has been made in improving the properties of these fibers through improvements in the polyacrylonitrile precursor fibers, optimization of the pyrolysis process, and a reduction of fiber diameter from 8 to 5 μm. This has produced an increase in breaking strain from 1% to more than 2%.

Pitch has attracted much attention in the U.S.A. and Japan as a source for producing carbon fibers. The high carbon yield, around 90%, together with the easy availability of pitch, seemed at first to offer the potential of a low-cost route for making carbon fibers. The difficulties of purifying the pitch and market forces seem to destine these fibers to primarily very high-modulus applications although the range of moduli is from 200 to 900 GPa. The difference in behavior of these fibers compared with that of PAN-based carbon fibers is due to a more graphitic structure, which leads both to higher moduli but also lower compressive strengths.

Alumina-based ceramic fibers are made from a variety of types of alumina. The alumina fibers possessing the highest Young's moduli are made from α-alumina, which is the most stable and crystalline form of aluminum oxide. The granular structure of α-alumina produces high-modulus but very brittle fibers. Other forms of alumina such as η and γ alumina, having much smaller grains usually associated with an amorphous silica phase, give fibers with lower Young's moduli and less brittle behavior.

Ceramic fibers, made from organo-silicide precursors that are pyrolyzed, are based on silicon associated with carbon, often oxygen, and sometimes nitrogen and titanium. Any crystalline phase is usually β-silicon carbide, which is often embedded in an amorphous matrix of silicon and oxygen, the latter element being introduced during pyrolysis. The most recent fibers of this type limit the oxygen content and hence the amorphous phase by modifying the pyrolysis conditions. The result is improved fiber properties.

4. STATISTICS OF FIBER REINFORCEMENT

Fibers are very fine filaments of matter and in any section of a composite structure there will be found thousands and most probably millions of fibers. Such large populations lend themselves to statistical analysis, and the inherent scatter in fiber properties ensures that a statistical approach to describing their properties is necessary.

It is not possible to give a single value for the strength of a fiber because they show considerable scatter and the mean strength decreases with fiber length. Figure 6 shows the scatter of strength of Toray's T300 carbon fibers. Both the scatter and the length dependence are due to the distribution of defects in the fibers, and the longer the fiber the greater chance there is of it containing a sizable defect and so being considerably weakened. Failure of the fiber as a function of applied load is therefore controlled by the random distribution of defects and requires a statistical treatment for its analysis.

The effect of a random distribution of a single type of defect on the strength of a solid has been described by Weibull, who likened the failure of the solid to the breaking of a chain in which the weakest link controls failure. If the probability of failure of a link under an applied stress σ is P_0, its probability of survival is $1 - P_0$, and the survival probability of the chain is $(1 - P_0)^n$ where n is the number of links. If P_n is the probability of the chain's failure we can write

$$1 - P_n = (1 - P_0)^n \tag{19}$$

or

$$P_n = 1 - \exp[n \ln(1 - P_0)]. \tag{20}$$

In the case of a body of volume V let us

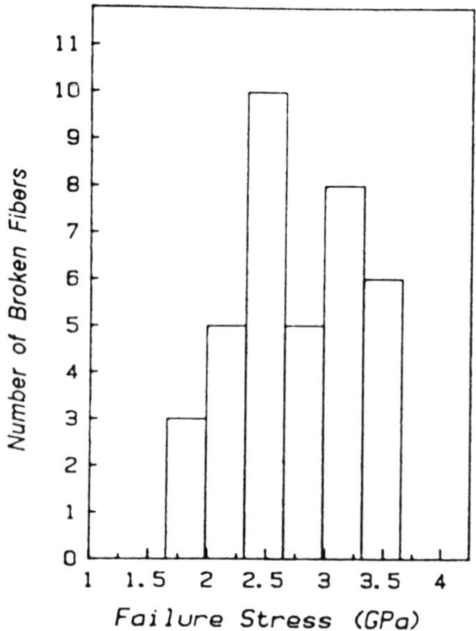

FIG. 6. The failure stress of carbon fibers shows considerable scatter.

consider it divided up into small volumes V_0 analogous to links in a chain. In this case n is analogous to V/V_0.

The quantity $\ln(1 - P_0)$ depends only on the stress if V_0 is considered constant, so that for the whole specimen the risk of failure is, from Eq. (20),

$$P_v = 1 - \exp\left[-\int_v f\left(\sigma, \frac{1}{V_0}\right)\right]dV.$$

Weibull put $f(\sigma,1/V_0)$ equal to $[(\sigma - \sigma_u)/\sigma_0]^m$ where σ is the applied stress, σ_u is the stress below which there is a zero probability of failure, and the two material parameters σ_0 and m are measures of the density and variability of the flaws in the materials.

In a straight tensile test in which the stress σ is constant throughout the body we can write

$$P_v = 1 - \exp\{-[(\sigma - \sigma_u)/\sigma_0]^m V\}. \tag{21}$$

This is the distribution function characterizing the scatter of strengths found with repeated tests on specimens of volume V and material constants σ_u, σ_0, and m. The parameter m is known as the Weibull shape parameter and characterizes the distribution of strengths. The greater the value of m the less is the scatter of results.

Writing the probability of survival $(1 - P_v)$ as P_s we obtain from Eq. (21), if σ_u is zero,

$$\ln\ln(1/P_s) = \ln V + m\ln\sigma - m\ln\sigma_0. \tag{22}$$

If we now consider a cylindrical fiber with diameter D that can vary with different fibers taken from the same bundle but all having the same lengths, we can write from Eq. (22)

$$\ln\ln(1/P_s) = m\ln\sigma + 2\ln D + \text{const.} \tag{23}$$

If $\ln\ln 1/P_s - 2\ln D$ is plotted as a function of $\ln\sigma$, the Weibull shape parameter may be measured from the gradient of the curve obtained, as is shown in Fig. 7 for carbon fibers. The fiber diameter was found not to vary significantly and so was considered a constant in plotting Fig. 7. In this way the distribution of fiber properties as described by Weibull statistics can be obtained from a series of tensile tests, usually not less than

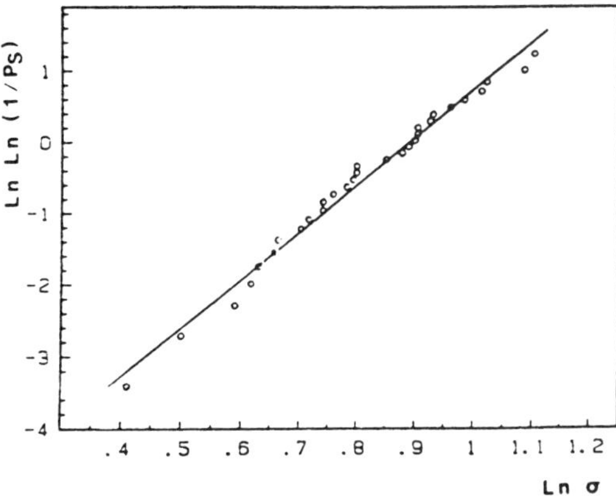

FIG. 7. A Weibull plot of the values shown in Fig. 6 reveals a straight line the gradient of which gives the Weibull modulus.

twenty-five depending on the scatter of results, with specimens of equal lengths.

As explained above, the average strength of fibers taken from a bundle decreases as a function of length, and this offers another approach to obtaining the Weibull shape parameter. If the median failure stress (P_s = 0.5), which is not necessarily the same as the mean stress, for a number of specimens is denoted by σ_{50} Eq. (22) can be written

$$\ln\ln\left(\frac{1}{0.5}\right) = \ln l + 2\ln D + m\ln\sigma_{50} + \text{const.} \tag{24}$$

If the mean fiber diameter from one lot of specimens to another is considered to be constant we can write

$$\ln\sigma_{50} = -(1/m)\ln l + \text{const.} \tag{25}$$

A plot of $\ln\sigma_{50}$ as a function of $\ln l$ for several gauge lengths gives a straight line curve with a gradient of $-1/m$.

Theoretically the values of m calculated by the two approaches should be identical; however, this is hardly ever the case. The difference is due to the difficulty of manipulating fine fibers during which defects may be created and, more significantly, weaker fibers are broken before they can be tested. This difficulty inevitably increases with increasing specimen length so that merely handling the fibers provides an unavoidable selection process and modifies the failure distribution curve. In addition, and although the approach based on Weibull statistics is extremely valuable and widely developed in analyzing fiber data, there are doubts as to whether the failure of fibers falls within the limits of the theory. This is because the fibers are very fine and it is not true that the same distribution of defects can be expected to exist in all small sections of the fibers.

The value of the Weibull shape parameter for most fibers is low, often around 5, which reflects the large dispersion found in their properties. This value of m is similar to that obtained with ceramics and can be compared with the values for m obtained for steel that give $m > 20$.

Because all fibers show great dispersion in their properties they do not all fail in the same manner. The mode of failure may involve slow crack propagation from surface or internal faults leading to eventual catastrophic failure. This is the case of many organic fibers, which may deform plastically, although very-high modulus organic fibers split on failure as a result of their high anisotropy. Other fibers such as carbon, glass, boron, and ceramic fibers show no plastic deformation and behave as purely brittle elastic bodies, their ultimate failure being determined by defect size as described by Griffith (1920) for elastic bodies.

5. ORGANIC FIBERS

It was not until around 1938 that the first truly synthetic organic fibers were produced from short molecules, and continuous glass fibers were also commercially produced just before this date. Both the polyamide and glass fibers were developed in the U.S.A., respectively by E. I. du Pont de Nemours and what became Owens Corning. Almost simultaneously with the work of Carothers at Du Pont, polyamide was produced in Germany by I. G. Farben.

5.1 Molecular Structure and Mechanical Properties

Since the beginning of production of polyamide, or nylon, fibers other organic fibers have been produced, and a trend toward the production of increasingly more rigid and complex molecules used in making the fibers is clearly discernible from Table 1. The aramid family of fibers made up of aromatic polyamides possesses remarkably high Young's moduli, more than twenty times that of conventional polyamide fibers. The more rigid molecule confers not only greater elastic modulus and failure stress but also better thermal resistance. This latter quality is a considerable advantage when these fibers are compared with other high-modulus organic fibers made from simpler molecules such as polyethylene made by straightening out the molecular structure and exploiting the covalent bonds in the backbone of the molecule.

5.2 Melt-Spun Fibers

Nylon (polyamide) and polyester (polyethylene terephthalate) fibers are extensively used for reinforcing rubber, and these flexible composites are used in fan belts, hoses, flexible drives, moving walkways, tires, and other applications. They account for 28% and 48%, respectively, of the total synthetic-fiber production. Both types of fibers are produced by melt spinning and are subsequently drawn to increase their Young's moduli. They are both semicrystalline with their molecular structures arranged as shown in Fig. 5 in microfibrils consisting of crystalline blocks of regularly folded molecules linked by less well-ordered molecules. Surrounding the microfibrils there are elongated but irregularly arranged molecules forming a mesomorphous phase. Crystallinity is often about 60% in polyamide fibers and somewhat less in the polyester fibers.

Neither fiber shows real elastic behavior even at low strains, and plastic deformation as well as creep occurs when the fiber, either by itself or in an elastomeric composite, is loaded. The properties required of these fibers, for example in a tire, are high strength, dimensional stability, fatigue resistance, thermal stability, and good bonding to the matrix. Tensile failure in both nylon and polyester fibers is similar, involving slow crack growth from the surface, in a radial direction. The crack opens as the material ahead of the crack deforms plastically. A point is reached at which the crack becomes unstable and rapid failure occurs. Almost identical fracture surfaces are obtained with both broken ends, one of which can be seen in Fig. 8. The composite structures for which organic

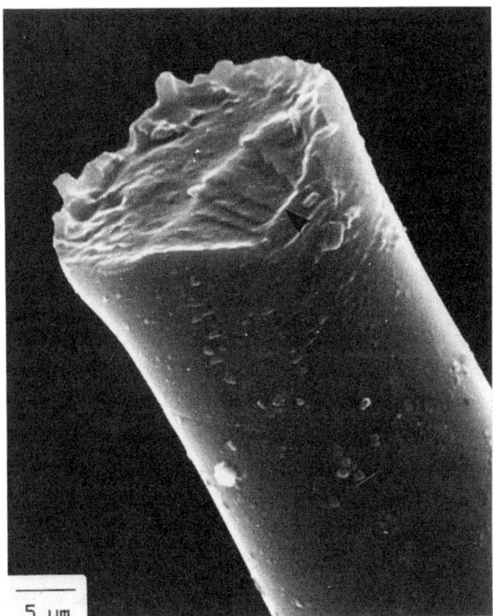

FIG. 8. The fracture morphologies of each end of a nylon or polyester fiber broken in tension or creep are very similar in appearance.

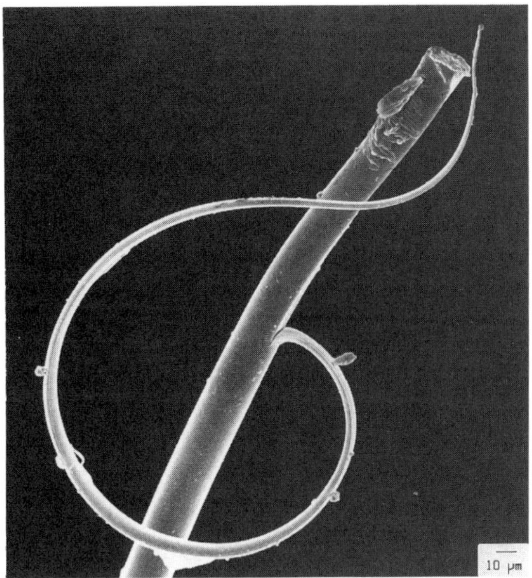

FIG. 9. Polyester and nylon fibers subjected to fatigue conditions fail with distinctive fracture morphologies.

fibers are used as reinforcements are often subjected to dynamic loading and fatigue is a major concern. Many organic fibers can fail by a distinct failure process leading to a characteristic fracture morphology involving longitudinal crack growth, which gradually reduces the load-bearing fiber cross section until failure occurs, as shown by Bunsell and Hearle (1974). This leads to a distinctive fracture morphology of which a particularly striking example is shown in Fig. 9. The loading conditions that lead to this type of failure are unusual, as a necessary criterion for fatigue failure is that the minimum cyclic load be nearly zero. Indeed, it is possible to avoid fatigue by increasing the overall loading pattern on the fiber and if this does not raise it into a regime of creep failure, the fiber will no longer break. The fatigue failure of these fibers seems to depend on the shear stresses set up between the intrafibrillar and interfibrillar amorphous zones due to differences in compliances of these regions.

Figure 10 illustrates the deformation of the molecular structures of a polyamide or polyester fiber during cyclic loading. Upon loading the tie molecules in the intrafibrillar amorphous zones are pulled slightly out of the crystalline zones so that on unloading they buckle, inducing shear stresses at the edge of the microfibril. These shear stresses lead to the breakage of molecules and a local deterioration of the fiber. Eventually the fatigue crack appears, and its development is influenced greatly by the anisotropic structure of the fiber.

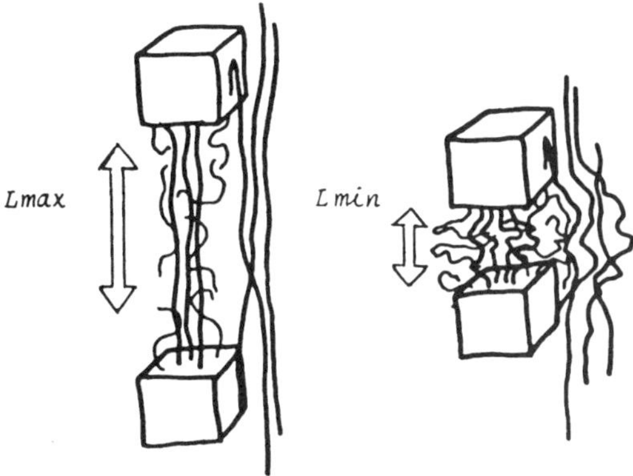

FIG. 10. During cyclic loading of nylon or polyester fibers, the tie molecules are slightly pulled out of the crystallites and buckle if the load is sufficiently reduced. This leads to the development of stresses between the microfibrils and the interfibrillar amorphous phase and the deterioration of the molecular structure.

5.3 Aramid Fibers

The aramid group of reinforced fibers, in the form of Kevlar from Du Pont de Nemours, has been available since 1972 and has been adopted for many composite structures. These fibers, which are made from a mesophase or liquid-crystal solution of the polymer in concentrated sulfuric acid, have remarkable tensile properties. Molecular poly(*p*-phenylene terephthalamide) (PPTA) alignment is produced when the mesophase solution passes through the spinneret. The shear stresses that are induced align the structure in much the same way as tree trunks are aligned during transport by river. Conventional fiber drawing is not necessary, although control of water content is important at this stage if optimum properties are to be obtained. The fibers are five times as strong as steel in air for the same weight, and this difference rises to thirty times when they are used in water, because of their low density and the buoyancy of the water. This is of great importance as these aramids find great use in cables, which are often impregnated with a resin to form a flexible composite structure, for marine and submarine applications. The highly anisotropic structure of aramid fibers results, however, in them being weak in all other directions. As a consequence failure of these fibers is nearly always highly fibrillar, as shown in Fig. 11. The structure of aramid fibers is very well ordered, and the basic molecular arrangement is shown in Fig. 12. Their highly anisotropic behavior means that the fibers are not used in primary loading structures subjected to compressive forces.

The processes that limit the compressive properties of aramid fibers confer on them, however, great tenacity, which means that they absorb much more energy than more brittle fibers when broken. This is because the fiber can deform plastically in compression. An elastic fiber can be bent to minimum radius of curvature at which the tensile stresses in the convex surface attain the breaking stress of the material. When an aramid fiber is bent slippage of the molecular structure occurs on and near the concave surface leading to the development of kink bands as can be seen from Fig. 13. The result of this plastic deformation is that the neutral stress axis is displaced from the fiber axis toward the convex surface. In this way the tensile stresses on the convex surface do not attain the failure stress of the fiber, which can be bent to a zero radius of curvature. This results in the fiber being difficult to cut as well as giving great tenacity and resistance to impact loading. At the present the aramid fibers, other than Kevlar, that are commercially available are the Arenka fiber

FIG. 11. Aramid "Kevlar" fibers are highly anisotropic and fibrillate on failure.

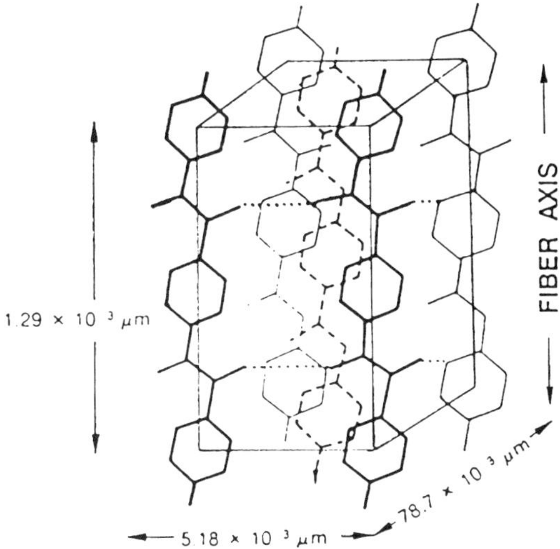

FIG. 12. The unit cell of the Kevlar aramid fiber showing the rigid PPTA molecules tied together transversely by weak secondary bonds.

from Akzo in Holland and the Technora fiber from Teijin in Japan. The Technora fiber, formally known as HM-50, from Teijin fits into the aramid family although it is not produced from a mesophase. It has an intermediate modulus.

The production of aramid fibers involves complex and expensive chemistry; however, the high-performance properties of the fibers are a result of the molecules being aligned parallel to the fiber's axis. Table 2 shows that although simpler polymers usually possess low mechanical properties they could, if all their molecules were aligned parallel to one another, give materials with the characteristics of high-performance materials.

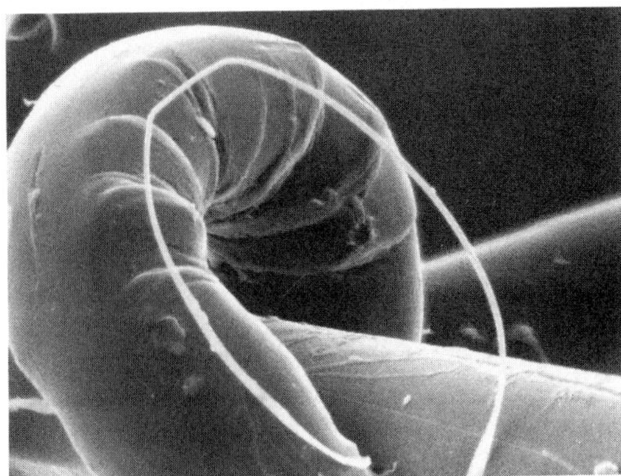

FIG. 13. Bending of aramid fibers leads to the development of kink or slip bands on the concave surface as a result of local plastic deformation.

Table 2. Ultimate strengths of various polymers.

Polymer	Density (g/cm³)	Ultimate strength (GPa)	Strength of commercial fiber (GPa)
Polyethylene	0.96	37	0.95
Polyamide 6	1.14	32	1.00
Polyvinylalcohol	1.28	29	1.00
Poly(*p*-phenylene terephthalamide)	1.43	30	2.65
Polyethylene terephthalate	1.37	28	1.00
Poly prepylene	0.91	17	0.95
Regenerated cellulose	1.50	18	0.55

5.4 Theoretical Performance of Organic Fibers

It is possible to calculate the theoretical maximum Young's modulus for a simple organic structure for the optimum case in which all the macromolecules are aligned in the direction of the applied load. Let us consider the molecule shown in Fig. 14. The force f applied to the molecule deforms the AB and BC bonds slightly and increases the valence angle $\widehat{ABC}$. Two force parameters can be defined relating the effort and the deformation produced, such that

$$K_a = (f/\Delta a)\sin(\theta/2) \tag{26}$$

and

$$K_\theta = fh/\theta.$$

These two parameters can be measured directly by infrared or Raman spectrometry, and for an olefin molecule we find that

$$K_a \approx 460 \text{ N m}^{-1}$$

and

$$K_\theta \approx 82 \times 20^{-20} \text{ N m}.$$

From the geometry of the molecule we see that

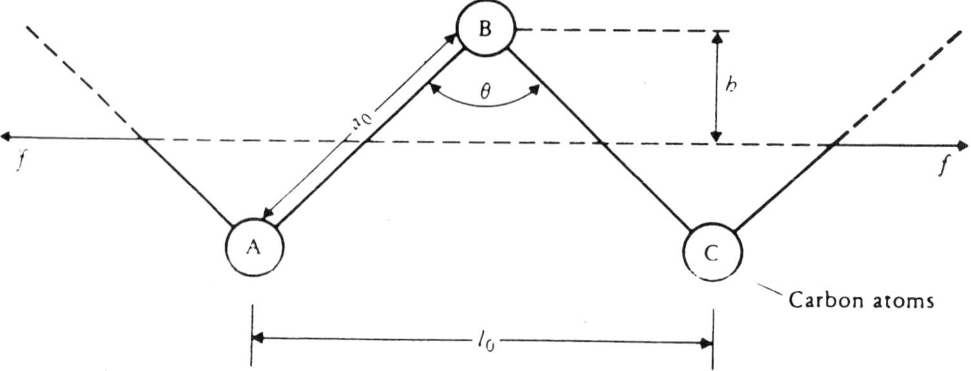

FIG. 14. Schematic representation of a part of a polymer chain being stretched by the force f; A, B, and C are main-chain carbon atoms.

$$l_0 = 2a \sin(\theta/2),$$

so that the deformation produced by the force f is

$$dl = 2[(a/2)\cos(\theta/2)d\theta + \sin(\theta/2)da] \tag{27}$$

and the strain

$$\epsilon = \frac{dl}{l_0} = \frac{(a/2)\cos(\theta/2)d\theta + \sin(\theta/2)da}{a\sin(\theta/2)}. \tag{28}$$

The couple fh producing a change of angle of $d\theta$ results in an angular compliance that, using Eq. (26), can be seen to be

$$\frac{d\theta}{f} = \frac{(a/2)\cos(\theta/2)}{K_\theta}. \tag{29}$$

The deformation of the AB and BC bonds is produced by the component of force, $f\cos(90° - \theta/2)$, so that the deformation compliance is

$$\frac{da}{f} = \frac{\sin(\theta/2)}{2K_a}. \tag{30}$$

The total compliance is obtained by replacing da and $d\theta$ in Eq. (28), using relations (29) and (30), and dividing by f so that

$$\frac{dl}{f} = a^2\frac{\cos^2(\theta/2)}{2K_\theta} + \frac{\sin^2(\theta/2)}{K_a}. \tag{31}$$

The Young's modulus E of the molecule can be calculated if its cross-sectional area A can be estimated, as

$$E = \frac{f/A}{dl/l_0}. \tag{32}$$

It has been estimated that A is of the order of 18×10^{-20} m^2 so that the theoretical maximum Young's modulus of a simple linear organic polymer is about 240 GPa. This is much greater than the moduli measured with most organic fibers because of their far from optimum molecular arrangements; however, ultrasonic measurements have revealed that, at a microscopic level, a simple polymer such as polyethylene possesses a modulus of 260 GPa. This observation has encouraged the optimization of the drawing processes of simple polymers with the aim of producing high-modulus fibers and films other than by producing increasingly complex and rigid molecular systems.

5.5 Ultrahigh-Performance Polyethylene Fibers

Two main manufacturing approaches exist for producing fibers by the above technique. The sol-gel approach, in which the polymer is dissolved in a solvent and then spun, is at present the only technique yielding commercially available fibers. High-modulus polyethylene fibers produced by DSM in Holland, Allied Chemicals in the U.S.A., and Toyobo in Japan have properties rivaling those of the aramid fibers with a lower specific gravity of 0.97. These fibers are limited in temperature to a maximum of 120 °C and suffer from creep. Figure 15 shows the behavior of PE fibers as a function of temperature described by Dessain *et al.* (1992). The abrupt change in slope is due to a solid-state phase change from orthorhombic to hexagonal. In the absence of an applied stress this phase change occurs around 140 °C, just below the melting point at 146 °C. However, the phase change can be induced at lower temperatures if a stress is applied. The level of stress that has to be applied to induce the phase change increases as the temperature is lowered until a critical point is reached around 5 °C below which the fiber fails before the phase change can occur.

The other manufacturing route is to draw directly from the melt as described by Bashir *et al.* (1984). This is potentially a cheaper technique but the Young's moduli of the fibers produced are not so high.

6. GLASS FIBERS

Glass has found many uses over the centuries from Roman times and earlier. A machine for producing glass fibers incorporating a type of spinneret was demonstrated at a British Association meeting in 1841. During the First World War glass fibers were produced in Germany to replace asbestos by touching a heat-softened glass rod to a spinning wheel, a technique that has been refined more recently for producing filaments of materials that are impossible to draw. In 1931 two American firms, Owen Illinois Glass Co. and Corning Glass Works, developed a method of spinning glass filaments from the melt through spinnerets. The two firms combined in 1938 to form the Owens Corning Fiberglass Corporation. Since that time extensive use of glass fibers has been made. Initially the glass fibers were destined for filters and textile uses; however, the development of heat-setting resins opened up the possibility of fiber-reinforced composites, and in the years following the Second World War the fiber took a dominant role in this type of material. Today, by far the greatest volume of composite materials is reinforced with glass fibers.

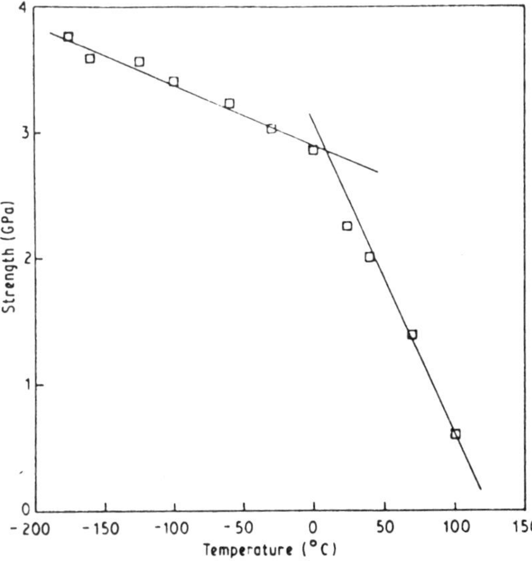

FIG. 15. The mean strength of ultrahigh-modulus polyethylene fibers as a function of temperature. The change in slope reveals a change in crystalline structure.

Several types of glass fiber exist but all are based on silica (SiO_2), which is melted and mixed with other elements to create specialty glasses (see GLASSES). Fibers of glass are produced by extruding molten glass through holes in a spinneret with diameters of 1 or 2 mm and then drawing the filaments by one of two techniques to produce fibers having diameters usually between 5 and 15 μm. The spinnerets usually contain several hundred holes so that a strand of glass fiber is produced.

The strength of glass fibers depends on the size of flaws, most usually at the surface, and as the fibers would be easily damaged by abrasion, either with other fibers or by coming into contact with machinery in the manufacturing process, they are coated with a size. The purpose of this coating is both to protect the fiber and to hold the strand together. The size may be temporary, usually a starch–oil emulsion, to aid handling of the fiber, which is then removed and replaced with a finish to help fiber-matrix adhesion in the composite. Alternatively the size may be of a type that has several additional functions, which are to act as a coupling agent and lubricant and to eliminate electrostatic charges.

Continuous glass fibers may be woven, as are textile fibers, made into a nonwoven mat in which the fibers are arranged in a random fashion, used in filament winding, or chopped into short fibers. In this last case the fibers are chopped into lengths of up to 5 cm and lightly bonded together to form a mat, or chopped into shorter lengths of a few millimeters for inclusion in molding resins.

The structure of all glass fibers is based on silica (SiO_2), which is shown in Fig. 16, mixed with other materials according to the properties desired. The structure is vitreous with no definite compounds being formed and no crystallization taking place, although at a very local level structures could be envisaged, as indicated by the dashed circles. Figure 16 shows how the silica forms a three-dimensional network, which is not regular. This open network is a result of the rapid cooling that takes place in fiber production, with the glass cooling from about 1500 °C to 200 °C in between 0.1 and 0.3 s. Despite this rapid rate of cooling there appear to be no appreciable residual stresses within the fiber and the structure is isotropic.

The most widely used glass for fiber-reinforced composites is called E glass. Glass fibers with superior mechanical properties are known as S or R glass with increased alumina content, and the compositions of a number of glasses produced in fiber form are shown in Table 3.

Glass is set to remain the most widely used reinforcement for general composites. Glass fibers are cheaper than most other relatively high-modulus fibers and, because of their flexibility, do not require very specialized machines or techniques to handle them. Their elastic modulus is, however, low when compared with other fibers that exist, and the specific gravity of glass, which for E-type glass is 2.54, is relatively high. The poor specific value of the mechanical properties of glass fibers means that they are not used for structures requiring light weight as well as high properties.

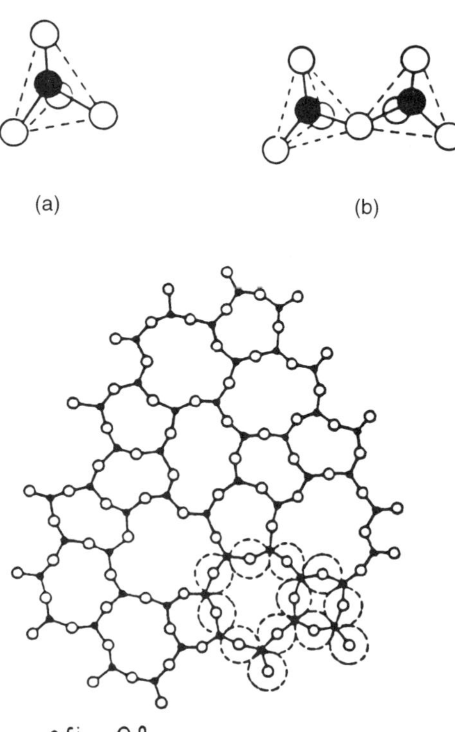

FIG. 16. The structure of glasses is a three-dimensional network based on silica but usually modified by the presence of other elements.

7. BORON AND SILICON-CARBIDE FIBERS ON A SUBSTRATE

The first fiber produced with a much increased Young's modulus compared with glass fibers was the boron fiber, made by chemical vapor deposition onto a substrate that was a tungsten wire. Boron, the fifth element in the periodic table, is the lightest element with which it was found practical to make fibers. The first boron fibers were produced in the U.S.A. and then in France and the U.S.S.R. at the beginning of the 1960s. They had remarkable properties, with a Young's modulus exceeding 400 GPa and quite extraordinary strength in compression.

Table 3. Typical compositions of various glasses used in fiber production; values given as percentages.

Glass type	E	S	R	A	C	D
SiO_2	54	65	60	72	65	74
Al_2O_3	15	25	25	0.7	4	
CaO	18		9	10	14	0.2
MgO	4	10	6	2.5	3	0.2
B_2O_3	8					
F	0.3					
Fe_2O_3	0.3					
TiO_2						0.1
NaO_2				14	8	1.2
K_2O	0.4				0.5	1.3
SO_3				0.8		
B_2O_3					5.5	23

This latter quality was enhanced by their large diameter of 140 μm. The manufacturing technique requires that each fiber be made separately and at a low speed. A mixture of boron trichloride (BCl_3) and hydrogen (H_2) heated to around 1000 °C passes over a continuous tungsten filament, having a diameter of around 13 μm, heated by electric current. Boron is deposited preferentially in the form of nodules onto the irregularities on the tungsten filament, which is produced by a wire-drawing process. The nodular structure of the fiber results in its having a surface that is not smooth, and the stress concentrations produced on the surface at the intersection of two nodules have been shown by Vega-Boggio and Vingsbo (1976) to be a factor in determining the fiber strength. Other defects associated with the core–mantle interface can also limit the strength of the fiber. Figure 17 shows a cross section of a boron fiber revealing its composite structure. The penetration of the small boron atoms into the tungsten core causes it to expand. This expansion is resisted by the boron mantle so that the core is put into compression and the mantle into tension. The cooling of the boron filament after its production induces a compressive stress into the surface.

The boron fiber shown in Fig. 17 has been additionally coated to protect it during the manufacture of a metal-matrix composite. Boron fibers proved to be a good reinforcement for light-metal alloys, particularly when the fiber was protected by a coating of SiC or B_4C. The American space shuttle is composed of a skeleton made up of more than 200 tubes of boron-fiber–reinforced aluminum, but the very high cost of the fiber has restricted wider use. Most of the applications originally foreseen for boron fibers have been taken over by carbon fibers, although they are used in some sports goods.

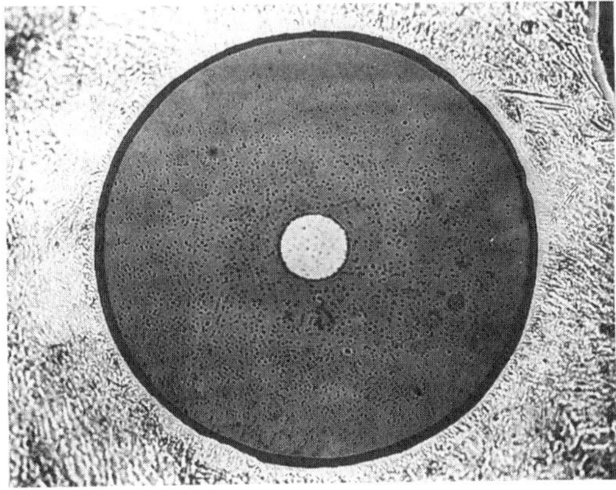

FIG. 17. Boron fibers are composed of a boron mantle deposited onto a tungsten core.

Silicon-carbide fibers, produced by a similar technique to that used to make boron fibers, remain a possible reinforcement for metal-matrix composites including titanium and intermetallic matrix composites. These fibers are based on materials that are readily available, and the substrate can either be tungsten wire or a carbon filament, which is cheaper. These SiC fibers have very good mechanical properties, with a Young's modulus of 412 GPa, and can be used at high temperature, around 1000 °C, although reactions between the fibers and the matrices they are reinforcing may limit their temperature range in a particular system. This production route for making fibers remains, however, very expensive and it is difficult to imagine a major development of this type of fiber.

8. CARBON FIBERS

Carbon fibers made by carbonizing fibers obtained from bamboo were the first filaments used in Edison's incandescent electric lamps, but they were extremely brittle and were rapidly replaced by tungsten wire. Many fibers can be converted into carbon fibers, as shown by Sitting (1980), the basic requirement being that the precursor fiber carbonizes rather than melts when heated. The route adopted in the U.S.A. in the 1950s and early 1960s to producing fibers from the next lightest element after boron was to use rayon, regenerated cellulosic fibers. This proved a slow process that did not yield spectacular results. The approach taken in Great Britain and Japan was to use polyacrylonitrile (PAN) fibers as precursors for making carbon fibers. This route, based on a modified form of acrylic textile fibers, was to prove successful and is the technique used today to make by far the greatest number of carbon fibers. The properties of carbon fibers made by this route depend on the temperature of pyrolysis used, so that it is possible to produce a family of carbon fibers with different structures. Figure 18 shows the routes adopted to make carbon fibers from both PAN and pitch precursors, as described by Toray Industries, the largest carbon-fiber manufacturer. Figures 19 and 20, taken from the results of Fitzer and Heine (1988), show how the strength of the fibers reaches a maximum at around 1500 °C whereas the elastic modulus increases continuously with increasing temperature up to around 2800 °C. The polyacrylonitrile molecule structure is linear, and conversion into carbon fiber requires its conversion into a three-dimensional structure and the elimination of all atoms except the carbon atoms. This is achieved by first heating the fiber to about 250 °C under a tensile load in air to produce a change in color from white to black and the cross linkage of the structure. Further heating in an inert atmosphere to above 1000 °C eliminates most of the elements other than carbon. Above 1500 °C the presence of elements other than carbon is negligible. The structure that results from the pyrolysis of PAN is highly anisotropic, with the basic structural units formed by the carbon-atom groups aligned parallel to the fiber axis. There is complete rotational disorder in the radial direction, and the relatively poor stacking of the carbon atoms means that a graphite structure is never achieved. For this reason it is incorrect to refer to carbon fibers made from PAN as graphite fibers. The structure contains pores, which account for the density of the fibers being less than that of densified carbon.

Table 4 shows the properties of a range of fibers used in composite materials. The values given for the carbon fibers are typical; however, there exists a considerable range of properties for these fibers.

An alternative route for making carbon fibers is from pitch obtained from either the residue of oil refining or the coking of coal. The high carbon yield of pitch, which approaches 90%, makes it an attractive and cheap source for making carbon-fiber precursors. Cost, however, is increased by the purification processes that are necessary before spinning. The pitch is converted into a mesophase or liquid-crystal solution, which is then spun, giving precursor fibers with aligned microstructures. From this point the carbon-fiber fabrication route is the same as that for the PAN-based fibers. However, the response to heating of the pitch-based precursors is not the same. There is no peak in strength at 1500 °C as this is due in the PAN-based fibers to the elimination of nitrogen, which is followed by the growth of carbon structural units resulting in a fall in strength. Pitch-based precursors heated to around

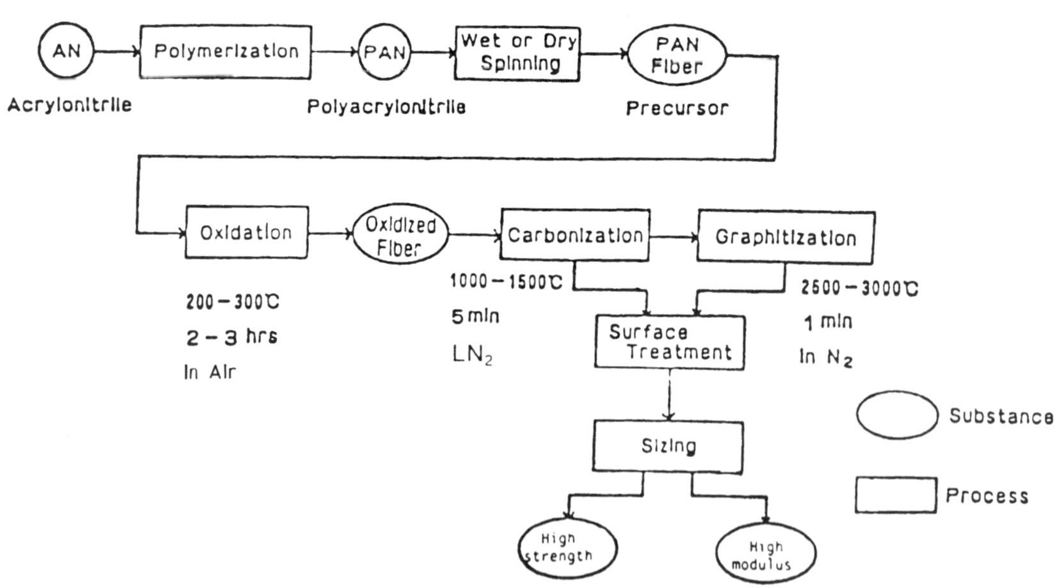

FIG. 18. Schematic flow diagram of the production of carbon fibers from PAN precursors (Toray information).

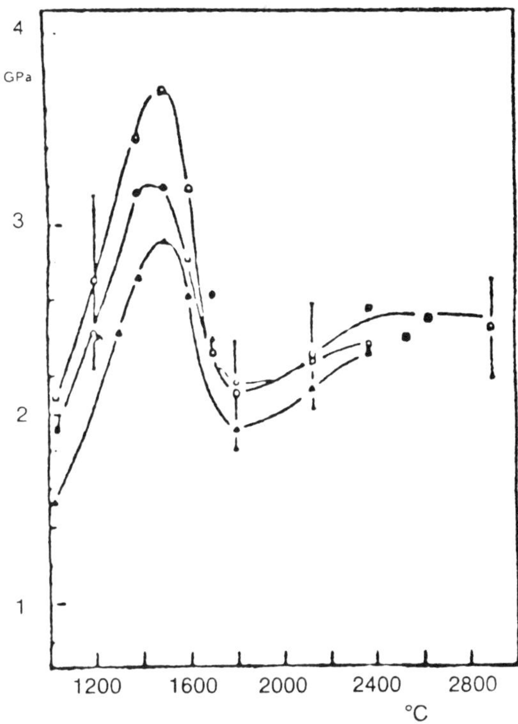

FIG. 19. The strength of carbon fibers produced from PAN depends on the heat-treatment temperature. The three curves were obtained with slightly different cross-linking temperatures but all show a maximum strength at 1500 °C (Vegga-Boggio and Vingsbo, 1977).

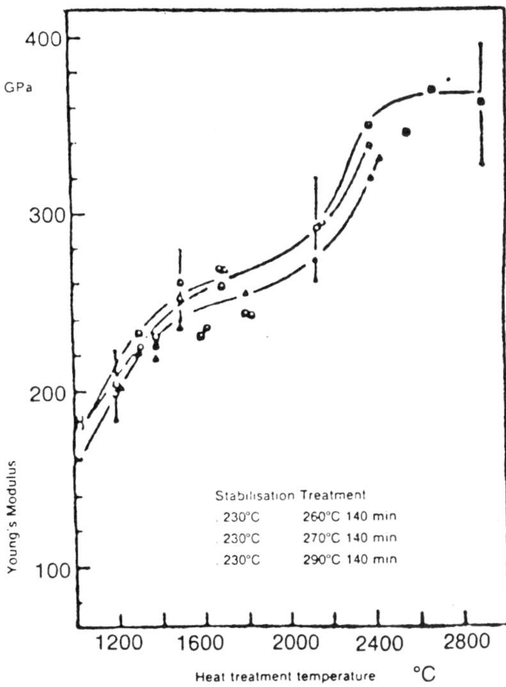

FIG. 20. The Young's modulus of carbon fibers produced from PAN increases with temperature (Fitzer and Heine, 1988).

2300 °C give fibers with Young's moduli as high as those obtained with PAN-based fibers at 2900 °C, and heating to these higher temperatures gives even greater stiffness. It is therefore more economical to produce high-modulus carbon fibers from pitch than it is from PAN. The properties are due to a less disordered, more graphitic microstructure of the pitch-based fibers, which, however, leads to lower compressive strength, as can be seen in Fig. 21.

9. ALUMINA FIBERS

Alumina is usually produced by heating aluminum hydroxide or other aluminum salts, such as ammonia alum, to above 1000 °C in air. During the heating process, the starting materials lose water and other fugitive components at 400–500 °C and thus become aluminum oxide (Al_2O_3) in chemical composition. They undergo several transformations of crystal structure, and finally become the most stable alumina called α-alumina. All other transitional aluminas are transformed into α-alumina at 1000–1100 °C. As a result of the dense packing and high regularity, α-alumina occurs as large crystal grains, generally larger than 0.1 μm; has a high degree of crystallinity, high density (3.97×10^{-3} kg m^{-3}), and a stable and inactive surface; and is strong and hard.

The first small-diameter continuous ceramic fiber, which became available in 1979 from Du Pont, consisted of almost pure α-alumina and was called Fiber FP, as described by Dhingra (1980). This fiber was made by blending the alumina in powder form with water and other compounds to form a mixture that could be spun. The filaments were then extruded through spinnerets, drawn, and fired in two steps up to 1300 °C to produce an α-alumina fiber. Properties of the fiber are shown in Table 4. It can be seen that the fiber had a low strain to fail-

Table 4. Comparison of fiber properties.

Fiber	Composition	Diameter (μm)	Specific gravity	Young's modulus E (GPa)	Strength $\bar{\sigma}$ (GPa)	Strain to failure (%)	Specific modulus E/ρ (GPa)	Specific strength σ/ρ (GPa)
Polyamide 66	PA66	20	1.2	<5	1	20	4	0.8
Polyester	PET	15	1.38	<18	0.8	15	13	0.6
Kevlar 49	PPTA	12	1.45	135	3	4.5	93	2.1
Polyethylene	PE	38	0.96	117	3	3.5	120	3.1
E glass	SiO_2	5–25	2.54	72	2	2.7	28	0.8
S glass	etc.	5–15	2.48	85	4.8	5.3	24	1.9
Boron-W	B, W	140	2.5	400	3.6	0.9	160	1.4
SiC–C	SiC, C	140	3.2	425	3.7	0.9	133	1.2
Short alumina silicate	97 Al_2O_3, 3SiO_2	3	3.3	300	2	0.7	91	0.6
Alumina	α-Al_2O_3	19	3.92	414	1.6	0.4	97	0.4
Alumina silicate	85 Al_2O_3, 15SiO_2	12–17	3.25	210	1.8	0.85	65	0.6
Saphikon	α-Al_2O_3	130	3.95	350	2.4	0.7	89	0.6
Nicalon	SiC, C, O, Si	14	2.55	220	3.0	1.4	86	1.18
Hi-Nicalon	SiC, C	14	2.74	270	2.8	1	99	1.02
Tyranno M	Si, Ti, C, O	8.5	2.37	190	3.3	1.7	80	1.39
High-strength carbon	C	5	1.82	230	4.8	2.1	126	2.64
High-modulus carbon	C	5	1.94	588	3.9	0.7	303	2.01

ure, which made it difficult to handle and limited its use. However, the FP fiber retained its strength well up to 1000 °C, showing almost no loss after 300 h when heated to this temperature in air. Above these temperatures, however, grain growth occurred making the fiber even more brittle and generally weakening it. The FP fiber is not produced commercially.

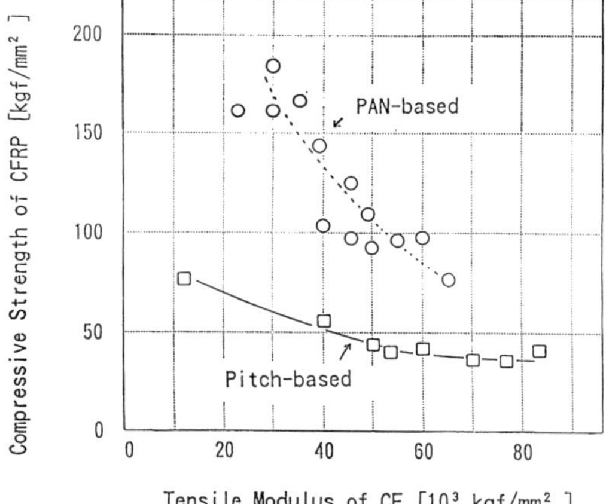

FIG. 21. The compressive strength of pitch-based carbon fibers is lower than that of PAN-based carbon fibers (Toray information).

In order to produce a fiber that could be a candidate for reinforcing high-temperature composites, Du Pont developed a fiber known as PRD-166. This fiber was described as being microcrystalline based on α-alumina but containing approximately 20 wt% of zirconia partially stabilized with yttrium oxide. The largest α-alumina grains were of the order of 0.5 μm. The zirconia showed two size populations with average sizes being 0.1 μm and 10 nm, which a study by small-angle x-ray diffraction has revealed to be in a tetragonal form.

The role of zirconia is to stabilize the structure, so improving strength and toughness. Stress concentrations associated with micropores produced a phase change in the zirconia from tetragonal to monoclinic. The associated increase in volume compressed the crack tip and impeded crack propagation. This resulted in an increase in strength although the fiber still remained brittle. The lower Young's modulus of the zirconia reduced slightly the rigidity of the PRD-166 fiber compared with the FP fiber. The PRD-166 fiber has not been made available commercially.

The properties of two other alumina fibers are shown in Table 4. Both the ICI Safimax fiber and the Sumitomo fiber contain silica, which inhibits grain growth, but in differing amounts. The greater the percentage of silica the greater the strength of the fiber but the lower the Young's modulus.

The γ-alumina fiber from Sumitomo Chemicals contains 15 wt% of silica and, as Table 4 shows, this leads to a much lower Young's modulus than that of the α-alumina fibers but a greater strain to failure and therefore easier handlability. The alumina is crystallized in the γ phase with a very fine grain size, a low density, and an active surface. The role of the silica is described as being to stabilize the alumina in the γ phase. These fibers typically show a Weibull modulus of less than 7. The fibers conserve their strength and stiffness at least up to 800 °C with little noticeable loss in tensile properties at 1000 °C. Above this temperature the tensile properties fall and at temperatures above 1150 °C the fiber deforms plastically. The fiber is observed to creep above 1000 °C. Up to 1100 °C the behavior is that of a material creeping by diffusion at grain boundaries. Above this temperature, softening of the silica seems to account for the rapid acceleration of the creep process.

The Saffil fiber is a staple fiber with a diameter of 3 μm produced by ICI since the early 1970s and consists of about 97% alumina and 3% silica. Such fibers are produced by solution spinning, which leads to little scatter in diameter. The solution consists of an aluminum salt, such as aluminum chlorohydroxide $AlC_x(OH)_y$, which contains hydrated species that eliminate water to link octahedral units by Al–O–Al bridges. As the process continues, a viscous solution is obtained, which can be spun into fibers. This is helped by the addition of organic polymers to control rheology and to lower the spinning temperature. Progressive increases in temperature burn off the organic material and convert the aluminum salt to alumina. The Saffil fiber was originally produced as a refractory felt.

The widest use of the Saffil-type fiber in composites is in the form of a mat that can be shaped into the form desired and then infiltrated with molten metal, usually aluminum alloy. It is probably the most successful fiber reinforcement for metal-matrix composites.

The Nextel series of fibers produced by 3-M are all based on mullite with varying percentages of boron oxide. The fibers are oval in cross section but a study by image analysis gives an average diameter of 11 μm. The Nextel 440 fiber has been found by x-ray diffraction to consist of two crystalline phases, one of mullite and the other of γ-alumina. The more crystalline Nextel 480 contains only mullite as a crystalline phase and much reduced boron oxide. The grain size in this latter fiber is 50 nm and the structure is found not to evolve up to 1200 °C except for the completion of the mullite crystallization. The manufacturer reports that the Nextel fibers conserve at least 75% of their properties up to 1000 °C and that they creep over 1000 °C.

The ceramic fibers based on alumina seem to be possible candidates for reinforcing light alloys with the observation, however, that their specific gravities are quite high. Such reinforcements permit metal structures to be made with enhanced properties, notably creep resistance at high temperatures. The phase change to α-alumina above about 1150 °C and subsequent grain growth would seem to exclude these fibers from most ceramic-matrix composite materials. A solution to this problem would be the use of continuous single-crystal fibers, which would contain no grain boundaries. Such fibers are produced in the form of continuous α-alumina fibers grown from the molten ceramic under the commercial name Saphikon. These fibers have diameters of around 100 μm.

10. CERAMIC FIBERS FROM ORGANIC PRECURSORS

The work of Yajima *et al.* (1987) in Japan was first published in Japan in the mid-1970s and gave rise to the first and so far the most successful fine ceramic fiber based on silicon carbide. This fiber is produced by Nippon Carbon under the name of Nicalon. The manufacture of this fiber involves the production of a polycarbosilane precursor fiber, which consists of cycles of six atoms arranged in a similar manner to the cubic structure of β-SiC.

The molecular weight of this polycarbosilane is low, around 1500, which makes drawing of the fiber extremely difficult. In addition, methyl groups (CH_3) in the polymer are not included in the Si–C–Si chain so that during pyrolysis the hydrogen is driven off, leaving a residue of free carbon. The precursor fibers are subjected to heating in air at about 300 °C to produce cross linking of the structure. This oxidation makes the fiber infusible but has the drawback of introducing oxygen into the structure, which remains after pyrolysis. The ceramic fiber is obtained by a slow increase in temperature in an inert atmosphere up to 1300 °C. The fiber that is obtained contains a majority of SiC but also significant amounts of free carbon and excess silicon combined with oxygen. The distribution of these elements is uniform across the fiber diameter, as has been shown by electron-microprobe analysis.

The strength and Young's moduli of Nicalon fibers tested in air or an inert atmosphere show little change up to 1000 °C. Above this temperature, both these properties fall with the greatest change being in the strength of the fiber.

Nicalon fibers as supplied now by the manufacturer are microcrystalline with typical grain sizes of the β-SiC crystals of around 2 nm, as is revealed by x-ray diffraction and transmission-electron microscopy. Heating above 1100 °C produces grain growth, which stabilizes with an average grain size of approximately 3 nm, at which point grain growth appears to stop. Under no load the fiber shrinks during such a heat treatment but the shrinkage stops with the stabilization of the grain size.

Simon and Bunsell (1984) have shown that when a load is applied to the fibers, it is found that a creep threshold stress exists above which creep occurs. The primary creep phase is clearly greatly influenced by the structural changes occurring. The ceramic grade 200 series Nicalon fibers, which evolved from earlier fibers, has been heat-treated in order to stabilize grain size at around 3 nm and improve the properties at high temperature.

The β-SiC crystals are embedded in an amorphous matrix consisting of Si, O, and C, which also contains aggregates of free carbon with sizes of around 2 nm. A typical composition by weight of a Nicalon fiber is 65% SiC, 15% free carbon, and an amorphous phase consisting of Si, O, and some carbon. The free carbon and the oxygen in the structure have been shown to play a determining role in the use of Nicalon fibers at high temperatures.

The existence of the creep stress thresholds is attributed to the presence of the particles of carbon in the structure, which impede its movement and rearrangement. In this case the carbon particles confer a useful property to the fiber, but their presence, together with the oxygen in the structure, finally limits the use of the fiber at high temperature. Degradation of Nicalon fibers is slower in air than in an inert atmosphere as a result of a silica coating being created at the fiber surface. This oxide coating is considered to slow the expulsion of the oxides of carbon produced in the interior of the fiber.

A later generation of these fibers known as Hi-Nicalon has been developed using the same precursor but using electron-beam radiation rather than oxidation to produce the cross linking of the precursor. The resulting fiber has a higher Young's modulus (~240 GPa) and higher strength.

Although the Nicalon fibers are the most widely available fine ceramic fiber based on silicon and carbon, the technology of this process of fiber production is being actively explored in many laboratories by a number of routes. The Tyranno fiber produced by Ube Industries is made by a process analogous to that used for the Nicalon fiber, but with significant differences. The manufacture of the Tyranno fiber begins with the production of polydimethylsilane, $(Si(CH_3)_2)_n$, as for the Nicalon fiber, which is then mixed with titanium alkoxide in a ratio of approximately 1:10, heated to 340 °C in nitrogen, and polymerized for 10 h (Gupta, 1988). The reaction product is concentrated at 310 °C for 2 h in a stream of nitrogen to produce polytitanocarbosilane. The molecular weight of this organo-metallic polymer is low, around 1500. The structure of the polytitanocarbosilane precursor is complex as a result of condensation of Si–H bonds and cross linking of titanium compounds occurring simultaneously. The precursor fiber is converted into a ceramic fiber by heating in air at about 180 °C and is then heat-treated in nitrogen at around 1000 °C. The resulting fiber contains a small amount of titanium, around 2% by weight, which is said to inhibit crystallization, and similar percentages of Si, O, C, and SiC as the Nicalon fiber. The Tyranno fiber has a diameter of around 11 μm.

The Tyranno fiber retains its strength up to 1200 °C, at least for short exposure times, although above this temperature the structure of the fiber changes to consist of β-SiC and crystallized silica.

A form of Tyranno fiber known as Tyranno LOX-E has been developed using the same electron-beam cross-linking process as mentioned for the Nicalon fiber. The Tyranno LOX-E fiber contains 4.7% oxygen.

A group of ceramic fibers that are being actively studied is based on a silicon carbide–silicon nitride system and is produced from polycarbosilazane precursors. Various routes for the manufacture of these fibers are possible, as has been shown by Lipowitz *et al.* (1987), and as yet these fibers are at the laboratory or early pilot-plant stage. The fibers can be produced by the pyrolysis of polysilazanes prepared from tri(*N*-methylamino) methylsilane polymerized by heating to 520 °C for several hours and then heated to between 1200 and 1500 °C for 2 h. During pyrolysis a weight loss of approximately 35% is reported to occur, and finally the Si_xC_2 ceramic fibers that are produced are described as being a shiny black in appearance.

This type of fiber was initially studied for the dielectric properties that it is possible to obtain with this class of material. The electrical resistivity of silicon carbide–silicon nitride is about 10^{12} times that obtained for graphite.

Other approaches to producing fibers based on Si–C–N have been studied and reported in the literature by researchers from Dow Corning. These fibers were prepared by melt spinning of amorphous thermoplastic polymers followed by a cross-linking step to gel the fibers and to make them infusible. Fibers consisting of Si–N–C were derived from hydridopolysilazane (HPZ) polymers prepared by reaction of trichlorosilane and hexamethyldisilazane. The fibers were pyrolyzed in nitrogen or argon at temperatures between 1100 and 1400 °C. The ceramic fibers produced were amorphous.

11. WHISKERS

Single ceramic crystals in the form of filaments are known as whiskers, and it has been known for many years that their mechanical properties can approach those that are theoretically possible. This is because of the lack of defects, which weaken other forms of matter. The requirement that the whiskers be free of defects means that they are extremely small, with diameters of around 1 μm and less and lengths usually not more than a few millimeters. Bigger diameters result in dramatic falls in mechanical properties.

Despite their small dimensions, whiskers have potential as reinforcements because of their aspect ratios of length to diameter, which are great enough to produce a reinforcing effect in plastic matrices. Difficulties of orienting the whiskers in a matrix and especially health hazards due to their size, which mean they can become lodged in the lungs if breathed in and so cause serious illness, have limited their use.

There are a great many possible types of whisker reinforcements. The most widely commercialized whisker type is based on silicon carbide.

12. CONCLUSION

Fibers exist in many forms and have evolved at a remarkable rate during the second half of the twentieth century. Their remarkable properties depend on their microstructure, which is often complex and anisotropic. In terms of strength fibers represent some of the strongest forms of ordinary matter, while their elastic moduli can be nearly as great as that of diamond. Such fibers can be used to reinforce all classes of solids whether they are rubber, plastics, light metals, glasses, or ceramics. The composite materials formed in this way are now and will increasingly become the engineering materials of the modern world.

GLOSSARY

Alumina: Aluminum oxide.

Aramid: A generic term for wholly aromatic polyamides.

Liquid Crystal: A solution in which the molecular structure is locally oriented.

Pitch: Residue of the oil-refining process or the coking process of oil.

Polyacrylonitrile (PAN): A polymer that can be spun to form fibers, which in turn can be converted to carbon fibers by pyrolysis.

Regenerated Fibers: Those artificial fibers that are made from naturally occurring cellulose.

Synthetic Fibers: Those of which their macromolecular structure is manmade.

Works Cited

Bashir, Z., Odell, J. A., Keller, A. (1984), *J. Mater. Sci.* **19,** 3713–3725.

Black, W. B., Preston, J. (1973), *High Modulus Wholly Aromatic Fibers,* New York: Marcel Dekker.

Bunsell, A. R. (Ed.) (1988), *Fibre Reinforcements for Composite Materials,* Amsterdam: Elsevier.

Bunsell, A. R., Hearle, J. W. S. (1974), *J. Appl. Polym. Sci.* **18,** 267–291.

Cox, H. L. (1952), *Brit. J. Appl. Phys.* **3,** 72–79.

Dessain, B., Moulaert, O., Kennings, R., Bunsell, A. R. (1992), *J. Mater. Sci.* **27,** 4515–4522.

Dhingra, A. K. (1980), *Philos. Trans. R. Soc. London, Ser. A* **294,** 411–417.

Fitzer, E., Heine, M. (1988). "Carbon Fiber Manufacture and Surface Treatment," in: A. R. Bunsell (Ed.), *Fibre Reinforcements for Composite Materials,* Amsterdam: Elsevier, Chap. 3, pp. 73–148.

Griffith, A. A. (1920), *Philos. Trans. R. Soc. London, Ser. A* **221,** 163–198.

Gupta, P. K. (1988), "Glass Fibers for Composite Materials," in: A. R. Bunsell (Ed.), *Fibre Reinforcements for Composite Materials,* Amsterdam: Elsevier.

Inglis, C. E. (1913), *Trans. Inst. Naval Architecture* **55,** 219–230.

Lipowitz, J., Freeman, H. A., Chem, R. T., Prack, E. R. (1987), *Adv. Ceram. Mater.* **2,** 121–128.

Moncrieff, R. W. (1979), *Man-made Fibres,* 6th ed., Guildford, U.K.: Butterworth.

Morton, W. E., Hearle, J. W. S. (1975), *Physics of Textile Fibres,* 2nd ed., London: Heinemann.

Simon, G., Bunsell, A. R. (1984), *J. Mater. Sci.* **19,** 3658–3670.

Sitting, M. (1980), *Carbon and Graphite Fibers: Manufacture and Applications,* Park Ridge, NJ: Noyes Dates Co.

Vega-Boggio, J., Vingsbo, O. (1976), *J. Mater. Sci.* **11,** 273–282.

Vegga-Boggio, J., Vingsbo, O. (1977), *J. Mat. Sci.* **12,** 2519–2524.

Yajima, S., Hasegawa, X., Hayashi, J., Iiuma, M. (1987), *J. Mater. Sci.* **13,** 2569–2576.

Further Reading

Bunsell, A. R. (1988), *Fibre Reinforcements for Composite Materials,* Amsterdam: Elsevier.

Moncrieff, W. E. (1979), *Man Made Fibres,* Guildford, U.K.: Butterworth.

Kelly, A., Macmillan, M. H. (1986), *Strong Solids,* Oxford, U.K.: Clarendon.

Lee, S. M. (Ed.) (1991), *International Encyclopedia of Composites,* New York: VCH.

Kelly, A. (Ed.) (1994), *Concise Encyclopedia of Composite Materials,* Oxford, U.K.: Pergamon-Elsevier.

Brook, R. J. (Ed.) (1991), *Concise Encyclopedia of Advanced Ceramic Materials,* Oxford, U.K.: Pergamon.

RELATIVITY

See GRAVITATION AND GENERAL RELATIVITY

RELATIVITY, SPECIAL

ROMUALDO TABENSKY, *Departamento de Física, Facultad de Ciencias Físicas y Matemáticas, University of Chile, Santiago, Chile*

	Introduction	369
1.	**Lorentz Transformations**	370
1.1	Derivation	370
1.2	Examples	371
2.	**Mathematical Tools**	372
2.1	Line Element and Poincaré Transformations	372
2.2	4-Vectors and Tensors	373
2.3	Fields	375
3.	**Relativistic Laws**	375
3.1	Particle Kinematics	375
3.2	Fermi-Walker Transport	375
3.3	Uniform Acceleration	376
3.4	Pure Boosts	376
3.5	Thomas Precession	376
3.6	Doppler Effect	377
3.7	Classical Interactions	378
4.	**Experimental Evidence**	380
	Glossary	382
	Works Cited	382
	Further Reading	382

INTRODUCTION

Understanding of Nature has always come slowly and hard. Relativity is no exception. It all began with Galileo Galilei's clear thinking and careful observation about four hundred years ago. He detected a feature shared by all laws of Nature to a high degree of precision. He detected a simple but general truth from among the immensely complex natural phenomena, thus showing the way to all future science. He observed that experiments performed on a ship, making its way through calm waters, produce results wholly independent of the ship's speed. Today we may imagine instead the more impressive situation where a stewardess is pouring a cup of coffee on a transparent aircraft speeding through midair, a simple chore when judged by the passengers, but nevertheless, a feat of high precision when judged by grounded observers. Both see two entirely different solutions to actual laws of nature that ought to exist within the equations of physics. The importance of Galileo's remark lies mainly in the fact that it provides a general requirement every true law of Nature ought to obey, thus narrowing the number of possibilities in the search for truth. However, it must be stressed that, after four hundred years we have as yet no real understanding of why Nature chose to behave in such a nice manner. The relativity principle states that every theory built for Nature will have to contain transformation laws for each and every variable used in it, so that solutions to a given phenomenon, as viewed by different observers, can be mapped into each other. The relativity principle requires no specific form for said transformations, only their existence. In particular, if coordinates are chosen to describe a theory, they will have to be accompanied by some relativistic transformations. Galileo found a consistent set of such transformations that implied adding the aircraft's constant velocity to all other velocities. Today we are aware, from high-energy experiments, of the existence of an upper limit to all velocities. Increasing energy, without limit, would have to be provided in order to accelerate particles toward the enormous, albeit finite, speed of light. If Galileo was right, the speed limit could easily be violated by an observer running in the opposite direction. Actually this speed limit has to be a universal invariant if no preferred reference exists. The existence of a limit to all speeds is not really understood, just accepted as an experimental fact. It is Einstein's merit to have foreseen this very lucidly, much before experiments made it obvious, thus clearing the way for logical reasoning that will be repeated here.

Newton, on the basis of Galileo's insights, created in 1687 his theory of mechanics, which dominated physics up to this century.

3-527-28138-X/96/$5.00 + .50

This was the first general predictive theory about motion ever known to humans. It was a tremendous success since it allowed people to foresee by calculation natural phenomena such as the motion of planets for centuries ahead, or tides on earth. Fantastic power was laid in people's hands. The belief spread that only details remained to be discovered. Notwithstanding, momentum was being built up in the area of electrical phenomena, culminating, in 1860, with the discovery of Maxwell's equations. Incompatibilities with Galileo's relativity appeared here for the first time.

1. LORENTZ TRANSFORMATIONS

1.1 Derivation

The hardest part pertaining to relativity is almost over; the mission is clear and the axioms known. What remains is mostly mathematical reasoning. The first task is to find a group of transformations for time and space coordinates that will preserve the speed of light. These transformations will have to come very close to Galileo's in some appropriate limit, since the latter proved themselves so accurate in Newton's mechanics. It is highly nontrivial that such transformations exist at all, and it is even more conspicuous that they come with enough room to accommodate three independent velocity parameters as well as three rotation parameters, indispensable for specifying all possible observers' inertial frames. The one-dimensional case is easier to grasp and will therefore be considered first: All possible motions are reduced to a single line, thus simplifying the reasoning considerably. Nonetheless the full transformations will be given in the next section.

Space-time coordinates (t,x) are identified with events. They label anything that happens on a point x in space at a time t. This mathematical idealization belongs entirely to Newtonian mechanics and is incorporated without change into relativity. It was Einstein who adopted the tactics to stay as close as possible to the successful Newtonian theory. Coordinates might eventually turn out to be a concept of approximate validity only; nonetheless, all accepted modern theories are fundamentally and inextricably based on them, this being the very reason so much care has been lavished in elucidating their transformation properties. The existence of space-time coordinates is a basic simplifying assumption that should be true for all observers if the relativity principle is to hold. Let our aircraft passengers set up their time coordinates with accurate clocks, and their space coordinates, by bouncing laser beams in an obvious manner. Since the speed of light is assumed to be the same in every frame, it is only natural to measure distances by means of a clock and light rays, thus reducing all space-time measurements to time measurements alone. A pair of real numbers (t',x') will thus be associated, by the passengers, with every event. Exactly the same is done by ground observers who measure (t,x) instead. Ambiguities arise since an origin has to be set in both cases; however, these may be dispelled by choosing any commonly agreed upon event to be labeled as (0,0) by both observers. Moreover, a common time unit has to be set for all inertial frames. It is plain that the situation in which different inertial observers use different units of time is perfectly possible on physical grounds and will show up in subsequent discussions. This point is subtle; it involves deciding when two clocks in relative motion are identical.

Homogeneity of space-time forces Lorentz transformations $(t,x) \leftrightarrow (t',x')$ to be linear, that is, given by

$$t = \alpha t' + \beta x', \qquad x = \gamma t' + \delta x'.$$

Demand now the invariance of the speed of light, chosen to be 1 in natural units, by setting $x = t$ or $x = -t$ whenever $x' = t'$ or $x' = -t'$ and get

$$t = \alpha t' + \beta x', \qquad x = \beta t' + \alpha x'.$$

The relative velocity V of the primed inertial frame relative to the unprimed one is determined by the locus of events $x' = 0$ and $x = Vt$. This sets $\beta = \alpha V$, thus

$$t = \alpha(t' + Vx'), \qquad x = \alpha(Vt' + x'). \tag{1}$$

Notice that these transformations contain the two independent parameters V and α, the latter being an overall scale involving a genuine velocity V. It was clear from the beginning that a scale factor would remain undetermined since no prescription has been given for comparing time units in the two inertial systems. It is plain, however, that no matter what the prescription is, it will assign a definite value α for each velocity V, so that $\alpha = \alpha(V)$. A naive attempt to set $\alpha = 1$, say, fails since a composition of two such transformations will no longer have $\alpha = 1$, as may be simply checked. The same calculation reveals that the composition of two such transformations, with velocities U and V, yields a new transformation given by

$$x = (1 + UV)\alpha(U)\alpha(V)\left(x' + \frac{U + V}{1 + UV}t'\right),$$
$$t = (1 + UV)\alpha(U)\alpha(V)\left(t' + \frac{U + V}{1 + UV}x'\right). \quad (2)$$

Two important facts emerge from these equations: first, that the relativistic addition of the velocities U and V,

$$\frac{U + V}{1 + UV}, \quad (3)$$

obtains, and second, that the composed transformation is of the same kind as the original one only if

$$\alpha\left(\frac{U + V}{1 + UV}\right) = (1 + UV)\alpha(U)\alpha(V). \quad (4)$$

Thus the unknown function $\alpha(V)$ satisfies a functional relation with a general solution easily checked to be

$$\alpha(V) = \frac{(1 + V)^{p}/(1 - V)^{p}}{\sqrt{1 - V^2}} \quad (5)$$

for arbitrary p. The solution was arrived at by solving Eq. (4) for infinitesimal values of U, thus reducing the problem to a differential one. The relativity transformations associated with any value of p are perfectly consistent; the one-dimensional world they define has, however, a preferred direction unless $p = 0$. This can be seen from just about any physical consideration described in the following. Nevertheless, right-left symmetry may be established directly by requiring Eq. (1) to remain unchanged under the simultaneous replacements $(x,V) \rightarrow (-x,-V)$, getting $\alpha(-V) = \alpha(V)$, which fixes $p = 0$. Lorentz transformations, as well as their inverse, are then arrived at:

$$t = \frac{t' + Vx'}{\sqrt{1 - V^2}}, \qquad x = \frac{x' + Vt'}{\sqrt{1 - V^2}},$$
$$t' = \frac{t - Vx}{\sqrt{1 - V^2}}, \qquad x' = \frac{x - Vt}{\sqrt{1 - V^2}}. \quad (6)$$

For sake of reference, add $y' = y$ and $z' = z$ in three dimensions. The fact that the two sets of equations differ solely by the sign of V means that observers in relative motion see each other with reversed velocities. This, of course, comes as no surprise; however, most other physical consequences have proved to be utterly shocking. So, newcomers, beware! A masterful account of how difficult all this was to assimilate is given by Miller (1981).

1.2 Examples

The Lorentz transformations will now be applied to the following illustrative experiment. Consider a driver D speeding trough a tunnel of length L with velocity V. At the end of the tunnel, the driver bounces back, reversing his velocity, becoming observer D^*. Figure 1 depicts the situation. Vertical lines OF and PE represent the left and right ends of the tunnel as viewed by a grounded observer G. OE is the driver's world line $x = Vt$. The primed frame is chosen to be the driver's rest frame. Line OT is defined by $t' = 0$ and thus corresponds to the driver's space axis. EF describes the driver returning with velocity $-V$. Every event in the figure is perfectly defined; so are their coordinates,

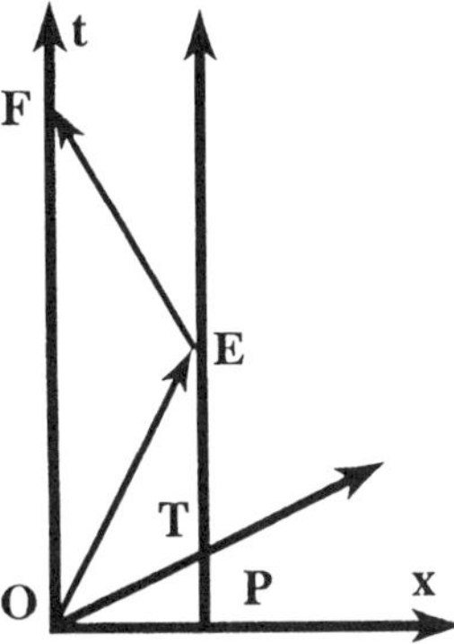

FIG. 1. A drive *OEF* crossing a tunnel with entrance *OF* and ending *PE*.

which are calculated from Lorentz transformations, with results displayed in Table 1. The driver's trip to the end of the tunnel E takes a time L/V as measured by G. We read $t' = L/\gamma V$ for the same event, which means that the driver's own watch recorded a time shortened by a factor γ^{-1} defined by Eq. (23). This striking relativistic result is called time dilation, since the driver's own time, his proper time, gets scaled up by a factor γ when viewed by ground observers. Very little additional effort is required in order to see that a stationary watch OF will measure more time than the driver's watch OEF; the scaling factor is γ once more. Care must be taken not to confuse this last number with the time coordinate of E in the driver's frame! So, if the driver left a grounded twin at the tunnel's left end O, he meets him again at F somewhat younger. Twins are almost never the same age since their histories do not match in every detail. The one who moves most stays younger, albeit only a very small fraction of a second in a lifetime. Now turn to event T with coordinates $(0,L/\gamma)$ in the driver's frame. T is simultaneous to O as perceived by the driver. Therefore $\gamma^{-1}L$ is the length of the tunnel when measured by the driver, in full correspondence to the length L measured by ground observers. So the tunnel appears shorter, by a factor γ^{-1}, a result known as Lorentz contraction. At high speeds the tunnel looks more like a door!

No discussion on Lorentz transformations can avoid mentioning the basic invariance

$$t^2 - x^2 = (t')^2 - (x')^2, \tag{7}$$

which can be checked from Eq. (6). This quadratic form plays a role similar to that played by $x^2 + y^2$ in the x,y Euclidean plane. First, and above all, it measures the square of proper time, which is nothing but the time displayed by a watch. This becomes plain when $x' = 0$ (the coordinate of the watch in its own rest frame) is set into the above equation. The same type of reasoning shows that the negative of the quadratic form yields the square of proper distance. Moreover, this invariance is actually a defining property for Lorentz transformations, since any linear transformation, connected to the identity, that satisfies Eq. (7) may be shown to be given by Eq. (6). This quadratic form is also very easy to generalize to more dimensions as shall shortly be done. From what has been just said the proper time $d\tau$ elapsed between two infinitesimally separated events (t,x) and $(t + dt,\ x + dx)$ is given by the all-important formula

$$d\tau^2 = dt^2 - dx^2.$$

This formula provides the time measured by a watch in arbitrary motion.

Table 1. Data Related to Fig. 1.

t	x		t'	x'
0	L	P	γLV	γL
LV	L	T	0	L/γ
L/V	L	E	$L/\gamma V$	0
$2L/V$	0	F	$2\gamma L/V$	$-2\gamma L$

2. MATHEMATICAL TOOLS

2.1 Line Element and Poincaré Transformations

Einstein's physical arguments opened the way to the use of very powerful geometrical tools suggested two years later by Minkowski. These are the perfect analog to Euclidean-space geometric tools, such as vectors and tensors, as well as scalar and vector products. Instead of the Euclidean distance between two adjacent points given by Pythagoras, the proper time $d\tau$ of two adjacent events separated by (dt,dx,dy,dz) is taken be given by

$$d\tau^2 = dt^2 - dx^2 - dy^2 - dz^2. \tag{8}$$

This all-important formula, known as the line element, provides the time measured by a clock in its arbitrary voyage through space. It has been already derived in two dimensions, and can be similarly derived in four dimensions as well (Pauli, 1958). Nevertheless the task shall be avoided for lack of space. The very existence of accurate timekeeping devices points to some basic undiscovered physics. However, experimental confirmation is strong. Modern atomic clocks are capable of detecting fractions of a second in a century, and their very calibration, from city to city, requires the use of Eq. (8). In fact, the most conspicuous difference from Galileo's relativity lies in that two watches departing from a single event and arriving at another, via different world lines, display different times, as obvious from the same Eq. (8).

Transformations $(t,x) \to (t',x')$, which preserve the line element, are known as Poincaré transformations; they constitute one of the most basic symmetry laws of nature. These transformations are exhausted by constant time and space displacements, as well as linear Lorentz transformations. An elementary proof is given by Weinberg (1972). It is simple to check, however, that the transformations

$$x'^{\mu} = \delta^{\mu} + \Lambda^{\mu}_{\nu}x^{\nu}, \tag{9}$$

with Λ^{μ}_{ν} constrained by

$$\eta_{\mu\nu} = \Lambda^{\alpha}_{\mu}\Lambda^{\beta}_{\nu}\eta_{\alpha\beta}, \tag{10}$$

preserve proper time $d\tau$. The following conventions have been adopted: Indices range from 0 to 3, repeated upper and lower indices imply summation, and finally, the Minkowski tensor $\eta_{\mu\nu}$ is defined by $\eta_{00} = 1$, $\eta_{11} = \eta_{22} = \eta_{33} = -1$ and zero otherwise. Poincaré transformations are thus parametrized by the four displacements δ^{μ}, and by Λ^{μ}_{ν}. This latter 4×4 matrix, once forced to satisfy Eq. (10), will constitute a six-parametric family, just right to accommodate three velocity components and three rotation angles necessary to define an arbitrary inertial frame. Even though the explicit solution to this quadratic equation is not particularly useful, it will nonetheless be provided. Write Eq. (10) in matrix form

$$\tilde{\Lambda}\eta\Lambda = \eta, \tag{11}$$

where Λ stands for the 4×4 matrix built out of Λ^{μ}_{ν}, the upper index conventionally chosen to be Λ's first index for matrix multiplication purpose, and the tilde denotes transposition. Check that a product of two solutions is also a solution. This suggests the possibility of building finite solutions by multiplying infinitesimal ones infinite times, a popular trick in Lie group theory. Therefore let $\Lambda = 1 + \Delta$ be an infinitesimal solution; then

$$\eta\tilde{\Delta} + \Delta\eta = 0, \tag{12}$$

or

$$\Delta_{\mu\nu} = -\Delta_{\nu\mu}, \tag{13}$$

which is a solvable constraint on $\Delta_{\mu\nu}$ defined by $\Delta_{\mu\nu} = \eta_{\mu\sigma}\Delta^{\sigma}_{\nu}$. It is therefore plausible that

$$\lim_{n\to\infty} (1 + \Delta/n)^n = \exp\Delta \tag{14}$$

satisfies Eq. (11) too, a fact that, in turn, can be rigorously shown by means of Eq. (12) written in the form $\tilde{\Delta} = -\eta^{-1}\Delta\eta$. This provides the set of all Lorentz transformations connected to the identity, known as proper Lorentz transformations. A more transparent form to this solution will be mentioned later.

2.2 4-Vectors and Tensors

The main task of relativity is the construction of invariant laws of nature. Tensors are the appropriate mathematical entities to be used in this endeavor, for they obey the simplest transformation properties. 4-vectors are simple tensors with components denoted by a^{μ} and pictured by arrows joining two events in space-time. They transform from one inertial frame to another according to

$$a'^{\mu} = \Lambda^{\mu}_{\nu}a^{\nu},$$

thus leaving the scalar product defined by

$$\boldsymbol{a}\cdot\boldsymbol{b} = \eta_{\mu\nu}a^{\mu}b^{\nu} = a^0b^0 - a^1b^1 - a^2b^2 - a^3b^3 \tag{15}$$

invariant. This product shares the same algebraic properties as the ordinary scalar product in three dimensions, as well as analogous geometric properties. The main difference lies in that there exist real 4-vectors with negative or zero squares. However, physical interpretation exists in all cases, namely, timelike 4-vectors ($\boldsymbol{a}\cdot\boldsymbol{a} > 0$) are those along which particles can travel at speeds less than that of light, whereas spacelike 4-vectors ($\boldsymbol{a}\cdot\boldsymbol{a} < 0$) correspond to higher, nonachievable speeds. Photons travel along lightlike vectors ($\boldsymbol{a}\cdot\boldsymbol{a} = 0$). The set of all lightlike 4-vectors, also called null vectors, that emerge from an event is known as the light cone. For any of these 4-vectors a frame exists where they take the following canonical values:

$$
\begin{aligned}
&\boldsymbol{a}\cdot\boldsymbol{a} > 0, \quad && \boldsymbol{a} = \delta(1,0,0,0),\\
&\boldsymbol{a}\cdot\boldsymbol{a} = 0, \quad && \boldsymbol{a} = \delta(1,1,0,0),\\
&\boldsymbol{a}\cdot\boldsymbol{a} < 0, \quad && \boldsymbol{a} = \delta(0,1,0,0).
\end{aligned} \tag{16}
$$

The scalar product measures a host of important quantities like proper time, proper lengths, angles, and even speeds, all of which can be understood with help of the canonical forms. A beautiful argument will now be provided to show that mathematical orthogonality, for example, describes physical simultaneity. Consider Fig. 2. EA is an observer that emits light at E, recapturing it at A after reflection off N. Since the speed of light is a universal constant, the time it takes for light to travel toward N equals the time it takes from N to A, as viewed by observer EA. M is the midpoint between E and A. M is therefore simultaneous to N. The squaring of $EN = EM + MN$ and $NA = MA - MN$ renders $EM = MA$ orthogonal to MN, because $EN^2 = NA^2 = 0$. So the set of simultaneous events to M is the hyperplane orthogonal to the observer's time axis. This reinforces the fact that simultaneity is highly observer dependent.

In order to define tensors, think of some linear map T whose component form, in some frame, is given by

$$a_T^\mu = T^\mu_\nu a^\nu,$$

where $\boldsymbol{a}_T = T(\boldsymbol{a})$. The same map T, in a primed frame, takes the values

$$T^{\mu\prime}_\nu = \Lambda^\mu_\sigma (\Lambda^{-1})^\tau_\nu T^\sigma_\tau.$$

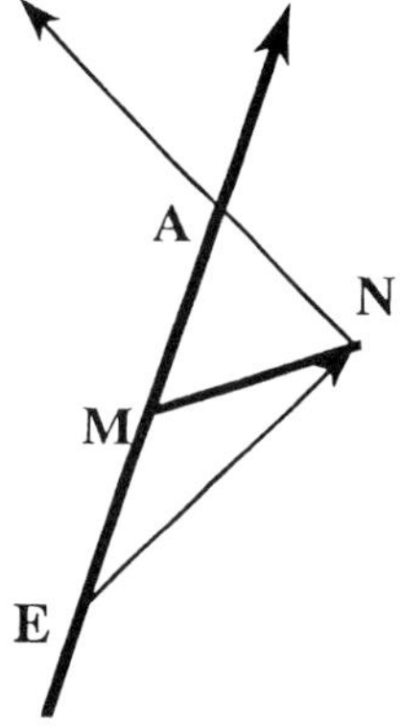

FIG. 2. Observer EMA bounces light ENA off N. M is the midpoint of EA.

So T in its upper index transforms via Λ coefficients, whereas in its lower index it transforms with the matrix inverse coefficients. The upper index is said to be contravariant, the lower covariant. The index position carries information about its transformation properties. A vector is a once contravariant tensor. Tensors may be many times contravariant as well as covariant. Homogeneous transformations are crucial, if only because the vanishing of a tensor is a frame-independent statement, and thus a powerful tool for writing laws in physics—so much so, that all known laws take precisely this form in some representation of the Lorentz group. Tensors obey too many useful properties to be listed here. For example, summation of an upper index with a lower one yields a new tensor that transforms as if the two indices never existed. The matrix inverse of a doubly covariant tensor is doubly contravariant and vice versa. Products of tensors produce higher-order tensors, and so on.

Please look at Eq. (10), which defines the Lorentz group. The right-hand side is the transform of $\eta_{\alpha\beta}$ considered as doubly covariant; the left-hand side is $\eta_{\mu\nu}$ again. These invariant tensors are very scarce and precious; their very existence allows for invariant physical and geometrical laws. Invariant tensors appear as coefficients in the equations of physics, thus preserving them in diverse frames. All invariant tensors are known to be algebraic combinations of $\eta_{\mu\nu}$, $\eta^{\mu\nu}$, δ^μ_ν, and $\epsilon_{\mu\nu\lambda\rho}$. The second doubly contravariant tensor is the matrix inverse of the first; the four times covariant Levi–Civita tensor is completely antisymetric with $\epsilon_{0123} = 1$. It should be recalled, however, that the Levi–Civita tensor transforms as such only when the determinant of Λ^μ_ν equals 1, which is true for proper transformations. These were the only tensors used in building laws of Nature until Dirac introduced spinors in 1932. Actually, spinors were known to mathematicians before that, a fact that in no way lessens Dirac's supreme achievement. Many a time it is harder to select physically relevant facts from the truly vast amount of mathematical results than it is to rediscover them. However, today it is possible to search successfully for all invariant tensors of any representation of the Lorentz group. The highly developed theory of Lie group representations provides the necessary tools (see

GROUP THEORY). The enormous literature, as well as profound theorems, may be retrieved from the excellent encyclopedic book of Barut and Rączka (1977). Most relevant facts about spinors are to be found in the outstanding book by Penrose and Rindler (1984).

2.3 Fields

Physical arguments shall be provided later in favor of a field description of nature. Meanwhile, pertinent definitions are in place. A field is a set of real or complex numbers assigned to each event x^μ. These numbers have diverse transformation laws. Fields may be scalar, vector, spinor, or tensor in character. A scalar field $\Phi(x^\mu)$ transforms as follows:

$$\phi'(x'^\mu) = \phi(x^\mu), \tag{17}$$

whereas a covariant vector field transforms as

$$A'_\nu(x'^\mu) = (\Lambda^{-1})^\lambda_\nu A_\lambda(x^\mu), \tag{18}$$

and so on. It can be easily checked that partial derivatives add an extra covariant index to any tensor field, with the result that

$$\phi_{,\mu} = \frac{\partial \phi}{\partial x^\mu} \tag{19}$$

is a covariant vector field if Φ is a scalar. The same holds for tensor fields. A relativistic partial differential equation can be immediately written, namely

$$\eta^{\mu\nu}\phi_{,\mu,\nu} = 0,$$

which is nothing but the standard wave equation. The simplest of all relativistic equations already correctly describes physical phenomena.

3. RELATIVISTIC LAWS

3.1 Particle Kinematics

Particles are described by curves in spacetime. The very same Newtonian concept is valid in relativity. However, a modified parametrization becomes useful. The fact that time and space coordinates mix under Lorentz transformations makes it convenient to treat them as uniformly as possible. So time and space coordinates will be given as functions of proper time, the only natural parameter in sight, that is $x^\mu = x^\mu(\tau)$.

The 4-velocity and 4-acceleration are in turn defined by

$$V^\mu = V^\mu(\tau) = \frac{dx^\mu}{d\tau}, \qquad a^\mu = a^\mu(\tau) = \frac{dV^\mu}{d\tau}, \tag{20}$$

which obey the much used properties

$$\boldsymbol{V}\cdot\boldsymbol{V} = 1, \qquad \boldsymbol{V}\cdot\boldsymbol{a} = 0. \tag{21}$$

The first relation comes from the meaning of proper time, while the second property follows by differentiating the first. These constraints owe their existence to the fact that the particle's world line contains only three arbitrary functions, just like in Newton's theory. V^μ and a^μ may be expressed in terms of the usual 3-velocity $\mathbf{v}$ and 3-acceleration $\mathbf{a}$ as follows

$$\boldsymbol{V} = \gamma(1,\mathbf{v}), \qquad a = \gamma^4[\mathbf{v}\cdot\mathbf{a}, \mathbf{a} + \mathbf{v}\times(\mathbf{v}\times\mathbf{a})], \tag{22}$$

where γ is defined by

$$\gamma = 1/\sqrt{1-\mathbf{v}^2}. \tag{23}$$

All this is obtained by plain differentiation and some algebraic rearrangement.

3.2 Fermi-Walker Transport

An accelerating observer faces the fundamental problem of how is he to define a time-independent direction in space. Prior to relativity, such a direction was simply given by a time-independent vector. Unfortunately, the same answer is inconsistent within relativity, simply because a space direction in the observer's rest frame stays orthogonal to the changing time axis $\boldsymbol{V}$. A typical example is the definition of a uniformly accelerated motion; the acceleration 4-vector cannot remain constant in time since $\boldsymbol{V}$ varies in Eq. (21). A propagation law along the world line is therefore mandatory.

It is reasonable to assume that said law will preserve all scalar products, and thus be, in effect, a Lorentz transformation. The transportation from τ to $\tau + d\tau$, along the world line, is an infinitesimal Lorentz transformation, which in turn was shown to be determined by an antisymmetric tensor like in Eq. (13). The only such tensor that can be built from V_μ and a_μ is a multiple of $V_\mu a_\nu - a_\mu V_\nu$. The transport law takes therefore the form

$$\frac{dp^\mu}{d\tau} = A(V^\mu a_\nu - a^\mu V_\nu)p^\nu.$$

A is determined by the obvious requirement that $\boldsymbol{V}$ itself should satisfy the equation, thus

$$\frac{Dp^\mu}{d\tau} = \frac{dp^\mu}{d\tau} + (V^\mu a_\nu - a^\mu V_\nu)p^\nu = 0. \qquad (24)$$

The first equality is known as the Fermi-Walker derivative, whereas its vanishing provides the so-called Fermi-Walker transport law. This derivative may be extended to tensors of all types.

3.3 Uniform Acceleration

The condition for uniform acceleration is now easily stated:

$$\frac{Da^\mu}{d\tau} = \frac{da^\mu}{d\tau} + \boldsymbol{a}\cdot\boldsymbol{a}V^\mu = 0.$$

The scalar $\boldsymbol{a}\cdot\boldsymbol{a}$ is clearly independent of time as may be verified by multiplying the equation by a_μ. The solution to this second-order differential equation is thus given in terms of exponentials as

$$V^\mu = M^\mu \exp(\tau g) + N^\mu \exp(-\tau g),$$

where $g^2 = -\boldsymbol{a}\cdot\boldsymbol{a}$, and $\boldsymbol{M}$ and $\boldsymbol{N}$ are null vectors with $2\boldsymbol{M}\cdot\boldsymbol{N} = 1$ in order to satisfy $\boldsymbol{V}\cdot\boldsymbol{V} = 1$. The canonical form $2\boldsymbol{M} = (1,1,0,0)$ and $2\boldsymbol{N} = (1,-1,0,0)$ gives the explicit solution

$$t = g^{-1}\sinh g\tau, \qquad x = g^{-1}\cosh g\tau,$$

where y and z coordinates vanish. Appropriate initial conditions have been set. Elimination of τ gives

$$x = \frac{c^2}{g}\sqrt{1 + \frac{g^2t^2}{c^2}} \approx \frac{c^2}{g} + \frac{gt^2}{2} \quad \text{for small } t,$$

where the speed of light c has been inserted. Notice that the low-speed approximation resembles Galileo's free-fall formula while $x \approx ct$ is true for large times. It thus takes an infinite time for a particle to achieve the speed of light when acted upon by a constant external force relative to the particle's frame. This reinforces the belief that it is impossible to accelerate matter above the speed of light.

3.4 Pure Boosts

Former discussions suggest the problem of looking for Lorentz transformations that map a 4-velocity $\boldsymbol{U}$ into a different one $\boldsymbol{V}$ without rotation. Said transformation Λ will boost velocities and do nothing else, and so it has to be built entirely out of $\boldsymbol{U}$ and $\boldsymbol{V}$; thus

$$\Lambda^\mu_\nu = AU^\mu U_\nu + BV^\mu V_\nu + CU^\mu V_\nu + DV^\mu U_\nu + E\delta^\mu_\nu.$$

The unknown coefficients are determined by requiring that Eq. (10) be satisfied, a calculation that is straightforward but long. It may be somewhat simplified by the use of two null vectors that can be formed out of $\boldsymbol{U}$ and $\boldsymbol{V}$, which either are eigenvectors of Λ, or are transformed into one another. Only the first case is connected to the identity. The end result turns out to be

$$\Lambda^\mu_\nu = \delta^\mu_\nu - \frac{1}{1+\gamma}[U^\mu U_\nu + V^\mu V_\nu + U^\mu V_\nu - (1+2\gamma)V^\mu U_\nu], \qquad (25)$$

with $\gamma = \boldsymbol{U}\cdot\boldsymbol{V}$. This formula can be obtained even more simply by exponentiating $\Delta^\mu_\nu = \Gamma(V^\mu U_\nu - U^\mu V_\nu)$ and fixing Γ by the requirement that $\boldsymbol{U}$ gets mapped to $\boldsymbol{V}$. Pure rotations may be obtained the same way. Composition of boosts and rotations leads to a general proper Lorentz transformation.

3.5 Thomas Precession

Thomas precession will now be considered. Take a particle whose world-line's tangent is $\boldsymbol{U}(\tau)$ and consider a comoving Fermi-Walker–transported position 4-vector $\boldsymbol{P}$

orthogonal to it. By definition, $\boldsymbol{U} \cdot \boldsymbol{P} = 0$. Let $\boldsymbol{V}$ denote a time-independent 4-vector that represents the time axis of an observer at rest. Transformation (25) when applied to $\boldsymbol{P}$ gives a 4-vector $\boldsymbol{Q}$ orthogonal to $\boldsymbol{V}$, which may be thought of as a position vector for the observer $\boldsymbol{V}$. $\boldsymbol{Q}$ can only rotate as time evolves. This rotation is called Thomas precession. In relativity, a frame that does not spin with respect to itself appears to rotate in the eyes of an inertial observer. The corresponding angular velocity will now be calculated. From $\boldsymbol{U} \cdot \boldsymbol{P} = 0$, it follows that $P^0 = \mathbf{u} \cdot \mathbf{P}$ when $\boldsymbol{U} = \gamma(1,\mathbf{u})$ and $\boldsymbol{P} = (P^0,\mathbf{P})$. Fermi-Walker transport for $\boldsymbol{P}$ takes the form

$$\frac{d\mathbf{P}}{d\tau} = \gamma^3 \mathbf{a} \cdot \mathbf{P},$$

after some algebra based on Eq. (24). $Q^\mu = \Lambda^\mu_\nu P^\nu$ is true by definition when Λ^μ_ν is given by Eq. (25). In the rest frame, that is, when $\boldsymbol{V} = (1,0,0,0)$ and $\boldsymbol{Q} = (0,\mathbf{Q})$, $\boldsymbol{Q}$ is given by

$$\mathbf{Q} = \mathbf{P} - \frac{\gamma}{\gamma + 1} (\mathbf{P} \cdot \mathbf{U})\mathbf{U},$$

and the inverted version by

$$\mathbf{P} = \mathbf{Q} + \frac{\gamma^2}{\gamma + 1} (\mathbf{U} \cdot \mathbf{Q})\mathbf{U}.$$

A rather long calculation finally gives

$$\frac{d\mathbf{Q}}{dt} = \frac{\gamma^2}{\gamma + 1} (\mathbf{a} \times \mathbf{U}) \times \mathbf{Q}, \tag{26}$$

where a change to the rest frame time was performed. It is now plain that $\boldsymbol{Q}$ precesses with the following angular velocity

$$\mathbf{\Omega}_{\mathrm{T}} = \frac{\gamma^2}{\gamma + 1} \mathbf{a} \times \mathbf{U}. \tag{27}$$

This is an entirely relativistic effect that helped historically in the understanding of spin-orbit coupling for atoms and nuclei (Jackson, 1975).

It is important to realize that Eq. (26) describes the rotation of a comoving frame and not the evolution of any real body, if only because said body should contract, stretch, and shear as well. The concept of a finite rigid body has been extensively discussed in the literature, with the overall conclusion that it cannot be defined in any satisfactory manner for relativity. Trautmann *et al.* (1964) give a masterful account of this problem. However, locally, a moving observer can nevertheless draw a congruence of world lines by dragging arrowheads of Fermi-Walker–transported position vectors. Said congruence describes particles of a locally rigid, nonspinning body. When this motion is viewed by an observer at rest, the body appears to rotate at an angular velocity

$$\mathbf{\Omega} = \frac{\gamma^2}{2} \mathbf{a} \times \mathbf{U}$$

instead, which differs appreciably from Eq. (27) for high speeds only.

3.6 Doppler Effect

This important effect consists in that an observer in motion relative to a light-emitting source perceives a shifted frequency. Let a source $\boldsymbol{U}$ emit light signals at frequency $\nu(\boldsymbol{U})$, which are received by an observer $\boldsymbol{V}$ at frequency $\nu(\boldsymbol{V})$, as depicted by Fig. 3. A and A' denote two infinitesimally separated bursts of light emitted by $\boldsymbol{U}$ and received by $\boldsymbol{V}$ at events B and B'. $AB = \boldsymbol{k}$ and $A'B' = \boldsymbol{k} + d\boldsymbol{k}$ are null vectors. A and A' are separated by proper time $d\tau(\boldsymbol{U})$, B and B' by $d\tau(\boldsymbol{V})$. Therefore $AA' = \boldsymbol{U}d\tau(\boldsymbol{U})$ and $BB' = \boldsymbol{V}d\tau(\boldsymbol{V})$. It is plain from Fig. 3 that

$$\boldsymbol{U}d\tau(\boldsymbol{U}) + (\boldsymbol{k} + d\boldsymbol{k}) - \boldsymbol{V}d\tau(\boldsymbol{V}) - \boldsymbol{k} = 0.$$

When this equation is multiplied by $\boldsymbol{k}$,

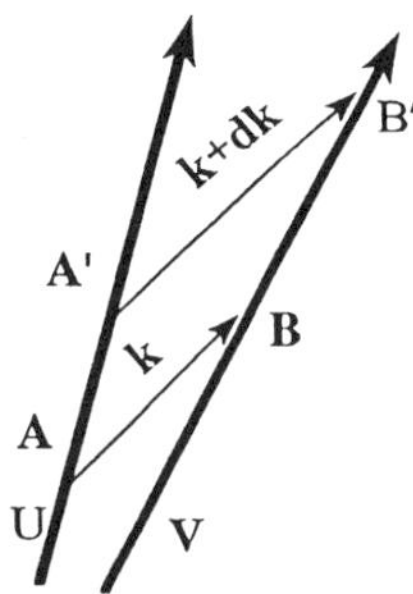

FIG. 3. Doppler effect. A source *O* emits bursts of light "*k*" at observer *V*.

$$\frac{d\tau(\boldsymbol{V})}{d\tau(\boldsymbol{U})} = \frac{\boldsymbol{U}\cdot\boldsymbol{k}}{\boldsymbol{V}\cdot\boldsymbol{k}}$$

obtains since $\boldsymbol{k}\cdot d\boldsymbol{k} = 0$ follows from $\boldsymbol{k}\cdot\boldsymbol{k} = 0$; simply because a frequency is the inverse of a period

$$\frac{\nu(\boldsymbol{U})}{\nu(\boldsymbol{V})} = \frac{\boldsymbol{U}\cdot\boldsymbol{k}}{\boldsymbol{V}\cdot\boldsymbol{k}} \tag{28}$$

holds too. This result when applied to a moving source with 3-velocity $\mathbf{U}$ and a light ray impinging the observer at rest $\boldsymbol{V}$ in the unitary direction $\mathbf{n}$ renders

$$\nu = \nu_0 \frac{1 - \mathbf{n}\cdot\mathbf{U}/c}{\sqrt{1 - \mathbf{U}^2/c^2}}. \tag{29}$$

Use of $\boldsymbol{U} = \gamma(1,\mathbf{U})$ and $\boldsymbol{V} = (1,0,0,0)$ as well as $\boldsymbol{k} \propto (1,\mathbf{n})$ has been made. ν_0 stands for the source frequency and ν for the observed frequency. The speed of light has been inserted for aesthetic reasons alone. In contrast to the Newtonian result of zero frequency shift for the transverse Doppler effect, that is when $\mathbf{n}\cdot\mathbf{U} = 0$, special relativity predicts a finite shift. Many other kinematical effects such as aberration of light, light bouncing off moving mirrors, and much more, are discussed in most textbooks (Misner *et al.*, 1972). In all cases, it is the clear physical statement, combined with the help of a world diagram, that reduces a problem to straightforward algebraic manipulation. Explicit Lorentz transformations are almost never necessary, just as Galileo transformations are seldom used in mechanics.

The really interesting and challenging issues deal with interactions. It is here that relativity provokes the most drastic changes, which we proceed to consider now.

3.7 Classical Interactions

Take two particles under interaction. Within the Newtonian scheme there will be an instantaneous force between them, so that the slightest change in any one of the particles will be immediately felt by the other. So there is energy propagating with infinite speed. It is simple to see that, in relativity, this fact implies acausal phenomena, like an observer manipulating his own history. This occurs because two simultaneous events in some frame have no definite causal relation in other frames; either event may happen earlier. So, if noncausal phenomena are ruled out, the very concept of force at a distance ceases to exist. Interactions have to be dealt with in an entirely different manner. It is the concept of a field that comes to rescue. Space is imagined filled with some undefined entity, described by various numbers assigned to each and every event. As a particle moves through this field, its motion is affected, and so is the field perturbed. Actually, this seems to be the only way to retain conservation laws, for if the interaction propagation speed is finite, then some form of energy will have to reside on neither particle, but in the field. Thorough discussions in this respect are given by Sudarshan and Makunda (1974), who prove no-go–related theorems. The field concept dominates today's fundamental theories of Nature. With it, a fantastically diverse array of natural phenomena can be succinctly summarized into a few equations. In the sequel this will be illustrated for electromagnetic phenomena.

Physicists have learned that basic laws are best described via variational principles. A real number, called action, is assigned to variables chosen to describe a system. The system evolves by minimizing the action. In particular, the variables chosen to describe a charge q are its world-line coordinates $x^\mu(\tau)$, and a covariant field $A_\mu(x^\nu)$. The action S for said system is taken to be

$$S = -m\int d\tau - q\int A_\mu dx^\mu - \frac{1}{16\pi}\int F^{\mu\nu}F_{\mu\nu}d^4x, \tag{30}$$

where

$$F_{\mu\nu} = A_{\nu,\mu} - A_{\mu,\nu} \tag{31}$$

is the electromagnetic field, m the mass, and q the charge. Recall that indices are raised with the inverse Minkowski tensor. The action is, above all, a relativistic invariant and will thus produce a relativistic theory. The first term in Eq. (30), when written in terms of time with the help of

$$d\tau = dt\sqrt{1 - \left(\frac{d\mathbf{x}}{dt}\right)^2},$$

is seen to become essentially Newton's kinetic energy for low speeds. The second term is a line integral along the charge's world line, which picks up values of the field at the charge's position only. Both terms, upon variation of the world line, provide the equations of motion that govern the way charges move in an electromagnetic field, namely

$$m\,\frac{dV^{\mu}}{d\tau} = qF^{\mu\nu}V_{\nu}, \tag{32}$$

with $\boldsymbol{V}$ given by Eq. (20). However, if only these two terms were present in the action, the field would not react to changes of the motion of the charge and no conservation laws could possibly exist. The last term renders the action invariant under translations because the field itself enters as a dynamical variable, and thus can be translated as a whole, rendering conservation of energy and momentum possible. Notice that the gauge replacement $A_{\mu} \rightarrow A_{\mu} + \Lambda_{,\mu}$ produces no effect on the motion of the charge, as seen from Eq. (32) or from the action itself. This suggests that only $F_{\mu\nu}$, defined by Eq. (31) and invariant under such transformations, should participate in the action via the last term of Eq. (30). S generates Maxwell's equations upon variation of the field. It is no exaggeration to say that the truly vast array of classical electromagnetic phenomena are a logical consequence of Eq. (30) alone.

Conservation of energy and momentum follows from the action S too. Standard theorems from mechanics show that the energy and momentum carried by particles turn out to be

$$E = \frac{mc^2}{\sqrt{1 - (\mathbf{v}/c)^2}}, \qquad \mathbf{P} = \frac{m\mathbf{v}}{\sqrt{1 - (\mathbf{v}/c)^2}}. \tag{33}$$

These, by use of Eq. (20), may be condensed into the formula

$$P^{\mu} = mV^{\mu},$$

from which the transformation laws of energy-momentum follow. The most dramatic difference from Newton's conservation laws lies in that relativity allows only four conserved quantities, whereas Newton's mechanics requires in addition mass conservation. Particle creation and annihilation is thus theoretically possible. Specific formulas differ, of course, too. Einstein's energy-momentum formulas are successfully verified on a day-to-day basis in all high-energy accelerator facilities. Hagedorn (1964) gives an extensive, first-class account of applications to collision physics.

A further illustration of the power of relativity in constructing laws of nature will be now given via the derivation of the equation of motion for charged particles that have an intrinsic spin and therefore a magnetic moment. Let $\boldsymbol{S}$ be a 4-vector intended to describe the intrinsic spin of some particle. When viewed from the particle's rest frame the spin is a space vector, therefore $\boldsymbol{S}\cdot\boldsymbol{U} = 0$ should hold. The size of the spin is preserved in time, so $\boldsymbol{S}\cdot\boldsymbol{S}$ should be time independent. If no electromagnetic field were present it is the Fermi-Walker derivative of $\boldsymbol{S}$ that should vanish, Therefore one expects this derivative to be linear in $F^{\mu\nu}$. By far the simplest equation that satisfies these requirements is

$$\frac{DS^{\mu}}{d\tau} = \frac{gq}{2m}\,(\delta^{\mu}_{\nu} - V^{\mu}V_{\nu})F^{\nu}_{\lambda}S^{\lambda}, \tag{34}$$

as may be instantly verified from the fact that the Fermi-Walker derivative of $\boldsymbol{V}$ vanishes. The extremely useful tensor within parentheses projects 4-vectors onto the hyperplane perpendicular to $\boldsymbol{V}$; it is there to insure $\boldsymbol{S}\cdot\boldsymbol{V} = 0$ for all terms. The arbitrary constant g is known as the Landé constant or simply as the g factor. Equation (34) was discovered by Bargmann, Michel, and Telegdi in 1959 (Jackson, 1975). They provide the theoretical foundation to the prominent $g - 2$ electron and muon experiments that have set new standards of precision in physics (Newman *et al.*, 1968).

When quantum ideas were incorporated into relativity a truly fantastic blend emerged. Dirac solved the problem of constructing a relativistic Schrödinger equation. In quantum mechanics, particles are no longer pictured as traveling along trajectories; probabilistic distributions over space are assumed instead. World lines are replaced by complex probability-amplitude fields called wave functions, and so action (30) is valid no more. Dirac realized that an electron was instead correctly described by spinor fields. These are mathematical objects more basic than tensors, which will not be covered here. However, much relevant information may be found in the superb work by Penrose and Rindler (1984).

4. EXPERIMENTAL EVIDENCE

Special relativity has been continuously tested with ever-increasing precision. This magical interplay between theoretical understanding and experimental search has led us to abandon without doubt the historical conception of absolute time, and with it, Galilean relativity.

All experiments performed at constant gravitational potential verify special relativity; however, gravitation sometimes plays a role in modifying the metric properties of special relativity in a measurable way. Einstein's general relativity provides the theoretical basis for gravitation. An example of the incidence of gravity is provided by the Hafele and Keating (1972) experiment. They flew highly stabilized cesium-beam clocks around the world, once eastward and once westward, on commercial jet flights. These clocks' recorded time was compared with the corresponding time recorded by the reference atomic time scale at the U.S. Naval Observatory, MEAN. The three clocks may be considered to start together and reunite after the flights' end. Thus they travel along three different world lines that emerge from one and fuse into one again at the end. The corresponding proper times can thus be unambiguously calculated from the flight data provided by the captains. Relative to the atomic time scale of the U.S. Naval Observatory, the eastward-going clock lost 59 ± 10 ns whereas the westward-going clock gained 273 ± 7 ns, in full correspondence to the predicted values of 40 ± 23 and 275 ± 21 ns, respectively. It must be stressed that these latter values include gravitational effects that are of the same order as the special relativistic effects. This experiment settles once and for all the famous twin paradox, and cleanly rules out absolute time.

A particularly vivid test of time dilation emerged as a by-product of the famous muon $g - 2$ experiment at CERN described by Bailey *et al.* (1977). Very fast muons were kept in a ring magnet at 3.098 GeV. Muons at rest have a lifetime of 2.197 11(8) μs whereas the fast muon's lifetime was measured to be 64.378(26) μs which agrees within 0.1% with the theoretical value of 64.435(9) μs for time dilation. So these particles live about 30 times longer just because they are fast.

Alväger *et al.* performed, in 1964, at the CERN Synchrotron also, an experiment designed to test in a very straightforward way Einstein's hypothesis on the constancy of the speed of light. They measured the speed of γ rays produced by the decay of fast neutral pions with energies over 6 GeV by timing over a known distance, finding no discrepancies with the accepted value for the speed of light. This result is all the more impressive since the pions themselves travel at almost the speed of light. It is appropriate to mention here another experiment performed by Brown *et al.* (1973) where the speed of 7-GeV γ rays was compared to the speed of visible light, detecting no variation within parts in a million. The speed of 11-GeV electrons was shown to coincide with the speed of light within parts per million too. So in effect this work has shown that the limiting velocity for electrons is approximately that of light.

Most relativistic experiments, however, were designed to test the isotropy of the speed of light simply because Galileo's relativity can admit at best only one inertial frame Σ where light propagates with equal speed in all directions. This is so because in any other frame S a preferred direction is automatically established by the velocity **u** of Σ as seen by S. Einstein's relativity circumvents this anisotropy by its peculiar velocity composition, so constructed as to insure said isotropy. In order to describe these experiments in a unified manner it is convenient to construct a formula for the time it takes light to travel from A to B along some specified path in the laboratory frame S within the context of Galileo's relativity. An elementary analysis to second order in u/c renders this time to be

$$t(A \to B) = \left(1 + \frac{\boldsymbol{u}\cdot\boldsymbol{u}}{2c^2}\right)\int_A^B \frac{ds}{c} - \frac{\boldsymbol{u}}{c}\cdot\int_A^B \frac{d\boldsymbol{s}}{c} + \frac{1}{2}\int_A^B \frac{ds}{c}\left(\frac{\boldsymbol{u}\cdot\boldsymbol{t}}{c}\right)^2, \tag{35}$$

where $d\mathbf{s}$ is the laboratory's arc element and $\mathbf{t}$ its unit tangent. Anisotropy shows up in the second and third terms. Experiments intended to detect anisotropy should ideally be designed to measure the second term, which is typically the largest. However, before the discovery of the Mössbauer effect in 1958, interference fringe counting was the only precision method available, which in turn required splitting a pencil of light along two different paths that eventually reunite to form the desired interference pattern. Unfortunately the second term vanishes for the overall time difference of said two paths since an integral of a vector line element along a closed loop is involved. Early experiments were therefore forced to measure the tiny third term.

The experiment that triggered doubts about firmly established space-time notions of the day was performed by Michelson-Morley (1887). It will now be outlined briefly as a prototype of early experiments. From Fig. 4 we see that a monochromatic beam I is split by a half-silvered mirror O. The right beam is reflected at mirror R, the transmitted beam is reflected at mirror U. Both reunite at E. Arms OR and OU are of the same length L. Σ, the ether of that time, sweeps across the laboratory frame S at velocity $\mathbf{u}$. The difference in arrival time at E, calculated from Eq. (35), is given by

$$t_R - t_U = (u^2L/c^3)\cos(2\theta). \tag{36}$$

The whole apparatus is then rotated 90° in order to detect fringe shifts with a fractional value given by

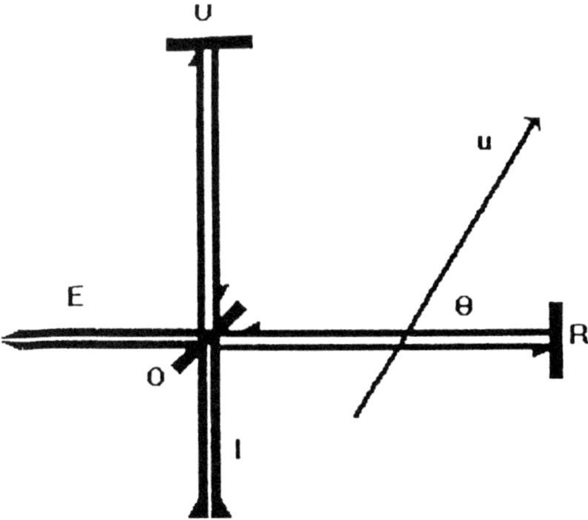

FIG. 4. Michelson-Morley interferometer.

$$(2u^2L/\lambda c^2)\cos(2\theta). \tag{37}$$

Strict equality of the arms' lengths is not important since L would get replaced by the average in the formula above. The fractional fringe shift, if observed at all, was no more than 0.02. The expected value from the Earth's estimated velocity is twenty times larger. Conceptually similar experiments, like the one due to Kennedy and Thorndike (1932), were performed over the years with ever-increasing precision. Let us just mention one of the latest and most precise performed by Brillet and Hall (1979). This is a one-arm Michelson-Morley experiment that rotates continuously. The arm is a Fabry-Pérot interferometer cavity whose length is measured in wavelengths of a servo-stabilized He-Ne laser, read out with extreme sensitivity as a frequency shift by heterodyning with another highly stable CH_4 laser. When time dependence was fitted by use of an equation almost identical to Eq. (37), no discrepancies of greater then a part in 10^6 existed.

A different type of experiment was achieved almost simultaneously by Champeney *et al.* (1963) and Turner and Hill (1964). We shall outline the former. A γ-ray source ^{57}Co and an absorber ^{57}Fe of exactly the same frequency are placed symmetrically and diametrically opposed on a fast-rotating horizontal disk. Diametrically opposed stationary proportional counters are placed close to the rim. Ideally all γ rays emitted toward the absorber will be absorbed when the disk rotates slowly. However, when the disk rotates fast the emitted frequency will no longer match the frequency seen by the absorber because of Doppler's effect. The relativistic shift is quadratic in u/c as given by Eq. (29) when applied in the absorber's instantaneous rest frame. Galileo's result is, however, much larger and first order in u/c predicting a frequency shift large enough to avoid Mössbauer resonant absorption. In order to calculate the frequency shift it is necessary to relate the time of absorption t_A to the time of emission t_E. Equation (35) gives

$$t_A = t_E + \left[1 + \tfrac{1}{2}\left(\frac{u}{c}\right)^2\right]\frac{D}{c} - \frac{u_\perp D}{c^2}\cos(\Omega t_E),$$

where D is the diameter of the rotating disk, Ω is its angular velocity, and $u_\perp$ is the component of the ether velocity **u** upon the disk. Recall that $\nu_E/\nu_A = dt_A/dt_E$ so that the relative frequency shift expected from Galileo's relativity is

$$\frac{\nu_A - \nu_E}{\nu_E} = -\frac{u_\perp D\Omega}{c^2}\sin(\Omega t_E).$$

The authors place a limit of only 1.6 ± 2.8 m/s for $u_\perp$ as compared with $2.99\,792\,458 \times 10^9$ m/s for the speed of light, which of course is a null result for anisotropy.

Highly sophisticated and technologically advanced experiments have been performed over the years validating relativity with ever-increasing accuracy. The literature can be traced from the following excellent review articles: Newman *et al.* (1978), Haugan and Will (1987), and Will (1992).

GLOSSARY

Boost: A Lorentz transformation that modifies velocities without rotations.

Event: A point in space at a given time.

Inertial Frame: System of reference where noninteracting particles preserve their velocities.

Light Cone: Set of all events with possible communication via light signals to a single event called vertex.

Line Element: Quadratic invariant in the space-time coordinate differentials that measures proper time.

Null Rays: Same as light rays.

Proper Length: Length measured in the object's rest frame.

Proper Time: Time measured by a watch in its rest frame.

World Line: Locus of events swept by a particle in motion.

Works Cited

Alväger, T., Farley, F. J. M., Kjellman, J., Wallin, I. (1964), *Phys. Lett.* **12,** 260–262.

Bailey, J., *et al.* (1977), *Nature* **268,** 301–305.

Barut, A. O., Rączka. R. (1977), *Theory of Group Representation and Applications,* Warsaw: Polish Scientific Publishers.

Brillet, A., Hall, J. L. (1979), *Phys. Rev. Lett.* **42,** 549–552.

Brown, B. C., Masek, G. E., Maung, T., Miller, E. S., Ruderman, H., Vernon, W. (1973), *Phys. Rev. Lett.* **30,** 763–766.

Champeney, D. C., Isaak, G. R., Khan, A. M. (1963), *Phys. Lett.* **7,** 241–243.

Hafele, J. C., Keating, R. C. (1972), *Science* **177,** 168–170.

Hagedorn, R. (1964), *Relativistic Kinematics,* New York: W. A. Benjamin.

Haugan, M. P., Will, C. M. (1987), *Phys. Today* **40** (5), 69–76.

Jackson, J. D. (1975), *Classical Electrodynamics,* New York: Wiley.

Kennedy, R. J., Thorndike, E. M. (1932), *Phys. Rev.* **42,** 400–418.

Michelson, A. A., Morley, E. W. (1887), *Am. J. Sci.* **34,** 333.

Miller, A. I. (1981), *Special Theory of Relativity,* Reading, MA: Addison-Wesley.

Misner, C. W., Thorn, K. S., Wheeler, J. A. (1972), *Gravitation,* New York: W. H. Freeman.

Newman, D., Ford, G. W., Rich, A., Sweetman, E. (1978), *Phys. Rev. Lett.* **40,** 1355–1358.

Pauli, W. (1958), *Theory of Relativity,* London: Pergamon Press.

Penrose, R., Rindler, W. (1984), *Spinors and Space-Time,* Cambridge, U.K.: Cambridge Univ. Press.

Sudarshan, E. C. G., Makunda, N. (1974), *Classical Dynamics: A Modern Perspective,* New York: Wiley.

Trautman, A., Pirani, F. A. E., Bondi, H. (1965), *Lectures in General Relativity,* 1964 Brandeis Summer Institute in Theoretical Physics, Vol. 1, Englewood Cliffs, NJ: Prentice-Hall.

Turner, K. C., Hill, H. A. (1964), *Phys. Rev.* **134,** B252–B256.

Weinberg, S. (1972), *Gravitation and Cosmology,* New York: Wiley.

Will, C. M. (1992), *Phys. Rev. D* **45,** 403–411.

Further Reading

Barut, A. O., Rączka, R. (1977), *Theory of Group Representation and Applications,* Warsaw: Polish Scientific Publishers.

Einstein, A. (1952), *The Principle of Relativity,* New York: Dover.

Galilei, Galileo (1960), *On Motion and On Mechanics,* Madison: The Univ. of Wisconsin Press.

Hagedorn, R. (1964), *Relativistic Kinematics,* New York: W. A. Benjamin.

Jackson, J. D. (1975), *Classical Electrodynamics,* New York: Wiley.

Miller, A. I. (1981), *Special Theory of Relativity,* Reading, MA: Addison-Wesley.

Misner, C. W., Thorn, K. S., Wheeler, J. A. (1972), *Gravitation,* New York: W. H. Freeman.

Panofsky, W. K. H. (1980), "Special Relativity

Theory in Engineering," in: H. Woolf (Ed.), *Some Strangeness in the Proportion*, Reading, MA: Addison-Wesley.

Pauli, W. (1958), *Theory of Relativity*, London: Pergamon Press.

Penrose, R., Rindler, W. (1984), *Spinors and Space-Time*, Cambridge, U.K.: Cambridge Univ. Press.

Shadowitz, A. (1968), *Special Relativity*, Toronto: N.B. Saunders.

Sudarshan, E. C. G., Makunda, N. (1974), *Classical Dynamics: A Modern Perspective*, New York: Wiley.

Synge, J. L. (1956), *Relativity: The Special Theory*, Amsterdam: North-Holland.

Trautman, A., Pirani, F. A. E., Bondi, H. (1965), *Lectures in General Relativity*, 1964 Brandeis Summer Institute in Theoretical Physics, Vol. 1, Englewood Cliffs, NJ: Prentice-Hall.

Weinberg, S. (1972), *Gravitation and Cosmology*, New York: Wiley.

REMOTE SENSING

RONALD J. HOLYER, *Remote Sensing Division, Naval Research Laboratory, Stennis Space Center, Mississippi, U.S.A.*

	Introduction	385
1.	**Remote-Sensing Basics**	389
1.1	Remote-Sensing Configurations	389
1.1.1	Passive Sensing	389
1.1.1.1	Emitted Radiant Energy	390
1.1.1.2	Reflected Solar Radiant Energy	391
1.1.2	Active Sensing	391
1.1.2.1	Microwave Wavelengths	392
1.1.2.2	Visible Light	393
1.1.3	Imaging and Nonimaging Systems	393
1.2	Atmospheric Windows	394
2.	**Signatures from Interactions of Electromagnetic Radiation with the Environment**	394
2.1	Mechanisms	395
2.1.1	Reflectance	395
2.1.2	Scattering	396
2.1.3	Emission and Absorption	396
2.2	Typical Signatures	397
2.2.1	Vegetation	397
2.2.2	Minerals	397
2.2.3	Oceans and Lakes	398
2.2.4	Sea Ice	399
2.2.5	Atmosphere	400
3.	**Representative Remote-Sensing Systems**	400
3.1	Operational Systems	401
3.1.1	Landsat	401
3.1.2	SPOT	402
3.1.3	TIROS Series	402
3.1.4	Geostationary Series	403
3.1.5	Defense Meteorological Satellite Program	403
3.1.6	European Remote-Sensing Series	404
3.2	Research Systems	405
3.2.1	Seasat	405
3.2.2	CZCS	406
3.2.3	TOPEX/POSEIDON	406
3.2.4	AVIRIS	407
4.	**Typical Applications**	407
4.1	Climate Change—the Ozone Hole	407
4.2	Geomorphology—Volcanology	408
4.3	Arctic Science—Sea-Ice Classification	408
4.4	Arctic Science—Ice Sheets	408
4.5	Earth's Biosphere—Global Biomass	409
4.6	Meteorology—Tropical Storms	409
4.7	Oceanography—Coastal Water Clarity	409
4.8	Land Use—Vegetation Classification	410
5.	**Future of Remote Sensing**	410
	Glossary	411
	Acronyms and Abbreviations	412
	Works Cited	413
	Further Reading	413

INTRODUCTION

"Remote sensing is the science and art of obtaining information about an object, area, or phenomenon through analysis of data acquired by a device that is not in contact with the object, area, or phenomenon under investigation" (Lillesand and Kiefer, 1994). The remotely detected signal can be of many forms. Distributions and characteristics of force, acoustic energy, and electromagnetic energy can be affected by objects and phenomena, leading to many candidates for exploitation by remote-sensing systems. Visual and auditory systems in humans and animals are the most common remote sensors. The photographic camera was the first artificial remote-sensing device. When man developed the ability to observe regions of the electromagnetic spectrum beyond the range of hu-

3-527-28138-X/96/$5.00 + .50

man vision and photographic sensitivity, the term "remote sensing" came into common usage to encompass the new field of observational science. Evelyn Pruitt, formerly of the Office of Naval Research, is widely recognized as the originator of the term. While the definition of remote sensing given above is broad enough to include such activities as seismic geophysical prospecting, underwater sonar and acoustic systems, astronomy, and even certain modern medical diagnostic procedures, common usage of the term remote sensing has come to apply to airborne or spaceborne systems sensing electromagnetic radiation reflected or emitted by the Earth/atmosphere system or objects located therein.

Numerous events, ranging from the role of aerial photographs in the Cuban missile crisis to television news presentations of weapons guidance videos from Desert Storm, have made the public aware that remote sensing is a key technology for intelligence and military applications. However, information concerning these applications of remote sensing is generally classified, placing military and intelligence remote-sensing systems and applications outside the scope of this article. This article is limited to airborne or spaceborne systems designed to obtain geophysical information by sensing electromagnetic radiation upwelling from the Earth/ocean/atmosphere system. These are described and illustrated with selected application examples from geomorphology, climate-change, meteorology, oceanography, land-use, and Arctic studies. Remote sensing has also been applied to exploration of other planets in our solar system. The properties of all of the major bodies in the solar system, with the exception of Pluto, have been remotely sensed from flyby or orbiting spacecraft. Planetary applications are not covered here. The interested reader is referred to O'Donnell and Sonett (1992); see also SPACE INSTRUMENTATION: PHYSICS.

An extensive history of remote sensing has been compiled in Chap. 2 of Colwell (1983). That chapter gives over a hundred references to early remote sensing, which would be an excellent starting point for readers interested in historical details. An overview of the history starts with balloons as the first remote-sensing platforms. In 1858 photographers Gaspard Felix Tournachon and Colonel Aime Laussedat, planning to produce topographic maps from photographs, ascended in balloons to take "bird's eye" photographs of Paris. Many years later in 1898, Laussedat succeeded in developing a mathematical analysis for converting overlapping perspective views into orthophotographic projections. This was probably the first quantitative rather than descriptive information extracted from remotely sensed data. Other pioneers in platform development included Julius Neubronner who, in 1903, designed and patented a breast-mounted aerial camera for carrier pigeons. The camera, which weighed only 70 grams, exposed 38-mm negatives automatically at 30-s intervals by a built-in timing mechanism. Along with balloons and pigeons, kites took their turn as remote-sensing platforms. English meteorologist E. D. Archibald, Russian government counselor R. Thiele, A. Batut of Paris, and American meteorologist G. R. Lawrence were among those taking aerial photographs from kites during the period 1880 to 1900. Lawrence devised camera systems weighing more than 450 kg, raised these to heights of over 1000 m, and took pictures as large as 1.35×2.4 m^2. Lawrence was in San Francisco on 18 April 1906, the day of the great earthquake, and was able to raise his camera to obtain aerial photographs of the disaster-stricken city.

The first aerial photographs taken from an airplane were collected in 1909. Aerial photography for military purposes blossomed in World War I. At first aerial reconnaissance was undertaken with ordinary cameras made to be used on the ground, but by the end of 1915, Lt. Col. J. T. C. More Brabazon, RAF, produced a camera specifically for aerial use. Photo reconnaissance has played an increasingly prominent role in every military conflict from that time until the present.

Rockets as remote-sensing platforms appeared at about the same time as airplanes. Alfred Maul led the way into rocketry for remote sensing. In 1907, Maul obtained a patent for a gyroscopically stabilized camera mount. By 1912, he was launching a 41-kg system, including a 20×25-cm^2 format camera and gyro, to altitudes of 800 m. In 1946, space pictures were taken from V-2 rockets launched from White Sands Proving Ground. The modern era of remote sensing from spaceborne platforms became possible in 1957 with the launch of Sputnik I, the first Earth-orbiting satellite. The first space photograph of the Earth was transmitted by Explorer 6 in 1959. By 1960 the Television and Infrared Observation Satellite (TIROS) was in space as the first operational Earth-observation satellite.

The history of remote sensing hinges on sensor technology developments as well as platforms. Detectors transform electromagnetic radiation incident on their surfaces into electrons or other detectable and recordable signals. Detectors can generally be classified on the basis of the physical processes by which the conversion from radiation to detectable output is made. The first "detectors" were iodized platings of silver on copper. The greatest difficulty was making the exposed image resistant to further light action. In 1839 Daguerre learned how to stabilize the exposed plates using sodium thiosulfate. In 1871, Maddox developed a sensitized gelatin emulsion of silver-halide grains so that the invisible image could be developed later, thus requiring that only the camera and not the developing apparatus must be carried aloft. Roll film appeared around 1897. Maxwell, in 1855, was the first to give serious consideration to the possibilities of color photography. By 1895, du Hauron had developed three-color separations. In 1924, Mannes and Godowsky obtained a patent on a multilayer color film. Kodachrome film was put on the market in 1935.

Extension of detector technology beyond the visible (VIS) spectrum began in response to a requirement for camouflage detection. A new color infrared (CIR) film incorporated sensitivities to green, red, and near-infrared (NIR) wavelengths in lieu of the usual blue, green, and red. The use of this film depended primarily on the differences in NIR reflectance between camouflage and natural green foliage. Kodak applied for a patent on CIR film in 1942. Although originally developed as a military reconnaissance film, CIR has found its greatest use in plant science for classification of vegetation types and detection of diseased or stressed vegetation. The extension to shorter-wavelength ultraviolet (UV) light occurred simultaneously with extension into the NIR. Some of the earlier work at UV wavelengths included the astronomer R. W. Wood who discovered "Wood's spot" on the moon using UV photographs at a wavelength of 323 nm.

Modern remote sensing has only been possible because film has been replaced by electro-optical sensors of many types. Thermal detectors rely on the increase of temperature in heat-sensitive materials resulting from absorption of radiation. The temperature rise results in some secondary variation of the material that can be monitored electrically. In the case of the bolometer, the secondary effect is a change in resistance. In thermocouples and thermopiles, the effect is a temperature-dependent voltage across a junction of two dissimilar metals (Seebeck effect). Thermal detectors require numerous incident photons to produce a resolvable change. Sensitivity of thermal detectors, therefore, tends to be low. Response time, i.e., the time delay before the photogenerated effect is observed in the output circuit, is also large for thermal detectors.

Quantum detectors employ direct interaction of the incident photon with the electron energy levels within the material to produce free charge carriers. Quantum detectors are of several types. Photoemissive detectors include phototubes, photomultipliers, Vidicons, and orthicons. The photoemissive mechanism involves absorption of photons from the incident radiation, which excite electrons within the material in such a manner that they are emitted through the surface barrier. Freed electrons are accelerated to an anode, and the resulting anode current is directly proportional to the incident photon flux.

Another effect, called photoconductivity, is a bulk effect wherein incident photons produce free charge carriers (electrons or holes) that cause the resistance of the material to vary inversely with the number of incident photons. This effect, first noticed in selenium in the mid-nineteenth century, has led to the use of semiconductors as detectors (see DETECTORS, SEMICONDUCTOR). Materials such as lead sulfide (PbS), indium antimonide (InSb), and mercury cadmium telluride (Hg:Cd:Te) fall in the photoconduction group. Photodiodes are a third type of solid-state detector. A photodiode is a material containing a *p-n* junction that modifies its electrical properties in response to incident light striking the junction. The effect of the incident light can be monitored as a photovoltage, or as photoconduction. Silicon (Si), indium arsenide (InAs), and indium antimonide (InSb) are common photodiode materials.

Early remote-sensing systems with electro-optical detectors scanned an image with a single detector. However, for high-resolution imaging, the required bandwidth of the detector becomes very large and the dwell time of the detector on a single image element becomes very small, producing poor sensitivity. The detector array, one- or two-dimensional arrangements of detectors integrated into a single device, has overcome some of this difficulty. A type of detector array called the charge-coupled device is especially prevalent in modern remote-sensing systems (see CHARGE-COUPLED DEVICES). Efficiency of these devices is quite high, and they can operate at megahertz rates, permitting high-resolution imaging in scanning instruments. Figure 1 illustrates the progression of remote-sensing hardware from single-element or very small detector arrays with a scanning mirror to a one-dimensional array providing cross-track coverage without a scanning mirror ("push-broom" sensor). One-dimensional arrays can also be used in the spectral dimension, combined with a scan mirror for cross-track coverage, to form an imaging spectrometer. The most recent development is a two-dimensional array of detectors spanning spectral and cross-track spatial dimensions, resulting in a "push-broom" spectrometer. Chen (1985) projects that silicon arrays sensitive over the 0.4- to 1.1-μm range will contain 10 000 elements arranged linearly, or 4000 $\times$ 4000 elements in two-dimensional arrays, by the year 2000. In the same time frame, Hg:Cd:Te arrays sensitive over the 1- to 30-μm range are projected to contain up to 8000 elements in linear arrays. Other detector technologies have been developed for radars, lidars, microwave radiometers, scatterometers, and altimeters. These specialized detectors will not be described in this article, but the data collected and applications of these sensors are reviewed.

Remote sensing is rarely employed without some form of coincident *in situ* data, which is commonly called ground truth. The term is not meant literally since "ground truth" can be "water truth" when investigating water features, or can be "air truth" collected in support of satellite-borne atmospheric sounders. Ground truth can even be collected remotely, as would be the case if aerial photography is used as ground truth for satellite imagery. Common uses of ground truth are to calibrate remote-sensor data, to verify information extracted remotely, or to aid in the interpretation of remotely sensed data. Ground truth is often difficult and time consuming to collect properly in accordance with statistical sampling design. Special instrumentation must often be developed for this purpose. The field of ground-truth collection is, therefore, a major subdiscipline within remote sensing.

Meteorologists began to utilize systematic

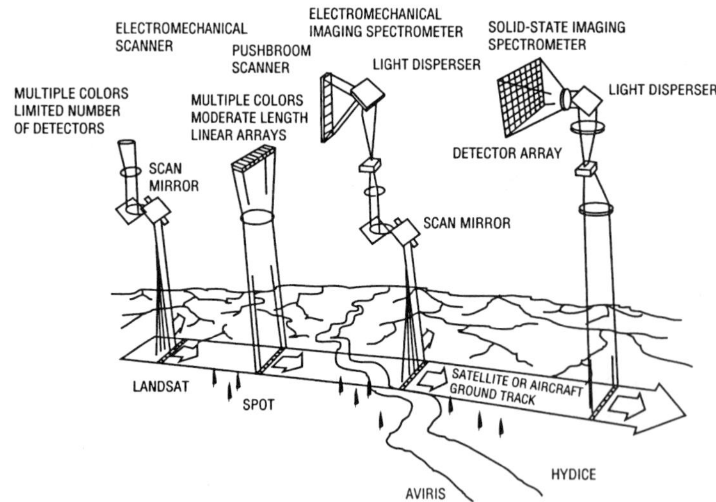

FIG. 1. Detector geometries used in remote-sensing systems.

orbital observations in 1960. TIROS I, the first meteorological satellite, and its successors were remarkably successful. Earth-science applications of satellite data became operational in 1972 when the Earth Resources Technology Satellite (ERTS-1), later renamed Landsat, was launched. Ocean scientists joined the satellite remote-sensing community in a major way in 1978 when the National Oceanographic and Atmospheric Administration (NOAA) Advanced Very High-Resolution Radiometer (AVHRR), Seasat, and Coastal Zone Color Scanner (CZCS) became operational. Seasat failed after three months, but the other two systems were very successful and the oceanographic community was firmly convinced of the value of remotely sensed data.

A plethora of remote-sensing systems has developed in recent years and is planned for the near future. Governments of many countries maintain archives of satellite and aircraft data. Commercial organizations have entered the remote-sensing field, providing remotely sensed data and products derived from remotely sensed data to commercial users and to the general public. Remote sensing has become a major area of applied physics. This article reviews the physics and applications that are the basis of this growing field. A very thorough two-volume work called *The Manual of Remote Sensing* has been produced by the American Society of Photogrammetry (Colwell, 1983). This manual of over 2000 pages aggregates much of the material on this topic into a single reference document.

1. REMOTE-SENSING BASICS

Energy is propagated by electromagnetic (EM) radiation from the source, directly through space, or indirectly by reflection or absorption and reradiation to the remote sensor. Electromagnetic radiation is made manifest only by its interaction with matter. Detection of this interaction with matter is the essence of remote sensing. J. C. Maxwell's concept of electromagnetic radiation was that a mathematically smooth wave motion existed in the magnetic and electric force fields, a concept he formulated as a set of differential equations expressing the interrelationship of electric and magnetic fields. Thus, in any region where there is a time-varying electric field, a magnetic field appears automatically in that same region as a conjugate partner. Similarly, a time-varying magnetic field is associated with a coincident electric field. When electromagnetic waves interact with matter the result will depend on both the magnetic and electric properties of the matter as well as characteristics of the waves such as wavelength or polarization.

Maxwell's wave formulation of electromagnetic radiation fails to account for certain significant interaction phenomena. These phenomena become more evident and consequential as the wavelength of the electromagnetic waves decreases. Max Planck offered another concept of electromagnetic radiation: the photon. The photon is a packet (or quantum) of radiant energy. Both the wave and photon formulations are important in remote sensing. Detector theory is generally photon based, but interactions of radiant energy with matter in the Earth's environment are often best modeled using wave theory.

Electromagnetic energy is characterized by its frequency or wavelength. Wavelength ranges are often identified by a name: visible light, x rays, microwaves, etc. The terminology associated with the electromagnetic spectrum is shown in Fig. 2. Every object is a source of electromagnetic radiation by a process called radiant emission. Likewise, every object reflects electromagnetic radiation from external sources. A remote-sensing system can sense the natural reflection and emission of an object or can supply its own energy source directed toward the object of interest. An example of a sensor with its own energy source is radar, which transmits a pulse of energy and then detects its reflection (the term echo is often used). Remote-sensing systems that supply their own source of electromagnetic radiation are called active systems; those that detect emitted or reflected natural radiation are called passive.

1.1 Remote-Sensing Configurations

1.1.1 Passive Sensing Passive remote-sensing systems observe either electromagnetic radiation emitted by a material, or solar energy reflected by the material. The intervening atmosphere is also an emitting,

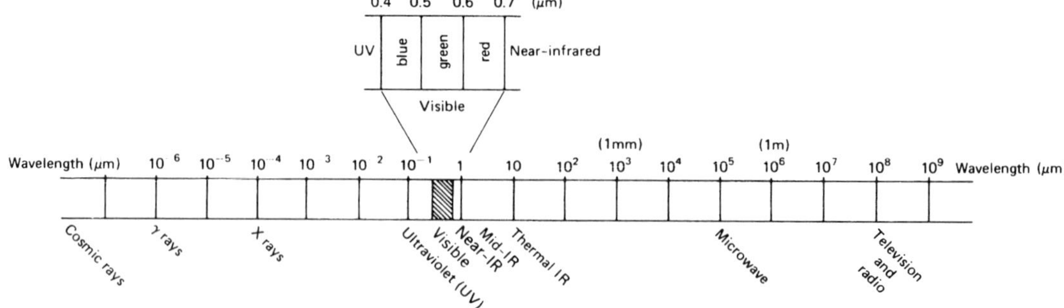

FIG. 2. The electromagnetic spectrum (from Lillesand and Kiefer, 1994, reprinted by permission of John Wiley & Sons, Inc.). Divisions of the spectrum (microwave, infrared, etc.) refer to ranges of wavelengths, not to single wavelengths as the labels in the figure might imply. The end points of ranges are imprecise and neighboring ranges may overlap.

scattering, and absorbing medium, confounding the interpretation of the received signal, or, on the positive side, making it possible to sense atmospheric parameters as well as to sense the Earth's surface.

1.1.1.1 Emitted Radiant Energy. Electromagnetic radiation is produced by transformations of energy from another form into electromagnetic waves (or photons, if you prefer). Electrical, chemical, thermal, magnetic, nuclear, and other forms of energy can be converted into electromagnetic energy by several mechanisms. Molecular processes occurring in nature that give rise to electromagnetic radiation are especially useful for remote sensing. At infrared (IR) and visible wavelengths, electromagnetic radiation is produced by molecular excitation followed by decay back to a lower-energy state. The energy loss required for decay to a lower-energy state is accomplished by radiation of electromagnetic energy. The frequency of the radiated (emitted) energy is directly related to the energy difference between the excited and subsequent decayed energy levels. If the source of the excitation is thermal energy (random motion of particles of matter), electronic, vibrational, or rotational excitation can result from particle collisions. The collisional process and subsequent decay are random, leading to emission over a wide spectral band. If an ideal source (called a blackbody) transforms heat energy into radiant energy at the maximum rate permitted by thermodynamic laws, then the spectral emittance is given by *Planck's formula,* which is displayed graphically in Fig. 3. Spectral emittance of a blackbody is seen to be a function of temperature. Spectral emittance is a maximum at the wavelength λ_m given by *Wien's displacement law,*

$$\lambda_m = A/T, \tag{1}$$

where $A = 2898\ \mu\text{m K}$ and T is temperature of the emitting material in kelvins. The total emitted energy integrated over all wavelengths, M, is given by the *Stefan–Boltzman law:*

$$M = \sigma T^4, \tag{2}$$

where $\sigma = 5.669 \times 10^{-8}\ \text{W m}^{-2}\ \text{K}^{-4}$. Natural materials do not emit radiation as ideal radiators (blackbodies), but rather emit at some reduced efficiency. The emission efficiency relative to a blackbody is called emissivity. Emissivity ranges from zero to one and it varies with wavelength for most materials. The variability of emissivity with wavelength provides a "signature" of a particular material that is useful for remote-object identification.

Since Planck's formula states that the amount and spectral distribution of energy are related to object temperature, emitted radiation can be used to measure remotely the temperature of an object. If the emissivity of the object is not unity, then emissivity must be known to accomplish the conversion from radiant emittance to temperature. However, emissivity is well known for many important natural materials. The remote sensing of temperature is common. Satellite-borne in-

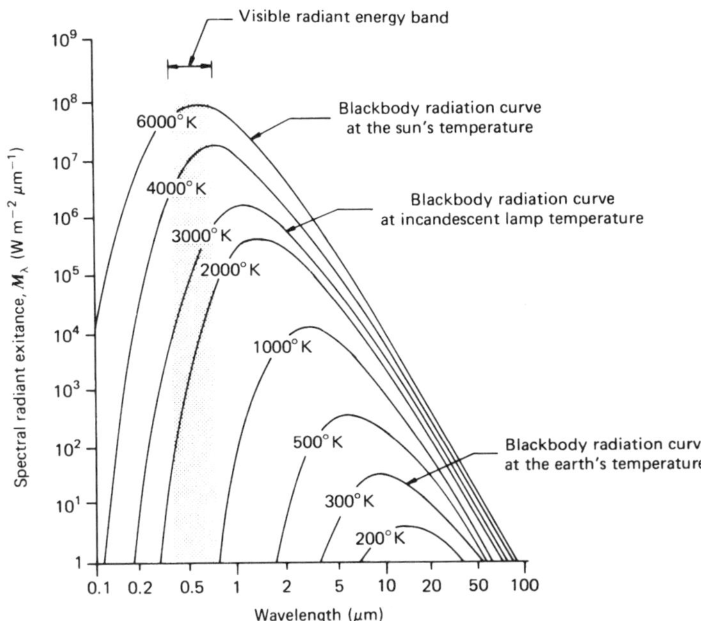

FIG. 3. Spectral distribution of energy radiated from blackbodies of various temperatures (from Lillesand and Kiefer, 1994, reprinted by permission of John Wiley & Sons, Inc.).

struments that measure sea-surface temperature are an example. Insertion of typical Earth temperatures (~300 K) into Eq. (1) results in a wavelength of maximum emission of approximately 10 μm. Wavelengths close to this value are the most commonly used for remote sensing of surface temperature because of the energy maximum, and, more importantly, because of high atmospheric transmittance at this wavelength. This portion of the electromagnetic spectrum is often labeled "thermal IR" in recognition of this fact.

Emitted energy within the Earth's environment is most dominant at the thermal-IR wavelengths, but the long-wave "tail" of the blackbody curve (Fig. 3) extends into the microwave portion of the spectrum where very low energy levels are emitted by natural materials. Microwave radiometers can detect these low-energy signals with adequate signal-to-noise ratios for useful passive sensing. Sea-ice age, soil moisture content, atmospheric water-vapor concentration, and precipitation rate are among the environmental parameters that can be sensed by operating passively at microwave wavelengths.

1.1.1.2 Reflected Solar Radiant Energy. The Sun is a blackbody radiator at a temperature of 6000 K. The solar temperature is determined from the location of the maximum in the spectrum and the relationship given in Eq. (1). The pure blackbody distribution of solar energy is modified by gaseous absorption in the outer layers of the Sun. The solar-energy spectrum measured at the Earth's surface has been further modified by absorption and scattering as the energy traversed the atmosphere. Figure 4 shows the solar-energy spectrum at the top of the atmosphere and at the Earth's surface. Direct sunlight and atmospherically scattered sunlight (sky light) provide ambient illumination for the remote sensing. Reflective remote sensing is limited to the UV, VIS, and NIR spectral ranges because of the presence of solar energy in these bands. The NIR wavelength range is sometimes referred to as reflective IR.

1.1.2 Active Sensing Active remote-sensing systems that supply their own energy source offer significant advantages. Active systems usually have both daytime and nighttime capability since they do not rely on an ambient source of illumination. Some active systems operate at wavelengths that penetrate clouds and rain, so as to provide all-weather sensing. An active energy source can be transmitted continuously or in short bursts or pulses. With a pulsed energy source, the travel time of the pulse from the

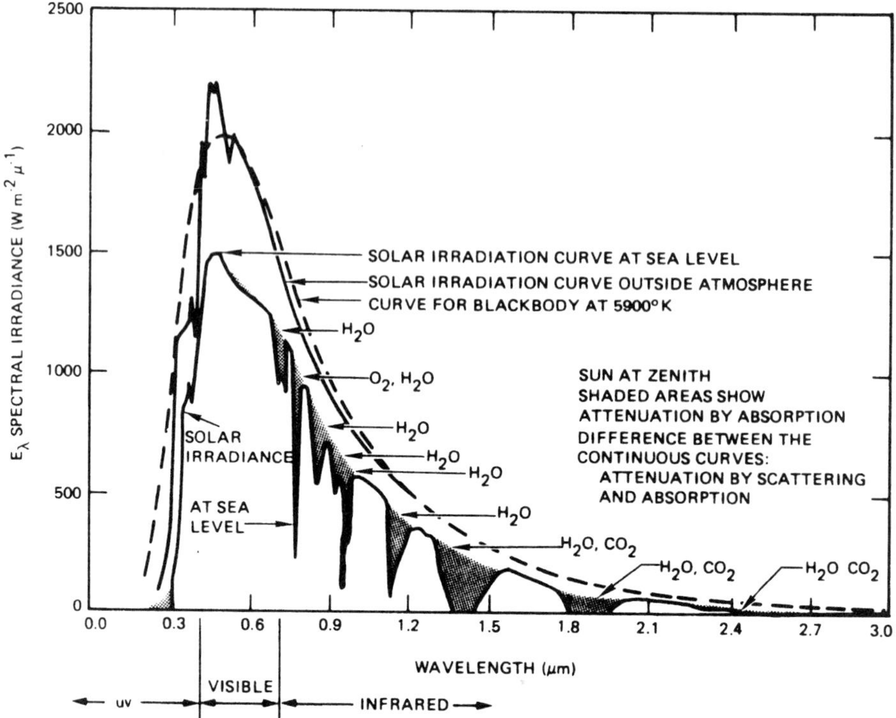

FIG. 4. Solar spectral irradiance at the Earth's surface (from Chahine *et al.*, 1983).

transmitter to the interacting object and back to the receiver can be measured, thereby providing object-distance information.

1.1.2.1 Microwave Wavelengths. Active microwave sensors include radars, scatterometers, and altimeters. Each of these produces microwave energy directed toward Earth by an antenna and, in turn, detects the reflected energy. The word "radar" is an acronym for radio detection and ranging. Radars transmit short pulses of microwave energy in the direction of interest and record the strength and origin of reflections of the pulse from objects within the system's field of view [see RADAR TECHNOLOGY or Skolnik (1980, 1990)]. Airborne radar remote sensing is done with a fixed antenna mounted below an aircraft and pointed to the side. Such systems, termed side-looking airborne radar (SLAR), obtain cross-track position of reflectors based on the time of pulse propagation. That is, reflections of the transmitted pulse from close objects return sooner than reflections from distant objects. Resolution of the reflector's along-track position is provided by aircraft forward motion between pulses. The SLAR produces continuous strips of imagery formed by encoding the strength of the returned pulse echoes as gray scales. SLAR systems were first developed in the early 1950s for military reconnaissance where terrain features were considered "clutter" that masked the targets of interest. With declassification and continued improvement, SLAR has become a powerful tool for acquiring natural-resource data.

Extension of SLAR to spaceborne platforms is not practical because of antenna size requirements. The field of view of a radar is inversely proportional to the antenna diameter. An antenna that would provide adequate spatial resolution on the Earth's surface from orbital altitudes would be too large to deploy in space. The synthetic-aperture radar (SAR) concept is required for spaceborne radar. The idea is that an array of small antennas can be made equivalent to a large antenna. The small antennas do not need to record signals simultaneously, but a large linear antenna can be "synthesized" by the along-track motion of a single small antenna. Radar pulses are transmitted and received from multiple locations along the track, and the returned signals are then coherently processed to produce the effect of a large antenna. SAR has the interesting property of providing ground resolution that is independent of sensor-to-target range, thus permitting high-resolution radar imaging from space. The SAR on the European Remote-Sensing Satellite (ERS-1) obtains radar images at 25-m ground resolution. The future Radarsat will provide 10-m resolution for numerous Arctic, land, and ocean applications.

A scatterometer is a nonimaging active microwave system. Scatterometers record the strength of the backscattered signal without invoking either real or synthetic-aperture image-formation processes. The field of view of the scatterometer is thus large. The observed phenomenon here is the backscatter efficiency, called the radar backscatter cross section, of the material in the field of view. A common use of scatterometry is to measure wind speed over the ocean surface, where radar backscatter cross section increases with increased surface roughness resulting from winds.

The radar altimeter is essentially a nonimaging radar. A pulse is transmitted vertically to the Earth's surface and the elapsed time until the echo is returned is measured. Travel time yields a calculated distance from the antenna to the reflecting surface. If precise orbital information is available to determine the spacecraft position, the distance estimate can be used to determine the elevation of the Earth's surface precisely to within a few centimeters. Surface elevation information is especially valuable over the oceans where variations in the Earth's gravity field and ocean phenomena such as tides and currents result in sea-surface elevation changes that can be measured by altimetry. Investigators have also measured trends in the thickness of the Greenland ice sheet by altimetric methods.

1.1.2.2 Visible Light. Active systems that emit UV, VIS, or NIR light have also been developed. The light source is typically a laser and the systems are called *lidar* (light detection and ranging). The laser is an excellent energy source because of its properties of collimation, monochromaticity, and amenability to short pulsing. Most present lidar systems are ground-based systems looking up into the atmosphere, but the Space Shuttle or Space Station would provide ideal platforms for Earth-viewing lidar experiments. The applications of a spaceborne lidar would include atmospheric species, cloud physics, cloud-top heights, stratospheric aerosols, and surface albedo. The lidar is usually operated in a pulsed mode that allows travel-time information as well as strength of the signal to be recorded. Travel times allow retrieval of vertical profiles of the returned signal.

1.1.3 Imaging and Nonimaging Systems Most remote sensors are imagers; that is, they produce two-dimensional arrays of information representing the spatial distribution of signal within the swath of the sensor. Cameras achieve imaging capability by using a two-dimensional recording medium and an optical element that focuses the angular distribution of scene brightness at the aperture onto the film in a manner that preserves geometric relationships. Figure 1 shows how scanning mirrors, detector arrays, and platform motion can be combined in various ways to achieve imaging. Because of imaging capabilities and the vantage point of high altitude, remotely sensed data are ideally suited for studying spatial distributions of geophysical information. In many applications there is no other way to obtain the data that airborne or spaceborne remote sensing supplies. For example, satellites provide global observations of sea-surface temperature every six hours at 1-km spatial resolution. There is no way that such a data set could be obtained *in situ* by ships and instrumented buoys. Oceanographers were unable to collect adequately sampled, spatially synoptic sea-surface data on regional scales prior to satellite remote sensing. Imaging sensors provide similar sampling benefits for land applications.

The imaging nature of remote-sensing systems dictates that remote-sensing and digital image-processing technologies have developed in concert. Without modern digital image storage, processing, and analysis systems, remote sensing would not reach its potential. Historically, remote-sensing data acquisition has always led data-analysis capabilities. That is, the ability to collect data far exceeds the capability to analyze it. This imbalance in favor of data acquisition continues today. With the rapidly decreasing cost of very capable image-processing workstations and powerful software, there are opportunities today for lower-budgeted organizations, even individuals, to participate in Earth-science investigations based on remotely sensed data, a pursuit that was restricted to well-funded government agencies just a decade ago.

A small number of remote-sensing systems are nonimaging. Nonimaging systems generally give reliable quantitative measures of the integrated intensity of electromagnetic radiation from all objects within their field of view. Often these measurements are made as a function of time and wavelength. The radar altimeter, for example, detects a return of energy integrated over a field of view as large as 10 km along a single line directly below the spacecraft. The altimeter performs very sophisticated time-gated processing of the signal returned from the field of view, producing satellite-to-surface distance measurements that are accurate to a few centimeters. Atmospheric sounders and lidar systems are other remote-sensing systems that are normally deployed in nonimaging configurations.

1.2 Atmospheric Windows

Remote sensing involves propagation of radiation through the atmosphere. Emissive sensing scenarios involve one pass through the atmosphere as the emitted radiation passes from the Earth to an airborne or spaceborne sensor. Reflective and active configurations require two passes through the atmosphere, one for illumination, the other for reflected radiation. For this reason, remote sensing of the Earth's surface is only possible at wavelengths where the atmosphere is relatively transparent. The atmosphere has numerous spectral bands where it is nearly opaque to electromagnetic radiation because of absorption and scattering. The primary absorbers are the triatomic molecules water vapor, carbon dioxide, and ozone. Rayleigh scattering occurs from molecules and other particles that are small with respect to wavelength of the radiation. The strength of Rayleigh scattering is inversely proportional to the fourth power of wavelength. Hence, there is a much stronger tendency for short wavelengths to be scattered by this mechanism. The "blue" sky is a manifestation of Rayleigh scattering. Another type of scattering is Mie scattering, which occurs when particles are of the same size as wavelength. Water or ice droplets and dust are major contributors to Mie scattering.

Figure 5 shows the transmission of a typical atmosphere over a vertical path from the Earth's surface to space. The plotted transmission represents the cumulative effects of the scattering and absorption mechanisms just described. Several spectral bands exist where the transmission of the atmosphere is fairly high. These are called atmospheric windows, and all remote sensing of the Earth's surface is done in one (or more) of these windows. The most commonly used atmospheric windows are the VIS and NIR window (0.4–2.5 μm), two thermal-IR windows (3–5 and 8–13 μm), and the microwave window (1 cm and longer wavelengths). At microwave wavelengths above 2 cm, the transmission of the atmosphere is essentially unity. The discussion of atmospheric windows assumes that the objective is to sense the Earth's surface. However, if the objective is to sense atmospheric phenomena, sensing must occur at wavelengths where there is significant interaction between the atmosphere and the ambient or transmitted energy. Much atmospheric remote sensing, therefore, occurs at wavelengths corresponding to atmospheric absorption bands and not in atmospheric windows.

2. SIGNATURES FROM INTERACTIONS OF ELECTROMAGNETIC RADIATION WITH THE ENVIRONMENT

Interaction of electromagnetic radiation with matter can trigger a number of mechanisms that affect the properties of the resulting wave. The incident radiation, upon interaction with a surface or object, can experience a number of changes—primarily of magnitude, direction, polarization, wavelength, and phase. The effect is a function of the characteristics of the matter and the characteristics of the incident radiation. Some mechanisms operate over a narrow range of wavelengths; other mechanisms are of a broad-band nature. When an electromagnetic wave is incident on an interface

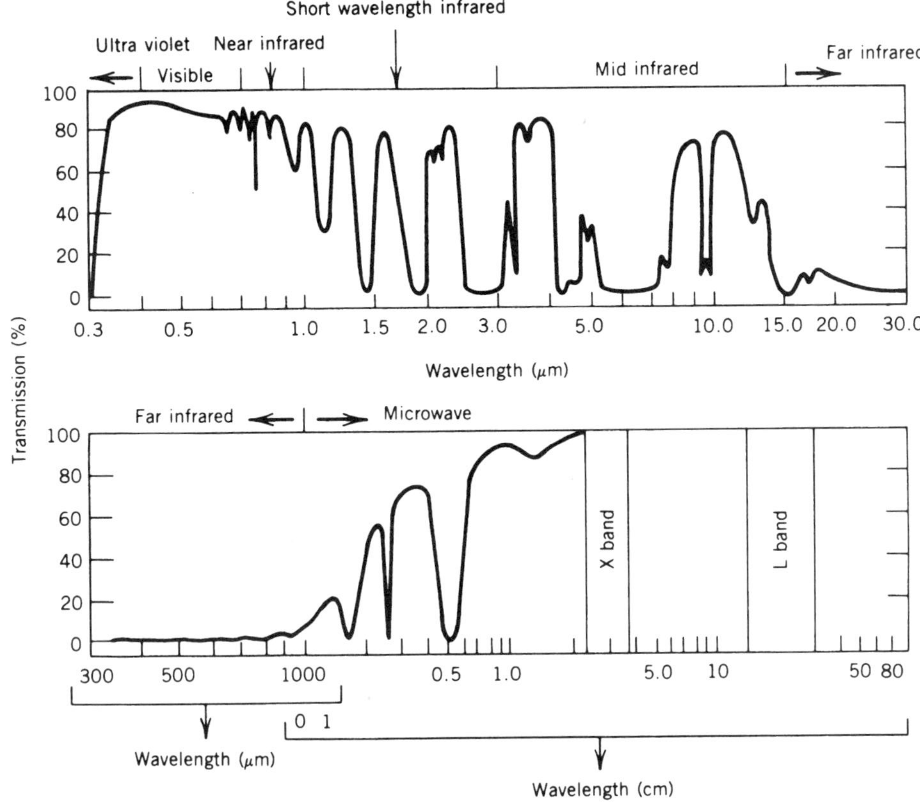

FIG. 5. Atmospheric spectral transmission over a vertical path (from Elachi, 1987, reprinted by permission of John Wiley & Sons, Inc.).

between two materials (in the case of remote sensing one of the materials is usually the atmosphere), some of the energy is reflected, some is absorbed, and some is transmitted through the interface as illustrated in Fig. 6. Reflected energy is commonly divided into two categories; specular (mirrorlike) and diffuse. The transmitted energy is usually absorbed in the bulk material and either reradiated by electronic or thermal processes or dissipated as heat. Conservation of energy requires that the sum of the reflected, absorbed, and transmitted energy is equal to the incident energy. Reflection, scattering, and transmission coefficients of a material are, therefore, not completely independent. These mechanisms are not limited to solid surfaces, but also function as EM radiation traverses the atmosphere or a waterbody.

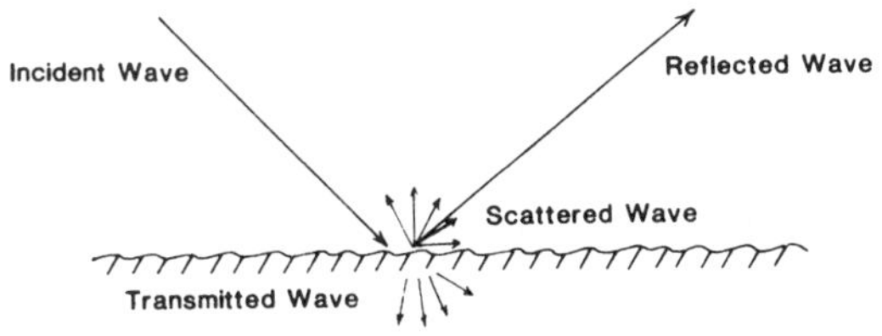

FIG. 6. Wave interaction with an interface (from Elachi, 1987, reprinted by permission of John Wiley & Sons, Inc.).

2.1 Mechanisms

2.1.1 Reflectance If the interface between a material and the atmosphere is smooth relative to the incident wavelength λ (i.e., $\lambda \gg$ surface roughness), the dominant effect is reflectance in the specular direction given by Snell's law. The reflection coefficient is a function of the complex index of refraction n and the incidence angle. Figure 7 shows the reflection coefficient of a material as a function of incidence angle, polari-

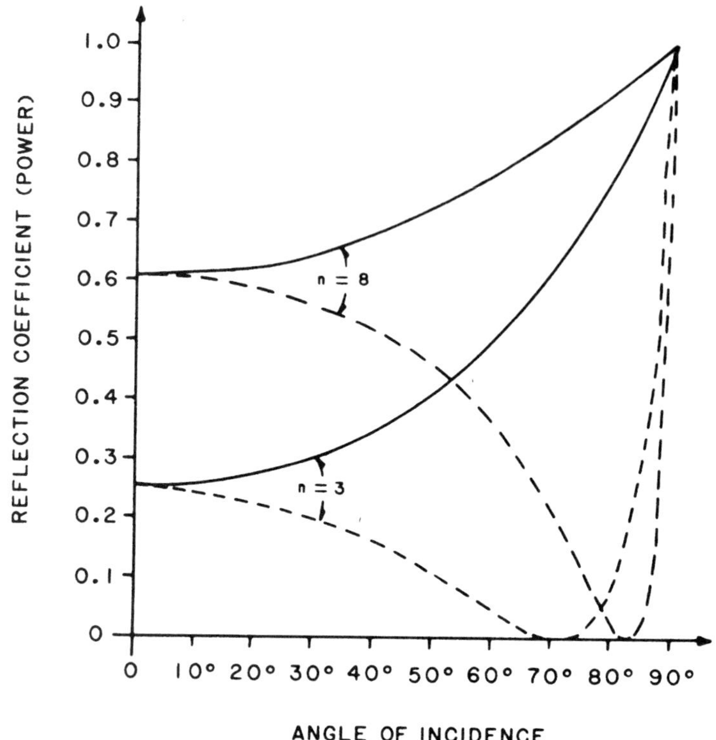

FIG. 7. Reflection coefficient of a half-space with two indices of refraction ($n = 3$ and $n = 8$) as a function of incidence angle. The solid line corresponds to horizontal polarization, the dashed curve to vertical polarization. (From Elachi, 1987, reprinted by permission of John Wiley & Sons, Inc.)

zation, and index of refraction. An interesting feature of the vertical polarization case in Fig. 7 is a reflection null at an angle given by the arctangent of the index of refraction. The angle of no reflection is called the Brewster angle. The index of refraction for a smooth material changes with wavelength, based on its dielectric, conductive, and magnetic properties. Consequently, reflectivity will be a function of wavelength leading to a spectral-reflectance signature that is unique to a material.

2.1.2 Scattering If the interface between two surfaces is rough, scattering is the dominant interaction mechanism. Rayleigh offers the following criterion of surface roughness:

$$h \geq \lambda/(8 \cos\theta), \quad (3)$$

where h is the height of surface variations above a plane in units of wavelength, λ is wavelength, and θ is the incidence angle. The importance of Eq. (3) is that wavelength determines roughness. What might appear rough to the human eye may be smooth to a radar. Scattered radiation has little directivity and is phase incoherent, and its amplitudes fluctuate in a random manner.

In addition to random scattering by surface irregularities, there is another type of scattering, called Bragg scattering, that arises from periodic structure on the interface. The physical principle is that of a diffraction grating. Constructive interference results in stronger backscatter in directions where scattering centers are uniformly spaced at multiples of one wavelength. Bragg scattering is commonly invoked to explain radar backscatter from the sea surface. Wind-generated ripples on the sea surface and radar energy both occur at centimeter scales, thus creating the necessary Bragg conditions.

2.1.3 Emission and Absorption Emission of electromagnetic radiation by objects has already been discussed in Sec. 1.1.1.1. In addition to the temperature and wavelength dependence expressed by Planck's law, emissivity, which is the emissive efficiency of a material relative to a blackbody, is a function of additional variables such as incidence angle, wavelength, and polarization. Kirchhoff's law states that emissivity is equal to absorptance. That is, good absorbers are good emitters. Materials often have unique emissive/absorptive signatures. Water is a good example of a strong spectral dependence in absorption. The absorption coefficient of water restricts remote sensing of the interior volume of a waterbody to a small spectral range in the visible portion of the spectrum. Outside of the visible range the absorption coefficient of water is so high that any electromagnetic energy entering the volume is absorbed near the surface and is not available for backscatter across the interface and into the atmosphere for subsequent detection. For this reason, radars and IR scanners, for example, observe only surface effects over water. Sea-surface temperatures measured remotely are sometimes called "skin temperatures" in recognition of this fact. The radiated skin temperature and subsurface bulk thermodynamic temperature below the surface are, fortunately, usually close. However, differences between skin and bulk temperatures of greater than 10 °C have been observed under unusual conditions.

2.2 Typical Signatures

Interaction mechanisms are different for each material and vary with the state of the material at the time of interaction. These differences create "signatures" for material identification, or produce relationships that can be exploited for observation of the state of the material (temperature, moisture content, age, etc.). Figure 8 shows generic reflectance curves for soil, vegetation, and water. Water is a low-reflectance material at visible wavelengths with no useful reflectance outside of the visible band. Vegetation is characterized by high reflectance in the NIR centered at 1 μm. Soil is characterized by a constant increase in reflectance with wavelength across the spectrum from 0.4 to 2.0 μm. Within these general categories high variability exists because of different species and condition of vegetation, type and moisture content of soil, and turbidity of water. Remote sensing systems are optimized to detect these differences, yielding detailed information about material and its state. Selected examples of signatures and remote-sensing systems designed to exploit them follow.

2.2.1 Vegetation Within the vegetation category, characterized by high NIR reflectance, there is considerable variability based upon species and stress caused by drought or disease. Figure 9 shows typical reflectance curves for deciduous and coniferous trees, which vary considerably in the NIR. Figure 9 also shows the spectral width (0.76–0.9 μm) of channel 4 of the Landsat Thematic Mapper (TM). TM band 4 is ideally placed for deciduous/coniferous discrimination.

2.2.2 Minerals Minerals also show considerable variability within their generic categories. Figure 10 shows reflectance curves for several mineral types. These curves cover the spectral range 2.0–2.5 μm, which corresponds to one of the atmospheric windows shown in Fig. 5. These differences should, therefore, be exploitable for remote classification of minerals in regions of sparse vegetation. The bandwidth of TM channel 7 is also shown in the figure. While it occupies the correct spectral range, TM band 7 is not

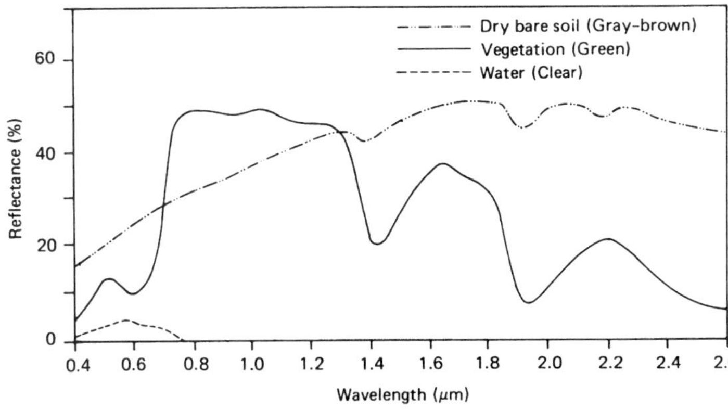

FIG. 8. Typical spectral-reflectance curves for vegetation, soil, and water (from Lillesand and Kiefer, 1994, reprinted by permission of John Wiley & Sons, Inc.).

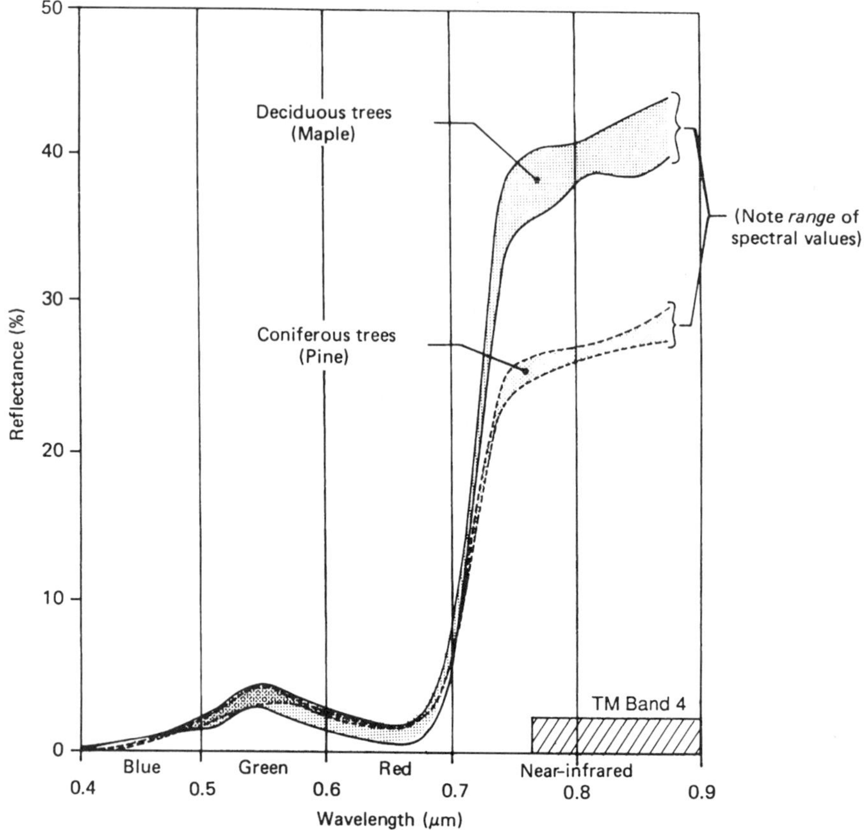

FIG. 9. Generalized spectral-reflectance envelopes for deciduous (broad-leaved) and coniferous (needle-bearing) trees (from Lillesand and Kiefer, 1994, reprinted by permission of John Wiley & Sons, Inc.).

expected to be very useful for mineral classification. The reason for this conclusion is that the differences between reflectance curves of the various minerals are contained in narrow-banded "fine structure." Thematic Mapper band 7 is clearly too broad to resolve the spectral differences. A hyperspectral sensor with numerous narrow bands would be required for mineral classification in the 2.0–2.5-μm region. Mineral classification was one of the primary applications that led the remote sensing community into hyperspectral systems in recent years.

2.2.3 Oceans and Lakes Biological constituents and optical properties of the water column have been retrieved by remote sensing of oceans and lakes. The parameter receiving most attention has been the concentration of the photosynthetic pigment chlorophyll-*a*. Figure 11 shows water-leaving spectral radiance of ocean waters with varying concentrations of chlorophyll-*a*. Note that radiance is inversely related to chlorophyll at blue wavelengths (0.4–0.5 μm) and directly related to chlorophyll at red wavelengths (0.6–0.7 μm). A "hinge point" exists at 0.52 μm where radiance is independent of chlorophyll. This behavior suggests that a good way to retrieve chlorophyll-*a* concentration of a waterbody from remote-sensing data would be to relate chlorophyll-*a* to the ratio of radiance at 0.45 μm to radiance at 0.55 μm. The ratio of two spectral channels is beneficial because changes in illumination level are canceled in the ratio. Figure 11 shows the spectral bandwidths for the CZCS. These bands are located to exploit the hinge point for normalization of illumination levels. Chlorophyll-*a* algorithms for CZCS are based on channel $\frac{1}{2}$ and channel $\frac{1}{3}$ ratios.

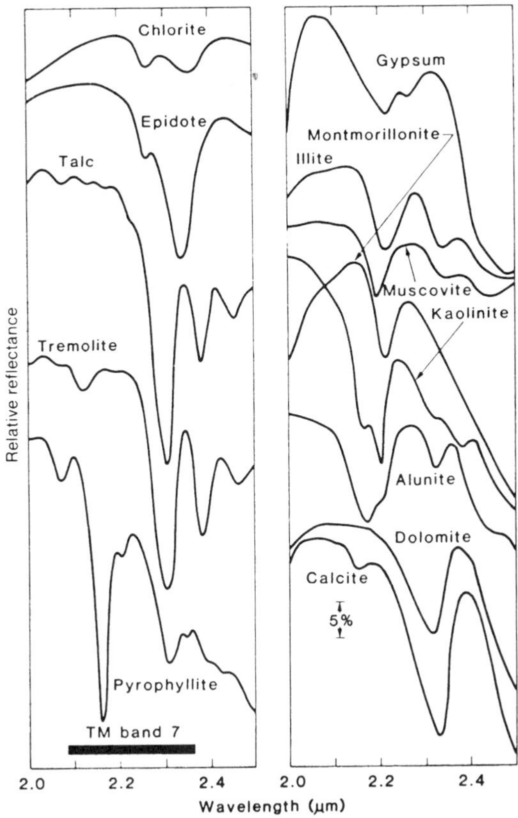

FIG. 10. Selected laboratory spectra of minerals, showing diagnostic absorptance and reflectance characteristics. The spectra are displaced vertically to avoid overlap. The bandwidth of the Landsat Thematic Mapper is also shown. (From Lillesand and Kiefer, 1994, reprinted by permission of John Wiley & Sons, Inc.)

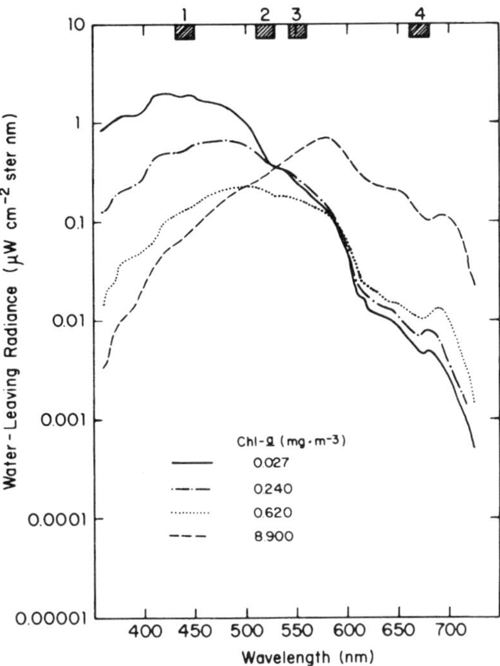

FIG. 11. Surface-leaving spectral radiance for varying concentrations of chlorophyll. Bandwidths of Coastal Zone Color Scanner bands 1–4 are also shown. (From Gordon *et al.*, 1985, reprinted by permission of Academic Press, Inc.)

2.2.4 Sea Ice Large ocean-to-atmosphere fluxes of heat and moisture occur in the Arctic. The Arctic is the birthplace of much of the weather in the Northern Hemisphere, and these fluxes play an important role in weather formation. The amount and thickness of ice cover is an important parameter in estimating heat and moisture fluxes in polar regions. Heat and moisture fluxes from sea ice into the atmosphere are much lower through ice cover than from open water. Ice thickness is difficult to measure directly, but it can be inferred from ice age, which can be measured remotely. Newly formed ice is smooth and has the same salt content as sea water. As ice ages, brine is leached from the ice and the ice becomes gradually "fresher." Snow accumulation on the ice surface also contributes fresh water to older ice. Melting and refreezing associated with the summer melt season and other dynamic processes produce morphological changes as the ice ages. The surface of multiyear ice is rough and porous and of low salinity. Changes in salinity of sea ice change its dielectric properties, which, in conjunction with roughness, change its emissivity, so that the emitted energy levels become a function of the age of the ice. Figure 12 is a plot of emitted energy (expressed as a brightness temperature) as a function of frequency in the microwave region for sea ice of various ages. The figure clearly shows that the microwave band around 30 GHz (called the K_a band) is a good frequency for classifying sea ice on the basis of age. The Special Sensor Microwave Imager (SSM/I) on the Defense Meteorological Satellite Program (DMSP) spacecraft images the Earth at four microwave frequencies, one of which is 37 GHz, which was selected in part for ice-classification purposes.

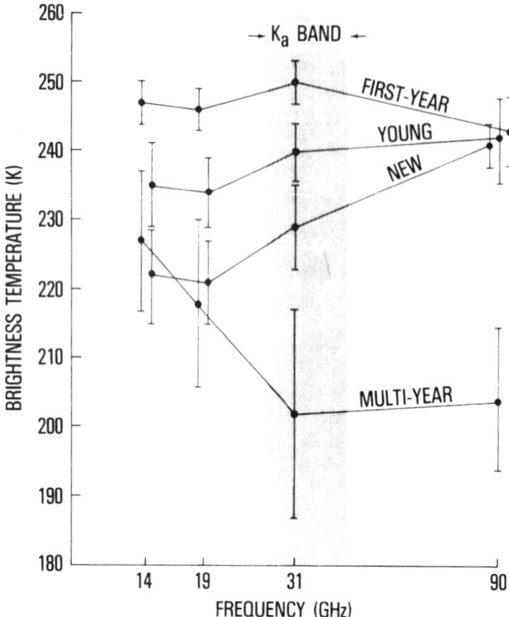

FIG. 12. Radiometric brightness temperature as a function of frequency between 14 and 90 GHz for four ice types. Shaded area shows the K_a band of frequencies. (From Troy *et al.*, 1981.)

The backscatter of sea ice in response to incident radar signals also varies with ice age. For *C*-band radar energy the backscatter for various ice types is given in Table 1. The relationship between ice age and radar backscatter varies with season. Table 1 contains the winter data, which are applied to an ice-classification example in Sec. 4.3.

Further reading on Arctic applications of remote sensing can be found in Massom (1991).

2.2.5 Atmosphere Radiance in absorption bands of atmospheric gases with a fixed concentration can be used to measure vertical profiles of atmospheric temperature. The atmospheric emitting constituent should be uniformly mixed in the atmosphere so that the emitted radiation can be considered a function of the temperature distribution only. The most important absorbing and emitting gaseous constituents in the Earth's atmosphere are water vapor, ozone, and carbon dioxide. Concentrations of water vapor and ozone are variable in time and space and, therefore, unsuitable for temperature sounding. However, the CO_2 band at 15 μm has been utilized for temperature sounding. Looking toward the Earth from space, a sensor can "see" only the upper region of the atmosphere if the atmospheric absorption coefficient at the sensed wavelength is high. The sensor could see progressively deeper into the atmosphere as the absorption coefficient is lowered. If the sensor is designed with several spectral bands with appropriately chosen wavelengths to give the desired range of atmospheric absorption coefficients, the result is an atmospheric vertical profiling (sounding) system. A common procedure for atmospheric sounders is to place their spectral bands along the "shoulder" of an absorption band where small changes in wavelength result in the desired changes in path absorption. Figure 13 shows nominal weighting functions for an eight-channel atmospheric sounder called Satellite Infrared Spectrometer (SIRS-B) on the Nimbus satellite. These sounding channels on the shoulder of the 15-μm absorption band permit temperature retrievals from various depths into the atmosphere. The weighting functions plotted in Figure 13 represent the relative contribution to the sensed atmospheric temperature in each channel as a function of altitude. Each channel is dominated by energy emitted in a different altitude range leading to an atmospheric-temperature profiling capability.

Table 1. ERS-1 Synthetic-Aperture Radar sea-ice classification table for winter (from Fetterer *et al.*, 1994). Backscatter is given in decibels relative to incident power density.

Ice type	Mean backscatter (dB)	Standard deviation (dB)
Multiyear	−10.0	1.01
First-year rough	−14.0	1.06
First-year smooth	−17.0	1.26
New ice/open water	−21.8	1.66

3. REPRESENTATIVE REMOTE-SENSING SYSTEMS

Kramer (1994) gives a survey and short description of 110 spaceborne and 150 airborne missions and sensors, and this listing does not include military or intelligence systems. These are too numerous even to list in this article. Several of the most common and

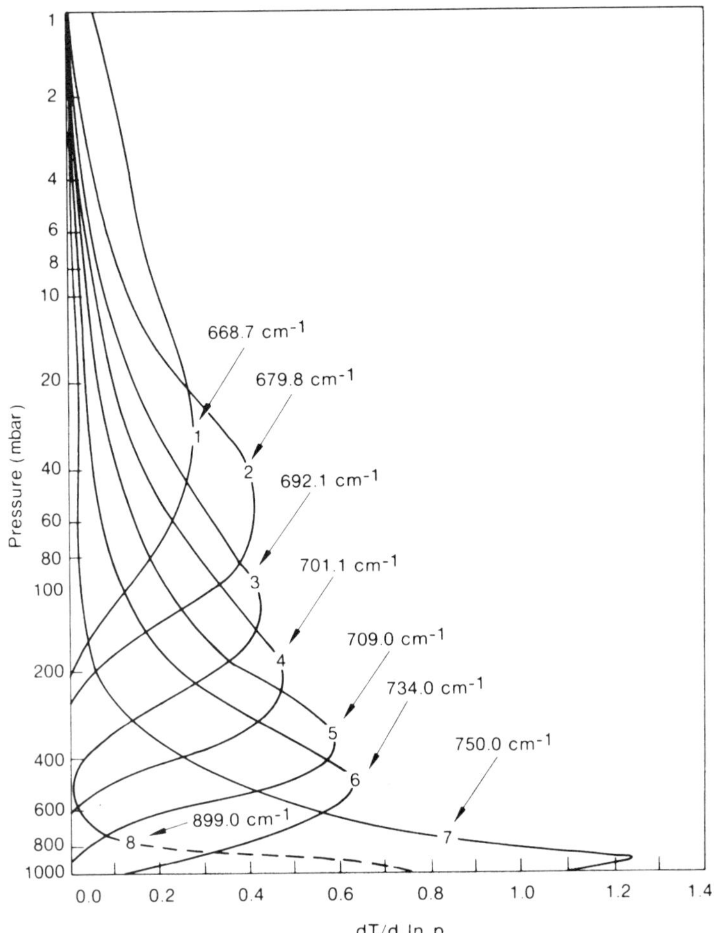

FIG. 13. Nimbus SIRS-B sensor weighting functions for each band in the 15-μm region (from Chen, 1985, reprinted by permission of Academic Press, Inc., Credit: D. Q. Wark, NASA).

representative operational and research satellite systems are described below.

3.1 Operational Systems

Operational systems are those that provide a long-term data source that is the basis of routine products for an end-user community. Satellite sensors that provide inputs for daily weather forecasts are an example. The operational algorithms that convert detector voltage to geophysical information are mature. Algorithms associated with operational systems have been previously validated by researchers, and have been reduced to practice in a form where the accuracy of the resulting parameters is well documented.

3.1.1 Landsat Landsat images are providing the basis for the mapping of detailed structure of the Earth's surface. Applications include land use, agriculture, forestry, geology, water resources, and mapping. Landsat is particularly suited for long-term monitoring of standing vegetation biomass, biological productivity, and the movement of fragile ecosystem boundaries. Landsat-1 was launched in July 1972 equipped with a multispectral scanner (MSS) as the primary sensor with 79-m resolution. Landsat-2 followed in 1975 and Landsat-3 in 1978. The second generation of the Landsat series began with Landsat-4 in 1982 followed by Landsat-5 in 1984. Landsats 4 and 5 extended the sensor suite with a Thematic Mapper (TM) with 30-m (120-m for thermal band) ground resolution. The Landsat series are in sun-synchronous orbit, so that they pass over at approximately 9:30 a.m. local time. The swath width is 185 km. Coverage of a given point on the Earth occurs every 18 days.

Table 2. Landsat MSS spectral bands

Channel no.	Wavelength range (nm)
4	500–600
5	600–700
6	700–800
7	800–1100
8	(10.4–12.6) × 10^3

Tables 2 and 3 give the Landsat MSS and TM spectral bands. (Not all satellites had all bands and the band-numbering system was not the same for all satellites.)

Landsat-6 with an Enhanced Thematic Mapper (ETM) was launched in October 1993, but failed to achieve its orbit. The High-Resolution Multispectral Imager (HRMSI) is the new sensor for Landsat-7, which is planned for a 1997 launch. Landsats 1–5 have collected 820 000 scenes, or 70 terabytes of data.

3.1.2 SPOT The *Systeme Pour l'Observation de la Terre* (SPOT) was conceived and designed by the French Centre National d'Etudes Spatiales (CNES) with participation from Sweden and Belgium. SPOT-1 was launched in 1986 with SPOT-2 and SPOT-3 following in 1990 and 1993, respectively. SPOT-1, -2, and -3 have identical orbits and sensor systems. The SPOT orbit repeats every 26 days, meaning that any given point on the Earth can be imaged from the same viewing angle at this interval. However, the pointable optics on SPOT enable off-nadir viewing so that coverage from different view angles can be obtained at 1-, 4-, and occasionally 5-day intervals. The primary SPOT sensor is the High-Resolution Visible (HRV) imaging system. The HRV operates in either of two modes:

1. a 10-m-resolution "panchromatic" (black and white) mode over the range 0.51 to 0.73 μm, or
2. a 20-m-resolution multispectral (CIR) mode over the ranges 0.50 to 0.59, 0.61 to 0.68, and 0.79 to 0.89 μm.

Table 3. Landsat TM Spectral Bands

Channel no.	Wavelength range (nm)
1	450–520
2	520–600
3	600–690
4	760–900
5	1550–1750
7	(2.08–2.35) × 10^3
6	(10.4–12.5) × 10^3

Because of the off-nadir viewing capability, stereoscopic imaging is possible with SPOT.

The SPOT program has been designed to provide long-term continuity of data through the continuing upgrade and launching of a series of satellite systems. SPOT-4 and -5 are planned for the future. SPOT-4, planned for a 1997 launch, will add a band covering the 1.58- to 1.75-μm portion of the spectrum. The new band is intended for vegetation monitoring and mineral discrimination. SPOT-4 plans also call for the addition of a separate, wide-field-of-view sensor called the Vegetation Monitoring Instrument (VMI), which will be useful in a range of applications where frequent, large-area coverage is important. The VMI will have 1-km spatial resolution and 2000-km swath width for daily global coverage. Spectral bands will be blue (0.43–0.47 μm), green (0.50–0.59 μm), red (0.61–0.68 μm), near-IR (0.79–0.89 μm), and mid-IR (1.58–1.75 μm).

3.1.3 TIROS Series TIROS was the first Earth-observation satellite. The third generation of TIROS satellites are still providing operational meteorological and oceanographic data, twenty-five years after the launch of TIROS-I in 1970. Purdom and Menzel (1995) present a review of the history, present status, and future capabilities of weather satellites. The first generation of TIROS satellites carried Vidicon television cameras operating at visible wavelengths for imaging clouds. Some of the first-generation satellites also carried a scanning infrared radiometer and an Earth radiation-budget sensor. The infrared radiometer was primarily for cloud-top temperatures. Satellites were launched into sun-synchronous circular orbits to provide favorable illumination for the cameras. The second generation of TIROS satellites carried a scanning radiometer (SR), the Very High-Resolution Radiometer (VHRR), and the vertical temperature-profile radiometer (VTPR). In addition to the cloud imaging and cloud-top temperatures, the second generation also measured surface temperatures, and vertical profiles of atmospheric temperature and water vapor. TIROS-N, placed into orbit in 1978, was the first of the third generation of satellites. Modern TIROS-N class satellites carry the AVHRR, which provides VIS, NIR, and multispectral IR imagery. Multispectral IR capability is provided for atmospherically corrected surface temperatures. The AVHRR is capable of measuring global sea-surface temperature with a bias of a few tenths of a degree, and a rms error of approximately 0.7 °C with respect to drifting-buoy data. Third-generation satellites also carry the TIROS Operational Vertical Sounder (TOVS), which consists of a High-Resolution Infrared Sounder (HIRS/2), a stratospheric sounding unit (SSU), and a microwave sounding unit (MSU). The TIROS series were the first satellites for detecting and tracking tropical storms. Table 4 lists pertinent specifications of the AVHRR sensor on the current TIROS satellites.

3.1.4 Geostationary Series A geostationary satellite is in a circular orbit with the orbital plane identical with the Earth's equatorial plane and at an altitude such that the orbital period is identical to the Earth's 23.93-h rotational period. The required geosynchronous period is obtained by placing the satellite at approximately 35 800 km altitude. Under this orbital configuration, the geostationary satellite is in a fixed position relative to the Earth's surface. Because the satellite is in the equatorial plane, its view of high latitudes is very oblique. Geostationary satellites are not generally useful at latitudes in excess of 55°. The main advantage afforded by geostationary orbit is the ability to image the same Earth scene continuously. Operational geostationary satellites transmit images to Earth every 30 minutes. This temporal sampling is ideal for tropical-storm tracking, and severe-weather forecasting. The 12-h sampling interval of a polar orbiting satellite is too long for these meteorological applications.

Table 4. AVHRR spectral bands and applications

Channel no.	Spectral band (μm)	Applications
1	0.58–0.68	Cloud mapping
2	0.725–1.0	Land/water discrimination, vegetative index
3	3.55–3.93	Water-vapor correction
4	10.3–11.3	Thermal mapping
5	11.5–12.5	Water-vapor correction

The first geostationary environmental satellite in 1966 was the Applications Technology Satellite (ATS), which was primarily a cloud imager. The Synchronous Meteorological Satellite (SMS) was launched in 1974 followed by the Geostationary Operational Environmental Satellites (GOES) in 1976. There are normally two GOES platforms, designated GOES East and GOES West, in operation at any given time. These are centered over the equator at 75° W and 135° W, giving combined coverage that reaches from the coast of Africa to the Western Pacific. The sensor on the GOES 1 to GOES 3 was a Visible and Infrared Spin-Scan Radiometer (VISSR). GOES 4 to 7 also carry a VISSR Atmospheric Sounder (VAS), which is both an imager and a sounder. The current GOES satellites (GOES I-M) carry a five-channel (four IR and one visible) imager and a 19-channel (18 IR and one visible) sounder. The spatial resolution of the current GOES is 4 km in the thermal channels and 1 km in the visible channel. Tables 5 and 6 list pertinent information on the current geostationary sensors.

3.1.5 Defense Meteorological Satellite Program The Defense Meteorological Satellite Program (DMSP) is the meteorological program of the U.S. Department of Defense, which originated in the mid-1960s. The evolution of technology in the DMSP satellites generally followed that of the civilian TIROS series. Current spacecraft carry an Operational Linescan System (OLS) as the primary system. The OLS has one visible band (0.4–1.1 μm) and one IR band (10.5–12.6 μm) and has either fine (0.56-km) or smoothed (2.7-km) resolution over a 3000-km swath. Day and night cloud-cover imagery is the OLS objective.

The DMSP satellite carries several other instruments, some of which are of interest in highly specialized applications. However, the Special Sensor Microwave Imager (SSM/I) has broad applications of general interest. The SSM/I is a passive microwave radiometer with four microwave frequency channels, each channel being recorded with both horizontal and vertical polarization for a total of eight channels. The SSM/I builds on a heritage of such instruments as the Scanning

Table 5. GOES I-M Imagery Products.

Product	Spectral bands[a] (μ)m 1 0.65	2 3.7	3 6.7	4 11	5 12	Comments
Clouds	X	X	X	X	x	Channels 2 and 4 are used in combination at night for low-level cloud and fog detection. Channels 4 and 5 are used in combination for cirrus detection.
Water			X	X	X	The 6.7-μm radiance depicts upper tropospheric water vapor. Channels 4 and 5 are used in combination for low-level water vapor depiction.
Surface Temperature		x		X	x	The visible channel is used for daytime cloud screening. Channels 2 and 4 are used in combination over cloud-free ocean whereas channels 4 and 5 are used over land.
Winds	X		X	X		Visible imagery is used for low-level cloud-tracked winds. IR window imgery is used primarily for upper-level cloud motion winds, whereas IR water-vapor imagery is used to provide mid-level winds.
Albedo and IR flux	X		x	X	X	Used for diurnal variation estimation.
Fires and smoke	X	X		x	x	Channels 2, 4, and 5 are used in combination to detect fires. Smoke is observed in visible channel noting it originates from a point source located by the IR channels.

[a]X denotes a primary spectral channel while x denotes a secondary spectral channel.

Multichannel Microwave Radiometer (SMMR) on Nimbus-7 and Seasat. Applications of these data include ocean-surface wind speed, ice coverage and age, areas and intensity of precipitation, cloud water content, and land-surface moisture. The swath width is 1394 km and the spatial resolution varies with channel as shown in the SSM/I characteristics listed in Table 7.

3.1.6 European Remote-Sensing Series

The European Remote-Sensing Satellite, ERS-1, launched by the European Space Agency (ESA) in 1991, is oriented toward ice

Table 6. GOES I-M sounder products.

Product	Resolution (km) Vert.	Hor.	Accuracy Abs.	Rel.	Comments
Temp					Multiple fields of view (FOV) are used for sounding in partly cloudy regimes. Surfaces skin temperature assumes cloud-free FOV.
Profile	3–5	50	2–3 K	1 K	
Land	...	10	2 K	1 K	
Sea	...	10	1 K	0.5 K	
Moisture					Motion is derived through sounding image animation.
Profile	2–4	50	30%	20%	
Total	...	10	20%	10%	
Motion	3 layers	50	6 m/s	3 m/s	
Cloud					Heights are specified using a combination of CO_2 slicing and the window channel temperature–height assignment methods.
Height	2 layers	10	50 mb	25 mb	
Amount	Total	10	15%	5%	
Ozone			30%		Motion is derived through animation of ozone signal derived by differencing ozone and IR window channel images.
Total	...	50	10 m/s	15%	
Motion	1 layer	50		5 m/s	
IR flux	Total	50	10 W/m^2	3 W/m^2	Derived from a linear combination of all IR channel radiances in 3.7–15 μm

Table 7. SSM/I sensor characteristics.

Ch.	Freq. (GHz)	λ (mm)	Pol.	Res. (km)	Environmental products
1	19.35	15.5	V/H	50	Ocean-surface wind, land-surface moisture
2	22.24	13.5	V	50	Columnar water-vapor content, ocean-surface wind
3	37.00	8.1	V/H	25	Rain, cloud water content, ice cover
4	85.50	3.5	V/H	12.5	Rain rate

and ocean monitoring and toward developing and promoting economic and commercial applications of remotely sensed data. Payload priority is given to active microwave instruments. ERS-1 carries a suite of four major instruments. The Active Microwave Instrumentation (AMI) operates in the *C* band and combines the functions of a SAR, a wave scatterometer, and a wind scatterometer. In the SAR mode, the AMI observes a minimum 80-km swath at 100- or 30-m resolution. The wave scatterometer samples each 100 km along track looking at a 5-km square. Wave information covers the wavelength range <100 to 1000 m in 12 logarithmically spaced steps. Wave directional resolution is 30°. Winds are observed at 50-km resolution over a 400-km swath. Wind accuracy is 2 m/s or 10% over the range 4 to 24 m/s. A K_u-band (13.5-GHz) radar altimeter provides significant wave height with 0.5-m accuracy in addition to precise height measurements ($1\sigma < 10$ cm). The Precise Range and Range Rate Experiment (PRARE) is an active tracking system designed to give accurate orbit information in support of the altimeter, as does a laser retroreflector for improved ground-based tracking of satellite position. The final instrument is the Along-Track Scanning Radiometer (ATSR). The ATSR is an IR instrument sensing in the same 3.7-, 10-, and 12-μm channels and the same ground resolution (1 km) as the AVHRR. However, the ATSR scans in a conical pattern to give two looks at each point on the ground, one look nearly vertical and the other at some considerable off-zenith angle. The conical scan limits the swath of the ATSR to 500 km. Multiple paths combined with multiple wavelengths lead to a claim of 0.5-K accuracy in sea-surface temperature over a $50 \times 50\text{-km}^2$ square that is 80% cloud covered. The ATSR also contains a nadir-looking, 2-channel microwave sounder to aid in water-vapor corrections to the IR data.

3.2 Research Systems

The progression in remote sensing systems is typically from aircraft prototype to spaceborne research system to operational satellite. There have been many research satellites that did not move directly into operational systems, but that were highly successful and generated large volumes of valuable data and are worth mentioning here. Data from research systems are generally available only to the science team responsible for development and operation of the sensor. However, some research data sets have been archived and made available to the public. Selection of research systems for inclusion in this section emphasizes those whose data are widely available.

3.2.1 Seasat Seasat, launched in June of 1978, was the first satellite system dedicated to oceanography. Seasat carried primarily microwave sensors to demonstrate that ocean properties such as ocean-surface temperature, wind, and current could be measured to useful accuracy and be mapped globally on an all-weather basis. Seasat carried five sensors including a radar altimeter (ALT), a SEASAT-A satellite scatterometer (SASS), a SAR, a visible and infrared radiometer (VIRR), and a Scanning Multichannel Microwave Radiometer (SMMR). Seasat failed after three months in orbit, but its data set was so good that most of the mission objectives could be met with the data in hand. Data from the Seasat mission are available from the National Aeronautics and Space Administration's Jet Propulsion Laboratory (NASA/JPL).

The ALT operated at 13.5 GHz, was nadir viewing, and illuminated a surface area of diameter ranging from 2.4 to 12 km depending on sea state. Precision measurements of sea-surface height and also estimates of significant wave height were retrieved. The SASS, transmitting at 14.6 GHz, measured surface wind speed. The SAR, imaging in the *L* band (1.275 GHz) looked to the side of the subsatellite track with a 100-km swath cen-

Table 8. Summary of Seasat results.

Sensor	Swath (km)	Res. (km)	Observable	Demonstrated accuracy (1σ)	Demonstrated range of observable
ALT	Nadir	2–20	Altitude	8 cm (precision)	. . .
			Significant wave height	10% or 0.5 m	0–10 m
			Wind speed	2 m/s	0–10 m/s
SASS	1200	50	Wind speed	1.3 m/s	4–26 m/s
			Wind direction	16°	0°–360°
SMMR	600	20–100	Sea-surface temperature	1.0 °C	10–30 °C
			Wind speed	2 m/s	0–25 m/s
			Atmospheric water vapor	10% or 0.2 g/cm^2	0–6 g/cm^2
SAR	100	25 m	Wavelength	12%	>100 m
			Wave direction	15°	0°–360°

tered 20° off nadir. The primary objective was to measure ocean-wave wavelength and direction, but the SAR also demonstrated iceberg detection and ice-lead mapping. The SMMR was carried primarily for sea-surface temperature, but it also measured water-vapor content in the atmosphere using various 6.6-, 10.7-, 18-, 21-, and 37-GHz vertically and horizontally polarized channels. While each of these sensors has been improved over the years and flown on later spacecraft, this combination of sensors on a single platform was unique and has not been duplicated since. Table 8 summarizes the Seasat results. Kirwan *et al.* (1983) is a special volume presenting the scientific results of Seasat.

3.2.2 CZCS The Coastal Zone Color Scanner (CZCS) is the first instrument devoted to the spaceborne measurement of ocean color. Table 9 shows the CZCS spectral bands and the water constituents that they were optimized to sense.

CZCS scans a swath of 1636 km with 0.8-km spatial resolution at nadir. Launched in October 1978 as a proof-of-concept mission, but remaining in service until June of 1986, CZCS produced approximately 86 000 ocean-color images. CZCS demonstrated the feasibility of measuring phytoplankton pigment concentrations from space. The sensor also demonstrated capabilities to monitor ocean dumping, noxious algal blooms, and red tides, and to perform habitat assessment. Based on the success of CZCS, a follow-on ocean-color sensor called the Sea-viewing Wide Field-of-view Sensor (SeaWiFS) is planned for a late 1995 launch.

Table 9. CZCS sensor summary.

Band	Wavelength (nm)	Band responds to
1	423–453	Chlorophyll absorption
2	510–530	Chlorophyll hinge point
3	540–560	Yellow substance, sediments
4	660–680	Chlorophyll absorption
5	700–800	Land vegetation, land/water discrimination, atmospheric aerosols
6	10.5–12.5 μm	Sea-surface temperature

3.2.3 TOPEX/POSEIDON The Topography Experiment for Ocean Circulation is a joint U.S. and French Earth-observation mission. Launched in 1992, TOPEX/POSEIDON as a dedicated altimetry mission is the heart of the World Ocean Circulation Experiment (WOCE). The research objective of TOPEX/POSEIDON is measurement of sea-surface topography for modeling of global changes in ocean circulation and sea level. This objective requires a combination of high altimetric precision and high orbital accuracy.

The spacecraft carries two altimeters and supporting sensors. The radar altimeter (ALT) is of Seasat heritage but operates at two frequencies (13.6 and 5.3 GHz), which allows correction for ionospheric path delays. The second altimeter is the Single-Frequency Solid-State Altimeter (SSALT). This experimental altimeter demonstrates the concept of low-power, low-weight, low–data-rate, and low-cost altimetry for future Earth-observing missions. Supporting sensors on TOPEX/POSEIDON include the TOPEX Microwave Radiometer (TMR), a multichannel microwave radiometer for water-vapor corrections. Accurate orbit positioning is assisted by Doppler Orbitography and Radiopositioning Integrated by Satellite (DORIS), a one-way microwave tracking system, a Laser Reflector Array (LRA) for ground-based laser tracking, and a GPS Demonstration Receiver (GPSDR), which uses signals from the GPS satellites combined with ground reference stations to achieve altitude accuracy of <10 cm.

3.2.4 AVIRIS The Airborne Visible/Infrared Imaging Spectrometer (AVIRIS) represents the current state of the art in hyperspectral systems. The term hyperspectral implies several hundred spectral bands. AVIRIS is a NASA-sponsored Earth-observing imaging spectrometer, which acquires data from a NASA ER-2 aircraft at altitudes up to 65 000 feet. This imaging spectrometer measures the total upwelling radiance from 0.4 to 2.45 μm in 224 spectral channels placed across the spectrum at 0.01-μm intervals. Data are collected as 11- by 100-km^2 images with 20-m spatial resolution. AVIRIS is the forerunner of several new hyperspectral systems. The Hyperspectral Digital Image Collection Experiment (HYDICE) is the next-generation airborne hyperspectral system. NASA has announced a spaceborne hyperspectral imager (HSI) for a summer 1996 launch. Hyperspectral systems (and even ultraspectral, meaning thousands of channels) are expected to be of growing importance in the next few years.

4. TYPICAL APPLICATIONS

4.1 Climate Change—the Ozone Hole

The Polar Ozone and Aerosol Monitor (POAM) and its follow-on, the Orbiting Ozone and Aerosol Monitor (OOAM), utilize proven visible solar-occultation technology and techniques from free-flying spacecraft to measure ozone and constituents important in ozone photochemistry in the stratosphere. The technique consists of measuring solar extinction (i.e., atmospheric absorption of sunlight) at nine wavelengths between 0.353 and 1.059 μm as the Sun rises and sets behind the Earth's atmosphere. From these measurements the abundance profiles of the absorbing gases and aerosols can be determined. The acquisition geometry is shown in Fig. 14.

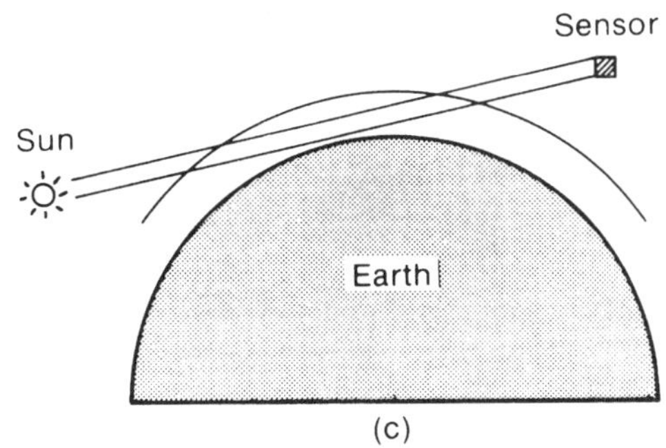

FIG. 14. Solar-occultation sounding method used in POAM-II retrievals of stratospheric ozone profiles (from Chen, 1985, reprinted by permission of Academic Press, Inc.).

The POAM-II sensor was developed by the Naval Research Laboratory and launched on the polar orbiting SPOT-3 satellite. In operation, the sensor calculates the position of the next sunrise/sunset event. The sensor/servo system then acquires and tracks the luminosity center of the Sun as it rises or sets. Nine coaligned telescopes with lens/filter/detector combinations measure solar energy at nine wavelengths. Field of view of the sensors is 0.01°, which corresponds to 0.6-km vertical resolution in the profiles. The instrument provides measurement from 10 to 60 km above the Earth's surface by observing 14 sunrise and 14 sunset events per day.

Plate 1 shows typical ozone-concentration trend measurements from POAM-II in the Southern Hemisphere (over Antarctica). The vertical axis in the figure is altitude in kilometers and the horizontal axis is date ranging from October 1993 to February 1995. The color-encoded ozone concentrations are in units of molecules per cubic centimeter. The figure then shows ozone concentration as a function of altitude and date. Notice the trend of the ozone maximum to lower altitudes in the period December through September, followed by ozone at depletion levels during September through December, and then a reestablishment of the ozone layer at higher altitudes in December, thus completing the cycle. Similar displays can also be produced of water vapor, NO_2, and other constituents. OOAM is planned for a July 1997 launch into a 57° inclination orbit so that its occultation measurements will be made predominantly at mid and low latitudes. OOAM in conjunction with POAM will provide excellent global coverage.

4.2 Geomorphology—Volcanology

Geomorphology is the study of land forms and landscapes, including the description, classification, origin, development, and history of planetary surfaces. Geomorphic processes occur on time scales ranging from seconds to hundreds of years and spatial scales from hundreds of meters to thousands of kilometers. Modern large-scale geomorphic analysis, sometimes called macrogeomorphology, makes extensive use of global observations from spacecraft that employ a variety of imaging and sensing systems. Remote sensing has revolutionized the interpretation of large-scale planetary features.

An excellent example of geomorphological structure observable from space is seen in the radar image of the Mauna Loa volcano shown in Plate 2. This image of the Mauna Loa volcano on the Big Island of Hawaii shows the capability of imaging radar to map lava flows and other volcanic structures. The large summit crater, called Mokuaweoweo Caldera, is clearly visible near the center of the image. Leading away from the caldera (toward the top right and lower center) are the two main rift zones shown here in orange. Rift zones are areas of weakness within the upper part of the volcano that are often ripped open as new magma approaches the surface at the start of an eruption. The most recent eruption of Mauna Loa was in 1984, when segments of the northeast rift zones were active. The South Kona District, known for cultivation of macadamia nuts and coffee, can be seen in the lower left as white and blue areas along the coast.

The image was acquired by the Spaceborne Imaging Radar-C/X-Band Synthetic-Aperture Radar (SIR-C/X-SAR) aboard the space shuttle Endeavour on 2 October 1994. The radar illumination is from the left of the image. Colors in the image are obtained using the following radar channels: Red represents the L band horizontally transmitted and received; green represents the L band horizontally transmitted, vertically received; blue represents the C band (horizontally transmitted, vertically received). Resulting color combinations in this radar image are caused by differences in surface roughness of the lava flows. Smoother flows, called pahoehoe flows, are depicted in red, and rougher flows, called a'a flows in volcanology terminology, are shown in yellow and white. Mauna Loa is one of 15 volcanoes worldwide that are being monitored by the scientific community as an "International Decade Volcano." Radar imagery is ideally suited for volcanology because it penetrates smoke and steam that frequently enshroud an active volcano and prohibit VIS or IR viewing.

The image shown in Plate 2 and others are available over Internet from JPL. Other organizations also maintain on-line archives of remotely sensed data, which can be downloaded by any individual with Internet access.

4.3 Arctic Science—Sea-Ice Classification

ERS-1 SAR imagery received at the Alaska SAR Facility (ASF) is routinely and automatically classified to create ice-type maps. Over 5700 SAR images have been classified and reside in the ASF product archive. The operational ice-type classification algorithm uses lookup tables containing parametric descriptions of the backscatter statistics of different sea-ice types (Fetterer *et al.,* 1994). There is a lookup table for each season, since backscatter is affected by seasonal environmental conditions. A winter lookup table (Table 1) is used when temperatures are below −10 °C. Plate 3 shows the raw ERS-1 SAR image and the operational ice-type classification. Ice-classification errors on the order of 6% have been reported for this ice-classification algorithm. (See also Sea Ice.)

4.4 Arctic Science—Ice Sheets

The ice sheets of Greenland cover 10% of the Earth's land area and contain over 75% of the world's fresh water. Judging from the historical record, future modulations of sea level by fluctuating ice sheet volume are almost assured. Therefore, it is important to monitor the ice sheets and understand the processes that might result in ice sheet growth or decay. Plate 4 shows the topography of the Greenland ice sheet as measured by the Geosat radar altimeter. Zwally *et al.* (1989) observe growth in ice sheet thickness of about 0.2 m/yr over the period 1978 to 1986. Growth of the ice sheet can be attributed to precipitation in excess of the long-term average. Increased precipitation in polar regions may be indicative of global warming.

4.5 Earth's Biosphere—Global Biomass

Plate 5 shows the first truly global view of the Earth's biosphere. The ocean areas within the figure are a composite of 31 352 CZCS scenes from November 1978 through June 1981. The values encoded as colors are chlorophyll concentration based on CZCS channel ratios. The violet/purple shades represent low chlorophyll concentration and the yellow/red shades denote high chlorophyll concentration. The color patterns in Plate 5 show the areas of high primary productivity to be in the polar regions and in coastal upwelling zones along the western boundaries of the continents.

The land portion of Plate 5 is produced from three years of AVHRR data from the NOAA-7 satellite. The parameter displayed over land is the Normalized Difference Vegetation Index (NDVI), which is produced by a ratio of channel sums and differences:

$$\mathrm{NDVI} = \frac{(\text{channel 2} - \text{channel 1})}{(\text{channel 2} + \text{channel 1})}. \tag{4}$$

The AVHRR NDVI integrated over the growing season has been found to be correlated with seasonal primary productivity (Prince, 1991). NDVI values calculated at the end of the growing season have been correlated with above-ground biomass (Kennedy, 1989). Details of the processing that transformed AVHRR channels 1 and 2 into the land-color patterns in Plate 5 are not available, but shades of green denote more lush vegetation and shades of yellow more sparse vegetation. Cracknell *et al.* (1994) have produced a journal special issue describing current global data sets for land from the AVHRR.

4.6 Meteorology—Tropical Storms

Satellite images of hurricanes are now common. During daylight hours tropical storms and their associated structure are apparent in visible-band images from sensors such as the AVHRR or GOES. During daytime and nighttime, the IR bands are useful, providing cloud heights and eye location to the trained observer. On occasion a situation arises where high-level outflow from the storm forms a cirrus-cloud layer over the storm that conceals the storm's center and structure from conventional VIS and IR viewing. In this case the SSM/I passive microwave radiometer on the DMSP satellite can be very beneficial. The microwave radiometer at 85 GHz with horizontal polarization penetrates the cirrus layer to reveal the interior structure of the storm. The 85H channel is sensitive to rain rate, which is a useful parameter exhibiting the eye of the storm and its feeder bands and other structure. Figure 15 is an 85H image of hurricane "Zelda" in the Philippine Sea on 4 November 1994. This is a well-developed hurricane with winds of 135 knots. In Fig. 15 the dark blotches are areas of heavy rain. Note how the "rain" image provided by the microwave sensor captures the eye and spiral banding of the storm.

4.7 Oceanography—Coastal Water Clarity

The CZCS system collected multispectral VIS/NIR data in five channels over the world's oceans for eight years. Algorithms were developed to convert the spectral radi-

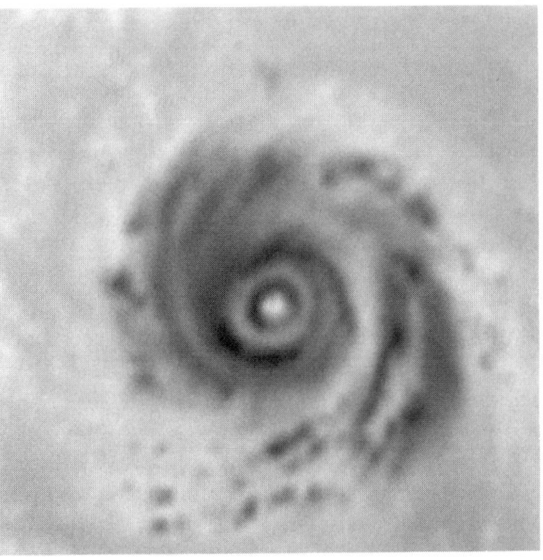

FIG. 15. SSM/I 85-GHz (horizontal polarization) image of hurricane "Zelda."

ances to geophysical parameters of interest to oceanographers. Chlorophyll concentration was the parameter of most general interest because of its association with biological processes. Plate 6 shows another parameter that was commonly derived from CZCS data. This parameter is diffuse attenuation coefficient, which is related to the depth of penetration of a diffuse (omnidirectional) light field into the ocean (see OPTICS, UNDERWATER). The diffuse attenuation coefficient is the reciprocal of the depth at which the light-field intensity is reduced by absorption and backscatter to 37% of its value at the surface. That is, if the attenuation coefficient is 0.1 m^{-1}, then the light field is reduced to 37% of its surface value at a depth of 10 m. Diffuse attenuation coefficient is, therefore, a quantitative way of expressing water clarity, i.e., lower values of the attenuation coefficient correspond to clearer water. Diffuse attenuation coefficient is a function of wavelength. Plate 6 shows attenuation at 0.49 μm. Note the "turbid" waters associated with the Mississippi River outflow and estuaries such as Mobile Bay. Clearest coastal waters along this section of coastline are in the Florida panhandle region.

Plate 6 is not a single image, but is a composite of several images collected over the month of October 1979. Cloud-free portions in individual images have been composited together to form this cloud-free monthly composite. Some artifacts of the compositing process are visible in the figure.

4.8 Land Use—Vegetation Classification

The most common applications of Landsat data are in agriculture, forestry, land use, habitat studies, and other fields where determination of ground-cover type and its condition are central to the application. Plate 7 is a Landsat TM scene from the USGS Batesville quad area of west-central Mississippi. Bands 4, 5, and 3 are used as red, green, and blue components, respectively, for this color rendition. This scene has been classified into eighteen land-use categories using classification techniques consisting of hybrid supervised/unsupervised methods. First, supervised training sites (where land-use type is known and supplied to the classifier) were used to create Level 1 classification categories. These Level 1 categories are water, wetlands, forestry, agriculture, and barren land. Then, the pixels in each of these classes are isolated and further classified using an unsupervised technique that accomplishes grouping (without *a priori* knowledge of what the subcategories represent) of pixels with similar spectral content. Performing the initial separation allows the unsupervised technique to focus on pixels that are in close proximity to each other in spectral space (see Lillesand and Kiefer, 1994, for more information on supervised and unsupervised classification). The hybrid classification was performed on the scene shown in Plate 7 using TM bands 2, 3, 4, 5, and 7 (see Table 3 for spectral ranges). Plate 8 is the classification result with a color assigned to each category.

5. FUTURE OF REMOTE SENSING

Remote sensing is a very effective way to achieve global sampling of geophysical and environmental data on the time and space scales required for global-change studies that are currently receiving much attention. NASA has made a major commitment to acquiring remotely sensed data as an integral part of global-change research by establishing the Mission to Planet Earth initiative. Major new programs like Mission to Planet Earth, combined with the traditional applications of remote sensing, should ensure steady growth for the remote-sensing field. Several current trends in the field are worth noting. These trends are expected to continue and to contribute to continued growth.

1. In the early days of remote sensing the perspective and approach were scene based—that is, obtain an image and extract from it information of interest, treating the scene as an independent piece of data. The trend today is toward analysis of data sets as opposed to scenes. High-volume batch analysis of global data sets leading to databases and statistical descriptions of phenomenology is becoming the new mindset in remote sensing.
2. In the past image-processing hardware was expensive and only government agencies or a few prosperous universities could support remote-sensing activities. Today very capable hardware and soft-

ware are priced where any organization, even some individuals, can enter this field. The trend toward drastically decreased funding requirements for analysis tools should attract many new individuals into remote sensing and should improve educational opportunities contributing toward the ultimate enhancement of the discipline.

3. Historically, remote-sensing data have also been expensive. Landsat scenes presently cost \$4400 to commercial users. However, Landsat data collected prior to 1985 are now available for \$250 per scene, allowing access to this data by a much wider user community. Many organizations are now making their archives of remotely sensed data, or products derived from remotely sensed data, available to the general public via the Internet and the World Wide Web at no cost. For example, Plate 2 was obtained from an Internet archive of SIR-C data that is available to the public. The future certainly holds increased low-cost access to ever-larger volumes of data.
4. Remote-sensing hardware and sensors are also expected to undergo dramatic improvements in the future. For example, NASA/JPL has just announced the second-generation solid-state imaging technology. The device, called the Active Pixel Sensor, is virtually a "camera on a chip." This technology is expected to replace charge-coupled devices because it is cheaper to produce (by a factor of approximately 3), uses less power, and is less susceptible to radiation damage in space.

Because of the factors listed above, remote sensing should be a growing field that will eventually become large enough to sustain profitable commercial enterprise. Initial attempts at commercialization of the Landsat program were not a resounding success, but as remote sensing grows, future commercialization or at least industry/government partnerships should become common. The greatest potential for commercial activity in remote sensing is probably not in the acquisition and distribution of data, but rather in "value added" enterprise—that is, analysis and interpretation of remotely sensed data to form derived products that are of interest to a specific user community.

Exploiters and preservers of the Earth's environment will both find remote sensing to be increasingly crucial to their objectives. Therefore, remote sensing and the future of our planet are inexorably linked.

GLOSSARY

Albedo: Ratio of the amount of electromagnetic energy reflected by a surface to the amount of energy incident upon it.

Aperture: The entrance for EM radiation into a sensing system.

Atmospheric Correction: A procedure that removes the effect of the intervening atmosphere between the scene and the remote sensor. Ideally, the result is data as they would have been sensed if the intervening space were a vacuum.

Backscatter Cross Section: The ratio of power return in a radar echo to power received by the target reflecting the signal.

Brightness Temperature: The temperature of a blackbody radiating the same amount of energy per unit area at the wavelength under consideration as the observed body.

Collimation: Formation of a beam of EM radiation that is nondiverging, i.e., the cross-sectional area of a collimated beam is independent of distance from the source.

Field of View: Angular extent observed by the optics.

Geomorphology: The study of land forms and landscapes, including the description, classification, origin, development, and history of planetary surfaces.

Geosynchronous: Satellite orbit with orbital plane coincident with the Earth's equatorial plane and orbital period the same as the Earth's rotational period. As a result the satellite is stationary over an equatorial point on the Earth. Also called geostationary orbit.

Hyperspectral: A remote-sensing system with hundreds of spectral bands.

Ice Lead: Linear open water or recently refrozen area within the Arctic ice pack.

Multispectral: A remote-sensing system with multiple spectral bands, the number ranging from two to tens of bands.

Nadir: Point on the ground where a line from the sensor to the Earth's center intersects the Earth's surface.

Orthographic Projection: Projection by parallel rays onto a plane at right angles to the rays.

Orthophotographic: Photographs that conform to an orthographic projection.

Pixel: Contraction of picture element. Image element corresponding to instantaneously imaged element of a scene.

Push-Broom: Term applied to a scanning technique where the across-track coverage is obtained by a linear array of detectors oriented perpendicular to the flight track. The field of view of each detector sweeps out a path parallel to the flight direction. By comparison, a conventional scanner moves the field of view of a single detector in the cross-track direction to obtain cross-track coverage of the scene.

Radiometer: Device for quantitatively measuring radiant energy.

Signature: Set of characteristics by which a material or an object can be identified.

Sounding: Data collected in profile, usually along a vertical atmospheric path.

Specular: Refers to a surface that is smooth with respect to the wavelength of incident energy. Reflection is at the same angle as incidence but on the opposite side of the surface normal.

Sun Synchronous: Satellite orbit with altitude and inclination such that the satellite passes over points with the same latitude at the same time each day.

Supervised Classification: Digital-information extraction technique in which the operator provides examples that the computer uses as a basis for assignment of pixels to categories.

Swath: The width of a strip of the Earth's surface viewed by the system.

Unsupervised Classification: Digital-information extraction technique in which the computer assigns pixels to categories without benefit of examples provided by the operator.

Zenith Angle: Angle between a vector and local vertical on the Earth's surface.

ACRONYMS AND ABBREVIATIONS

ALT	Radar Altimeter
AMI	Active Microwave Instrumentation
ASF	Alaska SAR Facility
ATS	Applications Technology Satellite
ATSR	Along-Track Scanning Radiometer
AVHRR	Advanced Very High-Resolution Radiometer
AVIRIS	Airborne Visible/Infrared Imaging Spectrometer
CIR	Color infrared
CNES	*Centre National d'Etudes Spatiales*
CZCS	Coastal Zone Color Scanner
DMSP	Defense Meteorological Satellite Program
DORIS	Doppler Orbitography and Radiopositioning Integrated by Satellite
EM	Electromagnetic
ERS	European Remote-Sensing Satellite
ERTS	Earth Resources Technology Satellite
ESA	European Space Agency
ETM	Enhanced Thematic Mapper
FoV	Field of view
GOES	Geostationary Operational Environmental Satellite
GPS	Global Positioning System
GPSDR	GPS Demonstration Receiver
HIRS	High-Resolution Infrared Sounder
HRMSI	High-Resolution Multispectral Imager
HRV	High-Resolution Visible
HSI	Hyperspectral Imager
HYDICE	Hyperspectral Digital Image Collection Experiment
JPL	Jet Propulsion Laboratory
LRA	Laser Reflector Array
MSS	Multispectral Scanner
MSU	Microwave Sounding Unit
NASA	National Aeronautics and Space Administration
NDVI	Normalized Difference Vegetation Index
NIR	Near-infrared portion of the EM spectrum
NOAA	National Oceanographic and Atmospheric Administration
OLS	Operational Linescan System
OOAM	Orbiting Ozone and Aerosol Monitor
POAM	Polar Ozone and Aerosol Monitor
PRARE	Precision Range and Range Rate Experiment
SAR	Synthetic-Aperture Radar

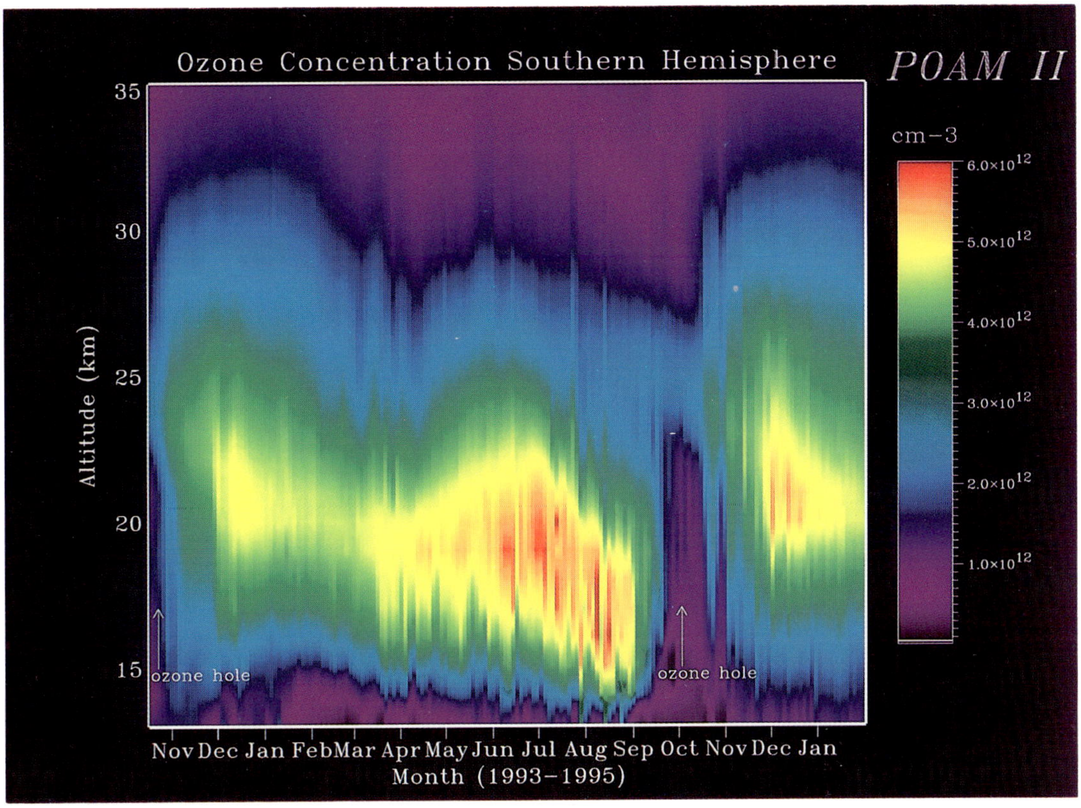

Plate 1. Altitude and time distribution of stratospheric aerosols measured by POAM-II (Credit: D. T. Chen, Naval Research Laboratory).

Plate 2. SIR-C synthetic-aperture radar image of Mauna Loa, Hawaii (Credit: NASA/JPL).

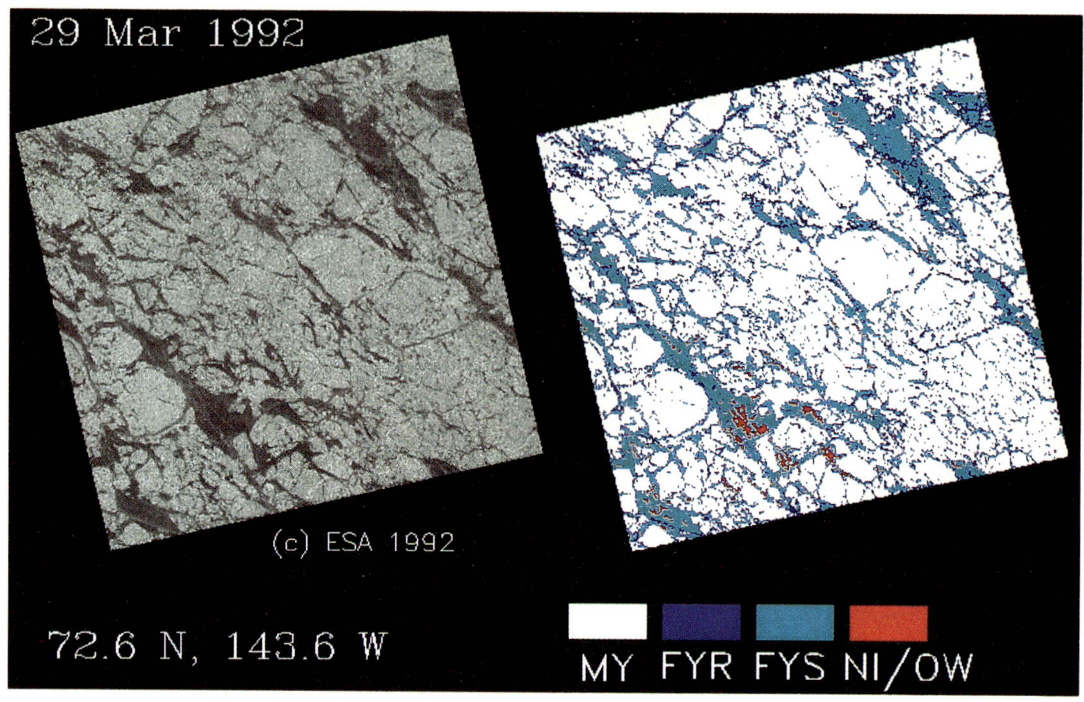

Plate 3. An ERS-1 SAR image from the Beaufort Sea at approximately 72 °N (left), with the winter lookup table given in Table 1 applied to produce an ice classification (right). The table-lookup algorithm found 67% multiyear (MY) ice, 16% first-year rough (FYR) ice, 15% first-year smooth (FYS) ice, and 2% new ice or open water (NI/OW). (From Fetterer *et al.,* 1994.)

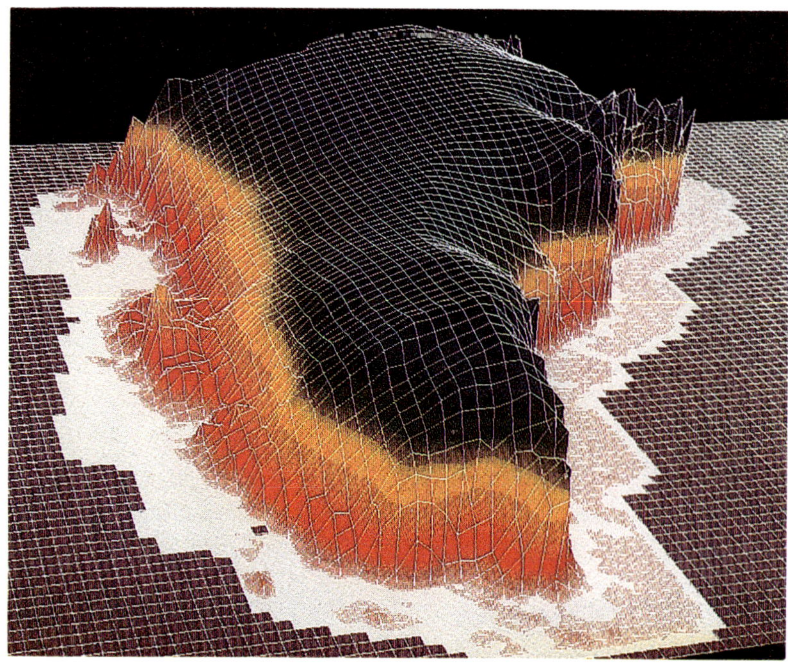

Plate 4. Surface topography of the Greenland ice sheet, as derived from Geosat radar altimetry (Credit: NASA/GSFC).

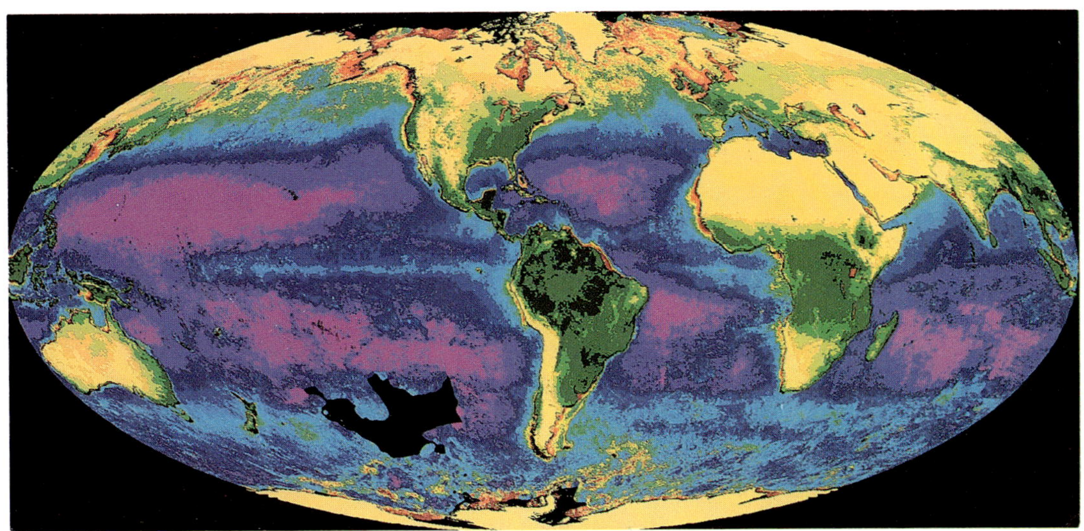

Plate 5. The global biosphere produced from composites of satellite data over a three-year period (Credit: C. J. Tucker, NASA/GSFC).

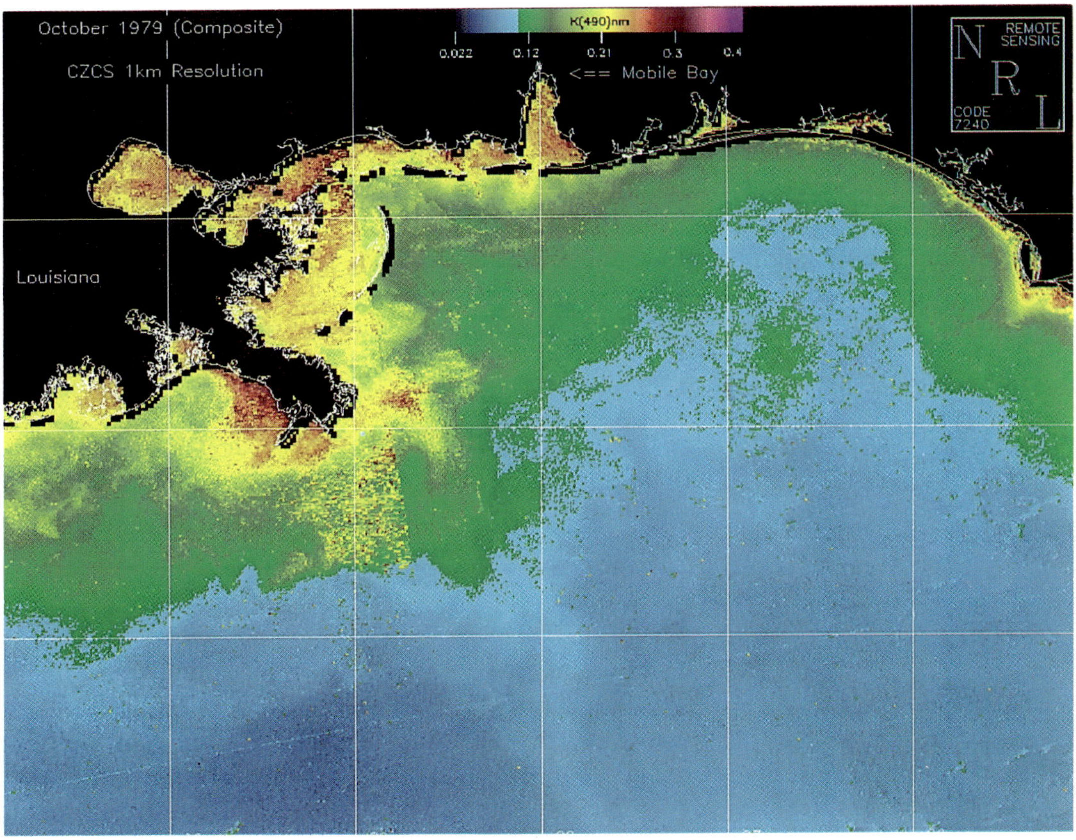

Plate 6. Image of water turbidity along the coast of the Gulf of Mexico from the Mississippi River Delta, Louisiana, to Appalachacola, Florida. The parameter displayed is the diffuse attenuation coefficient of the surface waters as determined from a ratio of CZCS spectral bands. Higher values of diffuse attenuation coefficient indicate more turbid water (shades of red/yellow) and lower values encoded as blue are clearer water. (Credit: R. A. Arnone, Naval Research Laboratory.)

Plate 7. Landsat TM image of the USGS Batesville quad area of Mississippi. Band 4 is driving the red gun, band 5 the green, and band 3 the blue of this color rendition. (Credit: Institute for Technology Development/Space Remote Sensing Center, Stennis Space Center, Mississippi.)

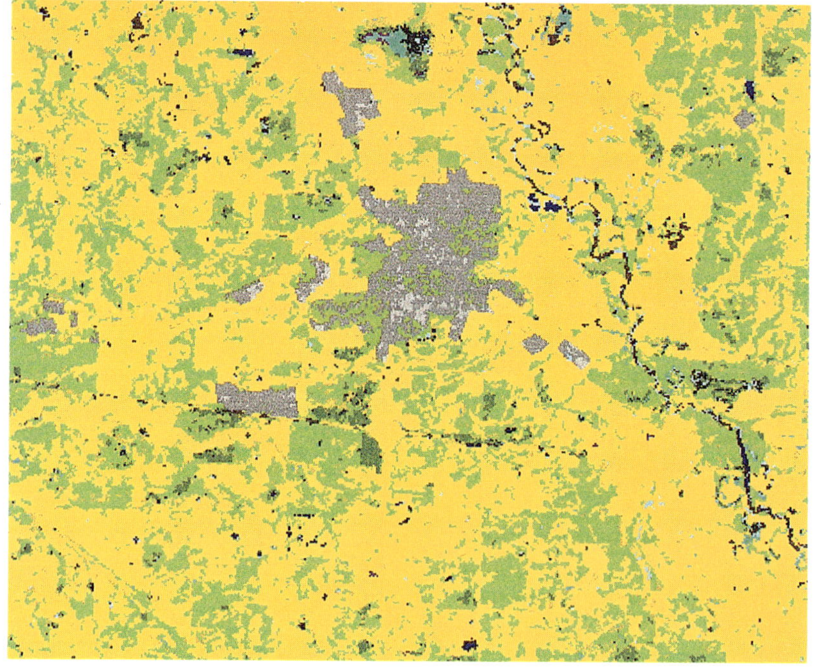

Plate 8. Results of classification of Plate 7 into eighteen land-use categories. The predominant categories are forestry subcategories (various shades of green), agriculture subcategories (various shades of yellow), and two urban subcategories (shades of gray). Other colors represent various types of barren land, wetlands, aquaculture, etc. (Credit: Institute for Technology Development/Space Remote Sensing Center, Stennis Space Center, Mississippi.)

RESEARCH AND DEVELOPMENT

ROBERT A. FROSCH, *John F. Kennedy School of Government, Harvard University, Cambridge, Massachusetts, U.S.A.*

	Introduction	415
1.	**Varieties and Terms**	415
2.	**Relationships among the Varieties of R&D**	416
3.	**Interactions among Kinds of Knowledge in R&D**	417
3.1	Processes of Implementation	418
4.	**Institutions in R&D**	419
4.1	Performers	419
4.2	Other Organizations	420
4.3	Industrial Policy	420
5.	**Funding of R&D**	420
6.	**Concluding Remarks**	421
	Works Cited	421
	Further Reading	421

INTRODUCTION

"Research and development" (R&D) refers to the processes by which new knowledge is discovered and new technology, products, and manufacturing processes are created. "Technology" is used here in the broad sense, to mean the capability and knowledge of "how to" make products and processes. ("A technology" or "the technology" frequently refers to both the body of knowledge and the tools and processes required to make particular kinds of products.) "Product" is also used here in the broadest possible sense, to include things and technical ideas, and abstract or nonhardware processes and products, including software.

1. VARIETIES AND TERMS

"Research" is frequently used to refer to the entire research and development process, but in its more specific sense refers to the search for new knowledge. When driven by interest in the subject and the knowledge itself, it is frequently referred to as "curiosity-driven" research, as "basic" research, or as "fundamental" research. The latter two terms imply that the new knowledge underlies knowledge obtained with other motivations, although this is not always the case.

The U.S. National Science Foundation (NSF) states (NSF, 1993, p. 94) that "the objective of basic research is to gain a more complete knowledge or understanding of the subject under study, without specific applications in mind." It adds that in industry basic research "does not have specific immediate commercial objectives, although it may be in fields of present or potential commercial interest."

"Applied research" is used for the search for new knowledge when there is a purpose in mind for the new knowledge to serve—particularly when it is wanted to help solve some specific problem, frequently a problem posed in an attempt to develop a new technology. The NSF (1993) describes it thus: "Applied Research. . . . to determine the means by which a specific, recognized need may be met." "Mission-oriented research" refers to research undertaken because the resulting knowledge is expected to help the work of the organization doing or sponsoring the research. "Mission-oriented research" is sometimes extended to include the idea of "mission-oriented basic research," in which the areas in which basic research is to be conducted are selected from subject areas in which new knowledge is believed to be likely to be of use in carrying out the mission purpose of the organization. It is usually used in the context of government agency research programs.

"Development" is used for the work necessary to create particular new products, processes for the manufacture of products, and software. NSF (1993) says ". . . systematic use of the knowledge or understanding

3-527-28138-X/96/$5.00 + .50

gained from research directed toward the production of useful materials, devices . . ."

"Exploratory development" has been used by the United States Department of Defense (DOD) in a sense nearly synonymous with "applied research," referring to the search for technological capabilities that may be applied to defense products.

"Advanced development" refers to work undertaken to demonstrate that a particular product or process can be created and function. Advanced development, while it demonstrates the necessary principles and frequently results in a working model, does not usually result in a product or process that has the cost, quality, reliability, and other properties required for replication and general use in the market. Advanced development is sometimes said to result in "proof of principle."

"Engineering development" refers to work that results in a product or process that has the appropriate manufacturability, quality, cost, reliability, and other properties required for routine use. Engineering development may include pilot production, the production of limited quantities of the product on a realistic, but not necessarily full-scale or full-rate, production system. Engineering development is sometimes said to result in "reduction to practice."

Engineering development may include a process of "test and evaluation" of the resulting product and/or manufacturing processes, to demonstrate that they do have the expected properties.

"Operational development" is sometimes used (e.g., in DOD) to refer to development work made necessary by product and manufacturing problems that develop in actual use.

"Innovation" refers to the whole set of processes used to bring a new product or process into the market. This generally includes business processes, as well as the technical processes of R&D.

Basic and applied research are generally considered to be in the domain of science, and more often involve scientists than engineers, while development is generally considered to be in the domain of engineering, and more often involves engineers than scientists. People educated as scientists and as engineers engage in both kinds of work.

2. RELATIONSHIPS AMONG THE VARIETIES OF R&D

The definitions of varieties of R&D described by the terms above cannot be made precise, but are approximate descriptions of activities that may be difficult to distinguish. It is frequently assumed that the terms describe a linear temporal sequence of stages in which basic research is followed by applied research, which is followed by advanced development and then engineering development, the sequence leading in an invariant orderly way to the introduction of new products or processes. This has been called the "linear model" of R&D. In actual practice, the R&D process is often far more intricate and complex.

There are cases in which discoveries in basic research lead to an understanding of possibilities for products or processes, following which there is an orderly progression through further applied research into various stages of development. Such cases, when based upon important and really new discoveries, often lead to revolutionary new products and processes. However, the development of a product may start at any stage of the presumed linear model. For example, what is called "continuous improvement" is normally a sequence of small, incremental changes made to an existing product or process to make it better in use, cheaper, more reliable, etc. These might be described as small engineering development changes, or as operational development changes, frequently undertaken without being formally described as development. It is not unusual for problems in making continuous improvements, or difficulties in engineering development, to lead to requirements for advanced development of technology, or to applied research to produce knowledge needed before the development process can successfully be completed. Sometimes problems arising in development or applied research may suggest gaps in fundamental knowledge, thus leading basic research in a new direction.

Many products and processes are based upon "craft knowledge," developed through years, even centuries or millennia, of practical experimentation, not upon systematically derived scientific knowledge. Systematic attempts are sometimes later mounted to create the scientific knowledge necessary to understand why and how these craft skills work, and what is the basis of the craft knowledge of how to do things. This is particularly the case when attempts are made to extend the use of craft knowledge beyond the domain in which it developed. Thus development may pull upon research.

Basic and applied research frequently depend upon tools that are the product of craft skills, or of much systematic development work. Even in the case when the fundamental idea for the tool arises in the basic research, it may be unusable until much development and engineering work has been done to make the tool useful.

Thus there are complex feedback and feedforward relationships among the stages of the presumed linear model.

3. INTERACTIONS AMONG KINDS OF KNOWLEDGE IN R&D

The complexity of R&D and of the R&D process, driven in part by the many possible relationships among the varieties of R&D, outlined above, is further increased by relationships among subjects and kinds of knowledge. In order to create any new product or process, it is necessary to merge knowledge and technology of various kinds. No physical product or process is exclusively electrical, electronic, chemical, or mechanical; real development always requires knowledge and technology from a variety of fields. For example, electronics must be sound from a heat-flow point of view and must be structurally supported by mechanical means; its manufacture requires chemistry. The development of new transportation vehicles may embody combustion chemistry controlled by electronics, carrying cargo by mechanical means, etc.

Thus development is likely to require a horizontal fusion of subjects and disciplines coming together into a new mix of technology and knowledge. Completely new product possibilities are sometimes created by the process of "technology fusion," as for example in the fusion of mechanical, electrical, electronic control, and computer technologies to create intelligent robotic automation. In addition to horizontal technology fusion, each of the individual subjects may require vertical extension of the underlying knowledge of the discipline to provide new solutions to new problems. All this may require support from other subjects, mathematical solutions for new chemistry, for example, or new means of doing structural or thermal computations for a regime not encountered before in previous products or processes.

The general nature of the requirements for contributions from other disciplines and technologies may be understood in advance, but the requirements are unlikely to be predictable in detail for any but the simplest developments. An ability to draw upon multiple sources of knowledge and technology is likely to be key for successful development, especially of complex products and processes. The history of applied research, and of development, is full of cases in which knowledge created by curiosity-driven basic research, not previously believed to have any connection with any application, has been crucial to the solution of applied and developmental problems. It is generally straightforward to predict what kinds of knowledge will be required for an application or a development; it has proved to be impossible to know in advance what knowledge and technology will be irrelevant. While apparently unconnected in a direct way to application, the body of basic knowledge derived from curiosity-driven research has proven to be essential to application, although sometimes with a long time lag.

The horizontal and vertical complexity of applied research and product and process development has resulted in most industrial R&D, and the R&D of mission-oriented government agencies, being strongly interdisciplinary. Academic research, not directly driven by application requirements, and thus generally in the curiosity-driven research area, has tended to be disciplinary in nature as a result of the traditional organization of scholarly subjects. However, increasing involvement of universities with industrial problems and sponsors, and the increasingly complicated relationships among the underlying disciplines, may change this.

One way of visualizing the complexity of a development process is to consider a matrix in which the rows are labeled with the various technological problems to be solved during a development project, while the columns are labeled with the kinds of knowledge expected to be required (or known from the history of a project to have been required) to complete the project successfully. Such a matrix display provides a map of the transform of knowledge and technology into product or process. Examples can deepen the understanding of the complex ways in which various kinds of knowledge and technology are the ingredients for development of products and processes.

3.1 Processes of Implementation

Since the process of developing a product or manufacturing process can be very complex, and since the nature of what is being developed may be forced by the facts of nature to change during the development, the relationship between the customer for the final development result and those who are carrying out the development is extremely important. (The immediate customer for the final result may be the ultimate market consumer or user, but it is likely that the proximate customer will be someone in a firm or government agency acting as a surrogate for the final user.) Experience suggests that it is extremely difficult to make the requirements for the development result sufficiently complete and unambiguous without also making them so constraining as to preclude the developer from finding or creating the best solution to the problem in terms of performance, cost, etc.

During applied research and development, it is often the case that the understanding of the real nature of the customer's problem may change, as well as the course of the development changing the understanding of possible solutions to the problem. The course of successful applications R&D is frequently marked by adjustments both in the understanding of the problem, and thus its redefinition, and in new understanding of the necessary knowledge and technology that must be discovered and created to solve the, possibly redefined, problem.

To deal with these changes that occur naturally in the course of applied research and development, reasonable dialog between customer and developer before, during, and after completion of a development can be an important determinant of success.

For the reasons just discussed, successful development projects are frequently the product of tightly knit interdisciplinary teams. In many industrial cases, the team must include business and marketing people if the real end requirements are to be reflected in the result of the development.

To enter the market, the product must be manufacturable. It is important that the process of development and design of the product include consideration, design, and development of the means by which it will be manufactured. The process of development and design that considers both together, with the appropriate timing, so that the product and its manufacturing means are well fitted together is referred to as "concurrent engineering." Concurrent engineering enriches the interdisciplinary development team by requiring that product development and design people and manufacturing development and design people work together as a team, creating further complexity and timing complications in the development process.

During the course of research and development, knowledge and technological understanding gained by participants in one part of the process may have to be learned and understood by those who must participate in other parts of the process. When the development has been completed, the results, whether operation of the manufacturing process, or use, maintenance, and operation of the product, will have to be understood and applied by operators and users. The process of moving the knowledge and technological "know-how" from one group of people to another is referred to as "technology transfer."

Technology transfer may occur through the reading of professional and scholarly papers or of patents and similar professional material, but most often, when successful, occurs through direct contact between people with knowledge of technology and its underlying bases and those who will use the knowledge in product and process application and use. In this sense, the continuing dialog between developers and users, referred to above, is the continuing mechanism of technology transfer—really a process in which those who need, and create the requirements for, the knowledge and technology and those who discover and create it continually educate each other.

In its broadest sense, the term technology transfer is used to describe the process by which knowledge and technology created in one organization, or part of an organization, for a particular purpose, is transferred to an entirely different organization, or part of an organization, to be used for some new and different development purpose. While this has sometimes been attempted by the transfer of documents, or simple teaching, when successful it usually involves protracted discussions and the temporary or permanent transfer of people between the organizations. The comment about interdisciplinary teams, made above with regard to development, also applies to technology transfer since it can be regarded as a logical portion of, or extension of, development.

Innovation, the total process of creating change in products, processes, or businesses, and commercialization, the process of ensuring that the results of development are a marketable product or process, may be thought of as the development and technology transfer process extended to the whole set of business or government requirements for introducing something new. The previous comments about the rich complexity and interdisciplinary team nature of the process continue to apply; the field of disciplines and knowledge must be broadened further to encompass all the business and social knowledge and technology required.

The nature of the market for a product or process is sometimes such that commercial introduction must be made by a specific time, or the opportunity for introduction will be missed. In such cases, the "window of opportunity" for introduction creates a pacing time constraint for the development, requiring that it be successfully completed by a given time. This may occur, for example, because of a race between competitors to introduce similar products or services, or because there is a limited time in which a product may be of use before there is a change to a completely new product or technology.

4. INSTITUTIONS IN R&D

4.1 Performers

There are a variety of institutions performing R&D. These include universities, industrial laboratories that are part of general businesses, private commercial laboratories performing R&D for outside customers, private nonprofit research institutions, government laboratories belonging to mission agencies, government laboratories chartered as national laboratories for particular purposes (and frequently operated by private operators under government contract), and private foundations performing R&D.

In the U.S., many universities are referred to as "research universities." In these universities, the faculty both teach and perform research. This is usually basic research, generally within the boundaries of one of the traditional academic disciplines such as physics, chemistry, or a branch of engineering, although increasing attention has been given to subjects, such as environmental issues, that require interdisciplinary work. At the time of this writing (1993), in many universities increasingly strong support ties with industrial and mission agency sponsors are changing the traditional practice in the direction of mission-oriented interdisciplinary research. Faculty members in many universities are increasingly involved in outside consulting work for private industry, or in entrepreneurial activities that involve them in the transfer of their knowledge into industrial technology, and in development work based upon their research. This trend is particularly strong in the biological and health-related sciences.

Industrial laboratories are characteristically engaged in product and process development, although many engage in applied research, and some in some basic research. Industrial R&D is characteristically multidisciplinary, since it is strongly pulled and motivated by product and process development requiring a variety of kinds of knowledge and technology.

Government laboratories belonging to mission agencies span the spectrum from a few that engage in mission-oriented basic research and applied research, to the majority, which are more like industrial laboratories dedicated to the developments required by their respective agencies. Their work is also characteristically multidisciplinary. In the United States, the Department of Defense operates the largest number of government laboratories, although many other departments and agencies have laboratories of their own.

The "national laboratories" in the United States are a special case. Created to deal with the technology required for nuclear weapons and related defense requirements, they are operated by contractors for the Department of Energy. They, too, are multidisciplinary and product and process oriented.

Private commercial laboratories are generally development oriented, performing specialized development and test work tailored to specifically defined customer needs.

4.2 Other Organizations

In addition to the R&D performing organizations, and the sponsors and funders of R&D (see Sec. 5), there are other kinds of organizations important to R&D and its participants, both performers and sponsors. Many of these serve as public markets for scientific, technological, and engineering information and knowledge, and as a means of public quality control on technical information.

These other organizations include a variety of technical societies, such as The American Physical Society, the American Chemical Society, the American Association for the Advancement of Science, and other scientific societies, and a variety of engineering societies. These serve as technical fora in which knowledge and technical information are exchanged at meetings and through technical journals published by the societies. There are also a variety of technical journals published by commercial publishers, and technical books published by university and commercial presses.

There are also trade associations of particular manufacturing sectors that serve, among other functions, as clearinghouses for technical information, and sometimes for the organization of cooperative R&D [e.g., the Electric Power Research Institute (EPRI)]. In addition, there are organizations that cut across industry and serve R&D in a functional manner, such as the Industrial Research Institute (IRI), a forum for the discussion of industrial R&D issues, especially issues of management of R&D.

In the U.S., the national academies (the National Academy of Science, the National Academy of Engineering, and the Institute of Medicine, and their operating arm, the National Research Council) play a role in providing scientific and technological advice to the federal government—frequently advice concerning R&D policies, budgets, and sponsorship.

4.3 Industrial Policy

In Europe and Japan, industrial R&D and government support of industrial R&D have been dominant. In contrast, in the period from the end of World War II until the present in the U.S., Defense Department needs and funding have tended to dominate the funding and performance of R&D. During the last decade in the U.S., there have been increasing calls for the use of the technology developed through this work to be available for industrial use. U.S. government policy has begun to move in the direction of opening government-owned technology and the capabilities and work of the national and government laboratories to use by industry. Mechanisms such as the Cooperative Research and Development Agreement (CRADA) have been developed to foster the transfer of technology from the government to industry.

At the same time, it has become clear that in many fields of technology, such as microelectronics and computer technology, industrial technology productive capability has surpassed that of the government for most purposes, and there is an increasing push to use more "civilian" technology for defense and other government purposes. Policy development in the U.S. in this area is moving very rapidly.

5. FUNDING OF R&D

An idea of the scale and performers of R&D may be obtained from Table 1.

In 1994, the split between federally and

Table 1. R&D in the U.S., European Community (EC), and Japan for 1990 (Alic *et al.*, 1992: Table 4.2, p. 92, Table 4.5, p. 100, Fig. 4.3, p. 98; OECD, 1992, Table 14, p. 22).

	Total (10^9 $)	Government (%)	Private (%)
U.S.	145.5	44[a]	56
EC	101.9	40.8	59.2
Japan	67.0	17.9	73.1

[a]DOD, Dept. of Energy, NASA = 60% of the 44%.

Table 2. U.S. DOD R&D funding in 1990 by category (Alic *et al.*, 1992, Table 4.7, p. 102).

Category	Fraction (%)
Applied research (research and exploratory development)	9
Advanced development	28
Engineering and operational systems development	55
Management and support	8

Table 3. U.S. DOD R&D funding in 1994 by category (AAAS, 1994, Table 1-5, p. 49).

Category	Fraction (%)
Basic research	18.4
Applied research	18.4
Development	59.1
Facilities	4.1

nonfederally funded R&D in the U.S. was roughly 43% to 57% (AAAS, 1994, Table 1-11, p. 55). The DOD funding in 1990 was split among categories of R&D as shown in Table 2.

In 1994, a slightly different categorization gives the figures in Table 3.

The proportions are different for industry-supported R&D, somewhat favoring development. For 1993, the AAAS gives numbers as in Table 4.

Table 4. U.S. industry-supported R&D by category (AAAS, 1994, pp. 23–24).

Category	Fraction (%)
Basic, or fundamental, research	about 5
Exploratory or applied research	nearly 25
Development	73

6. CONCLUDING REMARKS

The complexity of the R&D enterprise, as an intellectual and operating endeavor, as a system of performers and sponsors, and as a social system of people engaged in it will be apparent from the preceding. The complexity continues to evolve with the changing character of modern society.

Works Cited

AAAS (1994), *AAAS Report XVIII Research and Development FY 1994,* Washington, DC: Intersociety Working Group, American Association for the Advancement of Science.

Alic, J. A., Branscomb, L. M., Brooks, H., Carter, A. B., Epstein, G. L. (1992), *Beyond Spinoff: Military and Commercial Technologies in a Changing World,* Boston: Harvard Business School Press.

NSF (1993), *National Science Board, Science and Engineering Indicators—1993,* Washington, DC: U.S. Government Printing Office (NSB 93-1).

OECD (1992), *Main Science and Technology Indicators 1992/2,* Paris: OECD.

Further Reading

AAAS (1994), *AAAS Report XVIII Research and Development FY 1994,* Washington, DC: Intersociety Working Group, American Association for the Advancement of Science.

Alic, J. A., Branscomb, L. M., Brooks, H., Carter, A. B., Epstein, G. L. (1992), *Beyond Spinoff: Military and Commercial Technologies in a Changing World,* Boston: Harvard Business School Press.

Brooks, H. (1994), *The Relationship between Science and Technology,* Vol. 23, Pt. 5, *Research Policy,* Amsterdam: Elsevier, pp. 477–486.

Landau, R. Rosenberg, N. (Eds.), (1986), *The Positive Sum Strategy: Harnessing Technology for Economic Growth,* Washington, DC: National Academy Press.

NSF (1993), *National Science Board, Science and Engineering Indicators—1993,* Washington, DC: U.S. Government Printing Office (NSB 93-1).

OECD (1992), *Main Science and Technology Indicators 1992/2,* Paris: OECD.

RESISTORS

RONALD F. DZIUBA, *Electricity Division, National Institute of Standards and Technology, Gaithersburg, Maryland, U.S.A.*

	Introduction	423
1.	**Electrical Resistance**	424
1.1	Definition of Resistance	424
1.2	Unit of Resistance	425
1.2.1	Quantized Hall Resistance	425
1.3	Importance of Resistors	425
2.	**Resistor Fabrication**	426
2.1	Base Material	426
2.2	Resistive Element	427
2.3	Resistor Terminations and Terminals	427
2.4	Protective Enclosure	428
3.	**Types of Resistors**	428
3.1	Carbon Composition Resistors	428
3.2	Wirewound Resistors	428
3.3	Metal-Foil Resistors	429
3.4	Thin-Film Resistors	429
3.4.1	Metal Film	429
3.4.2	Metal-Oxide Film	430
3.4.3	Carbon Film	430
3.5	Thick-Film Resistors	430
4.	**Resistor Characteristics**	430
4.1	Accuracy	430
4.2	Stability	431
4.3	Power Rating	431
4.4	Temperature Coefficient of Resistance	431
4.5	Load Coefficient	432
4.6	Voltage Coefficient	432
4.7	Humidity Effects	432
4.8	Pressure Effects	432
4.9	Frequency Effects	432
5.	**Classification of Resistors**	433
5.1	Standard Resistors	433
5.2	Resistors for Electronic Circuits	433
5.3	Integrated-Circuit Resistors	433
5.4	High-Current Resistors	434
5.5	High-Voltage Resistors	434
	Glossary	434
	Works Cited	435
	Further Reading	435

INTRODUCTION

Electrical resistance is a fundamental property of a normal electrically conductive material or conductor by which it impedes the flow of electric current in a circuit, resulting in a voltage drop across the conductor and the generation of heat energy. The electrical resistance in a circuit, measured in ohms, is equal to the ratio of the voltage drop across the resistor measured in volts divided by the current measured in amperes. *Resistors* are a class of electrical components whose primary function is to introduce resistance in an electrical circuit for purposes of operation, protection, or control. They are used for voltage measurements, current measurements, setting voltage biases, controlling amplifier gains, setting time constants, matching and loading circuits, heat generation, temperature and humidity measurements, strain gauges, and other related functions.

The first resistors used by experimenters were constructed of iron, copper, or other pure-metal wires of arbitrary lengths and sizes. It soon became evident that a "standard" of resistance was needed to permit the intercomparison of results by different experimenters. In 1861 the British Association for the Advancement of Science appointed a committee to establish resistance standards (British Association for the Advancement of Science, 1913). The first resistance standard consisted of a coil of platinum–silver alloy wire sealed in a container filled with paraffin. It was known as the British Association unit and served as a standard for nearly two decades. Its disuse was a consequence of its large temperature coefficient of resistance (TCR). By 1900 wirewound resistance standards were constructed from manganin (Pe-

3-527-28138-X/96/$5.00 + .50

terson, 1956), a copper–manganese–nickel alloy having substantially a zero TCR near room temperature. The next major development in wire-drawn resistance alloys occurred around 1948 with the discovery of Evanohm® (Peterson, 1956) by the Wilbur B. Driver Co., an alloy of nickel, chromium, aluminum, and copper. Like manganin it also has a small TCR; however, its resistivity is 2.3 times that of manganin. Today manganin and Evanohm® are still the alloys of choice for the construction of wirewound resistors—manganin for low-value resistors (<100 Ω) and Evanohm® for high-value resistors up to 100 MΩ.

The origin of non-wirewound resistors dates back to the time of World War I with their use in communication equipment (Coursey, 1949). Frequency errors of wirewound resistors resulting from series inductance and shunt capacitance made them unsuitable for use in high-frequency amplifiers. A need also arose for multi-megohm resistors that were not available as wirewound resistors for use in radio receivers. Carbon composition resistors were developed to satisfy these needs. Mixtures of carbon and resin are altered to produce resistance materials having a wide variety of resistivities. For the construction of resistors, either the mixture is extruded as a solid rod or it can be deposited on a glass substrate. These resistors can be produced in various physical sizes, resistance values, and power ratings.

Later, film-type resistors supplanted carbon composition resistors for general purpose use because of better accuracies, stability, lower noise, and high-frequency performance. Film-type resistors are available with thin-film or thick-film resistive elements. The materials of choice for the construction of film-type resistors are pure metals, metal alloys, conductive oxides, and semiconductors. Film-type resistors have really come into their own in integrated-circuit and surface-mount technologies.

Around 1964 the metal-foil resistor was developed (Kaufman *et al.*, 1988). This new technology resulted in the production of resistors having smaller TCRs and better long-term stability than film-type resistors. The TCR, accuracy, and stability of these metal-foil resistors are approaching those of the best precision wirewound resistors.

The first section of this article defines resistance and the unit of resistance. It also describes the use of resistors in the measurement of other physical parameters. The following sections describe resistor fabrication, types of resistors, resistor characteristics, and classification of resistors according to use.

1. ELECTRICAL RESISTANCE

1.1 Definition of Resistance

The concept of electrical resistance is based on the experimental work of Georg S. Ohm, who in 1826 formulated the quantitative relationship between voltage and current in a dc electrical circuit. This relationship or "Ohm's law" expresses the fact that the magnitude of the current flowing in an electrical circuit depends directly on the electrical potential difference (voltage) and is inversely proportional to the property of the circuit known as the resistance (Wellard, 1960). In modern terminology his experimental equation for bulk metallic conductors can be expressed as

$$R = E/I,$$

where R is the resistance and E is the potential difference across the conductor when a current I flows. Resistance is a fundamental property of any conductor and can be expressed by the following integral expression:

$$R = \int \rho dl/A,$$

where l is the length, A the cross-sectional area, and ρ the resistivity of the conductor. For thin-film resistors it is convenient to define a quantity ρ_s, called the sheet resistance, which is equal to ρ divided by the film thickness d. A thin-film resistor consisting of a simple rectangle of length l (in the direction of current) and width w has a resistance of

$$R = (\rho/d)(l/w) = \rho_s(l/w).$$

The ratio l/w is usually equated to the number of squares of side w. Then ρ_s has the units ohms per square (Ω/□), since "squares" is a pure number. Thus, the value of a thin-film resistor is equal to ρ_s multiplied by the number of squares.

The power P (measured in watts) dissipated in a resistor may be expressed by any one of the following expressions:

$$P = EI = I^2R = E^2/R,$$

where electrical energy is irreversibly converted into heat.

1.2 Unit of Resistance

In 1862 the BA Committee adopted the name "ohm" for the unit of resistance (British Association for the Advancement of Science, 1913). The following year the committee adopted the "absolute" system of units, which provided compatibility between the mechanical units and electrical units. Throughout the years many different systems of units have been in use, which culminated in 1960 with the adoption of the International System of Units (SI), which since then has been accepted internationally (Harris, 1962). In SI units the ohm is defined in terms of the mechanical units of length, mass, and time—the meter (m), kilogram (kg), and second (s), respectively—and the unit of electric current, the ampere (A). The ohm (Ω) is given by the definition

$$1\ \Omega = 1\ \mathrm{m}^2\ \mathrm{kg}\ \mathrm{s}^{-3}\ \mathrm{A}^{-2}.$$

The ohm can be realized by means of a calculable capacitor whereby its impedance, determined from length measurements and the velocity of light, is compared with a resistance through a complex chain of ac and dc bridges.

Commercial, industrial, and scientific requirements for the long-term repeatability and worldwide consistency of measurements of resistance often exceed the accuracy with which the SI unit of the ohm can be readily realized. To meet these severe demands, it has become necessary to establish a "representation" of the ohm that has superior long-term reproducibility and constancy as compared with the present direct realization of the ohm.

1.2.1 Quantized Hall Resistance By international agreement, the representation of the ohm has been based on the quantum Hall effect (QHE) since 1 January 1990 (Hartland, 1992). The QHE occurs at the interface between two semiconducting layers in suitable MOSFET or heterostructure devices when placed in a large applied magnetic field greater than 1 T and cooled to a temperature of 4.2 K or below. Under these conditions, if a constant current of less than 100 μA flows between the source and drain electrodes, plateau regions of constant Hall voltage and hence Hall resistance occur as a function of magnetic field or gate voltage depending upon the type of semiconductor device. These plateaus as shown in Fig. 1 occur at values of the Hall resistance R_H given by

$$R_H(I) = V_H/I = R_K/I,$$

where V_H is the Hall voltage, I the current through the device, I refers to the ith plateau, and R_K is the von Klitzing constant and is believed to be equal to h/e^2, h being the Planck constant and e the elementary charge. The voltage V_x along the channel is essentially zero at a plateau site, as shown in Fig. 1. The value of the von Klitzing constant adopted internationally beginning 1 January 1990 is equal to

$$R_{K-90} = 25\ 812.807\ \Omega.$$

This conventional value is believed to be consistent with the SI ohm to within 0.005 Ω, corresponding to a relative uncertainty of 0.2 ppm. It is based on direct determinations of R_K using calculable capacitors and indirect measurements involving combinations of fundamental constants that yield h/e^2 independently of R_K. Still there is some small doubt concerning the equality of $R_K \equiv h/e^2$, which can only be resolved with more precise direct measurements of R_K. However, the ohm can be reproduced by means of the QHE to about 0.001 ppm.

1.3 Importance of Resistors

Resistors are important in the measurement of other electrical quantities. A large number of physical and chemical phenomena are investigated by means of measurements of voltages, and the measurement of voltage is customarily carried out by the measurement of resistance ratios. Electric

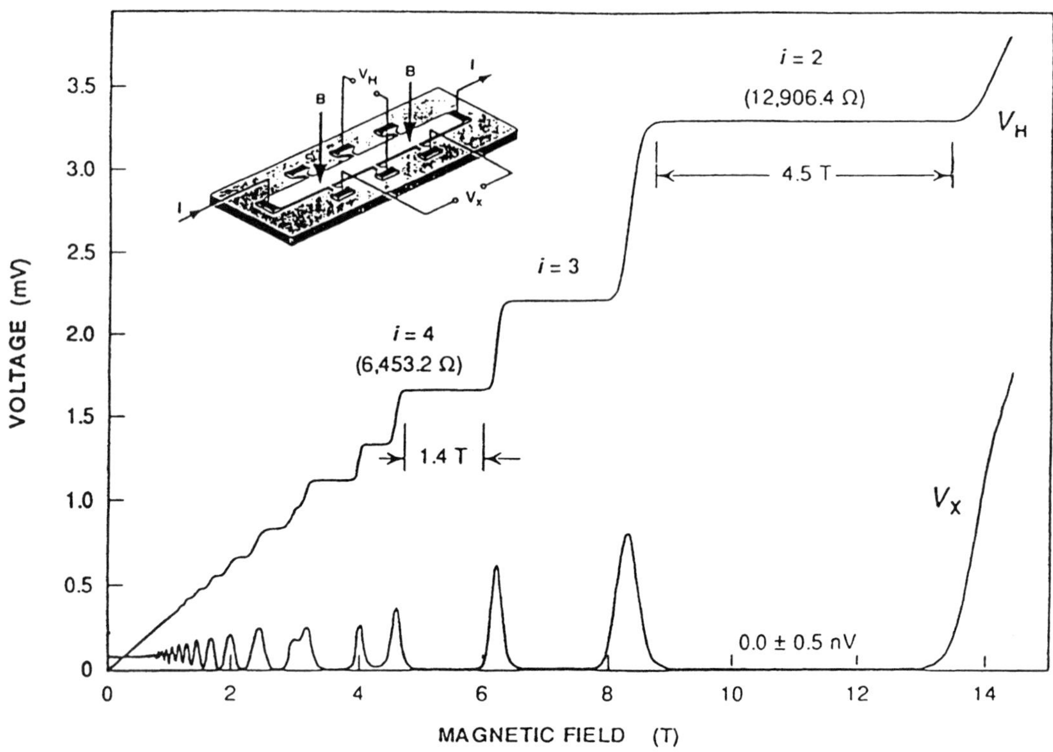

FIG. 1. The quantum Hall effect showing plateaus of constant Hall resistance with values $R_H(i)$.

current is readily most conveniently measured in terms of a voltage across a known resistor. Multimegohm resistors are used in the measurement of low currents extending downward to the picoampere range. Special low-valued resistors, called shunts, are used to measure high currents extending upward to the kiloampere range.

Resistors constructed out of pure metals such as copper or platinum have a high positive TCR and are used in the measurement of temperature. Thermistors are a special type of resistor with a very high negative or positive TCR and are also widely used to measure temperature. Changes in resistance with dimensions are utilized for measuring small displacements, and change in resistance with stress is utilized for the measurement of liquid pressures and force.

2. RESISTOR FABRICATION

The main factors involved in the selection of a resistor for a particular application are cost, resistance value, accuracy, power rating, time stability, frequency dependence, TCR, voltage coefficient of resistance (VCR), and the influence of other environmental factors such as humidity and pressure. Many of these factors can be minimized or optimized during the resistor fabrication process. The constructional features of a resistor can be separated into four general categories: base material, resistive element, terminations, and protective enclosure (Wellard, 1960).

2.1 Base Material

The base material for the resistive element must be a good insulator and have good mechanical stability over the range of ambient conditions to which the resistor is subjected. Common base materials for precision wirewound resistors are usually in the form of insulated metal cylinders, thin mica cards, or ceramic spools. For film-type resistors it is more important to consider matching the thermal-expansion coefficient of the base or "substrate" material with that of the closely adhering resistive element. Fused-sil-

Table 1. Typical electrical properties of some resistance materials.

Material	Composition (%)	ρ (20 °C) ($\mu\Omega$ cm)	TCR (20 °C) (ppm/K)
Carbon	C	3500	−500
Constantan	Cu 60, Ni 40	49	0 ± 20
Copper	Cu	1.7	+3900
Evanohm®[a]	Ni 75, Cr 20, Al 2.5, Cu 2.5	110	0 ± 5
Gold	Au	2.4	+3400
Manganin	Cu 84, Mn 12, Ni 4	48	0 ± 15
Nichrome®[b]	Ni 80, Cr 20	110	+140
Palladium–silver	Pd 60, Ag 40	42	+20
Platinum	Pt	10	+3000
Silver	Ag	1.6	+3800
Tantalum	Ta	13.5	+3100
Zeranin®[c]	Cu 92, Mn 7, Ge 1	43	0 ± 10

[a]Registered trademark, Wilbur B. Driver Company.
[b]Registered trademark, Driver Harris Company.
[c]Registered trademark, Isabellenhütte.

ica glass or alumina ceramic materials are generally used as the substrate material for thin-film, thick-film, and metal-foil resistors because of their excellent electrical and thermal properties.

2.2 Resistive Element

The main considerations in the design and construction of the resistive element are

1. the type of resistance material, and
2. the geometry of the resistance material as applied to the base material.

The material chosen for the resistive element depends on the type of resistor and its application. The main requirements are that the material have suitable resistivity and be sufficiently stable so that any changes in resistance value that occur during its operating life may be expected to fall below some prespecified value. Finally, the process used to construct the resistive element with the chosen material must be such that the final resistor can be made to meet its specifications at a reasonable cost. Tables 1 and 2 list electrical properties of some resistance materials used in the construction of wirewound and film-type resistors. Section 3 describes in more detail considerations of the geometry of the resistance material as applied to the base material.

2.3 Resistor Terminations and Terminals

Terminations are applied to the resistive element to define the exact value of resistance, while terminals provide the mechanical means by which the resistive element is connected into the electrical circuit. Termination is usually accomplished by soldering or welding lead wires to the ends of the resistive element. This is easily accomplished for wirewound elements but is more complicated for film-type elements. Termination for resistive films is commonly made by depos-

Table 2. Typical electrical properties of thin-film and thick-film resistance materials.

Type	Resistance range	TCR (ppm/K)	VCR (ppm/V)	ρ_s ($\Omega/\square$)
Carbon film	4 Ω to 20 MΩ	−250 to −400	<500	50 to 5 × 10^6
Cermet	10 Ω to 5 MΩ	−550 to +150	<100	10 to 10^5
Tin oxide	10 Ω to 100 kΩ	−280 to +140	<25	20 to 400
Nickel–chromium	15 Ω to 150 kΩ	−300 to +350	<1	25 to 300
Tantalum	15 Ω to 150 kΩ	−100 to +100	<1	25 to 300
Tantalum–nitride	15 Ω to 150 kΩ	−25 to +25	<1	25 to 300
Palladium, silver glaze	10 Ω to 1 MΩ	−250 to +250	<1	10 to 10^4

iting noble-metal (gold or silver) contact pads or using noble-metal conducting paint at the ends of the element. Leads then can be attached by compression, welding, or soldering. For carbon composition resistors, the leads are embedded within the resistive material.

2.4 Protective Enclosure

A resistor enclosure provides mechanical and environmental protection to the resistive element. The protective means usually takes one of three forms:

1. an enclosure usually of metal with insulating feedthroughs, ceramic, or glass,
2. a coating applied in liquid form and then cured, or
3. a molding in a suitable encapsulating material.

3. TYPES OF RESISTORS

3.1 Carbon Composition Resistors

The carbon composition resistor is perhaps the most widely used resistor in discrete circuits because of its low cost, high reliability, and small size. Basically, the resistive element is a mixture of carbon and a suitable binder. It is molded under high temperature and pressure into a cylinder with embedded solder-coated wire leads and an insulating plastic or ceramic jacket. Different resistance values are obtained by varying the carbon and filler content. Table 3 lists the standard color code (Wellard, 1960) for composition and some axial-type resistors. Different color bands are used to designate the resistance value and accuracy.

3.2 Wirewound Resistors

The base materials for precision wirewound resistors are usually in the form of insulated metal cylinders, thin mica cards, or ceramic spools. The metal cylinders are coated with an insulating enamel or varnish, and the resistance wire is wound in close thermal contact with the metal cylinder in order to dissipate readily any heat generated by the wire. These metal cylinders are ordinarily of brass, which has a coefficient of thermal expansion nearly the same as that of the wire resistance alloys. This is important to avoid small stresses in the wire introduced by temperature changes. Such stresses can change both the resistance value and its stability.

The mica-card winding form also provides a coefficient of thermal expansion compatible with the resistance wire to minimize temperature effects. The thin profile of these cards produces resistors with very thin cross section, thereby reducing inductance effects of wire loops and making the most efficient use of space. Most ceramic spools have poor heat conductivity and their temperature coefficients of thermal expansion are considerably smaller than for the resistance wires. If ceramic spools are used in the construction of precision resistors, the resistance wire is loosely wound on the spool and the structure sealed in close contact with a thermally con-

Table 3. Resistor color code.

Color of band	1st band, 1st significant figure	2nd band, 2nd significant figure	3rd band, multiplier	4th band, tolerance
Black	0	0	1	...
Brown	1	1	10	±1%
Red	2	2	10^2	±2%
Orange	3	3	10^3	...
Yellow	4	4	10^4	...
Green	5	5	10^5	...
Blue	6	6	10^6	...
Violet	7	7	10^7	...
Gray	8	8	10^8	...
White	9	9	10^9	...
Silver	...	...	10^{-2}	±10%
Gold	...	...	10^{-1}	±5%
None	...	...	...	±20%

ducting insulating fluid such as mineral oil or silicone fluid. The spools for general-purpose wirewound resistors are usually made out of ceramic, epoxy, steatite, or a vitreous material.

Almost all resistance alloys used for constructing wirewound resistors have either a copper or a nickel base (see Table 1). Manganin alloy, whose composition is approximately 84% Cu, 12% Mn, and 4% Ni, has a resistivity of about 48 $\mu\Omega$-cm and a thermal emf to copper of <3 μV/K, and is widely used to construct low-value resistors of nominal values below 100 Ω. Thermal emf refers to the Seebeck thermoelectric effect whereby a voltage is developed when the junctions of a circuit of two dissimilar metals are maintained at different temperatures (Harris, 1962). The TCR of manganin can be reduced to zero at a temperature near room temperature by proper heat treatment; however, the curvature of its resistance–temperature curve at room temperature is approximately -0.5 ppm/K^2. It is subject to surface oxidation and has a pressure coefficient of resistance (PCR) of approximately $+2.3$ ppb/hPa, where ppb refers to parts per billion (10^9).

For the construction of wirewound resistors of nominal values of 100 Ω or higher, Evanohm® is the usual alloy of choice. Its nominal composition is 75% Ni, 20% Cr, 2.5% Al, and 2.5% Cu. It has a resistivity of 110 $\mu\Omega$-cm and a thermal emf to copper of <1 μV/K. The resistance–temperature curve for Evanohm® is much flatter than that for manganin; its curvature at room temperature is approximately -0.05 ppm/K^2. Heat treatment of Evanohm® changes its temperature and time stability. A prolonged heat-treatment process of temperature cycling and soaking time is necessary in order to condition the alloy for zero TCR at a selected temperature and long-term stability. Its PCR is approximately -1.1 ppb/hPa and opposite in sign of the PCR for manganin.

The limitation of wirewound resistors at high-frequency operation is their inherent reactance that is developed as a result of the "coil" type of construction. A number of winding techniques have been developed to reduce the residual inductance and capacitance of wirewound resistors. The most common are the bifilar, thin-card, Ayrton–Perry, and reverse-pi winding methods (Wellard, 1960). The bifilar winding is accomplished by bending the wire back onto itself at its midpoint to form a long loop or "hairpin" with the two sides as close together as the wire insulation will permit. This technique reduces the inductance; however, it may result in a large capacitive effect if the loop is long. Consequently, this technique is used for short loops or low-valued resistors. Winding the resistance wire on a thin card made of mica or other insulating material produces low residual inductance due to the presence of currents in opposite direction separated only by the thickness of the card. Capacitive effects are small since the starting and finishing ends of the winding are at opposite ends of the card. The card form of construction is sometimes used with two windings in parallel, wound in opposite directions around the card, with one winding spaced between the turns of the other. This method, called the Ayrton–Perry winding, gives better cancellation of magnetic fields than the single winding. The reverse-pi method consists of windings in an even number of sections or "pi's" on a bobbin resistor form. As each section is filled, the direction of winding is reversed in the next section, so as to help cancel the inductive effects.

3.3 Metal-Foil Resistors

In the construction of this type of resistor, a layer of thin bulk metal (foil) several microns thick is cemented to a glass or ceramic substrate (Kaufman *et al.*, 1988). The foil is etched using a pattern designed for low residual reactance. The foil is compressed slightly by the substrate as a result of their unequal coefficients of expansion. The compressed foil has a negative TCR that cancels out the inherent positive TCR of the foil. As a result, the metal-foil resistor has a TCR close to zero. Since the foil has a thickness of several microns instead of the extremely thin film of a film resistor, the natural stability of the alloy is preserved, resulting in a resistor having better long-term stability.

3.4 Thin-Film Resistors

3.4.1 Metal Film Materials in thin-film form have higher resistivities than the bulk material because of the additional resistivity due to scattering of the conduction electrons at the boundary of the film (Maissel, 1970). The films are deposited on the substrate by either vacuum evaporation or cathode sputtering.

Nickel–chromium alloy is generally used in the manufacture of thin-film resistors, following logically from its use in bulk form for wirewound resistors. The properties of these films greatly depend on the film composition, which can vary over wide limits depending on deposition parameters such as substrate temperature, source temperature, vacuum, or rate of deposition. Good stable films with TCRs ranging from 5 to 100 ppm/K are obtained by selection. The films have poor stability under conditions of high humidity and therefore must be fully sealed against the entry of moisture.

Single-metal films are often used as resistive elements to avoid problems of composition control such as those occurring in the deposition of alloys such as nickel–chromium. Tantalum films are widely used because of their relatively high resistivities and low TCRs as compared with other pure-metal films and the tough, self-protective oxide that tantalum forms during heat treatment in oxygen. This oxide has excellent dielectric properties; this has led to the development of tantalum capacitors.

3.4.2 Metal-Oxide Film The most popular metal-oxide or cermet film (Maissel, 1970) used in the manufacture of resistors is chromium–silicon monoxide (Cr–SiO). This film is produced by the coevaporation of Cr and SiO and features high resistivity and stability with a TCR on the order of 100 ppm/K. The electrical resistivity can be varied over several orders of magnitude by varying the Cr content.

Tin-oxide (SnO_2) films (Maissel, 1970) used as resistive elements are formed by spraying the material in vapor form onto a heated glass or ceramic substrate. The reaction results in an extremely adherent, fully oxidized film. SnO_2 films have high resistivity and are very stable and rugged. They can withstand temperatures as high as 450 °C without deterioration. One interesting feature of SnO_2 films is that they possess a high degree of transparency. They have therefore found wide application in areas such as the manufacture of so-called "conducting glass" and as heating elements.

3.4.3 Carbon Film Carbon-film resistors (Grisdale *et al.*, 1951) have supplanted carbon composition resistors for general-purpose use because of lower cost, better tolerances, increased stability, lower noise, and better high-frequency performance. They are fabricated by pyrolytic decomposition of carbon on the surface of a cylindrical ceramic substrate. The characteristics of the resistors are sensitive to the deposition conditions, especially film thickness. During fabrication the resistance value of a carbon-film resistor is adjusted by a procedure known as spiraling, i.e., a thin grinding wheel is used to cut a groove through the film along a helical path to increase the resistance.

3.5 Thick-Film Resistors

Mixtures of metal, metal compounds, glass, and solvents are commonly used for the manufacture of resistive elements (Dummer, 1970). The usual method of applying these thick films on a flat or cylindrical alumina substrate is by dipping or rolling. The substrate with the liquid glaze (e.g., tantalum, tantalum nitride, glass, and a carrier) is fired at approximately 1000 °C to fuse the glass particles and bind the film to the substrate. This technology provides a resistive element that is impervious to environmental conditions without the need for an air-tight encapsulation. The inherent ruggedness of this glaze allows it to absorb higher voltage surges and overloads than a thin-film counterpart.

4. RESISTOR CHARACTERISTICS

A summary of key characteristics of different resistor types is listed in Table 4.

4.1 Accuracy

All resistors have a specified initial accuracy or tolerance. It expresses the maximum deviation in resistance from its nominal value in either percent or ppm. Carbon composition and some other film-type resistors use a color code to denote tolerances as listed in Table 3. Tolerances are typically 5% to 20% for carbon composition resistors, 0.5% to 10% for carbon-film resistors, 0.1%

Table 4. Typical resistor characteristics.

Characteristic	Wirewound (precision)	Metal foil	Metal film	Carbon film	Carbon composition
Resistance range	0.1 mΩ to 100 MΩ	0.5 Ω to 1 MΩ	1 Ω to 10 MΩ	1 Ω to 10 TΩ	1 Ω to 100 MΩ
Power dissipation	10 mW to 100 W	0.5 to 2 W	50 mW to 2 W	0.1 to 2 W	$\frac{1}{8}$ to 2 W
Tolerance	2 ppm to 1%	50 ppm to 1%	0.05% to 1%	0.5% to 10%	5%, 10%, 20%
Frequency limit	50 kHz	100 MHz	400 MHz	100 MHz	1 MHz
TCR (ppm/K)	±1 to ±20	0 ± 2.5	±20 to ±200	−200 to −1000	−800 to +1600
VCR (ppm/V)	0	<0.1	<1	−10	−50
Stability	Best	Good	Good	Moderate	Poorest
Advantages	Highest accuracy and stability	Low TCR, low VCR, high speed	Best frequency response	High resistance and low cost	Low cost and small size

to 1% for metal-film resistors, 0.01% to 1% for precision wirewounds, and 0.0001% (1 ppm) to 0.01% (100 ppm) for ultraprecision or standard resistors (Wellard, 1960).

4.2 Stability

Stability refers to the change in resistance value with time or environmental stress. Stresses include high- or low-temperature exposure, application of full rated power, short-time overload, moisture, soldering heat, or radiation exposure. Carbon composition resistors generally have the poorest stability; wirewounds are the most stable, followed by metal-foil, metal-film, and carbon-film resistors.

4.3 Power Rating

The power rating of a resistor is defined as the maximum specified wattage that can be continuously dissipated at a maximum specified temperature without damage to the resistor. The choice of materials in the construction of the resistor along with its physical size determines its power rating. An important consideration is the temperature rise of the resistor. If the resistor is mounted so that it is close to other heat-producing components or has restricted ventilation, the power dissipation of the resistor may have to be reduced to avoid reaching the maximum allowable temperature.

4.4 Temperature Coefficient of Resistance

All resistance materials exhibit some change in resistance with temperature, and the magnitude of this change is usually expressed in %/°C or ppm/°C. For general-purpose resistors over a limited temperature range, this change in resistance with temperature is treated as a linear function given by the relationship

$$R = R'[1 + \alpha_0(t - t')],$$

where R and R' are the resistance values at temperature t and reference temperature t', respectively, and α_0 is the mean TCR expressed in proportional parts per °C over the temperature range of operation (Harris, 1962). For most materials the resistance–temperature curve is not linear; therefore, for precision resistors, especially wirewounds, this relationship is more accurately expressed as

$$R = R'[1 + \alpha(t - t') + \beta(t - t')^2],$$

where α is the slope of the resistance–temperature curve at t' and β is the curvature at any temperature over the range of operation.

The TCRs of film-type resistors (Maissel, 1970) can span a wide range, as shown in Table 2. The main factors that cause this effect are the type of substrate, film thickness, film composition, film structure, and important parameters during deposition such as substrate temperature. Resistance in bulk materials results from electron scattering by phonons, impurities, and defects. In thin-film materials, surface boundaries cause additional scattering effects that influence the resistivities and TCRs of these films. Typically, thin films have lower TCRs compared with bulk materials primarily because of the high resistivity of thin films.

4.5 Load Coefficient

The load coefficient of a resistor is defined as the change in resistance caused by the production of joule heating (I^2R) in the resistor at the rate of one watt and is usually expressed in ppm/W. This change, however, is a function of time. When a current starts flowing in a resistor the resistance material (wire or film) very quickly takes up a temperature above that of the base material in which it is mounted. The amount of this change depends on the thermal contact between the resistance and base materials, thermal time constant of the resistance and base materials, and TCR of the resistance material, and also upon the changes in stress in the resistance material and its support. The load coefficient of a resistor is a rather indefinite quantity, varying with time of flow of current and with the environment of the resistor.

4.6 Voltage Coefficient

When voltage is applied to a carbon or film-type resistor there may be a decrease in resistance apart from changes caused by loading of the resistor (Wellard, 1960). This decrease in resistance with increase in voltage results from the breakdown of contact resistances between carbon particles in carbon composition resistors or the breakdown of boundary resistances between islands of material on substrates of film-type resistors. This change in resistance per applied volt is called the voltage coefficient of resistance. In carbon composition resistors, the change in resistance value due to applied voltage is usually −200 ppm/V or less. For carbon-film resistors, the VCR can vary from −10 to −50 ppm/V. Metal-film resistors have VCRs from −1 to −30 ppm/V depending on their wattage, whereas metal-oxide–film resistors have VCRs from −1 to −5 ppm/V. Wire-wound resistors do not exhibit a VCR.

4.7 Humidity Effects

Unless the resistive element is hermetically sealed in a glass or metal container with glass-to-metal seals, moisture can permeate a resistor and change its resistance value. Moisture can produce two reversible effects:

1. On the surface of a high-value resistor or its housing it can provide a path for leakage currents and thus lower the apparent resistance.
2. If moisture is absorbed through the insulator protecting the resistive element, such as the enamel coating of resistance wire, or the jacket of a carbon composition resistor, or the overglaze film of film-type resistors, the insulator will swell, resulting in pressure being exerted on the resistive element. This effect usually causes the resistance to increase and is directly proportional to the insulator thickness and inversely proportional to the thickness of the resistive element (Starr *et al.*, 1970). The increase can vary from 0.1% to 10% for a 50% change in relative humidity depending on the type of resistor.

4.8 Pressure Effects

Resistance materials are affected by changes in pressure (Meaden, 1965). These changes of resistance as a function of pressure are not well understood, and, depending on the material, resistors may have either a positive or negative pressure coefficient of resistance. Fortunately, for well-constructed resistors these changes of resistance with pressure are only significant for those of the highest quality. For pressure changes of 500 hPa, the change in resistance would be within 1 ppm.

4.9 Frequency Effects

A resistor may be represented to a first approximation by the circuit of Fig. 2, which shows a pure dc resistance R in series with

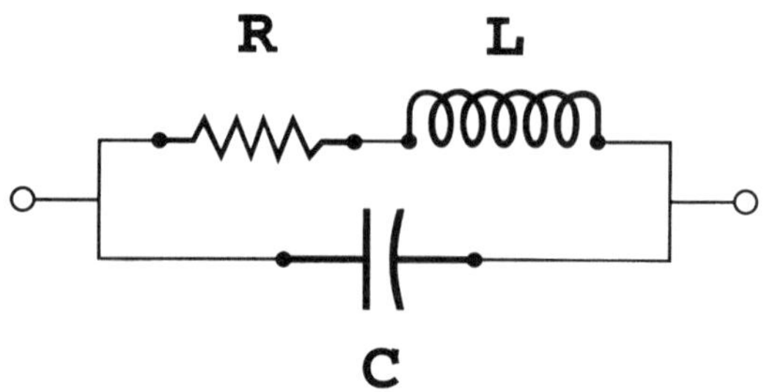

FIG. 2. Simplified model of a resistor.

an inductance L and this combination shunted by a capacitance C (Harris, 1962). The net equivalent inductance is given by

$$L' = L - CR^2.$$

The overall inductive effect of a resistor is reduced by its shunt capacitance. The time constant of a resistor is defined as

$$\tau = L'/R = L/R - CR.$$

The time constant of a resistor may be either positive or negative, depending on which of the terms, L/R or CR, predominates. For a well-designed ac resistor, the value of the time constant should be 10^{-7} or less. Low-valued resistors generally have a positive time constant, which means a net inductive effect, while high-valued resistors have a negative time constant or a net capacitive effect. The crossover point for wirewound resistors is around 100 Ω. Film-type resistors have the best high-frequency performance. The effective dc resistance of these resistors for most resistance values remains fairly constant up to 100 MHz and then decreases at higher frequencies (Hierholzer *et al.*, 1977).

For high-accuracy requirements other frequency effects need to be considered. These include dielectric losses, losses arising from eddy currents induced in nearby conductors, and losses due to skin effect. The skin effect (Harris, 1962) results from a decrease in the current density in the interior of the resistive element and an increase toward its surface as the frequency increases. The effective resistance increases and is a function of the resistivity of the material and the geometry of the resistive element.

5. CLASSIFICATION OF RESISTORS

The most common classification of resistors is according to their intended use and inherent performance.

5.1 Standard Resistors

Standard resistors (Harris, 1962) are used for calibration purposes in resistance measurements. They are usually commercially available only in decimal multiples from 0.01 mΩ to 10 TΩ. Standard resistors of nominal values to 100 MΩ are constructed of wire or strip resistance material. Manganin is the material of choice for resistors up to 10 Ω, and Evanohm® material is usually used in the construction of resistors above 10 Ω to 100 MΩ. Above 100 MΩ, the best available resistors are of the thin–metal-film construction. Standard resistors 100 Ω and below are constructed as four-terminal types to eliminate the problems of lead and contact resistances and to define the resistance more precisely. The important requirements of standard resistors are stability with time, low TCR, small thermal emfs against copper, well-defined terminations, and ruggedness. Standard resistors are adjusted to 0.01% or better from nominal value for wirewound resistors, and from 1% to 10% from nominal value for metal-film resistors.

5.2 Resistors for Electronic Circuits

The majority of resistors manufactured are intended for use in electronic circuits. These include resistors of all types of construction as listed in Sec. 3. For critical low-frequency circuits wirewounds are still in use; otherwise, metal-film resistors are used because of their excellent high-frequency performance. The other important requirements include small size, high reliability, and low cost. Resistor tolerances vary from 0.01% to 20% depending on their use. These resistors are intended for use in circuits involving potentials up to several hundred volts, but currents seldom over 10 to 100 mA.

A variety of thin- and thick-film resistor chips are available for hybrid microelectronic circuits (Hierholzer *et al.*, 1977). They measure as little as 1 mm by 1 mm, and have resistance values from a few ohms to 1 GΩ. A typical power rating is 0.25 W or less, with TCRs in the order of 20 to 200 ppm/K.

5.3 Integrated-Circuit Resistors

Integrated-circuit resistors (Fogiel, 1972) that lie in the range 20 Ω to 30 kΩ are generally formed by the diffusion of p-type material in an n-type epitaxy layer. Their tolerances, however, are poor, being in the order of ±25%. The tolerance of the ratio of two diffused resistors can be as low as ±2%. Diffused resistors are not desirable for higher values, since large chip areas are required. Above 30 kΩ, resistors can be formed from ion-implanted layers. These are quite reproducible, have a large dynamic range, and are linear.

5.4 High-Current Resistors

High-current resistors or shunts (Harris, 1962) are used for the measurement of direct current in the range from 10 A to 10 kA or higher. They are four-terminal resistors made from sheets or rods of manganin, using several elements in parallel for the lower-resistance, high-current values. They are designed for operation either in an air environment or mounted in an oil-filled container equipped with a stirring mechanism. The range of resistance is from about 10 μΩ to 10 mΩ. Because of the high power dissipation and the less well-defined thermal conditions, the accuracy obtainable is limited to 0.01% at the low-current values, and degrades to about 0.1% for high-current values.

5.5 High-Voltage Resistors

High-voltage resistors (Wellard, 1960) are designed to fulfill the special requirement for high-voltage, high-resistance units capable of dissipating moderate power. These resistors are rated from 5 to 20 kV, have a resistance range from 2 kΩ to 1 GΩ, and are rated from 5 to 20 W. These resistors are noninductive and are used primarily in high-voltage bleeder circuits, high-voltage dividers, and high-voltage networks.

GLOSSARY

Ayrton–Perry Winding: A resistor winding method where alternate winding layers are wound in opposite directions to reduce inductance by causing electrical fields to cancel.

Bifilar Winding: A resistor winding method in which a wire is folded in half into a long loop or "hairpin" with the two sides as close together as possible and then wound side by side.

Cermet: A resistance alloy consisting of a metal oxide, usually chromium–silicon monoxide (Cr–SiO).

Evanohm®: A resistance alloy of nickel, chromium, aluminum, and copper, developed by the Wilbur B. Driver Co. in 1948, having a low TCR and a resistivity 77.5 times that of copper.

Load Coefficient: The change in resistance caused by the production of joule heating in a resistor at the rate of one watt and usually expressed in ppm/W.

Manganin: A resistance alloy of copper, manganese, and nickel, developed around 1890, having a low TCR and a resistivity 27.8 times that of copper.

Ohm: The unit of resistance required to cause a voltage drop of 1 V with a flow of 1 A of current.

ppm: Parts per million (10^6), which is also 0.0001%.

Quantized Hall Resistance (QHR): A reference based on the QHE that defines a base resistance believed to be equal to the ratio of the fundamental constants h/e^2 and other quantized resistances equal to submultiples of h/e^2.

Quantum Hall Effect (QHE): An observed property at low temperatures in special microelectronic samples in which a two-dimensional electron gas exists and energy states are quantized when subjected to a strong magnetic field.

Resistance: The property of a conductor that causes it to impede the flow of electrons.

Resistance Alloy: A material, with properties of moderate to high resistivity, low temperature coefficient of resistance, and long-term stability, used in the construction of resistors.

Resistivity: The resistance per unit length of a material with uniform cross section.

Shunts: Special class of resistors of nominal values from about 10 μΩ to 10 mΩ designed to carry large currents in the range from 10 A to kA or higher.

Stability: The degree to which the resistance changes with time or environmental stress.

Standard Resistors: The highest quality of resistors used for calibration purposes in resistance measurements.

Temperature Coefficient of Resistance (TCR): The change in resistance per a degree change in temperature expressed in ppm/K or ppm/°C.

Thermal emf: The voltage generated when the junctions of a circuit of two dissimilar metals are maintained at different temperatures.

Voltage Coefficient of Resistance (VCR): The change in resistance as a function of voltage expressed as ppm/V.

Works Cited

British Association for the Advancement of Science (1913), *Reports of the Committee on Electrical Standards,* Oxford, U.K.: Cambridge Univ. Press.

Coursey, P. R. (1949), "Fixed Resistors for Use in Communication Equipment," *Proc. IEE* **96,** Pt. III, No. 41, 169–186.

Dummer, G. W. A. (1970), *Materials for Conductive and Resistive Functions,* New York: Hayden.

Fogiel, M. (1972), *Modern Microelectronics,* New York: Research and Education Association.

Grisdale, R. O., Pfister, A. C., van Roosbroeck, W. (1951), "Pyrolytic Film Resistors: Carbon and Borocarbon," *Bell System Tech. J.* 271–314.

Harris, F. K. (1962), *Electrical Measurements,* New York: Wiley.

Hartland, A., (1992), "The Quantum Hall Effect and Resistance Standards," *Metrologia* **29,** 175–190.

Hierholzer, E. L., Drexler, H. B. with Powers, J. H. (1977), "Resistors and Passive-Parts Standardization," in C. A. Harper (Ed.), *Handbook of Components for Electronics,* New York: McGraw-Hill.

Kaufman, M., Seidman, A. H., Sheneman, P. J. (1988), *Electronics Source Book for Technicians and Engineers,* New York: McGraw-Hill.

Maissel, L. I. (1970), "Thin-Film Resistors," in: L. I. Maissel, R. Glang (Eds.), *Handbook of Thin-Film Technology,* New York: McGraw-Hill, Chap. 18.

Meaden, G. T. (1965), *Electrical Resistance of Metals,* New York: Plenum.

Peterson, C. (1954), "Alloys for Precision Resistors," in: *Precision Electrical Measurements,* New York: Philosophical Library.

Starr, C. D., Schlenker, J., Graule, R. (1970), "Evaluating Moisture Resistance of Magnet Wire Enamels," *Insulation* **16** (2), 29–31.

Wellard, C. L. (1960), *Resistance and Resistors,* New York: McGraw-Hill.

Further Reading

Berry, R. W., Hall, P. M., Harris, M. T. (1968), *Thin-Film Technology,* New York: Van Nostrand Reinhold, Chap. 7.

Gore, T. S., Jr. (1989), "Resistors," in D. G. Fink, D. Christiansen (Eds.), *Electronics Engineers' Handbook,* New York: McGraw-Hill.

Marsten, J. (1962), "Resistors—A Survey of the Evolution of the Field," *Proc. IRE* **50,** 920–924.

Meeldijk, V. (1995), *Electronic Components—Selection and Application Guidelines,* New York: Wiley.

Prange, R. E., Girvin, S. M. (1990), *The Quantum Hall Effect,* New York: Springer-Verlag.

Thomas, J. L. (1948), "Precision Resistors and Their Measurement," National Bureau of Standards Circular 470, Washington, DC: U.S. Government Printing Office.

RESONANT TUNNELING

E. E. MENDEZ, *IBM Research Division, T. J. Watson Research Center, Yorktown Heights, New York, U.S.A.*

L. ESAKI, *University of Tsukuba, Tsukuba, Ibaraki, Japan*

1. **Introduction** 437
2. **Quantum-Mechanical Tunneling** 439
2.1 Tunneling and the WKB Approximation 439
2.2 Tunneling in Semiconductor Structures 439
2.3 Tunneling Current 441
3. **Resonant Tunneling in Semiconductor Heterostructures** 441
3.1 Resonant Tunneling Probability 441
3.2 Resonant Tunneling Current 443
3.3 Charge Accumulation and Bistability 444
3.4 Resonant Tunneling in Type I Heterostructures 445
3.5 Band-Structure Effects 446
3.6 Resonant Tunneling of Holes 447
3.7 Interband Resonant Tunneling 447
3.8 Magnetic-Field Effects 447
3.9 Low-Dimensionality Resonant Tunneling 448
4. **Applications of Resonant Tunneling** 450
4.1 High-Frequency Oscillators ... 450
4.2 Other Applications 451
5. **Summary and Prospect** 452

Glossary 452

Works Cited 453

Further Reading 454

1. INTRODUCTION

Tunneling phenomena are strictly quantum mechanical, with no counterpart in classical mechanics. Their consequences, however, are easily observable even at room temperature and are at the core of new applications in electronics. In this article we describe those effects, focusing on the most spectacular of them—resonant tunneling.

When subjected to a repulsive force, a classical particle moving in one dimension at an initial speed v is slowed down and the direction of its motion can eventually be reversed if the force is sufficiently strong. In this case, the particle is turned back at a point that depends on v and on the intensity of the force. However, in the realm of quantum mechanics, in which the position of an object is described by the probability of finding it there at a given time, there is a finite probability of finding the object beyond the classical turning point.

Using a mathematical language in which a force is defined in terms of the gradient of a potential energy, we can say that quantum mechanically it is possible for an object to tunnel through a potential barrier (repulsive potential) higher than the total energy of the object, provided the width and height of that barrier are finite. In spite of the counterintuitiveness of this concept, there are numerous phenomena in nature successfully explained by tunneling, e.g., the electric-field–induced ionization of hydrogen or the alpha decay of heavy nuclei. Moreover, the concept of tunneling is the foundation of inventions like the electron-field–emission and scanning tunneling microscopes, the Esaki diode, and the Josephson junction.

The fact that in quantum mechanics a particle is described by a wave function makes it possible to draw analogies between the concept of tunneling and other undulatory phenomena. One of these analogies can be found in the exponential attenuation of an electromagnetic wave propagating from a medium with index of refraction n_1 to an-

3-527-28138-X/96/$5.00 + .50

other of index $n_2 < n_1$. If the incident angle of the electromagnetic wave at the interface between the two media exceeds a critical value, the intensity of the wave in the second medium decays exponentially with the distance perpendicular to the interface. The wave is totally reflected if that medium is infinite, but if its thickness is comparable to the wavelength then a small fraction of the wave is transmitted, as first observed by Newton in the transmission of light between two glass plates separated by a thin air gap. The analogy with tunneling arises mathematically from the similarity between the equation for e.m. waves propagating through a thin slab and Schrödinger's wave equation for an object in a potential barrier.

Although the idea of tunneling was already used in the late 1920s to explain atomic- and nuclear-physics effects and was extensively studied in the 1950s and 1960s in relation with phenomena in solid-state physics, it has been with the advent of engineered semiconductor heterostructures and of the scanning tunneling microscope that the study of tunneling effects has reached unprecedented levels. Probably it was Iogansen (1964) who first suggested the possibility of resonant transmission of electrons in crystals (and, in particular, semiconductors) through a system of potential barriers. The development of new techniques for epitaxial growth of high-quality semiconductor films has led to the experimental demonstration of the phenomenon of *resonant tunneling*, by which under certain conditions an electron can pass unattenuated through two consecutive potential barriers confining a thin semiconductor region (see Fig. 1).

Resonant phenomena are not unique to tunneling through repulsive potentials. In fact, for certain energies an attractive potential (quantum well) can be transparent to particles normally scattered by it. A well-known example of such a phenomenon is the Ramsauer–Townsend effect, in which the cross section of electrons scattered by noble-gas atoms exhibits minima for certain resonant energies of the attractive atomic potential.

Resonant tunneling can be compared to the transmission of an electromagnetic wave through a Fabry–Pérot resonator. The equivalent of a Fabry–Pérot resonant cavity is formed by the semiconductor potential well defined by the two potential barriers, which are analogous to the Fabry–Pérot mirrors. The resonance condition is met when the energy of the electron coincides with one of the possible quantized energies of the semiconductor well (the normal modes of the Fabry–Pérot cavity).

The phenomenon of resonant tunneling could in principle be observed in multiple metal-insulator heterostructures, but it is the long mean free path of semiconductors that makes them ideal for the experimental study of the effect. It is manifested in the current-voltage characteristics of double-barrier heterostructures as regions of negative differential conductance, which are the foundation for new devices being explored for high-frequency electronics.

In this article we review the physical basis

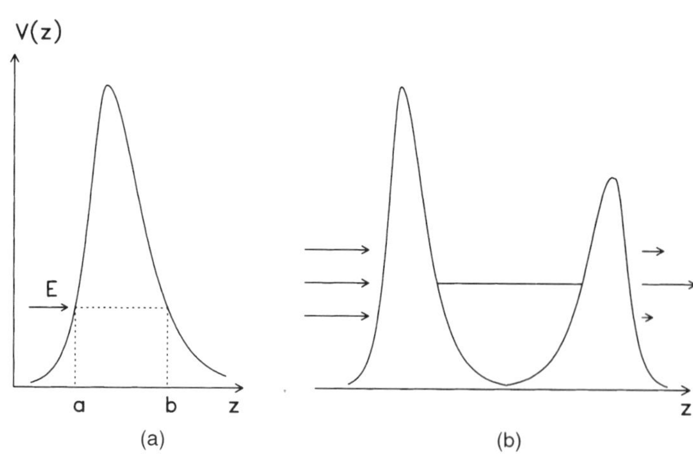

FIG. 1. (a) Generic potential profile, whose gradient is equal to a force acting on an electron of energy *E*. Points *a* and *b* are the classical turn-around points. Quantum mechanically, the electron has a small (but nonzero) probability of tunneling through the potential barrier and "reappearing" at point *b*. (b) A generic double-barrier potential profile, enclosing a quantum well. When the total energy of the electron coincides with a quasibound state of the well, tunneling becomes resonant and the tunneling probability is enhanced orders of magnitude over nonresonant tunneling.

of resonant tunneling, its main consequences, and its practical implications. First we introduce in detail the concept of tunneling, before describing resonant tunneling through a double-barrier structure. We focus on its consequences for heterostructures made of semiconductors and on concepts such as charge accumulation, tunneling time, and intrinsic bistability. We then discuss manifestations of resonant tunneling in several circumstances—for instance, under a magnetic field and under lateral confinement conditions. The final section is devoted to applications based on resonant tunneling, with emphasis on the physical parameters that enter into the optimization of device performance.

2. QUANTUM-MECHANICAL TUNNELING

Tunneling is a process in which a particle passes from one continuum of states to another (classically allowed regions) by traversing a potential barrier whose height is larger (forbidden region) than the total energy of the particle. The phenomenon is characterized by the probability that a particle incident on the barrier appear on the other side of it. For a free particle of mass m such a probability is calculated by solving Schrödinger's equation, which in one dimension (the z axis) reads

$$\left[-\frac{\hbar^2}{2m}\frac{d^2}{dz^2} + V(z) - E\right]\psi(z) = 0. \qquad (1)$$

Here $V(z)$ describes the potential barrier, and $\psi(z)$ is the quantum-mechanical wave function representing the particle of energy E. The module squared, $|\psi(z)|^2$, gives the probability of finding the particle at position z.

2.1 Tunneling and the WKB Approximation

In practice, there are only a few tunneling potentials for which Eq. (1) can be solved analytically. Although in most cases it is not difficult to find numerical solutions, to get a physical insight it is frequently advantageous to use simplifying approximations for calculating the tunneling probability, the most widely used of which is the semiclassical Wentzel–Kramers–Brillouin (WKB) approximation. This general method can be applied to the probability of tunneling through a barrier, as described in detail in quantum-mechanics textbooks, provided the change of the potential within a de Broglie wavelength is small relative to the particle's kinetic energy.

In the WKB approximation, the tunneling probability T through a potential $V(z)$ is given by

$$T = C\exp(-2\beta), \qquad (2)$$

when β, defined by

$$\beta(E) = \int_a^b \left\{\frac{2m}{\hbar^2}[V(z) - E]\right\}^{1/2} dz, \qquad (3)$$

is sufficiently large. Here a and b correspond to the classical turning points [see Fig. 1(a)]. The value of the prefactor C depends on whether the quantity $V(z) - E$ changes continuously or discontinuously at those turning points. If both changes are continuous, then $C = 1$; if both are discontinuous then C depends on E. The WKB approximation can be generalized to the cases where β is no longer large; the resulting expressions are then slightly more complicated than Eq. (2).

Tunneling of electrons in solids involves Bloch states and crystalline potentials, so that in principle the equations governing that process are much more complicated than those presented so far here. However, it is possible to use an effective-mass envelope-function approximation, in which the free-electron mass is replaced by an effective mass m^* and the complete potential $U(r)$ by the additional potential $V(r)$ imposed over the crystal lattice potential $U_0(r)$. The envelope wave-function solution $\psi(r)$ of Schrödinger's equation for such a reduced potential describes the envelope of the rapidly varying Bloch functions, provided $V(r)$ varies slowly on the scale of the lattice potential.

2.2 Tunneling in Semiconductor Structures

In a heterostructure of two semiconductors, say A and B, the composition varies on a distance of a few lattice constants. The potential $V(r)$ is provided by this discontinuity between the two materials. If a layer of B is imbedded in A, and if the conduction-band edge of B is higher in energy than that of A (see Fig. 2), then B acts as a potential barrier for electrons moving through the material interfaces. For a range of energies below the conduction-band offset between A and B, electrons tunnel through such a rectangular barrier.

Tunneling in semiconductors is not limited to heterostructures or to a single electronic band. In fact, under a strong electric field $\mathscr{E}$, electrons from the valence band can tunnel to empty states in the conduction band (interband tunneling) of the same material over a distance $E_g/e\mathscr{E}$, where E_g is the bandgap energy between the two bands and e is the electronic charge. (This is the mechanism for the Zener effect, which explains dielectric breakdown in solids.) If two spatial regions of a semiconductor are heavily p-type and n-type doped, respectively, so that electron and hole distributions are degenerate, both kinds of carriers have a significant probability of tunneling through to the other band. This is the basis of the Esaki or tunnel diode [Fig. 3(a)].

Another example of interband tunneling is offered by a heterostructure in which the barrier material B is such that the edge of its valence band is close in energy to that of the conduction band of material A, as indicated in Fig. 3(b). Inside the tunneling barrier, as in the p-n junction just discussed, the electron changes its character to holelike because of its proximity in energy to the valence band. In this case, it is also possible to use a WKB approximation to calculate the tunneling probability, provided the change from electronlike to holelike through the bandgap is taken into account. This can be done, e.g., by an analytical continuation into the gap of the two-band dispersion relation between energy and momentum.

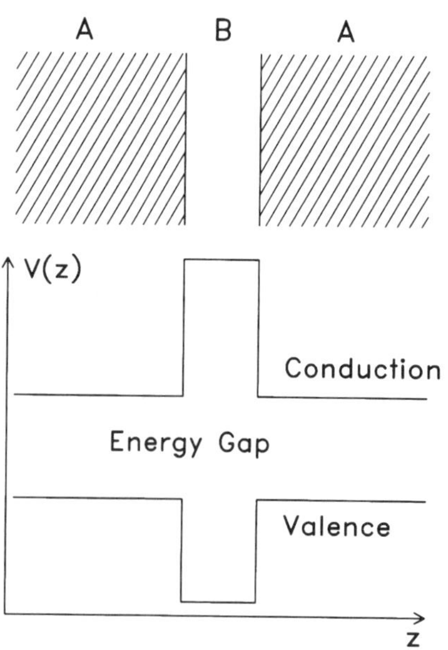

FIG. 2. Electronic potential energy along the direction z for a heterostructure consisting of a slab of material B in a material A whose conduction-band edge is higher in energy than that of B. Material B acts as a potential barrier to the motion of electrons across the interfaces between the two materials. On the other hand, when the conduction-band edge of B is lower in energy than that of A, B acts as a quantum well for electrons.

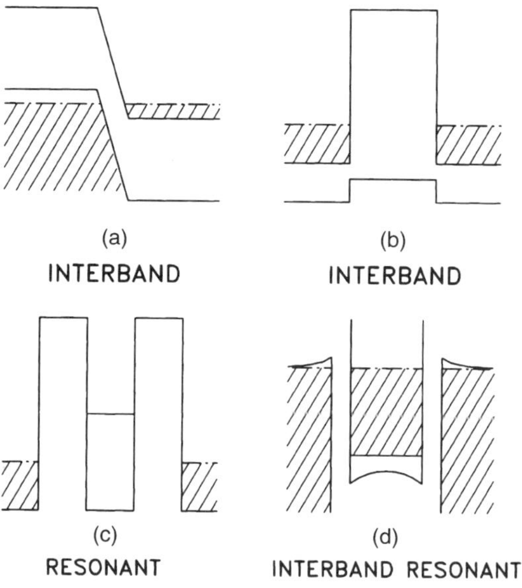

FIG. 3. Various potential profiles that can lead to negative differential resistance. (a) Interband tunneling, which can occur in a heavily doped p-n semiconductor junction. The so-called tunnel diode is based on this effect. (b) Interband tunneling through a single-barrier heterojunction. Because of the proximity in energy of the conduction and valence bands of the electrodes and the tunneling barrier, respectively, electrons from the electrodes have a hole character when tunneling through the barrier. (c) Resonant tunneling through two rectangular barriers. (d) Interband resonant tunneling, which can be seen as a particular case of (a) in which two p-n junctions are placed back to back with the n region in the center being narrow enough to exhibit quantum-size effects.

General to the various tunneling configurations outlined above are energy and momentum conservation laws. The total energy of the tunneling particle is conserved in the process, although, of course, its kinetic and potential energies are position dependent. In the simple case of a one-dimensional potential, transverse momentum $p_\perp$ ($p_\perp = \hbar k_\perp$, where $k_\perp$ is the transverse component of the wave vector) is also conserved, since there are no forces acting on the particle other than along the potential direction. As we will see later, a transverse magnetic field induces a force perpendicular to the field and the tunneling current, and under such perturbation transverse momentum is no longer conserved. Even in the absence of such a force, transverse momentum may not be conserved if a phonon is involved in the process (phonon-assisted tunneling) or if there is scattering induced by interface roughness. In practice, although weaker than the conservation of energy, the conservation of transverse momentum holds well, in a statistical sense. (That is, most electrons do not suffer scattering, which at the very least changes momentum direction.) In fact, the clearest macroscopic evidence of resonant tunneling is a direct consequence of the momentum-conservation law.

2.3 Tunneling Current

The macroscopic quantity of interest in tunneling structures is most frequently the electric current through the barrier. The tunneling current per unit area, J, is the product of the average electron velocity, the density of states, and the tunnel probability, and can be written as

$$J = \frac{e}{4\pi^3\hbar}\int dk_z d^2k_\perp f(E)T(E)\left(\frac{\partial E}{\partial k_z}\right), \tag{4}$$

where $f(E)$ is the Fermi–Dirac distribution for a total energy E, and $T(E)$ is the tunneling probability defined as the ratio between the incident and transmitted probability currents.

Under equilibrium, the net tunneling current is zero for any of the semiconductor structures mentioned above since the tunneling probability is the same for electrons moving from left to right as for those tunneling in the opposite direction. However, if a voltage is applied between the two end regions (electrodes) of A, a tunneling current will flow from the negative to the positive "electrode," expressed by

$$J = \frac{e}{4\pi^3\hbar}\int dE_z d^2k_\perp[f(E) - f(E + eV)]T(E_z). \tag{5}$$

The tunneling current can then be readily evaluated once the tunneling probability through the barrier is known.

3. RESONANT TUNNELING IN SEMICONDUCTOR HETEROSTRUCTURES

In the prototype heterostructure discussed above, the conduction-band edge of material B presented a potential barrier to electronic motion because of its higher position relative to that of material A. On the other hand, if the band edge of B is lower than that of A, layer B behaves as a quantum well, with one or more localized electronic levels at characteristic energies determined by the effective mass and the thickness of the layer (quantum-size effect). Resonant tunneling occurs in complex heterostructures comprising at least one quantum well and two barrier layers, like the one shown in Fig. 3(c).

3.1 Resonant Tunneling Probability

If an electron moves from the left to the right of such a rectangular double-barrier structure, its tunneling probability will be very small (of the order of the product of the tunneling probabilities for each individual barrier). On the other hand, when its energy coincides with one of the characteristic energies of the quantum well, its tunneling probability is enhanced drastically, to a value of the order of ratio between the transmission probabilities of the individual barriers. When such a resonance occurs we speak of resonant tunneling. If the two barriers are identical, the tunneling probability is unity, and the double-barrier system becomes totally transparent to the electron. The resonant process can be seen as a constructive interference in the well of the electron waves that tunnel through the two individual barriers.

The tunneling probability through a set of barriers and wells is frequently calculated using a transfer-matrix method (Kane, 1969; Tsu and Esaki, 1973). In such a formalism, each well/barrier is represented by a 2×2 matrix M, connecting the coefficients A and B of the general solution

$$\psi_e(z) = A \exp[ikz] + B \exp[-ikz] \tag{6}$$

at both sides of the well/barrier. The wave vector k in each region, defined by

$$\hbar^2k^2/2m = E - V, \tag{7}$$

is real in a well region ($E - V > 0$) and imaginary in a barrier ($E - V < 0$).

If an electron is incident from the left (region 1) on a set of wells/barriers consisting of n regions, only a transmitted wave will appear in region n. The transmission probability is then (Kane, 1969)

$$T = \frac{k_1 m_n}{k_n m_1} \frac{|A_n|^2}{|A_1|^2}. \tag{8}$$

The tunneling probability for a symmetrical double-barrier structure, calculated using the transfer-matrix method, is shown in Fig. 4(a), along with similar probabilities for triple and quintuple barriers. The sharp peaks in the transmission probability correspond to the resonant energies of the system. Near resonance, that probability is Lorentzian in shape, a result that is valid even for asymmetrical structures and that can be expressed by

$$T(E) = \frac{T_0}{1 + [(E - E_0)/\Delta E]^2}. \tag{9}$$

Here T_0, the maximum tunneling probability, is given by (Ricco and Azbel, 1984)

$$T_0 = 4T_LT_R/(T_L + T_R)^2, \tag{10}$$

where T_L and T_R are the individual tunneling probabilities of the left and right barriers, respectively, and ΔE is the half-width of the resonance, which is proportional to $T_L + T_R$. From Eq. (8) it follows immediately that $T_0 = 1$ when $T_L = T_R$ and that $T_0 = 4T_R/T_L$ when $T_R \ll T_L$.

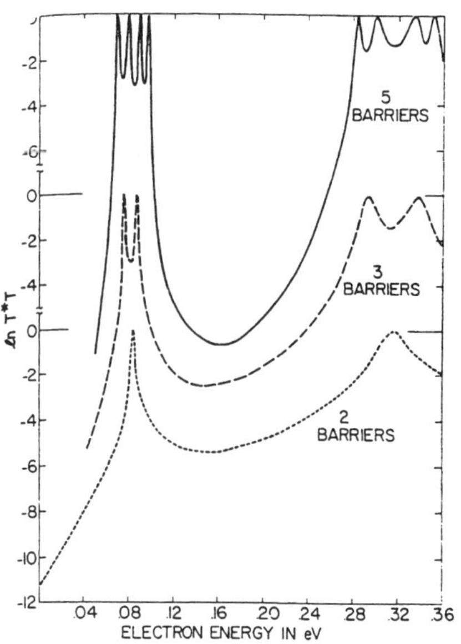

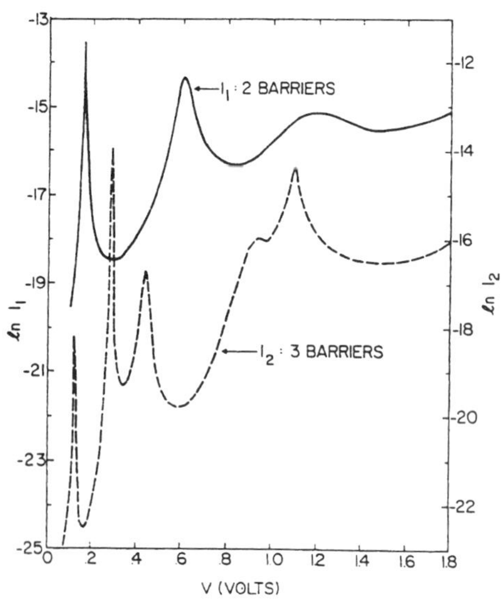

FIG. 4. (a) Probability (in logarithmic scale), calculated using the transfer-matrix method, for an electron of energy E to tunnel through two, three, or five rectangular barriers, each 0.3 eV high, 20 Å wide, and 50 Å apart from each other. (b) Calculated tunnel current density vs voltage applied to either a double- or a triple-barrier heterostructure similar to the ones described in (a). (After Tsu and Esaki, 1973.)

3.2 Resonant Tunneling Current

As for the case of a single barrier, once the tunneling probability of a double barrier is known the resonant tunneling current is calculated using Eq. (4). In semiconductor heterostructures, the two electrodes (from now on denoted emitter and collector, the former being at the lower potential) in practice have electron densities of the order of 10^{18} cm^{-3}, achieved through doping with n-type impurities. These electrons have a range of energies from 0 up to E_F, the Fermi energy. If one considers parabolic energy bands and zero temperature, the tunneling current density can be derived from Eq. (5) to be

$$J = \frac{em}{2\pi^2\hbar^3}\int_0^{E_F} dE_z(E_F - E_z)T(E_z), \tag{11}$$

where for simplicity we have assumed that the voltage drop between the emitter electrode and the quantum well is larger than the Fermi energy divided by the electronic charge. Using a Lorentzian lineshape like Eq. (9) for the transmission probability near resonance and assuming that $\Delta E \ll E_F$ (which in practice is always the case), it follows that in this approximation the current density is

$$J \simeq \frac{em\Delta E}{2\pi\hbar^3} T_0(E_F - E_0 + eV), \quad 0 \le E_0 - eV \le E_F;$$
$$J = 0 \text{ elsewhere.} \tag{12}$$

In other words, the current increases linearly with bias once the voltage aligns the localized state E_0 with the Fermi level, and it reaches a maximum at a bias such that E_0 coincides with the conduction-band edge. For higher voltages the tunneling current is ideally zero, as a consequence of conservation of transverse momentum. Although not exclusive to resonant tunneling, this abrupt drop in the current (and the corresponding negative differential resistance) is one of its defining characteristics (Fig. 5).

For nonparabolic bands and finite temperatures the expression for the current density is more complicated than Eq. (11) and its integration may have to be carried out numer-

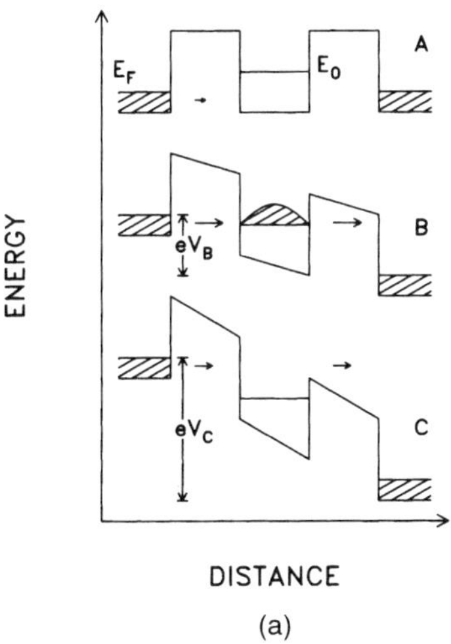

(a)

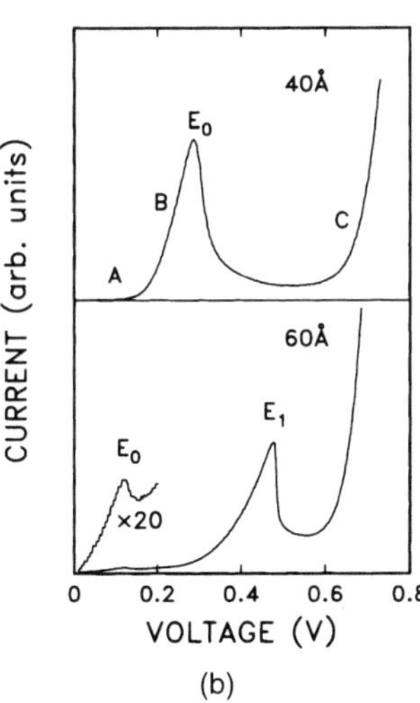

(b)

FIG. 5. (a) A double rectangular barrier under bias between two end electrodes, as exemplified by the semiconductor heterostructure n^+-GaAs/GaAlAs/GaAs/GaAlAs/n^+-GaAs. At very low voltage (A) the tunneling probability is very small and the current is negligible. When the external bias is such that the quantized state E_0 of the GaAs well lies between the Fermi level and the conduction-band edge of the emitter (B), resonant tunneling takes place and the current increases dramatically while charge accumulates in the well. When the bias pushes E_0 below the emitter's band edge, conservation of transverse momentum prevents electrons from tunneling and the current drops sharply, until at very high voltage (C) the right barrier effectively disappears and the current increases again. (b) Experimental current-voltage characteristics for two GaAs/GaAlAs double-barrier heterostructures with different quantum-well widths. A 40-Å well can accommodate only one bound state, giving rise to a single peak in its I-V characteristics, while a 60-Å well binds two discrete states.

ically, but that qualitative dependence of the tunneling current on voltage remains.

The negative differential resistance (NDR) characteristic of resonant tunneling provides information of two types. On one hand, the observation of NDR features implies a very high-quality heterostructure, free from materials defects and impurities and with sharp interfaces between its material components. (All these imperfections can act as scattering centers that break the $k_\perp$ conservation law.) On the other hand, the onset voltage of the resonant current and the peak voltages are directly related to the Fermi level of the electrode and the energies of the states in the quantum well, respectively.

Let us assume for a moment that when a voltage V is applied to a symmetric double-barrier heterostructure the voltage drops exclusively and uniformly in the undoped region. In this simple model, the peak voltage (times the electronic charge) corresponds to twice the quantum-state energy E_0, and the difference between the peak and onset voltages is twice the Fermi energy of the electrode. (The factor of 2 results from the fact that only half of the total voltage drops between the emitter and the center of the well.)

In practice, the peak voltages are directly related to the quantum energies, but the relation is not as simple as implied above. The reasons are multiple: part of the voltage drops at the electrodes, the drop is not uniform throughout the heterostructure, the energies of the quantum states are modified by the electric field, and there is an accumulation of electronic charge in the quantum well during the tunneling process. An exact calculation of the resonant voltage requires solving Schrödinger's equation as well as the electronic transport equations. Because of the complexity of the problem, a static approximation is frequently used, in which the accumulated charge in the well is ignored and the potential profile and energy levels at any voltage are calculated by self-consistently solving Poisson's and Schrödinger's equations.

3.3 Charge Accumulation and Bistability

Under steady-state resonant-tunneling conditions, charge builds up in the well proportionally to the tunneling current: $Q = \tau_R J$ (Goldman *et al.*, 1987). The proportionality constant, τ_R, represents the average time an electron spends in the well before tunneling out to the collector electrode. The lifetime of the electron in the well, τ_T, can be estimated semiclassically by calculating the probability per unit time for an electron of mass m and kinetic energy E to escape out of a well of width W through either the left or right barrier. The lifetime can then be written as $\tau_T^{-1} = \tau_L^{-1} + \tau_R^{-1}$, with

$$\frac{1}{\tau_T} \simeq \frac{\sqrt{2E/m}}{2W}(T_L + T_R). \tag{13}$$

Rigorously it is shown that $\tau_T = \hbar/2\Delta E$, where ΔE is the resonance half-width. From Eqs. (10) and (12) it then follows that the resonant-tunneling current is proportional to $T_L T_R/(T_L + T_R)$ and that the accumulated charge is proportional to $T_L/(T_L + T_R)$. In an asymmetric double-barrier heterostructure, the current is then determined by the tunneling probability of the less transparent barrier. On the other hand, the charge is controlled only by the collector barrier and is the largest for an opaque collector (relative to the emitter). The maximum charge accumulated is of the order of the two-dimensional charge in a well whose energy level is below the Fermi level by an amount E_F.

This analysis assumes a picture of resonant tunneling in two sequential steps, as opposed to a coherent process (Büttiker, 1988). An electron first tunnels from the emitter into the quasibound state in the quantum well, where, by scattering, it loses its original phase. Then the electron tunnels out of the well to the collector (Luryi, 1985). As far as the current and charge accumulation are concerned, however, the two pictures are in practice equivalent. The current peak is independent of scattering whenever the emitter Fermi energy is much larger than the resonance width ΔE (Weil and Vinter, 1987). The accumulated charge is in both cases proportional to τ_R because the integrated density of states is constant, as shown by integrating the wave function in the well and adding the contributions of all tunneling electrons (Gu and Gu, 1989).

In the discussion about the relation between the resonant voltages and the alignment of the quantum states with the emitter-band edge, it was implicitly assumed that the effective mass of the electron at the electrode and in the quantum well was the same. While this is frequently the case, there are instances in which such an assumption is not valid. As demonstrated experimentally, parallel-momentum conservation requires that, depending on whether the ratio m_w/m_b is greater or less than unity, the resonant voltage occurs when the quantum state is above or below the emitter band edge.

The accumulation of charge in the quantum well during the tunneling process has an interesting consequence: the appearance of bistability in the *I-V* characteristic, produced by two stable states at voltages around the NDR region (Eaves, 1990). The accumulated electrons create a space-charge layer, which alters the voltage distribution across the heterostructure for a given total voltage *V*. It is then possible to have two potential configurations, one in which the quantum state is still slightly above the emitter-band edge and the current is large, and another in which (for the same voltage) that state is below the emitter's edge and the current is low.

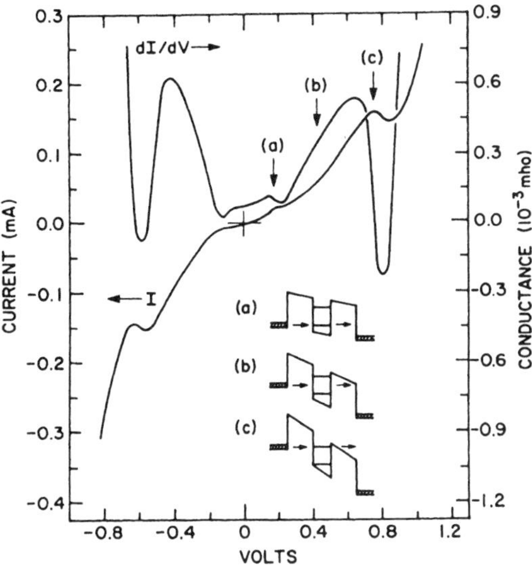

FIG. 6. First current-voltage characteristics that showed negative differential resistance due to resonant tunneling in a double-barrier semiconductor heterostructure. Although weak, the features (a) and (c) in the current (seen more clearly in the conductance) are consequences of resonant tunneling through the two quasibound states in a GaAs-GaAlAs heterostructure. (After Chang *et al.*, 1974.)

3.4 Resonant Tunneling in Type I Heterostructures

Resonant tunneling was first demonstrated experimentally in 1974, using (n^+)-GaAs-Ga_xAl_{1-x}As-GaAs-Ga_xAl_{1-x}As-(n^+)GaAs heterostructures (Chang *et al.*, 1974). The combination of two materials like GaAs and GaAlAs, whose lattice constants differ by less than 0.1% (even when $x = 1$), makes it possible the epitaxial deposition of alternate layers of them with the high degree of perfection necessary for the observation of resonant tunneling. The band gap of GaAlAs is larger than that of GaAs and its conduction-band edge is higher in energy. A GaAlAs slab acts then as a potential barrier to electrons in GaAs, and GaAlAs layers on both sides of a thin GaAs film confine its electrons in a quantum well. The barrier height increases linearly with the Al content, at an approximate rate of 0.1 eV for every 10% of Al.

The current-voltage characteristics of that first experiment showed two features, at 0.2 V and 0.8 V, associated with resonant tunneling through the two localized states in the GaAs well, as indicated in the inset of Fig. 6. The negative differential resistance (NDR) regions were observable only at liquid-nitrogen temperature (T = 77 K) or below, in a current background foreign to resonant tunneling and probably due to imperfections in the two semiconductors. Since then, dramatic advances in materials preparation by epitaxial techniques such as molecular beam epitaxy and metallorganic chemical vapor deposition have led to a drastic reduction of the leakage current, making possible the observation of resonant tunneling at room temperature and even of intrinsic bistability due to charge accumulation (Fig. 7).

Double-barrier heterostructures in other materials systems compatible for epitaxial growth (either lattice matched or strained layer) have also shown the characteristic NDR of resonant tunneling, e.g., GaAsP-GaAlAs, InGaAs-AlInAs, InGaAs-GaAs, etc. The observation of resonant tunneling has not been restricted to III-V compounds: Si-SiGe heterostructures have shown features evidencing the phenomenon, and there has even been a report of resonant tunneling in amorphous heterostructures.

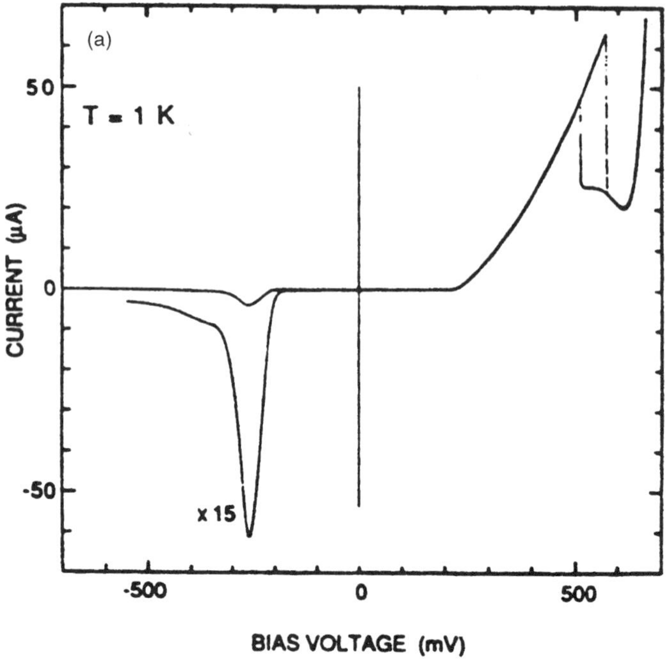

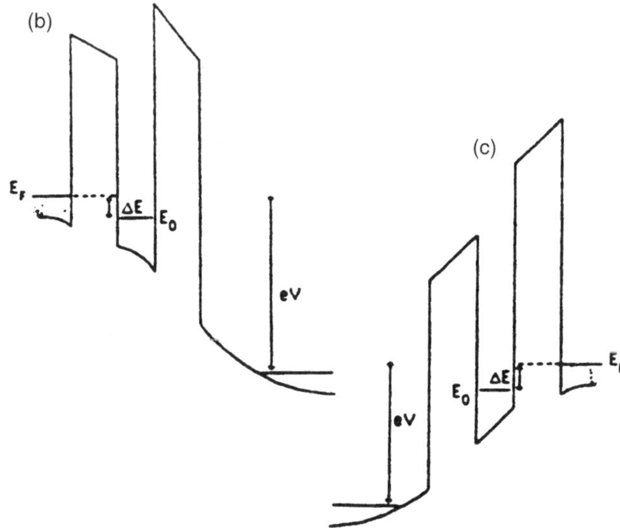

FIG. 7. (a) Current-voltage characteristics of an asymmetric double-barrier GaAs-GaAlAs heterostructure. (b) The tunneling current is much larger at forward bias because in this case, the two barriers become effectively more symmetrical and the resonant tunneling probability is larger than (c) at resonance under reverse bias. The large tunnel current and long lifetime in the well lead to a large charge accumulation in forward bias, which in turn gives rise to bistability. Under reverse bias, the accumulated charge is much smaller and no bistability is observed. (After Zaslavsky *et al.*, 1988.)

3.5 Band-Structure Effects

When studying tunneling phenomena, band-structure considerations beyond the conduction-band minima of the Brillouin zone are frequently ignored. This simplification is warranted in most cases, but there are circumstances in which close attention has to be paid to other critical points—for instance, when a high-Al composition ($x \geq 0.4$) is used in $Ga_{1-x}Al_xAs$-GaAs-$Ga_{1-x}Al_xAs$ resonant tunneling structures. Then, in the conduction band, the X point is close in energy to the Γ point and band mixing can occur. For $x > 0.45$, X is below Γ and the minimum-energy profile is a mixture of Γ (at GaAs) and X (at GaAlAs). This raises the question of which barrier determines the confined states in the quantum well—the one with the same symmetry or the lower one?

The answer has been provided by using the fact that hydrostatic pressure lowers the X point relative to Γ in GaAlAs. Starting with double-barrier $Ga_{1-x}Al_xAs$-GaAs-$Ga_{1-x}Al_xAs$ heterostructures with $x = 0.40$, it was found that pressure linearly shifted the resonant voltage to lower values, at a rate ($\simeq$1.8 meV/kbar) that remained constant well beyond the pressure (4 kbar) at which X and Γ cross. This result proved that the Γ barrier of GaAlAs (not necessarily the lowest barrier) determines the states of a quantum well at the Γ point of GaAs (Mendez *et al.,* 1986a). The uniform voltage shift was attributed to the slightly different pressure coefficients of GaAs and GaAlAs at Γ as well as to a pressure-induced increase of the electron effective mass.

When considering quantum-size effects at points of the Brillouin zone other than Γ, it is interesting to note that the X point in AlAs is lower in energy than in GaAs. In n^+GaAs-AlAs-GaAs-AlAs-n^+GaAs structures it is then possible to draw two potential profiles: one based on the Γ point, in which GaAs acts as a well and AlAs as a barrier, and another based on the X point, where AlAs is now a well while GaAs behaves as a barrier. In addition to the Γ-like quantum states in GaAs, X-like states are formed in AlAs, as evidenced by NDR features in the low-temperature I-V characteristics. Although the electrons from the n^+ electrode have a Γ character, their tunneling through X-point states still preserves transverse momentum (Mendez *et al.,* 1987).

3.6 Resonant Tunneling of Holes

The double-barrier potential profile that is at the center of resonant tunneling in semiconductor heterostructures is not exclusive to the conduction band. Many materials combinations, including the most familiar, GaAs-GaAlAs, exhibit a similar profile for the valence band (although with different barrier height) and holelike discrete states are formed in the quantum well. If the end electrodes are p doped, instead of n doped as considered in the above sections, then holes can tunnel resonantly through the quantum states, which can have either light- or heavy-hole character. Experimentally, both tunneling channels appear as NDR features of the I-V characteristics, although normally heavy-hole resonant currents are much smaller and resolved clearly only at low temperatures (Mendez *et al.,* 1985).

3.7 Interband Resonant Tunneling

As mentioned earlier, negative differential resistance is not unique to resonant tunneling. The I-V characteristics of tunnel diodes show it, and so do it the characteristics of a single barrier in which interband tunneling occurs, as shown in HgCdTe-HgTe heterostructures.

An interesting combination of resonant and interband tunneling occurs in structures based in the InAs-GaSb system, which is peculiar in that the conduction-band edge of InAs is lower in energy than the top of the valence-band edge of GaSb. As a result, in a heterostructure of the two materials there is an accumulation of electrons on the InAs side of the interface and a corresponding accumulation of holes on the GaSb side. If AlSb, which is also epitaxially compatible with InAs and GaSb, is included as part of a polytype heterostructure, a large number of combinations emerge (McGill and Collins, 1993). For instance, in a (p^+)GaSb-AlSb-InAs-AlSb-(p^+)GaSb heterostructure the GaSb electrodes are p type while the InAs quantum well contains free electrons arising from the charge-transfer process [Fig. 3(d)]. This structure can then be regarded as two tunnel diodes back to back, in which the common region (n-InAs) is so thin as to give rise to quantum-size effects and therefore to resonant tunneling. A symmetrical configuration, with the role of electrons and holes reversed, is provided by an interchange of InAs and GaSb.

The band discontinuity between two semiconductors is the origin of the rectangular profiles discussed above. By compositional grading of the materials it is possible to create other profiles, e.g., triangular or parabolic, either in the well or the barrier. Resonant tunneling associated with a parabolic well (in which the energy levels are equally separated) has been observed at a temperature of 7 K (Sen *et al.,* 1987).

3.8 Magnetic-Field Effects

A magnetic field H quantizes the electronic motion perpendicular to it into a set of discrete cyclotron orbits, with energies $E_n = (n + \frac{1}{2})\hbar\omega_c$ (Landau levels), where $\omega_c =$

eH/mc is the cyclotron frequency. When the field is parallel to the direction of tunneling, the current shows new NDR features at voltages below the main resonance, which, although generally weak in the I-V characteristics, are clearly resolved in plots of dI/dV vs V (Mendez *et al.*, 1986b). These field-induced structures are associated with the resonant tunneling of electrons through individual Landau levels in the quantum well. Since the energy separation between these levels is inversely proportional to the effective mass of the electron, from the voltage difference between consecutive field-induced resonances it is possible to determine the electronic effective mass.

In the presence of a magnetic field, the conservation of transverse momentum requires that the Landau-level index of electrons be conserved during the tunneling process. On the other hand, when the field is perpendicular to the current no Landau levels are formed in the well and a transverse force acts on the electrons, as a result of which the NDR is shifted (Ben Amor *et al.*, 1988). Since there is a correlation between the magnetic field and the lateral momentum gained by the electron, a plot of the voltage shift with field gives direct information on dispersion relation between energy and inplane momentum. This tool has been employed to map the nonparabolic dispersion of the valence band in GaAs (Hayden *et al.*, 1991) and even the anisotropy of holes in SiGe/Si heterostructures (Gennser *et al.*, 1991).

Much more pronounced than in conventional type I resonant tunneling (involving either electrons or holes) are the effects of a magnetic field in interband resonant tunneling, in which both electrons and holes participate. A system where the latter occurs is (p)GaSb-AlSb-InAs-AlSb-(p)GaSb, as described earlier. The small Fermi energy of holes in the GaSb electrodes, together with the large cyclotron energy of electrons in the InAs quantum well, contributes to the observation of distinct NDR structures associated with electron Landau levels even at modest magnetic fields [Fig. 8(b)]. For fields above 15 T the electronic spin splitting can be resolved, from which the Landé factor g—and its field dependence—has recently been determined (Mendez *et al.*, 1993).

The unique characteristics of that system allow us to probe by tunneling the two-dimensional electron gas in the InAs quantum well. This capability is especially interesting at high magnetic fields, under which a two-dimensional gas shows quantization of the in-plane Hall voltage and vanishing longitudinal resistance for certain fields (quantum Hall effect). In interband magnetotunneling, the zero-bias conductance exhibits a similar dependence: oscillatory behavior at low fields and zero conductance at high fields whenever the Fermi energy is between two Landau levels [Fig. 8(c)]. Because the quantum Hall effect relies on the existence of localized states that prevent transport of in-plane current, the manifestation of those states in "vertical" transport requires a better understanding of possible paths for the tunneling current, or maybe a deeper insight on the concept of localization.

3.9 Low-Dimensionality Resonant Tunneling

By quantizing the electronic energies into Landau levels, a parallel magnetic field effectively reduces the dimensionality of resonant-tunneling heterostructures. In the absence of the field, tunneling occurs between two three-dimensional regions through a two-dimensional well, whereas when the field is present tunneling is between one-dimensional electrodes via a fully quantized (zero-dimensional) well. A physical reduction of dimensionality can be achieved by lithographic means instead of by a magnetic field. In "quantum cylinders" (about 100 nm in diameter) made out of double-barrier structures, well-resolved multiple NDR have been observed at low temperatures, as a result of the quantization of the states in a GaInAs well in all three dimensions (Fig. 9). In this case, resonant tunneling takes place between two one-dimensional systems (the electrodes) via a zero-dimensional system (the quantum well).

By physically reducing the dimensions of a resonant-tunneling device, both its capacitance C and the charge accumulated in the well during the tunneling process are decreased. If the size is small enough, the energy necessary to charge the well with a single electron, $U_C = e^2/2C$, is larger than the thermal energy, kT, at temperatures around 0.1 K. The quantization of the charging en-

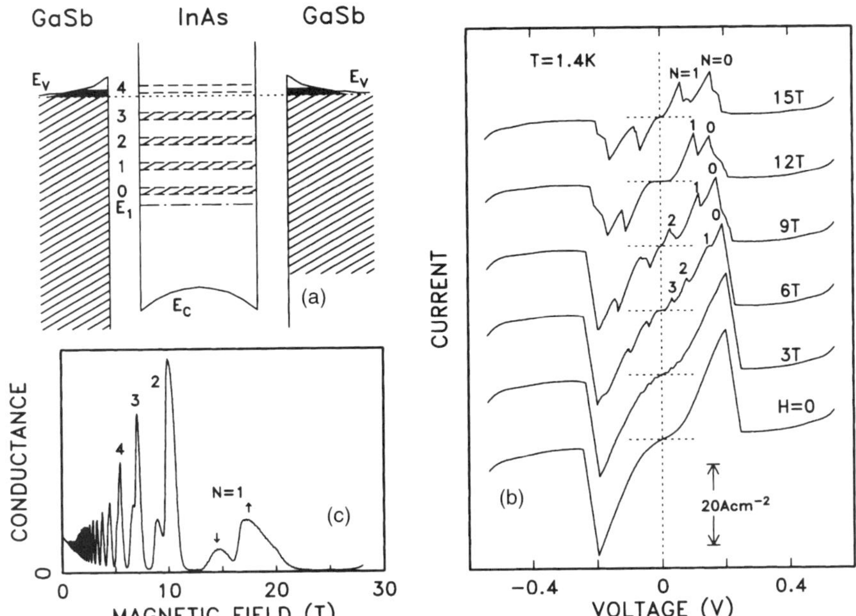

FIG. 8. (a) Band profile of a GaSb-AlSb-InAs-AlSb-GaSb heterostructure under a magnetic field H parallel to the materials interfaces. The InAs layer forms a quantum well for electrons while holes accumulate at the GaSb interfaces. The field creates a set of Landau levels in InAs whose separation is proportional to H. (b) Resonant tunneling is possible only when the energy of holes in the emitter coincides with that of an occupied Landau level in InAs, giving rise to a set of peaks in the I-V characteristics. The set of curves corresponds to a GaSb-AlSb-InAs-AlSb-GaSb heterostructure with 40-Å AlSb and 150-Å InAs layers. (c) Oscillations in the tunneling conductance, at $V = 0$, as a function of magnetic field, for the heterostructure of (b). In analogy with the quantum Hall effect, the conductance vanishes whenever the Fermi level lies in the gap between two Landau levels of the InAs well. (After Mendez *et al.,* 1991.)

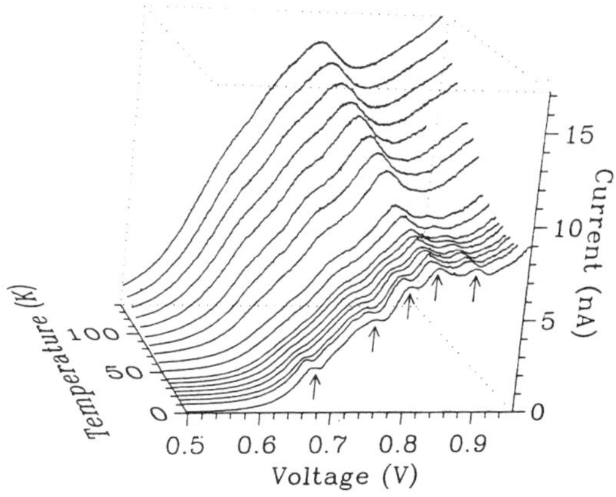

FIG. 9. Current-voltage characteristics of a 1000 Å quantum dot formed out of a GaInAs-GaAlAs double-barrier heterostructure. At high temperature, a negative differential resistance feature in the 0.7–0.8-eV range is due to resonant tunneling through the second-quantized state in the GaInAs well. At low temperatures new features are observable. They are evidence of size quantization in the plane of the interfaces. (After Reed *et al.,* 1988.)

ergy prevents tunneling for biases below a critical voltage $V_C = U_C/e$. This suppression of the tunneling current—normally denoted Coulomb blockade—has been observed below 1 K, coexisting with quantum-size effects (Su *et al.,* 1992).

The lateral dimensions of tunneling devices can be controlled by means of a Schottky-gate contact around an etched quantum cylinder. By applying a negative bias to the gate, relative to a conducting substrate, a depletion region is created around the quantum cylinder that has the effect of reducing the cylinder's cross section for the tunneling current. This geometry has allowed the observation of resonant tunneling through single impurities in quantum wells and of quantum-confinement effects controlled by a gate voltage (Dellow *et al.,* 1992; Guéret *et al.,* 1992).

Quantum wires also offer opportunities for resonant-tunneling effects in low-dimensionality systems. In an ingenious scheme

(Chang *et al.*, 1985), a thin heterostructure is grown on the edge of a conventional double-barrier structure so that the intersection of the two planes gives rise to a quantum wire. Resonant tunneling between two-dimensional electrodes through the one-dimensional well in the wire has been observed recently at $T = 0.3$ K (Dignam *et al.*, 1994).

Most experiments on resonant tunneling have utilized two different materials to create quantum well and barriers, but it is also possible to create a double-barrier configuration by in-plane modulation of the density of a two-dimensional (2D) electron gas. This modulation is achieved by two metal Schottky contacts on the surface, which, when properly biased, deplete of carriers the regions underneath them. Since the Fermi level is the same throughout the 2D plane, the conduction-band edge is raised at the depletion regions under the two contacts, creating a double-barrier configuration. An advantage of this lateral profile over the vertical one is the possibility of altering the barrier heights, and consequently the quantum states, with a change of the gate voltage. By varying such a voltage and keeping constant the bias between two Ohmic contacts (drain and source) to the 2D gas, it has been possible to observe negative differential resistance features in the drain-to-source *I-V* characteristic whenever the quantum states in the well between the two gates are aligned in energy with the conduction-band edge of the 2D gas in the electrodes (Ismail *et al.*, 1989).

4. APPLICATIONS OF RESONANT TUNNELING

The phenomenon of resonant tunneling in semiconductors not only provides experimental insight into a very fundamental concept of quantum mechanics, but also it lends itself to applications in electronics as a direct consequence of the negative differential resistance in the *I-V* characteristics of resonant-tunneling diodes discussed above. Any device that exhibits NDR can have gain, and in principle can be used as a reflection amplifier or as an oscillator. The former use, however, is not very practical because reflections at various parts of the transmission line can drive the circuit into oscillations. On the other hand, these instabilities can generate useful output power, and, in fact, they are the basis for high-frequency oscillators (Sollner *et al.*, 1983).

4.1 High-Frequency Oscillators

Compared to other devices with similar functions, resonant-tunneling diodes offer higher frequency and moderate power outputs. The cutoff frequency (above which the NDR disappears and the diode does not have gain), $f_{\max}$, and the maximum power (low frequency), w, are directly related to the electrical characteristics of the resonant-tunneling diodes through the following expressions:

$$f_{\max} = \frac{1}{2\pi RC}\sqrt{\frac{R}{R_s} - 1}. \qquad (14)$$

$$w = \tfrac{3}{16} I_{\max} V_{\max}. \qquad (15)$$

Here, $-R$ and C are, respectively, the negative resistance and capacitance that describe the device; R_s is the series resistance associated with the contact resistance, with the accumulation and depletion regions, and with the spreading resistance; and $I_{\max}$ and $V_{\max}$ are the maximum current and voltage excursions of the NDR region, respectively. In practice, $R \gg R_s$, so that Eq. (14) can be approximated by $f_{\max} = [2\pi\sqrt{RR_sC}]^{-1}$.

Since these parameters are related to physical and materials characteristics of the diodes, such as barrier height and width or doping level, by proper design it is possible to optimize the performance of resonant-tunneling oscillators. But because those parameters may affect performance in opposite ways, some design compromises are required. For instance, the peak current increases with decreasing barrier height and width. On the other hand, when the barrier height is reduced the thermionic current increases and so does the valley current, thus decreasing the peak-to-valley ratio and correspondingly increasing R. If the barriers become too narrow the capacitance increases and the cutoff frequency suffers as well. Low doping of the electrodes is favorable because it lowers the capacitance and increases $V_{\max}$ but has the detrimental effect of increasing the series resistance.

Band-structure parameters of the materials must be also considered for optimum design. There are cases in which a large separation in energy of critical points (e.g., the Γ and X points in InGaAs) reduces the likelihood of more than one tunneling path and enhances the peak-to-valley ratio. This (together with a reduced series resistance) explains the better performance of InGaAs-AlAs devices over those based on GaAs-AlAs. By taking advantage of all these criteria, frequencies as high as 700 GHz with an output power of 0.3 μW have been reached at room temperature in resonant-tunnel diodes based on InAs-AlSb (Fig. 10). At 100 GHz the output power was about 100 μW, which compares very favorably with the 1-μW power, at the same frequency, of the best tunnel diodes.

Traditional varistors or varactors, based on the voltage-dependent resistance or capacitance of nonlinear devices like *p-n* junctions or Schottky barriers, produce much larger power output than resonant-tunneling oscillators operating at their fundamental frequency [Eq. (14)], discussed above. However, the capability of generating higher harmonics in resonant-tunneling devices has increased their power output dramatically, up to 800 μW at 250 GHz. The variation of capacitance around zero voltage of a single, thick barrier has been exploited to generate 1 mW of power (Grönqvist *et al.*, 1991), a value comparable to that produced by Schottky varactors. By design optimization it may be possible to increase that power several times.

4.2 Other Applications

Although at the moment oscillators are the most mature devices based on resonant tunneling, by no means are they the only ones being explored. A resonant-tunneling diode can act as a switch by biasing it with a source resistance larger than its negative resistance. Their short switching time (2 ps) and low power dissipation compared with traditional semiconductor technologies could make these diodes attractive for logic circuits, but the implementation of logic operations with two-terminal devices introduces serious complications. Much more promising is the use of resonant-tunneling switches in sampling circuits, analog-to-digital converters, and multigigahertz shift registers for signal-processing applications.

Three-terminal devices based on resonant tunneling have also been proposed and demonstrated, for example, resonant-tunneling hot-electron or bipolar transistors. In the former, a double-barrier heterostructure replaces the emitter–base barrier of a conventional hot-electron transistor whereas in the latter it substitutes for the tunnel barrier at the emitter–base interface of a heterostructure bipolar transistor. In the first of these two schemes the electrons are injected into

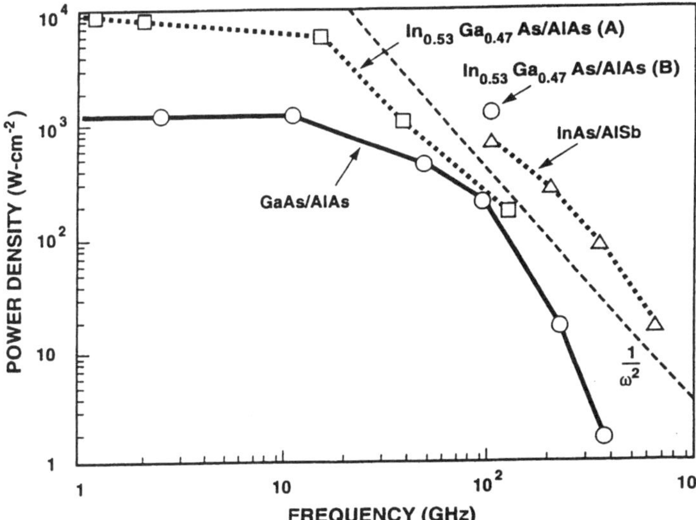

FIG. 10. A comparison of power density of resonant-tunneling oscillators based on different semiconductors. Because of the larger current densities of the InAs/AlSb pair, oscillations as high as 710 GHz with an output power of 0.3 μW have been achieved. (After Brown, 1992.)

the base by resonant tunneling, a feature that allows the device to act as a frequency multiplier or as an exclusive-NOR gate (Yokoyama *et al.,* 1985). Instead of a single transistor, MESFET technology requires eight devices for implementation of such a logic function.

The low-gain limitation of hot-electron transistors can be overcome by resonant-tunneling bipolar transistors, in which gains over 20 have been reported at 77 K, and up to 7 at room temperature, by placing the double-barrier heterostructure directly in the base (Capasso *et al.,* 1986). Recently, there has been a report of room-temperature integrated digital circuits with coexisting resonant-tunneling and double heterojunction bipolar transistors. The circuits, up to a 17-transistor full adder, required fewer components than traditional ones, thus allowing higher density and speed (Seabaugh *et al.,* 1993).

5. SUMMARY AND PROSPECT

The phenomenon of resonant tunneling reviewed here is quantum mechanical in nature, but it manifests itself in macroscopic effects even at room temperature, the most important of which is the presence of negative differential resistance in the current-voltage characteristics of semiconductor heterostructures with a double-barrier configuration. Such strong experimental evidence has made possible in the last ten years the study of resonant tunneling under a variety of circumstances and for a large group of semiconductors. By now our understanding of this phenomenon is quite complete, but its scientific interest remains unabated. An emerging area of fundamental research is the use of resonant tunneling as a spectroscopic tool to probe other quantum-mechanical systems—for example, two-dimensional gases at high magnetic fields, a topic only glimpsed here.

That the characteristic negative differential resistance of resonant tunneling is easily observed at room temperature makes some of the proposed applications good candidates for commercialization. Although it is hard to imagine that they will replace CMOS technology as the mainstream for low-power circuits, it is likely that they will find niche markets in high-frequency electronics.

The focus here has been on resonant tunneling through a double-barrier structure but the concept can be (and has been) generalized to multiple-barrier heterostructures. The ultimate case is when the barriers are so thin that there is strong coupling among the energy levels of the individual quantum wells. The electronic wave functions are no longer localized in individual wells but extended throughout the whole heterostructure. We speak then of a superlattice, by analogy with the periodic atomic lattice of a crystalline material (Esaki and Tsu, 1970; see also HETEROSTRUCTURES AND SUPERLATTICES, SEMICONDUCTOR). In contrast to the double-barrier case, electronic transport in superlattices is a much less explored subject both theoretically and experimentally. Because of the novelty of some of the phenomena that have already been predicted in the superlattice regime, the most outstanding of which is the possibility of generating electromagnetic radiation at ultrahigh frequencies (Bloch oscillations), we predict a much larger effort in this area in the next few years.

This article has dealt with semiconductors because these are the materials which have been hosts to resonant-tunneling effects so far. It is worth noting, however, that the resonant-tunneling concept for electrons in solids was first introduced in 1963 in the context of a proposed metal-insulator structure. Almost 30 years later, the first observation of room-temperature resonant tunneling in a metal-insulator ($CoSi_2$-CaF_2) double-barrier diode was reported. At the time of closing this review transistor action has just been demonstrated at 77 K in a resonant-tunneling device based on those two materials (Suemasu *et al.,* 1993). Although low materials quality still presents a challenge, these recent accomplishments are most welcome since metals have higher carrier concentrations and lower resistivities than semiconductors, both of which are favorable for high-speed operation.

GLOSSARY

Bistability: Condition of an electric circuit in which there are two stable voltage configurations for the same flow of current.

Bloch Oscillation: High-frequency electromagnetic oscillation predicted to occur when an electron moves in a periodic potential modulated by an external electric field.

Coulomb Blockade: Suppression of tunneling in a structure so small that the electrostatic energy to charge it with a single electron is larger than the thermal energy.

Cutoff Frequency: The highest frequency at which oscillations due to negative differential resistance are observed in a resonant-tunneling diode.

Heterostructure: A structure formed by two different materials. The term is normally employed in connection with semiconductor materials.

Interband Resonant Tunneling: A special case of resonant tunneling, in which the conduction and valence bands of two semiconductors are involved in the tunneling process.

Negative Differential Resistance: Electrical term that refers to a region in the current-voltage characteristic of an electronic device in which an increase in the applied voltage results in a decrease of the flow of current.

Potential Barrier: Mathematical concept used to describe a region of space where a repulsive force acts on a particle.

Quantum Well: Complementary to potential barrier, it refers to a region of space where an attractive force acts on a particle, or to a small region confined by two potential barriers.

Resonant Tunneling: A special case of tunneling, by which under certain conditions an electron has a high probability of traversing two consecutive potential barriers.

Tunneling: Quantum-mechanical phenomenon by which a moving particle can overcome a repulsive force and have a finite probability of reaching a region of space forbidden by classical mechanics.

WKB Approximation: Mathematical approximation, due to Wentzel, Kramers, and Brillouin, frequently used in quantum mechanics to calculate the transmission probability through a slowly varying potential barrier.

Works Cited

Ben Amor, S., Martin, K. P., Rascol, J. J. L., Higgins, R. J., Torabi, A., Harris, H. M., Summers, C. J. (1988), *Appl. Phys. Lett.* **53,** 2540–2542.

Brown, E. R. (1992), in: J. Shah (Ed.), *Hot Carriers in Semiconductor Nanostructures,* Boston: Academic, 469–498.

Büttiker, M. (1988), *IBM J. Res. Develop.* **32,** 63–75.

Capasso, F., Sen, S., Gossard, A. C., Hitchinson, A. L., English, J. H. (1986), *IEEE Electron Dev. Lett.* **7,** 573–576.

Chang, L. L., Esaki, L., Tsu, R. (1974), *Appl. Phys. Lett.* **24,** 593–595.

Chang, Y. C., Chang, L. L., Esaki L. (1985), *Appl. Phys. Lett.* **24,** 1324–1326.

Dellow, M. W., Beton, P. H., Langerak, C. J. G. M., Foster, T. J., Main, P. C., Eaves, L., Henini, M., Beaumont, S. P., Wilkinson, C. D. W. (1992), *Phys. Rev. Lett.* **68,** 1754–1757.

Dignam, M. M., Ashoori, R. C., Stormer, H. L., Pfeiffer, L. N., Baldwin, K. W., West, K. W. (1994), *Phys. Rev. B* **49,** 2269–2272.

Eaves, L. (1990), in: W. van Haeringen and D. Lenstra (Eds.), *Analogies in Optics and Micro Electronics,* Hingham, MA: Kluwer Academic Publishers, 227.

Esaki, L., Tsu, R. (1970), *IBM J. Res. Develop.* **14,** 61–65.

Gennser, U., Kesan, V. P., Syphers, D. A., Smith, III, T. P., Iyer, S. S., Yang, E. S. (1991), *Phys. Rev. Lett.* **67,** 3828–3831.

Goldman, V. J., Tsui, D. C., Cunningham, J. E. (1987), *Phys. Rev. B* **35,** 9387–9390.

Grönqvist, H., Kollberg, E., Rydberg, A. (1991), *Microwave Opt. Technol. Lett.* **4,** 33–35.

Gu, B.-Y., Gu, L. (1989), *Phys. Rev. B* **40,** 6124–6128.

Gueret, P., Blanc, N., Germann, R., Rothuizen, H. (1992), *Phys. Rev. Lett.* **68,** 1896–1899.

Hayden, R. K., Maude, D. K., Eaves, L., Valadares, E. C., Henini, M., Sheard, F. W., Hughes, O. H., Portal, J. C., Cury, L. (1991), *Phys. Rev. Lett.* **66,** 1749–1752.

Iogansen, L. V. (1964), *Soviet Phys. JETP* **18,** 146–150.

Ismail, K., Antoniadis, D. A., Smith, H. I. (1989), *Appl. Phys. Lett.* **55,** 589–591.

Kane, E. O. (1969), in: E. Burstein and S. Lundqvist (Eds.), *Tunneling Phenomena in Solids,* New York: Plenum.

Luryi, S. (1985), *Appl. Phys. Lett.* **47,** 490–492.

McGill, T. C., Collins, D. A. (1993), *Semicond. Sci. Technol.* **8,** S1–S5.

Mendez, E. E., Wang, W. I., Ricco, B., Esaki, L. (1985), *Appl. Phys. Lett.* **47,** 415–417.

Mendez, E. E., Calleja, E., Wang, W. I. (1986a), *Phys. Rev. B* **34,** 6026–6029.

Mendez, E. E., Esaki, L., Wang, W. I. (1986b), *Phys. Rev. B* **33,** 2893–2896.

Mendez, E. E., Wang, W. I., Calleja, E., Goncalves da Silva, C. E. T. (1987), *Appl. Phys. Lett.* **50,** 1263–1265.

Mendez, E. E., Ohno, H., Esaki, L., Wang, W. I. (1991), *Phys. Rev. B* **43,** 5196–5199.

Mendez, E. E., Nocera, J., Wang, W. I. (1993), *Phys. Rev. B* **47,** 13937–13940.

Reed, M. A., Randall, J. N., Aggarwal, R. J., Matyi, R. J., Moore, T. M., Wetsel, A. E. (1988), *Phys. Rev. Lett.* **60,** 535–537.

Ricco, B., Azbel, M. Ya. (1984), *Phys. Rev. B* **29,** 1970–1981.

Seabaugh, A. C., Taddiken, A. H., Beam III, E. A., Randall, J. N., Kao, Y.-C., Newell, B. (1993), *IEEE IEDM Tech. Digest,* 419–422.

Sen, S., Capasso, F., Gossard, A. C., Spah, R., Hatchinson, A., Chu, S. (1987), *Appl. Phys. Lett.* **51,** 1428–1430.

Sollner, T. C. L. G., Goodhue, W. D., Tannenwald, P. E., Parker, C. D., Peck, D. D. (1983), *Appl. Phys. Lett.* **43,** 588–590.

Su, B., Goldman, V. J., Cunningham, J. E. (1992), *Science* **255,** 313.

Suemasu, T., Kohno, Y., Suzuki, N., Watanabe, M., Asada, M. (1993), *IEEE IEDM Tech. Digest,* 553–556.

Tsu, R., Esaki, L. (1973), *Appl. Phys. Lett.* **22,** 562–564.

Weil, T., Vinter, B. (1987), *Appl. Phys. Lett.* **50,** 1281–1283.

Yokoyama, N., Imamura, K., Muto, S., Hiyamizu, S., Nishi, H. (1985), *Jap. J. Appl. Phys.* **24,** L853–854.

Zaslavsky, A., Goldman, V. J., Tsui, D. C., Cunningham, J. E. (1988), *Appl. Phys. Lett.* **53,** 1408–1410.

Further Reading

Bohm, D. (1951), *Quantum Theory,* Englewood Cliffs, NJ: Prentice-Hall.

Chang, L. L., Mendez, E. E., Tejedor, C. (1991), *Resonant Tunneling in Semiconductors,* New York: Plenum.

Duke, C. B. (1969), *Tunneling in Solids,* New York: Academic Press.

Esaki, L. (1975), "Long Journey into Tunneling," in: *Les Prix Nobel en 1973,* Stockholm: Imprimerie Royale, P. A. Norstedt & Söner.

Liu, H. C., Sollner, T. C. L. G. (1994), "High-Frequency Resonant-Tunneling Devices," in: R. A. Kiehl, T. C. L. G. Sollner (Eds.), *High Speed Heterostructure Devices,* Semiconductors and Semimetals, Vol. 3, Boston: Academic.

Mendez, E. E. (1988), "Physics of Resonant Tunneling in Semiconductors," in: E. E. Mendez, K. von Klitzing (Eds.), *Physics and Applications of Quantum Wells and Superlattices,* New York: Plenum.

Mendez, E. E. (1988), "Applications of Resonant Tunneling in Semiconductor Heterostructures," in: C. R. Leavens, R. Taylor (Eds.), *Interfaces, Quantum Wells and Superlattices,* New York: Plenum.

Price, P. J. (1992), "Electron Tunneling in Semiconductors," in: T. S. Moss (Ed.) *Handbook on Semiconductors,* Vol. 1, Amsterdam: North Holland.

Wolf, E. L. (1985), *Principles of Electron Tunneling Spectroscopy,* Oxford: Oxford University Press.

RESONATORS, ACOUSTIC

See FILTERS AND RESONATORS, ACOUSTIC

REVERSIBLE AND IRREVERSIBLE MATERIALS DEFORMATION

See SOLIDS, MECHANICAL PROPERTIES OF

RHEOLOGY

Elliot Kearsley, *10413 Englishman Drive, Rockville, MD 20852*

	Introduction	455
1.	**Constitutive Equations**	456
1.1	Constraints′	458
1.2	Continuum Mechanics	458
1.3	Linear Models	459
1.3.1	Newtonian Viscosity	460
1.3.2	Linear Elasticity	460
1.3.3	Linear Viscoelasticity	461
1.4	Nonlinear Models	463
1.4.1	Finite Elasticity	464
1.4.2	Differential Models	467
1.4.3	Models with Plastic Yield	469
1.4.4	Integral Models	470
1.4.5	Nonsimple Materials	471
2.	**Rheometry**	472
2.1	Viscometry	472
2.2	Strain-Energy Functions	474
2.3	Moduli of Linear Viscoelasticity	475
2.4	Normal Stresses	476
2.4.1	Rheogoniometry—Cone and Plate	477
2.4.1.1	Steady-State Properties	477
2.4.1.2	Time-Dependent Properties	478
2.4.2	Rheometry—Other Shearing Devices	478
2.4.2.1	Torsional Flow	478
2.4.2.2	Hole Pressure and Channel Flow	479
2.5	Complex Flows	479
2.5.1	Orthogonal Rheometry	479
2.5.2	Extensional Rheometry	480
2.5.3	Plastometer	481
3.	**Dimensionless Parameters**	482
	Glossary	482
	Works Cited	484
	Further Reading	484

INTRODUCTION

Rheology as a distinct science was born as recently as 1928. In that year, the word "rheology" was coined by E. C. Bingham of Lafayette College to denote the study of the deformation and flow of materials. The following year, an international organization, the Society of Rheology, was formed to promote the science. From the beginning, it was a highly interdisciplinary field and, although rheology is considered a branch of physics and the Society is a member society of the American Institute of Physics, the subject has close connections to the fields of physical chemistry, mathematics, materials science, and engineering.

For a more detailed explanation of the meaning of rheology, recall that, while the three laws of Newton's mechanics suffice to determine the response of point masses to forces, to predict the response of extended continuous bodies it is necessary to append equations associated with the particular material of the body that describe how the material distorts or changes shape under the influence of forces. Such equations, because they are intrinsic to the material, are called *constitutive equations* or sometimes, in the simplest cases, stress–strain laws. Rheology is concerned with the problems of formulating, establishing, and applying the mechanical constitutive equations of materials. These equations are mathematical models couched in the language of continuum mechanics, but it is understood that there is a microstructure and micromechanics underlying them (e.g., macromolecules of a polymer, grain boundaries of a metal, etc.).

Prior to the Second World War, development of rheology was sporadic and generally empirical, concerned mostly with linear theories. During the war, many practical problems with rheological content such as the evaluation of synthetic rubbers, the design of flame-thrower fluid, and the design of radar domes drew the attention of scientists. In the course of solving them, rheological instruments were designed, nonlinear phenomena were examined, and the development of the necessary basis in continuum mechanics was begun. At the time, much of this work was

3-527-28138-X/96/$5.00 + .50

classified secret, but publication after the war showed the progress that had been made, opening up the field and stirring up considerable interest in the subject. In the decades from 1950 to 1980, mathematicians and mechanicians developed a basis for describing nonlinear rheological phenomenology, while polymer physicists and physical chemists established some of the connections between the parameters of molecular chains and the mechanical behavior of associated materials. Currently, rheological interest is broadening to encompass liquid crystals, electrorheological and magnetorheological fluids, suspensions, and granular and composite materials. The Society of Rheology now has about 1300 members, mostly in the United States, in disciplines ranging through physics, chemistry, engineering, medicine, and biology. There is an international Biorheology Society and numerous national rheology societies and groups.

One can attempt to classify the activities of rheologists under five rubrics:

1. formulation of constitutive equations,
2. relating constitutive equations to micromechanics,
3. testing the validity of constitutive equations,
4. measuring rheological properties of materials, and
5. predicting and controlling the mechanical behavior of materials.

Technological applications of rheology are ubiquitous. To indicate just a few examples: extruding and molding of plastics, rubbers and metals; handling of cement and slurries; lubrication; specification and control of materials; well drilling and oil extraction; sealants and paints; antiearthquake mounts; and food processing.

1. CONSTITUTIVE EQUATIONS

The formulation of a constitutive equation must take into account some rather subtle geometrical considerations. The internal forces between a particle of material and its neighboring particles are to be related to the motion of the material but, intuitively, it must be motion relative to an imbedded convected coordinate system, that is, as "seen" by the moving particle rather than by a fixed observer. A variety of prescriptions have been proposed for formulating constitutive equations satisfying this and other fundamental requirements. A discussion of differences between some of them can be found in Huilgol (1975).

An exposition of rheology requires a different level of mathematical notation than that necessary for classical fluid mechanics. The use of tensor fields as the variables of constitutive equations is the conventional way to simplify the geometrical problems encountered. In this article, bold-faced letters are used to denote tensors (in the general sense, including vectors, pseudotensors, tensor fields, etc.) The transpose and inverse of second-order tensors are indicated by superscripts T and -1, respectively. The trace, determinant, divergence, and gradient of a tensor, say $\mathbf{A}$, are denoted by $\mathrm{tr}\mathbf{A}$, $\det\mathbf{A}$, $\nabla\cdot\mathbf{A}$, and $\nabla\mathbf{A}$, respectively. Scalar invariants of second-order tensors will be written as $I_i(\mathbf{A})$, $i = 1, 2, 3$, except that the argument may be dropped when it is obvious from context.

One postulates a *stress tensor*, $\boldsymbol{\sigma}$, at each point of a material body, which carries the local forces of interaction between contiguous parts. The stress tensor is a real, second-order tensor in three-dimensional space, symmetric (with rare exceptions when there are body moments or couple stresses). Thus, normally, only six of its nine components are independent. The tensor may be viewed as a linear operator that acts on a vector representing area to transform it to a vector representing a force acting on the area. The physical components of a stress tensor have the dimension of force per unit area.

To describe the change of shape of a body, a *strain tensor* is defined. Strain tensors are not unique in that any invertible function of a suitable strain tensor is also suitable, so that a wide range of choices is possible. The behavior of a material may be described using any one of various strain tensors, although the form of the constitutive equation will reflect that choice.

If each point of a material body is associated with a vector giving its position in space at some instant, we call the resulting vector field the *configuration* of the body at that instant. A particular configuration, the *reference configuration*, given by position vectors $\mathbf{X}$, the *material coordinates*, is used to label the points of the body. It is often convenient to choose a reference configuration to be stress free and/or an initial configuration at some zero of time. If the points $\mathbf{X}$ of the reference configuration trace out a path in time given by the vector function $\boldsymbol{\chi}$ and $\mathbf{x}$ is the configuration at time t, a vector field, then $\mathbf{x} = \boldsymbol{\chi}(\mathbf{X},t)$. The gradient of $\mathbf{x}$ with respect to $\mathbf{X}$ is a double tensor field, $\mathbf{F}(t)$, called the *deformation gradient* (often, simply *deformation*) describing the change of shape that is experienced in the neighborhood of each point of the material in going from configuration $\mathbf{X}$ to configuration $\mathbf{x}$. It is called a double tensor field because it depends on two configurations and acts as a tensor under coordinate transformation of either. The determinant of $\mathbf{F}$, the relative change in volume induced by the deformation, must be greater than zero if $\mathbf{F}$ is to be physically possible. (We do not allow a finite volume of material to contract to zero or to a negative volume.)

As a direct consequence of the polar decomposition theorem, it is apparent from these conditions that $\mathbf{F}$ may be factored into a product of a proper orthogonal tensor and a positive-definite symmetric tensor. In other words, any deformation of the material is locally equivalent to three stretchings in certain mutually perpendicular directions followed by a pure rotation (or, alternatively, rotation followed by stretching). Thus,

$$\mathbf{F} = \mathbf{RU} = \mathbf{VR}, \tag{1}$$

where $\mathbf{R}$ is a proper rotation and $\mathbf{U}$ and $\mathbf{V}$ are positive-definite, symmetric tensors called the *right stretch tensor* and *left stretch tensor*, respectively. It is worth noticing that $\mathbf{U}$ represents the stretches before rotation while $\mathbf{V}$ gives the stretches in their final orientation. Thus the principal stretches, that is, the eigenvalues of $\mathbf{U}$ and $\mathbf{V}$, are identical while their eigenvectors are not. The tensor $\mathbf{F}$ is not a suitable strain measure because its dependence on $\mathbf{R}$ (which represents a local rotation and not a true distortion of the material) is not just an expression of its orientation with respect to a reference configuration. The term $\mathbf{F}$ is said to be *observer dependent*. The stretch tensors, however, are suitable strain measures because for them $\mathbf{R}$ only specifies the change in orientation of the principal stretches and does not affect their values. But, in fact, the stretch tensors are seldom used as strain measures for the very practical reason that they are usually inconvenient to calculate.

Two tensors that are simply calculated from $\mathbf{F}$ and suitable for use as strain measures are $\mathbf{B}$ and $\mathbf{C}$, the so-called *left and right Cauchy–Green strain tensors*, respectively, defined as follows:

$$\mathbf{B} = \mathbf{V}^2 = \mathbf{F}\mathbf{F}^{\mathrm{T}}, \tag{2}$$

$$\mathbf{C} = \mathbf{U}^2 = \mathbf{F}^{\mathrm{T}}\mathbf{F}, \tag{3}$$

where $\mathbf{B}$ and $\mathbf{C}$ are symmetric, positive-definite tensors of rank two whose physical components are dimensionless. Of course, any other invertible tensor function of $\mathbf{U}$ or $\mathbf{V}$ is also a possible strain measure.

A relative strain tensor can be formed by choosing the reference configuration to be that at some arbitrary time, say τ, and then the gradient of $\mathbf{x}(t)$ with respect to $\mathbf{x}(\tau)$ is the *relative deformation gradient* $\mathbf{F}_\tau(t)$, the subscript τ indicating the time of the reference configuration. Deformation gradients compound by multiplication, thus,

$$\mathbf{F}_\tau(t) = \mathbf{F}_\xi(t)\mathbf{F}_\tau(\xi), \tag{4}$$

where ξ refers to a time intermediate to τ and t.

It is important not to confuse the tensor $\mathbf{F}$ of Eq. (1) with this relative tensor. It is conventional among rheologists to use the base symbol from tensors with respect to a fixed reference configuration to designate the analogous relative tensor but to attach a subscript to indicate the arbitrary reference configuration. Equation (4) offers several examples of this convention.

By analogy with Eq. (3), the *right relative Cauchy–Green strain tensor* is defined by

$$\mathbf{C}_\tau(t) = \mathbf{F}_\tau^{\mathrm{T}}(t)\mathbf{F}_\tau(t) = \mathbf{U}_\tau^2(t). \tag{5}$$

This notation can be extended to various other strain tensors; thus $\mathbf{U}_\tau(t)$ is a right relative stretch tensor, $\mathbf{B}_\tau(t)$ is a left relative Cauchy–Green strain tensor, etc. The Cauchy–Green strain tensors and their inverses are commonly used in the rheological treatment of flows. The various strain tensors are related by equations of the form

$$\mathbf{B}_\tau(t) = \mathbf{C}_t^{-1}(\tau). \tag{6}$$

The tensor $\mathbf{B}_\tau(t)$ is often called the *relative Finger tensor.*

Most rheological constitutive equations that are commonly used are so-called *simple materials,* for which the stress at a material point at any time can be calculated from the history of the deformation. In this context, the term *history* means the set of values of a property at the point for that time and all prior times. In short, a simple material is one for which the stress is a *functional* of the local deformation history. There is a geometrical restriction to be satisfied by this functional, the so-called *principle of material frame-indifference* (sometimes called *material objectivity*) which says, in loose terms, that the properties of the material described by the constitutive equation must be independent of the observer. It is for this reason that we rejected $\mathbf{F}$ as a strain measure. This restriction, when worked out, leads to a representation of a general simple material in the following form:

$$\boldsymbol{\sigma}(\mathbf{x},t) = \mathbf{F}\mathcal{F}^{t}_{\tau=-\infty}(\mathbf{C}(\mathbf{X},\tau))\mathbf{F}^{\mathrm{T}}, \tag{7}$$

where $\mathcal{F}(\cdot)$ is a tensor-valued functional of the history of its argument, in this case $\mathbf{C}(\mathbf{X},\tau)$ for τ ranging from $-\infty$ to t. The pre- and postmultiplications of the functional by deformation gradients in this equation are necessary to express the current stress in laboratory coordinates.

1.1 Constraints

The simple material of Eq. (7) is a useful theoretical concept, but it is much too general for practical applications. Somewhat more useful restricted models of ideal materials are defined in terms of *constraints,* which are conditions limiting the range of possible deformations. The constraints are themselves constitutive and must satisfy material objectivity. Of particular interest are *workless constraints,* maintained by implicit, internal forces that do no work in the course of any deformation allowed by the constraints. These forces of constraint do not directly affect the energetics. As a consequence, the stresses arising from the local history of constrained deformations are only part of the complete stress since they do not include that part of the stress associated with the workless constraints.

A very important example of a workless constraint, ubiquitous in rheology, is that of *incompressibility,* for which the deformation is isochoric, i.e., volume preserving. In this case, an unspecified isotropic stress is associated with maintaining the constraint and must be added to stress determined by local deformation history. As a result, only the deviatoric part of the stress is completely determined by the local deformation. Accordingly, Eq. (7) must then be modified to read

$$\boldsymbol{\sigma}(\mathbf{X},t) - p\mathbf{1} = \mathbf{F}\mathcal{F}^{t}_{\tau=-\infty}(\mathbf{C}(\mathbf{X},\tau))\mathbf{F}^{\mathrm{T}}, \quad \det\mathbf{C} = 1, \tag{8}$$

where p is an arbitrary pressure, $\mathbf{1}$ is the unit tensor, and $\mathbf{C}$ is restricted to be unimodular to assure an isochoric deformation. Of course, one does not expect to find real materials that are strictly incompressible any more than one expects to find ideal gases in thermodynamics. The model is an excellent approximation in many cases, however, when the energy associated with volume changes is ignorable compared with the energy of shearing, as is usually the case in stretchings and flows of elastomeric materials and of flowing liquids. Incompressibility is a widely used constraint of great importance. One of its most valuable aspects is that, for models of materials with this constraint, it is possible to write useful and exact expressions for the stress produced by certain simple, highly symmetric, inhomogeneous deformations.

If models are to be of practical use, further details of the form of a particular $\mathcal{F}$ will reflect other properties of the material such as isotropy or any anisotropy, its state of homogeneity, the limits and mode of the influence of past configurations, etc.

1.2 Continuum Mechanics

Such constitutive equations are useful only within the context of continuum mechanics. The general principles of continuum mechanics are contained in field equations expressing conservation of mass, the balances of linear momentum and moment of momentum, and, if thermodynamic effects are pertinent, appropriate forms of the first and second laws of thermodynamics. When these fields vary smoothly in the interior of a body of material, the equations are conveniently expressed in the form of differential equations. The mechanical field equations may be written as follows:

$$\dot{\rho} + \rho(\nabla \cdot \mathbf{v}) = 0 \quad \text{(conservation of mass)}, \tag{9}$$

$$\nabla \cdot \boldsymbol{\sigma} + \rho \mathbf{b} = \rho \dot{\mathbf{v}} \quad \text{(balance of momentum)}, \tag{10}$$

with $\dot{\mathbf{v}} = \partial\mathbf{v}/\partial t + (\mathbf{v}\cdot\nabla)\mathbf{v}$ and $\dot{\rho} = \partial\rho/\partial t + \mathbf{v}\cdot\nabla\rho$; and

$$\boldsymbol{\sigma} = \boldsymbol{\sigma}^{\mathrm{T}} \quad \text{(balance of moment of momentum)}, \tag{11}$$

where $\rho(\mathbf{X},t)$ is the density; a superposed dot indicates the *material derivative,* that is, the time derivative seen by a material particle; ∇ is the gradient operator; $\mathbf{v}$ is $\partial\mathbf{x}/\partial t$, the particle velocity; $\mathbf{b}$ is the body force, if any, acting on the particle; and Eq. (11) expresses the symmetry of the stress tensor in the absence of couple stresses as mentioned above. In addition, if thermal effects are pertinent, appropriate forms of the first and second laws of thermodynamics must be included with these principles. To complete the picture, to these general equations must be appended the constitutive equations that model some particular material and boundary conditions suitable to formulate a well-posed mathematical problem.

In a typical problem of continuum mechanics, the *surface tractions,* forces acting on the surface of the body, are given as well as the body forces (usually zero) or, alternatively, the surface is assigned a given shape or velocity. From the field equations and these boundary conditions, the deformation and stress in the interior are to be found. Usually, especially when there are nonlinearities present, there is no assurance that a solution exists or that it is unique even if it does. As a result, almost all known, nontrivial, exact solutions are so-called *inverse solutions* for which a tentative set of deformations is assumed, characterized by adjustable parameters and compatible with the given boundary conditions and constraints. The associated stress can then be calculated through the constitutive equation, and it is then only necessary to demonstrate that this stress can be made consistent with the field equations above by suitable adjustment of the parameters. This program can sometimes be carried out and an inverse solution found for certain constraints in highly symmetric geometries.

For incompressible materials, because Eqs. (9) and (11) are satisfied by assumption, only Eq. (10) need be considered. In that case, when there is sufficient symmetry to the problem, for a large class of constitutive equations and deformations, the arbitrary pressure term of the stress can be adjusted to satisfy Eq. (10). The adjective "controllable" is sometimes used to characterize this class of deformations.

To treat the great majority of rheological boundary-value problems for which these inverse methods are not sufficient, computational solutions must be sought through finite-element methods or other techniques for numerical solution of boundary-value problems. There is a large literature in the numerical analysis necessary for such computations because inverse solutions are rare. Furthermore, the recent amazing advances in computer simulation have made such methods often quite convenient, particularly where complex geometry and little symmetry are involved. Still, the solutions found by numerical methods usually do not have the convenient simplicity of inverse solutions.

1.3 Linear Models

The term "linear," as used here, is meaningless without an understanding of what variables depend linearly on what independent variables. For a linear stress-strain model, doubling the strain will cause a doubling of the stress. More generally, for a linear simple material, if a strain history is the sum of several simple strain histories then the resulting stress is the sum of the partial stresses associated with the simple histories. This is known as the *Boltzmann superposition principle.* This principle certainly cannot be generally valid because stresses compound additively, while strains do not. On the other hand, infinitesimal strains do compound additively (to first order in the strains) so that linear constitutive equations can be useful for treating small deformations. Furthermore, in the case of Newtonian fluids, the rate of strain (which compounds additively) rather than the strain history is relevant, so that, in this case too, there is no inherent rheological limitation to small deformations.

1.3.1 Newtonian Viscosity The *Newtonian fluid* obeys a classical constitutive equation in which the current rate of strain, rather than the whole strain history, determines the current stress. Remarkably, it is an approximation valid for many common liquids with compact molecules such as simple hydrocarbon oils or water over very large ranges of deformation rate. The concept of Newtonian fluids is so widely known and used in fluid mechanics that it is not really of much interest to the rheologist except as a model of the simplest common liquids. However, it is worthwhile here to formulate its constitutive equation in terms of deformation tensors in preparation for generalizing to models of more complex fluids.

The velocity gradient $\mathbf{L}$ is the rate of change of the deformation gradient:

$$\mathbf{L}(t) = \nabla\mathbf{v}(t) = \left.\frac{d\mathbf{F}_t(\tau)}{d\tau}\right|_{\tau=t}. \qquad (12)$$

The velocity gradient is not a materially objective quantity but it may be split into antisymmetric and symmetric parts, the rate of rotation of a particle, given by the *spin tensor* $\mathbf{W}$, and the rate of stretching, given by the *deformation-rate tensor* $\mathbf{D}$:

$$\mathbf{W} = \tfrac{1}{2}(\nabla\mathbf{v} - \nabla\mathbf{v}^{\mathrm{T}}), \quad \mathbf{D} = \tfrac{1}{2}(\nabla\mathbf{v} + \nabla\mathbf{v}^{\mathrm{T}}). \qquad (13)$$

The deformation rate is materially objective. The constitutive equation for an incompressible Newtonian fluid expresses the linear dependence of stress on the deformation rate, viz.,

$$\boldsymbol{\sigma} - p\mathbf{1} = 2\eta_0\mathbf{D}, \quad \mathrm{tr}(\mathbf{D}) = 0, \qquad (14)$$

where η_0 is a constant, the *Newtonian viscosity*, and p is the usual arbitrary pressure associated with incompressible media. Two characteristic properties of this model are common to all models that we would call fluids: The stress relaxes to an isotropic pressure when the strain rate goes to zero; further, any deviatoric component of stress induces an immediate nonzero rate of strain.

When Eq. (14) is put into Eq. (10), the classical Navier–Stokes equation for an incompressible Newtonian fluid results.

1.3.2 Linear Elasticity A purely elastic medium is one for which current stress is a function of current strain, that is, only the current configuration and an undeformed (that is, stress-free) configuration play a role in determining stress. *Classical infinitesimal elasticity* is based on a linear constitutive equation, an expression of *Hooke's law*. This approximation can be valid only when stretches are small and the effects of rotation unimportant. Nonetheless, it is very useful in describing brittle materials and metals in most common usages for which they are not subject to large deformations. There is a large engineering literature of various important problems in linear elasticity.

With $\mathbf{X}$ as the stress-free reference configuration, the *infinitesimal strain tensor* $\boldsymbol{\epsilon}$ is defined as

$$\boldsymbol{\epsilon} = \tfrac{1}{2}(\mathbf{F} + \mathbf{F}^{\mathrm{T}}) - \mathbf{1}. \qquad (15)$$

Then, Hooke's law for isotropic media can be written in the form

$$\boldsymbol{\sigma} = 2\mu\boldsymbol{\epsilon} + \lambda\,\mathrm{tr}(\boldsymbol{\epsilon})\mathbf{1}, \qquad (16)$$

where μ and λ are the classical *Lamé constants*. The familiar Young's modulus, E, and Poisson's ratio, ν, are related to them as follows:

$$\nu = \frac{\lambda}{2(\lambda + \mu)}, \quad E = \mu\frac{3\lambda + 2\mu}{\lambda + \mu}. \qquad (17)$$

By introducing the deviators $\mathbf{s}$ and $\mathbf{e}$ of $\boldsymbol{\sigma}$ and $\boldsymbol{\epsilon}$, respectively, Eq. (16) can be split into a scalar equation and a deviatoric equation:

$$\begin{aligned} &\mathbf{s} = 2\mu\mathbf{e}, \quad \mathrm{tr}(\boldsymbol{\sigma}) = (2\mu + 3\lambda)\,\mathrm{tr}(\boldsymbol{\epsilon}), \\ &\mathbf{s} = \boldsymbol{\sigma} - \tfrac{1}{3}\,\mathrm{tr}(\boldsymbol{\sigma})\mathbf{1}, \quad \mathbf{e} = \boldsymbol{\epsilon} - \tfrac{1}{3}\,\mathrm{tr}(\boldsymbol{\epsilon})\mathbf{1}. \end{aligned} \qquad (18)$$

An incompressible, isotropic, linear-elastic material has only one constitutive constant, an elastic modulus with dimension force/area. The constitutive equation can then be written in terms of the deviators:

$$\mathbf{s} = M\mathbf{e}, \qquad (19)$$

where M is the *shear modulus* and (in the linear approximation) $\mathbf{e}$ is equal to $\boldsymbol{\epsilon}$ since only isochoric deformations are allowed. The term modulus refers to a ratio of stress to strain, and the reciprocal of a modulus is a *compliance.*

The linear-elastic material is easily recognized as a solid because the stress deviator does not vanish no matter how long a shearing deformation is held.

1.3.3 Linear Viscoelasticity Between fluid and solid is *viscoelastic* matter which has a characteristic time and exhibits time-dependent responses. Viscoelastic materials show *stress relaxation* and *elastic recoil,* fluidlike and solidlike behavior, respectively. If a deformation is applied to a viscoelastic body and then held fixed, the stress will not remain fixed as with a purely elastic solid but will relax with time, perhaps to an isotropic pressure as with a fluid. On the other hand, if a deformation is produced by surface tractions and held for a time and then all applied tractions are suddenly released, the body will recoil as with an elastic solid, but will do so over a time span significant with respect to the characteristic time, perhaps, eventually, even returning to its original configuration. If the original configuration is not completely recovered, the residual deformation is called *permanent set.* Mechanical engineers sometimes distinguish between "viscoelastic" and "elasticoviscous" according to whether permanent set occurs, but this distinction is often unclear since relaxation is usually approximately exponential and it is generally difficult to tell whether one has waited long enough to establish permanent set. In practice, a more arbitrary definition of permanent set may be used.

If a traction is applied to a viscoelastic body and held constant, the body will *creep;* that is, the body will continue to deform for a time longer than its characteristic time—perhaps, if it is fluidlike, for as long as the traction is maintained.

A *Maxwell element* is a mechanical analog of a constitutive equation that exhibits simple exponential force relaxation. It consists of an elastic spring and a viscous dashpot connected in series fixed at one end and pulled by a force at the other so that the displacement is the sum of the responses of spring and dashpot. The characteristic time for the force at fixed displacement to relax exponentially to $1/e$ of its initial value, the quotient of the viscosity of the dashpot and the spring constant, is called the *relaxation time.*

A *Kelvin–Voigt element* connects the same components in parallel so that the force acting across the element is the sum of contributions from the spring and dashpot. Under a constant force, the displacement approaches a value determined by the spring constant. The displacement rate is a simple exponential with a characteristic time called the *retardation time.*

If a Kelvin–Voigt element is combined in series with another dashpot, a model is formed with both a retardation time and a relaxation time. When stress and strain deviators are substituted for the force and displacement in the equation for this mechanical analog, the result is the *Jeffreys model* with the following constitutive equation:

$$\boldsymbol{\sigma} + \lambda_{\text{rlx}} \frac{d\boldsymbol{\sigma}}{dt} = \eta\left[\mathbf{D} + \lambda_{\text{rtd}} \frac{d\mathbf{D}}{dt}\right], \tag{20}$$

where λ_{rlx} and λ_{rtd} are the relaxation time and retardation time, respectively, and η is a viscosity.

In a similar way, any number of these elements may be strung together into a passive, mechanical network analog of some particular, linear, constitutive equation. Substitution of stress and strain deviators for force and displacement in the equations for these networks leads to linear, viscoelastic constitutive equations in differential form describing isotropic, incompressible materials, valid for infinitesimal deformations. In general, such equations can be written as

$$P(\mathbf{s}) = Q(\mathbf{e}), \tag{21}$$

where P and Q are linear differential operators of the form

$$P = \sum_{r=0}^{p} p_r \frac{\partial^r}{\partial t^r}, \quad Q = \sum_{r=0}^{q} q_r \frac{\partial^r}{\partial t^r}, \tag{22}$$

in which the p_r and q_r are constitutive constants. (If the material is compressible, an analogous scalar equation on the isotropic parts of the stress and strain must be added to complete the picture.)

The differential equation, Eq. (21), may be seen to be of the same type as that which governs electrical networks of resistors and capacitors. Therefore, a similar analogy between linear viscoelasticity and passive electrical networks may be set up.

One advantage of constitutive equations formulated as linear differential equations is that an application of the Laplace transform to Eq. (21) removes the time variable and often allows the time-dependent, stress-analysis problem of a viscoelastic body to be replaced by an associated time-independent, linear-elastic problem whose boundary conditions are the transformed original boundary conditions. The solution of the viscoelastic problem is then found from the solution of the simpler elastic problem by inverse transformation.

A more common formulation of constitutive equations for linear-elastic media is in the form of integrals. Since the stress–strain relation is linear, one can superpose infinitesimal deformations to write a general stress equation for an incompressible, isotropic, viscoelastic material as follows:

$$\mathbf{s}(t) = \int_0^{\infty} \mathbf{e}(t - \xi) \frac{dG(\xi)}{d\xi} d\xi, \tag{23}$$

where $G(t)$ is the *stress-relaxation modulus*, the stress–strain ratio at time t as a result of a small constant strain applied at time zero on a body initially unstressed and at equilibrium. [The quantity $G(t)$ is a scalar since linearity implies that stress and strain are proportional.] Equation (23) expresses the stress as a summing up of contributions from configurations ranging from the present to the past.

The modulus $G(t)$ may be the stress-relaxation response of a mechanical network such as a set of Maxwell elements in parallel. Such a model would have a spectrum of relaxation times, each weighted by a modulus, so that $G(t)$ would be a sum of weighted exponential terms. On extending this idea to an infinite number of infinitesimally spaced relaxation times, a modulus with a continuous relaxation spectrum arises:

$$\begin{aligned} G(t) &= G_\infty + \int_0^{\infty} g(\tau) e^{-t/\tau} d\tau \\ &= G_\infty + \int_{-\infty}^{\infty} H(\ln\tau) e^{-t/\tau} d\ln\tau. \end{aligned} \tag{24}$$

The constant G_∞ has been added to allow a possible nonrelaxing elastic component. The function, $g(\tau)$, is called the *relaxation spectrum*. A logarithmic time scale is usually more practical and $H(\ln\tau)$, the *logarithmic relaxation spectrum*, is often encountered in the literature. The graph of any monotonically decreasing relaxation can be fitted with some appropriate relaxation spectrum. Much work has been done by polymer physicists in relating these spectra for polymeric materials to molecular structure.

If the strain history $\mathbf{e}(\tau)$ is zero from all past time up to time zero, Eq. (23) becomes

$$\mathbf{s}(t) = G(0)\mathbf{e}(t) + \int_0^{t} \mathbf{e}(t - \xi) \frac{dG(\xi)}{d\xi} d\xi, \tag{25}$$

where the first term on the right arises from the integration of the delta function associated with the jump of the stress-relaxation function at $\xi = 0$.

An equivalent constitutive relation can be formed from Eq. (25) by a change of dummy variables, $\tau = t - \xi$, and an integration by parts:

$$\mathbf{s}(t) = G(t)\mathbf{e}(0) + \int_0^{t} G(t - \tau) \frac{d\mathbf{e}(\tau)}{d\tau} d\tau. \tag{26}$$

Here, the initial term on the right is a consequence of a step discontinuity in strain history at $t = 0$.

An alternative form of the equations of an isotropic, incompressible, linear-viscoelastic medium might have been obtained by taking the current strain as determined by the history of the stress deviator, that is, by reversing the roles of stress and strain in the preceding. In that case, the equivalent of Eq. (25) would be

$$\mathbf{e}(t) = J(0)\mathbf{s}(t) + \int_{-\infty}^{t} \mathbf{s}(t - \xi) \frac{dJ(\xi)}{d\xi} d\xi, \tag{27}$$

where $J(t)$ is the *creep compliance*, the ratio of strain at time t to a constant stress that was applied to an undisturbed material at equilibrium at time zero.

From Eqs. (25) and (27), it is clear that the creep compliance and the stress-relaxation modulus are related. The relation connecting them is the following:

$$\int_0^{t} G(\tau) J(t - \tau) d\tau = 1. \tag{28}$$

The Laplace transform of this equation can be used to express the transform of either function in terms of the transform of the other, and, in principle, it allows one function to be calculated from the other.

It is often convenient to examine viscoelastic materials under steady-state sinusoidal excitation. If the deformation is small enough to justify linear analysis, the usual complex notation can be used to describe the response of the system. Thus, with the complex deformation history $\tilde{\mathbf{e}}(t) = \mathbf{e}_0 e^{i\omega t}$ and the corresponding complex stress $\tilde{\mathbf{s}}(t)$, the response is given by a complex modulus, $G^*(\omega)$, a function of frequency:

$$\tilde{\mathbf{s}}(t) = G^*(\omega)\tilde{\mathbf{e}}(t), \tag{29}$$

where $G^*(\omega)$ can be broken into a sum of real and imaginary parts each expressible in terms of the stress-relaxation modulus,

$$G^*(\omega) = G'(\omega) + iG''(\omega), \tag{30}$$

where $G'(\omega)$ and $G''(\omega)$, the *storage modulus* and *loss modulus*, respectively, are given by the following:

$$G'(\omega) = G(0) + \int_0^\infty \frac{dG(\tau)}{d\tau} \cos\omega\tau d\tau, \tag{31}$$

$$G''(\omega) = -\int_0^\infty \frac{dG(\tau)}{d\tau} \sin\omega\tau d\tau. \tag{32}$$

With the complex modulus, a useful equivalent to Eq. (21) can be written through use of the Fourier transform defined as follows:

$$\bar{g}(\omega) = \int_{-\infty}^{\infty} g(t)e^{-i\omega t}dt,\ g(t) = \frac{1}{2\pi}\int_{-\infty}^{\infty} \bar{g}(\omega)e^{i\omega t}d\omega, \tag{33}$$

where $\bar{g}(\omega)$ is the Fourier integral transform of $g(t)$. Then the Fourier-transformed, linear, viscoelastic, constitutive equation for an isotropic incompressible material takes the following compact form:

$$\bar{\mathbf{s}}(\omega) = G^*(\omega)\bar{\mathbf{e}}(\omega). \tag{34}$$

Of course, all these linear-viscoelastic equations can be generalized to describe compressible and/or anisotropic materials by decomposing the material properties into geometrically distinct parts and writing analogous equations for each. For example, compressible isotropic materials may be handled with the addition of another equation relating the traces of the stress and strain history.

1.4 Nonlinear Models

Linearity imposes very severe restrictions on a constitutive equation as is implied by the Boltzmann superposition principle. Consequently, the form of a linear constitutive equation is completely determined from little more than the choice of dependent and independent variables and the symmetries of the model. For materials with elasticity, linear constitutive equations can be valid only to infinitesimal deformations. On the other hand, nonlinear constitutive equations, freed from the restrictions of superposition, have great flexibility and can describe an infinite range of mechanical behaviors, often admitting phenomena quite unexpected to anyone familiar with only the linear equations of classical rheology. In fact, most materials do exhibit strongly nonlinear behavior: foods, biological materials, slurries and suspensions, soils, rocks, plastics, and rubbers, to name just a few. All elastic materials are nonlinear to some extent.

The following are some examples of the often surprising, inherently nonlinear phenomena that are observed.

1. *Poynting effect.* If a cylinder of elastic material, a length of wire or a rubber cylinder, is twisted about its axis by a force couple, the axis will lengthen. If the length is to be held fixed, a compressive axial force will be required and in certain configurations, buckling may occur.
2. *Weissenberg effect.* A phenomenon loosely related to the Poynting effect can be observed in fluids. If a rotating rod is lowered into a fixed cup of elastic fluid, the fluid will climb the rod in apparent defiance of gravity (see Fig. 1). This phenomenon is familiar to anyone who has used an eggbeater or cake mixer.
3. *Drag reduction* or *Toms effect.* The dissolution of minuscule amounts of high–molecular-weight polymer into fluids in turbulent flow will often produce amazing results. The rate of flow through a pipe

FIG. 1. The Weissenberg effect. The non-Newtonian fluid on the left (Separan AP-30) climbs the rotating rod while the Newtonian fluid (glycerol) on the right has a surface depression caused by inertia. Illustration taken from Tanner (1985), with permission from Clarendon Press.

between pressure reservoirs, for instance, can drastically increase. The details of the mechanism for this effect are still not entirely clear, but it is surely related to the nonlinearities induced by the solute.

4. *Die swell* or *Merrington effect.* An extrudate of molten plastic from a die of circular cross section will swell markedly as it leaves the die. In some cases, the diameter of the extrudate can swell to two or three times that of the die. In contrast, an extruded isothermal stream of Newtonian liquid at low Reynolds number may show a shrinking diameter; even at vanishing viscosity the swell is limited to about 15%. Die swell is very important in fiber spinning and other commercial processes.
5. *Tubeless siphon* or *Fano effect.* If a vertical tube is used to withdraw elastic liquid from a large reservoir, when the entrance of the tube is raised above the liquid level, the liquid may continue to flow up to the tube in a column supported by elastic forces (see Fig. 2).
6. *Flow through noncircular pipes.* An elastic liquid flowing through a straight tube of noncircular cross section will generally develop circulations in the plane of the cross section in addition to the flow along the axis. This would not be expected for a Newtonian fluid in isothermal laminar flow.

1.4.1 Finite Elasticity A purely elastic material is one with stress a simple function of strain, rather than a functional of strain history. If it is an isotropic material then, for purely geometrical reasons, the principal directions of the stress tensor and the strain tensor must coincide and the appropriate constitutive equation can be written as

$$\boldsymbol{\sigma} = \Lambda_0\mathbf{1} + \Lambda_1\mathbf{B} + \Lambda_2\mathbf{B}^2, \qquad (35)$$

where **B** is the left Cauchy–Green strain tensor with respect to a stress-free, undeformed state and the three scalars Λ_0, Λ_1, Λ_2 are scalar functions of the principal scalar invariants of **B**. Powers of **B** higher than the square need not appear in this equation because they can be reduced to polynomials of lower powers through the Hamilton–Cayley theorem.

Thermodynamic considerations imply a further condition on the constitutive equation for isothermal or adiabatic deformations of elastic media. In these cases, the work done in deforming the body from its undeformed state to any other equilibrium state of deformation is independent of the intervening deformations. Thus, a unique scalar quantity, the *strain energy,* can be associated with each state of deformation. Rigid rotations do not affect the strain energy; only the stretches contribute, and the strain energy of an isotropic, elastic material can be written as a function of the three principal scalar in-

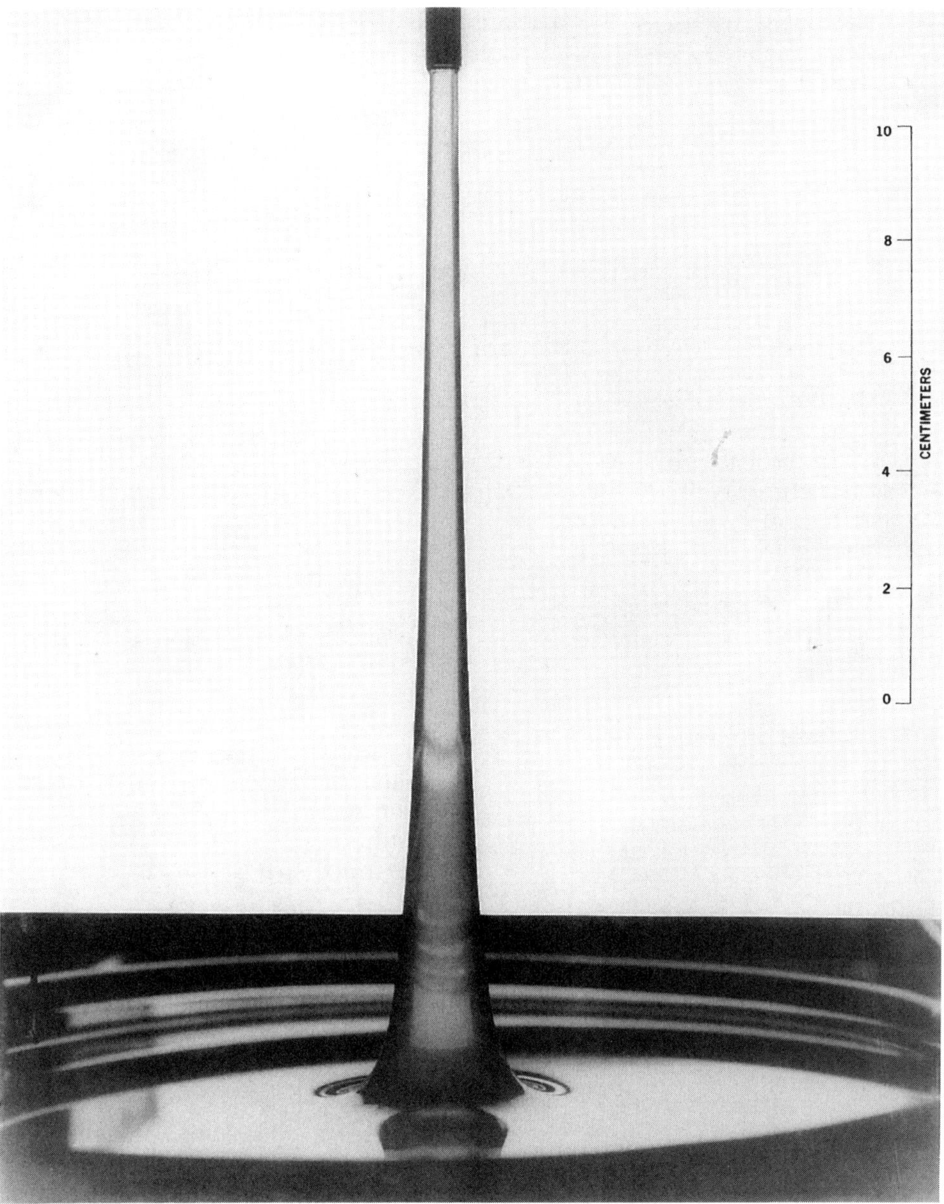

FIG. 2. The tubeless siphon or Fano effect. The siphon is begun with the tube below the surface of the elastic liquid. The liquid is sucked into the tube while the tube is raised above the surface, ultimately supporting a column of liquid. Properties of the liquid in extensional flow are related to the shape of the column. Illustration taken from Peng and Landel (1976) with permission.

variants of **B**. The result is called the *strain-energy function,* usually represented by $W(I_1,I_2,I_3)$. As a convention, the strain-energy function is written as energy per volume of material in the undeformed configuration. The mechanical properties of the material are completely contained in this function. For isothermal or adiabatic deformations, a purely elastic material has the following constitutive equation:

$$\boldsymbol{\sigma} = \frac{2}{\sqrt{I_3}}\mathbf{B}\,\frac{\partial W(I_1,I_2,I_3)}{\partial \mathbf{B}}\,. \tag{36}$$

The partial derivative in Eq. (36) can easily be expanded in terms of derivatives with respect to the scalar invariants and the equation then becomes

$$\boldsymbol{\sigma} = \frac{2}{\sqrt{I_3}}\left[\left(I_2\frac{\partial W}{\partial I_2} + I_3\frac{\partial W}{\partial I_3}\right)\mathbf{1} + \frac{\partial W}{\partial I_1}\mathbf{B} - I_3\frac{\partial W}{\partial I_2}\mathbf{B}^{-1}\right]. \tag{37}$$

Materials that are well modeled with finite elasticity usually are rubbery, polymeric materials when they are subjected to shearing deformations that are virtually isochoric. Therefore, it is particularly useful to add the constraint of incompressibility to Eq. (37) to produce the constitutive equation for an isotropic, incompressible, elastic material, viz.,

$$\boldsymbol{\sigma} - p\mathbf{1} = 2\frac{\partial W}{\partial I_1}\mathbf{B} - 2\frac{\partial W}{\partial I_2}\mathbf{B}^{-1}, \quad I_3 = 1, \tag{38}$$

where p is the usual arbitrary pressure associated with incompressibility and the strain energy $W(I_1,I_2)$ now may be taken as a function of only two invariants since I_3 is automatically 1 and thus fixed for an isochoric deformation.

In the particular case when the principal directions of strain are known, it is sometimes useful to replace Eq. (38) with relations for the principal stresses in terms of the principal stretches, the principal values of the stretch tensor, $\mathbf{V}$. The result is a set of equations known as the *Almansi form* of the equations of elasticity that is based on an expression of the strain energy in terms of stretches. Since the principal stresses are necessarily oriented parallel to the principal strains, we will identify each stress by a subscript of the corresponding stretch. With this notation, the Almansi form of Eq. (38) can be written

$$\begin{aligned}\sigma_\lambda - p &= \lambda\frac{\partial \bar{W}(\lambda,\mu,\nu)}{d\lambda},\\ \bar{W}(\lambda,\mu,\nu) &= W(I_1,I_2),\\ \lambda\mu\nu &= \sqrt{I_3} = 1,\end{aligned} \tag{39}$$

where σ_λ is the component of stress in the direction of principal stretch λ and the Almansi strain energy $\bar{W}$ is a symmetric function of the stretches λ, μ, and ν. Two similar equations are obtained by cycling the principal stretches and stresses in this equation.

Finite elasticity is used to describe large shearing motions of rubbery materials, usually polymeric, that show no significant permanent set. The constitutive mechanical properties of the material can be expressed as a strain-energy function. A plethora of particular forms of strain-energy functions have been proposed over the years, some based on models of the underlying molecular structure, others based on the phenomenology of various rubbery materials. The following is a small sampling of some common examples.

The simplest micromechanical model of the elasticity of elastomers is based on the entropic effects of a network of crosslinked flexible chains. It results in the *neo-Hookean model* for which stress is proportional to the left Cauchy–Green strain. The strain energy is then proportional to the first invariant, $I_1(\mathbf{B})$. An obvious generalization of this phenomenology results in the *Mooney–Rivlin model* for which the strain energy is a linear combination of the first and second invariants of $\mathbf{B}$,

$$W(I_1,I_2) = C_1 I_1(\mathbf{B}) + C_2 I_2(\mathbf{B}). \tag{40}$$

Here, C_1 and C_2 are constants. The neo-Hookean model is obtained by setting C_2 equal to zero. The Mooney–Rivlin model has been widely used as the simplest model that predicts behavior at least qualitatively similar to that observed with rubbery materials, but it has long been known that it is seriously inadequate for quantitative modeling.

Rubber elasticity is known to be associated predominantly with the entropy of deformation, and this entropy is often assumed to be a sum of independent contributions from each of the principal stretches. These ideas lead to the formulation of the *Valanis–Landel model,* for which the value of the strain energy is a sum over the values of a function of the principal stretches, that is,

$$\begin{aligned}\bar{W}(\lambda,\mu,\nu) &= f(\lambda) + f(\mu) + f(\nu),\\ \bar{W}(\lambda,\mu,\nu) &= W(I_1,I_2),\end{aligned} \tag{41}$$

where $f(\cdot)$ is a constitutive function of a principal stretch that we shall call the *V-L function.* In this case, Eq. (39), the Almansi form, is often useful for calculating stresses from deformations.

These models with strain energy in terms of principal stretches were considered by Mooney (1940) as special cases of a general expression for strain energy as a power series in the invariants. Valanis and Landel (1967) developed the model further and, subsequently, a number of specific forms for the V-L function, $f(\cdot)$, have been proposed.

Examples of strain energies for compressible elastic materials are somewhat rare. One example is that of Blatz and Ko (1962), used to describe deformations of sponge rubbers. It can be written in our notation as follows:

$$W(I_1,I_2,I_3) = \frac{\mu f}{2}\left[I_1 - 3 + \frac{1}{\alpha}(I_3^{-\alpha} - 1)\right] + \frac{\mu(1-f)}{2}\left[\frac{I_2}{I_3} - 1 + \frac{1}{\alpha}(I_3^{\alpha} - 1)\right],$$
$$\alpha = \nu/(1 - 2\nu), \quad 1 \geq f \geq 0, \tag{42}$$

where μ and ν are the shear modulus and Poisson's ratio, respectively, as observed at small deformations. The constant f is here an additional constitutive constant.

1.4.2 Differential Models Because the deformation rate $\mathbf{D}$ is a materially objective tensor, the most general nonlinear model for which stress is a function of deformation rate alone can be simply constructed by replacing the right-hand side of the model for a Newtonian fluid, Eq. (14), with a general isotropic tensorial function of $\mathbf{D}$,

$$\boldsymbol{\sigma} - p\mathbf{1} = 2\eta\mathbf{D} + 4\nu_2\mathbf{D}^2, \tag{43}$$

where η and ν_2 are material functions of the scalar invariants of $\mathbf{D}$. This is the *Reiner–Rivlin model.* It is said to be an inelastic model since it describes viscous behavior but does not describe recoil or many other phenomena associated with elasticity. The model is mostly of historical importance and is currently used only heuristically to model simple steady shearing flows. For such flows, the only nonzero scalar invariant of $\mathbf{D}$ is the second invariant, proportional to the square of the shear rate, κ^2, while $\mathbf{D}^2$ is equal to zero. The function η is a generalized, nonconstant viscosity, a monotonically increasing function of the shear rate. A common example of this form is the *power-law fluid,* in which η is proportional to a power of the shear rate, viz.,

$$\boldsymbol{\sigma} - p\mathbf{1} = C\kappa^{n-1}\mathbf{D}, \quad \kappa^2 = -I_2(\mathbf{D}), \tag{44}$$

where C is a material constant of suitable physical dimensions (depending on the dimensionless material constant n) and κ is the rate of shear with the dimension of reciprocal time. When n is set equal to 1, the equation becomes that of a Newtonian fluid, Eq. (14), with C set equal to η_0. The power-law fluid is useful because it can often be used to find exact solutions to geometrically simple problems and, for limited ranges of shear rate, it may fit data reasonably well. Usually, however, it is a poor representation of most rheologically complex fluids over wide ranges of shear rate (with the possible exception of slurries in simple, steady shearing flows).

To construct differential models of nonlinear viscoelasticity, it is necessary to define rates of change that are materially objective. The *Rivlin–Ericksen tensors* $\mathbf{A}_i(t)$, $i = 1, 2, \ldots$, are material time derivatives of the right relative strain, thus:

$$\mathbf{A}_n(t) \equiv \left.\frac{d^n\mathbf{C}_t(\tau)}{d\tau^n}\right|_{\tau=t}. \tag{45}$$

The higher derivatives can most easily be found with a recurrence relation for $\mathbf{A}_{n+1}$ in terms of $\mathbf{A}_n$, as follows:

$$\mathbf{A}_1(t) = 2\mathbf{D}(t),$$
$$\mathbf{A}_{n+1}(t) = \frac{d\mathbf{A}_n(t)}{dt} + \mathbf{v}\cdot\nabla\mathbf{A}_n(t) + \mathbf{A}_n(t)\mathbf{L}(t) + \mathbf{L}^{\mathrm{T}}(t)\mathbf{A}_n(t), \quad n = 1, 2, 3, \ldots, \tag{46}$$

where $\mathbf{L}$ is the velocity gradient. [See Eq. (12).] The nth-order Rivlin–Ericksen tensor has the physical dimensions of the nth power of reciprocal time.

The *nth-order Rivlin–Ericksen fluid* has the constitutive equation

$$\boldsymbol{\sigma} - p\mathbf{1} = \mathbf{f}(\mathbf{A}_1,\mathbf{A}_2,\ldots,\mathbf{A}_n), \tag{47}$$

where $\mathbf{f}(\cdot)$ is an isotropic tensor function of its arguments. There are representation theorems that may be used to develop $\mathbf{f}$ as a polynomial. For example, the stress of the second-order Rivlin–Ericksen fluid is an arbitrary isotropic tensor function of the first two Rivlin–Ericksen tensors that may be written out as the sum of terms in products of powers of the $\mathbf{A}_i$ with coefficients whose physical dimensions are those of stress times a power of time. Thus,

$$\begin{aligned}
\boldsymbol{\sigma} &= p\mathbf{1} + \boldsymbol{\sigma}_1 + \boldsymbol{\sigma}_2 + \boldsymbol{\sigma}_3 + \boldsymbol{\sigma}_4 + \boldsymbol{\sigma}_5 + \boldsymbol{\sigma}_6;\\
\boldsymbol{\sigma}_1 &= \alpha_1\mathbf{A}_1, \quad \boldsymbol{\sigma}_2 = \alpha_2\mathbf{A}_2 + \alpha_3\mathbf{A}_1^2,\\
\boldsymbol{\sigma}_3 &= \alpha_4(\mathbf{A}_1\mathbf{A}_2 + \mathbf{A}_2\mathbf{A}_1),\\
\sigma_4 &= \alpha_5\mathbf{A}_2^2 + \alpha_6(\mathbf{A}_1^2\mathbf{A}_2 + \mathbf{A}_2\mathbf{A}_1^2),\\
\boldsymbol{\sigma}_5 &= \alpha_7(\mathbf{A}_1\mathbf{A}_2^2 + \mathbf{A}_2^2\mathbf{A}_1),\\
\boldsymbol{\sigma}_6 &= \alpha_8(\mathbf{A}_1^2\mathbf{A}_2^2 + \mathbf{A}_2^2\mathbf{A}_1^2),
\end{aligned} \tag{48}$$

where the coefficients α_i are functions of the ten scalar invariants

$$\begin{aligned}
&\mathrm{tr}(\mathbf{A}_i),\ \mathrm{tr}(\mathbf{A}_i^2),\ \mathrm{tr}(\mathbf{A}_i^3), \quad i = 1, 2,\\
&\mathrm{tr}(\mathbf{A}_1\mathbf{A}_2),\ \mathrm{tr}(\mathbf{A}_1^2\mathbf{A}_2),\ \mathrm{tr}(\mathbf{A}_1\mathbf{A}_2^2),\ \mathrm{tr}(\mathbf{A}_1^2\mathbf{A}_2^2).
\end{aligned} \tag{49}$$

The representation of isotropic functions of sets of tensors is a highly technical subject, important to rheology and well reviewed in an article by Spencer (1971). The second-order Rivlin–Ericksen fluid in Eq. (48) is based on the representation of a general isotropic tensor function of only the first two $\mathbf{A}_i$, but it is already far too complicated to be of much use in applications.

A considerable simplification is achieved by restricting the stress to a power series in the $\mathbf{A}_i$ with constant coefficients. The resulting equations containing terms with coefficients of dimensions up to nth order in time are called simply *nth order fluids*. Thus, the second-order fluid has the constitutive equation

$$\boldsymbol{\sigma} = p\mathbf{1} + \eta_0\mathbf{A}_1 + \alpha_3\mathbf{A}_1^2 + \alpha_2\mathbf{A}_2, \tag{50}$$

where η_0 is the Newtonian viscosity and the α_i are constants. This is perhaps the simplest fluid that can be used to model the Weissenberg effect. Furthermore, an unexpected and useful feature of second-order fluids is that for any plane boundary-value problem determining a flow by boundary conditions on the velocity (rather than surface tractions), the flow appropriate for a Newtonian fluid is also a possible flow for a second-order fluid of the same viscosity: Only the stresses are different. This remarkable fact is known as the *Tanner–Giesekus theorem*, and it can be used to find inverse solutions to many simple plane flow problems by direct substitution of well-known Newtonian flows. The theorem is easily extended to parallel rectilinear flows and to potential flows such as occur in boundary layers. These solutions give insight into the mechanisms for many nonlinear phenomena but, unfortunately, the second-order fluid has some serious failings. It does not explain circulations in flow through noncircular pipes, for instance, and is certainly inadequate to describe stress relaxation.

The generalization of a linear differential model, valid for small deformations, to a nonlinear model valid for large deformations is not unique. Furthermore, the material time derivatives are ambiguous for finite deformations. Roughly speaking, this problem arises because the base vectors of the coordinate system convect differently for covariant and contravariant systems. In technical terms, raising and lowering indices does not generally commute with material time differentiation. As a result, the rate of a covariant tensor is not the same as the covariant form of the rate of the contravariant form of the tensor. Yet, either rate is objective. Indeed, so is any linear combination of them. Consequently, there is an infinity of materially objective time derivatives of constitutive tensors.

Three particular rates are so commonly used that they have been given names. If $\mathbf{S}$ is any generic objective tensor, $\mathbf{L}$ is the velocity gradient defined in Eq. (12), and $\mathbf{W}$ is the spin tensor defined in Eq. (13), then the three convected derivatives or rates of change of $\mathbf{S}$ are defined and named as follows:

$$\begin{aligned}
\frac{D^{\triangledown}\mathbf{S}}{Dt} &\equiv \frac{\partial\mathbf{S}}{\partial t} + (\mathbf{v}\cdot\nabla)\mathbf{S} - \mathbf{LS} - \mathbf{SL}^{\mathrm{T}} \text{ (Upper)},\\
\frac{D^{\triangle}\mathbf{S}}{Dt} &\equiv \frac{\partial\mathbf{S}}{\partial t} + (\mathbf{v}\cdot\nabla)\mathbf{S} + \mathbf{SL} + \mathbf{L}^{\mathrm{T}}\mathbf{S} \text{ (Lower)},\\
\frac{D^{0}\mathbf{S}}{Dt} &\equiv \frac{\partial\mathbf{S}}{\partial t} + (\mathbf{v}\cdot\nabla)\mathbf{S} + \mathbf{SW} - \mathbf{WS} \text{ (Jaumann)}.
\end{aligned} \tag{51}$$

The upper-convected derivative is the rate of a contravariant tensor as seen by an observer in a body-fixed frame of covariant base vectors but expressed in a space-fixed frame. Similarly, the lower-convected derivative relates to a covariant tensor. The Jaumann or *corotational derivative* is the average of these two, and it is the rate seen by an observer whose frame rotates with the material but does not stretch.

The *Oldroyd B-fluid* is a nonlinear form of the Jeffreys model. It is obtained by replacing time derivatives in Eq. (20) with upper-convected derivatives to get

$$\mathbf{s} + \lambda_{\text{rlx}} \frac{D^{\triangledown}\mathbf{s}}{Dt} = 2\eta\left[\mathbf{D} + \lambda_{\text{rtd}} \frac{D^{\triangledown}\mathbf{D}}{Dt}\right]. \tag{52}$$

This is a fairly simple, properly objective equation for a model that exhibits the Weissenberg effect. If the retardation time λ_{rtd} is set equal to zero, the equation becomes that of the *upper-convected Maxwell model.* This Maxwell model is one of the simplest properly objective viscoelastic equations, and it is widely used for calculating complex flows. It is known that the model is a poor representation of materials—for instance, it cannot describe shear-dependent viscosity—but, in many situations, it often shows qualitatively correct behavior.

Differential models are often favored when relating constitutive equations to micromechanics. For example, picturing a polymer solution as a dilute solution of "dumbbells," pairs of spherical particles joined by a Hookean spring and immersed in a Newtonian fluid, leads to the *dumbbell model.* This model is phenomenologically equivalent to the upper-convected Maxwell model with the addition that the material constants η, λ_{rlx}, and λ_{rtd} are expressed in terms of parameters of the micromechanics. An elaboration of this model based on the micromechanics of a less dilute solution is the *Giesekus model* for which a concentration dependence is added to the dumbbell model through the following equation:

$$\lambda \frac{D^{\triangledown}\mathbf{s}}{Dt} + \mathbf{s} + \frac{\alpha\lambda}{\eta}\mathbf{s}^2 = 2\eta\mathbf{D}, \quad 0 \leq \alpha \leq 1, \tag{53}$$

where α is a measure of the anisotropy induced by the flow due to the mean orientation of the dumbbells.

The *Leonov model* is based upon the idea that strain and strain rate may each be decomposed into elastic and irrecoverable parts. The irrecoverable strain rate is determined as a function of the elastic strain through a constitutive equation similar to Eq. (35) for stress. Then the elastic strain is determined through

$$\frac{D^{\triangledown}\mathbf{C}_e}{Dt} + \mathbf{D}_i\mathbf{C}_e + \mathbf{C}_e\mathbf{D}_i = 0, \tag{54}$$

where $\mathbf{C}_e$ and $\mathbf{D}_i$ are the recoverable strain and the irrecoverable strain rate, respectively. The stress is found through another equation on the elastic strain of the form of Eq. (35). The Leonov model thus requires two constitutive functions similar to strain-energy functions, the elastic strain energy and the *nonequilibrium potential* for determining the irrecoverable strain rate. As a result, there is the possibility of describing very complicated behavior. The simplest version proposed, however, contains only two material constants, a modulus and a relaxation time, and, for plane flows, it is equivalent to the Giesekus equation.

1.4.3 Models with Plastic Yield Many materials, such as metals or drilling muds, appear not to flow until the stress exceeds some minimum value. The *Bingham body* is a classical model of shearing in such materials. For this model, the constitutive equation takes different forms depending on whether or not the shear stress exceeds the *yield stress* σ_y, a constitutive constant. For simple shear, the Bingham body is defined by the following:

$$\begin{aligned} &\sigma_s - \sigma_y = \eta_p\kappa, \quad |\sigma_s| \geq \sigma_y; \\ &\kappa = 0, \quad |\sigma_s| \leq \sigma_y, \end{aligned} \tag{55}$$

where σ_s is the shear component of stress, σ_y is the yield stress, η_p is the plastic viscosity, a constitutive parameter, and κ is the rate of shear. A tensor form of Eq. (55) could easily be written, but that hardly seems worth while since the model is seldom used for analyzing any deformation other than simple shearing.

The Bingham body is simply a Newtonian fluid combined through the yield condition with a rigid body. Other such constitutive equations can be easily formed by combining solidlike equations (e.g., elasticity) valid for stresses below a yield stress with a fluid differential model (e.g., a power-law fluid) above the yield. A survey of such models is to be found in Bird *et al.* (1987). *Plasticity theories* also utilize this idea of a yield criterion distinguishing loading from unloading of stresses, but further supplement this with flow rules. The classical theory of plasticity as expounded in Hill (1950) has seen many subsequent elaborations and improvements in later years. These theories are particularly useful in describing metal-forming processes.

1.4.4 Integral Models In the case of linear viscoelasticity, for smooth histories at least, each differential model corresponds to an integral model. Which of these representations to use depends on the deformation history to be described. Discontinuous histories of stress or strain are usually more easily described by integral forms while sinusoidal histories are best treated with differential forms. For nonlinear viscoelasticity, this choice is seldom available and only the simplest nonlinear models may retain this convenience.

An early model useful for unvulcanized rubbers and molten polymers was based upon the micromechanics of a certain transient network with junctions that break and reform, the *Green–Tobolsky model,* with an integral constitutive equation:

$$\boldsymbol{\sigma}(t) = p\mathbf{1} + G\int_{-\infty}^{t} e^{(\tau-t)/\lambda}\mathbf{B}_t(\tau)d\tau. \tag{56}$$

This equation is equivalent to the upper-convected Maxwell model, Eq. (52), with λ_{rtd} set to zero and η equal to $G\lambda_{\mathrm{rlx}}$.

The *Lodge model* is based upon less restricted transient network micromechanics resulting in a somewhat more general integral form:

$$\boldsymbol{\sigma}(t) = p\mathbf{1} + \int_{-\infty}^{t} \mu(t-\tau)\mathbf{B}_t(\tau)d\tau,$$
$$\mu(\xi) = \sum_i \frac{G_i}{\lambda_i} e^{-\xi/\lambda_i}. \tag{57}$$

Here, the function $\mu(\cdot)$ is a sum of relaxation processes each with a relaxation time λ_i and a modulus G_i. These quantities are to be evaluated from fitting stress-relaxation curves or, possibly, are derived from more detailed micromechanical considerations.

Because stress relaxation and other mechanical behaviors typical of viscoelastic materials are clearly not equilibrium phenomena, classical thermodynamics (which is a thermodynamics of equilibrium states) is not sufficient to describe viscoelasticity. The *perfect elastic fluid* is a thermodynamically plausible, phenomenological model of viscoelasticity encompassing entropic as well as energetic effects. The model depends upon the assumption that adding a quantity to entropy can form a variable to be used in place of entropy in a caloric equation of state that will apply to these nonequilibrium viscoelastic phenomena. The caloric equation of state of classical thermodynamics is generalized by replacing entropy with a state function for nonequilibrium states that is the sum of entropy and a quantity called *entaxy.* Entaxy is a functional of the deformation history and the thermal history of the material, in contrast to the entropy, which is simply a function of the deformation and temperature at equilibrium (Bernstein *et al.,* 1964). Thus,

$$\Sigma(t) = \int_{-\infty}^{t} S\left(C_\tau(t), \int_t^\tau b_T(\xi)d\xi\right)b_T(\tau)d\tau,$$
$$\epsilon = f(v, s+\Sigma), \quad T = \frac{\partial f(v, s+\Sigma)}{\partial s}, \tag{58}$$

where $\Sigma(t)$ is the specific entaxy at time t, s is specific entropy, and v is the specific volume (the inverse of density, ρ). The function $S(\mathbf{C},\Delta t)$ is the rate of entaxy production, a function of the deformation from a past configuration and of the time interval as seen by the material. That time interval is dependent on b_T, the rate of a material clock, a positive function of temperature that is the inverse of a_T, the shift factor of time-temperature superposition (see Sec. 2.3, Moduli of Linear Viscoelasticity). The generalized caloric equation of state gives the specific internal energy ϵ as a function of specific volume v and $s + \Sigma$, the sum of entropy and entaxy. The absolute temperature T is the derivative of internal energy with respect to entropy, just as it is in equilibrium thermodynamics. The stress is given by a hydrostatic pressure, the derivative of internal energy with respect to specific volume, plus terms (not isotropic in general) that vanish at equilibrium. When a state is held until equilibrium is achieved, Σ goes to zero and classical thermodynamics is restored.

The stress equation for a perfect elastic fluid is written

$$\sigma(t) = p(v,T)\mathbf{1} + \rho T \int_{-\infty}^{t} \mathbf{F}_t(\tau) \times \frac{\partial S\left(\mathbf{C}_\tau(t), \int_\tau^t b_T(\xi)d\xi\right)}{\partial \mathbf{C}_\tau(t)} \mathbf{F}_t^{\mathrm{T}}(\tau) b_T(\tau) d\tau. \tag{59}$$

Entaxy production is assumed to have properties such that, while a material is held fixed and isothermal after deformation until thermodynamic equilibrium is established, the integral of Eq. (59) will monotonically relax, eventually to zero—the stress will approach a purely isotropic pressure. Thus, this is a model of a fluid. On the other hand, the integrand of Eq. (59) has similarities to the right-hand side of Eq. (36) for finite elasticity so that the model shows certain elastic properties when the material is not in equilibrium. Depending on the time scale of observation [see Deborah number, Eq. (80)], this model will display either elastic or viscous behavior. The material properties of a perfect elastic fluid are contained in the function $S(\cdot,\cdot)$, the function b_T, and the equilibrium pressure function, $p(v,T)$.

The perfect elastic fluid model is generally used to describe isothermal deformations of incompressible materials for which the thermodynamic details are largely irrelevant. In that case, Eq. (59) reduces to

$$\sigma(t) = p\mathbf{1} + \int_{-\infty}^{t} \left[\frac{\partial U(I_1, I_2, t-\tau)}{\partial I_1} \mathbf{B}_t(\tau) - \frac{\partial U(I_1, I_2, t-\tau)}{\partial I_2} \mathbf{B}_t^{-1}(\tau) \right] d\tau,$$

$$I_i = I_i(\mathbf{B}_t(\tau)). \tag{60}$$

This equation is called the *BKZ model* (for Bernstein, Kearsley, and Zapas; see Bernstein *et al.*, 1963, 1965) or sometimes the *Kaye-BKZ model.* The constitutive function $U(I_1, I_2, t - \tau)$ has the character of a time-dependent, weighted, strain-energy function of finite elasticity, and the integral on the right may be viewed as a weighted sum of the partial stresses arising from deformations from configurations at all past times τ to that at the present time, t. As a result, the material properties of the BKZ model are carried by the function U, which is unspecified. It must be measured or, possibly, derived from micromechanical considerations. A useful reference for applying this formulation to common deformations is found in Bernstein (1966).

The BKZ model with special forms of the function U arises often for simple micromechanically motivated models. The Green–Tobolsky and the Lodge models, Eqs. (56) and (57), are easily recognized as particular cases of Eq. (60). The *Doi–Edwards model* (1971) is based upon the idea of *reptation* proposed by de Gennes (1971) for the micromechanics of polymer melts. The chain molecules are visualized as confined by the surroundings to move within tubelike regions while the tubes themselves slowly change shape by diffusion. This rather elaborate picture turns out, at least in its simplest form, to generate a BKZ model.

Other integral models have been proposed, some of which may be viewed as generalizations of the BKZ model. For example, the derivatives of U in Eq. (60) may be replaced by arbitrary functions of the same variables, and these functions may be taken as factorable into a deformation-dependent part and time-interval–dependent part. Or, scalar invariants of the rate of deformation may be included as variables. Wagner and Stephenson (1979) propose a model in which an irreversibility is introduced by distinguishing "increasing" from "decreasing" deformation much as in plasticity theory. The form of the constitutive equation is modified according to this distinction. The resulting constitutive equation is often called the *Wagner model.*

1.4.5 Nonsimple Materials The constitutive equations above are all simple materials in the sense that stress at a point depends only on the history of deformation at that point. Such constitutive equations may be inadequate to describe complex materials for several reasons. The stress at a point may depend on nonlocal effects as might be expected for suspensions of stiff fibers. Or, there may be microstructure introducing internal degrees of freedom at a point, say the relative rotation of particles in a granular material or dislocation density in a lattice structure. Couple stresses may be required if electromagnetic interactions are to be included. The *Cosserat continuum* (Cosserat and Cosserat, 1909) is a model material for which each point of material, in addition to its position, carries vectors called *directors* that reflect micromechanical influences. A director might express some local mean orientation of particles in a suspension, for instance. If these directors affect the energetics and rheology of the material, the balance equations, Eqs. (9)–(11), must be modified accordingly. The Cosserat continuum and other models of this type are often referred to as *micropolar models.* Expositions of theories of this type are found in Green and Rivlin (1964), Toupin (1962), and Eringen (1962).

An important example of such nonsimple constitutive equations is the *Ericksen–Leslie model* of liquid crystals (Ericksen, 1961; Leslie, 1968). In this case, the director relates to the orientation of rodlike molecules. A constitutive equation for the asymmetric stress tensor is coupled with suitably modified balance equations and boundary conditions.

2. RHEOMETRY

To confirm the validity of constitutive equations as well as to realize their practical benefits, it is necessary to devise methods of observing and measuring the mechanical properties of target materials. Measurements of rheological properties of a material can only be made in the context of some assumptions about the form of a suitable constitutive equation. Ultimately, workable methods must be devised to produce well-defined deformations and to measure the forces required. Almost always, ideal measurements can only be approached in approximation. For example, infinite geometry may be modeled with finite apparatus by ignoring various end effects. Furthermore, the practicality of a given measurement technique is usually limited to a range of the properties of the test material. As a result, a great variety of apparatus have been devised, often for measurement of the same material property, but useful for different materials and circumstances. A practicing industrial processor of materials would like one simple test device that gives a number measuring "processability." Most technological processes are so complex, however, that no single measuring device can possibly be sufficient.

In the following discussion of rheometry, we shall often assume that inertial effects, surface-tension effects, and other nonrheological effects can be ignored, and we will say little about them. Nonetheless, sometimes these effects cannot be ignored. For instance, it is clear from Eq. (10) that the inertial term $\rho(\mathbf{v}\cdot\nabla)\mathbf{v}$ could have a big influence on high-velocity flows. In some situations, this term can produce complicating nonlinearities not directly related to the material rheology and then it must be taken into account. Fortunately, the range of deformation rate of interest for many materials and processes is such that rheological measurements at low–Reynolds-number flows are adequate and inertial effects can usually be ignored.

2.1 Viscometry

The rheology of Newtonian fluids conforming to Eq. (14) is completely determined by measuring the value of η_0, the constant viscosity. A variety of instruments have been devised for achieving this. The commonest of these is the *capillary viscometer,* for which the volume rate of laminar flow through the capillary under the influence of a known pressure difference is related to the viscosity through the *Hagen–Poiseuille law:*

$$\eta_0 = \pi R^4 \Delta P / 8QL, \tag{61}$$

where R is the radius of the capillary, ΔP is the difference in pressure across the capillary of length L, and Q is the volume flow per unit time. In practice, corrections to this equation are often needed, typically, a kinetic energy correction to account for the power expended in overcoming inertia and the Couette correction to account for end effects of the finite capillary through an addition to L of a fictitious corrective length. In gravity-driven capillary viscometers and in some other devices, it is actually the *kinematic viscosity,* η_0/ρ, which is measured.

A standard reference for the measurement of Newtonian viscosity by this and other classical methods is Barr (1931).

Nonlinear fluids that are modeled by a modification allowing viscosity to vary with shear rate in Eq. (14) are said to be non-Newtonian. If viscosity decreases with rate of shear, the material is said to be *pseudoplastic,* and if it increases, it is said to be *dilatant.* For some materials, such as certain suspensions and colloids, there is an internal structure when they are quiescent that is slowly broken up by shearing but, after a recovery period at rest, the fluid returns to its structured state. Such materials show a decrease in viscosity with time of shearing after rest, and they are called *thixotropic*. In rarer cases, the formation of a temporary structure is promoted by shearing and viscosity increases with time of shear. Such a material is said to be *rheopectic*.

In characterizing a non-Newtonian material, it is necessary to assign a shear rate to each measurement of viscosity. Furthermore, the Hagen–Poiseuille law is no longer valid since it is derived from the velocity profile of the flow, which differs here from the Newtonian case. Capillary viscometers can still be used, however, in the following way (based upon an equation for the gradient of the velocity). The volume flow rate Q is measured as a function of the pressure gradient, $a = \Delta P/L$. By calculation or graphical methods, the quantity κ, a rate of shear, is found from the following equation:

$$\kappa(a) \equiv \frac{1}{\pi a^2 R^3} \frac{d}{da} [a^3 Q(a)], \quad a \equiv \frac{\Delta P}{L}. \tag{62}$$

Finally, the viscosity as a function of shear rate is given by

$$\eta(\kappa) = R\Delta P/2L\kappa. \tag{63}$$

This procedure, viewed as a correction of the Hagen–Poiseuille law, is usually referred to as the *Rabinowitsch correction*. As a check on the validity of the method, measurements from capillaries of differing radii are compared. If they do not fall upon a single viscosity–rate-of-shear curve, problems with slip at the wall or other anomalies are to be suspected.

As the rate of shear is increased, even for Newtonian fluids, eventually it is found that laminar flow cannot be maintained, turbulence occurs, and the viscometer equations are no longer valid. But for many nonlinear liquids, other difficulties may interfere well before arriving at that point. The end corrections can be quite large for elastic fluids when considerable power is consumed in generating elastic energy. In that event, the Couette correction is inadequate and a procedure developed by Bagley (1957) may be used as an approximate correction for this entrance effect. As the fluid emerges from the tube, a swelling of the jet known as *die swell* (see above, Sec. 1.4) will usually occur and sometimes, at higher flow rates, an irregular unsteady extrusion known as *melt fracture*. These phenomena are presently only partially understood even though they are of great practical importance.

A *Couette viscometer* uses the laminar shearing flow between two coaxial cylinders to measure viscosity. In the most common, convenient form, the inner cylinder is rotated and the outer held fixed, but, at high Reynolds numbers, this arrangement has the drawback that Taylor vortices due to inertial forces may arise and disturb the flow. An instrument that overcomes this complication by exchanging the roles of the cylinders, thus stabilizing the flow, is sometimes called a *Searle viscometer*. In both cases, the following equation called the *Margules equation* is used to measure Newtonian viscosity:

$$\eta_0 = \frac{M}{4\pi L \Delta\Omega}\left(\frac{1}{R_i^2} - \frac{1}{R_o^2}\right), \tag{64}$$

where M is the torque acting on a cylinder, L is the length of the cylinders, $\Delta\Omega$ is the difference in angular velocity between the cylinders, and R_i and R_o are the radii of inner and outer cylinders, respectively. The Margules equation is based upon flow between infinite cylinders and, for very accurate measurements, end-effect corrections and other adjustments may be needed.

Couette viscometers may also be used to measure non-Newtonian viscosity by using a procedure analogous to the Rabinowitsch correction for capillaries, but it is somewhat more difficult to apply. Yang and Krieger (1978) consider various approximate methods of associating the measurement of viscosity with shear rate that are more practical and are often used.

To avoid the complication resulting from inhomogeneous shear rates for non-Newtonian fluids, it is convenient to measure viscosity from shearing flows between a rotating cone and a fixed plate. If the cone axis is perpendicular to the plate and the apex is just at the plate, the resulting flow is taken to be a homogeneous shearing in which fluid particles travel in circles about the cone axis. Actually, for nonlinear materials, this is an approximation true only for geometries with a small angle between the cone surface and the plane because, otherwise, normal stresses as well as inertial effects create transverse flows. For this reason, it is customary to restrict this angle to be less than 4°. With a *cone-and-plate viscometer,* for both Newtonian and non-Newtonian fluids, the viscosity and the rate of shear are found from simple equations, as follows:

$$\eta(\kappa) = \frac{3M\theta}{2\pi R^3 \Omega}, \quad \kappa = \frac{\Omega}{\theta}, \tag{65}$$

where κ is the rate of shear, M is the torque, θ is the angle between the surface of the cone and the plate, R is the radius, and Ω is the angular velocity between cone and plate.

There are many other viscometers designed to fit various practical or special technological needs. A variety of mechanisms can be used: timing the fall under gravity of a heavy ball immersed in the sample, measuring the damping of a vibrating reed or timing a disk in torsional oscillation in the test fluid, etc. A fairly complete listing of such devices as well as the classical viscometers can be found in Merrington (1949). These methods using complex flows can be used for relative viscometry of Newtonian liquids, where the linearity of the model allows comparisons of viscosity, but they are seldom practical for measuring material properties of non-Newtonian materials.

2.2 Strain-Energy Functions

The mechanical properties of a purely elastic material are contained in the strain-energy function: For an incompressible isotropic material, this is a function of two invariants of the deformation. The strain-energy function for a particular incompressible material is determined from measurements of force and displacement. [It can be seen from Eq. (38) that, to be precise, it is only the two derivatives of the strain energy with respect to the strain invariants that are experimentally accessible.] The ranges of the invariants that are physically possible are restricted by the requirement that the stretches be real positive quantities. Figure 3 illustrates the region of the $I_1(B)$–$I_2(B)$ plane containing physically meaningful sets of these two invariants. The region is enclosed by a cusp formed between two curves, loci of the values for simple extension and equibiaxial extension, respectively. The 45° line in Fig. 3 corresponds to values for pure shear. Strain energy is determined (up to an arbitrary constant of no physical significance—usually it is set to give zero strain energy in the undeformed state) by mapping the derivatives over this region. This is a laborious and demanding task at best. For some typical results, see Obata *et al.* (1970), who have measured the strain-energy derivatives of some typical rubbers as a function of degree of cross-linking and of temperature.

The most direct method of measuring the derivatives of the strain energy is through homogeneous, biaxial stretching of a sheet of material. Rivlin and Saunders (1951), in a classic description of various measurements of strain energy, describe such biaxial measurements in extensive detail. The method is useful, but it is tedious and, for practical reasons, it is not accurate for deformations close to the boundaries of the region of possible invariants shown in Fig. 3.

There are a number of inhomogeneous deformations that may be used to measure the strain-energy function of an incompressible material. Ericksen (1954) finds several classes of suitable deformations (sometimes called *controllable deformations*) for which the derivatives of the strain energy may be expressed in terms of the surface tractions needed to produce the deformations; that is, it is possible to produce them solely through the application of surface tractions no matter what the form of the strain energy. Torsion of a cylinder is an example. In this case, a distribution of normal force (because of the Poynting effect: see Sec. 1.4), in addition to a torque, must be applied to the ends of the cylinder to maintain the deformation. Of course, the actual amounts of traction required will depend upon the strain energy but, no matter what form the strain-energy function, some suitable torque and normal force will suffice. Penn and Kearsley (1976) demonstrate how measurements of torsion of an elastic cylinder can be used to evaluate the derivatives of the strain energy along the pure shear line of Fig. 3.

When the strain energy is of the Valanis–Landel type, the measurement task is much reduced. Then, the measurement of torque and normal force in torsion of a cylinder, or

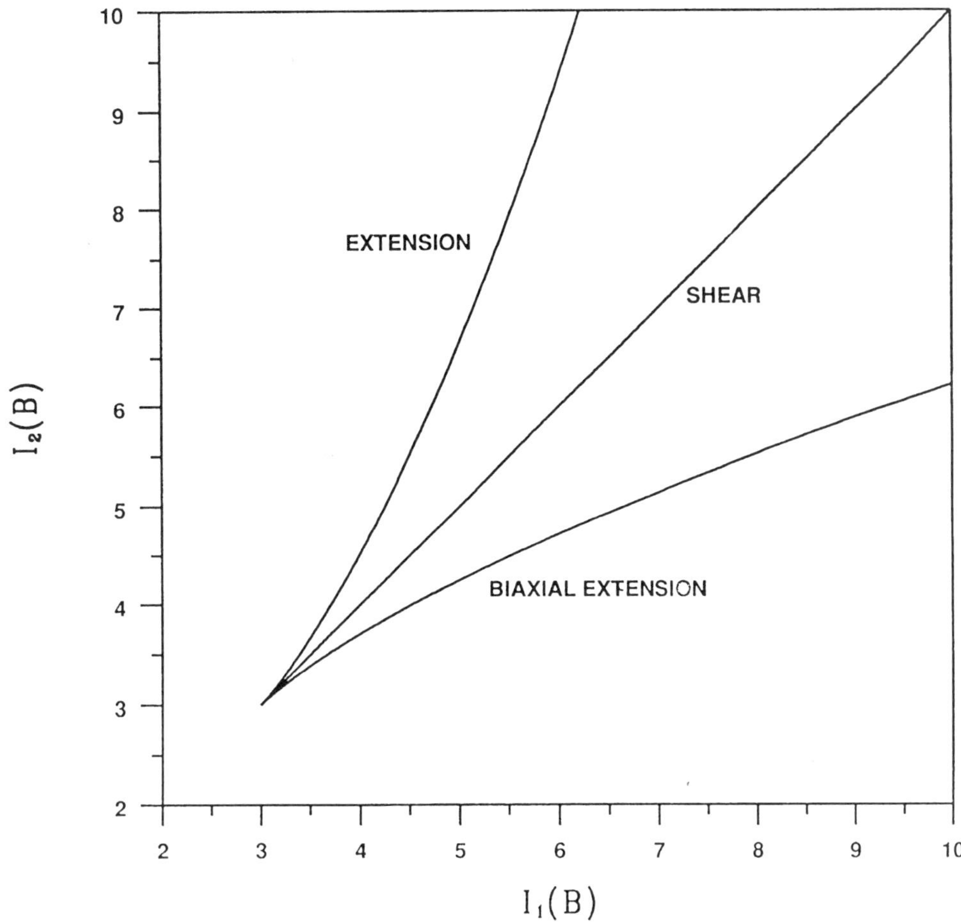

FIG. 3. Strain invariants of an incompressible material. Values of the invariants are confined to the cusped area between the lines for extension and biaxial extension because of the requirement that stretches be real and positive. The 45° line for which $I_1 = I_2$ is the locus of shear deformations.

the measurement of tractions for simple extension and equibiaxial extension, or any combination of two of these measurements is sufficient to determine the stress in any deformation (Kearsley, 1980; Kearsley and Zapas, 1980). These methods can be extremely useful because, in fact, most molecular models of rubber elasticity do predict a strain-energy function of Valanis–Landel form.

2.3 Moduli of Linear Viscoelasticity

Each of Eqs. (26), (27), and (29) formulates the mechanical properties of a linear-viscoelastic material. They indicate that the material may be characterized by measurements of stress relaxation or creep or sinusoidal deformation. For practical reasons, all three methods may be needed to cover a broad time scale of responses. For very short-time behavior, measurements of wave propagation through the material may be used. Once either $G(t)$ or $J(t)$ or $G^*(i\omega)$ is known, it is possible to calculate the stress corresponding to an arbitrary strain history, or the strain corresponding to an arbitrary stress history, by use of Boltzmann superposition principle.

For many polymeric materials, a plot of $\log J(t)$ vs $\log t$ measured at some fixed temperature T_0 can be superposed on the plot measured at another fixed temperature by translation along the log$t$ axis combined with a smaller translation along the log$J$ axis. The translation along the log$t$ axis is usually designated loga_T, where a_T is the reciprocal of the temperature-dependent rate of a material clock [see b_T of Eq. (59)]. Materials that fit this picture of a temperature-controlled, material clock are called *thermorheologically simple.* This curve-shifting procedure, known as *time-temperature superposition,* can then be used to draw a master curve of $J(t,T_0)$ at a reference temperature using superposed data from measurements at various temperatures. The shift factor a_T and the vertical shift are also plotted. In this way, extremely wide time scales can be plotted—so wide that it would be completely impractical to measure them at a single temperature. The resulting curve can be transposed to other temperatures by appropriate shifts taken from the plots.

2.4 Normal Stresses

In a certain class of flows, the strain history seen by a particle of material is independent of time in the sense that the strain history is independent of the chosen reference time, except for a possible rotation of the reference frame. The importance of this class is that for each such flow the stress for a general simple material modeled by a complicated functional as in Eq. (8) can be reduced to a set of functions of the first three Rivlin–Ericksen tensors. Such flows are called *motions with constant stretch history* (abbreviated MWCSH) because a material point experiences a deformation rate constant in time. They are often used in rheometry because of the simplicity of dealing with functions rather than functionals. A detailed analysis of the kinematics of MWCSH can be found in Huilgol (1975).

By far the commonest examples of MWCSH are steady laminar shear flows such as those occurring in viscometry. These flows constitute an important class called *viscometric flows.* A concise but readable treatment of the subject is found in Coleman *et al.* (1966). Viscometric flows are MWCSH with the added restriction that $\mathbf{L}^2 = (\nabla\mathbf{v})^2 = 0$. As a consequence, all Rivlin–Ericksen tensors of order greater than three are zero for these flows and, at each point of a flow, by rotation, a Cartesian coordinate system can be found in which the Rivlin–Ericksen tensors are as follows:

$$\mathbf{A}_1 = \begin{Vmatrix} 0 & \kappa & 0 \\ \kappa & 0 & 0 \\ 0 & 0 & 0 \end{Vmatrix}, \quad \mathbf{A}_2 = \begin{Vmatrix} 0 & 0 & 0 \\ 0 & 2\kappa^2 & 0 \\ 0 & 0 & 0 \end{Vmatrix},$$
$$\mathbf{A}_i = 0 \quad \text{when } i \geq 3. \tag{66}$$

The quantity κ and the rotation necessary to achieve this form will usually depend on position in the flow.

When Eq. (66) holds for all points of the flow in a fixed Cartesian coordinate system and κ is constant throughout the flow, the flow is a plane shearing flow with shear rate κ oriented so that the flow velocity is in direction 1, the shear gradient is in direction 2, and direction 3 is perpendicular to the other two. Locally, any viscometric flow of a simple material is equivalent to a steady plane shearing flow: Measurements of it can reveal no more rheological information than does steady plane shearing flow.

In viscometric flows, simple materials described by the functional of Eq. (8) are indistinguishable from the conceptually simpler Rivlin–Ericksen fluids described by the functions of Eq. (47) which, in light of Eq. (66), may be further simplified. Thus, for general simple materials in steady plane shearing flow a constitutive equation can be written as follows:

$$\boldsymbol{\sigma} = p\mathbf{1} + \eta(\kappa)\mathbf{A}_1 + [\psi_1(\kappa^2) + \psi_2(\kappa^2)]\mathbf{A}_1^2 - \tfrac{1}{2}\psi_1(\kappa^2) \tag{67}$$

where the $\mathbf{A}_i$ are in the form of Eq. (66). The rheological material properties determining material response in all viscometric flows are contained in three functions of shear rate: η is the viscosity, while ψ_1 and ψ_2 are called the *first normal-stress coefficient* and the *second normal-stress coefficient,* respectively, for reasons that will now be made clear.

From Eqs. (66) and (67), the components of stress are given by

$$\sigma_{12} = \sigma_{21} = \kappa\eta(\kappa),$$
$$\sigma_{11} = \kappa^2[\psi_1(\kappa^2) + \psi_2(\kappa^2)] + p,$$
$$\sigma_{22} = \kappa^2\psi_2(\kappa^2) + p, \quad \sigma_{33} = p, \tag{68}$$

all other components are zero.

Here, p is the arbitrary pressure associated with incompressibility and $\eta(\kappa)$ is the viscosity generating the shear stress. The only other nonzero components of stress are the diagonal components of the stress tensor, the *normal stresses*. Because p is arbitrary, the normal stresses themselves are not uniquely determined by shear rate, and it is customary to avoid the resulting ambiguity by considering differences of normal stresses. By convention, two fundamental normal-stress differences are defined, the *first and second normal-stress functions*, N_1 and N_2, each proportional to a corresponding normal-stress coefficient, ψ_1 and ψ_2, respectively, related as follows:

$$\begin{aligned} N_1 &\equiv \sigma_{11} - \sigma_{22} = \kappa^2\psi_1(\kappa^2), \\ N_2 &\equiv \sigma_{22} - \sigma_{33} = \kappa^2\psi_2(\kappa^2). \end{aligned} \tag{69}$$

The rheology of a simple material in a viscometric flow is completely characterized by the viscosity and the first and second normal-stress coefficients. Therefore, measuring these functions in any one viscometric flow characterizes a simple material for the whole class of flows.

For shearing flows that are not steady, direct analogs of the viscosity and normal-stress coefficients are often defined. Of course, these analogs are functions of time and they reflect the nonconstant stretch history at a particle. They apply to flows that are not MWCSH.

There are useful MWCSH other than viscometric flows. An example is steady extensional flow, that is, a pure stretching motion with constant rates of stretch. More examples are described in Huilgol (1975).

2.4.1 Rheogoniometry—Cone and Plate

As was indicated in Sec. 2.1 on viscometry, the cone-and-plate viscometer, when used with a shallow cone and at low rates of shear, is particularly convenient for measurement of nonlinear fluids because the deformation can then be taken as homogeneous and, in steady flows, viscometric. If means are supplied to measure the axial force on the cone and to vary the rate of rotation and the spacing between cone and plate, the device is called a rheogoniometer and it may be used to measure the normal-stress coefficients $\psi_1(\kappa^2)$ and $\psi_2(\kappa^2)$, as well as viscosity and, indeed, even certain mechanical properties of unsteady flows.

2.4.1.1 Steady-State Properties. In the commonest mode of operation of the rheogoniometer, the test fluid is supported in the space between cone and plate by the surface tension on a free boundary bridging from plate to cone rim. It is customary to assume that this surface is a segment of a sphere and that these surface-tension forces can be ignored with respect to the magnitudes of the normal-stress differences.

The viscosity is measured through Eq. (65) above; the first normal-stress difference, from

$$\psi_1(\kappa^2) = 2F/\pi\kappa^2R^2, \tag{70}$$

where κ is the rate of shear as given in Eq. (65), F is the axial force, and R is the radius.

Sources of error in applying Eq. (70) have been widely discussed. See Dealy (1982), for instance. Even when there is negligible transverse flow, inertial effects can influence the axial force at high shear rates. Corrections for such errors are possible. More serious are the edge effects that act on the free surface. For many materials, an indentation in the free surface appears at high shear rates, cutting deeply into the sample with increasing shear rate and completely invalidating the analysis. As a result of this and other difficulties, the method of Eq. (70) is necessarily restricted to measurements at rates of shear usually less than a few reciprocal seconds.

Measurements of the second normal-stress coefficient are even more difficult. Jackson and Kaye (1966) have assumed that a viscometric flow persists when the cone apex is displaced a distance d from contact with the plate. Upon making assumptions similar to those above, they arrive at the following equation:

$$\psi_2(\kappa^2) = -\frac{\Omega\theta}{\pi\kappa^2R}\left.\frac{\partial F}{\partial d}\right|_{d=0} - \kappa\frac{d\psi_1(\kappa^2)}{d\kappa}. \tag{71}$$

With $\psi_1(\kappa^2)$, calculated using Eq. (70), and from further data of axial force vs displacement at fixed shear rate, this equation can be used to measure the second normal-stress coefficient. However, the difficulties of accurate measurement coupled with the inaccuracies inherent in numerical or graphical evaluation of a derivative often make this method imprecise.

Another method of measuring the second normal-stress coefficient is possible with further modification of the apparatus. If means are provided to measure the pressure acting against the plate at points several distances from the apex of the cone, the logarithmic gradient of this pressure can be estimated. If the method is valid, this gradient will be a constant. Then ψ_2 can be found from this constant and ψ_1 through the following equation:

$$\psi_2(\kappa^2) = -\tfrac{1}{2}\left[\psi_1(\kappa^2) + \frac{1}{\kappa^2}\frac{dp_n(r)}{d\ln r}\right], \tag{72}$$

where $p_n(r)$ is the pressure against the plate at radial distance r.

With such measurements and with other methods, it has been found that, for polymeric materials, the second normal-stress coefficient is negative and of a magnitude only a fraction of the first normal-stress coefficient.

2.4.1.2 Time-Dependent Properties. For unsteady shearing, assuming effects due to inertia are ignorable, response functions may still be measured in principle using Eqs. (65) and (70) as in the steady shear case. Of course, in this case these responses are functions of time. For the most part, such measurements have been done on deformation histories termed *startup flow* or *cessation of steady shearing flow,* or on combinations of an oscillation or step displacement with a steady shearing. In practice, the actual measurements are usually confined to the corresponding analogs of viscosity and the first normal-stress difference since the second normal-stress difference is so difficult to measure. Usually, the cone-and-plate geometry is used.

Startup flow is achieved by suddenly shearing material that is at rest and in equilibrium, simultaneously observing the development of stress with time. The *shear-stress growth coefficient* and the *first normal-stress growth coefficient* are usually designated by $\eta^+(t,\kappa)$ and $\psi_1^+(t,\kappa)$, respectively, in an obvious generalization of the analogous steady-state response functions of Eq. (68). For many polymeric materials, in this flow a stress overshoot occurs in which the response initially grows, goes through a maximum, and then settles down to a steady-state value. The measurement is of considerable significance in relating the phenomenological rheology of the material to the micromechanics of entanglements of polymeric chains.

Cessation of steady shearing flow is achieved by suddenly stopping the shearing of a sample that had reached an equilibrium at some fixed rate of shear and recording the response as the stresses decay and the sample approaches equilibrium at rest. The corresponding response functions, the *shear-stress decay coefficient* and the *first normal-stress decay coefficient,* are defined in the obvious way and designated as $\eta^-(t,\kappa)$ and $\psi_1^-(t,\kappa)$, respectively.

Other time-dependent flows are sometimes measured with a cone-and-plate viscometer—for example, oscillatory shear or the combination of steady shear with a small sudden displacement or a small oscillation. It is also sometimes useful to do creeping shears and elastic recovery measurements.

2.4.2 Rheometry—Other Shearing Devices Because cone-and-plate rheometers produce a homogeneous rate of shear throughout the test sample, they are simple to analyze. Even in steady flow, however, the accurate measurement of ψ_2 is not easy. A number of other devices utilizing inhomogeneous shear are sometimes more or less useful. Only a few will be mentioned.

2.4.2.1 Torsional Flow. A *parallel-plate viscometer* shears the sample between a fixed plate and a parallel disk rotating with angular velocity ω. If the flow is assumed to be a slipping between plane-parallel rotating disks whose edges form a cylindrical boundary, that flow is viscometric. Because of inertia, the momentum balance of Eq. (10) cannot be exactly satisfied with such a flow, but with narrow spacing and at low shear rates it is very closely approximated. It is obvious that the shear rate is not homogeneous in this flow, and differentiation of data is needed to evaluate the viscometric coefficients. The viscosity is found from the following equation:

$$\eta(\kappa) = \frac{M}{2\pi R^3\kappa}\left[3 + \frac{d\log M(\kappa)}{d\log\kappa}\right], \quad \kappa = \frac{\omega R}{d}, \tag{73}$$

where M is the torque, R is the radius of the disk, and d is the spacing between plate and disk. Data that cannot be reduced to a unique curve when plotted as $M/(2\pi R^3)$ vs κ indicate that the flow is not as assumed.

If the thrust normal to the disk or plate can be measured as a function of ω, a combination of normal-stress coefficients can be measured. Then

$$\psi_1(\kappa^2) - \psi_2(\kappa^2) = \frac{F}{\pi R^2 \kappa^2}\left[2 + \frac{d \log F}{d \log \kappa}\right], \quad (74)$$

where F is the total force acting normally on the disk or plate.

2.4.2.2 Hole Pressure and Channel Flow. Prior to 1968, many rheological measurements relied upon determining the fluid thrust against a solid wall by reading pressure on a gauge connected to a small hole drilled in the wall. It was known that this method worked well for a Newtonian fluid where there is only a small error caused by inertia, which can be made ignorable by reducing the diameter of the flow hole. It was assumed that this would also apply to the errors for fluids with more complex rheology since experiments showed that for such fluids also, the pressure measurements approached a constant value in the limit for decreasing diameter of the hole. However, Broadbent *et al.* (1968) found that, for a viscoelastic fluid, there is a systematic error not caused by inertia that is intrinsic to the nonlinear material properties and cannot be eliminated by reducing the hole diameter. It is important to realize that the size of this error, sometimes called *hole pressure,* is often of the same order as that of the normal-stress differences and it cannot be ignored in rheological measurements. Indeed, we now know that early measurements of the second normal-stress difference that ignored this effect did not even have the correct sign. Tanner and Pipkin (1969) illustrated the mechanism for this hole pressure by an exact calculation of a second-order fluid in a plane flow over a deep narrow slot, the simplest example of the phenomenon. Measurements of normal stresses published prior to 1969 should not be relied on without checking to see if they are free of hole-pressure error. Once the mechanism for hole pressure was understood, the phenomenon was viewed as a possible means to measure normal-stress coefficients. Calculations suggest that measurements of hole pressure for shearing flow across a slotted trench relate directly to the first normal-stress coefficient; for shearing flow along a slotted trench, to the second normal-stress difference; and for a circular hole, to some linear combination of these. Discussion of the application of these ideas will be found in Lodge and Vargas (1983).

A direct method of observing the second normal-stress difference is possible with the flow under gravity of a viscoelastic fluid down a tilted open channel. If the channel has a flat bottom with perpendicular side walls, the shearing will vary from the walls to the center of the channel. For a Newtonian fluid, this does not affect the shape of the surface, which would be flat. For a viscoelastic fluid, however, the surface will curve in response to the adjustment of the height of the fluid across the flow to counterbalance the varying normal stress. In principle, for a negative second normal-stress coefficient this will produce an upward bulge, and for positive, a concave surface. In practice, this curvature can be measured and used to estimate the second normal-stress coefficient. It is found to be an upward bulge for polymeric fluids.

2.5 Complex Flows

In general, the three viscometric coefficients, functions of shear rate, are not sufficient to characterize completely the mechanical behavior of a material, even when restricted to the special case of MWCSH. Measurements of flows more complex than simple shearing may be required. For instance, in the flow between eccentric rotating discs discussed below, five coefficients, each a function of two variables, are needed for complete characterization. Other flows, such as extensional flow, also discussed below, cannot be viewed as superpositions of shearing. They are then completely unrelated to viscometric flows but closely related to technologically important flows.

2.5.1 Orthogonal Rheometry An interesting flow, which is possible between parallel disks that rotate with the same angular velocity on axes through their slightly displaced centers, is a steady flow with circular flow lines that is a MWCSH but it is not viscometric. The deformation experienced by a particle in this flow is a shearing that rotates with respect to the particle. The particle itself revolves and rotates in laboratory coordinates. Huilgol (1969) shows that, ignoring inertia, for simple materials in this flow the resulting stress is a constant and characterized by five material functions of two parameters, the angular velocity and the eccentricity of the centers. The *eccentric disk rheometer* or *Maxwell–Chartoff orthogonal rheometer* is a device to generate and measure this flow. Actually, for practical reasons, the apparatus drive one disk and allow the other to rotate freely on a low-friction bearing. There is no control of the cylindrical surface of the sample between disks. Tangential and possibly normal forces also are measured at one disk. Kearsley (1970) has pointed out that, for the flow that is assumed to occur, since there is no net torque on the disks, all the energy dissipated by the flow must be supplied by forces on the uncontrolled cylindrical surface. Thus, in fact, the apparatus must produce a flow at variance with the one assumed. Nevertheless, the devices seem to work satisfactorily for small eccentricities and rates of rotation such that inertial effects can be ignored. In this limit, the five functions are reduced to three expressing linear properties. The storage and loss moduli of Eqs. (31) and (32) can be measured in this way. If the x and y axes are taken in the plane of a disk, parallel and perpendicular, respectively, to the projection of the line connecting centers, then the dynamic moduli may be found from the following equations:

$$G'(\omega) = \frac{F_y h}{\pi d R^2}, \quad G''(\omega) = \frac{F_x h}{\pi \Omega d R^2}, \quad \omega = \Omega, \tag{75}$$

where G' and G'' are the storage and loss moduli, respectively, F_x and F_y are components of force acting tangentially on a disk, d is the offset of the disk axes, h is the thickness of the sample filling the space between disks, R is the radius of the disks, and Ω is the angular velocity of the disks, which is equal to ω, the frequency in the argument of the moduli. An important advantage of this method is that the dynamic moduli may be measured from steady forces and no phase-angle measurement is needed.

Two comparable devices are the *balance rheometer* or *Képès rheometer* and the *eccentric cylinder rheometer,* which operate on principles very similar to those of the eccentric disk rheometer. The former uses the flow between a rotating hemispherical cup and a concentric sphere rotating at the same angular velocity but about a slightly tilted axis. The latter uses the flow between a rotating cylindrical cup and a cylindrical bob rotating at the same angular velocity about a slightly displaced parallel axis. These devices are most useful for measuring dynamic linear properties in the limiting case of small deformation. The appropriate equations and a discussion of possible sources of error are to be found in Walters (1975).

2.5.2 Extensional Rheometry. *Extensional flows* are flows in which the material is stretched in simple extension. These flows are of great technological interest because they correspond to a commercially important process, spinning of fibers, but they are also of great interest to the polymer physicist because they have a strong effect in aligning polymer chains. More general flows occurring in commercial processing involve flow in contracting or expanding pipes or channels, flows that may be viewed as combining extensional and shearing flows. Sometimes the term *elongational flows* is reserved for these mixed flows because there is reason to believe that the extensional component will play a dominant role. A survey of the literature will show that much less is known about extensional flows than simple shearing flows. The reason for this is that there are many practical difficulties in generating and measuring well-defined extensional flows. Petrie (1979) gives a thorough survey of the literature up to the time of writing.

In simple extension, a cylinder of fluid is stretched along its axis while its surface is stress free. If the length of the cylinder increases exponentially in time, the local deformation rate is constant. When any transients die out, the history is effectively a MWCSH. In that event, the stress will achieve a steady-state value and a *tensile viscosity* or *Trouton viscosity,* $\eta_E(\alpha)$, can be defined as follows:

$$\eta_E(\alpha) = \lim_{t\to\infty} \frac{f(t)L_0 \exp(\alpha t)}{A_0\alpha}, \quad L(t) = L_0 \exp(\alpha t), \tag{76}$$

where $L(t)$ is the length of the cylinder, L_0 and A_0 are the length and cross section of the cylinder at time $t = 0$, $f(t)$ is the force of stretching, and α is a constant that determines the exponential of the stretching. It is apparent from this equation that the transients may never die out for some constitutive equations, which thus predict no limiting Trouton viscosity. A Newtonian fluid has a Trouton viscosity that is independent of shear rate: It is three times the Newtonian viscosity. Most polymeric fluids have Trouton viscosities that are strong functions of the stretching rate and that, at very low rates of deformation, appear to approach the Newtonian value of three times the zero shear viscosity. On the other hand, as the deformation rate is increased for these elastic fluids, the ratio of Trouton to shear viscosity can rise very rapidly and, at very high rates of shear, a steady state may be very difficult or impossible to reach. To measure Trouton viscosity, a controlled exponential stretching must be maintained and monitored until a steady state of stress sets in. This is extremely hard to achieve, particularly for low-viscosity fluids and high rates. Only very limited and scattered data will be found in the literature. Meissner (1972) has developed an instrument in which a strand of polymer melt is stretched between a pair of gearlike rollers rotating at constant angular velocity, thus producing a constant stretch rate while rapidly thinning. With this device he has drawn samples out to almost 100 times their original length and has made measurements at constant extension rates of several reciprocal seconds. More recently, a device is described by Tirtaatmadja and Sridhar (1993) with which they have measured Trouton viscosities at stretch rates of 5 s^{-1}.

It is much easier to produce high extension rates with instruments that need not produce a MWCSH: that is, with geometry for which a particle of fluid does not experience a constant history even if the flow is steady. Because high extension rates often occur in industrial processes, constant-force extensometers and various fiber-spinning devices of this type, capable of high stretch rates, have been used for technological measurements. An inconvenience of such measurements is that the results are very much a function of the particular history. Typically, in these cases, an apparent extensional viscosity is defined as the ratio of a local tensile stress and a stretching rate at some arbitrarily designated point on the threadline. Such measurements will not yield reliable values of η_E but they can give useful rough comparisons of materials. The tubeless siphon (mentioned in Sec. 1.4, above) has been used for this purpose. Oliver and Bragg (1974) developed a triple-jet device for determining extensional flow characteristics at very high stretching rates. Two converging jets at high velocity strike a central low-velocity jet causing it to stretch. The extensional strain rate is inferred from photographs. Results are somewhat sensitive to the flow conditions in the reservoirs.

Biaxial stretching of a sheet of polymeric material occurs in the commercially important processes of film blowing and blow molding and, for this reason, measurements of *biaxial extensional viscosity* are needed. Meissner (1985) has developed rotary clamps that may be arranged to produce anisotropic biaxial stretching of sheets of polymeric materials at constant rates. With this apparatus, biaxial analogs of the Trouton viscosities may be measured.

2.5.3 Plastometer The *plastometer* or *squeeze-film viscometer* is a simple device in which a cylindrical disk of test sample is squeezed between two parallel plates by a fixed load, usually a weight. The sample disk thins and flows radially under the constant load, while the spacing between the plates is monitored. The device can be used to measure viscosity but it is most interesting as a device to measure yield stress.

There is no general agreement on the meaning of the term "yield stress." The classical, theoretical meaning is "a stress below which no flow or, at least, no unrecoverable flow occurs." For example, σ_y of the Bingham body, Eq. (55), has a yield stress. Most people would agree that real materials actually undergo some flow at any stress: it is just a matter of the time scale necessary to perceive the effect. On the other hand, a pragmatist might define yield stress in terms of a particular measurement and time scale. In practice, many materials exhibit a yield stress, for all practical purposes, in the time scale of common usage.

The plastometer is used to measure the yield stress by observing the limiting spacing between parallel plates when flow ceases. The shear yield stress is then given by

$$\sigma_y = \frac{3\pi^{1/2}h_L^{5/2}F}{2V^{3/2}}, \quad V = \pi R_L^2 h_L, \tag{77}$$

where h_L is the limiting spacing, F is the loading force, R_L is the limiting radius of the disk, and V is the volume of the sample.

The actual flow of a material with yield stress in a plastometer is not completely understood. Covey and Stanmore (1981) give a survey of several widely differing solutions for a Bingham body.

3. DIMENSIONLESS PARAMETERS

The dimensional analysis of flows of rheologically complex fluids requires dimensionless parameters beyond the Reynolds number useful for Newtonian fluids. Indeed, the concept of a Reynolds number itself must be slightly generalized to accommodate the dependence of viscosity on rate of shearing. One way of doing this is as follows:

$$R_e = \rho UL/\eta(\kappa), \tag{78}$$

where R_e is the Reynolds number, U is a characteristic velocity, L is a characteristic length, and $\eta(\kappa)$ is the viscosity at the local rate of shear, κ. In this definition, Reynolds number may vary throughout the flow.

A *Weissenberg number, W,* is defined to represent a ratio of elastic to viscous effects:

$$W = \lambda_{\text{rlx}}U/L, \tag{79}$$

where λ_{rlx} is a typical relaxation time of the material, U is a typical fluid velocity, and L is a relevant length. In viscometric flows, W is a measure of the ratio of the normal force N_1 to the viscous force, $\kappa\eta(\kappa)$. More generally, it is meant to be a rough ratio of elastic forces to viscous forces.

A Deborah number, D_e, is defined as the ratio of a typical relaxation time of the material to a typical time scale of a measurement or of a residence time or flow time in a flow system. If this time scale is called T, the Deborah number is defined as

$$D_e = \lambda_{\text{rlx}}/T. \tag{80}$$

The name of this dimensionless number refers, whimsically, to the biblical prophetess Deborah who affirmed, according to one reading, "The mountains flow before the Lord." To a rheologist, this might seem a metaphorical statement that, if one's time scale were long enough, the most solid of materials would appear to flow.

Accordingly, the significance of the Deborah number is that in situations with high values of D_e the material tends to appear solidlike while at low D_e the material appears more fluid. Therefore, the appropriate constitutive equation for a situation often depends upon one's time scale as measured by the Deborah number.

Clearly, the definitions of such dimensionless parameters are far from unique, and the choice of the appropriate times, lengths, and velocities to be used in calculating them is open to considerable personal taste. But sensible, consistent definitions of these concepts have proved very useful for scaling and comparing materials and flows, especially in the treatment of technological phenomena.

GLOSSARY

Blow Molding: A process in which a bubble of molten material is shaped by inflation and solidification within a mold. Ordinary polyethylene gallon milk jugs are usually made in this way.

Body Force: A force from an external influence that acts on a particle of material, e.g., gravity. Body forces are vectors with dimensions of force per unit volume.

Body Moment: A force couple from an external influence that acts on a particle of material, e.g., the effect of a magnetic field acting on a magnetic dipole distribution.

Constitutive Equation: An equation relating to a particular material giving local physical properties of the material in terms of local influences. In this article, constitutive equations give mechanical properties in terms of deformation or deformation history, e.g., a stress–strain law.

Couple Stress: An interaction by which a force couple is exerted point-wise between contiguous surfaces of a material. Couple stresses are represented by an antisymmetric component of stress tensors.

Double Tensor: A tensor calculated from points in two different spaces that transforms as a tensor under coordinate transformations of either. For example, the deformation gradient $\mathbf{F}(t)$ is a double tensor because it depends tensorially on the coordinate system of the configuration $\mathbf{x}(t)$ at time t as well as that of reference configuration $\mathbf{X}$. (See Ericksen, 1960, p. 805.)

Elastomer: A polymeric substance capable of rubbery behavior in which large, essentially recoverable stretching can occur—for example, natural rubber.

Film Blowing: A process for forming polymer films in which a tube of molten plastic is extruded from an annular die, stretched axially and radially, and solidified by being drawn over a bubble of pressurized gas with rollers that flatten and pull the tube while sealing off the gas bubble.

Functional: A noun meaning a mathematical rule for calculating a quantity from a set of functions. It is a generalization of the idea of a function as a rule for calculating a quantity from a set of numbers. The functionals used in rheology are usually tensor valued, that is, they are rules for evaluating a tensor from histories of deformations.

Hamilton–Cayley Theorem: The theorem that every matrix satisfies its characteristic equation. Thus, for a second-order symmetric tensor in three-space such as stress or strain,

$$\mathbf{S}^3 - I_1(\mathbf{S})\mathbf{S}^2 + I_2(\mathbf{S})\mathbf{S} - I_3(\mathbf{S})\mathbf{1} = 0$$

where S is the tensor and the I_i are the principal scalar invariants. This theorem allows powers of these tensors to be expressed as a polynomial in lesser powers.

Isochoric: An isochoric deformation is a deformation without volume change. An incompressible material may undergo only isochoric deformations.

Polar Decomposition Theorem: A theorem of linear algebra that, as it applies to real tensors with strictly positive determinants, assures that they are equivalent to the product (in either order) of a proper orthogonal tensor and a symmetric, positive-definite tensor.

Principal Scalar Invariants: A set of scalar functions of the arguments of a matrix, invariant under orthogonal matrix transformations. For a second-order, symmetric tensor $\mathbf{S}$ in three-space, such as stress or strain, there are three scalar invariants as follows:

$$I_1(\mathbf{S}) \equiv \mathrm{tr}(\mathbf{S}), \quad I_2(\mathbf{S}) \equiv \tfrac{1}{2}[\mathrm{tr}^2(\mathbf{S}) - \mathrm{tr}(\mathbf{S}^2)], \quad I_3(\mathbf{S}) \equiv \det(\mathbf{S}).$$

Functions of these principal invariants—for instance, the principal values or eigenvalues—are also invariant. If the eigenvalues are available, the principal invariants may be written

$$I_1 = \lambda + \mu + \nu, \quad I_2 = \lambda\mu + \lambda\nu + \mu\nu, \quad I_3 = \lambda\mu\nu,$$

where λ, μ, and ν are the eigenvalues.

Shear: An isochoric deformation in which parallel surfaces of a material are displaced in their planes while remaining parallel. *Simple shear* is the isochoric deformation undergone by material sandwiched between two adherent, parallel, rigid plates that deform it by parallel displacement. *Pure shear* is a deformation without rotation composed of two orthogonally oriented stretches of reciprocal magnitudes. Since it is an isochoric deformation, the stretch in the third orthogonal direction must be unity. Simple shear may be decomposed locally into a pure shear plus a rigid rotation.

Stress Deviator: The deviatoric part of the stress tensor. The stress deviator has trace zero. The stress deviator $\mathbf{s}$ of stress $\boldsymbol{\sigma}$ is given by

$$\mathbf{s} = \boldsymbol{\sigma} - \tfrac{1}{3}\,\mathrm{tr}(\boldsymbol{\sigma})\mathbf{1}$$

Stretch: The stretch of an infinitesimal line segment is the ratio of the lengths before and after deformation. The eigenvalues of a stretch tensor give the stretches of lines oriented in the three, mutually perpendicular, principal directions of the tensor, the *principal stretches*.

Surface Traction: A vector or vector field, the force per unit area acting on a surface having, in general, both normal and tangential components. The surface may be a boundary of a body, an interface between bodies, or a notional surface in the interior of a body. A stress tensor may be thought of as a linear transformation that operates on a unit vector normal to a surface to produce a traction vector acting on the surface.

Works Cited

Bagley, E. B. (1957), *J. Appl. Phys.* **28,** 624–627.

Barr, G. (1931), *A Monograph of Viscometry,* New York: Oxford Univ. Press.

Bernstein, B. (1966), *Acta Mech.* **II/4,** 329–354.

Bernstein, B., Kearsley, E., Zapas, L. (1963), *Trans. Soc. Rheol.* **7,** 391–410.

Bernstein, B., Kearsley, E., Zapas, L. (1964), *J. Res. Natl. Bur. Stand. B* **68,** 103–113.

Bernstein, B., Kearsley, E., Zapas, L. (1965), *Trans. Soc. Rheol.* **9,** 27–39.

Bird, R. B., Dai, G. C., Yarusso, B. J. (1987), *Rev. Chem. Eng.* **1,** 1–70.

Blatz, P. J., Ko, W. L. (1962), *Trans. Soc. Rheol.* **6,** 223–251.

Broadbent, J. M., Kaye, A., Lodge, A. S., Vale, D. G. (1968), *Nature* **217,** 55–56.

Coleman, B. D., Markovitz, H., Noll, W. (1966), *Viscometric Flows of Non-Newtonian Fluids,* New York: Springer-Verlag.

Cosserat, E., Cosserat, F. (1909), *Théorie des Corps Déformables,* Paris: A. Hermann et Fils.

Covey, G. H., Stanmore, B. R. (1981), *J. Non-Newtonian Fluid Mech.* **8,** 249–260.

Dealy, J. M. (1982), *Rheometers for Molten Plastics,* New York: Van Nostrand Reinhold.

De Gennes, P. G. (1971), *J. Chem. Phys.* **55,** 572–579.

Doi, M., Edwards, S. F. (1978), *J. Chem. Soc. Faraday Trans. II* **74,** 1789–1829.

Ericksen, J. L. (1954), *Z. Angew. Math. Phys.* **5,** 466–486.

Ericksen, J. L. (1960), "Tensor Fields," in: S. Flügge (Ed.), *Handbuch der Physik,* Band III/1, Berlin: Springer-Verlag, pp. 793–858.

Ericksen, J. L. (1961), *Trans. Soc. Rheol.* **5,** 23–34.

Eringen, A. C. (1962), *Nonlinear Theory of Continuous Media,* New York: McGraw-Hill.

Green, A. E., Rivlin, R. S. (1964), *Arch. Rat. Mech. Anal.* **17,** 113–147.

Hill, R. (1950), *The Mathematical Theory of Plasticity,* New York: Oxford Univ. Press.

Huilgol, R. R. (1969), *Trans. Soc. Rheol.* **13,** 142–151.

Huilgol, R. R. (1975), *Continuum Mechanics of Viscoelastic Liquids,* New York: Wiley.

Jackson, R., Kaye, A. (1966), *Brit. J. Appl. Phys.* **17,** 1355–1360.

Kearsley, E. A. (1970), *J. Res. Natl. Bur. Stand. C* **74,** 19–20.

Kearsley, E. A. (1980), *J. Appl. Phys.* **51,** 4541–4542.

Kearsley, E. A., Zapas, L. J. (1980), *J. Rheol.* **24,** 483–500.

Leslie, F. M. (1968), *Arch. Rat. Mech. Anal.* **28,** 265–283.

Lodge, A. S., de Vargas, L. (1983), *Rheol. Acta* **22,** 151–170.

Meissner, J. (1972), *Trans. Soc. Rheol.* **16,** 405–420.

Meissner, J. (1985), *Chem. Eng. Commun.* **33,** 159–180.

Merrington, A. C. (1949), *Viscometry,* London: Edward Arnold.

Mooney, M. (1940), *J. Appl. Phys.* **11,** 582–592.

Obata, Y., Kawabata, S., Kawai, H. (1970), *J. Polym. Sci.* **8,** 903–919.

Oliver, D. R., Bragg, R. (1974), *Rheol. Acta* **13,** 830–835.

Peng, S. T. J., Landel, R. F. (1976), *J. Appl. Phys.* **47,** 4255–4260.

Penn, R. W., Kearsley, E. A. (1976), *Trans. Soc. Rheol.* **20,** 227–238.

Petrie, C. J. S. (1979), *Elongational Flow,* London: Pitman.

Rivlin, R. S., Saunders, D. W. (1951), *Philos. Trans. R. Soc. (London)* **A243,** 251–273.

Spencer, A. J. M. (1971), "Theory of Invariants," in: A. C. Eringen, (Ed.), *Continuum Physics,* Vol. I, New York: Academic Press, pp. 239–353.

Tanner, R. I., Pipkin, A. C. (1969), *Trans. Soc. Rheol.* **13,** 471–484.

Tanner, R. I. (1985), *Engineering Rheology,* Oxford, U.K.: Clarendon.

Tirtaatmadja, V., Sridhar, T. (1993), *J. Rheol.* **37,** 1081–1102.

Toupin, R. A. (1962), *Arch. Rat. Mech. Anal.* **11,** 385–414.

Valanis, K. C., Landel, R. F. (1967), *J. Appl. Phys.* **38,** 2997–3000.

Wagner, M. H., Stephenson, S. E. (1979), *J. Rheol.* **23,** 489–504.

Walters, K. (1975), *Rheometry,* London: Chapman and Hall.

Yang, T. M. T., Krieger, I. M. (1978), *J. Rheol.* **22,** 413–419.

Further Reading

Continuum Mechanics

Jaunzemis, W. (1967), *Continuum Mechanics,* New York: Macmillan.

Truesdell, C., Toupin, R. (1993), in: S. Flügge (Ed.), *Encyclopedia of Physics,* Vol. III/1, *The Clas-*

sical Field Theories, 2nd ed., New York: Springer-Verlag.

Truesdell, C., Noll, W. (1993), in: S. Flügge (Ed.), *Encyclopedia of Physics,* Vol. III/3, *The Non-linear Field Theories of Mechanics,* 2nd ed., New York: Springer-Verlag.

General Rheology

Barnes, H. A., Hutton, J. F., Walters, K. (1989), *An Introduction to Rheology,* Amsterdam: Elsevier.

Boger, D. V., Walters, K. F. (1993), *Rheological Phenomena in Focus,* Amsterdam: Elsevier.

Cheremisinoff, N. P. (1988), *Encyclopedia of Fluid Mechanics,* Vol. 7, *Rheology and Non-Newtonian Flows,* Houston: Gulf Publishing.

Joseph, D. D. (1990), *Fluid Dynamics of Viscoelastic Liquids,* New York: Springer-Verlag.

Middleman, S. (1968), *The Flow of High Polymers,* New York: Wiley.

Constitutive Equations

Bird, R. B., Armstrong, R. C., Hassager, O. (1987), *Dynamics of Polymeric Liquids,* Vol. 1, *Fluid Mechanics,* and Vol. 2, *Kinetic Theory,* 2nd ed., New York: Wiley.

Ferry, J. D. (1961), *Viscoelastic Properties of Polymers,* New York: Wiley.

Larson, R. G. (1988), *Constitutive Equations for Polymer Melts and Solutions,* Stoneham: Butterworths.

Schowalter, W. R. (1978), *Mechanics of Non-Newtonian Fluids,* Oxford, U.K.: Pergamon.

Tanner, R. I. (1985), *Engineering Rheology,* Oxford, U.K.: Clarendon.

Tschoegl, N. W. (1989), *The Phenomenological Theory of Linear Viscoelastic Behavior: An Introduction,* New York: Springer-Verlag.

Rheometry

Coleman, B. D., Markovitz, H., Noll, W. (1966), *Viscometric Flows of Non-Newtonian Fluids,* New York: Springer-Verlag.

Dealy, J. M. (1982), *Rheometers for Molten Plastics,* New York: Van Nostrand Reinhold.

Dealy, J. M., Wissbrun, K. F. (1990), *Melt Rheology and Its Role in Plastics Processing,* New York: Van Nostrand Reinhold.

Walters, K. (1975), *Rheometry,* London: Chapman and Hall.

Computational Mechanics

Crochet, M. J., Davies, A. R., Walters, K. F. (1984), *Numerical Simulation,* Amsterdam: Elsevier.

Gordon, G. V., Shaw, M. (1994), *Computer Programs for Rheologists,* Cincinnati: Hanser/Gardner.

Washizu, K. (1968), *Variational Methods in Elasticity and Plasticity,* Oxford, U.K.: Pergamon.

RHEOLOGY CONTROLLED BY MAGNETIC FIELDS

J. M. GINDER, *Research Laboratory, Ford Motor Company, Dearborn, Michigan, U.S.A.*

	Introduction	487
1.	**Ferrofluids**	490
1.1	Composition	490
1.2	Field-Dependent Viscosity	490
1.3	Applications of Ferrofluids	491
2.	**Magnetorheological Fluids**	491
2.1	Composition	491
2.2	Magnetic-Field–Induced Structure Formation	492
2.3	Steady-Shear Rheology	492
2.4	Oscillating-Shear Rheology	493
2.5	Magnetic Properties	494
2.6	Magnetostatic Models of Magnetorheology	494
3.	**Magnetic Powders**	496
3.1	Composition	496
3.2	Magnetic-Powder Rheology	497
4.	**Magnetorheological Elastomers**	497
4.1	Composition	497
4.2	Field-Dependent Viscoelastic Response	497
5.	**Applications of Magnetorheological Materials**	498
5.1	Magnetic-Circuit Considerations	498
5.2	Torque-Transfer Applications	498
5.3	Damping Applications	499
5.4	Process Applications	500
5.5	Outlook	500
	Glossary	501
	Works Cited	501
	Further Reading	502

INTRODUCTION

The formation of particle chains when magnetizable particles are exposed to a magnetic field is familiar to anyone who has experimented with iron filings and a permanent magnet. This ubiquitous phenomenon is also the basis for the sometimes dramatic field-induced changes in rheology exhibited by a variety of magnetorheological, or MR, materials (for a review of rheological phenomena, see RHEOLOGY). The application of a field to an initially random suspension of magnetizable particles results in the formation of chains, columns, or more complex solidlike structures aligned along the field direction, as shown in Fig. 1. Anisotropic interparticle forces induced by the applied field give rise to these structural and rheological changes. The formation of particle chains in an external field increases the electromagnetic energy stored in a MR suspension. The chains or columns are tilted or broken by applying a shear strain; as this deformation reduces the stored energy, a nonzero shear stress is required to impose the strain, as shown schematically in Fig. 2.

The rheology of many MR materials is commonly described by the Bingham plastic model (see, e.g., Macosko, 1994). Central to the Bingham model is the yield stress τ_y, which is a measure of the strength of solid materials. For applied stresses τ below τ_y, many materials are viscoelastic solids having nonzero shear moduli or stiffnesses G'. The shear stress required to produce a shear strain γ is then

$$\tau = G'\gamma, \quad \tau < \tau_y. \tag{1}$$

For applied stresses greater than τ_y, the material is able to flow. The Bingham equation relates the stress to the strain rate $\dot{\gamma}$ in this postyield condition:

$$\tau = \tau_y + \eta\dot{\gamma}, \quad \tau \geq \tau_y, \tag{2}$$

where η is the viscosity. These idealized pre-

3-527-28138-X/96/$5.00 + .50

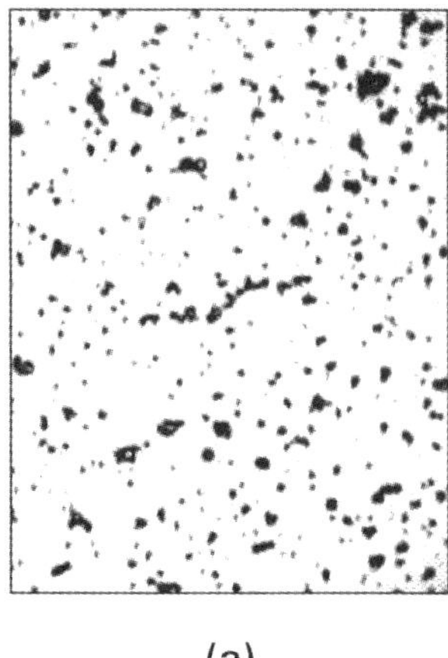

(a)

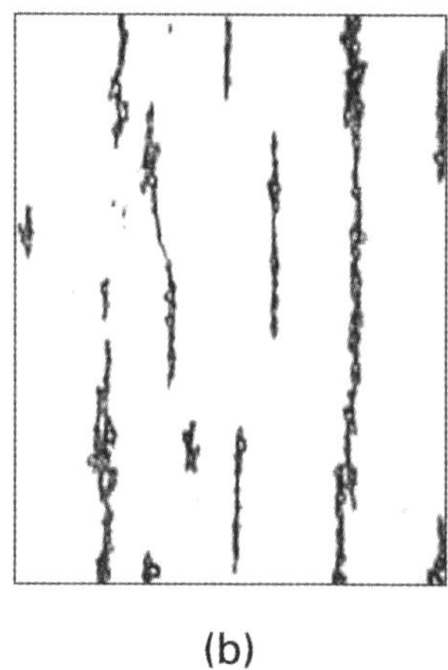

(b)

FIG. 1. Optical micrographs (300 × 400 μm) of a thin layer of a dilute suspension of a carbonyl-iron-based MR fluid: (a) zero applied field; (b) after the application of a ~20-kA/m field, showing the formation of particle chains aligned along the field direction.

and postyield behaviors are shown schematically in Fig. 3. While the scientific validity of the yield stress has been questioned, the Bingham model is undeniably quite useful in describing the flow of real materials like MR fluids. Magnetorheological phenomena involve field-dependent viscosities, yield stresses, or shear moduli.

The broad family of field-controllable magnetorheological materials includes ferrofluids, magnetorheological fluids, magnetic powders, and magnetorheological elastomers; the properties of these materials are summarized in Table 1. Ferrofluids, or colloidal magnetic fluids, typically contain magnetic particles less than 10 nm in size; consequently, the magnetorheological effect in ferrofluids is small, involving at most a tripling of the ferrofluid viscosity upon the application of a field (Rosensweig, 1985). Ferrofluids are of considerable utility in sealing, damping, and heat-transfer applications. Magnetorheological fluids are composed of micrometer-size magnetizable particles suspended at high concentration in nonmagnetic liquids. While similar in composition to ferrofluids, the larger particulates in MR fluids enable them to change reversibly from free-flowing liquids to solids having yield stresses of order 100 kPa on the application of a magnetic field. These changes enable the construction of a variety of electromechanical devices such as MR fluid clutches, brakes, and shock absorbers, several of which are presently under development for motion-control applications. Similar phenomena are already utilized in commercially available magnetic-powder clutches and brakes. Magnetorheological elastomers are obtained by dispersing micrometer-size magnetizable particles in a viscoelastic solid like a polymer gel or an elastomer. While these materials are solid under all circumstances, their modulus or stiffness can be varied by an applied field. These MR elastomers may find use in vibration-control applications.

Both the history and the physical properties of MR fluids are closely related to those of their electrical counterparts, the electrorheological (ER) fluids. ER fluids, which are

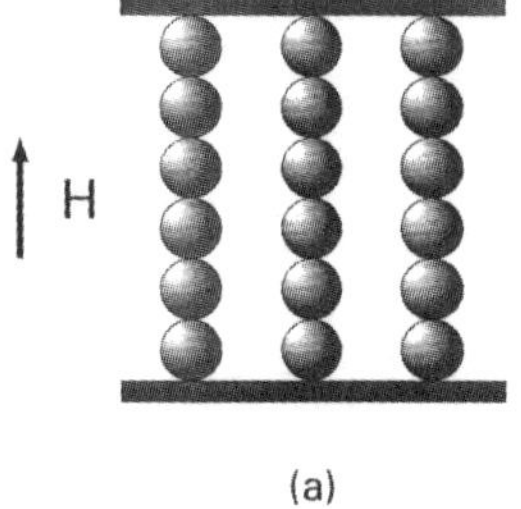

(a)

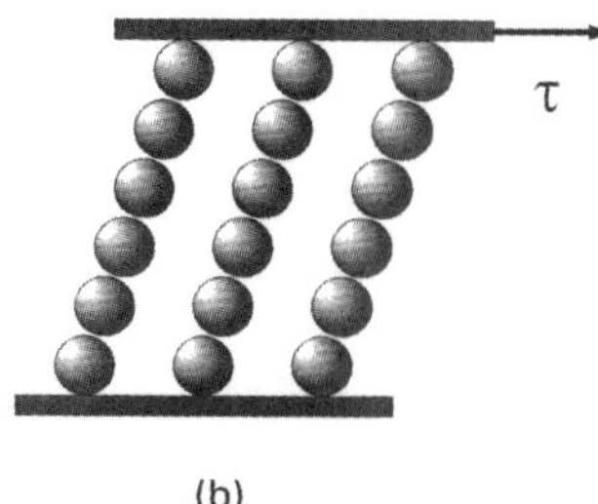

(b)

FIG. 2. Idealized structure of a MR fluid: (a) an applied field H induces the formation of chains or columns along the field; (b) a shear stress τ must be applied to deform the field-induced structure.

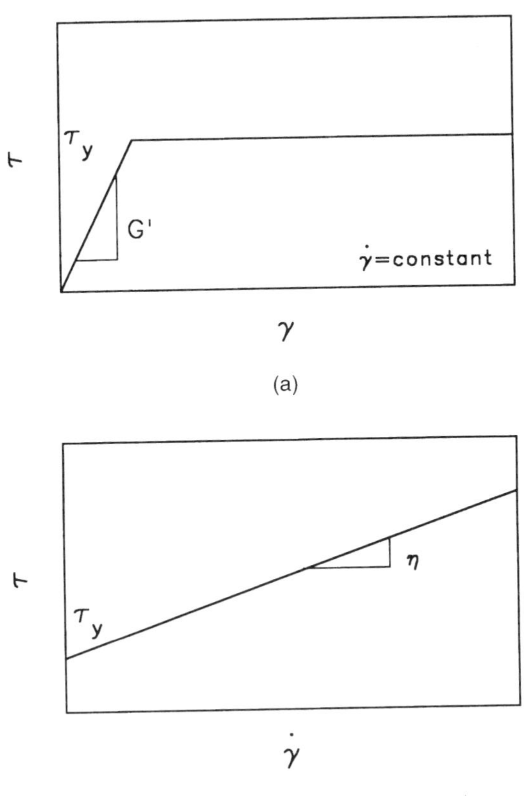

FIG. 3. Schematic rheological behavior of a Bingham solid in (a) preyield viscoelastic regime with shear modulus G' and (b) postyield plastic flow with viscosity η. The yield stress τ_y demarcates the pre- and postyield behavior.

suspensions of electrically polarizable particles in insulating liquids, undergo a transition to a solid state upon the application of a high electric field (see ELECTRORHEOLOGY). ER fluids were first developed by Willis Winslow in the 1940s (Winslow, 1947, 1949). By recognizing the analogy between the electrostatic polarization and magnetostatic magnetization of particles, Jacob Rabinow of the National Bureau of Standards was led to invent the first magnetorheological fluids and devices in the late 1940s (Rabinow, 1948, 1949, 1951). Rabinow's discovery soon inspired a flurry of activity on MR fluids and devices in the United States, England, the Soviet Union, and elsewhere (Parziale and Tilton, 1950; Blunden, 1951; Vorob'yeva, 1965). By the early 1950s it was recognized that dry magnetic powders would also function as field-controllable media for stress transfer (Grebe, 1952). Published activity in MR fluids began to decline in the late 1960s, while ER fluids began to enjoy increasing attention after the development of polymer particulate formulations in the late 1970s. The development of water-free ER fluids in the mid-1980s has led to significant improvements in their thermal and electrical performance. ER fluids having a relatively broad operating-temperature range and low electrical-power consumption are available in pilot-plant quantities from Bayer AG and Bridgestone. While these ER fluids perform well enough to permit the construction of practical devices like industrial dampers, their field-induced yield stresses are not high enough for applications in automotive torque transfer (Hartsock *et al.*, 1991). Moreover, their use requires potentially expensive high-voltage power supplies as well as the appropriate electrical wiring and connectors.

These limitations have led to a renaissance of interest in MR fluids and applications in this decade. A prototype rotary automotive shock absorber utilizing MR fluids was developed by a group at TRW (Pinkos *et al.*, 1993). MR fluids are useful in optical-surface finishing (Holton, 1993). MR-fluid–based devices are now being commercialized

Table 1. Composition and properties of magnetorheological materials.

	Ferrofluid	MR fluid	Magnetic powder	MR elastomer
Particulate material	Magnetite	Iron	Iron alloy	Iron
Particle size	2–10 nm	0.1–10 μm	10–50 μm	10–50 μm
Host material	Oils, water	Oils	Air	Elastomers
Volume fraction	0.02–0.2	0.1–0.5	0.5–0.6	0.1–0.5
Additives	Surfactants	Surfactants, thixotropes	Dispersants	...
Off viscosity (mPa-s)	2–200	100–1000	...	...
Magnetorheological changes	$\Delta\eta/\eta \sim 1$	$\tau_y \sim 100$ kPa	$\tau_y \sim 100$ kPa	$\Delta G' \sim 20$ kPa

by Lord Corporation for use in variable-resistance exercise equipment and seat vibration dampers. Because of their high yield stresses, wide operating-temperature range, and activation using electromagnets driven by low-voltage power supplies, other applications of MR fluids are likely to follow. In fact, magnetic-powder clutches and brakes have been used in motion-control and motor-testing applications for several decades; a magnetic-powder device even serves as the main clutch in an automotive power train.

Recent progress in the field of magnetorheological materials has been made in developing new MR-fluid formulations and inventing novel MR elastomers, as well as in understanding the connection between the magnetic properties of the constituents and the magnetorheological performance of the material. Questions about the long-term stability of these materials and the mass and power consumption of the devices that utilize them remain. This article is intended to review the present state of MR materials and device development as well as to provide an introductory guide for those who may wish to study or utilize these materials. Section 1 briefly reviews the composition, properties, and applications of ferrofluids. In light of the relatively weak magnetorheological effects that they possess, ferrofluids are not the main focus of this article; they nonetheless are of interest here as "proto-MR fluids." The composition, structural, rheological, and magnetic properties of magnetorheological fluids are treated in Sec. 2. Physical aspects of magnetic powders, i.e., dry MR fluids, are reviewed in Sec. 3. Section 4 treats the recently developed magnetorheological elastomers. Section 5 concludes with a discussion of the requirements for and opportunities in devices utilizing MR materials.

1. FERROFLUIDS

1.1 Composition

Ferrofluids, or colloidal magnetic fluids, are typically composed of single-domain particles of magnetite, Fe_3O_4, suspended with the aid of surfactants in polar liquids such as water or nonpolar liquids such as mineral oil. Ternary and higher ferrites, as well as metallic particulates of iron and cobalt, may also be utilized in ferrofluid formulations. The preparation of a stable ferrofluid involves a delicate balance between the attractive and repulsive interparticle forces as well as random Brownian forces on each particle (see, e.g., Rosensweig, 1985). The attractive forces are produced by magnetic dipole–dipole interactions and van der Waals forces caused by fluctuating electric dipoles. The repulsive forces are due to layers of adsorbed surfactant molecules on each particle (giving rise to so-called steric forces) as well as screened Coulomb repulsion if the particles are charged. If the particles are too small, their Brownian motion will greatly reduce the magnetic properties of the fluid. If they are too large, the attractive magnetic and van der Waals forces will cause the particles to agglomerate irreversibly. The choice of surfactant molecule necessary to stabilize the suspension depends on the nature of the suspending fluid and the surface chemistry of the particles, and is critical in preventing irreversible contact between particles. It is found that ferrofluids having particles with diameters between roughly 2 and 10 nm possess useful magnetic properties and acceptable stability (Berkovsky *et al.*, 1993).

1.2 Field-Dependent Viscosity

The relevance of ferrofluids in a discussion of magnetorheological phenomena is that, while ferrofluids remain fluid ($\tau_y = 0$) under all imposed magnetic fields, they exhibit viscosities that increase with applied field. Viscometric measurements on ferrofluids have revealed a field-dependent increase in viscosity (see, e.g., McTague, 1969). The origin of the viscosity enhancement is a single-particle effect caused by the magnetic restoring torque experienced by each suspended particle, $\mathbf{m} \times \mathbf{H}$, where $\mathbf{m}$ is the magnetic-dipole moment of the particle and $\mathbf{H}$ is the magnetic field. This torque favors the alignment of the permanent dipoles along the direction of the applied field and thus opposes the rotation of the particles in a shear flow, giving rise to an increase in the fluid viscosity. Viscosity enhancements $\Delta\eta(H)/\eta(0) \sim 2$ have been obtained (Rosensweig, 1985). Successful theories of the viscosity enhancement account for the effect of Brownian motion on the orientation of the dipoles (Shliomis, 1972).

1.3 Applications of Ferrofluids

The relatively small increases in ferrofluid viscosity with the application of a magnetic field have, to this author's knowledge, not been utilized in any applications. Ferrofluids are quite useful, however, because they are magnetic liquids: They experience magnetic-body forces in a magnetic-field gradient, and so the position of a drop of ferrofluid can be controlled by an applied field. Commercial applications of ferrofluids include rotary seals, magnetic bearings, motor dampers, and loudspeaker voice-coil dampers (Raj and Moskowitz, 1990). Monographs by Rosensweig (1985) and Berkovsky *et al.* (1993) provide excellent reviews of ferrofluid composition, properties, and applications.

2. MAGNETORHEOLOGICAL FLUIDS

2.1 Composition

The composition of MR fluids is similar to that of most ferrofluids: Magnetizable particles are suspended at high concentration in nonpolar or polar liquids with the aid of surfactant molecules. In distinct contrast to ferrofluids, the particles used in MR fluids are large enough to permit reversible particle aggregation in an applied magnetic field. Particle sizes in MR fluids may range from tens of nanometers to tens of micrometers. For magnetic particles above a critical size, which depends on particle shape and on the crystal structure of the constituent material, they consist of multiple magnetic domains. In the absence of an applied field, the particles then possess very small net dipole moments and thus experience little or no magnetic attractive forces. The application of a field causes domain-wall motion (Bean and Jacobs, 1960) that greatly increases the magnitude of the moments and thus the interparticle magnetic attraction. To obtain the largest moments and thus the largest magnetorheological effect, it is desirable to use particulates with high saturation magnetization. For example, magnetite has a saturation magnetization of 0.6 T, while iron has a saturation magnetization of 2.1 T, the largest of any element (see MAGNETIC MATERIALS). Materials for MR particulates can be produced by a number of techniques; one of the most widely used is so-called carbonyl iron, which is produced by the decomposition of iron carbonyl, $Fe(CO)_5$. Other processes to produce magnetic particulates include atomization, electrolysis, and grinding. The optimum particle-volume fraction in MR fluids is determined by a compromise between the benefits of heavy particulate loading—a reduction in sedimentation velocity and an increase in the MR effect—and a detrimental increase in zero-field viscosity. Typical volume fractions are between 0.1 and 0.5.

Since typical MR particulates are micrometers in size and are composed of metals or metal oxides having a mass density greater than that of most suspending fluids, they often suffer sedimentation under the prolonged action of gravity as well as centrifugation under the centrifugal forces encountered in rotary motion. Irreversible sedimentation or centrifugation is disastrous in practical applications of MR fluids and has been combated by several techniques. The most common is to add thixotropic agents such as stearates (Winslow, 1950) to promote the formation of a weakly associated network in the suspending fluid. The network possesses a nonzero yield stress in the absence of an applied field, locking the particles in place under quiescent conditions but permitting flow under low applied stresses. More recent approaches have involved the dispersal of highly anisotropic carbon fibers (Shtarkman, 1992) or nanometer-scale ceramic or organic particles (Shtarkman *et al.*, 1994), which also form thixotropic networks that entrain the magnetizable particles. The orientation of these rodlike fibers in shear flow may also reduce the rate of particle centrifugation. The price that is paid in these approaches is a sometimes substantial increase of the zero-field viscosity of the fluids. One promising alternative approach is to shrink the size of the magnetic particulates sufficiently to reduce the sedimentation velocity, which scales quadratically with the particle size, but not so much as to prohibit magnetorheological behavior. For example, researchers at BASF (Kormann *et al.*, 1994) have produced "nano-MRFs," ferrite-based MR fluids incorporating roughly 30-nm-diam particles suspended in polar liquids. These MR fluids appear to exhibit excellent stability against sedimentation and centrifugation and are not abrasive, but possess smaller yield stresses as a result of the reduced saturation magnetization of ferrites relative to iron.

2.2 Magnetic-Field–Induced Structure Formation

The application of a magnetic field to an isolated magnetizable particle induces the formation of a net magnetic-dipole moment **m** aligned with the field. In the low-field or linear regime, the magnitude of the moment is proportional to the magnetic field at the particle:

$$m = 4\pi a^3 \mu_0 \beta H, \tag{3}$$

where a is the particle radius, $\mu_0 = 4\pi \times 10^{-7}$ H/m is the permeability of free space, $\beta = (\mu_p - \mu_f)/(\mu_p + 2\mu_f)$, and μ_p (μ_f) is the relative permeability of the particle (fluid). In high fields in which the particle magnetization has saturated, the magnetization is independent of field and is given by

$$m = \frac{4\pi a^3}{3} \mu_0 M_s, \tag{4}$$

where M_s is the saturation magnetization of the particle. Pairs of particles experience an anisotropic force resulting from the dipole–dipole interaction energy

$$U = -\frac{m^2}{4\pi\mu_0 R^3}(3\cos^2\theta - 1), \tag{5}$$

where **R** is the vector separation between particles and θ is the angle between **m** and **R**. This anisotropic interaction, which leads to repulsive forces for particles aligned with moments head-to-head ($\theta \sim \pi/2$) and attractive forces for head-to-tail alignment ($\theta \sim 0$), promotes the formation of particle chains aligned along the field direction, as demonstrated in Fig. 1. This "pearl chaining" effect is a central feature of both MR and ER phenomena (Jones, 1995). The nature of the structures formed by this interaction has been of considerable interest; a number of optical studies of model MR-fluid systems have investigated the kinetics of chain formation and the eventual coalescence into columns or more complicated network structures (for example, Hayes, 1975; Fermigier and Gast, 1992; Liu *et al.*, 1995).

A lower bound on the response time for the magnetorheological effect is the time taken to form pairs of particles, the flocculation time τ_f. The flocculation time has been estimated from molecular-dynamics simulations and analytical models of ER fluids exposed to a stepwise increase in the field (Hass, 1993); it is predicted to be proportional to the suspending-fluid viscosity and inversely proportional to the square of the dipole moment on each particle. Utilizing the analogy between electrostatics and magnetostatics, electrorheological models can be translated to ones appropriate to magnetorheological phenomena. Following Hass (1993), an expression for the pair-formation time in a MR fluid in the linear regime can be obtained:

$$\tau_f \approx \frac{2\eta_f}{5\mu_0(\beta H)^2}\left[\left(\frac{\pi}{6\phi}\right)^{5/3} - 1\right], \tag{6}$$

where η_f is the viscosity of the suspending fluid and ϕ is the volume fraction of the suspended particles. As the field is increased, the pair-formation time decreases; at sufficiently high fields, the particle moments—and the rate of pair formation—saturate. By modifying Eq. (6) to treat saturated particles, the *minimum* pair-formation time in a MR fluid is $\tau_f(\text{sat}) \approx 18\eta_f[(\pi/6\phi)^{5/3} - 1]/5\mu_0 M_s^2$, or roughly 1 μs with $\mu_0 M_s = 2$ T, $\eta_f = 0.5$ Pa-s, and $\phi = 0.3$.

While no experimental response-time data are available on practical MR fluids, the few reports of transient MR-device performance indicate device response times of order 10 ms (Weiss *et al.*, 1993). It is likely that extrinsic factors such as inductive time constants, eddy-current effects, and hysteresis determine the structural response times of MR fluids. Moreover, a substantial magnetorheological stress cannot be obtained unless the particle network is strained. The rheological response time τ_{rheo} is thus inversely related to the shear-strain rate in steady-shear conditions: $\tau_{\text{rheo}} \sim \gamma_y/\dot{\gamma}$, where γ_y is the yield strain and $\dot{\gamma}$ is the shear-strain rate (Ginder and Ceccio, 1995).

2.3 Steady-Shear Rheology

Of central importance in both torque-transfer and damping applications is the rheological performance of MR fluids. The flows encountered in these applications typically involve shear of the working fluid. Steady-shear rheology is almost universally used to characterize the rheological performance of MR and ER fluids. Concentric-cylinder or parallel-plate rheometers are typically used: The shear stress transmitted through the fluid when it is sheared is measured. Concentric-cylinder rheometers are often preferred, since the variation in shear rate through the sample is small for small fluid gaps; the shear rate in parallel-plate rheometers varies linearly with distance from the axis of rotation (see, e.g., Macosko, 1994). Materials and measurement issues involved in designing and using rheometers for the study of MR fluids have been discussed by Janocha and Rech (1994).

Relatively few measurements of the steady-shear rheology of MR fluids are available in the literature. Parziale and Tilton (1950) measured the flux-density dependence of the shear stress transmitted through a carbonyl-iron–based MR fluid ($\phi \approx 0.5$) used in a parallel-disk MR clutch. Kordonsky and Shulman (1990) reported the rheological, magnetic, and thermal properties of several MR fluids. Lemaire and Bossis (1991) examined magnetite-based MR fluids, finding that the measured shear stresses depended on the roughness of the interface between the MR fluid and the test fixture. Weiss and co-workers (Weiss *et al.*, 1993; Weiss and Duclos, 1994) reported the shear-rate dependence of the shear stress in a proprietary MR fluid exposed to several applied magnetic fields (Fig. 4). The data are well fitted by the Bingham model [Eq. (2)] with a viscosity of order 0.3 Pa-s and a yield stress that increases sub-quadratically with applied field. A weak reduction of the yield stress with temperature from −50°C to +150 °C was also reported; Weiss *et al.* (1993) suggested that it was entirely consistent with the reduction in particle-volume fraction due to thermal expansion of the suspending fluid. Pinkos *et al.* (1993) inferred the apparent shear stress in a rotary damper incorporating a carbonyl-iron–based MR fluid; shear stresses approaching 50 psi, or over 300 kPa, were reported, although the nature or rate of the shear flow was not disclosed. Inverse MR fluids containing nonmagnetic particles, such as polystyrene microspheres, dispersed in ferrofluids have been studied by Kashevskii *et al.* (1989), Bossis and Lemaire (1991), and Popplewell *et al.* (1995). The latter measurements were found to be in good agreement with the predictions of a linear continuum model of MR fluids (Rosensweig, 1995a).

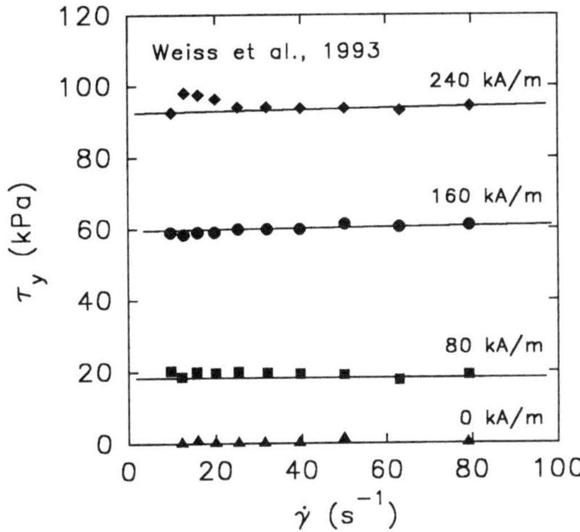

FIG. 4. Dependence of the shear stress of a MR fluid at 25 °C on shear rate and applied field (after Weiss *et al.*, 1993). The solid lines are fits by the Bingham model.

2.4 Oscillating-Shear Rheology

While many applications of MR fluids involve steady or unidirectional flows, some damping applications produce oscillatory flows. Oscillating-shear rheology is useful in defining the viscoelastic properties of complex fluids; in particulate systems, it can reveal qualitatively the nature and range of the interparticle interactions responsible for the viscoelastic behavior. These measurements may also be compared with theoretical predictions, which often are appropriate to small strain amplitudes and frequencies. They typically utilize sinuosoidal strain oscillations $\gamma^*(t) = \gamma_0 e^{i\omega t}$, where $i^2 = -1$, ω is the angular frequency of the oscillations, and t is the time, imposed on the sample; the complex stress transfer $\tau^* = \tau' + i\tau''$ is measured. The shear-storage modulus G', defined as τ'/γ_0, gives a measure of the elastic stiffness of the material. The loss modulus G'', given by τ''/γ_0, reflects viscous effects in the material. Weiss *et al.* (1994) have reported limited oscillating-shear rheological data on a proprietary MR fluid; the linear regime, in which the time dependence of the stress is linearly related to that of the applied strain, was found only for strain amplitudes below about 10^{-2}; the shear-storage moduli in this regime were as high as 2 MPa for applied fields of 160 kA/m. Magnetized MR fluids thus possess stiffnesses approaching those of common viscoelastic solids such as rubber.

2.5 Magnetic Properties

Knowledge of the magnetic properties of MR fluids can greatly assist the design of magnetic circuits for MR devices and may provide experimental tests for magnetostatic models of these materials. The variation of the flux density in a sample with applied field, $\mathbf{B} = \mathbf{B}(\mathbf{H})$ or the so-called *B-H* curve, can be measured using flux integration, vibrating-sample magnetometry, or other techniques. Carlson *et al.* (1994) reported the field dependence of the magnetization $M = B/\mu_0 - H$ for a MR fluid of unknown composition. A *B-H* curve derived from these data is shown in Fig. 5. The magnetic properties of MR fluids can be sensitive probes of the state of particle aggregation: Kordonsky and Shulman (1990) observed that the magnetization of a MR-fluid sample at a given magnetic field is larger when the particles have chained than when they are randomly distributed.

2.6 Magnetostatic Models of Magnetorheology

Models of magnetorheological phenomena are desirable both to satisfy basic scientific curiosity and to assist the development of MR fluids and devices. It was recognized early on that magnetically induced stresses were qualitatively responsible for the magnetorheological effect. More recent work has been inspired in part by considerable modeling and experimental activity on ER fluids. The connection between suspension microstructure and rheological behavior in MR fluids is likely to be very similar to that in ER fluids. Stress transfer in ER fluids occurs qualitatively through the tilting, snapping, and reformation of particle chains (Klingenberg and Zukoski, 1990). Simulations have shown how a field-dependent yield stress can result from these microstructural processes (Bonnecaze and Brady, 1992).

The many approaches used to model the electrostatic interactions in ER fluids can be directly transferred to model MR fluids so long as one is confined to the linear regime, i.e., for fields low enough not to produce saturation effects in the MR particulates. These models generally involve the calculation of the fields and forces in particle chains or three-dimensionally ordered structures. The restoring force produced by shearing the structure can then be calculated as a function of applied field and the permeability and volume fraction of the magnetizable particulates. These linear approaches predict that the shear stresses, yield stresses, and shear moduli are proportional to $\phi\mu_0 H^2$, where ϕ is the particle-volume fraction and H is the applied field.

The stresses measured in practical MR fluids increase with field more slowly than H^2. This serious discrepancy reflects the important roles of magnetic nonlinearity and saturation of the particles used in these fluids. These effects have been explicitly incorporated in a recent model of magnetorheology (Ginder and Davis, 1994). A nonlinear constitutive *B-H* relation was assumed, and the field distribution in an infinite linear chain of particles in a specified flux density was calculated using finite-element analysis (FEA) numerical techniques. The chain tension was then calculated from the Maxwell stress at the interface between particles and was used to infer the magnetically induced shear stress as a function of shear strain. The yield stresses predicted by this model for a saturation magnetization $\mu_0 M_s = 2$ T are shown in Fig. 6(a); they increase subquadratically with flux density and saturate at as

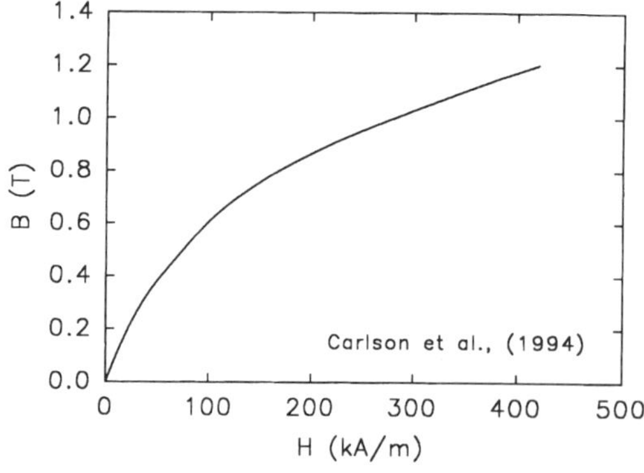

FIG. 5. *B-H* curve inferred from magnetization measurements on a MR fluid (after Carlson *et al.*, 1994).

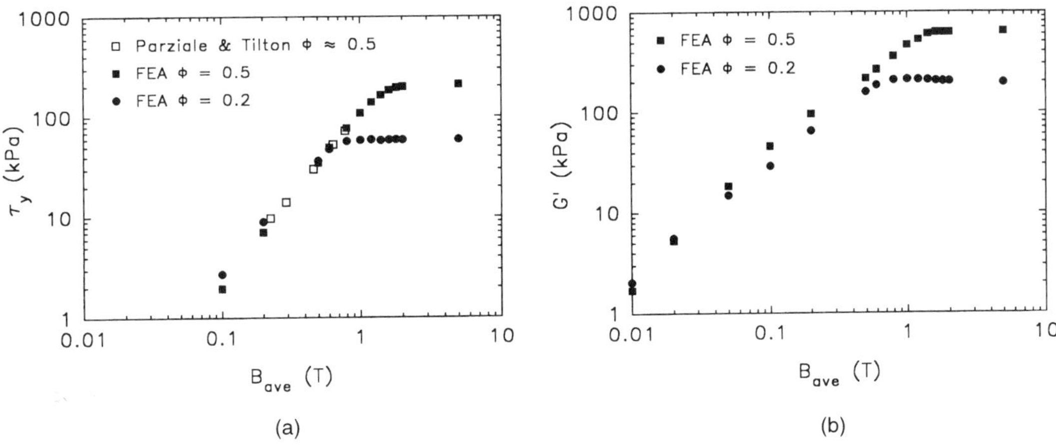

FIG. 6. Predictions of a finite-element analysis model for a MR fluid with $\mu_0 M_s = 2$ T: (a) flux-density dependence of the MR shear stress for volume fractions $\phi = 0.2$ (●) and 0.5 (■). The shear stresses measured in a MR fluid with $\phi \approx 0.5$ (□) (Parziale and Tilton, 1950) are also shown. (b) Variation of the shear modulus with flux density (after Ginder and Davis, 1994).

much as 210 kPa for $\phi = 0.5$. The predicted yield stresses compare quite favorably with experimental data from Parziale and Tilton (1950) on carbonyl-iron MR fluids with $\phi \approx 0.5$, also shown. While the shear moduli predicted by this model are as large as 600 kPa for $\phi = 0.5$ at saturation [Fig. 6(b)], they are far exceeded by the shear moduli measured on a proprietary MR fluid (Weiss *et al.*, 1994). This discrepancy is likely due to the fact that the model treats only the stresses directly resulting from magnetic interactions and not those associated with other interparticle-contact interactions—van der Waals forces, steric forces, and so on.

For applied fields too low for complete particle saturation, the consequences of saturation are still manifest in the magnetic saturation of the polar regions of each particle (von Guggenberg *et al.*, 1986; Jones and Saha, 1990; Ginder and Davis, 1994). The magnetic flux is channeled through the particles and is highest at the contact regions, causing this local saturation. A contour plot of the magnetization obtained by a FEA solution for a chain of particles (Fig. 7) reveals that the polar regions are indeed the first to saturate. Local saturation limits the increase of the field in the gaps between particles and thus reduces the quadratic dependence of the stresses on applied field or flux density. A simple analytical model (Ginder *et al.*, 1996) predicts that the yield stress in this regime is given by

$$\tau_y \approx \sqrt{6}\, \phi \mu_0 M_s^{1/2} H^{3/2}, \tag{7}$$

while the shear-storage modulus is given by

$$G' \approx 3\phi\mu_0 M_s H. \tag{8}$$

For high applied fields, the particles are completely saturated and so can be treated rigorously as dipoles. The results for the shear stress in simple dipole chains, in which the stresses scale quadratically with the dipole moment, can then be applied to calculate the yield stress at saturation (the maximum value of the yield stress). Results of FEA calculations for $\phi = 0.2$ and 0.5 (Fig. 8) also exhibit quadratic scaling with saturation magnetization. The FEA results are in agreement with the maximum (albeit not saturation-level) yield stresses reported for Lord and BASF MR fluids, as indicated in Fig. 8. The quadratic dependence of the maximum shear stress on the saturation magnetization of the particles underscores the desirability of MR particulates synthesized from elements or alloys having high M_s values. Iron–cobalt alloys having the largest known saturation magnetization of any material, $\mu_0 M_s \sim 2.4$ T, have recently been incorporated in a high-strength MR fluid (Carlson and Weiss, 1995).

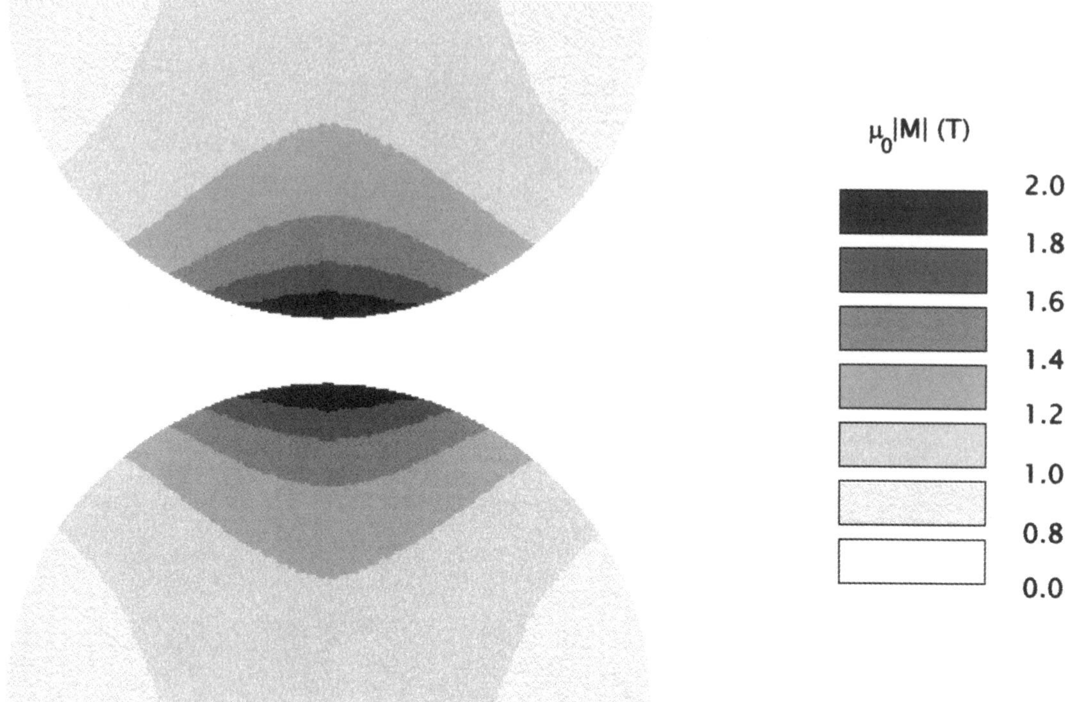

FIG. 7. Contour plot of the magnitude of the magnetization in a pair of particles embedded in an infinite chain modeled by finite-element analysis, showing the saturation of the magnetization near the contact points. The particles are assumed to have a saturation magnetization $\mu_0 M_s = 2$ T and are exposed to a flux density 0.4 T.

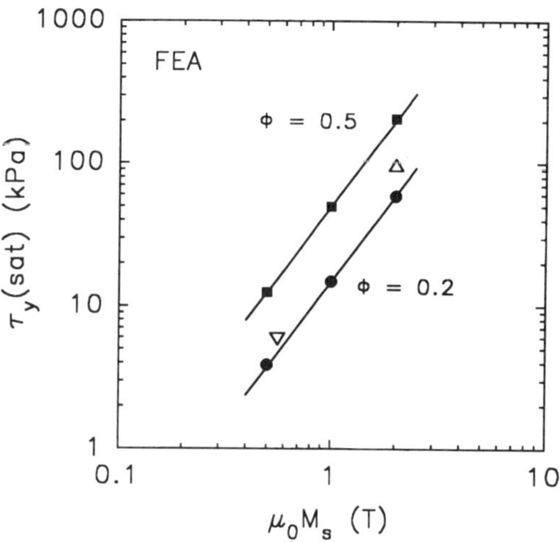

FIG. 8. Quadratic variation of the yield stress at saturation on the saturation magnetization of the particles for volume fractions $\phi = 0.2$ (●) and 0.5 (■), calculated using a FEA model (after Ginder and Davis, 1994). The largest shear stresses reported for MR fluids from Lord Corporation (△) (Weiss *et al.*, 1993) and BASF (▽) (Kormann *et al.*, 1994) are also shown.

3. MAGNETIC POWDERS

3.1 Composition

The magnetostatics of MR fluids are unaltered if the nonmagnetic suspending fluids are replaced by air to obtain dry magnetic powders. The highest-strength magnetic powders would thus involve the particulate materials with the highest saturation magnetization, iron or some of its alloys. Unfortunately, the major application of magnetic powders, in clutches and rotary brakes, is often associated with elevated temperatures that promote the eventual failure of these particles due to oxidation. Magnetic stainless steels such as 400-level steels, composed mainly of iron and chromium, are therefore used. Typical particle sizes in magnetic powders are from 10 to 100 μm; both spherically and irregularly shaped particles are used. Nonmagnetic powdered materials such as graphite, zinc oxide, or silicon dioxide may be added to act as lubricants or dispersants (Ramakrishnan and Pillai, 1980).

Magnetic powders are widely used in image development in xerography (Jones *et al.*, 1987), in which rotating drums housing arrays of permanent magnets are used to transport magnetizable particles to the photoreceptor drum on which the xerographic image is explosed. One-component xerography uses 10-μm polymeric toner particles loaded with both ferrite and pigment to form the image directly, while two-component xerography uses 100-μm ferrite or steel particles as a reusable transport medium to carry nonmagnetic 10-μm pigment particles to the photoreceptor drum.

3.2 Magnetic-Powder Rheology

The mechanics of noncohesive, isotropic powders is much more widely studied and better understood than that of magnetic powders, which are highly cohesive, anisotropic materials. The most extensive measurements of magnetic-powder rheology have been performed on magnetostabilized fluidized beds, in which a collection of magnetizable particles subjected to a stream of flowing gas or liquid is held in place by an imposed magnetic field (Rosensweig, 1995b). These beds were of interest for industrial chemical processes such as catalysis, molecular separation, and heat recovery. Several studies of the dependence of the yield stress on applied field and flow velocity have appeared (for example, Lee, 1983); as it happens, the yield stresses necessary for bed stabilization are several orders of magnitude smaller than those of interest in motion-control applications. Torque-transfer geometries have been used to characterize magnetic powders; a magnetic-powder disk clutch was used to infer the rheology of iron powders (Ramakrishnan and Pillai, 1980). The magnetically induced shear stresses were found to increase only weakly with increasing particle size and shear rate. Jones *et al.* (1987) have extended theories of powder mechanics to include magnetic interactions between particles. Such theories, first developed by Coulomb, use the conditions of static mechanical equilibrium to relate the shear and normal stresses in a powder. The application of magnetically induced stresses increases the effective normal stress and thus increases the shear stress and, in particular, its yield value.

4. MAGNETORHEOLOGICAL ELASTOMERS

4.1 Composition

Composites of magnetic particles in polymeric host media have been produced for some time, although usually for structural studies or dielectric applications. For example, the structure of a magnetostabilized bed of particles was investigated by polymerizing the suspending fluid, thereby "freezing" the particulate structure and permitting optical microscopy and magnetic measurements on the solidified specimens (Rosensweig *et al.*, 1981). Magnetic metallic particles have been aligned by magnetic fields and subsequently fixed by curing in silicone elastomers to produce transparent sheets whose conductivity along the field direction is extremely sensitive to pressure; these sheets can be used in touch-screen panels (Jin *et al.*, 1992). Highly anisotropic metallic magnetic particles have been dispersed and aligned in rigid epoxies to produce highly anisotropic dielectric materials for microwave frequencies (Behroozi *et al.*, 1990).

It has only recently been appreciated that the viscoelastic properties of magnetic-particle polymer composites can be varied by an applied field. A group at Toyota Central Research and Development has dispersed iron particles in liquid-silicone precursors, then cured the mixture to produce MR elastomers (Shiga *et al.*, 1992): viscoelastic solids whose storage and loss moduli are nonzero even in the absence of a field, but can be increased substantially, rapidly, and continuously by the application of a field. Other matrix materials may include ethylene–propylene, butadiene, and polyisoprene rubbers, as well as polyvinyl alcohol, polyacrylamide, and polystyrene gels. The preferred particle concentration ranges from 0.05 to 0.6 and can be varied spatially through the composite. After curing, these solids can be shaped or machined to the desired geometry and size.

4.2 Field-Dependent Viscoelastic Response

Shiga *et al.* (1992) measured the dynamic viscoelastic properties of several MR elastomers in an oscillating parallel-plate rheometer. One sample contained 150-μm electrolytic-iron particles dispersed randomly at $\phi \approx 0.27$ in room-temperature–curing silicone rubber. The shear-storage modulus of this material increased by 50% upon the application of a 320-Oe magnetic field. In another sample, 50-μm electrolytic iron was dispersed at $\phi \approx 0.2$ in an elevated-temperature–curable silicone-rubber precursor. A permanent magnet was used to orient the iron particles while the elastomer was cured at elevated temperature. The resulting anisotropic composite exhibited a shear-storage modulus of 7.8 kPa; the storage modulus increased to 21.5 kPa upon the application of a 740-Oe field.

5. APPLICATIONS OF MR MATERIALS

5.1 Magnetic-Circuit Considerations

Of critical importance in applications of MR materials—ferrofluids, MR fluids, magnetic powders, or MR elastomers—is the efficient delivery of the required magnetic field to the material. The production and transmission of the field are accomplished by a magnetic circuit that consists of a source of magnetomotive force, either a wire coil or a permanent magnet, and a flux path usually constructed of highly permeable materials like iron or steel (see MAGNETOSTATICS). The optimum MR-device design is that which will deliver the necessary field while consuming the least electrical power and weighing as little as possible, sometimes conflicting goals. Detailed magnetic-circuit design is greatly aided by numerical solutions of the magnetostatic field and flux density in the circuit that include the nonlinear B-H relationships of the circuit components. Several commercial numerical magnetostatic software programs based on finite-element and boundary-element techniques in two- and three-dimensional geometries are available for this purpose.

The approximate behavior of simple magnetic circuits can be understood by applying the fundamental laws of magnetostatics. Gauss's law, $\nabla \cdot \mathbf{B} = 0$, leads to the conservation of flux in the magnetic circuit. Ampére's law relates the line integral of the magnetic field along any contour (and in particular contours embedded in the magnetic circuit) to the current enclosed by it: $\nabla \times \mathbf{H} = \mathbf{J}$ or $\oint \mathbf{H} \cdot d\mathbf{l} = I_{tot}$, where J is the current density, I_{tot} is the total current contained by the contour, and $d\mathbf{l}$ is the differential line element on the contour. MR devices typically utilize electromagnets to allow control of the field in the MR material; in most cases the electromagnet is constructed around a toroidal coil of N turns such that $I_{tot} = NI$ where I is the current in the coil. If the coil has an average radius R_{coil} and a cross-sectional area A_{coil}, the electrical power consumed is $P \sim 8\rho R_{coil}(NI)^2/A_{coil}$, where ρ is the electrical resistivity of the wire. In this fashion, the electrical-power consumption can be related to the device shape and size as well as the fields in it. In typical magnetic-circuit designs, the constitutive relations between $\mathbf{B}$ and $\mathbf{H}$ are measured or provided for each material used in the magnetic circuit. The designer adjusts the magnetic-circuit geometry to obtain the desired device performance, power consumption, and packaging attributes.

5.2 Torque-transfer Applications

MR fluids and magnetic powders can be used to control the transfer of torque in clutches and rotary brakes. Magnetic-powder clutches and brakes are commercially available from a number of suppliers, including MagPower, Placid, Electroid, and T. B. Wood's Sons in the United States and Ogura and Mitsubishi Electric in Japan. These devices are used widely in manufacturing to control speed, torque, and tension, as well as in motor-testing applications. Magnetic-powder clutches are even used in automobiles: The main clutch on Subaru Justy vehicles equipped with continuously variable transmissions is a high–torque-capacity magnetic-powder clutch (Sakai, 1988). This application provides an example of the controllability afforded by the magnetorheological effect: The current to the clutch is varied continuously during engine acceleration and deceleration in order to reduce unwanted power-train transients. MR-fluid brakes are now being produced by Lord Corporation for use as variable-resistance devices on exercise machines. The smoother operating characteristics and improved heat-transfer capabilities of MR fluids are potential advantages over magnetic powders in these applications.

MR fluids and magnetic powders have

been utilized in two basic clutch or brake geometries, parallel disks and concentric cylinders, as shown schematically in Fig. 9. The torque-transfer characteristics of these geometries can be estimated if the MR fluids are assumed to obey a Bingham law with viscosity η and yield stress τ_y [Eq. (2)]. For a disk of radius R separated from each of the two opposing faces of its housing by a gap of width h, the total torque T assuming laminar flow is given by

$$T = \pi\eta R^3 \frac{R}{h} \Omega + \frac{4\pi}{3} R^3 \tau_y, \tag{9}$$

where Ω is the relative angular frequency of the disk and the housing. For concentric cylinders with a length L, inner-cylinder radius R, and gap h such that $h \ll R$, the torque is

$$T = 2\pi\eta L R^2 \frac{R}{h} \Omega + 2\pi L R^2 \tau_y. \tag{10}$$

In both of the above expressions, the first term on the right represents the contribution to the torque by Couette flow of a viscous fluid and is present at all times. The second term is due to the field-induced MR shear stress and is therefore controllable by an applied magnetic field. The rate of mechanical-power dissipation for both geometries is given by $P_{\text{diss}} = \Omega T$.

5.3 Damping Applications

Another important potential application of MR materials is in vibration damping. Shock absorbers and mounts are used in many automotive, aerospace, and industrial applications to reduce the vibrations transmitted from internal or external sources to operators, passengers, or sensitive equipment. Materials like MR fluids allow the damping force to be controlled according to external inputs from sensors, enabling significant improvements in damper performance. Researchers at TRW (Pinkos *et al.*, 1993), Lord Corporation (Carlson *et al.*, 1994, 1995), and Byelocorp Scientific (Kordonsky *et al.*, 1994) have utilized MR fluids in a variety of damping applications. A field-induced increase in damping force of as much as 3 kN, comparable to that required in many passenger-car automotive applications, was obtained in a shock-absorber prototype (Weiss *et al.*, 1993). Seat mounts for on- and off-road vehicles, engine mounts, and machinery mounts are other potential applications. A schematic cross-sectional view of a MR-fluid–based damper is shown in Fig. 10.

Like their passive counterparts, most MR-fluid dampers utilize fluid flow through an orifice or channel to produce a pressure drop ΔP and thus to generate a damping force. A MR-fluid valve is an electrically controllable orifice: The pressure drop across the valve is controllable by changes in the magnetic field

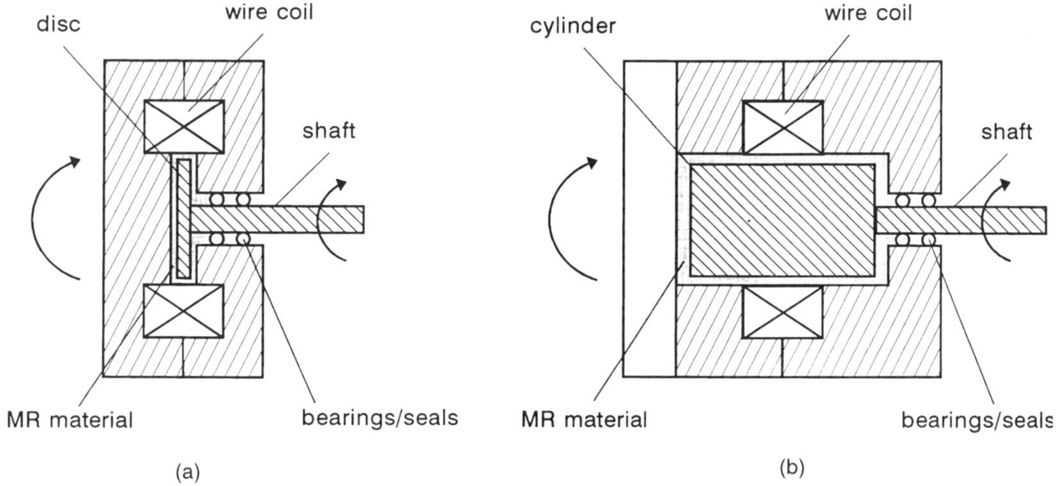

FIG. 9. Schematic cross-sectional view of (a) disk and (b) concentric-cylinder clutches.

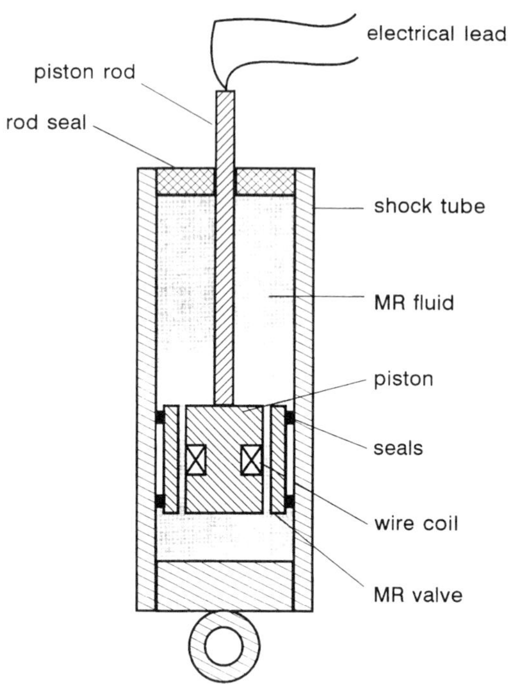

FIG. 10. Schematic cross-sectional view of a MR-fluid–based damper (after Carlson *et al.*, 1994).

in the channel. While ΔP depends on the detailed nature of the flow of solidified MR material in the channel, it is approximately given by

$$\Delta P \approx 12\eta \frac{Q}{wh^2}\frac{L}{h} + c\frac{L}{h}\tau_y, \tag{11}$$

where L, h, and w are the channel length, thickness, and width, Q is the volumetric-flow rate, and c is a numerical constant between 2 and 3. The damping force F generated in the channel is then the product of the pressure drop and the channel cross-sectional area: $F = hw\Delta P$. The first term on the right in Eq. (11) represents the pressure drop due to the (assumed) laminar Pouseille flow of the viscous MR fluid, which is proportional to the flow rate and is nominally independent of field. The second term on the right is the contribution to the pressure drop from the solidified MR fluid; it is essentially independent of flow rate but is strongly dependent on the magnetic field.

MR elastomers could potentially be used in damping and vibration isolation components such as engine mounts, although such applications have not yet appeared in the literature. Advantages of these materials include their solid-state character: No seals are required to contain fluids, and the suspended magnetizable particles do not sediment out over time.

5.4 Process Applications

Magnetorheological fluids may also find use in manufacturing and process applications. Of considerable recent interest is the use of MR-fluid–based polishing compounds in the deterministic finishing of optical components, an application being explored by Byelocorp Scientific, Inc. (Holton, 1993). Nonmagnetic abrasive particles are dispersed in an aqueous MR fluid; by applying a magnetic field, the abrasive grit is driven to the interface between the MR fluid and the optical component. The rate and spatial extent of the abrasive action can be controlled in real time by an electromagnet. Any nonferrous material may be polished by this technique.

5.5 Outlook

With the first MR-fluid devices now being commercialized, this is an exciting era in the field of magnetorheology. While the relative maturity of ferrofluid and magnetic-powder technology provides an "existence proof" for the viability of MR-fluid devices, the degree of acceptance and use of this technology in automotive and other industries will depend on a number of technical and economic factors. The not insubstantial mass and power consumption of many MR devices is an issue, although it should be noted that conventional electromechanical devices possess similar characteristics. The most efficient MR clutch designs embed wire coils inside the rotating clutch member, requiring slip rings that add cost and can fail over time. The minimum response time of MR devices has not been established. While inductive, eddy-current, and hysteretic effects conspire to slow device response, they can be combated with laminated magnetic-circuit materials and transient overvoltages. To take full advantage of the speed and controllability of the MR-device response, sensors are required as inputs to control hardware and software; while providing substantially improved performance, these sensors also add to the system cost. It might be possible to eliminate separate sensors and their associated expense if the MR material itself could be used as a strain or velocity sensor; this increase in material functionality might be achievable by using inductive or other effects. The abrasive effects of the magnetic particulates in MR fluids require much more attention than they have received; friction and wear may be particularly problematic in fluid seals. The toxicity and recyclability of MR fluids must be addressed.

Improvements in our understanding of the physics and chemistry of MR fluids should assist the development of new MR fluids. The stability of the fluids against sedimentation and centrifugation remains a central problem. The utilization of submicron magnetic particles produced by a variety of processes may improve stability. New surfactants may be identified to aid the stabilization and redispersal of the particles. Soluble polymers could be employed for fluid stabilization via the formation of a gel-like thixotropic network or by weakly flocculating the magnetizable particles. The nature of the structural deformations induced by applied stresses in MR fluids requires clarification. The assumption that isolated chains deform uniformly under an applied strain is undoubtedly a gross simplification of the situation in real, polydisperse systems. Structure deformation by fracture may be as relevant in MR fluids as it appears to be in magnetic powders. Models for the MR shear stress that account for not only the magnetostatic interactions between particles but also other colloidal-interaction forces are needed. The effects of magnetic hysteresis on the magnetic and rheological performance of MR fluids have not been investigated. The development of production devices and the emergence of new applications are powerful motivations for the continued pursuit of physical understanding of magnetorheological materials.

GLOSSARY

Magnetic Circuit: A circuit through which magnetic flux can be delivered. Usually consists of a source of magnetomotive force (either a wire coil or a permanent magnet) and a path containing iron or steel to transmit the flux.

Relative Permeability: The ratio of magnetic induction to magnetic field in a material, normalized by μ_0, the permeability of free space.

Saturation Magnetization: The maximum magnetization, or magnetic moment per unit volume, achievable in a material.

Thixotropy: A phenomenon in which a material exhibits time-dependent rheological properties.

Yield Stress: The stress at which a material deforms plastically or flows.

Works Cited

Bean, C. P., Jacobs, I. S. (1960), *J. Appl. Phys.* **31,** 1228–1230.

Behroozi, F., Orman, M., Reese, R., Stockton, W., Calvert, J., Rachford, F., Schoen, P. (1990), *J. Appl. Phys.* **68,** 3688–3693.

Berkovsky, B. M., Medvedev, V. F., Krakov, M. S. (1993), *Magnetic Fluids,* Oxford: Oxford Univ. Press.

Blunden, S. F. (1951), *Engineer* **191** (4961), 244–246.

Bonnecaze, R. T., Brady, J. F. (1992), *J. Rheol.* **36,** 73–115.

Bossis, G., Lemaire, E. (1991), *J. Rheol.* **35,** 1345–1354.

Carlson, J. D., Chrzan, M. J., James, F. O. (1994), U.S. Patent 5284330.

Carlson, J. D., Weiss, K. D. (1995), U.S. Patent 5382373.

Carlson, J. D., Chrzan, M. J., James, F. O. (1995), U.S. Patent 5398917.

Fermigier, M., Gast, A. P. (1992), *J. Colloid. Int. Sci.* **154,** 522–539.

Ginder, J. M., Davis, L. C. (1994), *Appl. Phys. Lett.* **65,** 3410–3412.

Ginder, J. M., Ceccio, S. L. (1995), *J. Rheol.* **39,** 211–234.

Ginder, J. M., Davis, L. C., Elie, L. D. (1996), in: W. Bullough (Ed.), *Proceedings of the 5th International Conference on Electrorheological Fluids and Magnetorheological Suspensions,* Singapore: World Scientific.

Grebe, O. (1952), *Elektrotech. Zeits.* **9,** 281–284.

von Guggenberg, P. A., Porter, A. J. II, Melcher, J. R. (1986), *IEEE Trans. Magn.* **MAG-22,** 614–619.

Hartsock, D. L., Novak, R. F., Chaundy, G. J. (1991), *J. Rheol.* **35,** 1305–1326.

Hass, K. C. (1993), *Phys. Rev. E* **47,** 3362–3373.

Hayes, C. F. (1975), *J. Colloid. Int. Sci.* **52,** 239–243.

Holton, W. C. (1993), *Photonics Spectra* (April), 29.

Janocha, H., Rech, B. (1994), *Rheology* (December), 199–203.

Jin, S., Tiefel, T. H., Wolfe, R., Sherwood, R. C., Mottine, J. J., Jr. (1992), *Science* **255,** 446–448.

Jones, T. B. (1995), *Electromechanics of Particles,* New York: Cambridge Univ. Press.

Jones, T. B., Saha, B. (1990), *J. Appl. Phys.* **68,** 404–410.

Jones, T. B., Whittaker, G. L., Sulenski, T. J. (1987), *Powder Technol.* **49,** 149–164.

Kashevskii, B. E., Kordonsky, V. I., Prokhorov, I. V. (1989), *Magnetohydrodynamics* **3,** 368–371.

Klingenberg, D. J., Zukoski, C. F. (1990), *Langmuir* **6,** 15–24.

Kordonsky, V. I., Shulman, Z. P. (1990), in: J. D. Carlson, A. F. Sprecher, H. Conrad (Eds.), *Proceedings of the Second International Conference on ER Fluids,* Lancaster, PA: Technomic, pp. 437–444.

Kordonsky, W. I., Gorodkin, S. R., Kolomentsev, A. V., Kuzmin, V. A., Luk'ianovich, A. V., Protasevich, N. A., Prokhorov, I. V., Shulman, Z. P. (1994), U.S. Patent 5353839.

Kormann, C., Laun, M., Klett, G. (1994), in: H. Borgmann, K. Lenz (Eds.), *Proceedings of the Fourth International Conference on New Actuators,* Bremen, Germany, 15–17 June, Bremen, Germany: AXON Technologie, pp. 271–274.

Lee, W.-K. (1983), *AIChE Symp. Ser.* **222,** 79.

Lemaire, E., Bossis, G. (1991), *J. Phys. D: Appl. Phys.* **24,** 1473.

Liu, J., Lawrence, E. M., Wu, A., Ivey, M. L., Flores, G. A., Javier, K., Bibette, J., Richard, J. (1995), *Phys. Rev. Lett.* **74,** 2828–2831.

Macosko, C. W. (1994), *Rheology,* New York: VCH.

McTague, J. P. (1969), *J. Chem. Phys.* **51,** 133–136.

Parziale, A. J., Tilton, P. D. (1950), *AIEE Trans.* **69,** 150–157.

Pinkos, A., Shtarkman, E., Fitzgerald, T. (1993), SAE International Congress and Exhibition, Detroit, MI, 1–5 March, SAE Technical Paper 930268.

Popplewell, J., Rosensweig, R. E., Siller, J. K. (1995), *J. Magn. Magn. Mater.,* in press.

Rabinow, J. (1948), *AIEE Trans.* **67,** 1308–1315.

Rabinow, J. (1949), SAE National Transportation Meeting, Cleveland, OH, 28–30 March, SAE Technical Paper 49-316.

Rabinow, J. (1951), U.S. Patent 2575360.

Raj, K., Moskowitz, R. (1990), *J. Magn. Magn. Mater.* **85,** 233–245.

Ramakrishnan, S., Pillai, K. P. P. (1980), *IEE Proc.* **127,** 81–88.

Rosensweig, R. E., Jerauld, G. R., Zahn, M. (1981), in: O. Brulin, R. K. T. Hsieh (Eds.), *Continuum Models of Discrete Systems,* Vol. 4, Amsterdam: North-Holland, pp. 137–144.

Rosensweig, R. E. (1985), *Ferrohydrodynamics,* Cambridge: Cambridge Univ. Press.

Rosensweig, R. E. (1995a), *J. Rheol.* **39,** 179–192.

Rosensweig, R. E. (1995b), *J. Electrostat.* **34,** 163–187.

Sakai, Y. (1988), SAE International Congress and Exhibition, Detroit, MI, 29 February–4 March, SAE Technical Paper 880481.

Shiga, M., Hirose, M., Okada, K. (1992), Japanese Patent laid-open No. 4-266970.

Shliomis, M. I. (1972), *Soviet Phys. JETP* **34,** 1291–1294.

Shtarkman, E. (1992), U.S. Patent 5167850.

Shtarkman, E. M., Starkovich, J. A., Davison, W. W., Peng, H.-H., Fitzgerald, T. J. (1994), U.S. Patent 5354488.

Vorob'yeva, T. M. (1965), *Electromagnetic Clutches and Couplings,* Oxford: Pergamon.

Weiss, K. D., Duclos, T. G., Carlson, J. D., Chrzan, M. J., Margida, A. J. (1993), SAE International Off-Highway and Powerplant Congress and Exhibition, Milwaukee, WI, 13–15 September, SAE Technical Paper 932451.

Weiss, K. D., Duclos, T. G. (1994), in: R. Tao, G. D. Roy (Eds.), *Electrorheological Fluids: Mechanisms, Properties, Technology, and Applications,* Proceedings of the Fourth International Conference on ER Fluids, Singapore: World Scientific, pp. 43–59.

Weiss, K. D., Carlson, J. D., Nixon, D. A. (1994), in: C. Rogers, G. Wallace (Eds.), *Proceedings of the Second International Conference on Intelligent Materials,* Lancaster, PA: Technomic, pp. 1300–1307.

Winslow, W. M. (1947), U.S. Patent 2417850.

Winslow, W. M. (1949), *J. Appl. Phys.* **20,** 1137–1140.

Winslow, W. M. (1950), U.S. Patent 2661825.

Further Reading

Colloidal Interactions and Stability

Russel, W. B., Saville, D. A., Schowalter, W. R. (1989), *Colloidal Dispersions,* Cambridge: Cambridge Univ. Press.

Rheological Techniques

Macosko, C. W. (1994), *Rheology,* New York: VCH.

Electrorheological Fluids

Havelka, K. O., Filisko, F. E. (Eds.) (1995), *Progress in Electrorheology: Science and Technology of Electrorheological Materials,* Proceedings of the American Chemical Society Symposium on Electrorheological Materials and Fluids, New York: Plenum.

Tao, R., Roy, G. D. (Eds.) (1994), *Electrorheological Fluids: Mechanisms, Properties, Technology, and Applications,* Proceedings of the Fourth International Conference on ER Fluids, Singapore: World Scientific Press.

U.S. Department of Energy Report DOE/ER/30172 (1993), *Electrorheological (ER) Fluids: A Research Needs Assessment,* Washington, DC: U.S. Department of Energy.

Magnetism and Magnetic Materials

Bozorth, R. M. (1951), *Ferromagnetism,* Toronto: van Nostrand.

Ferrofluids

Berkovsky, B. M., Medvedev, V. F., Krakov, M. S. (1993), *Magnetic Fluids,* Oxford: Oxford Univ. Press.

Rosensweig, R. E. (1985), *Ferrohydrodynamics,* Cambridge: Cambridge Univ. Press.

ROBOTICS

JOHN J. CRAIG, *Adept Technology, Inc., San Jose, California, U.S.A.*

	Introduction	505
1.	**Description of Position and Orientation**	506
2.	**Forward Kinematics of Manipulators**	507
3.	**Inverse Kinematics of Manipulators**	509
4.	**Velocities, Static Forces, Singularities**	510
5.	**Dynamics**	511
6.	**Trajectory Generation**	511
7.	**Manipulator Design and Sensors**	512
8.	**Position Control**	512
9.	**Force Control**	514
10.	**Computer Vision**	514
11.	**Locomotion**	515
12.	**Programming Robots**	515
13.	**Off-Line Programming and Simulation**	515
	Glossary	516
	Works Cited	517
	Further Reading	517

INTRODUCTION

The study of robotics concerns itself with the desire to synthesize some aspects of human function by the use of mechanisms, sensors, actuators, and computers. Obviously, this is a huge undertaking that seems certain to require a multitude of ideas from various "classical" fields. Various definitions have been used to describe the term *robot*—for example, the following from the Robot Association of America:

> A reprogrammable, multi-functional manipulator designed to move material, parts, tools, or special devices through various programmed motions for the performance of diverse tasks.

Perhaps the broadest definition covers any computer-controlled programmable machine.

The history of industrial automation is characterized by periods of rapid change in popular methods. Use of the *industrial robot* beginning in the 1960s, along with computer-aided design (CAD) systems and computer-aided manufacturing (CAM) systems, characterizes the latest trends in the automation of the manufacturing process. These technologies are leading industrial automation through another transition, the scope of which is still unknown.

Much of the present use of industrial robots is concentrated in rather simple, repetitive tasks that tend not to require high precision. In the 1980s and early 1990s relatively simple tasks like machine tending, material transfer, painting, and welding have been economically viable for robotization. Manufacturing market analysts predict that in the 1990s and into the next century industrial robots will become increasingly viable in applications that require more precision and sensory sophistication such as assembly tasks.

The predicted increase in the capabilities of industrial robots will cause a shift in which kinds of industries employ them. The automotive industry, where robots have been economically justified since the 1970s, will continue to be the leading user in the United States. However, the major growth of the U.S. robot population will occur in nonautomotive industries. Of lesser market size, but of great technological interest, are uses of robots in outer space and under the sea.

This article focuses on the mechanics and control of the most important form of the robot, the *mechanical manipulator*. Exactly what constitutes an industrial robot is sometimes debated. The distinction lies somewhere in the sophistication of the program-

3-527-28138-X/96/$5.00 + .50

mability of the device—if a mechanical device can be programmed to perform a wide variety of applications, it is probably an industrial robot. Machines that are for the most part relegated to one class of task are considered *fixed automation.* For the purposes of this article, the distinctions need not be debated as most topics are of a basic nature that applies to a wide variety of programmable machines.

By and large, the study of the mechanics and control of manipulators is not a new science, but merely a collection of topics taken from classical fields. Mechanical engineering contributes methodologies for the study of machines in static and dynamic situations. Mathematics supplies tools for describing spatial motions and other attributes of manipulators. Control theory provides tools for designing and evaluating algorithms to realize desired motions or force application. Electrical engineering techniques are brought to bear in the design of sensors and interfaces for industrial robots, and computer science contributes a basis for programming these devices to perform a desired task.

1. DESCRIPTION OF POSITION AND ORIENTATION

In the study of robotics we are constantly concerned with the location of objects in three-dimensional space. These objects are the links of the manipulator, the parts and tools with which it deals, and other objects in the manipulator's environment. At a crude but important level, these objects are described by just two attributes: their position and their orientation. Naturally, one topic of immediate interest is the manner in which we represent these quantities and manipulate them mathematically.

In order to describe the position and orientation of a body in space we will always attach a coordinate system, or *frame,* rigidly to the object. We then proceed to describe the position and orientation of this frame with respect to some reference coordinate system (see Fig. 1).

Since any frame can serve as a reference system within which to express the position and orientation of a body, we often think of *transforming* or *changing the description of* these attributes of a body from one frame to another. Study of robotics inevitably requires conventions and methodologies for dealing with the description of position and orientation, and the mathematics of manipulating these quantities with respect to various coordinate systems.

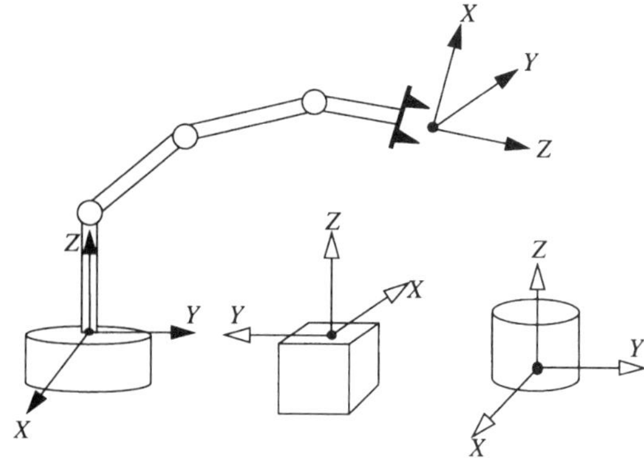

FIG. 1. Coordinate systems or "frames" are attached to the manipulator and objects in the environment.

We use the notation ${}^{A}P$ to represent a position P represented in a frame of reference A. This position is represented with a three-element vector whose individual elements are given subscripts x, y, and z:

$$ {}^{A}P = \begin{bmatrix} p_x \\ p_y \\ p_z \end{bmatrix}. \tag{1} $$

The leading superscript is used to indicate the frame of reference. A common problem is, given a point represented in one frame, to compute its value in a second reference frame:

$$ {}^{A}P = {}^{A}_{B}R\,{}^{B}P + {}^{A}P_{BORG}. \tag{2} $$

Equation (2) describes a general transformation mapping of a vector from its description in one frame to a description in a second frame. The symbol ${}^{A}_{B}R$ represents a *rotation matix* whose columns are the unit vectors of frame B written in frame A. The vector ${}^{A}P_{BORG}$ locates the origin of the reference system B written in the system A. Equation (2) is sometimes represented by the simpler notation

$$ {}^{A}P = {}^{A}_{B}T\,{}^{B}P, \tag{3} $$

where ${}^{A}_{B}T$ is called a *homogeneous transform* and has the structure

$$\begin{bmatrix} {}^{A}P \\ \hline 1 \end{bmatrix} = \left[\begin{array}{c|c} {}^{A}_{B}R & {}^{A}P_{BORG} \\ \hline 0 \quad 0 \quad 0 & 1 \end{array}\right] \begin{bmatrix} {}^{B}P \\ \hline 1 \end{bmatrix}. \tag{4}$$

That is, a "1" is added as the last element of the 4 × 1 vectors, and a row "[0 0 0 1]" is added as the last row of the 4 × 4 matrix. This construction allows the manipulation of position and orientation to be cast in the form of a matrix operator.

2. FORWARD KINEMATICS OF MANIPULATORS

Kinematics is the science of motion that treats motion without regard to the forces that cause it. Within the science of kinematics one studies the position, velocity, acceleration, and all higher-order derivatives of the position variables [with respect to time or any other variable(s)]. Hence, the study of the kinematics of manipulators refers to all the geometrical and time-based properties of the motion.

Manipulators consist of nearly rigid *links* that are connected with *joints,* which allow relative motion of neighboring links. These joints are usually instrumented with position sensors that allow the relative position of neighboring links to be measured. In the case of rotary or *revolute* joints, these displacements are called *joint angles.* Some manipulators contain sliding, or *prismatic,* joints in which the relative displacement between links is a translation, sometimes called the *joint offset.*

The number of *degrees of freedom* that a manipulator possesses is the number of independent position variables that would have to be specified in order to locate all parts of the mechanism. This is a general term used for any mechanism. For example, the well-known, so-called "four-bar linkage" (see Fig. 1) has only one degree of freedom (even though there are three moving members). In the case of typical industrial robots, because a manipulator is usually an open kinematic chain, and because each joint position is usually defined with a single variable, the number of joints equals the number of degrees of freedom.

At the free end of the chain of links that make up the manipulator is the *end-effector.* Depending on the intended application of the robot, the end-effector may be a gripper, welding torch, electromagnet, or other device. We generally describe the position of the manipulator by giving a description of the *tool frame,* which is attached to the end-effector, relative to the base *frame,* which is attached to the nonmoving base of the manipulator (see Fig. 2).

A very basic problem in the study of mechanical manipulation is that of *forward kinematics.* This is the static geometrical problem of computing the position and orientation of the end-effector of the manipulator. Specifically, given a set of joint angles, the forward kinematic problem is to compute the position and orientation of the tool frame relative to the base frame. Sometimes we think of this as changing the representation of manipulator position from a *joint-space* description into a *Cartesian-space* description. By Cartesian space we mean the space in which the position of a point is given with three numbers, and in which the orientation of a body is given with three numbers. It is sometimes called *task space* or *operational space.*

The forward kinematic formulation is generally accomplished as follows. A *link frame* is attached to each link following the so-called *Denavit–Hartenberg* convention (Denavit and Hartenberg, 1955). This convention

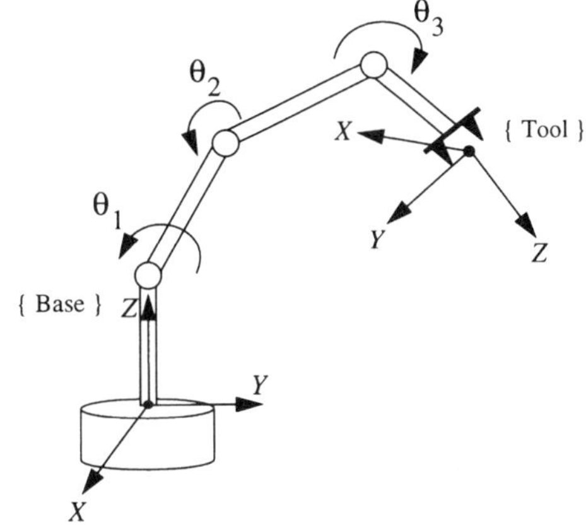

FIG. 2. Kinematic equations describe the tool frame relative to the base frame as a function of the joint variables.

specifies rules for locating link frames on links and defines four parameters that are sufficient to describe each link frame in terms of the preceding link frame. Figure 3 shows two successive link frames as well as the Denavit–Hartenberg parameters, a_i, α_i, d_i, and θ_i. The first two parameters, a_i and α_i, suffice to describe the relative spatial relationship between the lines that lie along successive joint axes. The parameter a_i, called the *link length,* is the length of the (usually unique) common perpendicular between the successive axes. The parameter α_i, called the *link twist,* is the angle between the two successive axes when projected onto a plane whose normal is the common perpendicular between the axes. The other two parameters, d_i and θ_i, have to do with the interconnection of links at joint i. The parameter d_i, called the *joint offset,* specifies the distance between the common perpendiculars describing the previous and following links at joint i. Finally, the parameter θ_i, called the *joint angle,* specifies the angle between successive common perpendiculars measured in a plane whose normal is joint axis i.

If the link frames have been attached to the links according to this convention, the following definitions of the link parameters are valid:

a_i = the distance from $\hat{Z}_i$ to $\hat{Z}_{i+1}$ measured along $\hat{X}_i$;

α_i = the angle between $\hat{Z}_i$ and $\hat{Z}_{i+1}$ measured about $\hat{X}_i$;

d_i = the distance from $\hat{X}_{i-1}$ to $\hat{X}_i$ measured along $\hat{Z}_i$; and

θ_i = the angle between $\hat{X}_{i-1}$ and $\hat{X}_i$ measured about $\hat{Z}_i$.

We usually choose $a_i > 0$ since it corresponds to a distance; however, α_i, d_i, and θ_i are signed quantities.

Given the Denavit–Hartenberg parameters for a mechanism, the transform that locates the ith frame in terms of the $(i - 1)$st frame is given by

$$
{}^{i-1}_{\ i}T = \begin{bmatrix} c\theta_i & -s\theta_i & 0 & a_{i-1} \\ s\theta_i c\alpha_{i-1} & c\theta_i c\alpha_{i-1} & -s\alpha_{i-1} & -s\alpha_{i-1}d_i \\ s\theta_i s\alpha_{i-1} & c\theta_i s\alpha_{i-1} & c\alpha_{i-1} & c\alpha_{i-1}d_i \\ 0 & 0 & 0 & 1 \end{bmatrix}, \tag{5}
$$

where $c\theta_i \equiv \cos\theta_i$, $s\theta_i \equiv \sin\theta_i$, and so on. Note that for any one link transformation, three of the four parameters are constants, and the fourth is variable. In the case of a revolute joint, θ_i is the joint variable, and in the case of a prismatic joint, d_i is the joint variable.

Finally, the forward kinematics of a manipulator is then easily found by concatenating the link transforms to compute

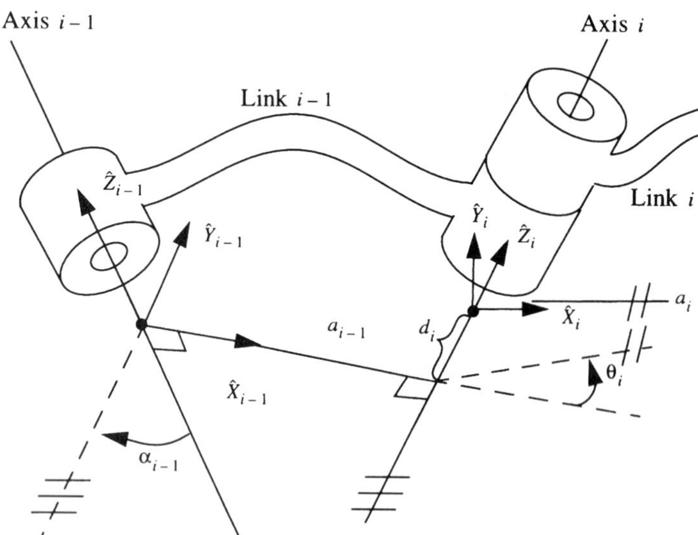

FIG. 3. Link frames are attached so that frame *i* is attached to link *i*, and has its *Z* axis aligned with the axis of rotation of joint *i*.

$$ {}^{0}_{N}T = {}^{0}_{1}T\,{}^{1}_{2}T\,{}^{2}_{3}T \cdots {}^{N-1}_{N}T. \quad (6) $$

This transformation, ${}^{0}_{N}T$, will be a function of all n joint variables. If the robot's joint position sensors are queried, the Cartesian position and orientation of the last link may be computed by ${}^{0}_{N}T$.

3. INVERSE KINEMATICS OF MANIPULATORS

The problem of *inverse kinematics* is posed as follows: Given the position and orientation of the end-effector of the manipulator, calculate all possible sets of joint angles that could be used to attain this given position and orientation (see Fig. 4). This is a fundamental problem in the practical use of manipulators.

The inverse kinematic problem is not as simple as the forward kinematics. Because the kinematic equations are nonlinear, their solution is not always easy or even possible in a closed form. Also, the questions of existence of a solution, and of multiple solutions, arise.

The existence or nonexistence of a kinematic solution defines the *workspace* of a given manipulator. The lack of a solution means that the manipulator cannot attain the desired position and orientation because it lies outside of the manipulator's workspace.

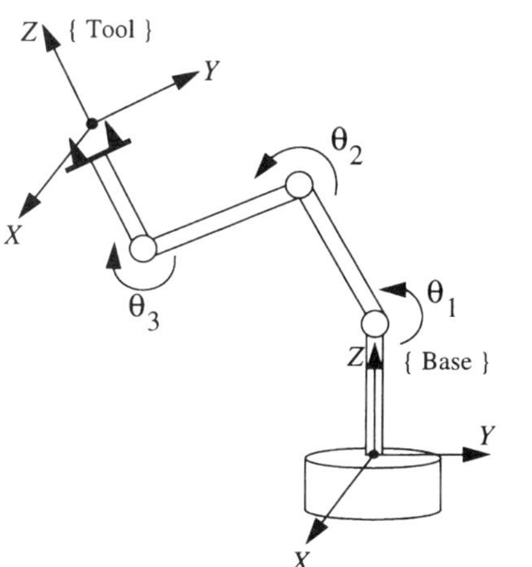

FIG. 4. For a given position and orientation of the tool frame, values for the joint variables can be calculated using the inverse kinematics.

Many industrial robots have six degrees of freedom, three of which form a *wrist* in which the last three axes of the robot are mutually orthogonal and intersect at a point. In these cases, the solution of the kinematic equations can be split into two problems: Solve for the first three joint variables in order to position the wrist point, and then solve for the final three joint angles to attain the desired orientation of the end-effector. The wrist point, or origin of the fourth link coordinate system, is given in the base system by

$$ {}^{0}P_{4ORG} = {}^{0}_{1}T\,{}^{1}_{2}T\,{}^{2}_{3}T\,{}^{3}P_{4ORG}, \quad (7) $$

where the left-hand side is given, and the right-hand side is a function of the joint variables of the first three degrees of freedom.

Solving a nonlinear set of equations such as Eq. (7) is difficult in general (Pieper and Roth, 1969; Tsai and Morgan, 1984). A sketch of a typical approach is as follows. Through algebraic manipulation, $n - 1$ of the n variables are eliminated, resulting in a transcendental equation in terms of the one remaining variable. Then, the trigonometric substitutions

$$ u = \tan(\theta/2), $$

$$ \cos\theta = \frac{1 - u^2}{1 + u^2}, $$

$$ \sin\theta = \frac{2u}{1 + u^2}, $$

are used to convert the problem from a transcendental equation into a polynomial in powers of u. Finally, the roots of the polynomial are found to solve for one variable, and the other variables are generally then easily found through back substitution. Multiple roots of the polynomial in u correspond to multiple physical configurations of the robot that all yield the same position of the end-effector. Since quartic (and lower-order) polynomials can be solved in closed form, mechanisms whose kinematic equations can be manipulated into polynomials of order 4 or lower are called *closed-form solvable.* It turns out that what may seem to be relatively minor or subtle changes to the design of the mechanical configuration of the linkages can have a dramatic effect on the order of this "characteristic polynomial." Robots with more than six degrees of freedom are in a whole other class of device characterized by an infinity of solutions for a given end-point positioning problem (Baker and Wampler, 1988).

4. VELOCITIES, STATIC FORCES, SINGULARITIES

In addition to dealing with static positioning problems, we may wish to analyze manipulators in motion (Hunt, 1978). Often in performing velocity analysis of a mechanism it is convenient to define a matrix quantity called the *Jacobian* of the manipulator. The Jacobian specifies a *mapping* from velocities in joint space to velocities in Cartesian space (see Fig. 5). The nature of this mapping changes as the configuration of the manipulator varies. At certain points, called *singularities,* this mapping is not invertible. An understanding of the phenomenon is important to designers and users of manipulators.

In the field of robotics, we generally speak of Jacobians that relate joint velocities to Cartesian velocities of the tip of the arm. For example,

$$\mathcal{V} = J(\Theta)\dot{\Theta}, \tag{8}$$

where Θ is the vector of joint angles of the manipulator and $\mathcal{V}$ is a vector of Cartesian velocities. This 6×1 Cartesian velocity vector is the 3×1 linear velocity vector and the 3×1 rotational velocity vector stacked together:

$$\mathcal{V} = \begin{bmatrix} v \\ w \end{bmatrix}. \tag{9}$$

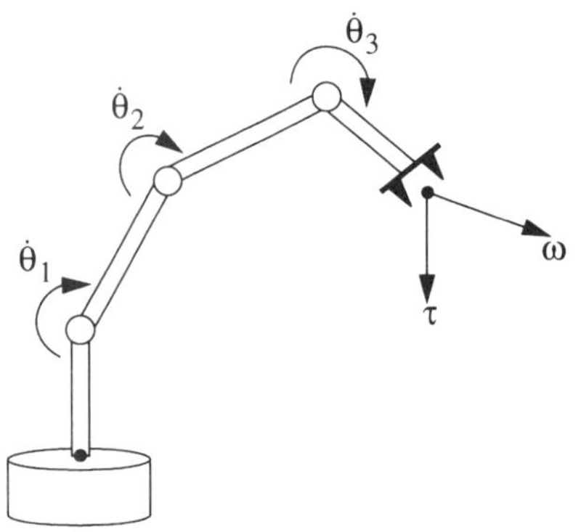

FIG. 5. The geometrical relationship between joint rates and velocity of the end-effector can be described in a matrix called the Jacobian.

Given that we have a linear transformation relating joint velocity to Cartesian velocity, a reasonable question to ask is: Is this matrix invertible? That is, is it nonsingular? If the matrix is nonsingular then we can invert it to calculate joint rates given Cartesian velocities:

$$\dot{\Theta} = J^{-1}(\Theta)\mathcal{V}. \tag{10}$$

For example, say we wish the hand of the robot to move with a certain velocity vector in Cartesian space. Using Eq. (10) we could calculate the necessary joint rates at each instant along the path.

Most manipulators have values of Θ where the Jacobian becomes singular. Such locations are called *singularities of the mechanism* or *singularities* for short. All manipulators have singularities at the boundary of their workspaces, and most have loci of singularities inside their workspaces—for example, in special configurations of the mechanism in which two revolute joints become aligned. In such configurations, the motion of either joint causes the same motion at the end point of the kinematic chain, and so it is as if the device has lost one degree of freedom.

It turns out that the Jacobian also appears in the relationship between static forces applied at the end-effector and static actuator torques (or forces) applied at the joints. It can be shown that

$$\tau = J^T\mathcal{F}, \tag{11}$$

where $\mathcal{F}$ is a 6×1 Cartesian force-moment vector acting at the end-effector, and τ is an $n \times 1$ vector of torques at the joints (for an n-jointed robot).

Manipulators do not always move through space; sometimes they are also required to contact a workpiece or work surface and apply a static force. In this case the problem arises: Given a desired contact force and moment, what set of *joint torques* is required to generate them? Once again, the Jacobian matrix of the manipulator arises quite naturally in the solution of this problem.

5. DYNAMICS

Dynamics is a huge field of study devoted to studying the forces required to cause motion. In order to make a manipulator accelerate from rest, glide at a constant end-effector velocity, and finally decelerate to a stop, a complex set of torque functions must be applied by the *joint actuators*. We use joint actuators as the generic term for devices that power a manipulator; for example, electric motors, hydraulic and pneumatic actuators, muscles, etc. The exact form of the required functions of actuator torque depends on the spatial and temporal attributes of the path taken by the end-effector as well as the mass properties of the links and payload, friction in the joints, etc. One method of controlling a manipulator to follow a desired path involves calculating these actuator torque functions using the dynamic equations of motion of the manipulator.

Using any of several techniques to derive the equations of motion for a mechanical manipulator (Luh *et al.*, 1980), it is found that these equations can be written in the form

$$\tau = M(\Theta)\ddot{\Theta} + V(\Theta,\dot{\Theta}) + G(\Theta), \tag{12}$$

where $M(\Theta)$ is the $n \times n$ *mass matrix* of the manipulator, $V(\Theta,\dot{\Theta})$ is an $n \times 1$ vector of centrifugal and Coriolis terms, $G(\Theta)$ is an $n \times 1$ vector of gravity terms, and τ is the $n \times 1$ vector of joint torques.

Each element of $M(\Theta)$ and $G(\Theta)$ is a complex function that depends on Θ, the position of all the joints of the manipulator. Each element of $V(\Theta,\dot{\Theta})$ is a complex function of both Θ and $\dot{\Theta}$. For an alternative formulation of manipulator dynamics, see Khatib (1985).

Another use of the dynamic equations of motion is in *simulation*. By reformulating the dynamic equations so that acceleration is computed as a function of actuator torque, it is possible to simulate how a manipulator would move under application of a set of actuator torques (see Fig. 6):

$$\ddot{\Theta} = M^{-1}(\Theta)[\tau - V(\Theta,\dot{\Theta}) - G(\Theta)]. \tag{13}$$

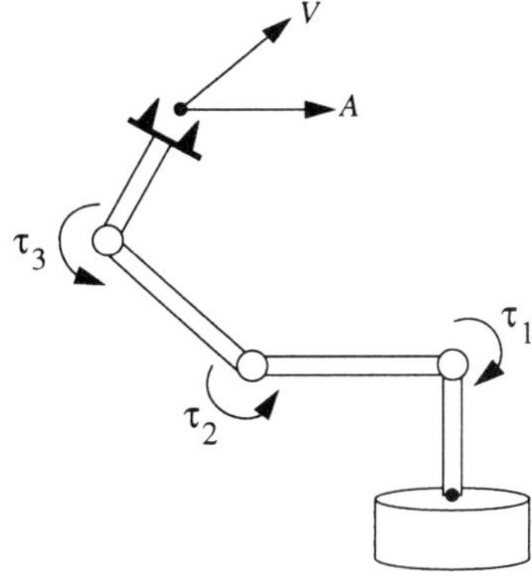

FIG. 6. The relationship between torques applied by the actuators and the resulting motion of the manipulator is embodied in the dynamic equations of motion.

We may then apply any of several known *numerical integration* techniques to integrate the acceleration to compute future positions and velocities.

6. TRAJECTORY GENERATION

A common way of causing a manipulator to move from here to there in a smooth, controlled fashion is to cause each joint to move as specified by a smooth function of time. Commonly, each joint starts and ends its motion at the same time, so that the manipulator motion appears coordinated. Exactly how to compute these motion functions is the problem of *trajectory generation* (see Fig. 7).

Often a path is described not only by a desired destination but also by some intermediate locations, or *via points*, through which the manipulator must pass en route to the destination. In such instances the term *spline* is sometimes used to refer to a smooth function that passes through a set of via points.

In order to force the end-effector to follow a straight line (or other geometric shape) through space, the desired motion must be converted to an equivalent set of joint motions. This *Cartesian trajectory generation* is a feature of most industrial robot control systems (Paul, 1981).

Whereas there is never a problem moving along smooth trajectories that are described

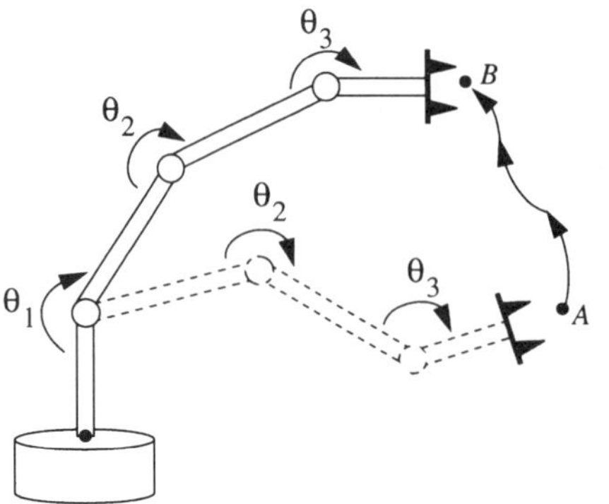

FIG. 7. In order to move the end-effector through space from point A to point B we must compute a trajectory for each joint to follow.

in terms of individual joint motions, when we impose an external path shape (such as straight-line) problems can arise.

Figure 8(a) illustrates that even though the initial and final positions of the robot are within the workspace, intermediate points may not be. A second type of problem with a straight-line path is shown in Fig. 8(b). In this case, the manipulator passes near a singularity that causes a very high joint velocity at joint 1 in order to maintain a constant tool tip velocity along the path. The closer the path comes to a singularity, the faster one or more joints are required to move. Finally, Fig. 8(c) indicates a third class of problem: Although all path points are reachable, they are not all reachable in the same kinematic solution. In this case, the path cannot be traversed. The problems illustrated in Fig. 8 are due to the realities of physical mechanisms, and are one cause of difficulty in programming robots to achieve useful tasks.

When a geometric model of the manipulator environment is available, advanced techniques can plan *collision-free paths* automatically (Lozano-Perez, 1987; Khatib, 1986). However, such functionality is presently not typically available in commercial robot systems.

7. MANIPULATOR DESIGN AND SENSORS

Although manipulators are in theory universal devices applicable to many situations, generally economics dictates that the intended task domain influence the mechanical design of the manipulator. Along with issues such as size, speed, and load capability, the designer must also consider the number of joints and their geometric arrangement. These considerations influence the manipulator's workspace size and quality, the stiffness of the manipulator structure (see Fig. 9), and other attributes.

Integral to the design of the manipulator are issues involving the choice and location of actuators, transmission systems, and internal position (and sometimes force) sensors.

8. POSITION CONTROL

Some manipulators are equipped with stepper motors or other actuators that can directly execute a desired trajectory. However, the vast majority of manipulators are driven by actuators that supply a force or a torque to cause motion of the links. In this case, an algorithm is needed to compute torques that will cause the desired motion. The problem of dynamics is central to the design of such algorithms but does not in itself constitute a solution. A primary concern of a *position control system* is to compensate

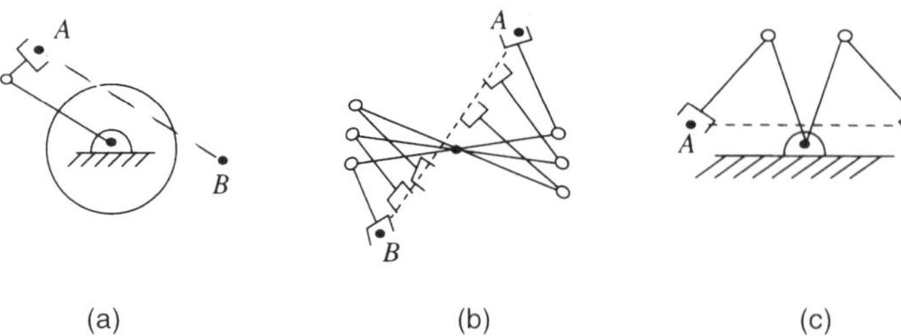

FIG. 8. Trajectory problem (a) intermediate points unreachable. Trajectory problem (b) high joint rates near singularity. Trajectory problem (c) no common solution for all path points.

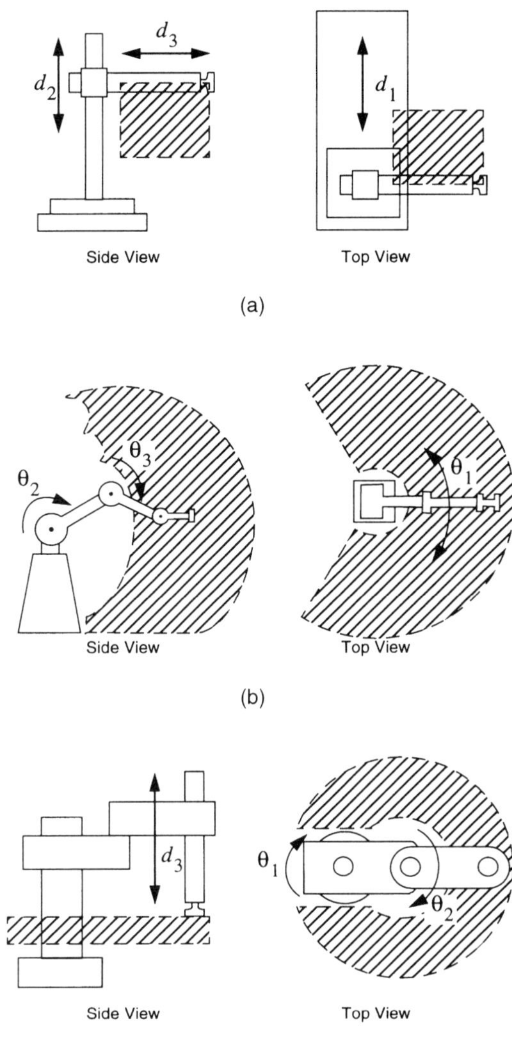

FIG. 9. Three common designs for robot mechanisms are Cartesian, Articulated, and Scara. Each type is characterized by a differently shaped workspace (as indicated) as well as other differentiating properties such as structural stiffness, positioning accuracy, etc.

automatically for errors in knowledge of the parameters of a system, and to suppress disturbances that tend to perturb the system from the desired trajectory (Franklin *et al.*, 1986). To accomplish this, position and velocity *sensors* are monitored by the *control algorithm,* which computes torque commands for the actuators (see Fig. 10). Many control systems are synthesized using linear approximate models of the dynamics of a

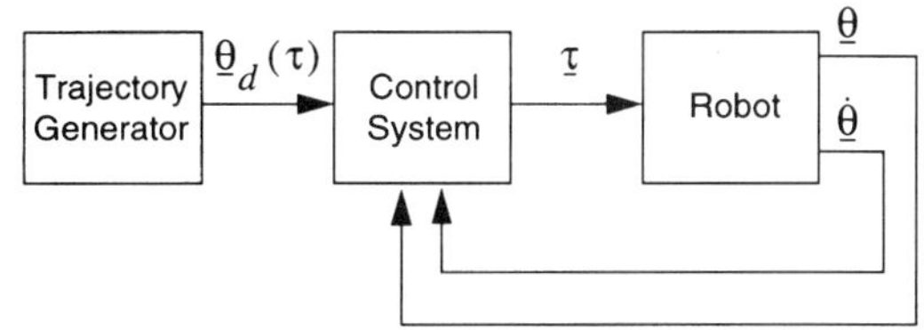

FIG. 10. In order to cause the manipulator to follow the desired trajectory, a position control system must be implemented. Such a system uses feedback from joint sensors to keep the manipulator on course.

manipulator. These linear methods are prevalent in current industrial practice.

Although control systems based on approximate linear models are popular in most current industrial robots, it is important to consider the complete nonlinear dynamics of the manipulator when synthesizing control algorithms. Some industrial robots now make use of *nonlinear control* algorithms in their controllers. These nonlinear techniques of controlling a manipulator promise better performance than do simpler linear schemes (Khosla, 1988; Khatib, 1987).

For a manipulator whose dynamics are given by

$$\tau = M(\Theta)\ddot{\Theta} + V(\Theta,\dot{\Theta}) + G(\Theta), \tag{14}$$

where $M(\Theta)$ is the $n \times n$ inertia matrix of the manipulator, $V(\Theta,\dot{\Theta})$ is an $n \times 1$ vector of centrifugal and Coriolis terms, and $G(\Theta)$ is an $n \times 1$ vector of gravity terms, a nonlinear *control law* computes torques to send to the actuators as

$$\tau = \alpha\tau' + \beta, \tag{15}$$

where τ is the $n \times 1$ vector of joint torques. We choose

$$\alpha = M(\Theta), \tag{16a}$$

and

$$\beta = V(\Theta,\dot{\Theta}) + G(\Theta), \tag{16b}$$

with the servo law

$$\tau' = \ddot{\Theta}_d + K_v\dot{E} + K_pE, \tag{17}$$

where

$$E = \Theta_d - \Theta, \quad (18)$$

where Θ_d is the vector of desired joint positions, usually a function of time, and K_v and K_p are diagonal constant matrices with values selected by the control system designer for good performance. Using the above relations it is quite easy to show that the closed-loop system is characterized by the error equation

$$\ddot{E} + K_v\dot{E} + K_pE = 0. \quad (19)$$

Note that this vector equation is decoupled since the matrices K_v and K_p are diagonal so that Eq. (19) could just as well be written on a joint-by-joint basis as

$$\ddot{e}_i + k_{vi}\dot{e} + k_{pi}e = 0. \quad (20)$$

This is ideal second-order performance, as a result of the nonlinear decoupling and linearizing control above. In practice this is difficult to achieve because of imperfect knowledge of the parameters that appear in Eq. (16). The problem of imprecise knowledge of system parameters can sometimes be dealt with by means of *adaptive control* (Craig *et al.,* 1986).

9. FORCE CONTROL

The ability for a manipulator to control forces of contact when it touches parts, tools, or work surfaces seems to be of great importance in applying manipulators to many real-world tasks. *Force control* is complementary to position control in that we usually think of one or the other as applicable in a certain situation (Mason, 1978). When a manipulator is moving in free space, only position control makes sense, since there is no surface to react against. When a manipulator is touching a rigid surface, however, position control schemes can cause excessive forces to build up at the contact or may cause contact to be lost with the surface when it was desired for some application. Since manipulators are rarely constrained by reaction surfaces in all directions simultaneously, using a mixed or *hybrid* control (Raibert and Craig, 1981) is required, with some directions controlled by a *position control law* and remaining directions controlled by a *force control law.*

In Fig. 11, the task of putting a peg into a hole generally requires some form of force sensing and control. Even a seemingly simple task like this must be broken into a succession of several distinct regimes, each with a corresponding control strategy. In free space approaching the surface, position control is used. While sliding across the surface searching for the hole, force control would be used in the Z direction to maintain contact. Once started in the hole, force control would typically be used in the X and Y directions to avoid excessive side forces that might lead to jamming or wedging.

10. COMPUTER VISION

Increasingly, robots are being used with various sensors so that they can better adapt to uncertainties in their environment. *Computer vision* encompasses a suite of techniques that are used to identify, locate, or inspect objects. Many such techniques operate

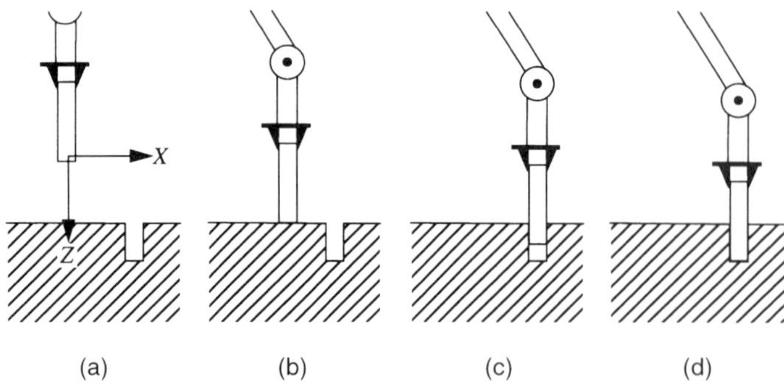

FIG. 11. In order for a manipulator to perform tasks that require contact with the environment (such as inserting a peg in a hole) the robot must be able to sense and control contact forces.

using a grayscale image acquired by a video camera. This image is then processed by various algorithms either to locate an object, to inspect an object for features, and/or to identify an object from a set of many possible objects. One simple technique uses *thresholding* to convert the grayscale image into strictly white or black regions, which are then analyzed for their size and shape. Often a *training phase* is done with the system first to train it to recognize the characteristics of the part to be identified. Another technique operates on the grayscale image to perform *edge detection* in order to convert the image into a set of lines or curves. These can then be matched to an expected set of edges in order to do object recognition, localization, or inspection. Some robot systems also use stereo vision (using two cameras) and solve a *correspondence* problem to identify features in each of the two views.

Computer vision techniques allow for more variability in the robot's workspace. For example, some arc-welding robots use a vision system to track the weld seam. Hence, if the parts have moved slightly, the operation can still proceed successfully. Robots without sensors must perform all operations by *dead reckoning,* which may lead to failures due to minor misalignments or changes in the environment. Similarly, robots use vision systems to see parts on conveyor belts, to find fiducial points on car bodies for alignment purposes, to locate a single object in a bin of parts, and so on.

11. LOCOMOTION

The vast majority of industrial robots are essentially analogous to a human arm mounted on the floor, or on a workbench. An important exception in current industrial practice is the AGV or *automatic guided vehicle.* AGVs are automatically guided carts that deliver parts in factories, or mail in some offices, or perform other operations in which mobility is required.

Although still rare, some robots have both mobility and one or more manipulators. There are also undersea and prototype outer-space robots that have both locomotion and manipulation abilities. Along with wheeled vehicles, there have been *legged* robots with one, two, four, or more legs (Raibert, 1986). These systems offer the promise of providing exploration or reconnaissance abilities in environments where it has traditionally been difficult for unmanned vehicles to go.

12. PROGRAMMING ROBOTS

A *robot programming language* serves as the interface between the human user and the industrial robot (Gruver and Soroka, 1988). Central questions arise such as: How are motions through space described easily by the programmer? How are multiple manipulators programmed so that they can work in parallel? How are sensor-based actions described in a language?

Robot manipulators differentiate themselves from fixed automation by being "flexible," which means programmable. Not only are the movements of manipulators programmable, but through the use of sensors and communications with other factory automation, manipulators can adapt to variations as the task proceeds.

The sophistication of the user interface is becoming extremely important as manipulators and other programmable automation are applied to more and more demanding industrial applications. The problem of programming manipulators encompasses all the issues of "traditional" computer programming, and so is an extensive subject in itself. Additionally, some particular attributes of the manipulator programming problem cause additional issues to arise.

13. OFF-LINE PROGRAMMING AND SIMULATION

An *off-line programming system* is a robot programming environment that has been sufficiently extended, generally by means of computer graphics, that the development of robot programs can take place without access to the robot itself (SILMA, 1995). A common argument raised in their favor is that an off-line programming system will not cause production equipment (i.e., the robot) to be tied up when it needs to be reprogrammed; hence, automated factories can stay in production mode a greater percentage of the time (see Fig. 12).

It also serves as a natural vehicle to tie

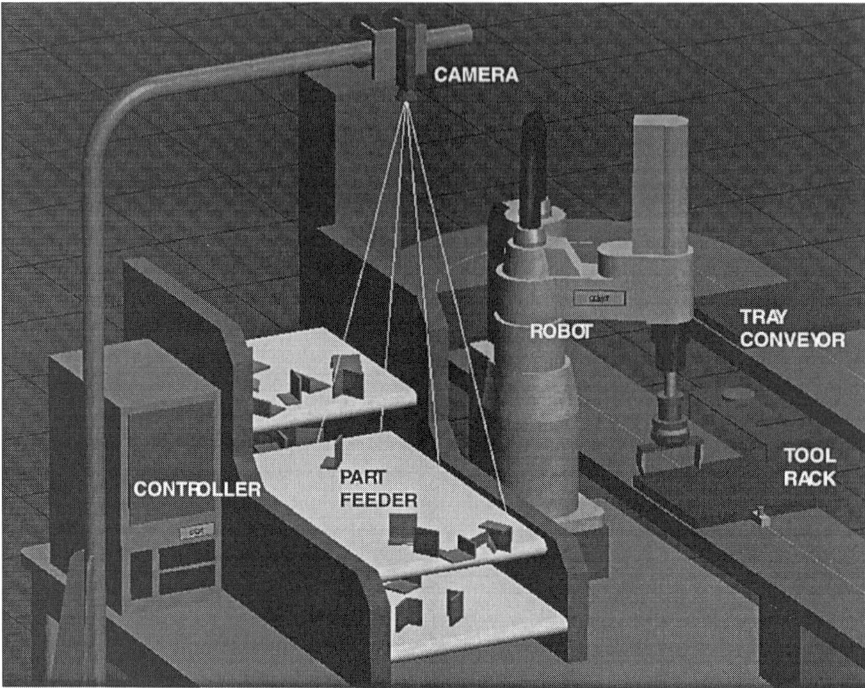

FIG. 12. Off-line programming (OLP) systems allow the programmer to interact with a 3-D simulated virtual world in order to program a robot.

computer-aided design (CAD) databases used in the design phase of a product to the actual manufacturing of the product. In some cases, this direct use of CAD data can dramatically reduce the programming time required for the manufacturing process.

GLOSSARY

Edge Detection: A technique used in computer vision in which changes in image intensity are linked together spatially to find image features that correspond to edges of objects in the image.

End-Effector: The gripper, tool, or other (generally active) object at the distal end of a manipulator.

Force Control: A style of controlling a manipulator such that the forces of contact (between the end-effector and other objects in the robot's environment) are directly controlled.

Homogeneous Transform: A 4×4 matrix containing both position and orientation information that specifies one coordinate system relative to a reference coordinate system.

Jacobian: Shortened name for the Jacobian matrix, a linear operator that maps joint velocities into linear and rotational velocities of the end point of a manipulator.

Kinematics: The study of position and motion (e.g., velocity, acceleration, etc.) without regard to the forces that cause the motion.

Mass Matrix: For an N-jointed manipulator, the $N \times N$ matrix that is a function of manipulator configuration and describes the mass distribution of all the linkages.

Off-Line Programming: The use of a computer language or system (generally augmented by the use of computer graphics) such that a robot may be programmed without access to the robot itself during programming.

Prismatic Joint: A joint between links that allows linear motion.

Revolute Joint: A joint between links that allows rotary motion.

Rotation Matrix: A matrix, usually of dimension 3 × 3, that describes the orientation of one coordinate system relative to a reference coordinate system.

Singularity: In robotics, a configuration of the mechanical linkage at which the Jacobian matrix loses full rank.

Works Cited

Baker, D., Wampler, C. (1988), "On the Inverse Kinematics of Redundant Manipulators," *Int. J. Rob. Res.* **7** (2), 3–21.

Craig, J., Hsu, P., Sastry, S. (1986), "Adaptive Control of Mechanical Manipulators," in: *Proceedings of the IEEE International Conference on Robotics and Automation, San Francisco,* Washington, DC: IEEE Computer Society Press.

Denavit, J., Hartenberg, R. S. (1955), "A Kinematic Notation for Lower-Pair Mechanisms Based on Matrices," *J. Appl. Mech.* **22,** 215–221.

Franklin, G., Powell, J. D., Emami-Naeini, A. (1986), *Feedback Control of Dynamic Systems,* Reading, MA: Addison-Wesley.

Gruver, W., Soroka, B. (1988), "Programming, High Level Languages," in: R. Dorf, S. Nof (Eds.), *The International Encyclopedia of Robotics,* New York: Wiley Interscience.

Hunt, K. (1978), *Kinematic Geometry of Mechanisms,* Cambridge, U.K.: Oxford Univ. Press.

Khatib, O. (1986), "Real-Time Obstacle Avoidance for Manipulators and Mobile Robots," *Int. J. Rob. Res.* **5** (1), 90–98.

Khatib, O. (1987), "A Unified Approach for Motion and Force Control of Robot Manipulators: The Operational Space Formulation," *IEEE J. Rob. Autom.* **RA-3** (1), 43–53.

Khosla, P. K. (1988), "Some Experimental Results on Model-Based Control Schemes," in: T. Pavlidis (Ed.), *Proceedings of the IEEE International Conference on Robotics and Automation, Philadelphia,* 3 vols., Washington, DC: IEEE Computer Society Press, p. 1380.

Lozano-Perez, T. (1987), "A Simple Motion Planning Algorithm for General Robot Manipulators," *IEEE J. Rob. Autom.* **RA-3** (3), 224–238.

Luh, J. Y. S., Walker, M. W., Paul, R. P. (1980), "On-Line Computational Scheme for Mechanical Manipulators," *J. Dyn. Syst. Meas. Control* **102** (2), 69–76.

Mason, M. (1978), "Compliance and Force Control for Computer Controlled Manipulators," M.S. Thesis, MIT AI Laboratory.

Paul, R. P. (1981), *Robot Manipulators: Mathematics, Programming, and Control,* Cambridge, MA: MIT Press.

Pieper, D., Roth, B. (1969), "The Kinematics of Manipulators under Computer Control," *Proceedings of the Second International Congress on Theory of Machines and Mechanisms, Zakopane, Poland,* Warsaw: Polytechnic Institute of Warsaw, Vol. 2.

Raibert, M., Craig, J. (1981), "Hybrid Position/Force Control of Manipulators," *J. Dyn. Syst. Meas. Control* **103** (2), 126–133.

Raibert, M. (1986), *Legged Robots That Balance,* Cambridge, MA: MIT Press.

SILMA Inc. (1995), "The CimStation User's Manual," Version 5.0, Available from SILMA Inc., 1601 Saratoga-Sunnyvale Rd., Cupertino, Calif., 95014.

Tsai, L., Morgan, A. (1984), "Solving the Kinematics of the Most General Six- and Five-Degree-of-Freedom Manipulators by Continuation Methods," Paper 84-DET-20, ASME Mechanisms Conference, Boston.

Further Reading

Craig, J. J. (1989), *Introduction to Robotics: Mechanics and Control,* 2nd ed., Reading, MA: Addison-Wesley.

Fu, K., Gonzalez, R., Lee, C. S. G. (1987), *Robotics: Control, Sensing, Vision, and Intelligence,* New York: McGraw-Hill.

Rivin, E. (1988), *Mechanical Design of Robots,* New York: McGraw-Hill.

ROCKS AND PHYSICAL GEOLOGY

See STRUCTURAL AND PHYSICAL GEOLOGY

SAFETY AND HEALTH IN THE WORKPLACE

See WORKPLACE HEALTH AND SAFETY

SATELLITES, ARTIFICIAL

STEPHEN J. PADDACK AND BRUCE A. CAMPBELL, *Advanced Missions Analysis Office, Goddard Space Flight Center, National Aeronautics and Space Administration, Greenbelt, Maryland, U.S.A.*

ENRICO P. MERCANTI, *Goddard Space Flight Center, National Aeronautics and Space Administration (Ret.), Annapolis, Maryland, U.S.A.*

SAMUEL WALTER MCCANDLESS, *User Systems, Inc., Chesapeake Beach, Maryland, U.S.A.*

	Introduction	519
1.	**Physical and Technical Principles**	520
1.1	Orbit Principles	520
1.1.1	Orbits Used by Artificial Satellites	523
1.2	Space Environment	524
1.2.1	Orbital Debris	526
2.	**Applications of Artificial Satellites**	527
2.1	Manned Spacecraft	527
2.2	Unmanned Satellites	527
2.2.1	Earth Observation	527
2.2.2	Space Environment	528
2.2.3	Planetary Exploration	529
2.2.4	Space Exploration	530
2.2.5	Commercial Satellites	531
2.2.6	Military Satellites	531
3.	**Design, Function, and Operation of Artificial Satellites**	532
3.1	Satellite Design Phases and Processes	532
3.1.1	Conceptual Study Phase	533
3.1.2	Mission Analysis Phase	533
3.1.3	Definition Phase	533
3.1.4	Design and Development Phase	534
3.1.5	Operations Phase	535
3.2	Satellite Subsystems and Functions	536
3.2.1	Power Generation and Distribution	537
3.2.2	Thermal Control	537
3.2.3	Attitude Determination and Control	537
3.2.4	Orbit Maintenance and Propulsion	537
3.2.5	Structure and Mechanisms	538
3.2.6	Communications, Command, Control, and Data Handling	538
3.3	Satellite Operations	538
3.4	Space Law	539
4.	**Artificial Satellite Manufacturing**	540
5.	**Economic Aspects of Artificial Satellites**	542
	Glossary	543
	Further Reading	543

INTRODUCTION

Artificial satellites are now associated with many aspects of everyday life. Communications satellites route thousands of long-distance telephone calls around the world continuously. Most of these satellites also distribute video signals for movie channels and news stations. The pictures of the weather conditions shown in these newscasts are also provided by satellites. Some of these remote-sensing satellites also aid in monitoring the world's resources (oil, forests and crops, water) as well as evaluating the environment and determining man's effect on it. Other artificial satellites are used for navigation around the globe, exploring the solar system, examining the universe around us, and even carrying people into space.

Artificial satellites may be grouped into two major categories: manned and unmanned. While many principles of design and operation are common between these two categories, the requirements associated with providing a suitable environment for persons to survive in space are unique to manned missions and significantly affect the configuration of the spacecraft. Within these

3-527-28138-X/96/$5.00 + .50

two major categories, there are a number of subcategories that can be defined, mostly in terms of the objective the satellite is to achieve.

Most manned missions to date have been limited to near-Earth operations of relatively short durations. However, Soviet/Russian space stations have been the home for many cosmonauts for extended periods, some up to and beyond a year. The Apollo missions to the Moon represent the furthest that man has ventured into space (about 250 000 km away) although the missions all lasted less than two weeks each.

Unmanned space missions are many and varied, calling for different designs of the satellites that perform them. Some major categories have already been mentioned: communications, remote sensing, navigation, and scientific. Within these categories, more specific definition of the uses can be made. For instance, remote sensing satellites include weather satellites, land-resource satellites, ocean- and atmosphere-sensing satellites, and even spy satellites. The types of scientific satellites run the gamut from Earth-orbiting space-environment sensors and astronomical telescopes to interplanetary craft.

Though these missions may vary widely, and the design of the spacecraft may depend heavily on its mission, a number of principles are common to just about all systems that operate in space. After discussing these common principles, some of the particular demands on the design and/or operation of artificial satellites are explored.

1. PHYSICAL AND TECHNICAL PRINCIPLES

Satellites are constrained to operate from orbits and must exist in the environment of space for periods of time dependent on their uses.

1.1 Orbit Principles

The path through space of an artificial satellite is caused primarily by gravitational forces. In the classical two-body problem the primary body is massive with respect to the satellite and no other forces act on the bodies. Given two bodies in space, a primary and a satellite, one of two things occurs. Either the bodies collide or there will be relative motion between the bodies that follows certain predictable paths. In most cases, the primary body is used as the center of reference and the path or trajectory of the satellite is referred to as its orbit. Prior to the 1950s the motion of bodies in space was observed by astronomers and analyzed and explained by celestial mechanicians and physicists interested in central force-field problems. Since the advent of rockets and artificial satellites, however, orbital mechanics has become a popular field of study and employment.

The names of the historical observers and theorists in this field include Copernicus, Brahe, Kepler, Galileo, and Newton. Copernicus made the bold statement, based on observations, that the Earth and the other planets revolve around the Sun. Brahe made the precise observations that allowed Kepler to derive the now well-known three laws of planetary motion. These laws state that a body in orbit around another 1) describes a conic section, 2) sweeps out equal areas in space in equal times, and 3) completes one orbit in a time, the orbital period, whose square is directly proportional to the cube of the semimajor axis of the orbit. Galileo performed many experiments on the motions of bodies, and observed moons in orbit around Jupiter using the first astronomical telescope. Newton used the findings of his predecessors to derive the fundamental equations of motions based on gravitational force and energy principles.

An orbit and the position of the orbiting body in the orbit are uniquely defined by a set of six orbital elements. Figure 1 shows how the size, shape, and position of an orbiting body are defined. The three of the orbital elements associated with these properties are the semimajor axis (a), eccentricity (e), and true anomaly (v).

Orbits are classified according to shape: circular, elliptical, parabolic, and hyperbolic, as shown in Fig. 2. Newtonian mechanics show that the kinetic- and potential-energy levels of an orbiting object in space require its being in one of the conic-section classes, thereby explaining the observations of Brahe and Kepler's first law.

Circular and elliptic orbits are called "closed" orbits because a body in closed or-

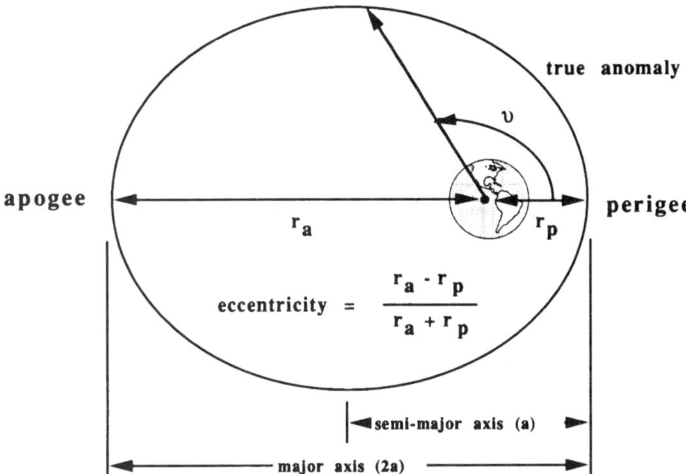

FIG. 1. Orbital elements defining the size and shape of an orbit, and position of a body in an orbit.

bit is constrained to revolve around its primary mass. A satellite in a closed Earth orbit, for example, remains revolving around its primary, the Earth. A body in parabolic orbit possesses the minimum energy required to escape from its primary. A body in hyperbolic orbit possesses more than the minimum energy required to escape. Satellites launched from the Earth with hyperbolic velocities with respect to the Earth are used for interplanetary space travel or to escape from the solar system.

The shape of an orbit is given by its eccentricity (e). The value of eccentricity for closed orbits is bounded to be equal to or greater than zero but less than unity. A circular orbit has an eccentricity of zero. Elliptical orbits have eccentricity values greater than zero but less than one. A parabolic orbit has an eccentricity of exactly unity. Hyperbolic orbits have eccentricities greater than unity. In summary,

$e = 0 \Rightarrow$ circular orbit,

$0 < e < 1 \Rightarrow$ elliptic orbits,

$e = 1 \Rightarrow$ parabolic orbit,

$e > 1 \Rightarrow$ hyperbolic orbit.

The orientation in space of an orbit is defined by the remaining three elements. These are the argument of perigee (ω); the right ascension of the ascending node (Ω); and the inclination (i). These take on meaning only when presented in a spatial reference frame. Figure 3 shows an inertially fixed geocentric

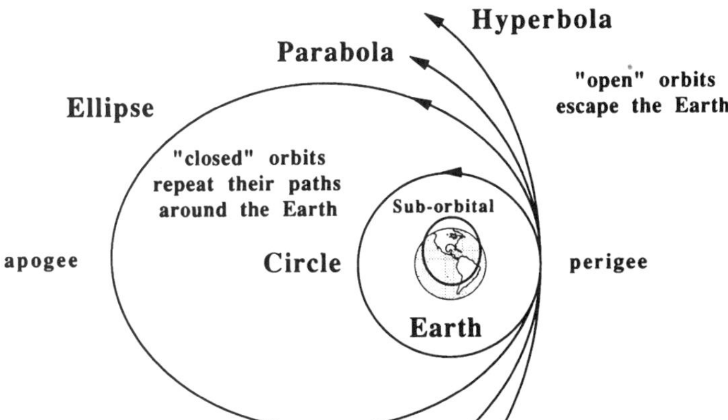

FIG. 2. Orbit types. Orbits around the Earth (or any other body) are conic sections.

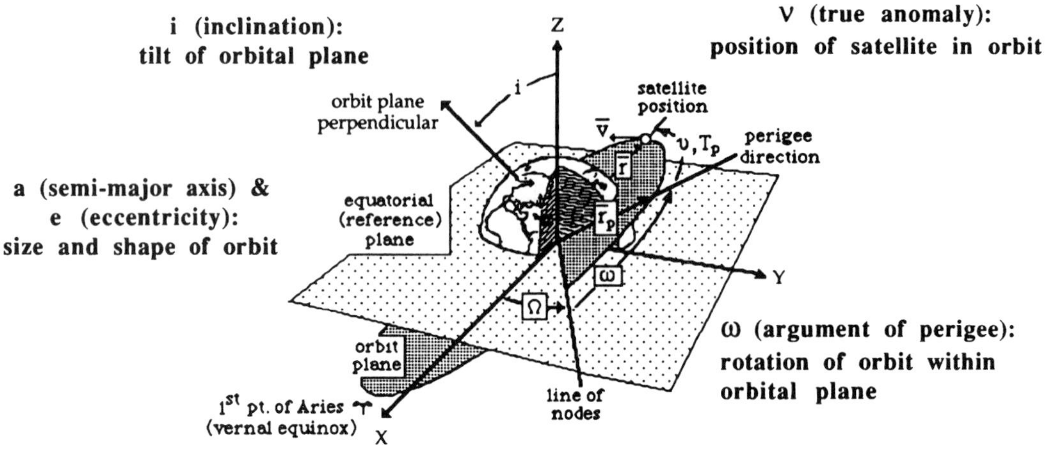

FIG. 3. Orbit description: the six "orbital elements" required to describe completely the orbit and position of a satellite in space.

Cartesian reference system where the fundamental (reference) plane is coincident with the Earth's equatorial plane and contains the X and Y axes. The positive X axis is directed to a specified point in space of astronomical significance. The position of the constellation Aries at a given time is often used. The Y axis is constructed perpendicular to the X axis. Finally, the Z axis is coincident with the Earth's spin axis with the positive direction defined as north. An orbit placed in this reference frame is also shown in the figure.

The inclination i is the angle between the plane of an orbit and the fundamental plane. The right ascension of the ascending node, Ω, is the angle in the fundamental plane measured from the positive X axis to the ascending nodal point (where an Earth-orbiting satellite pierces the equatorial plane when traveling from south to north). The right-hand rule is used for determining positive rotation. The argument of perigee, ω, is the angle measured in the plane of the orbit from the ascending node to the point of perigee.

These five parameters, a, e, i, Ω, ω, define the size, shape, and orientation of an orbit in space. The position of the satellite in the orbit can be given by a single angle that defines where the satellite is with respect to the point of perigee. This angle is called the true anomaly, v. The position of the satellite can be defined also by a time parameter, i.e., time from perigee, or the mean anomaly. The mean anomaly is based on the mean angular motion of the satellite.

In summary, the classical six orbital elements required to define completely an orbit and the position of a satellite in the orbit are

- a, semimajor axis;
- e, eccentricity;
- i, inclination;
- Ω, right ascension of the ascending node;
- ω, argument of perigee;
- v, true anomaly.

Kepler's second law is referred to as the areal law. It states simply that spatially equal areas are swept out in equal times. The most important thing this law reveals is that the speed of a satellite in an orbit will change as the orbital distance changes. Kepler's second law and its meaning to orbital velocities are shown in Fig. 4. Notice that, since its orbital distance does not change, the velocity of a satellite in a circular orbit is constant.

Kepler's third law is a mathematical statement that the time it takes for a body to complete one orbit is related to the size of the orbit. The time it takes to complete one orbit is called the orbital period, T. Later on,

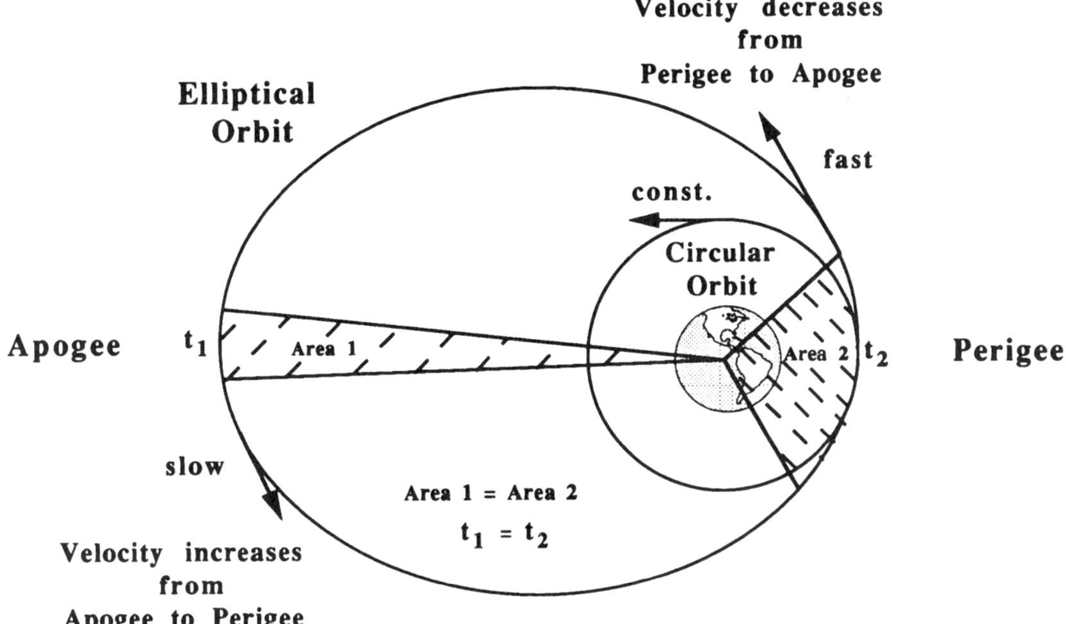

FIG. 4. Orbital velocities, illustrating Kepler's "areal" law, and its meaning for the velocity of a satellite in orbit around the Earth.

Newton was able to derive the constants associated with this relationship, which we can now describe as

$$T^2 = (4\pi^2/\mu)a^3,$$

in which μ represents the "gravitational parameter" for whatever body the satellite is orbiting. This law is constantly used by the orbit designer to satisfy the requirements of space-flight missions. One of the most obvious examples is the selection of an orbital distance giving an orbital period that precisely matches the rotation rate of the Earth, known as a "geosynchronous" orbit.

1.1.1 Orbits Used by Artificial Satellites The orbits chosen for different kinds of spacecraft depend on how the satellite is to be used. Some are low (altitude) Earth orbiting (LEO), some are geostationary (GEO), others are in high (altitude) Earth orbits (HEO), as depicted in Fig. 5. These terms, although often used in astronautics, are not uniquely defined except for the geostationary orbits. As a general rule, however, low Earth orbits are considered to be from the lowest altitude possible to allow continuing in orbit without reentering as a result of atmospheric drag to about 1000-km altitude.

Geostationary orbits are very specifically defined. As shown by Kepler's third law, close Earth-orbiting spacecraft have shorter orbital periods than spacecraft with larger orbits. It takes the Moon about 28 days to orbit the Earth once, but a spacecraft in LEO can complete one orbit in about an hour and a half. It is not difficult, then, to realize that there is a certain size orbit whose period is exactly one day, the geosynchronous orbit described earlier. If, in addition, the orbital plane of the spacecraft lies in the extended equatorial plane, and the orbit is circular (constant velocity), then the spacecraft will remain over a given longitudinal position on the Earth's equator. This special orbit is referred to as "geostationary" because a satellite in such an orbit remains fixed in the sky relative to any point on the Earth. Far from being stationary, the spacecraft is simply orbiting at an angular velocity that matches the Earth's rotational rate.

High Earth orbits, like low Earth orbits, are not uniquely defined. Rather, the term HEO is used to refer to orbits that are neither LEOs nor GEOs.

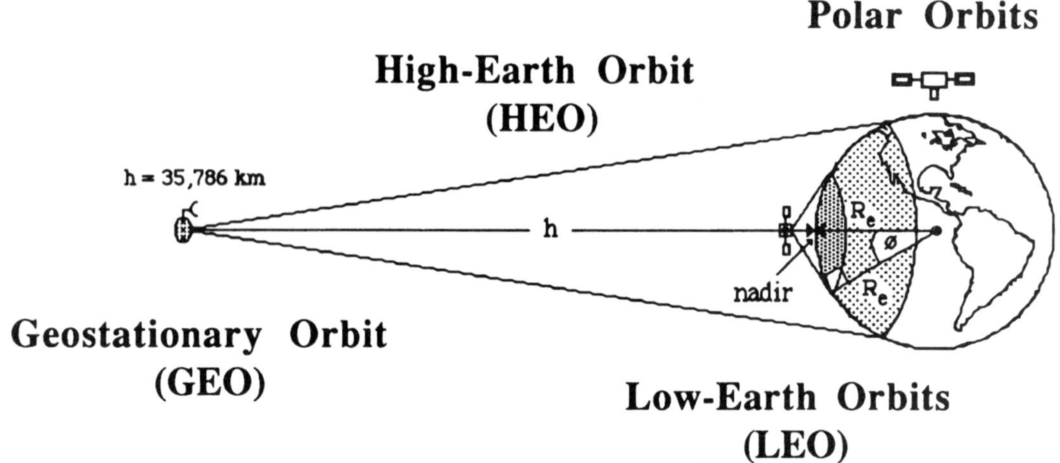

FIG. 5. Useful orbits with some common classifications of frequently used Earth orbits.

The following are some examples of space-flight missions requiring the various types of orbits. LEO orbits are used for any close Earth application such as Earth-resource monitoring and mapping or spy satellites. Because it takes more energy to achieve a higher orbit, all of the manned missions, with the exception of the Apollo flights to the Moon, have been and will continue to be exclusively in LEO orbits. The Hubble Space Telescope and other space-observing science satellites are in LEO orbits, since their fundamental requirement is to simply be out of the Earth's atmosphere to assure clarity of spacecraft sensor "vision." Some of these spacecraft are in highly inclined "polar" orbits that allow them to view the polar regions as well as the low-latitude regions.

GEO orbits are especially valuable for communications spacecraft and for global-weather spacecraft. Because the satellite remains fixed in the sky relative to any point on the Earth, it can be seen continuously by anyone in its field of vision using a fixed antenna. Figure 5 shows the large area of the Earth visible from a geostationary position, and a communications satellite can connect users thousands of miles apart. Besides providing the daily weather maps we see every night on the news, geostationary weather satellites can monitor developing storms far away from the home country and provide early warning of approaching danger.

High Earth orbits are used for scientific observations, navigation satellites, and also some communication satellites. These orbits can be highly elliptical, circular, or anything in between. An example of a science mission using high-eccentricity orbits would be those used for studying the magnetosphere, where it is important to take data close to the Earth as well as further from it. A Soviet astrodynamicist named Molniya developed a very special highly elliptical orbit that spends most of its time over the northern hemisphere. Such orbits are used by Russia for communications purposes, and a series of these satellites was named for the astrodynamicist. NASA has at least two spacecraft in very high circular orbits about half the distance to the Moon that can also be considered in HEO orbits.

Figure 5 shows that the area of the Earth that can be seen from a satellite changes with orbital distance. This is just another parameter that must be taken into account when designing an orbit for a particular application. The field of orbit design in astrodynamics combines celestial mechanics, engineering, and physics into a single discipline.

1.2 Space Environment

The environment in which satellites must operate is quite different from that of terrestrial systems. The Sun is the major environmental influence on the Earth and the space around it. The environment in the vicinity of the Earth is affected by two major products of the Sun's nuclear furnace: electromagnetic radiation, of which visible light is a part, and the outpouring of high-energy solar particles.

In the late 19th century, Stefan, Boltzmann, Wien, and Planck described the energy emitted by a blackbody as a result of its temperature. The Sun can be considered a blackbody at around 5750 K, which emits energy at essentially all wavelengths of the electromagnetic spectrum. However, the maximum energy output for such a blackbody occurs in the visible frequencies, toward which living things on Earth have developed and which the spacecraft designer uses to generate electrical power in space. The Earth also radiates electromagnetic energy, but because of its lower temperature (around 300 K) the Earth radiates its maximum energies in the infrared wavelengths. Both the Sun's and the Earth's energy must be taken into consideration in the design of satellites, especially in the area of thermal control.

In addition to the electromagnetic energy released by the Sun, a significant amount of particulate matter is emitted into space by the energy processes in and around the Sun. The Sun is composed mainly of charged particles (a plasma of loose protons and electrons), exposure to which can prove a hindrance or hazard for the operation of spacecraft and for the survival of humans in this environment. The Earth constantly encounters particulates streaming continuously from the Sun, which make up the relatively steady solar wind. However, occasional sharp increases in solar activity associated with disturbances such as sunspots and solar flares can produce a significant increase in particulate radiation that may be encountered by spacecraft in orbit around the Earth.

The environment in which spacecraft operate is affected in many ways by the electromagnetic and particulate radiations from the Sun. For example, the short-wavelength radiations from the Sun have enough energy to knock electrons out of the atmosphere, resulting in an ionized region, known as the ionosphere, that can modify and interfere with communication and remote sensing between Earth and space. The ionosphere is highly correlated with solar activity and varies dramatically geographically, seasonally, and with the day–night cycle. Radiation belts are also created, consisting of charged particles that become trapped within the Earth's geomagnetic field. These particles can affect spacecraft electronics and interfere with and threaten man's presence in space.

The Earth influences the space environment in other ways. The atmosphere extends, in continuously decreasing densities, far into space. Variations in temperature, atmospheric density, and the geomagnetic field result from complex and changing interactions between the Earth and the Sun, creating a dynamic environment in which spacecraft must operate. Figure 6 depicts the many factors involved in creating the space environment encountered by spacecraft in Earth orbit. The space-systems engineer must be familiar with this complex and fluctuating environment in order to determine how it is going to affect the spacecraft, both physically and operationally. Among other things, the effects mentioned influence thermal control, cause spacecraft charging, corrode materials, cause upsets in electronic subsystems, and create drag that modifies the spacecraft orbit. Man has also contaminated space with significant amounts of debris that create concern for the design of future missions. All of these considerations become increasingly important when humans are placed into the space environment. They are particularly susceptible to high-en-

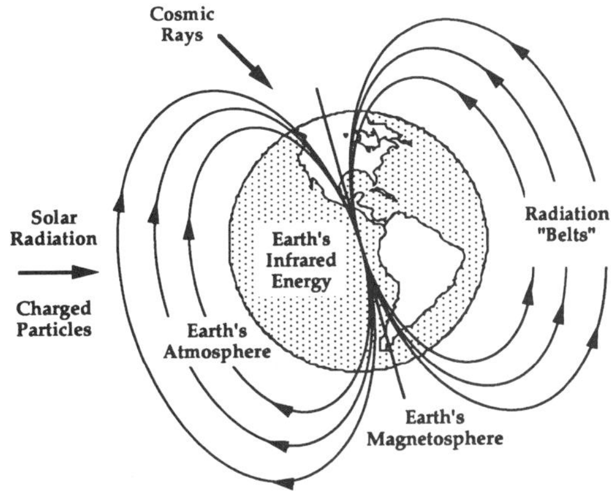

FIG. 6. Factors affecting the environment encountered by spacecraft in Earth orbit.

ergy particle radiations, which can cause cell damage and can be fatal if the radiation reaches high levels during a significant solar event.

1.2.1 Orbital Debris The world's rapidly escalating exploration of space is influencing the space environment in potentially threatening ways. The proliferation of discarded launch-vehicle parts and "dead" satellites locked in Earth orbits or traversing interplanetary space—collectively referred to as "space debris"—can cause enormous damage in a number of ways. Three of the most widely discussed dangers are the following:

1. A satellite or launch vehicle collides with a piece of debris. If the debris is not detectable and not tracked from space or the ground, the active satellite or rocket would have no knowledge it was on a collision course. Thus, if the two objects, each traveling in excess of 15 000 mph, collide there could be, at worst, the total destruction of both or, if the debris is of small mass, less but still substantial damage could ensue. The concern focuses on the geostationary orbits with their high population of unused communications and weather satellites and discarded rocket-stage hardware.
2. The orbit of the debris object decays. Gravitational forces could cause the debris to penetrate the near-Earth atmosphere and either burn up or crash on the Earth's surface at a location that may be highly populated. Because of the reentry velocity of the descending mass and its usually inoperative electric-signal–generating subsystems, it may not be detectable by ground surveillance instruments. Consequently, the incoming debris would impact the surface in much the same manner as a projectile would.
3. Contamination of other bodies. Because we do not know what life may exist on other planets, an interplanetary probe can, if it lands with a significant mass and at high speed, cause untold damage to the planet's environment, terrain, and possible life there. If the object carries contaminants from Earth, their spread on a distant celestial body could wreak havoc with humans or plant life.

These and other dangers that can result from space debris are prompting concern among international bodies such as the U.N. and countries heavily engaged in deploying satellites, shuttles, space stations, and interplanetary probes in space. This new awareness resulted in the first European Conference on Space Debris, held in Germany in April 1993. Organized by the European Space Agency (ESA), it was cosponsored by Italian, British, French, and German government space organizations. More than 100 presentations were made, 250 experts attended, and 17 countries were represented including the United States, Japan, and Russia. And certainly the U.N. had to have been involved, too. Conference organizer W. Flury stated that the purpose of the conference was to present space-debris–research results and define future research directions, identify methods of debris control, discuss needed regulations, and start considering global policy issues. He concluded, "Decision makers have to be convinced that debris mitigation measures have to be applied despite additional costs and the fact that immediate benefits to their projects may be very small."

Currently, worldwide efforts pertaining to space debris are focused on the following aspects:

1. The detection, measurement, and movements of large and particulate-sized manmade and naturally occurring objects to minimize placement of satellites in potentially dangerous domains of space. Techniques used or under consideration include radar, ground telescopes, space optics, and seismic instruments. The capture or collection of small debris particles using thin film and micropore foams is also under experimental investigation. Models are also being created and modeling techniques improved to define and predict more precisely the quantity of space debris, its location, and its motion as functions of time and position.
2. Study of the effects of impacts due to space debris on the performance and safety of rockets and satellites, with research into methods of protecting (shielding) these vehicles from damage actively in process. Manmade objects retrieved from space are now examined for space-debris effects or damage.
3. In the area of risk analysis, calculation of

the probability of collisions in the various orbits and in launch trajectories with special emphasis placed on geostationary orbits. Statistical databases, particularly of manmade abandoned or lost space segments, are under development to use in mapping safer anticollision routes for future flights.

An important risk factor pertains to nuclear-power sources in orbit that may malfunction and be abandoned in space, constituting an extremely hazardous environment should impact or even proximity occur.

4. In close conjunction with ongoing scientific and technological thrusts, study of the legal aspects involved with providers of space debris (liability, culpability, etc.), carried out internationally and by aerospace corporations, probably in much the same manner as it did when public use of aircraft soared. Policies to regulate space debris are now being formulated for later adoption by manufacturers, commercial users of space, governments, and the U.N.

2. APPLICATIONS OF ARTIFICIAL SATELLITES

2.1 Manned Spacecraft

The Soviets were first in achieving manned space flight with the launch of Air Force Major Yuri A. Gagarin in the 4706-kg (10 400-lb) Vostok 1 ("East") spacecraft on 12 April 1961 for a single orbit. Three weeks later, on 5 May 1961, the United States launched Navy Commander Alan B. Shepard, Jr., into space. The purpose behind the Mercury program was to gain basic data on the effects of space flight on human beings. Subsequent two-man Gemini flights were designed to evaluate the ability of performing the tasks in space required for a manned lunar landing. On 20 July 1969, only eight years after President Kennedy challenged the nation, astronauts Neil Armstrong and Edwin "Buzz" Aldrin descended to the surface of the Moon. After the lunar landings, using remaining Apollo equipment, the United States launched a space station, Skylab, and as an act of détente, conducted a joint space operation involving the docking of an Apollo capsule with a Soyuz spacecraft.

From 1975 to 1981 no American manned missions were launched, while design and construction of the next manned space vehicle, the Space Shuttle, proceeded. Since 12 April 1981, the shuttle fleet has flown missions carrying satellites, experiments, and crews of up to seven persons into low Earth orbit. With the exception of the *Challenger* explosion on 28 January 1986, the Space Shuttle has proven to be a capable and versatile space vehicle. Figure 7 shows a comparison of the various U.S. manned missions.

Soviet and American space programs paralleled each other in many ways. The one-man Vostok spacecraft was modified to carry a larger crew and was redesignated Voskhod ("Sunrise"). The first Soyuz ("Union") spacecraft was launched in April 1967 and has been used, with modifications, ever since. From 1971 to 1982 six successful Salyut space stations were launched and visited by a total of 29 crews. The current Russian space station, Mir ("Peace"), has been the home for cosmonauts for over a year, and several docking and laboratory modules have been added to the basic core.

The future of manned space flight is becoming more international, with Space Shuttle flights to the Russian "Mir" and with the involvement of Russia, the European Space Agency, and the Japanese in the International Space Station project.

2.2 Unmanned Satellites

In addition to manned spacecraft, the United States, the Soviet Union/Russia, and more recently many other countries have developed, have launched, and operate hundreds of unmanned spacecraft designed to perform military, national-interest, scientific, and commercial applications. The following sections highlight some of the application areas in which these spacecraft perform, and the evolution of spacecraft within each area.

2.2.1 Earth Observation Because of the unique vantage point of space, many satellites are designed to observe the Earth to collect data on the environment. TIROS 1 (Television Infrared Observation Satellite), the first weather satellite, was launched in April 1960 and returned over 22 500 pictures of the Earth. TIROS, NOAA (National Ocean-

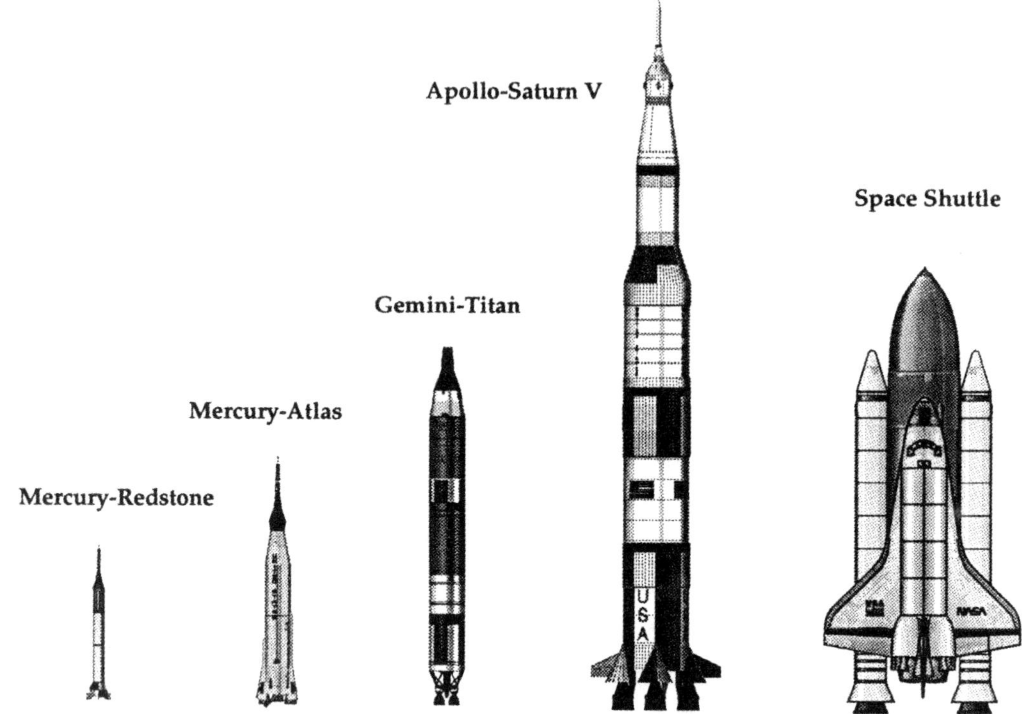

FIG. 7. U.S. manned launch vehicles. The figure, drawn approximately to scale, shows the relative sizes of the vehicles used for these missions. All of these vehicles have also been used to launch artificial satellites into orbit.

ographic and Atmospheric Administration), GOES (Geostationary Operational Environmental Satellite; Fig. 8), and the military DMSP (Defense Meteorological Satellite Program) systems monitor the Earth's weather, allowing timely prediction of climatic changes and their effects on the planet.

In September 1991, the Upper Atmosphere Research Satellite (UARS) was deployed from the Space Shuttle to conduct the first comprehensive studies of the stratosphere, mesosphere, and thermosphere regions of the atmosphere. The Total Ozone Mapping Spectrometer (TOMS), a NASA instrument that "hitched a ride" on U.S. Nimbus and Soviet/Russian Meteor weather satellites, discovered the "ozone hole" region of dangerously depleted ozone levels that appears over Antarctica each fall. Missions such as these allow researchers to map the short- and long-term changes occurring in the atmosphere and to determine what effects may be attributable to humans.

ERTS 1 (Earth Resources Technology Satellite) was launched in July 1972 into an orbit over the poles to gather information on geology, crops, population, and pollution, even gathering correlative information from ground-based instruments. ERTS evolved into the Landsat series of satellites, which continue to focus on discovering and monitoring Earth's resources in even the remotest areas of the world.

In June 1978, Seasat was launched to gather information on sea temperature, sea ice, wind speed and direction, and wave heights. Though its mission was cut short on account of an electrical failure, Seasat demonstrated the beneficial use of synthetic-aperture radar (SAR) in Earth-monitoring applications. A U.S./French satellite, TOPEX/POSEIDON, launched in August 1992, uses a radar altimeter to measure precisely the changes in global sea states to help determine the ocean's role in Earth's climate.

2.2.2 Space Environment James Van Allen used data from America's first satellite

FIG. 8. GOES. The Geostationary Operational Environmental Satellites are built by NASA for the National Oceanographic and Atmospheric Administration (NOAA) of the U.S. Department of Commerce.

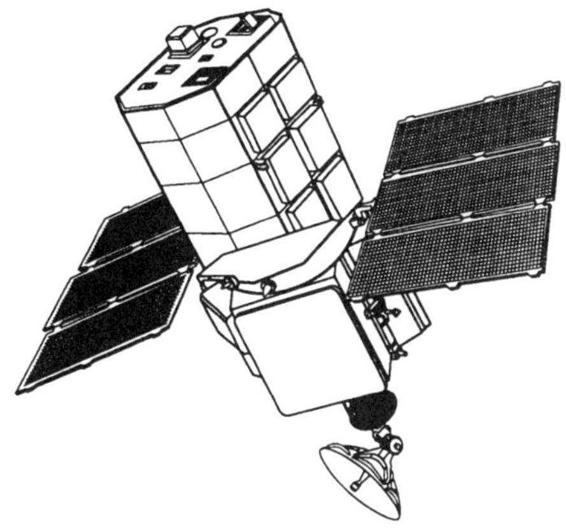

FIG. 9. SolarMax. This satellite was designed to help determine changes in the Earth's space environment due to changes in the Sun's activities.

(Explorer 1) to discover the "belts" of energetic particles encircling our world that now bear his name. Subsequent Explorer spacecraft allowed scientists to probe the solar wind and the Earth's magnetic field. The interaction between these phenomena was examined by the International Sun-Earth Explorer (ISEE) satellites, launched in 1977 and 1978, two in orbit around the Earth and one positioned in a special orbit between the Sun and the Earth. Pioneer and Voyager spacecraft, sent to visit the planets, also sampled the interplanetary environment. Now, that the Pioneer and Voyager craft are on their way out of our solar system, scientists continue to monitor their signals in hopes of gaining information on interstellar space.

The Sun has the greatest effect on the near-Earth space environment, and a series of Orbiting Solar Observatory (OSO) spacecraft launched between 1962 and 1975 were used to study solar flares and temperature differences on the Sun's surface. The Solar Maximum Mission (SMM or "SolarMax"; see Fig. 9) was launched in 1980 to observe the Sun during the maximum phase of its 11-year cycle. This spacecraft was the objective of a Space Shuttle rendezvous and EVA repair mission in 1984. Ulysses, a joint U.S.–European Space Agency spacecraft deployed from the Space Shuttle in October 1990, used a gravity assist from Jupiter to send it into a solar–polar orbit to allow mapping the solar environment in this previously unexplored region of our solar system.

Long-term effects of the space environment on spacecraft have been investigated with the Long Duration Exposure Facility (LDEF). It was released from the Space Shuttle in 1984 for what was planned as a one-year stay, but the *Challenger* accident delayed its recovery until 1990 when it was finally returned to Earth for analysis.

2.2.3 Planetary Exploration Planetary exploration began with flights to our nearest neighbor, the Moon. Before astronauts were sent there, several Ranger spacecraft were sent crashing into the lunar surface, sending back high-resolution photographs right up until impact. Lunar Orbiters photographed the Moon from orbit to locate possible landing sites, and Surveyor craft soft-landed on the Moon and sent back photographs and conducted lunar-soil sampling experiments.

Mariner 2 became the first spacecraft to fly by another planet, passing within 21 000 miles of Venus in December 1962. Subsequent Mariner and Pioneer spacecraft were sent toward Mercury, Venus, Mars, Jupiter, and Saturn—all the planets known to early astronomers! In 1976, two Viking spacecraft successfully landed on Mars, sending back pictures of the Martian landscape and conducting soil and atmospheric experiments. Spectacular pictures and a wealth of planetary information on Jupiter, Saturn, Uranus, and Neptune were sent back from the two Voyager spacecraft, now on their way out of our solar system. The Magellan spacecraft used a synthetic-aperture radar (SAR) to map more than 98% of the surface of the cloud-enshrouded planet Venus from September 1990 to September 1992. Galileo (Fig. 10), deployed in October 1989 from the Space Shuttle, used both Venus and the Earth for gravity-assist boosts to reach the planet Jupiter. Along the way, and despite a failed main-antenna deployment, Galileo flew by the asteroids Gaspra (October 1991) and Ida (August 1993), giving us our first close looks at these solar-system companions.

The Soviets sent numerous spacecraft to explore Venus, and two of their Venera craft successfully landed on the surface in March 1982. The landers were only able to transmit pictures and data for a short time until the harsh environment (temperatures that will melt lead and pressures about 100 times that on Earth) disabled the spacecraft. While the United States conducted the first flyby of a comet (Giaccobini-Zinner) in 1985, a flotilla of spacecraft from the Soviet Union, the European Space Agency, and Japan made somewhat more noteworthy flybys of Halley's comet in 1987.

2.2.4 Space Exploration The Orbiting Astronomical Observatory (OAO 2) was launched in December 1968 with a complement of ultraviolet instruments designed to examine the stars and discovered evidence of the existence of "black holes." High-Energy Astronomy Observatory (HEAO) spacecraft were launched between 1977 and 1979 to conduct x-ray, gamma-ray, and cosmic-ray surveys of the universe, and the International Ultraviolet Explorer (IUE) conducted observations in the ultraviolet (UV) wavelengths. Launched in November 1989, the Cosmic Background Explorer (COBE) carried instruments designed to search the infrared background of deep space for information about the "Big Bang" theory of the origin of our universe.

Deployed in April 1990 from the Space Shuttle, the Hubble Space Telescope (HST, Fig. 11) was the first of the "Great Observatories," facility-class satellites designed to offer scientists unparalleled opportunities to explore the universe. HST has a 2.4-m (94.5-in.) diameter mirror that can pick up objects 50 times fainter and 7 times more distant than present ground-based observatories, potentially expanding the universe visible to astronomers by 500-fold! Since it was designed for on-orbit servicing, in December 1993 astronauts were able to insert a set of corrective optics to offset the "spherical aberra-

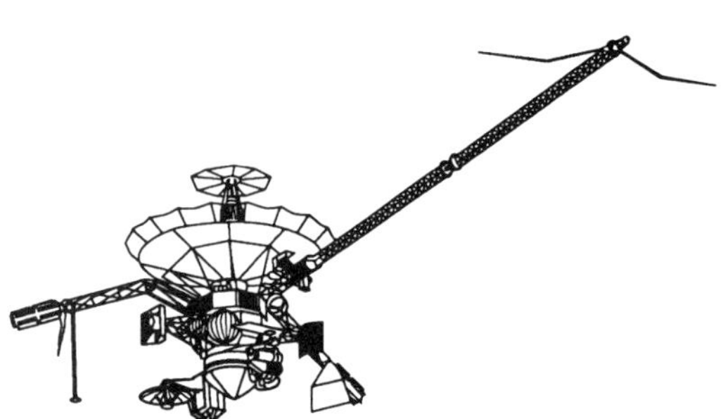

FIG. 10. Galileo. This spacecraft was designed to release a probe into the atmosphere of Jupiter and to remain in Jovian orbit to study the planet and its moons.

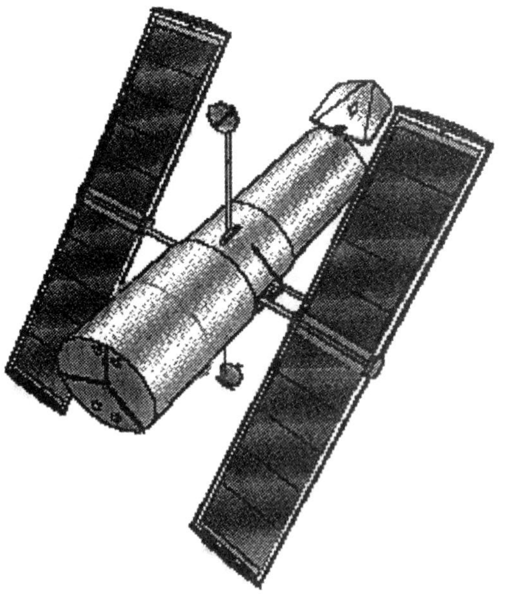

FIG. 11. Hubble Space Telescope. From its position high above the Earth's atmosphere, HST provides an unobstructed view of the universe.

tion" flaw discovered in the primary mirror that compromised HST's capabilities. The second great observatory, the Compton Gamma Ray Observatory (GRO), was deployed from the Space Shuttle in April 1992. GRO was designed to capture the relatively unimpeded gamma-ray emissions from cataclysmic cosmic events like supernovas, black holes, and even the remnants of the Big Bang. Other great observatory satellites may follow; however, the current trend is toward smaller, cheaper satellites.

2.2.5 Commercial Satellites By far, the largest commercial interest in space systems is in the relay of information or communications. Communication satellites had a humble start with Echo 1, launched in August 1960, which was simply a huge radio-reflecting balloon in low Earth orbit. By 1963, Syncom 1 demonstrated the capability and advantages of communicating via a satellite in a geostationary orbit. Today, over one hundred communication satellites, some similar to the one shown in Fig. 12, exist in this orbit, each capable of relaying thousands of telephone calls and television pictures simultaneously around the globe. In 1962, President Kennedy signed a bill creating the Communications Satellite Corporation (Comsat), a joint government–industry undertaking allowing transfer of U.S.-developed satellite communications technology for commercial use. The International Telecommunications Satellite Consortium (Intelsat), with Comsat representing the United States, promotes international use of these capabilities. Both organizations have been highly successful.

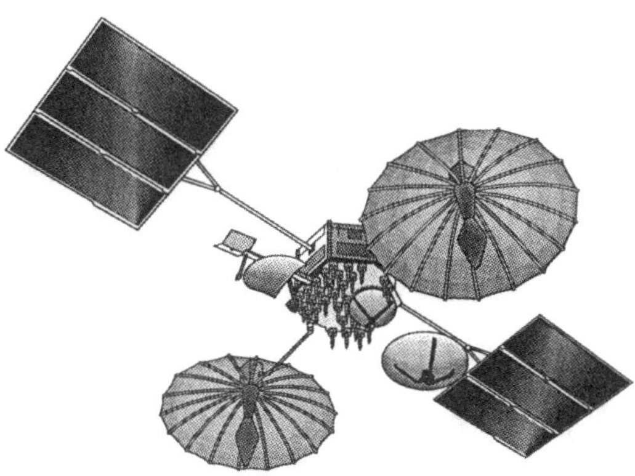

FIG. 12. Communications satellite. Though not a commercial satellite, NASA's Tracking and Data Relay Satellite (TDRS) displays the large antennas and solar arrays required to transmit and receive from geostationary orbit.

A similar arrangement exists involving dissemination of data gathered by the Landsat satellites for which demand by the commercial public has grown. The French SPOT (Satellite Probatoire de l'Observation de la Terre) satellite competes in this area, offering high-resolution photographs of points of interest on the Earth to news media, governments, and even military organizations worldwide.

2.2.6 Military Satellites The early experiments using captured V-2 rockets were conducted under the auspices of the military and led to the development of the Intercontinental Ballistic Missile (ICBM). As other uses of space became apparent, the military conducted their own experimental programs and developed systems to take advantage of the new "high ground" of space.

Today, the military routinely conducts force-enhancement missions such as communications, navigation, and remote environmental sensing using dedicated systems in space. Surveillance and monitoring missions are also routinely conducted, though the systems used are often classified. However, the worth of these systems became generally apparent during the war with Iraq, where coalition forces scored a decisive victory using these capabilities extensively.

Navigation satellites are probably the next most successful area in the commercial use of space systems. In this case, however, the private sector is taking advantage of a military satellite. The Global Positioning System (GPS, Fig. 13) and the earlier Navy Navigation System (Transit) satellites were designed for use by military forces to provide a worldwide, continuous positioning capability. As an afterthought, the U.S. government made these signals available for nonmilitary use, and the number of civilian users now far outweighs the military. Unlike the satellite communications industry, companies that provide GPS services benefit in that the cost associated with the satellites has already been borne by the U.S. government.

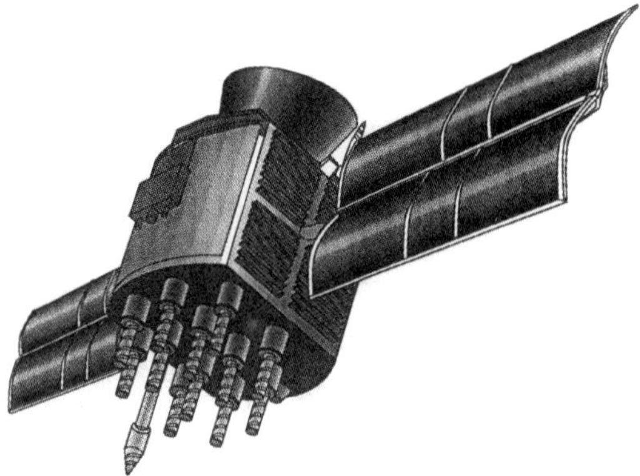

FIG. 13. Global Positioning System satellite. Though the primary mission of the constellation of GPS satellites is to provide continuous, very accurate navigation information to the U.S. military, the number of civilian and foreign users continues to grow.

3. DESIGN, FUNCTION, AND OPERATION OF ARTIFICIAL SATELLITES

3.1 Satellite Design Phases and Processes

The prerequisite for all satellite design phases and processes described is that a need exists for the mission under consideration to justify the time, effort, and sizable expense required to warrant its development. The need is often defined by individuals or groups of space scientists scattered all over the world. To further their research, they may desire the placement of instruments in space to acquire new, more, or better data on the basis of their analyses to date. Demands for artificial satellites also originate from federal military and civilian government agencies, foreign countries, and industry. Organizations such as international bodies, NASA advisory committees, the National Academy of Sciences, the National Space Council, and university research institutions also play important roles in defining needs for satellites—manned or unmanned.

Whatever the source of the initial concept for the mission, further development usually follows a structured approach. Figure 14 shows a typical mission-development approach used by NASA, which is easily compared with approaches taken by industry or other government agencies.

The approach uses defined phases associated with the level of detail expected at each stage of mission development. A number of reviews are used to ensure that the development is proceeding as expected and that all factors are such that the development can

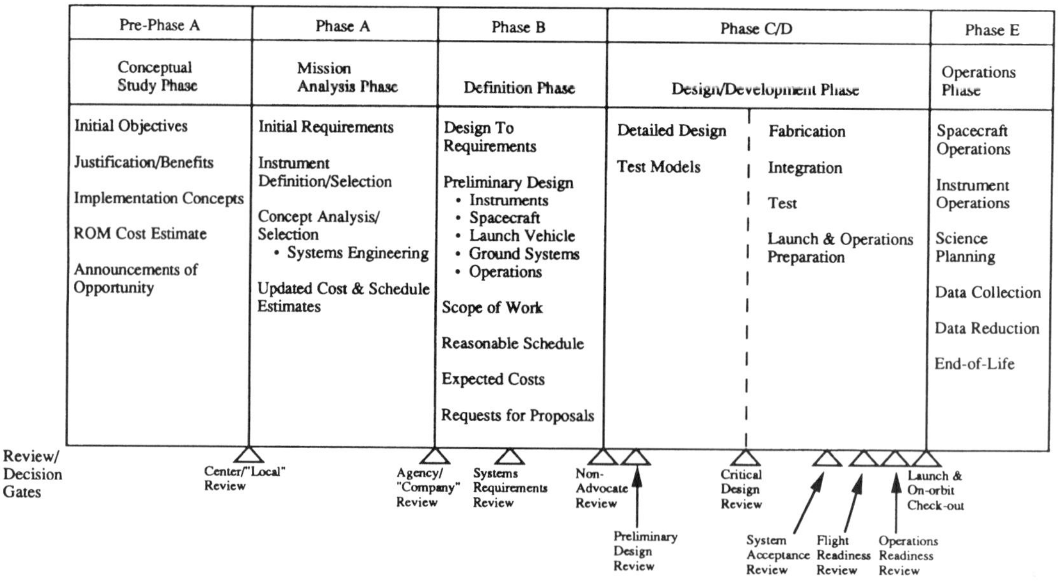

Pre-Phase A	Phase A	Phase B	Phase C/D		Phase E
Conceptual Study Phase	Mission Analysis Phase	Definition Phase	Design/Development Phase		Operations Phase
Initial Objectives Justification/Benefits Implementation Concepts ROM Cost Estimate Announcements of Opportunity	Initial Requirements Instrument Definition/Selection Concept Analysis/ Selection • Systems Engineering Updated Cost & Schedule Estimates	Design To Requirements Preliminary Design • Instruments • Spacecraft • Launch Vehicle • Ground Systems • Operations Scope of Work Reasonable Schedule Expected Costs Requests for Proposals	Detailed Design Test Models	Fabrication Integration Test Launch & Operations Preparation	Spacecraft Operations Instrument Operations Science Planning Data Collection Data Reduction End-of-Life

Review/ Decision Gates: Center/"Local" Review; Agency/ "Company" Review; Systems Requirements Review; Non-Advocate Review; Preliminary Design Review; Critical Design Review; System Acceptance Review; Flight Readiness Review; Operations Readiness Review; Launch & On-orbit Check-out

FIG. 14. Mission development chart, showing the various design "phases" with their components and the many reviews involved.

proceed to the next level. Considering the expense associated with such missions, and the usual irretrievability of the satellite once placed into orbit, these factors include budgetary and political as well as quality-assurance and technical elements. The following sections describe these phases and the level of detail expected, usual activities, and the reviews and/or decision gates associated with continuing development of a typical satellite mission.

3.1.1 Conceptual Study Phase This initial phase (sometimes referred to as Pre-Phase A) may be accomplished by the scientific/user community, but usually involves the assistance of an organization such as a NASA center or a spacecraft-manufacturing company that has experience in spacecraft design and development. The phase begins with formulation of a clear statement of the desired objectives of the mission. Associated with these is a justification for the mission describing the benefits that will be achieved. Using descriptions of the instruments or experiments that may gather the desired information, a number of different concepts may be formulated that can achieve the mission objectives and determine the feasibility of conducting the mission. These concepts include ground system and operations as well as launch-vehicle and spacecraft descriptions, but the level of detail is usually "back of the envelope" in scope. After the concept is formulated, a rough order-of-magnitude (ROM) cost for the mission is estimated, usually also using generic cost information as a basis. In addition, preliminary overall schedules and risk assessments are generated.

The results of this phase are reviewed at a "local" level, within the NASA Center or Headquarters Division, university, laboratory, or the like, to determine if the concept warrants further consideration outside the group. Within NASA, if the concept looks promising, an Announcement of Opportunity (AO) may be issued, which is the first step in defining and selecting the instruments or experiments that will fly on the mission if it is fully developed and launched.

3.1.2 Mission Analysis Phase The purpose of the Mission Analysis Phase (Phase A) is to translate the broad mission concepts and objectives into a feasible preliminary system design. This is a refinement and expansion of the Conceptual Study Phase work made with the intention of providing a concise, clear overview of the proposed system. Implementation is done in cooperation with flight ground systems, discipline engineering, and instrument specialists. Financial analysts utilize parametric cost models to derive top-level resource estimates for the overall mission and its component elements. The typical unmanned mission study has a duration of approximately nine months to two years. Study costs are usually one to two percent of total mission costs.

Key activities during this phase are system and subsystem trade studies (comparing different systems to identify the best candidate), analyses of performance requirements, identification of advanced-technology and long lead-time items, risk assessments, end-to-end system life-cycle versus cost studies, estimates of other costs, setting schedules, and selection of system and operational concepts. More specific items addressed include justification for the selection of a specific strategy and preliminary concepts on mission management (particularly if it is not an internal NASA effort).

Specific tasks to support this phase often include feasibility assessments, cost-estimate modeling, identification of spacecraft/ground data communications interfaces and satellite tracking-system interfaces, the design of information-dissemination systems for science data, system engineering for instrument design and their interfaces, definition of alternative strategies for the mission, study of new technology needs, and risk analyses.

3.1.3 Definition Phase The purpose of the Definition Phase (Phase B) is the refinement of the mission and system architecture and the system design created during the Mission Analysis Phase. It is designed to convert the previous phase preliminary system design into a mature final design and baseline for entering the Execution Phase where fabrication begins. Functional, operational, and performance requirements are defined. Established interface requirements and specifications are allocated down to the subsystem or major-component level. Firm cost estimates and schedules are prepared for transition to the subsequent Execution Phase (C/D). Once a detailed baseline configuration that satisfies all the mission and program requirements has been established, the Definition Phase design, costs, and schedules are submitted to NASA for approval. This phase has a duration of from one to two years. The costs associated with this study are usually four to eight percent of total mission costs.

Key activities during this phase are revalidation of mission requirements and system-operation concepts, conversion of mission requirements to much more specific project requirements, risk analyses, system and subsystem studies and trades, updating schedule and life-cycle costs, and system and subsystem concept selection and validation.

This phase covers the full range of technical, management, resource, facility, and procurement assumptions, ground rules, plans, procedures, and full documentation of the entire proposed flight/ground system. Alternative designs are analyzed to allow the choice of one optimum flight and ground-system approach considering technical performance, cost, risks, and schedules. The baseline design is totally defined as well as all major interfaces and system, subsystem, and component specifications.

Specific tasks conducted during this phase include telemetry assessments, analysis of instrument hardware, Space Shuttle interface analyses (as needed), acquiring knowledge of the space environment, determining TDRSS usage, assessment of make or buy decisions of components, setting up production plans, establishing facility requirements, making risk assessments, and offset planning, estimating cost, schedule, and resources, management planning and scheduling, mission requirements assessments, payload integration plans, planning satellite operations, and devising data-transfer and -processing systems.

3.1.4 Design and Development Phase Usually by this point a commitment has been made to develop and build the spacecraft and implement the mission. For this reason the Design Phase (Phase C) and the Development Phase (Phase D) are combined. The purpose of the Design/Development Phase (Phase C/D) is the actual implementation of the approved mission plan. Now is when the hardware is built, assembled, and tested. The successful end result of this phase is that the satellite will have achieved the desired orbit and is properly oriented in space with all space and ground systems functioning as expected to satisfy previously stipulated mission objectives.

Depending on the mission and/or the organization responsible for it, this phase is sometimes subdivided into separate Design/Development, Launch, Post-Launch checkout, and Operations phases. One example comprises the NOAA weather satellites where the spacecraft is built by either NASA or one of its contractors. (In parallel, the payload complement of meteorological and weather sensors derives from within the agency or is obtained from other contractors expert in this type of space instrumentation. Or, by virtue of a cooperative arrangement with other countries, an instrument may be supplied by a foreign government to fly on the U.S. satellite.) These sensors are then sent to the spacecraft manufacturer for integration and subsequent activities involving the entire space segment. After launch and on-orbit checkout of all systems, the control or Operations phase is transferred from NASA to NOAA.

Even if mission operations are not the responsibility of NASA, the agency may become involved if its shuttle is used to rendezvous with the satellite to perform repair, retrieval, or component- and personnel-exchange functions.

Another example of dividing this phase involves commercial or foreign satellites that are delivered to NASA or DOD sites (usually in Florida or California but sometimes on the shuttle) for Launch-phase purposes only.

The duration of this phase is extremely variable, often ranging from five to ten years depending on its complexity and whether the satellite is manned or unmanned. But subsequent copies of the same or moderately improved satellite system can be placed in orbit within much shorter time periods—sometimes in less than one year from the initial launch.

The major categories of this phase, listed below, are applicable to all mission elements including the space and ground systems:

1. design and development,
2. fabrication,
3. assembly,
4. integration,

5. environmental and other tests,
6. launch,
7. post-launch checkout, and
8. post-launch operations (may include retrieval, on-orbit change or repair, orbit changes, etc.).

Associated with these eight Design/Develop Phase categories are a host of peripheral efforts needed to assure success of the space mission or to determine what went wrong in the event of a failure. These efforts are applicable to the basic spacecraft and to its complement of life-support equipment and scientific instruments, in addition to the ground segment.

During the course of this phase, as the building, test, and assembly of the various hardware elements proceed, a concurrent effort is directed toward making the physical systems act together as a single entity. A partial list of these activities includes the following:

1. Preparing for shipping the fully assembled satellite to the launch site, installing it on the launch vehicle or shuttle, checking all systems for ground or air transit damage, and verifying correct electrical and mechanical interconnections with the launch vehicle and the shroud.
2. Because part of the Control Center is a diagnostic clinic for satellite anomalies in space, preparing and checking emergency short-term remedial control procedures that can be executed from the ground if needed. This effort also includes defining and validating fallback positions, which usually entail compromises in mission objectives.
3. Essentially the same as **2,** except that the activity is geared to anticipating and planning for unexpected changes in environmental conditions such as solar flares, micrometeorite fields, orbital debris, spacecraft outgassing that can blind sensors, and adverse weather conditions especially during the launch trajectory.
4. Establishing and agreeing upon priorities to exercise when anomalies occur in orbit. This is usually a contentious issue, with mission managers, spacecraft manufacturers, and scientists having conflicting priority requirements. For manned missions, the human element, naturally, takes precedence.
5. Planning for the uninterrupted flow of commands and data in the event one or more ground stations or relay satellites fail.

By this time, many of the difficult-to-resolve or -accomplish tasks and backup positions have become routine on prior missions and are not specifically discussed herein. Nonetheless, for every satellite flight, the issues resurface, are studied, and are satisfactorily planned for and resolved as failure modes occur.

The outputs for this phase comprise all aspects of the space and ground segments of the mission and their operation. Prior to proceeding to the Operations Phase, the launch must have been successful, the spacecraft should be deployed and in the correct orbit, the spacecraft and ground-systems communications should be in normal working order, and the spacecraft will have been operating successfully with the tracking stations or via a signal-relay satellite (e.g., TDRSS).

3.1.5 Operations Phase The purpose of the Operations Phase (Phase E) is routine operation of the satellite including data acquisition, processing, and distribution until or unless a problem develops or the mission is terminated. The conduct of this phase varies considerably depending on whether the mission is a principal investigator type, an Earth-observing–facility mission (e.g., Landsat) with many investigators and investigator organizations involved, or one transferred to another federal agency (e.g., NOAA) or the private or international sector. The duration of this phase is variable but can be five years or longer.

Key activities during this phase are command and control of the spacecraft and instruments and data-related functions such as transmission, capture, processing, archiving, and distribution. Special scientific operations may be authorized and conducted. Orbital changes may be required. Spacecraft or instrument components may be retrieved, repaired in orbit, or replaced if necessary and if possible.

During this phase routine tasks conducted include

- science-data reviews with scientists and government officials;

- flight assessments of hardware, software, science data, orbits, and mission objectives;
- flight-system monitoring;
- monitoring the acquisition, processing, dissemination, and archiving of mission data;
- compilation and production of a "lessons learned" document for the mission, which is used as a guide in the development of additional satellites in the series and for future new space missions.

3.2 Satellite Subsystems and Functions

Although satellites bear the imprint of specific designs to support particular missions, they all have certain elements in common, such as

1. a payload that captures an applications-specific objective, such as communications, remote sensing, or navigation, and
2. mission/payload-peculiar spacecraft-bus augmentations that provide special payload-related support functions possibly not offered by a standard spacecraft design. These mission/payload-peculiar support functions include such things as special data downlinks and antennas, data recorders, cryogenic coolers, data encryption, and other subsystems that are directly related to the unique needs of a particular payload.

However, no matter what the payload/mission, the satellite bus must consider the following in its design, and probably will incorporate integral subsystems that perform these functions:

1. power generation and distribution,
2. thermal control,
3. attitude determination and control,
4. orbit maintenance and propulsion,
5. structure and mechanisms,
6. communications, command, control, and data handling.

Each of these areas and the subsystems that perform them are discussed in the following sections. Figure 15 shows the dispo-

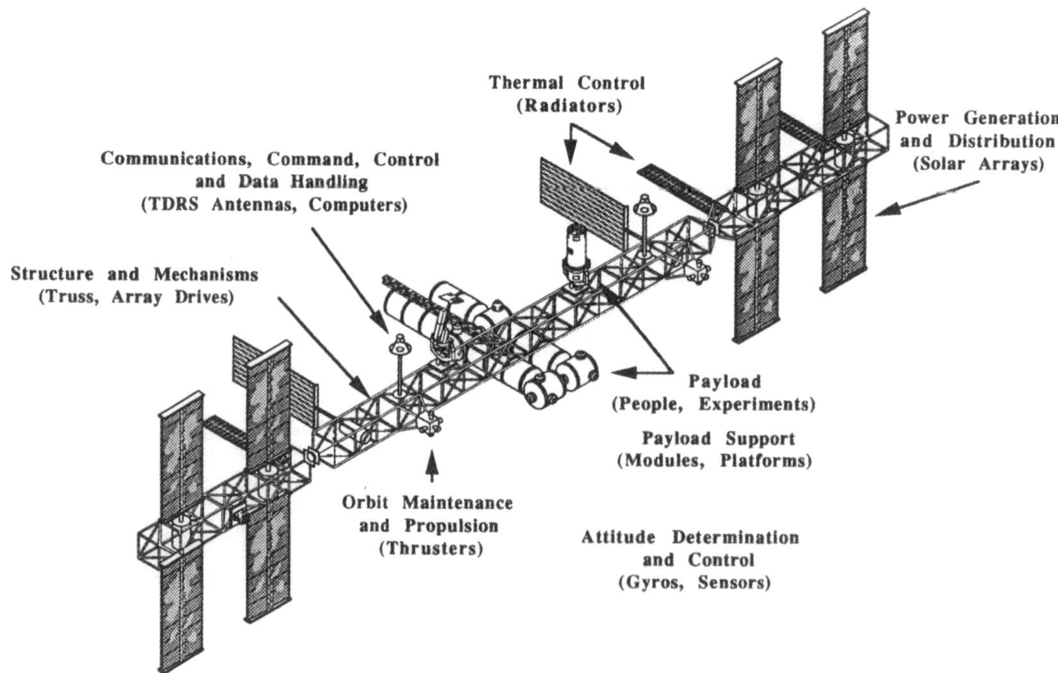

FIG. 15. Spacecraft components. All spacecraft have some common elements required to operate in space, as depicted in this early version of the international space station.

sition of the various components in one large satellite structure.

3.2.1 Power Generation and Distribution This subsystem provides the power used by the spacecraft and payload throughout the mission. Solar cells, which directly convert the Sun's light photons to electricity, are by far the most popular source of spacecraft power. Heat exchangers can use the Sun's radiated energy to heat a working fluid that can power a generator to create power. As a spacecraft operates further from the Sun, it may be necessary to use a different source of power such as nuclear-power generators. These systems have been used in deep space for distant planetary exploration, but have also been used in near-Earth orbits for high-power applications. Current space nuclear systems consist of radioisotope thermal generators (RTGs) that use radioactive decay to produce electrical power directly, and nuclear reactors that heat a working fluid to power a generator and produce electricity.

Power storage is achieved by using batteries to store energy. Nearly all solar-powered spacecraft use batteries to provide power during mission or orbit times when the Sun's energy is not available. Other chemical devices, such as the fuel cells used on the Space Shuttle, can be used to provide power over limited-duration time periods.

3.2.2 Thermal Control A spacecraft thermal-control subsystem keeps the temperature of the spacecraft and its mission-important elements within specified ranges. Thermal-control devices are categorized as being either passive or active. Passive devices shield, insulate, or change thermal-exposure characteristics. They include external coatings, thermal blankets, and louvers that selectively close or open spacecraft compartments to space. Active thermal-control devices include heaters, refrigerators, and heat pumps or pipes that use a working fluid to move heat energy from one spacecraft location to another.

3.2.3 Attitude Determination and Control Satellites employ a variety of techniques to control the satellite body axes to some reference orientation. Three-axis attitude-control systems use control torquing methods such as thrusters, spinning momentum wheels, and magnetic devices to maintain a satellite within an acceptable tolerance with respect to pointing references aligned with the center of the Earth, the direction of flight, and a third axis normal to these. Spin-stabilized systems rotate the entire body of the satellite to create a gyroscopically stable orientation with respect to the Earth. The spacecraft may incorporate a de-spun platform, which allows the payload to remain fixed with respect to a particular direction while the rest of the spacecraft spins for stability. Gravity-gradient control methods utilize the forces of gravity to orient the spacecraft.

All of these systems have their particular virtues and limitations. For example, spin-stabilized spacecraft produce a rotisserie effect with a beneficent thermal result. However, the spacecraft usually employ surface-mounted solar cells for power generation, and on a cylindrical spacecraft only about $\frac{1}{3}$ of these cells produce power at any time. Also, spin-stabilized satellites cannot achieve the accurate pointing control required by some remote sensors.

The control torquing methods are directed by a control system that determines the direction, amount of torque, and duration of actuation with respect to reference information provided by devices that can sense the location of the Earth, Sun, or stars. The systems designer must consider the mission/payload requirements and select the proper design from these fundamental approaches or some hybrid combination of these techniques.

3.2.4 Orbit Maintenance and Propulsion External forces, such as drag or solar-radiation pressure, cause the precise orbit desired by the mission to be perturbed. The orbit-maintenance subsystem uses thrusters periodically to restore the satellite to its desired orbit or, in many cases, to modify the orbital characteristics of a satellite (usually a change in the orbital period and associated altitude) to place the satellite into a different time- and space-sampling pattern with respect to specific Earth locales.

Most satellites utilize an integral propulsion system to transfer the satellite from an intermediary parking orbit, achieved by the launch vehicle, to the final desired orbital location. In a few cases, this system may be used to return the satellite to a maintenance orbit where it can be repaired or refueled or, at the end of life, it may be used to de-orbit the satellite from space. In many cases, an integral orbital-maintenance and propulsion system is designed having a variety of different thrusters associated with the relatively large energy requirements of orbital-plane changes and the much smaller energy requirements associated with orbital maintenance and spacecraft attitude control.

3.2.5 Structure and Mechanisms Spacecraft structural considerations must deal with the many physical configurations the craft experiences in achieving and operating in its orbit. The spacecraft must fit into the usually cramped launch-vehicle payload shroud, and survive the process of launch and orbital placement. On station, the spacecraft may need to deploy components into operational configurations, some requiring motors to keep them accurately pointed. Additionally, the structure must have sufficient strength and use the proper materials to maintain spacecraft structural integrity during mission lifetimes that often exceed 10 years of orbital stress produced by the space environment and by station-keeping and operational attitude changes.

3.2.6 Communications, Command, Control, and Data Handling Satellites send information to mission control locations about the spacecraft's state of health and operational status. Status information includes the on-or-off condition of payload and satellite subsystem elements, and health information verifies that all systems are operating within established tolerances of temperature, power, etc., or conversely that they are exceeding preset boundaries. On the basis of this information and the plan for normal or anomalous operation of the spacecraft, the ground control station will transmit commands and other information to the satellite for immediate implementation or for on-board storage for future use.

Most spacecraft are engaged in the collection or transfer of some form of information. In many cases, data rates and volumes are much greater than can be accommodated by the satellite health and status reporting system. When this happens, a separate system of transmitters, antennas, and on-board storage devices for later downlink to ground stations is needed. These special systems are extensive and expensive but are absolutely necessary for satellites that collect high-resolution remote-sensing data or act as a communications relay between widely displaced Earth locations.

Modern spacecraft use computers in nearly every aspect of the satellite system. A computer forms the heart of the attitude-control process, accepting information from the reference sensors and computing the instructions to the control torquing systems. Power-systems management is under the control and guidance of computers, and even the thermal-control system depends on computers to monitor and control component temperatures. Propulsion and orbit-maintenance systems utilize computers to determine the event time and the thrust time and direction for maneuvers, and data-handling and communication systems use computers for downlink timing control and for command authentication and execution. Finally, a master computer is required to determine the occurrence of any serious anomalous event that may affect the survival of the spacecraft, and decide the proper reaction to the event, including assuming control of spacecraft subsystems and instruments, to ensure a safe configuration until the problem can be corrected.

3.3 Satellite Operations

The mission objectives for most unmanned satellites require an operations system that provides a two-way ground-to-space communications link for the command and control of the spacecraft, its systems and subsystems, and the payload complement of instruments.

During the course of normal operations, instruments may have to be commanded into various electronic and physical configurations either routinely or in response to some anomalous situation. Typical commands include power on/partial/off; high/low-gain data-acquisition modes; sensor pointing; and activation of protective measures to prevent overheating or excessive cooling, radiation exposures, and micrometeorite damage. In some instances, the operational configuration of an instrument is varied in real time in response to computer- or scientist-based analysis of sensor data. If a computer is utilized to accomplish an operational change, it may be located on the spacecraft or at the ground control center.

The entire satellite may have to be commanded into different positional orientations during the course of normal operations to compensate for rotational limits of solar arrays (relative to the main body of the spacecraft) that may require a noon yaw turn maneuver to reposition the arrays ready for the next cycle of solar pointing. Pressurized inert gases (argon) that provide jet thrust for correcting naturally induced changes in orbital altitudes or inclinations are periodically released to accomplish these adjustments. As a function of spacecraft orientation or orbital position—both relative to the Earth—antennas and instrument probes may require routine pointing adjustments to facilitate ground communications or permit proper instrument operation.

In a limited number of cases, normal operations may require use of jet thrusters to position the spacecraft for a rendezvous with NASA's Space Shuttle for retrieval, repair, or equipment/personnel replacement purposes. Situations such as this involve complex and very precise orbital maneuvers and spacecraft orientations.

Whatever the operational mode, the main ingredient in its successful accomplishment is the communications link with the ground control center. Because of data-relay satellites, virtually round the clock communications are now possible. The NASA Tracking and Data Relay Satellite System (TDRSS) is also capable of relaying vast quantities of scientific data to NASA ground receiving stations, where it can be retransmitted via commercial communications satellites to users in any part of the world. U.S. weather satellites in geostationary Earth orbits have been retransmitting other NOAA-acquired meteorological data to anyone within the range of the antenna pattern—essentially acting simultaneously as data-relay devices. A typical ground station is shown in Figure 16.

In addition to the NASA control centers, ground data-receiving facilities, and TDRSS, other U.S. institutions (e.g., universities; local, regional, state, and federal agencies) and foreign countries have installed their own data-receiving stations. For example, Landsat data are now routinely acquired by more than a dozen countries, various universities, U.S. government agencies, and private organizations. These facilities can process imagery data suitable for use by Earth scientists and environmentalists worldwide.

FIG. 16. A typical ground station. Many ground stations require large, movable antennas in order to ensure successful two-way communications with the satellite. The cost of these stations and their operation cannot be ignored in initial mission planning.

3.4 Space Law

The placement and operation of a satellite in a space environment requires satisfying seven basic "laws" related to outer space of the U.N. Committee on the Peaceful Uses of Outer Space (COPUOS). Also considered part of international law are a number of U.N. requirements mostly pertaining to notifying that international body of launch intentions; operational, debris contribution and disposal plans; and untoward occurrences. Their exact titles are as follows:

1. Treaty on principles governing the activities of states in the exploration of outer space, including the Moon and other celestial bodies (1967).
2. Agreement on the rescue of astronauts, the return of astronauts, and the return of objects launched into outer space (1968).
3. Convention on international liability for damage caused by space objects (1972).
4. Convention on registration of objects launched into outer space (1975).
5. Agreement governing the activities of states on the Moon and other celestial bodies (1979).

6. Treaty banning nuclear-weapons tests in the atmosphere, in outer space, and under water (1963).
7. Convention on the prohibition of military or any other hostile use of environmental-modification techniques (1977).

It is to be noted that not all the 154 U.N. member countries subscribe to each of these laws. For example, there is no general agreement on the definition of outer space—one commonly accepted boundary defines this domain as the region 100 km or more from the Earth's surface. But several equatorial countries claim that the country's sovereignty extends from the surface to infinity, thereby establishing their boundaries to include geostationary satellites flying overhead.

The U.N. requires notice from the launching state if a satellite is no longer in orbit, particularly if it may reenter the atmosphere and crash. If damage to life or property in another country occurs, the launching state is liable. The country where the impact occurs must search for and return the debris to the launching state, but the latter can be billed for expenses incurred. In the event astronauts are involved, the country must care for them and promptly return them to the launching country. Satellite crew members may not be detained for interrogation purposes.

If a satellite causes interference of any type (electromagnetic, radioactivity, etc.) with another satellite, its owner is liable for correcting the situation. At times orbiting optical devices have been considered suspicious if the instrument can be focused on a hostile country for espionage purposes. Situations of this nature apply to all Earth-observing satellites, but their use remains legal.

With regard to interplanetary exploration, the U.N. prohibits any activity or landing that may affect the well-being of the target planet. Contaminants, radioactive sources, thermonuclear detonations, and electronic interferences fall in this category of restrictions, as do environmental modifications. In addition, no Earth state can establish a sovereign claim on any extraterrestrial body.

In the not too distant future, it is expected that countries that launch and then abandon satellites and other debris in space must take action, still to be precisely defined, to prevent their reentry to Earth or posing a collision problem for operational satellites in orbital flight.

In summary, international laws are intended to work in the best overall interests of mankind. Compliance does add to the cost of space exploration, but only in the near term. The long-range benefits of these agreements to space exploration far outweigh the immediate inconveniences and expenses.

4. ARTIFICIAL SATELLITE MANUFACTURING

Several major institutions manufacture satellites in the United States and abroad. NASA's Goddard Space Flight Center in Maryland has all the facilities required to design, manufacture, build, test, launch, and operate scientific spacecraft. Another similar U.S. government institution is the Naval Research Laboratory in Washington, DC. U.S. commercial manufacturers include Lockheed/Martin (a merger of Lockheed with Martin Marietta, which had previously acquired the satellite facilities of GE Astrospace, formerly RCA Astrospace), Loral, TRW, and Hughes, to name a few. Foreign manufacturers include Aerospatiale, Alcatel, and Matra Marconi (France), Daimler-Benz Aerospace (Germany), Alenia Spazio (Italy), and SPAR Aerospace (Canada). There are also several small satellite manufacturers such as Orbital Sciences Corp. and CTA among others. In addition, there are several academic facilities capable of manufacturing satellites such as the Johns Hopkins University's Applied Physics Lab and the California Institute of Technology's Jet Propulsion Laboratory.

In most cases, satellite builders will obtain many of the components used in the satellite from specialized component manufacturer companies. Such companies exist for solar arrays, batteries, attitude-control sensors and systems, and most other "standard" spacecraft components.

Despite the plethora of manufacturers, many of the procedures and processes of building satellites are quite similar. In order to build and test a spacecraft properly, some very specialized equipment and facilities are required. One of the most demanding requirements is cleanliness. Contaminants can cause valves to stick, electronics to short, moving parts to wear, and sensors to be blinded. For this reason, space components are manufactured using strictly controlled procedures to minimize introduction of foreign matter. The satellites themselves are assembled in "clean rooms" (Fig. 17), which maintain a clean environment (see AEROSOLS) in a large enough area to assemble a complete spacecraft. Before being brought into a clean room, space-flight components and even tools are cleaned, usually using an evaporative solvent. Within the clean room, air is constantly circulated and filtered to keep floating particulates down to a specified level. Personnel who work in the clean room are required to wear special clothing, including booties for their feet and hats over their hair to minimize contaminants from their bodies. Gloves are always used to keep body oils off flight components.

Once a spacecraft is assembled it is tested to ensure that it was assembled correctly, that it will survive the trip into space, and that it will operate correctly when it arrives on station. Comprehensive tests exercising all of the spacecraft and instrument systems are performed both in ambient conditions within the clean room, and under conditions simulating the anticipated space environment. In order to simulate the space environment, large thermal/vacuum chambers (Fig. 18) that can contain the entire spacecraft are used. These chambers use pumps and cryogenics to reach the almost zero pressure and extremely low temperatures that will be experienced by the satellite. Pass-through connectors must be built into the walls of the chamber to allow external test connections to the spacecraft for power and data transfer.

To ensure that a satellite will survive the forces encountered during the rocket flight into space, the entire spacecraft is placed on a vibration table and "shaken" in all three axes up to and slightly beyond the levels that will be experienced. In another room, the spacecraft is subjected to low-frequency sound waves at decibel levels simulating the noises generated within the rocket's payload shroud during launch. Other pre-launch tests include electromagnetic interference/electromagnetic compatibility (EMI/EMC) tests to ensure that no spacecraft component emits any spurious signals or will not perform correctly if subjected to any expected signals. This test requires a large dedicated room covered on the inside with anechoic materials to absorb these radiations and prevent reflected signals.

Throughout this process, quality-control inspectors keep watch to ensure that proper procedures are being followed. This includes ensuring that the parts used to construct the satellite are of the proper quality and have been tested to prove adequate performance. In most cases, components and subsystems

FIG. 17. Clean room. Notice the wall of filters used to keep the air in the room clean.

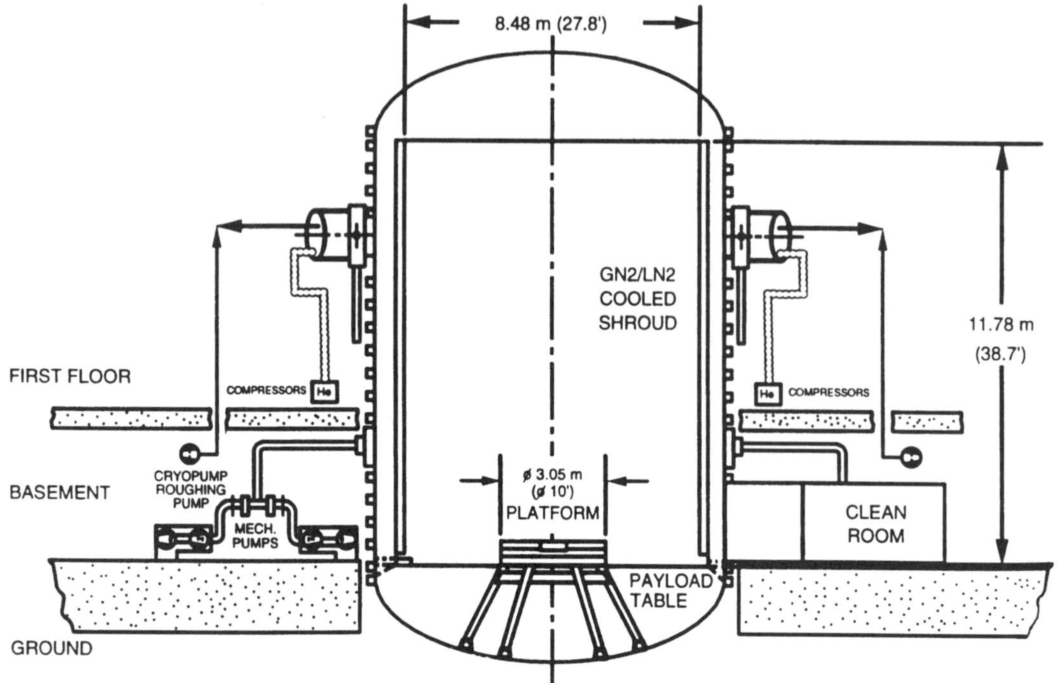

FIG. 18. Thermal vacuum chamber. Large chambers like this are required to accommodate a complete spacecraft. The chambers utilize pumps and cryogenics to expose the spacecraft to the expected environment it will experience in space.

have already been subjected to the same types of tests that the completed spacecraft will be subjected to after assembly.

5. ECONOMIC ASPECTS OF ARTIFICIAL SATELLITES

As mentioned several times previously, satellites represent an extremely large investment. A typical communications satellite may cost up to $\$50 \times 10^6$ to buy and as much or more to place into space. Satellites also require ground stations to communicate with and control the satellite. If users are involved, remote ground equipment is also required. The GPS system mentioned earlier consists of a constellation of 24 satellites to ensure continuous positioning capability worldwide. This system represents an investment of many billions of dollars to establish, and will continue to cost taxpayers hundreds of millions to maintain and replenish satellites as they fail. The Hubble Space Telescope cost over 10^9 dollars and is scheduled for repair missions from the Space Shuttle (which itself costs around $\$500 \times 10^6$ per flight) every two to three years. The Earth Observation System (EOS) is projected to cost over $\$7 \times 10^9$ for the constellation of satellites that will monitor the effects of man on the Earth's environment, and the International Space Station program will cost over $\$30 \times 10^9$.

As one can imagine, competition to supply hardware and services for these space endeavors among satellite manufacturers is intense. And while the rewards may be great, the risk is also great considering that, with the exception of communications satellites and other repeated satellite designs, each system is usually new and unproven. Since, in most cases, the satellite cannot be repaired once in space, systems must be robust and redundant or overdesigned. However, as this adds to weight and costs, the amount of robustness must be considered in design trade-offs. Usually, a particular launch vehicle is chosen at the start of design, and having to change to a larger vehicle because of increased weight may require some redesign that may cause the program to exceed funding limits. This could result in a delay of mission start, compromise of mission objectives, or even termination of the entire project, all of which have occurred in the past. Though launch vehicles are generally very reliable, insurance rates can be as high as 25% of the total cost of the satellite.

However, considering the benefits realized in communications, entertainment, navigation and national security, weather and environmental monitoring, and the general increase in knowledge of the Earth, the Universe, and our place in it, the costs associated with the use of artificial satellites are more than justified.

GLOSSARY

a: Orbit semimajor axis.

AMAO: Advanced Missions Analysis Office at NASA's Goddard Space Flight Center.

AO: Announcement of Opportunity.

COBE: Cosmic (Ray) Background Explorer.

Comsat: Communications Satellite Corporation.

COPUOS: U.N. Comm. on Peaceful Uses of Outer Space.

DMSP: Defense Meteorological Satellite Program.

DOD: Department of Defense.

e: Orbit eccentricity.

EMC: Electromagnetic compatibility.

EMI: Electromagnetic interference.

EOS: Earth Observation System.

ERTS: Earth Resources Technology Satellite.

ESA: European Space Agency.

EVA: Extravehicular activity.

GEO: Geostationary Earth orbit.

GOES: Geostationary Operational Environmental Satellite.

GPS: Global Positioning System.

GRO: Gamma Ray Observatory.

GSFC: Goddard Space Flight Center.

HEAO: High Energy Astronomy Observatory.

HEO: High-altitude Earth orbit.

HST: Hubble Space Telescope.

i: Orbit inclination.

ICBM: Intercontinental ballistic missile.

Intelsat: International Telecommunications Satellite Consortium.

ISEE: International Sun-Earth Explorer.

IUE: International Ultraviolet Explorer.

LDEF: Long Duration Exposure Facility.

LEO: Low-altitude Earth orbit.

NASA: National Aeronautics and Space Administration.

NOAA: National Oceanographic and Atmospheric Administration.

OAO: Orbiting Astronomical Observatory.

OSO: Orbiting Solar Observatory.

Phase A: Analysis phase of mission design.

Phase B: Definition phase of mission design.

Phase C/D: Execution phase of mission design.

Phase E: Operations phase of mission.

Pre-Phase A: Conceptual study phase of mission design.

ROM: Rough order of magnitude.

RTG: Radioisotope thermal generator.

SAR: Synthetic-aperture radar.

SMM: Solar Maximum Mission.

SPOT: Satellite Probatoire d' Observation de la Terre.

T: Orbit period.

TDRS: Tracking and Data Relay Satellite.

TDRSS: TDRS system.

TIROS: Television Infrared Observation Satellite.

TOMS: Total Ozone Mapping Spectrometer.

U.N.: United Nations.

UARS: Upper Atmosphere Research Satellite.

UV: Ultraviolet.

v: True anomaly.

ω: Argument of perigee.

Ω: Right ascension of the ascending node.

μ: Gravitational parameter.

Further Reading

Allahdadi, F. (Ed.) (1993), *Space Debris Detection and Mitigation,* SPIE Proceedings No. 1951, Bellingham, WA: SPIE.

Bate, R., Mueller, D., White, J. (1971), *Fundamentals of Astrodynamics,* New York: Dover.

Bilstein, R. (1989), *Orders of Magnitude, A History of the NACA and NASA, 1915–1990,* Washington, DC: U.S. Government Printing Office.

Clark, P. (1988), *The Soviet Manned Space Program,* New York: Orion Books.

Dutton, L. (1990), *Military Space,* London: Brassey's.

Feher, K. (1983), *Digital Communications, Satellite/Earth Station Engineering,* Englewood Cliffs, NJ: Prentice-Hall.

Flight Projects Directorate (1994), *Project Management Handbook,* Greenbelt, MD: NASA/Goddard Space Flight Center.

Flury, W. (Ed.) (1993), Special issue on Space Debris, *Adv. Space Res.* **13** (8).

Griffin, M., French, J. (1991), *Space Vehicle Design,* Washington, DC: American Institute of Aeronautics and Astronautics.

Jasentuliyana, N. (Ed.) (1992), *Space Law: Development and Scope,* Westport, CT: Praeger.

Kerrod, R. (1988), *The Illustrated History of NASA,* Anniversary Ed., New York: Gallery Books.

Logsdon, T. (1992), *The Navstar Global Positioning System,* New York: Van Nostrand Reinhold.

McElroy, J. (1986), *Space Science and Applications,* New York: IEEE.

National Research Council Committee on Space Debris (1995), *Orbital Debris: a Technical Assessment,* Washington, DC: National Academy Press.

Pisacane, V., Moore, R. (1994), *Fundamentals of Space Systems,* New York: Oxford Univ. Press.

Schnapf, A. (1985), *Monitoring Earth's Ocean, Land, and Atmosphere from Space—Sensors, Systems and Applications,* Washington, DC: American Institute of Aeronautics and Astronautics.

Simpson, J. (Ed.) (1994), *Preservation of Near-Earth Space for Future Generations,* Cambridge, England: Cambridge Univ. Press.

Tascione, T. (1984), *Introduction to the Space Environment,* Colorado Springs: U.S. Air Force Academy.

Wertz, J., Larson, W. (1991), *Space Mission Analysis and Design,* Boston: Kluwer.

Winkler, L. (1990), "Legal Aspects of Astronomy," *Griffith Observer* **54** (8), 2–9.

SCANNING ACOUSTICAL MICROSCOPY

BUTRUS T. KHURI-YAKUB, *Edward L. Ginzton Laboratory, Stanford University, Stanford, California, U.S.A.*

GLEN WADE, *Department of Electrical and Computer Engineering, University of California at Santa Barbara, Santa Barbara, California, U.S.A.*

ALBERT C. WEY, *Sonoscan, Inc., Bensenville, Illinois, U.S.A.*

	Introduction	545
1.	**Historical Perspectives**	546
2.	**Scanning Laser Acoustic Microscope (SLAM)**	546
2.1	SLAM Configurations	547
2.1.1	Solid-Surface Systems	547
2.1.2	Holography	547
2.1.3	Tomography	548
2.2	Systems Considerations	548
2.2.1	Laser-Beam Readout	548
2.2.2	Scanning Techniques	549
2.2.3	Detection Techniques	549
2.2.4	Configuration of Elements	551
2.2.5	Numerical Aperture	551
2.2.6	Resolution	552
2.2.7	Sensitivity	552
2.3	Applications	553
2.3.1	Biomedical	553
2.3.2	Nondestructive Evaluation	553
2.3.3	Materials Technology	553
2.3.4	Electronic Components	553
3.	**Scanning Acoustic Microscope (SAM)**	554
3.1	Principles of Operation	554
3.2	Transducer, Lens, and Test Media	556
3.2.1	Transducers	556
3.2.2	Buffer Rod	556
3.2.3	Lenses	556
3.2.4	Matching Layers	556
3.2.5	Operating Media	557
3.3	System Configurations	558
3.4	Scanning	559
3.5	Response of the SAM	560
3.6	Imaging Applications	562
3.6.1	Biological Applications	562
3.6.2	Materials Science	563
3.7	Metrology	565
3.7.1	Amplitude Only	565
3.7.2	Amplitude and Phase	566
4.	**Near-Field Scanning Acoustic Microscope (NFSAM)**	567
4.1	Tip NFSAM	568
4.2	Pinhole NFSAM	569
5.	**Conclusions**	569
6.	**Acknowledgments**	569
	Glossary	569
	Works Cited	570
	Further Reading	571

INTRODUCTION

Scanning electron microscopes, optical microscopes, scanning tunneling microscopes, and atomic-force microscopes are some of the imaging tools that have shown us, and taught us, a lot about the structure of materials in a myriad of fields and applications. Mostly, it is the surfaces of materials that have been the field of view of all these instruments, as none can show structure below the surface, especially in opaque materials. More importantly, the mechanical properties of the material and their variations are not the object of imaging of any of the above-named instruments. They measure changes in the index of refraction or topography of a material surface. Acoustic microscopes have been developed to answer a need for visualizing variations in the mechanical properties of a sample, especially below the surface. The scanning laser acoustic microscope, a transmission-imaging instrument, images the interior of samples by visualizing the shadow cast by material-property variations in the path of the acoustic beam. The scanning acoustic microscope, a reflection-imaging instrument; images the surface and near-surface variations in the mechanical properties of a sample. The near-field scanning acoustic microscope can be either a reflection or a transmission instru-

3-527-28138-X/96/$5.00 + .50

ment; however, its resolution is determined by the size of an aperture or a tip that is used to transmit and/or receive the signal. This article describes the principles of operation of these three instruments and shows some of the many applications where they have found a home.

1. HISTORICAL PERSPECTIVES

The ability to use sound for "seeing" has long been the dream of technologists throughout the world. A number of scientists and engineers have contributed ideas that have played a part in the realization of that dream. Some of the early thinkers in this regard include Leonardo da Vinci, Daniel Colladon, and Charles Sturm on passive sonar, and Lord Rayleigh, O. P. Richardson, Hiram S. Maxim, Paul Langevin, and M. C. Chilowski on active sonar (Lasky, 1977; Wade, 1987). S. J. Sokolov invented a variety of clever instruments to "see" with sound including a liquid–surface system and a bulk-diffraction system for detecting flaws in metal test pieces. In both systems the readout was accomplished by means of coherent light. He was also interested in acoustic microscopy and proposed an "ultrasonic microscope" (Sokolov, 1949) for using sound at 3 GHz where the wavelength would be short enough to make the instrument capable of resolving truly minute objects. Reimar Pohlman, a German contemporary of Sokolov, proposed the employment of a unique type of cell for instantaneous ultrasonic imaging where the image information is read out by reflected light (Pohlman, 1937).

SLAM was developed by Korpel, Kessler, and colleagues in the early 1970s (Korpel and Kessler, 1971; Korpel *et al.*, 1971). The research on SAM started with Calvin F. Quate and his colleagues in 1973 (Lemons and Quate, 1974). Since then, researchers throughout the world have joined in a sizable and effective international study of their potentialities.

Early demonstrations of near-field imaging were done using microwaves (Ash and Nichols, 1972), acoustics (Durr *et al.*, 1980), and optics (Pohl *et al.*, 1984). More recently, scanning tips such as those used in scanning tunneling and atomic-force microscopes have been used to image surfaces and variations in the mechanical properties of surfaces with atomic resolutions. These instruments are gaining in acceptance and applications not unlike the SLAM and SAM.

2. SCANNING LASER ACOUSTIC MICROSCOPE (SLAM)

SLAM works by introducing a continuous plane wave of ultrasound at frequencies up to several hundred megahertz into the bottom surface of a microscopic object as shown in Fig. 1. A fluid coupling (water in the figure) is required to bring the ultrasound to the object. As the ultrasound travels through the object, internal structural variations differentially attenuate and scatter it, resulting in a nonuniform wave pattern at the exit side of the object that corresponds to the localized acoustic properties. This pattern is duplicated onto a detector plate (labeled the coverslip in the figure), which is basically a metallized plastic block forming an optical mirror. The coverslip is placed in

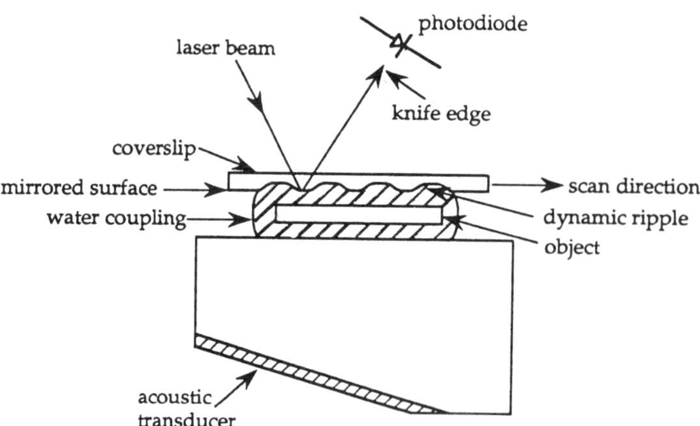

FIG. 1. Schematic diagram of the scanning laser acoustic microscope (SLAM).

close proximity to the sample and is acoustically coupled with fluid. The mirror becomes slightly distorted by the pressure of the ultrasound at each point. A laser beam is focused onto the bottom surface of the coverslip to detect the ultrasonic pattern by measuring the minute distortions of it. The reflected laser signal is converted into a video signal, and the acoustic image appears on a cathode-ray tube (CRT) screen. By rapid sweeping of the laser beam, each SLAM image is created in $\frac{1}{30}$ of a second, thus allowing true real-time inspection of the object (Kessler, 1974).

Key concepts associated with SLAM operation include point-by-point optical scanning, dynamic-ripple diffraction, real-time imaging, scanning laser-beam probe, and synchronously scanned television monitor. These concepts must be taken into account in analyzing SLAM's performance. They include classical laws of diffraction as applied to both light and sound and make use of mathematical tools worked out by such pioneer classicists as Rayleigh, Abbe, Michelson, Kirchhoff, and Fresnel. Also needed, however, in theoretical treatments of SLAM are more modern methods of systems analysis such as transfer-function theory, convolution, impulse response, sampling theory, and other methods associated with the communication-systems–theory point of view. First, we consider the different types of SLAM systems; then we address the issues relating to the performance of these systems.

2.1 SLAM Configurations

2.1.1 Solid-Surface Systems The laser beam in the SLAM impinges upon a solid surface that can support a moving ripple if struck by a continuous wave of ultrasound. The ultrasonic waves first are scattered by internal anomalies in the object. These waves then cast a nonuniform wave pattern onto the coverslip. As the waves strike the lower surface of the coverslip (see Fig. 1), they generate a moving ripple pattern on that surface. This is because the surface is dynamically distorted by particle displacements in the incident and reflected ultrasonic beams. The distortion pattern moves across the surface with velocity $V/\sin q$, where V is the sound velocity in the liquid and q the incident angle. SLAM uses a rapid-scanning laser beam to read out the information carried by the surface perturbation on the coverslip. The use of a scanning laser beam has a number of attractive features. For example, it makes it possible to discriminate between dynamically and statically scattered light because the former is Doppler shifted by the ultrasound frequency. Also, because the electrical output is at the ultrasonic frequency, electronic image processing in phase and amplitude is readily accomplished.

2.1.2 Holography The most essential characteristic of a hologram is that it records both phase and amplitude information. In contrast, SLAM, as ordinarily used, records only the amplitude. To perform holographic reconstruction, SLAM must be modified to acquire and utilize the phase information as well. When both the phase and the amplitude are employed properly in reconstructing the image, we obtain a hologram. A commercially available instrument that does this has been named HOLOSLAM.

To construct HOLOSLAM, a quadrature detector is incorporated into the circuitry of a conventional SLAM so that it can obtain the amplitude and phase information necessary for holographic reconstruction. Quadrature detection is accomplished by multiplying the detected signal with two coherent electronic reference signals, the phase of one being shifted by 90° with respect to that of the other. The two outputs thus obtained represent the real and imaginary parts of the complex amplitude of the ultrasound detected by the laser beam.

The two competing images (i.e., the real and virtual images), ordinarily present in conventional holographic reconstruction, do not both need to be reconstructed with this approach. The two bands of signal, representing respectively the positive and negative spatial-frequency components, are both processed by quadrature detectors. Because of this, HOLOSLAM has four output ports as shown in the schematic diagram of Fig. 2.

The entire operation is summarized as follows: A laser beam scans across the shiny dynamically rippled surface as in the case of SLAM. The angular displacement of the beam is detected by a knife-edge and photodiode combination, whose output is an electronic signal carrying the spatial information of the ultrasound wave field. This electronic

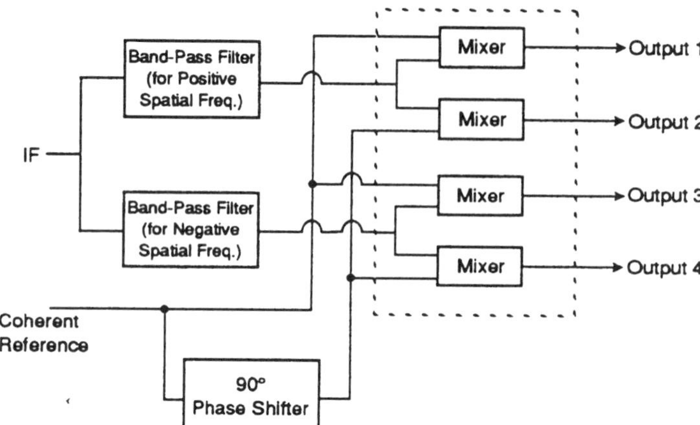

FIG. 2. Schematic diagram of the electronic circuitry for the HOLOSLAM.

signal is down-shifted to an intermediate frequency (IF). As indicated in the figure, the IF signal is then band-pass filtered by two filters to provide two signals corresponding respectively to the positive and negative spatial-frequency variations. The two signals are then processed by two quadrature detectors to produce four output signals in the video-frequency range. These output signals constitute the holographic data that are then used in straightforward fashion for holographic image reconstruction (Lin *et al.*, 1987).

2.1.3 Tomography To enhance the performance capabilities of the SLAM, tomographic principles as well as holographic principles are used separately or simultaneously. One technique for applying the former principles is called scanning tomographic acoustic microscopy (STAM). As in the case of HOLOSLAM, STAM utilizes digital signal processing of image information obtained with a modified SLAM. It is capable of producing authentic tomograms (cross-sectional images) and thus of overcoming the ambiguities associated with SLAM images when the microscopic objects are thick and structurally complex.

STAM is most suitable for the subsurface imaging of objects with planar structure. Using plane-wave insonification, the modified SLAM performs the data acquisition, after which digital processing of these acquired data accomplishes the image reconstruction. For planar geometry, STAM image reconstruction can be efficiently accomplished through the use of the so-called back-and-forth propagation (BFP) algorithm (Chiao and Lee, 1991). If this algorithm is used only once, that is, for a single viewing angle, the reconstruction is purely holographic. That is the case in HOLOSLAM. However, by applying the BFP algorithm over and over, that is, for a number of viewing angles, the BFP algorithm will produce a tomogram. Simulation and experimental results both show that STAM images can be significantly better, that is, more highly resolved, than images from either SLAM or HOLOSLAM.

Since the resolution improvement of STAM is directly related to the detection of the phase component of the sound field, STAM image reconstruction is sensitive to phase errors in the tomographic projections. If phase errors are present, the resolving power of the microscope may be reduced and unwanted artifacts may show up in the image. To correct for phase errors we can estimate the effect of these errors in the image reconstruction and calculate the correction to obtain the final image (Chiao and Lee, 1991). Operational results confirm the validity of this correction technique.

2.2 Systems Considerations

2.2.1 Laser-Beam Readout SLAM employs rapid optical scanning for detecting the acoustic fields point by point. This provides a number of important features. For example, since the sound-field parameters are ultimately translated into electrical signals, electronic filtering may be utilized to reject unwanted background such as noise. With these signals it is easy to choose between conventional imaging and holographic imaging simply by changing between linear detection and phase detection. Both lasers and laser-beam scanning devices are readily available, and the construction of an acoustic imaging system using them is relatively straightforward. Finally, with the appropriate devices and TV-compatible circuitry, real-time imaging is possible.

2.2.2 Scanning Techniques A key element in the SLAM scanning mechanism is an all solid-state acoustic-optic (A-O) light deflector. Bragg diffraction of laser light by acoustic waves produces horizontal deflection similar to that in a television screen and permits continuous scanning at rates up to 20 kHz or greater. The constant rate of angular change that characterizes a television scan permits the use of a wide optical aperture, leading to a small spot size.

The A-O light deflector is manufactured from high-quality flint glass as the interaction medium with indium-bonded lithium-niobate piezoelectric transducers for generating the acoustic waves. These waves are frequency modulated with a sawtooth function. A light beam entering the acoustic field at nearly normal incidence is deflected at an angle proportional to the acoustic frequency.

A configuration for up-shifted Bragg diffraction is illustrated schematically in Fig. 3. When Bragg-diffraction conditions are satisfied, the beam is deflected as if it were being reflected off the moving wave fronts of the sound. The phenomenon is reminiscent of the way Bragg-diffracted x rays are deflected off atomic planes in a crystal. A sound column acts, in essence, like a continuous distribution of "reflectors," and the optical rays that are reflected off successive crests add up constructively.

A frequency deviation of 40 MHz centered at 70 MHz will deflect a laser beam of wavelength 633 nm through an angle of 6.5 mrad centered at 11.4 mrad from the undeflected beam. Since the deflector utilizes phased-array beam steering, the deflected light will have nearly constant intensity across the deflection angle.

Resolution depends on the time–bandwidth product, which is determined by the length of the optical aperture and the bandwidth of the transducer. The actual aperture of the A-O deflector is 40 mm, which represents a transit time of 10 μs. With a 40-mm aperture and a transducer bandwidth of 40 MHz, 400 spots of resolution are achieved. The transit time is the time it takes to clear the sound from the optical aperture and is thus the switch time in a random-access, 400-spot system. A tradeoff exists between resolution and access time—the smaller the beamwidth, the faster the access time, with correspondingly less resolution.

2.2.3 Detection Techniques In current commercially available SLAMs, the amplitude of the scattered acoustic waves from the object is detected by a knife-edge and photodiode combination. Figure 1 illustrates the setup. A focused laser beam is directed toward the shiny surface upon which impinge the scattered acoustic waves, having been spatially modulated by the object. The angular position of the light reflected from that surface is modulated in turn by the dynamic activity of the corrugations produced by the acoustic waves. The photodiode, which detects the power carried by the light in the beam, is not, by itself, sensitive to the angular modulation of the beam. However, an appropriately positioned obstacle, such as a knife edge, will block a fraction of the light

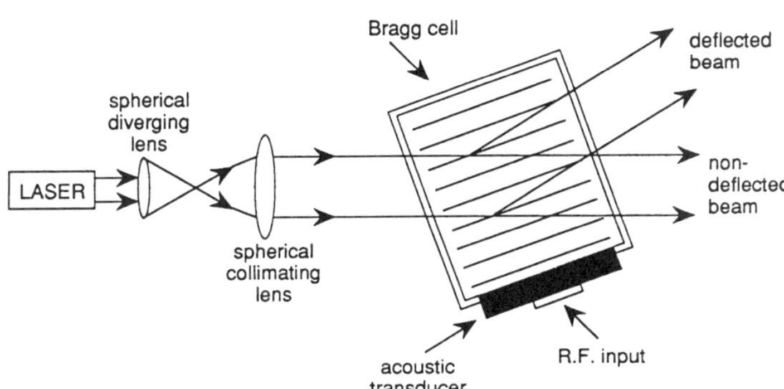

FIG. 3. Acoustic-optic light deflection using Bragg diffraction of laser light by acoustic waves.

depending upon the instantaneous position of the beam. Thus the knife edge demodulates the angular position fluctuations into corresponding intensity fluctuations. The knife-edge technique is very simple and is quite adequate in providing the detection capability needed by conventional SLAMs.

The transfer function for this approach, shown in Fig. 4, is antisymmetric in its response to spatial frequencies with motion in the scan direction. For negative spatial frequencies (corresponding to sine waves traveling on the surface in the opposite direction), the detected signal is opposite in phase to that for positive frequencies. For spatial frequencies perpendicular to the scan direction, the transfer function is symmetrical as shown (Rylander, 1982).

In ordinary SLAM operation, the insonifying wave is directed at an oblique angle of incidence against the object rather than at normal incidence. This is done because the scattering of a wave at normal incidence from an object generally produces a spatial-frequency spectrum on the shiny surface that would be centered at the origin and have its largest components there. From the transfer-function plot, we can see that such an object would not be effectively detected by the knife-edge system, at least as far as the waves traveling perpendicularly to the knife-edge direction (and hence in the scan direction) are concerned. For those waves the transfer function at the origin is zero. However, the spatial-frequency spectrum of the scattered waves may be easily biased about a finite, nonzero spatial frequency by employing obliquely (rather than normally) incident insonification. This is done in SLAM in order to avoid asymmetrical response about the spatial frequency corresponding to the bias; one sideband of the scattered spectrum is filtered out. This is quite suitable for real objects because of their generally symmetrical sidebands. Single-sideband detection with a synchronous detector can thus be used to reconstruct the images of such objects.

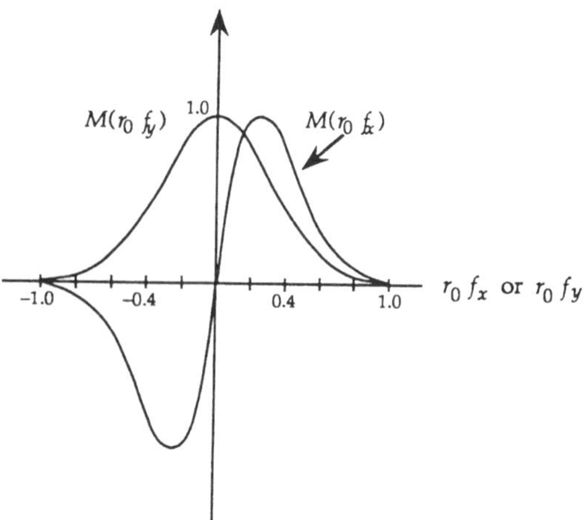

FIG. 4. Normalized transfer functions $M(r_0 f_y)$ and $M(r_0 f_x)$ of the knife-edge detector oriented perpendicular to the scan direction of the laser beam (i.e., in the x direction). The beam is assumed to have a Gaussian intensity profile. r_0 is the nominal spot radius and f_x, f_y are spatial frequencies.

The approach for image reconstruction must be somewhat different in HOLOSLAM and STAM. With these two microscopes the acquired signal constitutes the data to be digitally processed for holographic or tomographic image reconstruction. For example, the frequency spectrum corresponding to the acquired data is normalized by using inverse filtering or some other signal-processing technique to compensate for the nonuniform response of the knife-edge approach.

Finally, it should be pointed out that the peculiar shape of the knife-edge transfer function has a consequence in terms of resolution. In order to avoid the null response at the zero spatial frequency, the bandwidth of the spectrum of the detected acoustic wave field has to be limited so that it does not extend beyond the positive part of the transfer function. This, of course, brings about a limitation in the resolution of the reconstructed image.

Because of the above limitation, other, more complicated types of detection have been proposed for these microscopes. One such detection scheme uses a form of homodyne detector called the time-delay interferometric detector (TDID) (Rylander, 1982). The transfer function for the TDID is symmetric and has the maximum response for zero frequency. The detected acoustic wave field can have a greater bandwidth of the spatial-frequency spectrum with the TDID than with the knife-edge detector. Hence, higher resolution of the reconstructed image should be possible. The TDID is not currently being employed with these microscopes because the knife-edge detector is much simpler to implement.

2.2.4 Configuration of Elements Figure 5 shows all essential optical components of a typical scanning laser module in SLAM. A combination of a cylindrical lens, L1, and a spherical lens, L2, is used to shape the laser beam to make it suitable for the deflector aperture, which is 40 mm wide by 2 mm high. The laser beam is expanded horizontally by a 20× telescope (L1 and L2) to fill the 40-mm–wide aperture of the deflector. The lens L2 also vertically compresses the laser beam to 1 mm in height. After passing through the A-O deflector, the laser beam is compressed horizontally by a 0.125× telescope (L3 and L4) and vertically collimated by a lens, L3. Only the deflected laser beam is allowed to exit the scanning laser module, with a spot size of 2.5 mm in diameter. Finally, the beam is deflected vertically by a galvanometer-type mirror, driven by a current of appropriate wave form.

A knife-edge detection setup is employed to detect the periodic deflection of the probing optical beam caused by the acoustic surface perturbation of the coverslip. The knife edge lies in the Fourier-transform plane of the scanning laser module (i.e., the focal plane of L5) and therefore in the image plane of the scanner. This means that the laser beam always strikes the knife edge and the photodetector in the same place regardless of the scan angle. Conventional optical magnification may be inserted into the optical path in order to produce simultaneously a focused spot size appropriate to the resolution requirement for a given ultrasonic frequency and an optimal field of view.

2.2.5 Numerical Aperture SLAM's sensing system involves a mirror on the coverslip and a knife-edge detector. The characteristic for this system as a function of the angle of ultrasound incidence determines the numerical aperture (NA) of the system. The

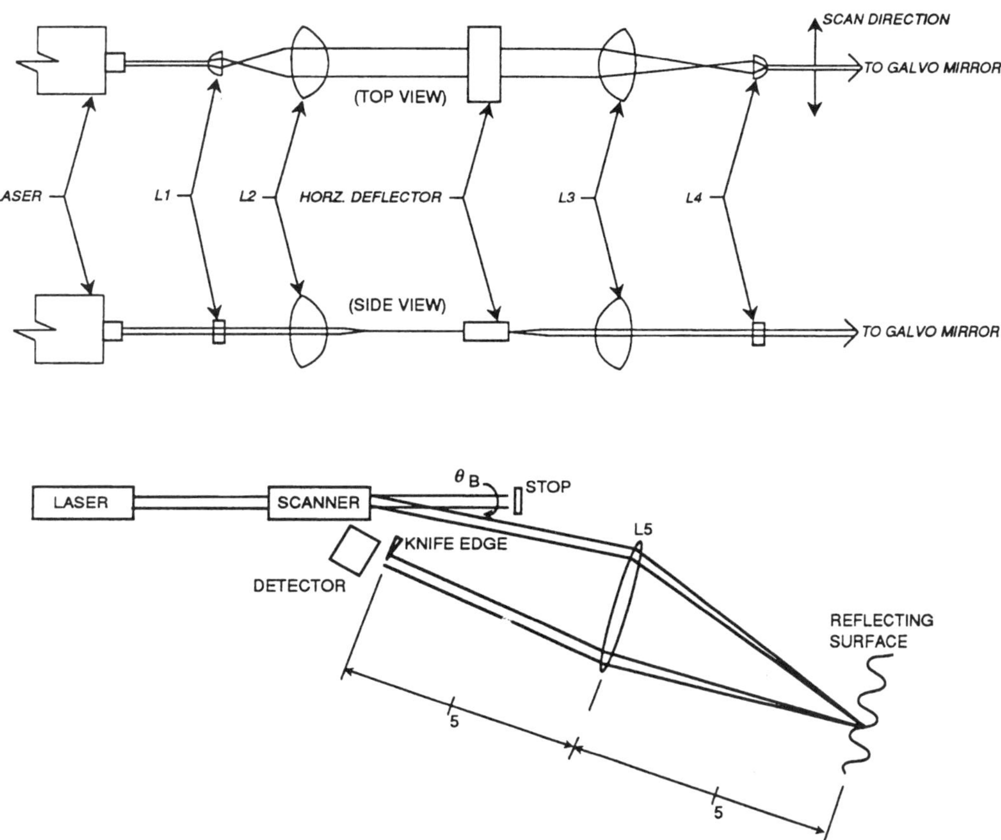

FIG. 5. The essential optical components of the scanning laser module in SLAM.

numerical aperture is defined as the sine of half the apex angle of the cone of rays admitted by the system. In SLAM, the numerical aperture is governed by the extent of the angular spectrum of sound diffracted by the specimen and detected by the coverslip. The coverslip in commercially available SLAMs is made of polycarbonate, in which the sound velocity is 2200 m/s, thus giving rise to a single critical angle at 43°. Assuming the "detectable" part of the spectrum to extend from the angle of incidence of the carrier, typically at 10°, to an angle somewhat lower than the critical angle of 43°, say 33°, we find for the numerical aperture of SLAM

$$\mathrm{NA} = \sin 33° - \sin 10° = 0.37.$$

2.2.6 Resolution The resolution R of SLAM is governed by the familiar relationship

$$R = \frac{\lambda/2}{\mathrm{NA}},$$

where l is the acoustic wavelength and NA, the numerical aperture. For example, at a frequency f of 100 MHz, with the sound velocity v in water at 1500 m/s and the numerical aperture at 0.37, we find for the acoustic resolution

$$R = \frac{v/2f}{\mathrm{NA}} = 20\ \mu\mathrm{m}.$$

Other ultrasonic frequencies available in a commercial SLAM are 10 and 30 MHz, which have acoustic resolutions of 200 and 67 μm, respectively.

2.2.7 Sensitivity In any imaging system the ultimate limitation on the detection of image information from weak signals is the inevitable existence of noise. Obviously no amount of amplification can make it possible to derive information from a signal whose power is substantially below the level of the noise entering the receiver. Laser-beam systems are inherently noisy because lasers suffer from high quantum noise. It is produced by the quantum limit on the possible accuracy of measurements involving an electromagnetic field. The quantum-noise power is given by $h\nu$ where h is Planck's constant and ν is the frequency of the laser beam.

For piezoelectric systems, the type of noise of greatest fundamental importance is the so-called "thermal noise." The thermal-noise power corresponding to a bandwidth B is given by the well-known expression kTB where k is Boltzmann's constant and T is the absolute temperature in kelvins. Thermal noise, in principle, can be eliminated simply by cooling all the elements of the system to absolute zero temperature. On the other hand, the expression for quantum noise shows that this noise is constant, and not amenable to reduction by physical means.

We can calculate the equivalent thermal-noise temperature of the readout process by equating the expressions for thermal noise and quantum noise and solving for T. Thus we get h/k for the noise temperature of the laser beam. If we enter this expression with the values of Planck's constant, Boltzmann's constant, and the frequency of the laser, we compute a temperature well in excess of 20 000 K. Hence the equivalent noise temperature associated with the readout process alone in the case of laser systems amounts to over 20 000 K.

From these calculations, we might guess that for ordinary circumstances the ultimate sensitivity of a laser system such as SLAM will be almost two orders of magnitude worse than that of a piezoelectric system such as SAM. Analytical study assuming ideal operating conditions for both systems has confirmed this expectation (Wang and Wade, 1974). The sensitivities achieved by these instruments in practice are, of course, not as good as those calculated in the study, the difference in the sensitivities actually achieved by the two systems being somewhat greater than the two orders of magnitude predicted.

The minimum detectable surface displacement is related to the sensitivity. It can be shown (Whitman and Korpel, 1969) that, with the knife-edge and photodiode detector, this displacement for SLAM is given by

$$\delta_{\min} = (2eB/\alpha P_0)^{1/2}(\lambda/2\pi),$$

where e is the charge on an electron, B the system bandwidth, a the responsivity of the photodiode, P_0 the laser power, and λ the optical wavelength. The conventional system uses a laser power of 10 mW and has a system bandwidth of 1 MHz, resulting in a calculated δ_{min} of 1.3×10^{-12} m. In practice, a minimum detectable displacement of 1×10^{-11} m has been demonstrated, which renders for SLAM an approximate 70 dB of dynamic range.

2.3 Applications

SLAM instruments are used in a wide variety of ways. Four of the chief categories of use are discussed below.

2.3.1 Biomedical SLAM provides a new capability for visualizing the structural characteristics of living tissue. Viable, unfixed, and unstained tissues retain much of their intrinsic, *in vivo* properties. The unique aspect of techniques made possible with SLAM is that they can differentiate elastic, viscoelastic, and density features of structure at the microscopic level and identify subtle changes in tissue architecture.

These changes are revealed by means of corresponding changes in the acoustic index of refraction, the acoustic absorption coefficient, and the acoustic impedance. This information is available in the acoustic micrographs and in the interferograms produced by SLAM.

Some of the studies undertaken using SLAM include embryological development of mice and chickens, physiology and pharmacology of hearts in organ culture, mammalian kidney tissue, metastatic adenocarcinoma in human liver, abnormal or diseased tissue, dentine and enamel components, collagen threads, muscle mechanics and wound maturation, bone pathologies, development of the organ structure of the fruit fly larva Drosophila, plant cell structure, and amphibian limb regeneration. SLAM has also been used to measure acoustic velocity in skeletal muscle, in rat tail tendon fibers, and in rat liver as a function of fat concentration.

2.3.2 Nondestructive Evaluation The most popular industrial application of SLAM is bond evaluation. Bonding, joining, and welding processes are extremely important in the manufacturing of a wide range of products. A variety of materials may be involved such as metals, ceramics, alloys, and polymers. Diffusion bonding, heat sealing, laminating, coating, and epoxy or adhesive bonding are a few examples of the many processes. Typical applications include the nondestructive evaluation of seam and spot welds, adhesive bonds, and laminated polymers.

SLAM has difficulty in scanning complex geometries. Nevertheless, special techniques have been worked out that are suitable for accommodating many complex shapes. Inspection of aircraft-engine turbine blades, for example, differs from the customary SLAM operation and requires a special fixture to control precisely the motion of the blade in the gap between transducer and coverslip. The interior of the blade consists of a network of channels for air cooling. Ribs that are bonded to each other separate the channels in each half of the blade. SLAM can distinguish between the bonded and unbonded ribs and can measure the percentage of bond for partially bonded ribs as well.

2.3.3 Materials Technology Another popular industrial application of SLAM is the nondestructive detection of voids, inclusions, cracks, and contamination in various materials. SLAM provides the means to visualize grain structure, grain boundaries, and impurities in metal specimens such as steel, copper, aluminum, and titanium alloy.

In structural applications, ceramics are used where high temperature and light weight are important. Silicon nitride, silicon carbide, and zirconia are receiving much attention for future engine applications. However, because of the inherent brittleness of ceramic materials, small defects are very critical to the structural integrity of the materials. SLAM has been very successful in nondestructive screening of ceramic samples.

Composite materials, with the combination of materials having different properties from the components and with manufactured anisotropy, have become a challenge for acoustic microscopists. SLAM can clearly define the relevant property distribution of materials by differential attenuation of ultrasound within the field of view. SLAM is used to detect fiber–matrix disbonding, matrix cracking, and interlaminar delamination in fiber-reinforced polymers, as well as in ceramic–matrix and metal–matrix composite samples that are subjected to tensile, fatigue, and impact tests.

2.3.4 Electronic Components Much attention has been given to the use of SLAM for nondestructive evaluation of microelectronics components, and for detecting cracks, disbonds, delaminations, voids, and porosity in these devices. SLAM has been extensively employed to detect disbond at molding-compound–paddle or molding-compound–die interfaces, and at the die attachment itself in plastic integrated-circuit packages such as plastic leaded chip carriers. SLAM is also used to optimize the process for consistent bond quality in tape automatic bonding interconnections. Other applications include, but are certainly not limited to, areas such as ceramic chip capacitors, substrates, surface-mount solder joints, thick film and ceramic dual in-line packages, or other hermetically sealed packages.

3. SCANNING ACOUSTIC MICROSCOPE (SAM)

3.1 Principles of Operation

The most common configuration is the reflection-type scanning acoustic microscope (SAM), as shown in Fig. 6. A piezoelectric material is attached to the flat end of a buffer rod to excite a plane-wave ultrasonic beam into the buffer rod. The other end of that rod has a spherical or cylindrical lens that focuses the sound beam, which is transmitted into the fluid. An electrical tone burst excites the transducer, thus transmitting a packet of ultrasonic energy that is focused by the lens to a diffraction-limited spot. Resolution is determined by the size of the focal spot, as it is with optical microscopes. However, because sound waves propagate at a velocity that is about five orders of magnitude slower than the speed of light, the resulting image resolution can be comparable to that of optical microscopes when operating in the gigahertz frequency range. A portion of the ultrasonic signal that is reflected by the sample and propagated back through the lens and onto the transducer, where it generates an electrical signal that is collected by the receiving electronics.

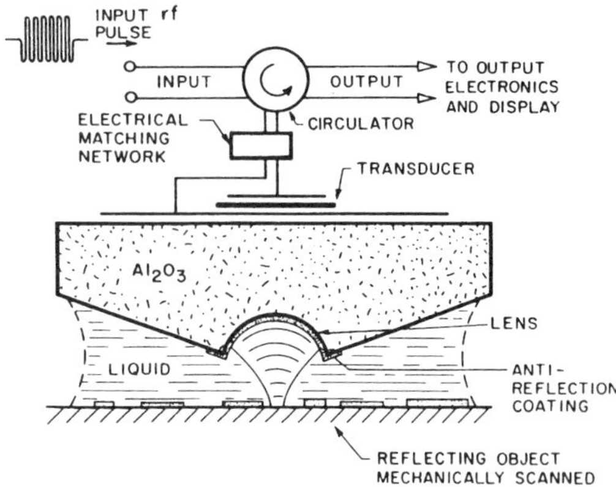

FIG. 6. Schematic diagram of a transducer–lens system for a scanning acoustic microscope (Quate, 1985).

The return signal's amplitude and/or phase is collected and used to modulate the intensity of the display monitor at a location corresponding to the location of the focal spot over the sample. By scanning either the transducer or the object, an acoustic image of the sample is formed on the display monitor. Typical scanning speeds are 20–30 s/frame for operating in the gigahertz frequency range, and for a field of view of up to 0.5×0.5 mm^2. For lower-frequency operation, a field of 5×5 cm^2 can be imaged in a few minutes.

Variations in the local mechanical properties of a sample or changes in the location of the surface with respect to the focus of the lens cause changes in the amplitude and phase of the reflected signal and appear as contrasted features in the image. Local variations can be induced by changes in orientation of the grains of a material, by the presence of a different material, or by defects at or below the surface of the sample.

SAM images show new features and do not blur as a lens is brought closer to a sample. This enhanced imaging capability is due to the physical interactions that take place between the ultrasonic field and the sample. When a focused ultrasonic beam in water impinges on a solid surface placed at its focus, part of the beam is reflected back, part is transmitted as longitudinal and shear waves into the bulk of the solid, and one part is converted into a surface wave that propagates on the surface. The energy in the surface wave leaks back into the water and away from the lens. As the lens is brought closer to a sample (defocused), the energy in the surface wave leaks back into the transducer. Thus, the reflected signal collected by the transducer is now made of the interference between the specular reflection from the surface of the sample and the mode-converted leaky surface wave. If there are defects in the sample, some of the energy that is propagated into the sample as longitudinal and shear waves can be reflected back and received by the transducer. Therefore, four types of acoustic signal paths can contribute signals at the transducer, and a variation in any of these signals changes the image.

Focused C-scan is the name of one variation of an acoustic microscope whose field of application is in imaging the internal structure of samples. A schematic of a focused C-scan imaging system is shown in Fig. 7. A focused transducer is excited with a broadband pulse, typically a half cycle at the frequency of operation, and the focus of the transducer is placed below the surface of sample at a plane where the defects are to be imaged. The preferred transducer has a broad bandwidth and short impulse response to separate in time the reflections from the interface from the reflection from within the sample. The surface of the sample and the near-surface region corresponding to the pulse duration are not inspected by this technique and are known as the "dead zone." To image other planes within the sample, the location of the focus is changed, and the sample is scanned again. The peak amplitude of the return echo modulates the intensity of the display monitor where the image is shown. Alternatively, several gates are used to measure the return signal at different locations, enabling imaging at more than one plane at a time. Yet another variation monitors the amplitude and time delay of the return signal from the back side of the sample, thereby generating a velocity map of the sample together with an image of total attenuation along the path of propagation.

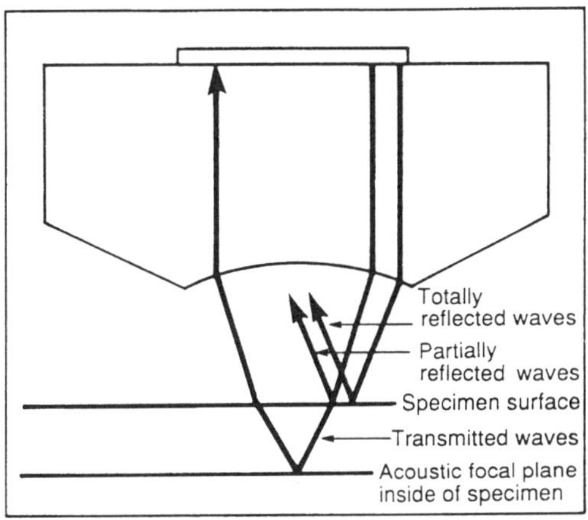

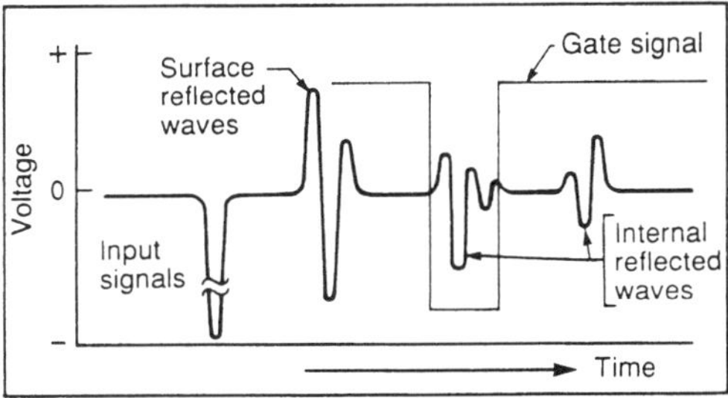

FIG. 7. Schematic diagram depicting the operation of a focused C-scan imaging system (courtesy of Olympus Corp.).

Thus differences between the C-scan and SAM are evident. The usual excitation of one is broad band (C-scan), and of the other (SAM) is narrow band. The transducer for the C-scan has a lens with a low *f*-number in order to enhance penetration into the sample, whereas for the SAM the *f*-number is low in order to excite surface waves on the sample and obtain the smallest focal spot for surface imaging. The domain of inspection of the C-scan system is the bulk of a sample, whereas the SAM does most of the work near the surface of the sample. Lastly, in a C-scan system, the longitudinal or shear wave is detected and interrogated for the presence of a defect. In a SAM, the detected signal is comprised of several components: the specular reflection, a mode-converted surface wave (or any near-surface mode that is allowed), and a possible reflection from a longitudinal or shear wave excited within the sample. It is important to note that at high frequencies the attenuation in samples reduces the penetration of the ultrasound beam, and limits the inspection to the surface of the sample.

It is also possible to defocus a transducer enough to separate the specular reflection temporally from the surface-wave reflection. This was first used in an acoustic microscope (Liang *et al.*, 1982) to make quantitative measurements of step heights and residual-stress variations. Later work (Yamanaka, 1982) used this phenomenon in a pulse-mode–excited system rather like the C-scan to measure variations of surface-wave velocity in ceramic samples. Another implementation (Gilmore *et al.*, 1986) used the surface-wave signal alone to image surfaces of samples for defects at or near the surface in a standard C-scan system; the spot size is about $0.2l_R$, where l_R is the surface-wave wavelength, at the 3-dB points because the surface wave is incident from a full circle around the focal spot. It is also possible to use the SAM with broadband excitation like a C-scan system to image solids and biological samples (Daft and Briggs, 1989).

3.2 Transducer, Lens, and Test Media

3.2.1 Transducers The transducer–lens system is at the heart of the SAM, and requires careful design to obtain good images and quantitative measurements. Various technologies and materials are available to make the transducer. Typically, a piezoelectric material such as zinc oxide or aluminum nitride is deposited in a vacuum chamber on the face of the buffer rod. Such a technology has proven capable of generating coherent ultrasonic waves in the frequency range of 30 MHz to 100 GHz. Thin metal bonding of a piezoelectric crystal, such as lithium niobate, has also been used for making ultrasonic transducers. Metal bonding has the advantage of allowing the use of materials with high piezoelectric coupling coefficients such as lithium niobate, which can be attached in different orientations to excite both longitudinal and shear waves. At frequencies below 50 MHz, it is possible to shape piezoelectric shells and use them to excite and focus the sound beam directly. The wide range of piezoelectric materials allows one to insure that the electrical impedance of the transducer is well matched to the electronics for optimum operation. Thin metal films, such as chrome–gold or aluminum, are used as the metallization defining the back contact and the top electrode of the transducer. The top electrode is used to define the active area of the transducer, and is chosen to reduce diffraction in the buffer rod and to give the desired illumination of the lens.

3.2.2 Buffer Rod Buffer rods are used as support for the transducer and to form the lens. Sapphire is the most widely used buffer-rod material at high frequencies because of its low attenuation (0.2 dB/cm at 1 GHz, with an f^2 dependence) and high ultrasonic velocity (11 100 m/s) compared with water (1500 m/s). The high velocity results in the reduction of spherical aberrations. However, sapphire has a high ultrasonic impedance ($Z_a = 44.3 \times 10^6$ kg/m^2 s) as compared with water ($Z_a = 1.5 \times 10^6$ kg/m^2 s); this impedance mismatch requires a matching layer ($Z_m = 8.15 \times 10^6$ kg/m^2 s) to improve the transmission from sapphire to water. Matching layers of borosilicate glass or carbon have been used for this purpose. These matching layers are deposited by physical sputtering or evaporation in a vacuum chamber. Fused quartz, silicon, aluminum and many other solids have also been used as buffer rods.

3.2.3 Lenses Lenses, spherical or cylindrical, may be made by grinding and polishing into the buffer-rod material. Chemical etching of materials such as silicon is another method of making lenses for acoustic microscopes. A typical 2-GHz lens in sapphire would have a radius of 50 μm and an aperture of 125 μm. At lower frequencies, both dimensions would increase proportionally. The lens surface must be free of scratches and defects, especially for high frequency operation (>1 GHz), where the wavelength of the ultrasonic beam in the water is a micron or less.

Fresnel lenses are yet a third type of lens being used in acoustic microscopes (Yamada *et al.*, 1986). Planar integrated circuit processing techniques are used to define these lenses in a process that allows making arrays of lenses, and consequently arrays of microscopes for scanning large areas.

3.2.4 Matching Layers A matching layer is usually attached to the lens to improve the transmission of the sound into the liquid. As in optics, the matching layer is a quarter wavelength thick, with the ideal impedance, which is the geometric mean between the impedances of the liquid and the buffer rod. The mechanical properties of several materials used in acoustic applications are summarized in Fig. 8. It is important to note that when the matching layer is deposited by a thin-film process, such as evaporation or sputtering, its mechanical properties will differ from those of the bulk material, requiring independent measurements to ascertain the mechanical parameters of the deposited material.

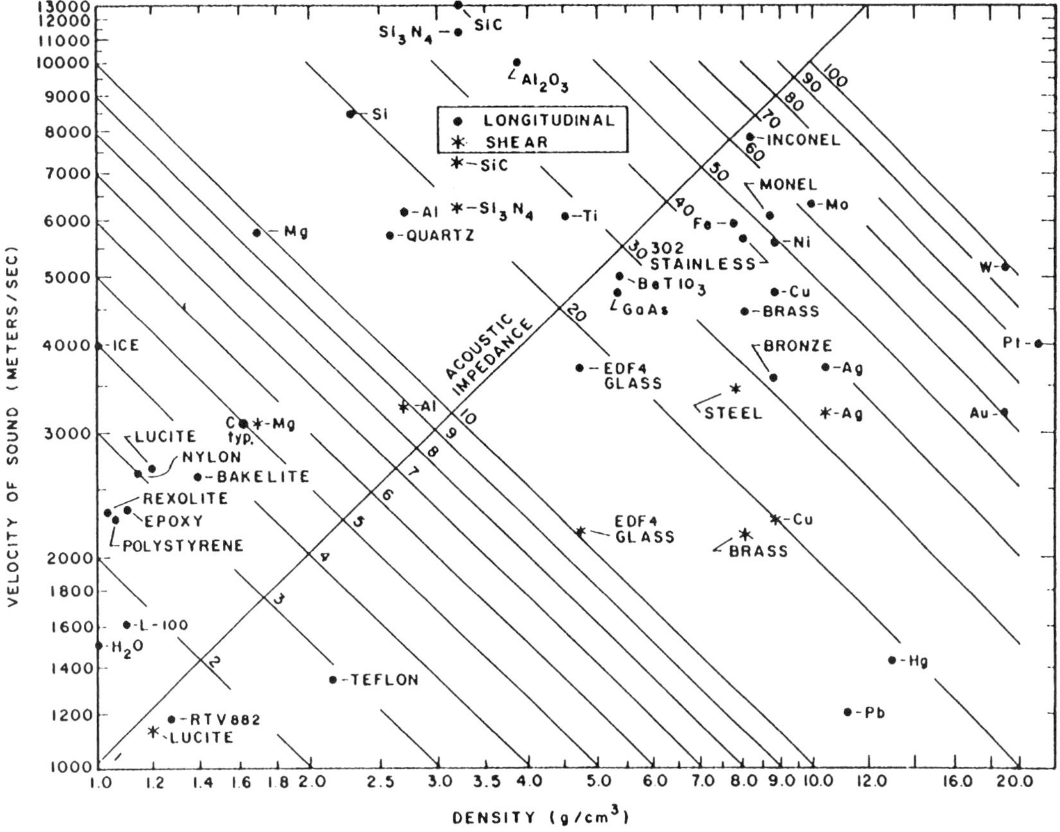

FIG. 8. Nomogram summarizing the acoustic properties of materials (after Eggelton, private communication).

3.2.5 Operating Media The properties of various liquids that are possible immersion media for acoustic microscopy are summarized in Table 1. Water is the usual medium in which samples are placed and tested, a natural and convenient choice. However, other liquids can offer advantages over water, particularly at very high frequencies (over 4 GHz) where attenuation in the water is prohibitive. Attenuation in water is

Table 1. Acoustic properties of selected liquids.

Liquid	Temperature	C (m/s)	$10^{17}\alpha/f^2$ (dB s^2/cm)	$10^{-5}Z$ (g/cm^2-s)
Water	25 °C	1495	191	1.5
Water	60 °C	1550	95	1.5
Methanol	30 °C	1088	262	0.866
Carbon disulfide	25 °C	1310	88 (3 GHz)	1.65
Mercury	24 °C	1449	50	16.7
Gallium	30 °C	2870	13.7	17.5
Nitrogen	77 K	850	120	0.68
Oxygen	90 K	900	86	1.0
Neon	27 K	600	201	0.72
Argon	87 K	840	132	1.2
Xenon	166 K	630	191	1.8
Helium	4.2 K	183	1966	0.023
Helium	1.95 K	227	610	0.033
Helium	0.4 K	238	15 (1 GHz)	0.035

2200 dB/cm at 1 GHz and at room temperature, and increases with the square of the frequency. For operation above 500 MHz the water is heated to about 60 °C where the attenuation is reduced to half its room-temperature value. The highest operating frequency achieved in water is 5 GHz with a resolution better than 2000 Å (Hadimioglu and Quate, 1983).

Liquid metals like gallium and indium are viable coupling fluids for acoustic imaging. As compared with water, liquid metals have high acoustic impedance and low attenuation, which are desirable features for increased signal level; however they also have larger velocities, which increase the wavelength and hence reduce the resolution. Liquid metals have been used very successfully (Attal and Cambon, 1978) to demonstrate that the improved impedance matches between the buffer-rod and liquid and between liquid and sample make it possible to penetrate deep into a sample. Features on silicon wafers have been imaged thus by looking through the back side of the wafer.

Air or gas at high pressure is another medium that has been used for imaging. Imaging in air affords the advantage of increased resolution because the speed of sound is five times smaller than in water, and it does not require wetting. One problem of working in air is the mismatch between the impedance of air and that of the sample and buffer rod. This mismatch is typically 4–6 orders of magnitude, which results in inefficient coupling of the sound into and out of the air, and which limits the instrument to imaging the topography of the sample only. Another problem is the large attenuation in the air, which is about 1.2 dB/cm at 1 MHz and which increases with the square of the frequency. These problems are somewhat reduced by operating at high pressure. An instrument operating at 500 MHz with argon gas at a pressure of 15 atm has been demonstrated (Wickramasinghe and Petts, 1982). This medium of operation has seen limited attention because of the risk of explosion in high-pressure vessels.

Cryogenic fluids such as liquid nitrogen and argon offer a medium of operation with low sound velocity and relatively low acoustic attenuation. Liquid nitrogen has been used for high-resolution imaging at frequencies up to 2.8 GHz. The best medium for high-resolution imaging by far is liquid helium at temperatures below 0.2 K, for it has very low sound velocity and negligible acoustic attenuation even at frequencies of tens of gigahertz. Imaging has been demonstrated at a frequency of 8 GHz with a resolution of 200 Å (Hadimioglu and Foster, 1984). Cryogenic liquids, like gases, allow very little coupling of the sound energy into a sample and are thus used mostly in surface-imaging instruments.

3.3 System Configurations

The first SAM was of the transmission type (Lemons and Quate, 1974). Two identical lenses were used, one to transmit the ultrasonic signal and the other to receive it after it had passed through a sample. This instrument proved difficult to align and operate, however, and yielded to the reflection-type microscope most widely used today. A typical schematic diagram for an amplitude-measuring SAM is shown in Fig. 9. The tone burst is derived by modulating a cw signal from a voltage-tuned oscillator (VTO) with a *pin* diode switch. The signal is applied to the transducer through a circulator, and the main pulse is gated out to avoid saturation of the receiver electronics. Heterodyne demodulation is used to detect the reflected pulse, which is then measured and stored in a digital computer via sample-and-hold (S/H) and analog-to-digital (A/D) conversion circuitry. The computer controls the mechanical scanner and stores the amplitude of the signal corresponding to a particular location over the sample.

Another class of systems measures both the amplitude and phase of the return signal from the sample. With the added information of phase, the systems carry out more quantitative measurements including inversion of the sample reflectance function, thin-film height, surface residual stress, image enhancement, surface roughness removal, multiple-frequency imaging, delamination thickness, and other image-processing applications (Liang *et al.*, 1985a). A schematic diagram of an amplitude- and phase-measuring instrument is shown in Fig. 10. (Reinholdtsen, 1989). The phase measurement is similar to that described in Sec. 2.1.2, the main difference being that the signal is mixed with a number of phase refer-

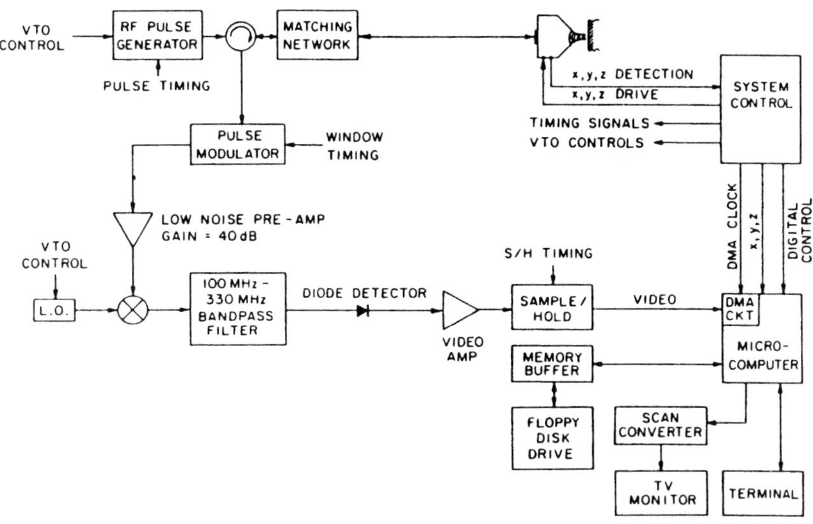

FIG. 9. Schematic diagram of a scanning acoustic microscope, where VTO stands for voltage-tuned oscillator (Lam, 1983).

ences before calculation of its magnitude and phase. This type of arrangement allows the measurement of amplitude with a sensitivity of 0.1% and phase with a sensitivity of 0.1°.

3.4 Scanning

At low frequencies, scanners made with dc or stepper motors are available with relocation accuracy of the order of 1 μm, and a field of view of tens of centimeters. In these systems, the scanning time is a few minutes for a scanning field of about 5 × 5 cm^2. At high frequencies, loudspeakers, or leaf springs of beryllium copper driven by piezoelectric stacks, are used for the fast-scanning direction, and universal motors with a belt drive are used for the slow-scan direction. The resulting frame rates are typically of the order of 20–30 s for a field of view of 500 × 500 μm^2.

One exciting implementation makes use of cylindrical arrays, which allow electronic scanning in one direction. A scanning speed of a couple of seconds per image for a scanning field of 3 × 3 cm^2 while operating at 25 MHz already has been demonstrated (Kubota *et al.,* 1988). SAMs operating at higher speeds and frequencies will indeed become a reality very soon.

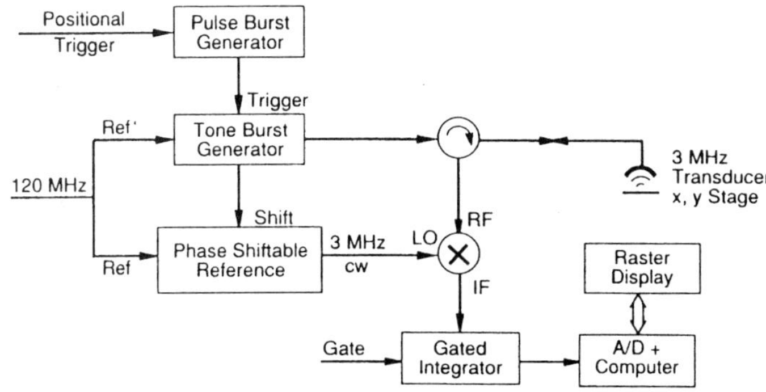

FIG. 10. Schematic diagram of an amplitude- and phase-measuring scanning acoustic microscope (Reinholdtsen, 1989).

3.5 Response of the SAM

In order to calculate the response of the SAM, it is necessary to consider the excitation, propagation, and interaction of the ultrasonic field with the sample. First, the propagation and focusing of the acoustic wave are presented. Next, the interaction of the focused beam with a sample is presented to explain the contrast mechanism of acoustic images.

The field distribution at the focus of the lens, called the point-spread function (PSF), determines the spatial resolution of the SAM. Several authors (Atalar, 1978; Wickramasinghe, 1979) have derived expressions for the PSF of an acoustic lens. These calculations are similar to those of optical lenses, and only the results of such calculations are presented in the following text.

In the system of Fig. 6, a quasi-cw excitation is applied to the transducer. The stress-field distribution inside the buffer rod undergoes large variations near the transducer, i.e., in the near field. Away from the transducer, the stress field varies slowly and decays as it propagates toward the lens. The near-field/far-field cutoff is known as the Fresnel length and is given by a^2/l, where a is the radius of the transducer and l is the wavelength of the sound in the buffer rod. Figure 11 shows the axial-field distribution at various normalized distances ($S = zl/a^2$) from the transducer. The lens can be placed at any location away from the transducer to affect its PSF.

As the ultrasonic wave reaches the lens, it is reflected in the buffer rod and refracted into the water according to Snell's law. The energy that remains within the buffer rod represents a loss in the energy available at the focus and a source of noise as it returns to the transducer through multiple reflections.

The energy that is transmitted into the water is focused at a distance equal to

$$F = R/(1 - V_W/V_S), \tag{1}$$

where V_W and V_S are the velocity of the wave in the water and solid, respectively. The above formula is valid for small angles or for lenses with large f-numbers; however, it is usually accepted as a measure of the location of the focus.

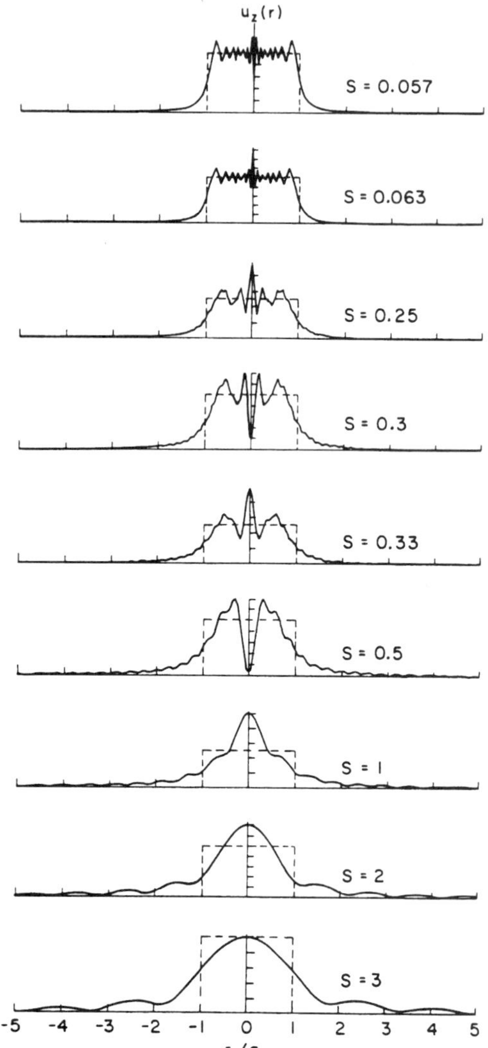

FIG. 11. Radial-field distribution due to plane piston transducer of radius a and at various distances ($S = zl/a^2$) from the transducer, for $ka = 100$ (Kino, 1987).

Figure 12 shows the PSF for a lens placed at different distances from the transducer. The specifications of the lens are f-number = 1.65, f = 40 MHz, transducer radius = 3.5 mm, and normalized distances S = 0.5, 1.0, and 10, respectively (Chou *et al.*, 1988). The importance of varying the parameter S can be seen by noting that in Fig. 11, the field distribution on the lens for S = 0.5 contains a null in the center of the lens, while for S = 1 the field is uniformly distributed over the lens and the phase has a parabolic dependence on radial distance. For S

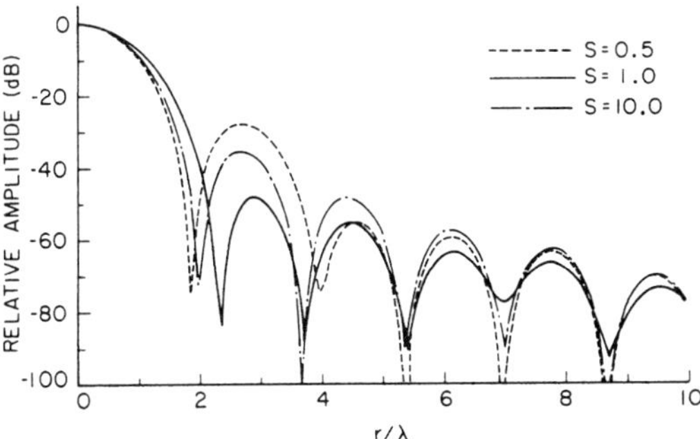

FIG. 12. Radial-field distribution at the focus of a lens with an f-number of 1.65, frequency of operation f = 40 MHz, where the lens is placed at different distances $S = z l/a^2$ from the transducer (Chou *et al.*, 1988).

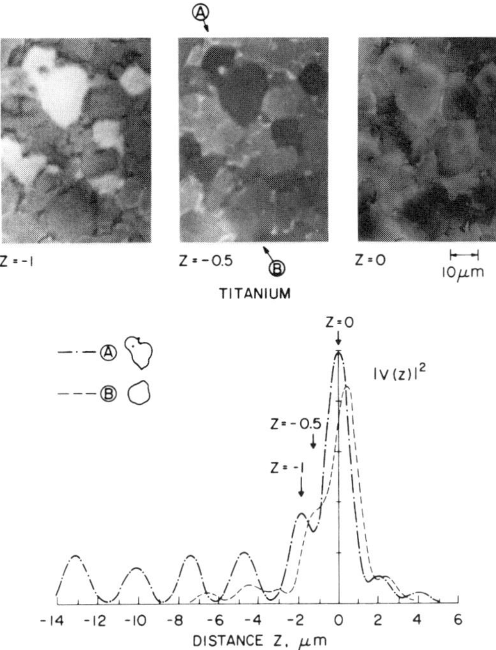

FIG. 13. Images of titanium sample along with corresponding $V(z)$ curves taken at three distances z between the lens and the object (Lam, 1983).

= 10, the field is almost constant in amplitude and phase over the lens. In Fig. 12, it is seen that the width of the PSF at the 3-dB points remains basically constant, while the sidelobes change by more than 20 dB. Thus, for surface imaging, it is best to use a lens located at S = 1 where the sidelobe level is lowest. For efficient surface-wave excitation, it is best to use a lens with the transducer located at S = 0.5, where the excitation off normal is most efficient.

Two important parameters of the focal spot are its lateral 3-dB width, which is equal to $f\# \times l$, and its 3-dB depth of field equal to $7.1 \times f\#^2 \times l$. Additionally, as a beam is focused within a sample its 3-dB width is the same as in water and its 3-dB depth of field is decreased by the ratio of the sound velocities of the sample to water.

Figure 13 shows three images of a sample of titanium taken at a frequency of 2 GHz and at three focus distances (Lam, 1983). At z = 0, the instrument is focused on the surface of the sample. Various grains in the crystal are clearly resolved because the amplitude of the reflection from each grain is different. This difference in the reflection can be due to differences in the mechanical impedance of the grains, which could be of different phases of the material or different orientations. Some of the dark spots are due to impurities or defects. The images that are taken when the sample is brought closer to the lens by 0.5 and 1.0 μm show not only improved contrast but also a contrast reversal; some of the features that were light in the image at focus are now dark and vice versa.

This contrast reversal was first noticed by several researchers, (Atalar *et al.* 1977; Weglen and Wilson, 1978) and explained theoretically and physically in an elegant fashion (Atalar, 1978; Wickramasinghe, 1979; Bertoni, 1984; Chou and Kino, 1987). Essentially, as the lens is brought closer to the sample, surface modes of propagation at the substrate–water interface are excited by the focused beam. These modes radiate their energy back into the water, where it is then collected by the lens and detected by the transducer. Often, it is the surface acoustic wave that is excited at the water–sample interface; such a "pseudosurface or leaky surface wave" is excited by rays arriving at the surface at the angle that satisfies Snell's law:

$$Q_R = \arcsin(V_W/V_R), \tag{2}$$

where V_R is the surface-wave velocity of the sample. The surface wave leaks back into the water after propagating at the interface, and is displaced from the specular reflection by a distance known as the Schoch displacement (Schoch, 1950), a phenomenon that is also well known in optics.

The contrast-reversal mechanism is interpreted as due to the interference between the specular reflection from the water–sample interface and the radiated signal from the surface mode. Any change in the amplitude or phase of either signal, i.e., due to the change in the distance between the lens and the sample, will effect a change in the output of the transducer. The variation of the output of the transducer as a function of distance between the lens and the sample is known as the $V(z)$ curve or the acoustic-material signature, and contains information about the mechanical properties of the sample and the illumination characteristic of the lens. Referring to any of the above-mentioned references, it is possible to write $V(z)$ as

$$V(z/\lambda) = C \int_{2}^{2\cos\theta_0} P^2(t)R(t)e^{-j2\pi(z/\lambda)t}dt, \qquad (3)$$

where $P(t)$ is the pupil function of the lens in the spatial-frequency domain, $R(t)$ the reflectance function of the water–sample interface (reflection coefficient as a function of angle), C a constant, and q_0 the maximum opening angle of the lens. Figure 14 shows a calculation of $V(z)$ of a silicon-nitride sample in water. The frequency of operation is 40 MHz, and the lens has an $f\# = 1.65$. Notice the typical character in the negative z direction (the sample is brought closer to the lens), which reflects the continuous addition of the surface wave and the specular reflections in and out of phase. The spacing between the nulls is measured to evaluate the surface-wave velocity (Kushibiki and Chubachi, 1985).

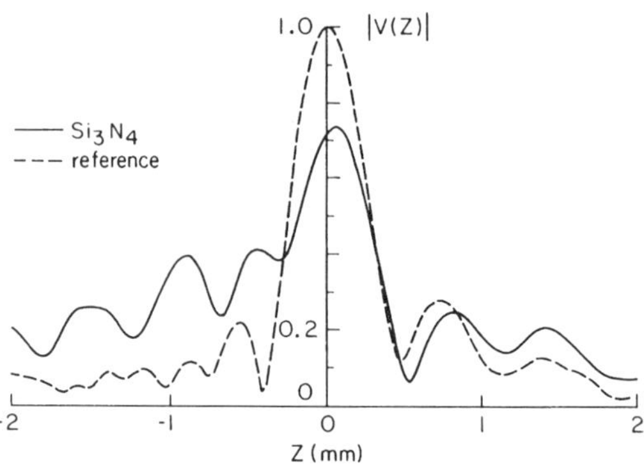

FIG. 14. $V(z)$ of a silicon-nitride sample and reference material identical to silicon nitride but that does not allow the propagation of surface waves. The difference between the two wave forms is the surface-wave excitation, leakage, and detection (Chou *et al.*, 1988).

The above equation shows that $V(z)$ is given by the Fourier transform of the product of the pupil function of the lens and the reflectance function of the sample. This equation expresses the power of the microscope as a quantitative tool for material characterization. Either the pupil function of a lens or the reflectance function of a sample can be measured by an inverse Fourier transform of a measured $V(z)$. For instance, the reflectance function of a water–air, water–lead, or water–Teflon interface has a constant amplitude and phase over most angles of interest and can be factored out in the above equation. Thus, an inverse Fourier transform of a measured $V(z)$ gives the pupil function of the lens. Once the pupil function is known, a measured $V(z)$ off of a sample can be inverted to evaluate the reflectance function of the water–sample interface. This method has been used (Liang *et al.*, 1985b) to characterize lenses and unknown samples.

3.6 Imaging Applications

Most of the early imaging work was directed toward imaging biological samples. SAMs are well suited for this application because they have a similar resolution to optical microscopes without the need for special sample preparation. In addition, biological samples can be imaged alive in water. Today, however, the major applications of SAMs are in the field of materials science for the examination of structural materials such as plastics, rocks, semiconductors, and metals. In this section we show images taken by SAMs at various frequencies and of different materials.

3.6.1 Biological Applications Figure 15, one of the first taken with an acoustic microscope, shows an acoustic micrograph of an unstained section of human lung tissue taken at a frequency of 600 MHz and with a magnification of about 200× (Lemons, 1975). The image shows very clearly several features of the human lung tissue—the air sacs, capillaries, cell walls, and blood vessels—and is typical in the detail it shows of variations in the mechanical properties of tissue. Much better resolution can be ob-

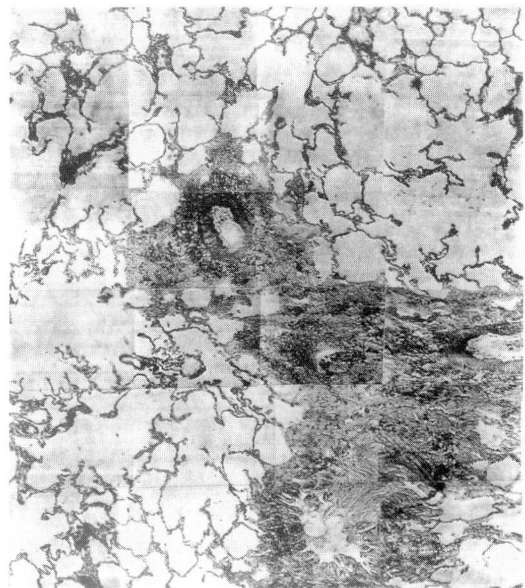

FIG. 15. Image of human lung tissue (Lemons, 1975).

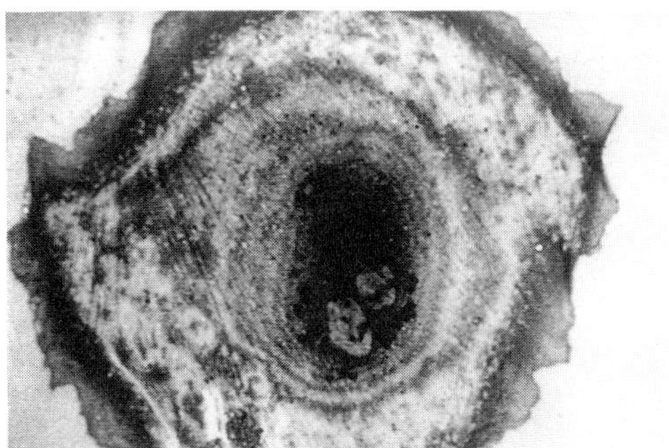

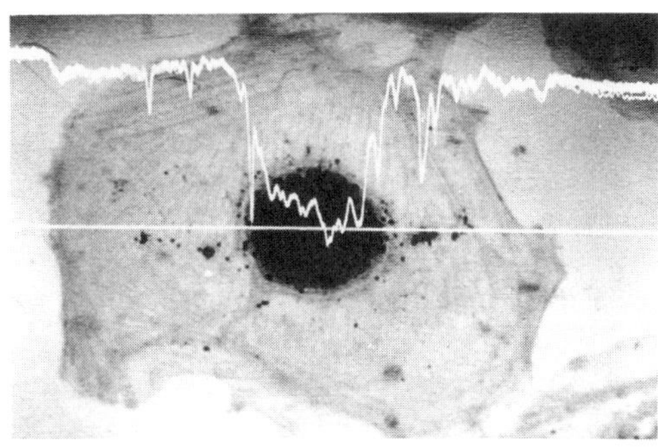

FIG. 16. Image of frog heart cells. Top image is a cell on a plastic substrate and bottom image is a cell on a glass substrate (courtesy of M. Hoppe, Leitz-Wetzlar).

tained by operating the acoustic microscope at a higher frequency, as seen in Fig. 16, which is an image of frog heart cells taken at a frequency of 1.6 GHz (Hoppe and Bereiter-Hahn, 1985). The top image is taken with the cells on a plastic substrate; the bottom image is taken on a glass substrate. The bottom image also shows a line scan of the amplitude variation over the middle of the cell. The images indicate that when a strong mismatch exists between the cells and the substrate, the image is dominated by the reflection from that interface. The image of the cell with the plastic substrate clearly shows more contrast from within the cell.

Figure 17 shows the image of a Myxobacterium taken at 8 GHz in liquid helium (Hadimioglu and Foster, 1984). This is the highest-resolution image taken to date, showing the details of the bacterium with a resolution of 200 Å. The contrast seen in the image corresponds to different levels of focus over the sample.

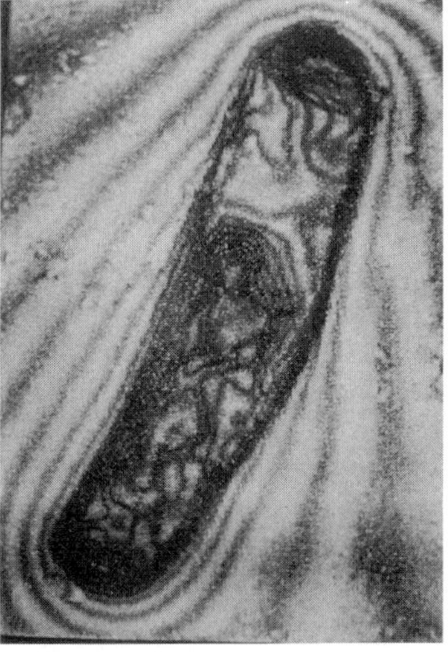

FIG. 17. Image of Myxobacterium taken with the scanning acoustic microscope operating in liquid helium (Hadimioglu and Foster, 1984).

3.6.2 Materials Science Figure 18 shows the image of an integrated circuit taken at a frequency of 1 GHz (courtesy of Olympus Corporation). The figure is a superposition of three images taken at focus levels of $z = 0$, -2, and -4 μm, and displays the

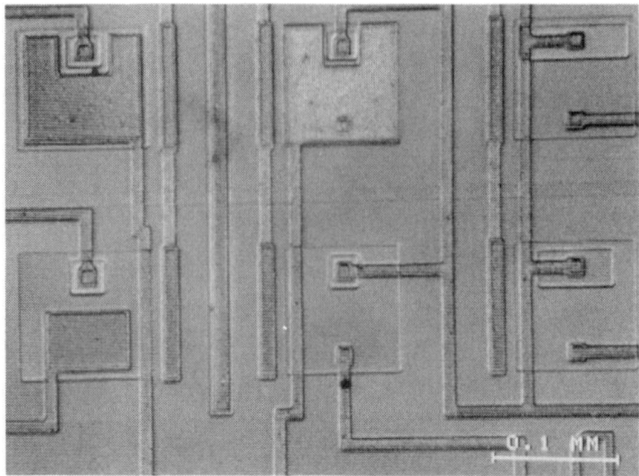

FIG. 18. Three superposed images of an integrated circuit taken at three focus distances (courtesy of Olympus Corp.).

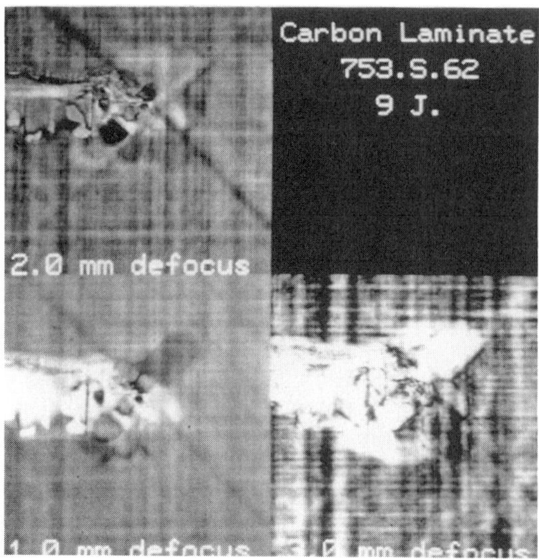

FIG. 20. Image of impact damage in a carbon-epoxy laminate. The sample was damaged with an impactor having energy 9 J, and the images were taken at three focus distances (Reinholdtsen, 1989).

various levels of the circuit with good contrasts; the outlines of the metallizations and diffusions are very clearly shown. Figure 19 shows a comparison between the optical and acoustic images of a chrome metallization on glass; the optical image shows the pattern with no visible defects, while the acoustic image shows clearly that there are poor-adhesion regions in the film. The two defocus images demonstrate the ability of the instrument to perform subsurface imaging.

Figure 20 shows a much lower-frequency image of impact damage in a carbon-epoxy laminate at 3 MHz, but the instrument is still a copy of its higher-frequency cousin. The image shows the layering of the many plies in the composite. It shows breaks in fibers, and it shows disbonds at various levels in depth of the sample. For these materials, acoustic microscopy is superior to dye-enhanced x-ray and focused C-scan imaging.

Figure 21 shows an image of a polished

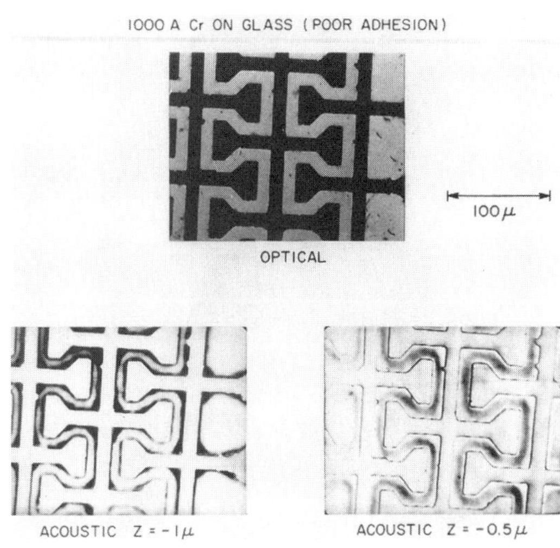

FIG. 19. A comparison between optic and acoustic images of poorly adhered chrome films on a glass substrate (courtesy of C. F. Quate).

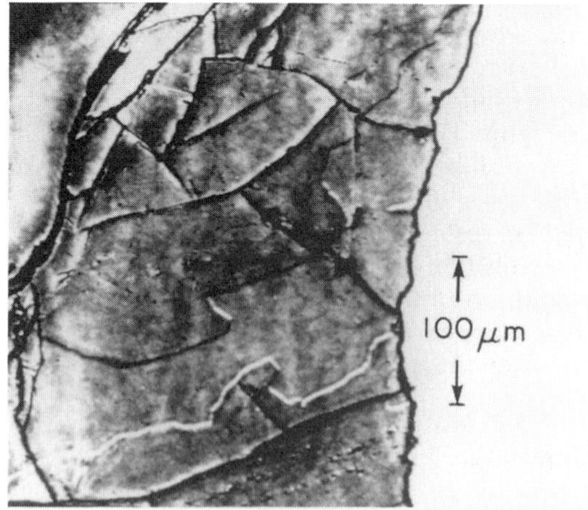

FIG. 21. Image of a granite sample showing cracks and grain boundaries (courtesy of G. A. D. Briggs).

thin section of granite. The image is taken at a frequency of 370 MHz, and shows clear details of cracking and grain boundaries in the rock (Briggs *et al.*, 1989). SEM and transmitted polarized-light images of the sample show less detail of the microcracks and the grain boundaries, illustrating yet again the power of the SAM over traditional imaging techniques.

Figure 22 (courtesy of Hitachi Construction Machinery Co.) shows a sample of a silicon-steel plate with and without residual stress. Localized differences of surface acoustic-wave velocities are observed with 6.5% tension deformation. The image was taken at a frequency of 600 MHz.

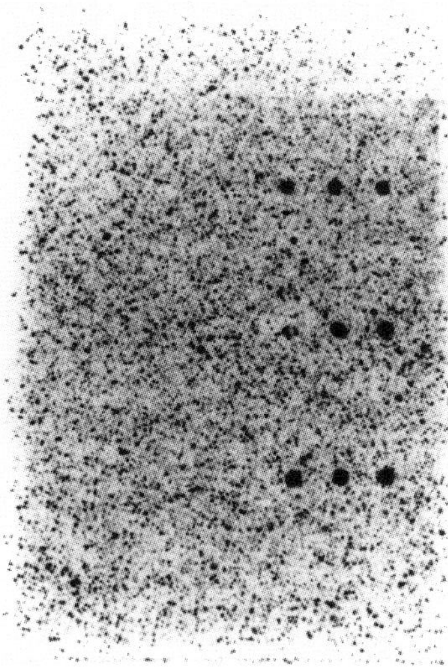

FIG. 23. C-scan image inside a René-95 specimen showing porosity and the ends of flat-bottomed holes (Gilmore, *et al.*, 1986).

Figure 23 is an example using a focused C-scan imaging system to look inside materials. A René-95 steel sample with 0.25-, 0.375-, and 0.5-mm electric-discharge–machined holes is imaged where the plane of focus is 3.1 mm below the surface and at the end of the holes. The sample is 24 × 35 mm and the imaging plane is the bond line between two pieces of the same material. The images show the holes along with a large amount of porosity because of the large depth of field of the lens (Gilmore *et al.,* 1986).

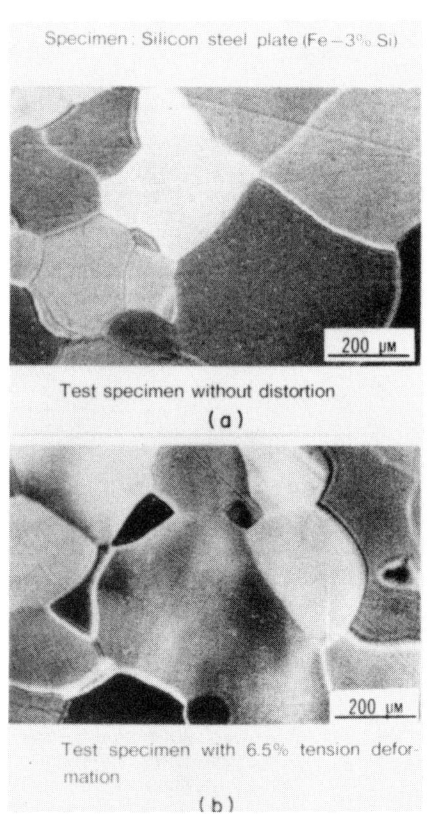

FIG. 22. Images of a silicon-steel plate showing differences due to residual stress (courtesy of J. S. Okagima, Hitachi Construction Company).

3.7 Metrology

3.7.1 Amplitude Only As mentioned earlier, it is possible to measure some of the mechanical properties of a sample through Fourier inversion of $V(z)$ curve. Using a SAM that measures the amplitude of $V(z)$, it is possible to calculate the surface-wave velocity and surface-wave attenuation. The most prominent work in this area is that of the group headed by Profs. Chubachi and Kushibiki at Tohoku University in Sendai, Japan. They use cylindrical lenses that excite surface waves in only one direction and make very accurate measurements of the surface-wave velocity and attenuation (Kushibiki and Chubachi, 1985). The distance between the nulls in the negative side of $V(z)$ is given by

$$D_Z = (l_R/\sin q_R)(1 + \cos q_R)/2, \qquad (4)$$

where D_z is the spacing between nulls and q_R is the critical angle for exciting the surface wave. The attenuation of the surface wave

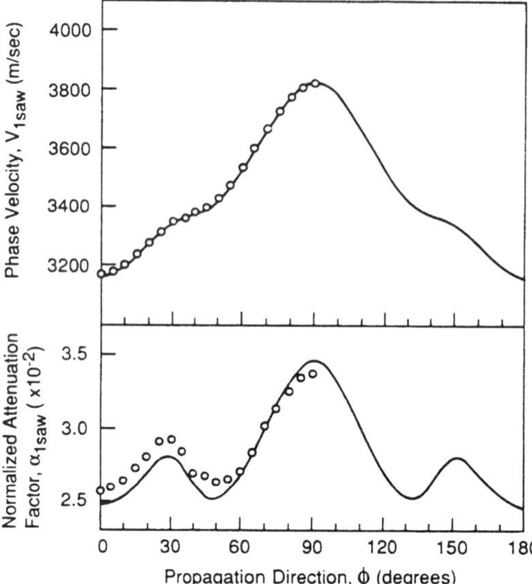

FIG. 24. Comparison of theoretical calculations and experimental measurements of the velocity and attenuation of a *Y*-cut α-quartz sample as a function of direction of propagation (Kushibiki and Chubachi, 1985).

can be deduced subsequently from the decay of the $V(z)$ curve as a function of z in the negative z direction (Weglein, 1985).

With a spherical lens, the surface wave is excited from all directions over a sample, and if the sample is anisotropic, variations in the surface-wave velocity will result in an average value for the surface-wave velocity. The cylindrical lens overcomes this problem and allows an accurate measure of anisotropy of the sample. A typical result of this work is shown in Fig. 24, where the surface-wave velocity of a sample of *Y*-cut α-quartz is shown as a function of angle (Kushibiki and Chubachi, 1985). Each point on the chart represents an independent $V(z)$ measurement and evaluation of the velocity at that angle. The agreement between theory and experiment is excellent and reflects an error of about 0.15%. The attenuation of the surface wave is also shown as a function of angle where the agreement with theory is also good.

3.7.2 Amplitude and Phase Some SAMs measure both amplitude and phase of the acoustic signal (Liang *et al.*, 1985; Reinholdtsen, 1989). The advantages offered by these instruments include enhanced height sensitivity, full $V(z)$ inversion, and increased signal processing ability for edge enhancement and subsurface-defect detection.

3.7.2.1 Reflectance-Function Inversion. Figure 25 shows the result of a full inversion of the reflectance function, using the Fourier-transform relationship between $V(z)$ and the reflectance function, of a sample of alu-

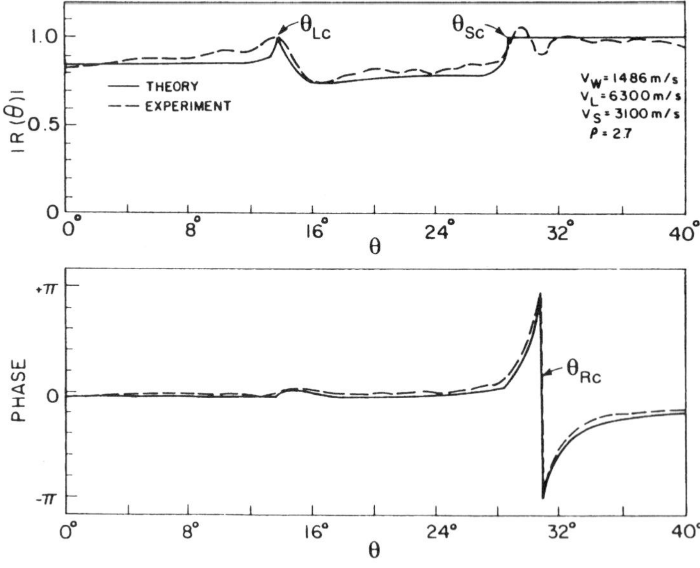

FIG. 25. Comparison of theoretical calculation and experimental measurements of the reflectance function on an aluminum sample. Here q_{Lc} is the longitudinal-wave critical angle, q_{Sc} is the shear-wave critical angle, and q_{Rc} is the surface-wave critical angle (Liang *et al.*, 1985).

minum, where q_{Lc} is the longitudinal-wave critical angle, q_{Sc} is the shear-wave critical angle, and q_{Rc} is the surface-wave critical angle. Because agreement between theory and experiment is excellent, it is possible to deduce the density and the longitudinal-wave, shear-wave, and surface-wave velocities of the sample. The ringing at the critical angles is due to the limited working distance, because the lens touches the sample for large values of defocus. The above results indicate that the best measurement system would then consist of a cylindrical lens and an amplitude- and phase-measuring instrument.

3.7.2.2 Residual-Stress Measurement. For a relatively large defocus distance, and with short-duration pulses, it is possible to separate, in the time domain, the specular reflection from the mode-converted surface wave. The phases of the two pulses can be compared to determine, very accurately, any variations of the surface-wave velocity, or the distance between the sample and lens. Such a configuration has been used to measure residual stress in a flat and homogeneous sample. A Pyrex disk 3.24 mm in thickness and 7.5 cm in diameter was treated to induce a circularly symmetric residual stress that decreased away from the center of the sample. An indentation-based measurement of the circumferential (solid circles) and radial (hollow circles) residual stresses is shown in Fig. 26. The figure also shows a measurement of the variation in the surface-wave velocity (phase of surface-wave reflection with respect to the specular reflection) in the sample at similar locations. The agreement between the two measurements is excellent. Because it is possible to measure a phase shift of 0.1°, a residual-stress change of less than 1 MPa can be measured.

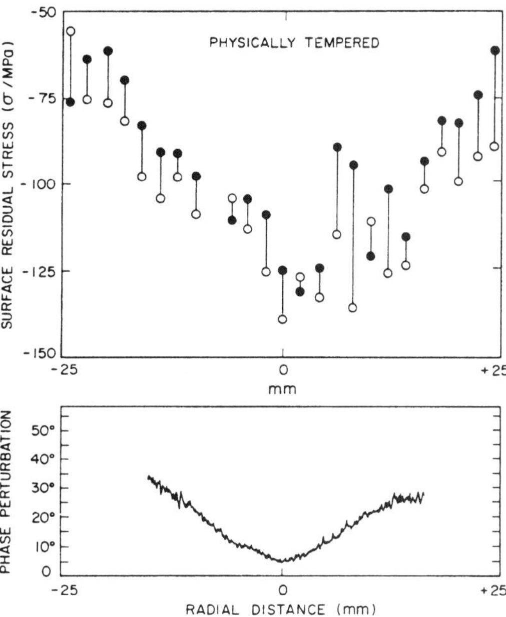

FIG. 26. Comparison of the measurement of residual stress in a Pyrex disk: top, by damage extent due to a Vickers indentor (courtesy of Marshall & Evans); bottom, by variation in surface-wave velocity (Liang *et al.*, 1985).

4. NEAR-FIELD SCANNING ACOUSTIC MICROSCOPE (NFSAM)

The transmission NFSAM is similar to the SLAM except that the sound field on the surface of a sample is detected with a tip or a pinhole that is much smaller than the wavelength. Subwavelength resolution is obtained for surface features, while subsurface features are imaged through their shadow on the surface of the sample. The reflection NFSAM is similar to the SAM; the sound can be incident through a tip or a pinhole, and variations in the reflected signal from the sample are used to image variations in the mechanical properties of a sample. In both transmission and reflection modes of operation, the signal-to-noise ratio in a NFSAM is quite small because the tips or pinholes are much smaller than the wavelength, and very small amounts of energy are collected or changed by the presence of the sample. Very simply, the sound emitted from a pinhole will propagate as a collimated beam for a distance roughly equal to the diameter of the pinhole, and thus define the near field. Beyond this distance, the beam will spread and decay rapidly in strength. Samples have to be placed in the near field in order to be imaged. The images obtained by the NFSAM represent variations in the mechanical properties of the sample with a spatial resolution equal to the size of the pinhole or tip.

The near-field scanning acoustic microscope (NFSAM) is a variation on the above two instruments where the resolution is determined by the physical dimensions of a tip or a pinhole and their distance from the

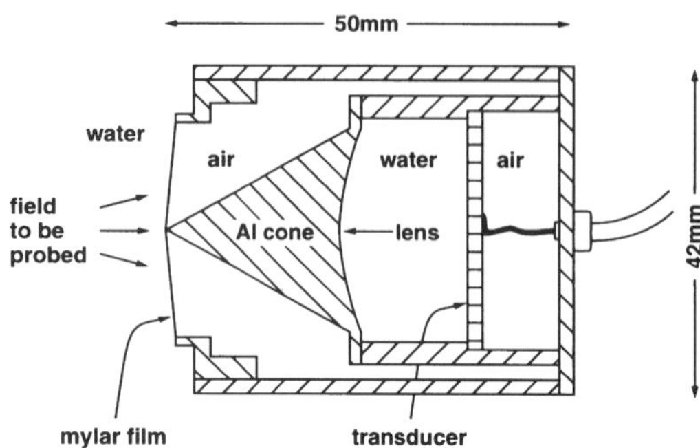

FIG. 27. Schematic diagram of a high-resolution acoustic probe. The aluminum–water interface acts as a positive lens (Durr *et al.*, 1980).

sample, rather than the dimensions of a focal spot. Two types of systems are distinguished, tip systems and pinhole systems.

4.1 Tip NFSAM

A schematic of the first tip NFSAM, reported by Durr *et al.* (1980) is shown in Fig. 27. The system is for a probe with a resolution of *l*/4. The probe consists of a metal cone whose tip receives the signal and determines the resolution. The back side of the metal cone is curved to transmit the received sound into water as a plane wave. The plane wave of sound is then detected with a flat piezoelectric transducer. This probe was used to demonstrate a resolution of about 75 μm while operating at a frequency of 4 MHz where the wavelength of sound in water is 375 μm. Zieniuk and Latuszek (1986) used a similar system with a sapphire cone to demonstrate *l*/10 resolution while operating at a frequency of 30 MHz. Guthner *et al.* (1989) used a resonant quartz-crystal tuning fork as a NFSAM. As the edge of the tuning fork was brought close to a sample, its resonant frequency and amplitude changed depending on the local properties of the sample. This system demonstrated imaging with resolution of less than 1 μm while operating at a frequency of 32 kHz.

Some impressive NFSAM images were obtained when coupling the sound in and out of atomic tips such as those used in scanning tunneling microscopy (STM) and atomic-force microscopy (AFM). Figure 28 shows a block diagram of the system of Uozumi and Yamamuro (1989). In this system a transducer is attached to the top side of a rod with a tip of atomic dimensions at the bottom. The transducer operated at a frequency of 1.44 MHz in a pulse-echo mode. The amplitude of the signal reflected from the atomic tip was measured, and variations in this signal were used to image a sample. This system was used to demonstrate imaging with a resolution of the order of 100 Å. Figure 29 shows a schematic of the system of Takata *et al.* (1989). In this system the tip was vibrated by a piezoelectric cylinder, and the transmitted acoustic signal was detected by a piezoelectric transducer on the back side of the sample. Operating at a frequency of 70 kHz, images of surface were obtained with a resolution comparable to those of the system of Uozumi and Yamamuro. In another version of a NFSAM, Kolosov and Yamanaka (1993) take advantage of the nonlinear relationship of force on an atomic tip and distance to the sample to image subsur-

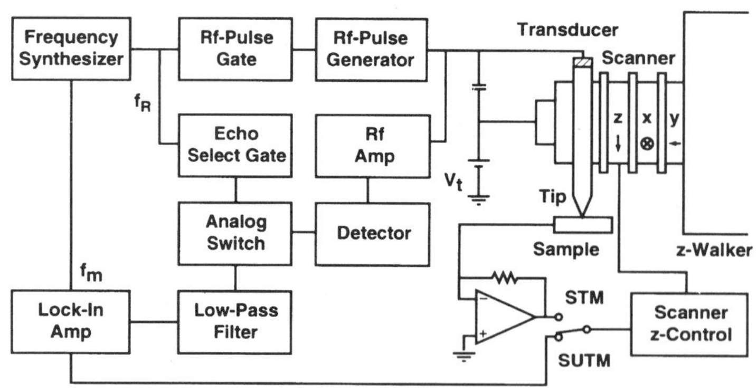

FIG. 28. Block diagram of the experimental setup of the scanning ultrasonic tip microscope (Uozumi and Yamamuro, 1989).

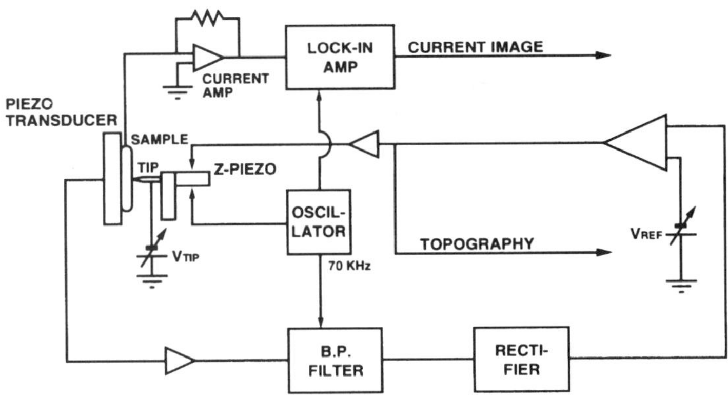

FIG. 29. Schematic diagram of the tunneling acoustic microscope (Takata *et al.*, 1989).

face features with a spatial resolution of the order of 100 Å.

4.2 Pinhole NFSAM

The first demonstration of subwavelength resolution was provided by Ash and Nicholls (1972) using electromagnetic radiation with a cavity with a pinhole. Khuri-Yakub *et al.* (1989) modified a SAM by inserting a shim with a pinhole at the focal plane of the lens to make a NFSAM. A schematic of this system is shown in Fig. 30. This instrument demonstrated imaging with an *l*/10 resolution while working at a frequency of 3 MHz. One important aspect of this system is that a simple modification to an existing SAM can be made to improve the resolution of surface images.

The NFSAM is still in its infancy compared with the other two instruments. No doubt, many applications for imaging variations of mechanical properties with atomic resolution will be available soon.

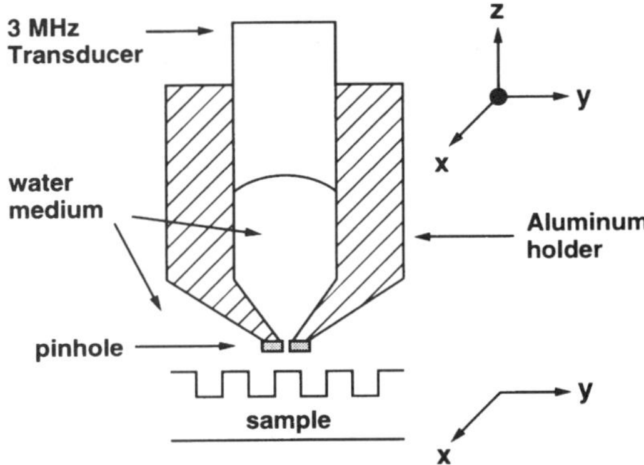

FIG. 30. Schematic diagram of a SAM lens with a pinhole (Khuri-Yakub *et al.*, 1989).

5. CONCLUSIONS

Acoustic microscopy is maturing into a very useful technology for imaging the variations of the mechanical properties of materials. Quantitative measurements are possible, and are gaining in credibility and acceptance by various communities for biological applications and material characterization. The field should continue to grow, and scanning acoustic microscopes will become traditional fixtures in materials-evaluation laboratories.

6. ACKNOWLEDGMENTS

The authors thank Andrew Briggs, Ching-Hua Chou, Babur Hadimioglu, Davis Kent, Larry W. Kessler, Gordon S. Kino, Hua Lee, Z. C. Lin, A. Meyyappan, and Calvin F. Quate for many useful discussions and comments. This work was supported by the National Science Foundation through its Center for High-Speed Image/Signal Processing at the University of California at Santa Barbara, and by the Department of Energy under Contract No. DE-FGO3-84-ER45157 at Stanford University.

GLOSSARY

Acoustic-Material Signature *V*(*z*): Output of a transducer–lens system as a function of the distance between the lens and the sample.

C-Scan Imaging System: Imaging system where the display shows defects in a plane parallel to the scanning plane.

Dead Zone: Near-surface region that is not inspected by a C-scan imaging system.

Defocus: Refers to moving a focused lens closer to or farther away from a sample.

Lamb Wave: A type of mechanical wave that is guided by a plate.

Piezoelectricity: A phenomenon where electric dipoles are generated in a solid by the application of mechanical stress, or a mechanical stress is generated as a result of the application of an electric field.

Point-Spread Function (PSF): The field distribution at the focal spot of a lens.

Rayleigh or Surface Wave: Mechanical wave that is trapped within a wavelength at the interface between a solid surface and vacuum. It is called leaky when it leaves the solid because of loading by a medium such as water.

SAM: Scanning acoustic microscope.

Works Cited

Ash, E. A., Nicholls, G. (1972), "Super-Resolution Aperture Scanning Microscope," *Nature* **237,** 510–512.

Atalar, A., Quate, C. F., Wickramasinghe, H. K. (1977), "Phase Imaging in Reflection with the Acoustic Microscope," *Appl. Phys. Lett.* **31,** 791–793.

Atalar, A. (1978), "An Angular Spectrum Approach to Contrast in Reflection Acoustic Microscopy," *J. Appl. Phys.* **49,** 5130–5139.

Attal, J., Cambon, G. (1978), "Improvements in the Resolution of the Reflective Scanning Acoustic Microscope Application to the Non destructive Evaluation of Microelectronics," in: J. deKlerk and B. R. McAvoy (Eds.), *Proceedings of IEEE Ultrasonics Symposium, 1978,* New York: IEEE, p. 212.

Bertoni, H. L. (1984), "Ray-Optical Evaluation of V(z) in the Reflection Acoustic Microscope," *IEEE Trans. Sonics Ultrason. Ind. Eng. Chem.* **31,** 105–116.

Briggs, A. (1992), *Acoustic Microscopy,* New York: Oxford Univ. Press.

Briggs, G. A. D., Daft, C. M. W., Fagan, A. F., Field, T. A., Lawrence, C. W., Montoto, M., Peck, S. D., Rodriguez, A., Scruby, C. B. (1989), in: H. Lee, G. Wade (Eds.), *Acoustical Imaging,* Vol. 17, New York: Plenum, pp. 1–16.

Chiao, R., Lee, H. (1991), *Int. J. Imaging Syst. Technol.* **3,** 334–353.

Chou, C. H., Kino, G. S. (1987), "The Evaluation of V(Z) in a Type II Reflection Microscope," *IEEE Trans. Ultrason. Ferroelec. Freq. Control* **34,** 341–345.

Chou, C. H., Khuri-Yakub, B. T., Kino, G. S. (1988), "Lens Design for Acoustic Microscopy," *IEEE Trans. Ultrason. Ferroelec. Freq. Control* **35,** 464–469.

Daft, C. M. W., Briggs, G. A. D. (1989), "Wide-band Acoustic Microscopy of Tissue," *IEEE Trans. Sonics Ultrason. Ind. Eng. Chem.* **36,** 258–263.

Durr, W., Sinclair, D. A., Ash, E. A. (1980), "High Resolution Acoustic Probe," *Electron. Lett.* **16,** 805–806.

Gilmore, R. S., Tam, K. C., Howard, D. R. (1986), *Acoustic Microscopy from 10 to 100 MHz for Industrial Applications,* General Electric Corporate Research and Development Report No. 86CRD015.

Güthner, P., Fischer, U. Ch., Dransfeld, K. (1989), *Appl. Phys. B* **48,** 89–92.

Hadimioglu, B., Quate, C. F. (1983), "Water Acoustic Microscopy at Suboptical Wavelengths," *Appl. Phys. Lett.* **43,** 1006–1007.

Hadimioglu, B., Foster, J. S. (1984), "Advances in Superfluid Helium Acoustic Microscopy," *J. Appl. Phys.* **56,** 1976.

Hoppe, M., Bereiter-Hahn, J. (1985), "Applications of Scanning Acoustic Microscopy: Survey and New Aspects," *IEEE Trans. Sonics Ultrason. Ind. Eng. Chem.* **32,** 289–301.

Kessler, L. W. (1974), *J. Acoust. Soc. Am.* **55,** 909–918.

Kessler, L. W., Korpel, A., Palermo, P. R. (1971), *Nature* **232,** 110–111.

Khuri-Yakub, B. T., Cinbis, C., Chou, C. H., Reinholdtsen, P. A. (1989), "Near-Field Scanning Acoustic Microscope," in: B. R. McAvoy (Ed.), *Proceedings of the IEEE Ultrasonics Symposium, 1989,* pp. 805–807.

Kino, G. S. (1987), *Acoustic Waves: Devices, Imaging, and Analog Signal Processing,* Englewood Cliffs, NJ: Prentice-Hall.

Kolosov, O., Yamanaka, K. (1993), *Jpn. J. Appl. Phys.* **32,** L1095–L1098.

Korpel, A., Kessler, L. W., Palermo, P. R. (1971), "An Acoustic Microscope Operating at 100 MHz," *Nature* **323,** 110–111.

Korpel, A., Kessler, L. W. (1971), in: A. Metherell (Ed.), *Acoustical Holography,* Vol. 3, New York: Plenum, pp. 23–43.

Kubota, J., Okada, H., Musha, Y., Takishita, Y., Iwasaki, A., Sasaki, S. (1988), "Electronic Scanning of 25 MHz Ultrasound for Imaging IC Packages," in: B. R. McAvoy (Ed.), *Proceedings of the IEEE Sonics and Ultrasonics Symposium,* New York: IEEE, p. 767–770.

Kushibiki J.-I., Chubachi, N. (1985), "Material Characterization by Line-Focus-Beam Acoustic Microscope," *IEEE Trans. Sonics Ultrason. Ind. Eng. Chem.* **32,** 189–212.

Lam, L. K. (1983), *Techniques of Acoustic Microscopy,* Ph.D. Thesis, Stanford University.

Lasky, M. (1977), *J. Acoust. Soc. Am.* **61,** 283–297.

Lemons, R. A., Quate, C. F. (1973), "A Scanning Acoustic Microscope," in: J. de Klerk (Ed.), *Proceedings of the IEEE Ultrasonics Symposium,* New York: IEEE, pp. 18–20.

Lemons, R. A., Quate, C. F. (1974), "Acoustic Microscope—Scanning Version," *Appl. Phys. Lett.* **24,** 163–165.

Lemons, R. A. (1975), *Acoustic Microscopy by Mechanical Scanning,* Ph.D. Thesis, Stanford University.

Liang, K. K., Bennett, S. D., Khuri-Yakub, B. T., Kino, G. S. (1982), "Surface Wave Velocity Measurements at 50 MHz," in: B. R. McAvoy (Ed.), *Proceedings of the IEEE Ultrasonics Symposium, 1982,* Vol. 2, New York: IEEE, p. 604.

Liang, K. K., Bennett, S. D., Khuri-Yakub, B. T., Kino, G. S. (1985a), "Precise Phase Measurements with the Acoustic Microscope," *IEEE Trans. Sonics Ultrason. Ind. Eng. Chem.* **32,** 266–273.

Liang, K. K., Kino, G. S., Khuri-Yakub, B. T. (1985b), "Material Characterization by the Inversion of the V(Z)," *IEEE Trans. Sonics Ultrason. Ind. Eng. Chem.* **32,** 213–224.

Lin, Z. C., Lee, H., Wade, G., Oravecz, M. G., Kessler, L. W. (1987), *IEEE Trans. Ultrason. Ferroelec. Freq. Control* **UFFC-34,** 293–300.

Pohl, D. W., Denk, W., Lang, M. (1984), *Appl. Phys. Lett.* **44,** 652–653.

Pohlman, R. (1937), *Z. Phys.* **107,** 497–507.

Quate, C. F. (1985), "Acoustic Microscopy," *Phys. Today* **38** (8), 34–42.

Reinholdtsen, P. (1989), *Image Processing for an Amplitude and Phase Acoustic Microscope,* Ph.D. Thesis, Stanford University.

Rylander, R. L. (1982), *A Laser-Scanned Ultrasonic Microscope Incorporating a Time-Delay Interferometric Detector,* Ph.D. Thesis, University of Minnesota.

Schoch, A. (1950), *Ergeb. Exakt. Naturwiss.* **23,** 127.

Sokolov, S. J. (1949), *D. K. Akad. Nauk SSSR* [Sov. Phys. Dokl.] **64,** 333–335.

Takata, K., Hasegawa, T., Hosaka, S., Hosoki, S. (1989), "Tunneling Acoustic Microscope," *Appl. Phys. Lett.* **55,** 1718–1720.

Uozumi, K., Yamamuro, K. (1989), "A Possible Novel Scanning Ultrasonic Tip Microscope," *J. Appl. Phys.* **28,** L1297–L1299.

Wade, G. (1987), in: H. W. Jones (Ed.), *Acoustical Imaging,* Vol. 15, New York: Plenum, pp. 1–28.

Wang, K., Wade, G. (1974), in: P. S. Green (Ed.), *Acoustical Holography,* Vol. 5, New York: Plenum, p. 239.

Weglein, R. D., Wilson, R. G. (1978), "Characteristic Material Signatures by Acoustic Microscopy," *Electron. Lett.* **14.**

Weglein, R. D. (1985), "Acoustic Micrometrology," IEEE Trans. Sonics Ultrason. *Ind. Eng. Chem.* **32,** 225–234.

Whitman, R. L., Korpel, A. (1969), *Appl. Opt.* **8,** 1567–1576.

Wickramasinghe, K. (1979), "Contrast and Imaging Performance in the Scanning Acoustic Microscope," *J. Appl. Phys.* **50,** 664–672.

Wickramasinghe, H. K., Petts, C. R. (1982), "Gas Medium Acoustic Microscopy," in: *Scanned Image Microscopy,* London: Academic.

Yamada, K., Shimizu, H., Minakata, M. (1986), "Planar Structure Focusing Lens for Operation at 200 MHz and its Application to the Reflection-Mode Acoustic Microscope," in: B. R. McAvoy (Ed.), *Proceedings of the IEEE Ultrasonics Symposium, 1986,* New York: IEEE, p. 745.

Yamanaka, K. (1982), "Broad Band Pulses of Leaky Surface Acoustic Waves in Acoustic Microscopy," in: B. R. McAvoy (Ed.), *Proceedings of the IEEE Ultrasonics Symposium, 1982,* New York: IEEE, Vol. 2, p. 609.

Zieniuk, J. K., Latuszek, A. (1986), "Ultrasonic Pin Scanning Microscope: A New Approach to Ultrasonic Microscopy," in: B. R. McAvoy (Ed.), *Proceedings of the IEEE Ultrasonic Symposium,* New York: IEEE, pp. 1037–1039.

Further Reading

Auld, B. A. (1973), *Acoustic Fields and Waves in Solids,* New York: Wiley-Interscience.

Briggs, A. (1992), *Acoustic Microscopy,* New York: Oxford Univ. Press.

Khuri-Yakub, B. T., Quate, C. F. (1992), *Selected Papers on Scanning Acoustic Microscopy,* Washington, DC: SPIE Optical Engineering Press.

Kino, G. S. (1987), *Acoustic Waves: Devices, Imaging, and Analog Signal Processing,* Englewood Cliffs, NJ: Prentice-Hall.

SCANNING TUNNELING MICROSCOPY

See TUNNELING MICROSCOPY

SCATTERING OF LIGHT

See RAMAN SCATTERING

SCHOTTKY BARRIERS

L. MAGAUD AND F. CYROT-LACKMANN, *Laboratoire d'Etudes des Propriétés Electroniques des Solides CNRS, Grenoble, France*

	Introduction	573
1.	**Description and Electrical Behavior**	575
1.1	Description of the Schottky Barrier	575
1.2	Schottky Barrier and Ohmic Contact at Equilibrium	575
1.3	Schottky Barrier and Ohmic Contact under an Applied Bias	576
1.3.1	Schottky Barrier under an Applied Bias	576
1.3.2	Ohmic Contact under an Applied Bias	577
2.	**Experimental Description**	578
2.1	General Trends	578
2.2	Phenomenological Models and Correlations	580
2.3	Schottky-Barrier Height Measurements	581
2.3.1	Capacitance Measurement, *C-V*	581
2.3.2	Current Measurement, *I-V*	581
2.3.3	Photoresponse, I-$h\nu$	582
2.3.4	Application of XPS to Schottky-Barrier Measurements	582
2.3.5	Ballistic-Electron Emission Microscopy (BEEM)	583
3.	**Theoretical Models**	584
3.1	Schottky Model	584
3.2	Bardeen Model	584
3.3	Defect Model (UDM)	585
3.4	Cowley–Sze Model	585
3.5	Metal-Induced Gap States Model (MIGS)	586
3.6	Correlations between Schottky-Barrier Height, Valence-Band Offset, Dangling-Bond Level, and Transition-Metal Impurity Level	587
3.6.1	Charge-Neutrality Level and the Semiconductor Dangling-Bond Level	587
3.6.2	Correlation with Valence-Band Offset	587
3.6.3	Correlation with Transition-Metal Impurity Levels	587
3.7	Discussion	588
3.7.1	Variation of the Schottky-Barrier Height with Metal Coverage	588
3.7.1.1	Small Coverages, $\theta < 1$ Monolayer (ML)	588
3.7.1.2	Thick Coverages, $\theta \gg 1$ ML	588
3.7.2	Comparison of MIGS and Defects Models	588
4.	**Microscopic Aspect: the Electronic-Structure Calculation Point of View**	589
5.	**Conclusion**	589
	Acknowledgment	590
	Glossary	590
	Works Cited	590
	Further Reading	591

INTRODUCTION

From an electronic-structure point of view, materials can be classified into two groups: metals and insulators. If the Fermi level falls in a partially filled band, states are available for conduction, even at 0 K, and the material is metallic. On the other hand, if all bands are either empty or fully occupied, the Fermi energy is greater than or equal to the last occupied-state energy, and no neighboring state is available for conduction at 0 K, the system is an insulator. Electrons have to jump from the last filled band (called the valence band) to the lowest empty band (the conduction band) to allow conduction. The last filled states and the first empty ones are separated by a region of forbidden energy: the insulator band gap. If the gap width is not too large (typically <1.4 eV) so

3-527-28138-X/96/$5.00 + .50

that for $T \neq 0$ K electrons can be excited into the conduction band, the system is called a semiconductor. When different materials are brought into contact, interesting properties may be expected. This is the case for some metal–semiconductor interfaces that, depending on the sign of the applied bias, let the current flow or not. They form one type of diode, called a Schottky diode.

The rectifying properties of metal–semiconductor interfaces were reported for the first time by Braun (1874). He attributed this anomalous conductivity to the existence of a highly resistive, thin interfacial layer. In 1938, Schottky (Schottky, 1938) explained the rectifying properties by the presence of a space-charge layer on the semiconductor side of the interface that is depleted of its carriers. For this reason, these metal–semiconductor interfaces were called Schottky barriers.

A model was proposed only 60 years after Braun first reported it, by Mott (1938) and Schottky (1940); it is known as the "Schottky model." The metal and the semiconductor Fermi levels align and no interface dipole is taken into account. This results in a linear dependence of the Schottky barrier height Φ_{bn} on the metal work function Φ_m (Schottky rule). For neutrality reasons, the space charge in the depletion layer is compensated by a charge of opposite sign at the interface, on the metal side. This model is far too simple since it neglects the role of surface or interface states. In fact, it was shown (Schottky 1939) by experiments on metal–selenium contacts that new models had to take these states into account.

Bardeen, in 1947, introduced the idea of Fermi-level pinning by states in the semiconductor band gap. Charge transfer at the interface occurs between the metal and the semiconductor surface states. This idea was developed by Heine (1965) and later by Tejedor *et al.* (1977) and Tersoff (1984), who replaced surface states by interface states since Heine demonstrated that surface states do not subsist at the interface. Their model is known as MIGS, "metal-induced gap states," or IDIS, "induced density of interface states."

In 1979, Spicer (Spicer *et al.*, 1979, 1988) proposed a different origin for the states involved in Fermi-level pinning. The metal deposition liberates energy, which creates defects, and the Fermi level is then pinned at the defect level. This UDM, "unified defect model," involves donor and acceptor defects and explains experimental differences observed for low metal coverages on *n*- and *p*-type semiconductors (photoemission measurements).

Usually, Schottky-barrier heights are determined by electrical measurements: capacitance-voltage (*C-V*) or current-voltage (*I-V*). These techniques do not allow the study of microscopic mechanisms involved in the Schottky-barrier determination. Furthermore, they are sensitive to the interface quality and great care must be taken in how they are applied. X-ray photoemission spectroscopy (XPS) allows the study of the evolution of the Fermi level with the metal coverage. The influence of the deposition temperature is difficult to follow because of the different modes of growth and especially island growth, which leaves uncovered areas on the semiconductor surface.

The discrepancy between model predictions and experimental results led many experimentalists to propose correlations between the Schottky-barrier height and the chemical or physical properties of its constituents. No systematic view emerges from the enormous amount of experimental data, and these correlations only apply to a given type of material. For example, in the transition metal-silicide–silicon interface, Φ_{bn} was related to the core-level shift of the transition metal, or the silicide heat of formation (Andrews and Phillips, 1975). Effective work functions have also been proposed to replace the metal work function in the Schottky rule (Freeouf, 1980; Freeouf and Woodall, 1981). For III-V semiconductors, Φ_{bn} has also been related to the heat of formation of the most stable metal–anion compound (Brillson, 1978).

Despite years of study, no general agreement exists on the mechanisms involved in the Schottky-barrier height determination. Furthermore, the large amount of experimental and theoretical data do not provide a unique description. It appears that several mechanisms compete to determine the Schottky-barrier height but that MIGS is the main one.

Further studies of the physical properties of Schottky barriers are needed from a fundamental point of view and for practical applications. Schottky barriers play an important role in the device industry for diode and transistor fabrication. They are preferred to *p-n* junctions when high switching rates are required. The understanding of the Schottky barrier might allow the development of barriers of predetermined height and hence barrier engineering. It would be especially interesting to obtain small barrier heights on an *n*-type semiconductor for infrared detector fabrication, for example.

This article is organized as follows: Schottky barriers, Ohmic contacts, and their electrical behavior under an applied bias are described in the first section. General remarks on the Schottky-barrier values (nondependence on the metal, the doping, . . .), experimental correlations, and methods of Schottky-barrier determination are presented in the second section. Theoretical models are discussed in the third section. The fourth section deals with the microscopic aspects of the interface and electronic-structure calculations.

The reader is referred to the articles SEMICONDUCTORS, ELEMENTAL, ELECTRONIC PROPERTIES and SEMICONDUCTORS, COMPOUND, ELECTRONIC PROPERTIES in this Encyclopedia for the definition of the physical quantities relevant to semiconductors.

1. DESCRIPTION AND ELECTRICAL BEHAVIOR

1.1 Description of the Schottky Barrier

The physical quantities involved in a Schottky barrier are described in Fig. 1 for the case of a metal on an *n*-type semiconductor. The Schottky-barrier height Φ_{bn} is the difference between the bottom of the conduction band and the Fermi level. At equilibrium, the Fermi levels must be the same in the metal and the semiconductor and charge transfer occurs. In the metal, the charges (Q_m) remain at the surface because of the high density of states ($\sim 10^{22}$ cm^{-3}). In the semiconductor, charges may have two contributions: a surface one (if surface or interface states exist) (Q_{ss}) and a space charge (Q_{sc}). The space charge is due to uncompensated ionized donors. Their density depends on the doping (10^{14}–10^{18} cm^{-3}) and is much lower than in the previous case so that the charge extends into the semiconductor and

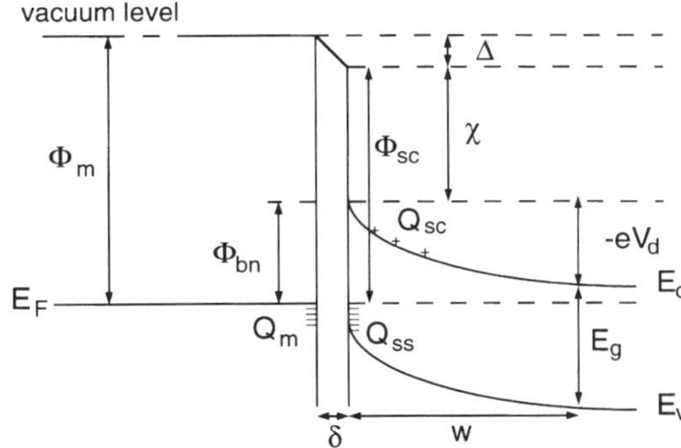

FIG. 1. Energy-band diagram for a metal–*n*-type semiconductor Schottky barrier. E_F is the Fermi level, Δ the interface dipole, E_g the gap width, E_v the top of the valence band, E_c the bottom of the conduction band, W the depletion-zone length, Φ_m the metal work function, Φ_{bn} the Schottky-barrier height, Φ_{sc} the semiconductor work function, V_d the diffusion potential, χ the semiconductor electronic affinity, Q_{sc} the space charge in the semiconductor, Q_{ss} the interface charge in the semiconductor, Q_m the interface charge in the metal, and δ the mean distance between Q_{ss} and Q_m.

creates a band bending. The length scales for these contributions are different: Surface charges are on ~ 1 nm while band bending occurs on ~ 100–1000 nm. The metal work function Φ_m is the difference between the vacuum level and the Fermi-level position. The semiconductor affinity χ is the difference between the vacuum level and the bottom of the conduction band. The term E_g is the gap width. For silicon, $E_g = 1.12$ eV and $\chi = 4.07$ eV; transition-metal work functions are around 4.5–5.5 eV. In this article, any level in the gap is labeled with respect to the top of the valence band. The interface dipole Δ and the Schottky-barrier height Φ_{bn} are expressed in terms of energies. Except when it is explicitly specified, an *n*-type semiconductor is always considered.

1.2 Schottky Barrier and Ohmic Contact at Equilibrium

The formation of the diode is explained (Fig. 2) in a simplified case where no interface states are taken into account (Rhoderick and Williams, 1988). We first consider an *n*-type semiconductor. If the metal work function is larger than the semiconductor one [Fig. 2(a)], electrons diffuse from the semiconductor to the metal and create a band

(a) $\Phi_m > \Phi_{sc}$ - n type

vacuum level

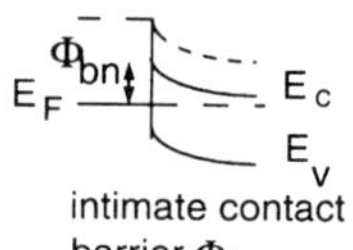

intimate contact
barrier Φ_{bn}

(b) $\Phi_m < \Phi_{sc}$ - n type

vacuum level

intimate contact
ohmic contact

(c) $\Phi_m > \Phi_{sc}$ - p type

vacuum level

intimate contact
ohmic contact

(d) $\Phi_m < \Phi_{sc}$ - p type

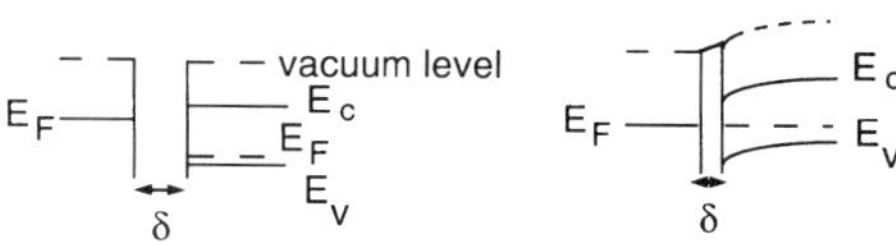

intimate contact
barrier Φ_{bp}

FIG. 2. Metal–semiconductor contact. (a) $\Phi_m > \Phi_{sc}$: *n*-type semiconductor, Schottky barrier. (b) $\Phi_m < \Phi_{sc}$: *n*-type semiconductor, Ohmic contact. (c) $\Phi_m > \Phi_{sc}$: *p*-type semiconductor, Ohmic contact. (d) $\Phi_m < \Phi_{sc}$: *p*-type semiconductor, Schottky barrier.

bending. Electrons have to pass a barrier Φ_{bn} between the semiconductor and the metal; Φ_{bn} is the Schottky-barrier height. In the other case ($\Phi_m < \Phi_{sc}$) [Fig. 2(b)], electrons diffuse from the metal to the semiconductor. They accumulate at the bottom of the conduction band which goes closer to or even below the Fermi level and create an accumulation layer. No barrier appears and the system has the resistivity of the semiconductor; it is an Ohmic contact. For an accumulation layer, the density of states is that of the valence or the conduction band, that is 10^{19} cm^{-3}. The corresponding band bending is then smaller than for the depletion layer. The behavior is reversed for metal on a *p*-type semiconductor interface [Figs. 2(c) and 2(d)].

This description is a simplified one since it ignores the role of surface states. If surface states exist, they contain a charge Q_{ss}; band bending already exists at the free semiconductor surface (see Sec. 3.2). The type of the contact (Ohmic or rectifying) cannot so easily be deduced from the difference between the metal and the semiconductor work functions, and an interface metal–*n*-type semiconductor with $\Phi_m < \Phi_{sc}$ can exhibit rectifying properties.

1.3 Schottky Barrier and Ohmic Contact under an Applied Bias

1.3.1 Schottky Barrier under an Applied Bias If $V_{sc\text{-}m} < 0$ [Fig. 3(a)], the band bending diminishes and the conduction band CB goes up by eV. The metal–semicon-

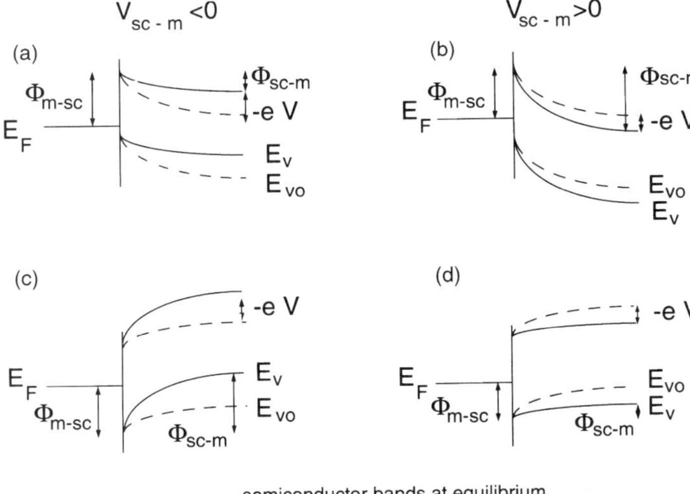

FIG. 3. Schottky barrier under an applied bias V_{sc-m}. (a) $V_{sc-m} < 0$, n-type semiconductor. (b) $V_{sc-m} > 0$, n-type semiconductor. (c) $V_{sc-m} < 0$, p-type semiconductor. (d) $V_{sc-m} > 0$, p-type semiconductor.

ductor barrier $\Phi_{m\text{-sc}}$ stays the same while the semiconductor–metal barrier $\Phi_{sc\text{-m}}$ decreases. The equilibrium is broken, and electrons diffuse from the semiconductor to the metal and create a current from the metal to the semiconductor. If the bias voltage is further increased, the flat-band regime is reached (for $V = V_d$, the diffusion potential) and the electrons can freely flow from the semiconductor to the metal.

If $V_{sc\text{-m}} > 0$ [Fig. 3(b)], the conduction band is lowered and the diffusion of the electrons from the semiconductor to the metal decreases. The metal–semiconductor barrier is unchanged. The contact is blocked. A p-type semiconductor exhibits opposite behavior [Figs. 3(c) and 3(d)].

1.3.2 Ohmic Contact under an Applied Bias For an Ohmic contact with an n-type semiconductor, electrons diffuse from the metal to the semiconductor and create an accumulation layer. The charge is an accumulation one; there is no zone without carriers and hence no resistive zone. The bias is applied to the whole semiconductor, which is more resistive than the metal. If $V_{sc\text{-m}} < 0$ [Fig. 4(a)], electrons go from the semicon-

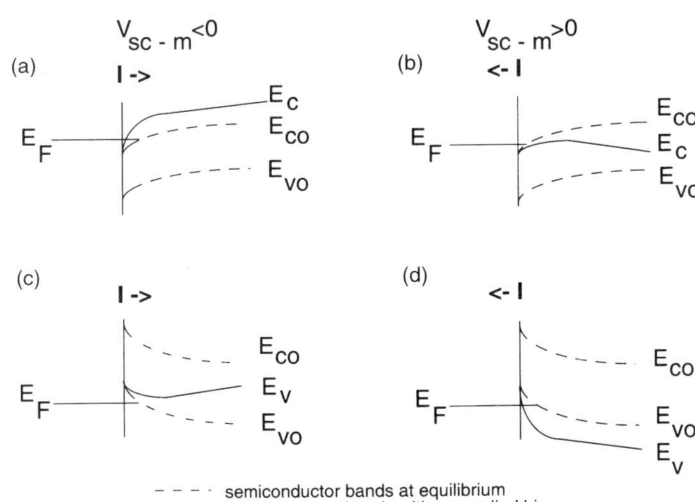

FIG. 4. Ohmic contact under an applied bias V_{sc-m}. (a) $V_{sc-m} < 0$, n-type semiconductor. (b) $V_{sc-m} > 0$, n-type semiconductor. (c) $V_{sc-m} < 0$, p-type semiconductor. (d) $V_{sc-m} > 0$, p-type semiconductor.

ductor to the metal. This flow creates a current from the metal to the semiconductor. If $V_{sc\text{-}m} > 0$ [Fig. 4(b)], the current goes from the semiconductor to the metal. Current can freely flow in the two directions. Interfaces with a p-type semiconductor exhibit a reverse behavior [Figs. 4(c) and 4(d)].

2. EXPERIMENTAL DESCRIPTION

Even though the study of the Schottky-barrier properties has a long history, it is difficult to draw a global view. Many experimental results are conflicting or only apply to very specific systems. We now review some general properties such as the nondependence of the barrier on the metal used or on the type of doping. We then briefly present a few correlations between the barrier height and chemical properties of its constituents. Part of the difficulty in the analysis of the results might come from the sensitivity of the measurements to experimental conditions. Different experimental methods are presented.

2.1 General Trends

Covalent or III-V semiconductor–metal Schottky barriers do not show much dependence on the nature of the metal. This is illustrated in Fig. 5 for Si and Fig. 6 for GaAs. In these figures, the plain line is the "metal-induced gap states" model prediction (see Sec. 3.5). Most of the values lie close to the mid-gap level. This is not true for barriers involving II-VI ionic semiconductors, which vary with the metal.

Table 1 shows examples of I-V measurements for different metal–semiconductor interfaces for both n- and p-type doping. In most cases, the sum of the barriers for n and p types is equal to the gap width:

$$\Phi_{bn} + \Phi_{bp} = E_g. \tag{1}$$

This means that the barrier is independent of the doping. The values tabulated correspond to thick diodes. Exceptions to this rule have been found for low metal coverages on III-V semiconductors (Al, Ga, In, Au/GaAs; Spicer *et al.*, 1988), which exhibit different Fermi-level positions for n- and p-type

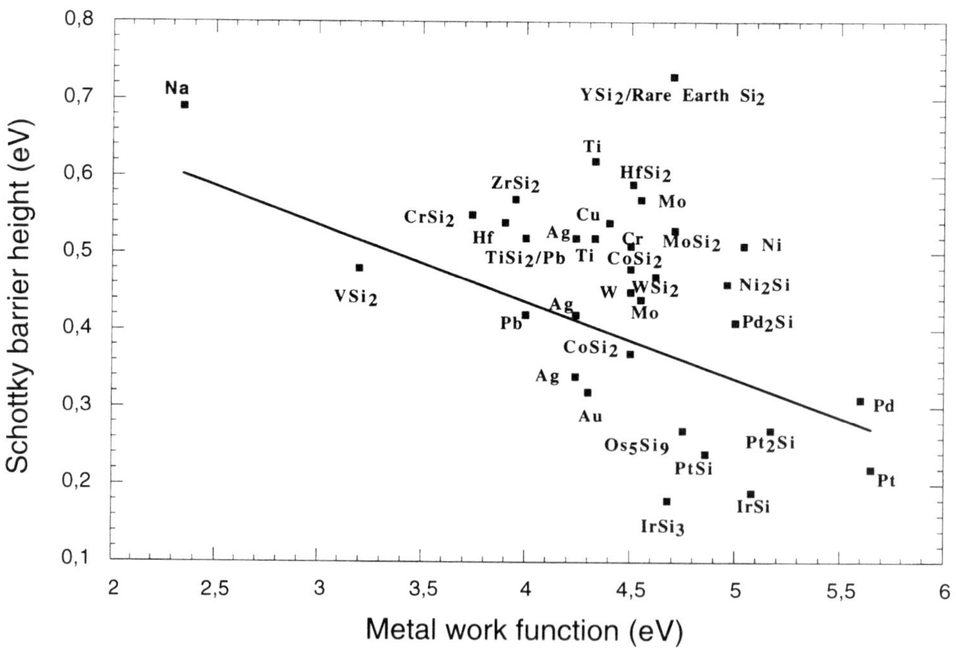

FIG. 5. $E_g - \Phi_{bn} = f(\Phi_m)$ for metal–Si interfaces. The straight line indicates the MIGS prediction.

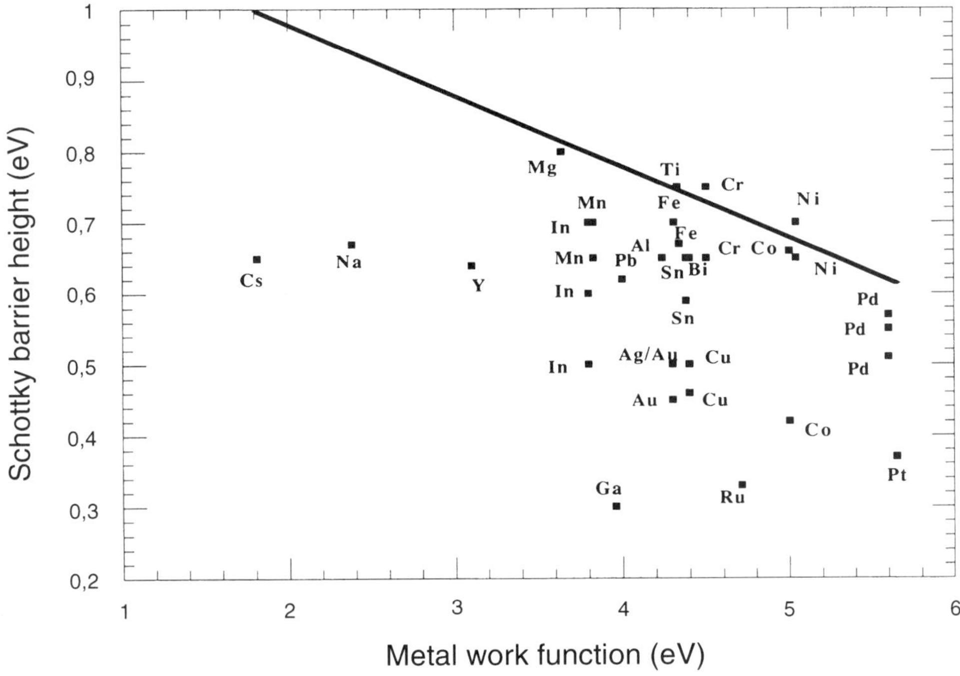

FIG. 6. $E_g - \Phi_{bn} = f(\Phi_m)$ for metal–GaAs interfaces. The straight line indicates the MIGS prediction.

doping. This is discussed in Sec. 3.3. The fact that the Schottky-barrier height does not depend on the semiconductor type of doping is the first meaning of "Fermi-level pinning." The Fermi level lies at a given position in the semiconductor band gap, whatever the doping. A second kind of pinning is usually evoked: a pinning of the Fermi level to its thick interface value for small metal coverages, generally before the deposition of one metal monolayer (1 ML). The Fermi level quickly moves from its value for isolated atoms (small coverage, $\theta \ll 1$ ML) to its saturated value for $\theta \sim 1$ ML. Calculations reproduce this behavior (Ortega *et al.*, 1988).

Table 1. I-V measurements for different metals on *n*- and *p*-type Si and GaAs.

Semiconductor	Metal	Φ_{bn}	Φ_{bp}	$\Phi_{bn} + \Phi_{bp}$
Si	Au	0.8	0.25	1.05
$E_g = 1.12$	Ag	0.67	0.47	1.14
	Cu	0.61	0.50	1.11
	Al	0.72	0.58	1.3
	Cr	0.61	0.50	1.11
	$ErSi_2$	0.39	0.70	1.09
GaAs	Ag	0.88	0.63	1.51
$E_g = 1.42$	Au	0.90	0.42	1.32
	Hg	0.72	0.68	1.4
	Cu	0.96	0.45	1.39
	Al	0.85	0.61	1.46
	Cr	0.77	0.56	1.33

Interfacial reactivity seems to have a small influence on the barrier-height determination. Ag, Au, Cu, and Pd on GaAs give nearly identical barriers (0.5, 0.55 eV) despite having completely different interface reactivities: No reaction exists at an Ag–GaAs interface while Pd on GaAs is highly reactive (Spicer *et al.*, 1988).

Other behaviors are not so clear. The interface geometry seems to have a small influence on the barrier height, but a well-known conflicting example exists: the $NiSi_2$–Si (111) interface exhibits two barriers, 0.65 and 0.75 eV, depending on whether or not $NiSi_2$ is rotated by 180° with respect to the silicon. This case will be further discussed in Sec. 4. There are also conflicting experimental results on the role of an interfacial oxygen or oxide layer.

These general trends apply to most metal–semiconductor interfaces. A huge amount of experimental data exists, and exceptions can be found in many cases.

2.2 Phenomenological Models and Correlations

Many correlations have been proposed between the Schottky-barrier height and the chemical properties of its constituents. Only a few will be discussed here; for a more detailed review, see Mönch (1990).

In the transition metal-silicide–silicon case, several relations were proposed between Φ_{bn} and the silicide bulk properties. Andrews and Phillips (1975) found a very good correlation of Φ_{bn} with the silicide heat of formation but no model was proposed. It corresponds to the straight line in Fig. 7. In this figure, more experimental points have been added and the relation is not so clear. Dependence on the transition-metal core-level shift has been proposed, too.

Propositions have been made to keep the Schottky rule and change the metal work function. In Tm–Si interfaces, Freeouf (1980) assumed the existence of a mixed phase close to the interface and replaced the transition-metal work function Φ_m by the geometrical mean, $(\Phi_m \Phi_{Si}^4)^{1/5}$, even though it does not correspond to the silicide work function. A similar approach was proposed for III-V and II-VI semiconductors. Anion microclusters at the interface were assumed to modify Φ_m so that the Schottky rule remains valid (Freeouf and Woodall, 1981), but no experimental evidence of their existence has been found.

New compounds can be formed at reactive metal–III-V or –II-VI semiconductor interfaces, even at room temperature, and Brillson (1978) related the barrier height with the heat of formation of the most stable metal–anion compound.

A common-anion rule has been proposed, too: Semiconductors with identical anions should exhibit very similar Schottky-barrier heights with the same metal, but this is contrary to what many experimental results give.

The idea of a correlation between the barrier height and bond strength (heat of formation) is very interesting but gives poor results. This might indicate that more subtle phenomena are involved in the barrier deter-

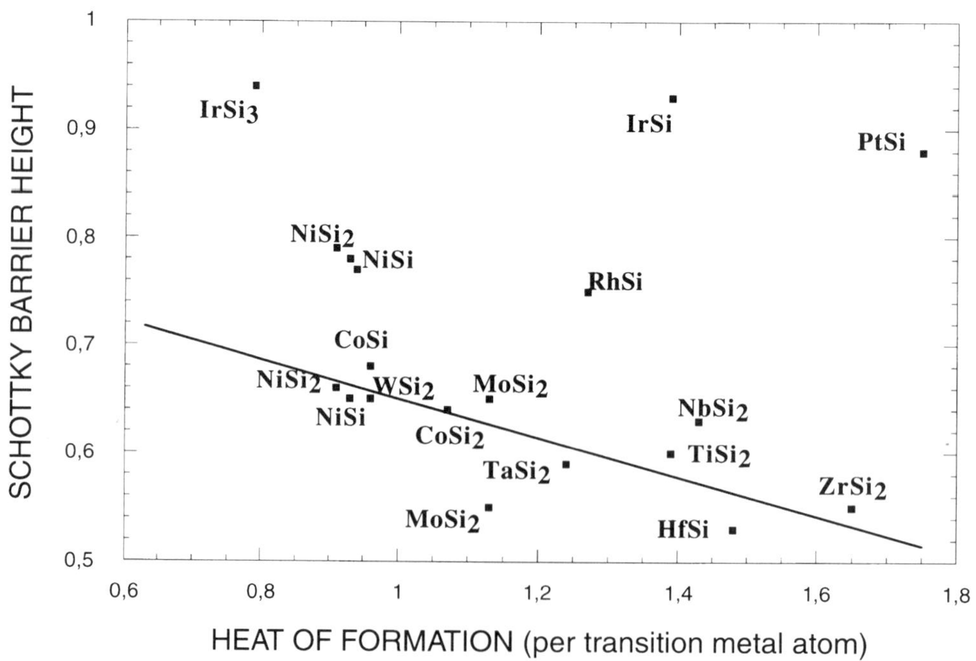

FIG. 7. $\Phi_{bn} = f(\Delta H_f)$ for transition-metal–silicon interfaces. The straight line indicates Andrews and Phillips correlation (Andrews and Phillips, 1975).

mination. It will be discussed along with the models.

2.3 Schottky-Barrier Height Measurements

Five techniques of Schottky-barrier height determination will now be presented (Rhoderick and Williams, 1988). They give the same information but face different types of problems. The combined use of at least two of them should give more confidence in the determined values.

2.3.1 Capacitance Measurement, *C-V* The relation between W, the depletion-zone width, and V_d, the diffusion potential, is given by

$$W = (2\epsilon V_d/eN_d)^{1/2}.$$

If a bias V is applied, $V = V_{\text{m-sc}} > 0$ (direct), the potential becomes $V_d - V$ and

$$W = [2\epsilon(V_d - V)/eN_d]^{1/2}. \qquad (2)$$

If V is modified, W is modified too and a differential capacitance can be defined:

$$C = \delta Q/\delta V. \qquad (3)$$

The charge associated with the band bending, $-e(V_d - V)$, is given by

$$Q_{sc} = [2\epsilon eN_d(V_d - V)]^{1/2}; \qquad (4)$$

combining with Eqs. (2) and (3) gives

$$C = (\epsilon eN_d/2)^{1/2}(V_d - V)^{-1/2} = \epsilon/W: \qquad (5)$$

C is the capacitance of a flat capacitor of width W. Equation (5) can be rewritten as

$$C^{-2}(V) = 2(V_d - V)/\epsilon eN_d; \qquad (6)$$

thus $C^{-2}(V)$ is linear in V. The term N_d can be determined from the slope, and the abscissa at the origin gives V_d. If N_d is not constant, $C^{-2}(V)$ is no longer linear.

In practice, the junction is reverse polarized and small oscillations δV are superposed. If no trap exists, the variation δQ corresponds to the variation of the free-carrier concentration at the edge W of the depletion zone on a width δW. Figure 8 shows a typical C^{-2}-V spectrum.

If traps exist, they contain part of the charge δQ. To get rid of them, one has to perform measurements at high frequency (>100 kHz). For such a frequency, only the semiconductor depletion charge changes. Another problem then arises: the series resistance contribution, which disappears at low frequency. A compromise must be found to minimize these two contributions.

2.3.2 Current Measurement, I-V If thermoionic emission is assumed to be the only current transport mechanism, the current density J in the forward direction with $V > 3kT/q$ is given by

$$J = J_0 \exp(eV/kT), \qquad (7)$$

where $J_0 = A^*ST^2 \exp(-\Phi_{bn}/kT)$ is the saturation current, A^* the effective Richardson constant, T the temperature, V the applied bias, and S the diode area. For simplicity, we do not introduce the Schottky-barrier lowering (due to image force), but it must be taken into account in barrier-height determination (more details can be found in Rhoderick and Williams, 1988). Very often, experimental measurements do not follow the

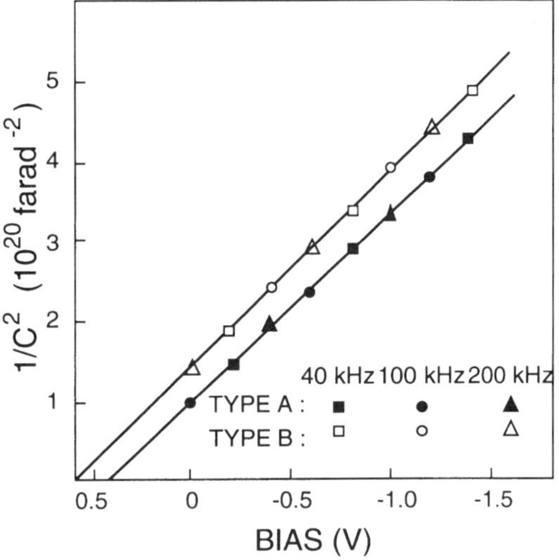

FIG. 8. Typical *C-V* characteristics for type A and B $NiSi_2$/Si (111) interfaces at three different frequencies (from Tung *et al.*, 1986). Type A: $NiSi_2$ and Si have the same crystallographic axis; type B: $NiSi_2$ is rotated by 180° with respect to Si.

linear relation between ln(J) and V. One of the reasons is that A^* and Φ_{bn} depend on the applied bias and that other current-transport mechanisms are involved. To take this effect into account, J is written as

$$J = J_0 \exp\left(\frac{eV}{nkT}\right) \quad \text{and then } n = \frac{e}{kT}\frac{\delta V}{\delta \ln(J)}, \tag{8}$$

Here n is the *ideality factor*. It must be close to 1 to deduce Φ_{bn} from the I-V characteristic ($n < 1.1$). If ln(J) is plotted against V, J_0 is the extrapolated value for $V = 0$. The value of n is deduced from the slope. Then Φ_{bn} is given by

$$\Phi_{bn} = kT \ln(A^* S T^2 / J_0). \tag{9}$$

The choice of A^* is not crucial; a change by a factor 3 results in a variation of kT for Φ. A typical example is given in Fig. 9.

2.3.3 Photoresponse, *I-hν* A monochromatic light is directed upon the metal surface (front illumination). When a photon arrives at the surface, an electron of the metal can take its energy $h\nu$. If this energy is large enough, the electron can go from the metal to the semiconductor conduction band. In the semiconductor, electrons will be accelerated because of the electric field in the space-charge zone. Electrons can cross the barrier if $h\nu > \Phi_{bn}$, and a photocurrent I_{ph} is induced. The process is described in Fig. 10. An electron with an energy E_1 receives the photon energy $h\nu$. It is excited to $E_2 = E_1 + h\nu$ and creates a hole of energy E_1. If $E_2 > \Phi_{bn}$, the electron can pass the barrier and is collected in the semiconductor. The hole, even if E_1 is smaller than E_v, will be pushed back to the metal. The photocurrent I_{ph} from the semiconductor to the metal is given by Fowler's formula:

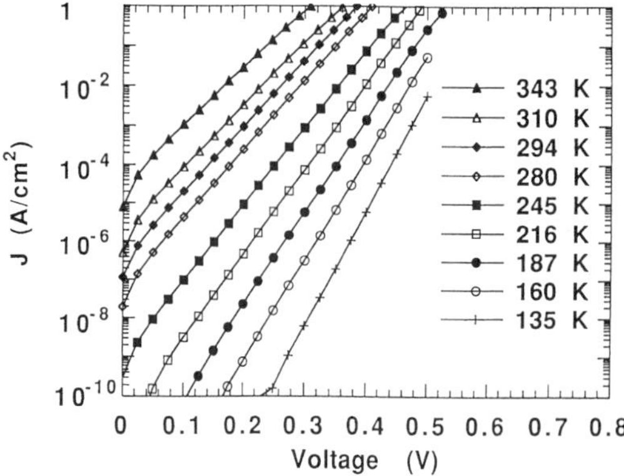

FIG. 9. Typical *I-V* characteristics for Al–*n*-GaAs samples for various temperatures. Ideality factors are between 1.01 and 1.03 (from Muret *et al.*, 1992).

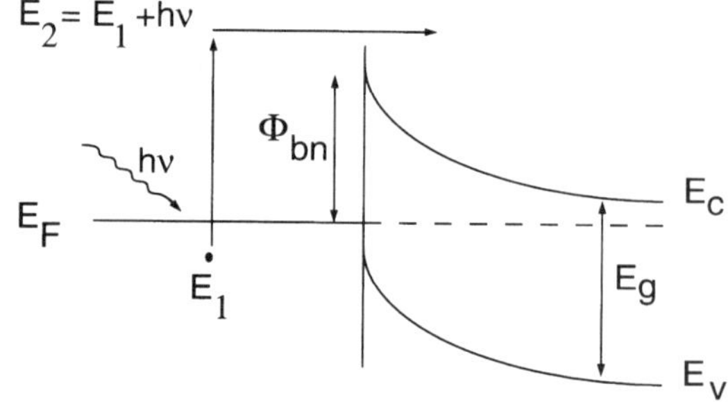

FIG. 10. Principle of photoresponse measurement.

$$I_{ph} = AT^2\{x^2/2 + \pi^2/6 - (e^{-x} - e^{-2x}/2^2 + e^{-3x}/3^2 - \cdots)\} \quad \text{for } x > 0. \tag{10}$$

Here A is a constant, T is the temperature, and $x = (h\nu - \Phi_{bn})/kT$. If $x > 6$, only the parabolic term is kept and

$$I_{ph} \sim (h\nu - \Phi_{bn})^2. \tag{11}$$

When $(I_{ph})^{1/2}$ is plotted against ν, a straight line should be obtained and Φ_{bn} is given by the extrapolated value at the energy axis. A typical spectrum is shown in Fig. 11. If the metal thickness is too large, the photocurrent will not be linear and absorption occurs in the metal. The thickness should be of the order of the electron mean free path (a few times 10 nm).

2.3.4 Application of XPS to Schottky-Barrier Measurements XPS has been extensively used for band bending and Schottky-barrier height determination for over a decade (Hollinger, 1987). It allows the examination of the first stages of Schottky-barrier formation. The principle is explained in Fig. 12.

Because of the small penetration length of XPS, bands can be considered as flat in the region of interest. The binding energy E_c of a semiconductor core level is measured for the free surface and for increasing metal cover-

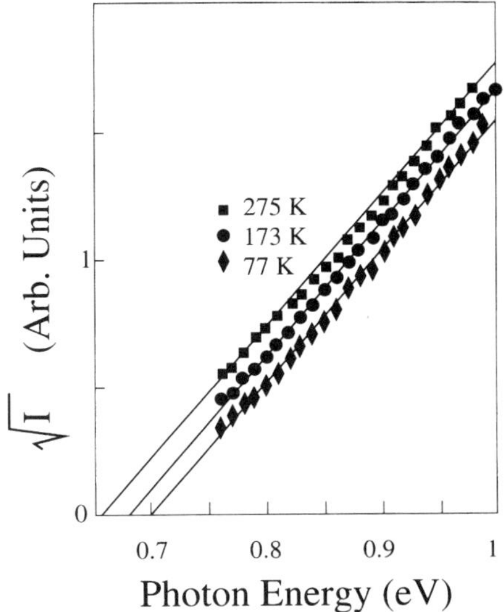

FIG. 11. Typical I-$h\nu$ spectra of $CoSi_2$–Si(n) (from Duboz *et al.*, 1989).

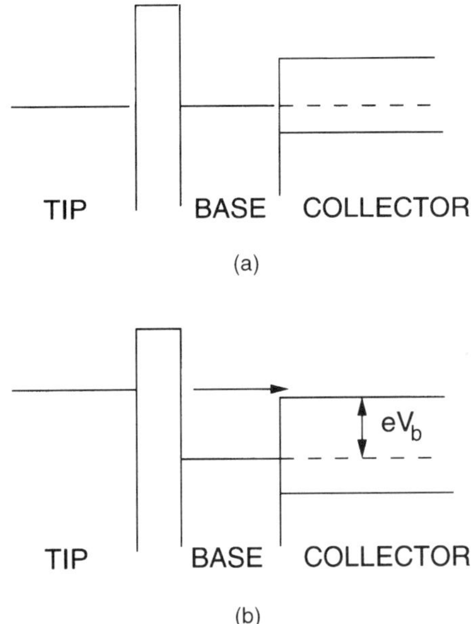

FIG. 13. Schematic energy-band diagram for the three-terminal BEEM experiment. Collector current I_c is measured between the base (metal) and the collector (semiconductor). (a) The energy-band diagram for zero tunnel bias, $V = 0$. (b) The energy-band diagram for tunnel bias greater than the barrier voltage, $eV > eV_b$. (From Kaiser and Bell, 1988.)

age. The core-level shift δ ($=E_{ci} - E_c$) reproduces the variation of the band bending. It is then possible to follow the evolution of the Schottky-barrier height $\Phi_{bn} = CB_i - E_{Fi} + \delta$ as a function of the metal coverage.

2.3.5 Ballistic-Electron Emission Microscopy (BEEM) This new technique for Schottky-barrier height determination was proposed by Kaiser and Bell (1988; see also Bell *et al.*, 1988). It is based on a scanning tunneling microscopy (STM) method and can be used for Schottky barriers, semiconductor heterojunctions, and other buried-interface studies. The principle of BEEM is described in Fig. 13.

A STM tip is positioned near the metal surface to allow electron tunneling between the tip and the metal (base). This injects ballistic electrons into the base. The tunnel current I_t is kept constant. If the metal width is smaller than the attenuation length (>10 nm), these ballistic electrons may propagate up to the interface. If the metal–tip bias V is smaller than V_b, electrons cannot enter the semiconductor. When V is increased above V_b, ballistic electrons enter the semiconductor (collector) and the collector current I_c increases drastically. Figure 14 shows a BEEM I_c-V spectrum for the Au–Si structure recorded by Kaiser and Bell. A theoretical treatment was developed by these authors, who proposed that for V just above V_b, the

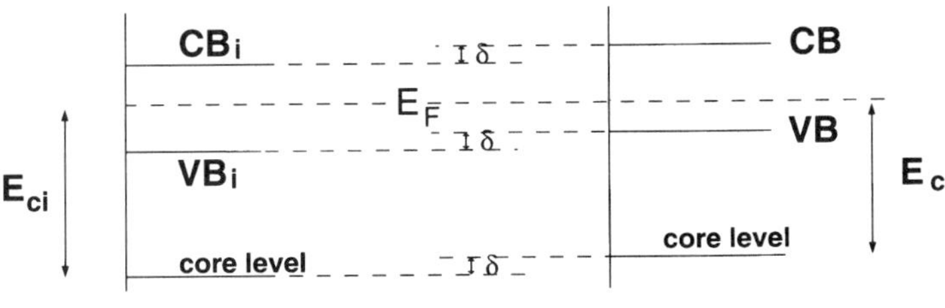

FIG. 12. Application of XPS to Schottky-barrier height measurement.

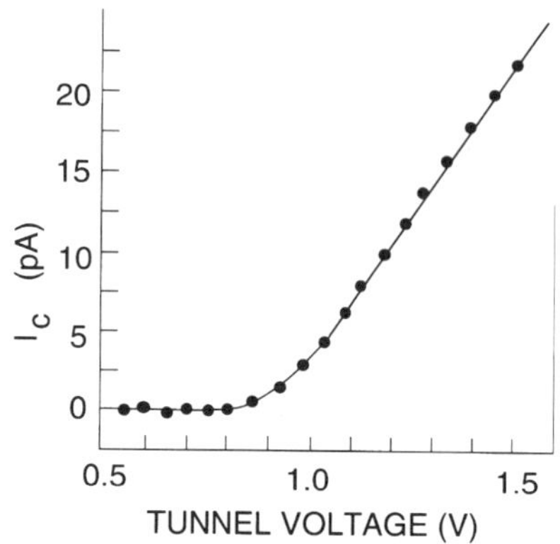

FIG. 14. BEEM I_c-V (dots) for the Au–Si structure recorded at a tunneling current $I_t = 0.87$ mA (from Kaiser and Bell 1988).

I_c-V spectrum behaves as $(V_b - V)^2$, but the power of this dependence is disputed.

This method allows the determination of the spatial variations of the Schottky-barrier height with nanometer-scale resolution.

3. THEORETICAL MODELS

The first model for Schottky barriers (Mott, 1938: Schottky, 1940) appeared more than 60 years after Braun's report and was soon shown incorrect by experiments. In 1947, the understanding of the semiconductor electronic structure allowed Bardeen to propose the idea of Fermi-level pinning by states in the semiconductor band gap, which is the key point of today's theories.

3.1 Schottky Model

This model (Fig. 15) (Schottky, 1940, 1942) assumes that no state exists in the band gap of the semiconductor so that no interface dipole can be induced. Only space charges exist in the semiconductor. At equilibrium, the Fermi levels must align in the metal and the semiconductor. Charge transfer occurs and creates an accumulation charge Q_m on the metal surface and a space charge Q_{sc} in the semiconductor. The whole system is neutral so that $Q_m = -Q_{sc}$.

As Fig. 15 shows,

$$\Phi_{bn} = \Phi_m - \chi. \tag{12}$$

The Schottky barrier depends linearly on the metal work function. It is a property of bulk metal and semiconductor. The same idea has been developed for band lineup in heterojunctions (the reader is referred to the article HETEROJUNCTIONS, SEMICONDUCTOR in this Encyclopedia for further information). This model gave good results for ionic semiconductors but does not agree with experiments on covalent and III-V semiconductors.

3.2 Bardeen Model

The Schottky model ignores the role of surface states at the metal–semiconductor interface. It is well known that a semiconductor free surface may show states in the gap, and Bardeen (1947) proposed a model where these states play a major role. They are filled up to some level Φ_0 and carry a charge Q_{ss} [Fig. 16(a)]. Because of the neutrality of the system, Q_{ss} must be compensated by a space charge Q_{sc} inside the semiconductor so that $Q_{ss} + Q_{sc} = 0$. This space charge is associated with band bending and depends on the density of surface states.

At the metal–semiconductor contact [Fig. 16(b)], charge transfer creates a supplementary charge Q_{ss}, compensated by the same charge of opposite sign on the metal side. If the density of surface states is high enough, the surface locally behaves like a metal

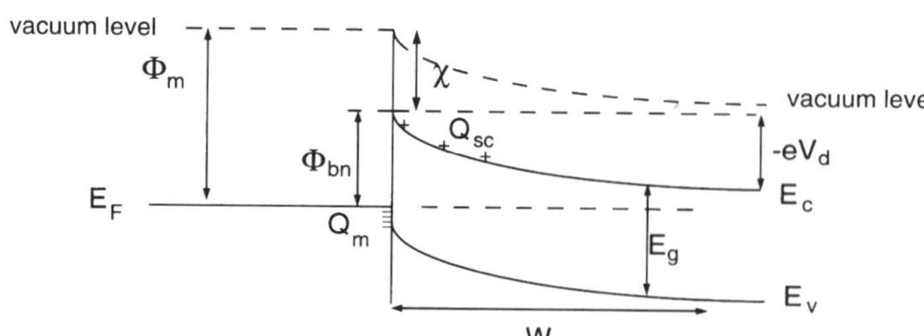

FIG. 15. Schottky-model description of the Schottky barrier for an n-type semiconductor. $\Phi_{bn} = \Phi_m - \chi$.

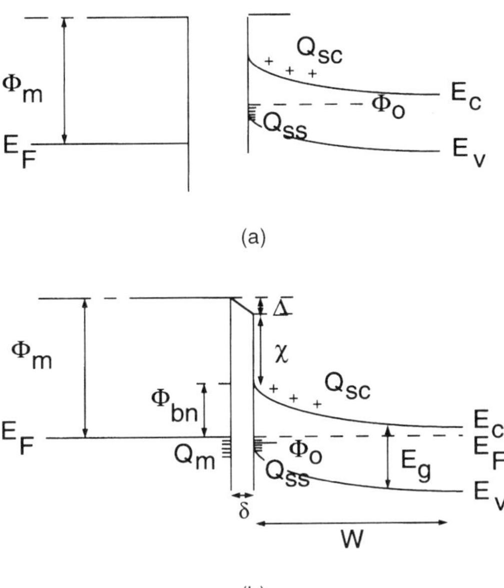

FIG. 16. Fermi-level pinning by states in the semiconductor band gap (Bardeen, Cowley–Sze, MIGS models). (a) Metal and semiconductor free surfaces. (b) Metal and semiconductor in contact.

whose Fermi level is Φ_0. The high density of surface states can accommodate all the charge transferred from the metal without noticeably moving E_F from Φ_0. The charge transfer creates a dipole Δ that tends to align E_F in the metal with Φ_0. The Schottky barrier is given by

$$\Phi_{bn} = E_g - \Phi_0. \quad (13)$$

The space-charge region in the semiconductor remains unchanged. The Schottky barrier is a property of the semiconductor surface and is completely independent of the metal. Here E_F is pinned near Φ_0 by the high density of surface states.

Bardeen is the originator of the idea of Fermi-level pinning by states in the semiconductor band gap. This is the key point of the models, but the nature of these states is still under discussion.

3.3 Defect Model (UDM)

Cleaved 110 surfaces of III-V semiconductors show no surface states in the band gap and the Bardeen model cannot be applied. Spicer *et al.* (1979) proposed a model where the Fermi level is pinned by defect levels in the band gap. The energy needed for the defect creation is provided by the metal deposition (condensation heat, etc.). The model applies to III-V–insulator and III-V–metal interfaces; this is why it is called the "unified defect model" (UDM). If Φ_1 is the defect level inside the semiconductor band gap, relative to the top of the valence band, and if the density of defects is high enough, charge transfer between the metal and the defect states creates a dipole that aligns E_F and Φ_1. E_F is pinned by these states and

$$\Phi_{bn} = E_g - \Phi_1. \quad (14)$$

When this model was first proposed, not much was known about defects, and Spicer first attributed their origin to anion or cation vacancies. Further studies of the GaAs system showed that antisites are the main defects. In GaAs, the key defect is the As_{Ga} antisite, which is a double donor with levels at 0.52 and 0.75 eV above the top of the valence band. These donors are partly compensated by acceptors, most likely Ga_{As} antisite. The Fermi-level position depends on the doping and the ratio $R = As_{Ga}/Ga_{As}$. In fact, $R > 1$ because GaAs crystals are grown from the As-rich side of the phase diagram. Interfaces with *n*- (*p*-) type GaAs are pinned at 0.75 (0.52) eV. (For further details, see Spicer *et al.*, 1988).

This model has been successfully applied to many III-V–metal systems. Zur *et al.* (1983) modeled the role of donor and acceptor defects on the Schottky-barrier height in two different cases: thick overlayer and thin overlayer (free semiconductor surface with defects). They found that for thick diodes Fermi levels cannot be very different (~0.1 eV) for *n*- and *p*-type semiconductors, but for submonolayer coverages, substantial differences may exist. The UDM explains why E_F can be pinned at different levels in the gap for *n*- and *p*-type semiconductors since it involves donor and acceptor defects.

3.4 Cowley–Sze Model

There is a general agreement on the pinning of the Schottky barrier by interface states, but the nature of these states is still controversial. The development proposed by

Cowley and Sze (1965) only assumes their existence and that their density D_s is constant between Φ_0, the filling level of surface states, and the final Fermi level [Fig. 16(b)]. ϕ_0 corresponds to a local charge neutrality of the surface. The interface charge due to metal–semiconductor charge transfer is

$$Q_{ss} = -eD_s(E_g - \Phi_0 - \Phi_{bn})\ \mathrm{C\ cm^{-2}}. \quad (15)$$

The calculation of the depletion width W of a semiconductor with a band bending $-eV_d$ is a classical electrostatic problem. It is done by solving Poisson's equation. Under the assumptions that for $x < W$, $\rho = -eN_d$, and for $x > W$, $\rho = 0$, $dV/dx = 0$, we find

$$W = (2\epsilon/eN_dV_d)^{1/2}; \quad (16)$$

here N_d is the donor density and is assumed to be constant in the whole depletion zone. At the interface, the charge per unit area associated with the band bending is given by

$$Q_{sc} = [2\epsilon_s N_d(\Phi_{bn} - \phi_n)]^{1/2}\ \mathrm{C\ cm^{-2}}. \quad (17)$$

The value of $\phi_n = E_c - E_f$ depends on the doping. The whole system is neutral so that $Q_m = -(Q_{sc} + Q_{ss})$. The dipole Δ at the interface can be obtained by Gauss's law on the surface of the metal and the semiconductor:

$$\Delta = -e\delta Q_m/\epsilon_i. \quad (18)$$

From Fig. 16(b), Δ can also be expressed as

$$\Delta = \Phi_m - \chi - \Phi_{bn}, \quad (19)$$

which comes from the fact that E_F is constant in the whole system. Combining relations (17), (18), and (19), one can express Φ_{bn} as

$$\Phi_{bn} = (1 + \alpha)^{-1}(\Phi_m - \chi) + [1 - (1 + \alpha)^{-1}] \times (E_g - \Phi_0) + V, \quad (20)$$

where $\alpha \doteq e^2D_s\delta/\epsilon_i\epsilon_0$. For $\delta \sim 0.4$–0.5 nm, $N_d < 10^{18}\ \mathrm{cm^{-3}}$, $\epsilon_i \sim 1$, $\epsilon_s \sim 12$, V can be neglected. If only the work function changes when the metal changes, the barrier varies with a slope $S = (1 + \alpha)^{-1}$. Equation (20) can be rewritten as

$$\Phi_{bn} = S(\Phi_m - \chi) + (1 - S)(E_g - \Phi_0). \quad (21)$$

If $D_s \to 0$, $S \to 1$, $\Phi_{bn} = \Phi_m - \chi$; this is the Schottky-model prediction. If $D_s \to \infty$, $S \to 0$, $\Phi_{bn} = E_g - \Phi_0$; this is the Bardeen-model prediction. The Schottky and Bardeen models are two limiting cases of this description. For the general case, $0 < S < 1$.

3.5 Metal-Induced Gap States Model (MIGS)

In 1965, Heine argued that real surface states cannot exist at the metal–semiconductor interface but that they are replaced by metal-induced states in the semiconductor band gap. The metallic wave-function tails decrease exponentially in the semiconductor and create a charge transfer. The semiconductor surface states go through a resonance (in opposition to Bardeen's theory). The model has been further developed by Tejedor *et al.* (1977), who introduced the "charge-neutrality level" (CNL) concept: Surface or interface states in the gap equally come from the valence and the conduction bands. States closer to the valence (conduction) band have a predominantly valence (conduction) character. For some level Φ_0, the valence- and the conduction-band contributions are equal. The term Φ_0 corresponds to a local neutrality of the surface and is the so-called "charge-neutrality level."

The charge transfer between the metal and the semiconductor creates a dipole that tends to set the metal Fermi level equal to Φ_0. If the density of interface states is high enough, the difference between the metal Fermi level and the semiconductor CNL is nearly completely screened by the charge transfer. The Schottky barrier is then given by Eq. (21) for small S. Common values of S are 0.1 for covalent and III-V semiconductors and 0.3 for ionic II-VI semiconductors.

The small dependence of the Schottky-barrier height on the metal used for a given semiconductor can be explained within this model. If the density of interface states is high enough, they can absorb a charge transfer due to a change in the metal work function without noticeably moving the Fermi level of the whole system.

Tersoff (1984) calculated the CNL for different semiconductors by coupling the states of the semi-infinite semiconductor complex band structure to the metallic wave-function tails. They are shown in Table 2. For silicon, $\Phi_0 = 0.36$ eV. This gives Schottky-barrier heights in the 0.6–0.8-eV range for all metal–n-Si interfaces, in good agreement with most experimental results. These values were obtained for a continuum of metallic states. If the metal DOS were made of discrete peaks for the energies in the band gap region, E_F would be pinned at one of these peaks, maybe far away from Φ_0.

The physical meaning of the CNL is fairly vague. Its value has been related to the semiconductor dangling-bond level (see Sec. 3.6.1, and Lannoo and Friedel, 1991). The existence and the role of the interface states have been confirmed by first-principles calculations (Yndurain, 1971; Louie *et al.*, 1977). They showed that a smooth density of states appears in the semiconductor band gap, close to the interface.

MIGS and IDIS are two different names for the same model. IDIS (induced density of interface states) applies for metal–semiconductor and semiconductor–semiconductor interfaces while MIGS is only for metal–semiconductor junctions.

3.6 Correlations between Schottky-Barrier Height, Valence-Band Offset, Dangling-Bond Level, and Transition-Metal Impurity Level

The charge-neutrality level is the basic key of the MIGS model, but its definition is rather vague. Lannoo and Friedel (1991) established the equivalence of the CNL to the semiconductor dangling-bond level and gave it a more physical meaning. Within this framework, the correlations between the Schottky-barrier height and various properties such as valence-band offset and transition-metal impurity levels can easily be explained.

Table 2. Charge-neutrality levels calculated by Tersoff (1984).

Sc	CNL	Sc	CNL
Si	0.36	GaSb	0.07
Ge	0.18	GaP	0.81
AlAs	1.05	InP	0.76
GaAs	0.7		
InAs	0.5		

3.6.1 Charge-Neutrality Level and the Semiconductor Dangling-Bond Level We first consider an ideal metal–covalent-semiconductor interface with no rearrangement of atoms (no relaxation nor reconstruction). Each dangling bond interacts with the metal states and becomes a resonant state (Lannoo and Friedel, 1991). If the resonance is symmetric, it is centered on the dangling-bond level E_{db}. There is one electron per dangling bond and so the Fermi level is equal to E_{db}. The surface is neutral, and E_{db} is the charge-neutrality level.

III-V semiconductor nonpolar surfaces exhibit two different dangling bonds, one for the anion (level E_a) and one for the cation (E_c), which go through a resonance because of the coupling with the metal wave functions. If the resonance is broad enough (the resonance width large compared with $E_c - E_a$) and the metal–anion and metal–cation couplings do not differ too much, E_F is pinned at the average dangling-bond level $(E_c + E_a)/2$. The CNL is then $(E_c + E_a)/2$, and E_F can be pinned in the gap even though the free surface does not show states in it.

3.6.2 Correlation with Valence-Band Offset The IDIS model is generally used to describe semiconductor heterojunctions. The same argument as above can be applied to band lineup, and one then finds that the two semiconductor dangling-bond levels must align. If two interfaces between the same metal M and two different semiconductors $SC1$ and $SC2$ have a difference in the Schottky-barrier heights $\Delta\Phi$ ($=\Phi_1 - \Phi_2$, Fig. 17), $\Delta\Phi$ is then the valence-band offset ΔEv of the heterojunction $SC1$–$SC2$.

3.6.3 Correlation with Transition-Metal Impurity Levels The vacuum level was first proposed as a reference level for transition-metal impurities in semiconductors. Tersoff and Harrison in 1987 argued that transition-metal impurities are pinned on the bulk cation vacancy level.

Delerue *et al.* (1988) with a defect molecule model and charge-dependent tight-binding calculations proposed that transition-metal impurity levels are pinned at the average dangling-bond level. Transition atom levels can then be used to determine Schottky-barrier height or valence-band offset.

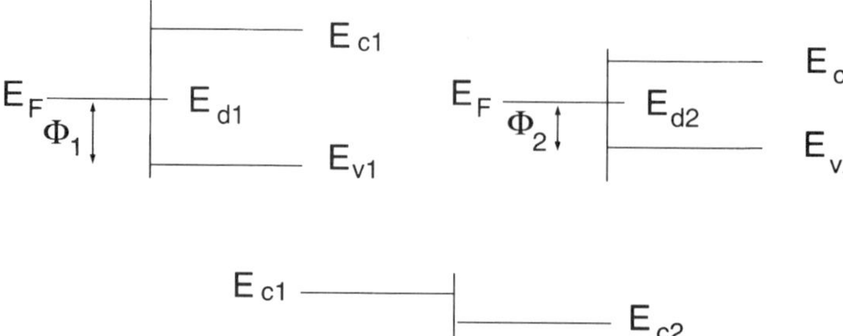

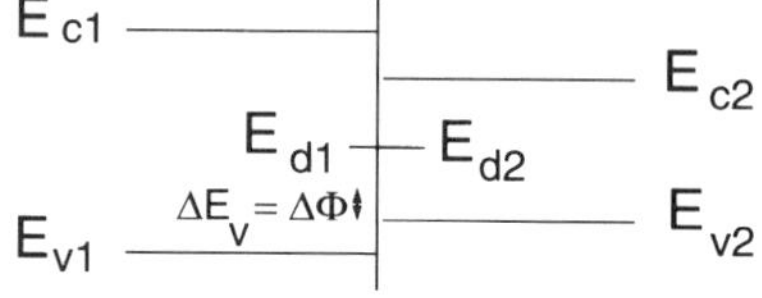

FIG. 17. Correlation between Schottky-barrier height and valence-band offset. E_{d1} (E_{d2}) is the dangling-bond level of semiconductor 1 (2).

3.7 Discussion

It is difficult to choose between the defects or the MIGS model because most of the time they give the same predictions. In the following section, we review some experimental evidence of the Schottky-barrier behavior in light of the two models.

3.7.1 Variation of the Schottky-Barrier Height with Metal Coverage

3.7.1.1 Small Coverages, θ < 1 Monolayer (ML). For very small metal coverages, $\theta < 1$ ML, some systems have been found to have two different pinning positions depending on the type of doping. This is the case for Cs, In, Ga, and Al on GaAs (Spicer *et al.*, 1988). In this case, metal atoms do not have their bulk metallic properties and the MIGS model does not apply. The semiconductor space charge must compensate the surface charge. The dipole length is of the order of the Debye length of the semiconductor, which is very large (1000 nm). The dipole potential (~1 eV) is obtained with a small interface density (10^{12} cm^{-2}) and a large distance. This density of interface states can be provided by defects; acceptors will pin the Fermi level for n-type semiconductor and donors for p-type.

3.7.1.2 Thick Coverages, θ ≫ 1 ML. The compensating charge to the interface charge in the semiconductor is provided by the metal. The screening length is about the Thomas–Fermi wavelength—that is, a few angstroms. A high density of interface states (10^{14} cm^{-2}) is then needed. This density is easily reached within the MIGS model where at least one dangling bond per surface atom is available. On the other hand, 10^{14} cm^{-2} is a very large quantity of defects, which cannot exist in many interfaces of high crystallographic quality.

Submonolayer measurements need careful analysis. Photovoltage effects can induce a difference in the pinning positions for n- and p-type semiconductors. The growth mode is important too. Island growth can create metallic islands for small coverages, and MIGS can exist. Another problem can come from the fact that part of the semiconductor surface remains free and the Fermi level is set at the semiconductor surface pinning position instead of the metal–semiconductor interface one.

3.7.2 Comparison of MIGS and Defects Models

The success and failure of MIGS and defects models are summarized in Table 3.

Schottky-barrier heights with Si and GaAs do not show linear variation in Φ_m (as predicted by the MIGS model) (Figs. 5 and 6), but the MIGS model description seems to be predominant. The electronegativity difference between the metal and the semiconductor influences the charge transfer at the interface and so the variations of E_F about Φ_0. In general, there must be a competition between the defect density and the MIGS density. When the MIGS pinning is not efficient enough, defects must play a role. Strain could have some effects too.

Schottky, Bardeen, MIGS, and defects are macroscopic models, based on macroscopic values, and therefore they cannot take into account the microscopic structure of the interface. This requires the use of complex

Table 3. Comparison of MIGS and defects models

	Defects	MIGS
Different pinning positions for n- and p-type semiconductor for small coverage	Yes—different defects are involved for n and p types	No—but experimental difference can come from the photovoltage effect
Pinning for nonpolar *III-V* semiconductor surfaces (no surface state in the gap)	Yes, no need of surface states	Yes—pinning to the average dangling-bond level if the resonance is broad enough
Pinning for small coverage $\theta <$ 1 ML	Yes	No—MIGS do not exist
Pinning for thick coverage	Yes, with a defect density of 10^{14} cm^{-2}, difficult to imagine	Yes—pinning by MIGS
Relative independence of the SBH with respect to the metal used	Yes—the energy given by the metal deposition creates native defects that are a bulk property of the semiconductor	Yes—if the MIGS density is high enough, they can absorb any supplementary charge due to a modification of the Fermi level in the metal
Band lineup in heterojunctions	No	Yes—IDIS model

electronic-structure calculation methods, which are briefly discussed in the following section.

4. MICROSCOPIC ASPECT: THE ELECTRONIC-STRUCTURE CALCULATION POINT OF VIEW

The microscopic geometry of the interface might have some importance in the Schottky-barrier height determinations. The strength of metal–semiconductor bonds at the interface and the positions of the neighbors are important parameters that are ignored in macroscopic models.

Electronic band-structure calculations within the local-density approximation take the system geometry into account. They are known to give very good results for bulk metals. Problems arise when they are applied to semiconductors. These methods are based on ground-state theory, and in principle, they cannot reproduce the semiconductor gap. They give gap widths about half the experimental values. In practice, the top of the valence band and even the bottom of the conduction band are well described: only the gap width is reduced.

Dealing with interfaces brings out another problem: the loss of translational symmetry. These methods only apply to periodic systems. A supercell technique is then needed: An infinity of interfaces are juxtaposed to restore periodicity.

As an example, let us consider the $NiSi_2$–Si (111) interface. The system exists with two orientations:

1. $NiSi_2$ and Si axes are identical; and
2. $NiSi_2$ is rotated by 180° with respect to Si.

This results in a change of the third neighbors of Ni interfacial atoms, and experimental barrier heights differ by 0.1–0.15 eV. Whether this difference is due to defects or geometry was an important question in the late 1980s. Even though results (linear muffin-tin orbitals method) from different authors disagree somewhat, they always give different barrier heights for the two orientations. This confirms most experimental results that the difference in barrier heights is more probably due to the difference in geometry than to defects.

Despite all the difficulties of *ab initio* electronic-structure calculation methods, they can give very interesting information on the role of the interface geometry because they allow a microscopic study. Other methods have been recently developed to take into account the energetics of the interface; for example, quantum molecular dynamics would allow the determination of the most stable geometry together with its electronic structure. The use of these methods for complicated systems such as Schottky barriers is a challenge for the next few years.

5. CONCLUSION

More than one century after the first report of the rectifying properties of the metal–semiconductor interface and despite the huge amount of experimental and theoretical work, the phenomena involved in the Schottky barrier are not yet completely understood. Correlations between the barrier height and some bulk properties of its constituents give interesting information; for example, the relation to the heat of formation indicates that interface bonds play a role. Macroscopic models highlight the role of the states in the semiconductor band gap at the interface. Whether these states are due to defects or are induced by the metal is not yet finally established, but MIGS seems to be predominant. Variations of the Schottky-barrier height with respect to the MIGS prediction can come from defects or strains, but the nature of the bonds and the interface geometry can play a role too. Electronic-structure calculations are needed to go beyond the models, up to the microscopic aspect of the interface.

The large spread in experimental results, which can even be contradictory, does not give a general view of the problem. This comes from the sensitivity of the barrier to the conditions of the experiment: temperature, reactivity, mode of growth. Measurement-related problems add further difficulties. Nonetheless, general trends can be drawn for most metal–semiconductor interfaces: The barrier height does not depend on the doping for thick interfaces. It is not very sensitive to the nature of the metal, and it reaches its thick-limit value for metal coverages as small as a few monolayers. The development of new techniques like BEEM gives access to new types of information which may explain many discrepancies.

More experimental and theoretical studies are thus needed to improve the understanding of the mechanism of the Schottky barrier.

ACKNOWLEDGMENT

The authors wish to thank P. Muret, O. Symko, and J. Y. Veuillen for very fruitful discussions.

GLOSSARY

Accumulation Zone: Region of the semiconductor where the top of the valence band (bottom of the conduction band) is closer to or above (below) the Fermi level.

Charge-Neutrality Level (CNL): In the semiconductor, surface states are filled up to this CNL. It corresponds to a local neutrality of the surface.

Depletion Zone, Space-Charge Region: Region in the semiconductor where ionized donors or acceptors (depending on the doping) are not compensated. No carriers subsist in this region.

Electronic Affinity: Energy difference between the bottom of the conduction band of a semiconductor and the vacuum level.

MIGS: "Metal-induced gap states," model that assumes Fermi-level pinning by metal-induced states in the semiconductor gap. It is applied to Schottky barriers and heterojunctions and is also called IDIS, "induced density of interface states."

Ohmic Contact: Metal–semiconductor contact that lets the current flow in both directions.

Schottky-Barrier Height Φ_b: Barrier electrons (holes) have to jump at a metal–semiconductor junction. For an *n*-type semiconductor, $\Phi_{bn} = E_c - E_F$ and for a *p*-type $\Phi_{bp} = E_F - E_v$.

Schottky Lowering: Barrier-height lowering due to image force effect.

UDM: "Unified defect model," model that assumes the Fermi-level pinning by native defects in the semiconductor.

Vacuum Level: Energy level of a resting electron.

Work Function: Energy needed to extract an electron from the Fermi level and take it to infinity (to the vacuum level).

Works Cited

Andrews, J. M., Phillips, J. C. (1975), *Phys. Rev. Lett.* **35,** 56–59.

Bardeen, J. (1947), *Phys. Rev.* **71,** 717–727.

Braun, F. (1874), *Ann. Phys. (Leipzig)* **153,** 556.

Brillson, L. J. (1978), *Phys. Rev. Lett.* **40,** 260–263.

Cowley, A. M., Sze, S. M. (1965), *J. Appl. Phys.* **36,** 3212–3220.

Delerue, C., Lannoo, M., Langer, J. M. (1988), *Phys. Rev. Lett.* **61,** 199–202.

Duboz, J. Y., Badoz, P. A., Arnaud d'Avitaya, F., Rosencher, E. (1989), *Phys. Rev. B* **40,** 10607–10610.

Freeouf, J. L. (1980), *Solid State Commun.* **33,** 1059–1061.

Freeouf, J. L., Woodall, J. M. (1981), *Appl. Phys. Lett.* **39,** 727–729.

Heine, V. (1965), *Phys. Rev.* **138,** A1689–A1696.

Hollinger, G. (1987), in: *Semiconductor Interfaces: Formation and Properties,* Springer Proceedings in Physics 22, New York: Springer-Verlag, pp. 210–231.

Kaiser, W. J., Bell, L. D. (1988), *Phys. Rev. Lett.* **60,** 1406–1409.

Bell, L. D., Kaiser, W. J. (1988), *Phys. Rev. Lett.* **61,** 2368–2371.

Lannoo, M., Friedel, P. (1991), in: *Atomic and Electronic Structure of Surfaces,* Springer Series in Surface Science 16, New York: Springer-Verlag, Chap. 7.

Louie, S. G., Chelikowsky, J. R., Cohen, M. L. (1977), *Phys. Rev.* **15,** 2154–2162.

Mönch, W. (1990), *Rep. Prog. Phys.* **53,** 221–278.

Mott, N. F. (1938), *Proc. Cambridge Philos. Soc.* **34,** 568–572.

Muret, P., Elguennouni, D., Missous, M., Rhoderick, E. H., Baptist, R., Pellissier, A. (1992), *Appl. Surf. Sci.* **56–57,** 341–347.

Ortega, J., Sanchez-Dehesa, J., Flores, F. (1988), *Phys. Rev. B* **37,** 8516–8518.

Rhoderick, E. H., Williams, R. H. (1988), in: *Metal Semiconductor Contacts,* Oxford: Clarendon, Chap. 5.

Schottky, W. (1938), *Naturwissenschaften* **26,** 843.

Schottky, W. (1939), *Z. Phys.* **113,** 367–414.

Schottky, W. (1940), *Phys. Z.* **41,** 570–573.

Schottky, W. (1942), *Z. Phys.* **118,** 539–592.

Spicer, W. E., Chye, P. W., Skeath, P. R., Su, C. Y., Lindau, I. (1979), *J. Vac. Sci. Technol.* **16,** 1427–1433.

Spicer, W. E., Kendelewicz, T., Newman, N., Cao, R., McCants, C., Miyano, K., Lindau, I. (1988), *Appl. Surf. Sci.* **33/34,** 1009–1029.

Tejedor, C., Flores, F., Louis, E. (1977), *J. Phys. C.* **10,** 2163–2177.

Tersoff, J. (1984), *Phys. Rev. Lett.* **52,** 465–468.

Tersoff, J., Harrison, W. (1987), *Phys. Rev. Lett.* **58,** 2367–2370.

Tung, R. T, Ng, K. K., Gibson, J. M., Levi, A. F. J. (1986), *Phys. Rev. B* **33,** 7077–7090.

Yndurain, Y. (1971), *J. Phys. C.* **4,** 2849–2858.

Zur, A., McGill, T. C., Smith, D. L. (1983), *Phys. Rev. B* **28,** 2060–2067.

Further Reading

Electrical Behavior, Schottky Lowering, Experimental Methods

Rhoderick, E. H., Williams, R. H. (1988), *Metal Semiconductor Contacts,* Oxford: Clarendon, Chap. 5.

Sze, S. M. (1981), *Physics of Semiconductor Devices,* New York: Wiley-Interscience.

Correlations between Barrier Height and Semiconductor Dangling-bond Level

Lannoo, M., Friedel, P. (1991), *Atomic and Electronic Structure of Surfaces,* Springer Series in Surface Science No. 16, New York: Springer-Verlag, Chap. 7.

Phenomenological Models and Correlations

Mönch, W. (1990), *Rep. Prog. Phys.* **53,** 221–278.

Comparison MIGS/Defects

Flores, F., Ortega, J. (1992), *Appl. Surf. Sci.* **56–58,** 301–310.

Spicer, W. E., Cao, R., Miyano, K., Kendelewicz, T., Lindau, I., Weber, E., Liliental-Weber, Z., Newman, N. (1988), *Appl. Surf. Sci.* **41/42,** 1–16.

CONTENTS OF PREVIOUS VOLUMES

Volume 1

03-A. Accelerators, Linear 1
03-A. Accelerators, Potential-Drop 27
11-D. Acoustic Properties of Liquids 43
02-A. Acoustical Instrumentation 63
07-C. Acoustical Tomography 89
07-E. Acoustics, Architectural 109
07-E. Acoustics, Linear 137
07-E. Acoustics, Nonlinear 183
07-E. Acoustics, Physiological 213
07-E. Acoustics, Psychological 261
07-D. Acoustics, Underwater 293
09-E. Aerodynamics 329
20-E. Aeronautics 379
17-B. Aerosols 415
19-E. Aerospace Medical Physics 443
18-C. Air-Pollution: Components, Causes, and Cures 489
01-C. Algebraic Methods 515
16-B. Aluminum 539
10-B. Amorphous Materials, Structure of 559
05-A. Amplifiers, Electronic 593
05-A. Amplifiers, Optical 607
01-C. Analytic Methods 623

Volume 2

01-E. Artificial Intelligence 1
20-E. Astronautics 25
20-A. Astronomical Telescopes 65
20-E. Astronomy 99
20-E. Astrophysics 117
07-D. Atmospheric Acoustics 145
20-E. Atmospheric Physics 181
20-D. Atmospheric Structure 201
04-C. Atomic Cooling and Trapping 225
04-C. Atomic Spectroscopy 245
04-D. Atoms 285
17-C. Auger Spectroscopy 297
18-A. Automotive Propulsion Systems 323
03-A. Betatrons 347
19-D. Biological Effects of Electromagnetic and Particle Radiation 365
19-D. Biological Effects of Sound and Ultrasound 403
19-D. Biomagnetism and Geomagnetic Field Detection by Organisms 421
19-C. Biomagnetism, Medical Aspects 453
18-C. Biomass Energy 473
19-E. Biomechanics of Living Organisms 489
19-E. Biomedical Engineering 513
19-C. Biomedical Uses of Radiation 549
19-E. Biophysics 571
19-D. Biorheology 609
02-A. Bus Systems and Computer Interfacing 639
07-E. Erratum: Acoustics, Physiological 653

Volume 3

02-C. Calibration and Maintenance of Test and Measuring Equipment 1
16-B. Carbon Materials 21
20-E. Cartography 41
17-D. Catalysis 67
01-E. Catastrophe Theory 85
14-D. Cathodoluminescence 121
19-E. Cellular Biomechanics 141
16-B. Ceramics, Material Properties of 169
08-D. Chaotic Phenomena 189
16-C. Characterization and Analysis of Materials 215
05-A. Charge-Coupled Devices 241
05-E. Charged-Particle Optics 273
17-C. Chemical Analysis 307
17-D. Chemical Kinetics 345
17-D. Chemical Reactions 377
17-D. Chemiluminescence 413
17-C. Chromatography 429
05-A. Circuit Elements 455
05-A. Circuits, Analog 479
05-A. Circuits, Digital 507
18-B. Coal 527
12-D. Cohesive Energy 559
12-D. Collective Phenomena in Solids 579

Volume 4

17-D. Combustion 1
16-B. Composite Materials 17
09-D. Compressible Flows 43
01-C. Computer-Aided Design in Electronics 71
01-C. Computer-Aided Manufacturing 89
01-C. Computer Databases 103
01-C. Computer Graphics 127
01-A. Computer Hardware 145
01-C. Computer Programming Languages 163
01-A. Computers 189
17-C. Conformational Analysis 229
02-D. Constants, Fundamental 243
16-B. Copper 267
08-D. Critical Phenomena 277
02-A. Cryogenics 311
16-D. Crystal Growth 335
10-B. Crystalline State 365
10-C. Crystallography 385
16-C. CVD (Chemical Vapor Deposition) 409
03-A. Cyclotrons 427
02-C. Data Acquisition 463
02-A. Detectors, Particle, Calorimetric 485
02-A. Detectors, Scintillation 503
02-A. Detectors, Semiconductor 513
17-C. Dialysis 533
13-D. Diamagnetism 557
16-B. Erratum: Carbon Materials 573

Volume 5

16-B. Diamond and Diamondlike Carbon 1
14-D. Dielectric Properties of Insulators 25
08-A. Diesel Engines 47
11-D. Diffusion and Ionic Conduction in Liquids 61
11-D. Diffusion in Thin Films 75
17-B. Disperse Systems 87
02-A. Display Technology 101
20-B. Earth, Interior Structure of 127
18-E. Electric Power Engineering 149
19-D. Electrical Conduction and Dielectric Behavior in Biological Systems 177
02-A. Electrical Instrumentation 201
17-E. Electrochemistry 223
05-E. Electrodynamics, Classical 259
14-D. Electroluminescence 295
02-A. Electromagnetic Quantities, Basic, Measurement of 327
05-D. Electromagnetic Radiation 345
05-D. Electromagnetic Wave Propagation 379
10-C. Electron Diffraction 405
12-D. Electron Level Splitting 431
10-C. Electron Microscopy 453
04-D. Electron Paramagnetic Resonance 475
04-D. Electron Scattering by Atoms and Molecules 499
15-D. Electron States in Zero-, One-, and Two-Dimensional Structures 531
12-D. Electron States: Localized 549
12-D. Electron Structure of Liquids 571
12-D. Electron Structure of Solids 595

Volume 6

05-A. Electronic Circuits 1
05-C. Electronic Components and Assemblies: Reliability and Testing 21
04-D. Electronic Structure of Atoms and Molecules 45
15-D. Electronic Structure of Surfaces 99
12-D. Electronic Transport in Solids 121
10-D. Electrorheology 145
05-A. Electrostatic Capacitative Energy Storage 155
05-E. Electrostatics 177
02-A. Ellipsometers 191
19-C. Energetics of Biological Processes 207
18-C. Energy Conversion 229
07-E. Engineering Acoustics 245
18-E. Environmental Health and Safety 271
11-D. Equations of State 291
12-D. Excitons 311
17-B. Explosives 327
06-B. Fiber Optics 359
12-D. Field Emission and Field Ionization 379
10-C. Field Ion Microscopy 393
09-E. Fluid Dynamics, Equilibrium 409
09-E. Fluid Dynamics, Non-equilibrium 437
01-C. Fourier & Other Mathematical Transforms 481
01-D. Fractal Geometry 501
16-B. Fullerenes 515
09-C. Fusion, Inertial Confinement 545
09-C. Fusion, Magnetic Confinement 575

Volume 7

09-A Fusion Technologies 1
20-E Galaxies and Cosmology 21
16-B Gallium Arsenide 43
01-C Game Theory 69
20-E Geodesy 89
20-D Geoelectromagnetism 109
01-C Geometrical Methods 125
20-E Geophysics 149
18-C Geothermal Energy 197
16-B Germanium 219
20-D Glaciers and Ice Sheets 239
16-B Glasses 251
15-D Grain Boundaries 271
16-B Graphite 289
03-D Gravitation and General Relativity 303
01-C Green's Functions 341
01-C Group Theory 365
08-A Heat Engines and Refrigerators 383
08-D Heat Transfer 405
11-D Heat-Pulse Propagation in Solids 417
13-D Heavy-Fermion Phenomena 435
15-D Heterostructures and Superlattices, Metallic 465
15-D Heterostructures and Superlattices, Semiconductor 477
02-C High-Pressure Techniques 495
07-C Holography, Acoustical 511
06-C Holography, Electron 531
06-C Holography, Optical 541
20-E Hydrology 563
19-C Imaging Techniques, Biomedical 581

Volume 8

05-A Inductors 1
01-E Information Science 21
12-D Insulators and Semiconductors, Conductivity in 41
10-B Insulators, Structure of 85
09-D Intense Particle Beams 103
14-D Interaction of Solids with Particles and Radiation 117
15-B Intercalation Compounds 133
02-A Interferometers and Interferometry 157
16-C Ion-Beam Modification of Materials 173
11-D Ionic Conduction and Diffusion in Solids 193
20-D Ionosphere 223
02-A Laboratory Instrumentation 247
06-D Laser Photochemistry 283
06-D Laser Physics 299
06-C Laser Technology 321
06-A Lasers, Dye, Technology and Engineering 331
06-A Laser, Free-Electron 353
06-A Lasers, Gas 371
06-C Lasers, Industrial Uses of 397
06-A Lasers, Semiconductor 425
06-A Lasers, Solid State 443
11-D Lattice Vibrations, Statistics of 465
06-A Light-Emitting Diodes 485
10-B Liquid Crystals, Structure of 515
10-B Liquids, Simple, Structure of 533
05-A Machines, Electrical 561
13-A Magnetic Devices 587

Volume 9

16-B Magnetic Materials ... 1
16-D Magnetic Ordering in Solids ... 15
04-D Magnetic Resonance and Quadrupole Resonance, Nuclear ... 31
19-C Magnetic Resonance Imaging ... 47
02-A Magnetic Resonance Spectrometers ... 71
13-A Magnetic Storage of Information ... 95
09-D Magnetohydrodynamics ... 111
13-A Magneto-optical Devices ... 157
20-B Magnetospheres of the Earth and Planets ... 187
05-E Magnetostatics ... 207
13-A Magnetostrictive Materials and Devices ... 229
05-A Magnets ... 245
02-C Maintenance and Calibration of Laboratory Instrumentation ... 261
01-E Manufacturing Engineering ... 279
20-E Marine Geophysics ... 305
02-A Mass and Density, Measurement of ... 335
16-D Materials Degradation ... 349
16-C Materials Preparation: Solids ... 365
16-C Materials Treatment ... 399
01-C Mathematical Modeling ... 417
09-A Measurement and Instrumentation of Flow ... 445
02-A Measurement of Magnetic Properties and Quantities ... 463
07-A Measurement, Acoustical ... 491
10-D Mechanical and Elastic Waves in Solids ... 515
10-D Mechanical Properties of Liquids ... 531
10-D Mechanical Properties of Solids ... 545
07-D Mechanical Vibration and Damping ... 561
03-D. Erratum: Gravitation and General Relativity ... 591
02-A. Erratum: Laboratory Instrumentation ... 591
06-A. Erratum: Lasers, Dye, Technology and Engineering ... 591

Volume 10

01-E Mechanics, Classical ... 1
19-C Medical Use of Lasers ... 33
01-A Memories ... 61
16-D Mesoscopic Systems ... 81
12-D Metal-Insulator Transitions ... 113
10-B Metallic Glasses ... 129
16-E Metallurgy, Physical ... 141
12-D Metals and Alloys, Conductivity in ... 163
10-B Metals and Alloys, Structure of ... 199
20-E Meteorology and Climatology ... 215
02-E Metrology ... 239
05-A Microelectronics ... 253
15-C Microlithography ... 281
01-A Microprocessors ... 297
02-A Microsensors ... 321
05-E Microwave Circuits ... 349
01-A Modulators and Demodulators, Electrical ... 379
01-A Modulators and Demodulators, Optical ... 393
04-D Molecular and Atomic Clusters ... 411
04-D Molecular and Atomic Collison Processes with Ion Beams ... 437
15-C Molecular Beam Epitaxy ... 471
04-C Molecular Spectroscopy ... 491
04-D Molecules ... 525
06-A Monochromators ... 553
01-C Monte-Carlo Methods ... 567

Volume 11

13-D Mößbauer Effect 1
02-C Muon Spin Rotation/Relaxation/Resonance 23
04-D Muonic, Mesonic, and Baryonic Atoms and Molecules 55
04-B Muonium and Positronium 79
07-E Music, Electronic 97
07-A Musical Instruments 129
16-B Nanophase Materials 173
18-B Natural Gas 201
05-C Network Theory, Electrical 221
01-C Networks, Computer 263
01-C Neural Networks 275
19-E Neurobiophysics 297
04-D Neutral Atomic and Molecular Collision Processes 323
10-C Neutron Diffraction 339
10-C Neutron Scattering 353
09-D Nonhomogeneous Flows 379
01-D Nonlinear Systems 417
18-C Nuclear Energy, Fission 429
18-B Nuclear Fuels and Isotopes 479
19-E Nuclear Medicine 513
03-D Nuclear Reactions 543
03-D Nuclear Structure 571
15-B. Erratum: Intercalation Compounds 601

Volume 12

18-C Nuclear Waste Management 1
19-B Nucleic Acids 19
20-E Oceanography 47
20-B Oil Shales and Tar Sands 77
01-C Operations Research 101
06-C Optical Communications 119
06-A Optical Components and Systems 157
01-A Optical Computing 177
06-A Optical Filters 195
02-A Optical Instrumentation 215
06-A Optical Interconnections 243
06-C Optical Microscopy 259
14-D Optical Properties of Solids 285
06-A Optical Scanners 337
01-A Optical Storage 369
19-C Optical Tomography 395
06-D Optics, Atmospheric 405
06-E Optics, Geometrical 435
06-E Optics, Linear 451
06-E Optics, Nonlinear 487
19-E Optics, Physiological 541
06-D Optics, Underwater 571

Volume 13

08-D Order-Disorder Transitions 1
05-A Oscillators 17
02-A Oscilloscopes, Analog and Digital 37
17-D Osmosis 59
16-C Packaging Technology 73
20-D Paleomagnetism 89
13-D Paramagnetism 101
03-A Particle Colliders 123
02-A Particle Detectors, Tracking ...141
12-D Particle Impact Phenomena ...175
03-E Particle Physics 193
03-A Particle Sources—Ion 213
03-D Particles, Elementary 223
01-E Patents and Intellectual Property 233
01-C Perturbation Methods 245
18-B Petroleum 269
11-D Phase Equilibria 297
11-D Phase Separation 323
11-D Phase Transitions: Renormalization and Scaling 343
11-D Phase Transitions, Structural 373
01-E Philosophy of Physics 389
01-E Philosophy of Technology 417
11-D Phonons in Crystal Lattices 439
14-A Photodetectors 459
12-D Photoemission and Photoelectron Spectra 477
14-D Photoluminescence 497
19-D Photosynthesis 513
14-A Photovoltaic Devices 533
01-F Physics and Engineering Organizations 559
Contents of Previous Volumes 589

Volume 14

04-A Physics and Technology of Ion and Electron Sources 1
01-E Physics and Technology, History of 35
01-C Physics Applications of Computers—Experimental 55
20-E Physics in Archaeology 69
01-E Physics Literature and Publications 87
01-F Physics Research Facilities 97
01-F Physics, Technology, and Society 107
14-A Piezoelectric Devices 129
14-A Piezoelectric Resonators and Applications 147
09-C Plasma Etching 171
09-E Plasma Physics 199
09-D Plasma Waves and Instabilities 237
20-E Plate Tectonics 273
13-B Plates and Films, Magnetic 297
10-D Point and Extended Defects in Crystals 317
02-A Polarimeters and Polarization Spectrometers 341
17-C Polarography, Voltammetry, and Related Techniques 371
12-D Polarons 383
16-D Polymer Dynamics 415
17-D Polymerization and Polymer Reactions 445
10-B Polymers: Crystal Structure and Morphology 477
12-B Polymers, Electrical and Electronic Properties 497
10-D Polymers, Mechanical Properties of 531
17-B Polymers, Molecular Properties of 549
14-B Polymers, Optical Properties of 569
12-C Positron-Annihilation Spectroscopy 607
Contents of Previous Volumes 633

Volume 15

05-E Power Electronics 1
05-A Power Transmission 25
01-C Probability Theory 51
10-B Properties and Applications of Amorphous Materials 71
19-A Prostheses and Artificial Limbs 115
19-B Prosthetic Materials 141
19-D Protein Dynamics 163
19-B Proteins and Enzymes 185
03-E Quantum Electrodynamics 215
01-C Quantum Logic 229
02-C Quantum Measurements 257
03-E Quantum Mechanics 275
06-E Quantum Optics 307
06-A Quantum Optoelectronic Devices 339
16-B Quartz 365
10-D Quasicrystals 377
05-A Radar Technology 399
16-D Radiation Damage in Crystals .429
02-A Radiation Detectors, Infrared ..459
02-A Radiation Detectors, Particle, Gamma, and X-Ray 473
04-D Radiation Interaction with Molecules 509
20-A Radio Telescopes 527
03-D Radioactivity 547
17-E Radiochemistry 565
14-D Raman Scattering 587
Contents of Previous Volumes 613

LIST OF RECOMMENDED UNITS AND SYMBOLS

RECOMMENDED UNITS AND CONVERSION FACTORS

The SI system provides six basic units: meter m, kilogram kg, second s, ampere A, kelvin K, candela cd, and mole mol.

Some important derived units are also allowed and bear special names, e.g.:

1 N (newton) = 1 kg m s^{-2}
1 J (joule) = 1 N m = 1 kg m^2 s^{-2}
1 W (watt) = 1 J s^{-1} = 1 kg m^2 s^{-3}
1 Pa (pascal) = 1 N m^{-2} = 1 kg m^{-1} s^{-2}

For mass, the gram g, or the metric ton t which equals 1000 kg, may be used instead of kilogram kg.

The so-called "long ton" (UK) and "short ton" (US) have been abandoned and will not be used in the *Encyclopedia of Applied Physics.*

For pressure, the bar (name and symbol alike) may be used, and for temperature, the degree celsius °C.

From all of these, decimal multiples or fractions can be derived:

Power of ten	Prefix	Symbol	Power of ten	Prefix	Symbol
10	deca	da	10^{-1}	deci	d
10^2	hecto	h	10^{-2}	centi	c
10^3	kilo	k	10^{-3}	milli	m
10^6	mega	M	10^{-6}	micro	μ
10^9	giga	G	10^{-9}	nano	n
10^{12}	tera	T	10^{-12}	pico	p
10^{15}	peta	P	10^{-15}	femto	f
10^{18}	exa	E	10^{-18}	atto	a

For mass, multiples or fractions are derived from g, not from kg (since the latter already contains a prefix), e.g., mg.

Units with the prefixes are considered as one entity and can, therefore, be raised to any power, e.g., cm^3.

The liter is now considered as synonymous with dm^3 (which does not hold in the older literature!). Please use the capital letter L as unit symbol (following a recent IUPAC recommendation). The use of the Ångström unit is discouraged; it should be replaced by fractional meters (1 Å = 100 pm = 0.1 nm).

SELECTED QUANTITIES, UNITS, AND SYMBOLS

Name of quantity[a]	Symbol[b]	SI unit[c]	Name of unit	Other units
Space and time				
length*	l	m	meter[d]	
breadth, width	b	m		
height	h	m		
radius	r	m		
thickness	d	m		
area	A,S	m^2		a (are) h (hectare)
volume	V	m^3		L, l(liter)[c]
plane angle	$\alpha,\beta,\gamma,$ ϑ,φ	1, rad	radian	° (degree) ′ (minute) ″ (second)
solid angle	ω,Ω	1, sr	steradian	
wavelength	λ	m		
wave number	σ,ν	m^{-1}		
time*	t	s	second	min (minute) h (hour) d (day)
frequency	ν, f	s^{-1}		Hz (hertz)[f]
relaxation time	τ	s		
velocity	u,v	$m\ s^{-1}$		km/h
acceleration	a	$m\ s^{-2}$		
Mechanics				
mass*	m	kg	kilogram	g (gram) t (tonne)[g]
(mass) density	ρ	$kg\ m^{-3}$		g/cm^3
momentum	$\mathbf{p}$	$kg\ m\ s^{-1}$		
angular momentum	$\mathbf{L}$	$kg\ m^2\ s^{-1}$		
force	$\mathbf{F}$	N	newton[f]	
moment of force	$\mathbf{M}$	N m		
weight	G, W	N		
pressure	p	Pa	pascal	bar (bar)[h]
energy	E, W	J	joule	W h (watt hour)[i] eV (electron volt)[j]
work	W,A	J		
power	P	W	watt	J/s, V A[k]
Molecular Physics and Thermodynamics				
thermodynamic temperature*	T	K	kelvin	°C (degrees Celsius)
Celsius temperature	ϑ,t			°C
number of entities	N			
Avogadro constant[l]	N_A,L	mol^{-1}	(particles) per mole	
Boltzmann constant[m]	k,k_B	$J\ K^{-1}$		
Planck constant[n]	h	J s		
(molar) gas constant[o]	R	$J\ mol^{-1}\ K^{-1}$		
(quantity of) heat	Q	J		
entropy[p]	S	$J\ K^{-1}$		
internal energy[p]	U	J		
Helmholtz function,[p] (Helmholtz) free energy, Helmholtz energy	F,A	J		
enthalpy[p]	H	J		
Gibbs function,[p] (Gibbs) free energy, Gibbs energy	G	J		
heat capacity[p]	C_p,C_V	$J\ K^{-1}$		

Name of quantity[a]	Symbol[b]	SI unit[c]	Name of unit	Other units
Chemical Physics				
amount of substance*	n	mol	mole	
relative atomic mass	A_r	1		
relative molecular mass	M_r	1		
atomic mass constant	m_u	kg		u (atomic mass unit)[q]
mass of a portion (of substance B)	$m_B, m(B)$	kg		g (gram)
molar mass (of substance B)	$M_B, M(B)$	$kg\ mol^{-1}$		
concentration (of substance B)	$c_B, c(B)$	$mol\ m^{-3}$		mol/L
mole fraction[r] (of substance B)	$\kappa_B, \kappa(B)$	1		
mass fraction (of substance B)	$\omega_B, \omega(B)$	1		%,‰,ppm,ppb
volume fraction (of substance B)	$\varphi_B, \varphi(B)$ $\phi_B, \phi(B)$	1		%,‰,ppm,ppb
mass concentration	ρ	$kg\ m^{-3}$		g/L
molality		$mol\ kg^{-1}$		mmol/kg
volume concentration[s]	σ	1		
molar volume	V_m	$m^3\ mol^{-1}$		L/mol
molar heat capacity	C_m	$J\ mol^{-1}\ K^{-1}$		
molar conductivity	Λ_m	$S\ m^2\ mol^{-1}$	(S: siemens)	
Faraday constant[t]	F	$C\ mol^{-1}$		
Electricity and Magnetism				
quantity of electricity[u]	Q	C	coulomb	
charge density	ρ	$C\ m^{-3}$		
electric potential	ϕ, V	V	volt	
electric potential difference, voltage	$U, \Delta\phi, \Delta V$	V		
electric dipole moment	$\mathbf{p}, \mathbf{p}_c$	C m		
electric current*	I	A	ampere	
electric current density	j	$A\ m^{-2}$		
electric field strength	$\mathbf{E}$	$V\ m^{-1}$		
electric displacement	$\mathbf{D}$	$C\ m^{-2}$		
capacitance	C	F	farad	
permittivity	ε	$F\ m^{-1}$		
relative permittivity	ε	1		
dielectric polarization	$\mathbf{P}$	$C\ m^{-2}$		
electric susceptibility	χ_c	1		
polarization (of a particle)	α	$m^2\ C\ V^{-1}$		
magnetic flux	Φ	Wb	weber	
magnetic flux density	$\mathbf{B}$	T	tesla	
magnetic field strength	$\mathbf{H}$	$A\ m^{-1}$		
permeability	μ	$H\ m^{-1}$, $N\ A^{-2}$	(H: henry)	
relative permeability	μ_r	1		
magnetization	M	$A\ m^{-1}$		
magnetic susceptibility	χ	1		
molar magnetic susceptibility	χ_m	$m^3\ mol^{-1}$		
(electrical) resistance	R	Ω	ohm	
(electrical) conductance	G	S	siemens	
(electrical) resistivity	ρ	Ω m		
(electrical) conductivity	κ, σ	$S\ m^{-1}$		
self-inductance	L	H	henry	
Radiation				
radiant energy	Q, W, Q_c	J		
luminous intensity*	I	cd	candela	
radiant intensity	I_c	$W\ sr^{-1}$, W		
emissivity, emittance	ε	1		

Name of quantity[a]	Symbol[b]	SI unit[c]	Name of unit	Other units
absorptance	α	1		
reflectance	ρ, R	1		
transmittance	τ	1		
absorption coefficient:				
linear (decadic)	a	m^{-1}		
molar (decadic)	ε	$m^2\ mol^{-1}$		
refractive index	n	1		
molar refraction	R_m	$m^3\ mol^{-1}$		
angle of optical rotation	α	1, rad		
Transport Properties				
flux of quantity X	J_X, J	(varies)		
mass flow rate	q_m, $\dot{m}$	$kg\ s^{-1}$		
volume flow rate	q_V, $\dot{V}$	$m^3\ s^{-1}$		
heat flow rate	Φ	W		
thermal conductivity	κ, k, λ	$W\ m^{-1}\ K^{-1}$		
coefficient of heat transfer	h	$W\ m^{-2}\ K^{-1}$		
thermal diffusivity	a	$m^2\ s^{-1}$		
diffusion coefficient	D	$m^2\ s^{-1}$		
thermal diffusion coefficient	D_T	$m^2\ s^{-1}$		
viscosity	η, μ	Pa s		
kinematic viscosity	ν	$m^2\ s^{-1}$		

[a]SI base quantities are marked by asterisks (*)
[b]Recommended by IUPAC
[c]SI base units as well as derived and supplementary units are listed; all are to be used with prefixes as needed
[d]do not use "metre"
[e]do not use "litre"; $1\ L = 10^{-3}\ m^3$
[f]$1\ Hz = 1\ s^{-1}$; $1\ N = 1\ kg\ m\ s^{-2}$
[g]Formerly metric ton; $1\ t = 10^3\ kg$
[h]$1\ bar = 10^5\ Pa$
[i]$1\ W\ h = 3.6 \times 10^3\ J$
[j]$1\ eV = 1.602\ 189 \times 10^{-19}\ J$
[k]$1\ W = 1\ J/s = 1\ VA$
[l]$N_A = 6.022\ 136\ 7 \times 10^{23}\ mol^{-1}$
[m]$k = 1.380\ 658 \times 10^{-23}\ J\ K^{-1}$
[n]$h = 6.626\ 075\ 5 \times 10^{-34}\ J\ s$
[o]$R = 8.314\ 510\ J\ mol^{-1}\ K^{-1}$
[p]Molar quantities can be distinguished from the quantity of a system by adding the subscript m; e.g., molar internal energy U_m, in $J\ mol^{-1}$
[q]$1\ u = 1.660\ 565\ 5 \times 10^{-27}\ kg$
[r]A more accurate, but rather uncommon, name is "amount-of-substance fraction"
[s]σ refers to the total volume of a mixture, whereas the volume fraction φ relates the volume of a substance to the volume of several components before mixing
[t]$F = 9.648\ 530\ 9 \times 10^4\ C\ mol^{-1}$
[u]Also called electric charge